ELSEVIER'S

DICTIONARY OF CHEMICAL ENGINEERING

VOLUME 1

CHEMICAL ENGINEERING AND LABORATORY EQUIPMENT

ADD: 1) ELS—'s dict of chem-eng—
2) DICT— of chem-eng—

ELSEVIER PUBLISHING COMPANY
335 JAN VAN GALENSTRAAT
P.O. BOX 211, AMSTERDAM, THE NETHERLANDS

ELSEVIER PUBLISHING CO. LTD.
BARKING, ESSEX, ENGLAND

AMERICAN ELSEVIER PUBLISHING COMPANY, INC.
52 VANDERBILT AVENUE
NEW YORK, NEW YORK 10017

LIBRARY OF CONGRESS CARD NUMBER: 68–54865
STANDARD BOOK NUMBER 444–40714–6

PRINTED IN THE NETHERLANDS

ELSEVIER'S

DICTIONARY OF CHEMICAL ENGINEERING

VOLUME 1

CHEMICAL ENGINEERING AND LABORATORY EQUIPMENT

IN SIX LANGUAGES

ENGLISH/AMERICAN - FRENCH - SPANISH
ITALIAN - DUTCH - GERMAN

COMPILED AND ARRANGED ON
AN ENGLISH ALPHABETICAL BASIS BY

W. E. CLASON

FORMERLY HEAD OF THE TRANSLATION DEPARTMENT
N.V. PHILIPS, GLOEILAMPENFABRIEKEN
EINDHOVEN (THE NETHERLANDS)

ELSEVIER PUBLISHING COMPANY

AMSTERDAM/LONDON/NEW YORK

1968

ELSEVIER'S

DICTIONARY OF CHEMICAL ENGINEERING

VOLUME 1
CHEMICAL ENGINEERING
AND LABORATORY EQUIPMENT

VOLUME 2
CHEMICAL ENGINEERING;
PROCESSES AND PRODUCTS

A complete catalogue listing all
Elsevier dictionaries will be sent upon request

PUBLISHER'S NOTE

By the most conservative estimates, there are at least two thousand basic languages in current use today. By some definitions, this figure could reach more than five thousand languages, not including a vast number of dialects. Added to this, the incredible current advances in engineering and science have rendered most existing technical dictionaries obsolete. The accumulated result has been a fast growing need for specialized multilingual dictionaries, particularly in the face of international scientific cooperation and exchange of knowledge.

To meet this need, Elsevier publishes a number of multilingual technical dictionaries relating to special fields of science and industry. Edited under authoritative auspices, they draw upon rich sources of knowledge.

In planning this dictionary, the author and publishers have been guided by the principles proposed by the United Nations Educational, Scientific and Cultural Organization (UNESCO). The aim is to ensure that each dictionary will fit into place in a pattern which may progressively extend over all interrelated fields of science and technology and cover all necessary languages.

For each language, there is an alphabetical list of words referring to corresponding numbers in the basic table. The system of thumb-indexing enables one to find any language at one. The binding, smooth paper and convenient size result in an enjoyable and valuable reference book.

PREFACE

In compiling this dictionary the author has taken the view that, in general, it is unnecessary to include purely chemical terms, such as natriumchloride or installations and processes where the name is derived from a proper noun. This basic principle was, however, not adhered to in the case of:

1. a chemical substance with more than one name
2. an installation or process also known under a name not derived from a proper noun.

In view of the great diversity of terms, the author has had to limit the subjects under abbreviations to the most important ones, whereby the selection was not always an easy matter.

Finally, for some items an attempt has been made to normalize the terminology in current use.

The author wishes to thank those both at home and abroad who have assisted him in his work. A special word of thanks is due to his wife, who took care of the entire manuscript, and checked the spelling and terminology with great accuracy.

September, 1968 W.E. Clason

ABBREVIATIONS

aam	acetic acid manufacture
aap	auxiliary apparatus
abs	absorption, adsorption etc.
act	acid treatment
agr	agricultural chemistry
aic	inorganic industrial chemistry
ana	analysis
app	general apparatus
bac	bacteriological chemistry
ble	bleaching
bre	brewing
brk	breaking and milling
boi	boilers and boiling
bui	building materials
bur	burners and burning
cen	centrifuges and centrifuging
cer	ceramics
chr	chromatography
cle	cleaning and degreasing
clo	cooling
coa	coating(s)
col	colloids
con	control and regulation
cor	corrosion
cov	conserving industry
cpt	component parts
cry	crystallography and crystallization
ctt	cold treatment
dai	dairy industry
dec	decorating and painting
dis	distillation
dry	drying
dye	dyeing and printing
elc	electrochemistry
eva	evaporation
exp	explosives
ext	extraction
fat	fats and oils
fea	fire extinguishers and extinguishing
fer	fertilizers
fil	filters and filtration
fmi	fermentation
fou	foundry
fue	fuels, petroleum industry
gen	general subjects
gla	glass
glu	glues
htt	high temperature treatment
hyg	hygienics
imp	impregnation
ins	instruments
iso	isotopes
iwt	industrial waste treatment
lab	laboratory apparatus
lea	leather industry
lub	lubrication
mat	sundry materials
mcy	medical and pharmaceutical chemistry
met	metals and alloys

min	mining and minerals
mit	military chemistry
mix	mixers and mixing
nue	nuclear chemistry
nit	nutrition including beverages
oic	organic industrial chemistry
pap	paper industry
per	perfumery
pho	photography
pla	plastics
pol	polishing
pre	pressurized apparatus and processes
rad	radiation chemistry
rub	rubber industry
sap	safety appliances
sep	separators and separating
sin	sugar industry
soa	soap industry
sor	sorting
spe	spectroscopy and spectrography
sto	storage
sua	sulfuric acid manufacture
tan	tanning industry
tes	testing apparatus and methods
tex	textile chemistry
tra	transport
vac	vacuum processes and apparatus
wel	welding and soldering

---

adj	adjective
f	feminine
GB	English, British usage
m	masculine
n	neuter
pl	plural
US	English, American usage
v	verb
f	Français
e	Español
i	Italiano
n	Nederlands
d	Deutsch

BASIC TABLE

A

1 ABLATIVE COATING, HEAT SHIELD sap
f couche *f* disparaissante en fondant
e capa *f* desaparecente por fusión
i strato *m* da defluire in seguito a fusione
n afsmeltlaag, hitteschild *n*
d Abbrennschicht *f*, Abschmelzschicht *f*, Hitzeschild *m*

2 ABRADER, ABRADING DEVICE aap
f meuleuse *f*
e amoladora *f*
i dispositivo *m* per abrasione, molatrice *f*
n slijpapparaat *n*
d Schleifapparat *m*

3 ABRASION MACHINE, ABRASION TESTER, ABRASION TESTING MACHINE aap,ins
f abrasimètre *m*, machine *f* d'abrasion, machine *f* pour essais d'abrasion
e abrasímetro *m*, máquina *f* para ensayos de abrasión
i abrasimetro *m*, macchina *f* per prova d'abrasione
n slijtagebeproevingsmachine, slijtagemachine
d Abnutzungsprüfapparat *m*, Abriebmaschine *f*, Abriebprüfungsmaschine *f*

4 ABRASION TESTING EQUIPMENT aap,ins
f installation *f* pour essais d'abrasion
e equipo *m* para pruebas de abrasión
i attrezzatura *f* per prova d'abrasione
n slijtagebeproevingsinstallatie
d Abriebprüfungsanlage *f*

5 ABRASIVE BELT, LINISHER aap,cpt
f bande *f* abrasive, ruban *m* émerisé
e cinta *f* abrasiva
i nastro *m* abrasivo
n slijpband
d Schleifband *n*

6 ABRASIVE CLOTH aap,mat
f toile *f* à l'émeri
e tela *f* abrasiva, tela *f* de esmeril
i tela *f* abrasiva
n schuurlinnen *n*
d Schmirgelleinen *n*

7 ABRASIVE FINISHING MACHINE pla,pol
f machine *f* de finissage par abrasion
e máquina *f* de acabado con abrasivos
i macchina *f* di finitura per abrasione
n afwerkslijpmachine
d Zurichtschleifmaschine *f*

8 ABRASIVE PAPER aap,mat
f papier *m* à l'émeri
e papel *m* abrasivo, papel *m* de esmeril
i carta *f* abrasiva
n schuurpapier *n*
d Polierpapier *n*, Sandpapier *n*, Schleifpapier *n*, Schmirgelpapier *n*

9 ABRASIVE WHEEL pol
f disque *m* abrasif
e rueda *f* abrasiva
i ruota *f* abrasiva
n slijpschijf
d Schleifscheibe *f*

10 ABSOLUTE-REST PRECIPITATION TANK sew
f récipient *m* de sédimentation à liquide tranquille
e tanque *m* de sedimentación de líquido en reposo
i serbatoio *m* di sedimentazione con liquido immobile
n bezinkingsvat *n* met stilstaande inhoud
d Ruhebecken *n*

11 ABSORBENT, ABSORBER abs
f absorbeur *m*
e absorbente *m*, medio *m* absorbente
i sostanza *f* assorbente
n absorbens *n*, absorberend medium *n*
d Absorptionsmittel *n*

12 ABSORBENT FABRIC, ABSORPTION FABRIC abs,mat
f tissu *m* absorbant
e tela *f* de absorción
i tessuto *m* assorbente
n absorberend weefsel *n*
d absorbierendes Gewebe *n*

13 ABSORBENT PAPER, ABSORPTION PAPER abs,mat
f papier *m* absorbant
e papel *m* secante
i carta *f* assorbente
n absorberend papier *n*
d absorbierendes Papier *n*

14 ABSORPTIOMETER abs,ins
f absorptiomètre *m*, mesureur *m* d'absorption
e medidor *m* de absorción
i absorciometro *m*, misuratore *m* d'assorbimento
n absorptiemeter
d Absorptionsmesser *m*, Absorptionsmessgerät *n*

15 ABSORPTION APPARATUS abs,ana,lab
f appareil *m* d'absorption
e aparato *m* de absorción
i apparecchio *m* d'assorbimento
n absorptietoestel *n*
d Absorptionsgerät *n*

16 ABSORPTION BOTTLE, ABSORPTION FLASK abs
f bouteille *f* d'absorption
e frasco *m* de absorción
i bottiglia *f* d'assorbimento
n absorptiefles
d Absorptionsflasche *f*

17 ABSORPTION BULB abs
f ballon *m* d'absorption
e bola *f* de absorción
i pallone *f* d'assorbimento
n absorptieballon, absorptiereservoir *n*
d Absorptionskugel *f*, Absorptionsküvette *f*

18 ABSORPTION CELL spe
f cellule *f* d'absorption
e cuba *f* de absorción
i vaschetto *m* d'assorbimento
n absorptiecel
d Absorptionsküvette *f*

19 ABSORPTION COLUMN, ABSORPTION TOWER abs
f colonne *f* d'absorption, tour *f* d'absorption
e columna *f* de absorción, torre *f* de absorción
i colonna *f* d'assorbimento, torre *f* d'assorbimento
n absorptiekolom, absorptietoren
d Absorptionskolonne *f*, Absorptionssäule *f*, Absorptionsturm *m*

20 ABSORPTION COLUMN PLATE, ABSORPTION TOWER PLATE abs
f plateau *m* de colonne d'absorption, plateau *m* de tour d'absorption
e plato *m* de columna de absorción, platillo *m* de torre de absorción
i piatto *m* di colonna d'assorbimento, placca *f* di torre d'assorbimento
n absorptiekolombodem, absorptiekolomschotel
d Absorptionskolonnenboden *m*, Passette *f*

21 ABSORPTION HYGROMETER ins
f hygromètre *m* à absorption
e higrómetro *m* de absorción
i igrometro *m* ad assorbimento
n absorptiehygrometer
d Absorptionshygrometer *n*

22 ABSORPTION JAR lab
f verre *m* d'absorption
e tarro *m* de absorción
i giarra *f* d'assorbimento
n absorptieglas *n*
d Absorptionsglas *n*

23 ABSORPTION LAGOON iwt
f bassin *m* à absorption
e pantano *m* de absorción
i bacino *m* d'assorbimento
n afvalwaterbassin *n*, afvalwaterpoel
d Abwasserteich *m*

24 ABSORPTION PIPETTE ana,lab
f pipette *f* d'absorption
e pipeta *f* de absorción
i pipetta *f* d'assorbimento
n absorptiepipet
d Absorptionspipette *f*

25 ABSORPTION PLANT abs
f installation *f* d'absorption
e equipo *m* de absorción
i impianto *m* d'assorbimento
n absorptie-installatie
d Absorptionsanlage *f*

26 ABSORPTION REFRIGERATING MACHINE abs,clo
f machine *f* frigorifique à absorption
e máquina *f* refrigeradora de absorción
i macchina *f* frigorifera ad assorbimento

n absorptiekoelmachine
d Absorptionskühlmaschine *f*

27 ABSORPTION REFRIGERATOR abs,clo
f frigidaire *m* à absorption
e refrigerador *m* de absorción
i frigorifero *m* ad assorbimento
n absorptiekoelkast
d Absorptionskühlschrank *m*

28 ABSORPTION SEPARATOR abs
f séparateur *m* à absorption
e separador *m* de absorción
i separatore *m* ad assorbimento
n absorptiescheider
d Absorptionsscheider *m*, Absorptionsseparator *m*

29 ABSORPTION TRAP abs,lab
f piège *m* à absorption
e trampa *f* de absorción
i trappola *f* ad assorbimento
n absorptie-opzet
d Absorptionsaufsatz *m*

30 ABSORPTION TUBE ana
f tube *m* absorbeur
e tubo *m* absorbente, tubo *m* de absorción
i tubo *m* d'assorbimento
n absorptiebuis
d Absorptionsröhre *f*

31 ACCELERATOR BATH pho
f bain *m* accélérateur
e baño *m* acelerador
i bagno *m* accelerante
n versnellingsbad *n*
d Beschleunigungsbad *n*

32 ACCELERATING BELT dec
f ruban *m* accélérateur
e cinta *f* aceleradora
i nastro *m* acceleratore
n versnellingsband
d Beschleunigungsband *n*

33 ACCESSORY APPARATUS aap
f appareils *pl* accessoires
e aditamientos *pl*, aparatos *pl* accesorios
i equipaggiamento *m* accessorio
n bijbehorende apparaten *pl*
d Zusatzgeräte *pl*

34 ACCUMULATOR, ENERGY ACCUMULATOR pre
f accumulateur *m* d'énergie
e acumulador *m* de energía
i accumulatore *m* d'energia
n energieaccumulator
d Kraftspeicher *m*

35 ACCUMULATOR, SECONDARY CELL, STORAGE CELL elc
f accumulateur *m*
e acumulador *m*
i accumulatore *m*
n accu, accumulator
d Akkumulator *m*, Sammler *m*

36 ACCUMULATOR COUPLE elc
f couple *m* d'accumulateur
e par *m* de acumulador
i coppia *f* d'accumulatore
n accuplatenpaar *n*
d Akkumulatorplattenpaar *n*

37 ACCUMULATOR ELEMENT elc
f bloc *m* de plaques d'accumulateur
e paquete *m* de placas de acumulador
i elementi *pl* d'accumulatore
n accuplatenblok *n*
d Akkumulatorplattenpaket *n*

38 ACCUMULATOR FRAME elc
f cadre *m* d'accumulateur
e cuadro *m* de acumulador
i telaio *m* d'accumulatore
n accuplaatraam *n*
d Akkumulatorrahmen *m*

39 ACCUMULATOR FRAME PLATE elc
f plaque *f* à cadre d'accumulateur
e placa *f* de cuadro de acumulador
i piastra *f* a telaio d'accumulatore
n accumassaplaat
d Rahmenplatte *f*

40 ACCUMULATOR GROUP elc
f faisceau *m* d'accumulateur, groupe *m* d'accumulateur
e grupo *m* de acumulador
i gruppo *m* d'accumulatore
n accuplatenstel *n*
d Akkumulatorplattensatz *m*

41 ACCUMULATOR PLATE elc
f plaque *f* d'accumulateur
e placa *f* de acumulador
i piastra *f* d'accumulatore
n accuplaat
d Akkumulatorplatte *f*

42 ACCUMULATOR SEPARATOR elc
f séparateur *m* d'accumulateur
e separador *m* de acumulador

i separatore *m* d'accumulatore
n accuseparator
d Akkumulatorscheider *m*

43 ACCUMULATOR TANK, STORAGE TANK — sto
f cuve *f* collectrice, récipient *m* d'emmagasinage
e tanque *m* colector
i serbatoio *m* polmone
n verzameltank
d Sammelbehälter *m*

44 ACETIFIER — aam
f récipient *m* acétificateur
e cuba *f* acetificadora
i serbatoio *m* acetificatore
n snelazijnkuip
d Schnellsäurer *m*

45 ACETIMETER, ACETOMETER — ana,ins
f acétimètre *m*, burette *f* acétimétrique
e acetímetro *m*
i acetimetro *m*
n acetometer, azijnmeter
d Acetometer *n*, Essigmesser *m*

46 ACETYLATION FLASK — lab
f matras *m* d'acétylation
e matraz *m* de acetilación
i matraccio *m* d'acetilazione
n acetyleerkolf
d Azetylierkolben *m*

47 ACETYLENE BLOWPIPE, ACETYLENE BURNER — bur,lab
f bec *m* à acétylène, chalumeau *m* à acétylène
e mechero *m* para acetileno, soplete *m* de acetileno
i becco *m* da acetilene, cannello *m* acetilenico
n acetyleenbrander
d Azetylenbrenner *m*

48 ACETYLENE CYLINDER — tra
f cylindre *m* à acétylène
e botella *f* de acetileno, cilindro *m* para acetileno
i bombola *f* per acetilene
n acetyleencilinder
d Azetylenzylinder *m*

49 ACETYLENE GENERATOR — aap
f générateur *m* d'acétylène
e generador *m* de acetileno
i generatore *m* d'acetilene
n acetyleengenerator
d Azetylenentwickler *m*, Azetylengenerator *m*

50 ACID AGEING ROOM (GB), ACID AGING ROOM (US) — tex
f chambre *f* de traitement par vapeur acide
e cámara *f* de vaporizado húmedo ácido
i camera *f* di trattamento per vapore acido
n zure-stoombehandelingsruimte
d Säuredampfungsraum *m*

51 ACID ANNEALING BATH — met
f bain *m* acide précédant le recuit
e baño *m* ácido antes del recocido
i bagno *m* precedente alla ricottura
n zuurbad *n* voor het gloeien
d Säurebad *n* vor dem Glühen

52 ACID BATH — elc
f bain *m* acide, bain *m* d'acide
e baño *m* ácido
i bagno *m* acido, bagno *m* d'acido
n zuurbad *n*
d Säurebad *n*

53 ACID CARBOY — tra
f bonbonne *f* à acide
e bombona *f* para ácidos, damajuana *f*
i damigiana *f* per acidi
n mandefles voor zuren
d Säureballon *m*

54 ACID COLUMN, ACID TOWER — abs
f colonne *f* d'acides, tour *f* d'acides
e columna *f* de ácidos, torre *f* de ácidos
i colonna *f* d'acidi, torre *f* d'acidi
n zuurkolom, zuurtoren
d Säurekolonne *f*, Säureturm *m*

55 ACID CONTAINER — aap
f récipient *m* d'acide
e depósito *m* de ácido, recipiente *m* para ácido
i serbatoio *m* d'acido
n zuurvat *n*
d Säuregefäss *n*

56 ACID COPPER PLATING BATH — elc
f bain *m* de cuivrage acide
e baño *m* de cobreado ácido
i bagno *m* di ramatura acido
n zuur koperbad *n*
d saures Kupferbad *n*

57 ACID EGG (US), cor
ACID ELEVATOR (US),
BLOW CASE (GB),
MONTEJUS
f monte-jus *m*
e elevador *m* de presión, montaácidos *m*, montajugos *m*
i montaacidi *m*, montaliquidi *m*
n drukvat *n*
d Druckbirne *f*, Druckfass *n*, Montejus *m*

58 ACID EMBOSSING BATH, elc
ACID ETCHING BATH
f bain *m* de décapage acide
e baño *m* de decapaje ácido
i bagno *m* per decapaggio acido
n zuur beitsbad *n*
d Säurebeizbad *n*

59 ACID FEEDER aap,cpt
f tube *m* d'amenée d'acide
e alimentador *m* de ácido
i alimentatore *m* d'acido
n zuurtoevoerbuis
d Säurezufuhr *f*

60 ACID FIXING BATH pho
f bain *m* de fixage acide
e baño *m* de fijador ácido
i bagno *m* di fissazione acido
n zuurfixeerbad *n*
d Säurefixierbad *n*

61 ACID FROSTING PLANT coa,gla
f installation *f* de matage à l'acide
e equipo *m* de esmerilado al ácido
i impianto *m* di smerigliatura all'acido
n zuurmatteerinstallatie
d Säuremattieranlage *f*

62 ACID FULLING MACHINE, tex
ACID MILLING MACHINE
f fouleuse *f* à l'acide
e batanadora *f* al ácido, máquina *f* de batanado ácido
i mulino *m* da folletare all'acido
n zure volmolen
d saure Walke *f*

63 ACID FURNACE BOTTOM, htt
ACID HEARTH
f sole *f* acide
e solera *f* ácida
i suola *f* acida
n zure ovenbodem
d saure Sohle *f*

64 ACID GATE VALVE (US), aap,cpt
ACID SLUICE VALVE (GB)
f soupape *f* à vanne pour acides
e válvula *f* corredera para ácidos
i valvola *f* a saracinesca per acidi
n zuurafsluiter
d Säureschieber *m*

65 ACID HOSE tra
f tube *m* flexible pour acides
e tubo *m* flexible para ácidos
i tubo *m* flessibile per acidi
n zuurslang
d Säureschlauch *m*

66 ACID LINING met
f garnissage *m* acide
e revestimiento *m* ácido
i rivestimento *m* acido
n zure voering
d saures Futter *n*

67 ACID MEASURING TANK ins
f récipient *m* de mesure pour acides
e tanque *m* medidor de ácido
i serbatoio *m* da misura d'acido
n zuurmeetvat *n*
d Säuremessgefäss *n*

68 ACID PICKLING TANK elc
f récipient *m* de décapage à l'acide
e tanque *m* para decapaje ácido
i vasca *f* di decapaggio con acido
n zuurbeitsvat *n*
d Säurebeizgefäss *n*

69 ACID PITCHER lab
f broc *m* pour acides
e garro *m* para ácidos
i bricco *m* per acidi
n zuurschenkkan
d Säurekanne *f*

70 ACID POLISHING PLANT gla
f installation *f* de polissage à l'acide
e equipo *m* de pulimento al ácido
i impianto *m* di pulitura all'acido
n zuurpolijstinstallatie
d Säurepolieranlage *f*

71 ACID PUMP aap
f pompe *f* à acides
e bomba *f* para ácidos
i pompa *f* d'acidi
n zuurpomp
d Säurepumpe *f*

72 ACID RECOVERY TOWER act
f colonne *f* de récupération d'acides
e columna *f* de recuperación de ácidos
i colonna *f* da ricupero d'acidi
n zuurterugwinningskolom
d Säurerückgewinnungssäule *f*, Vitriolkolonne *f*

73 ACID REFRACTORY mat
f réfractaire *m* acide
e refractario *m* ácido
i refrattario *m* acido
n zuur vuurvast materiaal *n*
d saurer feuerfester Stoff *m*

74 ACID RESISTING STONEWARE, ACIDPROOF STONEWARE mat
f grès *m* résistant aux acides
e gres *m* ácidorresistente
i gres *m* resistente all'acido
n zuurvast gres *n*
d säurefestes Steingut *n*

75 ACID ROOM, ACID STORE sto
f dépôt *m* d'acides, parc *m* d'acides
e almacén *m* de ácidos, pañol *m* de ácidos
i deposito *m* d'acidi
n zuurkamer
d Säureraum *m*

76 ACID SPOTTING TESTER tex
f appareil *m* d'essai de résistance à l'acide
e probador *m* de resistencia al ácido
i collaudatore *m* di resistenza all'acido
n zuurbestendigheidsbeproevingsapparaat *n*
d Säurefestigkeitsprüfgerät *n*

77 ACID STOP BATH pho
f bain *m* d'arrêt acide
e baño *m* de parada
i bagno *m* d'arresto acido, bagno *m* interruttore
n onderbreekbad *n*
d saures Unterbrechungsbad *n*

78 ACID STORAGE TANK, ACID TANK sto
f réservoir *m* d'emmagasinage d'acides
e tanque *m* de almacenamiento de ácidos
i serbatoio *m* da immagazzinamento d'acidi
n zuuropslagtank
d Säurebehälter *m*, Säurezisterne *f*

79 ACID TINNING BATH elc
f bain *m* d'étamage acide
e baño *m* de estañado ácido
i bagno *m* di stagnatura acido
n zuur tinbad *n*
d saures Zinnbad *n*

80 ACIDIMETER, ACIDOMETER ana,ins
f acidimètre *m*, acidomètre *m*, pèse-acide *m*
e acidímetro *m*, pesa-ácidos *m*
i acidimetro *m*, misuratore *m* dell'acidità
n acidimeter, zuurmeter
d Acidimeter *n*, Säuremesser *m*

81 ACIDPROOF MATERIAL mat
f matière *f* résistante aux acides
e material *m* ácidorresistente
i materiale *m* a prova d'acido, materiale *m* resistente all'acido
n zuurvast materiaal *n*
d säurebeständiges Material *n*, säurefestes Material *n*

82 ACIDPROOF METAL SHIELD aap
f écran *m* métallique résistant aux acides
e cubierta *f* metálica ácidorresistente
i schermo *m* metallico resistente all'acido
n zuurvast metaalscherm *n*
d säurebeständiger Metallschirm *m*

83 ACIDPROOF TRAVELLING BAND tra
f ruban *m* transbordeur résistant aux acides, ruban *m* transporteur résistant aux acides
e cinta *f* transportadora ácidorresistente
i nastro *m* trasportatore resistente all'acido
n zuurvaste transportband
d säurebeständiges Förderband *n*, säurebeständiges Transportband *n*

84 ACIDULATED BATH elc
f bain *m* acidulé
e baño *m* acidulado
i bagno *m* acidulato
n aangezuurd bad *n*
d angesäuertes Bad *n*

85 ACIERAGE BATH, coa
IRON PLATING BATH,
STEELING BATH
f bain *m* d'aciérage
e baño *m* de aceración, baño *m* de hierrado
i bagno *m* d'acciaiatura
n verijzeringsbad *n*, verstalingsbad *n*
d Stählungsbad *n*, Verstählungsbad *n*

86 ACROMETER, ins
OLEOMETER
f oléomètre *m*
e oleómetro *m*
i oleometro *m*
n oleometer
d Ölmesser *m*, Ölwaage *f*

87 ACTIVATED CARBON FILTER cle,fil
f filtre *m* à charbon actif
e filtro *m* de carbón activo
i filtro *m* a carbone attivo
n actieve-koolfilter *n*
d Aktivfilter *m*, Aktivkohlenfilter *m*

88 ACTIVATED SINTERING OVEN elc
f four *m* de frittage activé
e horno *m* de sinterización activada
i forno *m* di sinterizzazione attivata
n oven voor geactiveerd sinteren
d Ofen *m* für aktiviertes Sintern

89 ACTIVATED SLUDGE INSTALLATION iwt
f établissement *m* de traitement de boue activée
e instalación *f* de tratamiento de fango activado
i impianto *m* di trattamento di fanghiglia attivata
n installatie voor geactiveerd slib
d Belebtschlammanlage *f*

90 ACTIVATED SLUDGE TANK iwt
f réservoir *m* de boue activée
e tanque *m* de fango activado
i serbatoio *m* di fanghiglia attivata
n geactiveerd-slibtank
d Belebtschlammzisterne *f*

91 ACTIVE CAP dis
f calotte *f* active
e platillo *m* eficaz
i campanella *f* attiva
n werkzame klok
d tätige Glocke *f*

92 ACTIVE LAUNDRY nue
f laverie *f* de décontamination
e lavandería *f* de decontaminación
i lavanderia *f* di decontaminazione
n ontsmettingswasserij
d Entseuchungswäscherei *f*

93 ACTIVE PLATE abs
f plateau *m* actif
e plato *m* activo, platillo *m* activo
i piatto *m* attivo, placca *f* attiva
n werkzame plaat, werkzame schotel
d praktischer Boden *m*

94 ADAPTER dis,lab
f allonge *f*
e alargadera *f*
i allunga *f*
n allonge
d Destilliervorstoss *m*

95 ADAPTER, cpt
ADAPTOR,
ADOPTER
f élément *m* de centrage, manchon *m* de liaison, pièce *f* de liaison
e adaptador *m*, pieza *f* de ajuste, pieza *f* de conexión, racor *m*
i accessorio *m* d'aggiustamento, elemento *m* di centratura, manicotto *m* di riduzione, raccordo *m*
n passtuk *n*, verloopstuk *n*
d Einsatzstück *n*, Passstück *n*, Reduzierstück *n*

96 ADAPTER PLATE cpt
f plaque *f* de montage, plateau *m* porte-moule
e chapa *f* adaptadora, placa *f* de fijación, plato *m* adaptador
i piastra *f* d'adattamento, piastra *f* di base
n matrijsmontageplaat
d Werkzeugaufspannplatte *f*

97 ADAPTER RING (US), RETAINER RING (GB) cpt
f bague *f* pour douille
e anillo *m* adaptador
i anello *m* di riduzione, riduttore *m*
n pasring
d Haltering *m*, Passring *m*

98 ADAPTER WITH SINTER chr
f raccord *m* avec disque fritté
e racor *m* con disco sinterizado
i raccordo *m* con disco sinterizzato
n verloopstuk *n* met filterschijf
d Vorstoss *m* mit Filterplatte

99 ADHERENCE METER, ADHEROMETER, INKOMETER ins
f adhérencemètre *m*
e adherencímetro *m*, adherómetro *m*
i aderometro *m*
n adhesiemeter
d Adhäsionsmesser *m*

100 ADIABATIC CALORIMETER ins
f calorimètre *m* adiabatique
e calorímetro *m* adiabático
i calorimetro *m* adiabatico
n adiabatische calorimeter
d adiabatisches Kalorimeter *n*

101 ADJUSTABLE PITCH PROPELLER mix
f hélice *f* à pas variable
e hélice *f* de paso regulable
i elica *f* a passo variabile
n propeller met verstelbare spoed
d Propeller *m* mit einstellbarer Ganghöhe

102 ADJUSTABLE RANGE THERMOMETER ins
f thermomètre *m* à ajustement réglable
e termómetro *m* de ajuste regulable
i termometro *m* ad aggiustamento regolabile
n instelbare thermometer
d Einstellthermometer *n*

103 ADJUSTING SCREW cpt
f vis *f* de réglage
e tornillo *m* de ajuste, tornillo *m* regulador
i vite *f* di registrazione, vite *f* di regolazione
n regelschroef, stelschroef
d Adjustierschraube *f*, Regulierschraube *f*, Stellschraube *f*

104 ADMISSION PIPE cpt
f tube *m* d'admission
e tubo *m* de admisión
i tubo *m* d'ammissione
n inlaatpijp
d Einlassrohr *n*

105 ADSORPTION APPARATUS abs
f appareil *m* d'adsorption
e aparato *m* de adsorción
i apparecchio *m* d'adsorbimento
n adsorptietoestel *n*
d Adsorptionsgerät *n*

106 ADSORPTION COLUMN, ADSORPTION TOWER abs
f colonne *f* d'adsorption, tour *f* d'adsorption
e columna *f* de adsorción, torre *f* de adsorción
i colonna *f* d'adsorbimento, torre *f* d'adsorbimento
n adsorptiekolom, adsorptietoren
d Adsorptionskolonne *f*, Adsorptionssäule *f*, Adsorptionsturm *m*

107 AERATION ROOM iwt
f chambre *f* d'aérage, chambre *f* d'aération
e cámara *f* de aireación
i camera *f* d'aerazione
n aëratieruimte
d Belüftungsraum *m*

108 AERATION TANK iwt
f tank *m* d'aérage, tank *m* d'aération
e tanque *m* de aireación
i serbatoio *m* d'aerazione
n aëratietank
d Belüftungstank *m*

109 AERATION VALVE cpt
f clapet *m* d'aérage, clapet *m* d'aération, clapet *m* de ventilation
e válvula *f* de aireación, válvula *f* de ventilación
i valvola *f* di ventilazione
n ontluchtingsklep
d Belüftungsklappe *f*

110 AERATOR fou
f diviseur-aérateur *m*
e desintegrador *m* por aire

i disintegratore-aeratore *m*
n zandwolf
d Sandschleuder *f*

111 AERATOR hyg
f installation *f* d'aérage, installation *f* d'aération
e aireador *m*
i aeratore *m*
n ontluchter
d Belüftungsanlage *f*, Lüfter *m*

112 AERATOR PIPE iwt
f conduit *m* d'aération, tube *m* d'aération, tuyau *m* d'aération
e tubo *m* de aireación
i tubo *m* d'aerazione
n aëratiepijp
d Belüftungsrohr *n*

113 AEROGRAPH cer,tex
f pinceau *m* vaporisateur
e aerógrafo *m*, pistola *f* para pintar
i aerografo *m*, spruzzatore *m* per colori
n verfpatroonspuit
d Färbpistole *f*

114 AEROMETER ins
f aéromètre *m*
e aerómetro *m*
i aerometro *m*
n aërometer, dampdichtheidsmeter
d Aerometer *n*, Luftdichtemesser *m*, Luftmesser *m*

115 AEROSOL PACKING chr
f cartouche *m* d'aérosol
e cartucho *m* de aerosol
i cartoccio *m* d'aerosole
n aërosolpakking
d Aerosolpackung *f*

116 AFFINATION CENTRIFUGE cen,sin
f centrifugeur *m* d'affination
e centrífuga *f* de afinación
i centrifuga *f* d'affinazione
n raffineercentrifuge
d Affinationszentrifuge *f*, Raffinierzentrifuge *f*

117 AFTERCHROME DYEING BATH, tex
TOP CHROME BATH
f bain *m* de teinture postérieure au chrome
e baño *m* de postmordentado crómico
i bagno *m* di tintura posteriore al cromo
n nachromeerbad *n*
d Nachchromierungsbad *n*

118 AFTERCONDENSER pre
f postcondensateur *m*
e postcondensador *m*
i condensatore *m* posteriore
n nacondensor
d Nachkondensator *m*

119 AFTERCOOLER, clo
FINAL COOLER
f réfrigérateur *m* final
e postrefrigerador *m*
i frigorifero *m* posteriore, postrefrigeratore *m*
n nakoeler
d Nachkühler *m*

120 AFTERCOPPERING BATH tex
f bain *m* de traitement postérieur au sel de cuivre
e baño *m* de postratamiento con sal de cobre
i bagno *m* di trattamento posteriore con sale di rame
n nabehandelingsbad *n* met koperzout
d Nachbehandlungsbad *n* mit Kupfersalz

121 AFTERFRACTIONATING act,dis
COLUMN,
AFTERFRACTIONATING TOWER
f colonne *f* de fractionnement postérieur, tour *f* de fractionnement postérieur
e columna *f* de fraccionamiento posterior, torre *f* de fraccionamiento secundario
i colonna *f* di frazionamento posteriore, torre *f* di postfrazionamento
n nafractioneerkolom, nafractioneertoren
d Nachfraktionierungskolonne *f*, Nachfraktionierungsturm *m*

122 AFTERRIPENING BATH, pho
SECOND RIPENING BATH
f bain *m* de maturation en refonte
e baño *m* de maturación química
i bagno *m* di maturazione chimica
n chemisch rijpingsbad *n*
d chemisches Reifungsbad *n*, Nachreifungsbad *n*

123 AFTERTREATMENT ROOM tex
f chambre *f* de traitement postérieur
e cámara *f* de postratamiento
i camera *f* di trattamento posteriore

n nabehandelingskamer
d Nachbehandlungsraum *m*

124 AGARSLANT TUBE bac,lab
f tube *m* à culture à section ovale
e tubo *m* de cultivo de sección óvala
i tubo *m* di cultura a sezione ovale
n ovale cultuurbuis
d ovales Agarrohr *n*

125 AGATE MORTAR lab
f mortier *m* en agate
e mortero *m* de ágata
i mortaio *m* d'agata
n agaatmortier, agaatvijzel
d Achatmörser *m*

126 AGE-HARDENING PLANT, DISPERSION HARDENING PLANT, PRECIPITATION HARDENING PLANT met
f installation *f* de durcissement structural
e equipo *m* de endurecimiento estructural
i impianto *m* d'incrudimento strutturale
n structurele-hardingsinstallatie
d strukturelle Härtungsanlage *f*

127 AGEING OVEN (GB), AGING OVEN (US) htt
f étuve *f* de vieillissement
e estufa *f* de envejecimiento
i stufa *f* d'invecchiamento
n verouderingsstoof
d Alterungsschrank *m*

128 AGEING ROOM (GB), AGING ROOM (US) tex
f chambre *f* de traitement par vapeur
e cámara *f* de vaporizado húmedo
i camera *f* di trattamento al vapore
n stoombehandelingskamer
d Dampfungsraum *m*

129 AGEING TANK (GB), AGING TANK (US), SOURING TANK cer
f récipient *m* de mûrissage
e tanque *m* de curar
i serbatoio *m* di maturazione, serbatoio *m* d'omogenizzazione
n homogenisatietank
d Homogenisiergefäss *n*

130 AGITATED BATCH CRYSTALLIZER cry
f cristallisoir *m* intermittent à agitation
e cristalizador *m* agitado intermitente
i cristallizzatore *m* intermittente ad agitazione
n niet-continue kristallisator met roerwerk
d Chargenkristallisator *m* mit Rührwerk, diskontinuierlicher Rührwerkkristallisator *m*

131 AGITATED CLEANER, AGITATED CLEANING APPARATUS elc
f appareil *m* de dégraissage à agitation
e aparato *m* de desengrase con agitación
i apparecchio *m* di sgrassaggio con agitazione
n ontvettingsroerwerk *n*
d Flutreinigungsgerät *n*

132 AGITATED PAN DRIER dry
f bac *m* sécheur à agitation
e cuba *f* secadora agitada
i tino *m* essiccatore ad agitazione
n droogpan met roerwerk
d Trockenpfanne *f* mit Rührwerk

133 AGITATING ARM mix
f bras *m* agitateur
e brazo *m* agitador
i braccio *m* agitatore
n roerarm
d Rührarm *m*

134 AGITATING BRUSH mix
f goupillon *m* agitateur
e cepillo *m* agitador, cepillo *m* batidor
i pennello *m* agitatore
n roerborstel
d Rührbürste *f*

135 AGITATING PAN mix
f bac *m* d'agitation
e recipiente *m* de agitación
i marmitta *f* dell'agitatore
n mengbak, roerbak
d Rührkessel *m*, Rührpfanne *f*

136 AGITATION VAT mix
f récipient *m* d'agitation
e cuba *f* agitadora, cuba *f* de agitación, cuba *f* de agitadores

i serbatoio *m* d'agitazione
n roervat *n*
d Rührgefäss *n*

137 AGITATOR, STIRRER, STIRRING MILL mix
f agitateur *m*, cuve *f* à brasser, mélangeur *m*, pétrisseur *m*
e agitador *m*
i agitatore *m*
n roerapparaat *n*, roerder, roermachine, roerwerk *n*
d Agitator *m*, Mischer *m*, Rührer *m*, Rührwerk *n*

138 AGITATOR WITH SCRAPERS mix
f agitateur *m* à raclettes
e agitador *m* de arrastres
i agitatore *m* a raschietti
n roerapparaat *n* met schrapers
d schabendes Rührwerk *n*

139 AIR ACCUMULATOR, PRESSURE ACCUMULATOR pre
f accumulateur *m* d'air comprimé
e acumulador *m* de aire comprimido
i accumulatore *m* d'aria compressa
n persluchtaccumulator
d Drucksammler *m*

140 AIR AGITATOR mix
f agitateur *m* à air comprimé
e agitador *m* de aire comprimido
i agitatore *m* ad aria compressa
n roerwerk *n* met perslucht
d Druckluftrührwerk *n*

141 AIR ATOMIZING BURNER bur
f chalumeau *m* pulvérisateur à air comprimé
e mechero *m* de pulverización de aire comprimido
i bruciatore *m* atomizzatore con aria compressa
n verstuiverbrander met perslucht
d Zerstäuberbrenner *m* mit Pressluft

142 AIR-BALLASTED ACCUMULATOR pla,pre
f accumulateur *m* d'énergie à charge d'air
e acumulador *m* aerocargado
i accumulatore *m* d'energia con carico d'aria
n energieaccumulator met luchtbelasting
d Kraftspeicher *m* mit Luftbelastung

143 AIR BATH clo
f bain *m* d'air
e baño *m* de aire, estufa *f* de aire
i bagno *m* d'aria
n luchtbad *n*
d Luftbad *n*

144 AIR BLAST met,pre
f souffleuse *f*
e soplante *m*
i soffieria *f*
n luchtblaasinrichting
d Gebläse *n*

145 AIR BLEED HOLE cpt
f ventouse *f*
e agujero *m* de toma de aire
i orificio *m* di presa d'aria
n ontluchtingsopening
d Entlüftungsöffnung *f*

146 AIR BLEEDING VALVE cpt
f soupape *f* à désaération
e válvula *f* de escape del aire
i valvola *f* di spurgo, valvola *f* per evacuazione dell'aria
n ontluchtingsklep
d Entlüftungsventil *n*

147 AIR-BORNE SEALING GUN, HOT PATCHING GUN, PATCHING GUN cer,htt
f pistole *f* à fermer des pores
e pistola *f* para cerrar poros
i pistola *f* da chiudere pori
n poriëndichtend pistool *n*
d porendichtende Pistole *f*

148 AIR BRUSH (GB), AIR JET (US), AIR KNIFE (GB) pre
f lame *f* d'air, racle *m* à air comprimé
e pulverizador *m* de aire comprimido
i lama *f* d'aria
n luchtmes *n*
d Luftbürste *f*, Luftpinsel *m*

149 AIR CELL (US), AIR DEPOLARIZED CELL (GB) elc
f pile *f* à dépolarisation par air
e pila *f* de despolarización por aire
i pila *f* depolarizzata ad aria
n luchtzuurstofcel, luchtzuurstofelement *n*
d Luftsauerstoffelement *n*

150 AIR CHAMBER bre
f réservoir *m* à air
e depósito *m* de aire
i serbatoio *m* d'aria
n windketel
d Windkessel *m*

151 AIR CIRCULATION AUTOCLAVE pre
f autoclave *m* à circulation d'air
e autoclave *f* con circulación de aire
i autoclave *f* a circolazione d'aria
n autoclaaf met luchtcirculatie
d Umlaufautoklav *m*

152 AIR CLASSIFIER sor
f classeur *m* par air comprimé, trieuse *f* à air comprimé
e clasificador *m* por corriente de aire
i classificatore *m* con corrente d'aria
n persluchtsorteermachine
d Luftsetzmaschine *f*

153 AIR CLEANER, AIR FILTER cle
f filtre *m* d'air
e filtro *m* de aire, purificador *m* de aire
i filtro *m* d'aria
n luchtfilter *n*
d Luftfilter *m*, Luftreiniger *m*

154 AIR COCK cpt
f robinet *m* de dégagement d'air
e grifo *m* de aire, llave *f* de aire
i rubinetto *m* per l'aria
n ontluchtingskraan
d Abblasehahn *m*, Entlüftungshahn *m*

155 AIR-COMPRESSION PUMP pre
f pompe *f* à compression d'air
e bomba *f* de compresión de aire
i pompa *f* a compressione d'aria
n luchtcompressiepomp
d Luftverdichtungspumpe *f*

156 AIR COMPRESSOR pre
f compresseur *m* d'air
e compresor *m* de aire
i compressore *m* d'aria
n luchtcompressor
d Luftkompressor *m*, Luftverdichter *m*

157 AIR CONDENSER pre
f condensateur *m* d'air
e condensador *m* de enfriamiento por aire
i condensatore *m* ad aria
n luchtcondenser
d Luftkondensator *m*, Luftverdichter *m*

158 AIR CONDITIONER hyg
f appareil *m* de climatisation, appareil *m* de conditionnement de l'air
e acondicionador *m* del aire, aparato *m* climatizador
i condizionatore *m* dell'aria
n luchtconditioneerapparaat *n*, luchtverzorgingsapparaat *n*
d Klimatisierungsgerät *n*

159 AIR CONDITIONING CABINET hyg
f chambre *f* de climatisation, chambre *f* de conditionnement de l'air
e cabina *f* del aire acondicionado
i camera *f* di condizionamento dell'aria
n conditioneerkast, kast met regelbare temperatuur en vochtigheid
d Klimaschrank *m*

160 AIR CONDITIONING PLANT hyg
f établissement *m* de climatisation, établissement *m* de conditionnement de l'air
e equipo *m* climatizadora
i impianto *m* per condizionamento dell'aria
n luchtconditioneerinstallatie, luchtverzorgingsinstallatie
d Klimaanlage *f*, Klimatisierungsanlage *f*

161 AIR CONDUIT cpt
f conduite *f* d'air
e conducto *m* aerífero, conducto *m* de aire
i condotta *f* d'aria
n luchtleiding
d Luftleitung *f*

162 AIR COOLANT clo
f air *m* de refroidissement
e aire *m* de enfriamiento
i aria *f* di raffreddamento
n koellucht
d Kühlluft *f*

163 AIR-COOLED HEAT EXCHANGER clo
f échangeur *m* de chaleur à refroidissement par air
e cambiador *m* de calor enfriado al aire
i scambiatore *m* di calore con raffreddamento all'aria
n met lucht gekoelde warmte-uitwisselaar

d luftgekühlter Wärmeaustauscher *m*

164 AIR-COOLED OIL VAPOR DIFFUSION PUMP (US), AIR-COOLED OIL VAPOUR DIFFUSION PUMP (GB) vac
f pompe *f* à diffusion d'huile à refroidissement par air
e bomba *f* de difusión de aceite enfriada al aire
i pompa *f* a diffusione d'olio con raffreddamento all'aria
n met lucht gekoelde oliediffusiepomp
d luftgekühlte Öldiffusionspumpe *f*

165 AIR CUSHION sap
f coussin *m* d'air, matelas *m* d'air
e amortiguador *m* de aire
i cuscino *m* pneumatico
n luchtkussen *n*
d Luftkissen *n*

166 AIR DIFFUSER, DIFFUSER PLATE, FILTER SLAB iwt
f dalle *f* poreuse, diffuseur *m* d'air, plaque *f* diffuseuse, plaque *f* filtrante
e difusor *m* de aire, placa *f* filtrante
i diffusore *m* d'aria, piastra *f* filtrante, piastra *f* porosa
n filterplaat, luchtverdeelplaat
d Belüfterplatte *f*, Filterplatte *f*, Luftverteiler *m*, Verteilerplatte *f*

167 AIR DRAIN sap
f conduite *f* de dégagement d'air
e conducto *m* de escape de gases, conducto *m* de ventilación
i intercapedine *f* d'aerazione
n ontluchtingsleiding
d Zugkanal *m*

168 AIR DRIER dry
f sécheur *m* de l'air
e secador *m* del aire
i essiccatore *m* dell'aria
n luchtdroogapparaat *n*
d Luftdarre *f*, Lufttrockner *m*

169 AIR DUCT bre,cpt
f tuyau *m* à air
e conducto *m* de aire
i canale *m* per aria
n luchtkanaal *n*
d Luftkanal *m*

170 AIR ELUTRIATOR cle
f appareil *m* à élutrier l'air
e levigador *m* de aire
i elutriatore *m* dell'aria
n elueerapparaat *n*, luchtwasapparaat *n*
d Elueranlage *f*, Luftwäsche *f*

171 AIR EXHAUSTER, EXHAUSTING FAN hyg
f ventilateur *m* aspirant
e expulsor *m* de aire
i espulsore *m* d'aria
n luchtuitblaasinrichting
d Entlüftungsgebläse *n*

172 AIR EXIT, AIR OUTLET cpt
f sortie *f* d'air
e salida *f* de aire
i bocca *f* d'uscita dell'aria
n luchtuitlaat
d Luftauslass *m*

173 AIR-FILLED COOLING RING, AIR RING clo
f anneau *m* refroidisseur à air
e anillo *m* refrigerante con aire
i anello *m* refrigerante ad aria
n koelluchtring
d Kühlluftring *m*

174 AIR FILTER bre,fil
f filtre *m* à air
e filtro *m* de aire
i filtro *m* d'aria
n luchtfilter *n*
d Luftfilter *m*

175 AIR FLOW DRIER dry
f sécheur *m* à courant d'air
e secadora *f* por corriente de aire
i essiccatore *m* per corrente d'aria
n droogapparaat *n* met luchtstroom
d Luftstromtrockner *m*

176 AIR FLOW METER ins
f débitmètre *m* pour air
e flujómetro *m* para aire
i flussometro *m* per aria
n luchtstroommeter
d Luftstrommesser *m*

177 AIR FURNACE, AIR OVEN htt
f étuve *f* à air chaud
e estufa *f* a aire caliente
i forno *m* ad aria calda

n hete-luchtoven, oven met hete-luchtverwarming
d Heissluftofen *m*

178 AIR FURNACE (US), REVERBERATORY FURNACE (GB) htt
f four *m* réverbère
e horno *m* de reverbero
i forno *m* a riverbero
n vlamoven
d Flammofen *m*

179 AIR GAS GENERATOR, DRY BLAST PRODUCER fue
f générateur *m* de gaz à l'air
e generador *m* de gas de aire
i generatore *m* di gas d'aria
n luchtgasgenerator
d Aerogengasgenerator *m*, Luftgasgenerator *m*

180 AIR HARDENING PLANT, SELF-HARDENING PLANT met
f installation *f* de trempe à l'air
e equipo *m* de temple al aire
i impianto *m* di tempera all'aria
n installatie voor harding aan de lucht
d Lufthärtungsanlage *f*

181 AIR HEATER htt
f réchauffeur *m* de l'air
e calentador *m* del aire
i aerotermo *m*, riscaldatore *m* dell'aria
n luchtverhitter
d Lufterhitzer *m*

182 AIR HOLE cpt
f évent *m*, ventouse *f*
e ventosa *f*
i vuoto *m* d'aria
n luchtgat *n*, trekgat *n*
d Luftloch *n*, Zugloch *n*

183 AIR HUMIDIFIER hyg
f appareil *m* d'humectation de l'air, appareil *m* de mouillage de l'air
e humectador *m* del aire, humidificador *m* del aire
i umidificatore *m* dell'aria
n luchtbevochtiger
d Luftanfeuchtungsgerät *n*

184 AIR HUMIDITY INDICATOR, HYGROMETER (GB), MOIST-O-GRAPH (US) ins
f hygromètre *m*
e higrómetro *m*
i igrometro *m*
n hygrometer, vochtigheidsmeter
d Hygrometer *n*, Luftfeuchtigkeitsmesser *m*

185 AIR-HYDRAULIC PRESS pre
f presse *f* aérohydraulique
e prensa *f* aerohidráulica
i pressa *f* aeroidraulica
n lucht-waterdrukpers
d lufthydraulische Presse *f*

186 AIR-INFLATED GRINDER pol
f disque *m* à polir gonflable
e disco *m* dilatable para pulir
i disco *m* gonfiabile per smerigliare
n opblaasbare polijstschijf
d aufgeblasene Polierscheibe *f*

187 AIR INLET cpt
f entrée *f* de l'air
e entrada *f* del aire, toma *f* del aire
i bocca *f* d'entrata dell'aria, bocca *f* d'immissione
n luchtinlaat
d Lufteinlass *m*

188 AIR INTAKE VALVE, AIR SNIFFLER, MUSHROOM VALVE, POPPET VALVE cpt
f reniflard *m*
e válvula *f* de entrada de aire, válvula *f* de vástago
i valvola *f* a fungo, valvola *f* d'insufflazione
n snuifklep
d Schnarchventil *n*, Schnüffelventil *n*

189 AIR JACKET clo
f chemise *f* à circulation d'air
e camisa *f* de circulación de aire
i camicia *f* d'aria
n luchtmantel
d Luftkühlmantel *m*, Luftmantel *m*

190 AIR-KNIFE COATER coa
f machine *f* à enduire avec lame d'air, machine *f* à enduire avec recle à air comprimé
e máquina *f* para recubrir con chorro de aire dosificador
i spalmatrice *f* a lama d'aria
n kootmachine met luchtmes
d Luftbürstenauftragmaschine *f*, Schlitzdüsenauftragmaschine *f*

191 AIR LIFT fue,pre
f appareil *m* de levage à air comprimé, engin *m* de levage pneumatique
e elevador *m* de agua por aire comprimido
i pompa *f* per liquidi ad aria compressa
n pneumatische hefinrichting
d Druckluftheber *m*

192 AIR-LIFT AGITATOR mix
f agitateur *m* à air comprimé
e agitador *m* con aire comprimido, agitador *m* con presión de aire
i agitatore *m* a pressione d'aria
n pneumatisch roerapparaat *n*
d pneumatisches Rührwerk *n*

193 AIR-LIFT PUMP pre
f pompe *f* élévatoire à air comprimé
e bomba *f* elevadora de aire comprimido
i pompa *f* elevatrice a pressione d'aria
n persluchtpomp, luchtlift, mammoetpomp
d Druckluftwasserheber *m*, Mammutpumpe *f*

194 AIR LIQUEFYING PLANT ctt,gas
f installation *f* de liquéfaction d'air
e equipo *m* de licuación de aire
i impianto *m* di liquefazione d'aria
n installatie voor het vloeibaar maken van lucht
d Luftverflüssigungsanlage *f*

195 AIR LOCK sap
f écluse *f* pneumatique
e esclusa *f* de aire
i camera *f* d'equilibrio
n luchtsluis
d Gasschleuse *f*, pneumatische Schleuse *f*

196 AIR MONITOR rad,sap
f moniteur *m* pour l'air
e avisón *m* del aire, monitor *m* del aire
i avvisatore *m* dell'aria, monitore *m* dell'aria, segnalatore *m* dell'aria
n luchtmonitor, luchtverklikker
d Luftmonitor *m*, Luftwarngerät *n*

197 AIR NOZZLE cpt
f buse *f* d'air, tuyau *m* d'air
e tobera *f* de aire
i spruzzatore *m* d'aria, ugello *m* d'aria
n mondstuk *n* voor perslucht
d Luftdüse *f*

198 AIR-OPERATED DIAPHRAGM VALVE cpt,sap
f soupape *f* à membrane pneumatique
e válvula *f* neumática a membrana
i valvola *f* con diaframma a comando pneumatico
n pneumatische membraanklep
d luftgesteuertes Membranventil *n*

199 AIR PREHEATER htt
f pré-réchauffeur *m* de l'air
e precalentador *m* del aire
i preriscaldatore *m* dell'aria
n luchtvoorverhitter
d Luftvorwärmer *m*

200 AIR PRESSURE REDUCER pre
f détendeur *m* de pression de l'air
e reductor *m* de presión del aire
i riduttore *m* di pressione atmosferica, riduttore *m* di pressione dell'aria
n luchtdrukreduceerventiel *n*
d Luftdruckminderer *m*, Luftdruckreduzierventil *n*

201 AIR PUMP pre
f pompe *f* à air, pompe *f* pneumatique
e bomba *f* neumática
i pompa *f* d'aria, pompa *f* pneumatica
n luchtpomp
d Luftpumpe *f*

202 AIR RECEIVER pla,pre
f réservoir *m* à air comprimé
e depósito *m* de aire comprimido
i serbatoio *m* d'aria compressa
n persluchtaccumulator
d Pressluftspeicher *m*

203 AIR RELIEF COCK, VENT COCK sap
f robinet *m* d'évacuation d'air
e grifo *m* de escape de aire
i rubinetto *m* di sfogo
n ontluchtingskraan
d Entlüftungshahn

204 AIR SEAL htt
f fermeture *f* à air
e cierre *m* con aire
i chiusura *f* ad aria
n luchtafsluiting
d Luftabdichtung *f*

205 AIR SEPARATOR sor
f tamiseur *m* à air
e separador *m* de aire
i separatore *m* ad aria
n luchtseparator
d Windsichter *m*, Windsortierer *m*

206 AIR-SWEPT BALL MILL brk
f broyeur *m* à boules avec jet d'air dépoussiéreur
e molino *m* de bolas con barrido de aire, molino *m* pulverizador con extracción por chorro de aire
i mulino *m* a palle con getto d'aria classificatore
n luchtstroomkogelmolen
d Stromsichtermühle *f*

207 AIR TEST APPARATUS tes
f appareil *m* d'essai d'étanchéité
e aparato *m* de prueba de estanquieadad
i collaudatore *m* di tenuta stagna
n luchtdichtheidsbeproevings-apparaat *n*
d Luftdichtigkeitsprüfgerät *n*

208 AIR VALVE bre,cpt
f soupape *f* à air
e válvula *f* de aire
i valvola *f* per aria
n luchtklep
d Luftventil *n*

209 AIR VENT bre,cpt
f soupape *f* d'échappement
e válvula *f* de escape
i valvola *f* di scappamento
n uitlaatklep
d Auslassventil *n*

210 AIR VENT fou
f traînée *f* d'air
e respiro *m*, salida *f* de aire
i canale *m* d'aria
n luchtafvoerkanaal *n*
d Entlüftungsnut *f*, Entlüftungs-rille *f*

211 AIRLESS ELECTROSTATIC SPRAYING APPARATUS dec
f appareil *m* de pistolage électro-statique sans air
e aparato *m* de pintura electro-stática a pistolete sin aire
i apparecchio *m* di pittura elettrostatica con vaporizzatore senza aria
n elektrostatisch luchtvrij spuit-apparaat *n*
d elektrostatischer luftfreier Spritzapparat *m*

212 ALARM FLOAT sap
f flotteur *m* d'alarme
e flotador *m* avisador
i galleggiante *m* d'allarme
n alarmvlotter
d Alarmschwimmer *m*

213 ALARM VALVE sap
f soupape *f* d'alarme
e válvula *f* de alarma, válvula *f* de seguridad
i valvola *f* d'allarme
n alarmklep, waarschuwingsklep
d Alarmklappe *f*, Warnventil *n*

214 ALARM WHISTLE sap
f sifflet *m* d'alarme
e pito *m* de alarma, silbato *m* de alarma
i fischio *m* d'allarme
n alarmfluit
d Alarmpfeife *f*

215 ALBUMINIMETER ins
f albuminimètre *m*
e albuminímetro *m*
i albuminimetro *m*
n albuminometer
d Albuminimeter *n*, Eiweiss-messer *m*

216 ALCOHOLIMETER, ALCOHOLOMETER ins
f alcoolomètre *m*
e alcoholímetro *m*
i alcoolometro *m*
n alcoholmeter
d Alkoholometer *n*

217 ALCOHOLIZATION PRESS oic
f presse *f* d'alcoolisation
e prensa *f* de alcoholización
i pressa *f* d'alcoolizzazione
n verdringpers
d Verdrängerpresse *f*

218 ALCOVE, CANAL, CHANNEL gla
f canal *m* d'alimentation
e canal *m* de alimentación

i canale *m* d'alimentazione
n aanvoerkanaal *n*
d Speisekanal *m*

219 ALDEHYDE COLUMN dis
f colonne *f* de séparation d'aldéhyde
e columna *f* de separación de aldehido
i colonna *f* separatrice d'aldeide
n aldehydeafscheidingskolom
d Aldehydabscheidekolonne *f*

220 ALIGNING TRAY chr
f gabarit *m*
e bandeja *f* de alineación, soporte *m* de alineación
i sopporto *m* d'allineamento
n werksjabloon *n*
d Arbeitsschablone *f*

221 ALIQUOT PIPETTE lab
f pipette *f* à division aliquote
e pipeta *f* de división alícuota
i pipetta *f* a divisione aliquota
n pipet met aliquote verdeling
d Pipette *f* mit Aliquot-Unterteilung

222 ALKALI HYDROMETER, ALKALIMETER, POTASH BULB ins
f alcalimètre *m*
e alcalímetro *m*, pesaálcalis *m*
i alcalimetro *m*
n alkalimeter, loogzoutmeter
d Alkalimeter *n*, Kaliapparat *m*, Laugenmesser *m*

223 ALKALI SPOTTING TESTER tex
f appareil *m* d'essai de résistance à l'alcali
e probador *m* de resistencia al alcali
i collaudatore *m* di resistenza al alcali
n alkalibestendigheidsbeproevingsapparaat *n*
d Alkalibeständigkeitsprüfgerät *n*

224 ALKALINE BATH elc
f bain *m* alcalin
e baño *m* alcalino
i bagno *m* alcalino
n alkalisch bad *n*, basisch bad *n*
d alkalisches Bad *n*, basisches Bad *n*

225 ALKALINE CLEANING PLANT, ALKALINE DEGREASING PLANT cle,elc
f établissement *m* de dégraissage alcalin
e equipo *m* de desengrase alcalino
i impianto *m* di sgrassaggio alcalino
n alkalische ontvettingsinstallatie
d Reinigungsanlage *f* mit alkalischen Lösungsmitteln

226 ALKALINE CLEANING TANK cle,elc
f récipient *m* de dégraissage alcalin
e tanque *m* para desengrase alcalino
i serbatoio *m* di sgrassaggio alcalino
n tank voor het alkalisch ontvetten
d Behälter *m* für Reinigung mit alkalischen Lösungsmitteln

227 ALKALINE IMMERSION CLEANING PLANT elc
f installation *f* de dégraissage alcalin par immersion
e equipo *m* de desengrase alcalino por inmersión
i impianto *m* di sgrassaggio alcalino per immersione
n alkalische dompelontvettingsinstallatie
d alkalische Tauchreinigungsanlage *f*

228 ALKALINE PAINT STRIPPING BATH dec
f bain *m* alcalin de décapage
e baño *m* alcalino de decapaje
i bagno *m* alcalino di rimozione della vernice
n alkalisch afbijtbad *n*
d alkalisches Entlackungsbad *n*

229 ALKALINE SPRAY CLEANING MACHINE, SPRAY WASHER elc
f installation *f* de dégraissage alcalin au jet
e equipo *m* de desengrase alcalino al chorro
i impianto *m* di sgrassaggio allo spruzzo
n alkalische spuitreinigingsmachine
d alkalische Spritzreinigungsmaschine *f*

230 ALKALINE STORAGE BATTERY elc
f accumulateur *m* alcalin
e acumulador *m* alcalino
i accumulatore *m* alcalino

n alkalische accu
d alkalischer Akkumulator *m*

231 ALKALINITY DECREASING BATH gen
f bain *m* de réduction de l'alcalinité
e baño *m* de reducción de la alcalinidad
i bagno *m* di riduzione dell'alcalinità
n alkaliniteitsverlagingsbad *n*
d Alkalinitätsherabsetzungsbad *n*

232 ALKALINITY METER ins
f alcalinimètre *m*
e alcalinímetro *m*
i misuratore *m* d'alcalinità
n alkaliniteitsmeter
d Alkalinitätsmesser *m*

233 ALKALIPROOF PAPER pap
f papier *m* résistant aux alcalis
e papel *m* alcalinorresistente
i carta *f* resistente all'alcali
n alkalivast papier *n*
d alkalibeständiges Papier *n*

234 ALKYLATOR feu
f établissement *m* d'alkylisation
e equipo *m* de alkilización
i impianto *m* alchilatore
n alkyleerinstallatie
d Alkylieranlage *f*

235 ALLOY ANODE elc
f anode *f* alliée, anode *f* d'alliage
e ánodo *m* aleado
i anodo *m* legato
n gelegeerde anode, legeringsanode
d Legierungsanode *f*

236 ALUDEL aap,dis
f aludel *m*
e vasija *f* periforme con dos cuellos
i vaso *m* a forma di pera per collegamenti in serie
n aludel
d Aludel *m*

237 ALUDEL FURNACE htt
f four *m* à aludel
e horno *m* de vasija periforme
i forno *m* con vaso a forma di pera
n aludeloven
d Doppelofen *m*

238 ALUMINA MORTAR lab
f mortier *m* en alumine
e mortero *m* de alumina
i mortaio *m* d'allumina
n aluminamortier, aluminavijzel
d Tonerdemörser *m*

239 ALUMINIUM FACTORY (GB), ALUMINUM MILL (US) met
f aluminerie *f*
e fábrica *f* de aluminio
i fabbrica *f* d'alluminio
n aluminiumfabriek
d Aluminiumwerk *n*

240 ALUMINIZED SCREEN coa
f écran *m* aluminisé
e pantalla *f* aluminiada
i schermo *m* alluminizzato
n gealuminiseerd scherm *n*
d aluminisierter Schirm *m*

241 ALUMINIZING PLANT coa
f installation *f* d'aluminisation
e equipo *m* de aluminiación, equipo *m* de aluminización
i impianto *m* d'alluminiatura
n aluminiseerinstallatie
d Aluminisieranlage *f*

242 ALUMINOTHERMIC WELDING PLANT, THERMITE WELDING PLANT wel
f installation *f* de soudure aluminothermique
e equipo *m* de soldadura aluminotérmica
i impianto *m* di saldatura alla termite
n termietlasinstallatie
d Thermitschweissanlage *f*

243 ALUNDUM FILTER CRUCIBLE htt
f creuset *m* à filtre d'alundon
e crisol *m* con filtro de alundón
i crogiolo *m* con filtro d'alundo
n kroes met alundumfilter
d Tiegel *m* mit Alundumfilter

244 AMALGAM CELL, MERCURY CELL elc
f cellule *f* au mercure
e celda *f* de mercurio
i cella *f* a mercurio
n kwikcel
d Quecksilberzelle *f*

245 AMALGAM ELECTRODE elc
f électrode *f* à amalgame
e electrodo *m* de amalgama

i elettrodo *m* ad amalgama
n amalgaamelektrode
d Amalgamelektrode *f*

246 AMALGAMATING BARREL elc,met
f récipient *m* d'amalgamation
e barril *m* de amalgamación
i serbatoio *m* d'amalgamazione
n amalgameervat *n*
d Amalgamiergefäss *n*

247 AMALGAMATING BATH elc,met
f bain *m* d'amalgamation
e baño *m* de amalgamación
i bagno *m* d'amalgamazione
n amalgameerbad *n*
d Amalgamierbad *n*

248 AMALGAMATING PAN elc,met
f cuve *f* d'amalgamation
e cuba *f* de amalgamación, cubeta *f* de amalgamación
i siviera *f* d'amalgamazione
n amalgameerpan
d Amalgamierpfanne *f*

249 AMALGAMATING PLANT elc,met
f installation *f* d'amalgamation
e equipo *m* de amalgamación
i impianto *m* d'amalgamazione
n amalgameerinstallatie
d Amalgamieranlage *f*

250 AMALGAMATOR elc,met
f machine *f* d'amalgamation
e máquina *f* de amalgamación
i macchina *f* d'amalgamazione
n amalgameermachine
d Amalgamiermaschine *f*

251 AMMONIA ABSORPTION REFRIGERATING MACHINE abs,clo
f machine *f* frigorifique à ammoniaque
e máquina *f* refrigeradora de amoníaco
i macchina *f* frigorifera ad ammoniaca
n ammoniakkoelmachine
d Ammoniakkältemaschine *f*

252 AMMONIA CIRCULATING PUMP pre
f pompe *f* à circulation de l'ammoniaque
e bomba *f* de circulación del amoníaco
i pompa *f* a circolazione dell'ammoniaca
n ammoniakcirculatiepomp
d Ammoniakumlaufpumpe *f*

253 AMMONIA COMPRESSOR pre
f compresseur *m* à ammoniaque
e compresor *m* de amoníaco
i compressore *m* per ammoniaca
n ammoniakcompressor
d Ammoniakkompressor *m*, Ammoniakverdichter *m*

254 AMMONIA EVAPORATING COIL clo,eva
f serpentin *m* de l'évaporateur d'ammoniaque
e serpentín *m* del evaporador de amoníaco
i serpentina *f* dell'evaporatore d'ammoniaca
n ammoniakdampslang
d Ammoniakdampfschlange *f*

255 AMMONIA GENERATING PLANT gas
f installation *f* de génération d'ammoniaque
e equipo *m* de generación de amoníaco
i impianto *m* di generazione d'ammoniaca
n ammoniakontwikkelings-installatie
d Ammoniakentwicklungsanlage *f*

256 AMMONIA RECOVERY PLANT gas
f installation *f* de récupération d'ammoniaque
e equipo *m* de recuperación de amoníaco
i impianto *m* di ricupero d'ammoniaca
n ammoniakterugwinningsinstal-latie
d Ammoniakrückgewinnungs-anlage *f*

257 AMMONIA SCRUBBER fue
f établissement *m* de récupération des eaux ammoniacales
e equipo *m* de recuperación de las aguas amoniacales
i impianto *m* di ricupero delle acque ammoniacali
n ammoniakwasserij
d Ammoniakwäsche *f*

258 AMMONIA STILL dis
f établissement *m* de destillation d'ammoniaque
e equipo *m* de destilación de amoníaco
i impianto *m* di distillazione d'ammoniaca

n ammoniakdestilleerinrichting
d Ammoniakabtreibeapparat *m*, Ammoniakdestillieranlage *f*

259 AMPOULE (US), AMPULLA (GB) mcy,lab
f ampoule *f*
e ampolla *f*
i ampolla *f*
n ampul, ampulla
d Ampulle *f*

260 ANAEROBE APPARATUS bac,lab
f appareil *m* pour cultures anaérobiques
e aparato *m* anaeróbico
i apparecchio *m* anerobico
n anaërobeapparaat *n*
d Anaerobenapparat *m*

261 ANAEROBIC CULTURE DISH bac,lab
f boîte *f* pour cultures anaérobiques
e fuente *f* de cultivo anaeróbico
i capsula *f* di cultura anerobica
n anaërobische cultuurschotel
d Anaerobenschale *f*

262 ANALYTICAL BALANCE ana,lab
f balance *f* analytique, balance *f* pour analyses
e balanza *f* analítica, balanza *f* de análisis, balanza *f* de precisión
i bilancia *f* analitica, bilancia *f* di precisione, bilancia *f* per analisi
n analytische balans
d Analysenwaage *f*, chemische Waage *f*

263 ANALYZER ana,lab
f analysateur *m*
e analizador *m*
i analizzatore *m*
n analysator
d Analysator *m*

264 ANCHOR PADDLE IMPELLER mix
f roue *f* motrice à palettes en forme d'ancre
e rueda *f* móvil de paletas en forma de ancla
i girante *m* a pale in forma d'ancora
n ankervormige schoepenroerder
d ankerförmige Paddelrührstange *f*

265 ANCHOR PLATE, FOUNDATION WASHER cpt
f contreplaque *f* d'ancrage, plaque *f* de fondation
e placa *f* de anclaje, placa *f* de fundación
i placca *f* d'ancoraggio, placca *f* di fondazione
n ankerplaat
d Ankerplatte *f*

266 ANEROID MANOMETER, BELLOW-TYPE GAGE (US), BELLOW-TYPE GAUGE (GB) ins
f manomètre *m* anéroïde
e manómetro *m* aneroide
i manometro *m* aneroide
n aneroïdmanometer, doosmanometer, metaalmanometer
d Aneroidmanometer *n*, Deformationsdruckmesser *m*

267 ANGLE BAR, TURNING BAR pla
f barre *f* de retournement, tringle *m* de retournement
e barra *f* de rotación
i barra *f* di rotazione, barra *f* per volgere
n omkeerrol
d Wendestange *f*

268 ANGLE HEAD, ANGULAR EXTRUDER HEAD, OBLIQUE HEAD pla
f tête *f* de boudineuse à angle, tête *f* de boudineuse oblique
e cabezal *m* de budinadora en ángulo, cabezal *m* de extrusión oblicuo
i testa *f* di trafila ad angolo, testa *f* di trafila obliqua
n dwarse spuitkop, schuine spuitkop
d Schrägspritzkopf *m*

269 ANGLE PRESS pla
f presse *f* de moulage à double eff
e prensa *f* de moldear angular
i pressa *f* di stampaggio a ginocchiera
n pers met dwarse kop
d Winkelpresse *f*

270 ANGLE PRESS WITH VERTICAL AND HORIZONTAL RAM pla
f presse *f* de moulage à double effet à tête d'équerre
e prensa *f* de moldear angular con pistón vertical y horizontal
i pressa *f* di stampaggio a ginocchiera a pistone verticale e orizzontale

n pers met dwarse kop en verticale en horizontale plunjer
d Winkelpresse *f* mit vertikalem und wagerechtem Kolben

271 ANGLE THERMOMETER ins
f thermomètre *m* coudé
e termómetro *m* acodado
i termometro *m* a gomito
n thermometer met haaks kwikvat
d Winkelthermometer *n*

272 ANGLE VALVE, cpt
CORNER VALVE, MITER VALVE
f soupape *f* d'équerre
e válvula *f* angular, válvula *f* de escuadra
i valvola *f* a disco con seda a 45°
n hoekklep
d Eckventil *n*, Winkelventil *n*

273 ANGULAR HEAD, pla
OBLIQUE HEAD
f tête *f* d'angle, tête *f* oblique
e cabezal *m* angular, cabezal *m* oblicuo
i testa *f* ad angolo, testa *f* obliqua
n dwarse kop, schuine kop
d Schrägkopf *m*

274 ANGULAR HEAD EXTRUDER pla
f boudineuse *f* à tête d'angle, boudineuse *f* à tête oblique
e budinadora *f* con cabezal angular, budinadora *f* con cabezal oblicuo
i trafila *f* con testa ad angolo, trafila *f* con testa obliqua
n spuitmachine met dwarse kop, spuitmachine met schuine kop
d Schrägkopfspritzmaschine *f*

275 ANILINE FACTORY, dye
ANILINE WORKS
f fabrique *f* d'aniline
e fábrica *f* de anilina
i fabbrica *f* d'anilina
n anilinefabriek
d Anilinfabrik *f*

276 ANILINE POINT APPARATUS ins
f appareil *m* à point d'aniline
e aparato *m* de punto de anilina
i apparecchio *m* di punto d'anilina
n anilinepuntapparaat *n*
d Anilinpunktgerät *n*

277 ANILINE POINT ins
THERMOMETER
f thermomètre *m* pour l'appareil à point d'aniline
e termómetro *m* para el aparato de punto de anilina
i termometro *m* per l'apparecchio di punto d'anilina
n anilinepuntthermometer
d Anilinpunktthermometer *n*

278 ANION EXCHANGE COLUMN elc
f colonne *f* échangeuse d'anions
e columna *f* cambiadora de aniones
i colonna *f* scambiatrice d'anioni
n anionenuitwisselingskolom
d Anionenaustauschsäule *f*

279 ANNEALING BOX, htt,met
ANNEALING CAN, ANNEALING POT
f caisse *f* de recuit
e caja *f* de recocido
i cassetta *f* di ricottura
n gloeikist
d Glühkiste *f*

280 ANNEALING FURNACE, gla
COOLING FURNACE
f four *m* de refroidissement
e horno *m* de enfriamiento
i forno *m* di raffreddamento
n koeloven
d Kühlofen *m*

281 ANNEALING FURNACE, htt
HEATING STOVE
f four *m* à recuire
e horno *m* de recocido
i forno *m* di ricottura
n gloeioven
d Glühofen *m*

282 ANNULAR BASIN, sto
ANNULAR TANK
f bassin *m* annulaire
e pantano *m* anular
i bacino *m* anulare
n ringvormig bassin *n*
d ringförmiges Becken *n*

283 ANNULAR KILN, cer
HOFFMANN KILN, RING KILN
f four *m* annulaire
e horno *m* anular
i forno *m* anulare
n ringoven
d Ringofen *m*

284 ANNULAR VALVE cpt
f clapet *m* annulaire, soupape *f* annulaire
e válvula *f* anular

i valvola *f* anulare
n ringklep
d Ringventil *n*

285 ANODE elc
f anode *f*
e ánodo *m*
i anodo *m*
n anode
d Anode *f*

286 ANODE BASKET elc
f panier *m* porte-anode
e cesta *f* soporteánodo
i paniere *m* portanodo
n anodekorf
d Anodenkorb *m*

287 ANODE BUTT elc
f reste *m* d'anode
e residuo *m* de ánodo
i residuo *m* d'anodo
n anoderest
d Anodenstumpf *m*

288 ANODE COMPARTMENT elc
f boîte *f* de l'anode
e compartimiento *m* del ánodo
i cassetta *f* dell'anodo
n anodehuis *n*
d Anodengehäuse *n*

289 ANODE FRAME elc
f bâti *m* de l'anode
e bastidor *m* del ánodo
i telaio *m* dell'anodo
n anodefreem *n*
d Anodengestell *n*

290 ANODIC CLEANING PLANT elc
f établissement *m* de dégraissage anodique
e equipo *m* de desengrase anódico
i impianto *m* di sgrassaggio anodico
n anodische ontvettingsinstallatie
d anodische Reinigungsanlage *f*

291 ANODIC CLEANING TANK elc
f récipient *m* de dégraissage anodique
e tanque *m* de desengrase anódico
i serbatoio *m* di sgrassaggio anodico
n anodische ontvettingstank
d anodischer Reinigungsbehälter *m*

292 ANODIC OXIDATION BATH, coa,elc
ANODIC TREATMENT BATH,
ANODIZING BATH
f bain *m* d'oxydation anodique
e baño *m* de anodización, baño *m* de eloxación
i bagno *m* d'ossidazione anodica
n anodiseerbad *n*, eloxeerbad *n*
d Eloxierbad *n*

293 ANODIZING RACK coa,elc
f support *m* d'anodisation
e gancho *m* de anodización, percha *f* para anodizar
i sospensione *f* per ossidazione anodica
n anodiseerrek *n*
d Eloxiergestell *n*, Eloxierrahmen *m*

294 ANODIZING TANK coa,elc
f récipient *m* d'anodisation
e tanque *m* para anodizar
i serbatoio *m* per ossidazione anodica
n anodiseervat *n*
d Eloxiergefäss *n*

295 ANTECHAMBER, fue
PRECOMBUSTION CHAMBER
f antichambre *f* de combustion
e cámara *f* de precombustión
i camera *f* di precombustione
n voorkamer
d Vorkammer *f*

296 ANTI-FOAMING SPLASH HEAD dis,lab
f ampoule *f* de garde antimousse
e bola *f* antiespuma
i ampolla *f* antischiuma
n schuimvangopzet
d Schaumfängeraufsatz *m*

297 ANTI-SHOCK MOUNTING cpt
f dispositif *m* antichoc
e dispositivo *m* amortiguador
i guarnizione *f* antiurto
n schokdempend montage-element *n*
d stossdämpfendes Montageelement

298 ANTIMONY ELECTRODE elc
f électrode *f* d'antimoine
e electrodo *m* de antimonio
i elettrodo *m* d'antimonio
n antimoonelektrode
d Antimonelektrode *f*

299 APOTHECARY BALANCE lab
f trébuchet *m*
e balanza *f* para farmacia
i bilancia *f* per farmacia

n apothekersbalans
d Apothekerwaage *f*

300 APPARATUS app
f appareil *m*, appareillage *m*
e aparato *m*, aparatos *pl*
i apparato *m*, apparecchiatura *f*
n apparaat *n*, apparatuur, toestel *n*
d Apparat *m*, Apparatur *f*

301 APPARATUS FOR COLUMN CHROMATOGRAPHY, chr
COLUMN CHROMATOGRAPHY APPARATUS
f appareil *m* pour la chromatographie sur colonne
e aparato *m* para la chromatografía sobre columna
i apparecchio *m* per la chromatografia a colonna
n zuilenchromatograaf
d Säulenchromatographiegerät *n*

302 APPLICATION PRINTING MACHINE, tex
DIRECT PRINTING MACHINE
f imprimeuse *f* directe de textile
e máquina *f* estampadora directa de tejido
i stampatrice *f* diretta di tessuto
n directe textieldrukmachine
d direkte Textildruckmaschine *f*

303 APPLICATIONS LABORATORY lab
f laboratoire *m* de réalisation
e laboratorio *m* de aplicaciones
i laboratorio *m* per applicazioni
n ontwikkelingslaboratorium *n*
d Laboratorium *n* für Anwendungsmöglichkeiten

304 APPLICATOR, htt
HEATING INDUCTOR, WORK COIL
f inducteur *m* de chauffage
e inductor *m* de equipo de caldeo
i induttore *m* riscaldante
n verhittingsinductor, verwarmingsinductor, werkspoel
d Heizinduktor *m*, Heizschleife *f*

305 APPLICATOR, chr
THIN-LAYER SPREADER
f dispositif *m* d'étalement
e extendedor *m* de la capa
i distensore *m* della pellicola
n uitstrijker van de dunne laag
d Dünnschichtstreicher *m*

306 APPLICATOR ROLL pla
f cylindre *m* enducteur
e rodillo *m* de aplicación
i cilindro *m* trasportatore, rullo *m* trasferitore
n opbrengwals
d Auftragwalze *f*

307 APPLIED CHEMISTRY gen
f chimie *f* appliquée
e química *f* aplicada
i chimica *f* applicata
n toegepaste chemie
d angewandte Chemie *f*

308 APPLIED PRESSURE pre
f pression *f* de service
e presión *f* aplicada
i pressione *f* applicata
n bedrijfsdruk, toegepaste druk
d Betriebsdruck *m*

309 APPROACH (GB), cpt
APPROACHING PASSAGE (GB),
CHANNEL (US)
f canal *m* d'amenée
e canal *m* de acceso
i canale *m* d'entrata
n aanspuitkanaal *n*
d Zugangskanal *m*

310 APRON sap
f tablier *m*
e delantal *m*
i grembiale *m*
n schort
d Schürze *f*

311 APRON tra
f ruban *m* transbordeur, ruban *m* transporteur
e banda *f* transportadora
i nastro *m* trasportatore
n transportband
d Förderband *n*, Transportband *n*

312 APRON CONVEYER tra
f convoyeur *m* à courroie à section en auge
e correa *f* transportadora en forma de artesa, transportador *m* de banda articulada
i nastro *m* trasportatore a forma di truogoli, trasportatore *m* a nastro, trasportatore *m* a piastra e a cassette
n trogband, trogvormige transportband
d Plattenband *n*, Plattenförderer *m*, Trogbandförderer *m*

313 APRON CONVEYER FEEDER tra
f alimentateur *m* de ruban transporteur
e alimentador *m* de banda transportadora
i alimentatore *m* di nastro trasportatore
n bandtoevoer
d Bandaufgeber *m*

314 APRON CONVEYOR WITH CLOSED PANS tra
f transporteur *m* à palettes avec joues à recouvrement
e transportador *m* de bandejas
i trasportatore *m* a nastro articolato con cassette, trasportatore *m* a truogoli chiusi
n trogvormige transporteur met zijschotten
d Kastenbandförderer *m*, Muldenförderer *m* mit Seitenbrettern

315 AQUEOUS HOMOGENEOUS REACTOR nue
f réacteur *m* homogène aqueux
e reactor *m* homogéneo acuoso
i reattore *m* omogeneo acquoso
n homogene reactor met splijtstofoplossing
d wässrig-homogener Reaktor *m*

316 ARBOR (US), SPINDLE (GB) cpt,gen
f arbre *m*, axe *m*
e árbol *m*, eje *m*
i albero *m*, asse *m*
n as, spil
d Achse *f*, Spille *f*, Spindel *f*, Welle *f*

317 ARC FURNACE htt,met
f four *m* électrique à arc
e horno *m* eléctrico de arco
i forno *m* elettrico ad arco
n vlamboogoven
d Lichtbogenofen *m*

318 ARC FURNACE ELECTRODE htt,met
f électrode *f* de four électrique
e electrodo *m* de horno de arco
i elettrodo *m* di forno ad arco
n vlamboogovenelektrode
d Lichtbogenofenelektrode *f*

319 ARC IMAGE FURNACE, INDIRECT ARC FURNACE htt
f four *m* à arc indirect, four *m* à arc libre
e horno *m* de arco indirecto
i forno *m* ad arco indiretto, forno *m* ad arco libero
n indirecte vlamboogoven, vlamboogoven met indirecte verhitting
d indirekter Lichtbogenofen *m*

320 ARC WELDING PLANT wel
f installation *f* de soudure à l'arc électrique
e equipo *m* de soldadura por arco eléctrico
i impianto *m* di saldatura ad arco elettrico
n elektrische-booglasinstallatie
d Lichtbogenschweissanlage *f*

321 ARCH-TYPE DRIER dry
f séchoir *m* en arc
e secador *m* en forma de arco
i essiccatoio *m* ad arco
n boogvormige droogtunnel
d bogenförmiger Trockner *m*

322 ARCHIMEDEAN SCREW, PROPELLER PUMP tra
f vis *f* d'Archimède
e tornillo *m* de Arquímedes
i vite *f* d'Archimede
n schroef van Archimedes, schroefpomp, tonmolen
d Transportschnecke *f*

323 ARCHING OF A CYLINDER pla
f bombé *m* d'un cylindre
e comba *f* de un cilindro, corvadura *f* de un cilindro
i bombatura *f* di un cilindro
n welving van walscilinder
d Walzenbombage *f*, Walzenwölbung *f*

324 ARCHLESS KILN, SCOTCH KILN, SCOVE cer,htt
f four *m* à briques enfermées en pât
e horno *m* de ladrillos encerrados con barro
i forno *m* a mattoni inchiusi in pasta
n oven met in brij opgesloten vuurvaste stenen
d Ofen *m* mit in Schlicker eingeschlossenen feuerfesten Steinen

325 AREOMETER, HYDROMETER ins
f aréomètre *m*, hydromètre *m*
e areómetro *m*, hidrómetro *m*
i areometro *m*, idrometro *m*
n areometer, hydrometer, vochtmeter

d Aräometer *n*, Senkwaage *f*, Tauchwaage *f*

326 ARGON CHAMBER ins,nue
f chambre *f* d'ionisation à argon
e cámara *f* de ionización de argón
i camera *f* d'ionizzazione ad argon
n argonionisatievat *n*
d Argonisationskammer *f*

327 ARGON LIQUEFIER gas
f installation *f* de liquéfaction d'argon
e equipo *m* de licuación de argón
i impianto *m* di liquefazione di argon
n installatie voor het vloeibaar maken van argon
d Argonverflüssigungsanlage *f*

328 ARK cer
f cuve *f* d'argile hydratée
e cuba *f* de arcilla hidratada
i tino *m* d'argilla idratata
n vat voor water-kleimengsel
d Behälter *m* für Wasser-Tonerdemischung

329 ARM AGITATOR, mix
ARM MIXER, BLADE AGITATOR, BLADE MIXER
f agitateur *m* à palettes
e agitador *m* de paletas
i agitatore *m* a pale
n bladroerder
d Balkenrührwerk *n*, Paddelrührwerk *n*, Stabrührwerk *n*

330 ARRASTRE, brk,min
ROCKING MILL
f bocard *m* à pilons
e arrastre *m*
i mulino *m* a pilone, mulino *m* da pestare
n stampmolen
d Arrastra *f*, Pochmühle *f*

331 ARSENIC APPARATUS lab,mcy
f signaleur *m* d'arsenic
e indicador *m* de arsénico
i indicatore *m* d'arsenico
n arseenindicator, arsenicum-indicator
d Arsenanzeigegerät *n*

332 ARSENIC DETECTION ins,lab,mcy
APPARATUS,
ARSENIC DETERMINATION APPARATUS
f appareil *m* pour le dosage de l'arsenic
e aparato *m* dosificador de arsénico
i apparecchio *m* per dosare l'arsenico
n arseenbepalingsapparaat *n*, arsenicumbepalingsapparaat *n*
d Arsenbestimmungsgerät *n*

333 ARSENIC DISTILLATION ins,lab
APPARATUS
f appareil *m* à distiller l'arsenic
e aparato *m* de destilar el arsénico
i apparecchio *m* da distillare l'arsenico
n arseendestilleerapparaat *n*, arsenicumdestilleerapparaat *n*
d Arsendestillierapparat *m*

334 ARSINE GENERATOR gas,lab
f générateur *m* d'arsine
e generador *m* de arsino
i generatore *m* d'arsino
n arsinegenerator
d Arsenwasserstofferzeuger *m*, Arsinerzeuger *m*

335 ARTIFICIAL AGEING fou,pla
ROOM (GB),
ARTIFICIAL AGING ROOM (US)
f chambre *f* de vieillissement accéléré
e cámara *f* de envejecimiento acelerado
i camera *f* d'invecchiamento artificiale
n kunstmatige-verouderingsruimte
d Warmauslagerungsraum *m*

336 ARTIFICIAL ATMOSPHERE htt
FURNACE
f four *m* à atmosphère artificielle
e horno *m* de atmósfera controlada
i forno *m* ad atmosfera controllata
n oven met beheerste atmosfeer
d Schutzgasofen *m*

337 ARTIFICIAL SILK FACTORY tex
f usine *f* de rayonne
e fábrica *f* de rayón
i fabbrica *f* di raion, fabbrica *f* di seta artificiale
n kunstzijdefabriek
d Kunstseidefabrik *f*

338 ARTIFICIALLY AERATED iwt
TRICKLING FILTER
f lit *m* bactérien à ventilation forcée
e filtro *m* percolador de aireación artificial

i letto *m* percolatore aerato artificialmente
n kunstmatig geaëreerd druppelfilter *n*
d künstlich belüfteter Tropfkörper *m*

339 ASBESTOS CEMENT PIPE cpt
f tuyau *m* en amiante-ciment
e tubería *f* de cemento-asbesto
i tubo *m* di cemento-amianto, tubo *m* di fibrocemento, tubo *m* in eternit
n asbestcementbuis, eternietbuis
d Asbestzementrohr *n*, Eternitrohr *n*

340 ASBESTOS-COVERED ELECTRODE wel
f électrode *f* à revêtement d'amiante
e electrodo *m* recubierto de amianto
i elettrodo *m* a rivestimento d'amianto
n met asbest omhulde elektrode
d asbestumhüllte Elektrode *f*

341 ASBESTOS FILTER fil
f filtre *m* en amiante
e filtro *m* de amianto
i filtro *m* d'amianto
n asbestfilter *n*
d Asbestfilter *m*

342 ASBESTOS INSET cpt
f doublure *f* en amiante
e inserción *f* de amianto
i inserzione *f* d'amianto
n asbesttussenlaag
d Asbesteinlage *f*

343 ASBESTOS PACKING cpt
f garniture *f* d'amiante
e empaquedura *f* de amianto
i guarnizione *f* d'amianto
n asbestpakking
d Asbestdichtung *f*, Asbestpackung *f*

344 ASBESTOS WIRE GAUZE (GB), ASBESTOS WIRE GAZE (US) cpt
f toile *f* métallique à amiante
e tela *f* metálica amiantada
i rete *f* di filo metallico ed asbesto, reticella *f* amiantata
n asbestdraadnet
d Asbestdrahtnetz *n*, Drahtnetz *n* mit Asbesteinlage

345 ASBESTOS WOOL, COTTON ASBESTOS mat
f laine *f* d'amiante
e lana *f* de amianto
i lana *f* d'amianto
n asbestwol
d Asbestwolle *f*

346 ASCENDING TUBE bre
f jauge *f*
e tubo *m* de ascensión
i colonna *f* montante, tubo *m* ascend
n stijgbuis
d Steigrohr *n*

347 ASCENDING-WATER-TUBE BOILER bre
f chaudière *f* tubulaire verticale
e caldera *f* tubular vertical
i caldaia *f* tubolare verticale
n verticale buisketel
d Steilrohrkessel *m*

348 ASCENSION PIPE cpt,tra
f tuyau *m* de prise de gaz
e tubo *m* ascendente
i tubo *m* da presa di gas
n stijgbuis
d Steigrohr *n*

349 ASHLESS FILTER cle,cpt,fil
f filtre *m* sans cendres
e filtro *m* sin cenizas
i filtro *m* privo di ceneri, filtro *m* senza ceneri
n asvrij filter *n*
d aschenarmer Filter *m*

350 ASHLESS FILTER fil,lab,mat
f papier *m* filtre sans cendres
e papel *m* filtro sin cenizas
i carta *f* da filtro senza ceneri
n asvrij filtreerpapier *n*
d aschenarmes Filtrierpapier *n*

351 ASPHALT COATED PIPE cpt
f tube *m* asphalté
e tubo *m* asfaltado
i tubo *m* asfaltato
n geasfalteerde buis
d asphaltiertes Rohr *n*

352 ASPHALT COATING coa
f couche *f* d'asphalte
e revestimiento *m* de asfalto
i rivestimento *m* d'asfalto
n asfaltlaag
d Asphaltschicht *f*

353 ASPHALT DIP coa
f bain *m* d'asphalte
e baño *m* de asfalto
i bagno *m* d'asfalto
n asfaltbad *n*
d Asphalttauchbad *n*

354 ASPHALT LINING coa
f garnissage *m* d'asphalte
e guarnición *f* de asfalto
i guarnizione *f* d'asfalto
n asfaltvoering
d Asphaltfutter *n*

355 ASPIRATOR cle,pre
f aspirateur *m*
e aspirador *m*
i aspiratore *m*
n aanzuiger, aspirator
d Aspirator *m*, Saugapparat *m*

356 ASPIRATOR BOTTLE, FILTERING FLASK lab
f fiole *f* à filtration
e frasco *m* de filtración, pera *f* de aspiración
i bottiglia *f* a filtrazione
n aspirator
d Saugflasche *f*

357 ASSAY BALANCE ins
f balance *f* d'essai
e balanza *f* de ensayos
i bilancia *f* per la prova
n goudschaal
d Goldwaage *f*, Probierwaage *f*

358 ASSAY CRUCIBLE tes
f coupelle *f*
e crisol *m* de ensayo
i crogiolo *m* assaggiatore
n essaaikroes
d Probiertiegel *m*

359 ASSAY FLASK lab
f matras *m* d'essai
e matraz *m* de ensayo
i matraccio *m* d'assaggio
n essaaikolf
d Probierkolben *m*

360 ASSAY FURNACE, CUPEL FURNACE ana
f four *m* de coupellation
e horno *m* de copelación, horno *m* de ensayo
i forno *m* d'assaggio, forno *m* di coppellazione
n cupelleeroven, essaaieeroven
d Kupellierofen *m*

361 ATMIDOMETER, ATMOMETER, EVAPORIMETER ins
f atmidomètre *m*, atmomètre *m*, évaporimètre *m*
e atmidímetro *m*, evaporímetro *m*
i atmidometro *m*, evaporimetro *m*
n atmidometer, atmometer, verdampingsmeter
d Atmomesser *m*, Dunstmesser *m*, Verdampfungsmesser *m*, Verdunstungsmesser *m*

362 ATMOSPHERIC GAS BURNER bur
f brûleur *m* de gaz à tirage naturel
e quemador *m* de gas de tiro natural
i bruciatore *m* di gas a tiraggio naturale
n gasbrander met natuurlijke trek
d Gasbrenner *m* mit natürlichem Zug

363 ATOMIC ARC WELDING PLANT wel
f installation *f* de soudure à l'hydrogène atomique
e equipo *m* de soldadura por hidrógeno atómico
i impianto *m* di saldatura ad idrogeno atomico
n atomaire-waterstoflasinstallatie
d Arkatomschweissanlage *f*

364 ATOMIZER pre
f atomiseur *m*, pulvérisateur *m*, vaporisateur *m*
e atomizador *m*, nebulizador *m*, pulverizador *m*
i atomizzatore *m*, nebulizzatore *m*, polverizzatore *m*, spruzzatore *m*
n vernevelingsapparaat *n*, verstuiver
d Zerstäuber *m*

365 ATOMIZING NOZZLE cpt,pre
f buse *m* de pulvérisateur, gicleur *m* d'atomiseur
e tobera *f* de atomizador, tobera *f* de pulverizador
i ugello *m* di polverizzatore, ugello *m* di spruzzatore
n verstuivermondstuk *n*
d Sprühdüse *f*, Streudüse *f*, Zerstäuberdüse *f*

366 ATOMIZING SCRUBBER cle
f laveuse *f* à atomiseur
e lavador *m* atomizador
i lavatrice *f* spruzzatrice
n wasinstallatie met verstuiver
d Wäsche *f* mit Zerstäuber

367 ATTACHABLE CORE OF EXTRUDER HEAD pla
f mandrin *m* amovible de boudineuse
e macho *m* desmontable del cabezal de budinadora
i anima *f* smontabile della testa di trafila
n afneembare spuitkern
d Aufsteckdorn *m* für Spritzmaschinenkopf

368 ATTACHMENT, DISTILLATION HEAD dis
f rectificateur *m*, tête *f* à fractionner
e enganche *m*
i attacco *m*
n opzet
d Aufsatz *m*

369 ATTEMPERATOR con
f régulateur *m* de température
e atemperador *m*, regulador *m* térmico, serpentín *m* de enfriamiento
i attemperatore *m*
n temperatuurregelaar
d Temperaturregulator *m*

370 ATTRITION MILL brk
f broyeur *m* à friction
e molino *m* de frotamiento
i molazza *f* a sfregamento
n wrijfkollergang, wrijfmolen
d Reibkollergang *m*

371 AUSTENIZING HEAT TREATMENT FURNACE htt,met
f four *m* d'austénitisation
e horno *m* de austenitización
i forno *m* d'austenitizzazione
n austenitiseeroven
d Austenitisierungsofen *m*

372 AUTOCLAVE, POTHEATER pre
f autoclave *m*
e autoclave *f*
i autoclave *f*
n autoclaaf
d Autoklav *m*, Dampfkochtopf *m*, Druckkessel *m*

373 AUTOCLAVE PRESS pre
f presse *f* autoclave
e prensa *f* autoclave
i pressa *f* autoclave
n autoclaafpers
d Autoklavenpresse *f*, Kesselpresse *f*

374 AUTOCLAVE SHELL pre
f gaine *f* d'autoclave
e coraza *f* de autoclave
i involucro *m* d'autoclave
n autoclaafmantel
d Autoklavenmantel *m*

375 AUTOGENOUS WELDING PLANT wel
f installation *f* de soudure autogène
e equipo *m* de soldadura autógena
i impianto *m* di saldatura autogena
n autogene-lasinstallatie
d Autogenschweissanlage *f*

376 AUTOMATIC BURETTE lab
f burette *f* automatique
e bureta *f* automática
i buretta *f* automatica
n zelfregelende buret
d Selbstreglerbürette *f*

377 AUTOMATIC CAPILLARY PIPETTE bac,lab
f micropipette *f* à ajustage automatique
e micropipeta *f* de ajuste automático
i micropipetta *f* ad aggiustamento automatico
n automatische steekpipet
d automatische Stechpipette *f*

378 AUTOMATIC CHARGING EQUIPMENT tra
f appareillage *m* de chargement automatique
e aparatos *pl* de carga automática
i apparecchiatura *f* di caricamento automatico
n automatische chargeerinrichting
d automatische Beschickungsvorrichtung *f*

379 AUTOMATIC DISPENSING PIPETTE lab
f pipette *f* automatique à incliner sur flacon
e pipeta *f* de distribución automática
i pipetta *f* a distribuzione automatica
n pipet met automatische afloop
d Kippautomat *m*

380 AUTOMATIC DOSER tra
f distributeur-doseur *m*
e distribuidor *m* dosificador
i distributore-dosatore *m*

n doseerinrichting
d Dosiereinrichtung *f*, Zuteiler *m*

381 AUTOMATIC FIRE SPRINKLER, fea
SPRINKLER
f extincteur *m* automatique d'incendies
e rociador *m* extinguidor automático
i spruzzatore *m* automatico
n automatische sproeier
d automatische Wasserspritze *f*

382 AUTOMATIC OVERFLOW PIPETTE lab
f pipette *f* à débordement automatique
e pipeta *f* de rebosamiento automático
i pipetta *f* a fuoruscita automatica
n automatische overlooppipet
d automatische Pipette *f*

383 AUTOMATIC PICKLING EQUIPMENT elc
f automate *m* de décapage
e equipo *m* automático de decapaje
i impianto *m* automatico di decapaggio
n beitsautomaat
d Beizautomat *m*

384 AUTOMATIC RESPIRATOR sap
f masque *m* respiratoire automatique
e máscara *f* respiradora automática
i maschera *f* respiratrice automatica
n automatisch zuurstoftoestel *n*
d selbsttätiges Sauerstoffapparat *m*

385 AUTOMATIC SAMPLER CHANGER nue
f changeur *m* automatique d'échantillons
e cambiador *m* automático de muestras
i scambiatore *m* automatico di campioni
n automatische monsterwisselaar
d selbsttätiger Probenwechsler *m*

386 AUTOMATIC SOFTENING INSTALLATION, hyg
AUTOMATIC SOFTENING PLANT
f installation *f* d'adoucissement automatique
e equipo *m* ablandador automático
i impianto *m* d'addolcimento automatico
n automatische onthardings-installatie
d selbsttätige Enthärtungsanlage *f*

387 AUXILIARY ANODE elc
f anode *f* auxiliaire
e ánodo *m* auxiliar
i anodo *m* ausiliario
n bijanode, hulpanode
d Hilfsanode *f*, Zusatzanode *f*

388 AUXILIARY APPARATUS aap
f appareils *pl* auxiliaires
e aparatos *pl* auxiliarios
i apparecchiatura *f* ausiliaria
n hulpapparaten *pl*, hulpapparatuur, hulptoestellen *pl*
d Hilfsapparate *pl*, Hilfsgeräte *pl*

389 AUXILIARY ELECTRODE elc
f électrode *f* auxiliaire
e electrodo *m* auxiliar
i elettrodo *m* ausiliario
n hulpelektrode
d Hilfselektrode *f*

390 AWN CUTTER, bre
f ébarbeuse *f*
e desbarbadora *f*
i sbarbatrice *f*, spuntaresta *f*
n gerstkorter
d Entgranner *m*

391 AXIAL-FLOW AGITATOR mix
f agitateur *m* à flux axial
e agitador *m* de flujo axial
i agitatore *m* a flusso assiale
n roerwerk *n* met axiale stroming
d Rührwerk *n* mit Axialströmung

392 AXIAL HEAD (US) pla
STRAIGHT HEAD (GB)
f tête *f* droite
e cabezal *m* de extrusión recto
i testa *f* assiale
n axiale spuitkop
d Längsspritzkopf *m*

393 AZOTOMETER, ins
NITROMETER
f azotomètre *m*
e azotímetro *m*, nitrómetro *m*
i azotometro *m*
n azotometer
d Azotometer *n*

B

394 BACK CLOTH, RUNNER CLOTH, UNDERCLOTH mat,pla
f bande *f* ensouple, feuille *f* ensouple
e correa *f* acompañante
i nastro *m* d'accompagnamento
n meelopende baan, meelopende foelie
d Mitläuferband *n*, Mitläuferfolie *f*

395 BACK COVER rub
f couche *f* inférieure, sous-couche *f*
e capa *f* inferior
i coperchio *m* inferiore
n aangehechte onderlaag, ruglaag
d Rückseitenbelag *m*, Unterschicht *f*

396 BACK-PRESSURE PLATE rub
f plaque *f* à contrepression
e placa *f* de contrapresión
i piastra *f* a contropressione
n tegendrukplaat
d Gegendruckplatte *f*

397 BACK-PRESSURE VALVE cpt,sap
f clapet *m* de retenue, soupape *f* de retenue
e válvula *f* de contrapresión
i valvola *f* di ritegno
n terugslagklep
d Rückschlagklappe *f*, Rückschlagventil *n*

398 BACK-SIPHONAGE PREVENTER cpt
f barrière *f* de reflux, coupe-vide *m*
e interruptor *m* de sifonaje de retorno
i interruttore *m* del vuoto
n terugstroomonderbreker
d Rohrunterbrecher *m*, Rückflussverhinderer *m*

399 BACK-TANNING BATH tex
f bain *m* de traitement postérieur à l'acide tannique
e baño *m* de postratamiento con ácido tánico
i bagno *m* di trattamento posteriore all'acido tannico
n looizuurnabehandelingsbad *n*
d Gerbsäurenachbehandlungsbad *n*

400 BACK-UP ROLL, BACKING-UP ROLL gen
f cylindre *m* de support
e cilindro *m* reforzador
i cilindro *m* di sopporto
n steunwals
d Stützwalze *f*

401 BACKED FABRIC mat,pla
f tissu *m* doublé
e tejido *m* forrado con plástico
i tessuto *m* accoppiato
n gecacheerd weefsel *n*, weefsel *n* op plastiekfoelie
d kaschiertes Gewebe *n*

402 BACKFLOW VALVE, CHECK VALVE, REFLUX VALVE, STOP VALVE cpt
f soupape *f* d'arrêt
e válvula *f* de retención
i valvola *f* d'arresto, valvola *f* intercettante
n afsluitklep, sperklep
d Absperrventil *n*

403 BACKING ELECTRODE (GB), WELDING GROUND (US) wel
f contre-électrode *f*
e contraelectrodo *m*
i controelettrodo *m*
n onderlegelektrode
d Gegenelektrode *f*

404 BACKING PLATE pla
f contreplaque *f*, contreplateau *m*, plaque *f* intercalée, plaque *f* intermédiaire
e contraplaca *f*, placa *f* intermedia
i contropiastra *f*, piastra *f* intermedia
n basisplaat, steunplaat
d Grundplatte *f*, Zwischenplatte *f*

405 BACKING STRIP coa,htt
f bande *f* de renforcement
e contrachapa *f*, chapa *f* espaldar, pletina *f* de respaldo
i nastro *m* di rinforzo
n steunstrip
d Stützstreifen *m*

406 BACKWASH PIPE fue
f tube *m* de lavage par inversion de courant
e tubo *m* de lavado a contracorriente
i tubo *m* di desoliatura, tubo *m* di lavaggio ad inversione di flusso
n terugloopbuis
d Rückwäscherohr *n*

407 BACKWASHER tex
f dégraisseuse *f* de la laine
e alisadora *f* de la lana
i sgrassatrice *f* della lana
n wolontvettingstoestel *n*
d Wollentfettungsgerät *n*

408 BACTERIA BED, BACTERIA FILTER, BIOLOGICAL FILTER iwt
f filtre *m* bactérien, filtre *m* biologique
e filtro *m* bacteriano
i letto *m* batterico, letto *m* biologico
n bacteriefilter *n*
d biologischer Körper *m*

409 BACTERIA FILTER APPARATUS bac,fil,lab
f appareil *m* à filtration stérile
e aparato *m* de filtración bacteriana
i apparecchio *m* a filtrazione sterile
n steriel filtreerapparaat *n*
d Bakterienfiltriergerät *n*

410 BACTERIA PROPAGATION TANK bac
f récipient *m* de propagation de bactéries
e tanque *m* para la propagación de bacterias
i serbatoio *m* per la propagazione di batteri
n bacteriënbroedbak, bacteriënkweekbak
d Bakteriengärungsgefäss *n*, Bakterienzuchtbehälter *m*

411 BACTERICIDAL EQUIPMENT bac
f installation *f* bactéricide
e equipo *m* bactericida
i impianto *m* battericida
n bacteriëndodende installatie
d bakterientötende Anlage *f*

412 BACTERIOLOGICAL EQUIPMENT bac
f installation *f* bactériologique
e equipo *m* bacteriológico
i impianto *m* batteriologico
n bacteriologische inrichting
d bakteriologische Anlage *f*

413 BACTERIOLOGICAL PIPETTE bac,lab
f pipette *f* bactériologique
e pipeta *f* bacteriológica
i pipetta *f* batteriologica
n bacteriologiepipet
d Bakteriologiepipette *f*

414 BACTERIOLOGICAL TEST TUBE bac,lab
f tube *m* à essai bactériologique
e tubo *m* de ensayo bacteriológico
i tubo *m* da saggio batteriologico
n reageerbuis voor bacteriologie
d bakteriologisches Reagenzglas *n*

415 BAFFLE cpt,pla
f tôle *f* de séparation
e placa *f* de desviación
i deflettore *m*
n verdeelschuif
d Trennblech *n*

416 BAFFLE, BAFFLE PLATE abs,cpt
f chicane *f*, plaque *f* de déviation
e chicana *f*, placa *f* deflectora, placa *f* interceptora
i deflettore *m*, lamiera *f* di guida
n keerplaat, keerschot *n*
d Leitblech *n*, Prallplatte *f*, Stauscheibe *f*

417 BAFFLE COLUMN, BAFFLE TOWER dis
f colonne *f* à chicanes
e columna *f* de chicanas, columna *f* de deflectores
i colonna *f* a diflettori, colonna *f* a diaframmi
n zuil met keerschotten
d Traufenkolonne *f*

418 BAFFLE PLATE cpt,pla
f chicane *f*, plaque *f* de retenue
e placa *f* interceptora
i diaframma *m*
n zeefplaat
d Stauplatte *f*, Stauscheibe *f*

419 BAFFLE WALL gla
f barrage *f* de déviation
e pared *f* de desviación
i parete *f* di desviazione
n keerwand
d Prallwand *f*

420 BAG FILTER (GB), met,fil
SACK FILTER (US)
f filtre *m* à sac
e filtro *m* de bolsa
i filtro *m* a sacco
n zakfilter *n*
d Beutelfilter *m*, Taschenfilter *m*

421 BAG FILTRATION PLANT (GB),fil
SACK FILTRATION PLANT (US)
f installation *f* de filtrage en sacs
e equipo *m* de filtración en bolsas
i impianto *m* di filtrazione in sacchi
n zakfilterinstallatie
d Taschenfilteranlage *f*

422 BAG MOLD (US), pla
BAG MOULD (GB)
f moule *m* à sac en caoutchouc
e molde *m* con saco elástico
i stampo *m* a sacco di gomma
n matrijs met rubberzak
d Pressform *f* mit Gummisack

423 BAG SEALER, pla
BAG SEALING MACHINE
f ensacheuse *f*
e máquina *f* para cerrar sacos
i saldatrice *f* per sacchetti
n zakjessluitmachine
d Beutelschliessmaschine *f*

424 BAG-TYPE CONSTRUCTION elc
f montage *m* habillé
e montaje *m* tipo saco
i montaggio *m* a catodo rivestito
n constructie met gebonden buidel
d gewickelter Beutel *m*, Wickelbauweise *f*

425 BAGASSE ROLLER sin
f rouleau *m* à bagasse
e rodillo *m* para bagazo
i rullo *m* per bagassa
n bagassewals
d Bagassenwalze *f*

426 BAGGED ANODE elc
f anode *f* ensachée
e ánodo *m* en saquete, ánodo *m* enfundado
i anodo *m* imballato
n anode in zak
d Anode *f* im Beutel

427 BAGGING MACHINE tra
f machine *f* à ensacher
e máquina *f* para llenar sacos
i macchina *f* per insaccare
n zakkenvulmachine
d Sackfüller *m*

428 BAIT gla
f pontil *m*, tirant *m*
e barra *f* de conexión
i pontello *m*
n hechtijzer *n*, trekstang
d Hefteisen *n*

429 BAKING FURNACE, htt
GAS-EXPELLING FURNACE
f four *m* pour expulsion de gaz
e horno *m* para expulsión de gases
i forno *m* per espulsione di gas
n gasuitdrijfoven
d Gasaustreibeofen *m*

430 BAKING PRESS (US), pla
BLOCK PRESS (GB)
f presse *f* à blocs
e prensa *f* de bloques
i pressa *f* per blocchi
n blokpers
d Blockpresse *f*

431 BAKING STOVE, htt,tex
POLYMERIZING STOVE
f étuve *f* à polymériser
e estufa *f* de polimerización
i stufa *f* di polimerizzazione
n polymeriseerstoof
d Polymerisierofen *m*

432 BALANCE app,lab
f balance *f*
e balanza *f*
i bilancia *f*
n balans
d Waage *f*

433 BALANCE ARM, lab
BALANCE BEAM
f fléau *m* de balance
e astil *m* de balanza
i giogo *m* di bilancia
n balansarm
d Waagebalken *m*

434 BALANCE DESSICATOR lab
f récipient *m* à chlorure de calcium
e capa *f* para cloruro de calcio
i piattello *m* di bilancia per cloruro di calcio
n balansdroogmiddelhouder
d Waagentrockenmittelbehälter *m*

435 BALANCE PAN lab
f bassin *m* de balance, plateau *m* de balance
e platillo *m* de balanza
i piattello *m* di bilancia
n balansschaal
d Waagschale *f*

436 BALANCE RIDER, RIDER lab
f cavalier *m*
e caballero *m*, jinetillo *m*
i cavaliere *m*
n ruiter
d Reiter *m*

437 BALANCING RESERVOIR, COMPENSATING RESERVOIR, EQUALIZING RESERVOIR iwt
f réservoir *m* compensateur, réservoir *m* régulateur du débit
e depósito *m* de equilibrio
i serbatoio *m* compensatore, serbatoio *m* regolatore
n compensatievat *n*
d Ausgleichbehälter *m*

438 BALE DYEING TROUGH PACK DYEING TROUGH tex
f cuve *f* de teinture en balles
e cuba *f* de teñido en balas, cuba *f* de teñido en líos
i tino *m* di tintura in balle
n kuip voor het verven van balen stof
d Ballenfärbekübe *f*

439 BALE-OUT FURNACE htt,tex
f four *m* fixe
e horno *m* fijo
i forno *m* fisso
n vaste oven
d ortsfester Ofen *m*

440 BALL ANODE elc
f anode *f* à billes
e ánodo *m* de bolas
i anodo *m* a palle
n kogelanode
d Kugelanode *f*

441 BALL ANODE CAGE elc
f panier *m* pour anode à billes
e jaula *f* para ánodo de bolas
i gabbia *f* per anodo a palle
n kogelanodehouder
d Kugelanodenhalter *m*

442 BALL BURNISHING DRUM met,pol
f tambour *m* de polissage à billes
e tambor *m* de bruñido con bolas
i tamburo *m* di brunitura per palle
n kogelpolijsttrommel
d Kugelpoliertrommel *f*

443 BALL BURNISHING MACHINE pol
f machine *f* de polissage à billes
e máquina *f* de bruñido con bolas
i macchina *f* di brunitura con palle
n kogelpolijstmachine
d Kugelpoliermaschine *f*

444 BALL CHECK NOZZLE pla,rub
f tuyère *f* en fermeture à boule
e tobera *f* con bola obturadora
i bocchettone *m* con chiusura a sfera
n spuitkop met kogelafsluiting
d Kugelverschlussdüse *f*

445 BALL CHECK VALVE cpt
f soupape *f* de retenue à bille
e válvula *f* de retención a bola
i valvola *f* di ritegno a sfera
n kogelterugslagklep
d Kugelrückschlagventil *n*

446 BALL COCK cpt
f robinet *m* à bille
e grifo *m* de flotador, llave *f* de bola
i rubinetto *m* a galleggiante, rubinetto *m* sferico
n vlotterkraan
d Kugelhahn *m*, Schwimmerhahn *m*

447 BALL CONDENSER clo
f condensateur *m* sphérique
e condensador *m* esférico
i condensatore *m* sferico
n kogelcondensor
d Kugelkühler *m*

448 BALL CUP cpt
f coussinet *m* de bille, coussinet *m* de boulet
e asiento *m* de bola
i tazza *f* a sfera
n kogelklepzitting
d Kugelpfanne *f*, Kugelschale *f*

449 BALL CUP VALVE (GB), BALL VALVE (GB), GLOBE VALVE (US) cpt,sap
f clapet *m* à bille, soupape *f* à boulet
e válvula *f* de bola
i valvola *f* a sfera
n kogelklep, kogelventiel *n*
d Ballventil *n*, Kugelventil *n*

450 BALL GRINDER, BALL MILL, PEBBLE MILL — brk
f broyeur *m* à boules, broyeur *m* à galets, jarre *f* de broyage
e molino *m* de bolas
i mulino *m* a palle, mulino *m* a sassi
n kogelmolen
d Kugelmühle *f*

451 BALL-JOINT MANIPULATOR — nue
f manipulateur *m* à articulation à rotule
e manipulador *m* con articulación de rótula
i manipolatore *m* a giunto sferico
n kogelgewrichtmanipulator
d Kugelgelenkmanipulator *m*

452 BALLING DEVICE (US), PILL MACHINE (GB) — mcy
f pilulier *m*
e máquina *f* de hacer píldoras, pildorero *m*
i dispositivo *m* da fare pillole
n pillenmachine, pillenvormer
d Pillenmaschine *f*

453 BALLING FURNACE, BLACK ASH FURNACE, SODA FURNACE — aic
f four *m* de soude
e horno *m* de carbonatar, horno *m* de sosa
i forno *m* di soda
n sodaoven
d Sodaofen *m*

454 BANBURY MIXER, INTERNAL MIXER WITH FLOATING WEIGHT — mix,pla
f mélangeur *m* interne à piston, mélangeur *m* interne Banbury
e mezclador *m* interno Banbury, mezclador *m* interno con pistón
i mescolatore *m* Banbury
n banburymenger, zware kneedmachine
d Banbury-Mischer *m*, Gummikneter *m*, Innenmischer *m* mit Stempel, Stempelmischer *m*

455 BAND CONVEYER, BELT CONVEYER — tra
f transbordeur *m* à bande, transporteur *m* à bande
e transportador *m* de cinta
i trasportatore *m* a nastro
n bandtransportinrichting
d Bandförderer *m*

456 BAND DRIER, BELT DRIER — dry
f sécheur *m* à bandes transporteuses
e secadora *f* de cintas
i essiccatore *m* a nastri
n etagedroger
d Bandtrockner *m*

457 BAND FEED, BELT FEED — tra
f alimentateur *m* à bande
e alimentador *m* de cinta
i alimentatore *m* a nastro
n bandtoevoerinrichting
d Bandzuführer *m*

458 BAND FILTER, TRAVELLING BAND FILTER — cle,fil
f filtre *m* à bande transporteuse
e filtro *m* de cinta transportadora
i filtro *m* a nastro trasportatore
n lopende-bandfilter *n*
d Bandfilter *m*, Wandernutsche *f*

459 BAND HEATER, EXTRUDER BAND HEATER — pla,rub
f élément *m* de chauffage annulaire
e calentador *m* de cinta
i riscaldatore *m* a nastro avvolto
n ringvormig verwarmingselement
d Bandheizkörper *m*

460 BAND OVEN — nut
f four *m* à ruban transporteur
e horno *m* de correa
i forno *m* a nastro trasportatore
n bandoven
d Förderbandofen *m*

461 BAR AGITATOR — mix
f agitateur *m* à barres
e agitador *m* de barras
i agitatore *m* a barre
n stangenroerwerk *n*
d Stangenrührer *m*

462 BAR ANODE — elc
f anode *f* en barre
e ánodo *m* de barra
i anodo *m* in barra
n staafanode
d Stabanode *f*, Stangenanode *f*

463 BAR MILL, ROD GRINDER, ROD MILL — brk
f broyeur *m* à barres
e machacadora *f* de barras
i frantoio *m* a barre
n staafmolen
d Stabmühle *f*

464 BAR MOULD (GB), pla
COMPOSITE MOLD (US),
COMPOSITE MOULD (GB)
f moule *m* à empreintes différentes
e molde *m* compuesto, molde *m* de varias matrices
i stampo *m* ad impronte differenti
n gevarieerde matrijs
d Mehrfachform *f*

465 BAR ROTATING DYEING MACHINE tex
f machine *f* de teinture rotative à barres de support
e máquina *f* de teñido giratoria con barras de soporte
i macchina *f* di tintura rotativa con barre di sopporto
n draaiende verfmachine met steunstaven
d drehende Färbmaschine *f* mit Stützstäben

466 BAR SCREEN cpt
f grille *f* à barreaux
e emparillado *m* de barrotes, parrilla *f* de barrotes
i graticola *f* a barre
n staafrooster *n*
d Stabrost *m*, Stangenrost *m*

467 BARE REACTOR nue
f réacteur *m* sans réflecteur
e reactor *m* sin reflector
i reattore *m* nudo, reattore *m* senza riflettore
n reflectorloze reactor
d Reaktor *m* ohne Reflektor, unreflektierter Reaktor *m*

468 BARIUM MIXER mcy
f malaxeur *m* de bouillie opaque
e mezclador *m* de papilla de bario
i mescolatore *m* di pasta opaca
n bariumpapmenger
d Bariumbreimischer *m*

469 BARK, tex
BECK
f cuve *f*
e cuba *f*, tina *f*
i tino *m*
n kuip
d Küpe *f*

470 BARK MILL tan
f broyeur *m* d'écorce
e molino *m* de triturar casca
i mulino *m* di corteccia
n runmolen, schorsmolen
d Rindenmühle *f*

471 BARK TANNERY tan
f tannerie *f* à l'écorce
e teñería *f*
i conceria *f* alla scorza
n runlooierij
d Lohgerberei *f*

472 BARKING DRUM pap
f tambour *m* de décorticage
e tambor *m* descortezador
i tamburo *m* per scortecciare
n ontschorsingstrommel, schiltrommel
d Entrindungstrommel *f*

473 BARKING MACHINE pap
f machine *f* de décorticage
e máquina *f* descortezadora
i scortecciatrice *f*
n schilmachine
d Schälmaschine *f*

474 BARKOMETER, ins,tan
BARKTROMETER
f taninomètre *m*
e tanímietro *m*
i tanninometro *m*
n looizuurmeter
d Lohmesser *m*

475 BARLEY CLEANER, bre
BARLEY DRESSER
f nettoyeuse *f* d'orge, tarare *m*
e limpipadora *f* de cebada
i pulitrice *f* d'orzo
n gerstschoonder
d Gerstenputzmaschine *f*

476 BARLEY DRIER, bre
BARLEY KILN
f touraille *f* à sécher l'orge
e secadero *m* de cebada
i essiccatoio *m* per orzo
n gerstdroger
d Gerstendarre *f*, Gerstentrockner *m*

477 BARLEY FLOOR bre
f grenier *m* à orge
e granero *m* de cebada
i magazzino *m* per orzo
n gerstvloer
d Gerstenboden *m*

478 BARLEY GRADER, bre
BARLEY SORTER
f trieur-diviseur *m* d'orge
e clasificadora *f* de cebada
i classificatrice *f* d'orzo
n gerstsorteermachine
d Gerstensichter *m*, Gerstensortiermaschine *f*

479 BAROMETRIC LEG, BAROMETRIC PIPE cpt
f tube *m* barométrique, tube *m* d'eau de condensation
e tubo *m* barométrico
i tubo *m* barometrico
n condenswaterbuis
d Fallwasserrohr *n*

480 BARREL ATTEMPERATOR, COOLER KEG COIL bre
f serpentin *m* de fût intérieur
e serpentín *m* enfriador interior
i serpentina *f* raffreddatrice interiore
n vatkoeler
d Fasskühler *m*

481 BARREL FILLING MACHINE, RACKER, RACKING MACHINE bre
f soutireuse *f* à fûts
e llenadora *f* de barriles
i riempifusti *m*
n vatenvulmachine
d Fassabfüllmaschine *f*, Fässfüller *m*

482 BARREL MIXER mix
f malaxeur *m* à tambour, mélangeur *m* à tambour
e mezclador *m* de tambor
i mescolatrice *f* a bariletto, mescolatrice *f* a tamburo
n trommelmenger
d Trommelmischer *m*

483 BARREL PLATING PLANT elc
f établissement *m* de galvanoplastie au tonneau
e equipo *m* de galvanoplastia en tambor
i impianto *m* di galvanoplastica in tamburo, impianto *m* di rotogalvanostegia
n trommelgalvaniseerinstallatie
d Trommelgalvanisieranlage *f*

484 BARREL ROLLING DRUM, TUMBLING DRUM pol
f tonneau *m* de polissage
e tambor *m* de pulido
i tamburo *m* di pulitura
n polijsttrommel
d Poliertrommel *f*

485 BARRIER cpt
f barrière *f*
e barrera *f*
i barriera *f*
n poreuze wand
d poröse Wand *f*

486 BASE EXCHANGER iwt
f echangeur *m* de bases
e cambiador *m* de bases
i scambiatore *m* di basi
n basenuitwisselaar
d Basenaustauscher *m*

487 BASIC BOTTOM htt
f sole *f* basique
e solera *f* básica
i suola *f* basica
n basische bodem
d basischer Boden *m*

488 BASIC LINING met
f garnissage *m* basique
e revestimiento *m* básico
i rivestimento *m* basico
n basische voering
d basisches Futter *n*

489 BASIC PADDLE IMPELLER mix
f roue *f* motrice à palette normale
e rueda *f* móvil de paleta normal
i girante *m* a pala normale
n normale schoepenroerder
d normale Paddelrührstange *f*

490 BASIL BUFF pol
f disque *m* à polir en cuir de mouton
e disco *m* de polir a pellejo ovino
i disco *m* di pulitura a pelle di pecora
n polijstschijf van schapeleer
d Polierscheibe *f* aus Schafsleder

491 BASIN iwt
f bassin *m*
e alberca *f*, pantano *m*
i bacino *m*
n bassin *n*, bekken *n*
d Becken *n*

492 BASIN FLOOR iwt
f radier *m* d'un bassin
e fondo *m* de la alberca
i fondo *m* del bacino
n bassinbodem
d Beckensohle *f*

493 BASKET, CENTRIFUGE SIEVE sin
f panier *m* de centrifugeur
e cesta *f* de centrífuga
i paniere *m* di centrifuga
n centrifugezeef
d Schleudersieb *n*, Zentrifugensieb *n*

494 BASKET CENTRIFUGE sin
f centrifugeur *m* à panier
e centrífuga *f* de cesta
i centrifuga *f* a paniere
n trommelcentrifuge
d Trommelschleuder *f*
Trommelzentrifuge *f*

495 BASKET EVAPORATOR eva,pre
f évaporateur *m* à panier
e evaporador *m* de cesta
i evaporatore *m* a paniere
n trommelverdamper
d Trommelverdampfer *m*

496 BAT cer
f brique *f* porteuse
e ladrillo *m* portador
i mattone *m* portatore
n draagsteen
d Tragziegel *m*

497 BATCH gla
f composition *f*, mélange *m* vitrifiable
e mezcla *f*
i miscuglio *m*
n gemeng *n*
d Gemenge *n*

498 BATCH CENTRIFUGE mix
f centrifugeur *m* à lots
e centrífuga *f* de lotes
i centrifuga *f* ad intermittenza, centrifuga *f* per serie
n centrifuge met afgepaste hoeveelheden
d diskontinuierliche Schleuder *f*, Satzzentrifuge *f*

499 BATCH CHARGER, BATCH FEEDER gla
f alimentateur *m* de composition
e alimentador *m* de mezclas
i alimentatore *m* di miscugli
n gemengvoeder *n*
d Gemengespeiser *m*

500 BATCH CRYSTALLIZER cry
f cristallisoir *m* à lots, cristallisoir *m* intermittent
e cristalizador *m* intermitente
i cristallizzatore *m* intermittente
n niet-continue kristallisator
d Chargenkristallisator *m*, diskontinuierlicher Kristallisator *m*, Satzkristallisator *m*

501 BATCH DISTILLATION PLANT dis
f installation *f* de distillation discontue, installation *f* de distillation intermittente
e equipo *m* de destilación discontinua, equipo *m* de destilación intermitente
i impianto *m* di distillazione discontinua, impianto *m* di distillazione a lotti
n intermitterende destillatie-inrichting, niet-continue destillatie-inrichting
d absatzweise Destillationsanlage *f*, Blasendestillationsanlage *f*, intermittierende Destillationsanlage *f*

502 BATCH DYEING MACHINE tex
f machine *f* de teinture discontinue
e máquina *f* de teñido no continuo
i macchina *f* di tintura intermittente
n niet-continue verfmachine
d diskontinuierliche Färbmaschine *f*

503 BATCH FURNACE htt
f four *m* à charges, four *m* non-continu
e horno *m* de cargas per paquetes, horno *m* discontinuo
i forno *m* a lotti
n charge-oven, niet-continue oven
d Chargenofen *m*, Einsatzofen *m*

504 BATCH GRAVITY FILTER fil,sew
f filtre *m* à gravité intermittent
e filtro *m* de gravedad intermitente
i filtro *m* a gravità intermittente
n niet-continu open waterfilter *n*
d diskontinuierlicher offener Filter *m*

505 BATCH HOUSE gla
f salle *f* de composition
e sala *f* de mezclar
i sala *f* da mescolare
n gemengkamer
d Gemengehaus *n*, Gemengestube *f*

506 BATCH LEAF FILTER fil
f filtre *m* à feuilles intermittent
e filtro *m* de hojas intermitente
i filtro *m* a foglie intermittente
n niet-continu bladfilter *n*
d nichtkontinuierlicher Blattfilter *m*

507 BATCH MILL brk
f broyeur-frotteur *m* discontinu
e machacadora *f* mezcladora no continua
i frantoio *m* mescolatore intermittente

n niet-continue walsmolen
d diskontinuierlicher Kollergang *m*

508 BATCH MIXER gla
f tambour *m* mélangeur
e tambor *m* mezclador
i tamburo *m* mescolatore
n mengtrommel
d Mischtrommel *f*

509 BATCH MIXER mix
f mélangeur *m* à lots, mélangeur *m* intermittent
e mezclador *m* de lotes
i mescolatrice *f* ad intermittenza, mescolatrice *f* per serie
n menger met afgepaste hoeveelheden
d diskontinuierlicher Mischer *m*, Satzmischer *m*

510 BATCH-OFF MILL mix
f mélangeur *m* à cylindres
e mezclador *m* de cilindros
i mescolatore *m* a cilindri
n mengwals
d Mischwalze *f*

511 BATCH PAN (US), STOCK PAN (GB) mix,rub
f baquet *m* à mélange, cuve *f* à mélange
e cuba *f* para mezcla, tina *f* para mezcla
i marnetta *f*
n mengbak
d Mischwanne *f*

512 BATCH PILE gla
f bonne *f*
e montón *m*
i mucchio *m*
n gemenghoop
d Gemengehaufen *m*

513 BATCH REFINING KETTLE fat
f récipient *m* de désacidification
e cuba *f* de desacidificación
i serbatoio *m* per disacidificare
n ontzuringsketel
d Entsäuerungskessel *m*

514 BATCH-THROUGH CIRCULATION DRIER dry
f sécheur *m* à circulation intermittente
e secador *m* de circulación intermitente
i essiccatore *m* a circolazione intermittente
n niet-continue doorloopdroogmachine
d diskontinuierlicher Durchlauftrockner *m*

515 BATCH-TYPE FURNACE htt
f four *m* intermittent
e horno *m* discontinuo
i forno *m* discontinuo
n discontinue oven, oven met stationaire lading
d Stapelofen *m*

516 BATCHER gen
f machine *f* de dosage
e máquina *f* dosificadora
i macchina *f* dosatrice
n doseermachine
d Dosiermaschine *f*, Dosiervorrichtung *f*

517 BATCHING APPARATUS tex
f appareil *m* d'huilage
e aparato *m* de engrase
i apparecchio *m* d'oliaggio
n oliewolf
d Batschgerät *n*

518 BATH elc,gen
f bain *m*
e baño *m*
i bagno *m*
n bad *n*
d Bad *n*

519 BATH LINING elc
f revêtement *m* du bain
e revestimiento *m* del baño
i rivestimento *m* del bagno
n badbekleding
d Badauskleidung *f*

520 BATING VAT, PUERING VAT lea
f cuve *f* de chipage
e cuba *f* de maceración
i tino *m* di macerazione
n zweldervat *n*, zwelkuip
d Beizbrühegefäss *n*, Schwödbottich *m*

521 BATTERY, PRIMARY CELL elc
f élément *m* primaire, pile *f*
e elemento *m* primario, pila *f*
i elemento *m* primario, pila *f*
n batterij, primair element *n*, primaire cel
d Batterie *f*, Primärelement *n*

522 BATTERY CASE SEPARATOR, elc
SEPARATOR
f plaque *f* de séparation
e placa *f* de separación
i separatore *m*
n separator
d Scheider *m*, Separator *m*

523 BATTERY GRID elc
f grille *f* d'accumulateur
e rejilla *f* de acumulador
i griglia *f* d'accumulatore
n accurooster *n*
d Akkumulatorgitter *n*

524 BATTERY TESTER elc,tes
f pipette *f* pèse-acide
e probador *m* de acumuladores
i collaudatore *m* d'accumulatori
n accutester
d Akkumulatorenprüfer *m*,
Batterieprüfer *m*

525 BATTLEDORE, gla
PALLETTE
f batte *f*
e rafrentador *m*
i paletta *f*
n palet *n*
d Handbalett *n*

526 BAUDELOT COOLER, bre
DRIZZLING WORT COOLER,
IRRIGATION COOLER
f Baudelot *m*, refroidisseur *m* à
ruissellement
e enfriador *m* Baudelot
i raffreddatore *m* a pioggia
n sproeikoeler
d Berieselungskühler *m*,
Rieselkühler *m*

527 BAUXITE MINE met,min
f mine *f* de bauxite
e mina *f* de bauxita
i mina *f* di bauxite
n bauxietmijn
d Bauxitgrube *f*

528 BAYONET HEATER htt
f élément *m* de chauffage à
baïonnette
e elemento *m* de caldeo de
bayoneta
i elemento *m* riscaldatore a
baionetta
n bajonetverhittingselement *n*
d Bajonettheizelement *n*

529 BEAD CUTTING MACHINE gla
f machine *f* à couper des perles
e máquina *f* cortadora de perlas
i macchina *f* da tagliare perle
n parelsnijmachine
d Perlenschneidemaschine *f*

530 BEAKER lab
f vase *m*
e vaso *m*
i becher *m*, bicchiere *m*
n beker
d Becherglas *n*

531 BEAKER COVER GLASS, lab
WATCH GLASS
f verre *m* de montre
e vidrio *m* de reloj
i vetro *m* da orologio
n horlogeglaasje *n*
d Uhrgläschen *n*, Uhrglasschale *f*

532 BEAM HOUSE, tan
BEAM ROOM
f atelier *m* de chevalets
e taller *m* de caballetes
i officina *f* di cavalletti
n schaafboomwerkplaats
d Gerbenbankhaus *n*

533 BEATER rub
f batteur *m*
e máquina *f* batidora
i sbattitore *m* a frusta
n opklopper
d Schaumschläger *m*

534 BEATER, pap
BEATER ROLL
f cylindre *m* à couteaux
e cilindro *m* de batidora
i cilindro *m* a coltelli
n messenrol, walsrol
d Holländerwalze *f*, Messerwalze *f*

535 BEATER MIXER mix
f mélangeur *m* à battre,
mélangeur-batteur *m*
e batidora-mezcladora *f*
i mescolatore *m* sbattitore
n klutser, klutsmenger
d Schlagmischer *m*

536 BEATING ENGINE, pap
HOLLANDER
f hollandais *m*
e batidora *f*, holander *m* batidor
i mescolatore *m* olandese,
olandese *m*
n hollander, maalbak
d Holländer *m*

537 BEATING MACHINE, FOAM BEATER, FOAMING MACHINE, FROTHING MACHINE, WHISKING MACHINE — rub
f batteuse *f* à mousse, fouetteuse *f*
e batidora *f* mecánica
i frullatore *m*, produttore *m* di schiuma, sbattitore *m*
n klutsmachine, opklopmachine, schuimklopper
d Schaumschlagmaschine *f*, Schlagmaschine *f*

538 BECK, BARK — tex
f cuve *f*
e cuba *f*, tina *f*
i tino *m*
n kuip
d Küpe *f*

539 BED, SETTING — cer
f groupe *m* de cornues
e grupo *m* de retortas
i gruppo *m* di storte
n retortengroep
d Retortengruppe *f*

540 BED PLATE — cpt
f plaque *f* de base, plateau *m*
e placa *f* de asiento, placa *f* de base
i fondello *m*, piastra *f* di base
n grondplaat
d Grundplatte *f*

541 BEEHIVE COKE-OVEN, BEEHIVE FURNACE — htt
f four *m* à ruche
e horno *m* de colmena
i forno *m* ad alveare
n stolpvormige cokesoven
d Bienenkorbofen *m*

542 BEEHIVE KILN, ROUND KILN — htt
f étuve *f* à ruche
e estufa *f* de colmena
i stufa *f* ad alveare
n stolpvormige droogstoof
d Bienenkorbtrockenofen *m*

543 BEER HYDROMETER — bre,ins
f hydromètre *m* à bière
e hidrómetro *m* de cerbeza
i areometro *m* a birra
n bierareometer
d Bierwürzearäometer *n*

544 BEER STILL — bre,des
f colonne *f* à distiller le moût
e columna *f* de destilación del mosto
i colonna *f* da distillare il mosto
n wortdestilleerkolom
d Maischeentgeistungssäule *f*, Würzedestillierkolonne *f*

545 BEET SLICER — sin
f coupe-betteraves *m*
e rebañador *m* de remolachas
i affettatrice *f* di barbabietole
n bietensnijder
d Rübenschnitzelmaschine *f*

546 BEETLING MACHINE — tex
f moulin *m* à pilons
e máquina *f* de enfurtir, máquina *f* de estampar
i mulino *m* da pestare
n stampmolen
d Stampfkalander *m*

547 BEETROOT JUICE PUMP, SUGARBEET JUICE PUMP — sin
f pompe *f* à jus de betteraves
e bomba *f* de zumo de remolachas
i pompa *f* a succo di barbabietole
n ruwsappomp
d Rübensaftpumpe *f*

548 BEETROOT LIFT, SUGARBEET LIFT — sin,tra
f élévateur *m* de betteraves
e elevador *m* de remolachas
i elevatore *m* di barbabietole
n bietenophaler
d Rübenelevator *m*

549 BEETROOT PUMP, SUGARBEET PUMP — sin,tra
f pompe *f* à betteraves
e bomba *f* de remolachas
i pompa *f* a barbabietole
n bietenpomp
d Rübenpumpe *f*

550 BEETROOT SILO, SUGARBEET SILO — sin,tra
f silo *m* à betteraves
e silo *m* de remolachas
i silo *m* a barbabietole
n bietensilo
d Rübensilo *n*

551 BEETROOT WASHING MILL, SUGARBEET WASHER — sin
f laveuse *f* de betteraves
e lavadora *f* de remolachas

i lavatrice *f* di barbabietole
n bietenwasmolen
d Rübenwäsche *f*

552 BEKEN DUPLEX KNEADER mix,pla
f malaxeur *m* Beken
e amasador *m* con paletas que engranan
i impastatrice *f* a due operatori intersecanti
n bekenkneedmachine
d Beken-Kneter *m*

553 BELL (US), PLUNGER (GB) fou
f cloche *f*
e campana *f*
i campana *f*
n dompelklok
d Tauchglocke *f*

554 BELL (US), SOCKET (GB) cpt
f emboîtement *m*, manchon *m* élargi, manchon *m* mandriné
e enchufe *m*, tubo *m* abocarbado
i bicchiere *m*
n buismof, buisverwijding
d Muffenkelch *m*, Rohrmuffe *f*

555 BELL FURNACE htt
f four *m* à cloche
e horno *m* de campana
i forno *m* a campana
n klokoven
d Glockenofen *m*, Haubenofen *m*

556 BELL JAR lab
f cloche *f* en verre
e campana *f* de vidrio
i campana *f* di vetro
n glazen klok, stolp
d Glasglocke *f*

557 BELL-MOUTH INTAKE cpt
f trompe *f* d'entrée
e trompa *f* de entrada
i raccordo *m* d'imbocco, strombatura *f* d'imbocco
n inlooptromp
d Einlauftrompete *f*

558 BELL-SHAPED FUNNEL lab
f entonnoir *m* en forme de cloche
e embudo *m* en forma de campana
i imbuto *m* a campana
n klokvormige trechter
d Glockentrichter *m*

559 BELL-TYPE ANNEALING FURNACE htt
f four *m* de recuit à cloche
e horno *m* de recocido campaniforme
i forno *m* di ricottura a campana
n stolpgloeioven
d Haubenglühofen *m*

560 BELL-TYPE DIFFERENTIAL PRESSURE MANOMETER ins
f manomètre *m* différentiel type cloche
e manómetro *m* diferencial tipo campana
i manometro *m* differenziale tipo campana
n differentiële manometer met dompelklok
d Tauchglockendifferenzdruckmesser *m*

561 BELLOW-TYPE GAGE (US), BELLOW-TYPE GAUGE (GB), ANEROID MANOMETER ins
f manomètre *m* anéroïde
e manómetro *m* aneroide
i manometro *m* aneroide
n aneroïdmanometer, doosmanometer, metaalmanometer
d Aneroidmanometer *n*, Deformationsdruckmesser *m*

562 BELLOWS cpt,sap
f raccord *m* à soufflet
e fuelle *m* de dilatación
i compensatore *m* di dilatazione a soffietto
n harmonikabalg
d Faltenbalch *m*, Rohrdehnungsausgleicher *m*

563 BELT CONVEYER, BAND CONVEYER tra
f transbordeur *m* à bande, transporteur *m* à bande
e transportador *m* de cinta
i trasportatore *m* a nastro
n bandtransportinrichting
d Bandförderer *m*

564 BELT DRIER, BAND DRIER dry
f sécheur *m* à bandes transporteuses
e secadora *f* de cintas
i essiccatore *m* a nastri
n etagedroger
d Bandtrockner *m*

565 BELT FEED, tra
BAND FEED
f alimentateur *m* à bande
e alimentador *m* de cinta
i alimentatore *m* a nastro
n bandtoevoerinrichting
d Bandzuführer *m*

566 BELT FURNACE, met
CONVEYOR FURNACE
f four *m* à tapis
e horno *m* de correa transportadora
i forno *m* a nastro trasportatore
n transportbandoven
d Förderbandofen *m*

567 BENCH, gla
SIEGE
f banc *m*
e plaza *f*, solera *f*
i suola *f*
n bank
d Ofenbank *f*, Ofenherd *m*

568 BEND cpt
f coude *m*
e codo *m*
i gomito *m*
n bocht
d Krümmer *m*

569 BEND WITH VENT dis,lab
f coude *m* avec prise de vide
e codo *m* con racor tubular
i gomito *m* con raccordo tubolare
n boog met aangezette buis
d Bogen *m* mit Rohransatz

570 BENEFICIATION PLANT (US), min
ORE DRESSING PLANT (GB)
f installation *f* de traitement de minerais
e equipo *m* para beneficio de minerales
i impianto *m* di trattamento di minerali
n ertsbewerkingsinstallatie
d Erzaufbereitungsanlage *f*

571 BENT ADAPTER dis,lab
f tube *m* de recette
e tubo *m* acodado
i tubo *m* a gomito
n gebogen opzet
d gebogener Destilliervorstoss *m*

572 BENT BAR ANODE, elc
CURVED BAR ANODE
f anode *f* en barre incurvée
e ánodo *m* de barra curvada
i anodo *m* a barra ricurva
n gebogen staafanode
d gebogene Stabanode *f*, gebogene Stangenanode

573 BENZENE RECOVERY PLANT fue
f installation *f* de récupération du benzol
e equipo *m* de recuperación del benzol
i impianto *m* di ricupero del benzolo
n benzolterugwinningsinstallatie
d Benzolrückgewinnungsanlage *f*

574 BENZENE SCRUBBER, cle,pet
BENZENE WASHER
f laveur *m* de gaz au benzol
e lavador *m* de gas con benzol
i depuratore *m* di gas mediante benzolo
n benzol-gaswasser
d Benzolgaswäsche *f*

575 BERYLLIA-LINED FURNACE htt
f four *m* à garnissage d'oxyde de béryllium
e horno *m* de guarnición de berilia
i forno *m* a guarnizione d'ossido di berillio
n oven met berylliumoxydevoering
d Ofen *m* mit Berylliumoxydfutter

576 BERYLLIUM MODERATED nue
REACTOR,
BERYLLIUM REACTOR
f réacteur *m* modéré au béryllium
e reactor *m* moderado de berilio
i reattore *m* a berillio, reattore *m* moderato a berillio
n reactor met beryllium moderator, reactor met berylliumremstof
d berylliummoderierter Reaktor *m*, Berylliumreaktor *m*

577 BEWEL, gla
BOW, LADLE SHANK, SHANK
f brancard *m*
e maneral *m*
i portante *m*
n draagbeugel, handbeugel
d Tiegelschere *f*, Tragschere *f*

578 BICHROMATE BATH, cle,met
DICHROMATE BATH
f bain *m* au bichromate
e baño *m* de bicromato

i bagno *m* al bicromato
n bichromaatbad *n*
d Bichromatbad *n*

579 BILLET pla,rub
f poupée *f*
e cigarro *m*, paquete *m*
i billetta *f*, ceppo *m*, spezzone *m*
n pop
d Puppe *f*

580 BIMETALLIC INSTRUMENT ins
f instrument *m* bimétallique
e aparato *m* bimetálico
i strumento *m* bimetallico
n bimetaalinstrument *n*
d Bimetallinstrument *n*

581 BIMETALLIC STRIP ins
f bilame *m*
e lámina *f* bimetálica
i lamina *f* bimetallica
n bimetaalstrook
d Bimetallstreifen *m*

582 BIMETALLIC TUBE cpt
f tube *m* bimétallique
e tubo *m* bimetálico
i tubo *m* bimetallico
n bimetaalbuis
d Bimetallrohr *n*

583 BIN, tra
BUCKET
f benne *f*
e cajón *m*
i benna *f*
n bak, mand
d Kasten *m*, Kübel *m*

584 BIO-AERATION PLANT iwt
f installation *f* de bio-aération
e equipo *m* de bioaireación
i impianto *m* di bioaerazione
n biologische aëratie-installatie, slibactiveringsinstallatie
d biologische Aerationsanlage *f*, Schlammbelebungsanlage *f*

585 BIOLOGICAL CLEANING cle,iwt
PLANT
f installation *f* d'épuration biologique
e equipo *m* de depuración biológica
i impianto *m* di depurazione biologica
n biologische reinigingsinstallatie
d biologische Reinigungsanlage *f*

586 BIOLOGICAL FILTER, iwt
BACTERIA BED, BACTERIA FILTER
f filtre *m* bactérien, filtre *m* biologique
e filtro *m* bacteriano
i letto *m* batterico, letto *m* biologico
n bacteriefilter *n*
d biologischer Körper *m*

587 BIOLOGICAL HOLE nue
f trou *m* d'expériences biologiques
e cavidad *f* de experimentaciones biológicas
i cavità *f* di sperimentazioni biologiche
n holte voor biologische doeleinden
d Strahlenkanal *m* für biologische Zwecke

588 BIOLOGICAL SHIELD nue,sap
f écran *m* biologique
e pantalla *f* biológica
i schermo biologico
n biologisch scherm *n*
d biologischer Schirm *m*

589 BIOSTABILIZER bac
f biostabilisateur *m*, tambour *m* de fermentation
e bioestabilizador *m*
i biostabilizzatore *m*
n gisttrommel
d Gärtrommel *f*

590 BIPOLAR ELECTRODE, elc
SECONDARY ELECTRODE
f électrode *f* bipolaire
e electrodo *m* bipolar
i elettrodo *m* bipolare
n bipolaire elektrode
d Bipolarelektrode *f*

591 BIRD CAGE nue,tra
f cage *f* de transport
e jaula *f* de transporte
i gabbia *f* di trasporto
n transportkooi
d Transportkäfig *m*

592 BISCUIT, mat
BISCUIT PORCELAIN, BISQUE
f biscuit *m*, porcelaine *f* dégourdie
e bizcocho *m* de porcelana
i biscotto *m* di porcellana
n biscuit *n*
d Biskuit *n*

593 BISCUIT FIRING KILN cer,htt
f four *m* à biscuiter
e horno *m* de bizcochar
i forno *m* da biscottare
n biscuitoven
d Biskuitofen *m*

594 BITE, pla,rub
MILL OPENING, ROLL CLEARANCE
f écartement *m*, intervalle *m*, jeu *m*
e espacio *m* perjudicial
i giuoco *m*, interdistanza *f*
n walsafstand, walsspleet
d Spaltweite *f*, Walzenspalt *m*

595 BLACK ANNEALING OVEN, met
OPEN ANNEALING OVEN
f four *m* de recuit en noir
e horno *m* de recocido en negro
i forno *m* di ricottura in nero
n zwartgloeioven
d Schwarzglühofen *m*

596 BLACK ANODIZING PLANT coa
f installation *f* d'oxydation anodique en noir
e equipo *m* de anodización en negro
i impianto *m* d'ossidazione anodica in nero
n zwartanodiseerinstallatie
d Schwarzeloxieranlage *f*

597 BLACK ASH FURNACE, aic
BALLING FURNACE, SODA FURNACE
f four *m* de soude
e horno *m* de carbonatar, horno *m* de sosa
i forno *m* di soda
n sodaoven
d Sodaofen *m*

598 BLACKENING APPARATUS coa
f appareil *m* de noircissement
e aparato *m* de ennegrecimiento
i apparecchio *m* d'annerimento
n zwartingsapparaat *n*
d Schwärzungsgerät *n*

599 BLACKENING BATH pho
f bain *m* de noircissement
e baño *m* de ennegrecimiento
i bagno *m* d'annerimento
n zwartingsbad *n*
d Schwärzungsbad *n*

600 BLADE, cpt
ROTOR
f palette *f*, rotor *m*
e paleta *f*, rotor *m*
i paletta *f*, rotore *m*
n kneedas, rotor
d Knetkörper *m*, Rotor *m*

601 BLADE AGITATOR, mix
BLADE MIXER, ARM AGITATOR, ARM MIXER
f agitateur *m* à palettes
e agitador *m* de paletas
i agitatore *m* a palette
n bladroerder
d Balkenrührwerk *n*, Paddelrührwerk *n*, Stabrührwerk *n*

602 BLANK CARBURIZING OVEN, htt,met
PSEUDOCARBURIZING OVEN
f four *m* de pseudocémentation
e horno *m* de seudocementación
i forno *m* di pseudocementazione
n pseudocementeeroven
d Pseudozementierofen *m*

603 BLANK MOLD (US), gla
BLANK MOULD (GB), PARISON MOLD (US)
f moule *m* de paraison
e molde *m* de parisón
i stampo *m* di sbozzo greggio
n glasklompvorm
d Külbelform *f*

604 BLANK NITRIDING OVEN, htt,met
PSEUDONITRIDING OVEN
f four *m* de pseudonitruration
e horno *m* de seudonitruración
i forno *m* di pseudonitrurazione
n pseudonitreeroven
d Pseudonitrierofen *m*

605 BLANKET nue
f enveloppe *f* fertile
e envoltura *f* fértil
i mantello *m*
n mantel
d Mantel *m*

606 BLANKING MACHINE pla,rub
f machine *f* à découper
e máquina *f* troqueladora, prensa para recortar
i pressa *f* da tagliare
n uitstansmachine
d Stanzmaschine *f*

607 BLAST BURNER, bur,lab
BLOWLAMP
f brûleur *m* à haute pression
e quemador *m* de alta presión

i becco *m* ad alta pressione
n blaaslamp
d Hochdruckbrenner *m*

608 BLAST FURNACE STOVE, htt,met
COWPER STOVE, HOT BLAST STOVE
f appareil *m* régénérateur de vent chaud
e aparato *m* Cowper
i apparecchio *m* di Cowper, preriscaldatore *m* del vento
n stenen windverhitter
d Cowper *m*, steinerner Winderhitzer *m*

609 BLAST ROASTING OVEN, htt,met
ROAST-SINTERING OVEN
f four *m* à grillage de frittage, four *m* de frittage sur grille
e horno *m* de sinterización por tostación
i forno *m* ad arrostimento di sinterizzazione
n sinterroostoven
d Sinterröstofen *m*

610 BLEACH CHAMBER ble
f bain *m* de chlorure de chaux
e baño *m* de cloruro de calcio
i bagno *m* di cloruro di calcio
n bleekkalkbad *n*
d Chlorkalkbad *n*

611 BLEACHERY, ble,tex
BLEACHING PLANT
f blanchisserie *f*, établissement *m* de blanchiment
e equipo *m* de blanqueo
i impianto *m* da sbianca, impianto *m* di candeggio
n bleekinrichting, blekerij
d Bleichanlage *f*, Bleicherei *f*

612 BLEACHING BATH ble,tex
f bain *m* de blanchiment
e baño *m* de blanqueo
i bagno *m* di candeggio
n bleekbad *n*
d Bleichflotte *f*

613 BLEACHING BOILER ble,pap
f cuve *f* de blanchiment
e cuba *f* de blanqueo
i serbatoio *m* d'imbiancatura
n bleekkast
d Bleichkessel *m*

614 BLEACHING CABINET ble,pap
f armoire *f* de blanchiment
e armario *m* de blanqueo
i armadio *m* d'imbiancatura
n bleekkast
d Bleichkasten *m*

615 BLEACHING CLAY LAYER cle,fue
f couche *f* d'argile décoloratrice
e capa *f* de arcilla descoloradora
i strato *m* d'argilla decoloratrice
n ontkleuringskleilaag
d Entfärbungstonschicht *f*

616 BLEACHING-FIXING BATH pho
f bain *m* de blanchiment et fixage
e baño *m* de blanqueo y fijado
i bagno *m* d'imbianchimento e di fissaggio
n blekend fixeerbad *n*
d Bleichfixierbad *n*

617 BLEACHING HOLLANDER, ble,pap
POACHER, POTCHER
f hollandais *m* à blanchiment
e batidora *f* de blanqueo
i olandese *m* ad imbiancatura
n bleekhollander
d Bleichholländer *m*

618 BLEACHING MACHINE ble,tex
f machine *f* de blanchiment
e máquina *f* para blanqueo
i macchina *f* di candeggio
n bleekmachine
d Bleichmaschine *f*

619 BLEACHING TOWER ble,pap
f tour *f* de blanchiment
e torre *f* de blanqueo
i torre *f* da imbiancare
n bleektoren
d Bleichturm *m*

620 BLEND TANK mix,rub
f réservoir *m* de mélanges
e recipiente *m* de mezclas
i marnetta *f* per mescolanze
n mengselbak, mengseltrog
d Mischbehälter *m*

621 BLENDER, mix
MIXER
f mélangeur *m*
e mezclador *m*
i mescolatore *m*
n menger
d Mischer *m*

622 BLOCK MILL, brk
PAN MILL
f moulin *m* à blocs de pierre
e molino *m* con bloques de piedra

i mulino *m* a blocchi di petra
n stootblokmolen
d Stossblockmühle *f*

623 BLOCK MOLD (US), PRESSING MOULD (GB) gla
f moule *m* à presser
e molde *m* a prensar
i stampo *m* di pressa
n persvorm
d Pressform *f*

624 BLOCK PRESS (GB), BAKING PRESS (US) pla
f presse *f* à blocs
e prensa *f* de bloques
i pressa *f* per blocchi
n blokpers
d Blockpresse *f*

625 BLOCK PRINTING APPARATUS tex
f imprimeuse *f* à moule
e aparato *m* de estampado con molde
i stampatrice *f* a forma
n houtblokdrukmachine
d Holzmodelldruckmaschine *f*

626 BLOCK SKIVING MACHINE (US), BLOCK SLICING MACHINE (GB) pla
f découpeuse *f*, trancheuse *f*, tronçonneuse *f*
e máquina *f* cortadora de bloques
i taglierina *f*
n bloksnijmachine
d Blockschneidmaschine *f*

627 BLOCKING MACHINE (GB), STAMPING PRESS (US) pla
f presse *f* à marquer
e prensa *f* de embutir, prensa *f* de estampar
i pressa *f* per stampigliare
n stanspers
d Prägepresse *f*

628 BLOW CASE (GB), ACID EGG (US), ACID ELEVATOR (US), MONTEJUS cor
f monte-jus *m*
e elevador *m* de presión, monta-ácidos *m*, montajugos *m*
i montaacidi *m*, montaliquidi *m*
n drukvat *n*
d Druckbirne *f*, Druckfass *n*, Montejus *m*

629 BLOW-DOWN VALVE cpt,sap
f soupape *f* d'échappement, soupape *f* de purge
e válvula *f* de purga
i valvola *f* di scarico, valvola *f* di spurgo
n afblaasklep, spuiklep
d Abblaseventil *n*

630 BLOW GUN cle,pla
f pistole *f* à air comprimé
e pistola *f* de aire comprimida
i pistola *f* ad aria compressa
n blaasluchtpistool *n*, persluchtpistool *n*
d Pressluftpistole *f*

631 BLOW HEAD gla
f tête *f* de soufflage
e cabeza *f* de soplado
i testa *f* di soffiatura
n blaaskop
d Blaskopf *m*

632 BLOW MOLD (US), BLOW MOULD (GB) gla
f moule *m* à souffler
e molde *m* a soplar
i stampo *m* a soffiare
n blaasvorm
d Blasform *f*

633 BLOW-OFF PIPE cpt
f tubelure *f* d'évacuation, tuyau *m* d'échappement
e tubo *m* de descarga
i tubo *m* di scarico
n afblaaspijp, uitlaatpijp
d Abblasrohr *n*, Auspuffrohr *n*

634 BLOW-OUT PIPETTE lab
f pipette *f* à souffler
e pipeta *f* de descarga soplada
i pipetta *f* a soffiamento
n uitblaaspipet
d Ausblaspipette *f*

635 BLOW STICK, SHELL CORE pla,pre
f noyau-carapace *m*
e macha *f* en cáscara
i anima *f* a guscio
n holle kern, schaalkern
d Hohlkern *m*

636 BLOW-THROUGH VALVE, SNIFTING VALVE cpt
f reniflard *m*
e válvula *f* de purga
i valvola *f* di scarico
n snuifklep
d Schnarchventil *n*, Schnarrventil *n*, Schnüffelventil *n*

637 BLOWER pre,sap
f souffIerie *f*, souffleuse *f*
e soplante *f*
i soffiatore *m*, soffieria *f*
n aanjager, blower
d Gebläse *n*

638 BLOWER, clo
FAN, VENTILATOR
f ventilateur *m*
e ventilador *m*
i ventilatore *m*
n ventilator
d Lüfter *m*, Ventilator *m*

639 BLOWING IRON, gla
BLOWPIPE
f fêle *f* de souffleur, felle *f* de verrier
e caña *f* de vidriero
i canna *f* di soffiatore
n blaaspijp, glasblazerspijp
d Glasmacherpfeife *f*

640 BLOWING ROLLER tex
f rouleau *m* vaporisateur
e rodillo *m* vaporizador
i rullo *m* vaporizzatore
n stoombadrol
d Dampfbadrolle *f*

641 BLOWPIPE, lab
TORCH
f chalumeau *m*
e soplete *m*
i cannello *m* ferruminatorio
n blaaspijp
d Gebläselampe *f*, Lötrohr *n*

642 BLOWTORCH BURNER, htt,lab
OXYHYDROGEN BURNER
f chalumeau *m* oxhydrique
e soplete *m* oxídrico
i cannello *m* ossidrico
n knalgasbrander
d Knallgasgebläse *n*

643 BLUE ANNEALING OVEN htt
f four *m* de bleuissage
e horno *m* de pavonado
i forno *m* da scaldare al calore blu
n blauwgloeioven
d Blauglühofen *m*

644 BLUE DIP elc
f bain *m* bleu
e baño *m* azul
i bagno *m* azzurro
n kwikbeitsbad *n*
d Blaubrenne *f*

645 BLUE LINE BURETTE lab
f burette *f* à trait bleu
e bureta *f* con trazo azul
i buretta *f* a tratto azzurro
n buret met ingebrande blauwe emailstreep
d Schellbachbürette *f*

646 BLUE TONING IRON BATH pho
f bain *m* de virage bleu au fer
e baño *m* de viraje azul al hierro
i bagno *m* di viraggio azzurro al ferro
n ijzerblauwkleurbad *n*
d Eisenblautonbad *n*

647 BLUE WATER GAS gas
GENERATOR
f générateur *m* de gaz à l'eau
e generador *m* de gas de agua
i generatore *m* di gas d'acqua
n watergasgenerator
d Wassergaserzeuger *m*

648 BLUEING BATH (GB), tex
BLUING BATH (US)
f bain *m* d'azurage, bain *m* de bleutage
e baño *m* de azulado
i bagno *m* per azzurrare, bagno *m* per azzurreggiare
n blauwkuip
d Bläubad *n*, Bläuküpe *f*

649 BLUNGER cer
f mélangeur *m* d'argile
e mezclador *m* de pasta
i impastatrice *f* a coltelli per argilla
n kleikneder
d Tonmischer *m*

650 BOARDING MACHINE, lea
GRAINING MACHINE
f greneuse *f*
e graneladora *f*, máquina *f* de granelar
i macchina *f* da granire
n krispelmachine
d Krispelmaschine *f*

651 BOB, pol
POLISHING BOB
f touffe *m* polisseur
e bruñidor *m*
i brunitoio *m*
n polijstpop
d Schwabbel *f*

652 BOBBIN (US), elc
DOLLY (GB)
f aggloméré *m*
e aglomerado *m*
i impasto *m* depolarizzante
n buidel
d Beutel *f*, Puppe *f*

653 BODY, cpt
CHASE, FRAME
f bâti *m*, châssis *m*
e bastidor *m*, cuna *f*
i custodia *f*, telaio *m*
n freem *n*
d Chassis *n*, Gehäuse *n*,
Rahmen *m*

654 BODY APRON, sap
LEAD-RUBBER APRON
f tablier *m* opaque
e delantal *m* de protección
i grembiale *m* di protezione
n loodrubberschort
d Bleigummischürze *f*

655 BODY OF THE ROLL, cpt
ROLLER BODY
f corps *m* du cylindre
e cuerpo *m* del cilindro
i corpo *m* del cilindro
n walslichaam *n*
d Walzenkörper *m*

656 BOGIE KILN (GB), cer,gla,htt
CAR TUNNEL KILN (US),
FIXED-FIRE ZONE KILN,
TRUCK CHAMBER KILN
f étuve *f* à chariot
e estufa *f* de carrillo
i stufa *f* a carrello
n wagendroogstoof
d Wagentrockenofen *m*

657 BOGIE-TYPE FURNACE htt
f four *m* à chariot
e horno *m* de carrillo
i forno *m* a carrello
n wagenoven
d Beschickungswagenofen *m*,
Wagenofen *m*

658 BOILER boi
f chaudière *f*
e caldera *f*
i caldaia *f*
n ketel
d Kessel *m*

659 BOILER, boi
BOILING POT
f chaudron *m*
e caldero *m*
i bollitore *m*
n kookketel
d Kochkessel *m*

660 BOILER FITTINGS, boi,cpt
BOILER MOUNTINGS
f accessoires *pl* de chaudière
e accesorios *pl* de caldera
i accessori *pl* di caldaia
n appendages *pl*
d Kesselarmaturen *pl*

661 BOILER HOUSE, boi
BOILER ROOM
f salle *f* des chaudières
e sala *f* de las calderas
i sala *f* delle caldaie
n ketelhuis *n*
d Kesselhaus *n*

662 BOILER PACKING boi,cpt
f garniture *f* de chaudière
e empaquetadura *f* de caldera
i guarnizione *f* di caldaia
n ketelpakking
d Kesselpackung *f*

663 BOILER PLATE boi,cpt
f tôle *f* de chaudière
e chapa *f* de caldera
i lamiera *f* di caldaia
n ketelplaat
d Kesselplatte *f*

664 BOILER RING boi,cpt
f joint *m* de chaudière
e anillo *m* de caldera
i anello *m* di caldaia
n ketelpakkingring
d Kesselring *m*

665 BOILING CAPILLARY boi,ins
f tube *m* capillaire d'ébullition
e tubo *m* capilar de ebullición
i tubo *m* capillare d'ebollizione
n kookcapillair *n*
d Siedekapillare *f*

666 BOILING CHIP, boi,mat
BOILING STONE
f corps *m* solide d'ébullition
e cuerpo *m* sólido de ebullición
i solido *m* immerso d'ebollizione
n kooksteentje *n*
d Ansiedescheibe *f*, Porzellan-
scheibe *f*, Siedesteinchen *pl*

667 BOILING FLASK lab
f matras *m*
e matraz *m*

i matraccio *m*
n kookkolf
d Kochkolben *m*

668 BOILING PAN bre
f chaudière *f* à moût
e caldera *f* de ebullición
i caldaia *f* del luppolo
n kookpan
d Siedepfanne *f*

669 BOILING TUBE, CHEMICAL TEST TUBE lab
f tube *m* à essai chimique
e tubo *m* de ensayo químico
i tubo *m* di saggio chimico
n chemisch reageerbuisje *n*
d chemisches Reagenzglas *n*

670 BOILING WATER IRRADIATION LOOP nue
f circuit *m* fermé à irradiation d'eau bouillante
e circuito *m* cerrado de irradiación de agua hirviente
i circuito *m* chiuso ad irradiazione d'acqua bollente
n gesloten circuit *n* voor het bestralen van kokend water
d Kreislauf *m* zum Bestrahlen von siedendem Wasser

671 BOILING WATER REACTOR nue
f réacteur *m* à eau bouillante
e reactor *m* de agua hirviente
i reattore *m* ad acqua bollente
n kokend-waterreactor
d Siedereaktor *m*, Siedewasserreaktor *m*, Verdampferreaktor *m*

672 BOLOMETER ins
f bolomètre *m*
e bolómetro *m*
i bolometro *m*
n bolometer
d Bolometer *n*

673 BOLTING CLOTH sor
f bluteau *m*, blutoir *m*
e gasa *f* de cerner, tela *f* de tamiz
i buratto *m*, stamigna *f*
n zeefdoek
d Beuteltuch *n*, Siebtuch *n*

674 BOMB CALORIMETER ins
f bombe *f* calorimétrique, calorimètre *m* de Berthelot
e calorímetro *m* de combustión
i calorimetro *m* a bomba
n bom van Berthelot
d kalorimetrische Bombe *f*

675 BOND TESTING MACHINE glu,tes
f machine *f* pour essai de collage
e máquina *f* de ensayo de adherencia
i macchina *f* per prova d'adesione
n hechtkrachtbeproevingstoestel *n*
d Haftfestigkeitsmessgerät *n*

676 BONE CRUSHER brk
f broyeur *m* d'os
e molino *m* de huesos
i frantoio *m* per ossa
n beendermolen
d Knochenbrecher *m*

677 BONE GLUE FACTORY glu
f fabrique *f* de colle d'os
e fábrica *f* de cola de huesos, fábrica *f* de osteocola
i fabbrica *f* di colla d'ossa
n beenderlijmfabriek
d Knochenleimfabrik *f*

678 BONEBLACK FILTER fil,sin
f filtro *m* en charbon animal
e filtro *m* de carbón animal, filtro *m* de negro animal
i filtro *m* di nero d'ossa
n beenderkoolfilter *n*, beenzwartfilter *n*
d Knochenkohlefilter *m*

679 BOOK TYPE PRESS pla,rub
f presse *f* à charnière horizontale
e prensa *f* de bisagras horizontales
i pressa *f* a cerniera orizzontale
n horizontale scharnierpers, pianoscharnierpers
d horizontale Scharnierpresse *f*

680 BOOSTER EXPANDER liq
f machine *f* d'expansion à compresseur
e máquina *f* de expansión con compresor
i macchina *f* ad espansione con compressore
n expansiemachine met compressor
d Expansionsmaschine *f* mit Kompressor

681 BOOSTER FAN pre
f ventilateur *m* additionnel
e ventilador *m* reforzador

i ventilatore *m* di rinforzo
n aanjager
d Zusatzventilator *m*

682 BOOSTER PUMP pre
f pompe *f* auxiliaire
e bomba *f* reforzadora
i pompa *f* di rinforzo
n hulppomp
d Förderpumpe *f*, Hilfspumpe *f*, Zusatzpumpe *f*

683 BOOSTER RAM pre
f piston-plongeur *m* supplémentaire
e pistón *m* suplementario
i pistone *m* supplementare
n hulppersplunjer
d Zusatzkolben *m*

684 BOOT (US), gla
COVER STONE, POPETTE (GB)
f brique *f* de couverture
e ladrillo *m* de recubrimiento
i mattone *m* di copertura
n afdeksteen
d Abdeckstein *m*

685 BORAL SHIELD nue,sap
f écran *m* en boral
e pantalla *f* de boral
i schermo *m* di boral
n boralscherm *n*
d Boralschirm *m*

686 BORED PLATEN PRESS, rub
DRILLED PLATEN PRESS
f presse *f* à plateaux forés
e prensa *f* con platos taladrados
i pressa *f* con canali ricavati nei piani
n etagepers met geboorde verwarmingsplaten
d Plattenpresse *f* mit gebohrten Kanälen

687 BORED WELL, cpt
DRILLED WELL
f cylindre *m* foré
e cilindro *m* taladrado
i cilindro *m* forato
n geboorde wals, holle wals
d ausgebohrte Walze *f*, Hohlwalze *f*

688 BORON CARBIDE MORTAR lab
f mortier *m* en carbure de bore
e mortero *m* de carburo de boro
i mortaio *m* in carburo di boro
n boriumcarbide mortier, boriumcarbide vijzel
d Borkarbidmörser *m*

689 BORON CHAMBER ins,nue
f chambre *f* d'ionisation à bore
e cámara *f* de ionización de boro
i camera *f* d'ionizzazione a boro
n boriumionisatievat *n*
d Borionisationskammer *f*

690 BOSH gla
f cuve *f* de refroidissement
e cuba *f* de enfriamiento
i mastello *m* da raffreddamento
n koelkuip
d Kühlbottich *m*

691 BOTTLE ana,lab
f fiole *f*, flacon *m*
e frasco *m*
i boccetta *f*, fiasco *m*
n fles, flesje *n*
d Flasche *f*, Fläschen *n*

692 BOTTLE-CLEANING dai
MACHINE
f rince-bouteilles *m*
e máquina *f* lavabotellas
i sciacquatrice *f* di bottiglie
n flessenspoelmachine
d Flaschenspülmaschine *f*

693 BOTTLE SOAKING bre,nut
MACHINE,
SOAKING WHEEL
f roue *f* trempeuse
e remojadora *f* de botellas
i maceratrice *f*
n flessenreiniger
d Flaschenweiche *f*

694 BOTTLING DEPARTMENT bre
f bouteillerie *f*, canetterie *f*
e sala *f* de embotellado
i bottigliera *f*
n bottelarij
d Flaschenkellerei *f*

695 BOTTLING MACHINE bre
f machine *f* d'embouteillage
e máquina *f* de embotellar
i imbottigliatrice *f*
n flessenvulmachine
d Abfüllmaschine *f*

696 BOTTLING TANK bre
f cuve *f* de soutirage
e tanque *m* de presión
i serbatoio *m* a pressione
n bottelarijtank
d Drucktank *m*

697 BOTTOM BACKING PLATE cpt,pla
f appui *m* inférieur, plaquette *f* d'appui
e contraplaca *f* inferior
i piastra *f* di taglio inferiore
n onderste stansplaat, stansonderlegger
d Stanzunterlage *f*

698 BOTTOM CLAMP PLATE (US), REAR SHOE (GB) cpt
f plaque *f* inférieure
e placa *f* trasera de sujeción del molde
i piastra *f* di base mobile
n onderste opspanplaat
d Aufspannkörper *m* hinten

699 BOTTOM-DISCHARGE CAR, BOTTOM-DUMPING CAR, HOPPER CAR tra
f wagon *m* de déchargement par le fond
e vagón *m* de descarga por el fondo
i carro *m* a tramoggia
n onderlosser
d Bodenentlader *m*, Bodenentleerer *m*

700 BOTTOM ELECTRODE htt
f électrode *f* de sole
e electrodo *m* de solera
i elettrodo *m* di suola
n bodemelektrode
d Bodenelektrode *f*

701 BOTTOM FORCE PRESS, BOTTOM RAM PRESS pre
f presse *f* à piston inférieur
e prensa *f* de émbolo inferior
i pressa *f* a pistone inferiore
n pers met onderliggende plunjer
d Unterkolbenpresse *f*

702 BOTTOM HALF MOLD (US), BOTTOM HALF MOULD (GB) cpt
f demi-moule *m* inférieur
e parte *f* inferior del molde
i semistampo *m* inferiore
n onderste vormhelft
d unterer Formteil *m*

703 BOTTOM MOUNTING PLATE cpt
f plateau *m* inférieur pour fixation du moule
e plato *m* inferior para montaje del molde
i piatto *m* inferiore per il montaggio dello stampo
n onderste opspanplaat
d untere Aufspannplatte *f*

704 BOTTOM PLATE cpt
f plateau *m* inférieur
e placa *f* base, placa *f* de fondo
i piatto *m* inferiore
n grondplaat, onderplaat
d Bodenplatte *f*, Grundplatte *f*, Unterplatte *f*

705 BOTTOM RAM PRESS, UP-STROKE PRESS pla
f presse *f* à piston ascendant, presse *f* ascendante
e prensa *f* de plato inferior móvil
i pressa *f* ascendente
n pers met onderplunjer
d Unterdruckpresse *f*, Unterkolbenpresse *f*

706 BOTTOM ROLL pre
f cylindre *m* inférieur
e cilindro *m* inferior
i cilindro *m* inferiore
n onderste wals
d Unterwalze *f*

707 BOTTOMING BATH tex
f bain *m* de teinture préliminaire
e baño de teñido preliminar
i bagno *m* di tintura preliminare
n voorverfbad *n*
d Vorfärbebad *n*

708 BOURDON GAUGE (GB), CHUCK GAGE (US), PRESSURE GAUGE (GB) ins
f manomètre *m* de Bourdon
e manómetro *m* de Bourdon
i manometro *m* di Bourdon
n bourdonmanometer
d Bourdonfedermanometer *n*

709 BOW, BEWEL, LADLE SHANK, SHANK gla
f brancard *m*
e maneral *m*
i portante *m*
n draagbeugel, handbeugel
d Tiegelschere *f*, Tragschere *f*

710 BOWL, ROLL gen
f cylindre *m*
e cilindro *m*
i cilindro *m*
n wals
d Walze *f*

711 BOWL CALENDER (GB), ROLL CALENDER (US) pap
f calandre *f* à cylindres
e calandra *f* de cilindros
i calandra *f* a cilindri
n walskalander
d Walzenkalander *m*

712 BOX ANNEALING OVEN, CLOSE ANNEALING OVEN, POT ANNEALING OVEN htt,met
f four *m* de recuit en caisse
e horno *m* de recocido en caja
i forno *m* di ricottura in cassetta
n kistgloeioven
d Kastenglühofen *m*

713 BOX CARBURIZING OVEN, PACK CARBURIZING OVEN htt,met
f four *m* de cémentation en caisse
e horno *m* de cementación en caja
i forno *m* di cementazione in cassetta
n kistcementeeroven
d Kastenzementierofen *m*

714 BOX FEEDER cer
f appareil *m* alimentateur d'argile
e aparato *m* alimentador de arcilla
i apparecchio *m* alimentatore d'argilla
n kleitoevoerapparaat *n*
d Tonzufuhrgerät *n*

715 BOX FURNACE, MUFFLE FURNACE coa,htt
f four *m* à moufle
e horno *m* de mufla
i forno *m* a muffola
n moffeloven
d Muffelofen *m*

716 BOX MALTING PLANT bre
f malterie *f* à cases
e maltería *f* en cajas
i malteria *f* a casse
n kastenmouterij
d Kastenmälzerei *f*

717 BOX NEGATIVE elc
f plaque *f* à caisson
e placa *f* de casetones
i piastra *f* negativa
n doosplaat
d Kastenplatte *f*

718 BRANCH cpt
f branchement *m*, dérivation *f*
e derivación *f*, pieza *f* de injerto
i derivazione *f*, diramazione *f*
n aftakking, spruitstuk *n*
d Abzweigung *f*

719 BRASS PLATING PLANT coa,elc
f installation *f* de placage au laiton
e equipo *m* de latonado
i impianto *m* d'ottonatura elettrolitica
n vermessinginstallatie
d Vermessingunganlage *f*

720 BRAZIER htt
f brasero *m*
e brasero *m*
i braciere *m*
n kolenkomfoor *n*
d Kohlenpfanne *f*

721 BRAZING PLANT wel
f installation *f* de soudure forte
e equipo *m* de soldadura fuerte
i impianto *m* di basatura
n hardsoldeerinstallatie
d Hartlötanlage *f*

722 BREAK ROLL, BREAKING ROLL brk,nut
f cylindre-broyeur *m*
e cilindro *m* agramador
i cilindro *m* di frantumazione
n breekwals
d Brechwalze *f*

723 BREAKDOWN MILL, BREAKER, CRACKER, CRACKING MILL brk
f broyeur *m* à cylindres, concasseur *m* à cylindres
e machacadora *f*, trituradora *f*
i frantoio *m* a cilindri, rompitore *m* a cilindri
n breekwalswerk *n*
d Brecher *m*, Brecherwalzwerk *n*, Reisswalzwerk *n*

724 BREAKER pap
f pilon *m*
e machacadora *f* de trapos
i raffinatrice *f*, tagliastracci *m*
n lompenwolf
d Halbzeugholländer *m*, Lumpenzerreisser *m*

725 BREAKER PLATE, SPIDER pre, rub
f croisillon *m*, étoile *f*
e cruceta *f* esparcidora
i crocetta *f*, crociera *f* spalmatrice

n centerstuk *n* voor doorn
d Armkreuz *n*

726 BREAKER PLATE (US), STRAINER (GB) cpt,sor
f grille *f*
e criba *f*
i disco *m* forato
n geperforeerde plaat
d Lochscheibe *f*

727 BREAST ROLL pap
f cylindre *m* antérieur
e cilindro *m* anterior
i cilindro *m* anteriore
n borstwals
d Brustwalze *f*

728 BREAST WALL gla
f barrage *m* latéral
e pared *f* lateral
i parete *f* laterale
n zijwand
d Seitenwand *f*

729 BREEDER REACTOR nue
f réacteur *m* autorégénérateur, réacteur *m* régénérateur
e reactor *m* autorregenerador, reactor *m* criador
i reattore *m* convertitore, reattore *m* rigeneratore
n aanmaakreactor, kweekreactor
d Brüter *m*, Brutreaktor *m*

730 BREW-HOUSE, BREWING ROOM bre
f salle *f* de brassage
e sala *f* de cocimiento
i sala *f* di cottura
n brouwhuis *n*
d Sudhaus *n*

731 BREW KETTLE, COPPER bre
f boiler *m*
e caldera *f* de cocción
i bollitore *m*
n brouwketel
d Siedekessel *m*

732 BREWERY bre
f brasserie *f*
e cervecería *f*, fábrica *f* de cerveza
i birreria *f*, fabbrica *f* di birra
n brouwerij
d Brauerei *f*

733 BRICK DRIER bui
f four *m* de séchage de briques
e horno *m* para secar ladrillos, secadero *m* de ladrillos
i forno *m* per l'essiccamento dei mattoni
n baksteendroogoven
d Ziegeltrockenofen *m*

734 BRICK FIELD, BRICK YARD cer
f briqueterie *f*
e ladrillar *m*
i mattonaia *f*
n steenbakkerij, steenfabriek
d Ziegelei *f*

735 BRICK KILN htt
f four *m* à briques
e horno *m* de cocer ladrillos
i fornace *f* da mattoni
n steenoven
d Ziegelofen *m*

736 BRICK-LINED PROCESS EQUIPMENT gen
f installation *f* à garnissage de briques
e equipo *m* elaborador forrado de ladrillos
i equipaggiamento *m* a rivestimento di mattoni
n met baksteen gevoerde installatie
d Anlage *f* mit Ziegelauskleidung

737 BRICK PRESS cer
f presse *f* à briques
e prensa *f* ladrillera, prensa *f* para ladrillos
i pressa *f* per mattoni
n steenpers
d Ziegelpresse *f*

738 BRIDGE WALL gla
f barrage *m* fixe
e tornollamas *m*
i parete *f* di separazione
n scheidingswand
d Trennwandbrücke *f*

739 BRIDLE ROLLER, PINCH ROLLER pap,tex
f rouleau *m* de tension
e rodillo *m* cogedor, rodillo *m* de tensión
i rullo *m* di tensione
n trekrol
d Treibrolle *f*

740 BRIGHT ACID DIP elc
f bain *m* acide de brillantage
e baño *m* ácido de abrillantamiento
i bagno *m* acido di brillantaggio
n zuur glansbeitsbad *n*
d saure Glanzbrenne *f*

741 BRIGHT ANNEALING OVEN htt
f four *m* de recuit blanc
e horno *m* de recocido brillante
i forno *m* di ricottura antiossidante
n blankgloeioven
d Blankglühofen *m*

742 BRIGHT DIP elc
f bain *m* de brillantage
e baño *m* de abrillantamiento
i bagno *m* di brillantaggio
n glansbeitsbad *n*
d Glanzbrennbad *n*, Glanzbrenne *f*

743 BRIGHT PICKLING BATH elc
f bain *m* brillanteur de décapage
e baño *m* abrillantador de decapaje
i bagno *m* acido di decapaggio per brillantaggio
n glansbeitsbad *n*
d Glanzbeizbad *n*

744 BRIGHT PLATING PLANT elc
f installation *f* de brillantage
e equipo *m* de abrillantamiento
i impianto *m* di brillantaggio
n glanspleetinstallatie
d Glanzplattieranlage *f*

745 BRINE BASIN aic
f récipient *m* de saumure
e saladar *m*
i serbatoio *m* di salamoia
n pekelvat *n*
d Salzlakegefäss *n*, Solegefäss *n*

746 BRINE CURING VAT lea
f bain *m* de salage
e baño *m* de curado en salmuera
i bagno *m* di salamoia
n pikke bad *n*
d Solba *n*

747 BRINE UMP clo
f pom *f* à extraction de la saun re, pompe *f* d'eau salée
e boml *f* de extracción de la salm ra
i pom *f* di salamoia
n peke mp
d Salzl pumpe *f*, Solepumpe *f*

748 BRINE TANK bre
f machine *f* à glace
e generador *m* de hielo
i generatore *m* di ghiaccio
n ijsgenerator
d Eisgenerator *m*

749 BRIQUET MACHINE fue
f machine *f* à briqueter
e briqueteadora *f*
i macchina *f* di brichettaggio
n brikettenmachine
d Brikettiermaschine *f*

750 BRIX DEGREE METER ins,sin
f mesureur *m* du degré brix
e medidor *m* del grado brixio
i misuratore *m* del grado brix
n brixweger
d Brixgradmesser *m*

751 BRONZING BATH coa
f bain *m* de bronzage
e baño *m* de bronzeado
i bagno *m* di bronzatura
n bronsbad *n*
d Bronzierbad *n*

752 BROWN TONING BATH pho
f bain *m* de virage en brun
e baño *m* de viraje pardo
i bagno *m* di viraggio seppia
n bruinkleurbad *n*
d Brauntonbad *n*

753 BROWNING BATH coa
f bain *m* de brunissage
e baño *m* de pavonado pardo
i bagno *m* di brunitura
n bruineerbad *n*
d Braunierbad *n*

754 BROWNSTONE CELL elc
f pile *f* au bioxyde de manganèse
e pila *f* de arenisca parda
i pila *f* a perossido di manganese
n bruinsteencel
d Braunsteinzelle *f*

755 BRUSH FURNISHER te
f brosse *f* à teinture rotative
e cobilla *f* de teñido giratoria
i spazzola *f* di tintura rotativa
n draaiende verfborstel
d rotierende Färbbürste *f*

756 BRUSH ROLL de
f cylindre *m* brosseur
e cepillo *m* cilíndrico
i spazzola *f* cilindrica

n borstelrol, borstelwals
d Bürstenwalze *f*

757 BRUSH SCRAPER cpt
f racleur *m* à brosse
e cepillo *m* rascador
i raschiatore *m* a spazzola
n rakelborstel
d Bürstenabstreicher *m*

758 BRUSH SIFTER sor
f trieuse *f* à brosse
e tamizador *m* de cepillo
i vagliatura *f* a spazzola
n borstelzeef
d Bürstensieb *n*

759 BRUSH SPREADER, dec
BRUSH SPREADING MACHINE
f machine *f* à enduire à la brosse
e engomadora *f* por cepillo, máquina *f* de embadurnar por cepillo
i spalmatrice *f* a spazzola
n borstelstrijkmachine
d Bürstenstreichmaschine *f*

760 BRUSH SPREADER ROLL dec
f brosse *f* cylindrique pour enduction, rouleau *m* à brosse
e cepillo *m* cilíndrico para esparcir
i cilindro *m* spalmatore a spazzola
n borstelstrijkrol
d Streichbürstenwalze *f*

761 BRUSHING MACHINE dec
f métier *m* à brosser
e máquina *f* de cepillar
i spazzolatrice *f*
n borstelmachine
d Bürstenmaschine *f*

762 BUBBLE CAP dis
f calotte *f* à barbotage
e platillo *m* de burbujeo
i campanella *f* di gorgogliamento
n borrelklok
d Fraktionierbodenglocke *f*

763 BUBBLE CHAMBER ins
f chambre *f* à bulles
e cámara *f* de burbujas
i camera *f* di bolle
n bellenkamer, bellenvat *n*
d Blasenkammer *f*

764 BUBBLE COUNTER lab
f compte-bulle *m*, compteur *m* de bulles de gaz
e contador *m* de burbujas de gas
i contatore *m* di bolle di gas
n gasbellenteller
d Blasenzähler *m*, Gasblasenzähler *m*

765 BUBBLE GAGE (US), ins
BUBBLE GAUGE (GB)
f debitmètre *m* à bulles de gaz
e flujómetro *m* de burbujas de gas
i flussometro *m* a bolle di gas
n gasbellenstromingsmeter
d Blasenzähler *m*, Gasblasenströmungsmesser *m*

766 BUBBLE PLATE, cpt
BUBBLE TRAY
f plateau *m* de barbotage
e plato *m* de burbujas
i piatto *m* a campanelle, vassoio *m* di gorgogliamento
n borrelplaat, klokschotel
d Glockenboden *m*

767 BUBBLE PLATE COLUMN, abs
BUBBLE TRAY COLUMN
f colonne *f* à plateaux de barbotage
e columna *f* de burbujeo
i colonna *f* di gorgogliamento
n borrelplaatkolom, klokschotel-kolom
d Glockenbodenkolonne *f*

768 BUBBLE TOWER, dis
FRACTIONATING TOWER
f tour *f* de fractionnement
e torre *f* de borbotaje, torre *f* de burbujeo, torre *f* de fraccionación
i torre *f* di frazionamento
n fractioneertoren, rectificeerkolom
d Fraktionnierturm *m*

769 BUCKET, tra
BIN
f benne *f*
e cajón *m*
i benna *f*
n bak, mand
d Kasten *m*, Kübel *m*

770 BUCKET CONVEYOR, tra
BUCKET ELEVATOR
f élévateur *m* à godets
e elevador *m* de cangilones
i elevatore *m* a tazze
n emmerelevator
d Becherwerk *n*, Eimerelevator *m*

771 BUCKING KIER tex
f vaisseau *m* de lessivage
e tina *f* de colada
i tino *m* di lisciviazione
n beukvat *n*
d Beuchgefäss *n*

772 BUDDLE min
f auge *f* à laver, augette *f*, laveur *m*
e artesa *f*, lavadero *m*
i fosso *m* per lavaggio
n waskuil, wasvat *n*
d Erzbütte *f*, Schlämmgraten *m*, Schlämmtrog *m*

773 BUFF pol
f feutre *m* à polir
e disco *m* de paño pulidor
i disco *m* di panno per pulire
n zwabbelschijf
d Schwabbel *f*, Schwabbelscheibe *f*

774 BUFFING MILL pol
f moulin *m* d'émeulage
e molino *m* de abrillantado
i mulino *m* di molatura
n polijstmolen
d Poliermühle *f*

775 BUFFING WHEEL, POLISHING WHEEL pol
f disque *m* à polir
e disco *m* pulidor
i ruota *f* lucidatrice
n polijstschijf
d Polierscheibe *f*

776 BUILDING-UP BATH elc
f bain *m* d'épaississement
e baño *m* de recrecimiento
i bagno *m* di riporto galvanico a spessore
n opgroeibad *n*
d Verstärkungsbad *n*

777 BUILT-UP CONSTRUCTION pla
f construction *f* en coquille
e construcción *f* compuesta, construcción *f* desmontable
i costruzione *f* composta
n samengestelde matrijsopbouw
d Schachtelkonstruktion *f*

778 BULB lab
f ballon *m*
e bola *f*
i pallone *m*
n bolvormig vat *n*
d Ballon *m*

779 BULB BURETTE ana
f burette *f* à ampoule
e bureta *f* de bola
i buretta *f* a pallone
n bolburet
d Kugelbürette *f*

780 BULB-FORM DISTILLING COLUMN dis,lab
f colonne *f* à deux boules
e columna *f* de dos bolas
i colonna *f* a due ampolle
n destilleeropzet met twee bollen
d Destillationsaufsatz *m* mit zwei Kugeln

781 BULB PIPETTE lab
f pipette *f* à ampoule
e pipeta *f* de bola
i pipetta *f* a pallone
n bolpipet
d Kugelpipette *f*

782 BULKING INSTALLATION rub
f installation *f* pour stockage et homogénéisation
e instalación *f* para almacenar y homogeneizar
i installazione *f* per immagazzinare ed omogeneizzare
n bulkinstallatie, homogeniseer-installatie
d Homogenisieranlage *f*

783 BULL'S EYE, ELECTRODE RING htt,met
f entrée *f* d'électrode
e entrada *f* de electrodo
i entrata *f* d'elettrodo
n elektrode-ingang
d Elektrodenzugang *m*

784 BUNDLE DYEING VAT, HANK DYEING VAT, YARN DYEING VAT tex
f cuve *f* de teinture en échevaux
e cuba *f* de teñido en suspensión de las madejas
i tino *m* di tintura delle matasse
n strengenverfkuip
d Strähnenfärbküpe *f*

785 BUNDLE OF FUEL ELEMENTS, CLUSTER OF FUEL ELEMENTS, STRINGER nue
f faisceau *m* d'éléments combustibles
e haz *m* de elementos combustibles

i fascio *m* d'elementi combustibili
n brandstofelementenbundel, splijtstofelementenbundel
d Brennstoffelementenbündel *n*

786 BURETTE lab
f burette *f*
e bureta *f*
i buretta *f*
n buret
d Bürette *f*

787 BURETTE CLAMP, BURETTE HOLDER lab
f porte-burette *f*
e portabureta *f*
i morsetto *m* di buretta
n buretklem
d Bürettenklemme *f*

788 BURETTE FLOAT lab
f flotteur *m* de burette
e flotador *m* de bureta
i galleggiante *m* di buretta
n buretvlotter
d Bürettenschwimmer *m*

789 BURLING IRON, PICKING TOOL, TWEEZER tex
f outil *m* d'épluchage
e desmotadora *f*, despinzadora *f*
i utensile *m* di spuntatura
n nopijzer *n*
d Noppeisen *n*

790 BURNER bur,lab
f brûleur *m*, chalumeau *m*
e quemador *m*
i becco *m*
n brander
d Brenner *m*

791 BURNER BLOCK bur
f bloc *m* de brûleur
e bloque *m* de quemador
i blocco *m* di bruciatore
n branderblok *n*
d Brennerblock *m*

792 BURNER FIRING BLOCK, QUARL BLOCK bur,gla
f brique *f* réfractaire de brûleur
e ladrillo *m* refractario de quemador
i mattone *m* refrattario di bruciatore
n vuurvaste brandersteen
d feuerfester Brennerstein *m*

793 BURNER HOLE, BURNER PORT bur,gla
f lunette *f*, porte *f* de brûleur
e puerta *f* de quemador
i porta *f* di bruciatore
n branderpoort
d Bütte *f*

794 BURNER HOUSING bur,gla
f chambre *f* de brûleur
e casa *f* de quemador
i cassa *f* di bruciatore
n branderhuis *n*
d Brennergehäuse *n*

795 BURNER SLIDE bur,gla
f chariot *m* de brûleur
e carro *m* de quemador
i slitta *f* di bruciatore
n branderslede
d Brennerwagen *m*

796 BURNER TUBE bur,lab
f tube *m* de brûleur, tube *m* de chalumeau
e tubo *m* de quemador
i tubo *m* di becco
n branderbuis
d Brennerrohr *n*

797 BURNING-IN KILN cer,gla
f four *m* à émailler
e horno *m* de vidriado, horno *m* para esmaltar
i forno *m* di ricottura
n emailleeroven, inbrandoven
d Einbrennofen *m*, Emaillierofen *m*

798 BURNISHING APPARATUS pol
f appareil *m* de brunissage
e aparato *m* de bruñido
i apparecchio *m* di brunitura
n bruineerapparaat *n*
d Druckpoliergerät *n*

799 BURST CAN DETECTOR, BURST CARTRIDGE DETECTOR ins,nue,sap
f détecteur *m* d'éléments combustibles crevés
e detector *m* de elementos combustibles quebrados
i rivelatore *m* d'elementi combustibili scoppiettati
n detector van gebarsten splijtelementen
d Detektor *m* für zerrissene Brennstoffelemente

800 BUTANE CYLINDER gas
f cylindre *m* à butane
e cilindro *m* de butano
i cilindro *m* a butano
n butaancilinder, butaanfles
d Butanflasche *f*

801 BUTANE TANK gas
f récipient *m* à butane
e tanque *m* de butano
i serbatoio *m* a butano
n butaantank
d Butantank *m*

802 BUTTER TUB, nut
FIRKIN
f barillet *m* à beurre, tinette *f* à beurre
e baril *m* para mantèca
i bariletto *m* di burro
n botervaatje *n*, botervat *n*
d Butterfass *n*

803 BUTTERFLY VALVE cpt
f soupape *f* à papillon
e válvula *f* de mariposa
i valvola *f* a farfalla
n vlinderklep
d Drosselklappe *f*, Regelklappe *f*

804 BUTYROMETER, dai,ins
MILK TEST BOTTLE
f butyromètre *m*
e butirómetro *m*
i butirometro *m*
n butyrometer
d Butyrometer *n*, Milchprober *m*

805 BY-PASS dis
f dérivation *f*, déviation *f*
e desviación *f*
i derivazione *f* in parallelo
n omleiding
d Nebenschluss *m*, Umleitung *f*

C

806 CABINET DESICCATOR, CABINET DRIER, DRYING CABINET — dry,lab
f séchoir *m*
e armario *m* secador
i camera *f* d'essiccazione
n droogkast
d Trockenschrank *m*

807 CADMIUM CELL — elc
f pile *f* à cadmium
e pila *f* de cadmio
i pila *f* a cadmio
n cadmiumcel, cadmiumelement *n*
d Kadmiumelement *n*

808 CADMIUM-NICKEL STORAGE BATTERY — elc
f accumulateur *m* à cadmium-nickel
e acumulador *m* de cadmio-níquel
i accumulatore *m* a cadmio-nichel
n nikkelcadmiumaccu
d Nickel Kadmiumakkumulator *m*

809 CADMIUM PLATING BATH — elc
f bain *m* de cadmiage
e baño *m* para cadmiar
i bagno *m* di cadmiatura
n cadmiumbad *n*
d Kadmiumbad *n*

810 CADMIUM PLATING PLANT — elc
f installation *f* de cadmiage
e equipo *m* para cadmiar
i impianto *m* di cadmiatura
n vercadmiuminstallatie
d Kadmierungsanlage *f*

811 CADMIUM PLATING TANK — elc
f récipient *m* de cadmiage
e tanque *m* para cadmiar
i serbatoio *m* di cadmiatura
n vercadmiumtank
d Kadmierungsbehälter *m*

812 CAGE MILL — bre
f broyeur *m* à deux cages tournantes
e triturador *m* de dos jaulas girantes
i mulino *m* rotativo a due gabbie
n kooimolen
d Käfigmühle *f*

813 CAGE PRESS, FILTER PRESS — fil,pre
f filtre-presse *f*, presse *f* à filtre
e filtro-prensa *f*, prensa *f* filtradora
i filtropressa *f*, pressa *f* da filtro
n filterpers
d Filterpresse *f*, Seiherpresse *f*

814 CALANDRIA, STEAM BELT — sin
f chambre *f* à vapeur
e calandria *f*
i camera *f* a vapore
n stoomkamer
d Heizkammer *f*

815 CALCIMETER — ins
f calcimètre *m*
e calcímetro *m*
i calcimetro *m*
n calcimeter
d Calcimeter *n*, Kalzimeter *n*

816 CALCINING FURNACE, KILN, ROASTING FURNACE — cer
f four *m* de grillage
e horno *m* de tostación
i forno *m* d'arrostimento
n roostoven
d Röstofen *m*

817 CALCINING OVEN — htt,oic
f four *m* d'incinération
e horno *m* de incineración
i forno *m* d'incenerazione
n verassingsoven
d Veraschungsofen *m*

818 CALCIUM CHLORIDE TUBE — lab
f ampoule *f* à chlorure de calcium
e tubo *m* de cloruro cálcico
i tubo *m* a cloruro calcico
n chloorcalciumbuisje *n*
d Kalziumchloridrohr *n*

819 CALDWELL CRUCIBLE — lab
f creuset *m* porcelaine à fond perforé amovible
e crisol *m* de porcelana con fondo perforado removible
i crogiolo *m* in porcellana con fondo perforato smontabile
n porseleinen kroes met uitneembare zeefbodem

d Porzellantiegel *m* mit herausnehmbarem Siebboden

820 CALENDER pap,tex
f calandre *f*
e calandra *f*
i calandra *f*
n kalander, stofglanzer
d Kalander *m*

821 CALENDER BOWL, CALENDER ROLL pap,tex
f cylindre *m* de calandre
e cilindro *m* calandrador, cilindro *m* de calandra
i cilindro *m* di calandra
n kalanderwals
d Kalanderwalze *f*

822 CALENDER COATER pap,tex
f calandre *f* d'enduction
e calandra *f* de recubrimiento
i calandra *f* accoppiatrice
n bedekkingskalander
d Auftragkalander *m*

823 CALENDER ROLLERS pap,tex
f rouleaux *pl* moteurs de la calandre
e rodillos *pl* motrices de la calandra
i rulli *pl* d'avanzamento della calandra
n kalanderaandrijfrollen *pl*
d Kalanderantriebsrollen *pl*

824 CALENDER STACK (US), GLAZING ROLLS (GB) pla
f calandre *f* finisseuse
e calandra *f* de satinar
i calandra *f* spianatora
n satineerkalander
d Glättkalander *m*

825 CALENDER TRAIN pap,tex
f batterie *f* de calandres
e juego *m* de calandras, tren *m* de calandras
i serie *f* di calandre
n tandemkalander
d Kalanderreihe *f*

826 CALIBRATED PIPETTE lab
f pipette *f* calibrée
e pipeta *f* calibrada
i pipetta *f* tarata
n ijkpipet
d Vollpipette *f*

827 CALIBRATING BURETTE lab
f burette *f* à calibrage
e bureta *f* de calibración
i buretta *f* taratrice
n calibreerburet
d Kalibrierbürette *f*

828 CALIBRATING PIPETTE lab
f pipette *f* à calibrage
e pipeta *f* de calibración
i pipetta *f* taratrice
n calibreerpipet
d Kalibrierpipette *f*

829 CALIBRATION FURNACE tes
f four *m* d'étalonnage
e horno *m* de calibración, horno *m* de patrón
i forno *m* campione, forno *m* di taratura
n ijkoven
d Eichofen *m*

830 CALKING CHISEL (US), CALKING TOOL (US), CAULKING CHISEL (GB), CAULKING IRON (GB) cle
f ciseau *m* de calfat
e cincel *m* de calafate
i scalpello *m* da calafato, scalpello *m* da presellatura
n breeuwijzer *n*, kalfaatijzer *n*
d Kalfateisen *n*

831 CALKING MALLET (US), CAULKING MALLET (GB) cle
f maillet *m* de calfat
e mazo *m* de calafate
i maglio *m* di calafato, mazzuolo *m* di calafatura
n breeuwhamer, kalfaathamer
d Kalfathammer *m*

832 CALOMEL ELECTRODE, CALOMEL HALF CELL elc
f demi-cellule *f* au calomel, électrode *f* au calomel
e electrodo *m* de calomelano, semicelda *f* de calomelano
i elettrodo *m* al calomelano
n calomelelektrode
d Kalomelhalbzelle *f*

833 CALORIC VALUE REGULATOR con
f régulateur *m* du pouvoir calorique
e regulador *m* del poder calorífico

i regolatore *m* del potere calorifico
n verbrandingswaarderegelaar
d Heizwertregler *m*

834 CALORIFIER clo
f réchauffeur *m* à serpentin
e calentador *m* de serpentín
i riscaldatore *m* a serpentina
n labyrintkoeler
d Labyrintkühler *m*

835 CALORIMETER ins
f calorimètre *m*
e calorímetro *m*
i calorimetro *m*
n calorimeter
d Kalorimeter *n*

836 CALORIZATOR, sin
JUICE WARMER
f calorisateur *m*
e calorizador *m*
i calorizzatore *m*
n calorisator
d Kalorisator *m*, Saftwärmer *m*

837 CALORIZING PLANT coa
f installation *f* de calorisation
e equipo *m* de calorización
i impianto *m* di calorizzazione
n caloriseerinstallatie
d Kalorisieranlage *f*

838 CALUTRON nue,spe
f calutron *m*
e calutrón *m*
i calutrone *m*
n calutron *n*
d Calutron *n*

839 CAMBERED BOWL cpt
f cylindre *m* cambré
e cilindro *m* abombado
i cilindro *m* con fascia di curvatura
n gebombeerde wals
d bombierte Walze *f*

840 CAMPBELL-PRESSMANN DRYING APPARATUS, lab
LYOPHILIZER
f lyophilisateur *m*
e liofilizador *m*
i essiccatoio *m* per congelamento di Campbell e Pressmann
n vriesdroogapparaat *n* volgens Campbell-Pressmann
d Campbell-Pressmann-Gefriertrockenapparat *m*

841 CAN elc
f bac *m*
e recipiente *m*
i bicchiere *m*
n beker
d Becher *m*

842 CAN, nue
CASING, JACKET
f chemise *f*, gaine *f*
e funda *f*
i camicia *f*, guaina *f*, ricopertura *f*
n bus, huls, omhulling
d Hülle *f*, Hülse *f*

843 CAN, nut
TIN
f boîte *f*
e bote *m*, lata *f*
i barattolo *m*, scatola *f*
n blik *n*
d Dose *f*

844 CAN SEALING MACHINE (US), bre
TIN SEALING MACHINE (GB)
f sertisseuse *f*
e máquina *f* para tapar latas
i macchina *f* per chiudere barattoli, macchina *f* per chiudere scatole
n blikmachine
d Dosenverschlussmaschine *f*

845 CANAL, gla
ALCOVE, CHANNEL
f canal *m* d'alimentation
e canal *m* de alimentación
i canale *m* d'alimentazione
n aanvoerkanaal *n*
d Speisekanal *m*

846 CANDLE FILTER fil
f filtre *m* à bougie
e filtro *m* de bujía
i candela *f* filtrante
n kaarsfilter *n*
d Filterkerze *f*, Kerzenfilter *m*

847 CANE CRUSHER sin
f broyeuse *f*
e quebrantadora *f*
i macchina *f* di schiacciatura
n rietkneuzer
d Rohrquetschmaschine *f*

848 CANE DISINTEGRATOR, sin
CANE SHREDDER
f défibreur *m*
e desfibradora *f*

i sfibratore *m*
n rietmolen
d Rohrentfaserer *m*

849 CANE SUGAR PLANT sin
f fabrique *f* de sucre de canne
e equipo *m* de azúcar de caña
i impianto *m* di zucchero di canna
n rietsuikerfabriek
d Rohrzuckerfabrik *f*

850 CANISTER, sap
GAS MASK FILTER
f filtre *m* de masque anti-gaz
e filtro *m* de máscara antigás
i filtro *m* di maschera antigas
n gasmaskerfilter *n*
d Filterbüchse *f*, Filtereinsatz *m*

851 CANNED PUMP, nue,pre
CANNED ROTOR PUMP
f pompe *f* à roteur gainé
e bomba *f* de rotor hermético
i pompa *f* a rotore stagno
n pomp met ingekapselde rotor
d Pumpe *f* mit eingekapseltem Läufer

852 CANNERY cov
f conserverie *f*
e fábrica *f* de conservas
i conservificio *m*, stabilimento *m* conserve
n conservenfabriek
d Konservenfabrik *f*

853 CANNING MACHINE (US), bre
TINNING MACHINE (GB)
f emboîteuse *f*
e llenadora *f* de latas
i riempitrice *f* per barattoli
n blikvulmachine
d Dosenfüllmaschine *f*

854 CANNON POT gla
f creuset *m* petit
e crisol *m* pequeño
i crogiolo *m* piccolo
n kleine glaspot
d kleiner Glashafen *m*

855 CAPILLARY, ins
CAPILLARY TUBE
f tube *m* capillaire
e tubo *m* capilar
i tubo *m* capillare
n capillaire buis
d Kapillare *f*, Kapillarrohr *n*

856 CAPILLARY BURETTE, ins
CAPILLARY MICROBURETTE
f burette *f* capillaire
e bureta *f* capilar
i buretta *f* capillare
n capillaire buret
d Kapillarbürette *f*

857 CAPILLARY FILTER cle,fil
f filtre *m* capillaire
e filtro *m* capilar
i filtro *m* capillare
n capillair filter *n*
d Kapillarfilter *m*

858 CAPILLARY PIPETTE lab
f pipette *f* capillaire
e pipeta *f* capilar
i pipetta *f* capillare
n capillaire pipet
d Kapillarpipette *f*

859 CAPILLARY TUBING ins
f verre *m* tubulaire capillaire
e tubería *f* de vidrio capilar
i vetro *m* tubulare capillare
n capillair buisglas *n*
d kapillares Rohrglas *n*

860 CAPILLARY VISCOMETER, ins
CAPILLARY VISCOSIMETER
f viscosimètre *m* capillaire
e viscosímetro *m* capilar
i viscosimetro *m* capillare
n capillaire viscositeitsmeter
d Kapillarviskosimeter *n*

861 CAPILLOMETER ins
f mesureur *m* de capillarité
e capilaridímetro *m*, medidor *m* de capilaridad
i misuratore *m* di capillarità
n capillariteitsmeter
d Kapillaritätsmesser *m*

862 CAPSULE gen
f capsule *f*
e cápsula *f*
i capsula *f*
n capsule
d Kapsel *f*

863 CAPSULING MACHINE, dai
SEALING MACHINE
f machine *f* de capsulage
e máquina *f* tapadora
i capsulatrice *f*
n capsuleermachine
d Verschlussmaschine *f*

864 CAR TUNNEL KILN cer,gla,htt (US),
BOGIE KILN (GB), FIXED-FIRE ZONE KILN, TRUCK CHAMBER KILN
f étuve *f* à chariot
e estufa *f* de carrillo
i stufa *f* a carrello
n wagendroogstoof
d Wagentrockenofen *m*

865 CARBIDE DRUM tra
f fût *m* pour carbure
e bidón *m* de carburo
i fusto *m* per carburo
n carbidvat *n*
d Karbidtrommel *f*

866 CARBIDE ETCHING BATH coa
f bain *m* d'attaque à l'acide du carbure
e baño *m* de ataque al ácido del carburo
i bagno *m* d'incisione all'acqua-forte del carburo
n carbide-etsbad *n*
d Karbidätzbad *n*

867 CARBIDE FEEDER gas
f dispositif *m* d'alimentation du carbure
e dispositivo *m* de alimentación del carburo
i dispositivo *m* d'alimentazione del carburo
n carbidtoevoerinrichting *f*
d Karbideinfallvorrichtung *f*, Karbideinwurfvorrichtung *f*

868 CARBIDE FUNNEL gas
f entonnoir *m* de carbure
e embudo *m* de carburo
i tramoggia *f* d'introduzione del carburo
n carbidtrechter
d Karbideinwurftrichter *m*

869 CARBIDE FURNACE htt
f four *m* à carbure
e horno *m* para carburo de calcio
i forno *m* per carburo
n carbidoven
d Karbidofen *m*

870 CARBIDE-TO-WATER gas ACETYLENE GENERATOR
f générateur *m* d'acétylène à chute de carbure dans l'eau
e generador *m* de acetileno de caída de carburo en el agua
i generatore *m* d'acetilene a caduta di carburo nell'acqua
n acetyleengenerator met in water vallend carbid
d Einfallazetylenentwickler *m*, Einwurfazetylenentwickler *m*, Karbideinwurfentwickler *m*

871 CARBO-HYDROGEN gas GENERATOR,
OILGAS GENERATOR
f générateur *m* de gaz de pétrole
e generador *m* de gas de petróleo
i generatore *m* di gas di petrolio
n oliegasgenerator, vetgasgenerator
d Fettgasgenenerator *m*, Ölgasgenerator *m*

872 CARBO-NITRIDING PLANT, met
CYANIDE HARDENING PLANT
f installation *f* de cémentation au cyanure, installation *f* d'endurcissement au cyanure
e equipo *m* de cementación al cianuro, equipo *m* de endurecimiento al cianuro
i impianto *m* di cementazione al cianuro, impianto *m* d'indurimento al cianuro
n cyaanhardingsbad *n*
d Zyansalzhärtungsbad *n*

873 CARBON ANODE elc,met
f anode *f* en charbon
e ánodo *m* de carbón
i anodo *m* in carbone
n koolanode
d Kohlenanode *f*

874 CARBON-ARC WELDING wel PLANT
f installation *f* de soudure à l'arc au charbon
e equipo *m* de soldadura por arco de carbón
i impianto *m* di saldatura con elettrodi di carbone
n booglasinstallatie met koolelektrode
d Kohlelichtbogenschweissanlage *f*

875 CARBON BAG ELECTRODE elc
f électrode *f* de charbon en sac
e electrodo *m* de carbón de saco
i elettrodo *m* di carbone in sacco
n koolbuidelelektrode
d Kohlenbeutelelektrode *f*

876 CARBON-BLACK-FILLED cle NEOPRENE WASHER

f laveuse *f* de néoprène à remplissage de noir de fumée
e arandela *f* de neopreno rellenada con negro de carbón
i depuratore *m* di neoprene con riempimento di nerofumo
n met lampzwart gevulde neopreenwasinrichting
d mit Lampenschwarz gefüllte Neoprenwäsche *f*

877 CARBON COMBUSTION CELL, CARBON CONSUMING CELL elc
f pile *f* à charbon
e pila *f* de carbón
i pila *f* a carbone
n verbrandingscel
d Kohlenstoffelement *n*

878 CARBON-DIOXIDE COOLED REACTOR nue
f réacteur *m* à refroidissement par acide carbonique
e reactor *m* enfriado al ácido carbónico
i reattore *m* con raffreddamento ad acido carbonico
n met koolzuur gekoelde reactor
d kohlensäuregekühlter Reaktor *m*

879 CARBON-DIOXIDE DETERMINATION APPARATUS ins,lab
f appareil *m* pour le dosage de l'acide carbonique
e dosificador *m* del ácido carbónico
i dosatore *m* dell'acido carbonico
n koolzuurbepalingsapparaat *n*
d Kohlensäurebestimmungsapparat *m*

880 CARBON-DIOXIDE EXTINGUISHING PLANT fla
f installation *f* extinctrice à acide carbonique
e equipo *m* contraincendios de ácido carbónico
i impianto *m* estintore ad acido carbonico
n koolzuurblusinstallatie
d Kohlensäurelöschanlage *f*

881 CARBON-DIOXIDE FLASK sto,tra
f cylindre *m* à acide carbonique
e botella *f* de ácido carbónico
i bombola *f* d'acido carbonico
n koolzuurfles
d Kohlensäureflasche *f*

882 CARBON-DIOXIDE GENERATOR gas
f générateur *m* d'acide carbonique
e generador *m* de ácido carbónico
i generatore *m* d'acido carbonico
n koolzuurgenerator
d Kohlensäureentwickler *m*

883 CARBON-DIOXIDE RECORDER ins
f signaleur *m* d'acide carbonique
e registrador *m* de ácido carbónico
i registratore *m* d'acido carbonico
n koolzuurgehalteschrijver, koolzuurmeter
d Kohlensäuregehaltgeber *m*

884 CARBON ELECTRODE htt
f électrode *f* de charbon
e electrodo *m* de carbón
i elettrodo *m* di carbone
n koolelektrode
d Kohleelektrode *f*

885 CARBON-LINED FURNACE htt
f four *m* à revêtement de charbon
e horno *m* revestido de carbón
i forno *m* con rivestimento di carbone
n oven met koolvoering
d Ofen *m* mit Kohlenfutter

886 CARBON-MONOXIDE DETECTOR ins,sap
f détecteur *m* d'oxyde de carbone
e detector *m* de óxido de carbono
i rivelatore *m* di monossido di carbonio, rivelatore *m* d'ossido di carbonio
n koolmonoxydeverklikker
d Kohlenmonoxydanzeiger *m*

887 CARBON PACKING, GRAPHITE PACKING cpt
f garniture *f* de graphite
e empaquetadura *f* de grafito
i guarnizione *f* di grafite
n grafietpakking
d Graphitpackung *f*

888 CARBON RESIDUE RECORDER lab
f indicateur *m* de résidu de carbone
e indicador *m* de residuo de carbono
i indicatore *m* di residuo di carbonio
n koolstofresidu-indicator
d Kohlenstoffrestgeber *m*

889 CARBONRESISTOR FURNACE htt
f four *m* à résistances de charbon

e horno *m* de resistencias de carbón
i forno *m* a resistenze di carbone
n koolweerstandsoven
d Kohlewiderstandsofen *m*

890 CARBON RESTORATION PLANT, RECARBURIZATION PLANT met
f installation *f* de récarburation, installation *f* de régénération de pièces décarburisées
e equipo *m* de recarburación, equipo *m* de regeneración de piezas descarburadas
i impianto *m* di ricarburazione, impianto *m* di rigenerazione di pezzi decarburati
n heropkolingsinstallatie, installatie voor het regenereren van ontkoolde delen
d Aufkohlunganlage *f*, Rückkohlungsanlage *f*

891 CARBON TETRACHLORIDE EXTINGGUISHER fea,sap
f extincteur *m* à tétrachlorure de carbone
e extintor *m* de incendios de tetracloruro de carbono
i estintore *m* a tetracloruro di carbonio
n tetrablusapparaat *n*
d Tetralöscher *m*

892 CARBON TETRACHLORIDE GENERATOR gas
f générateur *m* de tétrachlorure de carbone
e generador *m* de tetracloruro de carbono
i generatore *m* di tetracloruro di carbonio
n tetrachloorkoolstofgenerator, tetragenerator
d Tetrachlorkohlenstoffgenerator *m*, Tetragenerator *m*

893 CARBONATATION PAN, CARBONATION PAN, SATURATION PAN sin
f cuve *f* de saturation
e cuba *f* de carbonatación
i tino *m* di carbonatazione
n carbonatatieketel
d Entkalkungsgefäss *n*, Scheidepfanne *f*

894 CARBONATATION TOWER, CARBONATION TOWER, SATURATION TOWER sin
f colonne *f* de saturation
e columna *f* de carbonatación, torre *f* de saturación
i colonna *f* di carbonatazione, colonna *f* di saturazione
n carbonatatiekolom
d Entkalkungskolonne *f*, Sättigungskolonne *f*, Saturationssäule *f*

895 CARBONIC ACID PUMP sin
f pompe *f* à acide carbonique
e bomba *f* de ácido carbónico
i pompa *f* ad acido carbonico
n koolzuurpomp
d Kohlensäurepumpe *f*

896 CARBONIZER, CARBONIZING MILL tex
f établissement *m* de carbonisage
e equipo *m* de carbonización
i impianto *m* di carbonizzazione
n carboniseerinstallatie
d Karbonisieranstalt *m*

897 CARBONIZING OVEN tex
f four *m* de carbonisage
e horno *m* de carbonización
i forno *m* di carbonizzazione
n carboniseeroven
d Karbonisierofen *m*

898 CARBOY, DEMIJOHN tra
f ballon *m*, bonbonne *f*, dame-jeanne *f*, tourie *f*
e castaña *f*, damajuana *f*
i damigiana *f*
n mandefles
d Korbflasche *f*

899 CARBOY FILTER tra
f filtre *m* de tourie
e filtro *m* de damajuana
i filtro *m* di damigiana
n mandeflesfilter *n*
d Korbflaschenfilter *m*

900 CARBOY HAMPER tra
f panier *m* de dame-jeanne
e cesto *m* de damajuana
i cesta *f* di damigiana
n flesmand
d Ballonkorb *m*

901 CARBOY TIPPER, INCLINATOR tra
f culbuteur *m* de tourie
e basculador *m* de damajuana
i ribaltatore *m* di damigiana
n mandefleskantelaar, mandefleskipper

d Korbflaschenkipper *m*

902 CARBURETOR, CARBURETTER, CARBURETTOR fue
f carburateur *m*
e carburador *m*
i carburatore *m*
n carburator
d Vergaser *m*

903 CARBURETTOR AIR HEATER fue
f réchauffeur *m* de l'air du carburateur
e calentador *m* del aire del carburador
i riscaldatore *m* dell'aria del carburatore
n carburatorluchtverhitter
d Ansaugluftvorwärmer *m*, Vergaserluftheizer *m*

904 CARBURETTOR AIR HORN, CARBURETTOR AIR SCOOP fue
f prise *f* d'air du carburateur
e toma *f* de aire del carburador
i presa *f* d'aria del carburatore
n luchtinlaat van de carburator
d Vergaserluftfänger *m*, Vergaserluftschippe *f*

905 CARBURETTOR CONTROL GEAR fue
f installation *f* de commande de carburateur
e equipo *m* de mando del carburador
i impianto *m* di comando del carburatore
n carburatorregelinrichting
d Vergasereinstellung *f*, Vergaserregulierung *f*

906 CARBURETTOR CONTROL ROD fue
f tirant *m* de commande du carburateur
e varilla *f* de mando del carburador
i barra *f* di comando del carburatore
n regelstaaf van de carburator
d Vergaserregulierstange *f*

907 CARBURETTOR FLOAT fue
f flotteur *m* du carburateur
e flotador *m* del carburador
i galleggiante *m* del carburatore
n carburatorvlotter
d Schwimmer *m*, Vergaserschwimmer *m*

908 CARBURETTOR FLOAT CHAMBER fue
f chambre *f* de flotteur
e cámara *f* de flotación
i vaschetta *f* di carburatore
n vlotterkamer
d Schwimmergehäuse *n*

909 CARBURETTOR MAIN JET fue
f gicleur *m* principal du carburateur
e surtidor *m* principal del carburador
i ugello *m* principale del carburatore
n hoofdsproeier van de carburator
d Vergaserdüse *f*

910 CARBURIZATION FURNACE htt,met
f four *m* de carburation
e horno *m* de carburación
i forno *m* di carburazione
n carbureeroven
d Karburierofen *m*

911 CARBURIZING BATH met
f bain *m* de cémentation
e baño *m* de cementación
i bagno *m* di cementazione
n cementeerbad *n*
d Zementierbad *n*

912 CARBURIZING CONTAINER met
f caisse *f* de cémentation
e caja *f* de cementación
i cassetta *f* di cementazione
n cementeerkast, cementeerkist
d Zementierkasten *m*

913 CARBURIZING FURNACE htt,met
f four *m* de cémentation
e horno *m* de cementación
i forno *m* di cementazione
n cementeeroven
d Zementierofen *m*

914 CARRIER PIN (GB), INSERT PIN (US), RETAINING PIN pla,pre
f broche *f* à prisonnier, ergot *m* de centrage, goupille *f* de fixation
e clavija *f* de retenida, clavija *f* de sujeción
i spina *f* portainserto
n fixatiepen
d Haltestift *m*

915 CARRIER PLATE chr
f plaque-support *m*
e placa *f* de soporte
i piastra *f* di sopporto
n dragerplaat
d Trägerplatte *f*

916 CARRIER ROLLER, pap
SUPPORTING ROLLER,
TABLE ROLLER
f rouleau *m* de support, rouleau *m* porteur
e rodillo *m* de soporte, rodillo *m* portador
i rullo *m* di sopporto, rullo *m* portatore
n draagrol, steunrol
d Stützrolle *f*, Tragrolle *f*

917 CASCADE AERATOR cle,iwt
f aérateur *m* en cascade
e aireador *m* de cascada, aireador *m* de escalones
i aeratore *m* in cascata
n cascadeaërator
d Kaskadenlüfter *m*

918 CASCADE COOLER clo
f refroidisseur *m* en cascade
e enfriador *m* en cascada
i raffreddatore *m* in cascata
n cascadekoeler
d Kaskadenkühler *m*

919 CASCADE DYEING MACHINE dye
f machine *f* de teinture en cascade
e máquina *f* de teñido en cascada
i macchina *f* di tintura in cascata
n cascadeverfmachine
d Kaskadenfärbmaschine *f*

920 CASCADE ISOTOPE SEPARATION PLANT nue
f installation *f* de séparation d'isotopes en cascade
e equipo *m* de separación de isótopos en cascada
i impianto *m* di separazione di isotopi in cascata
n cascade-isotopenscheidings-installatie
d Kaskadenisotopentrennungs-anlage *f*

921 CASCADE OF SEPARATING UNITS nue
f cascade *f* d'étapes de séparation
e cascada *f* de etapas de separación
i cascata *f* di stadi di separazione
n scheidingstrappencascade
d Trennungsstufenkaskade *f*

922 CASCADE STRAINER rub
f tamis *m* en cascade
e tamiz *m* en cascada
i staccio *m* in cascata
n cascadezeef
d Kaskadensieb *n*

923 CASCADE TRAY dis
f plateau *m* à cascade
e plato *m* de cascada
i piatto *m* a cascata
n cascadeschotel
d Kaskadenboden *m*

924 CASCADE WASHER pho
f installation *f* de lavage en cascade
e arandela *f* en cascada
i lavatrice *f* in cascata
n cascadewasinrichting
d Kaskadenwässerungsanlage *f*

925 CASE elc
f bac *m* multiple
e recipiente *m* múltiple
i contenitore *m* multiplo
n meervoudige bak
d Blockkasten *m*

926 CASE (GB), elc
JACKET (US)
f chemise *f*
e camisa *f*
i camicia *f*
n huls
d Batteriekarton *m*, Gehäuse *n*, Papphülse *f*

927 CASE HARDENING BOX met
f pot *m* de cémentation
e caja *f* de cementación
i cassa *f* di cementazione
n cementeerkist, inzetkist
d Zementierkasten *m*

928 CASE HARDENING FURNACE, htt,met
SURFACE CONVERTING FURNACE
f four *m* de trempe de surface
e horno *m* de temple de superficie
i forno *m* di tempera di superficie
n inzethardingsoven, oppervlak-hardingsoven
d Einsatzhärtungsofen *m*, Oberflächenhärtungsofen *m*

929 CASED TUBE gen
f tube *m* en acier à chemise métallique
e tubo *m* de acero de camisa metálica
i tubo *m* d'acciaio a camicia metallica
n stalen buis met metaalomhulsel
d metallbekleidetes Stahlrohr *n*

930 CASING, CAN, JACKET nue
f chemise *f*, gaine *f*
e funda *f*
i camicia *f*, guaina *f*, ricopertura *f*
n bus, huls, omhulling
d Hülle *f*, Hülse *f*

931 CASING (US), SHELL (GB) htt,sap
f carcasse *f*
e carcasa *f*
i carcassa *f*, mantello *m* di forno
n ovenmantel
d Blechmantel *m*

932 CASK tra
f baril *m*, fût *m*, tonneau *m*
e barril *m*, tonel *m*
i barile *m*, fusto *m*
n fust *n*
d Gebinde *n*, Tonne *f*

933 CASKET, COFFIN nue,tra
f caisse *f* de transport
e caja *f* de transporte
i bara *f*, cassa *f* di trasporto, cofano *m*
n transportkist
d Transportbehälter *m*

934 CAST ANODE elc
f anode *f* coulée
e ánodo *m* colado
i anodo *m* colato
n gegoten anode
d Gussanode *f*

935 CAST COATING coa
f revêtement *m* fondu
e revestimiento *m* fundido
i rivestimento *m* fuso
n gegoten bekleding
d Gussplattierung *f*

936 CAST IRON PIPE cpt
f tuyau *m* en fonte
e tubo *m* de fundición
i tubo *m* a ghisa
n gietijzeren buis
d Gusseisenrohr *n*, Gussrohr *n*

937 CAST STEEL GRATE boi
f grille *f* en acier coulé
e parrilla *f* en acero colado
i griglia *f* in acciaio colato
n gietijzeren rooster *n*
d Gussstahlrost *m*

938 CASTING LADLE, LADLE fon
f poche *f*
e cuchara *f*
i secchia *f* di colata, siviera *f*
n gietpan
d Pfanne *f*

939 CASTING MACHINE, COATING MACHINE pho
f machine *f* à émulsionner
e máquina *f* emulsionadora
i macchina *f* d'emulsionatura
n emulsiegietmachine
d Emulsionsauftragmaschine *f*

940 CASTLE TYPE MANIPULATOR ins
f manipulateur *m* à garniture de cylindres concentriques
e manipulador *m* de empaquetadura de cilindros concéntricos
i manipolatore *m* a scatola
n manipulator met concentrische cilinderpakking
d Manipulator *m* mit Koaxialzylinderpackung

941 CAT CRACKING PLANT, CATALYTIC CRACKING PLANT CATALYTIC CRACKING UNIT fue
f installation *f* de cracking catalytique, installation *f* de craquage catalytique
e equipo *m* de craqueo catalítico, equipo *m* para pirodesintegración catalítica
i impianto *m* per cracking catalitico, impianto *m* per piroscissione catalitica
n catalytische kraakinstallatie
d katalytische Krackanlage *f*

942 CATALASE TUBE bac,lab
f tube *m* de catalase
e tubo *m* de catalase
i tubo *m* di catalase
n catalasebuisje *n*
d Katalaserohr *n*

943 CATALYTIC REFORMING PLANT fue
f installation *f* de réformation catalytique
e equipo *m* de reformación catalítica
i impianto *m* di riformazione catalitica
n katalytische omvormingsinstallatie, katalytische reformeringsinstallatie
d katalytische Reformieranlage *f*

944 CATALYTIC SULFUR REMOVING PLANT (US), CATALYTIC SULPHUR REMOVING PLANT (GB) cle
f installation *f* de désulfuration catalytique
e equipo *m* de desulfuración catalítica
i impianto *m* da desolforazione catalitica
n catalytische ontzwavelingsinstallatie
d katalytische Entschwefelungsanlage *f*

945 CATCH-ALL, PULP SAVER, RECLAIMER, SAVE-ALL pap
f ramasse-pâte *m*
e recogedor *m* de pasta
i ricuperatore *m* di polpa
n vezelstofvanger
d Faserstoffänger *m*, Stoffänger *m*

946 CATHODE elc
f cathode *f*
e cátodo *m*
i catodo *m*
n katode
d Katode *f*

947 CATHODE COMPARTMENT elc
f boîte *f* de la cathode
e compartimiento *m* del cátodo
i cassetta *f* del catodo
n katodehuis *n*
d Katodengehäuse *n*

948 CATHODE FRAME elc
f bâti *m* de la cathode
e bastidor *m* del cátodo
i telaio *m* del catodo
n katodefreem *n*
d Katodengestell *n*

949 CATHODIC CLEANING PLANT elc
f établissement *m* de dégraissage cathodique
e equipo *m* de desengrase catódico
i impianto *m* di sgrassaggio catodico
n katodische ontvettingsinstallatie *f*
d katodische Reinigungsanlage *f*

950 CATHODIC CLEANING TANK elc
f récipient *m* de dégraissage cathodique
e tanque *m* de desengrase catódico
i serbatoio *m* di sgrassaggio catodico
n katodische ontvettingstank
d katodischer Reinigungsbehälter *m*

951 CATHODIC PICKLING PLANT elc
f installation *f* de décapage cathodique
e equipo *m* de decapaje catódico
i impianto *m* di decapaggio catodico
n katodische beitsinstallatie
d katodische Beizanlage *f*

952 CATHODIC PICKLING TANK elc
f récipient *m* de décapage cathodique
e tanque *m* de decapaje catódico
i serbatoio *m* di decapaggio
n katodische beitstank
d katodischer Beizbehälter *m*

953 CATHODIC PROTECTION BATH coa
f bain *m* de protection cathodique
e baño *m* de protección catódica
i bagno *m* di protezione catodica
n katodisch beschermingsbad *n*
d katodisches Schutzbad *n*

954 CATION ABSORBING FILTER fil
f filtre *m* absorbant les cations
e filtro *m* absorbedor de cationes
i filtro *m* assorbente di cationi
n kationenabsorberend filter *n*
d kationenabsorbierender Filter *m*

955 CATION EXCHANGE COLUMN elc
f colonne *f* échangeuse de cations
e columna *f* cambiadora de cationes
i colonna *f* scambiatrice di cationi
n kationenuitwisselingskolom
d Kationenaustauschsäule *f*

956 CATION EXCHANGER elc,pla
f échangeur *m* de cations
e canjeador *m* de cationes, intercambiador *m* de cationes

i scambiatore *m* di cationi
n kationenuitwisselaar
d Kationenaustauscher *m*

957 CATION PERMEABLE MEMBRANE ins
f membrane *f* perméable aux cations
e membrana *f* permeable a los cationes
i membrana *f* permeabile ai cationi
n kationendoorlatende membraan
d für Kationen durchlässige Membrane *f*

958 CATIONIC COLLECTOR elc
f collecteur *m* cationique
e colector *m* catiónico
i collettore *m* cationico
n kationische verzamelaar
d kationischer Sammler *m*

959 CAUL (US), PRESSURE EQUALIZING PLATE pla
f coussin *m* de pression
e almohada *f* compensadora de presión
i cuscino *m* di pressione
n drukegalisator, drukkussen *n*, drukverdeler
d Druckausgleichplatte *f*, Druckkissen *n*

960 CAULKING CHISEL (GB), CAULKING IRON (GB), CALKING CHISEL (US), CALKING TOOL (US) cle
f ciseau *m* de calfat
e cincel *m* de calafate
i scalpello *m* da calafato, scalpello *m* da presellatura
n breeuwijzer *n*, kalfaatijzer *n*
d Kalfateisen *n*

961 CAULKING MALLET (GB), CALKING MALLET (US) cle
f maillet *m* de calfat
e mazo *m* de calafate
i mazuella *f* di calafatuggio
n breeuwhamer, kalfaathamer
d Kalfathammer *m*

962 CAUSTIC DIP elc
f bain *m* de soude caustique
e baño *m* de hidróxido sódico
i bagno *m* d'idrossido sodico
n bijtende-sodabad *n*
d Ätznatronbad *n*

963 CAUSTIC SCRUBBER cle,nue
f récipient *m* de lavage caustique de gaz
e recipiente *m* de lavado cáustico de gas
i recipiente *m* di lavaggio caustico di gas
n gaswasser met natriumhydroxyde
d Gaswäsche *f* mit Natriumhydroxyd

964 CAUSTIC SODA CELL elc
f pile *f* à la soude
e pila *f* de sosa, pila *f* Lalande
i pila *f* ad idrossico sodico
n alkali-element *n*
d Alkalielement *n*

965 CAUSTIC SODA PICKLING EQUIPMENT elc
f installation *f* de décapage à soude caustique
e equipo *m* de decapaje per sosa cáustica
i impianto *m* di decapaggio per soda caustica
n beitsinstallatie met bijtende soda
d Natronätzanlage *f*

966 CAVITY PLUG (US), MOULD INSERT (GB) pla,pre
f empreinte *f* rapportée
e asiento *m* de la matriz
i tassello *m* d'impronta
n matrijsinzetstuk *n*
d Gesenkeinsatz *m*, Matrizeneinsatz *m*

967 CELL CAVITY elc
f cavité *f* de cellule, creuset-bassin *m*
e cavidad *f* de celda, cuba *f* crisol
i crogiolo *m*
n elektrolytruimte
d Herd *m*

968 CELL CONNECTOR elc
f connexion *f*
e conexión *f*
i connessione *f*
n celverbinder, elementverbinder
d Polbrücke *f*, Steg *m*

969 CELL DUST REVERSING PRECIPITATOR iwt
f dépoussiéreur *m* à chicanes
e pulviprecipitador *m* reversible
i precipitatore *m* di polvere ribaltabile
n omkeerbaar cellenfilter *n*
d Umkehrzellenentstäuber *m*

970 CELL FILTER bre,fil
f filtre *m* à chambres, filtre *m* cellulaire
e filtro *m* de cámaras, filtro *m* celular
i filtro *m* a camere, filtro *m* cellulare
n cellenfilter *n*, cellulair filter *n*
d Zellenfilter *m*

971 CELLULOSE FACTORY pap
f fabrique *f* de cellulose
e fábrica *f* de celulosa
i fabbrica *f* di cellulosa
n cellulosefabriek
d Zellstoffabrik *f*

972 CEMENT BATCHER bui
f appareil *m* de dosage de ciment
e dosificador *m* de cemento
i dosificatore *m* di cemento
n cementverdeler
d Zementbeschickungsgerät *n*

973 CEMENT BLENDING PLANT bui
f installation *f* à mélanger le ciment
e equipo *m* mezclador de cemento
i impianto *m* mescolatore di cemento
n cementmenginrichting
d Zementmischanlage *f*

974 CEMENT FACTORY bui
f cimenterie *f*
e cementería *f*, fábrica *f* de cemento
i fabbrica *f* di cemento
n cementfabriek
d Zementfabrik *f*

975 CEMENT GUN coa
f guniteuse *f*
e pistola *f* para enlucido de cemento
i pistola *f* spruzza cemento
n cementkanon *n*
d Zementkanone *f*

976 CEMENT KILN bui
f four *m* à ciment
e horno *m* giratorio de cemento
i forno *m* di cemento
n cementoven
d Zementbackofen *m*, Zementofen *m*

977 CEMENT MILL, CEMENT MIXER bui
f bétonnière *f*, malaxeur *m* à ciment
e hormigonera *f*, mezcladora *f* de cemento
i betoniera *f*
n cementmolen
d Zementmühle *f*

978 CEMENT MIXER, RUBBER SOLUTION MIXER rub
f malaxeur *m* à dissolution, mélangeur *m* à dissolution
e mezcladora *f* para soluciones
i agitatore *m* per soluzioni, mescolatore *m* per soluzioni
n solutiemolen
d Gummilöser *m*, Lösungskneter *m*, Lösungsmastikator *m*

979 CEMENTED SOCKET JOINT cpt
f raccord *m* par mandrinage collé
e unión *f* de enchufe cementado
i giunto *m* incollato a bicchiere
n gelijmde buismof
d Einsteckklebeverbindung *f*

980 CENTERLESS GRINDING MILL pol
f moulin *m* de rectification sans pointes
e molino *m* de rectificación sin puntas
i mulino *m* di rettificazione senza punte
n molen voor puntloos slijpen
d Mühle *f* für spitzenloses Schleifen

981 CENTRIFUGAL ATOMIZER, CENTRIFUGAL SPRAYER cen
f pulvérisateur *m* centrifuge
e pulverizador *m* centrífugo
i polverizzatore *m* centrifugo
n centrifugaalverstuiver
d Zentrifugalzerstäuber *m*

982 CENTRIFUGAL BLENDER cen,mix
f mélangeur *m* centrifuge
e mezclador *m* centrífugo
i mescolatore *m* centrifugo
n centrifugaalmenger
d Zentrifugalmischer *m*

983 CENTRIFUGAL BLOWER cen,pre
f soufflerie *f* centrifuge
e soplante *m* centrífugo
i soffieria *f* a forza centrifuga
n centrifugaal blaastoestel *n*
d Schleudergebläse *n*

984 CENTRIFUGAL CLARIFIER cen,cle
f appareil *m* de clarification centrifuge

e clarificador *m* centrífugo
i apparecchio *m* di chiarificazione centrifugo
n centrifugaal klaringstoestel *m*
d Klärschleuder *f*

985 CENTRIFUGAL CLASSIFIER sor
f classeur *m* centrifuge
e clasificador *m* centrífugo
i classificatore *m* centrifugo
n centrifugaalclasseerder
d Schleuderklassierer *m*, Zentrifugalklassierer *m*

986 CENTRIFUGAL COMPRESSOR cen,pre
f compresseur *m* centrifuge
e compresor *m* centrífugo
i compressore *m* centrifugo
n centrifugaalcompressor
d Kreiselverdichter *m*, Zentrifugalkompressor *m*

987 CENTRIFUGAL COUNTER-CURRENT CONTACTOR cen
f contacteur *m* chimique centrifuge à contrecourant
e contactor *m* químico centrífugo de contracorriente
i contatore *m* chimico centrifugo a controcorrente
n centrifugaal chemisch contactmakend apparaat *n* met tegenstroom
d chemischer Zentrifugalkontaktapparat *m* mit Gegenstrom

988 CENTRIFUGAL DISK ATOMIZER, SPINNING DISK ATOMIZER soa
f atomiseur *m* à disque rotatif, disque *m* atomiseur
e atomizador *m* de disco giratorio, disco *m* atomizador
i atomizzatore *m* a disco rotativo, disco *m* atomizzatore
n draaischijfverstuiver
d Fliehkraftzerstäuber *m*, Zerstäuberscheibe *f*

989 CENTRIFUGAL DRIER cen,dry
f essoreuse *f*
e secadora *f* centrífuga
i essiccatoio *m* centrifugo
n droogcentrifuge
d Trockenschleuder *f*, Zentrifugaltrockner *m*

990 CENTRIFUGAL DRUM cen
f tambour *m* centrifuge
e tambor *m* centrífugo
i tamburo *m* centrifugo
n centrifugetrommel
d Schleudertrommel *f*

991 CENTRIFUGAL DUST COLLECTOR cle,ga
f collecteur *m* centrifuge de la poussière
e colector *m* centrífugo del polvo
i collettore *m* centrifugo della polvere
n centrifugale stofverzamelaar
d Zentrifugalstaubsammler *m*

992 CENTRIFUGAL FAN, PADDLE WHEEL FAN cen,clo
f ventilateur *m* centrifuge
e ventilador *m* centrífugo, ventilador *m* de rueda de palas
i aspiratore *m* centrifugo, ventilatore *m* centrifugo
n centrifugaalventilator
d Zentrifugalventilator *m*

993 CENTRIFUGAL FILTER cen,fil
f filtre *m* centrifuge
e filtro *m* centrífugo
i filtro *m* centrifugo
n centrifugaal filter *n*
d Schleuderfilter *m*

994 CENTRIFUGAL-FLOW AGITATOR mix
f agitateur *m* à flux centrifuge
e agitador *m* de flujo centrífugo
i agitatore *m* a flusso centrifugo
n roerwerk *n* met centrifugale stroming
d Rührwerk *n* mit Zentrifugalströmung

995 CENTRIFUGAL LATEX CLARIFIER, LATEX CLARIFYING CENTRIFUGE rub
f centrifugeur *m* à clarifier le latex
e centrífuga *f* para clarificar el látex
i centrifuga *f* per chiarificare il lattice
n latexklaringscentrifuge
d Klärzentrifuge *f* für Kautschukmilch

996 CENTRIFUGAL LATEX CONCENTRATOR rub
f centrifugeur *m* à concentrer le latex

e concentrador *m* centrífugo para el látex
i centrifuga *f* per concentrare il lattice
n centrifuge voor het concentreren van latex
d Zentrifuge *f* zum Konzentrieren von Kautschukmilch

997 CENTRIFUGAL MILL brk,cen
f désintégrateur *m* centrifuge
e molino *m* centrífugo
i mulino *m* centrifugo
n centrifugaalmolen
d Schleudermühle *f*, Zentrifugalmühle *f*

998 CENTRIFUGAL PUMP cen,pre
f pompe *f* centrifuge
e bomba *f* centrífuga
i pompa *f* centrifuga
n centrifugaalpomp, centripomp
d Kreiselpumpe *f*, Schleuderpumpe *f*, Zentrifugalpumpe *f*

999 CENTRIFUGAL SCREEN, CENTRIFUGAL SIEVE sor,sta
f tamis *m* centrifuge
e criba *f* centrífuga, tamiz *m* centrífugo
i staccio *m* centrifugo, vaglio *m* centrifugo
n zeefcentrifuge
d Schleudersieb *n*, Zentrifugalsifter *m*

1000 CENTRIFUGAL SEDIMENTATION VESSEL cen
f cuve *f* de décantation centrifuge
e cuba *f* de sedimentación centrífuga
i serbatoio *m* di sedimentazione centrifugo
n bezinkvat *n* met centrifugewerking
d Zentrifugenabsetzbehälter *m*

1001 CENTRIFUGAL SEPARATOR cen
f séparateur *m* centrifuge
e separador *m* centrífugo
i separatore *m* centrifugo
n centrifugaalseparator
d Zentrifugalseparator *m*

1002 CENTRIFUGAL SIFTING MACHINE cen,sor
f trieuse *f* centrifuge
e cernedor *m* centrífugo
i stacciatura *f* centrifuga
n centrifugale sorteermachine
d Schleudersichter *m*

1003 CENTRIFUGAL-SPRAY COATER coa,pap
f pulvérisateur *m* centrifuge pour revêtir
e pulverizador *m* centrífugo para revestir
i polverizzatore *m* centrifugo per rivestire
n centrifugale laagverstuiver
d Zentrifugalbeschichtungszerstäuber *m*

1004 CENTRIFUGAL STILL cen,dis
f appareil *m* centrifuge de distillation
e aparato *m* centrífugo para destilar
i apparecchio *m* centrifugo da distillare
n centrifugaal destillatieapparaat *n*
d Zentrifugaldestillierapparat *m*

1005 CENTRIFUGAL-TYPE HUMIDIFIER hyg
f humecteur *m* centrifuge, humidificateur *m* centrifuge
e humectador *m* centrífugo, humidificador *m* centrífugo
i umidificatore *m* centrifugo
n centrifugaal bevochtigingstoestel *n*
d Zentrifugalanfeuchtapparat *m*, Zentrifugalbefeuchter *m*

1006 CENTRIFUGE cen
f centrifugeur *m*, machine *f* centrifuge
e centrífuga *f*, máquina *f* centrífuga
i centrifuga *f*, macchina *f* centrifuga
n centrifuge
d Zentrifuge *f*

1007 CENTRIFUGE CUP lab
f bécher *m* centrifugeur
e vaso *m* de centrífuga
i bicchiere *m* di centrifuga
n centrifugebeker
d Zentrifugenbecher *m*

1008 CENTRIFUGE SIEVE, BASKET cen
f panier *m* de centrifugeur
e cesta *f* de centrífuga
i paniere *m* di centrifuga
n centrifugezeef
d Schleudersieb *n*, Zentrifugensieb *n*

1009 CENTRIFUGE TUBE mcy
f tube *m* centrifugeur
e tubo *m* de centrífuga
i tubo *m* di centrifuga
n centrifugebuisje *n*
d Zentrifugenglas *n*

1010 CERAMIC FILTER cer,fil
f filtre *m* céramique
e filtro *m* cerámico
i filtro *m* ceramico
n keramisch filter *n*
d Tonfilter *m*

1011 CERAMIC FUELLED GAS-COOLED REACTOR nue
f réacteur *m* à refroidissement à gaz avec combustible céramique
e reactor *m* de enfriamiento por gas con combustible cerámico
i reattore *m* a raffreddamento a gas con combustibile ceramico
n met gas gekoelde keramiekbrandstofreactor
d gasgekühlter Keramikbrennstoffreaktor *m*

1012 CERAMIC FURNACE cer,htt
f four *m* céramique
e horno *m* cerámico
i forno *m* ceramico
n keramische oven
d Brennofen *m*

1013 CERAMIC REACTOR nue
f réacteur *m* céramique
e reactor *m* cerámico
i reattore *m* con combustibile ceramico
n keramiekreactor
d keramischer Reaktor *m*, Reaktor *m* mit keramischen Brennstoff- und Moderatorelementen

1014 CHAIN AGITATOR, CHAIN STIRRER mix
f agitateur *m* à chaînes
e mezclador *m* de cadenas
i mescolatore *m* a catene
n kettingroerwerk *n*
d Kettenrührwerk *n*

1015 CHAIN BALANCE lab
f balance *f* à chaîne
e balanza *f* de cadena
i bilancia *f* a catena
n kettingbalans
d Kettenwaage *f*

1016 CHAIN HYDROMETER ins
f hydromètre *m* à chaîne
e hidrómetro *m* de cadena
i idrometro *m* a catena
n hydrometer met kettingophanging
d Hydrometer *n* mit Kettenaufhängung

1017 CHALK TEST APPARATUS, LIME BAG TESTER tes
f appareil *m* d'essai à la craie
e aparato *m* de prueba a la creta
i collaudatore *m* alla creta
n krijtproefapparaat *n*
d Kreideprobeapparat *m*

1018 CHALKING MACHINE coa,rub
f machine *f* à talquer
e máquina *f* para entalcar
i intalcatrice *f*
n talkeermachine, talkmachine
d Pudermaschine *f*

1019 CHAMBER, LEAD CHAMBER sua
f chambre *f* de plomb
e cámara *f* de plomo
i camera *f* di piombo
n loden kamer
d Bleikammer *f*

1020 CHAMBER BURETTE lab
f burette *f* à tête sphérique
e bureta *f* de cabeza esférica
i buretta *f* a testa sferica
n bolburet
d Ballonburette *f*

1021 CHAMBER DRIER, CORRIDOR DRIER cer
f sécheur *m* à chambres
e secador *m* de cámaras
i essiccatore *m* a camere
n kamerdroogapparaat *n*
d Kammertrockner *m*

1022 CHAMBER FURNACE, RETORT FURNACE htt
f four *m* à cornues
e horno *m* de retortas
i forno *m* a storte
n retortenoven
d Retortenofen *m*

1023 CHAMBER KILN cer
f four *m* à chambres
e horno *m* de cámaras de cocción múltiple
i forno *m* a camere

n kameroven
d Kammerofen *m*

1024 CHAMBER OF A RESERVOIR, RESERVOIR CHAMBER cpt
f compartiment *m* de réservoir
e cámara *f* de recipiente
i vasca *f* di serbatoio
n reservoirkamer
d Behälterkammer *f*

1025 CHAMBER PRESS fil,pre
f filtre-presse *f* à chambres
e filtroprensa *m* de cámaras
i passafiltro *m* a camere
n kamerpers
d Kammerpresse *f*

1026 CHANGE CAN MIXER (US), PONY MIXER (GB) mix
f mélangeur *m* change-bidons
e mezclador *m* con cubas intercambiables
i mescolatore *m* a bidoni intercambiabili
n wisselvatmenger
d Mischer *m* mit auswechselbaren Kannen

1027 CHANNEL, ALCOVE, CANAL gla
f canal *m* d'alimentation
e canal *m* de alimentación
i canale *m* d'alimentazione
n aanvoerkanaal *n*
d Speisekanal *m*

1028 CHANNEL (US), APPROACH (GB), APPROACHING PASSAGE (GB) cpt
f canal *m* d'amenée
e canal *m* de acceso
i canale *m*
n aanspuitkanaal *n*
d Zugangskanal *m*

1029 CHANNEL FOR EXCESS RUBBER, OVERFLOW CHANNEL rub,sap
f couloir *m* de trop-plein, dégorgeoir *m*
e canaleta *f* para rebosamiento
i tubo *m* di troppo pieno, tubo *m* sfioratore
n overloopkanaal *n*
d Überlaufrinne *f*

1030 CHANNEL-TYPE AGITATOR mix
f agitateur *m* à canaux
e agitador *m* de canales
i agitatore *m* a canali
n kanalenroerwerk *n*
d Kanälenrührwerk *n*

1031 CHANNEL-TYPE INDUCTION OVEN htt,met
f four *m* électrique à induction à canal
e horno *m* eléctrico de inducción de canal
i forno *m* elettrico ad induzione a canale
n inductieoven met ringkanaal
d Rinneninduktionsofen *m*

1032 CHARCOAL FILTER fil
f filtre *m* à charbon de bois
e filtro *m* de carbón vegetal
i filtro *m* di carbone di legna
n houtskoolfilter *n*
d Holzkohlenfilter *m*

1033 CHARCOAL GAS GENERATOR gas
f générateur *m* de gaz de charbon de bois
e generador *m* de carbón vegetal
i generatore *m* di gas di carbone di legna
n houtskoolgasgenerator
d Holzkohlengasgenerator *m*

1034 CHARCOAL PILE, CHARCOAL STACK fue
f bûcher *m* de charbon de bois
e pila *f* de carbón vegetal
i carbonaia *f*
n houtskoolmijt
d Holzkohlenmeiler *m*

1035 CHARCOAL PIT fue
f charbonnière *f*, fosse *f* de charbon de bois
e fosa *f* de carbón vegetal
i pozzo *m* di carbone di legna
n houtskoolput
d Holzkohlengrube *f*

1036 CHARCOAL RETORT fue
f cornue *f* à charbon de bois
e retorta *f* de carbón vegetal
i storta *f* a carbone di legna
n houtskoolretort
d Holzkohlenretorte *f*

1037 CHARGE CAVITY pla,rub
f cavité *f* d'alimentation
e cavidad *f* de carga
i cavità *f* d'alimentazione
n vulruimte
d Füllraum *m*

1038 CHARGE METERING DEVICE pla,rub
f dispositif *m* de dosage de l'alimentation
e dispositivo *m* dosificador de la alimentación
i dispositivo *m* di regolazione dell'alimentazione
n voedingsdoseerapparaat *n*
d Speisemessvorrichtung *f*

1039 CHARGE RESISTANCE FURNACE (US), DIRECT RESISTANCE OVEN(GB) htt
f four *m* à chauffage direct par résistance
e horno *m* de calentamiento directo por resistencia
i forno *m* a riscaldamento diretto a resistenza
n weerstandsoven met directe verhitting
d direkter Widerstandsofen *m*

1040 CHARGING FUNNEL CHUTE cpt
f dispositif *m* de chargement
e tolva *f* de carga
i dispositivo *m* di carico
n vulinrichting
d Beschickungsvorrichtung *f* Füllvorrichtung *f*

1041 CHARGING HOLE sin
f trou *m* de chargement
e abertura *f* de carga
i apertura *f* di carico
n vulopening
d Beschickungsöffnung *f*

1042 CHARGING TRAY, LOADING TRAY pla,rub
f chargeur *m* à trous
e bandeja *f* de carga
i caricatore *m*
n vulplaat
d Fülltablett *n*, Siebplatte *f*

1043 CHASE pla
f cadre *m* de moule
e cuna *f* de molde
i telaio *m* di forma
n matrijsfreem *n*
d Matrizenrahmen *m*

1044 CHASE pla
f frette *f*
e alojamiento *m* del cojinete, cuna *f* del cojinete
i carcassa *f*, custodia *f*
n houder voor samengestelde matrijs
d Backe *f*

1045 CHASE, BODY, FRAME cpt
f bâti *m*, châssis *m*
e bastidor *m*, cuna *f*
i custodia *f*, telaio *m*
n freem *n*
d Chassis *n*, Gehäuse *n*, Rahmen *m*

1046 CHASE, MOLD CHASE (US), MOULD CHASE (GB) pla
f châssis *m* de moulage
e bloque *m* sujetador de la matriz, chasis *m* del molde
i custodia *f*
n matrijsfreem *n*
d Formrahmen *m*

1047 CHASER, CHASER MILL dec,mix
f moulin *m* mélangeur
e molino *m* mezclador
i mulino *m* mescolatore
n mengmolen
d Mischmühle *f*

1048 CHECK, GUIDE PLATE, STOCK GUIDE cpt,mix
f flasque *f* de guidage latérale
e carrillo *m* de un mezclador
i guancia *f* di un mescolatore
n keerbord *n*, materiaalgeleider van een mengwals
d Backe *f* einer Mischmaschine

1049 CHECK VALVE, BACKFLOW VALVE, REFLUX VALVE, STOP VALVE cpt
f soupape *f* d'arrêt
e válvula *f* de retención
i valvola *f* d'arresto, valvola *f* intercettante
n afsluitklep, sperklep
d Absperrventil *n*

1050 CHEESE BASKET dai,nu
f cagerotte *f*, clayon *m*
e cesto *m* de queso
i cesto *m* per formaggio
n kaaskorf
d Käsekorb *m*

1051 CHEESE DRAINER, CHEESE SIEVE dai,nut
f clisse *f*
e cañizo *m*
i casciaia *f*
n kaashorde, kaasmat
d Käsehorde *f*

1052 CHEESE FUNNEL, CREAM FUNNEL dai,lab
f entonnoir *m* à crème
e embudo *m* de crema
i imbuto *m* a panna
n roomtrechter
d Sahnetrichter *m*

1053 CHEESE KNIFE dai,nut
f coutelas *m*
e cuchillo *m* para cortar queso
i coltello *m* per tagliare formaggio
n kaasmes *n*
d Käsesäbel *m*, Käseschwert *n*

1054 CHEESE MOLD (US), CHEESE MOULD (GB), CHEESE TUB dai,nut
f jatte *f* à fromage, moule *m* à fromage
e encella *f*, molde *m* de queso
i forma *f* per il cacio, forma *f* per il formaggio
n kaasnap, kaasvorm
d Käsenapf *m*, Yärb *m*

1055 CHEESE PRESS dai,nut
f presse *f* à former le fromage
e prensa *f* de formación de queso
i pressa *f* da formare formaggio, torchio *m* da formaggio
n kaaspers
d Käsepresse *f*, Laibformpresse *f*

1056 CHEESE SCOOP dai,nut,tes
f sonde *f* à fromage
e catador *m*
i trapano *m* da formaggio
n kaasboor
d Käsestecher *m*

1057 CHEESE SCREEN-CUTTER dai,nut
f grille-coupeuse *f*
e rejilla *f* de corte
i griglia *f* per tagliare
n roostermes *n*
d Käseharfe *f*

1058 CHEMICAL COLORING BATH (US), CHEMICAL COLOURING BATH (GB) dec
f bain *m* de coloration chimique
e baño *m* de coloración química
i bagno *m* di colorazione chimica
n chemisch kleurbad *n*
d chemisches Farbbad *n*

1059 CHEMICAL ENGINEERING, INDUSTRIAL CHEMISTRY gen
f chimie *f* industrielle
e química *f* industrial
i chimica *f* industriale, ingegneria *f* chimica
n chemische technologie, industriële chemie
d chemische Technologie *f*, technische Chemie *f*

1060 CHEMICAL EXPOSURE METER pho
f exposimètre *m* chimique
e exposímetro *m* químico
i esposimetro *m* chimico
n chemische belichtingsmeter
d chemischer Belichtungsmesser *m*

1061 CHEMICAL GLASSWARE lab
f verrerie *f* laboratoire
e cristalería *f* química
i cristalleria *f* chimica
n laboratiumglaswerk *n*
d Laborglas *n*

1062 CHEMICAL PICKLING BATH, elc
f bain *m* de décapage chimique
e baño *m* de decapaje químico
i bagno *m* di decapaggio chimico
n chemisch beitsbad *n*
d chemisches Beizbad *n*

1063 CHEMICAL PLATING BATH, ELECTROLESS PLATING BATH coa
f bain *m* de placage chimique
e baño *m* de quimioplastia
i bagno *m* di placcatura chimica
n chemisch pleetbad *n*
d chemisches Plattierungsbad *n*

1064 CHEMICAL STONEWARE lab
f céramique *f* chimique
e cerámica *f* química
i ceramica *f* chimica
n chemisch aardewerk *n*
d Laborsteingut *n*

1065 CHEMICAL STRIPPING BATH, SOLUTION STRIPPING BATH cle
f bain *m* de dépouillage chimique
e baño *m* de eliminación química
i bagno *m* di degalvanizzazione chimica
n chemisch laagverwijderingsbad *n*
d chemisches Entplattierungsbad *n*

1066 CHEMICAL TEST TUBE, BOILING TUBE lab
f tube *m* à essai chimique

e tubo *m* de ensayo
i tubo *m* di saggio chimico
n chemisch reageerbuisje *n*
d chemisches Reagenzglas *n*

1067 CHILEAN MILL, EDGE RUNNER brk
f broyeur *m* à meules verticales
e molino *m* chileno, molino *m* de muelas verticales
i macina *f* a mole verticali, molazza *f*
n grote kollergang
d Kollergang *m*

1068 CHILLING ROLL soa
f cylindre *m* refroidisseur
e cilindro *m* refrigerador
i cilindro *m* refrigerante
n koelwals
d Kühlwalze *f*

1069 CHIMNEY gen
f cheminée *f*
e chimenea *f*
i camino *m*, fumaiuolo *m*
n schoorsteen
d Schlot *m*

1070 CHINA INDUSTRY cer
f industrie *f* de la porcelaine
e industria *f* de la porcelana
i industria *f* della porcellana
n porseleinindustrie
d Porzellanindustrie *f*

1071 CHLORIDOMETER ins
f chloridomètre *m*
e cloridómetro *m*
i cloridometro *m*
n chloridometer
d Chloridometer *n*

1072 CHLORIMETER, CHLOROMETER ins
f chlorimètre *m*
e clorímetro *m*
i clorometro *m*
n chloorgehaltemeter
d Chlorgehaltmesser *m*

1073 CHLORINATOR cle
f appareil *m* à chlore, chloreur *m*
e clorador *m*, clorizador *m*, clorurador *m*
i cloratore *m*
n chloreerapparaat *n*
d Chlorapparat *m*, Chlorgerät *n*

1074 CHLORINE CONTACT CHAMBER cle
f chambre *f* à contact du chlore
e cámara *f* de contacto del cloro
i camera *f* a contatto del cloro
n chloorcontactkamer
d Chlorkontaktkammer *f*

1075 CHLORINE DOSING APPARATUS cle
f appareil *m* de dosage de chlore
e aparato *m* dosificador de cloro
i apparecchio *m* dosificatore di cloro
n chloordoseringsapparaat *n*
d Chlordosierungsgerät *n*

1076 CHLORINE GENERATOR gas
f générateur *m* de chlore
e generador *m* de cloro
i generatore *m* di cloro
n chloorgenerator
d Chlorentwickler *m*

1077 CHLORITE BLEACHING BATH tex
f bain *m* de blanchiment à chlorite
e baño *m* de blanqueo de clorito
i bagno *m* di candeggio a clorito
n chlorietbleekbad *n*
d Chloritbleichbad *n*

1078 CHOKING CALORIMETER, THROTTLE CALORIMETER ins
f calorimètre *m* à étranglement
e calorímetro *m* de estrangulación
i calorimetro *m* a strozzatura
n smoorcalorimeter
d Drosselkalorimeter *n*

1079 CHROMATE COATING BATH coa
f bain *m* pour couche de chromate
e baño *m* para capa de cromato
i bagno *m* per strato di cromato
n chromaatlaagbad *n*
d Chromatschichtbad *n*

1080 CHROMATE DIP, CHROMIC ACID DIP elc
f bain *m* de décapage à l'acide chromique
e baño *m* de decapaje de ácido crómico
i bagno *m* di decapaggio all'acido cromico
n chromaatbeitsbad *n*
d Chromatbeize *f*

1081 CHROMATE DYEING BATH dye
f bain *m* de teinture à chromate

e baño *m* de teñido de cromato
i bagno *m* di tintura a cromato
n chromaatverfbad *n*
d Chromatfärbbad *n*

1082 CHROMATING BATH cle,met
f bain *m* de mordançage
e baño *m* de decapaje al bicromato
i bagno *m* di mordenzatura
n bichromaatbeitsbad *n*
d Bichromatbeizbad *n*

1083 CHROMATOCOIL chr
f chromatographe *m* à serpentins
e cromatógrafo *m* de serpentines
i cromatografo *m* a serpentine
n spiraalchromatograaf
d Spiralenchromatograph *m*

1084 CHROMATOGRAPH chr
f chromatographe *m*
e cromatógrafo *m*
i cromatografo *m*
n chromatograaf
d Chromatograph *m*

1085 CHROMATOGRAPHIC COLUMN chr
f tube *m* pour chromatographie sur colonne
e columna *f* cromatográfica
i colonna *f* cromatografica
n chromatograafkolom
d Chromatographiesäule *f*

1086 CHROMATOGRAPHIC JAR chr
f bocal *m* pour chromatographie, cuve *f* pour chromatographie
e cuba *f* de cromatografía
i tino *m* di cromatografia
n scheidingskamer
d Trennkammer *f*

1087 CHROME PRINTING BATH tex
f bain *m* d'impression chromique
e baño *m* de impresión crómica
i bagno *m* di stampatura cromica
n chroomdrukbad *n*
d Chromdruckbad *n*

1088 CHROMIUM PLATING BATH elc
f bain *m* de chromage
e baño *m* de cromado
i bagno *m* di cromatura
n chroombad *n*, verchroombad *n*
d Chrombad *n*, Verchromungsbad *n*

1089 CHROMIUM PLATING PLANT elc
f installation *f* de chromage
e equipo *m* de cromado
i impianto *m* di cromatura
n verchromingsinstallatie
d Verchromungsanstalt *f*

1090 CHROMIUM PLATING TANK elc
f récipient *m* de chromage
e tanque *m* de cromado
i serbatoio *m* di cromatura
n verchroomtank
d Verchromungsbehälter *m*

1091 CHROMIZING OVEN coa
f four *m* de cémentation par le chrome
e horno *m* de cromización
i forno *m* di cromizzazione
n chromeeroven
d Inkromierofen *m*

1092 CHUCK GAGE (US), BOURDON GAUGE (GB), PRESSURE GAUGE (GB) ins
f manomètre *m* de Bourdon
e manómetro *m* de Bourdon
i manometro *m* di Bourdon
n bourdonmanometer
d Bourdonfedermanometer *n*

1093 CHURN, CHURNING VESSEL nut
f baratte *f*
e mantequera *f*
i zangola *f*
n karn, karnton, karnvat *n*
d Butterfass *n*

1094 CHURN DASHER, CHURN STAFF nut
f baratton *m*, batte-beurre *m*
e palo *m* batador
i pestello *m* di zangola
n karnpols, karnstok
d Butterstempel *m*, Butterstössel *m*

1095 CHURNER, CHURNING MACHINE dai,nut
f baratte *f* mécanique
e mantequera *f* mecánica
i zangola *f* meccanica
n karnmachine
d Buttermaschine *f*

1096 CHUTE, SLIDE iwt
f couloir *m*
e canalón *m*
i canaletto *m*
n goot
d Rutsche *f*

1097 CIRCULAR TANK sto
f récipient *m* circulaire
e tanque *m* circular
i serbatoio *m* circolare
n ronde tank
d kreisförmiger Behälter *m*

1098 CIRCULATING CYCLONE EVAPORATOR eva
f évaporateur *m* cyclone
e evaporador *m* ciclónico
i evaporatore *m* ciclonico
n cycloonverdamper
d Umlaufverdampfer *m*

1099 CIRCULATING FUEL REACTOR nue
f réacteur *m* à combustible circulant, réacteur *m* de circulation
e reactor *m* de combustible circulante
i reattore *m* a circolazione del combustibile
n reactor met circulerende splijtstof
d Kreislaufreaktor *m*, Reaktor *m* mit umlaufendem Spaltstoff

1100 CIRCULATING MACHINE tex
f machine *f* de teinture à circulation du colorant
e máquina *f* de teñido con circulación del colorante
i macchina *f* di tintura a circolazione del colorante
n verfmachine met circulerende verfstof
d Färbmaschine *f* mit Farbstoffumlauf

1101 CIRCULATING PUMP pre
f pompe *f* de circulation
e bomba *f* de circulación
i pompa *f* di circolazione
n circulatiepomp
d Umlaufpumpe *f*, Umwälzpumpe *f*

1102 CIRCULATION LOOP gen,nue
f circuit *m* de circulation
e circuito *m* de circulación
i circuito *m* di circolazione
n circulatiecircuit *n*
d Kreislaufleitung *f*

1103 CIRCULATOR (US), WORT CIRCULATING PUMP (GB) bre
f circulateur *m*
e bomba *f* de prefiltración
i pompa *f* per prefiltrato
n voorlooppomp
d Vorlaufpumpe *f*

1104 CISTERN, COLLECTING TANK, RECEIVING TANK sto
f récipient *m* collecteur, réservoir *m* à eau
e recipiente *m* de colección, recipiente *m* de agua
i serbatoio *m* collettore, serbatoio *m* d'acqua, serbatoio *m* di raccolta
n verzameltank, watertank
d Sammelbehälter *m*, Wasserbehälter *m*

1105 CLADDING coa
f doublage *m*, placage *m*
e chapado *m*, plaqueado *m*
i placcatura *f*
n bemanteling, plateren *n*
d Plattierung *f*

1106 CLAMPING FRAME cpt
f serre-flan *m*
e bastidor *m* de tensado
i telaio *m* di fissaggio
n montagefreem *n*, opspanfreem *n*
d Halterahmen *m*, Einspannrahmen *m*

1107 CLAMPING PLATE (US), MOUNTING PLATE (GB) cpt,pla
f plateau *m* de fixation, plateau *m* porte-moule
e placa *f* de sujeción
i piastra *f* portastampo
n opspanplaat
d Aufspannplatte *f*

1108 CLARIFICATION BASIN, CLARIFIER iwt
f bassin *m* de clarification, clarificateur *m*
e clarificadora *f*
i bacino *m* di chiarificazione, vasca *f* di chiarificazione
n klaringsbassin *n*, klaringsbekken *n*
d Klärbecken *n*

1109 CLARIFICATION PLANT, SEWAGE DISPOSAL PLANT, SEWAGE PLANT iwt
f installation *f* de clarification, station *f* d'épuration des eaux d'égout, usine *f* de purification des eaux d'égout
e equipo *m* de clarificación,

equipo *m* de utilización de aguas negras
i impianto *m* di chiarificazione, impianto *m* d'epurazione delle acque di fogna
n afvalwaterreinigingsinstallatie
d Abwasserreinigungsanlage *f*, Kläranlage *f*

1110 CLARIFICATION TANK iwt,sto
SEDIMENTATION TANK, SETTLING TANK
f bac *m* de décantation, récipient *m* de clarification
e tanque *m* de clarificación, tanque *m* de sedimentación
i serbatoio *m* di chiarificazione, serbatoio *m* di sedimentazione
n bezinkvat *n*, klaringstank
d Absetzbehälter *m*, Klärbehälter *m*

1111 CLARIFYING BATH pho
f bain *m* de clarification
e baño *m* clarificador
i bagno *m* di chiarificazione
n klaringsbad *n*
d Klärbad *n*

1112 CLARIFYING TUB, bre
FILTER VAT, LAUTER TUB, STRAINING VAT
f cuve *f* de filtration, cuve-filtre *m*
e cuba *f* de filtración
i tino *m* di filtrazione
n klaringskuip
d Klärbottich *m*, Läuterbottich *m*

1113 CLASSIFIER sor
f classeur *m*
e clasificador *m*
i classificatore *m*
n classificeerder
d Klassierer *m*

1114 CLASSIFIER WITH CONICAL TROUGHS, sor
SLOUGHING-OFF BOX
f classeur *m* à bacs coniques
e clasificador *m* de tolvas cónicas
i classificatore *m* a vasche coniche
n classificeerder met conische bakken
d Spitzkasten *m*

1115 CLASSIFYING CRYSTALLIZER cry,sor
f cristallisoir *m* classeur
e cristalizador *m* clasificador
i cristallizzatore *m* classificatore
n classificator-kristallisator
d Klassierkristallisator *m*

1116 CLAY BLANKET sin
f couche *f* de terrage
e colchón *m* de barro, pantalla *f* de arcilla
i strato *m* d'argilla
n kleidek *n*
d Tondecke *f*

1117 CLAY BLUNGER, cer,mix
CLAY MIXER
f malaxeur *m* d'argile
e amasadora *f* de arcilla
i mescolatore *m* d'argilla
n kleimenger
d Tonmischer *m*

1118 CLAY CRUCIBLE htt
f creuset *m* en argile
e crisol *m* de arcilla, crisol *m* de barro
i crogiolo *m* d'argilla
n kleikroes
d Tontiegel *m*

1119 CLAY KNEADING MACHINE (GB), mix
CLAY TEMPERING MACHINE
f machine *f* à pétrir l'argile
e amasadora *f* de arcilla
i macchina *f* gramolatrice d'argilla
n kleikneedmachine
d Tonknetmaschine *f*

1120 CLAY-LINED CRUCIBLE htt
f creuset *m* à revêtement en argile
e crisol *m* revestido de arcilla
i crogiolo *m* a rivestimento in argilla
n met klei gevoerde smeltkroes
d tongefutterter Tiegel *m*

1121 CLAY MILL cer,brk
f broyeur *m* d'argile
e molino *m* de arcilla
i mulino *m* d'argilla
n kleimolen
d Tonmühle *f*

1122 CLAY PIT cer
f argiliaire *f*, glaisière *f*
e barrera *f*, cantera *f* de arcilla
i cava *f* d'argilla, marniera *f*
n kleigroeve, kleikuil
d Tongrube *f*

1123 CLAY POT gla
f creuset *m* en argile
e crisol *m* en arcilla
i crogiolo *m* in argilla
n kleipot
d Tonerdehafen *m*

1124 CLAY PRESS cer
f presse *f* d'argile
e prensa *f* de arcilla
i pressa *f* d'argilla
n kleipers
d Tonpresse *f*

1125 CLAY SHREDDER cer
f déchireuse *f* d'argile
e desmenuzadora *f* de arcilla
i trinciatrice *f* d'argilla
n kleihakmachine
d Tonzermalmer *m*

1126 CLEAN REACTOR nue
f réacteur *m* virginal
e reactor *m* virginal
i reattore *m* vergine
n maagdelijke reactor, onvergiftigde reactor
d jungfräulicher Reaktor *m*

1127 CLEANER BLADE fil
f racleur *m*
e rascador *m*
i raschiera *f*
n afstrijkmes *n*
d Streichblech *n*

1128 CLEANING DOCTOR (GB), COLOR DOCTOR (US), COLOUR DOCTOR (GB) tex
f racloir *m*
e rascador *m*
i raschiatoio *m*
n schraapmes *n*
d Streichmesser *n*

1129 CLEANING MACHINE, DRESSING MACHINE bre,cle
f nettoyeuse *f*
e limpiadora *f*
i pulitrice *f*
n poetsmachine
d Putzmaschine *f*, Reinigungsmaschine *f*

1130 CLEANING PLANT cle,elc
f établissement *m* de dégraissage
e equipo *m* de desengrase
i impianto *m* di sgrassaggio
n ontvettingsinstallatie *f*
d Reinigungsanlage *f*

1131 CLEANING TANK cle,elc
f récipient *m* de dégraissage
e tanque *m* de desengrase
i serbatoio *m* di sgrassaggio
n ontvettingstank
d Reinigungsbehälter *m*

1132 CLEANSER, CUP SCRAPER, RAKER cle,rub
f curette *f* de tasse à latex
e limpiador *m* para recipiente de látex
i pulitore *m* per tazza di lattice
n cupwisser
d Latexbecherwischer *m*

1133 CLEAR WATER TANK, SERVICE RESERVOIR hyg
f réservoir *m* d'eau potable
e tanque *m* de agua dulce, tanque *m* de agua potable
i serbatoio *m* d'acqua filtrata, serbatoio *m* d'acqua potabile
n drinkwatertank
d Reinwasserbehälter *m*, Trinkwasserbehälter *m*

1134 CLEARING BATH pho
f bain *m* de clarification
e baño *m* clarificador
i bagno *m* chiarificatore
n klaringsbad *n*
d Klärungsbad *n*

1135 CLIMBING-FILM EVAPORATOR eva
f évaporateur *m* à couche montante
e evaporador *m* de capa ascendente
i evaporatore *m* a strato di salita
n verdamper met stijgende laag
d Aufsteigschichtverdampfer *m*, Kletterverdampfer *m*

1136 CLINICAL MAXIMUM THERMOMETER mcy
f thermomètre *m* médical
e termómetro *m* clínico
i termometro *m* clinico
n koortsthermometer
d Fieberthermometer *n*

1137 CLINKERING ZONE cer
f zone *f* de frittage, zone *f* de vitrification
e zona *f* de aglutinación, zona *f* de sinterización
i zona *f* di sinterizzazione, zona *f* di vetrificazione
n klinkerzone
d Sinterzone *f*

1138 CLOSE ANNEALING OVEN, htt,met
BOX ANNEALING OVEN,
POT ANNEALING OVEN
f four *m* de recuit en caisse
e horno *m* de recocido en caja
i forno *m* di ricottura in cassetta
n kistgloeioven
d Kastenglühofen *m*

1139 CLOSED CIRCUIT, gen
CLOSED CYCLE
f circuit *m* fermé, cycle *m* fermé
e ciclo *m* cerrado, circuito *m* cerrado
i ciclo *m* chiuso, circuito *m* chiuso
n gesloten circuit *n*, kringloop
d Kreislauf *m*

1140 CLOSED CONDUIT, tra
PIPE LINE
f tuyautage *m*, tuyauterie *f*
e conducto *m* cerrado, tubería *f*
i tubazione *f*
n pijpleiding
d geschlossene Leitung *f*

1141 CLOSED-CUP FLASH-POINT TESTER tes
f vérificateur *m* du point d'inflammation en vase fermé
e probador *m* del punto de deflagración en copa cerrada
i collaudatore *m* del punto d'infiammazione in coppa chiusa
n vlampunttester in gesloten vat
d Flammpunktprüfer *m* in geschlossenem Behälter

1142 CLOSED FILTER fil
f filtre *m* fermé
e filtro *m* cerrado
i filtro *m* chiuso
n gesloten filter *n*
d geschlossener Filter *m*

1143 CLOSED FRAME PRESS pla,rub
f presse *f* à bâti fermé
e prensa *f* de bastidor de caja, prensa *f* de chasis cerrado
i pressa *f* a telaio chiuso
n pers met gesloten freem
d geschlossene Rahmenpresse *f*

1144 CLOSED POT, gla
COVERED POT, HOODED POT
f creuset *m* couvert, creuset *m* fermé
e crisol *m* cerrado, crisol *m* cubierto
i crogiolo *m* coperto
n gesloten haven, gesloten pot, overdekte haven, overdekte pot
d geschlossener Hafen *m*, überdeckter Hafen *m*

1145 CLOSED RESERVOIR sto
f réservoir *m* couvert
e tanque *m* cubierto
i serbatoio *m* coperto
n overdekte tank
d überdeckter Behälter *m*

1146 CLOSED TRICKLING FILTER, fil
COVERED TRICKLING FILTER
f lit *m* bactérien couvert
e filtro *m* de escurrimiento cubierto, filtro *m* percolador cubierto
i letto *m* percolatore coperto
n afgedekt bacteriënfilter *n*
d geschlossener Filter *m*

1147 CLOTH BUFF pol
f roue *f* de polissage en drap
e disco *m* pulidor de paño
i lucidatrice *f* a tessuto
n polijstschijf met doek
d Tuchschwabbelscheibe *f*

1148 CLOTH FILTER fil
f filtre *m* de tissu
e filtro *m* de tela
i filtro *m* di tessuto
n weefselfilter *n*
d Gewebefilter *m*

1149 CLOTH ROLLS pol
f rouleaux *pl* de polissage drapés
e rollos *pl* pulidores cubiertos de tela
i rulli *pl* pulitori coperti di stoffa
n met stof beklede polijstrollen *pl*
d stoffbekleidete Polierrollen *pl*

1150 CLOTH TUBE FILTER, fil
STOCKING-TYPE FILTER
f filtre *m* à manches
e filtro *m* de mangas
i filtro *m* a maniche
n slangenfilter *n*, zakkenfilter *n*
d Schlauchfilter *m*

1151 CLOUD AND POUR POINT INDICATOR fue,ins
f signaleur *m* du point de trouble et de congélation
e indicador *m* del punto de turbiedad y de fluidez crítica
i indicatore *m* della temperatura di cristallizzazione del punto di scorrimento

n troebelings- en vloeipuntindicator
d Trübungs- und Stockpunktgeber *m*

1152 CLOUD POINT INDICATOR fue,ins
f signaleur *m* du point de congélation
e indicador *m* del punto de turbiedad
i indicatore *m* della temperatura di cristallizzazione
n troebelingspuntindicator
d Trübungspunktgeber *m*

1153 CLOUDBURSTING PLANT (US), cle
STEEL BALL PEENING PLANT (GB)
f installation *f* de traitement par jet de balles en acier
e equipo *m* de tratamiento por chorro con perdigonos de acero
i impianto *m* di trattamento con pallottole d'acciaio
n kogelstraalinstallatie
d Kugelstrahlanlage *f*

1154 CLUSTER LATTICE nue
f réseau *m* à faisceau
e red *f* de haz
i reticolo *m* a fascio
n bundelrooster *n*
d Bündelgitter *n*

1155 CLUSTER OF FUEL ELEMENTS, nue
BUNDLE OF FUEL ELEMENTS, STRINGER
f faisceau *m* d'éléments combustibles
e haz *m* de elementos combustibles
i fascio *m* d'elementi combustibili
n brandstofelementenbundel, splijtstofelementenbundel
d Brennstoffelementenbündel *n*

1156 COAGULATION CYLINDER, dai,nut
CURDLING CYLINDER
f cylindre *m* de caillage, cylindre *m* de coagulation
e cilindro *m* de coagulación, cilindro *m* de cuajamiento
i cilindro *m* d'accagliamento, cilindro *m* di coagulazione
n coaguleercilinder, stremcilinder
d Gerinnezylinder *m*, Koagulierzylinder *m*

1157 COAGULATION TANK col
f récipient *m* de coagulation
e tanque *m* de coagulación
i serbatoio *m* di coagulazione
n coaguleerbak
d Koagulationsbehälter *m*

1158 COAL ASH FURNACE lab
f four *m* à déterminer la teneur en cendres
e horno *m* para determinar el porcentaje de cenizas
i forno *m* da determinare il tenore di ceneri
n asgehaltebepalingsoven
d Aschengehaltbestimmungsofen *m*

1159 COAL GAS GENERATOR gas
f générateur *m* de gaz d'éclairage, générateur *m* de gaz de houille
e generador *m* de gas de hulla, generador *m* de gas del alumbrado
i generatore *m* di gas di carbone, generatore *m* di gas illuminante
n lichtgasgenerator
d Leuchtgasgenerator *m*

1160 COARSE BAR SCREEN, sor
COARSE RACK, COARSE SCREEN, TRASH RACK
f grille *f* grossière
e criba *f* gruesa
i griglia *f* grossa, griglia *f* grossolana
n grove zeef
d Grobrechen *m*

1161 COARSE FILTER, fil
ROUGHING FILTER
f préfiltre *m*
e filtro *m* grueso
i filtro *m* grosso, filtro *m* sgrossatore
n grof filter *n*
d Grobfilter *m*

1162 COATING coa
f couche *f*, recouvrement *m*, revêtement *m*
e capa *f*, recubrimiento *m*, revestimiento *m*
i ricoprimento *m*, rivestimento *m* strato *m*
n bedekking, bekleding, laag
d Beschichtung *f*, Schicht *f*, Überzug *m*

1163 COATING KNIFE (US), de
DOCTOR KNIFE (GB)
f racle *f*
e cuchilla *f* dosificadora, rasqueta *f*

i raschiatore *m*
n strijkmes *n*
d Rakel *f*, Streichmesser *n*

1164 COATING MACHINE, pho
CASTING MACHINE
f machine *f* à émulsionner
e máquina *f* emulsionadora
i macchina *f* d'emulsionatura
n emulsiegietmachine
d Emulsionsauftragmaschine *f*

1165 COATING PISTOL, coa
SPRAY GUN, SPRAYING PISTOL
f pistolet *m* de pulvérisation
e pistola *f* pulverizadora
i pistola *f* a spruzzo
n spuitpistool *n*
d Spritzpistole *f*

1166 COB MILL cer
f moulin *m* à battement
e molino *m* de percusión
i mulino *m* a percussione
n slagmolen
d Schlagmühle *f*

1167 COCK, cpt
TAP
f robinet *m*
e grifo *m*
i rubinetto *m*
n kraan
d Hahn *m*

1168 COCK MANIFOLD, lab
TAP MANIFOLD
f rampe *f* à robinets
e rampa *f* de grifos
i rampa *f* di rubinetti
n kranenstraat
d Hahnbrücke *f*

1169 COFFIN, nue,tra
CASKET
f caisse *f* de transport
e caja *f* de transporte
i bara *f*, cassa *f* di trasporto, cofano *m*
n transportkist
d Transportbehälter *m*

1170 COIL, dis
WORM
f serpentin *m*
e serpentín *m*
i serpentina *f*
n spiraalkoeler
d Rohrschlange *f*

1171 COILED TUBE CONDENSER clo
f condensateur *m* à serpentin
e condensador *m* de serpentín
i condensatore *m* a serpentina
n condensor met spiraalbuis
d Kondensator *m* mit Kühlschlange

1172 COKE BOX cle,sua
f boîte *f* à coke
e caja *f* de coque
i cassetta *f* a coke
n cokesbak
d Kokskasten *m*

1173 COKE OVEN fue
f four *m* de coke
e horno *m* de coque
i forno *m* di coke
n cokesoven
d Koksofen *m*

1174 COKE-OVEN GASGENERATOR gas
f générateur *m* de gaz de cokerie, générateur *m* de gaz de four à coke
e generador *m* de gas de coquería, generador *m* de gas de horno de coque
i generatore *m* di gas di cokeria, generatore *m* di gas di forno a coke
n cokesovengasgenerator
d Kokereigasgenerator *m*, Koksofengasgenerator *m*

1175 COKE PERCOLATOR, clo
COKE SCRUBBER, COKE-TRAY AERATOR
f ruisseleur *m* à coke
e aireador *m* de batea de coque, depurador *m* de coque
i irrigatore *m* a coke, percolatore *m* a coke
n cokessproeier
d Koksrieseler *m*

1176 COLD ACCUMULATOR ctt
f accumulateur *m* de froid
e acumulador *m* de frío
i accumulatore *m* di freddo
n koudeaccumulator
d Kältesammler *m*, Kältespeicher *m*

1177 COLD CLEAN REACTOR nue
f réacteur *m* froid virginal
e reactor *m* frío virginal
i reattore *m* freddo vergine
n koude maagdelijke reactor, koude onvergiftigde reactor
d kalter jungfräulicher Reaktor *m*

1178 COLD ENAMELLING PLANT coa
f installation *f* d'émaillage à froid
e equipo *m* de esmaltado en frío
i impianto *m* di smaltatura a freddo
n koudemailleerinstallatie
d Kaltemaillierungsanlage *f*, Tauchemaillierungsanlage *f*

1179 COLD EXTRACTION APPARATUS, COLD SOXHLET ext
f appareil *m* pour extraction avec des solvants refroidis
e aparato *m* de extracción con solventes enfriados
i apparecchio *m* d'estrazione con solventi raffreddati
n koudextraheerapparaat *n*
d Kaltextraktionsapparat *m*

1180 COLD FLOW TESTER tes
f machine *f* d'essai d'écoulement à froid
e aparato *m* para ensayo de escurrimiento en frío
i apparecchio *m* per la prova di scorrimento a freddo
n koudevloeibeproevingsapparaat *n*
d Prüfgerät *n* für kalten Fluss

1181 COLD GALVANIZING BATH elc
f bain *m* de zingage à froid
e baño *m* de cincado en frío
i bagno *m* di zincatura a freddo
n koudverzinkbad *n*
d Kaltverzinkungsbad *n*

1182 COLD LIQUID CHILLER clo
f appareil *m* de réfrigération à liquide
e refrigerador *m* de líquido
i raffreddatore *m* a liquido
n koelvloeistofapparaat *n*
d Kühlflüssigkeitsgerät *n*

1183 COLD MASHING VAT, MASH TUB, MASH TUN bre
f cuve *f* d'empâtage, cuve-matière *f*
e cuba *f* de mezcla, tina *f* de braceado
i tino *m* di miscela
n beslagkuip
d Maischbottich *m*

1184 COLD MOLD (US), COLD MOULD (GB) pla
f moule *m* froid
e molde *m* frío
i stampa *f* fredda
n koude vorm
d kalte Form *f*

1185 COLD REACTOR nue
f réacteur *m* froid
e reactor *m* frío
i reattore *m* freddo
n koude reactor
d kalter Reaktor *m*

1186 COLD RINSE (GB), COLD RINSE BATH (GB), COLD SWILL (US), COLD SWILL BATH (US) elc
f bain *m* de rinçage à froid
e baño *m* de enjuague en frío
i bagno *m* di sciacquatura a freddo
n koud spoelbad *n*
d Kaltspülbad *n*

1187 COLD-TEST JAR lab
f tube *m* à congélation
e tubo *m* de congelación
i tubo *m* a congelazione
n stolpuntvat *n*
d Stockpunktglas *n*

1188 COLD-VULCANIZING MACHINE rub
f machine *f* à vulcaniser à froid
e máquina *f* para vulcanizar en frío
i macchina *f* per vulcanizzazione a freddo
n koudvulcaniseermachine
d Kaltvulkanisiermaschine *f*

1189 COLD WELDING PLANT wel
f installation *f* de soudure à froid
e equipo *m* de soldadura en frío
i impianto *m* di saldatura a freddo
n koudwelinstallatie
d Kaltschweissanlage *f*

1190 COLDFINGER (US), IMMERSION CONDENSER, SUBMERGED CONDENSER pre
f condensateur *m* plongeur, doigt *m* réfrigérant, réfrigérant *m* à immersion
e condensador *m* de inmersión
i condensatore *m* ad immersione
n dompelcondensor
d Einhängekühler *m*, Tauchkondensator *m*

1191 COLLAR cer,ht
f col *m* de cornue

e cuello *m* de retorta
i collo *m* di storta
n retorthals
d Retortenhals *m*

1192 COLLECTING ELECTRODE fil
f électrode *f* de précipitation
e electrodo *m* de precipitación
i elettrodo *m* di precipitazione
n neerslagelektrode
d Niederschlagselektrode *f*

1193 COLLECTING ELECTRODE, coa
PASSIVE ELECTRODE
f électrode *f* passive
e electrodo *m* inerte, electrodo *m* pasivo
i elettrodo *m* passivo
n passieve elektrode
d passive Elektrode *f*

1194 COLLECTING ELECTRODE fil
RAPPER
f secoueur *m* d'électrode de précipitation
e sacudidor *m* de electrodo de precipitación
i scossatore *m* d'elettrodo di precipitazione
n neerslagelektrodeschudder
d Niederschlagselektrodenrüttler *m*

1195 COLLECTING TANK, sto
CISTERN, RECEIVING TANK
f récipient *m* collecteur, réservoir *m* à eau
e recipiente *m* de colección, recipiente *m* de agua
i serbatoio *m* collettore, serbatoio *m* d'acqua, serbatoio *m* di raccolta
n verzameltank, watertank
d Sammelbehälter *m*, Wasserbehälter *m*

1196 COLLECTING TRAP dis,lab
f collecteur *m* de fractions
e colector *m* de fracciones
i collettore *m* di frazioni
n fractieverzamelaar
d Fraktionssammler *m*

1197 COLLECTING TROUGH iwt,sto
f goulotte *f* collectrice, rigole *f* collectrice
e pilón *m* colector
i doccione *m* collettore
n verzamelgoot
d Sammelrinne *f*

1198 COLLECTING TUN, bre,sto
COLLECTING VESSEL
f cuve *f* guilloire
e cuba *f* de reposo, tina *f* de reposo
i tino *m* di raccolta
n verzameltank
d Sammelbottich *m*

1199 COLLECTOR, iwt
INTERCEPTING SEWER
f collecteur *m*
e alcantarilla *f* interceptadora, canal *m* colector
i canale *m* collettore, collettore *m* di fognatura
n verzamelkanaal *n*
d Abwassersammler *m*, Sammelkanal *m*

1200 COLLECTOR PLATES elc
f plaques *pl* collectrices
e placas *pl* colectoras
i piastre *pl* conduttrici
n contactplaten *pl*
d Kontaktplatten *pl*

1201 COLLOID MILL col
f broyeur *m* pour colloïdes, moulin *m* colloïdal
e molino *m* coloidal
i mulino *m* colloidale
n colloïdmolen, colloïdwals
d Kolloidmühle *f*, Kolloidwalze *f*

1202 COLONY COUNTER bac,lab
f compteur *m* de colonies
e contador *m* de colonias
i contatore *m* di colonie
n kolonieteller
d Kolonienzähler *m*

1203 COLOR BOX (US), tex
COLOUR BOX (GB)
f cuve *f* à colorant
e cuba *f* de colorante, pastera *f*
i tino *m* a colorante
n kleurstoftrog
d Farbstofftrog *m*

1204 COLOR COMPARISON TUBE lab
(US),
COLOUR COMPARISON TUBE (GB)
f tube *m* à comparaison de couleurs
e tubo *m* comparador de colores
i tubo *m* comparatore di colori
n kleurvergelijkingsbuis
d Farbenvergleichrohr *n*

1205 COLOR DOCTOR (US), COLOUR DOCTOR (GB), CLEANING DOCTOR (GB) — tex
- f racloir *m*
- e rascador *m*
- i raschiatoio *m*
- n schraapmes *n*
- d Streichmesser *n*

1206 COLOR MILL (US), COLOUR MILL (GB), PAINT MILL — dec
- f moulin *m* à couleurs, moulin *m* à molette
- e molino *m* para colores
- i mulino *m* a colori
- n verfmolen
- d Farbmühle *f*

1207 COLOR SCREEN (US), COLOUR SCREEN (GB) — fil,pho
- f filtre *m* des couleurs
- e pantalla *f* de los colores
- i filtro *m* dei colori, schermo *m* colorato
- n kleurenfilter *n*
- d Farbfilter *m*, Lichtfilter *m*

1208 COLOR SHOP (US), COLOUR SHOP (GB) — tex
- f atelier *m* de préparation du colorant
- e taller *m* de preparación del colorante
- i officina *f* di preparazione del colorante
- n kleurstofaanmakerij
- d Farbstoffküche *f*

1209 COLORIMETER — ins
- f colorimètre *m*
- e colorímetro *m*
- i colorimetro *m*
- n colorimeter
- d Kolorimeter *n*

1210 COLORIMETER TUBE — ins,lab
- f tube *m* de colorimètre
- e tubo *m* de colorímetro
- i tubo *m* di colorimetro
- n colorimeterbuis
- d Kolorimeterzylinder *m*

1211 COLORING BARREL (US), COLOURING DRUM (GB) — pla
- f tonneau *m* de coloration
- e tambor *m* de coloreado, tambor *m* de teñido
- i buratto *m* di colorazione
- n droogkleurtrommel
- d Einfarbtrommel *f*

1212 COLUMN, TOWER — gen
- f colonne *f*, tour *f*
- e columna *f*, torre *f*
- i colonna *f*, torre *f*
- n kolom, toren, zuil
- d Kolonne *f*, Säule *f*, Turm *m*

1213 COLUMN CHROMATOGRAPHY APPARATUS, APPARATUS FOR COLUMN CHROMATOGRAPHY — chr
- f appareil *m* pour la chromatographie sur colonne
- e aparato *m* per la cromatografía sobre columna
- i apparecchio *m* per la cromatografia a colonna
- n zuilenchromatograaf
- d Säulenchromatographiegerät *n*

1214 COLUMN PRESS, PILLAR PRESS — pre
- f presse *f* à colonnes
- e prensa *f* de columnas
- i pressa *f* a colonne
- n kolompers, zuilenpers
- d Säulenpresse *f*

1215 COLUMN UNIT — dis,lab
- f segment *m* de colonne
- e segmento *m* de columna
- i segmento *m* di colonna
- n kolomsegment *n*
- d Kolonnenschuss *m*

1216 COMB RACK — coa,htt
- f support *m* d'accrochage
- e gancho *m* de asimiento
- i telaio *m* di sospensione
- n ophangrek *n*
- d Einhängestell *n*

1217 COMBINATION CELL — elc
- f pile *f* avec une électrode métallique et une électrode non-métallique
- e pila *f* con un electrodo metálico y un electrodo no metálico
- i pila *f* con un elettrodo metallico ed un elettrodo non metallico
- n element *n* met een metaal- en een niet-metaalelektrode
- d Element *n* mit einer Metall- und einer Nichtmetallelektrode

1218 COMBINED BATH, TONING AND FIXING BATH — pho
- f bain *m* viro-fixateur

e baño *m* de fijador virador
i bagno *m* di viraggio e di fissaggio
n combinatiebad *n*, kleurend fixeerbad *n*
d Tonfixierbad *n*

1219 COMBINED VACUUM PUMP/ COMPRESSOR vac
f pompe *f* à vide et compresseur *m* combiné
e bomba *f* de vacío y compresor *m* combinado
i pompa *f* a vuoto e compressore *m* combinato
n vacuümpomp-compressor-combinatie
d Vakuumpumpe-Verdichter-zusammenbau *m*

1220 COMBUSTION BOAT lab
f godet *m* à fusion, nacelle *f* à fusion
e cápsula *f* de combustión, navicella *f* de fusión
i navicella *f* a fusione
n gloeischuitje *n*, verbrandings-schuitje *n*
d Einsetzer *m*, Glühschiffchen *n*, Verbrennungsschiffchen *n*

1221 COMBUSTION BOAT SHIELD lab
f cartouche *m* de chargement, gaine *f* de nacelle à fusion
e funda *f* de navicella de fusión
i camicia *f* di navicella a fusione
n gloeischuitjeomhulling, verbrandingsschuitjeomhulling
d Glühschiffchenhülle *f*, Muffel *m* für Verbrennungsschiffchen

1222 COMBUSTION CHAMBER fue,htt
f chambre *f* de combustion
e cámara *f* de combustión
i camera *f* di combustione
n verbrandingskamer
d Verbrennungskammer *f*

1223 COMBUSTION CRUCIBLE htt
f creuset *m* à combustion
e crisol *m* de combustión
i crogiolo *m* a combustione
n verbrandingskroes
d Verbrennungstiegel *m*

1224 COMBUSTION FURNACE, INCINERATOR htt
f four *m* crématoire, four *m* d'incinération
e horno *m* de análisis, horno *m* de combustión
i forno *m* d'incenerazione
n verbrandingsoven
d Verbrennungsofen *m*

1225 COMBUSTION TUBE lab
f tube *m* de combustion
e tubo *m* de combustión
i tubo *m* di combustione
n verbrandingsbuis
d Verbrennungsrohr *n*

1226 COMMERCIAL TANK elc
f cuve *f* commerciale
e cuba *f* comercial
i tino *m* industriale
n produktievat *n*
d Produktionsgefäss *n*

1227 COMMINUTOR, DISINTEGRATOR, SCREEN SHREDDER, TRITURATOR pap
f appareil *m* de dilacération, dilacérateur *m*
e trituradora *f*
i frantumatrice *f*, trituratore *m*
n fijnmaalmolen
d Rechengutzerkleinerer *m*, Rechenwolf *m*

1228 COMMUNICATING TUBES, COMMUNICATING VESSELS ins
f vases *pl* communicants
e tubos *pl* comunicantes
i vase *pl* comunicanti
n communicerende vaten *pl*
d kommunizierende Röhren *pl*

1229 COMPACTING MILL (GB), COMPACTOR MILL (US) pre
f moulin *m* compresseur
e molino *m* compactador
i mulino *m* di compressione
n comprimeermolen
d Komprimiermühle *f*

1230 COMPARISON SPECTROSCOPE ins,spe
f spectroscope *m* comparateur
e espectroscopio *m* comparador
i spettroscopio *m* comparatore
n vergelijkende spectroscoop
d vergleichendes Spektroskop *n*

1231 COMPARISON TUBE ins
f tube *m* de comparaison
e tubo *m* de testigo
i tubo *m* di paragone
n vergelijkbuis
d Vergleichröhre *f*

1232 COMPENSATING RESERVOIR, iwt
BALANCING RESERVOIR,
EQUALIZING RESERVOIR
f réservoir *m* compensateur, réservoir *m* régulateur du débit
e deposito *m* de equilibrio
i serbatoio *m* compensatore, serbatoio *m* regolatore
n compensatievat *n*
d Ausgleichbehälter *m*

1233 COMPOSITE MOLD (US), pla
COMPOSITE MOULD (GB),
BAR MOULD (GB)
f moule *m* à empreintes différentes
e molde *m* compuesto, molde de varias matrices
i stampo *m* ad impronte differenti
n gevarieerde matrijs
d Mehrfachform *f*

1234 COMPOSITE PLATE coa,elc
f dépôt *m* composé
e depósito *m* compuesto
i deposito *m* composto
n samengesteld neerslag *n*
d Mehrschichtenüberzug *m*, Verbundüberzug *m*

1235 COMPOUND MICROSCOPE ins
f microscope *m* composé
e microscopio *m* compuesto
i microscopio *m* composto
n samengestelde microscoop
d zusammengesetztes Mikroskop *n*

1236 COMPOUND ROLLS brk
f cylindres *pl* composés
e cilindros *pl* compuestos
i cilindri *pl* composti
n walsenparen *pl*
d Verbundwalzen *pl*

1237 COMPOUND TANNING BATH tan
f bain *m* de tannage mixte
e baño *m* de curtimiento mixto
i bagno *m* di concia mista
n gemengd looibad *n*
d gemischtes Gerbbad *n*

1238 COMPOUNDER-EXTRUDER rub
f boudineuse-mélangeuse *f*
e budinadora *f* mezcladora
i mescolatore-trafila *m*
n meng-spuitmachine
d Mischungsstrangpresse *f*

1239 COMPOUNDING ROOM mix
f salle *f* de mélange
e sala *f* de mezclar
i sala *f* per la preparazione delle mescolanze
n mengkamer
d Mischkammer *f*, Mischraum *m*

1240 COMPRESSED AIR pre
INSPIRATOR
f injecteur *m* d'air comprimé
e inyector *m* de aire comprimida
i iniettore *m* d'aria compressa
n persluchtinblazer
d Pressluftansauger *m*

1241 COMPRESSED AIR PIPE pre
f tube *m* pour pressions élevées
e tubo *m* para presiones elevadas
i tubo *m* per pressioni elevate
n hoge-drukbuis
d Hochdruckrohr *n*

1242 COMPRESSION MOLDING pre,rub
PRESS (US),
COMPRESSION MOULDING
PRESS (GB)
f presse *f* de moulage à compression
e prensa *f* de moldeo por compresión, prensa *f* para estampar a compresión
i pressa *f* di stampaggio per compressione, pressa *f* per stampatura a compressione
n compressievormpers
d Druckformpresse *f*

1243 COMPRESSION clo
REFRIGERATION MACHINE
f machine *f* de réfrigération par compression
e máquina *f* de refrigeración por compresión
i macchina *f* di refrigerazione a compressione
n compressiekoelmachine
d Kompressionskältemaschine *f*

1244 COMPRESSOR pre
f compresseur *m*
e compresor *m*
i compressore *m*
n compressor
d Kompressor *m*, Verdichter *m*

1245 COMPRESSOR JACKET clo
f chemise *f* de compresseur
e chaquete *m* de compresor
i camicia *f* di compressore
n compressormantel
d Kompressormantel *m*

1246 CONCENTRATION CELL elc
f pile *f* de concentration
e pila *f* de concentración
i pila *f* di concentrazione
n concentratiecel
d Konzentrationselement *n*

1247 CONCENTRIC CYLINDRICAL SHELL SILO sin
f silo *m* à cylindres concentriques
e silo *m* de cilindros concéntricos
i silo *m* a cilindri concentrici
n silo met concentrische cilinders
d Zylinderschalensilo *n*

1248 CONCRETE BIOLOGICAL SHIELD nue
f écran *m* biologique en béton
e pantalla *f* biológica de hormigón
i schermo *m* biologico di calcestruzzo
n biologisch betonscherm *n*
d biologische Betonabschirmung *f*

1249 CONCRETE HOPPER bui
f benne *f* à béton
e tolva *f* de hormigón
i benna *f* del calcestruzzo
n betonhopper
d Betonkübel *m*

1250 CONCRETE SHIELD nue,sap
f carapace *f* en béton
e blindaje *m* de hormigón
i schermo *m* in calcestruzzo
n betonpantser *n*
d Betonpanzer *m*

1251 CONCURRENT CENTRIFUGE, FLOW-THROUGH CENTRIFUGE cen,iso
f centrifugeur *m* a courant continu
e centrífuga *f* de corriente continua
i centrifuga *f* a corrente continua
n doorstroomcentrifuge
d Durchlaufzentrifuge *f*

1252 CONDENSATION COLUMN gen
f colonne *f* de condensation
e columna *f* de condensación
i colonna *f* di condensazione
n condensatiekolom
d Kondensationssäule *f*

1253 CONDENSED-MILK FACTORY dai
f fabrique *f* du lait condensé
e fábrica *f* de la leche condensada
i fabbrica *f* del latte condensato
n fabriek van gecondenseerde melk
d Kondensmilchfabrik *f*

1254 CONDENSER, CONDENSOR cpt
f condensateur *m*
e condensador *m*
i condensatore *m*
n condenser
d Kondensator *m*

1255 CONDENSER TUBE clo
f tube *m* de condensateur
e tubo *m* condensador, tubo *m* del condensador
i tubo *m* di condensatore
n condenserbuis
d Kondensatorrohr *n*

1256 CONDENSING COIL, COOLING COIL, REFRIGERATING COIL clo
f serpentin *m* de réfrigération
e serpentín *m* refrigerador
i serpentina *f* refrigerante
n koelslang
d Kühlschlange *f*, Kühlschlauch *m*

1257 CONDENSING VESSEL cpt
f séparateur *m* d'eau de condensation
e depósito *m* de condensación
i marmitta *f* di condensazione
n condenspot
d Kondensationstopf *m*, Kondenstopf *m*

1258 CONDITIONING OVEN tex
f étuve *f* de conditionnement
e estufa *f* de acondicionamiento
i stufa *f* di condizionamento
n conditioneerstoof
d Konditionierofen *m*

1259 CONDITIONING ZONE gla
f zone *f* de réglage de la température
e zona *f* de regulación de la temperatura
i zona *f* di regolazione della temperatura
n temperatuurregelingszone
d Temperaturregelungszone *f*

1260 CONDUCTIVITY CELL elc,mea
f cellule *f* à mesurer la conductivité
e célula *f* de medida de la conductividad

i cella *f* da misurare la conduttività
n geleidendheidsmeetcel
d Leitfähigkeitsmesszelle *f*

1261 CONE, FUNNEL, HOPPER — cpt
f entonnoir *m*
e embudo *m*
i imbuto *m*, tramoggia *f*
n trechter
d Trichter *m*

1262 CONE CLASSIFIER — sor
f classeur *m* à cône
e clasificador *m* de cono
i classificatore *m* a cono
n kegelclassificeerder
d Kegelklassierer *m*

1263 CONE CRUSHER — brk
f concasseur *m* à cône rotatif
e machacadora *f* de cono giratorio
i rompitore *m* a cono rotativo
n roterende-kegelbreker
d Drehkegelbrecher *m*

1264 CONE DRUM, CONICAL PELLETING DRUM — mcy
f tambour *m* conique pour former des comprimés
e tambor *m* cónico para formación de comprimidos
i tamburo *m* conico da formare compresse
n conische tabletteertrommel
d konische Tablettiertrommel *f*

1265 CONE IMPELLER — mix
f roue *f* motrice conique
e rueda *f* móvil cónica
i girante *m* conico
n conische roerder
d konische Rührstange *f*

1266 CONE MIXER, DOUBLE CONE MIXER — mix
f mélangeur *m* à double cône
e mezclador *m* de doble cono
i mescolatore *m* a doppio cono
n dubbele-kegelmenger
d Doppelkonusmischer *m*

1267 CONE WHEEL, TAPERED WHEEL — pol
f disque *m* conique
e disco *m* ahusado
i disco *m* conico
n conische schijf
d Kegelscheibe *f*

1268 CONFORMING ANODE — elc
f anode *f* préformée
e ánodo *m* preformado
i anodo *m* preformato
n omsluitende anode
d umschliessende Anode *f*

1269 CONICAL SCREEN CENTRIFUGE — cen
f centrifugeur *m* à écran conique
e centrífuga *f* de pantalla cónica
i centrifuga *f* a schermo conico
n centrifuge met conisch scherm
d Zentrifuge *f* mit konischem Schirm

1270 CONNECTING COCK — cpt,lab
f robinet *m* de communication, robinet *m* de connexion
e grifo *m* de comunicación, grifo *m* de conexión
i rubinetto *m* di comunicazione, rubinetto *m* di connessione
n verbindingskraan
d Verbindungshahn *m*

1271 CONNECTING CONDUIT — tra
f conduite *f* de communication
e tubería *f* de conexión
i condotta *f* di collegamento
n verbindingsleiding
d Verbindungsleitung *f*

1272 CONNECTING PIPE — tra
f tuyau *m* de communication, tuyau *m* de raccordement
e tubo *m* de conexión
i condotto *m* di raccordo, tubo *m* di collegamento
n aansluitbuis, verbindingsbuis
d Anschlussrohr *n*, Verbindungsrohr *n*

1273 CONSISTOMETER, PLASTIMETER, PLASTOMETER, RHEOMETER — ins
f consistomètre *m*, plastomètre *m*
e consistómetro *m*, plastómetro *m*
i consistometro *m*, plastometro *m*
n consistometer, plastometer
d Konsistenzmesser *m*, Plastometer *n*

1274 CONTACT AERATOR, SUBMERGED CONTACT AERATOR — iwt
f lit *m* immergé
e aireador *m* de contacto
i letto *m* sommerso
n contactaërator
d Emscherfilter *m*, Tauchkörper *m*

1275 CONTACT BED, CONTACT FILTER, FILL AND DRAW CONTACT BED — iwt
f lit *m* de contact
e lecho *m* de contacto
i letto *m* di contatto
n contactbed *n*
d Füllkörper *m*

1276 CONTACT CONDENSER, TUBULAR CONDENSER — pre
f condensateur *m* tubulaire
e condensador *m* tubular
i condensatore *m* tubolare
n buiscondensor
d Röhrenverdichter *m*

1277 CONTACT CONDUCTOR, CONTACT ELECTRODE — elc
f conducteur *m* de contact, électrode *f* de contact
e conductor *m* de contacto, electrodo *m* de contacto
i conduttore *m* di contatto
n stroomgeleider
d Kontaktelektrode *f*

1278 CONTACT PLATING PLANT — coa
f installation *f* de dépôt par contact
e equipo *m* de depósito por contacto
i impianto *m* di deposito per contatto
n contactbedekkingsinstallatie
d Kontaktplattierungsanlage *f*

1279 CONTACT SURFACE — iwt
f surface *f* de contact, surface *f* utile
e superficie *f* de contacto, superficie útil
i superficie *f* di contatto, superficie *f* utile
n contactoppervlak *n*, werkzaam oppervlak *n*
d Benetzungsfläche *f*

1280 CONTAINER, CONTAINMENT VESSEL, SAFETY CONTAINER — nue
f récipient *m* de sûreté
e recipiente *m* de seguridad
i involucro *m* del reattore, involucro *m* di contenimento
n veiligheidsvat *n*
d Sicherheitsbehälter *m*

1281 CONTAINER, JAR — elc
f bac *m*
e recipiente *m*
i contenitore *m*
n bak
d Zellengefäss *n*, Zellenkasten *m*

1282 CONTINUOUS BELT MIXER, CONTINUOUS RIBBON MIXER — mix
f mélangeur *m* à bande continue
e mezclador *m* de cinta continua
i mescolatore *m* a nastro continuo
n eindloze-bandmenger
d kontinuierlicher Bandmischer *m*

1283 CONTINUOUS CENTRIFUGAL FILTER, CONVEYOR-DISCHARGE CENTRIFUGE — cen,fil
f centrifugeur *m* à décharge par bande transporteuse, filtre *m* centrifuge continu
e centrífugo *m* de descarga por banda transportadora, filtro *m* centrífugo continuo
i centrifuga *f* a scarico per nastro trasportatore, filtro *m* centrifugo continuo
n centrifuge met lopende-band-ontlading, continu centrifugaal filter *n*
d kontinuierlicher Zentrifugal-filter *m*, Zentrifuge *f* mit Laufbandentladung

1284 CONTINUOUS COUNTER-CURRENT DECANTER — sep
f décanteur *m* continu à contrecourant
e decantador *m* continuo de contracorriente
i apparecchio *m* di decantazione continuo a controcorrente
n continu decanteerapparaat *n* met tegenstroom
d kontinuierlicher Gegenstrom-dekantierapparat *m*

1285 CONTINUOUS COUNTER-CURRENT LIQUID-LIQUID EXTRACTION PLANT — nue,sep
f installation *f* continue à contre-courant d'extraction par partage liquide-liquide
e equipo *m* continuo de contra-corriente de extracción por repartición líquido-líquido
i impianto *m* continuo a controcorrente per ripartizione liquido-liquido
n continue vloeistof-vloeistof-extractie-installatie met tegenstroombedrijf

d kontinuierliche Flüssigkeit-
Flüssigkeitextraktionsanlage *f*
in Gegenstrombetrieb

1286 CONTINUOUS DISTILLATION PLANT dis
f installation *f* de distillation continue
e equipo *m* de destilación continua
i impianto *m* di distillazione continua
n continue destilleerderij
d kontinuierliche Destillieranlage *f*

1287 CONTINUOUS DYEING MACHINE dye
f machine *f* de teinture continue
e máquina *f* de teñido continuo
i macchina *f* di tintura continua
n continue verfmachine
d kontinuierliche Färbmaschine *f*

1288 CONTINUOUS ELECTRODE htt
f électrode *f* continue
e electrodo *m* continuo
i elettrodo *m* continuo
n continue elektrode
d kontinuierliche Elektrode *f*

1289 CONTINUOUS FILTER, fil,iwt
PERCOLATING FILTER
f filtre *m* percolateur
e filtro *m* percolador
i filtro *m* percolatore
n doorzijgfilter *n*, percolator
d Seiher *m*, Sickerfilter *m*

1290 CONTINUOUS FLASH EVAPORATOR, eva
FLASH ROTARY VACUUM EVAPORATOR
f évaporateur *m* rotatif sous vide
e vacuoevaporador *m* giratorio
i evaporatore *m* rotativo sotto vuoto
n roterende vacuümverdamper
d Rotationsvakuumverdampfer *m*

1291 CONTINUOUS-FLOW CALORIMETER ins
f calorimètre *m* à courant continu
e calorímetro *m* de flujo continuo
i calorimetro *m* a flusso continuo
n stromingscalorimeter
d Durchflusskalorimeter *n*

1292 CONTINUOUS FURNACE htt
f four *m* continu
e horno *m* continuo
i forno *m* continuo
n continue oven
d Durchlaufofen *m*

1293 CONTINUOUS GRAIN IRRADIATION PLANT nue,nut
f installation *f* d'irridiation continue du blé, installation *f* d'irridiation continue de céréales
e equipo *m* de irradiación continua de cereales
i impianto *m* d'irradiazione continua del grano
n continue graanbestralingsinstallatie
d kontinuierliche Getreidebestrahlungsanlage *f*

1294 CONTINUOUS KILN, htt
CONTINUOUS STOVE
f étuve *f* continue
e estufa *f* continua
i stufa *f* continua
n doorloopdroogoven
d Durchlauftrockenofen *m*

1295 CONTINUOUS MERCURY STILL dis,lab
f appareil *m* à distiller le mercure en continue
e aparato *m* continuo de destilar mercurio
i apparecchio *m* continuo da destillare mercurio
n continu kwikdestilleerapparaat *n*
d Apparat *m* zur kontinuierlichen Quecksilberdestillation

1296 CONTINUOUS PASTEURISATION TOWER dai
f tour *f* de pasteurisation continue
e torre *f* de pasteurización continua
i torre *f* di pastorizzazione continua
n continue pasteuriseertoren
d kontinuierlicher Pasteurisierturm *m*

1297 CONTINUOUS PICKLING PLANT elc
f installation *f* de décapage continu
e equipo *m* de decapaje continuo
i impianto *m* di decapaggio continuo
n continue beitserij
d kontinuierliche Beizanlage *f*

1298 CONTINUOUS REHEATING FURNACE, htt,met
PUSHER-TYPE FURNACE
f four *m* poussant
e horno *m* empujador
i forno *m* á spinta
n doorschuifoven
d Durchstossofen *m*, Stossofen *m*

1299 CONTINUOUS SAND PLANT fou
f sablerie *f* continue
e equipo *m* continuo de arenas
i impianto *m* continuo per la lavorazione delle terre
n continue zandbereidings-installatie
d kontinuierliche Sandaufbereitungs-anlage *f*

1300 CONTINUOUS TANK FURNACE gla
f four *m* à bassin continu
e horno *m* de cuba continuo
i forno *m* a tino continuo
n continue wanoven
d Dauerwannenofen *m*

1301 CONTINUOUS-THROUGH CIRCULATION DRIER dry
f sécheur *m* à circulation continue
e secador *m* de circulación continua
i essiccatoio *m* a circolazione continua
n continue doorloopdroogmachine
d kontinuierlicher Durchlauf-trockner *m*

1302 CONTINUOUS VERTICAL RETORT htt,met
f cornue *f* verticale à charge continue
e retorta *f* vertical de carga continua
i storta *f* verticale a carica continua
n verticale retort met doorlopende lading
d Vertikalretorte *f* mit fortlaufender Beschickung

1303 CONTINUOUS VULCANIZER, rub
CONTINUOUS VULCANIZING MACHINE
f machine *f* à vulcaniser en continue
e máquina *f* de vulcanización continua
i vulcanizzatore *m* continuo
n continue vulcaniseermachine
d kontinuierliche Vulkanisier-maschine *f*

1304 CONTINUOUS WATER STILL dis,lab
f appareil *m* à eau distillée à marche continue
e aparato *m* continuo de destilar agua
i apparecchio *m* continuo da distillare acqua
n continu waterdestilleerapparaat *n*
d kontinuierlicher Wasserde-stillierapparat

1305 CONTROLLED TIPPING GROUND (GB), hyg
SANITARY LAND FILL PLOT (US)
f terrain *m* de décharge contrôlée
e terreno *m* de descarga controlada
i terreno *m* di scarico controllato
n aangewezen vuilbegraafplaats, aangewezen vuilnisstortplaats
d Grundstück *n* für geordnete Ablage

1306 CONTROLLED-VOLUME PUMP, pre
METERING PUMP
f pompe *f* à capacité variable
e bomba *f* de gasto regulado
i pompa *f* dosatrice
n pomp met regelbare capaciteit
d Pumpe *f* mit regelbarer Kapazität

1307 CONVERTER, gen
CONVERTOR
f récipient *m* de réaction
e recipiente *m* de reacción
i recipiente *m* di reazione
n reactievat *n*
d Reaktionsgefäss *n*

1308 CONVERTER, met
CONVERTOR
f convertisseur *m*
e convertidor *m*
i convertitore *m*
n convertor, peer
d Birne *f*, Konverter *m*

1309 CONVERTER, bre
FOREMASHER, PONYMASHER, PREMASHER
f hydrateur *m*
e premezclador *m*
i premescolatore *m*
n voorbeslagkuip
d Vormaischer *m*

1310 CONVERTER LINER coa
f revêtement *m* de convertisseur
e revestimiento *m* de convertidor
i rivestimento *m* di convertitore
n convertorvoering
d Konverterfutter *n*

1311 CONVEYOR BELT tra
f bande *f* transporteuse, ruban *m* transporteur
e banda *f* transportadora, cinta *f* transportadora
i nastro *m* trasportatore
n transportband
d Förderband *n*, Transportband *n*

1312 CONVEYOR FURNACE, BELT FURNACE met
f four *m* à tapis
e horno *m* de correa transportadora
i forno *m* a nastro trasportatore
n transportbandoven
d Förderbandofen *m*

1313 CONVEYORIZED PICKLING EQUIPMENT elc
f installation *f* de décapage continue
e equipo *m* de decapaje con transportador
i impianto *m* di decapaggio continuo
n continue beitsinstallatie
d kontinuierliche Beizanlage *f*

1314 COOLED ABSORPTION TOWER, COOLER ABSORBER abs,clo
f tour *f* d'absorption à refroidissement
e enfriador *m* absorbente, torre de absorción de enfriamiento
i torre *f* d'assorbimento raffreddata
n gekoelde absorptietoren
d Kühlabsorptionsturm *m*

1315 COOLER KEG COIL, BARREL ATTEMPERATOR bre
f serpentin *m* de fût intérieur
e serpentín *m* enfriador interior
i serpentina *f* raffreddatrice interiore
n vatkoeler
d Fasskühler *m*

1316 COOLING BED, HOT BED clo,met
f refroidisseur *m*
e lecho *m* de enfriamiento
i piano *m* di raffreddamento, raffreddatore *m*
n koelbed *n*
d Kühlbett *n*, Warmlager *n*

1317 COOLING COIL clo
f serpentin *m* de refroidissement
e serpentín *m* de enfriamiento
i serpentina *f* di raffreddamento
n koelslang, koelspiraal
d Kühlschlange *f*

1318 COOLING FIN clo
f ailette *f* de refroidissement
e aleta *f* de enfriamiento
i aletta *f* di raffreddamento
n koelvin
d Kühlrippe *f*

1319 COOLING FIXTURE, COOLING JIG, SHRINKAGE BLOCK clo,pla
f conformateur *m*, jauge *f* conformatrice
e platilla *f* de contracción
i conformatore *m*, controforma *f* di raffreddamento
n krimpblok *n*
d Abkühlvorrichtung *f*, Schrumpfblock *m*

1320 COOLING FURNACE, ANNEALING FURNACE gla
f four *m* de refroidissement
e horno *m* de enfriamiento
i forno *m* di raffreddamento
n koeloven
d Kühlofen *m*

1321 COOLING JACKET clo
f chemise *f* d'eau, enveloppe *f* de refroidissement
e camisa *f* de enfriamiento
i camicia *f* d'acqua, camicia *f* di raffreddamento
n koelwatermantel
d Kühlwassermantel *m*

1322 COOLING PIT, COOLING POND nue
f bassin *m* refroidisseur de radioactivité
e depósito *m* enfriador de radiactividad
i bacino *m* di decadimento, vasca di decadimento
n afstervingsbekken *n*
d Aktivitätsverminderungsbecken

1323 COOLING PRESS soa
f presse *f* réfrigérante
e prensa *f* de enfriamiento
i pressa *f* di raffreddamento
n koelpers
d Kühlpresse *f*

1324 COOLING ROLL MILL soa
f appareil *m* réfrigérant à cylindres
e cilindros *pl* de enfriamiento
i cilindri *pl* raffreddatori
n koelwals
d Kühlwalze *f*

1325 COOLING TOWER clo
f récupérateur *m* à cheminée, tour *f* de refroidissement
e torre *f* de enfriamiento
i torre *f* di raffreddamento
n koeltoren
d Gradierwerk *n*, Kühlturm *m*

1326 COOLING ZONE cer
f zone *f* de refroidissement
e zona *f* de enfriamiento
i zona *f* di raffreddamento
n koelzone
d Abkühlungszone *f*

1327 COOLSHIP, bre,clo
SHALLOW COOLER
f bac *m* refroidisseur
e bandeja *f* de enfriamiento
i rinfrescatoio *m*
n koelkuip
d Kühlschiff *n*

1328 COP DYEING APPARATUS dye
f appareil *m* de teinture en cannette
e aparato *m* de teñido en canilla
i apparecchio *m* di tintura in fuso di filo
n spoelverfapparaat *n*
d Garnwickelfärbgerät *n*

1329 COPPER, bre
BREW KETTLE
f boiler *m*
e caldera *f* de cocción
i bollitore *m*
n brouwketel
d Siedekessel *m*

1330 COPPER BIT, wel
SOLDERING IRON
f fer *m* à souder, soudoir *m*
e soldador *m*
i arnese *m* per saldare, saldatoio *m*
n soldeerbout
d Lötkolben *m*

1331 COPPER GAUZE FILTER cle,fil
(GB),
COPPER GAZE FILTER (US)
f filtre *m* en toile de cuivre
e filtro *m* en tela de cobre
i filtro *m* in tela di rame
n kopergaasfilter *n*
d Kupfergazefilter *m*

1332 COPPER PLATING BATH elc
f bain *m* de cuivrage
e baño *m* de cobreado
i bagno *m* di ramatura
n koperbad *n*
d Kupferbad *n*

1333 COPPER STRIP TESTER tes
f appareil *m* d'essai à ruban de cuivre
e probador *m* con tiro de cobre
i collaudatore *m* a nastro di rame
n koperstriptester
d Kupferstreifenprüfgerät *n*

1334 CORE cer,fou
f noyau *m*
e macho *m*, noyo *m*
i anima *f*
n kern
d Kern *m*

1335 CORE OF EXTRUDER HEAD, pla
EXTRUDER-HEAD CORE
f âme *f* de tête de boudineuse
e ánima *f* de cabezal de budinadora
i anima *f* di testa di trafila
n spuitkopdoorn
d Spritzdorn *m*

1336 CORE PIN, pla,pre
HOLE FORMING PIN
f broche *f* à trou lisse, broche-noyau *f*
e clavija *f* de macho, pasador *m*
i spina *f* d'anima
n kernstift
d Lochstift *m*

1337 CORE PRESSURE VESSEL nue
f récipient *m* à pression du coeur de réacteur
e recipiente *m* de presión del núcleo
i recipiente *m* a pressione del nocciolo
n drukvat *n* voor de actieve zone, reactorhartdrukvat *n*
d Druckgefäss *n* der aktiven Zone, Spaltzonedruckgefäss *n*

1338 CORED MOLD (US), CORED MOULD (GB) pla
f moule *m* à canaux
e molde *m* de canales
i stampo *m* a canali
n vorm met luchtkanalen
d Form *f* mit Luftkanälen

1339 CORELESS INDUCTION OVEN htt
f four *m* à creuset à induction
e horno *m* de crisol por inducción
i forno *m* a crogiolo ad induzione
n kernloze inductieoven
d Induktionstiegelofen *m*

1340 CORNER VALVE, ANGLE VALVE, MITER VALVE cpt
f soupape *f* d'équerre
e válvula *f* angular, válvula *f* de escuadra
i valvola *f* a disco con seda a 45°
n hoekklep
d Eckventil *n*, Winckelventil *n*

1341 CORRIDOR DRIER, CHAMBER DRIER cer
f sécheur *m* à chambres
e secador *m* de cámaras
i essiccatore *m* a camere
n kamerdroogapparaat *n*
d Kammertrockner *m*

1342 CORRODING POT met
f cuve *f* de corrosion
e vasija *f* de corrosión
i serbatoio *m* di corrosione
n mestpot
d Gärtopf *m*

1343 CORRUGATED ROLL, FLUTED ROLL rub
f cylindre *m* cannelé
e cilindro *m* acanalado
i cilindro *m* scanalato
n geribbelde wals, riffenwals
d geriffelte Walze *f*, Riffelwalze *f*

1344 COSSETTES CONVEYOR sin,tra
f transporteur *m* de cossettes
e transportador *m* de rebanadas
i trasportatore *m* di ritagli
n snijdseltransporteur
d Schnitzelförderer *m*

1345 COTTAGE STEAMER tex
f chambre *f* de traitement à vapeur
e cámara *f* de tratamiento al vapor
i camera *f* di trattamento al vapore
n stoomkamer
d Dampfkammer *f*

1346 COTTON ASBESTOS, ASBESTOS WOOL mat
f laine *f* d'amiante
e lana *f* de amianto
i lana *f* d'amianto
n asbestwol
d Asbestwolle *f*

1347 COUCHING PRESS pap
f presse *f* à coucher
e prensa *f* húmeda
i pressa *f* umida
n koetspers
d Gautschpresse *f*

1348 COUCHING ROLL pap
f cylindre *m* de presse à coucher
e cilindro *m* de prensa húmeda
i cilindro *m* di pressa umida
n koetswals
d Gautschwalze *f*

1349 COUNTER ELECTRODE spe
f contre-électrode *f*
e contraelectrodo *m*
i controelettrodo *m*
n tegenelektrode
d Gegenelektrode *f*

1350 COUNTER RESERVOIR stc
f réservoir *m* d'équilibre
e recipiente *m* equilibrador
i serbatoio *m* d'equilibrio, serbatoio *m* d'estremità
n tegendrukvat *n*
d Gegenbehälter *m*

1351 COUNTERCURRENT CENTRIFUGE cen,isc
f centrifugeur *m* à contrecourant
e centrifuga *f* de contracorriente
i centrifuga *f* a controcorrente
n tegenstroomcentrifuge
d Gegenstromzentrifuge *f*

1352 COUNTERCURRENT CONDENSER pr
f condensateur *m* à contrecourant
e condensador *m* de contracorrien
i condensatore *m* a controcorrent
n tegenstroomcondensor
d Gegenstromkondensator *m*

1353 COUNTERCURRENT CONTACT CONDENSER pr
f condensateur *m* tubulaire à contrecourant

e condensador *m* tubular de contracorriente
i condensatore *m* tubolare a controcorrente
n tegenstroombuiscondensor
d Gegenstromröhren-kondensator *m*

1354 COUNTERCURRENT COOLER bre,clo
f refroidisseur *m* à contrecourant
e enfriador *m* de contracorriente
i raffreddatore *m* a controcorrente
n tegenstroomkoeler
d Gegenstromkühler *m*

1355 COUNTERCURRENT DECANTER sep
f décanteur *m* à contrecourant
e decantador *m* de contracorriente
i recipiente *m* di decantazione a controcorrente
n tegenstroomdecanteerapparaat *n*
d Gegenstromdekantierapparat *m*

1356 COUNTERCURRENT DISTILLATION PLANT dis
f installation *f* de distillation à contrecourant
e equipo *m* de destilación de contracorriente
i impianto *m* di distillazione a controcorrente
n tegenstroomdestilleerinstallatie
d Gegenstromdestillieranlage *f*

1357 COUNTERFLOW COOLING TOWER clo
f tour *f* de refroidissement à contrecourant
e torre *f* de enfriamiento de contracorriente
i torre *f* di raffreddamento a controcorrente
n tegenstroomkoeltoren
d Gegenstromkühlturm *m*

1358 COUNTERFLOW FILTER fil
f filtre *m* à contrecourant
e filtro *m* de contracorriente
i filtro *m* a controcorrente
n tegenstroomfilter *n*
d Gegenstromfilter *m*

1359 COUNTERPRESSURE RACKER, ISOBAROMETRIC RACKER bre
f soutireuse *f* isobarométrique
e llenadora *f* de contrapresión, llenadora *f* isobarométrica
i isobarometro *m*
n tegendrukvulmachine
d Gegendruckfüller *m*, isobarometrischer Fassabfüller *m*

1360 COUPLE elc
f couple *m*
e par *m*
i coppia *f*
n platenpaar *n*
d Plattenpaar *n*

1361 COVER GLASS lab
f couvre-objet *m*
e cristal *m* protector, cubreobjetos *m*
i vetrino *m* per microscopio
n dekglaasje *n*
d Deckgläschen *n*

1362 COVER ROLLER, OVERPRINTING ROLLER tex
f rouleau *m* de surimpression
e rodillo *m* de sobreimpresión
i rullo *m* di soprastampatura
n overdrukrol
d Aufdruckrolle *f*

1363 COVER STONE, BOOT (US), POPETTE (GB) gla
f brique *f* de couverture
e ladrillo *m* de recubrimiento
i mattone *m* di copertura
n afdeksteen
d Abdeckstein *m*

1364 COVERED POT, CLOSED POT, HOODED POT gla
f creuset *m* couvert, creuset *m* fermé
e crisol *m* cerrado, crisol *m* cubierto
i crogiolo *m* coperto
n gesloten haven, gesloten pot, overdekte haven, overdekte pot
d geschlossener Hafen *m*, überdeckter Hafen *m*

1365 COVERED SLUDGE DRYING BED iwt
f lit *m* de séchage de boue couvert
e lecho *m* secador de fango cubierto
i letto *m* d'essiccamento del fango coperto
n overdekt slibbed *n*
d überdecktes Schlammbett *n*

1366 COVERED TRICKLING FILTER, CLOSED TRICKLING FILTER fil
f lit *m* bactérien couvert

e filtro *m* de escurrimiento cubierto, filtro *m* percolador cubierto
i letto *m* percolatore coperto
n afgedekt bacteriënfilter *n*
d geschlossener Filter *m*

1367 COWPER STOVE, BLAST FURNACE STOVE, HOT BLAST STOVE htt,met
f appareil *m* régénérateur de vent chaud
e aparato *m* Cowper
i apparecchio *m* di Cowper, preriscaldatore *m* del vento
n stenen windverhitter
d Cowper *m*, steinerner Winderhitzer *m*

1368 CRABBING MACHINE tex
f machine *f* à fixer le tissu
e máquina *f* fijadora del tejido
i macchina *f* per fissare la lucentezza del tessuto
n brandbok
d Einbrennmaschine *f*, Krabbmaschine *f*

1369 CRACKER, CRACKING MILL, BREAKDOWN MILL, BREAKER brk
f broyeur *m* à cylindres, concasseur *m* à cylindres
e machacadora *f*, trituradora *f*
i frantoio *m* a cilindri, rompitore *m* a cilindri
n breekwalswerk *n*
d Brecher *m*, Brecherwalzwerk *n*, Reisswalzwerk *n*

1370 CRACKER ROLL brk
f cylindre-broyeur *m*
e cilindro *m* triturador
i cilindro *m* rompitore
n breekwals
d Brechwalze *f*

1371 CRACKING PLANT fue
f installation *f* de cracking, installation *f* de craquage
e equipo *m* de cracking, equipo *m* de craqueo, equipo *m* de pirodesintegración
i impianto *m* di cracking, impianto *m* di piroscissione
n kraakinstallatie
d Krackanlage *f*

1372 CRANE MANIPULATOR, RECTILINEAR MANIPULATOR ins,nue
f manipulateur *m* rectiligne
e manipulador *m* rectilíneo
i manipolatore *m* rettilineo
n coördinatenmanipulator
d Koordinatenmanipulator *m*

1373 CREAM FUNNEL, CHEESE FUNNEL dai,lab
f entonnoir *m* à crème
e embudo *m* de crema
i imbuto *m* a panna
n roomtrechter
d Sahnetrichter *m*

1374 CREAMING TANK rub
f récipient *m* de crémage
e recipiente *m* de desnate
i serbatoio *m* per scrematura
n oproomtank
d Aufrahmgefäss *n*

1375 CROSS-BEATER MILL brk
f moulin *m* à croisillon percuteur
e molino *m* batidor de martillos
i mulino *m* a martelli
n kruisslagmolen
d Schlagkreuzmühle *f*

1376 CROSS-EXTRUDER HEAD, TRANSVERSAL EXTRUDER HEAD pla,rub
f tête *f* d'équerre de boudineuse
e cabezal *m* de budinadora transversal
i testa *f* di trafila ad angolo
n dwarse spuitkop
d Querspritzkopf *m*

1377 CROSS-FIRED FURNACE, SIDE-FIRED FURNACE gla
f four *m* à feu de coupage
e horno *m* de fuego cruzado
i forno *m* a fuoco incrociato
n kruisvuuroven
d Kreuzfeuerofen *m*

1378 CROSS-FLOW TRAY dis
f plateau *m* à une seule passe
e plato *m* de corrientes cruzadas
i piatto *m* a flusso trasversale
n kruisstroomschotel
d Kreuzstromboden *m*

1379 CROSS-HEAD EXTRUDER pla,rub
f boudineuse *f* à tête d'équerre
e budinadora *f* de cabezal transversal
i trafila *f* con testa ad angolo
n spuitmachine met dwarse spuitkop
d Querkopfspritzmaschine *f*

1380 CROWN gla
f toît *m* du four
e techo *m* del horno
i tetto *m* del forno
n ovendak *n*
d Ofendach *n*

1381 CRUCIBLE htt
f creuset *m*
e crisol *m*
i crogiolo *m*, crogiuolo *m*
n kroes, smeltkroes
d Schmelztiegel *m*, Tiegel *m*

1382 CRUCIBLE COVER lab
f couvercle *m* pour creuset
e cobertera *f* de crisol
i coperchio *m* di crogiolo
n kroesdeksel *n*
d Tiegeldeckel *m*

1383 CRUCIBLE FURNACE, CRUCIBLE MELTING FURNACE, POT FURNACE htt
f four *m* à creuset
e horno *m* de crisol
i forno *m* a crogiolo
n kroesoven, smeltkroesoven
d Tiegelofen *m*, Tiegelschmelzofen *m*

1384 CRUSH ROLLER tex
f rouleau *m* égalisateur
e rodillo *m* de igualación
i rullo *m* da uguagliare
n egaliseerrol
d Egalisierrolle *f*

1385 CRUSHER brk,pla
f broyeur *m*, concasseur *m*
e quebrantador *m*
i frantoio *m*
n breekapparaat *n*, breekmachine
d Brecher *m*

1386 CRUSHING ROLL brk,pla
f broyeur *m* à cylindres
e trituradora *f* de cilindros
i macinatore *m* a cilindri
n breekmachine, breekwals
d Walzenbrecher *m*

1387 CRUSHING TEST APPARATUS tes
f appareil *m* d'essai de compression longitudinale
e aparato *m* de prueba de compresión longitudinal
i collaudatore *m* di compressione longitudinale
n compressiebeproevingsapparaat *n* in lengterichting
d Kompressionsprüfapparat *m* in Längsrichtung

1388 CRUTCHER mix,soa
f mélangeur *m* de savon
e mezclador *m* de jabón
i mescolatore *m* di sapone
n zeepmengvat *n*
d Seifenmischer *m*

1389 CRYOMETER ins
f cryomètre *m*
e criómetro *m*
i criometro *m*
n cryometer
d Kryometer *n*

1390 CRYOSCOPE ins
f cryoscope *m*
e crioscopio *m*
i crioscopio *m*
n cryoscoop
d Kryoskop *n*

1391 CRYPTOMETER ins
f cryptomètre *m*
e criptómetro *m*
i criptometro *m*
n dekkrachtmeter
d Deckfähigkeitsmesser *m*, Deckkraftmesser *m*

1392 CRYSTALLIZATION TROUGH lab
f cuve *f* de cristallisation
e vaso *m* de cristalización
i serbatoio *m* di cristallizzazione
n kristallisatievat *n*
d Kristallisationsgefäss *n*

1393 CRYSTALLIZER cry
f cristallisoir *m*
e cristalizador *m*
i cristallizzatore *m*
n kristallisatieapparaat *n*
d Kristallisator *m*

1394 CRYSTALLIZING EVAPORATOR, EVAPORATOR CRYSTALLIZER cry
f cristallisoir *m* à évaporation
e cristalizador *m* de evaporación
i cristallizzatore *m* ad evaporazione
n verdampingskristallisatie-apparaat *n*
d Verdampfungskristallisator *m*

1395 CUBE MIXER mix
f mélangeur *m* cubique
e mezclador *m* cúbico
i mescolatore *m* cubico
n kubusmenger
d Würfelmischer *m*

1396 CUBING ROLL cer
f cylindre *m* de cubage
e cilindro *m* de cubicaje
i cilindro *m* di cubatura
n kubusvormende wals
d Würfelwalze *f*

1397 CUFF HEATER BAND, HEATER BAND htt,pla
f bande *f* de chauffage
e banda *f* de caldeo
i banda *f* riscaldatrice
n verwarmingslint *n*
d Erhitzungsband *n*

1398 CULTURE APPARATUS bac
f appareil *m* de culture
e aparato *m* de cultivo
i apparecchio *m* di cultura
n cultuurtoestel *n*
d Kulturgerät *n*

1399 CULTURE BOTTLE bac,lab
f flacon *m* de culture
e frasco *m* de cultivo
i matraccio *m* di cultura
n cultuurkolf
d Kulturkolben *m*

1400 CULTURE DISH, CULTURE PLATE, PETRI DISH bac
f boîte *f* de Pétri
e fuente *f* de cultivo
i capsula *f* di cultura, capsula *f* di Petri
n cultuurschotel
d Kulturschale *f*, Petri-Schale *f*

1401 CULTURE FLASK bac
f fiole *f* de culture, matras *m* de culture
e matraz *m* de cultivo
i matraccio *m* di cultura
n cultuurkolf
d Kulturkolben *m*

1402 CUP SCRAPER, CLEANSER, RAKER cle,rub
f curette *f* de tasse à latex
e limpiador *m* para recipiente de látex
i pulitore *m* per tazza di lattice
n cupwisser
d Latexbecherwischer *m*

1403 CUP WHEEL pol
f meule *f* boisseau
e muela *f* acopada, muela *f* de cubeta
i mola *f* a tazza
n komschijf
d Becherscheibe *f*, Topfscheibe *f*

1404 CUPEL ana,met
f coupelle *f*
e copela *f*
i coppella *f*
n copelleerschaal, kapel
d Kapelle *f*, Kupelle *f*

1405 CUPEL FURNACE ASSAY FURNACE ana,lab
f four *m* de coupellation
e horno *m* de copelación, horno *m* de ensayo
i forno *m* d'assaggio, forno *m* di coppellazione
n cupelleeroven, essaaieeroven
d Kupellierofen *m*

1406 CUPOLA FURNACE htt
f cubilot *m*
e cubilote *m*
i cubilotto *m*
n koepeloven
d Kupolofen *m*

1407 CURD SUSPENSION TUBE dai,nut
f tube *m* à suspension de caillebotte
e tubo *m* de dispersión de naterón, tubo *m* de suspensión de requesón
i tubo *m* a suspensione di ricotta
n kwarksuspensiebuis, wrongelsuspensiebuis
d Quarksuspensionsrohr *n*

1408 CURDLING CYLINDER, COAGULATION CYLINDER dai,nut
f cylindre *m* de caillage, cylindre *m* de coagulation
e cilindro *m* de coagulación cilindro *m* de cuajamiento
i cilindro *m* d'accagliamento, cilindro *m* di coagulazione
n coaguleercilinder, stremcilinder
d Gerinnezylinder *m*, Koagulierzylinder *m*

1409 CURDLING VAT dai,nut
f cuve *f* de caillage
e cuba *f* de cuajamiento
i tino *m* d'accagliamento
n kaastobbe, stremvat *n*, wrongelvat *n*
d Quarkbildungsfass *n*

1410 CURING BAG, WATER BAG rub
f sac *m* à cuisson, sac *m* à eau

e cámara *f* vulcanizadora de agua
i camera *f* per vulcanizzazione ad acqua
n hulpband, vulcanisatiekern, waterzak
d Wasserheizschlauch *m*

1411 CURING FURNACE, CURING OVEN htt,pla
f four *m* de moulage
e horno *m* de moldeo
i forno *m* d'indurimento
n uithardingsoven
d Härtungsofen *m*

1412 CURING OVEN, VULCANIZING OVEN rub
f étuve *f* de vulcanisation
e estufa *f* de vulcanización
i riscaldatore *m* di vulcanizzazione
n vulcaniseerstoof
d Vulkanisierofen *m*

1413 CURING PAN, VULCANIZING BOILER rub
f chaudière *f* de vulcanisation
e caldera *f* de vulcanización
i caldaia *f* di vulcanizzazione
n vulcaniseerketel
d Vulkanisierkessel *m*

1414 CURING PRESS VULCANIZING PRESS rub
f presse *f* de vulcanisation
e prensa *f* de vulcanización
i pressa *f* di vulcanizzazione
n vulcaniseerpers
d Vulkanisierpresse *f*

1415 CURING ROOM nut
f chambre *f* de conservation
e cámara *f* de curación
i camera *f* da conservare
n conserveerruimte
d Haltbarmachungsraum *m*

1416 CURING VESSEL nut
f cuve *f* de conservation
e cuba *f* de curación, cuba *f* de curado
i tino *m* di conservazione
n conserveervat *n*
d Haltbarmachungsgefäss *n*

1417 CURVED BAR ANODE, BENT BAR ANODE elc
f anode *f* en barre incurvée
e ánodo *m* de barra curvada
i anodo *m* a barra ricurva
n gebogen staafanode
d gebogene Stabanode *f*, gebogene Stangenanode *f*

1418 CURVED BLADE PROPELLOR mix
f hélice *f* a palettes courbées
e hélice *f* de palas curvadas
i elica *f* a palette ricurve
n propeller met gewelfde bladen
d Propeller *m* mit gewölbten Flügeln

1419 CUT-OFF DISK, CUTTING WHEEL, SLITTING WHEEL cpt
f disque *m* de tronçonnage
e disco *m* de cortar
i disco *m* per tranciare
n doórslijpschijf
d Trennscheibe *f*

1420 CUTTER, CUTTING DEVICE sep
f dispositif *m* de découpage
e dispositivo *m* de corte
i taglierina *f*
n snijinrichting
d Schneidvorrichtung *f*

1421 CUTTER, CUTTING MACHINE soa
f coupeuse *f*
e cortadora *f*
i taglierina *f*
n snijmachine
d Schneidmaschine *f*

1422 CYANIDATION PLANT, CYANIDING PLANT met
f installation *f* d'extraction d'or par cyanuration
e equipo *m* de beneficio por cianuración
i impianto *m* di cianurizzazione
n cyanide-uitlooginstallatie
d Zyanidlaugerei *f*

1423 CYANIDE COPPER PLATING BATH elc
f bain *m* de cuivrage au cyanure de potassium
e baño *m* de cobreado con cianura de potasio
i bagno *m* di ramatura al cianuro di potassio
n cyaankalikoperbad *n*
d Zyankalikupferbad *n*

1424 CYANIDE HARDENING PLANT, CARBO-NITRIDING PLANT met

f installation *f* de cémentation au cyanure, installation *f* d'endurcissement au cyanure
e equipo *m* de cementación al cianuro, equipo *m* de endurecimiento al cianuro
i impianto *m* di cementazione al cianuro, impianto *m* d'indurimento al cianuro
n cyaanhardingsbad *n*
d Zyansalzhärtungsbad *n*

1425 CYCLONE FILTER dai,fil
f filtre *m* à cyclone
e filtro *m* de ciclón
i filtro *m* a ciclone
n cycloonfilter *n*
d Zyklonfilter *m*

1426 CYCLONE THICKENER, HYDROCYCLONE iwt
f épaississeur-cyclone *m*
e espesador *m* ciclónico
i concentratore *m* ciclone
n cycloonverdikker, hydrocycloon
d Hydrozyklon *m*, Zykloneindicker *m*

1427 CYLINDER DRIER, DRYING CYLINDER dry,pap
f cylindre *m* sécheur
e cilindro *m* secador
i cilindro *m* essiccante
n droogwals
d Trockenwalze *f*

1428 CYLINDRICAL GRADER sor
f machine *f* de triage à cylindre
e cilindro *m* seleccionador
i cilindro *m* selettore
n sorteercilinder
d Sortierzylinder *m*

1429 CYLINDRICAL SCREEN CENTRIFUGAL FILTER fil
f filtre *m* centrifuge à tamis cylindrique
e filtro *m* centrífugo de tamiz cilíndrico
i filtro *m* centrifugo a vaglio cilindrico
n filtercentrifuge met cilindrische zeef
d Filterzentrifuge *f* mit Zylindersieb

1430 CYLINDRICAL SCREEN FEEDER cer
f alimentateur *m* à écran cylindrique
e alimentador *m* de pantalla cilíndrica
i alimentatore *m* a griglia cilindrica
n rolschermtoevoerinrichting
d Rollschirmspeiser *m*

1431 CYSTINE BRIDGE, DISULFIDE BRIDGE (US), DISULPHIDE BRIDGE (GB) lab
f pont *m* cystinique, pont *m* disulfure
e puente *m* cisteínico, puente *m* disulfuro
i ponte *m* bisulfurico, ponte *m* cistinico
n disulfidebrug
d Cystinbrücke *f*, Disulfidbrücke *f*

D

1432 DAIRY INDUSTRY dai
f industrie *f* laitière
e industria *f* lechera
i industria *f* casearia,
industria *f* dei prodotti di latte
n zuivelindustrie
d Milchwirtschaft *f*,
Molkereibetrieb *m*

1433 DAIRY STERILIZER dai
f stérilisateur *m* de lait
e esterilizador *m* de leche
i dispositivo *m* di sterilizzazione
di latte
n melksteriliseerapparaat *n*
d Milchsterilisator *m*

1434 DAMP AIR BLOWER hyg
f souffleuse *f* d'air humide
e soplante *m* humificador
i soffiatore *m* d'aria umida
n vochtige-luchtblower
d Feuchtluftgebläse *n*

1435 DAMPER pap
f appareil *m* mouilleur,
humidificateur *m*
e aparato *m* humectador,
humedecedor *m*
i apparecchio *m* umettatore
n bevochtiger
d Anfeuchtapparat *m*, Feuchter *m*

1436 DAMPER, con,fue
REGISTER, SMOKE SLIDE
VALVE
f registre *m*
e registro *m*
i paratoia *f*, registro *m*,
serranda *f*
n regelschuif
d Rauchschieber *m*,
Regelschieber *m*, Schieberplatte *f*

1437 DAMPING CYLINDER pap
f cylindre *m* mouilleur
e cilindro *m* humectador
i cilindro *m* umettatore
n bevochtigingswals
d Feuchtwalze *f*

1438 DANCER ROLL (US), pla
LOOSE ROLLER (GB)
f rouleau *m* fou
e rodillo *m* loco, rodillo *m* tensor
i rullo *m* folle
n spanrol
d Losrolle *f*, Spannrolle *f*

1439 DANDY (US), htt,met
RUNNING-OUT FIRE (GB)
f four *m* de premier affinage
e horno *m* de primer afino
i forno *m* di prima affinazione
n eerste raffineeroven
d Frischereiofen *m*

1440 DANDY ROLL pap
f cylindre *m* filigraineur
e cilindro *m* afiligranador
i cilindro *m* per filigrana,
cilindro *m* sgocciolatore
n filigramwals, watermerkwals
d Wasserzeichenwalze *f*

1441 DARK-GLOWING OVEN htt,met
f four *m* à recuit bleu
e horno *m* de recocido azul
i forno *m* a ricottura azzurra
n blauwgloeioven
d Blauglühofen *m*

1442 DAUBING APPARATUS coa
f appareil *m* de revêtement en
matière réfractaire
e aparato *m* de recubrimiento
de materia refractaria
i apparecchio *m* di rivestimento
con strato refrattario
n apparaat *n* voor het aanbrengen
van een vuurvaste laag
d Apparat *m* zur feuerbeständiger
Beschichtung

1443 DAY TANK gla
f bassin *m* journalier, bassin *m*
périodique
e cuba *f* periódica
i tino *m* periodico
n dagwan
d Tageswanne *f*

1444 DAYLIGHT OPENING (GB), pla,rub
MOLD OPENING (US)
f distance *f* entre plateaux,
lumière *f*
e abertura *f* entre platos, luz *f*
de la prensa
i apertura *f*, luce *f*
n daglichtopening

d lichte Einbauhöhe *f*, lichte Öffnungsweite *f*

1445 DAYLIGHT PRESS (GB), MULTI-PLATEN PRESS (US) pla,rub
f presse *f* à plateaux multiples
e prensa *f* de platos múltiples
i pressa *f* a piani
n etagepers
d Etagenpresse *f*, Plattenpresse *f*

1446 DEACIDIFICATION TANK cle
f récipient *m* de désacidification
e tanque *m* de desacidificación
i serbatoio *m* di rimozione dell'acido
n ontzuringstank
d Entsäuerungsgefäss *n*

1447 DEACIDIFYING TOWER gen
f tour *f* de désacidification
e torre *f* de desacidificación
i torre *f* di rimozione dell'acido
n ontzuringskolom
d Entsäuerungssäule *f*

1448 DEAD-WEIGHT PRESSURE GAGE (US), DEAD-WEIGHT PRESSURE GAUGE (GB), PISTON MANOMETER ins
f manomètre *m* à piston
e manómetro *m* de pistóne
i manometro *m* a pistone
n zuigermanometer
d Kolbendruckmesser *m*, Kolbenmanometer *n*

1449 DE-AERATING GRID hyg
f grille *f* à désaérer
e parrilla *f* desaireadora
i griglia *f* di disaerazione
n ontluchtingsrooster *n*
d Entlüftungsgitter *n*

1450 DE-AERATOR boi
f désaérateur *m*
e desaireador *m*
i disaeratore *m*
n ontluchter
d Entlüfter *m*

1451 DE-AERATOR EXTRACTION PUMP boi
f pompe *f* aspirante du désaérateur
e bomba *f* de extracción del desaireador
i pompa *f* aspirante del disaeratore
n ontluchterzuigpomp
d Entlüftersaugpumpe *f*

1452 DE-AIRING MACHINE cer
f machine *f* de désaération
e máquina *f* de desaireación
i macchina *f* di disaerazione
n ontluchtingsmachine
d Entlüftungsmaschine *f*

1453 DEALUMINIZING TANK cle
f récipient *m* de dégagement d'aluminium
e tanque *m* de desaluminación
i serbatoio *m* di disgregazione d'alluminio
n aluminiumafscheidingstank
d Aluminiumabscheidungsgefäss *n*

1454 DEARSENICATOR sua
f dispositif *m* d'extraction de l'arsenic
e dispositivo *m* de extracción del arsénico
i dispositivo *m* da eliminare l'arsenico
n arseenextractor
d Arsenextraktor *m*

1455 DEASPHALTING APPARATUS fue
f appareil *m* de désasphaltage
e aparato *m* desasfaltador
i apparecchio *m* di rimozione dell'asfalto
n asfaltextraheerapparaat *n*
d Entasphaltierungsvorrichtung *f*

1456 DEBENZOLIZATION PLANT fue
f installation *f* de débenzolage
e equipo *m* de desbenzoilación
i impianto *m* di rimozione del benzolo
n benzolextraheerapparaat *n*
d Entbenzolungsanlage *f*

1457 DEBISMUTHIZING PLANT met
f installation *f* d'élimination du bismuth
e equipo *m* de eliminación del bismuto
i impianto *m* da eliminare il bismuto
n bismutonttrekkingsinstallatie
d Wismutbeseitigungsanlage *f*

1458 DEBUTANIZER fue
f appareil *m* d'élimination du butane
e desbutanizador *m*
i apparecchio *m* di rimozione del butano
n butaanextractor
d Entbutanisierungsapparat *m*

1459 DECANNING PLANT, DECLADDING PLANT, DEJACKETING PLANT nue
f installation *f* à dégainer
e equipo *m* descamisador
i impianto *m* di spogliatura
n ontmantelingsinstallatie
d Enthülsungsanlage *f*

1460 DECANTER gen,lab
f décanteur *m*
e decantador *m*, garrafa *f*
i recipiente *m* di decantazione
n decanteervat *n*
d Dekantierapparat *m*, Dekantiergefäss *n*

1461 DECARBONIZATION PLANT iwt
f installation *f* de décarbonisation
e equipo *m* de descarbonización
i impianto *m* di decarbonizzazione, impianto *m* per eliminazione dell'acido carbonico
n koolzuurverwijderingsinstallatie
d Entkarbonisierungsanlage *f*

1462 DECARBOXYLATION VESSEL oic
f cuve *f* de décarboxylation
e cuba *f* de descarboxilación
i tino *m* di decarbossilazione
n decarboxyleerinstallatie
d Decarboxylieranlage *f*

1463 DECATIZING TANK, ROLL BOILING TANK tex
f récipient *m* de décatissage
e cuba *f* de decatizar
i serbatoio *m* di decatissaggio
n afglanstank, decanteertank
d Dekatiergefäss *n*

1464 DECHLORINATING PLANT hyg,iwt
f installation *f* de déchloration
e equipo *m* de descloración, equipo *m* de descloridación
i impianto *m* di declorazione
n ontchloringsinstallatie
d Entchlorungsanlage *f*

1465 DECINORMAL CALOMEL ELECTRODE elc
f électrode *f* au calomel décinormale
e electrodo *m* de calomelano decinormal
i elettrodo *m* al calomelano decinormale
n decinormale calomelelektrode
d Zehntelnormal-Kalomel-halbzelle *f*

1466 DECKLE pap
f moule *m* de papier à main
e molde *m* de papel a mano
i forma *f* di carta a mano
n handvorm
d Auflaufrahmen *m*, Deckel *m*

1467 DECOBALTER met
f dispositif *m* d'élimination du cobalt
e dispositivo *m* de remoción del cobalto
i dispositivo *m* di rimozione del cobalto
n kobaltafscheidingsapparaat *n*
d Entkobaltierungsgerät *n*

1468 DECOCTION APPARATUS ext
f appareil *m* de décoction
e aparato *m* de decocción
i apparecchio *m* di decozione
n afkookapparaat *n*
d Abkochapparat *m*, Absudapparat *m*

1469 DECOLORIZATION VESSEL fue,gla,tex
f cuve *f* de décoloration
e cuba *f* de descoloración
i tino *m* di decolorazione
n ontkleuringsvat *n*
d Entfärbungsgefäss *n*

1470 DECONTAMINATION ROOM, DETOXIFICATION ROOM hyg,nue
f chambre *f* de décontamination, chambre *f* de détoxication
e cámara *f* de descontaminación, cámara *f* de detoxificación
i camera *f* di decontaminazione, camera *f* di disintossicazione
n ontsmettingsruimte
d Dekontaminierungsraum *m*, Entgiftungsraum *m*

1471 DECORATION FIRING KILN, ENAMEL KILN, ENAMELLING FURNACE cer,htt
f four *m* à émailler
e horno *m* de esmaltado, horno *m* de esmaltar
i forno *m* a smaltatura, forno *m* per smaltare
n emailleeroven
d Emaillierofen *m*

1472 DECORTICATING MILL, DECORTICATOR pap,tan
f machine *f* à décortiquer

e máquina *f* descortizadora
i macchina *f* per scorticare
n ontschorsmachine, schilmachine
d Entrindemaschine *f*, Schälmaschine *f*

1473 DEENAMELLING PLANT met
f installation *f* d'élimination de l'émail
e equipo *m* de desesmaltado
i impianto *m* di rimozione dello smalto
n emailverwijderingsinstallatie
d Entemaillierungsanlage *f*

1474 DEEP-ETCH TESTING APPARATUS, MACROETCH TESTING APPARATUS tes
f appareil *m* d'essai de macro-attaque
e aparato *m* de prueba de ataque profondo, aparato *m* de prueba de macroataque
i collaudatore *m* a macroincisione all'acquaforte, collaudatore *m* ad attacco chimico profondo
n diepetsbeproevingsapparaat *n*
d Tiefätzprüfgerät *n*, Tiefbeizprüfgerät *n*

1475 DEEP-FREEZE PLANT, LOW TEMPERATURE REFRIGERATION PLANT clo
f installation *f* de réfrigération à basse température
e equipo *m* de refrigeración de baja temperatura
i impianto *m* congelatore, impianto *m* di refrigerazione a bassa temperatura
n diepvriesinstallatie
d Tiefkühlanlage *f*

1476 DEFECATING PLANT sin
f installation *f* de défécation
e equipo *m* de defecación
i impianto *m* di defecazione
n defecatie-installatie, nasatureerinstallatie
d Scheidesaturationsanlage *f*, Schlammsaftsaturationsanlage *f*

1477 DEFERRIFICATION TANK, DEFERRIZATION TANK hyg,iwt
f récipient *m* de déferrisation
e tanque *m* de desferrización
i serbatoio *m* per eliminare il ferro
n ontijzeringstank
d Enteisenungsgefäss *n*

1478 DEFIBERIZER, DEFIBRATOR pap
f défibreur *m*
e desfibrador *m*
i sfibratore *m*
n defibrator
d Zerfaserer *m*

1479 DEFINING BARREL, DEFLASHING BARREL cle,pol
f tonneau *m* d'ébarbage
e baril *m* de rebabado
i barile *m* di sbavatura
n ontgratingsvat *n*
d Entgratungsgefäss *n*

1480 DEFLAGRATING SPOON lab
f cuillère *f* de déflagration
e cucharilla *f* de deflagración
i cucchiaio *m* di deflagrazione
n deflagratielepel
d Abbrennlöffel *m*

1481 DEFLAGRATION PLANT gas
f installation *f* de déflagration
e equipo *m* de deflagración
i impianto *m* di deflagrazione
n deflagreerinstallatie
d Deflagrieranlage *f*

1482 DEFLAGRATION TUBE exp,lab
f tube *m* déflagrateur
e tubo *m* deflagrador
i tubo *m* deflagratore
n ontbrandingsbuis
d Verpuffungsröhrchen *n*

1483 DEFLOCCULATION TANK sin
f récipient *m* de défloculation
e tanque *m* de desfloculación
i serbatoio *m* da separare sotto forma di fiocchi
n ontvlokkingsinstallatie
d Entflockungsanlage *f*

1484 DEFROSTING PLANT nut
f installation *f* à décongeler
e equipo *m* de descongelación
i impianto *m* decongelatore
n ontdooiinstallatie
d Entfrostungsanlage *f*

1485 DEGASIFYING APPARATUS gas
f dispositif *m* de dégazage
e dispositivo *m* de desgasificación
i apparecchio *m* di degassazione
n ontgasser, ontgassingsapparaat *n*
d Entgaser *m*, Entgasungsgerät *n*

1486 DEGASIFYING PLANT, GAS EXPELLER PLANT gas

f installation *f* dégazeuse
e equipo *m* de desgasificación
i impianto *m* degassificatore
n ontgassingsinstallatie
d Entgasungsanlage *f*

1487 DEGATING DEVICE pre
f appareil *m* de décarottage
e aparato *m* de quitar los bebederos
i apparecchio *m* da staccare la materozza
n aanspuitingverwijderings-apparaat *n*
d Entgratungsgerät *n*

1488 DEGREASING BATH cle
f bain *m* de dégraissage
e baño *m* de desengrase, baño *m* desengrasante
i bagno *m* di sgrassaggio
n ontvettingsbad *n*
d Entfettungsbad *n*

1489 DEGUMMING BATH tex
f bain *m* de dégommage
e baño *m* de desengomado, baño *m* desgomante
i bagno *m* di sgommatura
n ontgommingsbad *n*
d Degummierungsbad *n*, Entbastungsbad *n*

1490 DEHUMIDIFIER, DEHUMIDIFYING APPARATUS hyg
f appareil *m* éliminateur de l'humidité
e aparato *m* de deshumidificación
i apparecchio *m* da eliminare l'umidità, deumidificatore *m*
n ontvochtingsapparaat *n*
d Entfeuchter *m*, Entfeuchtungsgerät *n*

1491 DEHYDRATION PLANT, DEWATERING PLANT dry,hyg
f installation *f* de déshydratation, installation *f* de drainage
e equipo *m* de deshidratación
i impianto *m* d'asciugamento, impianto *m* di drenaggio, impianto *m* di prosciugamento
n ontwateringsinstallatie
d Entwässerungsanlage *f*

1492 DEHYDROGENATION PLANT oic
f installation *f* de déshydrogénation
e equipo *m* de deshidrogenación
i impianto *m* di deidrogenazione
n dehydreerinstallatie, waterstof-extraheerinstallatie
d Dehydrierungsanlage *f*, Wasserstoffextrahieranlage *f*

1493 DEHYDROHALOGENATION PLANT fue,oic
f installation *f* de déshydrohalogénisation
e equipo *m* de deshidrohalogenización
i impianto *m* di deidroalogenizzazione
n waterstofhalogeenextraheer-installatie
d Wasserstoffhalogenextraktions-anlage *f*

1494 DEIONIZATION VESSEL elc
f cuve *f* de désionisation
e cuba *f* de desionización
i tino *m* di deionizzazione
n desionisatievat *n*
d Entionisierungsgefäss *n*

1495 DEJACKETING PLANT, DECANNING PLANT, DECLADDING PLANT nue
f installation *f* à dégainer
e equipo *m* descamisador
i impianto *m* di spogliatura
n ontmantelingsinstallatie
d Enthülsungsanlage *f*

1496 DELAY TANK nue
f réservoir *m* à diminution de radioactivité
e tanque *m* de diminución de radiactividad
i serbatoio *m* a diminuzione di radioattività
n versterftank
d Abklingbehälter *m*

1497 DELAYING BELT dec
f ruban *m* de ralentissement
e cinta *f* de retraso
i nastro *m* di rallentamento
n vertragingsband
d Verzögerungsband *n*

1498 DELIMING TANK tan
f récipient *m* de déchaulage
e tanque *m* de descalcificación
i serbatoio *m* di decalcificazione
n ontkalkingsvat *n*
d Entkalkungsgefäss *n*

1499 DELIVERY, DELIVERY PIPING tra
f conduite *f* de refoulement
e tubería *f* de descarga
i condotta *f* di scarico

n afvoerleiding
d Ableitung *f*

1500 DELIVERY FLASK lab
f matras *m* à verser
e matraz *m* de vertido
i matraccio *m* per versare
n schenkkolf
d Schenkkolben *m*

1501 DELIVERY VALVE, cpt
DISCHARGE VALVE
f soupape *f* de refoulement
e válvula *f* de descarga, válvula *f* de impulsión
i valvola *f* premente
n drukklep
d Druckventil *n*

1502 DELIVERY VALVE, cpt
HEAD VALVE
f soupape *f* de manoeuvre
e válvula *f* de impulsión
i valvola *f* di mandata
n stuurklep
d Steuerventil *n*

1503 DELUSTERING PLANT tex
f installation *f* de délustrage
e equipo *m* de deslustrar
i impianto *m* da delucidare
n ontglansinstallatie
d Entglänzungsanlage *f*

1504 DEMANGANIZING PLANT, hyg
MANGANESE REMOVAL PLANT
f installation *f* de démanganisation
e equipo *m* para desmanganizar
i impianto *m* di demanganizzazione
n mangaanverwijderingsinstallatie
d Entmanganungsanlage *f*

1505 DEMIJOHN, tra
CARBOY
f ballon *m*, bonbonne *f*, dame-jeanne *f*, tourie *f*
e castaña *f*, damajuana *f*
i damigiana *f*
n mandefles
d Korbflasche *f*

1506 DEMINERALIZING PLANT, cle
DESALTING PLANT
f installation *f* de déminéralisation, installation *f* de dessalement
e equipo *m* de desmineralización
i impianto *m* di demineralizzazione
n installatie voor het verwijderen van minerale zouten
d Entmineralisierungsanlage *f*, Entsalzungsanlage *f*

1507 DEMULSIFICATION ins,lab
NUMBER INDICATOR
f indicateur *m* de l'indice de démulsification
e indicador *m* del índice de desmulsificación
i indicatore *m* del numero di demulsificazione
n demulgeergetalindicator
d Entemulgierungszahlgeber *m*

1508 DEMULSIFYING APPARATUS rub
f appareil *m* à démulsifier
e desemulsor *m*, desmulsificador *m*
i apparecchio *m* demulsificante
n apparaat *n* voor het breken van emulsies, demulgeerapparaat *n*
d Demulgator *m*, Entemulgierungsgerät *n*, Vorrichtung *f* zum Brechen von Emulsionen

1509 DENATURATION PLANT oic
f installation *f* de dénaturation
e equipo *m* de desnaturalización
i impianto *m* di denaturazione
n denatureerinstallatie
d Denaturierungsanlage *f*

1510 DENICKLEFICATION BATH cle
f bain *m* de dénickelage
e baño *m* de desniquelificación
i bagno *m* di rimozione del nichel
n ontnikkelingsbad *n*
d Entnickelungsbad *n*

1511 DENITRIDING TOWER, sua
GLOVER TOWER
f tour *f* de dénitruration
e torre *f* de desnitruración
i torre *f* di denitrurazione
n denitreertoren
d Denitrierturm *m*

1512 htt,met
DENSE SINTERING OVEN
f four *m* de frittage à densité maxi[illegible]
e horno *m* de sinterización de densidad máxima
i forno *m* di sinterizzazione a densità massima
n oven voor het sinteren met maximale dichtheid
d Dichtsinterofen *m*

1513 DENSENER clo,fou,gla
f refroidisseur *m* de forme
e enfriador *m* de forma
i conchiglia *f* esterna
n aangepast koelstuk *n*
d Kokille *f*, profilierte Schreckpla[illegible]

1514 DENSIFIER, SLUDGE CONCENTRATOR SLUDGE THICKENER iwt
f récipient *m* épaississeur de boue
e tanque *m* concentrador de fango
i serbatoio *m* concentratore di fango
n slibindikker
d Schlammeindickbehälter *m*, Schlammeindicker *m*

1515 DENSIMETER, DENSOMETER ins
f densimètre *m*
e densímetro *m*
i densimetro *m*
n densimeter, dichtheidsmeter
d Densimeter *n*, Dichtemesser *m*

1516 DENSITOMETER ins,pho
f densitomètre *m*
e densitómetro *m*
i densitometro *m*
n densitometer, zwartingsmeter
d Densitometer *n*, Schwärzungsmesser *m*

1517 DENSITY BOTTLE, PYCNOMETER, PYKNOMETER ins
f pycnomètre *m*
e picnómetro *m*
i picnometro *m*
n pyknometer
d Pyknometer *n*

1518 DENSITY PUMP pre
f pompe *f* à colonnes de densités différentes
e bomba *f* de columnas de densidades diferentes
i pompa *f* a densità differenti
n pomp met verschillende kolomdichtheden
d Pumpe *f* mit verschiedenen Kolbendichten

1519 DENUDER elc
f décomposeur *m*
e descomposidor *m*
i descompositore *m*
n loogcel
d Amalgamzersetzer *m*, Zersetzungszelle *f*

1520 DEODORIZER, DEODORIZING APPARATUS, ODOR REMOVAL APPARATUS (US), ODOUR REMOVAL APPARATUS (GB) iwt
f appareil *m* de désodorisation, appareil *m* d'élimination de l'odeur
e aparato *m* de desodorización
i apparecchio *m* di deodorizzazione, apparecchio *m* d'eliminazione dell'odore
n stankverdrijvingsapparaat *n*
d Desodoriergerät *n*, Geruchsbeseitigungsgerät *n*

1521 DEOXIDIZING PLANT met
f installation *f* de désoxydation
e equipo *m* de desoxidación
i impianto *m* di disossidazione
n reductie-installatie
d Entoxydierungsanlage *f*, Reduktionsanlage *f*

1522 DEOXYGENATION PLANT, OXYGEN REMOVAL PLANT iwt
f installation *f* de désoxygénation
e equipo *m* de desoxigenación
i impianto *m* di disossigenazione
n zuurstofverwijderingsinstallatie
d Sauerstoffbeseitigungsanlage *f*, Sauerstoffentzuganlage *f*

1523 DEPARAFFINAZATION PLANT, DEWAXING PLANT fue
f installation *f* de déparaffinage
e equipo *m* de desparafinización, equipo *m* desparafinador
i impianto *m* di deparaffinazione, impianto *m* per eliminare la paraffina
n ontparaffineringsinstallatie
d Entparaffinierungsanlage *f*

1524 DEPASSIVE ANODE, DEPOLARIZED ANODE elc
f anode *f* dépolarisée
e ánodo *m* depolarizado
i anodo *m* depolarizzato
n gedepolariseerde anode
d depolarisierte Anode *f*, entpassivierte Anode *f*

1525 DEPHENOLIZING PLANT iwt
f installation *f* de déphénolage
e equipo *m* de desfenolización
i impianto *m* di defenolizzazione
n fenolextraheerinstallatie
d Entphenolierungsanlage *f*

1526 DEPHLEGMATOR dis
f déflegmateur *m*
e desflegmador *m*
i deflegmatore *m*
n deflegmator
d Dephlegmiersäule *f*, Dephlegmierturm *m*

1527 DEPHOSPHORIZING TANK met
f récipient *m* de déphosphoration
e tanque *m* desfosforante
i impianto *m* di defosforazione
n fosforverdrijvingsvat *n*
d Entphosphorungsgefäss *n*

1528 DEPICKLING PLANT tan
f installation *f* de ramollissement
e equipo *m* de remoción de la salazón
i impianto *m* di rammollimento
n weekmakingsinstallatie
d Weichmachungsanlage *f*

1529 DEPILATION TANK, UNHAIRING TANK lea
f récipient *m* de dépilage, récipient *m* d'épilage
e tanque *m* de depilación
i serbatoio *m* di depilazione
n blootvat *n*, ontharingskuip
d Enthaarungsgefäss *n*

1530 DEPITCHING TANK, DETARRING TANK tex
f cuve *f* de dégoudronnage
e cuba *f* de despintado, cuba *f* de desteñido
i botte *f* di scratamento
n ontteringskuip
d Entpechungsküpe *f*

1531 DEPOLYMERIZATION VESSEL oic
f cuve *f* de dépolymérisation
e cuba *f* de despolimerización
i tino *m* di depolimerizzazione
n depolymerisatievat *n*
d Depolymerisationsgefäss *n*

1532 DEPOSITING-OUT TANK, LIBERATOR TANK elc
f cuve *f* de dépôt total, cuve *f* de libération
e cuba *f* de depósito total, cuba *f* de liberación
i cella *f* d'esaurimento
n afwerkcel
d Entkupferungsbad *n*, Entmetallisierungsbad *n*

1533 DESALINATION MEMBRANE cle
f membrane *f* de dessalage
e membrana *f* de desalinación
i membrana *f* di desalinizzazione
n ontziltingsmembraan *n*
d Entsalzungsmembrane *f*

1534 DESALINATION SHIP cle
f vaisseau *m* de dessalage
e nave *f* de desalinación
i nave *f* di desalinizzazione
n ontziltingsvaartuig *n*
d Entsalzungsschiff *n*

1535 DESALTER, DESALTING APPARATUS cle
f dispositif *m* de dessalage, dispositif *m* de dessalaison, dispositif *m* de dessalement
e dispositivo *m* de desalación, dispositivo *m* de desalinación
i dispositivo *m* di desalinizzazione
n ontziltingsapparaat *n*, ontzoutingsapparaat *n*
d Entsalzungsgerät *n*

1536 DESCALING PLANT met
f installation *f* de décalaminage, installation *f* de détartrage
e equipo *m* de descascarillado
i impianto *m* di disincrostazione, impianto *m* di scagliatura
n oxydehuidverwijderings-installatie
d Entzünderungsanlage *f*

1537 DESENSITATION BATH phc
f bain *m* désensibilisateur
e baño *m* desensibilizador
i bagno *m* desensibilizzatore
n desensibilisatiebad *n*
d Desensibilisatorbad *n*

1538 DESICCATOR, ESSICCATOR dry,lab
f dessiccateur *m*, exsiccateur *m*
e desecador *m*, secador *m*
i essiccatoio *m*, essiccatore *m*
n drooginrichting, exsiccator
d Exsikkator *m*, Trockengefäss *n*

1539 DESICCATOR PLATE dry,lab
f disque *m* du dessiccateur, plaque *f* du dessiccateur
e plato *m* del desecador
i piatto *m* del dessiccatore
n exsiccatorplaat
d Exsikkatoreinsatz *m*, Exsikkatorplatte *f*

1540 DESILICONIZATION TANK, SILICA REMOVAL TANK iw
f récipient *m* de désiliciage, récipient *m* de désiliciation
e tanque *m* de desiliconización
i serbatoio *m* da rimozione della silice
n zandvanger
d Entkieselungsgefäss *n*, Sandfäng

1541 DESILTER iwt
f débourbeur *m*
e aparato *m* de desentarquinamiento, desenturbiador *m*
i separatore *m* di fango, separatore *m* di melma
n slibafscheider
d Schlammabscheider *m*

1542 DESILVERING BATH pho
f bain *m* de désargenture
e baño *m* de desplatación
i bagno *m* per rimozione dell'argento
n ontzilveringsbad *n*
d Entsilberungsbad *n*

1543 DESIZING TANK tex
f récipient *m* de désencollage
e tanque *m* de desapresto, tanque *m* de desencolado
i serbatoio *m* di sbozzimatura
n ontslichtingsvat *n*
d Entschlichtungsgefäss *n*

1544 DESLUDGING PLANT, SLUDGE REMOVAL PLANT iwt
f installation *f* d'enlèvement de boue, installation *f* d'évacuation de boue
e equipo *m* de desenlodamiento
i impianto *m* di scarico di fango
n ontslibbingsinstallatie
d Entschlammungsanlage *f*

1545 DESTATICISATION BATH sap
f bain *m* enlevant l'électricité statique
e baño *m* desestatizador
i bagno *m* eliminatore dell'elettricità statica
n bad *n* voor het opheffen van statische elektriciteit
d Bad *n* zur Beseitigung der statischen Elektrizität

1546 DESTRUCTIVE DISTILLATION PLANT, DRY DISTILLATION PLANT dis
f installation *f* de distillation destructive, installation *f* de distillation sèche
e equipo *m* de destilación destructiva, equipo *m* de destilación seca
i impianto *m* di distillazione a secco, impianto *m* di distillazione distruttiva
n droge-destillatie-installatie
d Trockendestillationsanlage *f*, Zersetzungsdestillationsanlage *f*

1547 DESTRUCTIVE TESTING APPARATUS tes
f appareil *m* d'essai destructif
e aparato *m* de prueba destructiva
i collaudatore *m* di rottura, collaudatore *m* distruttivo
n destructieve-beproevingsapparaat *n*
d zerstörendes Prüfgerät *n*

1548 DESTRUCTOR, INCINERATOR hyg,iwt
f destructeur *m*, incinérateur *m*
e destructor *m*, horno *m* de incineración
i forno *m* per incenerire
n destructor, verbrandingsoven
d Verbrennungsofen *m*

1549 DESTRUCTOR PLANT hyg
f installation *f* d'incinération de cadavres
e equipo *m* incinerador de cadáveres
i impianto *m* da incenerire cadaveri
n destructiebedrijf *n*, destructorinstallatie, kadaververbrandingsinstallatie
d Kadaververwertungsanlage *f*, Tierkörperbeseitigungsanlage *f*

1550 DESULFURATION TANK (US), DESULPHURATION TANK (GB) met
f récipient *m* de dessoufrage, récipient *m* de désulfuration
e tanque *m* de desulfuración, tanque *m* para desazufrar
i serbatoio *m* di desolforazione
n ontzwavelingsvat *n*
d Entschwefelungsgefäss *n*

1551 DESUPERHEATER pre
f réfrigérateur *m* de vapeur surchauffé
e desrecalentador *m*
i raffreddatore *m* di vapore surriscaldato
n oververhitte-stoomkoeler
d Heissdampfkühler *m*

1552 DETERGENT BATH tex
f bain *m* de lavage
e baño *m* detergente
i bagno *m* detergente
n wasbad *n*
d Waschflotte *f*

1553 DETINNING BATH elc
f bain *m* de désétamage
e baño *m* de desestañado

i bagno *m* da ricuperare lo stagno
n tinterugwinningsbad *n*
d Entzinnungsbad *n*, Zinnrück-gewinnungsbad *n*

1554 DETINNING DRUM elc
f tambour *m* de désétamage
e tambor *m* desestañador
i tamburo *m* da ricuperare lo stagno
n tinterugwinningstrommel
d Entzinnungstrommel *f*

1555 DETINNING PLANT elc
f installation *f* de désétamage
e equipo *m* de desestañado
i impianto *m* da ricuperare lo stagno
n tinterugwinningsinstallatie
d Entzinnungsanlage *f*

1556 DETONATOR exp
f amorce *f*, détonateur *m*
e detonador *m*, fulminante *m*
i detonatore *m*, involucro *m* del detonante
n slaghoedje *n*
d Zündkapsel *f*

1557 DETOXIFICATION ROOM, hyg,nue
DECONTAMINATION ROOM
f chambre *f* de décontamination, chambre *f* de détoxication
e cámara *f* de descontaminación, cámara *f* de detoxificación
i camera *f* di decontaminazione, camera *f* di disintossicazione
n ontsmettingsruimte
d Dekontaminierungsraum *m*, Entgiftungsraum *m*

1558 DETRITUS TANK, iwt
GRIT TANK
f récipient *m* dessableur
e tanque *m* de decantación de detritos
i serbatoio *m* dissabbiatore
n zandvanger
d Sandfänger *m*

1559 DEVELOPING BATH pho
f bain *m* de développement
e baño *m* de revelado
i bagno *m* di sviluppo, bagno *m* rivelatore
n ontwikkelbad *n*
d Entwicklungsbad *n*

1560 DEVELOPING TANK pho
f cuve *f* de développement
e tanque *m* de revelado
i sviluppatrice *f*, truogolo *m* per sviluppo
n ontwikkeltank
d Entwicklungstank *m*, Entwicklungstrog *m*

1561 DEVIL htt
f pot *m* à feu
e hornillo *m* portátil
i caldano *m*
n vuurduvel, vuurpot
d Feuertopf *m*

1562 DEVITRIFICATION PLANT cle
f installation *f* de dévitrification
e equipo *m* de desvitrificación
i impianto *m* di devitrificazione
n ontglazingsinstallatie
d Entglasungsanlage *f*

1563 DEVULCANIZER, rub
DIGESTER FOR RECLAIMING,
REGENERATION AUTOCLAVE,
REGENERATION BOILER
f autoclave *m* de régénération, chaudière *f* à régénération, dévulcaniseur *m*
e caldera *f* de regeneración, desvulcanizador *m*, digestor *m* para regeneración
i caldaia *f* di rigenerazione, devulcanizzatore *m*, digestore *m* per rigenerare
n regenereerautoclaaf, regenereerketel
d Digestor *m*, Entvulkanisierungs-gefäss *n*, Regenerierkessel *m*

1564 DEWATERING CHANNEL, tr
DISCHARGE CHANNEL,
DRAIN CHANNEL
f canal *m* de vidange
e canal *m* de escurrimiento
i canale *m* di vuotamento
n afvoerkanaal *n*
d Entleerungskanal *m*

1565 DEWATERING PLANT, hy
DEHYDRATION PLANT
f installation *f* de déshydratation, installation *f* de drainage
e equipo *m* de deshidratación
i impianto *m* d'asciugamento, impianto *m* di drenaggio, impianto *m* di prosciugamento
n ontwateringsinstallatie
d Entwässerungsanlage *f*

1566 DEWAXING PLANT, fu
DEPARAFFINAZATION PLANT

f installation *f* de déparaffinage
e equipo *m* de desparafinización, equipo *m* desparafinador
i impianto *m* di deparaffinazione, impianto *m* per eliminare la paraffina
n ontparaffineringsinstallatie
d Entparaffinierungsanlage *f*

1567 DEWING MACHINE tex
f machine *f* mouilleuse
e máquina *f* de humectación, máquina *f* de rociado
i macchina *f* umidificatrice
n invochtmachine
d Anfeuchtmaschine *f*

1568 DEZINCIFICATION BATH, cor,met
DEZINKING BATH
f bain *m* de dégagement du zinc, bain *m* de dézincification
e baño *m* de descincado
i bagno *m* di dezincatura, bagno *m* di rimozione dello zinco
n ontzinkingsbad *n*
d Entzinkungsbad *n*

1569 DIAL FEED PRESS (GB), pla,rub
ROTARY PRESS (US)
f presse *f* à barillet, presse *f* à tourelle revolver
e prensa *f* revólver
i pressa *f* a giostra, pressa *f* rotativa
n carousselpers, roterende pers
d Karussellpresse *f*, Revolverpresse *f*

1570 DIALYSER, gen
DIALYSIS APPARATUS
f dialyseur *m*
e dializador *m*
i dializzatore *m*
n dialysator
d Dialysator *m*

1571 DIAMOND DISK gla
f disque *m* diamanté
e disco *m* adiamantado
i disco *m* con diamanti
n diamantschijf
d Diamantscheibe *f*

1572 DIAMOND WHEEL pol
f meule *f* diamantée
e muela *f* adiamantada
i mola *f* con diamanti
n diamantslijpschijf
d Diamantschleifscheibe *f*

1573 DIAPHRAGM elc
f diaphragme *m*
e diafragma *m*
i diaframma *m*
n diafragma *n*
d Diaphragma *n*

1574 DIAPHRAGM, cpt
MEMBRANE
f diaphragme *m*, membrane *f*
e diafragma *m*, membrana *f*
i diaframma *m*, membrana *f*
n diafragma *n*, membraan *n*
d Diaphragma *n*, Membran *f*, Membrane *f*

1575 DIAPHRAGM PUMP, pre
MEMBRANE PUMP
f pompe *f* à diaphragme, pompe *f* a membrane
e bomba *f* de diafragma, bomba *f* de membrana
i pompa *f* a diaframma, pompa *f* a membrana
n diafragmapomp, membraanpomp
d Balgpumpe *f*, Membranpumpe *f*

1576 DIAPHRAGM VALVE cpt
f soupape *f* à membrane
e válvula *f* reguladora de membrana
i valvola *f* a membrana
n membraanafsluiter
d Membranschleuse *f*, Membranventil *n*

1577 DIATOMITE FILTER fil,iwt
f filtre *m* à diatomées, filtre *m* au kieselgur
e filtro *m* de harina fósil, filtro *m* de trípoli
i filtro *m* a diatomee, filtro *m* alla farina fossile
n kiezelgoerfilter *n*
d Diatomitfilter *m*, Kieselgur-Anschwemmfilter *m*

1578 DIAZOTIZING BATH pho
f bain *m* diazoïque
e baño *m* de diazotaje
i bagno *m* di diazotizzazione
n diazobad *n*
d Diazobad *n*

1579 DICHROMATE BATH, cle,met
BICHROMATE BATH
f bain *m* au bichromate
e baño *m* de bicromato
i bagno *m* al bicromato
n bichromaatbad *n*
d Bichromatbad *n*

1580 DICHROMATE CELL elc
f pile *f* à bichromate
e pila *f* de bicromato
i pila *f* a bicromato
n chroomzuurelement *n*
d Bichromatelement *n*, Dichromatelement *n*

1581 DIE APPROACH (GB), DIE CHANNEL (US) pla,rub
f canal *m* d'écoulement
e canal *m* de la tobera
i canale *m* dell'ugello
n extrusiekanaal *n*, spuitkanaal *n*
d Düsenkanal *m*

1582 DIE BLOCK cpt,pla
f porte-filière *m*
e portaboquilla *m*, portatobera *m*
i portafiliera *m*
n matrijsklem
d Werkzeughalter *m*

1583 DIE BLOCK pla,rub
f contreforme *f*, forme *f*
e portamatriz *m*, portamolde *m*, portaestampa *m*
i piastro *m* dello stampo
n matrijsblok *n*
d Formblock *m*

1584 DIE LAND (GB), ORIFICE LAND (US) pla,rub
f partie *f* de filière à section constante
e canal *m* de laminado, parte *f* final del canal de la tobera
i zona *f* a sezione costante
n spuitkanaal *n* met constante doorsnede
d Bügelkanal *m*, Bügellänge *f*, Führungskanal *m*

1585 DIE PLATE, DIE PLATEN pla,rub
f plateau *m* porte-moule
e portamolde *m*
i portafiliera *m*
n matrijsopspanplaat
d Formhalteflansch *m*

1586 DIE RELIEF (GB), ORIFICE RELIEF (US) pla,rub
f embouchure *f* des lèvres de la filière, sortie *f* de la filière
e ensanchamiento *m* de la salida de la tobera, salida *f* de la boquilla
i allargamento *m* della sezione d'uscita, bocca *f* della filiera, rilascio *m*
n extrusieprofiel *n*, mondstukopening, spuitmatrijs
d Düsenaustritt *m*, erweiterte Austrittsöffnung *f*

1587 DIE-SPINNING NOZZLE, SPINNERET, SPINNING NOZZL[E] tex
f filière *f*
e tobera *f* de hilatura
i ugello *m* di filiera
n spindop, spinkop
d Spinndüse *f*

1588 DIELECTRIC DRIER, HIGH-FREQUENCY DRIER dry
f étuve *f* par pertes diélectriques
e estufa *f* por pérdidas dieléctric[as]
i stufa *f* a perdite dielettriche
n droogoven met capacitieve verwarming
d dielektrischer Trockner *m*, Hochfrequenztrockner *m*

1589 DIFFERENTIAL AERATION CELL cor
f élément *m* d'aération différentielle
e célula *f* de aireación diferencial
i cella *f* d'aerazione differenzial[e]
n ontluchtingscel
d Belüftungselement *n*

1590 DIFFERENTIAL CALORIMETER ins
f calorimètre *m* différentiel
e calorímetro *m* diferencial
i calorimetro *m* differenziale
n differentiële calorimeter
d Differenzialkalorimeter *n*

1591 DIFFERENTIAL FLOTATION PLANT, SELECTIVE FLOTATION PLAN[T] met
f installation *f* de flottation sélective
e equipo *m* de flotación selectiva
i impianto *m* di flottazione selettiva
n selectieve flotatie-installatie
d selektive Flotationsanlage *f*

1592 DIFFERENTIAL MANOMETER, DIFFERENTIAL PRESSURE GAGE (US), DIFFERENTIAL PRESSURE GAUGE (GB) ins
f manomètre *m* différentiel
e manómetro *m* diferencial
i manometro *m* differenziale

n differentiële manometer
d Differentialmanometer *n*

1593 DIFFRACTION GRATING spe
f réseau *m* de diffraction
e red *f* difractora, rejilla *f* de difracción
i reticolo *m* di diffrazione
n diffractierooster *n*
d Beugungsgitter *n*

1594 DIFFUSER gen
f diffuseur *m*
e diffundidor *m*, difusor *m*
i diffusore *m*
n diffusietoestel *n*, diffusor
d Diffusionsapparat *m*, Diffusor *m*

1595 DIFFUSER PLATE, AIR DIFFUSER, FILTER SLAB iwt
f dalle *f* poreuse, diffuseur *m* d'air, plaque *f* diffuseuse, plaque *f* filtrante
e difusor *m* de aire, placa *f* filtrante
i diffusore *m* d'aria, piastra *f* filtrante, piastra *f* porosa
n filterplaat, luchtverdeelplaat
d Belüfterplatte *f*, Filterplatte *f*, Luftverteiler *m*, Verteilerplatte *f*

1596 DIFFUSION BARRIER nue
f barrière *f* à diffusion
e barrera *f* de difusión
i barriera *f* antidiffusiva
n diffusiewand
d Diffusionswand *f*

1597 DIFFUSION BATTERY sin
f batterie *f* de diffusion
e batería *f* de difusión
i batteria *f* di diffusione
n diffusiebatterij
d Diffusionsbatterie *f*

1598 DIFFUSION CELL sep
f cellule *f* à diffusion
e célula *f* de difusión
i cella *f* a diffusione
n diffusiecel
d Diffusionszelle *f*

1599 DIFFUSION COLUMN sin
f colonne *f* de diffusion
e columna *f* de difusión
i colonna *f* di diffusione
n diffusiekolom
d Diffusionssäule *f*

1600 DIFFUSION CONDENSATION PUMP, DIFFUSION PUMP pre
f pompe *f* à diffusion
e bomba *f* de difusión
i pompa *f* a diffusione
n diffusiepomp
d Diffusionspumpe *f*

1601 DIFFUSION SINTERING OVEN htt,met
f four *m* de frittage à diffusion
e horno *m* de sinterización de difusión
i forno *m* di sinterizzazione a diffusione
n diffusiesinteroven
d Diffusionssinterofen *m*

1602 DIFFUSION ZONE htt,met
f zone *f* de diffusion
e zona *f* de difusión
i zona *f* di diffusione
n diffusiegebied *n*
d Diffusionsgebiet *n*

1603 DIGESTER pap
f digesteur *m* de cellulose
e digestor *m* de celulosa
i digestore *m* di cellulosa
n cellulosekookketel
d Zellstoffkocher *m*

1604 DIGESTER, DIGESTER TANK, SLUDGE DIGESTION TANK iwt
f digesteur *m*, vase *m* de digestion
e tanque *m* digestor
i vasca *f* di digestione
n digereertank
d Faulbehälter *m*, Schlammfaulbehälter *m*

1605 DIGESTER FOR RECLAIMING, DEVULCANIZER, REGENERATION AUTOCLAVE, REGENERATION BOILER rub
f autoclave *m* de régénération, chaudière *f* à régénération, dévulcaniseur *m*
e caldera *f* de regeneración, desvulcanizador *m*, digestor *m* para regeneración
i caldaia *f* di rigenerazione, devulcanizzatore *m*, digestore *m* per rigenerare
n regenereerautoclaaf, regenereerketel
d Digestor *m*, Entvulkanisierungsgefäss *n*, Regenerierkessel *m*

1606 DIGESTION CHAMBER, SLUDGE DIGESTION COMPARTMENT iwt
f chambre *f* de digestion, chambre *f* de putréfaction de boue
e cámara *f* de digestión
i camera *f* di digestione di fango
n digereerkamer
d Faulraum *m*, Schlammfaulraum *m*

1607 DILATOMETER ins
f dilatomètre *m*
e dilatómetro *m*
i dilatometro *m*
n dilatometer
d Dilatometer *n*

1608 DIP, DIPPING BATH coa
f bain *m* d'immersion, bain *m* au trempé
e baño *m* de inmersión
i bagno *m* d'immersione
n dompelbad *n*
d Tauchbad *n*

1609 DIP BRAZING PLANT wel
f installation *f* de soudure forte par immersion
e equipo *m* de soldadura fuerte por inmersión
i impianto *m* di brasatura per immersione
n hardsoldeerinstallatie met dompelbad
d Immersionshartlötungsanlage *f*

1610 DIP ENAMELLING APPARATUS, WET ENAMELLING APPARATUS coa
f appareil *m* d'émaillage au trempé
e aparato *m* de esmaltado por inmersión
i apparecchio *m* di smaltatura ad immersione
n dompelemailleerapparaat *n*
d Nassemaillierungsgerät *n*

1611 DIP PIPE, INVERTED SIPHON sew
f siphon *m* renversé, siphon *m* submergé
e sifón *m* invertido
i sifone *m* rovesciato
n grondduiker
d Dücker *m*, Düker *m*

1612 DIP TANK dec
f récipient *m* d'immersion
e tanque *m* de inmersión
i serbatoio *m* d'immersione
n dompeltank
d Tauchgefäss *n*

1613 DIPPING BASKET ACETYLENE GENERATOR gas
f générateur *m* d'acétylène à contact
e generador *m* de acetileno por inmersión
i generatore *m* d'acetilene per immersione
n contactontwikkelaar van acetyleen
d Berührungsentwickler *m*, Tauchgenerator *m*

1614 DIPPING BATH, LIQUORING BATH tex
f bain *m* d'immersion
e baño *m* de inmersión
i bagno *m* d'immersione
n dompelbad *n*
d Tauchbad *n*

1615 DIPPING FORM coa
f forme *f* pour le trempé
e molde *m* para inmersión
i conformatore *m* ad immersione, stampo *m* per immersione
n dompelvorm
d Tauchform *f*

1616 DIPPING REFRACTOMETER ins
f réfractomètre *m* à immersion
e refractómetro *m* de inmersión
i rifrattometro *m* ad immersione
n dompelrefractometer
d Eintauchrefraktometer *n*

1617 DIPPING TANK tex
f bac *m* de trempage
e tanque *m* para inmersión
i serbatoio *m* per immersione
n dompeltank
d Einlegegefäss *n*, Tauchgefäss *n*

1618 DIPPING VESSEL cer,coa
f cuve *f* d'immersion
e cuba *f* de inmersión
i tino *m* d'immersione
n dompelbak
d Tauchgefäss *n*

1619 DIRECT ARC FURNACE, SMOTHERED ARC FURNACE htt
f four *m* à arc direct
e horno *m* de arco directo
i forno *m* ad arco diretto

n directe vlamboogoven, vlamboogoven met directe verhitting
d direkter Lichtbogenofen *m*

1620 DIRECT CONTACT CONDENSER pre
f condensateur *m* à contact
e condensador *m* de mezcla
i condensatore *m* a contatto
n mengcondensor
d Mischkondensator *m*

1621 DIRECT CYCLE INTEGRAL BOILING REACTOR nue
f réacteur *m* à eau bouillante à refroidissement direct et échangeur de chaleur incorporé
e reactor *m* de agua hirviente de enfriamiento directo y cambiador de calor encerrado
i reattore *m* ad acqua bollente a raffreddamento diretto e scambiatore di calore incorporato
n direct gekoelde kokend-water-reactor met ingebouwde warmte-uitwisselaar
d direkt gekühlter Siedewasser-reaktor *m* mit eingebautem Wärmeaustauscher

1622 DIRECT FIRED DRIER dry
f sécheur *m* à chauffage direct par gaz de combustion
e secador *m* de calentamiento directo por gas de combustión
i essiccatore a riscaldamento diretto per gas di combustione
n direct door verbrandings-gassen verhitte drooginrichting
d Rauchgastrockner *m*

1623 DIRECT FIRED FURNACE gla
f four *m* à chauffage direct
e horno *m* de calentamiento directo
i forno *m* a riscaldamento diretto
n oven met directe verhitting
d direkt geheizter Ofen *m*

1624 DIRECT PRINTING MACHINE, APPLICATION PRINTING MACHINE tex
f imprimeuse *f* directe de textile
e máquina *f* estampadora directa de tejido
i stampatrice *f* diretta di tessuto
n directe textieldrukmachine
d direkte Textildruckmaschine *f*

1625 DIRECT RESISTANCE OVEN (GB), CHARGE RESISTANCE FURNACE (US) htt
f four *m* à chauffage direct par résistance
e horno *m* de calentamiento directo por resistencia
i forno *m* a riscaldamento diretto a resistenza
n weerstandsoven met directe verhitting
d direkter Widerstandsofen *m*

1626 DIRECT ROTARY DRIER dry
f sécheur *m* rotatif à action directe
e secador *m* giratorio de acción directa
i essiccatoio *m* rotativo ad azione diretta
n direct-werkende roterende droger
d unmittelbar arbeitender Drehtrockner *m*

1627 DIRECT VISION SPECTROSCOPE ins,spe
f spectroscope *m* à vision directe
e espectroscopio *m* de visión directa
i spettroscopio *m* a vista diretta
n rechtzichtspectroscoop
d Geradesichtspektroskop *n*

1628 DIRT TRAP cle
f nid *m* à crasses, piège *m* à crasses
e nido *m* de escorias
i fermascorie *m*
n slakvanger
d Schlackenfang *m*

1629 DISAPPEARING-FILAMENT PYROMETER ins
f pyromètre *m* à disparition de filament
e pirómetro *m* de filamiento desvanecedor
i pirometro *m* a filamento scomparente
n gloeidraadpyrometer
d Glühfadenpyrometer *n*

1630 DISC, DISK gen
f disque *m*
e disco *m*
i disco *m*
n schijf
d Scheibe *f*

1631 DISCHARGE BASIN iwt
f bassin *m* de décharge
e alberca *f* de descarga, depósito *m* de descarga
i bacino *m* di scarico
n lozingsbekken *n*
d Entlastungsbecken *n*

1632 DISCHARGE CHANNEL, DEWATERING CHANNEL, DRAIN CHANNEL tra
f canal *m* de vidange
e canal *m* de escurrimiento
i canale *m* di vuotamento
n afvoerkanaal *n*
d Entleerungskanal *m*

1633 DISCHARGE CHANNEL, EFFLUENT TROUGH iwt
f rigole *f* de décharge
e arroyo *m* de descarga
i canaletto *m* di scarico
n afvloeigoot
d Abflussrinne *f*

1634 DISCHARGE CHANNEL, RELIEF SEWER iwt
f canal *m* de décharge, émissaire *m* de dérivation
e canal *m* de descarga
i canale *m* di scarico
n ontlastingskanaal *n*
d Entlastungskanal *m*

1635 DISCHARGE COCK, DRAIN COCK, PURGE COCK cpt
f robinet *m* de purge, robinet *m* de vidange
e grifo *m* de descarga, grifo *m* de purga
i rubinetto *m* di scarico
n aftapkraan
d Ablasshahn *m*, Entleerungshahn *m*

1636 DISCHARGE HEAD, EXTRUDER DISCHARGE HEAD pla
f tête *f* de boudineuse
e cabezal *m* de budinadora
i bocchettone *m* di trafila
n spuitkop
d Spritzkopf *m*

1637 DISCHARGE HOLE sin
f orifice *m* de décharge
e orificio *m* de descarga
i orificio *m* di scarico
n aftapopening
d Ausspeicherungsöffnung *f*

1638 DISCHARGE HOPPER, OUTLET HOPPER sap
f entonnoir *m* à trop-plein
e gargante *m* de rebose
i imbuto *m* di sfioro
n overlooptrechter
d Ablauftrichter *m*, Schütttrichter *m*

1639 DISCHARGE NOZZLE cpt
f ajutage *m* d'écoulement
e tobera *f* de descarga, tobera *f* de escape
i ugello *m* per erogazione
n uitstromingsmondstuk *n*
d Ausflussdüse *f*, Strahldüse *f*

1640 DISCHARGE PRINTING APPARATUS tex
f appareil *m* d'impression au rongeant
e aparato *m* de estampado por corrosión
i apparecchio *m* di stampatura al corrosivo
n beitsdrukapparaat *n*
d Ätzdruckgerät *n*

1641 DISCHARGE TROUGH sto
f reverdoir *m*
e bandeja *f* colectora
i vasca *f* di raccolta
n opvangvat *n*
d Sammelmulde *f*

1642 DISCHARGE VALVE, DELIVERY VALVE cpt
f soupape *f* de refoulement
e válvula *f* de descarga, válvula *f* de impulsión
i valvola *f* premente
n drukklep
d Druckventil *n*

1643 DISCHARGE VALVE, OUTLET VALVE, UNLOADING VALVE cpt
f soupape *f* de décharge
e válvula *f* de descarga
i valvola *f* di scarico
n uitlaatklep
d Ablaufventil *n*, Auslassventil *n*

1644 DISH lab
f bol *m*, cuvette *f*
e cápsula *f*, taza *f*
i bacinella *f*, capsula *f*
n schaal
d Schale *f*

1645 DISINFECTION ROOM hyg
f chambre *f* de désinfection
e cámara *f* para desinfectar
i camera *f* di disinfezione
n ontsmettingsruimte
d Desinfektionsraum *m*, Entseuchungsraum *m*

1646 DISINTEGRATING MILL, DISINTEGRATOR brk
f casse-pierres *m*, désintégrateur *m*
e desintegrador *m*, molino *m* desintegrador
i disintegratore *m*, mulino *m* disintegratore
n desintegrator, slingermolen
d Desintegrator *m*, Schleudermühle *f*

1647 DISINTEGRATING MILL, DISINTEGRATOR, OPENER, WILLEY, WILLOW tex
f broyeur *m* à couronnes dentées, loup *m* batteur, ouvreuse *f*
e abridora *f*, batidora *f*, batuar *m*, diabla *f*
i apritoio *m*, lupo *m*
n duivel, snar, wolf
d Klopfwolf *m*, Reisswolf *m*

1648 DISINTEGRATOR, COMMINUTOR, SCREEN SHREDDER, TRITURATOR pap
f appareil *m* de dilacération, dilacérateur *m*
e trituradora *f*
i frantumatrice *f*, trituratore *m*
n fijnmaalmolen
d Rechengutzerkleinerer *m*, Rechenwolf *m*

1649 DISINTEGRATOR, RAG CHOPPER, RAG DEVIL pap
f effileuse *f*, effilocheuse *f*
e desguince *m*, deshilachadora *f*, lobo *m* de trapos
i lupo *m*, tagliastracci *m*
n lompensnijder, lompenwolf
d Hadernschneider *m*, Lumpenschneider *m*

1650 DISK ANODE elc
f anode *f* circulaire plate
e ánodo *m* de disco
i anodo *m* a disco
n schijfanode
d Rundplattenanode *f*

1651 DISK ATOMIZER pre
f pulvérisateur *m* à disque
e nebulizador *m* de disco, pulverizador *m* de disco
i polverizzatore *m* a disco
n schijfverstuiver
d Plattenzerstäuber *m*, Scheibenzerstäuber *m*

1652 DISK-BOWL CENTRIFUGE cen
f centrifugeur *m* à assiette
e centrífuga *f* de platillo
i centrifuga *f* a piatto
n schotelcentrifuge
d Tellerzentrifuge *f*

1653 DISK DRIER, ROTARY SHELF DRIER dry
f sécheur *m* à plateaux
e secador *m* de platos
i essiccatore *m* a piatti
n draaitafeldroogapparaat *n*
d Tellertrockner *m*

1654 DISK FEEDER cer,tra
f plaque *f* distributrice
e plato *m* distribuidor
i disco *m* distributore
n verdeelschijf
d Verteilerscheibe *f*

1655 DISK FILTER fil
f filtre *m* à disques
e filtro *m* de discos
i filtro *m* a dischi
n schijffilter *n*
d Scheibenfilter *m*, Spaltölfilter *m*

1656 DISK MILL brk
f broyeur *m* à disques
e molino *m* de discos
i mulino *m* a dischi
n schijfmolen
d Scheibenmühle *f*

1657 DISK MIXER mix,pla,rub
f mélangeur *m* à disque
e mezclador *m* de disco
i mescolatore *m* a disco
n schijfmengmachine
d Scheibenmischer *m*

1658 DISK SCREEN iwt
f grille *f* à disque, râteau *m* à disque
e rejilla *f* de disco
i griglia *f* a disco
n schijfrooster *n*
d Scheibenrechen *m*

1659 DISK-TYPE ROTARY VACUUM FILTER fil
f filtre *m* à vide à disque rotatif

e filtro *m* de vacío de disco giratorio
i filtro *m* a vuoto a disco rotativo
n vacuümfilter *v* met roterende schijf
d Vakuumdrehscheibenfilter *m*

1660 DISK VALVE cpt
f soupape *f* à disque, soupape *f* à siège plan
e válvula *f* de disco
i valvola *f* a disco, valvola *f* a sede piana
n schijfklep
d Scheibenventil *n*, Tellerventil *n*

1661 DISK WHEEL pol
f roue *f* à disque
e rueda *f* de disco
i ruota *f* a disco
n schijfrad *n*
d Scheibenrad *n*

1662 DISPENSING BURETTE ana
f burette *f* à compensation
e bureta *f* de compensación
i buretta *f* di compensazione
n vulburet
d Abfüllbürette *f*

1663 DISPENSING SCOOP lab
f cuillère *f* de dosage
e cucharón *m* de dosificación
i cucchiaio *m* dosatore
n doseerlepel
d Dosierlöffel *m*

1664 DISPERSING APPARATUS gen
f appareil *m* à dispersion
e aparato *m* de dispersión, aparato *m* dispersador
i apparecchio *m* per dispergere
n dispergeerapparaat *n*
d Dispergiervorrichtung *f*

1665 DISPERSION FUEL ELEMENT nue
f élément *m* combustible à dispersion
e elemento *m* combustible de dispersión
i elemento *m* combustibile a dispersione
n dispersiebrandstofelement *n*, dispersiesplijtstofelement *n*
d Dispersionsbrennstoffelement *n*

1666 DISPERSION HARDENING PLANT, met
AGE-HARDENING PLANT, PRECIPITATION HARDENING PLANT,
f installation *f* de durcissement structural
e equipo *m* de endurecimiento estructural
i impianto *m* d'incrudimento strutturale
n structurele-hardingsinstallatie
d strukturelle Härtungsanlage *f*

1667 DISPERSION MIXER pla
f mélangeur *m* à dispersion
e amasador *m* dispersador
i impastatrice *f* a dispersione
n dispersiekneedmachine
d Dispersionskneter *m*

1668 DISPLACEMENT CELL elc
f pile *f* à échange d'ions
e pila *f* de cambio de iones
i pila *f* a scambio d'ioni
n element *n* met ionenuitwisseling
d Element *n* mit Ionenaustausch

1669 DISSOLVING DRUM gen
f tambour *m* de dissolution
e tambor *m* de disolución
i tamburo *m* di dissoluzione
n oplostrommel
d Lösetrommel *f*, Trommellöser *m*

1670 DISTILLATION COLUMN, dis
DISTILLATION TOWER, DISTILLING COLUMN, STILL HEAD
f colonne *f* à fractionner, colonne *f* de distillation
e columna *f* de destilación
i colonna *f* di distillazione
n destilleerkolom
d Destillierkolonne *f*, Destilliersäule *f*

1671 DISTILLATION CONDENSER, dis
DISTILLATION RECEIVER
f condensateur *m* de distillation
e condensador *m* de destilación, destiladora *f*
i condensatore *m* di distillazione
n destilleercondensor
d Destillationskondensator *m*

1672 DISTILLATION FLASK dis,lab
f ballon *m* à distiller, matras *m* de distillation
e matraz *m* de destilación
i matraccio *m* di distillazione
n destilleerkolf
d Destillationskolben *m*, Destillierkolben *m*, Fraktionierkolben *m*

1673 DISTILLATION FURNACE dis,met
f four *m* de distillation
e horno *m* de destilación
i forno *m* di distillazione
n destilleeroven
d Destillationsofen *m*

1674 DISTILLATION HEAD, ATTACHMENT dis
f rectificateur *m*, tête *f* à fractionner
e enganche *m*
i attacco *m*
n opzet
d Aufsatz *m*

1675 DISTILLATION PLANT, DISTILLERY dis
f distillerie *f*
e destilería *f*
i distilleria *f*
n branderij, stokerij
d Brennerei *f*

1676 DISTILLATION RANGE DETERMINATION APPARATUS dis,fue,ins
f appareil *m* de détermination de la gamme de distillation
e aparato *m* de determinación de la gama de destilación
i apparecchio *m* di determinazione della gamma di distillazione
n apparaat *n* voor het bepalen van het destilleerverloop
d Apparat *m* zur Bestimmung des Siedeverlaufs

1677 DISTILLER, DISTILLING APPARATUS, STILL dis
f alambic *m*
e alambique *m*
i alambicco *m*, lambicco *m*
n alambiek, destilleertoestel *n*
d Brennapparat *m*, Destillierapparat *m*

1678 DISTILLING CONNECTING TUBE, VERTICAL RECOVERY BEND dis,lab
f coude *m* à prises parallèles
e codo *m* de tomas paralelas
i gomito *m* a prese parallele
n destillatieoverloop
d Destillierbrücke *f*

1679 DISTILLING UDDER, MULTI-LIMB VACUUM RECEIVER ADAPTER dis,lab
f séparatuer *m* de Bredt, séparateur *m* de Pauly
e separador *m* de Bredt
i separatore *m* di Bredt
n destilleeruier
d Destilliereuter *m*, Destilliervorlage *f* nach Bredt

1680 DISTRIBUTING ROLL, WAVER pap
f cylindre-broyeur *m*
e cilindro *m* distribuidor
i cilindro *m* distributore, cilindro *m* macinatore
n distributiewals
d Reibwalze *f*, Verreibwalze *f* Verteilerwalze *f*

1681 DISTRIBUTOR gen
f distributeur *m*, répartiteur *m*
e distribuidor *m*
i distributore *m*, ripartitore *m*
n verdeler
d Verteiler *m*

1682 DISULFIDE BRIDGE (US), DISULPHIDE BRIDGE (GB), CYSTINE BRIDGE lab
f pont *m* cystinique, pont *m* disulfure
e puente *m* cisteínico, puente *m* disulfuro
i ponte *m* bisulfurico, ponte *m* cistinico
n disulfidebrug
d Cystinbrücke *f*, Disulfidbrücke *f*

1683 DIVISION WALL htt
f cloison *f*, paroi *f* de séparation
e pared *f* divisoria
i muro *m* divisorio
n scheidsmuur
d Trennwand *f*

1684 DOBBIN cer,dry
f disque *m* de support rotatif
e disco *m* de soporte giratorio
i disco *m* di sopporto rotativo
n draaischijf
d Drehscheibe *f*

1685 DOCTOR pap
f racle *f*, racloir *m*
e rascador *m*
i raschietto *m*
n afstrijker, afstrijkmes *n*
d Abstreifer *m*, Abstreifmesser *n*, Schaber *m*

1686 DOCTOR, SPONGE ELECTRODE elc
f électrode *f* à éponge

e electrodo *m* macrocelular
i elettrodo *m* macrocellulare
n tampon
d Schwammelektrode *f*

1687 DOCTOR BACK cle
f support *m* de racloir
e soporte *m* de rascador
i sopporto *m* di raschietto
n afstrijkmesdrager
d Abstreiferträger *m*

1688 DOCTOR BATH, SPONGE PLATING BATH elc
f bain *m* de revêtement électrolytique avec électrode à éponge
e baño *m* de revestimiento galvánico con electrodo macrocelular
i bagno *m* di rivestimento galvanico con elettrodo macrocellulare
n tamponeerbad *n*
d Schwammelektrodenbad *n*

1689 DOCTOR BLADE cle
f lame *f* de racloir
e cuchilla *f* rascadora
i lama *f* di raschietto
n afstrijkmeslemmet *n*
d Abstreiferklinge *f*

1690 DOCTOR BLADE HOLDER cle
f support *m* de lame de racloir
e soporte *m* de cuchilla rascadora
i sopporto *m* di lama di raschietto
n afstrijkmeslemmetdrager
d Abstreiferklingenträger *m*

1691 DOCTOR KNIFE (GB), COATING KNIFE (US) dec
f racle *f*
e cuchilla *f* dosificadora, rasqueta *f*
i raschiatore *m*
n strijkmes *n*
d Rakel *f*, Streichmesser *n*

1692 DOCTOR ROLL pla
f cylindre *m* docteur
e cilindro *m* de dosificación
i cilindro *m* dosatore
n doseerwals
d Dosierwalze *f*

1693 DOCTOR TESTER fue,tes
f dispositif *m* de détermination du soufre
e aparato *m* de determinación del azufre
i apparecchio *m* da determinare lo zolfo
n zwavelbepalingstoestel *n*
d Doktorprüfgerät *n*, Schwefelbestimmungsgerät *n*

1694 DOCTOR TREATMENT TANK fue
f récipient *m* de traitement au plombite
e tanque *m* de tratamiento doctor, tanque *m* para la desodorización de petróleo
i serbatoio *m* di raffinazione per eliminare cattivi odori
n doctortank, plumbiteertank
d Doktorgefäss *n*

1695 DOFFER tex
f cylindre *m* d'abattage, cylindre *m* peigneur
e cilindro *m* descargador de la carda
i cilindro *m* spogliatore
n afnemer, kamrol, strijkwals
d Kammwalze *f*

1696 DOGHOUSE gla
f cuvette *f* d'enfournement, ouverture *f* d'enfournement, ouvreau *m*
e abertura *f* de enhornamiento
i apertura *f* di prelievo
n inleggat *n*
d Einlegevorbau *m*, Vorbau *m*

1697 DOLLY (GB), BOBBIN (US) elc
f aggloméré *m*
e aglomerado *m*
i impasto *m* depolarizzante
n buidel
d Beutel *f*, Puppe *f*

1698 DOLLY, MOP, POLISHING LAP pol
f disque *m* en lisière de tissu, meule *f* flexible en chiffon
e bruñidor *m* de trapos
i brunitoio *m* a pezzi di tessuto
n lappenschijf
d Schwabbelscheibe *f*

1699 DOLLY TUB, KIEVE met
f cuve *f* à rincer l'or
e cuba *f* de agitación
i serbatoio *m* d'agitazione
n goudwaspan
d Schlämmfass *n*

1700 DOME BOTTOM sin
f fond *m* en coupole
e fondo *m* de cúpula
i fondo *m* a cupola
n koepelbodem
d Kugelschale *f*

1701 DOSING DEVICE iwt
f dispositif *m* de chasse,
dispositif *m* de dosage,
dispositif *m* de répartition
e dispositivo *m* dosificador
i dispositivo *m* di dosaggio
n doseringsapparaat *n*
d Beschickungsvorrichtung *f*,
Dosiervorrichtung *f*

1702 DOSING FEEDER (GB), cpt,pla
WEIGHT FEEDER (US)
f dispositif *m* de dosage
e alimentador *m* ponderal,
dispositivo *m* de dosificación
i dosatore *m*
n gewichtsdoseerinrichting
d Dosiervorrichtung *f* nach
Gewicht

1703 DOT WELDING PLANT wel
f installation *f* de soudure
d'apport
e equipo *m* de soldadura de aporto
i impianto *m* di saldatura d'apporto
n oplasinstallatie
d Aufschweissanlage *f*,
Auftragschweissanlage *f*

1704 DOUBLE ACTION PRESS pre
f presse *f* à double action
e prensa *f* de doble acción
i pressa *f* a doppia azione
n tweezijdige pers
d zweistufige Presse *f*

1705 DOUBLE ARM KNEADER mix,pla
f malaxeur *m* à deux bras
e amasador *m* de doble brazo
i impastatrice *f* a due operatori
n dubbele-armkneedmachine
d Doppelarmkneter *m*

1706 DOUBLE BEAT VALVE cpt
f soupape *f* à double siège
e válvula *f* de campana, válvula *f*
de doble asiento
i valvola *f* a due sede
n evenwichtsklep
d Gleichgewichtsventil *n*

1707 DOUBLE BLADE AGITATOR, mix
DOUBLE BLADE MIXER
f agitateur *m* à deux palettes,
mélangeur *m* à deux palettes
e agitador *m* de dos paletas,
mezclador *m* de dos paletas
i agitatore *m* a due pale,
mescolatore *m* a due pale
n dubbel roerwerk *n*
d doppelschauflicher Mischer *m*,
doppelschaufliches Rührwerk *n*

1708 DOUBLE BREWHOUSE bre
OUTFIT
f salle *f* double de brassage
e sala *f* doble de cocimiento
i reparto *m* doppio di cottura
n dubbel brouwhuis *n*
d doppeltes Sudwerk *n*

1709 DOUBLE CLAD VESSEL nue
f récipient *m* à deux chemises
e recipiente *m* de doble chapado
i serbatoio *m* a due camicie
n dubbelbekleed vat *n*
d doppelt bekleidetes Gefäss *n*

1710 DOUBLE CONE DRUM mix,pla
MIXER
f mélangeur *m* à tambour
biconique
e mezclador *m* de tambor de doble
cono
i mescolatore *m* a tamburo biconico
n dubbelconische kanteltrommel
d Doppelkegeltrommelmischer *m*

1711 DOUBLE CONE FILTER fil
f filtre *m* à double cône
e filtro *m* de doble cono
i filtro *m* a doppio cono
n dubbelconusfilter *n*
d Doppelkegelfilter *m*

1712 DOUBLE CONE MIXER, mix
CONE MIXER
f mélangeur *m* à double cône
e mezclador *m* de doble cono
i mescolatore *m* a doppio cono
n dubbele-kegelmenger
d Doppelkonusmischer *m*

1713 DOUBLE CONTINUOUS htt,met
REHEATING FURNACE
f four *m* poussant double
e horno *m* empujador doble
i forno *m* a spinta doppio
n dubbele doorschuifoven
d Doppelstossofen *m*

1714 DOUBLE FILTER, bre,fil
TWO STAGE FILTER
f double-filtre *m*

e filtro *m* doble
i filtro *m* doppio
n dubbelfilter *n*
d Doppelfilter *m*

1715 DOUBLE FLOOR KILN, TWO FLOOR KILN bre
f touraille *f* à deux plateaux
e tostadero *m* de dos platos
i essiccatoio *m* a due griglie
n tweevloereneest
d Zweihordendarre *f*

1716 DOUBLE FORCE PRESS (US), DOUBLE RAM PRESS (GB) pla
f presse *f* à double piston
e prensa *f* de doble émbolo, prensa *f* de doble pistón
i pressa *f* a doppio effetto
n dubbele plunjerpers
d Doppelkolbenpresse *f*

1717 DOUBLE MOTION AGITATOR mix
f agitateur *m* contrerotatif
e agitador *m* doble
i agitatore *m* controrotante
n roerwerk *n* met tegen elkaar in werkende propellers
d Doppelrührwerk *n*

1718 DOUBLE MOTION PADDLE mix
f palette *f* à double effet
e paleta *f* de doble efecto
i pala *f* a doppio effetto
n roerspaan met samengestelde beweging
d Schaufel *f* mit Doppelbewegung

1719 DOUBLE NABEN KNEADER, FISHTAIL TYPE KNEADER mix,pla
f mélangeur *m* à ailerons de requin
e amasador *m* con aletas, mezclador *m* con brazo cola de pez
i impastatrice *f* a doppia elica
n kantelmengmachine in deltavorm
d Fischschwanzkneter *m*

1720 DOUBLE PIPE CONDENSER pre
f condensateur *m* à double tube
e condensador *m* de doble tubo
i condensatore *m* a doppio tubo
n dubbele-buiscondensor
d Doppelrohrkondensator *m*

1721 DOUBLE PIPE CRYSTALLIZER cry
f cristallisoir *m* à double conduite
e cristalizador *m* de tubería doble
i cristallizzatore *m* di doppia condotta
n kristallisatieapparaat *n* met dubbel buizenstel
d Kristallisationsgerät *n* mit Doppelleitung

1722 DOUBLE PIPE EXCHANGER clo,htt
f échangeur *m* de chaleur à double tube
e cambiador *m* de calor de doble tubo
i scambiatore *m* di calore a doppio tubo
n dubbele-buiswarmte-uitwisselaar
d Doppelrohrwärmeaustauscher *m*

1723 DOUBLE ROLLER CRUSHER brk
f broyeur *m* à deux cylindres
e quebrantador *m* de dos cilindros
i frantoio *m* a due cilindri
n tweewalsenbreekmachine
d Zweiwalzenbrecher *m*

1724 DOUBLE SCREW EXTRUDER pla
f boudineuse *f* à deux vis
e prensa *f* de extrusión de doble husillo
i estrusore *m* a due viti
n dubbelschroefextrusiepers
d Doppelschneckenpresse *f*

1725 DOUBLE SHAFT KNEADER WITH HELICALLY ARRANGED BLADES mix,pla
f malaxeur *m* à palettes hélicoïdales
e amasador *m* con paletas helicoidales
i estrusore *m* bialbero a dischi mescolanti
n schijvenkneedmachine
d Scheibenkneter *m*

1726 DOUBLE SIDE GATE pla,pre
f entrée *f* bilatérale
e entrada *f* bilateral
i entrata *f* bilaterale
n tweezijdige aanspuitpoort
d zweiseitiger Angusssteg *m*

1727 DOUBLE TEXTURE PROOFING CALENDER, DOUBLING CALENDER tex
f calandre *f* pour le doublage de tissus
e calandra *f* para doblar tejidos
i calandra *f* per doppiatura di tessuti

n doubleerkalander
d Gewebedublierungskalander *m*

1728 DOUBLE TUBE COOLER clo
f réfrigérant *m* à tubes concentriques
e enfriador *m* de tubos concéntricos
i raffreddatore *m* a tubi concentrici
n concentrische-buizenkoeler
d Koaxialröhrenkühler *m*

1729 DOUBLE WALLED TUBE gen
f tube *m* à double paroi
e tubo *m* de doble pared
i tubo *m* a doppia parete
n dubbelwandige buis
d Doppelrohr *n*

1730 DOWN-DRAFT KILN (US), cer
DOWN-DRAUGHT KILN (GB)
f four *m* à flammes descendantes
e horno *m* de llamas descendentes
i forno *m* a fiamme discendenti
n oven met naar beneden gerichte vlam
d Ofen *m* mit niedergehender Flamme

1731 DOWN-STROKE PRESS pre
f presse *f* à piston descendant
e prensa *f* de plato superior móvil
i pressa *f* discendente
n pers met bovenliggende plunjer
d Oberkolbenpresse *f*

1732 DOWN-TAKE PIPE (GB), sew
DRAFT TUBE (US), SOIL PIPE
f tube *m* de chute, tube *m* de descente
e tubería *f* descendente, tubo *m* de bajada
i tubo *m* di caduta, tubo *m* discendente
n valbuis
d Abfallrohr *n*, Fallrohr *n*

1733 DOWNCOMER, met
DOWNTAKE
f prise *f* de gaz latérale
e toma *f* de gas lateral
i presa *f* di gas laterale
n hoogovengasafvoer
d Gichtgasabzugsrohr *n*

1734 DRAFT GAGE (US), ins
DRAUGHT GAUGE (GB)
f indicateur *m* du tirage
e indicador *m* del tiro
i indicatore *m* del tiraggio
n trekmeter
d Zugmesser *m*

1735 DRAFT REGULATOR (US),
DRAUGHT REGULATOR (GB)
f régulateur *m* du tirage
e regulador *m* del tiro
i regolatore *m* del tiraggio
n trekregelaar
d Zugregler *m*

1736 DRAG-OUT RINSE (GB), elc
DRAG-OUT RINSE BATH (GB),
DRAG-OUT SWILL (US),
DRAG-OUT SWILL BATH (US)
f bain *m* de pré-rinçage
e baño *m* de enjuague preliminar
i bagno *m* di sciacquatura preliminare
n voorspoelbad *n*
d Sparspülbad *n*, Vorspülbad *n*

1737 DRAIN CHANNEL, tra
DEWATERING CHANNEL,
DISCHARGE CHANNEL
f canal *m* de vidange
e canal *m* de escurrimiento
i canale *m* di vuotamento
n afvoerkanaal *n*
d Entleerungskanal *m*

1738 DRAIN COCK, cpt
DISCHARGE COCK, PURGE COCK
f robinet *m* de purge, robinet *m* de vidange
e grifo *m* de descarga, grifo *m* de purga
i rubinetto *m* di scarico
n aftapkraan
d Ablasshahn *m*, Entleerungshahn *m*

1739 DRAIN TUBE iwt
f drain *m*, tube *m* d'écoulement
e desaguadero *m*, tubo *m* de drenaje
i tubo *m* di drenaggio, tubo *m* di dreno, tubo *m* di fognatura
n afvoerbuis
d Abflussrohr *n*, Entwässerungsrohr *n*, Kanalisationsrohr *n*

1740 DRAIN VALVE cpt
f soupape *f* de vidange
e válvula *f* de drenaje, válvula *f* purgadora
i valvola *f* di spurgo
n afvoerklep
d Ablassventil *n*

1741 DRAINING BOARD sew
f égouttoir *m*
e goterón *m*
i asse *f* a gocciolare
n druipplank
d Ablaufbrett *n*, Tropfbrett *n*

1742 DRAW-OFF TRAY dis
f plateau *m* de soutirage
e plato *m* de descarga
i piatto *m* di prelievo
n aftapschotel
d Abnahmeboden *m*, Entnahmeboden *m*

1743 DRAW ROLL, HAUL-OFF ROLL (GB), TAKE-UP ROLL (US) pla
f cylindre *m* de tirage, cylindre récepteur
e cilindro *m* de tracción
i cilindro *m* di tirata
n afneemwals, aftrekwals
d Abzugswalze *f*

1744 DRAWBAR gla
f bloc *m* réfractaire de tirage
e bloque *m* refractario para estirar
i blocco *m* refrattario per tirare
n trekblok *n*
d Ziehblock *m*

1745 DREDGER, DREDGING APPARATUS cer,coa
f poudreuse *f*, saupoudroir *m*
e aparato *m* de espolvereo, polverera *f*
i spargitore *m* di polvere
n bestrooiingsapparaat *n*
d Aufpudergerät *n*, Aufstreusieb *n*

1746 DRENCHING PLANT lea
f installation *f* de trempage
e equipo *m* de remojado
i impianto *m* di rammollimento
n weekinstallatie
d Weichanlage *f*

1747 DRESSER tex
f encolleuse *f*
e adobador *m*, encoladora *f*, zurrador *m*
i macchina *f* ad imbozzimare
n slichtmachine
d Schlichtmaschine *f*

1748 DRESSING MACHINE, CLEANING MACHINE bre,cle
f nettoyeuse *f*
e limpiadora *f*
i pulitrice *f*
n poetsmachine
d Putzmaschine *f*, Reinigungsmaschine *f*

1749 DRESSING PLANT, FINISHING PLANT, SIZING PLANT te:
f installation *f* d'apprêtage
e equipo *m* de apresto
i impianto *m* d'apprettatura
n appreteerinrichting
d Appreturanlage *f*

1750 DRESSING PLANT, ORE CONCENTRATION PLANT mi
f installation *f* d'enrichissement des minerais
e equipo *m* de enriquecimiento de los minerales
i impianto *m* d'arricchimento dei minerali
n ertsconcentreerinstallatie
d Erzanreicherungsanlage *f*

1751 DRESSING ROLL te
f cylindre *m* d'apprêtage
e cilindro *m* de apresto
i cilindro *m* d'apprettatura
n appreteerwals
d Appreturwalze *f*

1752 DRIER, DRYER, DRYING APPARATUS dr
f séchoir *m*
e desecador *m*, secador *m*
i essiccatoio *m*, essiccatore *m*
n droger, droogapparaat *n*
d Trockenapparat *m*, Trockner *m*

1753 DRILLED PLATEN PRESS, BORED PLATEN PRESS ru
f presse *f* à plateaux forés
e prensa *f* con platos taladrados
i pressa *f* con canali ricavati ne piani
n etagepers met geboorde verwarmingsplaten
d Plattenpresse *f* mit gebohrten Kanälen

1754 DRILLED WELL, BORED WELL c
f cylindre *m* foré
e cilindro *m* taladrado
i cilindro *m* forato
n geboorde wals, holle wals
d ausgebohrte Walze *f*, Hohlwalze *f*

1755 DRIP COOLER clo
f réfrigérateur *m* à gouttes
e enfriador *m* de gotas
i raffreddatore *m* a gocce
n druppelkoeler
d Tropfenkühler *m*

1756 DRIP-PROOF BURNER bur
f brûleur *m* abrité
e quemador *m* resguardado
i bruciatore *m* protetto contro lo stillicidio
n spatdichte brander
d tropfdichter Brenner *m*

1757 DRIP-TIP CONDENSER lab
f réfrigérant *m* à biseau d'écoulement
e enfriador *m* con punta para escurrir
i raffreddatore *m* con punta stillicidia
n druppelende koeler
d Kühler *m* mit Abtropfspitze

1758 DRIP TUBE sap
f collecteur *m* de gouttes
e colector *m* de gotas
i collettore *m* di gocce
n druppelvanger
d Tropfenfänger *m*

1759 DRIZZLING WORT COOLER, BAUDELOT COOLER, IRRIGATION COOLER bre
f Baudelot *m*, refroidisseur *m* à ruissellement
e enfriador *m* Baudelot
i raffreddatore *m* a pioggia
n sproeikoeler
d Berieselungskühler *m*, Rieselkühler *m*

1760 DROP ELIMINATOR cle,hyg
f éliminateur *m* de gouttes, rideau *m* de chicanes, rideau *m* d'essuyage
e eliminador *m* de gotas
i sgocciolatore *m*
n druppelvanger
d Tropfenabscheider *m*

1761 DROP POINT INDICATOR fue,ins
f signaleur *m* du point de goutte
e indicador *m* del punto de goteo
i indicatore *m* del punto di goccia
n druppelpuntindicator
d Tropfpunktgeber *m*

1762 DROP TEST APPARATUS, FALLING-WEIGHT APPARATUS exp,tes
f mouton *m*, mouton-pendule *m*
e probador *m* de caída de peso
i collaudatore *m* a caduta
n valhamer
d Fallhammer *m*

1763 DROP VALVE cpt
f soupape *f* à siège conique, soupape *f* renversée
e válvula *f* de mandada de asiento cónico, válvula *f* invertida
i valvola *f* a sede conica
n kegelklep
d Kegelventil *n*, Pilzventil *n*

1764 DROPPING BOTTLE lab
f flacon *m* compte-gouttes
e frasco *m* cuentagotas
i boccetta *f* a contagocce, contagocce *m*
n druppelflesje *n*
d Tropfflasche *f*

1765 DROPPING ELECTRODE elc
f électrode *f* à gouttes
e electrodo *m* de gotas
i elettrodo *m* a gocce
n druppelelektrode
d Tropfelektrode *f*

1766 DROPPING FUNNEL lab
f ampoule *f* à décanter, entonnoir *m* compte-gouttes
e embudo *m* de goteo
i imbuto *m* a contagocce
n druppeltrechter
d Tropftrichter *m*

1767 DROPPING GLASS lab
f compte-gouttes *m*
e cuentagotas *m*
i contagocce *m*
n druppelaar
d Tropfenspender *m*, Tropfenzähler *m*, Tropfglas *n*

1768 DROPPING POINT APPARATUS ins
f appareil *m* à point de goutte
e aparato *m* de punto de goteo
i apparecchio *m* a punto di sgocciolamento
n druppelpuntapparaat *n*
d Tropfpunktapparat *m*

1769 DROSS FILTER fil,fou
f filtre *m* à crasses
e filtro *m* de escorias
i filtro *m* fermascorie
n zeefplaat
d Siebkern *m*, Siebplatte *f*

1770 DROSSING KETTLE met
f récipient *m* de décrassage, récipient *m* d'écrémage
e recipiente *m* de despumación
i serbatoio *m* per rimozione della feccia
n afschuimketel
d Schlickerkessel *m*

1771 DROSSING PLANT met
f installation *f* de décrassage
e equipo *m* de despumación
i impianto *m* per rimozione della feccia
n afschuiminstallatie
d Schlickeranlage *f*

1772 DRUM tra
f tambour *m*
e tambor *m*
i tamburo *m*
n drum, trommel
d Trommel *f*

1773 DRUM CONCENTRATOR min
f concentrateur *m* à tambour
e concentrador *m* de tambor
i concentratore *m* a tamburo
n trommelconcentrator
d Anreicherungstrommel *f*

1774 DRUM DRIER, SINGLE DRUM DRIER dry
f séchoir *m* à tambour
e secador *m* de cilindro, secador *m* de tambor
i essiccatore *m* a cilindro, essiccatore *m* a tamburo
n trommeldroger
d Röhrentrockner *m*, Trommeltrockner *m*, Walzentrockner *m*

1775 DRUM DYEING MACHINE dye
f machine *f* de teinture à tambour
e máquina *f* de teñido de tambor
i macchina *f* di tintura a tamburo
n trommelverfmachine
d Trommelfärbmaschine *f*

1776 DRUM FILTER, ROTARY FILTER fil
f filtre *m* à tambour
e filtro *m* giratorio, tambor *m* filtrante
i filtro *m* a tamburo
n trommelfilter *n*
d Trommelfilter *m*

1777 DRUM FURNACE met
f four *m* à tambour
e horno *m* cilíndrico
i forno *m* a tamburo
n trommeloven
d Trommelofen *m*

1778 DRUM GRIT-WASHER iwt
f tambour *m* laveur de sable
e tambor *m* lavador de arena
i tamburo *m* per lavaggio di sabbia
n zandwastrommel
d Sandwaschtrommel *f*

1779 DRUM KILN bre
f tambour *m* tourailleur, touraille *f* cylindrique
e tostadero *m* cilíndrico
i essiccatoio *m* a tamburo
n trommeleest
d Trommeldarre *f*

1780 DRUM MALTING PLANT bre
f malterie *f* à tambours
e maltería *f* de tambores
i malteria *f* a tamburi
n trommelmouterij
d Trommelmälzerei *f*

1781 DRUM MIXER mix
f mélangeur *m* à tonneau
e mezclador *m* cilíndrico, mezclador *m* de tambor
i mescolatore *m* a tamburo
n trommelmenger
d Mischtrommel *f*, Trommelmischer *m*

1782 DRUM REFLUX dis
f ballon *m* de reflux
e depósito *m* de reflujo
i barilozzo *m* di riflusso
n terugstroomballon
d Rücklaufbehälter *m*

1783 DRUM SCREEN, ROTARY SCREEN, SCREENING DRUM so
f tambour *m* cribleur, tambour *m* tamiseur
e cuba *f* cilíndrica, cuba *f* de tambor
i staccio *m* a tamburo, vaglio *m* rotativo
n zeeftrommel
d Siebtrommel *f*, Trommelreche[n *m*,] Trommelsieb *n*

1784 DRUM SLASHER te
f encolleuse *f* à tambour
e encoladora *f* de cilindro
i macchina *f* da imbozzimare a tamburo
n trommelslichtmachine
d Trommelschlichtmaschine *f*

1785 DRUM STRAINER bre
f filtre-tambour *m*
e tambor *m* filtrante
i filtro *m* a tamburo
n trommelfilter *n*
d Trommelfilter *m*

1786 DRUM-TYPE MAGNETIC SEPARATOR sor
f séparateur *m* magnétique à tambour
e separador *m* magnético de tambor
i separatore *m* magnetico a tamburo
n magnetische trommelscheider
d magnetischer Trommelscheider *m*, magnetischer Trommelseparator *m*

1787 DRUM WARMER gen
f chauffe-tambour *m*
e calentador *m* de tambor
i riscaldatore *m* a tamburo
n trommelwarmer
d Trommelwärmer *m*

1788 DRUM WASHER pap
f laveuse *f* à tambour
e lavador *m* de tambor
i lavatrice *f* a tamburo
n trommelwasser
d Trommelwäsche *f*

1789 DRUMMING SHOP lea
f atelier *m* à battre le cuir
e taller *m* de batanar
i officina *f* di bottalatura
n lederklopperij
d Lederklopfwerkstatt *f*

1790 DRY ACETYLENE GENERATOR gas
f générateur *m* d'acétylène sec
e generador *m* de acetileno seco
i generatore *m* d'acetilene secco
n droge acetyleengenerator
d Trockenvergaser *m*

1791 DRY BLAST PRODUCER, AIR GAS GENERATOR fue
f générateur *m* de gaz à l'air
e generador *m* de gas de aire
i generatore *m* di gas d'aria
n luchtgasgenerator
d Aerogengasgenerator *m*, Luftgasgenerator *m*

1792 DRY BOX nue
f boîte *f* de protection
e caja *f* de protección
i cassa *f* di protezione
n veiligheidskast
d Schutzkammer *f*

1793 DRY CELL elc
f pile *f* sèche
e pila *f* seca
i pila *f* a secco
n droge cel, droog element *n*
d Trockenelement *n*

1794 DRY CLEANING SHOP cle
f atelier *m* de nettoyage à sec
e taller *m* de limpieza en seco
i officina *f* di lavaggio a secco
n chemische wasserij
d chemische Reinigungsanstalt *f*

1795 DRY DISTILLATION PLANT, DESTRUCTIVE DISTILLATION PLANT dis
f installation *f* de distillation destructive, installation *f* de distillation sèche
e equipo *m* de destilación destructiva, equipo *m* de destilación seca
i impianto *m* di distillazione a secco, impianto *m* di distillazione distruttiva
n droge destillatie-installatie
d Trockendestillationsanlage *f*, Zersetzungsdestillationsanlage *f*

1796 DRY ENAMELLING APPARATUS coa,met
f appareil *m* d'émaillage au poudré
e aparato *m* de esmaltado con polvo
i apparecchio *m* di smaltatura a spolvero
n poederemailleerapparaat *n*
d Puderemaillierungsgerät *n*

1797 DRY END pap
f groupe *m* de séchage, section *f* sécheuse
e sección *f* de secado
i batteria *f* d'essiccatori, gruppo *m* d'essiccatori
n droogsectie
d Trockenpartie *f*

1798 DRY GAS CLEANING PLANT cle,gas
f installation *f* de dépoussiérage à sec du gaz, installation *f* d'épurage à sec du gaz

e equipo *m* de depuración seca del gas
i impianto *m* d'epurazione a secco del gas
n droge-gasreinigingsinstallatie
d trockene Gasreinigungsanlage *f*

1799 DRY ICE PRODUCER ctt
f appareil *m* producteur de neige carbonique, générateur *m* de glace sèche
e aparato *m* productor de hielo seco, generador *m* de nieve carbónica
i apparecchio *m* prodottore d'anidride carbonica solida, apparecchio *m* prodottore di ghiaccio asciutto
n droog-ijsapparaat *n*, koolzuur-sneeuwapparaat *n*
d Kohlensäureschneeentwickler *m*, Trockeneisentwickler *m*

1800 DRY ICE EXTINGUISHER fea
f extincteur *m* à neige carbonique
e extintor *m* de nieve carbónica
i estintore *m* ad anidride carbonica solida
n koolzuursneeuwblusser
d Kohlensäureschneelöscher *m*

1801 DRY PAN brk
f moulin *m* à rouleaux pour traitement sec
e molino *m* de rodillos para trabajar en seco
i mulino *m* a rulli per lavorare al secco
n droge kollergang
d Trockenkollergang *m*

1802 DRY POWDER EXTINGUISHER fea
f extincteur *m* à poudre
e extintor *m* de polvo seco
i estintore *m* a polvere secca
n droogblusser
d Trockenlöscher *m*

1803 DRY SPRAYER dec
f pistolet *m* pour peinture, pulvérisateur *m* pour couleurs
e atomizador *m* para colores, pulverizador *m* para colores
i polverizzatore *m* per colori
n verfspuit
d Farbspritzpistole *f*

1804 DRYING CABINET, CABINET DESICCATOR, CABINET DRIER dry,lab
f séchoir *m*
e armario *m* secador
i camera *f* d'essiccazione
n droogkast
d Trockenschrank *m*

1805 DRYING CHAMBER, DRYING LOFT, DRYING ROOM dry,pa
f chambre *f* de séchage, séchoir
e cámara *f* secadora, secadero *m*
i essiccatoio *m*
n droogkamer
d Darre *f*, Trockenkammer *f*

1806 DRYING CHEST te
f plaques *pl* creuses de séchage à chauffage à vapeur
e placas *pl* huecas secadoras con calentamiento de vapor
i piastre *pl* cave essiccatrici con riscaldamento a vapore
n met stoom verhitte holle droogplaten *pl*
d dampfgeheizte Trockenhohlplatten *pl*

1807 DRYING CYLINDER, CYLINDER DRIER dry,pa
f cylindre *m* sécheur
e cilindro *m* secador
i cilindro *m* essiccante
n droogwals
d Trockenwalze *f*

1808 DRYING DRUM d
f tambour *m* sécheur
e tambor *m* de secado
i tamburo *m* essiccante
n droogtrommel
d Trockentrommel *f*

1809 DRYING JAR dry,l
f colonne *f* à dessécher
e columna *f* secadera
i colonna *f* essiccatrice
n droogcilinder, droogtoren
d Trockenturm *m*, Trockenzylinder *m*

1810 DRYING KILN bre,d
f touraille *f* séchoir
e secadero *m*
i essiccatoio *m*
n droogeest
d Trockendarre *f*

1811 DRYING KILN d
f four *m* à sécher
e horno *m* de secado
i forno *m* essiccatore
n droogoven
d Darrofen *m*, Trockenofen *m*

1812 DRYING PAN, dry
PAN DRIER
f bac *m* sécheur
e cuba *f* secadora
i tino *m* essiccatore
n droogpan
d Trockenpfanne *f*

1813 DRYING PLANT bre
f installation *f* de tourailles
e equipo *m* de tostaderos
i impianto *m* d'essiccatoio
n eestafdeling
d Darrenanlage *f*

1814 DRYING PLANT dry
f installation *f* de séchage
e equipo *m* de secado
i impianto *m* d'essiccamento
n drooginstallatie
d Trockenanlage *f*

1815 DRYING STOVE dry,met
f étuve *f*
e estufa *f*
i stufa *f*
n droogoven, droogstoof
d Wärmeschrank *m*

1816 DRYING TOWER, dry,gas
TOWER DRIER
f tour *f* de séchage
e torre *f* de secado
i torre *f* d'essiccamento
n droogtoren
d Trockenturm *m*

1817 DRYING TRAY, dry
TRAY DRIER
f bac *m* de séchage
e cubeta *f* secadora
i recipiente *m* essiccatore
n droogbak
d Trockenschale *f*

1818 DRYING TUBE lab
f tube *m* absorbeur, tube *m* sécheur
e tubo *m* secador
i tubo *m* essiccatore
n droogbuis
d Trockenrohr *n*

1819 DUAL-FLOW TRAY dis
f plateau *m* à double sens d'écoulement
e plato *m* de dos flujos
i piatto *m* a passaggio contemporaneo di liquido e vapore
n tweestromenschotel
d zweiflutiger Boden *m*

1820 DUCT, cpt
GAS PASSAGE
f caniveau *m* de ventilation
e conducto *m* de gas
i condotto *m* di gas
n gaskanaal *n*
d Gaskanal *m*

1821 DUMP TANK nue
f réservoir *m* de vidange
e tanque *m* vaciador
i vasca *f* di raccolta
n opslagtank
d Ablassbehälter *m*

1822 DUMPING FLOOR bre
f plateau *m* basculant
e plato *m* basculante
i griglia *f* ribaltabile
n kantelvloer
d Kipphorde *f*

1823 DUMPING FLOOR KILN, bre
f touraille *f* à plateau basculant
e tostadero *m* de plato basculante
i essiccatoio *m* a griglia ribaltabile
n kantelvloereest
d Kipphordendarre *f*

1824 DUROMETER, ins,tes
HARDNESS TESTER,
SCLEROMETER
f duromètre *m*, scléromètre *m*
e durómetro *m*
i durometro *m*
n hardheidsbeproevingsapparaat *n*, hardheidsmeter
d Härteprüfer *m*

1825 DUST CARTRIDGE, fil,sap
DUST FILTER
f filtre *m* à poussière
e filtro *m* de polvo
i filtro *m* per polvere
n stoffilter *n*
d Staubfilter *m*

1826 DUST CATCHER, sap
DUST COLLECTOR
f collecteur *m* de poussière, dépoussiéreur *m*
e captador *m* de polvo, colector *m* de polvo, recogedor *m* de polvo
i collettore *m* della polvere
n stofvanger
d Staubabscheider *m*, Staubfänger *m*

1827 DUST CHAMBER sap
f chambre *f* à poussière
e cámara *f* de polvo
i camera *f* a polvere
n stofkamer
d Staubkammer *f*

1828 DUST COLLECTING TUBE ins,lab
f tube *m* échantillonneur pour des poussières
e tubo *m* probador de polvos
i tubo *m* collaudatore di polveri
n stofverzamelbuis
d Staubsammelrohr *n*

1829 DUST IMPINGER TUBE ins,lab
f tube *m* à choc pour le dosage des poussières
e tubo *m* de colisión para dosificación de polvos
i tubo *m* ad urto per dosaggio di polveri
n stootbuis voor stofbepaling
d Prallrohr *n* zur Staubbestimmung

1830 DUST RESPIRATOR sap
f masque *m* anti-poussière
e máscara *f* antipolvo
i maschera *f* antipolvere
n stofmasker *n*
d Staubmaske *f*

1831 DUST SETTLING CHAMBER cle,gas
GRAVITY SETTLING CHAMBER, SETTLING CHAMBER
f chambre *f* à poussière
e cámara *f* de polvo
i camera *f* a polvere
n stofkamer
d Staubkammer *f*

1832 DYE BATH dye
f bain *m* de teinture
e baño *m* de tinte
i bagno *m* di tintura
n verfbad *n*
d Färbebad *n*, Färbeflotte *f*

1833 DYE BEAKER lab
f bécher *m* à colorant
e vaso *m* de colorante
i bicchiere *m* a colorante
n verfbeker
d Färbebecher *m*

1834 DYE HOUSE dye
f teinturerie *f*
e tintoreria *f*
i tintoria *f*
n ververij
d Färberei *f*

1835 DYEING MACHINE dye
f appareil *m* de teinture
e máquina *f* de teñido
i apparecchio *m* di tintura
n verfapparaat *n*
d Färbeapparat *m*

1836 DYEING VAT dy
f cuve *f* de teinture
e recipiente *m* de teñido
i serbatoio *m* di tintura
n verfkuip
d Färbekūpe *f*

1837 DYEOMETER dye,in
f tintomètre *m*
e tintómetro *m*
i tintometro *m*
n kleurbadcolorimeter
d Farbbadkolorimeter *n*

E

1838 EARTHENWARE, STONEWARE cer
f argile *f* cuite, terre *f* cuite
e gres *m* cocido, loza *f* de barro
i terracotta *f*, terraglia *f*
n aardewerk *n*
d Steingut *n*

1839 EASING AIR FEEDHOLE cer,htt
f entrée *f* d'air thermorégulatrice
e entrada *f* de aire termorreguladora
i entrata *f* d'aria termoregolatrice
n temperatuur-regelende luchtinlaat
d Temperaturregelungsluftzufuhr *f*

1840 EBONITE BUSHING coa,sap
f manchon *m* en ébonite
e casquillo *m* de ebonita, manguito *m* de ebonita
i camicia *f* d'ebanite
n ebonietmof
d Ebonitmuffe *f*

1841 EBONITE LINING coa
f garnissage *m* en ébonite, revêtement *m* en ébonite
e guarnición *f* de ebonita, revestimiento *m* de ebonita
i guarnizione *f* d'ebanite, rivestimento *m* d'ebanite
n ebonietbekleding, ebonietvoering
d Ebonitfutter *n*

1842 EBULLIOMETER ins
f ébulliomètre *m*
e ebullómetro *m*
i ebulliometro *m*
n ebulliometer, kookpuntmeter
d Ebulliometer *n*, Siedepunktmesser *m*

1843 EBULLIOSCOPE ins
f ébullioscope *m*
e ebulloscopio *m*
i ebullioscopio *m*
n ebullioscoop
d Ebullioskop *n*

1844 ECCENTRIC PRESS pla,rub
f presse *f* à excentrique
e prensa *f* de excéntrica, prensa *f* excéntrica
i pressa *f* eccentrica
n excenterpers
d Exzenterpresse *f*

1845 ECHELON DIFFRACTION GRATING ins
f réseau *m* à échelons
e red *f* de difracción en escalones
i reticolo *m* a gradini
n trappenrooster *n*
d Stufengitter *n*

1846 ECONOMIZER, SMOKE GAS PREHEATER prc
f économiseur *m*
e calentador *m* del agua de alimentación, economizador *m*
i economizzatore *m*
n economiser, rookgasvoorverwarmer, voedingswatervoorverwarmer
d Ekonomiser *m*, Rauchgasvorwärmer *m*, Sparanlage *f*, Speisewasservorwärmer *m*

1847 EDGE FILTER fil
f filtre *m* à bord
e filtro *m* de canto
i filtro *m* ad orlo
n randfilter *n*
d Randfilter *m*

1848 EDGE INSULATOR elc
f isolateur *m* latéral
e aislador *m* lateral
i isolatore *m* laterale
n afstandhouder
d Randisolator *m*

1849 EDGE ROLLER, EDGE RUNNER brk
f meule *f* verticale
e muela *f* vertical, rueda *f* vertical
i mola *f* verticale
n kollergangsteen, verticale maalsteen
d Kollerstein *m*, Läufer *m*

1850 EDGE RUNNER, CHILEAN MILL brk
f broyeur *m* à meules verticales
e molino *m* chileno, molino *m* de muelas verticales
i macina *f* a mole verticali, molazza *f*

n grote kollergang
d Kollergang *m*

1851 EDUCTION PIPE tra
f tube *m* d'écoulement
e tubo *m* de escape, tubo *m* eductor
i tubo *m* d'efflusso
n afvoerbuis
d Abzugsrohr *n*, Auslassrohr *n*

1852 EDUCTION VALVE cpt
f soupape *f* d'écoulement
e válvula *f* de educción, válvula *f* de escape
i valvola *f* d'efflusso
n afvoerklep
d Abzugsventil *n*, Auslassventil *n*

1853 EDUCTOR pre
f appareil *m* d'extraction
e eductor *m*
i apparecchio *m* d'estrazione
n extraheertoestel *n*
d Auszuggerät *n*, Extrahiergerät *n*, Extraktionsgerät *n*

1854 EDULCORATION PLANT cle
f installation *f* d'édulcoration
e equipo *m* de purificación por medio de lavado
i impianto *m* edulcorante
n uitwasinstallatie
d Auswaschanlage *f*

1855 EFFERVESCENT FERMENTATION TANK bre,fmi
f installation *f* de fermentation effervescente
e tanque *m* de fermentación efervescente
i impianto *m* di fermentazione effervescente
n gistingsinstallatie met schuimvorming
d Schaumgärungsanlage *f*

1856 EFFLUENT COLLECTING BASIN pap
f bassin *m* collecteur de l'effluent
e depósito *m* colector del efluente
i vasca *f* collettrice dell'effluente
n afvoerverzamelbekken *n*
d Abflussbecken *n*

1857 EFFLUENT MONITOR ins,nue
f moniteur *m* de radioactivité d'effluent
e monitor *m* de radiactividad de efluente
i avvisatore *m* di radioattività d'effluente
n afvoermonitor
d Abflussmonitor *m*

1858 EFFLUENT PIPE iwt
f tube *m* de sortie
e tubo *m* de salida
i tubo *m* di scarico
n afvoerpijp
d Abflussrohr *n*

1859 EFFLUENT TROUGH, DISCHARGE CHANNEL iwt
f rigole *f* de décharge
e arroyo *m* de descarga
i canaletto *m* di scarico
n afvloeigoot
d Abflussrinne *f*

1860 EFFUSIOMETER ins
f effusiomètre *m*
e efusiómetro *m*
i effusiometro *m*
n effusiometer, gasdichtheidsmeter, uitstromingsmeter
d Ausströmungsmesser *m*, Effusiometer *n*, Gasdichtemesser

1861 EIKONOMETER ins
f iconomètre *m*
e iconómetro *m*
i iconometro *m*
n iconometer
d Ikonometer *n*

1862 EJECTING DEVICE (GB), PIECE KNOCKOUT (US) pla,rub
f dispositif *m* de démoulage, dispositif *m* éjecteur
e dispositivo *m* de eyección, expulsor *m*, eyector *m*
i dispositivo *m* d'eiezione, dispositivo *m* d'espulsione
n uitstootinrichting
d Auswerfer *m*, Auswurfvorrichtung

1863 EJECTION BAR (GB), EJECTION PIN (GB), KNOCKOUT PIN (US) rub
f barre *f* d'éjection, goupille *f* éjectrice, tige *f* d'éjecteur
e barra *f* de eyección, perno *m* eyector
i barra *f* d'espulsione, spina *f* d'espulsione, tirante *m* d'espulsione
n uitstootpin, uitwerperstift
d Ausdrückstift *m*, Ausstosser *m*, Auswerfer *m*, Auswurfstift *m*

1864 **EJECTION FRAME,** pla
EJECTOR FRAME
f plaque-support *m* d'éjecteur
e bastidor *m* de eyección
i travi *pl* d'espulsione
n uitwerperfreem *n*
d Ausdrückrahmen *m*,
Auswerferrahmen *m*

1865 **EJECTION RAM,** pla,rub
EJECTOR RAM
f piston *m* d'éjection
e pistón *m* de eyección
i pistone *m* espulsore
n uitwerperplunjer
d Ausdrückkolben *m*,
Auswerferkolben *m*

1866 **EJECTION SPRING,** rub
EJECTOR SPRING
f ressort *m* d'éjecteur
e muelle *m* de eyector
i molla *f* d' espulsore
n uitwerpveer
d Auswerferfeder *f*,
Auswurffeder *f*

1867 **EJECTION TIE BAR** pla,pre
f tige *f* de rappel
e vástago *m* expulsor
i tirante *m* dell'estrattore
n uitwerperstang
d Ausdrückstange *f*

1868 **EJECTOR** cpt
f éjecteur *m*
e eyector *m*
i eiettore *m*
n ejecteur
d Ejektor *m*

1869 **EJECTOR** iwt
f éjecteur *m* à air comprimé
e eyector *m* de aire comprimida
i eiettore *m* ad aria compressa
n persluchtstuwer
d Druckluftheber *m*

1870 **EJECTOR CONNECTING BAR,** pla,pre
EJECTOR ROD
f broche *f* d'éjecteur
e espiga *f* de expulsión
i asta *f* d'estrazione
n uitwerperplunjer
d Ausdrückbolzen *m*

1871 **EJECTOR FRAME GUIDE** pla,pre
f guide *m* de plaque-support
e guía *f* de bastidor
i guida *f* di travi d'estrazione
n uitwerperfreemgeleiding
d Ausdrückrahmenführung *f*

1872 **EJECTOR PIN PLATE (GB),** pla,rub
KNOCKOUT PIN PLATE (US)
f plaque *f* d'éjection
e placa *f* de eyector
i piastra *f* d'estrazione
n uitwerpplaat
d Ausdrückplatte *f*,
Auswerferplatte *f*

1873 **EJECTOR PUMP** pre
f pompe *f* à éjecteur
e bomba *f* de eyector
i pompa *f* ad eiettore
n ejectorpomp, zuigstraalpomp
d Ejektorpumpe *f*,
Saugstrahlpumpe *f*

1874 **ELASTOMETER** ins,rub
f élastomètre *m*
e elastómetro *m*
i elastometro *m*, elaterometro *m*
n elasticiteitsmeter, elaterometer
d Elastizitätsmesser *m*,
Elastometer *n*

1875 **ELECTRIC CONVECTION FURNACE,** htt
ELECTRIC CONVERTER
f radiateur *m* électrique à convection
e radiador *m* eléctrico de convección
i stufa *f* elettrica a convezione
n elektrische convectiekachel
d elektrischer Konvektionsofen *m*

1876 **ELECTRIC DEHYDRATOR** dry
f dessicateur *m* électrique
e desecador *m* eléctrico
i essiccatore *m* elettrico
n elektrische droogkast
d elektrischer Dörrschrank *m*,
Elektrodörrschrank *m*

1877 **ELECTRIC DESALTER** nut
f dispositif *m* de dessalage électrique
e dispositivo *m* de desalación eléctrico
i dispositivo *m* di desalinizzazione elettrico
n elektrisch ontziltingsapparaat *n*,
elektrisch ontzoutingsapparaat *n*
d elektrisches Entsalzungsgerät *n*

1878 ELECTRIC FURNACE htt
f four *m* électrique
e horno *m* eléctrico
i forno *m* elettrico
n elektrische oven
d Elektroofen *m*

1879 ELECTRIC SOLDERING IRON wel
f fer *m* à souder électrique
e soldador *m* eléctrico
i saldatore *m* elettrico
n elektrische soldeerbout
d elektrischer Lötkolben *m*

1880 ELECTRIC STEAM RISER boi,htt
f chaudière *f* électrique à vapeur
e caldera *f* eléctrica de vapor
i generatore *m* elettrico di vapore
n elektrische stoomketel, elektroketel
d Elektrodampfkessel *m*

1881 ELECTRIC WATER HEATER htt
f chaudière *f* électrique à eau chaude
e caldera *f* eléctrica para agua caliente
i caldaia *f* elettrica per acqua calda
n elektrisch heetwaterreservoir *n*
d Elektroheisswasserkessel *m*

1882 ELECTRICAL FILTER, ELECTRICAL PRECIPITATOR fil
f électrofiltre *m*, précipitateur *m* électrostatique
e electrofiltro *m*
i elettrofiltro *m*, precipitatore *m* elettrico
n elektrische vanger, elektrofilter *n*
d Elektrofilter *m*

1883 ELECTROANALYSIS FLASK ana,elc
f matras *m* d'électroanalyse
e matraz *m* de electroanálisis
i matraccio *m* d'elettroanalisis
n elektroanalysekolf
d Elektroanalysekolben *m*

1884 ELECTROBURNER bur
f brûleur *m* électrique
e electroquemador *m*
i bruciatore *m* elettrico
n elektrobrander
d Elektrobrenner *m*

1885 ELECTROCOATING APPARATUS, ELECTRODEPOSITION APPARATUS, ELECTROFORESIS APPARATUS, ELECTROPAINTING APPARATUS dec
f appareil *m* d'électrodéposition
e aparato *m* de revestimiento electroforético
i apparecchio *m* di rivestimento elettroforetico
n elektroforese-lakapparaat *n*
d elektroforetischer Lackapparat

1886 ELECTRODE elc
f électrode *f*
e electrodo *m*
i elettrodo *m*
n elektrode
d Elektrode *f*

1887 ELECTRODE RING, BULL'S EYE htt,met
f entrée *f* d'électrode
e entrada *f* de electrodo
i entrata *f* d'elettrodo
n elektrode-ingang
d Elektrodenzugang *m*

1888 ELECTRODE SALT BATH elc
f four *m* à bain de sel à électrodes
e horno *m* de baño salino de electrodos
i forno *m* a bagno salino ad elettrodi
n elektrodenzoutbadoven
d Elektrodensalzbadofen *m*

1889 ELECTRODE SCREEN sap,wel
f blindage *m* d'électrodes, écran *m* d'électrodes
e blindaje *m* de electrodos, pantalla *f* de electrodos
i schermo *m* d'elettrodi
n elektrodenscherm *n*
d Elektrodenschirm *m*

1890 ELECTRODE STEAM GENERATOR, ELECTRODE STEAM RAISER, ELECTRODE WATER HEATER htt
f chaudière *f* à eau chaude à électrodes, chaudière *f* à vapeur à électrodes
e caldera *f* de electrodos
i caldaia *f* ad elettrodi
n elektrodenketel
d Elektrodenkessel *m*

1891 ELECTRODECANTATION DEVICE cle
f dispositif *m* de décantation électrique

e dispositivo *m* de electrodecantación
i dispositivo *m* di decantazione elettrica
n elektrisch decanteerapparaat *n*
d elektrisches Dekantiergerät *n*

1892 ELECTRODEPOSITION BATH elc
f bain *m* pour dépôts électrolytiques
e baño *m* para depósitos electrolíticos
i bagno *m* per deposizioni elettrolitiche
n galvanobad *n*
d elektrolytisches Abscheidungsbad *n*

1893 ELECTRODIALYSIS VESSEL elc,oic
f cuve *f* d'électrodialyse
e cuba *f* de electrodiálisis
i tino *m* d'elettrodialisi
n elektrodialysevat *n*
d Elektrodialysegefäss *n*

1894 ELECTRODIPPING PLANT dec
f installation *f* électrique de vernissage au trempé
e equipo *m* eléctrico barnizador de inmersión
i impianto *m* elettrico di verniciatura per immersione
n elektrische dompellak-installatie
d Elektrotauchlackieranlage *f*

1895 ELECTRODISPERSION VESSEL col,met
f cuve *f* d'électrodispersion
e cuba *f* de electrodispersión
i tino *m* d'elettrodispersione
n elektrodispersievat *n*
d Elektrodispersionsgefäss *n*

1896 ELECTRODISSOLUTION TANK elc
f récipient *m* d'électrodissolution
e tanque *m* de electrodisolución
i serbatoio *m* d'elettrodissoluzione
n tank voor het elektrolytisch oplossen
d elektrolytisches Auflösungsgefäss *n*

1897 ELECTRO-EROSION METAL WORKING PLANT, met
SPARK METAL WORKING PLANT
f installation *f* d'usinage par électro-érosion,
installation *f* d'usinage par étincelage
e equipo *m* de trabajo por electro-erosión
i impianto *m* di lavorazione a scintillamento
n inrichting *f* voor elektro-erosieve behandeling
d Elektroerosionsbearbeitungsanlage *f*

1898 ELECTRO-EXTRACTION PLANT, met
ELECTROWINNING PLANT
f installation *f* d'extraction par voie électrolytique
e equipo *m* de electroextracción, equipo *m* de extracción por vía electrolítica
i impianto *m* d'estrazione per via elettrolitica
n elektrolytische winnings-installatie
d elektrolytische Metallgewinnungsanlage *f*

1899 ELECTROFACING BATH coa,elc
f bain *m* de recouvrement électrolytique en métal dur
e equipo *m* de recargue galvánico
i impianto *m* di rivestimento elettrolitico con metallo duro
n elektrolytische hardmetaal-bekledingsinstallatie
d elektrolytische Hartmetall-bekleidungsanlage *f*

1900 ELECTROFORMED MOLD (US), elc,pre
ELECTROFORMED MOULD (GB)
f moule *m* à usinage par électro-érosion
e molde *m* maquinado por electroerosión
i stampa *f* a lavorazione per elettroerosione
n matrijs gemaakt door middel van vonkerosie
d Matrize *f* angefertigt durch elektroerosive Bearbeitung

1901 ELECTROFORMING BATH elc
f bain *m* d'électroformage
e baño *m* para electroformación
i bagno *m* per galvanoplastica
n galvanoplastiekbad *n*
d Galvanoplastikbad *n*

1902 ELECTROGALVANIZING BATH coa,elc
f bain *m* de galvanisation électrolytique, bain *m* de zingage électrolytique
e baño *m* de galvanización electrolítica
i bagno *m* di galvanizzazione elettrolitica
n elektrolytisch galvaniseerbad *n*
d galvanisches Verzinkungsbad *n*

1903 ELECTROGILDING BATH elc
f bain *m* de dorage électrolytique
e baño *m* de doradura electrolítica
i bagno *m* di doratura elettrolitica
n elektrolytisch verguldingsbad *n*
d elektrolytisches Vergoldungsbad *n*

1904 ELECTROGRANODIZING BATH coa
f bain *m* de phosphatage électrolytique préventif
e baño *m* de fosfatización electrolítica preventiva
i bagno *m* di fosfatizzazione elettrolitica preventiva
n beschermend elektrolytisch fosfateerbad *n*
d elektrolytisches Schutzphosphatierungsbad *n*

1905 ELECTROLESS PLATING BATH, CHEMICAL PLATING BATH coa
f bain *m* de placage chimique
e baño *m* de quimioplastia
i bagno *m* di placcatura chimica
n chemisch pleetbad *n*
d chemisches Plattierungsbad *n*

1906 ELECTROLYTIC BATH elc
f bain *m* électrolytique
e baño *m* electrolítico
i bagno *m* elettrolitico
n elektrolysebad *n*
d Elektrolysebad *n*

1907 ELECTROLYTIC CELL elc
f cellule *f* électrolytique
e celda *f* electrolítica
i cella *f* elettrolitica
n elektrolytische cel
d elektrolytische Zelle *f*

1908 ELECTROLYTIC CLEANING BATH elc,cle
f bain *m* de dégraissage électrolytique
e baño *m* de desengrase electrolítico
i bagno *m* di sgrassaggio elettrolitico
n elektrolytisch ontvettingsbad *n*
d elektrolytisches Reinigungsbad *n*

1909 ELECTROLYTIC CLEANING PLANT elc
f établissement *m* de dégraissage électrolytique
e equipo *m* de desengrase electrolítico
i impianto *m* di sgrassaggio elettrolitico
n elektrolytische ontvettingsinstallatie
d elektrolytische Reinigungsanlage *f*

1910 ELECTROLYTIC CLEANING TANK elc
f récipient *m* de dégraissage électrolytique
e tanque *m* para desengrase electrolítico
i serbatoio *m* di sgrassaggio elettrolitico
n tank voor het elektrolytisch ontvetten
d Behälter *m* für elektrolytische Reinigung

1911 ELECTROLYTIC FURNACE elc,ht
f four *m* électrolytique, pot *m*
e horno *m* electrolítico
i forno *m* elettrolitico
n elektrolytische oven
d Elektrolytofen *m*

1912 ELECTROLYTIC LEAD REFINERY elc,me
f installation *f* de raffination électrolytique du plomb
e equipo *m* de afinado electrolítico del plomo
i impianto *m* di raffinazione elettrolitico del piombo
n elektrolytische loodraffineerder
d elektrolytische Bleivergütungsanlage *f*

1913 ELECTROLYTIC PARTING BATH, ELECTROLYTIC SEPARATING BATH, ELECTROPARTING BAT el
f bain *m* de séparation électrolytique
e baño *m* de separación electrolítica

i bagno *m* di separazione elettrolitica
n elektrolytisch metaalscheidingsbad *n*
d elektrolytisches Metallscheidungsbad *n*

1914 ELECTROLYTIC PICKLING BATH elc
f bain *m* de décapage électrolytique
e baño *m* de decapaje electrolítico
i bagno *m* di decapaggio elettrolitico
n elektrolytisch beitsbad *n*
d elektrolytisches Beizbad *n*

1915 ELECTROLYTIC REDUCTION TANK elc,met
f récipient *m* de réduction électrolytique
e tanque *m* de reducción electrolítica
i serbatoio *m* di riduzione elettrolitica
n elektrolytische-reductietank
d elektrolytisches Reduktionsgefäss *n*

1916 ELECTROLYTIC REFINING PLANT, ELECTROREFINING PLANT elc,met
f raffinerie *f* électrolytique
e refinería *f* electrolítica
i raffineria *f* elettrolitica
n elektrolytische raffineerderij
d elektrolytische Raffinerie *f*, elektrolytische Vergütungsanlage *f*

1917 ELECTROLYTIC STRIPPING BATH elc
f bain *m* de dépouillage électrolytique
e baño *m* de eliminación electrolítica
i bagno *m* di degalvanizzazione elettrolitica
n elektrolytisch laagverwijderingsbad *n*
d elektrolytisches Entplattierungsbad *n*

1918 ELECTROLYZER elc
f électrolyseur *m*
e electrolizador *m*
i elettrolizzatore *m*
n elektrolytische cel
d elektrolytische Zelle *f*

1919 ELECTROLYZER CASCADE elc,min
f cascade *f* d'électrolyseurs
e cascada *f* de electrolizadores
i cascata *f* d'elettrolizzatori
n cascade van elektrolytische cellen
d elektrolytische Zellenkaskade *f*

1920 ELECTROMAGNETIC SEPARATOR sor
f séparateur *m* électromagnétique
e separador *m* electromagnético
i separatore *m* elettromagnetico
n elektromagnetische afscheider
d elektromagnetischer Scheider *m*

1921 ELECTROMETALLISATION PLANT elc
f installation *f* de métallissation électrolytique
e equipo *m* de metalización electrolítica
i impianto *m* di metallizzazione elettrolitica
n elektrolytische metalliseerinstallatie
d elektrolytische Metallisierungsanlage *f*

1922 ELECTROMETRIC TITRATION APPARATUS lab
f appareil *m* de titrage potentiométrique
e aparato *m* de valoración electrométrica
i apparecchio *m* di titolazione elettrometrica
n elektrometrisch titreerapparaat *n*
d elektrometrisches Massanalysegerät *n*

1923 ELECTRON MICROSCOPE ins
f microscope *m* électronique
e microscopio *m* electrónico
i microscopio *m* elettronico
n elektronenmicroscoop
d Elektronenmikroskop *n*

1924 ELECTROPLATING BASE elc
f métal *m* de base
e metal *m* de base
i metallo *m* di fondo
n grondmetaal *n*
d Grundmetall *n*

1925 ELECTROPLATING BATH elc
f bain *m* de galvanoplastie
e baño *m* de galvanoplastia
i bagno *m* di galvanoplastica

n galvanostegiebad *n*
d Elektroplattierungsbad *n*, Galvanostegiebad *n*

1926 ELECTROPLATING GENERATOR elc
f dynamo *m* pour galvanoplastie
e dínamo *m* para galvanoplastia
i dinamo *m* per galvanoplastica
n gelijkstroomgenerator voor galvanostegie
d Gleichstromgenerator *m* für Elektroplattierung, Gleichstromgenerator *m* für Galvanostegie

1927 ELECTROPLATING VAT elc
f cuve *f* de galvanoplastie
e cuba *f* de galvanoplastia
i serbatoio *m* di galvanoplastica
n galvanostegievat *n*
d Elektroplattierungsgefäss *n*, Galvanostegiegefäss *n*

1928 ELECTROPOLISHING BATH elc,pol
f bain *m* de polissage électrolytique
e baño *m* de pulido electrolítico
i bagno *m* di pulitura elettrolitica
n elektrolytisch glansbad *n*, elektrolytisch polijstbad *n*
d Elektropolierbad *n*

1929 ELECTROREFINING BATH elc
f bain *m* d'électroraffinage
e baño *m* de electroafino
i bagno *m* d'elettroraffinazione
n elektrolytisch zuiveringsbad *n*
d elektrolytisches Raffinationsbad *n*

1930 ELECTROSTATIC GAS CONDITIONING APPARATUS gas
f installation *f* de dépoussiérage électrostatique de gaz
e equipo *m* de purificación electrostática de gas
i impianto *m* di purificazione elettrostatica di gas
n elektrostatische gasreinigingsinstallatie
d elektrostatische Gasreinigungsanlage *f*

1931 ELECTROSTATIC PRECIPITATOR fil
f dépoussiéreur *m* électrostatique, filtre *m* électrostatique
e filtro *m* electrostático, precipitador *m* electrostático
i filtro *m* elettrostatico
n elektrofilter *n*
d Elektrofilter *m*

1932 ELECTROSTATIC SEPARATOR sor
f trieuse *f* électrostatique
e clasificador *m* electrostático
i classificatore *m* elettrostatico
n elektrostatische sorteermachine
d elektrostatische Sortiermaschine *f*

1933 ELECTROSTATIC SPRAYING PLANT coa
f installation *f* de pistolage électrostatique
e equipo *m* de chorreado electrostático
i impianto di spruzzatura elettrostatica
n elektrostatische spuitinstallatie
d elektrostatische Spritzanlage *f*

1934 ELECTROTHERMAL OVEN htt
f four *m* électrothermique
e horno *m* electrotérmico
i forno *m* elettrotermico
n elektrowarmteoven
d Elektrowärmeofen *m*

1935 ELECTROTYPING SHELL elc
f coquille *f*
e coquilla *f*
i riporto *m* metallico
n schil
d Galvano *m*

1936 ELEMENT elc
f bloc *m* de plaques
e paquete *m* de placas
i elementi *pl*
n platenblok *n*
d Plattenpaket *n*

1937 ELEVATOR (US), SILO (GB) nut,tra
f silo *m*
e silero *m*, silo *m*
i silo *m*
n silo
d Schachtspeicher *m*, Silo *m*, Zellenspeicher *m*

1938 ELEVATOR KILN cer
f étuve *f* à élévateur
e estufa *f* de elevador
i stufa *f* ad elevatore
n hefstoof
d Hebetrockenofen *m*

1939 ELIMINATOR BAFFLE, ELIMINATOR PLATE cle,hyg
f chicane *f* d'essuyage
e placa *f* eliminadora de gotas
i piastra *f* di sgocciolamento
n druppelvangerplaat
d Tropfenabscheideblech *n*, Tropfenabscheideplatte *f*

1940 ELLIPTICAL ANODE, OVAL SECTION ANODE elc
f anode *f* elliptique
e ánodo *m* elíptico
i anodo *m* ellittico
n ovale anode
d Knüppelanode *f*

1941 ELLIPTICAL RETORT, RHENISCH RETORT htt
f cornue *f* elliptique
e retorta *f* elíptica
i storta *f* ellittica
n ovale retort
d ovale Retorte *f*

1942 ELUTION APPARATUS, ELUTOR sin
f appareil *m* d'élution
e aparato *m* de elución
i apparecchio *m* d'eluzione
n uitwasapparaat *n*
d Auswaschgerät *n*

1943 ELUTRIATING APPARATUS, ELUTRIATOR min,sor
f appareil *m* d'élutriation
e aparato *m* de levigación, elutriador *m*
i apparecchio *m* di lavaggio a controcorrente, elutriatore *m*
n elueerapparaat *n*, opslibapparaat *n*
d Abschlämmgerät *n*, Schlämmgerät *n*

1944 ELUTRIATION TANK min,sor
f récipient *m* d'élutriation
e cuba *f* de elutriación
i serbatoio *m* d'elutriazione
n elueertank, opslibtank
d Abschlämmgefäss *n*, Schlämmgefäss *n*

1945 EMBOSSER, EMBOSSING CALENDER lea,pap,tex
f calandre *f* à gaufrer, calandre *f* de grainage, gaufreuse *f*
e calandra *f* engofradora, calandra *f* de gofrado, calandra *f* de relieve
i calandra *f* per goffratura, goffratrice *f*
n gaufreerkalander, preegkalander, prentkalander
d Gaufrierkalander *m*, Prägekalander *m*

1946 EMBOSSING ANVIL rub
f table *f* de gaufrage
e mesa *f* de engofrado, yunque *m* de embutir
i tavola *f* di goffratura
n gaufreertafel, preegtafel, prenttafel
d Prägetisch *m*

1947 EMBOSSING APPARATUS pho
f appareil *m* graveur en relief, appareil *m* de repoussage
e grabador *m* en relieve, repujador *m*
i apparecchio *m* da lavoro in rilievo
n reliëfdrukapparaat *n*
d Tiefprägegerät *n*

1948 EMBOSSING DIE pla,rub
f matrice *f* de gaufrage
e troquel *m*
i stampo *m* per goffratura
n gaufreerstempel, preegstempel, prentstempel
d Prägestempel *m*

1949 EMBOSSING PLATE lea,pap,tex
f plaque *f* de gaufrage
e placa *f* de engofrado
i piastra *f* di goffratura
n gaufreerplaat, preegplaat, prentplaat
d Prägeplatte *f*

1950 EMBOSSING ROLLER lea
f rouleau *m* de gaufrage
e rodillo *m* de engofrado, rodillo *m* estampador, rodillo *m* repujador
i rullo *m* di goffratura
n gaufreerrol, preegrol, prentrol
d Prägerolle *f*

1951 EMERY GRINDER, EMERY GRINDING MACHINE pol
f concasseur *m* d'émeri
e machacador *m* de esmeril
i rompitore *m* di smeriglio
n amarilmolen
d Schmirgelmühle *f*

1952 EMERY POLISHING MACHINE aap,pol
f machine *f* à meule d'émeri
e máquina *f* de muela de esmeril

i smerigliatrice *f*
n amarilslijpmachine, slijpmachine
d Schleifmaschine *f*, Schmirgelmaschine *f*, Schmirgelschleifmaschine *f*

1953 EMERY ROLL pol
f cylindre *m* à émeri
e cilindro *m* de esmeril
i cilindro *m* a smeriglio
n amarilcylinder
d Schmirgelwalze *f*

1954 EMERY ROLLER tex
f rouleau *m* émerisé
e rodillo *m* de liga, rodillo *m* avanzador de la tela
i rullo *m* smerigliato
n amarilrol, met amaril bedekte drijfrol
d Schmirgelrolle *f*

1955 EMERY WHEEL pol
f disque *m* à émeri, meule *f* à émeri
e disco *m* de esmeril, muela *f* de esmeril
i disco *m* a smeriglio, mola *f* a smeriglio
n amarilschijf
d Schmirgelscheibe *f*

1956 EMULSIFIER, EMULSIFYING APPARATUS pho,pla,rub
f émulgateur *m*, émulsionneuse *f*
e aparato *m* para emulsionar, emulsionador *m*, emulsor *m*
i apparecchio *m* per emulsionare
n emulgator, emulgeerapparaat *n*
d Emulgator *m*, Emulgiergerät *n*

1957 EMULSION HARDENING BATH pho
f bain *m* de durcissement de la émulsion
e baño *m* de endurecimiento de la emulsión
i bagno *m* d'indurimento dell'emulsione
n emulsiehardingsbad *n*
d Emulsionshärtungsbad *n*

1958 EMULSION SPRAYER pho
f pulvérisateur *m* d'émulsion
e rociador *m* de emulsión
i spruzzatore *m* d'emulsione
n emulsieverstuiver
d Emulsionzerstäuber *m*

1959 EMULSION TROUGH pho,pla,rub
f auge *f* à émulsion
e tolva *f* de emulsión
i serbatoio *m* d'emulsione
n emulsietrog
d Emulsionstrog *m*

1960 ENAMEL KILN, ENAMELLING FURNACE, DECORATION FIRING KILN cer,htt
f four *m* à émailler
e horno *m* de esmaltado, horno *m* de esmaltar
i forno *m* a smaltatura, forno *m* per smaltare
n emailleeroven
d Emaillierofen *m*

1961 ENAMEL LINING coa
f revêtement *m* d'émail
e guarnición *f* de esmalte
i guarnizione *f* di smalto
n emailbekleding, emailvoering
d Emailbedeckung *f*, Emailfutter *n*

1962 ENAMELLING PLANT coa
f émaillerie *f*
e equipo *m* para esmaltar
i impianto *m* da smaltare
n emailleerderij
d Emaillierereі *f*, Emaillierwerk

1963 ENCAPSULATED FUEL UNIT (US), SEALED-IN FUEL ELEMENT (GB) nue
f élément *m* combustible scellé
e elemento *m* combustible encamisado
i elemento *m* combustibile incamiciato
n ingekapseld splijtstofelement *n*
d eingehülstes Brennstoffelement *n*

1964 ENCLOSED-SCALE THERMOMETER lab
f thermomètre *m* à échelle protégée
e termómetro *m* de escala protegida
i termometro *m* a scala protetta
n thermometer met ingebouwde schaal
d Einschlussthermometer *n*

1965 END CELL elc
f élément *m* de réduction, élément *m* de régulation
e elemento *m* de reducción, elemento *m* de regulación
i elemento *m* di regolazione
n regelcel, schakelcel
d Endzelle *f*

1966 END RUNNER MILL brk,lab
f mortier *m* mécanique
e mortero *m* mecánico
i mortaio *m* meccanico
n mechanische vijzel
d mechanischer Mörser *m*

1967 END WALL gla
f pignon *m*
e pared *f* de frente
i parete *f* di fronte
n kopwand
d Stirnwand *f*

1968 ENDLESS SCREW, cpt
WORM
f vis *f* sans fin
e tornillo *m* sin fin
i vite *f* senza fine
n schroef zonder eind, worm
d endlose Schraube *f*, Schnecke *f*

1969 ENDLESS WIRE, pap
ENDLESS WIRE SCREEN
f toile *f* sans fin
e tela *f* sin fin
i staccio *m* continuo, tela *f* senza fine
n koperdoek *n* zonder einde, langzeef
d Metalltuch *n*

1970 ENDURANCE TESTING tes
APPARATUS,
FATIGUE TESTING APPARATUS
f appareil *m* d'essai de résistance à la fatigue
e aparato *m* de prueba de fatiga
i collaudatore *m* di fatica
n vermoeiingsbeproevings-apparaat *n*
d Dauerfestigkeitsprüfgerät *n*

1971 ENERGY ACCUMULATOR, pre
ACCUMULATOR
f accumulateur *m* d'énergie
e acumulador *m* de energía
i accumulatore *m* d'energia
n energieaccumulator
d Kraftspeicher *m*

1972 ENFLEURAGE PLANT per
f installation *f* d'enfleurage
e equipo *m* de enfleuraje
i impianto *m* d'estrazione di profumo
n enfleurage-installatie
d Parfümextraktionsanlage *f*

1973 ENRICHED AIR, htt
ENRICHED BLAST
f air *m* enrichi d'oxygène
e viento *m* enriquecido con oxígeno, viento *m* oxigenado
i aria *f* arricchita da ossigeno
n met zuurstof verrijkte lucht
d angereicherte Luft *f*

1974 ENRICHED REACTOR, nue
ENRICHED URANIUM REACTOR
f réacteur *m* enrichi
e reactor *m* enriquecido
i reattore *m* ad uranio arricchito
n verrijkte reactor
d angereicherter Reaktor *m*, Reaktor *m* mit angereichertem Brennstoff

1975 ENRICHING SECTION dis
f zone *f* d'enrichissement
e zona *f* de enriquecimiento
i zona *f* d'arricchimento
n aanrijkzone, verrijkingszone
d Anreicherungszone *f*

1976 ENTRAINMENT FILTER fil
f filtre *m* d'entraînement
e filtro *m* de arrastre
i filtro *m* di trascinamento
n meesleepfilter *n*
d Mitreissfilter *m*

1977 ENTRAINMENT TUBE boi,des
f tube *m* d'entraînement
e tubo *m* de arrastre
i tubo *m* di trascinamento
n meeneembuis
d Hochreissrohr *n*

1978 ENTRANCE OPENING tra
f orifice *m* d'arrivée, orifice *m* d'entrée
e abertura *f* de entrada
i apertura *f* d'entrata, bocca *f* d'entrata
n ingangsopening
d Eintrittsöffnung *f*

1979 ENVELOPE KILN, cer
SHUTTLE KILN, TOP-HAT KILN
f four *m* à recuire avec calotte protectrice
e horno *m* cerámico con caperuza de protección
i forno *m* ceramico con cappa di protezione
n kapselbakoven, kokerbakoven
d Kapselbrennofen *m*

1980 EPICYCLIC MIXER, PLANET WHEEL MIXER — mix
f mélangeur *m* épicycloïdal
e mezclador *m* de engrenajes planetarios
i mescolatore *m* a ruota epicicloidale
n planeetwielmenger
d Planetenrührwerk *n*

1981 EPITHERMAL REACTOR — nue
f réacteur *m* épithermique
e reactor *m* epitérmico
i reattore *m* epitermico
n epitherme reactor
d epithermer Reaktor *m*

1982 EQUALIZING RESERVOIR, BALANCING RESERVOIR, COMPENSATING RESERVOIR — iwt
f réservoir *m* compensateur, réservoir *m* régulateur du débit
e deposito *m* de equilibrio
i serbatoio *m* compensatore, serbatoio *m* regolatore
n compensatievat *n*
d Ausgleichbehälter *m*

1983 EQUILIBRIUM BALL VALVE — cpt
f soupape *f* de vidange à flotteur
e válvula *f* de manguito
i valvola *f* di scarico a galleggiante
n uitlaatklep met vlotter
d Schwimmausflussventil *n*

1984 EQUILIBRIUM CHAMBER — ins
f chambre *f* d'équilibre
e cámara *f* de equilibrio
i camera *f* d'equilibrio
n evenwichtskamer
d Gleichgewichtskammer *f*

1985 EQUILIBRIUM FLASH EVAPORATION PLANT — dis,fue
f installation *f* d'évaporation par détente en équilibre
e equipo *m* de evaporación instantánea en equilibrio
i impianto *m* d'evaporazione instantanea in equilibrio
n evenwichtsdestillatie-inrichting
d Gleichgewichtsdestillations-anlage *f*

1986 ERLENMEYER FLASK, FLAT-BOTTOMED CONICAL FLASK — lab
f matras *m* d'Erlenmeyer
e matraz *m* de Erlenmeyer
i bevuta *f*, matraccio *m* d'Erlenmeyer
n erlenmeyerkolf
d Erlenmeyerkolben *m*

1987 ESSICCATOR, DESICCATOR — dry,lab
f dessiccateur *m*, exsiccateur *m*
e desecador *m*, secador *m*
i essiccatoio *m*, essiccatore *m*
n drooginrichting, exsiccator
d Exsikkator *m*, Trockengefäss *n*

1988 ESTERIFICATION TANK — fat
f récipient *m* d'estérification
e tanque *m* de esterificación
i serbatoio *m* d'esterificazione
n veresteringstank
d Veresterungsgefäss *n*

1989 ETCHING BATH — elc,met
f bain *m* d'attaque chimique, bain *m* de décapage
e baño *m* de ataque químico, baño *m* mordentador
i bagno *m* d'attacco chimico, bagno *m* di decapaggio
n etsbad *n*
d Ätzbad *n*

1990 ETCHING PITS, PITS — elc,met
f piqûres *pl*
e hoyos *pl* de ataque químico, picaduras *pl*
i figure *pl* d'attacco
n putjes *pl*
d Narben *pl*, Poren *pl*

1991 ETCHING TEST APPARATUS — tes
f appareil *m* d'essai pour le décapage
e aparato *m* de prueba por el ataque químico
i apparecchio *m* di prova per il decapaggio
n beitstestapparaat *n*
d Ätzprobiergerät *n*

1992 ETHER EXTRACTION APPARATUS — lab,oic
f appareil *m* extracteur d'éther
e aparato *m* extractor de éter
i apparecchio *m* estrattore d'etere
n etherextractietrechter
d Schütteltrichter *m* zum Ausäthern

1993 ETHERIFICATION PLANT — oic,tex
f installation *f* d'éthérification

e equipo *m* de eterificación
i impianto *m* d'eterificazione
n ethervormingsinstallatie
d Ätherbildungsanlage *f*

1994 ETHYLATION PLANT oic
f installation *f* d'éthylation
e equipo *m* de etilación
i impianto *m* d'etilazione
n ethyleerinrichting
d Ethylieranlage *f*

1995 EUDIOMETER ins
f eudiomètre *m*
e eudiómetro *m*
i eudiometro *m*
n eudiometer
d Eudiometer *n*

1996 EVAPORATING DISH eva,lab
f capsule *f* d'évaporation
e cápsula *f* de evaporación
i capsula *f* d'evaporazione
n indampschaaltje *n*
d Abdampfschale *f*,
Verdampfschale *f*

1997 EVAPORATING PAN dry,eva
f cuve *f* d'évaporation
e cuba *f* de evaporación
i tino *m* d'evaporazione
n verdampvat *n*
d Verdampfungsgefäss *n*

1998 EVAPORATING PLANT dry,eva
f installation *f* d'évaporation
e equipo *m* evaporador
i impianto *m* d'evaporazione
n verdampinstallatie
d Verdampferanlage *f*

1999 EVAPORATION APPARATUS dry,eva
f appareil *m* d'évaporation
e aparato *m* de evaporación
i apparecchio *m* d'evaporazione
n indampapparaat *n*, inkook-
apparaat *n*
d Einkochapparat *m*

2000 EVAPORATION AREA eva,htt
f aire *f* d'évaporation
e área *f* de evaporación
i area *f* d'evaporazione
n verdampingsoppervlak *n*
d Verdunstungsfläche *f*

2001 EVAPORATION PONDS, SALINE, SALT PANS eva,sal
f marais *pl* salants, saline *f*
e salina *f*
i salina *f*
n zouttuin
d Meersaline *f*, Salzgarten *m*

2002 EVAPORATIVE CENTRIFUGE cen
f centrifugeur *m* à évaporation
e centrífuga *f* de evaporación
i centrifuga *f* ad evaporazione
n verdampingscentrifuge
d Verdampfungszentrifuge *f*

2003 EVAPORATOR, VAPORIZER dry,eva
f évaporateur *m*
e desecador *m*, evaporador *m*,
vaporizador *m*
i evaporatore *m*
n verdamper, verdamplichaam *n*
d Verdampfer *m*

2004 EVAPORATOR COIL eva
f serpentin *m* d'évaporateur
e serpentín *m* de evaporador
i serpentina *f* d'evaporatore
n verdamperkoelbuizen *pl*
d Kühlschlangensystem *n*

2005 EVAPORATOR CONDENSER clo
f condensateur *m* de l'évaporateur
e condensador *m* del evaporador
i condensatore *m* dell'evaporatore
n verdampercondensor
d Verdampferkondensator *m*

2006 EVAPORATOR CRYSTALLIZER, CRYSTALLIZING EVAPORATOR cry
f cristallisoir *m* à évaporation
e cristalizador *m* de evaporación
i cristallizzatore *m* ad evaporazione
n verdampingskristallisatie-
apparaat *n*
d Verdampfungskristallisator *m*

2007 EVAPORATOR PREHEATER eva
f préchauffeur *m* d'évaporateur
e prerrecalentador *m* de
evaporador
i preriscaldatore *m* d'evaporatore
n verdampervoorverhitter
d Verdampfervorheizanlage *f*

2008 EVAPORATOR TOWER eva
f tour *f* d'évaporation
e torre *f* de evaporación
i torre *f* d'evaporazione
n verdampingstoren
d Verdampfersäule *f*

2009 EVAPORATOR TUBE dry,eva
f tube *m* d'évaporateur
e tubo *m* de evaporador
i tubo *m* d'evaporatore
n verdamperbuis
d Verdampferrohr *n*

2010 EVAPORIMETER, ATMIDOMETER, ATMOMETER ins
f atmidomètre *m*, atmomètre *m*, évaporimètre *m*
e atmidímetro *m*, evaporímetro *m*
i atmidometro *m*, evaporimetro *m*
n atmidometer, atmometer, verdampingsmeter
d Atmomesser *m*, Dunstmesser *m*, Verdampfungsmesser *m*, Verdunstungsmesser *m*

2011 EVENER ROLL, SPIKE ROLL tex
f cylindre *m* égaliseur
e cilindro *m* igualador, cilindro *m* de puntas
i cilindro *m* a denti
n getande rol
d Zahnrolle *f*

2012 EXCESS-ACETYLENE BURNER bur
f brûleur *m* réducteur
e quemador *m* reductor
i bruciatore *m* riducente
n reducerende brander
d Reduktionsbrenner *m*

2013 EXCESS-OXYGEN BURNER bur
f brûleur *m* oxydant
e quemador *m* oxidante
i bruciatore *m* ossidante
n oxyderende brander
d Oxydationsbrenner *m*

2014 EXCESS PRESSURE REGULATOR sap,pre
f régulateur *m* de surpression
e regulador *m* de sobrepresión
i regolatore *m* di soprapressione
n overdrukregelinrichting
d Überdruckregler *m*

2015 EXCESS RUBBER OVERFLOW, SPEW, SPUE rub
f trop-plein *m*
e rebosamiento *m* del molde
i troppo pieno *m*
n overloop
d Austrieb *m*, Formaustrieb *m*

2016 EXCHANGER gen
f échangeur *m*
e canjeador *m*, intercambiador *m*
i scambiatore *m*
n uitwisselaar
d Austauscher *m*

2017 EXHAUST DUCT cpt
f conduite *f* de désaération
e conducto *m* de exhaustación, tubería *f* de aspiración
i condotta *f* di sfiato
n ontluchtingsleiding
d Entlüftungsleitung *f*

2018 EXHAUST GAS ANALYZER ana
f analyseur *m* des gaz d'échappement
e analizador *m* de los gases de escape
i analizzatore *m* dei gas di scappamento
n uitlaatgasanalyseapparaat *n*
d Auspuffgasprüfgerät *n*

2019 EXHAUST HOSE cpt
f tuyau *m* d'échappement
e tubo *m* de escape
i tubo *m* di scarico
n afblaasslang
d Ausblaseschlauch *m*

2020 EXHAUST MANIFOLD cpt
f tube *m* collecteur du gaz d'échappement
e colector *m* del gas de escape
i collettore *m* del gas di scappamento
n uitlaatgassenleiding
d Auspuffleitung *f*

2021 EXHAUST MUFFLER (US), EXHAUST SILENCER (GB), MUFFLER (US), SILENCER (GB cpt
f pot *m* de détente, silencieux *m* d'échappement
e silenciador *m* del escape
i marmitta *f* di scarico, silenziatore *m*
n knalpot
d Auspufftopf *m*, Schalldämpfer *m*

2022 EXHAUST PIPE, OUTLET VENT, VENT PIPE cpt
f tuyau *m* d'évent
e tubo *m* de escape
i sfiatoio *m*, tubo *m* di sfiato
n afblaaspijp
d Abblasrohr *n*, Dunstrohr *n*

2023 EXHAUST PORT cpt
f lumière *f* d'échappement
e lumbrera *f* de escape, orificio *m* de escape
i luce *f* di scarico, portello *m* di scarico
n uitlaatopening
d Auspufföffnung *f*, Auspuffschlitz *m*

2024 EXHAUST STACK hyg
f carneau *m*, cheminée *f* de décharge
e chimenea *f* de descarga
i camino *m* di scarico
n afvoerschoorsteen
d Abzugschornstein *m*

2025 EXHAUST VALVE sap
f soupape *f* d'échappement
e válvula *f* de escape
i valvola *f* di scappamento
n uitlaatklep
d Auspuffventil *n*

2026 EXHAUSTER gas,hyg
f dispositif *m* d'aspiration, ventilateur *m* aspirant
e dispositivo *m* de evacuación, exhaustor *m*
i aspiratore *m*
n afzuiginrichting, exhauster
d Absaugevorrichtung *f*, Exhauster *m*

2027 EXHAUSTING FAN, AIR EXHAUSTER hyg
f ventilateur *m* aspirant
e expulsor *m* de aire
i espulsore *m* d'aria
n luchtuitblaasinrichting
d Entlüftungsgebläse *n*

2028 EXPANDER, EXPANSION MACHINE pre
f machine *f* à expansion
e máquina *f* de expansión
i macchina *f* ad espansione
n expansiemachine
d Expansionsmaschine *f*

2029 EXPANDING MANDRIL rub
f mandrin *m* extensible
e mandril *m* extensible
i mandrino *m* contraibile, mandrino *m* estensibile
n kalibreerdoorn, uitzetbare doorn, uitzetbare opspanstang
d ausziehbarer Dorn *m*, Expansionsdorn *m*, Spreizdorn *m*

2030 EXPANDING ROLLER (US), STRETCHER ROLLER (GB) rub
f galet *m* tendeur, rouleau *m* déplisseur
e ensanchador *m*, rodillo *m* extensor, rodillo *m* tensor
i rullo *m* di tensione, rullo *m* estensore, stenditore *m*
n spanrol, strekrol
d Breithalter *m*, Spannrolle *f*

2031 EXPANSION JOINT cpt
f joint *m* de dilatation
e junta *f* de dilatación
i giunto *m* di dilatazione
n expansiekoppeling
d Dehnungsfuge *f*, Dehnungsstoss *m*, Dilatationsstoss *m*, Expansionsverbindung *f*

2032 EXPELLER fat
f presse *f* hélicoïdale
e almazara *f*, prensa *f* de tornillo
i pressa *f* elicoidale
n schroefpers
d Schneckenpresse *f*

2033 EXPERIMENTAL PLANT, PILOT PLANT tes
f installation *f* d'essai, installation *f* expérimentale
e equipo *m* experimental
i impianto *m* pilota, impianto *m* sperimentale
n proefinstallatie
d Versuchsanlage *f*

2034 EXPLOSION BOMB lab
f bombe *f* calorimétrique
e bomba *f* calorimétrica
i bomba *f* calorimetrica
n calorimeterbom
d Kalorimeterbombe *f*

2035 EXPLOSION VALVE sap
f clapet *m* d'explosion
e válvula *f* antiexplosión
i valvola *f* d'esplosione
n explosieklep
d Explosionsklappe *f*

2036 EXPLOSIVE PRESSING MACHINE met,pre
f machine *f* à presser par explosion
e máquina *f* de prensado por explosión
i macchina *f* da pressare per esplosione

n explosiepers
d Explosionspresse *f*

2037 EXPLOSIVE TESTER tes
f appareil *m* d'essai d'explosion
e probador *m* de explosivos
i apparecchio *m* di prova d'esplosione
n springstoftestapparaat *n*
d Sprengstoffprüfgerät *n*

2038 EXPLOSIVE VAPOR DETECTOR (US), EXPLOSIVE VAPOUR DETECTOR (GB) ins,sap
f détecteur *m* de vapeurs explosives
e detector *m* de vapores explosivos
i rivelatore *m* di vapori esplosivi
n explosieve-gassenverklikker
d Explosivgasanzeiger *m*

2039 EXPLOSIVES FACTORY exp
f fabrique *f* d'explosifs
e fábrica *f* de explosivos
i fabbrica *f* d'esplosivi
n springstoffenfabriek
d Sprengstoffabrik *f*

2040 EXTENDED-SURFACE COIL, FIN-TYPE HEATING COIL, FINNED-TUBE HEATING COIL, SPIRAL FIN-TUBE HEATING COIL cle,hyg
f réchauffeur *m* d'air en tubes ailetés
e calentador *m* de aire con tubos de aletas
i riscaldatore *m* d'aria con tubi ad alette
n ribbenbuisluchtverhitter
d Bandrippenrohrheizregister *n*

2041 EXTENSOMETER ins
f extensomètre *m*
e extensímetro *m*
i estensimetro *m*
n rekmeter
d Dehnungsmesser *m*

2042 ESTINGUISHER, FIRE EXTINGUISHER fea
f extincteur *m*
e apagador *m*, extintor *m* de incendios
i estintore *m*
n blusapparaat *n*, brandblusapparaat *n*
d Feuerlöscher *m*, Löscher *m*

2043 EXTRACTION APPARATUS, EXTRACTOR ext
f appareil *m* d'extraction
e aparato *m* de extracción
i apparecchio *m* d'estrazione
n extractieapparaat *n*, extraheerapparaat *n*
d Extrahiergerät *n*, Extraktionsgerät *n*

2044 EXTRACTION BATTERY ext
f batterie *f* d'appareils d'extraction
e batería *f* de aparatos de extracción
i gruppo *m* d'apparecchi d'estrazione
n extractiegroep, extraheerbatteri
d Extrahiergruppe *f*, Extraktionsgruppe *f*

2045 EXTRACTION COLUMN ext
f colonne *f* d'extraction
e columna *f* de extracción
i colonna *f* d'estrazione
n extractiekolom, extraheerkolom
d Extrahiersäule *f*, Extraktionssäule *f*

2046 EXTRACTION CUP WITH SIPHON ext
f panier *m* en verre pour extraction
e cesto *m* extractor con sifón
i cesta *f* estrattrice con sifone
n extractie-inzetstuk *n* met sifon
d Extraktionseinsatz *m* mit Heberrohr

2047 EXTRACTION FLASK ext
f fiole *f* à extraction
e matraz *m* de extracción
i matraccio *m* ad estrazione
n extractiekolf
d Extraktionskolben *m*

2048 EXTRACTION FUNNEL, SEPARATING FUNNEL lab
f ampoule *f* à décanter, entonnoir *m* à séparation
e embudo *m* de extracción
i imbuto *m* separatore
n scheitrechter
d Scheidetrichter *m*

2049 EXTRACTION RACK lab
f support *m* d'extraction
e bastidor *m* de extracción
i cremagliera *f* d'estrazione
n extractierek *n*, extraheerrek *n*

d Extrahiergestell *n*,
Extraktionsgestell *n*

2050 EXTRACTION THIMBLE, FILTER CUP lab
f cartouche *m* à extraction, manchon *m* poreux
e capa *f* porosa de extracción
i inviluppo *m* poroso
n poreuse huls
d Extraktionseinsatz *m*, poröse Hülle *f*

2051 EXTRACTIVE DISTILLATION COLUMN dis,ext
f colonne *f* de distillation extractive
e columna *f* de destilación extractiva
i colonna *f* di distillazione estrattiva
n extractieve-destillatiekolom
d Extraktivdestillationssäule *f*

2052 EXTRACTIVE DISTILLATION PLANT dis,ext
f installation *f* de distillation extractive
e equipo *m* de destilación extractiva
i impianto *m* di distillazione estrattiva
n extractieve-destillatie-inrichting
d Extraktivdestillationsanlage *f*

2053 EXTRACTOR bre
f corps *m* du plongeur
e aparato *m* de tiraje
i apparecchio *m* a spina
n aansteekhuis *n*
d Anstichapparat *m*

2054 EXTRUDED ANODE elc
f anode *f* extrudée
e ánodo *m* extruído
i anodo *m* estruso
n gespoten anode
d stranggepresste Anode *f*

2055 EXTRUDER, EXTRUDING MACHINE, FORCING MACHINE pla
f boudineuse *f*
e budinadora *f*
i trafila *f*
n spuitmachine
d Spritzmaschine *f*, Strangpresse *f*

2056 EXTRUDER BAND HEATER, BAND HEATER pla,rub
f élément *m* de chauffage annulaire
e calentador *m* de cinta
i riscaldatore *m* a nastro avvolto
n ringvormig verwarmings-element *n*
d Bandheizkörper *m*

2057 EXTRUDER CORE, EXTRUSION CORE pla,rub
f âme *f* de tête de boudineuse, poinçon *m*
e alma *f* de cabezal de budinadora
i spina *f* di trafila
n spuitkopdoorn
d Spritzdorn *m*

2058 EXTRUDER DIE, EXTRUSION DIE pla,rub
f filière *f* de boudineuse
e boquilla *f* de budinadora
i bocchettone *m* di trafila
n spuitmatrijs, spuitmondstuk *n*
d Spritzmatrize *f*, Spritzmundstück *n*

2059 EXTRUDER DISCHARGE HEAD, DISCHARGE HEAD pla
f tête *f* de boudineuse
e cabezal *m* de budinadora
i bocchettone *m* di trafila
n spuitkop
d Spritzkopf *m*

2060 EXTRUDER-HEAD CORE, CORE OF EXTRUDER HEAD pla
f âme *f* de tête de boudineuse
e ánima *f* de cabezal de budinadora
i anima *f* di testa di trafila
n spuitkopdoorn
d Spritzdorn

2061 EXTRUSION MANDRIL pla,rub
f poinçon *m* d'extrusion
e mandril *m* de budinadora
i mandrino *m* di trafila
n spuitdoorn
d Spritzdorn *m*

2062 EXTRUSION PLASTOMETER ins,pla
f plastomètre *m* d'extrusion
e plastómetro *m* de extrusión
i plastometro *m* d'estruzione
n extrusieplastometer
d Ausflussplastometer *n*

2063 EXTRUSION PLUNGER pla,rub
f piston *m* d'extrusion
e pistón *m* de budinadora
i pistone *m* di trafila
n spuitplunjer
d Strangpresskolben *m*

2064 EYE, gla
POT FURNACE EYE
f trou *m* d'admission de gaz combustibles
e mirilla *f*
i orificio *m* per gas combustibili
n ovengat *n*
d Ofenloch *n*

F

2065 FABRIC DUST COLLECTOR cle,fil,gas
f collecteur *m* de poussière en tissu
e colector *m* de polvo de tela
i collettore *m* di polvere in tessuto
n stofverzamelaar uit textiel
d Stoffstaubsammler *m*

2066 FABRIC EXPANDER ROLL tex
f cylindre-étireur *m* de tissus
e cilindro *m* estirador de tejidos
i cilindro *m* per stirare tessuti
n weefselstrekwals
d Gewebestreckwalze *f*

2067 FACTORY, MILL, WORKS gen
f établissement *m*, fabrique *f*, manufacture *f*, usine *f*
e fábrica *f*
i fabbrica *f*, stabilimento *m*
n fabriek
d Fabrik *f*, Werk *n*

2068 FADEOMETER, FADING TESTER ins,tes
f fadéomètre *m*
e fadeómetro *m*, medidor *m* de descoloración
i fadeometro *m*, misuratore *m* di decolorazione
n fadeometer, lichtechtheidsbeproevingsapparaat *n*
d Fadeometer *n*, Lichtechtheitsprüfgerät *n*

2069 FALLING-BALL VISCOSIMETER (GB), FALLING-SPHERE VISCOSIMETER (US) ins
f viscosimètre *m* à chute de bille
e viscosímetro *m* de bola descendente
i viscosimetro *m* a sfera cadente
n viscosimeter met vallende kogel
d Kugelfallviskosimeter *n*

2070 FALLING-FILM COLUMN, FALLING-FILM TOWER dis
f colonne *f* à film
e columna *f* de película
i colonna *f* a film fluente
n zuil met dalende film
d Fallfilmkolonne *f*

2071 FALLING-FILM STILL dis
f appareil *m* de distillation à couche descendante
e aparato *m* de destilación de película bajante
i apparecchio *m* di distillazione con pellicola descendente
n destilleerinstallatie met dalende vloeistoffilm
d Destillieranlage *f* mit herabsinkender Flüssigkeitsschicht, Fallfilmdestillationsapparat *m*

2072 FALLING-WEIGHT APPARATUS, DROP TEST APPARATUS exp,tes
f mouton *m*, mouton-pendule *m*
e probador *m* de caída de peso
i collaudatore *m* a caduta
n valhamer
d Fallhammer *m*

2073 FALSE BOTTOM, STRAINER bre
f faux fond *m*
e falso fondo *m*
i fondo *m* falso, fondo *m* filtrante
n bodem van de klaringskuip
d Läuterboden *m*, Seihboden *m*, Senkboden *m*

2074 FAN soa
f palette *f*
e paleta *f*
i paletta *f*
n schoep
d Schaufel *f*

2075 FAN, BLOWER, VENTILATOR clo
f ventilateur *m*
e ventilador *m*
i ventilatore *m*
n ventilator
d Lüfter *m*, Ventilator *m*

2076 FAN GATE pla,pre
f entrée *f* en éventail
e entrada *f* de abanico
i entrata *f* in ventaglio
n waaieraanspuitpoort
d Fächeranguss *m*

2077 FANTAIL BURNER, FISHTAIL BURNER bur,lab
f bec *m* de gaz en éventail

e mechero *m* de mariposa
i bruciatore *m* a fiamma in forma di ventaglio
n waaierbrander
d Fächerbrenner *m*

2078 FANTAIL DIE pla
f filière *f* plate
e boquilla *f* en cola de pez
i filiera *f* a ventaglio
n conisch verlopende spleetmatrijs
d Fischschwanzdüse *f*

2079 FAST BURNING FUSE exp
f mèche *f* à combustion rapide
e mecha *f* de combustión rápida
i miccia *f* a combustione rapida
n snelle lont
d Schnellzündschnur *f*

2080 FAT HARDENING PLANT, fat
FAT HYDROGENATION PLANT
f installation *f* d'hydrogénation d'huile
e equipo *m* de hidrogenación de grasa
i impianto *m* d'idrogenazione di grasso
n vethardingsinstallatie
d Fetthärtungswerk *n*, Ölhärtungsanlage *f*

2081 FAT-LIQUORING DRUM tan
f tambour *m* pour le graissage du cuir, tambour *m* pour l'huilage du cuir
e tambor *m* de impregnación del cuero con aceite
i tamburo *m* d'impregnazione del cuoio con olio
n vetlooitrommel
d Fettgerbebottich *m*

2082 FAT RECOVERY PLANT, fat,oic
GREASE RECOVERY PLANT
f installation *f* de récupération de graisse
e equipo *m* de recuperación de grasa
i impianto *m* di ricupero di grasso
n vetterugwinningsinstallatie
d Fettrückgewinnungsanlage *f*

2083 FAT TRAP, cle,fat
GREASE SEPARATOR,
GREASE TRAP
f tôle *f* à retenir la graisse
e interceptor *m* de grasa
i eliminatore *m* di grasso
n vetvanger
d Fettabscheider *m*, Fettfänger *m*

2084 FATIGUE TESTING APPARATUS, tes
ENDURANCE TESTING APPARATUS
f appareil *m* d'essai de résistance à la fatigue
e aparato *m* de prueba de fatiga
i collaudatore *m* di fatica
n vermoeiingsbeproevingsapparaat *n*
d Dauerfestigkeitsprüfgerät *n*

2085 FAURE PLATE, elc
PASTED PLATE
f plaque *f* à oxyde rapporté, plaque *f* Faure
e placa *f* de óxido empastado
i piastra *f* ad ossido riportato
n roosterplaat
d pastierte Platte *f*

2086 FEED, gen
FEEDER
f alimentateur *m*, dispositif *m* d'alimentation
e alimentador *m*, dispositivo *m* de alimentación
i alimentatore *m*, dispositivo *m* d'alimentazione
n aanvoerinrichting, voedingsapparaat *n*
d Speisevorrichtung *f*, Zufuhr *f*

2087 FEED, pla
LOAD, MOLD CHARGE (US),
MOULD CHARGE (GB)
f chargement *m*
e carga *f*
i carica *f*
n vulling
d Beschickung *f*

2088 FEED BUSH, pla
SPRUE BUSH
f buse *f* du pot d'alimentation, busette *f*, cheminée *f* du moule, douille *f* de carotte
e manguito *m* del bebedero
i bocchello *m*
n aanspuitbus
d Angussbuchse *f*

2089 FEED-CHECK VALVE boi,pre
f soupape *f* d'arrêt de l'alimentation
e válvula *f* de retorno de alimentación, válvula *f* reguladora de cierre automático

i valvola *f* di ritegno dell'alimentazione
n voedingsafsluitklep
d Speiseabsperrventil *n*

2090 FEED CYLINDER pla
f cylindre *m* d'injection, pot *m* d'injection
e cilindro *m* de inyección
i cilindro *m* d'iniezione
n vulcilinder
d Füllzylinder *m*

2091 FEED FUNNEL, FEED HOPPER pla
f entonnoir *m* d'alimentation, trémie *f* d'alimentation, trémie *f* de chargement
e embudo *m* de alimentación, tolva *f* de alimentación
i tramoggia *f* d'alimentazione, tramoggia *f* di carica
n voedingstrechter, vultrechter
d Beschickungstrichter *m*, Einfülltrichter *m*

2092 FEED HOPPER WITH SLOT DIE pla
f fente *f* de coulée
e tolva *f* de alimentación con ranura de colada
i filiera *f* di colata
n gietapparaat *n* met spleet-opening
d Schlitzgiesser *m*

2093 FEED MILL pla
f cylindre *m* réchauffeur
e cilindro *m* precalentador
i cilindro *m* preriscaldatore
n opwarmwals
d Vorwärmwalze *f*

2094 FEED ORIFICE (GB), GAT (US) pla
f entrée *f* d'injection
e entrada *f*
i apertura *f* d'entrata
n aanspuitopening
d Anschnitt *m*

2095 FEED PIPE boi,pre
f tube *m* d'alimentation
e tubo *m* alimentador
i tubo *m* d'alimentazione
n toevoerbuis, voedingsbuis
d Speiserohr *n*, Zuflussrohr *n*

2096 FEED PUMP pre
f pompe *f* d'alimentation
e bomba *f* alimentadora
i pompa *f* d'alimentazione
n voedingspomp
d Beschickungspumpe *f*

2097 FEED REGULATOR, FEED TIMER gen
f régulateur *m* de l'alimentation
e regulador *m* de la alimentación
i regolatore *m* dell'alimentazione
n voedingsregelaar
d Speiseregler *m*

2098 FEED ROLLER pla,rub,sin
f rouleau *m* alimentateur
e rodillo *m* alimentador
i rullo *m* alimentatore
n toevoerrol, voedingsrol
d Speiserolle *f*, Zufuhrrolle *f*

2099 FEED SLIDE pla
f vanne *f* doseuse
e compuerta *f* de dosificación
i cassetto *m* dosatore
n doseerschuif
d Dosierschieber *m*

2100 FEED VALVE, FILLING VALVE cpt
f soupape *f* de remplissage
e válvula *f* de llenado
i valvola *f* di riempimento
n vulklep
d Füllventil *n*

2101 FEED-WATER HEATER boi,pre
f réchauffeur *m* de l'eau d'alimentation
e recalentador *m* del agua de alimentación
i preriscaldatore *m* dell'acqua d'alimentazione
n voedingswatervoorwarmer
d Speisewasservorwärmer *m*

2102 FEEDING HEAD gen
f tête *f* de remplissage
e boca *f* de alimentación
i testa *f* d'alimentazione
n vulopening
d Einfüllkopf *m*, Füllöffnung *f*

2103 FEELER CONTROL, TRACER CONTROL con
f commande *f* à palpeur
e accionamiento *m* por palpador
i azionamento *m* con tastatore
n aftastbesturing
d Fühlersteuerung *f*

2104 FELT DRIER pap
f feutre *m* sécheur
e secador *m* de fieltro
i essiccatore *m* a feltro
n droogvilt *n*
d Filztrockner *m*

2105 FELT FABRIC DUST COLLECTOR cle,fil,gas
f collecteur *m* de poussière en feutre
e colector *m* de polvo de fieltro
i collettore *m* di polvere in feltro
n vilten stofverzamelaar
d Filzstaubsammler *m*

2106 FELTING MILL tex
f machine *f* de feutrage
e máquina *f* de enfieltrado, máquina *f* de fieltramiento
i macchina *f* di feltratura
n viltmachine *f*
d Filzmaschine *f*

2107 FELTING SCREEN, PULP MOLD (US), PULP MOULD (GB) pap,pla
f moule *m* à grille d'aspiration
e molde *m* de succión
i telaio *m* d'aspirazione
n pulppersmal
d Saugform *f*

2108 FEMALE MOLD (US), FEMALE MOULD (GB), NEGATIVE DIE pla,rub
f moule *m* fémelle, moule *m* négatif
e molde *m* hembra, molde *m* negativo
i stampo *m* femmina
n negatieve mal
d Negativform *f*

2109 FERMENTATION CAP bre,fmi
f coiffe *f* de fermentation
e capacete *m* de fermentación
i cappa *f* di fermentazione
n fermentatiekap, gistkap
d Gärkappe *f*, Mostkappe *f*, Süssmostkappe *f*

2110 FERMENTATION CELLAR, FERMENTING CELLAR bre,fmi
f cave *f* de fermentation
e bodega *f* de fermentación
i cantina *f* di fermentazione
n gistkelder
d Gärkeller *m*

2111 FERMENTATION INDUSTRY fmi
f industrie *f* de fermentation
e industria *f* de fermentación
i industria *f* di fermentazione
n fermenteringsindustrie
d Gärungsgewerbe *n*

2112 FERMENTATION TANK, FERMENTER fmi
f cuve *f* de fermentation
e cuba *f* de fermentación
i tino *m* di fermentazione
n gistkuip
d Gärbottich *m*, Gärungsküpe *f*

2113 FERMENTATION TANK, RAG FERMENTATION TANK, ROTTING TANK, STUFF VAT pap
f récipient *m* de pourrissage
e pudridor *m*
i serbatoio *m* di putrefazione
n ontsluitingstank, rottingsvat *n*
d Faulbütte *f*

2114 FERMENTATION TUBE lab
f tube *m* de fermentation
e tubo *m* de fermentación
i tubo *m* di fermentazione
n gistingsbuisje *n*
d Gärröhrchen *n*

2115 FERMENTATION VAT pap
f cuve *f* de macération
e macerador *m*
i tino *m* di macerazione
n lompenbak, stamptrog
d Mazerationsbütte *f*

2116 FERMENTATION VESSEL cer
f récipient *m* de pourrissage
e tina *f* de putrefacción
i tino *m* di putrefazione
n rottingskuip
d Maukgefäss *n*

2117 FERMENTING TANK, FERMENTING TUB, FERMENTING VAT bre
f minion *m*
e vasija *f* de fermentación
i tino *m* di fermentazione
n gistvat *n*
d Gärbütte *f*

2118 FERMENTING TUB ATTEMPERATOR bre,fmi
f refroidisseur *m* de cuve de fermentation
e serpentín *m* de enfriamiento

i serpentina *f* di raffreddamento
n gistkuipkoeler
d Gärbottichkühler *m*

2119 FERTILIZER FACTORY fer
f manufacture *f* d'engrais chimiques
e fábrica *f* de abonos
i fabbrica *f* di concimi chimici
n kunstmestfabriek
d Kunstdüngerfabrik *f*

2120 FESTOON DRIER (GB), pho,pla
LOOP DRIER (US)
f accrocheuse-sécheuse *f*, séchoir *m* à feston, séchoir *m* à guirlandes, séchoir *m* à plis supportés
e secador *m* de cinta colgante, secador *m* de festón, secador *m* por tendido ondulado
i essiccatore *m* a festoni
n droogtunnel met guirlande-inrichting
d Hängetrockner *m*, Schleifentrockner *m*

2121 FESTOON STEAMER tex
f appareil *m* de traitement de plis supportés à vapeur
e aparato *m* de tratamiento de festones con vapor
i apparecchio *m* di trattamento di festoni con vapore
n stoombehandelingsapparaat *n* met guirlande-inrichting
d Schleifendampfbehandlungsanlage *f*

2122 FETTLING TOOL, cer
SCRAPING TOOL
f couteau *m* racleur
e cuchillo *m* rascador
i coltello *m* da raschiare
n afschrapmes *n*
d Abschrotmesser *n*

2123 FIBER DUSTING PLANT (US), coa
FIBRE DUSTING PLANT (GB)
f installation *f* de flocage
e equipo *m* para cubrir con fibras
i impianto *m* distributore di fibre
n vezelverstuivingsinstallatie
d Beflockungsanlage *f*

2124 FIBER RECOVERING PLANT (US), pap
FIBRE RECOVERING PLANT (GB)
f récupérateur *m* de fibres
e recuperador *m* de fibras
i ricuperatore *m* di fibre
n vezelvanger
d Fasernfänger *m*

2125 FILL AND DRAW CONTACT BED, iwt
CONTACT BED, CONTACT FILTER
f lit *m* de contact
e lecho *m* de contacto
i letto *m* di contatto
n contactbed *n*
d Füllkörper *m*

2126 FILLER PLATE (US), pla
LOADING SHOE (GB)
f lunette *f*, plaque *f* de remplissage
e placa *f* independiente de carga
i piastra *f* di carica
n vulplaat
d Füllplatte *f*

2127 FILLER ROD, pla,wel
WELDING ROD
f baguette *f* d'apport
e varilla *f* de aportación
i bacchetta *f* di rapporto
n oplasstaaf
d Aufschweissstab *m*

2128 FILLING MACHINE gen
f machine *f* à remplir, remplisseuse *f*
e máquina *f* de rellenar
i macchina *f* da riempire
n vulmachine
d Abfüllmaschine *f*

2129 FILLING RINGS, dis
RASCHIG RINGS
f anneaux *pl* de Raschig, anneaux *pl* de remplissage
e anillos *pl* de Raschig, anillos *pl* llenadores
i anelli *pl* di Raschig, anelli *pl* di riempimento
n raschigringen *pl*, vulringen *pl*
d Füllringe *pl*, Raschigringe *pl*

2130 FILLING SHOVEL gla
f palette *f*
e pala *f*
i pala *f*
n gemengschop
d Schaufel *f*

2131 FILLING SPACE (US), pla
LOADING CHAMBER (GB), LOADING SPACE (US), LOADING WELL (GB)

f chambre *f* de charge, chambre *f* de compression
e cámara *f* de carga, cámara *f* de compresión
i camera *f* di carica, camera *f* di compressione
n vulruimte
d Füllraum *m*

2132 FILLING VALVE, FEED VALVE cpt
f soupape *f* de remplissage
e válvula *f* de llenado
i valvola *f* di riempimento
n vulklep
d Füllventil *n*

2133 FILM BADGE ins,nue,sap
f film-témoin *m*, plaquette *f* pelliculaire
e película *f* dosimétrica
i pellicola *f* dosimetrica
n filmplak
d Filmdosimeter *n*, Filmplakette *f*

2134 FILM CASTING MACHINE pho,pla
f machine *f* de coulée des pellicules
e máquina *f* de colada de películas
i macchina *f* da colare pellicole
n filmgietmachine
d Filmgiessmaschine *f*

2135 FILM-PACK-TYPE COOLING TOWER clo
f tour *f* de refroidissement à couche d'eau
e torre *f* de enfriamiento con capa de agua
i torre *f* di raffreddamento a strato d'acqua
n koeltoren met waterfilm
d Wasserschichtkühlturm *m*

2136 FILM-TYPE ABSORBER abs
f absorbeur *m* à couche mince
e absorbedor *m* de película
i assorbitore *m* a strato sottile
n filmabsorptieapparaat *n*
d Dünnschichtabsorber *m*

2137 FILM-TYPE EVAPORATOR eva
f évaporateur *m* à couche mince
e evaporador *m* de película
i evaporatore *m* a strato sottile
n filmverdamper
d Dünnschichtverdampfer *m*

2138 FILTER fil,gen
f filtre *m*
e filtro *m*
i filtro *m*
n filter *n*
d Filter *m*

2139 FILTER AREA, FILTER SURFACE fil
f aire *f* de filtrage, surface *f* du filtre
e área *f* filtradora, superficie *f* del filtro
i area *f* filtrante, superficie *f* del filtro
n filteroppervlak *n*
d Filterfläche *f*, Filteroberfläche *f*

2140 FILTER BAG fil
f sac *m* pour filtrer
e manga *f* para filtrar
i sacco *m* da filtrare
n filtreerzak
d Filterbeutel *m*, Filtersack *m*

2141 FILTER BED sew
f couche *f* filtrante
e capa *f* filtrante, lecho *m* filtrante, lecho *m* percolador
i letto *m* filtrante
n filterbed *n*, filtreerlaag
d Filterbett *n*, Filterschicht *f*

2142 FILTER BLOCK fil
f bloc *m* multicellulaire pour nettoyage des filtres
e bloque *m* multicelular para drenaje de filtros
i blocco *m* multicellulare per drenaggio di filtri
n multicellulair filterreinigings-blok *n*
d multizellularer Filterreinigungs block *m*

2143 FILTER CAKE sin
f gâteau *m* filtré, tourteau *m* de filtre
e torta *f* de cachaza, torta *f* del filtroprensa
i focaccia *f* di filtro
n filterkoek
d Filterkuchen *m*

2144 FILTER CAKE GRID sin
f treillis *m* de gâteau filtré
e parrilla *f* de torta de cachaza
i telaio *m* di focaccia di filtro
n filterkoekraam *n*
d Filterkuchenrahmen *m*

2145 FILTER CELL sin
f cellule *f* de filtre
e célula *f* de filtro
i cellula *f* di filtro
n filtercel
d Filterzelle *f*

2146 FILTER CHAMBER fil
f chambre *f* de filtration
e cámara *f* de filtración
i camera *f* di filtrazione, camera *f* di filtro
n filtreerkamer
d Filtrierkammer *f*

2147 FILTER CLOTH fil
f étamine *f* à filtrer
e tela *f* filtrante, tela *f* para filtrar
i tessuto *m* di filtro
n filterdoek *n*
d Filtertuch *n*

2148 FILTER CONE fil
f cône *m* de filtre
e cono *m* de filtro
i cono *m* di filtro
n filterconus, filterkegel
d Filtereinsatz *m*, Filterkegel *m*

2149 FILTER CONTROLLER, FILTER REGULATOR con,fil
f régulateur *m* de débit du filtre
e regulador *m* de presión del filtro
i regolatore *m* di pressione del filtro
n filterdrukregelaar
d Filterdruckregler *m*

2150 FILTER CUP, EXTRACTION THIMBLE lab
f cartouche *m* à extraction, manchon *m* poreux
e capa *f* porosa de extracción
i inviluppo *m* poroso
n poreuse huls
d Extraktionseinsatz *m*, poröse Hülle *f*

2151 FILTER CYLINDER fil
f tube *m* poreux à filtrer
e tubo *m* poroso para filtrar
i tubo *m* poroso da filtrare
n poreuze filtreerbuis
d poröses Filtrierrohr *n*

2152 FILTER ELEMENT, FILTER PLATE fil
f plaque *f* poreuse
e placa *f* porosa
i piastra *f* porosa
n poreuze plaat
d Filtereinsatz *m*

2153 FILTER FELT fil
f feutre *m* à filtrer
e fieltro *m* filtrador
i feltro *m* da filtrare
n filtreervilt *n*
d Filtrierfilz *m*

2154 FILTER FRAME fil
f cadre *m* de filtre
e bastidor *m* de filtro, marco *m* de filtro
i portafiltro *m*
n filterraam *n*
d Filterrahmen *m*

2155 FILTER FUNNEL, FILTERING FUNNEL fil
f entonnoir *m* filtreur
e embudo *m* filtrador
i imbuto *m* filtratore
n filtreertrechter
d Filtriertrichter *m*

2156 FILTER GATE, RING GATE, RING PENCIL GATE pla
f attaque *m* annulaire
e entrada *f* anular
i entrata *f* anulare
n ringaansnijding
d Ringanschnitt *m*

2157 FILTER GAUZE (GB), FILTER GAZE (US) fil,pap
f toile *f* filtrante
e gasa *f* filtrante
i garza *f* filtrante
n filtreergaas *n*
d Filtergaze *f*

2158 FILTER HOUSING cle,fil,gas
f boîte *f* de filtre
e caja *f* de filtro
i cassa *f* di filtro
n filterhuis *n*
d Filtergehäuse *n*

2159 FILTER LAYER, FILTERING LAYER fil
f couche *f* filtrante
e capa *f* del filtro
i strato *m* del filtro
n filterlaag
d Filterschicht *f*

2160 FILTER MANTLE fil
f chemise *f* de filtre
e camisa *f* de filtro
i camicia *f* di filtro
n filtermantel
d Filtermantel *m*

2161 FILTER PLANT, FILTRATION PLANT fil
f installation *f* de filtration
e equipo *m* de filtración
i impianto *m* di filtrazione
n filterinstallatie, filtreerinstallatie
d Filteranlage *f*, Filtrieranlage *f*

2162 FILTER PRESS, CAGE PRESS fil,pre
f filtre-presse *f*, presse *f* à filtre
e filtroprensa *f*, prensa *f* filtradora
i filtropressa *f*, pressa *f* da filtro
n filterpers
d Filterpresse *f*, Seiherpresse *f*

2163 FILTER RUFFLE fil
f manchon *m* de filtre
e manguito *m* de filtro
i manicotto *m* di filtro
n filtermanchet
d Filtermanschette *f*

2164 FILTER SCREEN, FILTERING SCREEN fil
f tamis *m* filtrant
e criba *f* filtradora
i vaglio *m* filtrante
n filtreerzeef, zeeffilter *n*
d Filtriersieb *n*, Siebfilter *m*

2165 FILTER SLAB, AIR DIFFUSER, DIFFUSER PLATE iwt
f dalle *f* poreuse, diffuseur *m* d'air, plaque *f* diffuseuse, plaque *f* filtrante
e difusor *m* de aire, placa *f* filtrante
i diffusore *m* d'aria, piastra *f* filtrante, piastra *f* porosa
n filterplaat, luchtverdeelplaat
d Belüfterplatte *f*, Filterplatte *f*, Luftverteiler *m*, Verteilerplatte *f*

2166 FILTER STICK fil
f baguette *f* filtrante
e varilla *f* de filtro
i bacchetta *f* di filtro
n filterstaafje *n*
d Filterstäbchen *n*

2167 FILTER TAP bre,fil
f robinet *m* de filtration
e grifo *m* de extracción, grifo *m* de filtración
i becco *m* di filtrazione, rubinetto *m* di filtrazione
n kraan van de klaringskuip
d Abläuterhahn *m*

2168 FILTER THICKENER fil,sir
f filtre *m* épaississant
e filtro *m* espesador
i filtro *m* ispessitore
n filterindikker, indikfilter *n*
d Eindickfilter *m*, Filtereindicker

2169 FILTER TUBE fi
f tube *m* filtrant
e tubo *m* filtrador
i tubo *m* filtrante
n filtreerbuis
d Filterrohr *n*

2170 FILTER UNIT fi
f groupe *m* de filtres, groupe *m* filtrant
e grupo *m* de filtros
i aggregato *m* di filtri, aggregato filtrante
n filteraggregaat *n*
d Filteraggregat *n*

2171 FILTER VAT, CLARIFYING TUB, LAUTER TUB, STRAINING VAT br
f cuve *f* de filtration, cuve-filtre
e cuba *f* de filtración
i tino *m* di filtrazione
n klaringskuip
d Klärbottich *m*, Läuterbottich *m*

2172 FILTERING BASIN fil,hyg,se
f dépôt *m* de filtration
e depósito *m* de filtración
i bacino *m* di filtrazione
n filtreerbekken *n*
d Filtrierbecken *n*

2173 FILTERING CRUCIBLE la
f creuset *m* filtrant
e crisol *m* filtrador
i crogiolo *m* filtrante
n filtreerkroes
d Filtriertiegel *m*

2174 FILTERING DIAPHRAGM, FILTERING MEMBRANE f
f membrane *f* de filtre

e membrana *f* de filtro
i membrana *f* di filtro
n filtermembraan *n*
d Filtermembrane *f*

2175 FILTERING FLASK, ASPIRATOR BOTTLE lab
f fiole *f* à filtration
e frasco *m* de filtración, pera *f* de aspiración
i bottiglia *f* a filtrazione
n aspirator
d Saugflasche *f*

2176 FILTERING JAR fil,lab
f vase *m* à filtration
e vaso *m* de filtración
i vasetto *m* a filtrazione
n filtreerbeker
d Filtriergefäss *n*

2177 FIN clo
f ailette *f*
e aleta *f*
i aletta *f*
n rib, vin
d Rippe *f*

2178 FIN-TYPE HEATING COIL, FINNED-TUBE HEATING COIL, EXTENDED-SURFACE COIL, SPIRAL FIN-TUBE HEATING COIL cle,hyg
f réchauffeur *m* d'air en tubes ailetés
e calentador *m* de aire con tubos de aletas
i riscaldatore *m* d'aria con tubi ad alette
n ribbenbuisluchtverhitter
d Bandrippenrohrheizregister *n*

2179 FINAL BATH pho
f bain *m* final
e baño *m* final
i bagno *m* finale
n eindbad *n*
d Schlussbad *n*

2180 FINAL BLEACHING BATH tex
f bain *m* de blanchiment final
e baño *m* de blanqueo final
i bagno *m* di candeggio finale
n nableekbad *n*
d Nachbleichbad *n*

2181 FINAL BOILING POINT RECORDER ins
f enregistreur *m* de point d'ébullition final
e registrador *m* de punto de ebullición final
i registratore *m* di punto d'ebollizione finale
n eindkookpuntschrijver
d Endsiedepunktgeber *m*, Siedeendegeber *m*

2182 FINAL CLEANING TANK tex
f récipient *m* de nettoyage final
e tanque *m* de limpieza final
i serbatoio *m* di pulitura finale
n nareinigingstank
d Nachreinigungsgefäss *n*

2183 FINAL COOLER, AFTERCOOLER clo
f réfrigérateur *m* final
e postrefrigerador *m*
i frigorifero *m* posteriore, postrefrigeratore *m*
n nakoeler
d Nachkühler *m*

2184 FINAL FILTER, SUBSEQUENT FILTER fil
f filtre *m* finisseur
e filtro *m* final
i filtro *m* finale
n eindfilter *n*
d Nachfilter *m*

2185 FINAL SETTLING TANK, SECONDARY SEDIMENTATION TANK iwt
f décanteur *m* secondaire, récipient *m* de clarification finale
e tanque *m* de clarificación final
i serbatoio *m* di sedimentazione finale
n laatste klaringskuip
d Nachklärgefäss *n*

2186 FINAL SINTERING OVEN htt,met
f four *m* de frittage final
e horno *m* de sinterización final
i forno *m* di sinterizzazione finale
n afsinteroven
d Fertigsinterofen *m*

2187 FINE GRINDING MILL brk
f moulin *m* de concassage excessif
e molino *m* de molienda fina, molino *m* de trituración fina
i mulino *m* di macinazione fina
n fijnmolen
d Feinmühle *f*

2188 FINE SAND FILTER fil
f filtre *m* à sable fin
e filtro *m* de arena fina
i filtro *m* a sabbia fina
n fijnzandfilter *n*
d Feinsandfilter *m*

2189 FINE SIEVE sor
f tamis *m* fin
e tamiz *m* fino
i vaglio *m* fino
n fijne zeef
d Feinsieb *n*

2190 FINENESS TESTER pap,tes
f appareil *m* à examiner la finesse du broyage
e probador *m* del grado de finura
i collaudatore *m* del grado di finezza
n fijnheidstestapparaat *n*, maalgraadmeter
d Mahlgradprüfer *m*

2191 FINERY met
f four *m* d'affinage
e horno *m* de afino
i forno *m* d'affinazione
n fijnoven
d Feinofen *m*

2192 FINISHING BOX tex
f case *f* d'apprêt, case *f* d'apprêtage
e caja *f* de aprestado
i cassa *f* d'apprettatura
n appreteerbak
d Appretierkasten *m*

2193 FINISHING PLANT gen
f installation *f* de finition
e acabaduría *f*, equipo *m* de acabado
i impianto *m* di finitura
n nabewerkingsinstallatie
d Endbearbeitungsanlage *f*, Nachbearbeitungsanlage *f*

2194 FINISHING PLANT lea
f installation *f* de finissage
e equipo *m* de acabado
i impianto *m* di finissaggio
n afwerkingsinstallatie
d Aufbereitungsanlage *f*

2195 FINISHING PLANT, DRESSING PLANT, SIZING PLANT tex
f installation *f* d'apprêtage
e equipo *m* de aprestado
i impianto *m* d'apprettatura
n appreteerinrichting
d Appretieranlage *f*

2196 FINISHING SHOP ger
f atelier *m* de finition
e taller *m* de acabado
i officina *f* di finitura
n afwerkhal
d Zurichterei *f*

2197 FINNED JACKET cl
f enveloppe *f* à ailettes
e camisa *f* con aletas
i camicia *f* ad alette
n mantel met koelvinnen
d Kühlrippenmantel *m*

2198 FINNED TUBE cl
f tube *m* à ailettes
e tubo *m* de aletas
i tubo *m* ad alette
n ribbenbuis
d Rippenrohr *n*

2199 FIRE ALARM sa
f avertisseur *m* d'incendie
e avisador *m* de incendios
i avvisatore *m* antincendio
n brandmelder
d Feuermelder *m*

2200 FIRE COPPER br
f chaudière *f* à feu nu
e caldera *f* de fuego directo
i caldaia *f* a fuoco diretto
n brouwketel met directe verhitting
d Feuerpfanne *f*

2201 FIRE ENGINE HOSE, FIRE HOSE fe
f tuyau *m* à incendie
e manguera *f* de incendios
i tubo *m* per pompieri
n brandslang, brandspuitslang
d Feuerlöschschlauch *m*

2202 FIRE EXTINGUISHER, EXTINGUISHER fe
f extincteur *m*
e apagador *m*, extintor *m* de incendios
i estintore *m*
n blusapparaat *n*, brandblusapparaat *n*
d Feuerlöscher *m*, Löscher *m*

2203 FIRE EXTINGUISHING ROSE, SPARGER fe
f arroseur *m* extincteur d'incend

e rociadera *f* extinguidora de incendios
i spruzzatore *m* antincendio
n brandblussproeier
d Feuerlöschbrause *f*

2204 FIRE GILDING PLANT coa
f installation *f* de dorure au feu
e equipo *m* de doradura al fuego
i impianto *m* di doratura al fuoco
n thermische verguldinstallatie
d Feuervergoldungsanlage *f*

2205 FIRE HYDRANT fea
f bouche *f* d'eau, bouche *f* d'incendie
e boca *f* de contraincendios
i bocca *f* da incendio, idrante *m* antincendio
n brandkraan
d Hydrant *m*

2206 FIRE KILN bre
f touraille *f* à feu nu
e tostadero *m* de fuego directo
i essiccatoio *m* a fuoco diretto
n eest met directe verhitting
d Feuerdarre *f*

2207 FIRE-OFF DIP elc
f bain *m* de pré-décapage
e baño *m* de decapaje preliminar
i bagno *m* di decapaggio preliminare
n voorbeitsbad *n*
d Gelbbrenne *f*

2208 FIRE REFINING FURNACE htt,met
f four *m* de raffinage au feu
e horno *m* de afino al fuego
i forno *m* d'affinazione al fuoco
n thermische veredelingsoven
d thermischer Vergütungsofen *m*

2209 FIRE STILL dis
f appareil *m* de distillation à chauffage direct
e aparato *m* de destilación de calentamiento directo
i apparecchio *m* di distillazione a riscaldamento diretto
n destilleerapparaat *n* met directe verhitting
d Destillieranlage *f* mit direkter Heizung

2210 FIRE WALL sap
f mur *m* coupe-feu, mur *m* mitoyen
e cortafuego *m*
i muro *m* tagliafiamma, paratia *f* parafiamma
n brandmuur
d Brandmauer *f*

2211 FIREWORKS FACTORY exp
f fabrique *f* de feu d'artifice
e fábrica *f* de fuegos artificiales
i fabbrica *f* di fuochi artificiali
n vuurwerkfabriek
d Feuerwerkfabrik *f*

2212 FIRING OVEN cer
f four *m* de cuisson
e horno *m* de cocción
i forno *m* di cottura
n bakoven
d Brennofen *m*

2213 FIRKIN, nut
BUTTER TUB
f barillet *m* à beurre, tinette *f* à beurre
e baril *m* para manteca
i bariletto *m* di burro
n botervaatje *n*, botervat *n*
d Butterfass *n*

2214 FISHTAIL BURNER, bur
FANTAIL BURNER
f bec *m* de gaz en éventail
e mechero *m* de mariposa
i bruciatore *m* a fiamma in forma di ventaglio
n waaierbrander
d Fächerbrenner *m*

2215 FISHTAIL TYPE mix,pla
KNEADER,
DOUBLE NABEN KNEADER
f mélangeur *m* à ailerons de requin
e amasador *m* con aletas, mezclador *m* con brazo cola de pez
i impastatrice *f* a doppia elica
n kantelmengmachine in delta-vorm
d Fischschwanzkneter *m*

2216 FITTINGS aap
f accessoires *pl*, garniture *f*
e accesorios *pl*, aditamientos *pl*
i accessori *pl*, raccordi *pl*
n appendages *pl*, uitrusting
d Armatur *f*, Ausrüstungsgegenstände *pl*

2217 FIVE-BOWL CALENDER, pap,pla
FIVE-ROLL CALENDER
f calandre *f* à cinq cylindres

e calandra *f* de cinco cilindros
i calandra *f* a cinque cilindri
n vijfwalsenkalander
d Fünfwalzenkalander *m*

2218 FIVE ROLLER MILL bre
f concasseur *m* à cinq cylindres
e molino *m* de cinco cilindros
i mulino *m* a cinque cilindri
n schrootmolen met vijf walsen
d Fünfwalzenmühle *f*

2219 FIXED-BED CRACKING PLANT fue
f installation *f* de craquage à lit fixe
e equipo *m* de craqueo en lecho fijo, equipo *m* de pirodesintegración en lecho fijo
i impianto *m* di piroscissione in letto fisso
n kraakinstallatie met vast opgestelde katalysator
d Festbettkrackanlage *f*

2220 FIXED-BED ION EXCHANGER elc
f échangeur *m* d'ions à lit fixe
e cambiador *m* de iones de lecho fijo
i scambiatore *m* d'ioni a letto fisso
n ionenuitwisselaar met vast harsbed
d Festbettionenaustauscher *m*

2221 FIXED-DIE PLATE (GB), STATIONARY PLATEN (US) pla
f plaque *f* fixe du moule
e placa *f* fija del molde
i piastra *f* d'impronta fissa
n vaststaande matrijsplaat
d feststehende Formplatte *f*

2222 FIXED-EXPANSION MACHINE pre
f machine *f* à expansion constante
e máquina *f* de expansión constante
i macchina *f* ad espansione costante
n constante-expansiemachine
d konstante Expansionsmaschine *f*

2223 FIXED-FIRE ZONE KILN, BOGIE KILN (GB), CAR TUNNEL KILN (US), TRUCK CHAMBER KILN cer,gla,htt
f étuve *f* à chariot
e estufa *f* de carrillo
i stufa *f* a carrello
n wagendroogstoof
d Wagentrockenofen *m*

2224 FIXED-PAN MILL brk
f broyeur *m* à cuve fixe
e molino *m* de cuba fija
i mulino *m* a tino fisso
n molen met stilstaande bodem
d Kollergang *m* mit feststehendem Teller

2225 FIXED PLATE pla
f matrice *f* fixe, plaque *f* à fixation
e matriz *f* fija, placa *f* base
i piastra *f* fissa
n vaste matrijsplaat, vaststaand matrijsgedeelte *n*
d Formplatte *f*, Stammplatte *f*

2226 FIXING BATH ph
f bain *m* de fixage
e baño *m* de fijado
i bagno *m* di fissaggio
n fixeerbad *n*
d Fixierbad *n*

2227 FLAME ANNEALING SHOP htt,me
f atelier *m* de recuit au chalumea
e taller *m* de recocido con soplete
i officina *f* di ricottura alla fiam
n vlamgloeiwerkplaats
d Flammenglühwerkstatt *f*

2228 FLAME CLEANING SHOP cle,me
f atelier *m* de décalaminage au chalumeau
e taller *m* de limpia con soplete
i officina *f* di pulitura alla fiamma
n werkplaats voor hamerslagverwijdering aan de vlam
d Flammentzunderungswerkstatt

2229 FLAME COLORING TESTER (US), FLAME COLOURING TESTER ana,te
f appareil *m* d'essai à la flamme
e probador *m* a la llama
i collaudatore *m* alla fiamma
n vlamkleuringsapparaat *n*
d Flammenfärbungsgerät *n*

2230 FLAME GUARD sa
f pare-flamme *m*
e parallamas *m*
i parafiamma *m*
n vlammenscherm *n*, vuurschern
d Flammenschutz *m*

2231 FLAME HARDENING PLANT, SHORTERIZING PLANT, TORCH HARDENING PLANT htt,m

f installation *f* de trempe au chalumeau
e equipo *m* de temple superficial a la llama
i impianto *m* di tempera alla fiamma
n vlamhardingsinstallatie
d autogene Flammenhärtungsanlage

2232 FLAME PLATING SHOP coa
f atelier *m* de placage à la flamme
e taller *m* de chapeado a la llama
i officina *f* di placcatura alla fiamma
n vlampleetwerkplaats
d Flammplattierungswerkstatt *f*

2233 FLAME SCALING SHOP cle,met
f atelier *m* de décalaminage au chalumeau
e taller *m* de descascarillado con soplete
i officina *f* di disincrostazione alla fiamma
n oxydehuidverbrandingswerkplaats
d thermische Entrostungswerkstatt *f*

2234 FLAME SPRAYING BURNER bur,coa
f brûleur *m* à métalliser par jet
e quemador *m* para metalizar por chorro
i bruciatore *m* per metallizzare con getto
n metaalspuitbrander
d Metallspritzbrenner *m*

2235 FLAME TRAP sap
f retenue-flamme *m*
e apagallamas *m*
i colpo *m* di ritorno di fiamma, rompifiamma *m*
n vlaminslagbeveiliging
d Flammenrückschlagsicherung *f*

2236 FLAME TREATING APPARATUS bur,pla
f appareil *m* de traitement superficiel à la flamme
e aparato *m* de tratamiento superficial a la llama
i apparecchio *m* di trattamento superficiale alla fiamma
n oppervlakbehandelingsbrander
d Oberflächenbehandlungsbrenner *m*

2237 FLAME TUBE BOILER, FLUE BOILER, TUBULAR BOILER boi
f chaudière *f* tubulaire
e caldera *f* de tubo de llama
i caldaia *f* tubolare
n vlampijpketel
d Flammrohrkessel *m*

2238 FLAP VALVE, TILTING DISK CHECK VALVE cpt
f clapet *m* de fermeture, soupape *f* à clapet
e válvula *f* de charnela, válvula *f* de mariposa
i valvola *f* a cerniera, valvola *f* di chiusura
n deurklep, scharnierende klep
d Klappenventil *n*, Verschlussklappe *f*

2239 FLARE, LIGHT BOMB, STAR SHELL mit
f obus *m* éclairant
e cohete *m* luminoso
i razzo *m* luminoso
n lichtgranaat, lichtkogel
d Leuchtbombe *f*

2240 FLARE BED htt
f canal *m* de gaz combustibles
e canal *m* de gases combustibles
i canale *m* di gas combustibili
n kanaal *n* voor verbrandingsgassen
d Verbrennungsgasenkanal *m*

2241 FLARING CUP brk
f meule *f* conique
e muela *f* de taza cónica
i mola *f* conica
n conische maalsteen
d Kegelmahlstein *m*

2242 FLASH-AGEING BATH (GB), FLASH-AGING BATH (US) tex
f bain *m* de traitement rapide par vapeur
e baño *m* de vaporizado húmedo rápido
i bagno *m* di trattamento al vapore rapido
n snelstoombad *n*
d Schnelldampfbad *n*

2243 FLASH BOILER boi,pre
f chaudière *f* à vaporisation rapide
e caldera *f* de vaporización rápida
i caldaia *f* a vaporizzazione rapida

n ketel met snelle verdamping
d Schnelldampfkessel *m*

2244 FLASH DISTILLATION PLANT dis
f installation *f* de distillation à évaporation instantanée
e equipo *m* de destilación con evaporación instantánea
i impianto *m* di distillazione equilibrata in corrente di vapore
n afdampinstallatie
d Kurzwegdestillationsanlage *f*

2245 FLASH DRIER dry
f sécheur *m* rapide
e secador *m* instantáneo
i essiccatore *m* istantaneo
n sneldroger
d Schnelltrockner *m*

2246 FLASH EVAPORATION PLANT dis
f établissement *m* d'évaporation avec réservoirs à basse pression
e equipo *m* de evaporación con cubas de baja presión
i impianto *m* d'evaporazione con serbatoio a bassa pressione
n verdampinstallatie met lage-druktanks
d Verdampfungsanlage *f* mit Niederdruckbehältern

2247 FLASH GROOVE pla
f rainure *f* d'échappement
e ranura *f* de escape
i scanalatura *f* di scarico
n overloopgroef
d Abflussnute *f*, Austriebnute *f*, Stoffabflussnute *f*

2248 FLASH LAND (GB), FLASH RIDGE (GB), FLASH SURFACE (US), LAND SURFACE (US) pla
f bord *m* d'appui, surface *f* d'appui
e borde *m* de rebaba, superficie *f* de rebaba
i appoggio *m*, battuta *f*
n afknijprand
d Abquetschfläche *f*, Abquetschrand *m*

2249 FLASH LATHE pla
f tour *m* à ébavurer
e pulidora *f*
i pulitrice *f*
n afbraambank
d Abgratbank *f*

2250 FLASH MOLD (US), FLASH MOULD (GB) pla
f moule *m* à couteau, moule *m* à déversage, moule *m* à échappement
e matriz *f* de desbordamiento, molde *m* de rebaba
i stampo *m* a coltello, stampo *m* con bava
n afknijpmatrijs, overloopmatrijs, snijpersvorm
d Abquetschform *f*, Überlaufform

2251 FLASH POINT CRUCIBLE lab
f creuset *m* ouvert pour point d'éclair, creuset *m* ouvert pour point d'inflammation
e crisol *m* abierto para punto inflamador
i crogiolo *m* scoperto per punto d'infiammabilità
n vlampuntkroes
d Flammpunkttiegel *m*

2252 FLASH POINT RECORDER ins
f enregistreur *m* du point d'éclair, enregistreur *m* du point d'inflammation
e registrador *m* del punto inflamador
i registratore *m* del punto d'infiammabilità
n vlampuntschrijver
d Flammpunktgeber *m*

2253 FLASH POINT TESTER fue
f vérificateur *m* du point d'éclair, vérificateur *m* du point d'inflammation
e probador *m* del punto inflamador
i collaudatore *m* del punto d'infiammabilità
n vlampuntbeproevingsapparaat *n*
d Flammpunktgerät *n*

2254 FLASH RING, VERTICAL FLASH RING pl
f gorge *f* circulaire
e anillo *m* de rebaba, ranura *f* de rebaba
i bava *f* verticale
n afknijpring
d Abquetschring *m*

2255 FLASH ROASTER me
f four *m* de grillage de concentré pulvérulents
e horno *m* de tostación de concentrados pulverizados

i forno *m* d'arrostimento alla fiamma
n vlamroostoven
d Flammenröstofen *m*

2256 FLASH ROASTING PLANT htt,min
f installation *f* de grillage de concentrés pulvérulents
e equipo *m* de tostación de concentrados pulverizados
i impianto *m* d'arrostimento di concentrati polverizzati
n vlamroostinstallatie
d Flammröstanlage *f*

2257 FLASH ROTARY VACUUM EVAPORATOR, eva
CONTINUOUS FLASH EVAPORATOR
f évaporateur *m* rotatif sous vide
e vacuoevaporador *m* giratorio
i evaporatore *m* rotativo sotto vuoto
n roterende vacuümverdamper
d Rotationsvakuumverdampfer *m*

2258 FLASHBACK CHAMBER gas,sap
f chambre *f* intermédiaire remplie d'eau
e cámara *f* intermedia llena de agua
i camera *f* intermedia a riempimento d'acqua
n met water gevulde tussenruimte van een acetyleengenerator
d wassergefüllter Zwischenraum *m*

2259 FLASK lab
f matras *m*
e matraz *m*
i matraccio *m*
n kolf
d Kolben *m*

2260 FLAT tex
f chapeau *m* de carde
e chapón de carda
i cappello *m* di carda
n kaardbeslag *n*
d Krempelgarnitur *f*

2261 FLAT ANODE elc
f anode *f* à bande plate
e ánodo *m* de cinta plana
i anodo *m* a nastro piano
n platte strookanode
d Flachstreifenanode *f*

2262 FLAT-BED GLAZER, pho
GLAZING PRESS
f presse *f* plate à glacer
e prensa *f* de glaseado
i pressa *f* da lucidare
n hoogglanspers
d Hochglanzpresse *f*

2263 FLAT BLADE PROPELLER mix
f hélice *f* à palettes planes
e hélice *f* de paletas planas
i elica *f* a pale piane
n propeller met platte bladen
d Propeller *m* mit flachen Flügeln

2264 FLAT-BOTTOMED CONICAL FLASK, lab
ERLENMEYER FLASK
f matras *m* d'Erlenmeyer
e matraz *m* de Erlenmeyer
i bevutta *f*, matraccio *m* d'Erlenmeyer
n erlenmeyerkolf
d Erlenmeyerkolben *m*

2265 FLAT-FIN TUBE HEATING COIL cle,hyg
f réchauffeur *m* d'air en tubes à ailettes planes
e calentador *m* de aire con tubos de aletas planas
i riscaldatore *m* d'aria con tubi ad alette piane
n lamellenbuisluchtverhitter
d Lamellenrohrheizregister *n*

2266 FLAT HEAD EXTRUDER pla
f boudineuse *f* à tête plate
e budinadora *f* con cabezal plano
i trafila *f* con testa piatta
n spuitmachine met platte spuitkop
d Flachkopfspritzmaschine *f*

2267 FLAT-PLATE PRINTING MACHINE tex
f machine *f* d'impression à plaque
e máquina *f* estampadora con placa
i macchina *f* stampatrice a zinco con incisione debole
n drukplaatverfmachine
d Druckplattenfärbmaschine *f*

2268 FLAT SHEETING DIE, pla
SLOT DIE
f filière *f* plate
e boquilla *f* de ranura ancha, tobera *f* de ranura ancha
i filiera *f* piana, testa *f* piana
n spleetmatrijs
d Breitschlitzdüse *f*

2269 FLAT STOPCOCK lab
f robinet *m* plat

e grifo *m* plano
i rubinetto *m* piatto
n platte kraan
d Flachhahn *m*

2270 FLAYING HOUSE lea
f écorcherie *f*, équarrissoir *m*
e desolladero *m*, despellejadura *f*
i casa *f* degli scorticatori, scorticatoio *m*
n vilderij
d Abdeckerei *f*

2271 FLEECE pla
f toison *f*, voile *f*
e velo *m*
i velo *m*
n vliesmat
d Vlies *n*

2272 FLESHING KNIFE lea
f écharnoir *m*
e descarnador *m*, luneto *m* raspador
i coltello *m* scarnaficatore, scarnatoio *m*
n vleesmes *n*
d Schabeisen *n*, Scherdegen *m*, Schereisen *n*

2273 FLESHING MACHINE lea
f écharneuse *f*, machine *f* à écharner
e máquina *f* descarnadora
i macchina *f* scarnaficatrice
n ontvleesmachine
d Entfleischmaschine *f*, Schabmaschine *f*

2274 FLEXIBLE MOLD (US), FLEXIBLE MOULD (GB) pla
f moule *f* flexible
e molde *m* flexible
i stampo *m* flessibile
n flexibele vorm
d flexibele Form *f*

2275 FLEXIMETER, FLEXOMETER ins
f flexomètre *m*
e flexímetro *m*
i flessometro *m*
n flexometer
d Flexometer *n*

2276 FLOAT cpt
f flotteur *m*
e flotador *m*
i galleggiante *m*
n vlotter
d Schwimmer *m*

2277 FLOAT, TROUGH min,sep
f auge *f* de sédimentation
e cuba *f* de sedimentación
i truogolo *m* di sedimentazione
n bezinktrog
d Ablagerungsmulde *f*

2278 FLOAT BOWL (US), FLOAT CHAMBER (GB) fue
f chambre *f* du flotteur, cuve *f* à niveau constant
e cámara *f* del flotador
i camera *f* del galleggiante
n vlotterkamer
d Schwimmergehäuse *n*

2279 FLOAT CARBURETTER fue
f carburateur *m* à niveau constant
e carburador *m* de nivel constante
i carburatore *m* a galleggiante
n vlottercarburator
d Schwimmervergaser *m*

2280 FLOAT STOP VALVE cpt
f soupape *f* d'arrêt à flotteur
e válvula *f* de cierre de flotador
i valvola *f* di chiusura a galleggiante
n sperklep met vlotter
d Schwimmabsperrventil *n*

2281 FLOAT VALVE cpt
f soupape *f* à flotteur
e válvula *f* de flotador
i valvola *f* a galleggiante
n vlotterklep
d Schwimmerventil *n*

2282 FLOATER gla
f barrage *m* mobile, flotteur *m*
e flotador *m*
i galleggiatore *m*
n vlotter
d Schwimmer *m*

2283 FLOATER HOLE gla
f trou *m* de flotteur
e orificio *m* de flotador
i orificio *m* di galleggiante
n vlotteropening
d Schwimmeröffnung *f*

2284 FLOATING BLANKET, SLUDGE BLANKET FILTER iwt
f filtre *m* à voile de boue
e colchón *m* de fango
i filtro *m* con sospensione di fango
n drijvend filter *n*
d schwebender Filter *m*

2285 FLOATING CHASE pla
f châssis *m* mobile
e cuna *f* móvil
i telaio *m* mobile
n beweegbaar matrijsdeel *n*
d beweglicher Formrahmen *m*

2286 FLOATING COVER fue,sap
f couverture *f* flottante
e cubierta *f* flotante
i coperta *f* flottante
n drijvende bedekking
d Schwimmdecke *f*

2287 FLOATING COVER iwt
f cloche *f* à gaz flottante
e campana *f* de gas flotadora
i copertura *f* galleggiante per la presa del gas
n drijvend gasdak *n*
d schwimmende Gasdecke *f*

2288 FLOATING DIE pla,rub
f matrice *f* à fixation élastique
e matriz *f* montada elasticamente
i matrice *f* fissata elasticamente
n elastisch bevestigde matrijs
d elastisch angebrachte Matrize *f*

2289 FLOATING KNIFE pla
f racle *f* en air, racle *f* pneumatique
e cuchilla *f* dosificadora neumática, cuchilla *f* flotante
i raschiatore *m* libero
n luchtrakel
d Luftrakel *f*

2290 FLOATING NOZZLE pla
f buse *f* à ressort, buse *f* élastique
e tobera *f* deslizable
i ugello *m* mobile
n elastische spuitneus
d elastische Düse *f*, Schiebedüse *f*

2291 FLOATING PLATE pla
f plateau *m* flottant
e placa *f* flotante, placa *f* intermedia móvil
i piastra *f* intermedia libera
n zwevende tussenplaat
d Schwebeplatte *f*, Schwebetisch *m*

2292 FLOATING PUNCH pla
f poinçon *m* mobile
e macho *m* flotante
i punzone *m* libero
n beweegbare stempel
d beweglicher Stempel *m*

2293 FLOCCULATION TANK col
f récipient *m* de floculation
e tanque *m* de floculación
i serbatoio *m* di flocculazione
n uitvloktank
d Ausflockungsbehälter *m*

2294 FLOCK SPRAYING GUN coa,dec
f pistolet *m* pulvérisateur de flocons
e pistola *f* pulverizadora de borras
i pistola *f* a spruzzo di fiocchi
n vlokkenspuitpistool *n*
d Flockenspritzpistole *f*

2295 FLOOR PLOUGH (GB), FLOOR PLOW (US) bre
f retourneur *m*
e removedor *m* para era
i mescolatore *m* per aia
n moutploeg
d Tennenwender *m*

2296 FLOTATION MACHINE met,min
f machine *f* de flottation
e máquina *f* de flotación
i macchina *f* di flottazione
n flotatiemachine
d Schwimmaufbereitungsmaschine *f*

2297 FLOTATION PLANT gen
f installation *f* de flottation
e equipo *m* de flotación
i impianto *m* di flottazione
n flotatie-installatie, schuim-installatie
d Flotationsanlage *f*

2298 FLOTATION SAVE-ALL iwt
f arrête-matière à flottation, ramasse-pâte *m* de flottation
e recuperador *m* de colas de flotación
i ricuperatore *m* di materiale sospeso per flottazione
n flotatiestofvanger
d Flotationstoffänger *m*

2299 FLOTATION TANK met,min
f récipient *m* de flottation
e tanque *m* de flotación
i serbatoio *m* di flottazione
n flotatietank
d Schwimmaufbereitungsgefäss *n*

2300 FLOUR MILL, GRIST MILL met
f moulin *m* à blé

e molino *m* harinero
i mulino *m* da grano
n korenmolen
d Getreidemühle *f*

2301 FLOUROMETER ins
f appareil *m* à déterminer la teneur en matière de remplissage
e medidor *m* de la proporción de la materia de relleno
i polverimetro *m*
n vulstofgehaltemeter
d Füllstoffgehaltmesser *m*

2302 FLOW COATING PLANT dec
f installation *f* d'arrosage de vernice
e equipo *m* rociador de barniz
i impianto *m* da annaffiare vernice
n lakgietinstallatie
d Lackgiessanlage *f*

2303 FLOW COMPARTMENT, SEDIMENTATION CHAMBER, SETTLING COMPARTMENT iwt
f chambre *f* de sédimentation
e cámara *f* de sedimentación
i camera *f* di sedimentazione
n bezinkruimte
d Absetzraum *m*, Klärraum *m*

2304 FLOW INDICATOR ins
f indicateur *m* de courant, indicateur *m* d'écoulement
e indicador *m* del gasto
i indicatore *m* della corrente
n stromingsindicator
d Strömungsanzeiger *m*

2305 FLOW RECORDER, VOLUME RECORDER ins
f appareil *m* enregistreur de volume
e registrador *m* de volumen
i registratore *m* di portata, registratore *m* di quantità, registratore *m* di volume
n volumeopnemer
d Mengenschreiber *m*

2306 FLOW TABLE tes
f table *f* d'essai d'écoulement
e mesa *f* de ensayo de flujo, mesa *f* de prueba de gasto
i tavola *f* di prova di flusso
n schokproefapparaat *n*, uitbreidproefapparaat *n*, vloeiproefapparaat *n*
d Fliessprobegerät *n*

2307 FLOW-THROUGH CENTRIFUGE, CONCURRENT CENTRIFUGE cen,iso
f centrifugeur *m* à courant continu
e centrífuga *f* de corriente continua
i centrifuga *f* a corrente continua
n doorstroomcentrifuge
d Durchlaufzentrifuge *f*

2308 FLOWMETER ins
f débitmètre *m*
e flujómetro *m*, gastómetro *m*
i flussometro *m*
n stromingsmeter
d Strömungsmesser *m*

2309 FLOWMETER WITH FLOATING INDICATOR, GAPMETER ins
f débitmètre *m* à flotteurs
e flujómetro *m* con indicadores flotantes
i flussometro *m* a galleggianti
n rotameter
d Rotamesser *m*

2310 FLUE fue
f conduite *f* de fumée
e canal *m* de humo
i condotto *m* di fumo
n rookgassenkanaal *n*
d Rauchgaskanal *m*

2311 FLUE BOILER, FLAME TUBE BOILER, TUBULAR BOILER boi
f chaudière *f* tubulaire
e caldera *f* de tubo de llama
i caldaia *f* tubolare
n vlampijpketel
d Flammrohrkessel *m*

2312 FLUE GAS ANALYZER ana,gas
f analyseur *m* des fumées
e analizador *m* de los gases de combustión
i analizzatore *m* dei fumi, analizzatore *m* dei gas di combustione
n rookgasanalysator
d Rauchgasprüfer *m*

2313 FLUE GAS PURIFIER cle,ga
f dépoussiéreur *m* des fumées
e depurador *m* de los gases de combustión
i depuratore *m* dei fumi, depuratore *m* dei gas di combustione

n rookgasreiniger
d Rauchgasreiniger *m*

2314 FLUFFING TABLE lea
f table *f* de ponçage
e mesa *f* de apomazar
i tavola *f* da pomiciare
n puimtafel
d Bimssteintisch *m*

2315 FLUID BED, FLUIDIZED BED, MOVING BED iwt,sew
f lit *m* fluide, lit *m* fluidifié, lit *m* suspendu
e lecho *m* fluidizado, lecho *m* flúido
i letto *m* fluidificato, letto *m* fluidizzato
n wervelbed *n*
d Fliessbett *n*, Wirbelbett *n*

2316 FLUID CATALYTIC CRACKING PLANT fue
f installation *f* de craquage catalytique fluide
e equipo *m* de craqueo con catalizador flúido, equipo *m* de pirodesintegración con catalizador flúido
i impianto di piroscissione con catalizzatore fluido
n kraakinstallatie met gefluïdiseerde katalysator
d Fliessbettkrackanlage *f*, Wirbelschichtkrackanlage *f*

2317 FLUID-ENERGY MILL, JET IMPACT MILL, JET MILL brk
f broyeur *m* à jet
e triturador *m* de chorro
i frantoio *m* a getto
n luchtstraalmolen
d Strahlmühle *f*

2318 FLUID INTERFACE DETECTOR nue
f détecteur *m* de la couche intermédiaire entre fluides
e detector *m* de la zona interfacial entre flúidos
i rivelatore *m* dello strato di separazione tra fluidi
n vloeistofgrensvlakdetector
d Flüssigkeitsgrenzflächen-detektor *m*

2319 FLUID MIXING TANK nue
f cuve *f* mélangeuse
e tanque *m* mezclador
i serbatoio *m* mescolatore
n mengtank
d Mischbehälter *m*

2320 FLUIDIZED BED COATING APPARATUS coa
f appareil *m* de revêtement à lit fluidisé
e aparato *m* de revestimiento de lecho fluidificado
i apparecchio *m* di rivestimento a letto fluidizzato
n wervelsinterapparaat *n*
d Wirbelsinterapparat *m*

2321 FLUIDIZED BED REACTOR, FLUIDIZED REACTOR nue
f réacteur *m* à lit fluidisé, réacteur *m* à tourbillonnement du combustible
e reactor *m* con combustible fluidificado
i reattore *m* a letto fluidizzato
n fluïdisatiereactor
d Fliessbettreaktor *m*, Wirbelschichtreaktor *m*

2322 FLUIDIZED FUEL BED fue,htt
f lit *m* de combustible fluidisé
e carga *f* de combustible fluidificado
i carica *f* di combustibile fluidificato
n vloeibare-brandstoflaag
d flüssige Brennstofflage *f*

2323 FLUIDIZED-SOLIDS REACTOR nue
f réacteur *m* à combustible solide fluidisé
e reactor *m* con sustancias de desintegración casifluidas, reactor *m* de combustible sólido fluidificado
i reattore *m* a combustibile solido fluidizzato
n reactor met gefluïdiseerde vaste brandstof
d Reaktor *m* mit quasiflüssigem Brennstoff

2324 FLUOBORATE-LEAD BATH, LEAD-FLUOBORATE BATH elc
f bain *m* au fluoborate de plomb
e baño *m* de fluoborato de plomo
i bagno *m* al fluoborato di piombo
n loodfluoboraatbad *n*
d Bleifluoboratbad *n*

2325 FLUORESCENCE MICROSCOPE ins
f microscope *m* à fluorescence
e microscopio *m* de fluorescencia

i microscopio *m* a fluorescenza
n fluorescentiemicroscoop
d Fluoreszenzmikroskop *n*

2326 FLUORIDATION PLANT hyg
f installation *f* de fluorisation
e equipo *m* de fluorización
i impianto *m* di fluorizzazione
n fluorideringsinstallatie *f*
d Fluorierungsanlage *f*

2327 FLUORIMETER, ins
FLUOROMETER
f fluoromètre *m*
e fluorímetro *m*
i fluorimetro *m*
n fluorescentiemeter, fluorimeter
d Fluoreszenzmesser *m*, Fluorometer *n*

2328 FLUORINE CELL gas
f générateur *m* de fluor
e generador *m* de flúor
i generatore *m* di fluoro
n fluorgenerator
d Fluorgenerator *m*

2329 FLUOSILICATE-LEAD BATH, elc
LEAD-FLUOSILICATE BATH
f bain *m* au fluosilicate de plomb
e baño *m* de fluosilicato de plomo
i bagno *m* al fluosilicato di piombo
n loodfluosilicaatbad *n*
d Bleifluosilikatbad *n*

2330 FLUTED ANODE, elc
RIBBED ANODE
f anode *f* cannelée
e ánodo *m* acanalado
i anodo *m* scanalato
n geribbelde strookanode
d Riffelanode *f*

2331 FLUTED FILTER, fil,lab
FOLDED FILTER
f filtre *m* à plis, filtre *m* au papier plissé
e filtro *m* acanalado, filtro *m* de pliegues
i filtro *m* a pieghe, filtro *m* pieghevole
n vouwfilter *n*
d Faltenfilter *m*

2332 FLUTED FUNNEL, lab
RIBBED FUNNEL
f entonnoir *m* cannelé
e embudo *m* acanalado
i imbuto *m* scanalato
n geribbelde filter *n*
d Rippentrichter *m*

2333 FLUTED ROLL, rub
CORRUGATED ROLL
f cylindre *m* cannelé
e cilindro *m* acanalado
i cilindro *m* scanalato
n geribbelde wals, riffelwals
d geriffelte Walze *f*, Riffelwalze *f*

2334 FLUX BLOCK gla
f brique réfractaire de bassin
e ladrillo *m* refractario de cuba
i mattone *m* refrattario di tino
n vuurvaste wannensteen
d feuerbeständiger Wannenstein *m*

2335 FLUX TRAP REACTOR, nue
HIGH FLUX REACTOR
f réacteur *m* à flux concentré, réacteur *m* à flux élevé
e reactor *m* de flujo concentrado reactor *m* de flujo elevado
i reattore *m* a flusso concentrato, reattore *m* a flusso elevato
n hoge-flux reactor
d Hochflussreaktor *m*

2336 FLYWHEEL PUMP pre
f pompe *f* à volant
e bomba *f* de volante
i pompa-volano *f*
n vliegwielpomp
d Schwungradpumpe *f*

2337 FOAM BEATER, rub
FOAMING MACHINE,
FROTHING MACHINE, BEATING MACHINE, WHISKING MACHINE
f batteuse *f* à mousse, fouetteuse
e batidora *f* mecánica
i frullatore *m*, produttore *m* di schiuma, sbattitore *m*
n klutsmachine, opklopmachine, schuimklopper
d Schaumschlagmaschine *f*, Schlagmaschine *f*

2338 FOAM EXTINGUISHER fea
f extincteur *m* à mousse
e extintor *m* de espuma
i estintore *m* a schiuma
n schuimblusser
d Schaumlöscher *m*

2339 FOAM METER soa
f moussemètre *m*
e espumómetro *m*
i misuratore *m* di schiuma
n schuimapparaat *n*
d Schaumapparat *m*

2340 FOAM SCRAPER, sin
FOAM SEPARATOR
f séparateur *m* d'écume
e paleta *f* despumadora
i separatore *m* di schiuma
n schuimscheider
d Schaumabscheider *m*

2341 FOLLOWER cpt,pla,pre
f élément *m* interchangeable, élément *m* rapport
e pieza *f* auxiliar
i pezzo *m* ausiliario, tassello *m*
n inzetstuk *n*
d Einsatz *m*

2342 FOOD INDUSTRY nut
f industrie *f* alimentaire
e industria *f* alimentaria, industria *f* de la alimentación
i industria *f* alimentare
n levensmiddelenindustrie, voedingsmiddelenindustrie
d Nährungsmittelindustrie *f*

2343 FOOT VALVE cpt
f soupape *f* d'aspiration
e válvula *f* de aspiración, válvula *f* de pie
i valvola *f* di fondo
n zuigklep
d Saugventil *n*

2344 FORCE (US), pla,pre
PUNCH (GB), TOP FORCE (GB)
f poinçon *m*
e molde *m* macho, punzón *m*
i punzone *m*
n positief matrijsdeel *n*, stempel
d Stempel *m*

2345 FORCE PLATE (US), pla
PLUNGER RETAINER PLATE (GB)
f plaque *f* porte-poinçon
e placa *f* de alojamiento de los machos
i piastra *f* portapunzone
n bevestigingsplaat
d Einsatzfutter *n*

2346 FORCE PLUG pla
f poinçon *m* fixé par vis
e macho *m* atornillado
i punzone *m* d'avvitamento
n stempelinzetstuk *n*
d aufgeschraubter Stempeleinsatz *m*

2347 FORCE PUMP pre
PRESSURE PUMP
f pompe *f* élévatoire, pompe *f* foulante
e bomba *f* impelente
i pompa *f* a pressione, pompa *f* premente
n perspomp
d Druckpumpe *f*

2348 FORCE SIDE PART (US), pla
MOVABLE MOULD PART (GB)
f partie *f* mobile du moule
e semimolde *m* móvil
i semistampo *m* mobile
n bewegende matrijshelft
d beweglicher Werkzeughalter *m*

2349 FORCED-CIRCULATION BOILER, boi,pre
FORCED-FLOW BOILER
f chaudière *f* à circulation forcée
e caldera *f* de circulación forzada
i caldaia *f* a circolazione forzata
n ketel met gedwongen circulatie
d Kessel *m* mit erzwungenem Umlauf, Zwangsumlaufkessel *m*

2350 FORCED-CIRCULATION EVAPORATOR eva,pre
f évaporateur *m* à circulation forcée
e evaporador *m* de circulación forzada
i evaporatore *m* a circolazione forzata
n verdamper met gedwongen circulatie
d Umwälzverdampfer *m*, Verdampfer *m* mit erzwungenem Umlauf, Zwangsumlaufverdampfer *m*

2351 FORCED-DRAFT COOLING TOWER (US), clo
INDUCED-DRAUGHT COOLING TOWER (GB)
f tour *f* de refroidissement à tirage forcé
e torre *f* de enfriamiento de tiro forzado
i torre *f* di refrigerazione a tiraggio forzato
n koeltoren met kunstmatige trek
d Kühlturm *m* mit künstlichem Zug

2352 FORCED-DRAFT FURNACE (US), htt
FORCED-DRAUGHT FURNACE (GB)
f four *m* à tirage forcé
e horno *m* de tiro forzado
i forno *m* a tiraggio forzato

n oven met kunstmatige trek
d Ofen *m* mit künstlichem Zug, Saugzugofen *m*

2353 FORCING MACHINE, EXTRUDER, EXTRUDING MACHINE pla
f boudineuse *f*
e budinadora *f*
i trafila *f*
n spuitmachine
d Spritzmaschine *f*, Strangpresse *f*

2354 FORE VACUUM PUMP (GB), FOREPUMP (US) vac
f pompe *f* à vide préliminaire élevé
e bomba *f* de vacío preliminar elevado
i pompa *f* a vuoto preliminare elevato
n voorvacuümpomp
d Vorvakuumpumpe *f*

2355 FORECOOLER clo
f pré-refroidisseur *m*, refroidisseur *m* préliminaire
e enfriador *m* preliminar, preenfriador *m*
i preraffreddatore *m*, raffreddatore *m* preliminare
n voorkoeler
d Vorkühler *m*

2356 FOREHEARTH, RECEIVER htt
f avant-creuset *m*
e antecrisol *m*
i avancrogiolo *m*
n voorhaard
d Vorherd *m*

2357 FOREMASHER, CONVERTER, PONYMASHER, PREMASHER bre
f hydrateur *m*
e premezclador *m*
i premescolatore *m*
n voorbeslagkuip
d Vormaischer *m*

2358 FORMER, LATEX FORMER rub
f forme *f* de trempage
e molde *m* para inmersión
i stampo *m* ad immersione
n dompelvorm
d Tauchform *f*

2359 FORMING DIE pla
f moule *m* de formage
e molde *m* de conformado
i punzone *m*
n mal
d Formwerkzeug *n*

2360 FOUNDATION WASHER, ANCHOR PLATE cpt
f contreplaque *f* d'ancrage, plaque *f* de fondation
e placa *f* de anclaje, placa *f* de fundación
i placca *f* d'ancoraggio, placca *f* di fondazione
n ankerplaat
d Ankerplatte *f*

2361 FOUNTAIN, LAUNDER, RUNNER THROUGH (US), SPOUT (GB), TAPPING SPOUT (GB) fon
f canal *m* de coulée, chenal *m* de coulée
e canal *m* de colada
i canale *m* di colata
n gietgoot, schenk, tapgoot
d Abstichrinne *f*, Förderinne *f*, Rutsche *f*

2362 FOUR-BOWL CALENDER, FOUR-ROLL CALENDER pap,pla
f calandre *f* à quatre cylindres
e calandra *f* de cuatro cilindros
i calandra *f* a quattro cilindri
n vierwalsenkalander
d Vierwalzenkalander *m*

2363 FOUR-ROLLER MILL bre
f concasseur *m* à quatre cylindres
e triturador *m* de cuatro cilindros
i frantoio *m* a quattro cilindri
n schrootmolen met vier walsen
d Vierwalzenmühle *f*

2364 FOURDRINIER MACHINE pap
f machine *f* à papier à table plate, machine *f* à papier de Fourdrin
e máquina *f* de papel de tela metálica sin fin
i macchina *f* di carta a tela metallica senza fine
n langzeefmachine
d Langsiebpapiermaschine *f*

2365 FRACTION COLLECTOR di
f récepteur *m* de fractions
e recipiente *m* de fracciones
i raccoglitore *m* di frazioni
n fractioneerontvanger
d Fraktioniervorlage *f*

2366 FRACTIONAL CRYSTALLIZATION VESSEL cry
f cuve *f* de cristallisation fractionnée
e cuba *f* de cristalización fraccionada
i tino *m* di cristallizzazione frazionata
n vat *n* voor gefractionneerde kristallisatie
d Umkristallisationsgefäss *n*

2367 FRACTIONAL DISTILLATION PLANT dis
f installation *f* de distillation fractionnée
e equipo *m* de destilación fraccionada
i impianto *m* di distillazione frazionata
n gefractioneerde-destillatie-inrichting
d fraktionierte Destillationsanlage *f*

2368 FRACTIONAL PRECIPITATION TANK sep
f récipient *m* de précipitation fractionnée
e tanque *m* de precipitación fraccionada
i serbatoio *m* di precipitazione frazionata
n gefractioneerde-destillatie-inrichting
d Gefäss *n* für fraktionierte Fällung

2369 FRACTIONATING COLUMN dis
f colonne *f* de fractionnement
e columna *f* de fraccionamiento
i colonna *f* di frazionamento
n fractioneerkolom
d Fraktioniersäule *f*

2370 FRACTIONATING FLASK dis,lab
f ballon *m* de distillation fractionnée, cornue *f* de distillation fractionnée
e cucúrbita *f* de destilación fraccionada
i cucurbita *f* per distillazione frazionata
n fractioneerkolf
d Fraktionierkolben *m*, Fraktionskolben *m*

2371 FRACTIONATING OIL VAPOR DIFFUSION PUMP (US), FRACTIONATING OIL VAPOUR DIFFUSION PUMP (GB) vac
f pompe *f* à diffusion d'huile à fractionnement
e bomba *f* de difusión de aceite fraccionadora
i pompa *f* a diffusione d'olio a frazionamento
n fractionerende oliediffusiepomp
d Fraktionieröldiffusionspumpe *f*

2372 FRACTIONATING TOWER, BUBBLE TOWER dis
f tour *f* de fractionnement
e torre *f* de borbotaje. torre *f* de burbujeo, torre *f* de fraccionación
i torre *f* di frazionamento
n fractioneertoren, rectificeerkolom
d Fraktionierturm *m*

2373 FRAME, BODY, CHASE cpt
f bâti *m*, châssis *m*
e bastidor *m*, cuna *f*
i custodia *f*, telaio *m*
n freem *n*
d Chassis *n*, Gehäuse *n*, Rahmen *m*

2374 FRAME FILTER fil
f filtre *m* à cadres
e filtro *m* de telas
i filtro *m* a tele
n raamfilter *n*
d Rahmenfilter *m*

2375 FRAME PLATE, MASS-TYPE PLATE, PLATED FRAME PLATE elc
f plaque *f* à cadre, plaque *f* à masse
e placa *f* de cuadro, placa *f* de masa
i piastra *f* a massa, piastra *f* a telaio
n massaplaat
d Rahmenplatte *f*

2376 FRAME PRESS, STRAIN PLATE PRESS pla,rub
f presse *f* à cadre
e prensa *f* de cuadro
i pressa *f* a telaio
n raampers, trekplatenpers
d Rahmenpresse *f*

2377 FREE-BEATING MILL pap
f moulin *m* à battement
e molino *m* de batido libre, molino *m* de pisones libre
i mulino *m* battitore
n slagmolen
d Schlagmühle *f*

2378 FREEZE-DRYING APPARATUS clo,dry
f sécheur *m* par congélation
e secador *m* por congelación
i apparecchio *m* di deidrazione a freddo
n vriesdroogapparaat *n*
d Gefriertrocknungsapparat *m*

2379 FREEZING POINT RECORDER, SOLIDIFICATION POINT RECORDER ins
f enregistreur *m* du point de solidification
e registrador *m* del punto de solidificación
i registratore *m* del punto di solidificazione
n stolpuntschrijver
d Erstarrungspunktschreiber *m*

2380 FRICTION CALENDER rub
f calandre *f* à friction
e calandra *f* de friccíon
i calandra *f* a frizione
n frictioneerkalander
d Friktionskalander *m*

2381 FRICTION PRESS, FRICTION SCREW PRESS, SCREW PRESS fat,pre
f balancier *m*, presse *f* à vis
e prensa *f* de husillo, prensa *f* de tornillo
i pressa *f* a vite
n frictiepers, schroefpers, spindelpers
d Spindelpresse *f*

2382 FRITTED DISK FUNNEL fil
f entonnoir *m* conique à plaque filtrante
e embudo *m* cónico de placa filtrante
i imbuto *m* conico a piastra filtrante
n kegelvormige glasfiltertrechter
d kegelförmiger Glasfiltertrichter *m*

2383 FRITTED GAS-DISPERSION TUBE, FRITTED GAS-DISTRIBUTION TUBE fil
f tube *m* à distribution de gaz à plaque frittée, tube *m* de barbotage à plaque frittée
e tubo *m* de distribución de gas con placa sinterizada
i tubo *m* di distribuzione di gas a piastra sinterizzata
n gasverdelingsbuis met gesinterde plaat
d Gaseinleitungsrohr *n* mit gesinterter Platte

2384 FRITTED GLASS FILTER FUNNEL fil
f entonnoir *m* cylindro-conique à plaque filtrante
e embudo *m* cilindro-cónico de placa filtrante
i imbuto *m* cilindro-conico a piastra filtrante
n glasfiltertrechter
d Glasfilternutsche *f*

2385 FRITTING OVEN, SINTERING OVEN htt,me
f four *m* de frittage
e horno *m* de sinterización
i forno *m* di sinterizzazione
n sinteroven
d Sinterofen *m*

2386 FROG SUIT nue,sa
f vêtements *pl* de sûreté
e vestimentas *pl* de seguridad
i scafandro *m*
n beschermende kleding
d Hautanzug *m*, Schutzkleidung *f*

2387 FRONT SHOE (GB), TOP CLAMP PLATE (US) pl
f plaque *f* d'entrée, plaque *f* supérieure
e placa *f* delantal de sujeción del molde
i fondello *m*, piastra *f* di base fissa
n bovenste opspanplaat
d Aufspannkörper *m* hinten

2388 FROSTING PLANT gl
f installation *f* de matage
e equipo *m* de esmerilado
i impianto *m* di smerigliatura
n matteerinstallatie
d Mattieranlage *f*

2389 FROTH FLOTATION PLANT me
f installation *f* de flottation par écume
e equipo *m* de flotación por espum
i impianto *m* di flottazione media schiuma
n schuimflotatie-installatie
d Schaumflotationsanlage *f*

2390 FUEL BED fue,ht
f lit *m* de combustible

e carga *f* de combustible
i carica *f* di combustibile
n brandstoflaag
d Brennstofflage *f*, Brennstoffschicht *f*

2391 FUEL CELL, GAS CELL elc
f pile *f* à gaz
e pila *f* de gas
i pila *f* a gas
n brandstofelement *n*
d Brennstoffelement *n*

2392 FUEL CHANNEL nue
f canal *m* de combustible
e canal *m* de combustible
i canale *m* di combustibile
n brandstofkanaal *n*, splijtstofkanaal *n*
d Brennstoffkanal *m*

2393 FUEL ELEMENT CORE nue
f noyau *m* d'un élément combustible
e núcleo *m* de un elemento combustible
i anima *f* di un elemento combustibile
n brandstofelementkern, splijtstofelementkern
d Brennstoffelementkern *m*, Spaltstoffelementkern *m*

2394 FUEL-FIRED FURNACE fue,htt
f four *m* chauffé au mazout
e horno *m* caldeado con aceite combustible
i forno *m* riscaldato con olio combustibile
n stookolieoven
d Heizölofen *m*

2395 FUEL LATTICE nue
f réseau *m* à éléments combustibles
e red *f* de elementos combustibles
i reticolo *m* ad elementi combustibili
n brandstofrooster *n*
d Brennstoffgitter *n*

2396 FUEL PIN nue
f goupille *f* de combustible
e tornillo *m* de combustible
i ago *m* di combustibile
n brandstofstift, splijtstofstift
d Brennstoffstift *m*, Spaltstoffstift *m*

2397 FUEL ROD nue
f barre *f* de combustible
e barra *f* de combustible
i barra *f* di combustibile
n brandstofstaaf, splijtstofstaaf
d Brennstoffstab *m*, Spaltstoffstab *m*

2398 FUGITOMETER dye,ins
f fugitomètre *m*
e fugitómetro *m*
i fugitometro *m*
n fugitometer
d Fugitometer *n*

2399 FULL WIDTH SCOURING MACHINE tex
f laveuse *f* de grande largeur
e lavadora *f* a la ancha
i lavatrice *f* di gran larghezza
n breedwasmachine
d Breitwaschmaschine *f*

2400 FULLING MILL, MILLING MACHINE tex
f fouleuse *f*
e batanadora *f*
i mulino *m* da folletare
n volmolen
d Walke *f*

2401 FUME CLOSET, FUME HOOD lab
f hotte *f*
e campana *f* de humo, humero *m*
i cappa *f* per fumi, fumaiolo *m*
n afzuigkap
d Abzug *m*, Rauchfang *m*

2402 FUME CUPBOARD lab
f sorbonne *f*
e campana *f* de humos
i cappa *f* di ventilazione
n zuurkast
d Abzug *m*, Abzugschrank *m*

2403 FUMIGATION ROOM nut
f chambre *f* de fumigation
e cámara *f* de fumigación
i camera *f* di fumigazione
n rookhok *n*
d Räucherraum *m*

2404 FUNNEL, CONE, HOPPER cpt
f entonnoir *m*
e embudo *m*
i imbuto *m*, tramoggia *f*
n trechter
d Trichter *m*

2405 FURNACE htt
f four *m*

e horno *m*
i forno *m*
n oven
d Ofen *m*

2406 FURNACE BOTTOM, HEARTH — htt
f sole *f* de four
e solera *f* de horno
i suola *f* di forno
n haardbodem
d Herdsohle *f*

2407 FURNACE CHARGE — htt
f charge *f* d'un four
e carga *f* de horno
i carica *f* di forno
n ovencharge, ovenlading
d Ofencharge *f*, Ofeneinsatz *m*, Ofengut *n*

2408 FURNACE HEARTH — gla,met
f âtre *m*
e hogar *m* de horno
i focolare *m* di forno
n ovenhaard
d Ofenherd *m*

2409 FURNACE LINING — htt
f garnissage *m* de four
e guarnición *f* de horno
i guarnizione *f* di forno
n ovenvoering
d Ofenfutter *n*

2410 FURNISHING ROLLER — tex
f rouleau *m* de teinture
e rodillo *m* tintador
i rullo *m* tintore
n verfrol
d Färbrolle *f*

2411 FUSED ELECTROLYTE CELL — elc
f pile *f* à électrolyte fondu
e pila *f* de electrólito fundido
i pila *f* ad elettrolito fuso
n gesmolten-elektrolytcel
d Hochtemperaturelement *n*

2412 FUSED ELECTROLYTE CONTAINER, SHELL — elc
f cuve *f* à électrolyte fondu
e cuba *f* de electrólito fundido
i tino *m* ad elettrolito fuso
n gesmolen-elektrolytbak
d Wanne *f* für geschmolzenen Elektrolyt

2413 FUSIBLE ARC WELDING ELECTRODE — wel
f électrode *f* fusible pour soudage à l'arc
e electrodo *m* fusible para soldadura ad arco
i elettrodo *m* fusibile per saldatura ad arco
n afsmeltende booglaselektrode
d Abschmelzelektrode *f*

2414 FUSING OVEN — htt
f four *m* de fusion
e horno *m* de fusión
i forno *m* di fusione
n smeltoven
d Schmelzofen *m*

2415 FUSION CASTING MOLD (US), FUSION CASTING MOULD (GB) — cer
f moule *m* de coulée à fusion
e molde *m* de colada de fusión
i forma *f* di colata a fusione
n smeltgietvorm
d Schmelzgussform *f*

2416 FUSION CONES, PYROMETRIC CONES, SEGER CONES — tes
f cônes *pl* de Seger, cônes *pl* pyrométriques
e conos *pl* de Seger, conos *pl* pirométricos
i coni *pl* di Seger, coni *pl* pirometrici
n segerkegels *pl*, smeltkegels *pl*
d Schmelzkegel *pl*, Segerkegel *pl*

2417 FUSION WELDING PLANT — wel
f installation *f* de soudure par fusion
e equipo *m* de soldadura por fusión
i impianto *m* di saldatura per fusione
n smeltlasinstallatie
d Schmelzschweissanlage *f*

G

2418 GABLE WALL gla
f barrage *m* d'alimentation
e pared *f* alimentadora
i parete *f* alimentatrice
n voedingswand
d Speisewand *f*

2419 GAGE (US), GAUGE (GB), MANOMETER ins
f manomètre *m*
e manómetro *m*
i manometro *m*
n manometer
d Manometer *n*

2420 GAGE COCK (US), GAUGE COCK (GB), PET COCK (US), TEST COCK (GB) boi
f robinet *m* d'essai, robinet *m* de jauge, robinet *m* de niveau d'eau
e grifo *m* de prueba, grifo *m* indicador del nivel
i rubinetto *m* di prova
n proefkraan, waterpeilkraan
d Probierhahn *m*, Wasserstandshahn *m*

2421 GAGE GLASS (US), GAUGE GLASS (GB), GLASS GAGE (US), GLASS GAUGE (GB) boi
f indicateur-tube *m*, tube *m* indicateur d'eau
e tubo *m* de nivel, tubo *m* de vidrio indicador
i indicatore *m* di livello a tubo di vetro
n peilglas *n*
d Wasserstandsglas *n*

2422 GAGE ROLLS (US), GAUGING ROLLS (GB) pla
f cylindres *pl* de calibrage
e cilindros *pl* de calibración
i cilindri *pl* calibratori
n kalibreerwalsen *pl*
d Kalibrierwalzen *pl*

2423 GALACTOMETER, LACTODENSITOMETER, LACTOMETER ins
f galactomètre *m*, pèse-lait *m*
e galactómetro *m*
i galattometro *m*
n galactometer, melkmeter
d Galaktometer *n*, Milchwaage *f*

2424 GALVANIC CELL elc
f cellule *f* galvanique
e celda *f* galvánica
i cella *f* galvanica
n galvanisch element *n*, galvanische cel
d galvanisches Element *n*

2425 GALVANIZING BATH coa
f bain *m* de galvanisation, bain *m* de zingage
e baño *m* de cincado, baño *m* de galvanización
i bagno *m* di galvanizzazione, bagno *m* di zincatura
n galvaniseerbad *n*, zinkbad *n*
d Galvanisierbad *n*, Zinkbad *n*

2426 GALVANIZING PLANT coa
f installation *f* de zingage
e equipo *m* de cincado
i impianto *m* di zincatura
n verzinkingsinstallatie
d Verzinkungsanlage *f*

2427 GALVANIZING TANK coa
f récipient *m* de zingage
e tanque *m* de cincado
i serbatoio *m* di zincatura
n verzinkingstank
d Verzinkungsgefäss *n*

2428 GANGUE met,min
f gangue *f*
e ganga *f*
i ganga *f*
n ganggesteente *n*, gangmineraal *n*
d Ganggestein *n*, Gangmineral *n*

2429 GAP TYPE PRESS, GOOSENECK PRESS, JAW TYPE PRESS, OPEN SIDE PRESS, SWANNECK PRESS rub
f presse *f* à col de cygne, presse *f* ouverte sur un côté
e prensa *f* abierta por un lado, prensa *f* acodillada, prensa *f* en cuello de cisne
i pressa *f* a collo di cigno
n muilpers, zwanehalspers
d einseitig offene Presse *f*, Maulpresse *f*, Schwanenhalspresse *f*

2430 GAPMETER, FLOWMETER WITH FLOATING INDICATOR — ins
f débitmètre *m* à flotteurs
e flujómetro *m* con indicadores flotantes
i flussometro *m* a galleggianti
n rotameter
d Rotamesser *m*

2431 GARBAGE GRINDER — iwt
f broyeur *m* d'ordures, délacérateur *m*
e triturador *m* de basuras
i laceratore *m* d'immondizie
n afvalmolen, vuilnismolen
d Müllwolf *m*, Müllzerkleinerer *m*

2432 GAS ANALYZER — ana,lab
f analyseur *m* de gaz
e analizador *m* de gas
i analizzatore *m* di gas
n gasanalysetoestel *n*
d Gasanalyseapparat *m*

2433 GAS-AND-PRESSURE-AIR BURNER — bur
f brûleur *m* à gaz et à air comprimé
e quemador *m* de gas y de aire comprimida
i bruciatore *m* a gas ed ad aria compressa
n gas-persluchtbrander
d Gas-Pressluftbrenner *m*

2434 GAS BALANCE — ins
f balance *f* pour gaz
e balanza *f* para gas
i bilancia *f* per gas
n gasbalans
d Gaswaage *f*

2435 GAS BOTTLE, GAS CYLINDER — lab,tra
f cylindre *m* à gaz
e botella *f* de gas, cilindro *m* de gas
i bombola *f* di gas
n gascilinder, gasfles
d Gaszylinder *m*

2436 GAS BURETTE — lab
f burette *f* à gaz
e bureta *f* de gas
i buretta *f* a gas
n gasburet
d Gasbürette *f*

2437 GAS BURNER — bur,lab
f bec *m* à gaz, brûleur *m* à gaz
e mechero *m* de gas, quemador *m* de gas
i becco *m* a gas, bruciatore *m* a ga
n gasbrander
d Gasbrenner *m*

2438 GAS CALORIMETER — ins
f calorimètre *m* à gaz
e calorímetro *m* de gas
i calorimetro *m* a gas
n gascalorimeter
d Gaskalorimeter *n*

2439 GAS CARTRIDGE FIRE EXTINGUISHER — fea
f extincteur *m* à bonbonne à gaz
e apagador *m* con botella de gas
i estintore *m* con bombola a gas
n brandblusser met gaspatroon
d Gaspatronenfeuerlöscher *m*

2440 GAS CELL, FUEL CELL — el(
f pile *f* à gaz
e pila *f* de gas
i pila *f* a gas
n brandstofelement *m*
d Brennstoffelement *n*

2441 GAS CLEANER, GAS CONDITIONER, GAS PURIFIER — cle,gas,la
f épurateur *m* de gaz
e depurador *m* de gas
i depuratore *m* di gas
n gasreiniger
d Gasreiniger *m*

2442 GAS COLLECTOR — ga
f collecteur *m* de gaz
e colector *m* de gases
i collettore *m* di gas
n gasopvanginrichting
d Gassammler *m*

2443 GAS-COOLED HEAVY WATER REACTOR — nu
f réacteur *m* à eau lourde à refroidissement au gaz
e reactor *m* de agua pesada refrigerado por gas
i reattore *m* ad acqua pesante raffreddato a gas
n met gas gekoelde zwaar-water-reactor
d gasgekühlter Schwerwasser-reaktor *m*

2444 GAS COOLED REACTOR, GAS REACTOR — n
f réacteur *m* à refroidissement a

e reactor *m* refrigerado por gas
i reattore *m* a gas, reattore *m* raffreddato a gas
n met gas gekoelde reactor
d gasgekühlter Reaktor *m*

2445 GAS COOLER clo,gas
f réfrigérant *m* de gaz, refroidisseur *m* de gaz
e refrigerador *m* de gas
i raffreddatore *m* di gas
n gaskoeler
d Gaskühler *m*

2446 GAS DETECTOR sap
f indicateur *m* de grisou, signaleur *m* de grisou
e gasoscopio *m*, grisuómetro *m*, grisuscopio *m*
i indicatore *m* di grisu
n mijngasverklikker
d Grubengasanzeiger *m*, Schlagwetteranzeiger *m*

2447 GAS DRIER, GAS DRYING APPARATUS dry,gas
f sécheur *m* de gaz
e secador *m* de gas
i essiccatore *m* di gas
n gasdroger
d Gastrockner *m*

2448 GAS ELECTRODE gas
f électrode *f* à gaz
e electrodo *m* de gas
i elettrodo *m* a gas
n gaselektrode
d Gaselektrode *f*

2449 GAS EXHAUSTER gas
f extracteur *m* de gaz
e extractor *m* de gases
i estrattore *m* di gas
n gasafzuiger
d Gassauger *m*

2450 GAS EXPELLER PLANT, DEGASIFYING PLANT gas
f installation *f* dégazeuse
e equipo *m* de desgasificación
i impianto *m* degassificatore
n ontgassingsinstallatie
d Entgasungsanlage *f*

2451 GAS-EXPELLING FURNACE, BAKING FURNACE cer,htt
f four *m* pour expulsion de gaz
e horno *m* para expulsión de gases
i forno *m* per espulsione di gas
n gasuitdrijfoven
d Gasaustreibeofen *m*

2452 GAS FILTER fil
f filtre *m* de gaz
e filtro *m* de gases
i filtro *m* di gas
n gasfilter
d Gasfilter *m*

2453 GAS-FIRED FURNACE htt
f four *m* à gaz
e horno *m* de gas
i forno *m* a gas
n gasoven
d Gasofen *m*

2454 GAS FLOWMETER ins
f débitmètre *m* à gaz
e flujómetro *m* de gases
i flussometro *m* a gas
n gasstromingsmeter
d Gasströmungsmesser *m*

2455 GAS GENERATOR gas
f gazogène *m*, générateur *m* de gaz
e gasógeno *m*, generador *m* de gas
i gassogeno *m*, generatore *m* di gas
n gasgenerator
d Gasgenerator *m*

2456 GAS-GRAPHITE REACTOR nue
f réacteur *m* à gaz-graphite
e reactor *m* de gas-grafito
i reattore *m* a gas-grafite
n gas-grafietreactor
d Gas-Graphitreaktor *m*

2457 GAS JACKET clo
f gaine *f* de gaz
e camisa *f* de circulación de gas
i camicia *f* di gas
n gasmantel
d Gasumhüllung *f*

2458 GAS-LIQUID CHROMATOGRAPH chr
f chromatographe *m* gaz-liquide
e cromatógrafo *m* gas-líquido
i cromatografo *m* gas-liquido
n gas-vloeistofchromatograaf
d Gas-Flüssigkeit-Chromatograph *m*

2459 GAS MANOMETER ins
f manomètre *m* à gaz
e manómetro *m* de gas
i manometro *m* a gas
n gasmanometer
d Gasmanometer *n*

2460 GAS MANTLE, INCANDESCENT MANTLE gas
f manchon *m* de bec à incandescence
e manguito *m* incandescente para gas
i reticella *f* ad incandescenza
n gasgloeilichtkousje *n*, gloeikousje *n*
d Glühstrumpf *m*

2461 GAS MASK sap
f masque *f* anti-gaz
e máscara *f* antigas
i maschera *f* antigas
n gasmasker *n*
d Gasmaske *f*

2462 GAS MASK FILTER, CANISTER sap
f filtre *m* de masque anti-gaz
e filtro *m* de máscara antigás
i filtro *m* di maschera antigas
n gasmaskerfilter *n*
d Filterbüchse *f*, Filtereinsatz *m*

2463 GAS METER ins
f compteur *m* de gaz
e gasómetro *m*
i contatore *m* di gas
n gasmeter
d Gasmesser *m*

2464 GAS MIXTURE SEPARATION PLANT gas
f installation *f* de séparation de mélange gazeux
e equipo *m* separador de mezcla gaseosa
i impianto *m* separatore di miscugli di gas
n gasscheidingsinstallatie
d Gaszerlegungsanlage *f*

2465 GAS PASSAGE, DUCT cpt
f caniveau *m* de ventilation
e conducto *m* de gas
i condotto *m* di gas
n gaskanaal *n*
d Gaskanal *m*

2466 GAS PICKLING PLANT cle
f installation *f* de décapage au gaz
e equipo *m* de decapaje con gas
i impianto *m* di decapaggio al gas
n gasbeitsinstallatie
d Gasbeizerei *f*

2467 GAS PIPETTE lab
f pipette *f* à gaz
e pipeta *f* de gas
i pipetta *f* a gas
n gaspipet
d Gaspipette *f*

2468 GAS PIPING tra
f conduite *f* de gaz
e tubería *f* de gas
i condotta *f* di gas
n gasleiding
d Gasleitung *f*

2469 GAS PRODUCING OVEN ga
f four *m* gazogène
e horno *m* generador de gas
i forno *m* generatore di gas
n cokesgasoven
d Gaswerkofen *m*

2470 GAS REGENERATOR ga
f régénerateur *m* de gaz
e regenerador *m* de gas
i rigeneratore *m* di gas
n gasregenerator
d Gasregenerator *m*

2471 GAS REGULATOR, PRESSURE GOVERNOR gas,sa
f régulateur *m* de la pression du
e regulador *m* de la presión del g
i regolatore *m* della pressione de gas
n gasdrukregelaar
d Gasdruckregler *m*

2472 GAS RETORT gas,h
f cornue *f* à gaz
e retorta *f* de gas
i storta *f* a gas
n gasretort
d Gasretorte *f*

2473 GAS SAMPLER te
f échantillonneur *m* de gaz
e muestrador *m* de gas
i apparecchio *m* di campionatura di gas
n gasmonsternemer
d Gasprobenehmer *m*

2474 GAS SAMPLING TUBE ana,g
f tube *m* échantillonneur de gaz
e tubo *m* muestrador de gas
i tubo *m* campione di gas
n gasmonsterbuis
d Gassammelrohr *n*

2475 GAS SAMPLING VALVE c
f soupape *f* d'introduction de l'échantillon

e válvula *f* de introducción de la muestra gaseosa
i valvola *f* d'introduzione del campione di gas
n gasmonsteruitgever
d Gasprobegeber *m*

2476 GAS SCRUBBER cle
f laverie *f* de gaz
e lavador *m* de gas, lavagases *m*
i lavatore *m* di gas
n gaswasserij
d Gaswäsche *f*, Gaswäscherei *f*

2477 GAS SEPARATOR fue
f séparateur *m* de gaz
e separador *m* de gas
i separatore *m* di gas
n gasafscheider
d Gasabscheider *m*

2478 GAS-SOLID CHROMATOGRAPH ana
f chromatographe *m* gaz-solide
e cromatógrafo *m* gas-sólido
i cromatografo *m* gas-solido
n gas-vaste-stof-chromatograaf
d Gas-Feststoff-Chromatograph *m*

2479 GAS STREAM REGULATOR con
f flacon *m* de Mariotte, régulateur *m* du débit de gaz
e regulador *m* de la corriente gaseosa
i regolatore *m* della corrente di gas
n gasstroomregelaar
d Gasstromregler *m*

2480 GAS-TIGHT CASING, GAS-TIGHT ENVELOPE gen
f enveloppe *m* étanche à l'air
e envoltura *f* hermética
i involucro *m* a tenuta di gas
n gasdicht omhulsel *n*
d gasdichte Umhüllung *f*

2481 GAS WASHING BOTTLE lab
f flacon *m* laveur de gaz
e frasco *m* lavador de gas
i boccetta *f* lavatrice di gas
n gaswasfles, wasfles
d Gaswaschflasche *f*, Waschflasche *f*

2482 GAS WASHING PLANT cle,gas
f installation *f* de lavage de gaz
e equipo *m* lavagases
i impianto *m* di lavaggio di gas
n gaswasinstallatie
d Gaswäscheanlage *f*

2483 GAS WEIGHING BALLOON ana
f pèse-gaz *m*
e frasco *m* de pesada de gas
i pesagas *m*
n gasweegflesje *n*
d Gaswägepipette *f*

2484 GAS WELDING PLANT wel
f installation *f* de soudure à flamme de gaz
e equipo *m* de soldadura por llama de gas
i impianto *m* di saldatura per fiamma di gas
n gasvlamlasinstallatie
d Gasflammschweissanlage *f*

2485 GAS WORKS gas
f usine *f* à gaz
e fábrica *f* de gas
i fabbrica *f* di gas
n gasfabriek
d Gasanstalt *f*, Gaswerk *n*

2486 GASHOLDER, GASOMETER sto
f gazomètre *m*
e gasómetro *m*
i gassometro *m*
n gashouder
d Gasbehälter *m*, Gasometer *m*

2487 GASIFYING CHAMBER fue
f chambre *f* gazéificatrice
e cámara *f* de gasificación
i camera *f* di gasificazione
n vergassingsruimte
d Vergaserraum *m*

2488 GASKET cpt
f garniture *f* d'étanchéité, rondelles *pl* de joints
e guarnición *f* de empaquetadura
i guarnizione *f* di cilindro, guarnizione *f* di flangia
n cilinderpakking, flenspakking
d Dichtungsring *m*, Dichtungsscheibe *f*

2489 GASKET GROOVE cpt
f gorge *f* pour la garniture
e rebajo *m* para la guarnición
i alloggiamento *m* per la guarnizione, gola *f* per la guarnizione
n pakkingsgroef
d Packungsrille *f*

2490 GASKET RING, JOINT RING, PACKING RING cpt
f bague *f* de garniture

e arandela *f* de guarnición
i anillo *m* di guarnizione
n pakkingring
d Dichtungsring *m*, Manschette *f*

2491 GASOLINE SEPARATOR (US), fue
GASOLINE TRAP (US), PETROL SEPARATOR (GB), PETROL TRAP (GB)
f séparateur *m* d'essence
e separador *m* de gasolina
i separatore *m* di benzina
n benzineafscheider
d Benzinabscheider *m*

2492 GASSING APPLIANCE, tex
SINGEING APPARATUS
f dispositif *m* de flambage, dispositif *m* de gazage
e dispositivo *m* de chamusquina, dispositivo *m* de gaseado
i dispositivo *m* di bruciatura, dispositivo *m* di gasatura
n afzengapparaat *n*, flambeer-apparaat *n*, zengapparaat *n*
d Flambierapparat *m*, Gasiergerät *n*, Senggerät *n*, Versengapparat *m*

2493 GATE gla
f brique *f* de fermeture
e ladrillo *m* de cierre
i mattone *m* di chiusura
n afsluitsteen
d Abschliessstein *m*

2494 GATE (US), pla
FEED ORIFICE (GB)
f entrée *f* d'injection
e entrada *f*
i apertura *f* d'entrata
n aanspuitopening
d Anschnitt *m*

2495 GATE, cpt
GATE VALVE, SLUICE VALVE
f obturateur *m*, robinet-vanne *m*
e válvula *f* de compuerta, válvula *f* de paso directo
i saracinesca *f*
n afsluitschuif
d Absperrschieber *m*, Abzugsschieber *m*

2496 GATE, pla
INLET
f canal *m* secondaire, entonnoir *m* de coulée, entrée *f*
e entrada *f*
i entrata *f*
n aanspuitkanaal *n*, aanspuitpoort
d Angusssteg *m*, Steg *m*

2497 GATE AGITATOR, mi
GATE MIXER
f agitateur *m* à grille tournante, malaxeur *m* à cadre
e agitador *m* de rejilla, mezclador *m* de rejilla
i agitatore *m* con asta rotante, mescolatore *m* a griglia
n roostermenger, roosterroer-werk *n*
d Gitterrührwerk *n*, Rostmischer

2498 GATE AND CULL pl
f culot *m*
e mazarota *f* y resto *m* de transferencia
i materozza *f*
n aanspuiting
d Anguss *m*

2499 GATE CUTTER pl
f coupe-carotte *m*, pince *f* coupante
e cortador *m* de bebederos, tenazas *pl*
i tronchettino *m*, tronchetto *m*
n aanspuitingsafsnijder, kniptang
d Abkneifzange *f*, Angussabschneider *m*

2500 GATE MARK p
f trace *f* de la carotte
e marca *f* de entrada
i zona *f* d'iniezione
n aanspuitplaats
d Angussstelle *f*

2501 GATE PADDLE IMPELLER m
f roue *f* motrice à palettes ouvertes
e rueda *f* móvil de paletas abiertas
i girante *m* a palette aperte
n roerder met open schoepen
d Rührstange *f* mit durchlochten Paddeln

2502 GATE VALVE, c
STRAIGHT-WAY VALVE
f soupape *f* droite
e válvula *f* de corredera, válvul de paso directo
i valvola *f* a saracinesca
n doorlaatklep, in-en uitlaatklep
d Durchgangsventil *n*

2503 GATHERING HOLE g
f ouvreau *m*, trou *m* de cueillage

e abertura *f* de presa
i apertura *f* di prelievo
n aftapopening
d Aushebeloch *n*

2504 GAUGE (GB), GAGE (US), MANOMETER — ins
f manomètre *m*
e manómetro *m*
i manometro *m*
n manometer
d Manometer *n*

2505 GAUGE COCK (GB), GAGE COCK (US), PET COCK (US), TEST COCK (GB) — boi
f robinet *m* d'essai, robinet *m* de jauge, robinet *m* de niveau d'eau
e grifo *m* de prueba, grifo *m* indicador del nivel
i rubinetto *m* di prova
n proefkraan, waterpeilkraan
d Probierhahn *m*, Wasserstandshahn *m*

2506 GAUGE GLASS (GB), GAGE GLASS (US), GLASS GAGE (US), GLASS GAUGE (GB) — boi
f indicateur-tube *m*, tube *m* indicateur d'eau
e tubo *m* de nivel, tubo *m* de vidrio indicador
i indicatore *m* di livello a tubo di vetro
n peilglas *n*
d Wasserstandsglas *n*

2507 GAUGING ROLLS (GB), GAGE ROLLS (US) — pla
f cylindres *pl* de calibrage
e cilindros *pl* de calibración
i cilindri *pl* calibratori
n kalibreerwalsen *pl*
d Kalibrierwalzen *pl*

2508 GAUZE STRAINER (GB), GAZE STRAINER (US) — fil
f tamis *m* métallique
e tamiz *m* metálico
i staccio *m* in garza metallica
n gaaszeef, metaalgaaszeef
d Gazesieb *n*

2509 GAY-LUSSAC TOWER, NITRIDING TOWER — sua
f tour *f* de nitruration
e torre *f* de nitruración
i torre *f* di nitrurazione
n nitreertoren
d Nitrierturm *m*

2510 GEAR-SHAPED ANODE, RIPPLE-ROUND ANODE — elc
f anode *f* crénelée, anode *f* dentelée
e ánodo *m* dentado
i anodo *m* dentato
n tandanode
d Zahnanode *f*

2511 GEAR-TYPE SPINNING PUMP — tex
f pompe *f* à extrusion à engrenages, pompe *f* de filature à engrenages
e bomba *f* hiladora de engrenajes
i pompa *f* di filatura ad ingranaggi
n tandradspinpomp
d Zahnradspinnpumpe *f*

2512 GELATINE STOVE — lab
f étuve *f* à gélatine
e estufa *f* de gelatina
i stufa *f* a gelatina
n gelatinestoof
d Gelatineofen *m*

2513 GELATINIZATION PLANT — oic
f installation *f* de gélatinisation
e equipo *m* de gelatinización
i impianto *m* di gelatinizzazione
n gelatineerinstallatie
d Gelatinierungsanlage *f*

2514 GENERAL LABORATORY BALANCE — lab
f trébuchet *m*
e balanza *f* universal de labor
i bilancia *f* universale di laboratorio
n precisiebalans, universele laboratoriumbalans
d Apothekerwaage *f*, Universallaboratoriumswaage *f*

2515 GENERATOR — gen
f générateur *m*
e generador *m*
i generatore *m*
n generator, ontwikkelaar
d Entwickler *m*, Generator *m*

2516 GENTLEMAN, ROTARY SOLDERING TABLE — wel
f table *f* rotative pour soudure
e mesa *f* giratoria para soldadura
i tavola *f* rotativa per saldatura
n roterende soldeertafel
d drehender Löttisch *m*

2517 GERM-PROOF FILTER bac,fil
f filtre *m* stérilisant
e filtro *m* de bacterias
i filtro *m* di batterie, filtro *m* sterilizzatore
n bacteriefilter *n*, steriel filter *n*
d Bakterienfilter *m*, EK-Filter *m*, Entkeimungsfilter *m*

2518 GERMAN NOZZLE, PARABOLIC NOZZLE pla
f buse *f* parabolique
e tobera *f* parabólica
i ugello *m* parabolico
n parabolisch mondstuk *n*
d Paraboldüse *f*

2519 GERMICIDAL LAMP bac,hyg
f lampe *f* germicide
e lámpara *f* germicida
i lampada *f* germicida
n kiemdodende lamp
d keimtötende Lampe *f*

2520 GERMINATING BOX bre
f germoir *m*
e caja *f* de germinación, germinador *m*
i germinatoio *m*
n kiemkast
d Keimkasten *m*

2521 GERMINATING DRUM bre
f tambour-germeur *m*
e tambor *m* de germinación
i germinatoio *m* a tamburo
n kiemtrommel
d Keimtrommel *f*

2522 GERMINATING FLOOR, GROWING FLOOR, MALTING FLOOR bre
f aire *f* de germination, germoir *m*
e era *f* de germinación
i aia *f* di germinazione
n kiemvloer
d Keimboden *m*, Keimtenne *f*, Malztenne *f*

2523 GILDING BATH coa
f bain *m* de dorure
e baño *m* de doradura
i bagno *m* di doratura
n goudbad *n*
d Goldbad *n*

2524 GLASS BEAD lab
f bille *f* en verre, perle *f* en verre
e abalorio *m*, cuenta *f* de vidrio, perla *f*. de vidrio
i perla *f* di vetro
n glasparel
d Glasperle *f*

2525 GLASS CRUCIBLE, GLASS MELTING POT, GLASS POT gla
f creuset *m* de verrerie
e crisol *m* de vidriería
i crogiolo *m* di vetreria
n glashaven, glaspot
d Glashafen *m*

2526 GLASS ELECTRODE ana
f demi-cellule *f* en verre, électrode *f* de verre
e electrodo *m* de vidrio, semicelda *f* de vidrio
i elettrodo *m* di vetro
n glaselektrode
d Glashalbzelle *f*

2527 GLASS FILTER fil
f filtre *m* en verre
e filtro *m* de vidrio
i filtro *m* di vetro
n glasfilter *n*, glazen filter *n*
d Glasfilter *m*

2528 GLASS FILTER CRUCIBLE fil,lab
f creuset *m* filtrant en verre
e crisol *m* filtrante de vidrio
i crogiolo *m* filtrante in vetro
n glasfilterkroes
d Glasfiltertiegel *m*

2529 GLASS FILTER DISK fil,lab
f plaque *f* filtrante
e placa *f* filtrante
i piastra *f* filtrante
n glasfilterplaat
d Glasfilterplatte *f*

2530 GLASS FURNACE, TANK FURNACE gla
f four à bassins
e horno *m* de cubas
i forno *m* a tini
n wanoven
d Glaswannenofen *m*, Wannenofen *m*

2531 GLASS GAGE (US), GLASS GAUGE (GB), GAGE GLASS (US), GAUGE GLASS (GB) boi
f indicateur-tube *m*, tube *m*, indicateur d'eau
e tubo *m* de nivel, tubo *m* de vidrio indicador
i indicatore *m* di livello a tubo di vetro

n peilglas *n*
d Wasserstandsglas *n*

2532 GLASS-LINED TANK sto
f récipient *m* vitrifié
e tanque *m* vidriado
i serbatoio *m* invetriato
n tank met glasvoering
d Behälter mit Glasfutter

2533 GLASS MOLD (US), GLASS MOULD (GB) gla
f moule *f* de verrerie
e prensa *f* de moldear vidrio
i stampo *m* di vetreria
n glasvorm
d Glasform *f*

2534 GLASS MORTAR lab
f mortier *m* en verre
e mortero *m* de vidrio
i mortaio *m* di vetro
n glazen mortier
d Glasmörser *m*

2535 GLASS PIPING tra
f tuyauterie *f* en verre
e tubería *f* de vidrio
i tubatura *f* di vetro
n glazen buisleiding
d Glasrohrleitung *f*

2536 GLASS PLATE lab
f plaque *f* en verre
e placa *f* de vidrio
i lastra *f* di vetro, piastra *f* di vetro
n glasplaat
d Glasplatte *f*

2537 GLASS ROD gla
f tige *f* en verre
e varilla *f* de vidrio
i asta *f* di vetro
n glasstaaf
d Glasstab *m*

2538 GLASS SPOON lab
f cuiller *f* de verre
e cuchara *f* de vidrio
i cucchiaio *m* di vetro
n glazen lepel
d Glaslöffel *m*

2539 GLASS STOPPER lab
f bouchon *m* en verre
e tapón *m* de cristal esmerilado
i tappo *m* di vetro
n glazen stop
d Glasstopfen *m*

2540 GLASS-STOPPERED BOTTLE lab
f flacon *m* bouché à l'émeri
e frasco *m* con tapón esmerilado
i boccetta *f* a tappo
n stopfles
d Stopfflasche *f*

2541 GLASS TANK gla
f bassin *m* de verrerie
e cuba *f* de vidriería
i tino *m* di vetreria
n glaswan
d Glaswanne *f*

2542 GLASS TUBE gla,lab
f tube *m* en verre
e tubo *m* de vidrio
i tubo *m* di vetro
n glazen buis
d Glasröhre *f*

2543 GLASS TUBING gla,lab
f verre *m* tubulaire
e vidrio *m* tubular
i vetro *m* tubolare
n buisglas *n*
d Röhrenglas *n*

2544 GLASS WOOL fil,lab
f laine *f* de verre
e lana *f* de vidrio
i lana *f* di vetro
n glaswol
d Glaswolle *f*

2545 GLASSED PADDLE IMPELLER mix
f roue *f* motrice à palettes vitrifiées
e rueda *f* móvil de paletas forradas de vidrio
i girante *m* a pale con rivestimento di vetro
n met glas bedekte schoepen-roerder
d glasierte Paddelrührstange *f*

2546 GLAZE (GB), LINING (US) cer,gla
f brasque *m*
e enlucido *m* refractario
i rivestimento *m* refrattario
n vuurvaste bekleding
d Tiegelglasur *f*

2547 GLAZE FIRING OVEN, GLOST OVEN cer
f four *m* de glaçure
e horno *m* de esmaltar, horno *m* de vidriar

i forno *m* di smaltatura
n glazuuroven
d Glasurofen *m*

2548 GLAZED POT gla
f creuset *m* vitrifié
e crisol *m* vidriado
i crogiolo *m* vetriato
n verglaasde pot
d verglaster Hafen *m*

2549 GLAZING APPARATUS pol
f appareil *m* d'émoussage
e aparato *m* de embotamiento
i apparecchio *m* smussatore
n ontruwingsapparaat *n*
d Schmirgelgerät *n*

2550 GLAZING APPARATUS (GB), fou,htt
RELINING APPARATUS (US)
f appareil *m* de brasquage
e aparato *m* de enlucido de refractario
i apparecchio *m* da rivestire
n bekledingsapparaat *n*
d Glasurbrandgerät *n*

2551 GLAZING CALENDER pap,pla
f calandre *f* à satiner, satineuse *f*
e calandra *f* de lustrar, calandra *f* de satinar, satinadora *f*
i calandra *f* da rasare, rasatrice *f*
n satineerkalander
d Glanzkalander *m*, Satinierkalander *m*

2552 GLAZING CYLINDER, pap
GLAZING ROLL
f cylindre *m* satineur
e cilindro *m* satinador
i cilindro *m* rasatore
n satineerwals
d Glanzwalze *f*, Satinierwalze *f*

2553 GLAZING MACHINE, pho
HIGH GLAZING MACHINE
f glaceuse *f*, machine *f* à glacer
e máquina *f* de glaseado
i lucidatrice *f*
n hoogglansmachine
d Hochglanzmaschine *f*

2554 GLAZING MILL pol
f moulin *m* de polissage préliminaire
e molino *m* de pulido preliminar
i mulino *m* di lucidatura preliminare
n voorpolijstmolen
d Vorpoliermühle *f*

2555 GLAZING PLANT lea
f installation *f* de cirage, installation *f* de parage
e equipo *m* de abrillantado
i impianto *m* di lucidatura
n glansinrichting
d Glanzstossanlage *f*

2556 GLAZING PRESS, pho
FLAT-BED GLAZER
f presse *f* plate à glacer
e prensa *f* de glaseado
i pressa *f* da lucidare
n hoogglanspers
d Hochglanzpresse *f*

2557 GLAZING ROLLS (GB), pla
CALENDER STACK (US)
f calandre *f* finisseuse
e calandra *f* de satinar
i calandra *f* spianatora
n satineerkalander
d Glättkalander *m*

2558 GLOBE VALVE (US), cpt,sap
BALL CUP VALVE (GB),
BALL VALVE (GB)
f clapet *m* à bille, soupape *f* à boulet
e válvula *f* de bola
i valvola *f* a sfera
n kogelklep, kogelventiel *n*
d Ballventil *n*, Kugelventil *n*

2559 GLORY HOLE gla
f four *m* auxiliaire
e horno *m* auxiliar
i forno *m* ausiliario
n hulpoven
d Einbrennofen *m*

2560 GLOSSIMETER, ins,pap
GLOSSMETER
f brillancemètre *m*
e brillancímetro *m*
i misuratore *m* della brillantezza
n glansmeter
d Glanzmesser *m*

2561 GLOVE BOX nuc
f boîte *f* de protection à gants
e cámara *f* de guantes
i camera *f* a guanti, scatola *f* a guanti
n handschoenkast
d Handschuhbox *f*, Schutzkammer mit eingebauten Handschuhen

2562 GLOVER TOWER, su
DENITRIDING TOWER
f tour *f* de dénitruration

e torre *f* de desnitruración
i torre *f* di denitrurazione
n denitreertoren
d Denitrierturm *m*

2563 GLUE COOKER glu
f bouilleur *m* de col
e caldera *f* para cola
i caldaia *f* per colla
n lijmketel, lijmkookketel
d Leimsiedekessel *m*

2564 GLUE COOKING PLANT glu
f fabrique *f* de colle
e fábrica *f* de cola
i fabbrica *f* di colla
n lijmkokerij, lijmziederij
d Leimsiederei *f*

2565 GO-DEVIL, cle
PORCUPINE
f hérisson *m*
e raspatubos *m*
i raschiatore *m* per tubi
n buiskrabber, buisschraper
d Molch *m*, Rohrwischer *m*

2566 GOB BURNER gla
f brûleur *m* de gouttes
e quemador *m* de gotas
i bruciatore *m* di sbozzi
n glaspostbrander
d Postenbrenner *m*

2567 GOB FEEDER gla
f alimentateur *m* de gouttes
e alimentador *m* de gotas
i alimentatore *m* di sbozzi
n glasdruppeltoevoer, glasportietoevoer, glaspost-toevoer
d Glaspostenzufuhr *f*

2568 GOGGLES sap
f lunettes *m* de protection
e gafas *pl* protectores
i occhiali *pl* di protezione
n veiligheidsbril
d Schutzbrille *f*

2569 GOLD DETERMINATION FLASK ana,lab
f matras *m* d'essayeur
e matraz *m* de ensayo de oro
i matraccio *m* di prova d'oro
n goudkolfje *n*
d Goldkölbchen *n*

2570 GOLD DIP BATH elc
f bain *m* de dorure par immersion
e baño *m* de doradura de inmersión
i bagno *m* di doratura per immersione
n dompelverguldbad *n*
d Tauchvergoldungsbad *n*

2571 GOLD PLATING BATH coa
f bain *m* de dorure
e baño *m* de plaqueado al oro
i bagno *m* di placcatura all'oro
n goudpleetinstallatie
d Goldplattieranlage *f*

2572 GOLD REDUCTION PLANT min
f installation *f* d'extraction d'or
e equipo *m* de extracción de oro
i impianto *m* d'estrazione d'oro
n goudextractie-installatie, gouduitsmelterij
d Goldausschmelzerei *f*

2573 GOLD TONING BATH pho
f bain *m* à trichlorure d'or
e baño *m* de tricloruro de oro
i bagno *m* a tricloruro d'oro
n goudbad *n*
d Goldtrichloridbad *n*

2574 GOLDBEATING WORKSHOP coa
f atelier *m* de battement d'or
e taller *m* de batidor de oro
i officina *f* di battitura d'oro
n goudslagerij
d Blattgoldschlägerei *f*

2575 GOOSENECK PRESS, rub
GAP TYPE PRESS, JAW TYPE PRESS, OPEN SIDE PRESS, SWANNECK PRESS
f presse *f* à col de cygne, presse *f* ouverte sur un côté
e prensa *f* abierta por un lado, prensa *f* acodillada, prensa *f* en cuello de cisne
i pressa *f* a collo di cigno
n muilpers, zwanehalspers
d einseitig offene Presse *f*, Maulpresse *f*, Schwanenhalspresse *f*

2576 GOUGING TEST APPARATUS tes
f appareil *m* à essayer l'émail
e probador *m* de esmalte
i collaudatore *m* di smalto
n emailbeproevingsapparaat *n*
d Emailprüfgerät *n*

2577 GRADER, sor
GRADING MACHINE
f trieuse *f*

e máquina *f* clasificadora
i cernitrice *f*
n sorteermachine
d Sortiermaschine *f*

2578 GRADING TABLE sor
f table *f* de triage
e mesa *f* de clasificación
i tavola *f* di cernita
n sorteertafel
d Sortiertisch *m*

2579 GRADUATE, PHARMACEUTICAL GRADUATE lab
f éprouvette *f* graduée de pharmacie
e frasco *m* graduado, vaso *m* graduado
i bicchiere *m* graduato
n apothekersmaatglas *n*
d Apothekermensur *f*

2580 GRADUATED CYLINDER, MEASURING CYLINDER lab
f cylindre *m* gradué, éprouvette *f* graduée
e cilindro *m* graduado
i cilindro *m* graduato
n maatcilinder, maatglas *n*
d Messzylinder *m*

2581 GRADUATED FLASK lab
f matras *m* gradué
e matraz *m* graduado
i matraccio *m* graduato
n maatkolf
d Messkolben *m*

2582 GRADUATED PIPETTE (GB), MEASURING PIPETTE (US) lab
f pipette *f* graduée
e pipeta *f* graduado
i pipetta *f* graduata
n meetpipet
d Messpipette *f*

2583 GRAIN CLEANING MACHINE, SMUT MILL nut
f tarare *m*
e máquina *f* para limpiar cereales
i macchina *f* per eliminare il carbonchio dei cereali
n graanreinigingsmachine
d Getreidereinigungsmaschine *f*

2584 GRAIN-OUT PLANT, GRAINING-OUT PLANT, SALTING-OUT PLANT soa
f installation *f* de relargage
e equipo *m* de salado
i impianto *m* di salatura
n uitzoutinstallatie
d Aussalzanlage *f*

2585 GRAIN-OUT VESSEL, GRAINING-OUT VESSEL, SALTING-OUT VESSEL soa
f cuve *f* de relargage
e cuba *f* de salado
i tino *m* di salatura
n uitzoutvat *n*
d Aussalzgefäss *n*

2586 GRAIN ROLL brk
f rouleau *m* de grenage
e rodillo *m* de graneo
i rullo *m* di granitura
n korrelwals
d Kornwalze *f*

2587 GRAINING BOARD lea
f marguérite *f*, paumelle *f*
e tablero *m* de granulado
i legno *m* da granice
n krispelhout *n*
d Krispelholz *n*

2588 GRAINING MACHINE, BOARDING MACHINE lea
f greneuse *f*
e graneladora *f*, máquina *f* de granelar
i macchina *f* da granire
n krispelmachine
d Krispelmaschine *f*

2589 GRAINING PLANT sin
f installation *f* de grenage
e equipo *m* de formación del grano
i impianto *m* di formazione del granello
n greininstallatie
d Kornmacherei *f*

2590 GRAINING ROLL lea
f cylindre *m* de crépissage, cylindre *m* de grenure
e cilindro *m* de granulación
i cilindro *m* da granire
n krispelwals
d Krispelwalze *f*

2591 GRAINING ROLLER cer,dec
f rouleau *m* à marbrer
e rodillo *m* de granelar
i rullo *m* da venare
n aderrol, marmerrol
d Maserierrolle *f*

2592 GRAINS DRIER bre
f séchuer *m* des drêches
e secadero *m* de orujo
i essiccatoio *m* di trebbie
n bosteldroger
d Trebertrockner *m*

2593 GRAINS REMOVER bre
f dédrêcheur *m*
e máquina *f* para sacar el orujo, removedor *m*
i macchina *f* per scaricare le trebbie
n bostelploeg
d Austrebermaschine *f*

2594 GRAINS STIRRER, RAKE bre
f piocheur *m*
e removedora *f*
i mescolatore *m* delle trebbie
n bostelhark
d Aufhackmaschine *f*

2595 GRAINS TANK bre
f silo *m* à drêches
e silo *m* de orujo
i silo *m* per trebbie
n bostelsilo
d Treberkasten *m*

2596 GRANARY bre
f grange *f*
e granero *m*
i granaio *m*
n graanopslagplaats
d Kornkammer *f*

2597 GRANT, UNDERBACK bre
f bac *m* reverdoir, reverdoir
e recipiente *m* colector
i recipiente *m* di raccolta, vasca *f* di raccolta
n lekgoot
d Läutergrand *m*, Läuterrinne *f*

2598 GRANT, UNDERBACK, UNDERLET bre
f batterie *f* de filtration
e batería *f* de filtración
i batteria *f* di chiarificazione
n serie klaringskuipen
d Läuterbatterie *f*, Würzegrand *m*

2599 GRANULATING PIT brk
f puits *m* à granulation
e pozo *m* para granulación
i pozzo *m* a granulazione
n granulatieput
d Granuliergrube *f*

2600 GRANULATOR sin
f granulateur *m*
e granulador *m*
i granulatore *m*
n granulateur
d Granulateur *m*

2601 GRAPHITE ELECTRODE htt
f électrode *f* de graphite
e electrodo *m* de grafito
i elettrodo *m* di grafite
n grafietelektrode
d Graphitelektrode *f*

2602 GRAPHITE LATTICE nue
f réseau *m* d'éléments combustibles modéré au graphite
e red *f* de elementos combustibles moderada con grafito
i reticolo *m* ad elementi combustibili moderato al grafite
n met grafiet gemodereerd splijtstofrooster *n*
d graphitmoderiertes Brennstoffgitter *n*

2603 GRAPHITE MODERATED REACTOR nue
f réacteur *m* modéré au graphite
e reactor *m* moderado con grafito
i reattore *m* moderato al grafite
n met grafiet gemodereerde reactor
d graphitmoderierter Reaktor *m*

2604 GRAPHITE MOLD (US), GRAPHITE MOULD (GB) fou
f moule *m* en graphite
e molde *m* de grafito
i forma *f* in grafite
n grafietvorm
d Graphitform *f*

2605 GRAPHITE PACKING, CARBON PACKING cpt
f garniture *f* de graphite
e empaquetadura *f* de grafito
i guarnizione *f* di grafite
n grafietpakking
d Graphitpackung *f*

2606 GRAPHITE PEBBLES nue
f billes *pl* de graphite
e bolas *pl* de grafito
i sfere *pl* di grafite
n grafietkogels *pl*
d Graphitkugeln *pl*

2607 GRATE fue
f grille *f*

e parrilla *f*
i griglia *f*
n rooster *n*
d Feuerrost *m*, Rost *m*

2608 GRATE MAGNET sep
f grille *f* magnétique
e pantalla *f* magnética
i griglia *f* magnetica
n magnetisch rooster *n*, roostermagneet
d magnetischer Rost *m*, Rostmagnet *m*

2609 GRATELESS GENERATOR fue
f générateur *m* sans grille
e generador *m* sin parrilla
i generatore *m* senza griglia
n roosterloze generator
d rostloser Generator *m*

2610 GRATING SPECTROSCOPE ins,spe
f spectroscope *m* à réseau
e espectroscopio *m* de retículo
i spettroscopio *m* a reticolo
n traliespectroscoop
d Gitterspektroskop *n*

2611 GRAVEL FILTER, fil,sew
GRAVEL PACKED FILTER,
GRAVELER (US),
GRAVELLER (GB)
f filtre *m* de gravier
e filtro *m* de grava
i filtro *m* di ghiaia
n grindfilter *n*
d Kiesfilter *m*

2612 GRAVITOMETER, lab
SPECIFIC GRAVITY BOTTLE
f gravitomètre *m*
e gravitómetro *m*
i gravitometro *m*
n gravitometer
d Gravitometer *n*

2613 GRAVITY CELL elc
f pile *f* à différence de poids spécifique
e pila *f* de densidad
i pila *f* a differenza di peso specifico
n element *n* met twee elektrolyten met verschillend soortelijk gewicht
d Zweischichtenelement *n*

2614 GRAVITY CONVEYOR tra
f goulotte *f*, transporteur *m* par gravité
e transportador *m* por gravedad
i trasportatore *m* a gravità
n rollentransportband
d Rollenbahn *f*

2615 GRAVITY FILTER, fil
OPEN FILTER
f filtre *m* à gravité, filtre *m* ouvert
e filtro *m* abierto, filtro *m* de gravedad
i filtro *m* a gravità, filtro *m* aperto, filtro *m* scoperto
n open filter *n*
d offener Filter *m*

2616 GRAVITY SETTLING CHAMBER, cle,gas
DUST SETTLING CHAMBER,
SETTLING CHAMBER
f chambre *f* à poussière
e cámara *f* de polvo
i camera *f* a polvere
n stofkamer
d Staubkammer *f*

2617 GREASE GUN lab
f pistolet *m* graisseur
e jeringa *f* de grasa
i pistola *f* a grasso
n vetspuit
d Fettspritze *f*

2618 GREASE RECOVERY PLANT, fat
FAT RECOVERY PLANT
f installation *f* de récupération de graisse
e equipo *m* de recuperación de grasa
i impianto *m* di ricupero di grasso
n vetterugwinningsinstallatie
d Fettrückgewinnungsanlage *f*

2619 GREASE SEPARATOR, cle,fat
GREASE TRAP, FAT TRAP
f tôle *f* à retenir la graisse
e interceptor *m* de grasa
i eliminatore *m* di grasso
n vetvanger
d Fettabscheider *m*, Fettfänger *m*

2620 GREEN HOUSE cer
f salle *f* d'examination
e sala *f* de inspección
i sala *f* di verifica
n controleruimte, inspectieruimte
d Kontrollraum *m*

2621 GRID, pla,pre
INSULATING GRID
f plaque *f* isolante

e parrilla *f* aislante
i griglia *f* isolante
n isolatierooster *n*
d Isolierrost *m*

2622 GRID TRAY dis
f plateau *m* à grille
e plato *m* de enrejado
i piatto *m* a griglia
n roosterschotel
d Gitterboden *m*

2623 GRID-TYPE HUMIDIFIER hyg
f humecteur *m* à grille, humidificateur *m* à grille
e humectador *m* de pantalla, humidificador *m* de pantalla
i umidificatore *m* a griglia
n bevochtigingstoestel *n* met rooster
d Gitteranfeuchtapparat *m*, Gitterbefeuchter *m*

2624 GRIFFIN FLOW TUBE lab
f tube *m* de pissette à soupape
e redoma *f* con válvula de retención
i spruzzetta *f* a valvola di ritegno
n spuitfles met terugslagklep
d Spritzflaschenrohr *n* mit Rückschlagventil

2625 GRINDER pap
f défibreur *m*
e desfibrador *m*
i sfibratore *m*
n houtslijpmachine
d Holzschleifmaschine *f*, Zerfaserer *m*

2626 GRINDERY pol
f atelier *m* de meulage
e taller *m* de amoladura
i officina *f* di molatura
n slijperij
d Schleiferei *f*

2627 GRINDING ELEMENTS brk
f corps *pl* de broyage, éléments *pl* broyeurs
e elementos *pl* de moliendo
i corpi *pl* macinanti
n maalkogels *pl*
d Mahlkörper *pl*

2628 GRINDING MACHINE pol
f meuleuse *f*
e máquina *f* de amolar
i macchina *f* di molatura
n slijpmachine
d Schleifmaschine *f*

2629 GRINDING MACHINE, POLISHING MACHINE pol
f machine *f* à polir, polissoir *m*
e máquina *f* de pulir
i lucidatrice *f*
n polijstmachine
d Poliermaschine *f*

2630 GRINDING MILL, MILL brk
f broyeur *m*, moulin *m*
e molino *m*
i frantumatore *m*, mulino *m*
n breekmolen, maalmolen
d Mühle *f*

2631 GRINDING ROLLER pol
f cylindre *m* à aiguiser
e cilindro *m* esmerilador
i cilindro *m* da affilare
n slijpwals
d Schleifwalze *f*

2632 GRINDING WHEEL pol
f disque *m* polisseur
e disco *m* de pulir
i disco *m* lucidatore
n polijstschijf
d Polierscheibe *f*

2633 GRINDING WHEEL pol
f roue *f* à meuler
e muela *f*
i mola *f*
n slijpschijf
d Schleifscheibe *f*

2634 GRIST BOX, SILO, STORAGE BIN bre
f trémi *f* à malt, wagon *m* à farine
e tolva *f* de harina
i cassone *m* per macinato
n schrootkaar
d Schrotrumpf *m*

2635 GRIST MILL, FLOUR MILL met
f moulin *m* à blé
e molino *m* harinero
i mulino *m* da grano
n korenmolen
d Getreidemühle *f*

2636 GRIST MILL, MALT CRUSHER, MALT GRINDING MILL, MALT MILL bre
f concasseur *m* de malt

e triturador *m* de malta
i frantoio *m* per malto
n schrootmolen
d Malzbrecher *m*, Malzmühle *f*, Malzspaltapparat *m*, Schrotmühle *f*

2637 GRIT BLASTING MACHINE, SHOT BLASTING MACHINE — cle
f machine *f* de grenaillage
e máquina *f* de chorreado con granalla
i macchina *f* di granigliatura
n gritstraalmachine
d Abstrahlmaschine *f*, Schleuderstrahlmaschine *f*

2638 GRIT CHAMBER, GRIT COLLECTOR — sew
f collecteur *m* de sable, filtre *m* de sable
e colector *m* de arena, filtro *m* de arena
i filtro *m* di sabbia, raccoglitore *m* di sabbia
n zandvanger
d Sandfang *m*, Sandräumer *m*

2639 GRIT TANK, DETRITUS TANK — iwt
f récipient *m* dessableur
e tanque *m* de decantación de detritos
i serbatoio *m* dissabbiatore
n zandvanger
d Sandfänger *m*

2640 GRIZZLY — brk,sor
f cribon *m* à barres parallèles
e cribón *m* de rejillas
i vaglio *m* a griglie
n roosterzeef
d Gittersieb *n*, Grubenrost *m*

2641 GROOVED CYLINDER (US), GROOVED ROLL (GB) — pla,rub
f cylindre *m* cannelé
e cilindro *m* acanalado
i cilindro *m* scanalato
n riffelwals
d Riffelwalze *f*

2642 GROOVED TOOL FOR MOLDING TEST SPECIMEN (US), GROOVED TOOL FOR MOULDING TEST SPECIMEN(GB) — pla
f moule *m* d'éprouvettes étalon
e molde *m* para obtención de probetas
i stampo *m* per provette normalizzate
n proefstaalmatrijs
d Rippenwerkzeug *n*

2643 GROUND JOINT, MALE JOINT — gla,lab
f joint *m* en verre rodé, joint *m* male, rodage *m* émeri
e junta *f* esmerilada
i giuntura *f* molata
n geslepen verbinding
d Aussenschliff *m*, Schliffhülse *f*

2644 GROUNDING MACHINE, STAINING MACHINE — pa
f machine *f* à teinter
e máquina *f* de tinción
i macchina *f* di tintura
n grondeermachine
d Fonciermaschine *f*, Grundiermaschine *f*

2645 GROWING FLOOR, GERMINATING FLOOR, MALTING FLOOR — br
f aire *f* de germination, germoir
e era *f* de germinación
i aia *f* di germinazione
n kiemvloer
d Keimboden *m*, Keimtenne *f*, Malztenne *f*

2646 GUARD RING — cer,te
f anneau *m* directif
e anillo *m* directivo
i anello *m* direttivo
n leiring, vlamrichter
d Führungsring *m*

2647 GUIDE BUSHING — c
f douille *f*, guidage *m*
e casquillo *m* guía, manguito *m* de guía
i bussola *f* di guida
n leibus, leisleuf
d Führungsbüchse *f*

2648 GUIDE PLATE, CHECK, STOCK GUIDE — cpt,m
f flasque *f* de guidage latérale
e carillo *m* de un mezclador
i guancia *f* di un mescolatore
n keerbord *n*, materiaalgeleider van een mengwals
d Backe *f* einer Mischmaschine

2649 GUIDE ROLL — p
f galet *m*, rouleau *m* guide
e rodillo *m* guía
i rullo *m* di guida
n leirol, leiwals
d Leitwalze *f*

2650 GUNCOTTON FACTORY exp
f fabrique *f* de coton-poudre, fabrique *f* de fulmicoton
e fábrica *f* de algodón explosivo, fábrica *f* de fulmicotón
i fabbrica *f* di cotone fulminante, fabbrica *f* di fulmicotone
n schietkatoenfabriek
d Schiessbaumwollefabrik *f*

2651 GUNPOWDER CHEST exp
f caisson *m* à poudre, coffre *m* à poudre
e caja *f* de pólvora
i cassetta *f* a polvere
n kruitkist
d Pulverkiste *f*

2652 GUNPOWDER FACTORY exp
f fabrique *f* de poudre, poudrerie *f*
e fábrica *f* de pólvora
i fabbrica *f* di polvere da sparo, fabbrica *f* di polvere pirica
n kruitfabriek
d Schiesspulverfabrik *f*

2653 GUNPOWDER HOUSE exp
f poudrière *f*
e polvorín *m*
i polveriera *f*
n kruithuis *n*
d Pulvermagazin *n*

2654 GUNPOWDER VAT exp
f baril *m* à poudre
e barril *m* de pólvora
i barile *m* di polvere
n kruitton, kruitvat *n*
d Pulverfass *n*

2655 GUTTER sew
f gouttière *f*
e canalón *m*
i canaletto *m*
n goot
d Gosse *f*, Rinne *f*

2656 GYPSUM PLASTER MOLD (US), fou
GYPSUM PLASTER MOULD (GB)
f moule *m* en plâtre
e molde *m* de yeso
i forma *f* in gesso
n gipsvorm
d Gipsform *f*

2657 GYRATORY CRUSHER brk
f concasseur *m* giratoire
e quebrantadora *f* giratoria
i frantoio *m* giratorio
n kegelbreker
d Kegelbrecher *m*, Kreiselbrecher *m*

H

2658 HACKLING MACHINE, HECKLING MACHINE, SCUTCHING MACHINE tex
f peigneuse *f*
e máquina *f* peinadora, máquina *f* rastrilladora
i macchina *f* di pettinatura
n hekelmachine
d Hechelmaschine *f*

2659 HAIR-PIN FURNACE, U-TYPE FURNACE cer
f four *m* à épingle à cheveux
e horno *m* de herradura
i forno *m* a forcella
n omloopoven
d Umkehrofen *m*

2660 HALF CELL, HALF ELEMENT, SINGLE ELECTRODE SYSTEM elc
f demi-cellule *f*, hémi-cellule *f*
e semicelda *f*
i semicellula *f*
n halve cel
d Halbzelle *f*

2661 HALF-CORN CYLINDER bre
f cylindre *m* de trieur
e cilindro *m* de máquina seleccionadora
i cilindro *m* di macchina selettrice
n trieurcilinder
d Trieurzylinder *m*

2662 HALF-STUFF HOLLANDER pap
f hollandais *m* à défilé, hollandais *m* à pâte effilachée
e batidora *f* de pasta para producto semiacabado
i mescolatore *m* olandese a pasta semiallestita
n halfgoedhollander, halfstof-hollander
d Halbzeugholländer *m*

2663 HALIDE LEAK DETECTOR gas,sap
f détecteur *m* de fuite d'halogénure
e detector *m* de fuga de halogenuro
i rivelatore *m* di fuga d'alogenuro
n halogenidelekzoeker
d Halogenidlecksucher *m*

2664 HALOGENATION PLANT aic
f installation *f* d'halogénation
e equipo *m* de halogenación
i impianto *m* d'alogenazione
n halogeneerinstallatie
d Halogenierungsanlage *f*

2665 HAMMER CRUSHER brk
f concasseur *m* à marteaux
e quebrantador *m* de martillos
i frantoio *m* a martelli
n hamerbreekmachine
d Hammerbrecher *m*

2666 HAMMER MILL brk,nut
f broyeur *m* à marteaux
e triturador *m* de martillos
i mulino *m* a martelli
n hamermolen
d Hammermühle *f*

2667 HAND MOLD (US), HAND MOULD (GB) pap
f moule *m* à main
e molde *m* a mano
i forma *f* a mano
n handvorm
d Handform *f*

2668 HAND MOLD (US), HAND MOULD (GB), LOOSE MOLD (US), PORTABLE MOUL (GB) pla
f moule *m* à main
e molde *m* a mano
i stampo *m* a mano
n handmatrijs
d Handform *f*

2669 HAND MOULDING MACHINE (US), HAND MOULDING MACHINE (C fou,pl
f machine *f* à mouler à main
e máquina *f* de moldear a mano
i macchina *f* formatrice a mano
n handvormmachine
d Handformmaschine *f*

2670 HAND-RABBLED FURNACE ht
f four *m* à brassage manuel, four *m* à ringardage manuel
e horno *m* de rableado a mano
i forno *m* a raschiatura a mano, forno *m* a rimescolamento a ma
n oven met handroerwerk

d Handfortschaufelungsofen *m*

2671 HANDLER, TAN VAT, TANNING PIT lea
f fosse *f* à tan
e foso *m* para curtido, noque *m*
i fossa *f* da conciare le pelli
n looikuil
d Lohgrube *f*

2672 HANGING FILTER, INSERTION FILTER ext,fil
f filtre *m* à insérer
e filtro *m* de inserción
i filtro *m* d'inserzione
n inzetfilter *n*
d Einsatzfilter *m*

2673 HANK DYEING VAT, BUNDLE DYEING VAT, YARN DYEING VAT tex
f cuve *f* de teinture en échevaux
e cuba *f* de teñido en suspensión de las madejas
i tino *m* di tintura delle matasse
n strengenverfkuip
d Strähnenfärbküpe *f*

2674 HARD CHROMIUM PLATING BATH, HARD PLATING BATH coa,elc
f bain *m* de chromage dur
e baño *m* de cromado duro
i bagno *m* di cromatura dura
n hardverchromingsbad *n*
d Hartverchromungsbad *n*

2675 HARD FACING PLANT coa,met
f installation *f* de rechargement dur
e equipo *m* de recargue duro, equipo *m* de revestimiento duro
i impianto *m* di rivestimento duro
n hardmetaalbekledingsinstallatie
d Hartmetallauftraganlage *f*

2676 HARDENING BATH pho
f bain *m* de durcissement
e baño *m* reforzador
i bagno *m* induritore
n hardingsbad *n*
d Härtebad *n*

2677 HARDENING FURNACE htt,met
f four *m* de trempe
e horno *m* de temple
i forno *m* di tempera
n hardingsoven
d Härtungsofen *m*

2678 HARDENING SHOP htt,met
f atelier *m* de trempage
e taller *m* de temple
i bottega *f* di tempera
n harderij
d Härterei *f*

2679 HARDNESS DROP TESTER, SCLEROSCOPE ins,tes
f scléroscope *m*
e escleroscopio *m*
i durometro *m* per caduta, scleroscopio *m*
n hardheidsbepalingsapparaat *n* met vallende kogel, scleroscoop
d Fallhärteprüfer *m*

2680 HARDNESS TESTER, DURIMETER, SCLEROMETER ins,tes
f duromètre *m*, scléromètre *m*
e durómetro *m*
i durometro *m*
n hardheidsbeproevingsapparaat *n*, hardheidsmeter
d Härteprüfer *m*

2681 HAUL-OFF ROLL (GB), DRAW ROLL, TAKE-UP ROLL (US) pla
f cylindre *m* de tirage, cylindre récepteur
e cilindro *m* de tracción
i cilindro *m* di tirata
n afneemwals, aftrekwals
d Abzugswalze *f*

2682 HEAD VALVE, DELIVERY VALVE cpt
f soupape *f* de manoeuvre
e válvula *f* de impulsión
i valvola *f* di mandata
n stuurklep
d Steuerventil *n*

2683 HEADS COLUMN dis
f tête *f* de rectification d'aldéhyde
e columna *f* de separación de aldehido
i testa *f* di rettificazione d'aldeide
n aldehydeafscheidingskolom
d Aldehydabscheidesäule *f*

2684 HEAP LEACHING PLANT cle,min
f installation *f* de lixiviation en amas
e equipo *m* de lixiviación en montones
i impianto *m* di lisciviazione in mucchi
n massale-uitlogingsinstallatie

d Haufenlaugungsanlage *f*

2685 HEAP LEACHING TANK cle,min
f récipient *m* de lessivation en amas, récipient *m* de lixiviation en amas
e tanque de lixiviación en montones
i serbatoio *m* di lisciviazione in mucchi
n massa-uitloogtank
d Haufenauslaugebehälter *m*

2686 HEARTH, FURNACE BOTTOM htt
f sole *f* de four
e solera *f* de horno
i suola *f* di forno
n haardbodem
d Herdsohle *f*

2687 HEARTH FURNACE htt,met
f four *m* à sole
e horno *m* de solera
i forno *m* a suolo
n haardoven
d Herdofen *m*

2688 HEARTH JACKET htt
f blindage *m* du creuset
e blindaje *m* del crisol
i schermatura *f* del crogiolo
n haardpantser *n*
d Gestellpanzer *m*, Herdpanzer *m*

2689 HEAT ACCUMULATOR htt
f accumulateur *m* de chaleur
e acumulador *m* de calor
i accumulatore *m* di calore
n warmteaccumulator
d Wärmespeicher *m*

2690 HEAT ALARM INDICATOR, THERMAL ALARM INDICATOR sap
f dispositif *m* d'alarme thermique
e termoavisador *m*
i dispositivo *m* d'allarme termico
n thermisch alarmapparaat *n*
d wärmebetätigte Alarmvorrichtung *f*

2691 HEAT CHAMBER, HOT-AIR CHAMBER, SPROUT CHAMBER bre
f chambre *f* de chaleur, cochon *m*
e cámara *f* de aire caliente, cámara *f* de calefacción
i camera *f* di riscaldamento
n kleine hel
d Darrsau *f*, Sau *f*, Wärmekammer *f*

2692 HEAT DETECTOR con,sap
f thermostat *m* indirect
e termóstato *m* indirecto
i termostato *m* indiretto
n indirecte thermostaat
d indirekter Thermostat *m*

2693 HEAT DISTORTION TESTING APPARATUS cer,con,te
f appareil *m* d'essai de distorsion thermique
e probador *m* de termotorsión
i collaudatore *m* di distorsione termica
n vormvastheidsbeproevings-apparaat *n*
d Formbeständigkeitsprüfgerät *n*

2694 HEAT EXCHANGER ge
f échangeur *m* de chaleur
e cambiador *m* de calor, intercambiador *m* térmico, termocanjeador *m*
i scambiatore *m* di calore
n warmteuitwisselaar
d Wärmeaustauscher *m*

2695 HEAT EXCHANGER LEAK DETECTION PROBE nue,sa
f sonde *f* détectrice de fuites d'échangeur de chaleur
e sonda *f* detectora de pérdidas de cambiador de calor
i sonda *f* rivelatrice di perdite di scambiatore di calore
n lekzoeker voor warmte-uitwisselaar
d Wärmeaustauscherlecksucher *m*

2696 HEAT INSULATING SLEEVE fou,h
f manchon *m* isolant
e manguito *m* aislante
i manicotto *m* isolante
n isolerende tapmanchet
d wärmeisolierender Trichter-einsatz *m*

2697 HEAT-JACKETED DRUM h
f tambour *m* chauffant à double enveloppe
e tambor *m* de calefacción de doble pared
i tamburo *m* di riscaldamento a doppia camicia
n dubbelwandige verwarmings-trommel
d doppelwandige Heiztrommel *f*

2698 HEAT PUMP p
f pompe *f* à chaleur, thermopom

e bomba *f* calorífica, bomba *f* de calor
i pompa *f* a calore
n warmtepomp
d Wärmepumpe *f*

2699 HEAT REGENERATOR gen
f régénérateur *m* de chaleur
e regenerador *m* de calor, termorregenerador *m*
i preriscaldatore *m* a ricupero
n warmteregenerator
d Wärmeregenerator *m*

2700 HEAT-SEALING PLANT, HIGH-FREQUENCY WELDING PLANT pla
f installation *f* de soudure à chaud
e equipo *m* de sellado en caliente
i impianto *m* di sigillatura a caldo
n warmlasinstallatie
d Heissiegelanlage *f*

2701 HEAT SHIELD, ABLATIVE COATING sap
f couche *f* disparaissante en fondant
e capa *f* desaparecente por fusión
i strato *m* da defluire in seguito a fusione
n afsmeltlaag, hitteschild *n*
d Abbrennschicht *f*, Abschmelzschicht *f*, Hitzeschild *m*

2702 HEAT STERILIZATION PLANT bac,hyg
f installation *f* de thermostérilisation
e equipo *m* de termoesterilización
i impianto *m* di sterilizzazione a caliente
n thermische steriliseerinstallatie
d Thermosterilisationsanlage *f*

2703 HEAT TINTING OVEN dec,htt
f four *m* de coloration thermique
e horno *m* de termocoloración
i forno *m* di colorazione termica
n thermische-kleuroven
d Wärmetönungsofen *m*

2704 HEAT TRANSFER RIG clo,nue
f ensemble *m* de transfert de chaleur
e aparejo *m* de termotransferencia
i attrezzatura *f* di trasmissione di calore
n warmteoverdrachtsaggregaat *n*
d Wärmeübertragungsaggregat *n*

2705 HEAT TREATMENT BATH htt,met
f four *m* à bain
e horno *m* de baño
i forno *m* a bagno
n warmtebehandelingsbad *n*
d Wärmebehandlungsbad *n*

2706 HEAT TREATMENT PLANT htt,met
f installation *f* de traitement thermique
e equipo *m* para termotratamiento
i impianto *m* di trattamento termico
n warmtebehandelingsinstallatie
d Wärmebehandlungsanlage *f*

2707 HEATER BAND, CUFF HEATER BAND htt,pla
f bande *f* de chauffage
e banda *f* de caldeo
i banda *f* riscaldatrice
n verwarmingslint *n*
d Erhitzungsband *n*

2708 HEATER TUNNEL (US), HEATING TUNNEL (GB) cer,pla
f tunnel *m* de chauffage
e túnel *m* de calentamiento
i tunnel *m* di riscaldamento
n verhittingstunnel, verwarmingstunnel
d Erwärmungstunnel *m*, Heiztunnel *m*

2709 HEATING CABLE htt
f câble *m* chauffant
e cable *m* de calefacción
i cavo *m* riscaldatore
n verwarmingskabel
d Heizkabel *n*

2710 HEATING CHAMBER, HEATING ROOM htt
f chambre *f* de chauffage
e cámara *f* de calefacción, sala *f* de calefacción
i camera *f* di riscaldamento
n opwarmkamer
d Wärmekammer *f*

2711 HEATING CHANNEL (GB), HEATING FLUE, HEATING PASSAGE (US) htt,pla
f canal *m* de chauffage
e canal *m* de caldeo
i canale *m* di riscaldamento
n verwarmingskanaal *n*
d Heizkanal *m*

2712 HEATING COIL, STEAM COIL pre
f serpentin *m* de chauffage
e serpentín *m* de calentamiento
i serpentina *f* di riscaldamento
n verwarmingsspiraal
d Heizschlange *f*

2713 HEATING ELEMENT htt
f élément *m* de chauffage
e elemento *m* de calentamiento
i elemento *m* di riscaldamento
n verhittingselement *n*
d Heizelement *n*

2714 HEATING INDUCTOR, APPLICATOR, WORK COIL htt
f inducteur *m* de chauffage
e inductor *m* de equipo de caldeo
i induttore *m* riscaldante
n verhittingsinductor, verwarmingsinductor, werkspoel
d Heizinduktor *m*, Heizschleife *f*

2715 HEATING JACKET htt
f chemise *f* de chauffage
e chaqueta *f* calentadora
i camicia *f* di riscaldamento
n verhittingsmantel
d Heizmantel *m*

2716 HEATING PIN pla,pre
f goupille *f* de chauffage
e clavija *f* de calefacción
i spina *f* di riscaldamento
n verwarmingspen
d Heizstift *m*

2717 HEATING PLATE, WARMING PLATE lab
f plaque *f* chauffante, plateau *m* chauffant
e chapa *f* caliente, placa *f* calentadora
i piastra *f* riscaldante
n verwarmingsplaat
d Heizplatte *f*

2718 HEATING SHOE htt,rub
f sabot *m* de chauffage
e zapata *f* de calentamiento
i ganascia *f* di riscaldamento
n verhittingsblok *n*, verhittingsschoen
d Heizschuh *m*

2719 HEATING STOVE, ANNEALING FURNACE htt
f four *m* à recuire
e horno *m* de recocido
i forno *m* di ricottura
n gloeioven
d Glühofen *m*

2720 HEATING TUBE lab
f four *m* grenade
e tubo *m* calefactor
i tubo *m* di riscaldamento
n verhittingsbuis
d Heizgranate *f*

2721 HEATING TUBE nue
f tube *m* de chaleur
e tubo *m* de calor
i tubo *m* di calore
n verhittingspijp
d Heizrohr *n*

2722 HEAVY AGGREGATE CONCRETE SHIELD, LOADED CONCRETE SHIELD nue,sa
f écran *m* en béton chargé
e pantalla *f* de hormigón cargado
i schermo *m* in calcestruzzo caricato
n scherm *n* van verzwaard beton
d Schwerbetonschirm *m*

2723 HEAVY HYDROGEN PLANT nu
f installation *f* d'hydrogène lourd
e equipo *m* de deuterio, equipo *m* de hidrógeno pesado
i impianto *m* d'idrogeno pesante
n zware-waterstofinstallatie
d Schwerwasserstoffanlage *f*

2724 HEAVY WATER BOILING REACTOR nu
f réacteur *m* à eau lourde bouillante
e reactor *m* de agua pesada hirviente
i reattore *m* ad acqua pesante in ebollizione
n kokend-zwaar-waterreactor
d Schwerwassersiedereaktor *m*

2725 HEAVY WATER MODERATED REACTOR, HEAVY WATER REACTOR nu
f réacteur *m* à eau lourde, réacteur *m* modéré en eau lourd
e reactor *m* de agua pesada, reactor *m* moderado de agua pesada
i reattore *m* ad acqua pesante, reattore *m* moderato ad acqua pesante
n reactor met zwaar water als moderator,

reactor met zwaar water als remstof, zwaar-waterreactor
d schwerwassermoderierter Reaktor *m*, Schwerwasserreaktor *m*

2726 HEAVY WATER PLANT nue
f installation *f* d'eau lourde
e equipo *m* de agua pesada
i impianto *m* d'acqua pesante
n zwaar-waterinstallatie
d Schwerwasseranlage *f*

2727 HECKLING MACHINE, HACKLING MACHINE, SCUTCHING MACHINE tex
f peigneuse *f*
e máquina *f* peinadora, máquina *f* rastrilladora
i macchina *f* di pettinatura
n hekelmachine
d Hechelmaschine *f*

2728 HELICAL CONVEYOR, SCREW CONVEYOR, WORM CONVEYOR tra
f transporteur *m* hélicoïdal
e transportador *m* helicoidal
i trasportatore *m* elicoidale
n transportworm
d Förderschnecke *f*

2729 HELICAL ELEMENT htt
f élément *m* en hélice
e elemento *m* helicoidal
i elemento *m* elicoidale
n spiraalelement *n*
d Spiralenheizelement *n*

2730 HELICAL PLATE elc
f plaque *f* à rosettes
e placa *f* de rosetas
i piastra *f* a rosette
n rozetplaat
d Grossoberflächenplatte *f*

2731 HELICOIDAL AGITATOR mix
f agitateur *m* hélicoïdal
e agitador *m* helicoidal
i agitatore *m* elicoidale
n schroefroerder
d Schraubenrührwerk *n*

2732 HELICOIDAL MIXER mix
f malaxeur *m* hélicoïdal
e mezclador *m* helicoidal
i mescolatore *m* elicoidale
n schroefmenger
d Mischschnecke *f*

2733 HELIUM CRYOSTAT ins
f cryostat *m* à hélium
e crióstato *m* de helio
i criostato *m* ad elio
n heliumcryostaat
d Heliumkryostat *m*

2734 HELIUM DIVING BELL sap
f cloche *f* à plongeur remplie d'hélium
e campana *f* de buzo llenada de helio
i scafandro *m* a riempimento d'elio
n met helium gevulde duikerklok
d heliumgefüllte Taucherglocke *f*

2735 HELIUM LEAK DETECTOR nue
f détecteur *m* de pertes d'hélium
e detector *m* de pérdidas de helio
i rivelatore *m* di perdite d'elio
n heliumdetector
d Heliumleckfinder *m*, Heliumlecksucher *m*

2736 HELIUM LIQUEFIER PLANT gas
f installation *f* de liquéfaction d'hélium
e equipo *m* de licuación de helio
i impianto *m* per liquefazione d'elio
n installatie voor het vloeibaar maken van helium
d Heliumverflüssigungsanlage *f*

2737 HELIUM PRESSURE TANK nue
f récipient *m* d'hélium sous pression
e tanque *m* de presión de helio
i serbatoio *m* a pressione d'elio
n heliumdrukvat *n*
d Heliumdruckgefäss *n*

2738 HELIX PRESSURE GAGE (US), HELIX PRESSURE GAUGE (GB) ins
f manomètre *m* hélicoïdal
e manómetro *m* helicoidal
i manometro *m* elicoidale
n schroefmanometer
d Schraubenmanometer *n*

2739 HEMATIMETER, HEMATOMETER, YEAST COUNTING APPARATUS bre,ins
f hématimètre *m*
e hematimetro *m*
i ematimetro *m*
n gisttelraam *n*
d Hämatimeter *n*, Hefezählvorrichtung *f*

2740 HERBICIDE FACTORY, WEEDKILLER FACTORY agr,cle
f usine *f* d'herbicides

e fábrica *f* de herbicidas
i fabbrica *f* d'erbicidi
n fabriek van onkruidverdelgende middelen
d Fabrik *f* von Unkrautvertilgungsmitteln

2741 HERMETIC SEAL, HERMETICAL CLOSURE gen
f bouchage *m* hermétique
e cierre *m* hermético
i chiusura *f* ermetica
n hermetische sluiting
d hermetischer Verschluss *m*

2742 HESSIAN CRUCIBLE, TRIANGULAR CRUCIBLE htt
f creuset *m* triangulaire
e crisol *m* triangular
i crogiolo *m* triangolare
n driehoekige smeltkroes
d dreieckiger Schmelztiegel *m*

2743 HETEROGENEOUS REACTOR nue
f réacteur *m* hétérogène
e reactor *m* heterogeneo
i reattore *m* eterogeneo
n heterogene reactor
d heterogener Reaktor *m*

2744 HIGH-ACTIVITY WASTE CONTAINER iwt,nue
f récipient *m* de déchets de haute activité
e recipiente *m* de desechos fuertemente radiactivos
i serbatoio *m* di rifiuti altamente radioattivi
n houder voor sterk radioactief afval
d Sammelbehälter *m* für Abfall mit hoher Aktivität

2745 HIGH BACKING-PRESSURE PUMP vac
f pompe *f* à vide à contrepression élevée
e bomba *f* de vacío de alta contrapresión
i pompa *f* a vuoto ad alta contropressione
n vacuümpomp met hoge tegendruk
d Vakuumpumpe *f* mit hohem Gegendruck

2746 HIGH FLUX REACTOR, FLUX TRAP REACTOR nue
f réacteur *m* à flux concentré, réacteur *m* à flux élevé
e reactor *m* de flujo concentrado, reactor *m* de flujo elevado
i reattore *m* a flusso concentrato, reattore *m* a flusso elevato
n hoge-flux reactor
d Hochflussreaktor *m*

2747 HIGH-FREQUENCY DRIER, DIELECTRIC DRIER dry
f étuve *f* par pertes diélectriques
e estufa *f* por pérdidas dieléctricas
i stufa *f* a perdite dielettriche
n droogoven met capacitieve verwarming
d dielektrischer Trockner *m*, Hochfrequenztrockner *m*

2748 HIGH-FREQUENCY DRYING PLANT dry
f installation *f* de séchage par haute fréquence
e equipo *m* de secado por alta frecuencia
i impianto *m* d'essiccazione per alta frequenza
n hoogfrequentdrooginrichting
d Hochfrequenztrockenanlage *f*

2749 HIGH-FREQUENCY FURNACE htt
f four *m* à haute fréquence
e horno *m* de alta frecuencia
i forno *m* ad alta frequenza
n hoogfrequente oven
d Hochfrequenzofen *m*

2750 HIGH-FREQUENCY PREHEATER, PREHEATER BY DIELECTRIC LOSSES htt,rub
f appareil *m* de préchauffage à haute fréquence, préchauffeur *m* pour radiovulcanisation
e precalentador *m* de alta frecuencia
i preriscaldatore *m* ad alta frequenza
n hoogfrequente voorverwarmingsinstallatie
d Hochfrequenzvorwärmer *m*

2751 HIGH-GLAZING MACHINE, GLAZING MACHINE pho
f glaceuse *f*, machine *f* à glacer
e máquina *f* de glaseado
i lucidatrice *f*
n hoogglansmachine
d Hochglanzmaschine *f*

2752 HIGH-PRESSURE ACCUMULATOR pre
f accumulateur *m* à haute pression

e acumulador *m* de alta presión
i accumulatore *f* ad alta pressione
n hoge-drukaccumulator
d Hochdruckakkumulator *m*

2753 HIGH-PRESSURE BOILER boi
f chaudière *f* à haute pression
e caldera *f* de alta presión
i caldaia *f* ad alta pressione
n hoge-drukketel
d Hochdruckkessel *m*

2754 HIGH-PRESSURE COMPRESSOR pre
f compresseur *m* à haute pression
e compresor *m* de alta presión
i compressore *m* ad alta pressione
n hoge-drukcompressor
d Hochdruckkompressor *m*, Hochdruckverdichter *m*

2755 HIGH-PRESSURE CYLINDER pre
f cylindre *m* à haute pression
e cilindro *m* de alta presión
i cilindro *m* ad alta pressione
n hoge-drukcilinder
d Hochdruckzylinder *m*

2756 HIGH-PRESSURE EQUILIBRIUM STILL dis
f appareil *m* de distillation en équilibre par pression élevée
e aparato *m* de destilación en equilibrio por alta presión
i apparecchio *m* di distillazione in equilibrio per alta pressione
n hoge-druk-evenwichtsdestilleer-apparaat *n*
d Hochdruckgleichgewichts-destillierapparat *m*

2757 HIGH-PRESSURE GAS PLANT gas
f installation *f* de production de gaz à pression élevée
e equipo *m* de producción de gas de alta presión
i impianto *m* di fabbricazione di gas ad alta pressione
n installatie voor de vervaardiging van gas onder hoge druk
d Hochdruckgasherstellungs-verfahren *n*

2758 HIGH-PRESSURE HEAT EXCHANGER htt
f échangeur *m* de chaleur à pression élevée
e intercambiador *m* de calor de alta presión
i scambiatore *m* di calore ad alta pressione
n hoge-drukwarmteuitwisselaar
d Hochdruckwärmeaustauscher *m*

2759 HIGH-PRESSURE HEAT-TRANSFER RIG nue
f ensemble *m* de transfert de chaleur à pression élevée
e aparejo *m* de termotransferencia de alta presión
i attrezzatura *f* di trasmissione di calore ad alta pressione
n hoge-druk-warmteoverdracht-aggregaat *n*
d Hochdruckwärmeübertragungs-aggregat *n*

2760 HIGH-PRESSURE HYDROGENATION PLANT fat
f installation *f* d'hydrogénation sous haute pression
e equipo *m* de hidrogenación de alta presión
i impianto *m* d'idrogenazione ad alta pressione
n hoge-drukhydreringsinstallatie
d Hochdruckhydrierungsanlage *f*

2761 HIGH-PRESSURE PLANT gen,pre
f installation *f* pour pressions élevées
e equipo *m* para altas presiones
i impianto *m* per alte pressioni
n hoge-drukinstallatie
d Hochdruckanlage *f*

2762 HIGH-PRESSURE WATER LOOP clo,nue
f circuit *m* fermé à eau sous pression élevée
e circuito *m* cerrado de agua de alta presión
i circuito *m* chiuso ad acqua ad alta pressione
n gesloten circuit *n* voor water onder hoge druk
d Hochdruckwasserkreislauf *m*

2763 HIGH-SPEED CENTRIFUGE, ULTRACENTRIFUGE cen
f centrifugeur *m* à vitesse élevée, ultracentrifugeur *m*
e centrífuga *f* de alta velocidad, supercentrífuga *f*, ultra-centrífuga *f*
i centrifuga *f* ad alta velocità, ultracentrifuga *f*
n snelle centrifuge, ultracentrifuge
d Schnellschleuder *f*, Schnell-zentrifuge *f*, Ultrazentrifuge *f*

2764 HIGH-TEMPERATURE DYEING ROOM dye
f salle *f* de teinture à température élevée
e sala *f* de teñido de alta temperatura
i sala *f* di tintura ad alta temperatura
n hoge-temperatuurverfruimte
d Hochtemperaturfärbraum *m*

2765 HIGH-TEMPERATURE FURNACE htt
f four *m* à température élevée
e horno *m* de alta temperatura
i forno *m* ad alta temperatura
n hoge-temperatuuroven
d Hochtemperaturofen *m*

2766 HIGH-TEMPERATURE GAS COOLED REACTOR nue
f réacteur *m* à haute température à refroidissement au gaz
e reactor *m* de alta temperatura refrigerado por gas
i reattore *m* a gas ad alta temperatura
n met gas gekoelde hoge-temperatuurreactor
d gasgekühlter Hochtemperatur-reaktor *m*

2767 HIGH-TEMPERATURE HELIUM COOLED REACTOR nue
f réacteur *m* à haute température à refroidissement à hélium
e reactor *m* de alta temperatura refrigerado por helio
i reattore *m* ad elio ad alta temperatura
n met helium gekoelde hoge-temperatuurreactor
d heliumgekühlter Hochtemperatur-reaktor *m*

2768 HIGH-VACUUM DIFFUSION PUMP vac
f pompe *f* à diffusion à vide élevé
e bomba *f* de difusión de alto vacío
i pompa *f* a diffusione ad alto vuoto
n hoogvacuümdiffusiepomp
d Hochvakuumdiffusionspumpe *f*

2769 HIGH-VACUUM DISTILLATION PLANT dis
f installation *f* de distillation sous vide élevé
e equipo *m* de destilación en vacío elevado
i impianto *m* di distillazione a vuoto spinto
n hoogvacuümdestillatie-inrichting
d Hochvakuumdestillationsanlage *f*

2770 HIGH-VACUUM STILL dis
f appareil *m* de distillation sous vide élevé
e aparato *m* de destilación en alto vacío
i apparecchio *m* di distillazione a vuoto spinto
n hoogvacuümdestillatieapparaat *n*
d Hochvakuumdestilliergerät *n*

2771 HIGH-VACUUM STOPCOCK cpt,vac
f robinet *m* à vide élevé
e grifo *m* de alto vacío
i rubinetto *m* a vuoto spinto
n hoogvacuümkraan
d Hochvakuumhahn *m*

2772 HIGH-VELOCITY THERMOCOUPLE, SUCTION PYROMETER gla,ins
f pyromètre *m* pour mesurer la température de gaz circulants à vitesse élevée
e pirómetro *m* para medida de la temperatura de gases circulantes con alta velocidad
i pirometro *m* per misurare la temperatura di gas circolanti ad alta velocità
n pyrometer voor temperatuur-meting aan snel stromende gassen
d Pyrometer *n* für Temperatur-messung an schnellströmende Gase

2773 HINGED FOLLOWER MOLD (US), HINGED FOLLOWER MOULD (GB) pl
f moule *m* à charnières
e molde *m* abatible
i stampo *m* a guance cernierate
n uitklapbare matrijs
d Klappenform *f*

2774 HINGED NOSE PIECE pl
f tête *f* à charnières, tête *f* articulée
e cabezal *m* articulado
i testa *f* a cerniere, testa *f* a squadra per trafila
n scharnierende spuitkop
d Spritzmaschinengelenkkopf *m*

2775 HOB met
f plateau *m* de four
e placa *f* de horno
i piastra *f* di forno
n ovenplaat
d Ofenplatte *f*

2776 HOB pla,rub
f poinçon *m*
e punzón *m*
i conio *m*, mozzo *m*
n matrijzenstempel, persstempel
d Kaltsenkstempel *m*, Pfaff *m*, Prägestempel *m*

2777 HOBBING PRESS pla,pre
f presse *f* à forcer
e prensa *f* hidráulica para mandrilar, prensa *f* para acuñar
i coniatrice *f*
n nerfpers
d Prägepresse *f*

2778 HOFFMANN KILN, ANNULAR KILN, RING KILN cer
f four *m* annulaire
e horno *m* anular
i forno *m* anulare
n ringoven
d Ringofen *m*

2779 HOLD-DOWN PLATE pla
f serre-flan *m*
e pisador *m*
i premilastra *f*
n plooihouder, verend klemraam *n*
d Niederhalter *m*

2780 HOLDING BASIN, STORING BASIN iwt,sto
f bassin *m* d'emmagasinement
e estanque *m* de almacenamiento
i bacino *m* d'immagazzinamento
n vergaarbekken *n*, voorraadsbekken *n*
d Speicherbecken *n*

2781 HOLDING FURNACE met
f four *m* de maintien de la température
e horno *m* mantenedor de la temperatura
i forno *m* d'attesa
n sudderoven, warmhoudoven
d Warmhalteofen *m*

2782 HOLDING TANK iwt,sto
f bac *m* intercepteur, récipient *m* d'interception
e tanque *m* colector
i serbatoio *m* collettore
n verzameltank
d Auffangbehälter *m*

2783 HOLDING VAT bre
f cuve *f* d'attente
e tina *f* de espera
i tino *m* di purificazione
n wachttank
d Wartebottich *m*

2784 HOLE FORMING PIN, CORE PIN pla,pre
f broche *f* à trou lisse, broche-noyau *f*
e clavija *f* de macho, pasador *m*
i spina *f* d'anima
n kernstift
d Lochstift *m*

2785 HOLLANDER, BEATING ENGINE pap
f hollandais *m*
e batidora *f*, holander *m* batidor
i mescolatore *m* olandese, olandese *m*
n hollander, maalbak
d Holländer *m*

2786 HOLLANDER BEATER ROLL pap
f cylindre *m* porte-lames
e cilindro *m* batidor, cilindro *m* portacuchillos
i cilindro *m* portacoltelli
n hollanderwals, messenwals
d Holländerwalze *f*, Messerwalze *f*

2787 HOLLOW MOLD (GB), HOLLOW MOULD (US) pla
f moule *m* de coulée
e molde *m* de colada
i forma *f* di colata, stampo *m* di colata
n gietvorm
d Giessform *f*, Gussform *f*

2788 HOMEGENEOUS REACTOR nue
f réacteur *m* homogène
e reactor *m* homogéneo
i reattore *m* omogeneo
n homogene reactor
d homogener Reaktor *m*

2789 HOMOGENEOUS SHIELD nue
f écran *m* homogène
e pantalla *f* homogénea
i schermo *m* omogeneo
n homogeen scherm *n*
d homogener Schirm *m*

2790 HOMOGENEOUS SOLUTION-TYPE REACTOR, WATER BOILER REACTOR nue
f réacteur *m* à solution homogène
e reactor *m* de solución homogénea
i reattore *m* omogeneo acquoso
n homogene oplossingsreactor
d homogener Lösungsreaktor *m*, Wasserkocherreaktor *m*

2791 HOMOGENIZER tex
f appareil *m* d'homogénéisation, homogénéisateur *m*
e aparato *m* de homogenización, homogenizador *m*
i apparecchio *m* d'omogeneizzazione, omogeneizzatore *m*
n homogenisator
d Homogenisator *m*, Homogenisierapparat *m*

2792 HONE, WHETSTONE pol
f affiloir *m*, queue *f*, queux *m*
e piedra *f* aguzadera, piedra *f* de afilar
i cote *f*, pietra *f* per affilare
n aanzetsteen, wetsteen
d Absetzstein *m*, Wetzstein *m*

2793 HOOD gen
f capot *m*, capuchon *m*, dôme *m*
e campana *f*, caperuza *f*
i cappa *f*, cappuccio *m*
n kap
d Haube *f*, Kappe *f*

2794 HOOD met
f hotte *f*
e campana *f*
i cappa *f*
n afzuigkap
d Abzughaube *f*

2795 HOODED POT, CLOSED POT, COVERED POT gla
f creuset *m* couvert, creuset *m* fermé
e crisol *m* cerrado, crisol *m* cubierto
i crogiolo *m* coperto
n gesloten haven, gesloten pot, overdekte haven, overdekte pot
d geschlossener Hafen *m*, überdeckter Hafen *m*

2796 HOP BACK, HOP STRAINER bre
f panier *m* à houblon
e colador *m* de lúpulo
i filtro *m* di luppolo
n hopzeef
d Hopfenseiher *m*

2797 HOP EXTRACTING APPARATUS bre
f extracteur *m* de houblon, laveur *m* de houblon
e extractor *m* de lúpulo
i estrattore *m* per luppolo
n hopuitloogapparaat *n*
d Hopfenentlauger *m*

2798 HOP MILL bre
f effeuilleuse *f* de houblon
e molino *m* de lúpulo
i mulino *m* di luppolo
n hopmolen
d Hopfenzerreissmaschine *f*

2799 HOP OAST bre
f touraille *f* de houblon
e tostadero *m* de lúpulo
i essiccatoio *m* di luppolo
n hopeest
d Hopfendarre *f*

2800 HOP STORAGE bre,sto
f magasin *m* de houblon
e almacén *m* de lúpulo
i magazzino *m* per luppolo
n hopopslagplaats
d Hopfenlager *n*

2801 HOPPER fou,tra
f trémie *f*
e tolva *f*
i tramoggia *f*
n aanvoertrechter
d Speisetrichter *m*

2802 HOPPER gen
f benne *f*
e tolva *f*
i benna *f*
n hopper, vultrechter
d Kübel *m*

2803 HOPPER, CONE, FUNNEL cpt
f entonnoir *m*
e embudo *m*
i imbuto *m*, tramoggia *f*
n trechter
d Trichter *m*

2804 HOPPER-BOTTOMED TANK sto
f récipient *m* à fond pyramidal, récipient *m* de décantation pyramidal

e tanque *m* con fondo de tolva
i serbatoio *m* a tramoggia
n tank met trechtervormige bodem
d Behälter *m* mit trichterförmigem Boden

2805 HOPPER CAR, BOTTOM-DISCHARGE CAR, BOTTOM-DUMPING CAR tra
f wagon *m* de déchargement par le fond
e vagón *m* de descarga por el fondo
i carro *m* a tramoggia
n onderlosser
d Bodenentlader *m*, Bodenentleerer *m*

2806 HOPPER DRIER dry
f sécheur *m* à trémie
e secadero *m* de tolva
i essiccatoio *m* a tramoggia
n droogtrechter
d Trichtertrockner *m*

2807 HOPPER VIBRATOR gen
f vibrateur *m* de benne
e vibrador *m* de tolva
i vibratore *m* di tramoggia
n hopperschudder
d Kübelschüttelvorrichtung *f*

2808 HORIZONTAL EXTRUDER HEAD pla,rub
f tête *f* droite de boudineuse
e cabezal *m* horizontal de budinadora
i testa *f* orizzontale di trafila
n horizontale spuitkop
d Längsspritzkopf *m*

2809 HORIZONTAL INJECTION PRESS pla,pre
f presse *f* à injection horizontale
e prensa *f* de inyección horizontal
i pressa *f* ad iniezione orizzontale
n horizontale spuitgietmachine
d Horizontalspritzgussmaschine *f*

2810 HORIZONTAL KILN bre
f touraille *f* horizontale
e tostadero *m* horizontal
i essiccatoio *m* orizzontale
n horizontale eest
d Plandarre *f*

2811 HORIZONTAL RETORT gas
f cornue *f* horizontale
e retorta *f* horizontal
i storta *f* orizzontale
n horizontale retort
d Horizontalretorte *f*

2812 HORIZONTAL RETORT OVEN htt
f four *m* à cornues horizontales, four *m* à retortes horizontales
e horno *m* de retortas horizontales
i forno *m* a storte orizzontali
n horizontale retortenoven
d Horizontalretortenofen *m*

2813 HORIZONTAL RIBBON MIXER mix
f mélangeur *m* horizontal à bandes
e mezclador *m* horizontal de bandas
i mescolatore *m* orizzontale a nastri
n mengmachine met lintvormige schoepen
d Horizontalbandmischer *m*

2814 HORIZONTAL VULCANIZER rub
f vulcanisateur *m* horizontal
e vulcanizador *m* horizontal
i caldaia *f* di vulcanizzazione orizzontale
n horizontale vulcaniseerketel
d horizontaler Vulkanisierkessel *m*

2815 HORSESHOE AGITATOR, HORSESHOE MIXER mix
f agitateur *m* en forme d'ancre
e agitador *m* en forma de herradura de caballo
i agitatore *m* in forma d'ancora
n ankerroerwerk *n*
d Ankerrührwerk *n*

2816 HOSE, HOSEPIPE gen
f flexible *m*, tuyau *m* flexible
e manga *f*, manguera *f*, tubo *m* flexible
i tubo *m*
n slang
d Schlauch *m*

2817 HOT-AIR CABINET dry
f chambre *f* à air chaud
e cámara *f* de aire caliente
i camera *f* ad aria calda
n hete-luchtkast
d Heissluftschrank *m*

2818 HOT-AIR CHAMBER, HEAT CHAMBER, SPROUT CHAMBER bre
f chambre *f* de chaleur, cochon *m*

e cámara *f* de aire caliente, cámara *f* de calefacción
i camera *f* di riscaldamento
n kleine hel
d Darrsau *f*, Sau *f*, Wärmekammer *f*

2819 HOT-AIR DOUBLE-WALL STERILIZER hyg,lab
f appareil *m* de stérilisation à double paroi à air chaud
e esterilizador *m* de doble pared con aire caliente
i sterilizzatore *m* a doppia parete con aria calda
n dubbelwandig steriliseerapparaat *n* met hete lucht
d doppelwandiges Sterilisiergerät *n* mit Heissluft

2820 HOT-AIR DRYING PLANT dry
f installation *f* de séchage à air chaud
e equipo *m* secador de aire caliente
i impianto *m* essiccatore ad aria calda
n hete-luchtdrooginrichting
d Heisslufttrockenanlage *f*

2821 HOT-AIR KILN bre
f touraille *f* à air chaud
e tostadero *m* de aire caliente
i essiccatoio *m* ad aria calda
n hete-luchteest
d Heissluftdarre *f*

2822 HOT-AIR STERILIZER hyg,lab
f appareil *m* de stérilisation à air chaud
e esterilizador *m* de aire caliente
i sterilizzatore *m* ad aria calda
n hete-luchtsteriliseerapparaat *n*
d Heissluftsterilisiergerät *n*

2823 HOT BED, COOLING BED clo,met
f refroidisseur *m*
e lecho *m* de enfriamiento
i piano *m* di raffreddamento, raffreddatore *m*
n koelbed *n*
d Kühlbett *n*, Warmlager *n*

2824 HOT BLAST STOVE, BLAST FURNACE STOVE, COWPER STOVE htt,met
f appareil *m* régénérateur de vent chaud
e aparato *m* Cowper
i apparecchio *m* di Cowper, preriscaldatore *m* del vento
n stenen windverhitter
d Cowper *m*, steinerner Winderhitzer *m*

2825 HOT CLEAN REACTOR nue
f réacteur *m* chaud virginal
e reactor *m* caliente virginal
i reattore *m* caldo vergine
n hete maagdelijke reactor
d heisser jungfräulicher Reaktor *m*

2826 HOT COMPACTING PRESS, HOT PRESSING MACHINE met,pre
f machine *f* de compression à chaud
e maquina *f* de compresión en caliente
i macchina *f* di compressione a caldo
n warmpersmachine
d Warmpressmaschine *f*

2827 HOT DESEAMER cle,htt
f décriqueuse *f* à chaud
e acanalador *m* por llama, escarpador *m* con soplete
i scriccatore *m* alla fiamma
n afvlammachine
d Flämmaschine *f*

2828 HOT-DIP COATING BATH coa
f bain *m* chaud de revêtement par immersion
e baño *m* caliente de revestimiento por inmersión
i bagno *m* caldo di rivestimento per immersione
n warm dompelmetalliseerbad *n*
d warmes Eintauchmetallisierungsbad *n*

2829 HOT-DIP GALVANIZING BATH elc,htt
f bain *m* chaud de galvanisation
e baño *m* caliente de galvanización
i bagno *m* galvanico caldo
n bad *n* voor thermisch verzinken
d Zinkbadofen *m*

2830 HOT-DIP TINNING BATH elc
f bain *m* chaud d'étamage
e baño *m* caliente de estañado
i bagno *m* caldo di stagnatura
n bad *n* voor thermisch vertinnen
d Zinnbadofen *m*

2831 HOT ENAMELLING OVEN coa,htt
f four *m* d'émaillage à chaud
e horno *m* de esmaltado en caliente
i forno *m* di smaltatura a caldo
n warmemailleeroven
d Puderemaillierofen *m*

2832 HOT FINISHING FURNACE htt,pol
f four *m* de finissage à chaud
e horno *m* de acabado en caliente
i forno *m* di finitura a caldo
n warmveredelingsoven
d Warmvergütungsofen *m*

2833 HOT FLOOR cer
f carreau *m* sécheur
e pavimento *m* secador
i pavimento *m* essiccatore
n droogvloer
d Trockenboden *m*

2834 HOT-FLUE DRYING PLANT tex
f installation *f* de séchage à air chaud
e equipo *m* secador de aire caliente
i impianto *m* essiccatore ad aria calda
n hete-luchtdrooginstallatie
d Heisslufttrockenanlage *f*

2835 HOT GALVANIZING PLANT coa
f installation *f* de galvanisation à chaud
e equipo *m* de galvanización en caliente
i impianto *m* di galvanizzazione a caldo
n thermische verzinkinstallatie
d Feuerverzinkungsanlage *f*

2836 HOT-GAS WELDING PLANT pla,wel
f installation *f* de soudure à gaz chaud
e equipo *m* de soldadura con gas caliente
i impianto *m* di saldatura ad aria calda
n hete-luchtlasinstallatie
d Heissgasschweissanlage *f*

2837 HOT-OIL PROCESSING ROOM tex
f salle *f* de teinture à huile chaude
e sala *f* de teñido con aceite caliente
i sala *f* di tintura ad olio caldo
n hete-olieverfruimte
d Heissölfärbraum *m*

2838 HOT PATCHING GUN, cer,htt
AIR-BORNE SEALING GUN,
PATCHING GUN
f pistole *f* à fermer des pores
e pistola *f* para cerrar poros
i pistola *f* da chiudere pori
n poriëndichtend pistool *n*
d porendichtende Pistole *f*

2839 HOT PREPARATION PLANT, cer
STEAM TEMPERING PLANT
f installation *f* de traitement de l'argile à la vapeur
e equipo *m* de tratamiento de la arcilla al vapor
i impianto *m* di trattamento dell'argilla al vapore
n stoombehandelingsinstallatie voor klei
d Dampfbehandlungsanlage *f* für Ton

2840 HOT REACTOR nue
f réacteur *m* chaud
e reactor *m* caliente
i reattore *m* caldo
n hete reactor
d heisser Reaktor *m*

2841 HOT RINSE (GB), elc
HOT RINSE BATH (GB), HOT SWILL (US), HOT SWILL BATH (US)
f bain *m* de rinçage à chaud
e baño *m* de enjuague en caliente
i bagno *m* di sciacquatura a caldo
n warm spoelbad *n*
d Heissspülbad *n*, Warmspülbad *n*

2842 HOT ROLLER, pap
STEAM-HEATED CYLINDER
f tambour *m* sécheur
e tambor *m* secador
i tamburo *m* essiccatore
n droogtrommel
d Dampftrommel *f*, Trockentrommel *f*

2843 HOT TINNING PLANT coa
f installation *f* d'étamage à chaud
e equipo *m* de estañado en caliente
i impianto *m* di stagnatura a caldo
n thermische vertinningsinstallatie
d Feuerverzinnungsanlage *f*

2844 HOT WATER FUNNEL lab
f entonnoir *m* à eau chaude, entonnoir *m* à filtration chaude
e embudo *m* de agua caliente, embudo *m* de filtrar en caliente
i imbuto *m* a filtrazione calda, imbuto *m* ad acqua calda
n warmwatertrechter
d Heisswassertrichter *m*

2845 HOT WORT RECEIVER, bre
SETTLING TANK
f bac *m* fermé
e tina *f* de sedimentación

i tino *m* di sedimentazione
n gesloten koelbak
d Setzbottich *m*

2846 HOUSING gen
f boîte *f*, boîtier *m*, enceinte *f*
e caja *f*, carcasa *f*
i carcassa *f*, involucro *m*
n huis *n*
d Gehäuse *n*

2847 HUMIDIFIER, hyg
MOISTENING APPARATUS
f humecteur *m*, humidificateur *m*
e humectador *m*, humidificador *m*
i umidificatore *m*
n bevochtigingstoestel *n*
d Anfeuchtapparat *m*, Befeuchter *m*

2848 HUMIDIFYING CHAMBER, tex
MOISTENING CHAMBER
f chambre *f* d'humidification
e cámara *f* de humidificación
i camera *f* d'umidificazione
n bevochtigingsruimte
d Befeuchtungskammer *f*

2849 HUMIDITY CABINET lab
f boîte *f* à atmosphère humide
e armario *m* humectador
i involucro *m* ad atmosfera umida
n vochtkast
d Feuchtigkeitskasten *m*

2850 HUMIDITY CONTROLLER cor,hyg
f régulateur *m* de l'humidité
e higrorregulador *m*
i regolatore *m* dell'umidità
n vochtigheidsregelinrichting
d Feuchtigkeitsregler *m*

2851 HUMIDITY DRIER cer,dry
f sécheur *m* à air chaud humide
e secadero *m* de aire caliente húmedo
i essiccatoio *m* ad aria calda umida
n droger met vochtige lucht
d Trockner *m* mit feuchter Luft

2852 HUMMELER, bre
AWN CUTTER
f ébarbeuse *f*
e desbarbadora *f*
i sbarbatrice *f*, spuntaresta *f*
n gerstkorter
d Entgranner *m*

2853 HUTCH min
f compartiment *m* de tamis
e compartimiento *m* de criba
i compartimento *m* di vaglio
n zeefafdeling
d Siebabteil *n*

2854 HYDRATION PLANT pap
f installation *f* de hydratation
e equipo *m* de hidratación
i impianto *m* d'idratazione
n hydratatie-installatie
d Hydratationsanlage *f*

2855 HYDRATOR pap
f appareil *m* d'hydratation, hydrateur *m*
e hidratador *m*
i apparecchio *m* d'idratazione
n hydratatieapparaat *n*
d Hydratierapparat *m*

2856 HYDRAULIC ACCUMULATOR sto
f accumulateur *m* hydraulique
e acumulador *m* hidráulico, hidroacumulador *m*
i accumulatore *m* idraulico
n hydraulische accumulator
d hydraulischer Akkumulator *m*

2857 HYDRAULIC CLASSIFIER sor
f classeur *m* hydraulique, trieuse hydraulique
e clasificador *m* hidráulico, hidroclasificador *m*
i classificatore *m* idraulico
n hydraulische classificator
d hydraulischer Erztrenner *m*, hydraulischer Klassierapparat *m*

2858 HYDRAULIC DESCALING cle,fou
APPARATUS
f appareil *m* de décalaminage par jet d'eau
e aparato *m* de decapaje hidráulico
i apparecchio *m* di scagliatura idraulica
n afspuitapparaat *n*
d Abspritzgerät *n*

2859 HYDRAULIC EXTRUDER, pla,rub
RAM EXTRUDER (GB),
STUFFER (US)
f expulseur *m* à pression hydraulique, presse *f* à filer
e prensa *f* de extrusión con émbolo hidráulica, prensa *f* hidráulica
i espulsatore *m* a pressione idraulica, estrusore *m* a pistone
n hydraulische spuitpers, plunjerspuitmachine

d hydraulische Strangpresse *f*, Kolbenstrangpresse *f*

2860 HYDRAULIC PRESS, HYDROPRESS pla,pre,rub
f presse *f* hydraulique
e hidroprensa *f*, prensa *f* hidráulica
i pressa *f* idraulica
n hydraulische pers
d hydraulische Presse *f*

2861 HYDRAULIC RAM pre
f bélier *m* hydraulique
e ariete *m* hidráulico
i ariete *m* idraulico
n waterram
d hydraulischer Widder *m*, Stossheber *m*

2862 HYDRAULIC SEPARATOR, HYDROSEPARATOR sor
f caisse *f* pointue, séparateur *m* hydraulique
e hidroseparador *m*, separador *m* hidráulico
i separatore *m* idraulico
n hydraulische scheider, hydraulische separator
d Hydroseparator *m*

2863 HYDRAULIC TESTING APPARATUS sap
f appareil *m* d'essai hydraulique
e aparato *m* de prueba idráulica
i apparecchio *m* di prova idraulica
n waterdrukbeproevingsapparaat *n*
d Wasserdruckprobiergerät *n*

2864 HYDRAULIC VULCANIZING PRESS rub
f presse *f* hydraulique à vulcaniser
e prensa *f* hidráulica para vulcanizar
i pressa *f* idraulica per vulcanizzare
n hydraulische vulcaniseerpers
d hydraulische Vulkanisierpresse *f*

2865 HYDROCYCLONE, CYCLONE THICKENER iwt
f épaisseur-cyclone *m*
e espesador *m* ciclónico
i concentratore *m* ciclone
n cycloonverdikker, hydrocycloon *n*
d Hydrozyklon *m*, Zykloneindicker *m*

2866 HYDRODESULFURIZATION TANK (US), HYDRODESULPHURIZATION TANK (GB) fue
f récipient *m* de hydrodésulfuration
e tanque *m* de hidrodesulfuración
i serbatoio *m* d'idrodesolforazione
n installatie voor hydrogenerend ontzwavelen
d Wasserstoffentschwefelungsanlage *f*

2867 HYDROEXTRACTOR, WHIZZER dry
f essoreuse *f* centrifuge, hydro-extracteur *m*, toupie *f* mécanique
e extractor *m* centrífugo, hidroextractor *m*
i estrattore *m* centrifugo, idroestrattore *m*
n droogcentrifuge
d Zentrifugaltrockenmaschine *f*

2868 HYDROFORMING PLANT fue
f installation *f* d'hydroréformation
e equipo *m* de hidroreformación
i impianto *m* di riformazione con un catalizzatore d'ossido di molibdeno
n hydroformeerinstallatie, installatie voor dehydrerend aromatiseren
d Reformieranlage *f* unter Wasserstoffdruck

2869 HYDROGEN ANNEALING PLANT htt
f installation *f* de recuit à l'hydrogène
e equipo *m* de recocido al hidrógeno
i impianto *m* di ricottura all'idrogeno
n gloeifrisinstallatie in waterstof
d Glühfrischanlage *f* in Wasserstoff

2870 HYDROGEN BOMB mit,nue
f bombe *f* à hydrogène
e bomba *f* de hidrógeno
i bomba *f* ad idrogeno
n waterstofbom
d Wasserstoffbombe *f*

2871 HYDROGEN ELECTRODE elc
f électrode *f* à hydrogène
e electrodo *m* de hidrógeno
i elettrodo *m* ad idrogeno
n waterstofelektrode
d Wasserstoffelektrode *f*

2872 HYDROGEN-TREATING TANK coa
f cuve *f* de traitement à l'hydrogène
e cuba *f* de tratamiento al hidrógeno
i tino *m* di trattamento all'idrogeno
n waterstofbehandelingsvat *n*
d Wasserstoffbehandlungsgefäss *n*

2873 HYDROGEN WELDING PLANT wel
f installation *f* de soudure à l'hydrogène
e equipo *m* de soldadura al hidrógeno
i impianto *m* di saldatura all'idrogeno
n waterstoflasinstallatie
d Wasserstoffschweissanlage *f*

2874 HYDROGENATION PLANT fat,fue
f installation *f* d'hydrogénation
e equipo *m* de hidrogenación
i impianto *m* d'idrogenazione
n hydreerinstallatie, hydrogeneringsinstallatie
d Hydrieranlage *f*, Hydrogenierungsanlage *f*

2875 HYDROGENATOR fat
f autoclave *m* à hydrogénation
e autoclave *f* para hidrógeno, tanque *m* hidrogenador
i autoclave *f* da idrogenare
n hydreerautoclaaf, hydrogeneringsketel, vethardingsketel
d Härtungsautoklav *m*, Härtungskessel *m*

2876 HYDROLYSIS PLANT iwt
f installation *f* d'hydrolyse
e equipo *m* de hidrólisis
i impianto *m* d'idrolisi
n hydrolyse-installatie
d Hydrolyseanlage *f*

2877 HYDROLYZING TANK, PRIVY VAULT, SEPTIC TANK iwt
f récipient *m* septique
e tanque *m* séptico
i serbatoio *m* settico, vasca *f* a flusso di liquame
n vuilnistank, vuilwatertank
d Abwasserfaulraum *m*, Faulgrube *f*, Faulraum *m*

2878 HYDROMETER, AREOMETER ins
f aréomètre *m*, hydromètre *m*
e areómetro *m*, hidrómetro *m*
i areometro *m*, idrometro *m*
n areometer, hydrometer, vochtmeter
d Aräometer *n*, Senkwaage *f*, Tauchwaage *f*

2879 HYDROMETER SET ins
f jeu *m* d'aréomètres, trousse *f* d'aréomètres
e juego *m* de hidrómetros
i gruppo *m* d'idrometri
n hydrometergroep
d Aräometersatz *m*

2880 HYDROSTATIC SLUDGE REMOVAL PLANT iw
f installation *f* d'enlèvement de boue par pression hydrostatique
e equipo *m* de remoción de fango por presión hidrostática
i impianto *m* di scarico di fango per pressione idrostatica
n hydrostatische ontslibbingsinstallatie
d hydrostatische Entschlammunganlage *f*

2881 HYDROSTATIC TESTING APPARATUS se
f appareil *m* d'essai hydrostatiq
e aparato *m* de prueba hidrostática
i apparecchio *m* di prova idrostatica
n hydrostatisch beproevingsapparaat *n*
d hydrostatisches Prüfgerät *n*

2882 HYDROTOMETER i
f hydrotomètre *m*
e hidrotómetro *m*
i idrotometro *m*
n hydrotometer
d Hydrotometer *n*

2883 HYDROXYLATION PLANT
f installation *f* d'hydroxylation
e equipo *m* de hidroxilación
i impianto *m* d'idrossilazione
n hydroxyleerinstallatie
d Hydroxylierungsanlage *f*

2884 HYGROMETER (GB), AIR HUMIDITY INDICATOR, MOIST-O-GRAPH (US)
f hygromètre *m*
e higrómetro *m*
i igrometro *m*

n hygrometer, vochtigheids-
meter
d Hygrometer *n*, Luftfeuchtig-
keitsmesser *m*

2885 HYGROSTAT ins
f hygrostat *m*
e higróstato *m*
i igrostato *m*
n hygrostaat
d Hygrostat *m*

2886 HYPERSENSITIZATION BATH pho
f bain *m* de hypersensibilisation
e baño *m* de hipersensibilización
i bagno *m* d'ipersensibilizzazione
n hypersensibilisatiebad *n*
d Hypersensibilisierungsbad *n*

2887 HYPO BATH, pho
SODIUM THIOSULPHATE BATH
f bain *m* à hyposulfite de sodium
e baño *m* de hiposulfito sódico
i bagno *m* ad iposolfito di sodio
n hypobad *n*
d Fixiernatronbad *n*, Hypobad *n*

2888 HYPOCHLORITE BLEACHING tex
TANK
f récipient *m* de blanchiment à
l'hypochlorite
e tanque *m* de blanqueo con
hipoclorito
i serbatoio *m* di candeggio con
ipoclorito
n hypochlorietbleektank
d Hypochloritbleichgefäss *n*

I

2889 ICE BAG mcy
f vessie *f* à glace
e bolsa *f* de hielo machacado
i vescica *f* di ghiaccio
n ijszak
d Eisbeutel *m*

2890 ICE BOX (US), ICE CHEST (GB), ICE SAFE (GB) clo
f armoire *f* à glace, armoire *f* frigorifique, glacière *f*
e armario *m* frigorífico, nevera *f*
i ghiacciaia *f*
n ijskast
d Eisschrank *m*

2891 ICE CALORIMETER ins
f calorimètre *m* à glace
e calorímetro *m* de hielo
i calorimetro *m* a ghiaccio
n ijscalorimeter
d Eiskalorimeter *n*

2892 ICE CELLS clo
f mouleaux *pl* à glace, mouleaux *pl* frigorifiques
e moldes *pl* frigoríficos, moldes *pl* para hielo
i celle *pl* per ghiaccio, celle *pl* refrigeratrici
n koelcellen *pl*
d Eiszellen *pl*, Kühlzellen *pl*

2893 ICE FACTORY clo
f manufacture *f* de glace
e fábrica *f* de hielo
i fabbrica *f* di ghiaccio
n ijsfabriek
d Eisfabrik *f*

2894 IGNITER, IGNITION DEVICE gen
f dispositif *m* d'allumage
e dispositivo *m* de ignición, encendedor *m*, inflamador *m*
i accenditore *m*
n ontsteekinrichting, ontsteker
d Zünder *m*, Zündvorrichtung *f*

2895 IGNITION ACCUMULATOR elc
f accumulateur *m* d'allumage
e acumulador *m* encendedor
i accumulatore *m* d'accensione
n ontstekingsaccu
d Zündakkumulator *m*

2896 IGNITION CAPSULE, INCINERATING DISH la
f nacelle *f* pour incinérations
e cápsula *f* de incineración
i capsula *f* d'incenerazione
n verassingsschaaltje *n*
d Veräschungsschale *f*

2897 IGNITION FLAME bu
f flamme *f* d'allumage, veilleuse
e llama *f* encendedora
i fiamma *f* d'infiammazione
n ontstekingsvlam, waakvlam
d Zündflamme *f*

2898 IGNITION TEST TUBE la
f tube *m* à recuire
e tubo *m* de recocido
i tubo *m* di ricottura
n gloeibuisje *n*
d Glühröhrchen *n*

2899 IHRIGIZING OVEN m
f four *m* d'imprégnation au silici
e horno *m* de silicionación
i forno *m* da insilare
n siliciumimpregneeroven
d Silinierofen *m*

2900 IMBIBITION MATRIX p
f matrice *f* pour tirage par imbibition
e matriz *f* para copia por imbibición
i matrice *f* da stampare per imbibizione
n opzuigmatrijs voor kleur-afdrukken
d Einsaugekopiermatrize *f*

2901 IMBIBITION ROLLER p
f rouleau *m* d'imbibition
e rodillo *m* de imbibición
i rullo *m* d'imbibizione
n drenkrol
d Tränkrolle *f*

2902 IMBIBITION TANK, SOAKING TANK
f récipient *m* d'imbibition
e tanque *m* de imbibición
i serbatoio *m* d'imbibizione
n drenkvat *n*
d Tränkgefäss *n*

2903 IMMERSION CONDENSER, pre
COLDFINGER (US),
SUBMERGED CONDENSER
f condensateur *m* plongeur, doigt *m* réfrigérant, réfrigérant *m* à immersion
e condensador *m* de inmersión
i condensatore *m* ad immersione
n dompelcondensor
d Einhängekühler *m*, Tauchkondensator *m*

2904 IMMERSION FILTER TUBE fil
f filtre *m* plongeur
e filtro *m* de inmersión
i filtro *m* ad immersione
n dompelfilter *n*
d Eintauchnutsche *f*

2905 IMMERSION HEATER htt
f thermo-plongeur *m*
e termoinmersor *m*
i elemento *m* riscaldatore ad immersione
n dompelaar, dompelelement *n*
d Tauchsieder *m*

2906 IMMERSION PLATING BATH elc
f bain *m* de dépôt au trempé, bain *m* de dépôt par immersion
e baño *m* de depósito por inmersión
i bagno *m* di deposito per immersione
n chemisch neerslagbad *n*
d Ansiedebad *n*, Eintauchplattierungsbad *n*

2907 IMMERSION PUMP pre
f pompe *f* plongeante
e bomba *f* de inmersión
i pompa *f* ad immersione
n onderwaterpomp
d Tauchpumpe *f*, Unterwasserpumpe *f*

2908 IMMERSION PYROMETER ins
f pyromètre *m* à immersion
e pirómetro *m* de inmersión
i pirometro *m* ad immersione
n dompelpyrometer
d Eintauchpyrometer *n*

2909 IMMERSION THERMOSTAT ins
f thermostat *m* à immersion
e termóstato *m* de inmersión
i termostato *m* ad immersione
n dompelthermostaat
d Eintauchthermostat *m*

2910 IMPACT BURNER bur
f brûleur *m* à impact
e quemador *m* de choque
i bruciatore *m* ad urto
n stootbrander
d Auftreffbrenner *m*, Prallbrenner *m*, Stossbrenner *m*

2911 IMPACT CLEANING DEVICE cle
f nettoyeuse *f* à jet de sable
e dispositivo *m* de limpieza por chorro de arena
i pulitrice *f* a getto di sabbia
n zandstraalreiniger
d Sandstrahlputzer *m*

2912 IMPACT DRIER dry
f sécheur *m* à jets d'air chaud
e secador *m* por choques de aire caliente
i essiccatore *m* per getti d'aria calda
n drooginrichting met warmeluchtstoten
d Trockenvorrichtung *f* mit Heissluftanprall

2913 IMPACT MILL, brk
IMPACT PULVERIZER
f broyeur *m* à percussion, casse-gueuses *m*
e molino *m* de golpe, triturador *m* por choque
i mulino *m* ad urto, mulino *m* con pestelli in croce
n slagkruismolen
d Schlagkreuzmühle *f*

2914 IMPACT MOLDING MACHINE (US), pla
IMPACT MOULDING MACHINE (GB)
f presse *f* à chocs, presse *f* à frappes
e prensa *f* de moldear por impacto
i pressa *f* a percussione
n slagpers
d Schlagpresse *f*

2915 IMPACT PENDULUM, rub,tes
SHOCK PENDULUM
f appareil *m* à marteau-pendule, mouton *m* pendulaire, mouton-pendule *m*
e péndulo *m* para prueba de choque
i maglio *m* a pendolo, pendolo *m* per prova d'urto
n slagarm, slagpendel, slingerslagapparaat *n*
d Pendelschlagwerk *n*, Schlagpendel *m*

2916 IMPACT SCREEN sor
f tamis *m* à frappes
e criba *f* de percusión
i vaglio *m* a percussione
n stootzeef
d Stossieb *n*

2917 IMPACT SEPARATOR, IMPINGEMENT SEPARATOR, INERT SEPARATOR sor
f séparateur *m* à frappes
e separador *m* por choques
i separatore *m* a percussione
n stootseparator
d Stossabscheider *m*

2918 IMPACT TESTING MACHINE tes
f appareil *m* d'essai au choc
e aparato *m* de prueba de resistencia al choque
i macchina *f* per prove d'urto
n slagproefapparaat *n*
d Schlagprobegerät *n*

2919 IMPACT WHEEL MIXER mix
f broyeur *m* centrifuge
e pulverizador *m* por impacto
i mulino *m* centrifugo
n turboverstuiver
d Turbozerstäuber *m*

2920 IMPELLER cen,mix
f roue *f* motrice
e rueda *f* móvil
i girante *m*, ruota *f*
n roterende roerder
d drehende Rührstange *f*

2921 IMPELLER BLADE cen,mix
f palette *f* de roue motrice
e paleta *f* de rueda móvil
i paletta *f* di girante, paletta *f* di ruota
n roerderblad *n*
d Rührstangenflügel *m*

2922 IMPELLER FLOTATION MACHINE mix
f machine *f* à flottation par agitation
e máquina *f* de flotación por agitación
i macchina *f* a flottazione per agitazione
n flotatiemachine met roerwerk
d Rührwerkflotationsmaschine *f*

2923 IMPERFORATE BASKET CENTRIFUGE cen
f centrifugeur *m* sans tamis, essoreuse *f* à tambour non perforé
e centrífuga *f* sin tamiz
i centrifuga *f* senza staccio
n zeefloze centrifuge
d sieblose Schleuder *f*

2924 IMPREGNATING MACHINE imp,pla
f machine *f* à imprégner et à enduire
e máquina *f* de impregnar con resina
i impregnatrice *f* di resina
n harsimpregneermachine
d Beharzungs- und Imprägniermaschine *f*

2925 IMPREGNATING PAN imp
f cuve *f* d'imprégnation
e cuba *f* de impregnación
i serbatoio *m* d'impregnazione
n impregneerketel
d Imprägnierkessel *m*

2926 IMPREGNATING PLANT imp
f installation *f* d'imprégnation
e equipo *m* de impregnación
i impianto *m* d'impregnazione
n impregneerinstallatie
d Imprägnieranlage *f*

2927 IMPREGNATION BATH imp
f bain *m* d'imprégnation
e baño *m* de impregnación
i bagno *m* d'impregnazione
n impregneerbad *n*
d Imprägnierbad *n*, Tränkbad *n*

2928 IMPREGNATION TANK imp,ru
f cuve *f* d'imprégnation
e tanque *m* de impregnación
i serbatoio *m* d'impregnazione
n impregneertank
d Imprägnierungsgefäss *n*, Tränkkessel *m*

2929 INACTIVE CAP di
f calotte *f* inactive
e platillo *m* ineficaz
i campanella *f* inattiva
n onwerkzame klok
d untätige Glocke *f*

2930 INCANDESCENT MANTLE, GAS MANTLE ga
f manchon *m* de bec à incandesce
e manguito *m* incandescente para gas
i reticella *f* ad incandescenza
n gasgloeilichtkousje *n*, gloeikousje *n*
d Glühstrumpf *m*

2931 INCENDIARY BOMB exp
f bombe *f* incendiaire
e bomba *f* incendiaria
i spezzone *m* incendiario
n brandbom
d Brandbombe *f*

2932 INCINERATING DISH, IGNITION CAPSULE lab
f nacelle *f* pour incinérations
e cápsula *f* de incineración
i capsula *f* d'incenerazione
n verassingsschaaltje *n*
d Veräschungsschale *f*

2933 INCINERATOR, COMBUSTION FURNACE htt
f four *m* crématoire, four *m* d'incinération
e horno *m* de análisis, horno *m* de combustión
i forno *m* d'incenerazione
n verbrandingsoven
d Verbrennungsofen *m*

2934 INCINERATOR, DESTRUCTOR hyg,iwt
f destructeur *m*, incinérateur *m*
e destructor *m*, horno *m* de incineración
i forno *m* per incenerire
n destructor, verbrandingsoven
d Verbrennungsofen *m*

2935 INCLINABLE PRESS pla
f presse *f* inclinable
e prensa *f* inclinable
i pressa *f* inclinabile
n hellende pers, kantelpers
d schrägeinstellbare Presse *f*

2936 INCLINATOR, CARBOY TIPPER tra
f culbuteur *m* de tourie
e basculador *m* de damajuana
i ribaltatore *m* di damigiana
n mandefleskantelaar, mandeflеskipper
d Korbflaschenkipper *m*

2937 INCLINED RETORT OVEN htt
f four *m* à cornues inclinées
e horno *m* de retortas inclinadas
i forno *m* a storte inclinate
n hellende retortenoven
d Schrägretortenofen *m*

2938 INCLINED SEAT VALVE cpt
f soupape *f* à siège incliné
e válvula *f* de asiento inclinado
i valvola *f* a sede inclinata
n klep met schuine zitting
d Schrägsitzventil *n*

2939 INCLINED TWO-ROLL CALENDER pap
f calandre *f* inclinée à deux cylindres
e calandra *f* inclinada de dos cilindros
i calandra *f* inclinata a due cilindri
n overhellende tweewalsenkalander
d überhängender Zweiwalzenkalander *m*

2940 INCREASER cpt
f raccord *m* de réduction
e cono *m* reductor, pieza *f* cónica de unión
i manicotto *m* di riduzione
n verloopstuk *n*
d Übergangsrohr *n*

2941 INCUBATOR bac
f étuve *f* à incubation
e estufa *f* bacteriológica, incubador *m*
i stufa *f* d'incubazione
n broedstoof, incubator
d Brutschrank *m*

2942 INCUBATOR THERMOMETER ins
f thermomètre *m* de contrôle pour étuves
e termómetro *m* de estufa bacteriológica
i termometro *m* di stufa d'incubazione
n broedstoofthermometer
d Brutschrankthermometer *n*

2943 INDENTER tes
f pénétrateur *m*
e penetrador *m*
i penetratore *m*
n indringlichaam *n*
d Eindruckkörper *m*

2944 INDICATOR DIAGRAM ins
f diagramme *m* d'indicateur
e diagrama *m* de indicador
i diagramma *m* tracciato dall'indicatore
n indicatordiagram *n*
d Indikatordiagramm *n*

2945 INDICATOR ELECTRODE elc
f électrode *f* indicatrice
e electrodo *m* indicador

i elettrodo *m* indicatore
n indicatorelektrode
d Indikatorelektrode *f*

2946 INDICATOR VAPORIZER chr
f vaporisateur *m* d'indicateur
e vaporizador *m* de indicador
i vaporizzatore *m* d'indicatore
n indicatorverstuiver
d Indikatorzerstäuber *m*

2947 INDIRECT ARC FURNACE, ARCH IMAGE FURNACE htt
f four *m* à arc indirect, four *m* à arc libre
e horno *m* de arco indirecto
i forno *m* ad arco indiretto, forno *m* ad arco libero
n indirecte vlamboogoven, vlamboogoven met indirecte verhitting
d indirekter Lichtbogenofen *m*

2948 INDIRECT COOLER, SURFACE COOLER clo
f refroidisseur *m* superficiel
e enfriador *m* superficial
i raffreddatore *m* superficiale, refrigerante *m* superficiale
n oppervlakkoeler *m*
d Oberflächenkühler *m*

2949 INDIRECT CYCLE INTEGRAL BOILING REACTOR nue
f réacteur *m* à eau bouillante à refroidissement indirect et échangeur de chaleur incorporé
e reactor *m* de agua hirviente de enfriamiento indirecto y cambiador de calor encerrado
i reattore *m* ad acqua bollente a raffreddamento indiretto e scambiatore di calore incorporato
n indirect gekoelde kokend-water-reactor met ingebouwde warmte-uitwisselaar
d indirekt gekühlter Siedewasser-reaktor *m* mit eingebautem Wärmeaustauscher

2950 INDIRECT RESISTANCE FURNACE (GB), RESISTOR FURNACE (US) htt
f four *m* à chauffage indirect par résistance
e horno *m* de calentamiento indirecto por resistencia
i forno *m* a riscaldamento indiretto a resistenza
n weerstandsoven met indirecte verhitting
d indirekter Widerstandsofen *m*

2951 INDIRECT ROTARY DRIER dry
f sécheur *m* rotatif indirect
e secadero *m* giratorio indirecto
i essiccatoio *m* rotativo indiretto
n indirecte droogcentrifuge
d indirekte Trockenzentrifuge *f*

2952 INDUCED DRAFT FURNACE (US), INDUCED DRAUGHT FURNACE (GB) htt
f four *m* à tirage induit par aspiration
e horno *m* de tiro por aspiración
i forno *m* a tiraggio indotto
n oven met kunstmatige trek
d Ofen *m* mit induziertem Zug

2953 INDUCED DRAUGHT COOLING TOWER (GB), FORCED DRAFT COOLING TOWER (US) clo
f tour *f* de refroidissement à tirage forcé
e torre *f* de enfriamiento de tiro forzado
i torre *f* di refrigerazione a tiraggio forzato
n koeltoren met kunstmatige trek
d Kühlturm *m* mit künstlichem Zug

2954 INDUCED ROLL SEPARATOR sep
f séparateur *m* à cylindre induit
e separador *m* de cilindro inducido
i separatore *m* a cilindro indotto
n separator met geïnduceerde wals
d Induktionswalzenseparator *m*

2955 INDUCTION BRAZING PLANT wel
f installation *f* de soudure forte à induction
e equipo *m* de soldadura fuerte por corriente inducido
i impianto *m* di brasatura ad induzione
n inductieve hardsoldeerinstallatie
d Induktionshartlötungsanlage *f*

2956 INDUCTION CHANNEL FURNACE htt
f four *m* à induction à canal
e horno *m* de inducción con canal
i forno *m* ad induzione a canale
n inductieoven met ringkanaal, inductieringoven
d Induktionsrinnenofen *m*

2957 INDUCTION FURNACE htt
f four *m* à induction
e horno *m* de inducción
i forno *m* ad induzione
n inductieoven
d Induktionsofen *m*

2958 INDUCTION FURNACE WITH SUBMERGED CHANNEL htt
f four *m* à induction à canal submergé
e horno *m* de inducción con canal sumergido
i forno *m* ad induzione a canale sommerso
n inductieoven met ringkanaal
d Induktionsrinnenschmelzofen *m*

2959 INDUCTION-TYPE MAGNETIC SEPARATOR sor
f séparateur *m* électromagnétique à induction
e separador *m* magnético por inducción
i separatore *m* magnetico ad induzione
n inductieve magnetische scheider
d induktiver Magnetscheider *m*

2960 INDUSTRIAL CHEMISTRY, CHEMICAL ENGINEERING gen
f chimie *f* industrielle
e química *f* industrial
i chimica *f* industriale, ingegneria *f* chimica
n chemische technologie, industriële chemie
d chemische Technologie *f*, technische Chemie *f*

2961 INERT CELL elc
f pile *f* non-amorcée
e pila *f* seca recargable con agua
i pila *f* a liquido stabilizzato
n vulelement *n*
d Füllelement *n*, Lagerelement *n*

2962 INERT-GAS METAL-ARC WELDING PLANT wel
f installation *f* de soudure à l'arc en atmosphère inerte
e equipo *m* de soldadura por arco en atmosfera de gas inerte
i impianto *m* di saldatura in gas inerte
n booglasinstallatie in inerte gasatmosfeer
d Lichtbogenschweissanlage *f* unter Schutzgas

2963 INERT GAS PURIFICATION PLANT gas,nue
f installation *f* d'épuration de gaz inertes
e equipo *m* depurador de gases inertes
i impianto *m* di depurazione di gas inerti
n edelgassenzuiveringsinstallatie
d Edelgasreinigungsanlage *f*

2964 INERT SEPARATOR, IMPACT SEPARATOR, IMPINGEMENT SEPARATOR sor
f séparateur *m* à frappes
e separador *m* por choques
i separatore *m* a percussione
n stootseparator
d Stossabscheider *m*

2965 INERTIAL DUST COLLECTOR cle,gas
f collecteur *m* de poussière à inertie
e colector *m* de polvo de inercia
i collettore *m* di polvere ad inerzia
n stofverzamelaar volgens het traagheidsprincipe
d Staubsammler *m* mit Trägheitswirkung

2966 INFILTRATION DEVICE iwt
f appareil *m* à infiltration
e aparato *m* de infiltración
i apparecchio *m* ad infiltrazione
n infiltratieapparaat *n*
d Infiltrationsgerät *n*

2967 INFILTRATION SLOT sew
f fente *f* d'infiltration
e ranura *f* de infiltración
i fessura *f* di drenaggio, scanalatura *f* di drenaggio
n sijpelsleuf
d Sickerschlitz *m*

2968 INFLAMMABLE VAPOR DETECTING EQUIPMENT (US), INFLAMMABLE VAPOUR DETECTING EQUIPMENT (GB) sap
f installation *f* détectrice de vapeurs inflammables
e equipo *m* detector de vapores inflamables
i impianto *m* rivelatore di vapori infiammabili
n verklikkerinstallatie voor ontvlambare gassen
d Überwachungsanlage *f* für feuergefährliche Gase

2969 INFLOW TUBE cpt,sew
f tuyau *m* d'admission, tuyau *m* d'entrée
e tubo *m* de afluencia, tubo *m* de llegada
i tubo *m* d'afflusso
n inlaatbuis
d Einlaufrohr *n*, Eintrittsrohr *n*, Zuflussrohr *n*

2970 INFRARED BATH dry,lab
f bain *m* à rayons infrarouges
e baño *m* de rayos infrarrojos
i bagno *m* a raggi infrarossi
n infraroodbad *n*
d Infrarotkessel *m*

2971 INFRARED DRIER dry
f sécheur *m* à l'infrarouge
e secadero *m* por rayos infrarrojos
i essiccatore *m* per raggi infrarossi
n infrarooddroger
d Infrarottrockner *m*

2972 INFRARED DRYING TUNNEL, dry
INFRARED TUNNEL OVEN
f tunnel *m* de séchage par rayons infrarouges
e galería *f* de secado por rayos infrarrojos
i tunnel *m* essiccatore per raggi infrarossi
n infrarooddroogtunnel
d Infrarottrockentunnel *m*

2973 INFRARED FILTER pho
f filtre *m* pour infrarouge
e filtro *m* por rayos infrarrojos
i filtro *m* per raggi infrarossi
n infraroodfilter *n*
d Infrarotfilter *m*

2974 INFRARED GAS ana,lab
ANALYZER
f analyseur *m* de gaz par lumière infrarouge
e analizador *m* de gases por luz infrarroja
i analizzatore *m* di gas per luce infrarossa
n infrarood-gasanalyseapparaat *n*
d Infrarot-Gasanalysegerät *n*

2975 INFRARED GAS OVEN dec
f four *m* à gaz à rayonnement infrarouge
e horno *m* de gas de radiación infrarroja
i forno *m* a gas a radiazione infrarossa
n infraroodgasoven
d Infrarotgasofen *m*

2976 INFRARED PREHEATER, pla,rub
PREHEATER BY INFRARED RADIATION
f préchauffeur *m* par rayons infrarouges
e precalentador *m* por rayos infrarrojos
i preriscaldatore *m* per raggi infrarossi
n infraroodvoorverwarmer
d Infrarotvorwärmer *m*

2977 INFRARED SENSITIZER BATH ph
f bain *m* sensibilisateur pour l'infrarouge
e baño *m* sensibilizador para el infrarrojo
i bagno *m* sensibilizzatore per l'infrarosso
n infraroodsensibilisatiebad
d Infrarot-Sensibilisatorbad *n*

2978 INFRASIZER min,sc
f microclasseur *m*
e microclasificador *m*
i microclassificatore *m*
n microclassificeerapparaat *n*
d Mikroklassierer *m*

2979 INITIAL BOILING POINT i
RECORDER
f enregistreur *m* de la températ initiale d'ébullition
e registrador *m* de punto de ebullición inicial
i registratore *m* di punto d'ebollizione iniziale
n beginkookpuntschrijver
d Siedebeginngeber *m*

2980 INJECTION CONDENSER, p
JET CONDENSER
f condensateur *m* à injection
e condensador *m* de inyección
i condensatore *m* a getto
n waterstraalcondensor
d Einspritzkondensator *m*

2981 INJECTION CYLINDER
f pot *m* d'injection
e cilindro *m* de inyección, cilindro *m* inyector
i cilindro *m* d'iniezione
n injectiecilinder
d Spritzzylinder *m*

2982 INJECTION MOLD (US), INJECTION MOULD (GB) pla
f moule *m* à injection
e molde *m* para inyección
i stampo *m* per iniezione
n spuitgietmatrijs
d Spritzgussform *f*

2983 INJECTION MOLDING MACHINE (US), INJECTION MOULDING MACHINE (GB), INJECTION PRESS pla
f machine *f* à injection, presse *f* à injection
e inyectora *f*, máquina *f* de moldeo por inyección
i pressa *f* d'iniezione
n spuitgietmachine
d Spritzgussmaschine *f*

2984 INJECTION MOLDING NOZZLE (US), INJECTION MOULDING NOZZLE (GB) pla
f buse *f* de presse à injection
e tobera *f* para prensa de inyección
i ugello *m* di pressa ad iniezione
n spuitgietmondstuk *n*
d Spritzdüse *f*, Spritzgussdüse *f*

2985 INJECTION PISTON, INJECTION PLUNGER (US), INJECTION RAM (GB) pla
f piston *m* d'injection
e émbolo *m* de inyección, pistón *m* de inyección
i pistone *m* d'iniezione
n injectieplunjer, spuitgietplunjer
d Spritzgusskolben *m*, Spritzkolben *m*

2986 INJECTION PUMP, SATURATOR bre
f saturateur *m*
e saturador *m*
i pompa *f* di saturazione, saturatore *m*
n injecteerpomp
d Imprägnierungspumpe *f*

2987 INJECTOR pre
f injecteur *m*
e inyector *m*
i iniettore *m*
n injector
d Injektor *m*

2988 INKOMETER, ADHERENCE METER, ADHEROMETER ins
f adhérencemètre *m*
e adherencímetro *m*, adherómetro *m*
i aderometro *m*
n adhesiemeter
d Adhäsionsmesser *m*

2989 INLET, GATE pla
f canal *m* secondaire, entonnoir *m* de coulée, entrée *f*
e entrada *f*
i entrata *f*
n aanspuitkanaal *n*, aanspuitpoort
d Angusssteg *m*, Steg *m*

2990 INLET OPENING sew
f orifice *m* d'adduction, orifice *m* d'admission, orifice *m* d'introduction
e abertura *f* de admisión
i apertura *f* d'introduzione
n inlaatopening
d Einflussöffnung *f*, Zuflussöffnung *f*

2991 INLET PIPE sew
f tuyau *m* d'admission, tuyau *m* d'alimentation
e tubo *m* de admisión, tubo *m* de entrada, tubo *m* de llegada
i tubo *m* d'afflusso
n inlaatpijp
d Einlassrohr *n*, Zuflussrohr *n*

2992 INLET VALVE cpt
f soupape *f* d'admission
e válvula *f* de admisión
i valvola *f* d'ingresso
n inlaatklep
d Einlassschieber *m*, Eintrittsventil *n*

2993 INSECTICIDE FACTORY hyg
f manufacture *f* d'insecticides
e fábrica *f* de insecticidas
i fabbrica *f* d'insetticidi
n insekticidefabriek
d Insektenvertilgungsmittelfabrik *f*, Schädlingsbekämpfungsmittelwerk *n*

2994 INSECTPROOF CONTAINER tra
f récipient *m* insectifuge
e recipiente *m* insectífugo
i recipiente *m* insettifugo
n tegen insekten beschermde houder
d insektenbeständiger Behälter *m*

2995 INSERT pla
f insertion *f*, prisonnier *m*
e enganche *m*, inserción *f*, pieza *f* insertada

i inserto *m*, pezzo *m* riportato
n ingegoten stuk *n*, inzetdeel *n*, inzetstuk *n*
d Eingussteil *m*, Einpressteil *m*, Einspritzteil *m*

2996 INSERT PIN (US), CARRIER PIN (GB), RETAINING PIN — pla,pre
f broche *f* à prisonnier, ergot *m* de centrage, goupille *f* de fixation
e clavija *f* de retenida, clavija *f* de sujeción
i spina *f* portainserto
n fixatiepen
d Haltestift *m*

2997 INSERT PYROMETER — ins
f pyromètre *m* à insertion
e pirómetro *m* de inserción
i pirometro *m* ad inserzione
n insteekpyrometer
d Einsatzpyrometer *n*

2998 INSERTION FILTER, HANGING FILTER — ext,fil
f filtre *m* à insérer
e filtro *m* de inserción
i filtro *m* d'inserzione
n inzetfilter *n*
d Einsatzfilter *m*

2999 INSIDE DIE — pla,pre,rub
f matrice *f* intérieure
e matriz *f* interior
i matrice *f* interiore
n binnenmatrijs
d innere Matrize *f*

3000 INSOLUBLE ANODE — elc
f anode *f* insoluble
e ánodo *m* insoluble
i anodo *m* insolubile
n onoplosbare anode
d unlösliche Anode *f*

3001 INSPISSATION VESSEL — sin
f cuve *f* épaississeuse
e cuba *f* espesadora
i serbatoio *m* d'inspessimento
n indikvat *n*
d Eindickgefäss *n*

3002 INSTALLATION, PLANT — gen
f établissement *m*, installation *f*
e equipo *m*, instalación *f*
i impianto *m*, installazione *f*
n inrichting, installatie
d Anlage *f*

3003 INSTRUMENTATION — gen
f instrumentation *f*
e instrumentación *f*
i strumentazione *f*
n apparatuur, instrumentatie
d Apparatur *f*, Instrumentausrüstung *f*

3004 INSULATING GRID, GRID — pla,pre
f plaque *f* isolante
e parrilla *f* aislante
i griglia *f* isolante
n isolatierooster *n*
d Isolierrost *m*

3005 INSULATION BOILER — boi
f chaudière *f* isolée
e caldera *f* aislada
i caldaia *f* isolata
n geïsoleerde ketel
d isolierter Kessel *m*

3006 INTAKE BASIN — iw
f bassin *m* d'arrivée
e alberca *f* de entrada
i bacino *m* d'ingresso
n opvangbekken *n*
d Einlaufbecken *n*

3007 INTAKE MAIN, INTAKE PIPE — boi,cp
f conduite *f* de prise
e tubería *f* de toma
i tubazione *f* di presa
n aftapleiding
d Entnahmeleitung *f*

3008 INTAKE SCREEN — iw
f grille *f* d'entrée
e rejilla *f* de admisión
i griglia *f* d'imbocco, griglia *f* d'ingresso
n opvangrooster *n*
d Einlaufrost *m*

3009 INTAKE SLUICE VALVE — bo
f vanne *f* de prise
e válvula *f* esclusa de entrada
i saracinesca *f* di presa
n schuifafsluiter
d Entnahmeschieber *m*

3010 INTAKE STRAINER — iw
f crépine *f* d'entrée
e colador *m* de entrada
i colatoio *m* d'ingresso
n opvangnet *n*
d Einlaufseiher *m*

3011 INTENSIFYING BATH pho
f bain *m* de renforcement
e baño *m* intensificador
i bagno *m* di rinforzo
n versterkbad *n*
d Verstärkungsbad *n*

3012 INTERCEPTING SEWER, COLLECTOR iwt
f collecteur *m*
e alcantarilla *f* interceptadora, canal *m* colector
i canale *m* collettore, collettore *m* di fognatura
n verzamelkanaal *n*
d Abwassersammler *m*, Sammelkanal *m*

3013 INTERCEPTOR, SINK TRAP, SIPHON sew
f siphon *m* inodore
e sifón *m* contra mal olor
i intercettore *m*, sifone *m* contro mal odore
n sifon, stankafsluiter, zwanehals
d Geruchverschluss *m*

3014 INTERCOOLER clo
f réfrigérant *m* intermédiaire
e interenfriador *m*, interrefrigerador *m*
i refrigeratore *m* intermedio
n tussenkoeler
d Zwischenkühler *m*

3015 INTERFERENCE FILTER pho
f filtre *m* interférentiel
e filtro *m* interferencial
i filtro *m* interferenziale
n interferentiefilter *n*
d Interferenzfilter *m*

3016 INTERMEDIATE HEAT EXCHANGER nue
f échangeur *m* de chaleur intermédiaire
e cambiador *m* de calor intermedio
i scambiatore *m* di calore intermedio
n tussenwarmte-uitwisselaar
d Zwischenwärmeaustauscher *m*

3017 INTERMITTENT CHLORINATION PLANT cle
f installation *f* de chlorination intermittente
e equipo *m* de clorinación intermitente
i impianto *m* di clorinazione intermittente
n niet-continue chloreerinstallatie
d Stosschlorungsanlage *f*

3018 INTERMITTENT OVEN htt
f four *m* intermittent
e horno *m* intermitente
i forno *m* intermittente
n niet-continue oven
d diskontinuierlicher Ofen *m*

3019 INTERMITTENT SOIL FILTER, SOIL FILTER fil,iwt
f filtre *m* par le sol, sol *m* filtrant
e filtro *m* natural
i filtro *m* naturale
n natuurlijk filter *n*
d Bodenfilter *m*

3020 INTERNAL COMBUSTION ENGINE gen
f moteur *m* à combustion interne
e motor *m* de combustión interna
i motore *m* a combustione interna, motore *m* endotermico
n verbrandingsmotor
d Brennkraftmaschine *f*, Verbrennungsmotor *m*

3021 INTERNAL MIXER mix,pla,rub
f mélangeur *m* fermé, mélangeur *m* interne
e mezclador *m* cerrado, mezclador *m* interno
i mescolatore *m* chiuso, mescolatore *m* interno
n gesloten menger, inwendige menger
d Innenmischer *m*

3022 INTERNAL MIXER WITH FLOATING WEIGHT, BANBURY MIXER mix,pla
f mélangeur *m* interne à piston, mélangeur *m* interne Banbury
e mezclador *m* interno Banbury, mezclador *m* interno con pistón
i mescolatore *m* Banbury
n banburymenger, zware kneedmachine
d Banbury-Mischer *m*, Gummikneter *m*, Innenmischer *m* mit Stempel, Stempelmischer *m*

3023 INTERNAL ROTARY-DRUM FILTER fil
f filtre *m* à tambour rotatif interne
e filtro *m* de tambor giratorio interno

i filtro *m* a tamburo rotativo interno
n filter *n* met interne roterende trommel
d Filter *m* mit Innendrehtrommel

3024 INTERNAL VIBRATOR cle
f pervibrateur *m*, pervibreur *m*
e vibrador *m* interno
i vibratore *m* interno
n inwendige trilinrichting
d Innenrüttler *m*

3025 INTROFIER BATH imp
f bain *m* accélérateur de l'imprégnation
e baño *m* acelerador de la impregnación
i bagno *m* acceleratore dell'impregnazione
n impregneerversnellingsbad *n*
d Durchtränkungsbeschleunigungsbad *n*, Imprägnierbeschleunigungsbad *n*

3026 INVERSE AGITATION TANK pho
f boîte *f* de développement à agitation par renversement
e cubeta *f* de revelado de agitación por inversión
i recipiente *m* di sviluppo ad agitazione per inversione
n tuimelontwikkeltank
d Kippdose *f*, Stülpentwicklungsdose *f*

3027 INVERTED MOLD (US), pla,rub
INVERTED MOULD (GB)
f moule *m* inversé
e molde *m* invertido
i stampo *m* rovescio
n omgekeerde matrijs
d umgekehrte Form *f*

3028 INVERTED SIPHON, sew
DIP PIPE
f siphon *m* renversé, siphon *m* submergé
e sifón *m* invertido
i sifone *m* rovesciato
n grondduiker
d Dücker *m*, Düker *m*

3029 INVERTED TRUNCATED-CONE BOTTOM sin
f fond *m* conique
e fondo *m* cónico
i fondo *m* conico
n kegelbodem
d Kegelboden *m*

3030 IODINE SCRUBBER gas,nue
f laverie *f* de gas à iode
e lavagases *m* de iodio
i lavatore *m* di gas ad iodio
n jodiumgaswasserij
d Jodgaswäsche *f*

3031 ION EXCHANGE UNIT, sew
IONIC COLUMN
f colonne *f* échangeuse d'ions
e columna *f* intercambiadora de iones
i colonna *f* scambiatrice d'ioni
n ionenuitwisselkolom
d Ionenaustauschsäule *f*, Ionitfilter *m*

3032 ION EXCHANGER gen,pla,sew
f échangeur *m* d'ions
e intercambiador *m* de iones
i scambiatore *m* d'ioni
n ionenuitwisselaar
d Ionenaustauscher *m*

3033 IONIZATION CHAMBER ins
f chambre *f* d'ionisation
e cámara *f* de ionización
i camera *f* d'ionizzazione
n ionisatiekamer, ionisatievat *n*
d Ionisationskammer *f*

3034 IONIZATION GAGE (US), ins
IONIZATION GAUGE (GB),
IONIZATION MANOMETER
f manomètre *m* à ionisation
e manómetro *m* de ionización
i manometro *m* ad ionizzazione
n ionisatiemanometer
d Ionisierungsmanometer *n*

3035 IONIZATION GAGE HEAD (US), ins
IONIZATION GAUGE HEAD (GB)
f tête *f* de manomètre à ionisation
e cabeza *f* de manómetro de ionización
i testa *f* de manometro ad ionizzazione
n ionisatiemanometerkop
d Ionisationsmanometerkopf *m*

3036 IRON-NICKEL STORAGE BATTERY ele
f accumulateur *m* à fer-nickel
e acumulador *m* de ferro-níquel
i accumulatore *m* a ferro-nichel
n ijzernikkelaccu, nikkelijzeraccu
d Nickel-Eisenakkumulator *m*

3037 IRON PLATING BATH, ACIERAGE BATH, STEELING BATH coa
f bain *m* d'aciérage
e baño *m* de aceración, baño *m* de hierrado
i bagno *m* d'acciaiatura
n verijzeringsbad *n*, verstalingsbad *n*
d Stählungsbad *n*, Verstählungsbad *n*

3038 IRON REMOVING APPARATUS sew
f appareil *m* de déferrisation, déferriseur *m*
e aparato *m* desferrizador
i apparecchio *m* per separazione del ferro
n ontijzeringsapparaat *n*
d Enteisener *m*

3039 IRRIGATION COOLER, BAUDELOT COOLER, DRIZZLING WORT COOLER bre
f Baudelot *m*, refroidisseur *m* à ruissellement
e enfriador *m* Baudelot
i raffreddatore *m* a pioggia
n sproeikoeler
d Berieselungskühler *m*, Rieselkühler *m*

3040 IRRIGATOR, PERCOLATOR clo
f asperseur *m*
e regadera *f*
i irrigatore *m*
n sproeier
d Rieseler *m*

3041 ISOBAROMETRIC RACKER, COUNTERPRESSURE RACKER bre
f soutireuse *f* isobarométrique
e llenadora *f* de contrapresión, llenadora *f* isobarométrica
i isobarometro *m*
n tegendrukvulmachine
d Gegendruckfüller *m*, isobarometrischer Fassabfüller *m*

3042 ISOMERIZER fue
f dispositif *m* d'isomérisation
e isomerizador *m*
i isomerizzatore *m*
n isomeriseertoestel *n*
d Isomerisiergerät *n*

3043 ISOTOPE PRODUCTION UNIT iso
f département *m* de production d'isotopes
e instalación *f* para producir isótopos
i impianto *m* di produzione d'isotopi
n isotopenproduktie-inrichting
d Isotopenproduktionswerkstatt *f*

3044 ISOTOPE SEPARATION PLANT iso
f installation *f* de séparation des isotopes
e equipo *m* de separación de los isótopos
i impianto *m* di separazione degl'isotopi
n isotopenscheidingsinstallatie
d Isotopentrennungsanlage *f*

3045 ISOTOPIC LABORATORY iso
f laboratoire *m* isotopique
e laboratorio *m* para isótopos
i laboratorio *m* per isotopi
n isotopenlaboratorium *n*
d Isotopenlaboratorium *n*

J

3046 J-BOX tex
f compensateur *m* Gantt
e caja *f* en J
i cassa *f* in J
n aflegvat *n*
d Ablegegefäss *n*, Kontinugefäss *n*

3047 JACK-LEG (US), sew
OVERFLOW PIPE (GB)
f tuyau *m* de trop-plein
e tubo *m* de rebosamiento
i tubo *m* di troppo pieno
n overlooppijp
d Überlaufrohr *n*

3048 JACKET, nue
CAN, CASING
f chemise *f*, gaine *f*
e funda *f*
i camicia *f*, guaina *f*, ricopertura *f*
n bus, huls, omhulling
d Hülle *f*, Hülse *f*

3049 JACKET (US), elc
CASE (GB)
f chemise *f*
e camisa *f*
i camicia *f*
n huls
d Batteriekarton *m*, Gehäuse *n*, Papphülse *f*

3050 JACKETED COIL CONDENSER clo
f réfrigérant *m* à serpentin chemisé
e enfriador *m* de serpentín en chaquetado
i raffreddatore *m* a serpentina a camicia
n ingekapselde spiraalkoeler
d kombinierter Flächen-Spiralkühler *m*

3051 JACKETED DISTILLING RECEIVER dis,lab
f récepteur *m* chemisé
e recipiente *m* enchaquetado
i raccoglitore *m* a camicia
n ingekapselde ontvanger
d ummantelte Vorlage *f*

3052 JACKETED HOT-AIR PRESS rub
f presse *f* de vulcanisation à chemise à l'air chaud
e prensa *f* de vulcanizar con camisa de aire caliente
i pressa *f* di vulcanizzazione con camicia d'aria calda
n vulcaniseerpers met hete luchtmantel
d Heissluftmantelvulkanisier-presse *f*

3053 JACKETED PRESSURE VESSEL pr
f récipient *m* à pression à chemise
e cuba *f* de presión con camisa
i tino *m* a pressione a camicia
n ommanteld drukvat *n*
d Manteldruckgefäss *n*

3054 JACKETED SEPARATORY FUNNEL lab,se
f ampoule *f* à décanter à jaquette isolante, ampoule *f* à décanter chemisée
e embudo *m* de separación enchaquetado
i imbuto *m* separatore a camicia
n ingekapselde scheitrechter
d ummantelter Scheidetrichter *m* ummantelter Tropftrichter *m*

3055 JACKETED TROUGH p
f cuve *f* à double chemise
e artesa *f* de doble pared
i vasca *f* a doppi a parete
n dubbelwandige trog
d doppelwandiger Trog *m*

3056 JAMB WALL g
f piédroit *m*
e pared *f* separadora
i parete *f* separatrice
n tussenwand
d Zwischenwand *f*

3057 JAPANNING KETTLE d
f chaudière *f* de vernissage
e caldera *f* de barnizado, caldera *f* de charol
i caldaia *f* da laccare
n lakketel
d Lackkessel *m*

3058 JAPANNING OVEN d
f four *m* de vernissage
e hornillo *m* japonés, horno *m* de barnizado, horno *m* de charol

i forno *m* da laccare
n lakoven
d Lackierofen *m*

3059 JAPANNING PLANT dec
f installation *f* de vernissage à la laque
e equipo *m* de barnizado con laca, equipo *m* de charol
i impianto *m* da laccare
n japanlakkerij, lakinstallatie
d Lackmalerei *f*

3060 JAR lab
f récipient *m*, vase *m*
e vasija *f*, vaso *m*
i vaso *m*
n pot, vat *n*
d Gefäss *n*, Stutzen *m*

3061 JAR, CONTAINER elc
f bac *m*
e recipiente *m*
i contenitore *m*
n bak
d Zellengefäss *n*, Zellenkasten *m*

3062 JAR MILL lab
f moulin *m* à boules de laboratoire
e molino *m* de bolas de laboratorio
i mulino *m* a palle di laboratorio
n laboratoriumkogelmolen
d Laboratoriumkugelmühle *f*

3063 JAW cpt
f mâchoire *f*, mordache *f*, mors *m*
e maníbula *f*, mordaza *f*
i ganascia *f*, guancia *f*
n schuifkern
d Backe *f*

3064 JAW BREAKER, JAW CRUSHER bre
f concasseur *m* à mâchoires
e machacadora *f* de madníbulas, machacadora *f* de mordazas
i frantoio *m* a ganasce
n bekkenbreker, kaakbreker
d Backenbrecher *m*, Maulbrecher *m*

3065 JAW TYPE PRESS, GAP TYPE PRESS, GOOSENECK PRESS, OPEN SIDE PRESS, SWANNECK PRESS rub
f presse *f* à col de cygne, presse *f* ouverte sur un côté
e prensa *f* abierta por un lado, prensa *f* acodillada, prensa *f* en cuello de cisne
i pressa *f* a collo di cigno
n muilpers, zwanehalspers
d einseitig offene Presse *f*, Maulpresse *f*, Schwanenhalspresse *f*

3066 JERKING MACHINE, JERKING TABLE min
f machine *f* à secousses
e máquina *f* sacudidora
i macchina *f* a scosse
n schudmachine
d Rüttelmaschine *f*

3067 JET, NOZZLE pre
f buse *f*, gicleur *m*, tuyère *f*
e abertura *f*, boquilla *f*, tobera *f*
i imboccatura *f*, spruzzatore *m*, ugello *m*
n sproeier, straalmondstuk *n*, straalpijp
d Düse *f*

3068 JET BLOWER pre
f soufflante *f* à jet
e soplante *m* de chorro
i soffiatore *m* a getto
n straalaanjager, straalblower
d Strahlgebläse *n*

3069 JET CARBURETTER, SPRAY CARBURETTER fue
f carburateur *m* à giclage
e carburador *m* de inyección
i carburatore *m* ad iniezione
n injectiecarburator
d Düsenvergaser *m*

3070 JET CONDENSER, INJECTION CONDENSER pre
f condensateur *m* à injection
e condensador *m* de inyección
i condensatore *m* a getto
n waterstraalcondensor
d Einspritzkondensator *m*

3071 JET DEFLECTOR fue,pre
f déflecteur *m* du jet
e deflector *m* del chorro
i diflettore *m* del getto
n straaldeflector
d Strahldeflektor *m*

3072 JET DRIER dry
f sécheur *m* à jets d'air
e secadero *m* de chorros de aire
i essiccatoio *m* a getti d'aria
n luchtstraaldroger
d Luftstrahltrockner *m*

3073 JET IMPACT MILL, brk
JET MILL, FLUID-ENERGY MILL
f broyeur *m* à jet
e triturador *m* de chorro
i frantoio *m* a getto
n luchtstraalmolen
d Strahlmühle *f*

3074 JET NOZZLE pre
f buse *f* à jet, tuyère *m* à jet
e tobera *f* de chorro
i ugello *m* a getto
n straalmondstuk *n*
d Strahldüse *f*

3075 JET PUMP pre
f pompe *f* à jet
e bomba *f* de chorro
i pompa *f* a getto
n straalpomp
d Strahlpumpe *f*

3076 JET-TYPE REFLUX RATIO HEAD dis,lab
f tête *f* de reflux avec gicleur
e cabezal *m* de reflujo con tobera
i testa *f* di riflusso con ugello
n opzet voor terugstroomregelaar met straalopening
d Aufsatz *m* für Rückflussregler mit Düse, Kolonnenkopf *m* mit Düsenrücklaufregler

3077 JET WASHER cle
f laveur *m* à jet
e lavador *m* de chorro
i lavatrice *f* a getto
n straalwasser
d Strahlwäscher *m*

3078 JIG, min,sor
JIG TABLE, JIGGER
f crible *m* vibrant
e criba *f* vibradora
i crivello *m* oscillante
n deintoestel *n*, zeeftoestel *n*
d Siebsetzmaschine *f*

3079 JIG, tex
JIGGER
f jigger *m*, machine *f* de teinture au large
e jigger *m*, máquina *f* de teñido al ancho
i jigger *m*, macchina *f* di tintura al largo
n breedverfmachine, jigger
d Breitfärbemaschine *f*

3080 JIGGING CONVEYOR, tra
OSCILLATING CHUTE, SHAKING SHOOT
f couloir *m* à secousses, gouttière *f* à secousses
e canal *m* de sacudidas
i scivolo *m* crivellatore
n schudgoot
d Schüttelrinne *f*

3081 JOCKEY POT, gla
MONKEY POT
f creuset *m* pour verre spécial
e crisol *m* para vidrio special
i crogiolo *m* per vetro speziale
n pot voor speciaal glas
d Sonderglashafen *m*

3082 JOINT, cpt
PACKING
f bourrage *m*, garniture *f*
e empaquetadura *f*
i guarnizione *f*
n pakking
d Packung *f*

3083 JOINT RING, cpt
GASKET RING, PACKING RING
f bague *f* de garniture
e arandela *f* de guarnición
i anillo *m* di guarnizione
n pakkingring
d Dichtungsring *m*, Manschette *f*

3084 JOINT SEALING, cpt
SCREEVE, SEALING RING
f anneau *m* de fermeture, cordon *m* d'étanchéité
e anillo *m* de cierre, anillo *m* de estanqueidad, tapa-juntas *m*
i anello *m* di chiusura, cordone *m* di sigillatura
n dichtingsgroef, dichtingsring
d Dichtungsrille *f*, Dichtungsring

3085 JOLLEY, cer
POTTER'S WHEEL
f tour *m* de potier
e torno *m* para moldear loza hueca
i tornio *m* da vasaio
n pottenbakkersschijf
d Töpferscheibe *f*

3086 JUICE BOILER, boi,fmi,si
REBOILER
f récipient *m* de rébullition
e rehervidor *m*
i serbatoio *m* a far bollire di nuovo
n opkoker, opkookvat *n*
d Aufkocher *m*, Aufkochgefäss *n*

087 JUICE TRAP sin
f collecteur *m* de jus
e colector *m* de zumo
i collettore *m* di sugo
n sapvanger
d Saftfänger *m*

088 JUICE WARMER, CALORIZATOR sin
f calorisateur *m*
e calorizador *m*
i calorizzatore *m*
n calorisator
d Kalorisator *m*, Saftwärmer *m*

3089 JUMPER elc
f cavalier *m*
e puente *m*
i ponte *m*
n uitschakeldraad
d U-Verbindung *f*

K

3090 KAMPOMETER ins
f campomètre *m*
e campómetro *m*
i campometro *m*
n campometer
d Kampometer *n*

3091 KATHAROMETER ins
f catharomètre *m*
e catarómetro *m*
i catarometro *m*
n katharometer
d Katharometer *n*

3092 KETONIZATION PLANT oic
f installation *f* de cétonisation
e equipo *m* de cetonización
i impianto *m* di chetonizzazione
n ketoniseerinstallatie
d Ketonisieranlage *f*

3093 KETTLE htt
f marmite *f*
e marmita *f*
i marmitta *f*, pentola *f*
n ketel
d Kessel *m*

3094 KIBBLER brk
f égrugeoir *m*, moulin *m* à égruger
e trituradora *f*
i mulino *m* da sgrossare
n grofmolen
d Grobmühle *f*, Schrotmühle *f*

3095 KIER ble
f bouilleur *m* à lessivage
e caldera *f* de blanquear
i bollitore *m* di candeggio
n bleekketel
d Beuchkessel *m*, Bleicherei-kocher *m*, Kocher *m*

3096 KIEVE, met
DOLLY TUB
f cuve *f* à rincer l'or
e cuba *f* de agitación
i serbatoio *m* d'agitazione
n goudwaspan
d Schlämmfass *n*

3097 KILN bre
f touraille *f*
e tostadero *m*
i essiccatoio *m*
n eest
d Darre *f*

3098 KILN, ce
CALCINING FURNACE,
ROASTING FURNACE
f four *m* de grillage
e horno *m* de tostación
i forno *m* d'arrostimento
n roostoven
d Röstofen *m*

3099 KILN CAR cer,htt,tr
f chariot *m* d'étuve
e carrillo *m* de estufa
i carrello *m* di stufa
n droogstoofwagen
d Trockenofenwagen *m*

3100 KILN FAN br
f ventilateur *m* de touraille
e ventilador *m* de tostadero
i ventilatore *m* all'essiccatoio
n eestventilator
d Darrenventilator *m*

3101 KILN FLOOR br
f plateau *m* de touraillage
e piso *m*, plato *m*
i graticcio *m*
n eestvloer
d Abdarrhorde *f*, Darrboden *m*, Horde *f*

3102 KILN FURNACE br
f foyer *m* de la touraille
e horno *m* del tostadero
i forno *m* dell'essiccatoio
n eesthaard
d Darrofen *m*

3103 KILN TURNER br
f retourneur *m* de touraille
e removedor *m* de tostadero
i mescolatore *m* d'essiccatoio
n eestomgooier
d Darrwender *m*

3104 KJELLDAHL FLASK l
f ballon *m* de Kjelldahl, matras *m* de Kjelldahl
e matraz *m* de Kjelldahl
i matraccio *m* di Kjelldahl
n destructiebuis volgens Kjelldal
d Kjelldahl-Kolben *m*

3105 KNEADER mix,p
f malaxeur *m*, pétrisseur *m*
e amasadora *f* mecánica

i impastatore *m*, masticatore *m*
n kneedmachine
d Anteigwerk *n*, Knetmaschine *f*

106 KNEADER ARM, KNEADER BLADE — mix,pla
f palette *f*
e brazo *m*, paleta *f*
i braccio *m*, operatore *m*, paletta *f*
n kneedarm
d Knetarm *m*, Knetschaufel *f*

107 KNIFE COATER (US), RUBBER SPREADER (GB) — rub
f machine *f* à enduire avec racle sur rouleau
e máquina *f* engomadora
i spalmatrice *f* a lama libera
n strijkmachine
d Rakelstreichmaschine *f*, Walzenstreichmaschine *f*

108 KNIFE HOLDER — sin
f porte-couteaux *m*
e sujetacuchillas *m*
i portacoltelli *m*
n messenblok *n*
d Messerkasten *m*

109 KNOCKOUT PIN (US), EJECTION BAR (GB), EJECTION PIN (GB) — rub
f barre *f* d'éjection, goupille *f* éjectrice, tige *f* d'éjecteur
e barra *f* de eyección, perno *m* eyector
i barra *f* d'espulsione, spina *f* d'espulsione, tirante *m* d'espulsione
n uitstootpin, uitwerperstift
d Ausdrückstift *m*, Ausstosser *m*, Auswerfer *m*, Auswurfstift *m*

3110 KNOCKOUT PIN PLATE (US), EJECTOR PIN PLATE (GB) — pla,rub
f plaque *f* d'éjection
e placa *f* de eyector
i piastra *f* d'estrazione
n uitwerpplaat
d Ausdrückplatte *f*, Auswerferplatte *f*

3111 KNOTTER — pap,tex
f épurateur *m* de pâte
e colector *m* de ramas
i annodatore *m*
n knopenvanger
d Knotenfänger *m*

3112 KONIMETER, KONOMETER — ins
f conimètre *m*
e conímetro *m*
i conimetro *m*
n konimeter, stofgehaltemeter
d Konimeter *n*, Staubmesser *m*

3113 KRYPTON LIQUEFYING PLANT — gas
f installation *f* de liquéfaction de crypton
e equipo *m* de licuación de criptón
i impianto *m* di liquefazione di cripton
n installatie voor het vloeibaar maken van krypton
d Kryptonverflüssigungsanlage *f*

3114 KYANIZING PLANT — imp
f installation *f* de kyanisation
e equipo *m* de kianización
i impianto *m* d'impregnazione del legname con sublimato corrosivo
n kyaniseerinstallatie
d Kyanisieranlage *f*

L

3115 LABELLING MACHINE lab,mcy
f machine *f* à étiqueter
e etiquetadora *f*, rotuladora *f* mecánica
i macchina *f* da etichettare
n etiketteermachine
d Etikettiermaschine *f*

3116 LABORATORY APPARATUS lab
f appareils *pl* de laboratoire
e aparatos *pl* de laboratorio
i apparecchiatura *f* di laboratorio
n laboratoriumapparaten *pl*
d Laboratoriumgeräte *pl*, Laborgeräte *pl*

3117 LABYRINTH CONDENSER clo
f condensateur *m* à labyrinthe
e condensador *m* de laberinto
i condensatore *m* a labirinto
n labyrintcondensor
d Labyrinthkondensator *m*, Labyrinthverdichter *m*

3118 LABYRINTH HEATING APPARATUS htt
f appareil *m* de chauffage à labyrinthe
e aparato *m* de calefacción de laberinto
i apparecchio *m* di riscaldamento a labirinto
n labyrintverwarmingstoestel *n*
d Labyrinthheizungsapparat *m*

3119 LABYRINTH JOINT cpt
f joint *m* à labyrinthe
e junta *f* de laberinto
i giunto *m* a labirinto
n labyrintverbinding
d Labyrinthverbindung *f*

3120 LABYRINTH PACKING cpt
f garniture *f* à labyrinthe
e empaquetadura *f* de laberinto
i guarnizione *f* a labirinto
n labyrintpakking
d Labyrinthpackung *f*

3121 LABYRINTH SEAL cpt
f fermeture *f* à labyrinthe
e cierre *m* de laberinto, cierre *m* laberíntico
i chiusura *f* a labirinto
n labyrintsluiting
d Labyrinthverschluss *m*

3122 LACHRYMATORY BOMB (GB), gas LACRIMATORY BOMB (US), TEAR GAS BOMB
f bombe *f* lacrymogène
e bomba *f* de gas lacrimógeno
i spezzone *m* lacrimogeno
n traangasbom
d Tränengasbombe *f*

3123 LACHRYMATORY GAS GENERATOR (GB), LACRIMATORY GAS GENERATOR (US), TEAR GAS GENERATOR mi
f générateur *m* de gaz lacrymogè
e generador *m* de gas lacrimógen
i generatore *m* di gas lacrimoger
n traangasgenerator
d Tränengasgenerator *m*

3124 LACQUER FACTORY, VARNISH FACTORY de
f fabrique *f* de laque, fabrique *f* de vernis
e fábrica *f* de barniz, fábrica *f* laca
i fabbrica *f* di lacca, fabbrica *f* di vernice
n lakfabriek, vernisfabriek
d Firnisfabrik *f*, Lackfabrik *f*

3125 LACQUER SPRAYING GUN d
f pistolet *m* au laque
e pistola *f* rociadora de laca
i pistola *f* a lacca
n lakspuit
d Lackpistole *f*, Lackspritze *f*

3126 LACQUERING PLANT d
f installation *f* de laquage
e equipo *m* de lacado
i impianto *m* di laccare
n lakinstallatie
d Lackieranlage *f*

3127 LACTODENSITOMETER, LACTOMETER, GALACTOME i
f galactomètre *m*, pèse-lait *m*
e galactómetro *m*
i galattometro *m*
n galactometer, melkmeter
d Galaktometer *n*, Milchwaage *f*

3128 LADLE
f louche *f*, puisoir *m*
e cuchara *f*
i siviera *f*

n scheplepel
d Schöpfkelle *f*

3129 LADLE, fon
CASTING LADLE
f poche *f*
e cuchara *f*
i secchia *f* di colata, siviera *f*
n gietpan
d Pfanne *f*

3130 LADLE, lab
SCOOP
f cuiller *f* à pot
e cangilón *m*
i cucchiaia *f*
n scheplepel
d Löffel *m*, Schöpfkelle *f*

3131 LADLE SHANK, gla
BEWEL, BOW, SHANK
f brancard *m*
e maneral *m*
i portante *m*
n draagbeugel, handbeugel
d Tiegelschere *f*, Tragschere *f*

3132 LAKE COLLECTOR fil
f collecteur *m* de masse de ségrégation
e colector *m* de masa de segregación
i collettore *m* di massa di segregazione
n zijgermassaverzameltank
d Seigerungsmassesammler *m*

3133 LAMBDA PIPETTE (US), lab
ULTRAMICROPIPETTE (GB)
f ultramicropipette *f*
e ultramicropipeta *f*
i ultramicropipetta *f*
n microlitertransferpipet
d Mikrolitertransferpipette *f*

3134 LAMELLAR COOLER clo
f refroidisseur *m* à lames
e enfriador *m* lamelar
i raffreddatore *m* a lamelle, raffreddatore *m* a lamine
n lamellenkoeler
d Lamellenkühler *m*

3135 LAMINAR-FLOW FILTER fil
f filtre *m* à écoulement laminaire
e filtro *m* de flujo laminar
i filtro *m* a flusso laminare
n filter *n* met laminaire stroming
d Laminarströmungsfilter *m*

3136 LAMINATING WEB pla
f strate *m* continu
e banda *f* continua para estratificados
i sopporto *m* impregnato da stampaggio
n geïmpregneerde baan
d Pressbahn *f*

3137 LAMINATOR rub
f laminoir *m* pour combiner des feuilles
e laminador *m* para combinar hojas
i laminatoio *m* per combinare foglie
n lamineerwals
d Lagenschichtwalze *f*, Schichtpressstoffwalze *f*, Verbundschichtwalze *f*

3138 LANCASHIRE BOILER pre
f chaudière *f* à deux tubes de fumée
e caldera *f* de dos hornos cilíndricos interiores
i caldaia *f* a due focolari interni
n vlampijpketel met boven-en binnenvuurgang
d Zweiflammrohrkessel *m*

3139 LANCE CUTTING PLANT, htt,met
OXYGEN CUTTING PLANT
f installation *f* d'oxycoupage
e equipo *m* de corte con lanzo, equipo *m* de oxicorte
i impianto *m* d'ossitaglio
n zuurstoflanssnijinstallatie
d Brennschneidanlage *f*

3140 LAND, pla
LAND AREA
f surface *f* de compression
e superficie *f* de presión
i superficie *f* totale di stampaggio
n afknijpvlak *n*
d Abquetschfläche *f*

3141 LAND AREA pla
f surface *f* d'appui
e superficie *f* de contacto
i superficie *f* d'appoggio
n contactoppervlak *n*
d Kontaktfläche *f*

3142 LAND SURFACE (US), pla
FLASH LAND (GB), FLASH RIDGE (GB), FLASH SURFACE (US)
f bord *m* d'appui, surface *f* d'appui

e borde *m* de rebaba, superficie *m* de rebaba
i appoggio *m*, battuta *f*
n afknijprand
d Abquetschfläche *f*, Abquetschrand *m*

3143 LAPPING DISK pol
f disque *m* de rodage
e disco *m* de repasado, disco *m* de tribopulidura
i disco *m* di lappatura
n lepschijf
d Läppscheibe *f*

3144 LAPPING MACHINE, LAPPING MILL pol
f machine *f* de rodage
e máquina *f* de repasado, molino *m* de tribopulidura
i macchina *f* di lappatura, smerigliatrice *f*
n lepmolen
d Läppmaschine *f*

3145 LATCH PLATE pla
f plateau-support *m* de prisonniers
e placa *f* de retención
i piastra *f* portainserti
n fixatieplaat voor in te persen metaaldelen
d Halteplatte *f*

3146 LATENSIFICATION BATH, LATENT IMAGE INTENSIFICATION BATH pho
f bain *m* de renforcement de l'image latente
e baño *m* reforzador de la imagen latente
i bagno *m* rinforzatore dell'immagine latente
n versterkbad *n* van het latente beeld
d Verstärkungsbad *n* des latenten Bildes

3147 LATERAL DELIVERY PIPE cpt
f tuyau *m* latéral de décharge
e tubo *m* lateral de descarga
i tubo *m* laterale di scarico
n zijafvoerbuis
d Seitenabflussrohr *n*

3148 LATEX CHANNEL rub
f rigole *f* d'écoulement du latex
e canal *m* de evacuación del látex
i canale *m* d'evacuazione del lattice
n latexafvoergoot
d Latexabflussrille *f*

3149 LATEX CLARIFYING CENTRIFUGE, CENTRIFUGAL LATEX CLARIFIER rub
f centrifugeur *m* à clarifier le latex
e centrífuga *f* para clarificar el látex
i centrifuga *f* per chiarificare il lattice
n latexlakringscentrifuge
d Klärzentrifuge *f* für Kautschukmilch

3150 LATEX FORMER, FORMER rub
f forme *f* de trempage
e molde *m* para inmersión
i stampo *m* ad immersione
n dompelvorm
d Tauchform *f*

3151 LATEXOMETER ins,rub
f latexomètre *m*
e latexómetro *m*
i areometro *m* di lattice, lattimetro
n latexometer
d Kautschukmilcharäometer *n*, Latexometer *n*

3152 LATTICE spe
f réseau *m*
e retículo *m*
i reticolo *m*
n rooster *n*
d Gitter *n*

3153 LAUGHING GAS GENERATOR, NITROUS OXIDE GENERATOR gas,mcy
f générateur *m* de gaz hilarant, générateur *m* de protoxyde d'azote
e generador *m* de gas hilarante
i generatore *m* di gas esilarante, generatore *m* d'ossido nitroso
n lachgasgenerator
d Lachgasgenerator *m*, Stickstoffoxydulgenerator *m*

3154 LAUNDER, FOUNTAIN, RUNNER THROUGH (US), SPOUT (GB), TAPPING SPOUT (GB) fon
f canal *m* de coulée, chenal *m* de coulée
e canal *m* de colada
i canale *m* di colata
n gietgoot, schenk, tapgoot
d Abstichrinne *f*, Förderrinne *f*, Rutsche *f*

155 LAUTER TUB, bre
CLARIFYING TUB, FILTER VAT,
STRAINING VAT
f cuve *f* de filtration, cuve-filtre *m*
e cuba *f* de filtración
i tino *m* di filtrazione
n klaringskuip
d Klärbottich *m*, Läuterbottich *m*

156 LAY-AWAY PIT, tan
LAYER
f puits *m* à tan
e pozo *m* para la casca
i pozzo *m* a tannino
n runkuil
d Lohgrube *f*

157 LAYER gen
f couche *f*
e capa *f*
i strato *m*
n laag
d Schicht *f*

158 LAYER LATTICE spe
f réseau *m* à couches
e retículo *m* de capas
i reticolo *m* a strati
n lagenrooster *n*
d Schichtgitter *n*

159 LEACH PIT tan
f puits *m* à lessive
e pozo *m* de lejía
i pozzo *m* a lisciva
n loogkuil
d Laugegrube *f*

160 LEACHING BATH, tan
LEECHING BATH
f bain *m* à lessive
e baño *m* de lejía
i bagno *m* di lisciva
n uitloogbad *n*
d Auslaugebad *n*

161 LEACHING PLANT, met
LIXIVIATION PLANT
f installation *f* de lessivage,
installation *f* de lixiviation
e equipo *m* de lixiviación
i impianto *m* di lisciviazione
n uitlooginstallatie
d Auslaugeanlage *f*, Laugerei *f*

162 LEACHING TANK met
f récipient *m* de lessivage,
récipient *m* de lixiviation
e tanque *m* de lixiviación
i serbatoio *m* di lisciviazione
n uitloogtank
d Auslaugegefäss *n*

3163 LEAD ANNEALING BATH met
f bain *m* de recuit en plomb fondu
e baño *m* de recocido en plomo
fundido
i bagno *m* di ricottura in piombo
fuso
n uitgloeibad *n* van gesmolten lood
d Ausglühbad *n* aus geschmolzenem
Blei

3164 LEAD ANODE elc,met
f anode *f* en plomb
e ánodo *m* de plomo
i anodo *m* in piombo
n loodanode
d Bleianode *f*

3165 LEAD ATOMIZER coa
f pulvérisateur *m* de plomb
e pulverizador *m* de plomo
i polverizzatore *m* di piombo
n loodverstuiver
d Bleizerstäuber *m*

3166 LEAD BATH htt
f bain *m* de plomb
e baño *m* de plomo
i bagno *m* di piombo
n loodbad *n*
d Bleibad *n*

3167 LEAD BATH FURNACE htt
f four *m* à bain de plomb
e horno *m* de baño de plomo
i forno *m* a bagno di piombo
n loodbadoven
d Bleibadofen *m*

3168 LEAD BLOCK exp,tes
f bloc *m* de plomb
e bloque *m* de plomo
i blocco *m* di piombo
n loodblok *n*
d Bleiblock *m*, Bleimörser *m* nach
Trauzl

3169 LEAD BUCKLE cer
f disque *m* perforé en plomb
e disco *m* perforado de plomo
i disco *m* perforato di piombo
n geperforeerde loodschijf
d perforierte Bleischeibe *f*

3170 LEAD CHAMBER, sua
CHAMBER
f chambre *f* de plomb
e cámara *f* de plomo

i camera *f* di piombo
n loden kamer
d Bleikammer *f*

3171 LEAD CRUCIBLE met
f creuset *m* à fondre le plomb
e crisol *m* de fundir el plomo
i crogiolo *m* da fondere il piombo
n loodsmeltkroes
d Bleischmelztiegel *m*

3172 LEAD DESILVERING PLANT met
f installation *f* de désargentation du plomb
e equipo *m* de desplatación del plomo
i impianto *m* di rimozione dell'argento dal piombo
n loodontzilveringsinstallatie
d Bleientsilberungsanlage *f*

3173 LEAD-FLUOBORATE BATH, elc
FLUOBORATE-LEAD BATH
f bain *m* au fluoborate de plomb
e baño *m* de fluoborato de plomo
i bagno *m* al fluoborato di piombo
n loodfluoboraatbad *n*
d Bleifluoboratbad *n*

3174 LEAD-FLUOSILICATE BATH, elc
FLUOSILICATE-LEAD BATH
f bain *m* au fluosilicate de plomb
e baño *m* de fluosilicato de plomo
i bagno *m* al fluosilicato di piombo
n loodfluosilicaatbad *n*
d Bleifluosilikatbad *n*

3175 LEAD HARDENING BATH met
f bain *m* de trempe en plomb
e baño *m* de temple en plomo
i bagno *m* di tempera in piombo
n loodhardingsbad *n*
d Bleihärtungsbad *n*

3176 LEAD IMPROVING FURNACE, met
SOFTENING FURNACE
f four *m* d'affinage du plomb
e horno *m* de afino del plomo
i forno *m* da affinare il piombo
n loodveredelingsoven
d Bleivergütungsofen *m*, Seigerofen *m*

3177 LEAD JACKET, coa,cpt
LEAD SHEATH
f armure *f* de plomb, gaine *f* de plomb
e envuelta *f* de plomo, vaina *f* de plomo
i camicia *f* di piombo, guaina *f* di piombo, rivestimento *m* di piombo
n loodmantel
d Bleihülle *f*, Bleimantel *m*

3178 LEAD KETTLE, met
LEAD POT
f pot *m* de plombier
e cazo *m* de plomero
i pentola *f* di stagnaio
n loodgieterspot, loodpot
d Bleitopf *m*

3179 LEAD-LINED STEEL TANK st
f récipient *m* en acier à revêteme[nt] intérieur de plomb
e cuba *f* de acero con revestimiento interno de plomo
i tino *m* d'acciaio con rivestiment[o] interno di piombo
n tank met loodvoering
d Bleifutterbehälter *m*

3180 LEAD LINING co
f garnissage *m* en plomb, revêtement *m* en plomb
e guarnición *f* de plomo, revestimiento *m* de plomo
i guarnizione *f* di piombo, rivestimento *m* di piombo
n loodbekleding, loodvoering
d Bleihärtebad *n*, Patentierbad *n*

3181 LEAD MELTING FURNACE, ht
LEAD SMELTER
f four *m* à fondre le plomb
e horno *m* de fundir el plomo
i forno *m* di fusione per piombo
n loodsmeltoven
d Bleischmelzofen *m*

3182 LEAD PATENTING BATH, ht
LEAD QUENCHING BATH
f bain *m* de trempe au plomb
e baño *m* de temple al plomo
i bagno *m* di patentamento, bagno *m* di tempera al piombo
n loodafschrikbad *n*
d Bleihärtebad *n*, Patentierbad *n*

3183 LEAD PRESS m
f presse *f* à plomb
e prensa *f* de plomo
i pressa *f* a piombo
n loodpers
d Bleipresse *f*

3184 LEAD PUMP m
f pompe *f* à plomb
e bomba *f* de plomo
i pompa *f* a piombo
n loodpomp
d Bleipumpe *f*

3185 LEAD-RUBBER APRON, sap
BODY APRON
f tablier *m* opaque
e delantal *m* de protección
i grembiale *m* di protezione
n loodrubberschort *n*
d Bleigummischürze *f*

3186 LEAD SHOT met
f grains *pl* de plomb
e granalla *f* de plomo
i graniglia *f* di piombo, pallini *pl* per tarare
n hagel, loodschroot *n*
d Ausgleichschrot *n*, Tarierschrot *n*

3187 LEAD-SILVER ANODE elc
f anode *f* en alliage de plomb et argent
e ánodo *m* de aleación de plomo y plata
i anodo *m* in lega di piombo ed argento
n loodzilveranode
d Bleisilberanode *f*

3188 LEAD SOLUBILITY TESTING APPARATUS tes
f appareil *m* d'essai de la solubilité de plomb
e probador *m* de la solubilidad de plomo
i collaudatore *m* della solubilità di piombo
n loodoplosbaarheidsproef-apparaat *n*
d Bleilöslichkeitsprüfgerät *n*

3189 LEAD STORAGE BATTERY elc
f accumulateur *m* à plomb
e acumulador *m* de plomo
i accumulatore *m* a piombo
n loodaccu
d Bleiakkumulator *m*

3190 LEAF FILTER fil
f filtre *m* à feuilles
e filtro *m* de hojas
i filtro *m* a foglie
n bladfilter *n*
d Blattfilter *m*

3191 LEAF PADDLE IMPELLER mix
f roue *f* motrice à palettes en feuilles
e rueda *f* móvil de palas en hojas
i girante *m* a palette in foglie
n bladvormige-schoepenroerder
d Rührstange *f* mit blattförmigen Paddeln

3192 LEAF PRESS tex
f presse *f* à satiner
e prensa *f* de satinar
i pressa *f* da rasare
n glanspers, satineerpers
d Glanzpresse *f*, Satinierpresse *f*

3193 LEAK DETECTION ASSEMBLY, nue,sap
LEAK DETECTION SYSTEM
f système *m* de détecteurs de fuites
e sistema *m* detector de pérdidas
i rivelatori *pl* di perdite
n lekzoekersysteem *n*
d Lecksuchersystem *n*

3194 LEAR, gla
LEER, LEHR
f four-tunnel *m* à refroidissement
e horno *m* de túnel de enfriamiento
i forno *m* a canale a raffreddamento
n kanaalkoeloven
d Kanalkühlofen *m*

3195 LEATHER GASKET, cpt
LEATHER PACKING
f garniture *f* en cuir
e empaquetadura *f* de cuero
i guarnizione *f* di cuoio
n lederpakking
d Lederpackung *f*

3196 LEATHER GLAZING MACHINE lea
f machine *f* à polir le cuir
e máquina *f* de abrillantar el cuero
i macchina *f* da lucidare il cuoio
n lederglansmachine
d Lederglanzmaschine *f*

3197 LEATHER WHEEL pol
f disque *m* de cuir
e disco *m* de cuero
i disco *m* di cuoio
n lederschijf
d Lederscheibe *f*

3198 LEATHER WORKS lea
f fabrique *f* de cuir
e fábrica *f* de cuero
i fabbrica *f* di cuoio
n lederfabriek
d Lederfabrik *f*

3199 LEPTOMETER ins
f leptomètre *m*
e leptómetro *m*
i leptometro *m*
n leptometer
d Leptometer *n*

3200 LEVEL CONTROLLER con
f régulateur *m* de niveau
e regulador *m* de nivel
i regolatore *m* di livello
n niveauregelaar
d Niveauregler *m*, Pegelregler *m*

3201 LEVEL INDICATOR ins
f indicateur *m* de niveau
e indicador *m* de nivel
i indicatore *m* di livello
n niveauaanwijzer
d Niveauindikator *m*, Pegelgeber *m*

3202 LEVELLING BOTTLE, TUBULATURE BOTTLE lab
f flacon *m* d'équilibre de niveau
e frasco *m* de equilibrio de nivel
i bottiglia *f* d'equilibrio di livello
n niveauflesje *n*
d Niveauflasche *f*

3203 LEVELLING BULB ins,lab
f vase *m* à niveau
e vaso *m* de nivel
i vaschetto *m* a livello
n niveauballon
d Niveaugefäss *n*

3204 LEVER gen
f levier *m*
e leva *f*, palanca *f*
i leva *f*
n hefboom
d Heber *m*

3205 LEVER SAFETY VALVE cpt,sap
f soupape *f* de sécurité de levier
e válvula *f* de seguridad de palanca
i valvola *f* di sicurezza di leva
n hefboomveiligheidsklep
d Hebersicherheitsventil *n*

3206 LEVIGATION TANK cle
f récipient *m* de lévigation
e tanque *m* de levigación
i serbatoio *m* di levigazione
n selectieve-bezinkingstank
d Abschlämmgefäss *n*

3207 LIBERATOR TANK, DEPOSITING-OUT TANK elc
f cuve *f* de dépôt total, cuve *f* de libération
e cuba *f* de depósito total, cuba *f* de liberación
i cella *f* d'esaurimento
n afwerkcel
d Entkupferungsbad *n*, Entmetallisierungsbad *n*

3208 LIGHT BOMB, FLARE, STAR SHELL mit
f obus *m* éclairant
e cohete *m* luminoso
i razzo *m* luminoso
n lichtgranaat, lichtkogel
d Leuchtbombe *f*

3209 LIGHT FILTER spe
f filtre-lumière *m*
e filtro *m* de luz
i filtro *m* di luce
n lichtfilter *n*
d Lichtfilter *m*

3210 LIGHT WATER REACTOR, WATER MODERATED REACTOR, WATER REACTOR nue
f réacteur *m* à eau légère
e reactor *m* de agua liviana
i reattore *m* ad acqua leggera
n licht-waterreactor, waterreactor
d Leichtwasserreaktor *m*, Wasserreaktor *m*

3211 LIME BAG TESTER, CHALK TEST APPARATUS tes
f appareil *m* d'essai à la craie
e aparato *m* de prueba a la creta
i collaudatore *m* alla creta
n krijtproefapparaat *n*
d Kreideprobeapparat *m*

3212 LIME BURNING KILN, LIME KILN htt
f four *m* à chaux
e horno *m* de cal
i forno *m* di calce
n kalkbrandoven, kalkoven
d Kalkbrennofen *m*

3213 LIME COATING coa
f revêtement *m* en chaux
e revestimiento *m* de cal
i rivestimento *m* di calce
n kalklaag
d Kalkschicht *f*

3214 LIME DEFECATION VESSEL, LIMING TANK sin
f récipient *m* de défécation, récipient *m* de saturation
e alcalizador *m*, tanque *m* de defecación
i serbatoio *m* di defecazione, serbatoio *m* di saturazione
n defecatiepan, saturatiepan
d Scheidepfanne *f*

3215 LIME DESULFURIZING PLANT (US), LIME DESULPHURIZATION PLANT (GB) met
f installation *f* désulfuratrice par la chaux
e equipo *m* de desulfuración con cal
i impianto *m* di desolforazione con calce
n ontzwavelingsinstallatie met kalk
d Entschwefelungsanlage *f* mit Kalk

3216 LIME KILN COOLING ZONE sin
f zone *f* de refroidissement d'un four à chaux
e zona *f* de enfriamiento de un horno de cal
i zona *f* di raffreddamento di un forno per calce
n afkoelzone van een kalkoven
d Abkühlungszone *f* eines Kalkofens

3217 LIME KILN FIRE ZONE sin
f zone *f* de feu d'un four à chaux
e zona *f* de fuego de un horno de cal
i zona *f* di fuoco di un forno per calce
n vuurzone van een kalkoven
d Feuerzone *f* eines Kalkofens

3218 LIME KILN HEATING ZONE sin
f zone *f* de chauffage d'un four à chaux
e zona *f* de calentamiento de un horno de cal
i zona *f* di riscaldamento di un forno per calce
n verwarmingszone van een kalkoven
d Erwärmungszone *f* eines Kalkofens

3219 LIME MIXER sin
f chauleur *m*
e cuba *f* mezcladora de cal
i tino *m* mescolatore di calce
n chauleur, kalkmengbak
d Kalkmischwanne *f*

3220 LIME MUD COLLECTOR, LIME SLUDGE COLLECTOR sin
f collecteur *m* de boue de chaux
e colector *m* de lodo de cal
i collettore *m* di fango di calce
n kalkneerslagcollector
d Scheideschlammsammler *m*

3221 LIME PIT tan
f puits *m* de pelainage, puits *m* de plainage, puits *m* de plamage
e pozo *m* de pelado, pozo *m* de pelambre, pozo *m* de pelote
i pozzo *m* di pelatura
n kalkput, plootput
d Äschergrube *f*

3222 LIME TANK tan
f récipient *m* de pelainage, récipient *m* de plainage, récipient *m* de plamage
e cuba *f* de pelado, cuba *f* de pelambre, cuba *f* de pelote
i tino *m* di pelatura
n kalkvat *n*
d Äscherfass *n*

3223 LIME-WATER SOFTENING PLANT hyg,sew
f installation *f* de désacidification par l'eau de chaux
e equipo *m* de desacidificación por agua de cal
i impianto *m* di disacidificazione a calce
n ontzuringsinstallatie met kalkwater
d Kalkwasserentsäuerungsanlage *f*

3224 LINER elc
f séparateur *m*
e separador *m*
i separatore *m*
n voering
d Scheider *m*

3225 LINER fou,pla
f bande *f* de garnissage
e banda *f* de guarnición
i striscia *f* di riempimento
n opvulstrip
d Einlegestreifen *m*

3226 LINER, SHIM gen
f chemise *f*
e chapa *f*
i spessore *m*
n blikvoering, busje *n*
d Blecheinlage *f*

3227 LINING coa,htt,met,pla
f chemise *f*, garnissage *m*, revêtement *m*
e camisa *f*, guarnición *f*, revestimiento *m*
i camicia *f*, guarnizione *f*, rivestimento *m*
n bekleding, voering
d Auskleidung *f*, Futter *n*

3228 LINING (US), cer,gla
GLAZE (GB)
f brasque *m*
e enlucido *m* refractario
i rivestimento *m* refrattario
n vuurvaste bekleding
d Tiegelglasur *f*

3229 LINISHER, aap,cpt
ABRASIVE BELT
f bande *f* abrasive, ruban *m* émerisé
e cinta *f* abrasiva
i nastro *m* abrasivo
n slijpband
d Schleifband *n*

3230 LINT DOCTOR tex
f contre-racloir *m*
e contrarrascador *m*
i controraschietto *m*
n tegenstrijker
d Gegenrakel *m*, Gegenschaber *m*

3231 LINTEL GIRDER, cpt
MANTLE RING
f marâtre *f*
e madrasta *f*
i maratra *f*
n draagring
d Tragkranz *m*

3232 LIP lab
f bec *m*
e pico *m*
i becco *m*, canna *f*
n schenktuit
d Ausguss *m*

3233 LIP, fou,met
SPOUT
f bec *m* de coulée
e pico *m* de colada
i becco *m* di colata
n schenktuit
d Ausguss *m*, Giessschnauze *f*

3234 LIQUATION VAT, met
SEGREGATION VAT
f cuve *f* de liquation, cuve *f* de ségrégation
e cuba *f* de licuación, cuba *f* de segregación
i tino *m* di liquazione
n segregatievat *n*, zijgervat *n*
d Ausschwitzgefäss *n*, Seigerungsgefäss *n*

3235 LIQUEFYING COLUMN gas
f colonne *f* de liquéfaction, colonne *f* de rectification
e columna *f* de licuación, columna *f* de rectificación
i colonna *f* di liquefazione, colonna *f* di rettificazione
n kolom voor het vloeibaar maken van lucht, luchtontledings-installatie, rectificeerkolom
d Luftzerlegungssäule *f*, Rektifiziersäule *f*, Verflüssigungssäule *f*

3236 LIQUEFYING PLANT gas
f installation *f* de liquéfaction
e equipo *m* de licuación
i impianto *m* di liquefazione
n installatie voor het vloeibaar maken van gassen
d Verflüssigungsanlage *f*

3237 LIQUID-BATH FURNACE htt
f four *m* à bain liquide
e horno *m* de baño líquido
i forno *m* a bagno liquido
n smeltbadoven
d Schmelzbadofen *m*

3238 LIQUID CARBURIZING BATH met
f bain *m* de cémentation
e baño *m* de cementación
i bagno *m* di cementazione
n cementeerbad *n*
d Zementationsbad *n*

3239 LIQUID FLOWMETER ins
f débitmètre *m* à liquides
e flujómetro *m* de líquidos
i flussometro *m* a liquidi
n vloeistofstromingsmeter
d Flüssigkeitsströmungsmesser *m*

3240 LIQUID HONING PLANT, pol
VAPOR BLASTING PLANT (US),
VAPOUR BLASTING PLANT (GB)
f installation *f* de honage au jet de vapeur
e equipo *m* de rectificación al chorro de vapor
i impianto *m* di rettificatura al getto di vapore
n hooninstallatie met dampstraal
d Honanlage *f* mit Dampfstrahl

3241 LIQUID HYDROGEN LOOP clo,nue
f circuit *m* fermé à hydrogène liquide
e circuito *m* cerrado de hidrógeno líquido
i circuito *m* chiuso ad idrogeno liquido
n gesloten circuit *n* voor vloeibare waterstof

d Kreislauf *m* mit flüssigem Wasserstoff

3242 LIQUID JET PUMP vac
f pompe *f* à jet de liquide
e bomba *f* de chorro de líquido
i pompa *f* a getto di liquido
n vloeistofstraalpomp
d Flüssigkeitsstrahlpumpe *f*

3243 LIQUID-LIQUID EXTRACTION PLANT nue,sep
f installation *f* d'extraction par partage liquide-liquide
e equipo *m* de extracción por repartición líquido-líquido
i impianto *m* d'estrazione per ripartizione liquido-liquido
n vloeistof-vloeistof-extractie-installatie
d Flüssig-Flüssig-Extraktions-anlage *f*

3244 LIQUID-METAL COOLED REACTOR nue
f réacteur *m* à refroidissement par métal fondu
e reactor *m* de enfriamiento por metal líquido
i reattore *m* a metallo liquido
n met vloeibaar metaal gekoelde reactor
d flüssigmetallgekühlter Reaktor *m*, Reaktor *m* mit Flüssigmetall-kühlung

3245 LIQUID-METAL FUEL REACTOR nue
f réacteur *m* à combustible métallique fondu
e reactor *m* de combustible metálico líquido
i reattore *m* con combustibile in metallo liquido
n reactor met splijtstof uit vloeibaar metaal
d Reaktor *m* mit Flüssigmetall-brennstoff

3246 LIQUID-METAL PUMP nue
f pompe *f* de métal liquide
e bomba *f* de metal líquido
i pompa *f* di metallo liquido
n vloeibaar-metaalpomp
d Flüssigmetallpumpe *f*

3247 LIQUID-PHASE CRACKING PLANT fue
f installation *f* de craquage en phase liquide
e equipo *m* de pirólisis en fase líquida
i impianto *m* di piroscissione in fase liquida
n kraakinstallatie volgens het vloeistoffazeproces
d Krackanlage *f* nach dem Flüssigphaseverfahren

3248 LIQUID-PHASE HYDROGENATION PLANT fat
f installation *f* d'hydrogénation en phase liquide
e equipo *m* de hidrogenación en fase líquida
i impianto *m* d'idrogenazione in fase liquida
n hydreerinstallatie volgens het vloeistoffazeproces
d Hydrieranlage *f* nach dem Flüssigphaseverfahren

3249 LIQUID-PHASE SINTERING OVEN htt
f four *m* de frittage en présence de la phase liquide
e horno *m* de sinterización con la fase líquida
i forno *m* di sinterizzazione con la fase liquida
n smeltsinteroven
d Schmelzsinterofen *m*

3250 LIQUID RESIDUE BURNER iwt
f brûleur *m* pour résidus de déchets liquides
e quemador *m* para desechos líquidos residuales
i bruciatore *m* per rifiuti liquidi residui
n brander voor vloeibare afvalresten
d Brenner *m* für flüssige Abfallreste

3251 LIQUID RESIDU FURNACE iwt
f four *m* pour résidus de déchets liquides
e horno *m* para desechos líquidos residuales
i forno *m* per rifiuti liquidi residui
n oven voor vloeibare afvalresten
d Ofen *m* für flüssige Abfallreste

3252 LIQUID SEAL, LIQUID SEALING cpt
f fermeture *f* hydraulique, joint *m* hydraulique
e cierre *m* de líquido, junta *f* hidráulica, obturador *m* de líquido
i chiusura *f* idraulica, giunta *f* idraulica

n gasslot *n*, vloeistofslot *n*
d Flüssigkeitsverschluss *m*, hydraulischer Abschluss *m*

3253 LIQUID-SLAG GAS GENERATOR gas
f gazogène *m* à cendres liquides
e gasógeno *m* de cenizas líquidas
i gasogeno *m* a fusione di ceneri
n slakkengasgenerator
d Abstichgenerator *m*

3254 LIQUOR FINISHING BATH coa,met
f bain *m* d'étamage préliminaire
e baño *m* de estañado preliminar
i bagno *m* di stagnatura preliminare
n voortinbad *n*
d Vorzinnbad *n*

3255 LIQUORING BATH, DIPPING BATH tex
f bain *m* d'immersion
e baño *m* de inmersión
i bagno *m* d'immersione
n dompelbad *n*
d Tauchbad *n*

3256 LIXIVIATION PLANT, LEACHING PLANT met
f installation *f* de lessivage, installation *f* de lixiviation
e equipo *m* de lixiviación
i impianto *m* di lisciviazione
n uitlooginstallatie
d Auslaugeanlage *f*, Laugerei *f*

3257 LOAD, FEED, MOLD CHARGE (US), MOULD CHARGE (GB) pla
f chargement *m*
e carga *f*
i carica *f*
n vulling
d Beschickung *f*

3258 LOADED CONCRETE SHIELD, HEAVY AGGREGATE CONCRETE SHIELD nue,sap
f écran *m* en béton chargé
e pantalla *f* de hormigón cargada
i schermo *m* in calcestruzzo caricato
n scherm *n* van verzwaard beton
d Schwerbetonschirm *m*

3259 LOADING APERTURE sin,tra
f orifice *m* de chargement
e orificio *m* de carga
i orificio *m* di caricamento
n vulopening
d Füllöffnung *f*

3260 LOADING CHAMBER (GB), LOADING SPACE (US), LOADING WELL (GB), FILLIN[G] SPACE (US) pl[a]
f chambre *f* de charge, chambre de compression
e cámara *f* de carga, cámara *f* de compresión
i camera *f* di carica, camera *f* di compressione
n vulruimte
d Füllraum *m*

3261 LOADING DEVICE ge[n]
f dispositif *m* de chargement
e cargador *m*, dispositivo *m* de carga
i dispositivo *m* di caricamento
n laadinrichting
d Ladevorrichtung *f*

3262 LOADING FUNNEL tr[a]
f entonnoir *m* de chargement, trémie *f* d'alimentation
e embudo *m* de carga, tolva *f* de carga
i imbuto *m* di caricamento, tramoggia *f* di caricamento
n laadtrechter, vultrechter
d Fülltrichter *m*

3263 LOADING PLATE (US), RETAINER PLATE (GB), TRANSFER CHAMBER pl[a]
f plaque *f* contenant la chambre de transfert
e placa *f* intermedia de alojamiento de la cámara de plastificación
i piastra *f* di carica
n tussenplaat, vulplaat
d Füllplatte *f*

3264 LOADING PLUNGER pla,ru[b]
f piston *m* chargeur
e pistón *m* cargador
i pistone *m* caricatore
n vulplunjer
d Füllkolben *m*

3265 LOADING SHOE (GB), FILLER PLATE (US) p[la]
f lunette *f*, plaque *f* de remplissage
e placa *f* independiente de carg[a]
i piastra *f* di carica
n vulplaat
d Füllplatte *f*

3266 LOADING TRAY, CHARGING TRAY — pla,rub
f chargeur *m* à trous
e bandeja *f* de carga
i caricatore *m*
n vulplaat
d Fülltablett *n*, Siebplatte *f*

3267 LOADING VESSEL, WEIGHTING VESSEL — tex
f cuve *f* à charger
e cuba *f* de cargado
i tino *m* di carcia
n verzwaringskuip
d Beschwerungsküpe *f*

3268 LOAM KNEADER, LOAM MILL, LOAM MIXER — mix
f malaxeur *m* pour terre glaise
e amasador *m* de tierra arcillosa
i impastatrice *f* per terra argillosa, impastatrice *f* per terra grassa
n leemkneedmachine
d Lehmknetmaschine *f*

3269 LOCKING RING — cpt,pla
f anneau *m* de fermeture, bague *f* de serrage
e aro *m* de cierre
i anello *m* di chiusura
n klemring, sluitring
d Klemmring *m*, Spannring *m*

3270 LONG-DURATION TESTER — tes
f appareil *m* d'essai à longue durée
e aparato *m* de ensayo de larga duración
i apparecchio *m* di prova di lunga durata
n duurproefapparaat *n*
d Dauerprüfgerät *n*

3271 LONG-STROKE PRESS — pla
f presse *f* à longue course
e prensa *f* de larga carrera
i pressa *f* a lunga corsa
n pers met lange plunjerslag
d Presse *f* mit grossem Hub, Presse *f* mit hohem Hub

3272 LONGITUDINAL-ARCH KILN — cer
f four *m* annulaire à parois longitudinales
e horno *m* anular de paredes longitudinales
i forno *m* anulare con pareti longitudinali
n ringoven met longitudinale tussenwanden
d Ringofen *m* mit Längswänden

3273 LONGITUDINAL-FINNED FUEL ELEMENT — nue
f élément *m* combustible à ailettes longitudinales
e elemento *m* combustible a aletas longitudinales
i elemento *m* combustibile ad alette longitudinali
n van langsribben voorzien splijtelement *n*
d Brennstoffelement *n* mit Längsberippung

3274 LOOP DRIER (US), FESTON DRIER (GB) — pho,pla
f accrocheuse-sécheuse *f*, séchoir *m* à feston, séchoir *m* a guirlandes, séchoir *m* à plis supportés
e secador *m* de cinta colgante, secador *m* de festón, secador *m* por tendido ondulado
i essiccatore *m* a festoni
n droogtunnel met guirlande-inrichting
d Hängetrockner *m*, Schleifentrockner *m*

3275 LOOSE MOLD (US), HAND MOLD (US), HAND MOULD (GB), PORTABLE MOULD (GB) — pla
f moule *m* à main
e molde *m* a mano
i stampo *m* a mano
n handmatrijs
d Handform *f*

3276 LOOSE PUNCH — pla,pre
f poinçon *m* mobile
e punzón *m* móvil
i punzone *m* mobile
n losse stempel
d beweglicher Stempel *m*

3277 LOOSE ROLLER (GB), DANCER ROLL (US) — pla
f rouleau *m* fou
e rodillo *m* loco, rodillo *m* tensor
i rullo *m* folle
n spanrol
d Losrolle *f*, Spannrolle *f*

3278 LOSS OF HEAD INDICATOR — ins
f indicateur *m* de perte de charge
e indicador *m* de pérdida de carga
i indicatore *m* di perdita di carico
n drukverliesindicator
d Druckverlustanzeiger *m*

3279 LOW-PRESSURE ACCUMULATOR pre
f accumulateur *m* à basse pression
e acumulador *m* de baja presión
i accumulatore *m* a bassa pressione
n lage-drukaccumulator
d Niederdruckakkumulator *m*

3280 LOW-PRESSURE VAPORIZER pre
f évaporateur *m* à basse pression
e evaporador *m* de baja presión
i evaporatore *m* a bassa pressione
n lage-drukverdamper
d Niederdruckverdampfer *m*

3281 LOW-TEMPERATURE CARBONISATION FURNACE fue
f four *m* de semicarbonisation
e horno *m* de carbonización de baja temperatura
i forno *m* di carbonizzazione a bassa temperatura
n oven voor het carboniseren bij lage temperatuur, zweeloven
d Schwelofen *m*

3282 LOW-TEMPERATURE CARBONISATION PLANT fue
f installation *f* de semicarbonisation
e equipo *m* de carbonización de baja temperatura
i impianto *m* di carbonizzazione incompleta, impianto *m* di distillazione a bassa temperatura
n carboniseerinstallatie met lage temperatuur, zweelinstallatie
d Schwelanlage *f*, Tieftemperaturverkokungsanlage *f*

3283 LOW-TEMPERATURE CARBONIZATION SHAFT fue
f puits *m* de semicarbonisation
e pozo *m* de carbonización de baja temperatura
i pozzo *m* di carbonizzazione a bassa temperatura
n carboniseerschacht met lage temperatuur, zweelschacht
d Schwelschaft *m*

3284 LOW-TEMPERATURE HYDROGENATION PLANT fat
f installation *f* d'hydrogénation à basse température
e equipo *m* de hidrogenación de baja temperatura
i impianto *m* d'idrogenazione a bassa temperatura
n hydreerinstallatie met lage temperatuur
d Tieftemperaturhydrieranlage *f*

3285 LOW-TEMPERATURE REFRIGERATION PLANT, DEEP-FREEZE PLANT clo
f installation *f* de réfrigération à basse température
e equipo *m* de refrigeración de baja temperatura
i impianto *m* congelatore, impianto *m* di refrigerazione a bassa temperatura
n diepvriesinstallatie
d Tiefkühlanlage *f*

3286 LOWER PLATEN pla
f plateau *m* inférieur
e mesa *f* inferior
i piano *m* inferiore
n onderste persplaat
d unterer Pressentisch *m*

3287 LUCITE WINDOW ins,spe
f fenêtre *f* en lucite
e ventana *f* de lucita
i finestra *f* di lucite
n lucietvenster *n*
d Luzitfenster *n*

3288 LUMINOUS-FLAME BURNER bur
f brûleur *m* lumineux
e quemador *m* de llama luminosa
i bruciatore *m* luminoso
n diffusiebrander
d Diffusionsbrenner *m*

3289 LUMP nue
f bloc *m*
e bloque *m*, cartucho *m*, trozo *m*
i blocco *m*
n blok *n*
d Block *m*, Brocken *m*

3290 LUSTRING PLANT tex
f installation *f* d'apprêt à lustre
e equipo *m* de acabado lustrado
i impianto *m* di lucidatura
n glansappreteerinstallatie
d Glanzappreturanlage *f*

3291 LYE BATH gen
f bain *m* de lessive
e baño *m* de lejía
i bagno *m* di lisciva
n loogbad *n*
d Laugebad *n*

292 LYE COLUMN, pap
LYE TOWER
f colonne *f* à lessive
e columna *f* de lejía
i colonna *f* a lisciva
n loogkolom
d Laugesäule *f*

3293 LYOPHILIZER, lab
CAMPBELL-PRESSMANN
DRYING APPARATUS
f lyophilisateur *m*
e liofilizador *m*
i essiccatoio *m* per congelamento
di Campbell e Pressmann
n vriesdroogapparaat *n* volgens
Campbell-Pressmann
d Campbell-Pressmann-Gefrier-
trockenapparat *m*

M

3294 MACERATION VAT tan
f cuve *f* de macération
e cuba *f* de maceración
i tino *m* di macerazione
n maceratievat *n*, uittrekvat *n*, zweedvat *n*
d Aufquellgefäss *n*, Mazeriergefäss *n*

3295 MACROETCH TESTING APPARATUS, DEEP-ETCH TESTING APPARATUS tes
f appareil *m* d'essai de macro-attaque
e aparato *m* de prueba de ataque profondo, aparato *m* de prueba de macroataque
i collaudatore *m* a macroincisione all'acquaforte, collaudatore *m* ad attacco chimico profondo
n diepetsbeproevingsapparaat *n*
d Tiefätzprüfgerät *n*, Tiefbeizprüfgerät *n*

3296 MAGNESIUM BOMB exp
f bombe *f* incendiaire à magnésium
e bomba *f* incendiaria de magnesio
i spezzone *m* incendiario a magnesio
n magnesiumbrandbom
d Magnesiumbrennbombe *f*

3297 MAGNESIUM CASING INCENDIARY BOMB exp
f bombe *f* incendiaire à chemise de magnésium
e bomba *f* incendiaria de camisa de magnesio
i spezzone *m* incendiario a camicia di magnesio
n brandbom met magnesiummantel
d Brandbombe *f* mit Magnesium-mantel

3298 MAGNESIUM CELL elc
f pile *f* au magnésium
e pila *f* de magnesio
i pila *f* al magnesio
n magnesiumelement *n*
d Magnesiumelement *n*

3299 MAGNESIUM GALVANIC ANODE elc
f anode *f* à magnésium pour protection galvanique
e ánodo *m* de magnesio para protección galvánica
i anodo *m* a magnesio per protezione galvanico
n magnesiumanode
d Magnesiumanode *f*

3300 MAGNETIC AGITATOR, MAGNETIC BAR STIRRER, MAGNETIC STIRRER lab,mi
f agitateur *m* à bras magnétique
e agitador *m* de brazo magnético
i agitatore *m* a braccio magnetico
n roerwerk *n* met magnetische roerarm
d Rührwerk *n* mit magnetischem Rührarm

3301 MAGNETIC CLARIFIER c
f clarificateur *m* magnétique, épurateur *m* magnétique
e clarificador *m* magnético, epurador *m* magnético
i apparecchio *m* di depurazione, chiarificatore *m* magnetico
n magnetisch klaringsapparaat *n* magnetisch zuiverapparaat *n*
d magnetischer Klärapparat *m*

3302 MAGNETIC COOLANT SEPARATOR c
f séparateur *m* magnétique d'impuretés du liquide réfrigérant
e separador *m* magnético de impurezas del líquido refrigerante
i separatore *m* magnetico di impurezze del liquido refrigerante
n magnetisch apparaat *n* voor he afscheiden van koelvloeistof-onzuiverheden
d magnetischer Apparat *m* zum Ausscheiden von Kühlflüssig-keitsverunreinigungen

3303 MAGNETIC FILTER
f filtre *m* magnétique
e filtro *m* magnético
i filtro *m* magnetico
n magneetfilter *n*
d Magnetfilter *m*

304 MAGNETIC FIRE DETECTOR fea
f indicateur *m* magnétique d'incendie
e detector *m* magnético de incendio
i rivelatore *m* magnetico d'incendio
n magnetische brandverklikker
d magnetischer Feuerdetektor *m*, magnetischer Feuermelder *m*

305 MAGNETIC GAS COOLING APPARATUS clo,gas
f appareil *m* magnétique refroidisseur de gaz
e aparato *m* magnético enfriador de gas
i apparecchio *m* magnetico raffreddatore di gas
n magnetisch gaskoelingsapparaat *n*
d magnetischer Gaskühlungsapparat *m*

306 MAGNETIC OIL FILTER fat,fil
f filtre *m* d'huile magnétique
e filtro *m* de aceite magnético
i filtro *m* d'olio magnetico
n magnetisch oliefilter *n*
d magnetischer Ölfilter *m*

307 MAGNETIC ORE SEPARATOR min,sep
f séparateur *m* magnétique de minerais
e separador *m* magnético para minerales
i separatore *m* magnetico di minerali
n magnetische ertsscheider
d magnetischer Erzscheider *m*

308 MAGNETIC SEPARATOR sep
f séparateur *m* magnétique
e separador *m* magnético
i separatore *m* magnetico
n magnetische scheider
d Magnetscheider *m*

309 MAIN CONDUIT, cpt,tra
MAIN PIPE LINE, PRINCIPAL CONDUIT
f conduite *f* maîtresse, conduite *f* principale
e conducto *m* principal, tubería *f* principal
i condotta *f* maestra, condotta *f* principale
n hoofdleiding
d Hauptleitung *f*, Stammleitung *f*

3310 MAIN HEAD pla,pre
f corps *m* de presse, plateau *m* supérieur, traverse *f*
e cabezal *m* superior fijo
i testa *f*, traversa *f*
n juk *n*
d Querhaupt *n*

3311 MALE DIE, pla,rub
POSITIVE DIE
f moule *m* mâle
e molde *m* macho, molde *m* positivo
i stampo *m* maschio
n positieve mal
d Patrize *f*, Positivform *f*

3312 MALE JOINT, gla,lab
GROUND JOINT
f joint *m* en verre rodé, joint *m* male, rodage *m* émeri
e junta *f* de vidrio esmerilado macha
i giuntura *f* molata
n geslepen verbinding
d Aussenschliff *m*, Schliffhülse *f*

3313 MALT BIN, bre
MALT HOPPER
f case *f* à malt, trémie *f* à mouture
e caja *f* de malta, tolva *f* de malta
i tramoggia *f* di malto
n moutkast
d Malzkasten *m*, Malzrumpf *m*

3314 MALT CLEANER, bre
MALT CLEANING MACHINE, MALT DEGERMINATING MACHINE
f dégermeuse *f*, nettoyeuse *f*
e desgerminadora *f*, limpiadora *f* de malta
i pulitrice *f* per radichette
n moutpoetsmachine
d Malzentkeimungsmaschine *f*, Malzputzmaschine *f*

3315 MALT CRUSHER, bre
MALT GRINDING MILL, MALT MILL, GRIST MILL
f concasseur *m* de malt
e triturador *m* de malta
i frantoio *m* per malto
n schrootmolen
d Malzbrecher *m*, Malzmühle *f*, Malzspaltapparat *m*, Schrotmühle *f*

3316 MALT EXTRACT BATH bre
f bain-marie *m* pour analyse de malt

e baño-maria *m* para análisis de malta
i bagno-maria *m* per analisi di malto
n moutanalysebad *n*
d Maischapparat *m*, Maischbad *n*

3317 MALT HOUSE, MALTING PLANT — bre
f malterie *f*
e maltería *f*
i malteria *f*
n mouterij
d Mälzerei *f*

3318 MALT KILN — bre
f touraille *f* de malt
e tostadero *m* de malta
i essiccatoio *m* di malto
n mouteest
d Malzdarre *f*

3319 MALT LOFT — bre
f grenier *m* à malt
e almacén *m* de malta
i magazzino *m* di malto
n moutzolder
d Malzboden *m*

3320 MALT PLOUGH (GB), MALT PLOW (US) — bre
f charrue *f* à malt
e arado *m* para malta
i aratro *m* per malto verde
n moutploeg
d Malzpflug *m*

3321 MALT POLISHER, MALT POLISHING APPARATUS — bre
f polisseuse *f* de malt
e máquina *f* para pulir malta, pulidora *f* de malta
i lucidatrice *f* di malto
n moutpolijstmachine
d Malzabreibeapparat *m*, Malzpoliermaschine *f*

3322 MALT SILO — bre
f silo *m* de malt
e silo *m* de malta
i silo *m* di malto
n moutsilo
d Malzsilo *n*

3323 MALT TURNER — bre
f retourneur *m* de malt
e removedor *m* de malta
i mescolatore *m* di malto
n moutomgooier
d Malzwendeapparat *m*

3324 MALTING FLOOR, GERMINATING FLOOR, GROWING FLOOR — br
f aire *f* de germination, germoir
e era *f* de germinación
i aia *f* di germinazione
n kiemvloer
d Keimboden *m*, Keimtenne *f*, Malztenne *f*

3325 MAMMOTH PUMP — pr
f pompe *f* mammouth
e bomba *f* mamut
i pompa *f* mammut
n mammoetpomp
d Mammutpumpe *f*

3326 MAN-HOLE — bc
f trou *m* d'homme
e agujero *m* de hombre
i passo *m* d'uomo
n mangat *n*
d Mannloch *n*

3327 MANDREL — el
f noyau *m* de support pour électroformage
e núcleo *m* de soporte para electroformación
i anima *f* di sopporto per galvanoplastica
n steunkern voor galvanoplastiek
d Stützkern *m* für Galvanoplastik

3328 MANDREL, MANDRIL — pla,pr
f mandrin *m*, poinçon *m*
e mandril *m*
i mandrino *m*, maschio *m*
n kern
d Dorn *m*

3329 MANDREL CARRIER — pla,pr
f porte-poinçon *m*
e portamandril *m*
i anello *m* a razze
n kernhouder
d Dornhalterung *f*

3330 MANGANESE REMOVAL PLANT, DEMANGANIZING PLANT — hy
f installation *f* de démanganisat
e equipo *m* para desmanganizar
i impianto *m* di demanganizzazi
n mangaanverwijderingsinstallat
d Entmanganungsanlage *f*

3331 MANGLE — cer,d
f sécheur *m* à air chaud vertical avec chaînes sans fin

e secador *m* de aire caliente vertical con cadenas sin fin
i essiccatoio *m* ad aria calda verticale con catene senza fine
n verticale hete-luchtdroger met eindloze kettingen
d aufrechter Heizlufttrockner *m* mit endlosen Ketten

3332 MANIFOLD cpt
f branchement *m*, distributeur *m*
e distribuidor *m*
i collettore *m*
n spruitstuk *n*, verdeelstuk *n*
d Verteiler *m*

3333 MANIPULATOR ins,nue
f manipulateur *m*
e manipulador *m*
i manipolatore *m*
n manipulator
d Manipulator *m*

3334 MANOCRYOMETER ins
f manocryomètre *m*
e manocriómetro *m*
i manocriometro *m*
n manocryometer
d Manokryometer *n*

3335 MANOMETER, GAGE (US), GAUGE (GB) ins
f manomètre *m*
e manómetro *m*
i manometro *m*
n manometer
d Manometer *n*

3336 MANTLE gen
f chemise *f*, enveloppe *f*, gaine *f*
e camisa *f*, envolvente *f*, vaina *f*
i camicia *f*, guaina *f*, involucro *m*, mantello *m*
n mantel, omhulling
d Hülle *f*, Mantel *m*

3337 MANTLE FILTER fil
f filtre *m* à chemise
e filtro *m* de camisa
i filtro *m* a camicia
n mantelfilter *n*
d Mantelfilter *m*

3338 MANTLE RING, LINTEL GIRDER cpt
f marâtre *f*
e madrasta *f*
i maratra *f*
n draagring
d Tragkranz *m*

3339 MANUAL REGULATOR con
f dispositif *m* de réglage manuel
e dispositivo *m* de regulación manual
i regolatore *m* a mano
n handregelapparaat *n*
d Handregler *m*

3340 MARGARINE FACTORY dai
f margarinerie *f*
e fábrica *f* de margarina
i fabbrica *f* di margarina
n margarinefabriek
d Margarinefabrik *f*

3341 MARINE COATING sap
f couche *f* de protection résistance à l'eau de mer
e capa *f* de protección resistente al agua de mar
i strato *m* di protezione resistente all'acqua di mare
n tegen zeewater bestendige deklaag
d seewasserfeste Deckschicht *f*

3342 MARKING TEMPLATE chr
f gabarit *m* de marcage
e plantilla *f* marcador
i sagoma *f* a tracciare
n merkplaat
d Beschriftungsschablone *f*

3343 MARSH GAS DETECTOR, METHANE DETECTOR, MINE GAS DETECTOR sap
f détecteur *m* de gaz de mine, détecteur *m* de grisou
e detector *m* de grisú
i rivelatore *m* di grisu
n mijngasdetector, mijngasverklikker
d Grubengasanzeiger *m*, Grubengasdetektor *m*

3344 MARVER gla
f mabre *m* de paraison, marbre *m*
e mármol *m*
i lastra *f* di marmo, marmo *m*
n walsplaat
d Marbel *m*, Marbeltisch *m*

3345 MASH AGITATOR, MASH MACHINE, MASH STIRRER bre
f agitateur *m* de trempe, vageur *m* de trempe
e agitador *m* de maisch, agitador *m* de mezcla
i agitatore *m* di miscela
n beslagroerwerk *n*
d Maischrührwerk *n*

3346 MASH COPPER bre
f chaudière *f* à trempe
e caldera *f* de masa de cebada
i caldaia *f* di saccarificazione
n kookketel
d Maischekessel *m*, Maischepfanne *f*

3347 MASH FILTER bre
f filtre *m* à moût
e filtro *m* de mosto
i filtro *m* per miscela
n beslagfilter *n*
d Maischefilter *m*

3348 MASH FILTER FRAME bre
f cadre *m* de filtre à moût
e marco *m* de filtro-prensa para mosto
i cornice *f* per filtro di miscela
n beslagfilterraam *n*
d Maischefilterrahmen *m*

3349 MASH FILTER PLATE bre
f plateau *m* de filtre à moût
e placa *f* de filtro-prensa para mosto
i piastra *f* di filtro per miscela
n beslagfilterplaat
d Maischefilterplatte *f*

3350 MASH TUB, bre
MASH TUN, COLD MASHING VAT
f cuve *f* d'empâtage, cuve-matière *f*
e cuba *f* de mezcla, tina *f* de braceado
i tino *m* di miscela
n beslagkuip
d Maischbottich *m*

3351 MASH TUN TAPS bre
f batterie *f* de filtration
e batería *f* de extracción, batería *f* de filtración
i batteria *f* di filtrazione
n filterbatterij
d Abläuterbatterie *f*

3352 MASK SPRAYING APPARATUS dec
f appareil *m* de pistolage à masque
e aparato *m* para pintar con estarcido
i apparecchio *m* di pittura a maschera
n sjabloonspuitapparaat *n*
d Schablonenspritzapparat *m*

3353 MASKED VALVE cpt
f reniflard *m* encastré
e válvula *f* con deflector de vástago, válvula *f* empotrada
i valvola *f* a fungo incassato
n verzonken snuifklep
d versunkenes Schnarchventil *n*

3354 MASS DIFFUSION COLUMN sep
f colonne *f* à diffusion massique
e columna *f* de difusión másica
i colonna *f* a diffusione di massa
n kolom voor massadiffusie
d Massendiffusionssäule *f*

3355 MASS DIFFUSION SCREEN sep
f écran *m* à diffusion massique
e pantalla *f* de difusión másica
i schermo *m* a diffusione di massa
n scherm *n* voor massadiffusie
d Massendiffusionsschirm *m*

3356 MASS SPECTROGRAPH spe
f spectrographe *m* de masse(s)
e espectrógrafo *m* de masa
i spettrografo *m* di massa
n massaspectrograaf
d Massenspektrograph *m*

3357 MASS-TYPE PLATE, elc
FRAME PLATE, PLATED FRAME PLATE
f plaque *f* à cadre, plaque *f* à masse
e placa *f* de cuadro, placa *f* de masa
i piastra *f* a massa, piastra *f* a telaio
n massaplaat
d Rahmenplatte *f*

3358 MASTER FORM elc
f forme *f* maîtresse
e forma *f* patrón
i forma *f* madre
n moedervorm
d Urmodell *n*

3359 MASTER GAUGE (GB), ins
REFERENCE GAGE (US)
f rapporteur *m*
e calibre *m* de referencia, galga *f* patrón
i calibro *m* campione, calibro *m* di riscontro
n standaardkaliber *n*
d Prüflehre *f*

3360 MASTER-SLAVE nu
MANIPULATOR
f manipulateur *m* asservi
e manipulador *m* piloto-operador
i manipolatore *m* asservito
n op afstand bediende manipulator
d magische Hände *pl*, Parallelmanipulator *m*

3361 MASTICATOR, pla,rub
PLASTICATOR,
PLASTIFICATOR
f masticateur *m*, plastificateur *m*
e masticador *m*, plastificador *m*
i masticatore *m*, plastificatore *m*
n plasticeermachine
d Mastikator *m*, Plastifiziermaschine *f*, Plastikator *m*

3362 MATCHED METAL DIES pla,pre
f moule *m* en deux parties rigides, moule *m* et contre-moule *m* rigides
e molde *m* y contramolde *m* rígidos
i stampo *m* a controstampo rigido, stampo *m* accoppiato
n verwarmde stijve mallen *pl*
d starre Formen *pl*, starre Formhälften *pl*

3363 MATERIAL TESTING REACTOR nue
f réacteur *m* d'essai des matériaux
e reactor *m* de ensayo de materiales
i reattore *m* per prove di materiali
n reactor voor materiaalonderzoek
d Materialprüfreaktor *m*, Materialprüfungsreaktor *m*

3364 MATRIX elc
f matrice *f*
e matriz *f*
i matrice *f*
n matrijs
d Matrize *f*

3365 MATRIX pla
f matrice *f*
e matriz *f*
i matrice *f*
n matrijs
d Mater *f*, Matrize *f*

3366 MATTE DIP elc
f bain *m* de matage
e baño *m* de mateado
i bagno *m* di mordenzatura
n matbeitsbad *n*
d Mattbrenne *f*

3367 MEASURING CYLINDER, lab
GRADUATED CYLINDER
f cylindre *m* gradué, éprouvette *f* graduée
e cilindro *m* graduado
i cilindro *m* graduato
n maatcilinder, maatglas *n*
d Messzylinder *m*

3368 MEASURING FLASK lab
f ballon *m* de mesure, ballon *m* jaugé, matras *m* gradué
e matraz *m* aforado
i matraccio *m* graduato, pallone *m* di misura
n maatkolf
d Messkolben *m*

3369 MEASURING INSTRUMENT ins
f appareil *m* de mesure
e instrumento *m* de medida
i strumento *m* di misura
n meetinstrument *n*
d Messgerät *n*

3370 MEASURING NOZZLE ins
f buse *f* de mesure, tuyère *f* de mesure
e tobera *f* de aforo, tobera *f* de medida
i ugello *m* di misura
n meetmondstuk *n*
d Messdüse *f*

3371 MEASURING PIPETTE (US), lab
GRADUATED PIPETTE (GB)
f pipette *f* graduée
e pipeta *f* graduada
i pipetta *f* graduata
n meetpipet
d Messpipette *f*

3372 MEASURING SCOOP lab
f cuiller *f* de dosage, louche *f* de dosage
e cuchara *f* de dosificación
i cucchiaio *m* di dosificazione
n maatlepel
d Messlöffel *m*, Messschaufel *f*

3373 MECHANICAL AGITATOR mix
f agitateur *m* mécanique
e agitador *m* mecánico
i agitatore *m* meccanico
n mechanisch aangedreven roerwerk *n*
d mechanisch betätigtes Rührwerk *n*

3374 MECHANICAL CLASSIFIER sor
f classeur *m* mécanique
e clasificador *m* mecánico
i classificatore *m* meccanico
n classificeermachine
d Klassiermaschine *f*

3375 MECHANICAL FLUE GAS PURIFIER cle,gas
f dépoussiéreur *m* mécanique des fumées

e depurador *m* mecánico de los gases de combustión
i depuratore *m* meccanico dei fumi, depuratore *m* meccanico dei gas di combustione
n mechanische rookgasreiniger
d mechanischer Rauchgasreiniger *m*

3376 MECHANICAL REFINING DEVICE met,min
f dispositif *m* de raffinage mécanique
e dispositivo *m* de afinación mecánico
i dispositivo *m* d'afino meccanico
n mechanisch raffineerapparaat *n*
d mechanisches Aufbereitungsgerät *n*

3377 MECHANICAL STABILITY TESTER tes
f stabilomètre *m* à latex
e comprobador *m* mecánico de estabilidad
i collaudatore *m* della stabilità
n mechanische stabiliteitsmeter
d mechanisches Stabilitätsprüfgerät *n*

3378 MECHANICAL TESTING APPARATUS tes
f appareil *m* d'essai mécanique
e comprobador *m* mecánico
i collaudatore *m* meccanico
n mechanisch beproevingsapparaat *n*
d mechanisches Prüfgerät *n*

3379 MECHANICAL THICKENER gen
f épaississeur *m* mécanique
e espesador *m* mecánico
i espessitore *m* meccanico
n mechanische indikker
d mechanischer Eindicker *m*

3380 MEDIUM-PRESSURE ACCUMULATOR pre
f accumulateur *m* à moyenne pression
e acumulador *m* de media presión
i accumulatore *m* a media pressione
n middeldrukaccumulator
d Mitteldruckakkumulator *m*

3381 MELTER,
MELTING AREA gla
f zone *f* de fusion
e zona *f* de fusión
i zona *f* di fusione
n smeltruimte
d Schmelzraum *m*

3382 MELTING CRUCIBLE htt
f creuset *m* de fusión
e crisol *m* de fusión
i crogiolo *m* di fusione
n smeltkroes
d Schmelztiegel *m*

3383 MELTING END gla
f bassin *m* de fusion, zone *f* de fusion
e parte *f* de fusión
i zona *f* di fusione
n smeltruimte
d Schmelzraum *m*

3384 MELTING FURNACE,
SMELTING FURNACE gla,met
f four *m* de fusion
e horno *m* de fusión
i forno *m* di fusione
n smeltoven
d Schmelzofen *m*

3385 MELTING POINT APPARATUS ins
f indicateur *m* de point de fusion
e indicador *m* de punto de fusión
i indicatore *m* di punto di fusione
n smeltpuntbepalingsapparaat *n*
d Schmelzpunktgeber *m*

3386 MELTING POINT RECORDER ins
f enregistreur *m* de point de fusion
e registrador *m* de punto de fusión
i registratore *m* di punto di fusione
n smeltpuntindicator
d Schmelzpunktschreiber *m*

3387 MELTING POINT TESTING APPARATUS,
SOFTENING POINT TESTING APPARATUS tes
f appareil *m* de détermination du point de ramollissement du bitume
e comprobador *m* del punto de reblandecimiento del betún
i collaudatore *m* del punto di rammollimento del bitume
n apparaat *n* voor het bepalen van het verwekingspunt van bitumen
d Erweichungspunktbestimmungsgerät *n* für Bitumen

3388 MELTING POINT TUBE la
f tube *m* pour déterminer le point de fusion
e tubo *m* de determinación del punto de fusión
i tubo *m* da determinare il punto di fusione

n smeltpuntbepalingsbuis
d Schmelzpunktbestimmungsrohr *n*

3389 MELTING POT gla
f creuset *m* à verre fondu
e crisol *m* de vidrio fundido
i crogiolo *m* di vetro fuso
n glaspot, smeltpot
d Glashafen *m*

3390 MELTING POT met
f calebasse *f*
e cazo *m* de plomero, olla *f* para fundir
i ramaiuolo *m*
n smeltpot
d Schmelztopf *m*

3391 MEMBRANE, DIAPHRAGM cpt
f diaphragme *m*, membrane *f*
e diafragma *m*, membrana *f*
i diaframma *m*, membrana *f*
n diafragma *n*, membraan
d Diaphragma *n*, Membran *f*, Membrane *f*

3392 MEMBRANE FILTER fil
f filtre *m* à membrane
e filtro *m* de membrana
i filtro *m* a membrana
n membraanfilter *n*
d Membranfilter *m*

3393 MEMBRANE PUMP, DIAPHRAGM PUMP pre
f pompe *f* à diaphragme, pompe *f* à membrane
e bomba *f* de diafragma, bomba *f* de membrana
i pompa *f* a diaframma, pompa *f* a membrana
n diafragmapomp, membraanpomp
d Balgpumpe *f*, Membranpumpe *f*

3394 MERCERIZATION PLANT, STEEPING PLANT tex
f installation *f* de mercerisage
e equipo *m* de mercerización
i impianto *m* di mercerizzazione
n merceriseerinstallatie
d Mercerieranlage *f*, Mercerisieranlage *f*

3395 MERCERIZER, MERCERIZING TANK, STEEPING VAT tex
f récipient *m* de mercerisage
e tanque *m* de mercerización
i serbatoio *m* di mercerizzazione
n merceriseertank
d Merceriergefäss *n*, Mercerisiergefäss *n*

3396 MERCURATION VESSEL, MERCURIZATION VESSEL oic
f cuve *f* de mercuration
e cuba *f* de mercuración
i tino *m* di mercurazione
n mercuriseervat *n*
d Merkuriergefäss *n*

3397 MERCURIFICATION PLANT met
f installation *f* d'extraction du mercure
e equipo *m* de mercurificación
i impianto *m* d'estrazione del mercurio
n kwikwinningsinstallatie
d Quecksilbergewinnungsanlage *f*

3398 MERCURY CATHODE elc
f cathode *f* de mercure
e cátodo *m* de mercurio
i catodo *m* di mercurio
n kwikkatode
d Quecksilberkatode *f*

3399 MERCURY CELL, AMALGAM CELL elc
f cellule *f* au mercure
e celda *f* de mercurio
i cella *f* a mercurio
n kwikcel
d Quecksilberzelle *f*

3400 MERCURY CHECK VALVE cpt,lab
f soupape *f* à pression de mercure
e válvula *f* de presión de mercurio
i valvola *f* a pressione di mercurio
n kwikdrukklep
d Quecksilberdruckventil *n*

3401 MERCURY COLUMN ins
f colonne *f* de mercure
e columna *f* de mercurio
i colonna *f* di mercurio
n kwikkolom
d Quecksilbersäule *f*

3402 MERCURY DIFFUSION PUMP, MERCURY VAPOR DIFFUSION PUMP (US), MERCURY VAPOUR DIFFUSION PUMP (GB) vac
f pompe *f* à diffusion de mercure
e bomba *f* de difusión de mercurio
i pompa *f* a diffusione di mercurio
n kwikdiffusiepomp
d Quecksilberdiffusionspumpe *f*

3403 MERCURY FILTER fil,lab
f filtre *m* à mercure
e filtro *m* de mercurio
i filtro *m* a mercurio
n kwikfilter *n*
d Quecksilberfilter *m*

3404 MERCURY GAGE (US), ins
MERCURY GAUGE (GB)
f manomètre *m* à mercure
e manómetro *m* de mercurio
i manometro *m* a mercurio
n kwikmanometer
d Quecksilbermanometer *n*

3405 MERCURY OXIDE CELL, elc
OXIDE OF MERCURY CELL
f pile *f* à oxyde de mercure-zinc
e pila *f* de óxido de mercurio-cinc
i pila *f* ad ossido di mercurio-zinco
n kwikcel
d Quecksilberoxydelement *n*

3406 MERCURY PURIFICATION cle,lab
APPARATUS
f appareil *m* à purifier le mercure
e aparato *m* purificador de mercurio
i apparecchio *m* di depurazione di mercurio
n kwikzuiveringsapparaat *n*
d Quecksilberreinigungsapparat *m*

3407 MERCURY SAFETY VALVE cpt
f soupape *f* d'arrêt à mercure
e válvola *f* de retención de mercurio
i valvola *f* di ritegno a mercurio
n kwikterugslagklep
d Quecksilberrückschlagventil *n*

3408 MERCURY SEAL cpt,mix
f fermeture *f* à mercure
e cierre *m* de mercurio
i chiusura *f* a mercurio
n kwikafsluiting
d Quecksilberverschluss *m*

3409 MERCURY-SEAL AGITATOR mix
f agitateur *m* à fermeture de mercure
e agitador *m* de cierre de mercurio
i agitatore *m* a chiusura a mercurio
n roerwerk *n* met kwikafsluiting
d Rührwerk *n* mit Quecksilberverschluss

3410 MERCURY VAPOR JET vac
DIFFUSION PUMP (US),
MERCURY VAPOUR JET
DIFFUSION PUMP (GB)
f pompe *f* à éjection et diffusion de vapeur de mercure
e bomba *f* de eyección y difusión de vapor de mercurio
i pompa *f* ad iniezione e diffusione di vapore di mercurio
n kwikdampstraalpomp
d Quecksilberdampfstrahlpumpe *f*

3411 MESH cpt,sor
f maille *f*
e malla *f*
i maglia *f*
n maas
d Masche *f*

3412 MESH SCREEN, cle,sep,sor
MESH SIEVE
f tamis *m* à mailles
e tamiz *m* de mallas
i staccio *m* a maglie
n mazenzeef
d Maschensieb *n*

3413 METAFILTRATION CASCADE fil
f filtre *m* à cascades
e filtro *m* de cascadas
i filtro *m* a cascate
n cascadefilter *n*
d Kaskadenfilter *m*

3414 METAL ARC WELDING wel
PLANT
f installation *f* de soudure à électrodes métalliques
e equipo *m* de soldadura con electrodos metálicos
i impianto *m* di saldatura ad elettrodi metallici
n lasinstallatie met metaalelektrod
d Schweissanlage *f* mit Metallelektroden

3415 METAL BATH htt
f bain *m* de métal liquide
e baño *m* de metal líquido
i bagno *m* a metallo liquido
n gesmolten metaalbad *n*, metaalbad *n*
d Metallbad *n*, Schmelzbad *n*

3416 METAL ELECTRODE wel
f électrode *f* métallique
e electrodo *m* metálico
i elettrodo *m* metallico
n metaalelektrode
d Metallelektrode *f*

3417 METAL PLATING BATH coa
f bain *m* à métallisation
e baño *m* de metalización
i bagno *m* di metalizzazione
n metalliseerbad *n*
d Metallisierbad *n*

3418 METAL SHIELD sap
f blindage *m*
e blindaje *m*
i schermatura *f*
n pantser *n*
d Panzer *m*

3419 METAL SPRAYING GUN coa
f pistolet *m* pulvérisateur de métal
e pistola *f* pulverizadora de metal
i pistola *f* a spruzzo di metallo
n metaalspuitpistool *n*, metaalverstuiver
d Metallspritzpistole *f*

3420 METALLIC COATING coa
f revêtement *m* métallique
e revestimiento *m* metálico
i rivestimento *m* metallico
n metaalbedekking
d Metallauskleidung *f*

3421 METALLIC-FUELLED GAS-COOLED REACTOR nue
f réacteur *m* à combustible métallique à refroidissement par gaz
e reactor *m* de combustible metálico enfriado al gas
i reattore *m* a combustibile metallico con raffreddamento a gas
n met gas gekoelde metalliekebrandstofreactor
d gasgekühlter Reaktor *m* mit metallischem Brennstoff

3422 METALLIC LINING coa
f garnissage *m* métallique
e guarnición *f* metálica
i guarnizione *f* metallica
n metaalvoering
d Metallfutter *n*

3423 METALLIC MIRROR coa
f miroir *m* métallique
e espejo *m* metálico
i specchio *m* metallico
n metaalspiegel
d Metallspiegel *m*

3424 METALLIC PACKING cpt
f garniture *f* métallique
e empaquetadura *f* metálica
i guarnizione *f* metallica
n metaalpakking
d Metallpackung *f*

3425 METALLIC SILVER BAR ELECTRODE elc
f électrode *f* d'argent en baguette
e electrodo *m* de plata en barra
i elettrodo *m* d'argento in bacchetta
n zilverstaafelektrode
d Silberstabelektrode *f*

3426 METALLIC TISSUE gen
f toile *f* métallique
e tela *f* metálica
i tessuto *m* metallico
n metaalweefsel *n*
d Metallgewebe *n*

3427 METALLIC-TONING BATH pho
f bain *m* de virage métallique
e baño *m* de viraje metálico
i bagno *m* di viraggio metallico
n metaalkleurbad *n*
d Metalltonbad *n*

3428 METALLIZATION PLANT, METALLIZING PLANT coa
f installation *f* de métallisation
e equipo *m* de metalización
i impianto *m* di metallizzazione
n metalliseerinstallatie
d Metallisieranlage *f*

3429 METERING PUMP, CONTROLLED-VOLUME PUMP pre
f pompe *f* à capacité variable
e bomba *f* de gasto regulado
i pompa *f* dosatrice
n pomp met regelbare capaciteit
d Pumpe *f* mit regelbarer Kapazität

3430 METERING TANK pla
f récipient *m* jauge
e tanque *m* de dosificación
i serbatoio *m* di misura
n doseertank
d Dosiergefäss *n*

3431 METHANE DETECTOR, MARSH GAS DETECTOR, MINE GAS DETECTOR sap
f détecteur *m* de gaz de mine, détecteur *m* de grisou
e detector *m* de grisú
i rivelatore *m* di grisu
n mijngasdetector, mijngasverklikker

d Grubenanzeiger *m*, Grubengas-
detektor *m*

3432 METHANIZATION PLANT oic
f installation *f* de méthanisation
e equipo *m* de metanización
i impianto *m* di metanizzazione
n methaniseerinstallatie
d Methanisieranlage *f*

3433 METHANOMETER ins
f méthanomètre *m*
e metanómetro *m*
i metanometro *m*
n methaangehaltemeter
d Methangehaltmesser *m*

3434 METHYLATION PLANT oic
f installation *f* de méthylation
e equipo *m* de metilación
i impianto *m* di metilazione
n methyleerinstallatie
d Methylieranlage *f*

3435 MICROBALANCE lab
f microbalance *f*
e microbalanza *f*
i microbilancia *f*
n microbalans
d Mikrowaage *f*

3436 MICROBURETTE lab
f microburette *f*
e microbureta *f*
i microburetta *f*
n microburet
d Mikrobürette *f*

3437 MICROMANIPULATOR bac,ins
f micromanipulateur *m*
e micromanipulador *m*
i micromanipolatore *m*
n micromanipulator
d Mikromanipulator *m*

3438 MICROMETER, ins
SCREW GAGE (US), SCREW
GAUGE (GB)
f micromètre *m* à vis
e micrómetro *m* de tornillo
i micrometro *m* a vite
n schroefmicrometer
d Schraublehre *f*, Schraubmikro-
meter *n*

3439 MICRONIZER MILL brk
f moulin *m* à tourbillonnement
e molino *m* pulverizador a
tamaños micrométricos
i mulino *m* polverizzatore a
dimensioni micrometriche
n wervelmolen
d Wirbelmühle *f*

3440 MICROSCOPE ins
f microscope *m*
e microscopio *m*
i microscopio *m*
n microscoop
d Mikroskop *n*

3441 MICROSPECTROSCOPE spe
f microspectroscope *m*
e microespectroscopio *m*
i microspettroscopio *m*
n microspectroscoop
d Mikrospektroskop *n*

3442 MICROSTRAINER fil
f micro-tamis *m*
e microtamiz *m*
i microstaccio *m*
n microzeef
d Feinsieb *n*, Mikrosieb *n*

3443 MIGRATION TUBE ins,lab
f tube *m* à déterminer la migratio
des ions
e tubo *m* para determinar la
migración de los iones
i tubo *m* per determinare la
migrazione degl'ioni
n ionenmigratiebuis
d Ionenwanderungsrohr *n*

3444 MILK-BOTTLE FILLING dai
MACHINE
f soutireuse *f* de bouteilles à lait
e llenadora *f* de botellas de leche
i imbottigliatrice *f* di latteria
n melkflessenvulmachine
d Milchflaschenfüllmaschine *f*

3445 MILK PASTEURIZER da
f pasteurisateur *m* de lait
e pasteurizador *m* de leche
i apparecchio *m* da pastorizzare
il latte
n melkpasteuriseerapparaat *n*
d Milchpasteurisierapparat *m*

3446 MILK-POWDER COLLECTOR da
f collecteur *m* de poudre de lait
e colector *m* de polvo de leche
i collettore *m* di polvere di latte
n melkpoederverzamelaar
d Milchpulversammler *m*

3447 MILK-POWDER DRYING da
ROOM
f sécherie *f* du poudre de lait

e sala *f* de secado del polvo de leche
i sala *f* d'essiccazione della polvere di latte
n melkpoederdroogkamer
d Milchpulvertrockenkammer *f*

3448 MILK TANK dai
f récipient *m* de lait
e tanque *m* de leche
i serbatoio *m* di latte
n melktank
d Milchbehälter *m*

3449 MILK TEST BOTTLE, dai,ins
BUTYROMETER
f butyromètre *m*
e butirómetro *m*
i butirometro *m*
n butyrometer
d Butyrometer *n*, Milchprober *m*

3450 MILL, gen
FACTORY, WORKS
f établissement *m*, fabrique *f*, manufacture *f*, usine *f*
e fábrica *f*
i fabbrica *f*, stabilimento *m*
n fabriek
d Fabrik *f*, Werk *n*

451 MILL, brk
GRINDING MILL
f broyeur *m*, moulin *m*
e molino *m*
i frantumatore *m*, mulino *m*
n breekmolen, maalmolen
d Mühle *f*

452 MILL OPENING, pla,rub
BITE, ROLL CLEARANCE
f écartement *m*, intervalle *m*, jeu *m*
e espacio *m* perjudicial
i giuoco *m*, interdistanza *f*
n walsafstand, walsspleet
d Spaltweite *f*, Walzenspalt *m*

453 MILL ROOM bre
f salle *f* de concassage
e sala *f* de trituración
i sala *f* dei mulini per malto
n schrootmalerij
d Schroterei *f*

454 MILLING MACHINE, tex
FULLING MILL
f fouleuse *f*
e batanadora *f*
i mulino *m* da folletare
n volmolen
d Walke *f*

3455 MILLING MACHINE, brk,min
ORE BREAKER
f concasseur *m* de minerais
e triturador *m* de minerales
i frantoio *m* per minerali
n ertsbreker
d Erzbrecher *m*

3456 MINE GAS DETECTOR, sap
MARSH GAS DETECTOR,
METHANE DETECTOR
f détecteur *m* de gaz de mine, détecteur *m* de grisou
e detector *m* de grisú
i rivelatore *m* di grisu
n mijngasdetector, mijngasverklikker
d Grubengasanzeiger *m*, Grubengasdetektor *m*

3457 MINERAL OIL PLANT fue
f installation *f* d'huile minérale
e equipo *m* de aceite mineral
i impianto *m* d'olio minerale
n aardolie-installatie
d Erdölanlage *f*

3458 MINERAL OIL REFINERY, fue
MINERAL OIL REFINING PLANT
f raffinerie *f* d'huile minérale
e refinería *f* de aceite mineral
i raffineria *f* d'olio minerale
n aardolieraffinaderij
d Erdölanlage *f*

3459 MIRROR MICROSCOPE ins
f microscope *m* à miroir
e microscopio *m* de espejo
i microscopio *m* a specchio
n spiegelmicroscoop
d Spiegelmikroskop *n*

3460 MIST ELIMINATOR dis
f antibrouillard *m*
e separador *m* de gotas
i separatore *m* di gocce
n druppelafscheider
d Tröpfchenabscheider *m*

3461 MIST EXTRACTOR fat
f piège *m* de gouttes d'huile
e extractor *m* de neblina de aceite
i estrattore *m* di nebbia d'olio
n spattenvanger
d Öltropfenfalle *f*

3462 MITER VALVE, cpt
ANGLE VALVE, CORNER VALVE
f soupape *f* d'équerre
e válvula *f* angular, válvula *f* de escuadra
i valvola *f* a disco con seda a 45°

n hoekklep
d Eckventil *n*, Winkelventil *n*

3463 MIXED-BED DEIONIZER elc
f installation *f* de désionisation à lit mixte
e equipo *m* de desionización de lecho mixto
i impianto *m* di deionizzazione a letto misto
n desionisatie-installatie met gemengd bed
d Mischbettentionisieranlage *f*

3464 MIXED-FLOW AGITATOR mix
f agitateur *m* à flux composés
e agitador *m* de flujos compuestos
i agitatore *m* a flussi composti
n roerwerk *n* met samengestelde stromingen
d Rührwerk *n* mit zusammengesetzten Strömungen

3465 MIXED-GAS BURNER bur
f brûleur *m* à gaz mixte
e quemador *m* de gas mixto
i bruciatore *m* a gas misto
n menggasbrander
d Mischgasbrenner *m*

3466 MIXED PHASE CRACKING PLANT fue
f installation *f* de craquage en phase mixte
e equipo *m* de pirólisis en fase mixta
i impianto *m* di piroscissione in fase mista
n kraakinstallatie volgens het gemengde-fazeproces
d Krackanlage *f* nach dem Gemischtphaseverfahren

3467 MIXER, BLENDER mix
f mélangeur *m*
e mezclador *m*
i mescolatore *m*
n menger
d Mischer *m*

3468 MIXING BASIN iwt,sew
f bassin *m* d'homogénéisation
e alberca *f* de homogenización
i bacino *m* di mescolazione, vasca *f* di mescolazione
n mengbekken *n*
d Mischbecken *n*

3469 MIXING BLADE mi
f palette *f* mélangeuse
e paleta *f* mezcladora
i paletta *f* mescolatrice
n mengarm
d Mischarm *m*

3470 MIXING CHAMBER mi
f salle *f* de mélange
e sala *f* de mezclado
i camera *f* di mescolatura
n mengkamer
d Mischraum *m*

3471 MIXING CONDENSER pr
f condensateur *m* à mélange
e condensador *m* mezclador
i condensatore *m* mescolatore
n mengcondensor
d Mischverdichter *m*

3472 MIXING CYLINDER, SHAKING CYLINDER lab,mi
f cylindre *m* mélangeur, éprouve *f* mélangeuse
e cilindro *m* mezclador
i cilindro *m* mescolatore
n mengmaatcilinder, mengmaatgl
d Mischzylinder *m*

3473 MIXING DRUM mi
f mélangeur *m* à tambour
e tambor *m* mezclador
i tamburo *m* mescolatore
n mengtrommel
d Mischtrommel *f*

3474 MIXING MILL, MIXING ROLLERS, OPEN MIL ROLLER MILL pla,ru
f malaxeur *m* à cylindres, mélangeur *m* ouvert
e amasadora *f* de cilindros, mezclador *m* abierto, mezclad de cilindros
i mescolatore *m* a cilindri, mescolatore *m* aperto
n mengwals, open mengwals
d offene Mischwalze *f*, Walzenmischer *m*

3475 MIXING NOZZLE f
f buse *f* de mélange, tuyère *f* d mélange
e tobera *f* mezcladora
i ugello *m* mescolatore
n mengmondstuk *n*
d Mischdüse *f*

3476 MIXING TANK mix
f cuve *f* de mélange
e cuba *f* de mezclado
i tino *m* di mescolazione
n mengvat *n*
d Mischbottich *m*

3477 MOBILOMETER ins,pla
f mobilomètre *m*
e mobilómetro *m*
i mobilometro *m*
n mobilometer
d Mobilometer *n*

3478 MODERATED REACTOR nue
f réacteur *m* à modération, réacteur *m* modéré
e reactor *m* de moderación, reactor *m* moderado
i reattore *m* a moderazione, reattore *m* moderato
n gemodereerde reactor
d moderierter Reaktor *m*

3479 MOIST CHAMBER bac
f chambre *f* humide
e cámara *f* húmeda
i camera *f* umida
n vochtkamer
d Feuchtkammer *f*

3480 MOIST-O-GRAPH (US), AIR HUMIDITY INDICATOR, HYGROMETER (GB) ins
f hygromètre *m*
e higrómetro *m*
i igrometro *m*
n hygrometer, vochtigheids-meter
d Hygrometer *n*, Luftfeuchtigkeits-messer *m*

3481 MOISTENING APPARATUS, HUMIDIFIER hyg
f humecteur *m*, humidificateur *m*
e humectador *m*, humidificador *m*
i umidificatore *m*
n bevochtigingstoestel *n*
d Anfeuchtapparat *m*, Befeuchter *m*

3482 MOISTENING CHAMBER, HUMIDIFYING CHAMBER tex
f chambre *f* d'humidification
e cámara *f* de humidificación
i camera *f* d'umidificazione
n bevochtigingsruimte
d Befeuchtungskammer *f*

3483 MOISTURE-PROOF CONTAINER gen,tra
f récipient *m* à l'épreuve de l'humidité
e recipiente *m* humidífugo
i recipiente *m* idrofugo
n vochtwerend vat *n*
d feuchtigkeitsgeschützter Behälter *m*

3484 MOISTURE TRAP dry
f collecteur *m* d'humidité
e colector *m* de humedad
i collettore *m* d'umidità
n vochtvanger
d Feuchtfänger *m*

3485 MOLASSES PUMP sin
f pompe *f* à mélasse
e bomba *f* de melaza
i pompa *f* a melassa
n melassepomp
d Melassepumpe *f*

3486 MOLASSES TANK sin
f récipient *m* de mélasse
e tanque *m* de melaza
i serbatoio *m* di melassa
n melassetank
d Melassebehälter *m*

3487 MOLD (US), MOULD (GB) elc
f moule *m*
e molde *m*
i forma *f*
n afdruk, stijp
d Gussform *f*

3488 MOLD CAVITY (US), MOULD CAVITY (GB) pla,pre
f cavité *f* de moule
e cavidad *f* de molde
i cavità *f* di stampo
n vormholte
d Formhohlraum *m*

3489 MOLD CHARGE (US), MOULD CHARGE (GB), FEED, LOAD pla
f chargement *m*
e carga *f*
i carica *f*
n vulling
d Beschickung *f*

3490 MOLD CHASE (US), MOULD CHASE (GB), CHASE pla
f châssis *m* de moulage
e bloque *m* sujetador de la matriz, chassis *m* del molde
i custodia *f*
n matrijsfreem *n*
d Formrahmen *m*

3491 MOLD CLAMP (US), pla,pre
MOULD CLAMP (GB)
f serrage *m* de moule, verrouillage *m* de moule
e pinza *f* para molde
i dispositivo *m* di chiusura di stampo, morsetto *m* di stampo
n matrijsvergrendeling
d Formenschluss *m*, Formverschluss *m*

3492 MOLD CLEANING GUN (US), pla
MOULD CLEANING GUN (GB)
f sableuse *f* pour nettoyage du moule
e máquina *f* para limpieza del molde
i macchina *f* per sabbiatura dello stampo
n vormzandstraler
d Formsandstrahlputzmaschine *f*

3493 MOLD INSERT (US), pla,pre
MOULD INSERT (GB)
f noyau
e inserción *f* en el molde
i tassello *m*
n inzetdeel *n*, inzetstuk *n*
d Formeneinsatz *m*

3494 MOLD OPENING (US), pla,rub
DAYLIGHT OPENING (GB)
f distance *f* entre plateaux, lumière *f*
e abertura *f* entre platos, luz *f* de la prensa
i apertura *f*, luce *f*
n daglichtopening
d lichte Einbauhöhe *f*, lichte Öffnungsweite *f*

3495 MOLD PART (US), pla,rub
MOULD PART (GB)
f partie *f* du moule
e parte *f* del molde
i parte *f* dello stampo
n matrijsgedeelte *n*
d Formteil *m*

3496 MOLD RETAINING FLANGE (US), pla,pre
MOULD RETAINING FLANGE (GB)
f bride *f* porte-moule, plateau *m* porte-moule
e brida *f* de sujeción del molde
i flangia *f* portastampo
n matrijsbevestigingsflens
d Formhalteflansch *m*

3497 MOLD WITH CONICAL SPLITS (US), pla,pre
MOULD WITH CONICAL SPLITS (GB)
f moule *m* à coquilles coniques
e molde *m* de cojinetes
i stampo *m* a guance
n konische matrijs
d Kegelform *f*

3498 MOLDING PRESS (US), pla,pre,rub
MOULDING PRESS (GB)
f presse *f* à compression, presse *f* de moulage
e prensa *f* de moldeo por compresión
i pressa *f*
n kunststofpers, pers
d Kunststoffpresse *f*, Presse *f*

3499 MOLDING SHOP (US), pla,pre
MOULDING SHOP (GB)
f atelier *m* de moulage
e taller *m* de moldeo
i officina *f* delle presse, reparto *m* presse
n perserij
d Presserei *f*

3500 MOLECULAR AIR PUMP (GB), vac
MOLECULAR-DRAG PUMP (US
f pompe *f* moléculaire
e bomba *f* de arrastre molecular
i pompa *f* molecolare
n moleculaire-luchtpomp
d Molekularluftpumpe *f*

3501 MOLECULAR DISTILLATION PLANT, dis
MOLECULAR STILL
i installation *f* de distillation moléculaire
e equipo *m* de destilación molecul
i impianto *m* di distillazione molecolare
n moleculaire-destillatie-installatie
d Kurzwegdestillieranlage *f*, Molekulardestillieranlage *f*

3502 MOLECULAR LAYER, co
MONOMOLECULAR LAYER,
UNIMOLECULAR LAYER
f couche *f* monomoléculaire
e capa *f* monomolecular
i strato *m* monomolecolare
n monomoleculaire laag
d Monomolekularschicht *f*

3503 MOLECULAR SIEVE ch
f tamis *m* moléculaire

e tamiz *m* molecular
i staccio *m* molecolare
n moleculaire zeef
d Molekularsieb *n*

3504 MONITOR nue,sap
f moniteur *m*
e monitor *m*
i avvisatore *m*, monitore *m*, segnalatore *m*
n monitor, verklikker
d Monitor *m*, Überwachungsgerät *n*, Warngerät *n*

3505 MONKEY POT, JOCKEY POT gla
f creuset *m* pour verre spécial
e crisol *m* para vidrio special
i crogiolo *m* per vetro speziale
n pot voor speciaal glas
d Sonderglashafen *m*

3506 MONOBATH pho
f monobain *m*
e monobaño *m*
i monobagno *m*
n monobad *n*
d Einbad *n*

3507 MONTEJUS, ACID EGG (US), ACID ELEVATOR (US), BLOW CASE (GB) cor
f monte-jus *m*
e elevador *m* de presión, montaácidos *m*, montajugos *m*
i montaacidi *m*, montaliquidi *m*
n drukvat *n*
d Druckbirne *f*, Druckfass *n*, Montejus *m*

3508 MOP, DOLLY, POLISHING LAP pol
f disque *m* en lisière de tissu, meule *f* flexible en chiffon
e bruñidor *m* de trapos
i brunitoio *m* a pezzi di tessuto
n lappenschijf
d Schwabbelscheibe *f*

3509 MORDANTING BATH pho
f bain *m* de mordançage
e baño *m* de mordentación
i bagno *m* di mordenzatura
n beitsbad *n*
d Beizbad *n*

3510 MORTAR lab
f mortier *m*
e mortero *m*
i mortaio *m*
n mortier, vijzel
d Mörser *m*, Reibschale *f*

3511 MORTAR BOX, MORTAR TROUGH bui
f auge *f* à mortier
e artesa *f* de mortero
i truogolo *m* di malta
n speciebak, speciekuip
d Mörtelkelle *f*

3512 MORTAR MILL brk,bui
f moulin *m* de mortier
e molino *m* de mortero
i mulino *m* di malta
n speciemolen
d Mörtelmühle *f*

3513 MOTHER BLANK, STARTING-SHEET BLANK elc
f plaque-support *m* de feuille de départ
e placa-soporte *m* de hoja de arranque
i piastra *f* di sopporto di lamierino catodico
n beginmatrijs
d Mutterblechkatodenblech *n*

3514 MOTHER CASK aam
f tonneau-mère *m*
e barril-madre *m*
i fustomadre *m*
n kaamvat *n*, moederkuip
d Essigbildner *m*, Mutterfass *n*

3515 MOULD INSERT (GB), CAVITY PLUG (US) pla,pre
f empreinte *f* rapportée
e asiente *m* de la matriz
i tassello *m* d'impronta
n matrijsinzetstuk *n*
d Gesenkeinsatz *m*, Matrizeneinsatz *m*

3516 MOUNTING PLATE gen
f plateau *m* inférieur
e placa *f* base de fijación
i piastra *f* di base
n montageplaat
d Säulenplatte *f*

3517 MOUNTING PLATE (GB), CLAMPING PLATE (US) cpt,pla
f plateau *m* de fixation, plateau *m* porte-moule
e placa *f* de sujeción
i piastra *f* portastampo
n opspanplaat
d Aufspannplatte *f*

3518 MOUNTING PLATEN pla,pre
f plateau *m*, table *f* de la presse
e mesa *f* de la prensa, plato *m*
i piano *m* della pressa
n opspanplaat, opspantafel
d Pressentisch *m*

3519 MOVABLE MOULD PART (GB), pla
FORCE SIDE PART (US)
f partie *f* mobile du moule
e semimolde *m* móvil
i semistampo *m* mobile
n bewegende matrijshelft
d beweglicher Werkzeughalter *m*

3520 MOVABLE PLATEN, pla,pre
MOVING DIE PLATE
f plaque *f* mobile du moule
e plato *m* móvil del molde
i piastra *f* mobile dello stampo
n beweegbare matrijsplaat, beweegbare plaat
d bewegliche Formplatte *f*

3521 MOVING BED, hyg
FLUID BED, FLUIDIZED BED
f lit *m* fluide, lit *m* fluidifié, lit *m* suspendu
e lecho *m* fluidizado, lecho *m* flúido
i letto *m* fluidificato, letto *m* fluidizzato
n wervelbed *n*
d Fliessbett *n*, Wirbelbett *n*

3522 MOVING-BED CATALYTIC fue
CRACKING PLANT
f installation *f* de craquage à lit mobile
e equipo *m* de pirólisis sobre lecho móvil
i impianto *m* di piroscissione a letto mobile
n kraakinstallatie met bewegend bed
d Wanderbett-Krackanlage *f*

3523 MOVING FIRE KILN cer
f four *m* à feu mobile
e horno *m* de fuego móvil
i forno *m* a fuoco mobile
n oven met zich verplaatsend vuur
d Ofen *m* mit fortschreitendem Feuer

3524 MUCK DRUM, boi
MUD DRUM
f collecteur *m* de boue
e colector *m* de fango
i collettore *m* di fango
n slibvanger
d Schlammfänger *m*

3525 MUD BOX sew
f caisse *f* à dépôt de boue
e caja *f* de fango
i cassa *f* di fango
n slibbak
d Schlammkasten *m*

3526 MUD HOG (US), brk
SWING HAMMER CRUSHER (GB)
f broyeur *m* à marteau oscillant rotatif
e machacadora *f* de martillo oscilante giratorio
i frantoio *m* a martello oscillante rotativo
n breker met draaiende zwaaiham
d Brecher *m* mit rotierendem Schwinghammer

3527 MUD PUMP se
f pompe *f* de boue
e bomba *f* de fango
i pompa *f* di fango
n slikpomp
d Schlammpumpe *f*

3528 MUD VALVE si
f soupape *f* de purge
e válvula *f* de lodo
i valvola *f* di spurgo
n modderklep
d Schlammventil *n*

3529 MUD WATER PUMP si
f pompe *f* de purge
e bomba *f* de lodo
i pompa *f* di spurgo
n modderwaterpomp
d Schlammwasserpumpe *f*

3530 MUFFLE htt,s
f moufle *f*
e mufla *f*
i muffola *f*
n moffel
d Muffel *f*

3531 MUFFLE FURNACE, coa,h
BOX FURNACE
f four *m* à moufle
e horno *m* de mufla
i forno *m* a muffola
n moffeloven
d Muffelofen *m*

3532 MUFFLE TUNNEL c
f tunnel *m* à moufler

e túnel *m* de moflar
i tunnel *m* a muffola
n moffeltunnel
d Muffeltunnel *m*

3533 MUFFLER (US), EXHAUST MUFFLER (US), EXHAUST SILENCER (GB), SILENCER (GB) cpt
f pot *m* de détente, silencieux *m* d'échappement
e silenciador *m* del escape
i marmitta *f* di scarico, silenziatore *m*
n knalpot
d Auspufftopf *m*, Schalldämpfer *m*

3534 MULLER, MULLER MIXER, MULLING MACHINE, SAND MILL brk
f frotteur *m*
e amasador *m*
i molazza-mescolatore *m*
n koldermolen, kollergang
d Kollergang *m*

3535 MULTI-CAVITY MOLD (US), MULTI-IMPRESSION MOULD (GB) pla,pre
f moule *m* à empreintes multiples
e molde *m* de varias posiciones, molde *m* múltiple
i stampo *m* multiplo
n meervoudige matrijs
d Mehrfachform *f*

3536 MULTI-COIL CONDENSER clo
f réfrigérant *m* à serpentins multiples
e enfriador *m* de serpentines múltiples
i raffreddatore *m* a serpentine multiple
n koeler met meer spiralen
d Mehrfachspiralkühler *m*

3537 MULTI-LIMB VACUUM RECEIVER, ADAPTER, DISTILLING UDDER dis,lab
f séparateur *m* de Bredt, séparateur *m* de Pauly
e separador *m* de Bredt
i separatore *m* di Bredt
n destilleeruier
d Destilliereuter *m*, Destilliervorlage *f* nach Bredt

3538 MULTI-PLATEN PRESS (US), DAYLIGHT PRESS (GB) pla,rub
f presse *f* à plateaux multiples
e prensa *f* de platos múltiples
i pressa *f* a piani
n etagepers
d Etagenpresse *f*, Plattenpresse *f*

3539 MULTI-PURPOSE REACTOR nue
f réacteur *m* mixte, réacteur *m* universel
e reactor *m* de multiuso
i reattore *m* a molteplici scopi
n reactor voor verschillende doeleinden
d Mehrzweckreaktor *m*

3540 MULTI-REGION REACTOR, MULTI-ZONE REACTOR nue
f réacteur *m* multizone
e reactor *m* de múltiples zonas
i reattore *m* a più regioni
n meerzonenreactor
d Mehrzonenreaktor *m*

3541 MULTI-TUBE HEAT EXCHANGER htt
f échangeur *m* de chaleur à plusieurs tubes
e cambiador *m* de calor multitubular
i scambiatore *m* di calore a tubi multipli
n meerbuizenwarmteuitwisselaar
d Mehrrohrenwärmeaustäuscher *m*

3542 MULTIPLE-EFFECT EVAPORATOR pre
f évaporateur *m* à effet multiple
e evaporador *m* de efecto múltiple
i evaporatore *m* a più stadi, evaporatore *m* ad effetto multiplo
n multipel-effect-verdamper
d Mehrkörperverdampfer *m*, Mehrstufenverdampfer *m*

3543 MULTIPLE ELECTRODE, POLYELECTRODE elc
f électrode *f* multiple, polyélectrode *f*
e electrodo *m* múltiple, polielectrodo *m*
i elettrodo *m* multiplo
n meervoudige elektrode
d Mehrfachelektrode *f*

3544 MULTIPLE-HEARTH FURNACE htt,met
f four *m* à étages
e horno *m* multiplaza
i forno *m* a ripiani
n etageoven
d Etagenofen *m*

3545 MULTIPLE-HEARTH ROASTER bur
f four *m* de frittage à étages
e horno *m* de tostación multiplaza
i forno *m* d'arrostimento a ripiani
n etageroostoven
d Etagenröstofen *m*

3546 MULTIPLE-PLATE FREEZER nut
f installation *f* de congélation à plusieurs étages
e equipo *m* de congelación de planos superpuestos
i impianto *m* frigorifero a piani multipli
n vriesinstallatie met etages
d Mehrplattengefrieranlage *f*

3547 MULTIPLE-SCREW EXTRUDER pla,pre
f boudineuse *f* à plusieurs vis
e extrusora *f* de varios husillos
i estrusore *m* multivite
n extrusiemachine met meer schroeven
d Mehrschneckenmaschine *f*

3548 MULTIPLE-STAGE CENTRIFUGAL PUMP pre
f pompe *f* centrifuge multicellulaire
e bomba *f* centrífuga de múltiples etapas, bomba *f* centrífuga multigradual
i pompa *f* centrifuga multipla
n centrifugaalpomp met dubbele inlaat
d Mehrstufenkreiselpumpe *f*

3549 MULTIPLE-STAGE COMPRESSOR pre
f compresseur *m* à plusieurs étages
e compresor *m* de múltiples etapas
i compressore *m* multiplo
n compressor met meervoudige circulatie
d mehrstufiger Verdichter *m*

3550 MULTIPLYING GAGE (US), MULTIPLYING GAUGE (GB) ins
f manomètre *m* multiplicateur
e manómetro *m* multiplicador
i manometro *m* moltiplicatore
n vermenigvuldigingsmanometer
d Vervielfachungsmanometer *n*

3551 MUSHROOM BURNER bur
f brûleur *m* champignon
e quemador *m* champiñon
i bruciatore *m* a testa a fungo
n paddestoelbrander
d Pilzbrenner *m*

3552 MUSHROOM VALVE, AIR INTAKE VALVE, AIR SNIFFLER, POPPET VALVE cp
f reniflard *m*
e válvula *f* de entrada de aire, válvula *f* de vástago
i valvola *f* a fungo, valvola *f* d'insufflazione
n snuifklep
d Schnarchventil *n*, Schnüffelventil *n*

3553 MUSTARD GAS DETECTOR, YPERITE DETECTOR ins,sa
f détecteur *m* de gaz moutarde, détecteur *m* d'ypérite
e detector *m* de gas mostaza, detector *m* de iperita
i rivelatore *m* d'iprite
n mosterdgasdetector, mosterdgasverklikker, yperietverklikke
d Senfgaswarngerät *n*, Yperitanzeigegerät *n*

3554 MUSTARD GAS GENERATOR, YPERITE GENERATOR gas,mi
f générateur *m* d'ypérite
e generador *m* de iperita
i generatore *m* d'iprite
n mosterdgasgenerator
d Senfgasgenerator *m*

N

3555 NAIL TEST APPARATUS exp
f appareil *m* d'essai au clou
e aparato *m* de prueba al clavo
i collaudatore *m* al chiodo
n spijkerbeproevingsapparaat *n*
d Nagelprüfgerät *n*

3556 NAKED-FLAME BURNER bur
f brûleur *m* à feu libre
e quemador *m* de llama desnuda
i bruciatore *m* a fuoco libero
n open-vlambrander
d Brenner *m* mit offener Flamme

3557 NAPALM BOMB, exp,mit
OIL BOMB
f bombe *f* napalm
e bomba *f* de aceite, bomba *f* napalm
i bomba *f* napalm
n napalmbom
d Napalmbombe *f*

3558 NAPHTALENE PLANT oic
f installation *f* de naphtaline
e equipo *m* de naftalina
i impianto *m* di naftalina
n naftaline-installatie
d Naphtalinanlage *f*

3559 NAPHTOLATION BATH tex
f bain *m* de naphtolage
e baño *m* de naftolización
i bagno *m* di naftolizzazione
n naftoleerinstallatie
d Naphtolgrundieranlage *f*, Naphtolieranlage *f*

3560 NATURAL DRAFT COOLING TOWER (US), clo
NATURAL DRAUGHT COOLING TOWER (GB)
f tour *f* de réfrigération à tirage naturel
e torre *f* de enfriamiento de tiro natural
i torre *f* di raffreddamento a tiraggio naturale
n koeltoren met natuurlijke trek
d Kühlturm *m* mit natürlichem Zug, Kühlturm *m* mit Selbstzug

3561 NATURAL DRAFT FURNACE (US), htt
NATURAL DRAUGHT FURNACE (GB)
f four *m* à tirage naturel
e horno *m* de tiro natural
i forno *m* a tiraggio naturale
n oven met natuurlijke trek
d Ofen *m* mit natürlichem Zug, Ofen *m* mit Selbstzug

3562 NATURAL GAS PIPELINE gas
f conduite *f* de gaz naturel
e gasoducto *m* de gas natural, tubería *f* de gas natural
i condotta *f* di gas naturale
n aardgasleiding
d Erdgasleitung *f*

3563 NATURAL GAS PLANT gas
f installation *f* de gaz naturel
e equipo *m* de gas natural
i impianto *m* di gas naturale
n aardgasinstallatie
d Erdgasanlage *f*

3564 NATURAL GAS WELL gas
f dépôt *m* de gaz naturel
e pozo *m* de gas natural
i giacimento *m* di gas naturale
n aardgasbel, aardgasbron
d Erdgasquelle *f*

3565 NATURAL SEASONING PLANT bui
f installation *f* de séchage du bois à l'air
e equipo *m* de cura de la madera al aire
i impianto *m* di stagionamento naturale
n houtdrooginstallatie aan de lucht
d Lufttrocknungsanlage *f*, natürliche Ablagerungsanlage *f*

3566 NATURAL URANIUM REACTOR nue
f réacteur *m* à uranium naturel
e reactor *m* de uranio natural
i reattore *m* ad uranio naturale
n reactor met natuurlijk uranium
d Natururanreaktor *m*, Reaktor *m* mit natürlichem Uran

3567 NEEDLE LUBRICATOR lub
f lubrificateur *m* à aiguille
e lubricador *m* de aguja
i lubrificatore *m* a spillo, lubrificatore *m* ad ago
n naaldsmeerapparaat *n*
d Nadelschmiergerät *n*

3568 NEEDLE NOZZLE pla,rub
f gicleur *m* à aiguille
e boquillo *m* de aguja, tobera *f* de aguja
i ugello *m* ad ago
n naaldmondstuk *n*
d Nadeldüse *f*

3569 NEEDLE VALVE cpt
f soupape *f* à pointeau
e válvula *f* cónica, válvula *f* de aguja
i valvola *f* a spillo, valvola *f* a spina conica
n naaldklep
d Nadelventil *n*

3570 NEGATIVE, NEGATIVE MATRIX elc
f matrice *f* négative, négatif *m*
e matriz *f* negativa, negativa *f*
i matrice *f* negativa
n negatieve matrijs
d Negativ *n*, Patrize *f*

3571 NEGATIVE DIE, FEMALE MOLD (US), FEMALE MOULD (GB) pla,rub
f moule *m* fémelle, moule *m* négatif
e molde *m* hembra, molde *m* negativo
i stampo *m* femmina
n negatieve mal
d Negativform *f*

3572 NEGATIVE ELECTRODE elc
f électrode *f* négative
e electrodo *m* negativo
i elettrodo *m* negativo
n negatieve elektrode
d negative Elektrode *f*

3573 NEGATIVE PLATE elc
f plaque *f* négative
e placa *f* negativa
i piastra *f* negativa
n negatieve plaat
d negative Platte *f*

3574 NEGATIVE TERMINAL elc
f borne *f* négative, pôle *m* négatif
e borna *f* negativa, polo *m* negativo
i morsetto *m* negativo
n negatief aansluitcontact *n*, negatieve pool
d negativer Pol *m*

3575 NEON LIQUEFIER gas
f installation *f* de liquéfaction de néon
e equipo *m* de licuación de neón
i impianto *m* di liquefazione di neon
n installatie voor het vloeibaar maken van neon
d Neonverflüssigungsanlage *f*

3576 NEPHELOMETER ins
f néphélomètre *m*
e nefelómetro *m*
i nefelometro *m*
n nefelometer, troebelheidsmeter
d Nephelometer *n*, Trübungsmesser

3577 NEPHELOMETRIC ANALYZER, TURBIDIMETRIC ANALYZER ana
f analyseur *m* néphélométrique
e analizador *m* nefelométrico
i analizzatore *m* nefelometrico
n nefelometrisch analyseerapparaat *n*
d nepholometrisches Analysiergerät *n*

3578 NERVE GAS GENERATOR gas,mit
f générateur *m* de gaz énervant
e generador *m* de gas enervante
i generatore *m* di gas da apoplessia nervosa
n zenuwgasgenerator
d Nervengasgenerator *m*

3579 NESTED ELECTRODE elc
f paquet *m* d'électrodes
e electrodos *pl* en haz
i pacchetto *m* d'elettrodi
n elektrodenpakket *n*
d Elektrodenpaket *n*

3580 NET, SECTION elc
f batterie *f*, système *m* multiple
e batería *f*, sistema *m* múltiple
i batteria *f*
n serieschakeling van cellen
d Badgruppe *f*

3581 NEUTRALIZATION BATH pho
f bain *m* de neutralisation
e baño *m* de neutralización
i bagno *m* di neutralizzazione
n neutraliseerbad *n*
d Neutralisierungsbad *n*

3582 NEUTRALIZATION VESSEL gen
f cuve *f* de neutralisation
e cuba *f* de neutralización
i tino *m* di neutralizzazione
n neutraliseervat *n*
d Neutralisiergefäss *n*

583 NEUTRON-ABSORBING GIASS PLATE nue,sap
f plaque *f* de verre à absorption de neutrons
e placa *f* de vidrio absorbedora de neutrones
i piastra *f* di vetro ad assorbimento di neutroni
n neutronen absorberende glasplaat
d neutronenabsorbierende Glasplatte *f*

584 NICKEL ANODE elc
f anode *f* de nickel
e ánodo *m* de níquel
i anodo *m* di nichel
n nikkelanode
d Nickelanode *f*

585 NICKEL-CADMIUM BATTERY elc
f accumulateur *m* nickel-cadmium
e acumulador *m* níquel-cadmio
i accumulatore *m* nichel-cadmio
n nikkelcadmiumaccu
d Nickel-Kadmiumakkumulator *m*

586 NICKEL-CLAD STEEL TANK sto
f récipient *m* en acier à revêtement de nickel
e tanque *m* de acero chapado de níquel
i serbatoio *m* in acciaio con rivestimento di nichel
n met nikkel beklede stalen tank
d Stahlgefäss *n* mit Nickelauskleidung

587 NICKEL FLASHING BATH elc
f bain *m* de nickelage rapide
e baño *m* de niquelado rápido
i bagno *m* di nichelatura rapida
n snel nikkelbad *n*
d Schnellnickelbad *n*

588 NICKEL PELLET, NICKEL SHOT cle,met
f grain *m* de nickel
e granalla *f* de níquel, pelotilla *f* de níquel
i pallottola *f* di nichel
n nikkelhagel
d Nickelschrot *m*

589 NICKEL PLATING BATH elc
f bain *m* de nickelage
e baño *m* de niquelado
i bagno *m* di nichelatura
n nikkelbad *n*
d Nickelbad *n*

3590 NICKEL STRIKING BATH elc
f bain *m* à couche d'attaque en nickel, bain *m* d'amorçage de nickelage
e baño *m* primario de niquelado
i bagno *m* d'attivazione di nichelatura
n basisnikkellaagbad *n*, voorvernikkelbad *n*
d Vorvernickelbad *n*

3591 NIP gla
f écart *m* de cylindres
e distancia *f* entre los cilindros, pasada *f* de cilindros
i distanza *f* tra cilindri
n walsafstand
d Walzenabstand *m*

3592 NIP ROLLS, NIPPER ROLLS pla,rub
f rouleaux *pl* pinceurs
e rodillos *pl* de compresión, rodillos *pl* estriadores
i rulli *pl* premitori
n drukrollen *pl*
d Haltewalze *f*

3593 NITER BATH (US), NITRE BATH (GB) htt
f bain *m* de salpètre
e baño *m* de salitre
i bagno *m* di salnitro
n salpeterbad *n*
d Salpeterbad *n*

3594 NITER BED (US), NITRE BED (GB) fer
f dépôt *m* de salpètre
e nitrera *f*, salitral *m*
i giacimento *m* di salnitro
n salpeterafzetting
d Salpeterablagerung *f*

3595 NITER FACTORY (US), NITRE FACTORY (GB) fer
f nitrière *f*
e nitrería *f*, salitrería *f*
i fabbrica *f* di salnitro
n salpeterfabriek
d Salpeterfabrik *f*

3596 NITER HOG (US), NITRE REWORKING VESSEL (GB) sua
f cuve *f* à récupération de nitre
e cuba *f* de recuperación de nitro
i tino *m* a ricupero di nitro
n salpeteropwerkvat *n*
d Salpeterwiederaufarbeitungsgefäss *n*

3597 NITRALIZING BATH htt
f bain *m* de nitrate de sodium fondu
e baño *m* de nitrato de sodio fundido
i bagno *m* di nitrato di sodio fuso
n gesmolten-natriumnitraatbad *n*
d geschmolzenes Natriumnitratbad *n*

3598 NITRATION PLANT sua,oic
f installation *f* de nitration
e equipo *m* de nitración
i impianto *m* di nitrazione
n nitratie-installatie
d Nitrationsanlage *f*

3599 NITRATOR oic
f appareil *m* de nitruration
e nitruradora *f*
i apparecchio *m* di nitrurazione
n nitreerapparaat *n*
d Nitriergerät *n*

3600 NITRIC ACID CONCENTRATOR aic
f appareil *m* concentrateur d'acide nitrique
e concentrador *m* de ácido nítrico
i concentratore *m* d'acido nitrico
n concentreerapparaat *n* voor salpeterzuur
d Salpetersäurekonzentrierapparat *m*

3601 NITRIDING BOX met
f caisse *f* de nitruration
e caja *f* de nitruración
i cassetta *f* di nitrurazione
n nitreerkist
d Nitrierkiste *f*

3602 NITRIDING FURNACE met
f four *m* de nitruration
e horno *m* de nitruración
i forno *m* di nitrurazione
n nitreeroven
d Nitrierofen *m*

3603 NITRIDING TOWER, GAY-LUSSAC TOWER sua
f tour *f* de nitruration
e torre *f* de nitruración
i torre *f* di nitrurazione
n nitreertoren
d Nitrierturm *m*

3604 NITRIFICATION PLANT, NITRIFYING PLANT bio
f installation *f* de nitrification
e equipo *m* de nitrificación
i impianto *m* di nitrificazione
n nitrificeerinstallatie
d Nitrifizieranlage *f*

3605 NITROGEN ARC WELDING PLANT, NITROGEN SHIELD ARC WELDING PLANT w
f installation *f* de soudure à arc en azote
e equipo *m* de soldadura a arco en nitrógeno
i impianto *m* di saldatura ad arco in azoto
n booglasinstallatie in stikstofatmosfeer
d Lichtbogenschweissanlage *f* unter Stickstoff

3606 NITROGEN FACTORY ga
f fabrique *f* d'azote
e fábrica *f* de nitrógeno
i fabbrica *f* d'azoto
n stikstoffabriek
d Stickstofffabrik *f*

3607 NITROGEN FIXATION PLANT ga
f installation *f* de fixage de l'azote
e equipo *m* de fijación del nitróg
i impianto *m* di fissazione dell'azoto
n stikstofbindingsinstallatie
d Stickstoffbindungsanlage *f*

3608 NITROGEN LIQUEFIER g
f installation *f* de liquéfaction d'azote
e equipo *m* de licuación de nitrógeno
i impianto *m* di liquefazione d'azoto
n installatie voor het vloeibaar maken van stikstof
d Stickstoffverflüssigungsanlage

3609 NITROGENIZING PLANT aic,o
f installation *f* de nitrogénation
e equipo *m* de nitrogenación
i impianto *m* da nitrogenare
n nitrogeneerinstallatie
d Nitrogenieranlage *f*

3610 NITROGLYCERINE FACTORY e
f fabrique *f* de nitroglycérine
e fábrica *f* de glonoina, fábrica de nitroglicerina
i fabbrica *f* di nitroglicerina
n nitroglycerinefabriek
d Glonoinanlage *f*, Nitroglyzerinanlage *f*, Sprengölanlage *f*

3611 NITROMETER, ins
AZOTOMETER
f azotomètre *m*
e azotímetro *m*, nitrómetro *m*
i azotometro *m*
n azotometer
d Azotometer *n*

3612 NITROUS OXIDE gas,mcy
GENERATOR,
LAUGHING GAS GENERATOR
f générateur *m* de gaz hilarant, générateur *m* de protoxyde d'azote
e generador *m* de gas hilarante
i generatore *m* di gas esilarante, generatore *m* d'ossido nitroso
n lachgasgenerator
d Lachgasgenerator *m*, Stickstoffoxydulgenerator *m*

3613 NODULIZING FURNACE htt,met
f four *m* à former des nodules
e horno *m* de nodular
i forno *m* da formare noduli
n noduleeroven
d Ofen *m* zur Knötchenbildung

3614 NOMOGRAPH ins
f nomographe *m*
e nomógrafo *m*
i nomografo *m*
n nomograaf
d Nomograph *m*

3615 NON-DESTRUCTIVE tes
TESTING APPARATUS
f appareil *m* d'essai non-destructif
e aparato *m* de prueba indestructiva
i collaudatore *m* non distruttivo
n apparaat *n* voor niet-destructieve beproeving
d zerstörungsfreies Prüfgerät *n*

3616 NON-FUSING ARC WELDING wel
ELECTRODE
f électrode *f* réfractaire pour soudure *f* à l'arc
e electrodo *m* refractario para soldadura por arco
i elettrodo *m* refrattario per saldatura ad arco
n niet-afsmeltende booglaselektrode
d nichtabschmelzende Elektrode *f*

3617 NON-LINED CONSTRUCTION, elc
UNWRAPPED CONSTRUCTION
f montage *m* sans habillage
e montaje *m* sin recubrimiento
i montaggio *m* senza rivestimento
n constructie zonder voering
d wickellose Bauweise *f*

3618 NON-RETURN VALVE cpt
f soupape *f* auto-retour, soupape *f* de retenue
e válvula *f* de retenido, válvula *f* de ritención
i valvola *f* di ritegno, valvola *f* di retenuta
n terugslagklep
d Rückschlagventil *n*

3619 NORMAL CALOMEL elc
ELECTRODE,
STANDARD CALOMEL ELECTRODE
f demi-cellule *f* au calomel normale, électrode *f* au calomel normale
e electrodo *m* de calomelano normal, semicelda *f* de calomelano normal
i elettrodo *m* al calomelano normale
n normale calomelelektrode
d Normalkalomelhalbzelle *f*

3620 NORMAL ELEMENT, elc
STANDARD CELL
f pile *f* étalon
e pila *f* patrón
i pila *f* campione
n normaalelement *n*
d Normalelement *n*

3621 NORMALIZATION FURNACE met
f four *m* à normaliser
e horno *m* normalizador
i forno *m* di normalizzazione
n normaliseeroven
d Normalisierofen *m*

3622 NOSE, gla
REFINER
f zone *f* d'affinage
e zona *f* de afino
i zona *f* d'affinazione
n raffineerruimte
d Raffinierraum *m*

3623 NOZZLE, pre
JET
f buse *f*, gicleur *m*, tuyère *f*
e abertura *f*, boquilla *f*, tobera *f*
i imboccatura *f*, spruzzatore *m*, ugello *m*
n sproeier, straalmondstuk *n*,straalpijp
d Düse *f*

3624 NOZZLE ADAPTER pla,pre
f tête *f* du pot
e portaboquilla *m*
i portaugello *m*
n aanspuitbusbevestiging, spuitmatrijshouder
d Angussbuchsenhalter *n*, Düsenpasstück *n*

3625 NOZZLE BLOCK pla,pre
f porte-buse *m*
e bloque *m* de boquilla, portaboquilla *m*
i portaugello *m*
n mondstukbevestiging
d Düsenblock *m*

3626 NULL ELECTRODE elc
f électrode *f* zéro
e electrodo *m* cero
i elettrodo *m* zero
n nulelektrode
d Nullelektrode *f*

3627 NUTRIENT BOTTLE bac,lab
f flacon *m* à nutrition
e frasco *m* de nutrición
i boccetta *f* di nutrizione
n voedingsbodemfles
d Nährbodenflasche *f*

3628 NYLON GAUZE FILTER (GB), NYLON GAZE FILTER (US) cle,fil,tex
f filtre *m* à gaze de nylon
e filtro *m* de gasa nilónica
i filtro *m* a garza di nailon
n nylongaasfilter *n*
d Nylongazefilter *m*

O

3629 OAST HOUSE bre
f salle *f* de séchage du houblon
e sala *f* de secado del lúpulo
i sala *f* d'essiccazione del luppolo
n hopdroogkamer
d Hopfentrockenkammer *f*

3630 OBLIQUE HEAD, ANGLE HEAD, ANGULAR EXTRUDER HEAD pla
f tête *f* de boudineuse à angle, tête *f* de boudineuse oblique
e cabezal *m* de budinadora en ángulo, cabezal *m* de extrusión oblicuo
i testa *f* di trafila ad angolo, testa *f* di trafila obliqua
n dwarse spuitkop, schuine spuitkop
d Schrägspritzkopf *m*

3631 OBLIQUE HEAD, ANGULAR HEAD pla
f tête *f* d'angle, tête *f* oblique
e cabezal *m* angular, cabezal *m* oblicuo
i testa *f* ad angolo, testa *f* obliqua
n dwarse kop, schuine kop
d Schrägkopf *m*

3632 OBNOXIOUS GAS ELIMINATOR sew
f éliminateur *m* de gaz d'odeur repoussante
e eliminador *m* de gases insalubles
i eliminatore *m* di gas puzzolenti
n afvoerinrichting voor kwalijkriekende gassen
d Abfuhr *f* für übelriechende Gase

3633 OBSERVATION PIPE, OBSERVATION TUBE gen
f tuyau *m* d'observation
e tubo *m* de observación
i tubo *m* d'osservazione
n waarnemingsbuis
d Beobachtungsrohr *n*

3634 OBSERVATION WINDOW gen
f fenêtre *f* d'observation
e ventana *f* de observación
i finestra *f* d'osservazione
n waarnemingsvenster *n*
d Beobachtungsfenster *n*

3635 ODOR REMOVAL APPARATUS (US), ODOUR REMOVAL APPARATUS (GB), DEODORIZER, DEODORIZING APPARATUS iwt
f appareil *m* de désodorisation, appareil *m* d'élimination de l'odeur
e aparato *m* de desodorización
i apparecchio *m* di deodorizzazione, apparecchio *m* d'eliminazione dell'odore
n stankverdrijvingsapparaat *n*
d Desodoriergerät *n*, Geruchsbeseitigungsgerät *n*

3636 OENOMETER, VINOMETER ins
f oenomètre *m*, pèse-vin *m*, vinomètre *m*
e enómetro *m*
i enometro *m*
n vinometer, wijnmeter
d Weinmesser *m*, Weinwaage *f*

3637 OIL BARREL, OIL DRUM tra
f tonneau *m* à huile
e bidón *m* de aceite
i barile *m* da olio
n olievat *n*
d Ölfass *n*

3638 OIL BATH oic
f bain *m* d'huile
e baño *m* de aceite
i bagno *m* d'olio
n oliebad *n*
d Ölbad *n*

3639 OIL BOMB, NAPALM BOMB exp,mit
f bombe *f* napalm
e bomba *f* de aceite, bomba *f* napalm
i bomba *f* napalm
n napalmbom
d Napalmbombe *f*

3640 OIL BURNER bur
f brûleur *m* à huile combustible, brûleur *m* à mazout

e quemador *m* de aceite combustible
i bruciatore *m* ad olio combustibile
n oliebrander
d Ölbrenner *m*

3641 OIL CATCHER, cle
OIL COLLECTOR
f collecteur *m* d'huile
e colector *m* de aceite
i raccoglitore *m* d'olio
n olievanger
d Ölfang *m*, Ölfänger *m*

3642 OIL CHANNEL lub
f rainure *f* d'huile
e pata *f* de araña, ranura *f* de engrase
i scanalatura *f* d'olio
n oliegroef
d Ölrille *f*

3643 OIL-COOLED APPARATUS clo
f appareil *m* à refroidissement à l'huile
e aparato *m* enfriado con aceite
i apparecchio *m* raffreddato ad olio
n met olie gekoeld apparaat *n*
d ölgekühlter Apparat *m*

3644 OIL COOLER clo
f refroidisseur *m* à l'huile
e refrigerante *m* de aceite
i raffreddatore *m* all'olio
n oliekoeler
d Ölkühler *m*

3645 OIL COOLER clo
f refroidisseur *m* de l'huile
e enfriador *m* del aceite
i raffreddatore *m* dell'olio
n oliekoeler
d Ölkühler *m*

3646 OIL CRUSHER, fat
OIL MILL
f huilerie *f*, moulin *m* à huile
e almazara *f*, molino *m* de aceite
i oleificio *m*
n oliemolen
d Ölmühle *f*

3647 OIL DIFFUSION PUMP vac
f pompe *f* à diffusion d'huile
e bomba *f* de difusión de aceite
i pompa *f* a diffusione d'olio
n oliediffusiepomp
d Öldiffusionspumpe

3648 OIL EJECTOR PUMP pre
f pompe *f* à éjecteur d'huile
e bomba *f* de eyector de aceite
i pompa *f* ad eiettore d'olio
n olie-ejectorpomp, oliezuigstraalpomp
d Ölejektorpumpe *f*, Ölsaugstrahlpumpe *f*

3649 OIL EXPELLER, fat,pre
OIL PRESS
f pressoir *m* d'huile
e prensa *f* para extraer aceites
i pressa *f* per l'estrazione d'olio
n oliepers
d Ölpresse *f*

3650 OIL FILTER, cle,fil
OIL PURIFIER
f filtre *m* d'huile
e filtro *m* de aceite
i filtro *m* d'olio
n oliefilter *n*
d Ölfilter *m*

3651 OIL-FINISHING TANK tan
f récipient *m* de finissage à l'huile
e tanque *m* de acabado en aceite
i serbatoio *m* di finitura all'olio
n tank voor het nabehandelen in olie
d Behälter *m* zum Nachbehandeln in Öl

3652 OIL-FIRED FURNACE htt
f four *m* chauffé au mazout
e horno *m* de combustión de petróleo
i forno *m* ad olio combustibile
n met olie gestookte oven
d ölgeheizter Ofen *m*

3653 OIL-HARDENING BATH htt,met
f bain *m* de trempe en huile
e baño *m* de temple en aceite
i bagno *m* di tempera in olio
n oliehardingsbad *n*
d Ölhärtungsbad *n*

3654 OIL JACKET clo
f chemise *f* d'huile
e chaqueta *f* de aceite
i camicia *f* d'olio
n oliemantel
d Ölmantel *m*

3655 OIL MIST DETECTOR lub
f signaleur *m* de brouillard d'huile
e detector *m* de niebla de aceite
i rivelatore *m* di nebbia d'olio

n olieneveldetector
d Ölnebelanzeiger *m*

656 OIL PRESSURE RELIEF VALVE cpt,sap
f soupape *f* régulatrice de la pression de l'huile
e válvula *f* de descarga de la presión del aceite
i valvola *f* di regolazione della pressione dell'olio
n oliedrukregelklep
d Öldruckregelventil *n*

657 OIL PUMP lub,pre
f pompe *f* d'huile
e bomba *f* de aceite, bomba *f* de lubricación
i pompa *f* d'olio
n oliepomp
d Ölpumpe *f*

658 OIL REMOVAL FILTER fil
f filtre *m* déshuileur
e filtro *m* desaceitador
i filtro *m* sgrassatore
n olievangfilter *n*
d Entölungsfilter *m*

659 OIL REMOVAL PLANT, OIL SEPARATION PLANT cle
f installation *f* de déshuilage
e equipo *m* desaceitador
i impianto *m* d'eliminazione d'olio, impianto *m* di separazione d'olio
n olieverwijderingsinstallatie
d Entölungsanlage *f*

660 OIL RING, SCRAPER RING lub
f anneau *m* de lubrification
e anillo *m* de lubricación, anillo *m* engrasador
i anello *m* di lubrificazione
n smeergroef
d Schmierrille *f*

661 OIL SEAL cpt
f fermeture *f* à huile
e cierre *m* de aceite
i chiusura *f* ad olio
n olieafsluiting
d Ölverschluss *m*

662 OIL SEPARATOR cle
f séparateur *m* d'huile
e separador *m* de aceite
i separatore *m* d'olio
n olieafscheider
d Ölabscheider *m*

3663 OIL TANK fat
f récipient *m* d'huile
e tanque *m* de aceite
i serbatoio *m* d'olio
n olietank
d Ölbehälter *m*

3664 OIL TESTING CENTRIFUGE cen,fat,lub,tes
f centrifugeur *m* à examiner l'huile
e centrífuga *f* para probar aceites
i centrifuga *f* collaudatrice d'olio
n olieonderzoekcentrifuge
d Ölprüfungszentrifuge *f*

3665 OIL VAPOR-JET PUMP (US), OIL VAPOUR-JET PUMP (GB) vac
f pompe *f* à éjection de vapeur d'huile
e bomba *f* de eyección de vapor de aceite
i pompa *f* ad iniezione di vapore d'olio
n oliedampstraalpomp
d Öldampfstrahlpumpe *f*

3666 OILGAS GENERATOR, CARBO-HYDROGEN GENERATOR fue,gas
f générateur *m* de gaz de pétrole
e generador *m* de gas de petróleo
i generatore *m* di gas di petrolio
n oliegasgenerator, vetgasgenerator
d Fettgasgenerator *m*, Ölgasgenerator *m*

3667 OILING APPARATUS tex
f appareil *m* d'ensimage
e aparato *m* de ensaimado
i apparecchio *m* d'oliatura
n smoutapparaat *n*
d Schmalzapparat *m*

3668 OINTMENT MILL mcy
f mélangeur *m* d'onguent
e mezclador *m* de pomada, mezclador *m* de ungüente
i mescolatore *m* di pomata, mescolatore *m* d'unguento
n zalfmachine
d Salbenmischer *m*

3669 OLEOMETER, ACROMETER ins
f oléomètre *m*
e oleómetro *m*
i oleometro *m*
n oleometer
d Ölmesser *m*, Ölwaage *f*

3670 OLIVE OIL PRESS fat
f pressoir *m* d'olives
e almazara *f*
i pressa *f* d'olive
n olijvenpers
d Olivenpresse *f*

3671 ONE-DAYLIGHT PRESS, pla,pre,rub
SINGLE-DAYLIGHT PRESS
f presse *f* à un étage
e prensa *f* de un solo hueco
i pressa *f* ad un'apertura
n één-etagepers
d Einetagenpresse *f*

3672 ONE-FLOOR KILN, bre
SINGLE-FLOOR KILN
f touraille *f* à un plateau
e tostadero *m* de un plato
i essiccatoio *m* ad un graticcio
n eest met één vloer
d Einhordendarre *f*

3673 ONE-FLUID CELL elc
f pile *f* à un liquide
e pila *f* de electrólito único, pila *f* de un líquido
i pila *f* ad un solo elettrolito
n element *n* met gemeenschappelijke elektrolyt
d Flüssigkeitselement *n*

3674 ONE-MARK PIPETTE lab
f pipette *f* à un seul repère
e pipeta *f* de una traza
i pipetta *f* ad una traccia
n volpipet
d Vollpipette *f*

3675 OPEN ANNEALING OVEN, met
BLACK ANNEALING OVEN
f four *m* de recuit en noir
e horno *m* de recocido en negro
i forno *m* di ricottura in nero
n zwartgloeioven
d Schwarzglühofen *m*

3676 OPEN-CUP FLASH-POINT TESTER tes
f vérificateur *m* du point d'inflammation à coupe ouverte
e probador *m* del punto imflamador en copa abierta
i collaudatore *m* del punto d'infiammabilità a coppa aperta
n vlampuntapparaat *n* met open schaal
d Flammpunktgerät *n* im offenen Tiegel

3677 OPEN FILTER, fi
GRAVITY FILTER
f filtre *m* à gravité, filtre *m* ouve
e filtro *m* abierto, filtro *m* de gravedad
i filtro *m* a gravità, filtro *m* aperto, filtro *m* scoperto
n open filter *n*
d offener Filter *m*

3678 OPEN FIRE KILN br
f touraille *f* à feu direct
e tostadero *m* de fuego directo
i essiccatoio *m* a fuoco diretto
n direct verhitte eest
d Rauchdarre *f*

3679 OPEN-FRONT ECCENTRIC PRESS pr
f presse *f* à excentrique à col de cygne
e prensa *f* excéntrica a cuello de cisne
i pressa *f* con eccentrico a collo di cigno
n zwanehalsexcenterpers
d Einständerexzenterpresse *f*

3680 OPEN MILL, pla,ru
MIXING MILL, MIXING ROLLERS, ROLLER MILL
f malaxeur *m* à cylindres, mélangeur *m* ouvert
e amasadora *f* de cilindros, mezclador *m* abierto, mezclado de cilindros
i mescolatore *m* a cilindri, mescolatore *m* aperto
n mengwals, open mengwals
d offene Mischwalze *f*, Walzenmischer *m*

3681 OPEN POT gl
f creuset *m* ouvert
e crisol *m* abierto
i crogiolo *m* scoperto
n open pot
d offener Hafen *m*

3682 OPEN SAND FILTER iv
f filtre *m* ouvert à sable
e filtro *m* abierto de arena
i filtro *m* aperto a sabbia
n open zandfilter *n*
d offener Sandfilter *m*

3683 OPEN SIDE PRESS, ru
GAP TYPE PRESS, GOOSENEC PRESS, JAW TYPE PRESS, SWANNECK PRESS

f presse *f* à col de cygne, presse *f* ouverte sur un côté
e prensa *f* abierta por un lado, prensa *f* acodillada, prensa *f* en cuello de cisne
i pressa *f* a collo di cigno
n muilpers, zwanehalspers
d einseitig offene Presse *f*, Maulpresse *f*, Schwanenhals-presse *f*

3684 OPEN STEAM OVEN htt,rub
f étuve *f* de vulcanisation à vapeur libre
e estufa *f* de vulcanización a vapor libre
i stufa *f* di vulcanizzazione a vapore libero
n vulcaniseerstoof met vrije stoom
d Freidampfvulkanisationskessel *m*

3685 OPENER, tex
DISINTEGRATING MILL, DISINTEGRATOR, WILLEY, WILLOW
f broyeur *m* à couronnes dentées, loup *m* batteur, ouvreuse *f*
e abridora *f*, batidora *f*, batuar *m*, diablo *m*
i apritoio *m*, lupo *m*
n duivel, snar, wolf
d Klopfwolf *m*, Reisswolf *m*

3686 OPENING SURFACE pla,pre
f surface *f* d'ouverture d'un moule
e superficie *f* de abertura de un molde
i superficie *f* di chiusura di un stampo
n open oppervlak *n* van een matrijs
d Öffnungsfläche *f* eines Werkzeugs

3687 OPERATING PLATFORM, gen
WORKING PLATFORM
f plate-forme *f* de manoeuvre
e plataforma *f* de servicio
i piattaforma *f* di servizio
n bedieningsbordes *n*
d Arbeitsbühne *f*

3688 OPTICAL PYROMETER ins
f pyromètre *m* optique
e pirómetro *m* óptico
i pirometro *m* ottico
n optische pyrometer
d optisches Pyrometer *n*

3689 ORE BIN, min
ORE BUNKER, ORE POCKET
f accumulateur *m* à minerais, silo *m* à minerais
e bolsa *f* de mineral, tolva *f* de mineral
i silo *m* di minerale
n ertsbunker
d Erzbunker *m*, Erztasche *f*

3690 ORE BREAKER, brk,min
MILLING MACHINE
f concasseur *m* de minerais
e triturador *m* de minerales
i frantoio *m* per minerali, rompitore *m*
n ertsbreker
d Erzbrecher *m*

3691 ORE CALCINING FURNACE min
f four *m* à calciner minerais
e horno *m* de calcinar minerales
i forno *m* da calcinare minerali
n ertscalcineeroven
d Erzbrennofen *m*

3692 ORE CONCENTRATION PLANT, min
DRESSING PLANT
f installation *f* d'enrichissement des minerais
e equipo *m* de enriquecimiento de los minerales
i impianto *m* d'arricchimento dei minerali
n ertsconcentreerinstallatie
d Erzanreicherungsanlage *f*

3693 ORE CRUSHER brk,min
f concasseur *m* de minerais
e bocarte *m*, trituradora *f* de minerales
i rompitore *m*, spaccatore *m*
n ertsbreker
d Erzbrecher *m*

3694 ORE DRESSING PLANT (GB), min
BENEFICIATION PLANT (US)
f installation *f* de traitement de minerais
e equipo *m* para beneficio de minerales
i impianto *m* di trattamento di minerali
n ertsbewerkingsinstallatie
d Erzaufbereitungsanlage *f*

3695 ORE GRADING MACHINE, min
ORE SIZING MACHINE
f machine *f* à classifier minerais, trieuse *f*
e máquina *f* para clasificar minerales

i macchina *f* per classificare minerali
n ertsclassificeermachine, ertssorteermachine
d Erzklassierungsmaschine *f*

3696 ORE HEARTH, SCOTCH HEARTH met,min
f four *m* de fusion des minerais de plomb
e horno *m* de fundición para minerales de plomo
i forno *m* per fusione dei minerali di piombo
n loodertssmeltoven
d Bleierzschmelzofen *m*

3697 ORE LEACHING PLANT min
f installation *f* de lixiviation de minerais
e equipo *m* de lixiviación de minerales
i impianto *m* da lisciviare minerali
n ertsuitlooginstallatie
d Erzauslaugeanlage *f*

3698 ORE ROASTING OVEN min
f four *m* à griller les minerais
e horno *m* de tostación de minerales
i forno *m* d'arrostimento di minerali
n ertsroostoven
d Erzröstofen *m*

3699 ORE SCREENER min,sor
f tamis *m* à minerais
e cribador *m* de minerales
i vaglio *m* di minerali
n ertszeef
d Erzsieb *n*

3700 ORE SINTERING OVEN min
f four *m* de frittage de minerais
e horno *m* de sinterización de minerales
i forno *m* da sinterizzare minerali
n ertssinteroven
d Erzsinterofen *m*

3701 ORE SMELTING HEARTH htt,met,min
f sole *f* de fusion de minerais
e solera *f* de fundición de minerales
i suola *f* di fusione di minerali
n ertssmelthaard
d Erzschmelzherd *m*

3702 ORE WASHERY mir
f laverie *f* de minerais
e lavadero *m* de minerales
i laveria *f* di minerali
n ertswasserij
d Erzwäsche *f*

3703 ORGANIC COOLED REACTOR nu
f réacteur *m* à réfrigérant organique
e reactor *m* de enfriador orgánic
i reattore *m* a refrigerante organico
n reactor met organisch koelmidc
d organisch gekühlter Reaktor *m*, Reaktor *m* mit organischem Kühlmittel

3704 ORGANIC MODERATED REACTOR nu
f réacteur *m* à modérateur organique
e reactor *m* de moderador orgáni
i reattore *m* a moderatore organi
n reactor met organische moderator, reactor met organische remstof
d organisch moderierter Reaktor Reaktor *m* mit organischem Moderator

3705 ORIFICE ge
f orifice *m*, ouverture *f*
e abertura *f*, orificio *m*
i apertura *f*, orificio *m*
n opening
d Öffnung *f*

3706 ORIFICE LAND (US), DIE LAND (GB) pla,r
f partie *f* de filière à section constante
e canal *m* de laminado, parte *f* final del canal de la tobera
i zona *f* a sezione costante
n spuitkanaal *n* met constante doorsnede
d Bügelkanal *m*, Bügellänge *f*, Führungskanal *m*

3707 ORIFICE MIXER m
f mélangeur *m* à disques perfor
e mezclador *m* de discos perfor
i mescolatore *m* a dischi perforati
n menger met geperforeerde schijven
d Lochscheibenmischer *m*

3708 ORIFICE PLATE, RETAINING RING pla,pre
f diaphragme *m* de mesure
e anillo *m* de retención
i diaframma *m* di misura
n orifice-meter, stuwring voor drukverhoging
d Messblende *f*, Stauring *m*

3709 ORIFICE RELIEF (US), DIE RELIEF (GB) pla,rub
f embouchure *f* des lèvres de la filière, sortie *f* de la filière
e ensanchamiento *m* de la salida de la tobera, salida *f* de la boquilla
i allargamento *m* della sezione d'uscita, bocca *f* della filiera, rilascio *m*
n extrusieprofiel *n*, mondstukopening, spuitmatrijs
d Düsenaustritt *m*, erweiterte Austrittsöffnung *f*

3710 OSCILLATING CHUTE, JIGGING CONVEYOR, SHAKING SHOOT tra
f couloir *m* à secousses, gouttière *f* à secousses
e canal *m* de sacudidas
i scivolo *m* crivellatore
n schudgoot
d Schüttelrinne *f*

3711 OSCILLATING CONVEYOR tra
f transportateur *m* oscillant
e transportador *m* oscilante
i trasportatore *m* oscillante
n trilband
d Schwingförderband *n*

3712 OSCILLATING MILL brk
f broyeur *m* oscillant, concasseur *m* oscillant
e machacador *m* oscilante
i frantoio *m* oscillante
n trilkogelmolen
d Schwingmühle *f*

3713 OSCILLATING SCREEN sor
f tamis *m* oscillant
e criba *f* oscilante
i vaglio *m* oscillante
n trilzeef
d Schwingsieb *n*

3714 OSCILLATING TABLE min
f table *f* à secousses
e mesa *f* oscilante
i tavola *f* oscillante
n schudtafel
d Schüttelhorde *f*

3715 OSMOMETER ins
f osmomètre *m*
e osmómetro *m*
i osmometro *m*
n osmometer
d Osmometer *n*

3716 OSMOTIC DIFFUSION PUMP vac
f pompe *f* à diffusion osmotique
e bomba *f* de difusión osmótica
i pompa *f* a diffusione osmotica
n osmotische diffusiepomp
d osmotische Diffusionspumpe *f*

3717 OSMOTIC MEMBRANE lab
f membrane *f* osmotique
e membrana *f* osmótica
i membrana *f* osmotica
n osmotisch membraan *n*
d osmotische Membrane *f*

3718 OSMOTIC SEPARATOR sep
f paroi *f* osmotique
e pared *f* osmótica
i parete *f* osmotica
n osmotische wand
d osmotische Wand *f*

3719 OUTFALL, OUTLET sew
f déchargeoir *m*, dégorgeoir *m*
e tubo *m* de salida
i sbocco *m*, scolatoio *m*
n uitlaat
d Ablauf *m*, Auslauf *m*

3720 OUTLET HOPPER, DISCHARGE HOPPER sap
f entonnoir *m* à trop-plein
e gargante *m* de rebose
i imbuto *m* di sfioro
n overlooptrechter
d Ablauftrichter *m*, Schütttrichter *m*

3721 OUTLET VALVE, DISCHARGE VALVE, UNLOADING VALVE cpt
f soupape *f* de décharge
e válvula *f* de descarga
i valvola *f* di scarico
n uitlaatklep
d Ablaufventil *n*, Auslassventil *n*

3722 OUTLET VENT, EXHAUST PIPE, VENT PIPE cpt
f tuyau *m* d'évent
e tubo *m* de escape

i sfiatoio *m*, tubo *m* di sfiato
n afblaaspijp
d Abblasrohr *n*, Dunstrohr *n*

3723 OUTSIDE DIE pre,rub
f matrice *f* extérieure
e matriz *f* exterior
i bocchettone *m* esterno
n buitenmatrijs
d äussere Matrize *f*

3724 OVAL SECTION ANODE, ELLIPTICAL ANODE elc
f anode *f* elliptique
e ánodo *m* elíptico
i anodo *m* ellittico
n ovale anode
d Knüppelanode *f*

3725 OVEN htt,met,min
f four *m*
e horno *m*
i forno *m*
n oven
d Ofen *m*

3726 OVEN PLANT htt
f installation *f* des fours
e equipo *m* de los hornos
i impianto *m* dei forni
n oveninstallatie
d Ofenanlage *f*

3727 OVEN-TYPE FURNACE, SEMI-MUFFLE-TYPE FURNACE htt
f four *m* à chambres
e horno *m* de cámaras
i forno *m* a camere
n kameroven
d Kammerofen *m*

3728 OVERFALL, OVERFLOW, SPILLWAY sew
f déversoir *m*
e vertedero *m*
i sfioratore *m*, stramazzo *m*
n overloop
d Überfall *m*, Überlauf *m*

3729 OVERFEED FUEL BED htt
f lit *m* de combustible alimenté d'en haut
e lecho *m* de combustible alimentado por la parte alta
i letto *m* di combustione con alimentazione dall'alto
n brandstoflaag met bovenvoeding
d Brennstoffschicht *f* mit Oberfeuerung

3730 OVERFLOW CAVITY pre,rub
f chambre *f* à déversage
e cavidad *f* de rebosamiento
i pozzetto *m* sfioratore
n overloopruimte
d Überlaufraum *m*

3731 OVERFLOW CHANNEL, CHANNEL FOR EXCESS RUBBER rub,sa
f couloir *m* de trop plein, dégorgeoir *m*
e canaleta *f* para rebosamiento
i tubo *m* di troppo pieno, tubo *m* sfioratore
n overloopkanaal *n*
d Überlaufrinne *f*

3732 OVERFLOW CONDUIT se
f conduite *f* de trop-plein
e tubería *f* de rebosamiento
i condotta *f* di troppo pieno
n overloopleiding
d Überlaufleitung *f*

3733 OVERFLOW CONE rub,sa
f embout *m*
e cono *m* de descarga
i cono *m* di troppo pieno
n overloopkegel
d Überlaufkonus *m*

3734 OVERFLOW OPENING, OVERFLOW ORIFICE sa
f orifice *m* de trop-plein
e orificio *m* de rebosamiento
i orificio *m* di troppo pieno
n overloopopening
d Ausflussöffnung *f*, Austriebsöffnung *f*

3735 OVERFLOW PIPE (GB), JACK-LEG (US) se
f tuyau *m* de trop-plein
e tubo *m* de rebosamiento
i tubo *m* di troppo pieno
n overlooppijp
d Überlaufrohr *n*

3736 OVERFLOW TRAP pre,ru
f sortie *f* de trop-plein
e lumbrera *f* de rebosamiento
i luce *f* di troppo pieno
n overloopuitlaat
d Überlaufauslass *m*

3737 OVERFLOW VALVE sa
f soupape *f* de trop-plein
e válvula *f* de derrame, válvula de rebosamiento

i valvola *f* di troppo pieno
n overloopklep
d Überlaufventil *n*

3738 OVERPRINTING ROLLER, COVER ROLLER tex
f rouleau *m* de surimpression
e rodillo *m* de sobreimpresión
i rullo *m* di soprastampatura
n overdrukrol
d Aufdruckrolle *f*

3739 OXIDATION AND REDUCTION CELL elc
f pile *f* à oxydation et réduction
e pila *f* de oxidación y reducción
i pila *f* d'ossidazione e riduzione
n oxydatie-reductie-element *n*
d Oxydations-Reduktionsbatterie *f*

3740 OXIDATION APPARATUS, OXIDIZER gen
f appareil *m* d'oxydation
e oxidador *m*
i apparecchio *m* d'ossidazione
n oxydeerapparaat *n*
d Oxydationsapparat *m*, Oxydator *m*

3741 OXIDATION ZONE gen
f zone *f* d'oxydation
e zona *f* de oxidación
i zona *f* d'ossidazione
n oxydatiegebied *n*
d Oxydationsgebiet *n*

3742 OXIDE COATING coa
f couche *f* d'oxyde
e capa *f* de óxido
i strato *m* ossidico
n oxydelaag
d Oxydschicht *f*

3743 OXIDE OF MERCURY CELL, MERCURY OXIDE CELL elc
f pile *f* à oxyde de mercure-zinc
e pila *f* de óxido de mercurio-cinc
i pila *f* ad ossido di mercurio-zinco
n kwikcel
d Quecksilberoxydelement *n*

3744 OXY-ACETYLENE BLOWPIPE, OXY-ACETYLENE BURNER wel
f chalumeau *m* oxyacétylénique
e soplete *m* oxiacetilénico
i cannello *m* ossiacetilenico
n acetyleen-zuurstofbrander
d Azetylen-Sauerstoffbrenner *m*

3745 OXY-ACETYLENE CYLINDER wel
f cylindre *m* à oxyacétylène
e cilindro *m* de oxiacetileno
i cilindro *m* ad ossiacetilene
n acetyleen-zuurstofcilinder
d Azetylen-Sauerstoffzylinder *m*

3746 OXY-ACETYLENE WELDING PLANT wel
f installation *f* de soudure oxyacétylénique
e equipo *m* de soldadura oxiacetilénica
i impianto *m* di saldatura ossiacetilenica
n autogene lasinstallatie
d autogene Schweissanlage *f*

3747 OXY-ARC CUTTING ELECTRODE sep
f électrode *f* pour oxycoupage
e electrodo *m* para oxicorte
i elettrodo *m* per ossitaglio
n boogelektrode met zuurstofkanaal
d Sauerstoffschneidelektrode *f*

3748 OXY-ARC CUTTING PLANT sep
f installation *f* d'oxycoupage à l'arc
e equipo *m* de oxicorte por arco
i impianto *m* d'ossitaglio per arco, impianto *m* di tranciatura ad arco ad ossigeno
n vlamboogsnijinstallatie met zuurstoftoevoer
d elektrische Sauerstoffschneidanlage *f*

3749 OXY-CALORIMETER, OXYGEN BOMB CALORIMETER ins
f oxycalorimètre *m*
e oxicalorímetro *m*
i ossicalorimetro *m*
n zuurstofcalorimeter
d Kalorimeterbombe *f*, Sauerstoffkalorimeter *n*

3750 OXYGEN BOMB TEST APPARATUS tes
f appareil *m* d'essai à la bombe à oxygène
e comprobador *m* con la bomba de oxígeno
i collaudatore *m* con bomba ad ossigeno
n bierer-davisbeproevingsapparaat *n*
d Sauerstoffbombenalterungsprüfgerät *n*

3751 OXYGEN CUTTING PLANT, htt,met
LANCE CUTTING PLANT
f installation *f* d'oxycoupage
e equipo *m* de corte con lanzo, equipo *m* de oxicorte
i impianto *m* d'ossitaglio
n zuurstoflanssnijinstallatie
d Brennschneidanlage *f*

3752 OXYGEN CYLINDER tra
f cylindre *m* à oxygène
e botella *f* de oxígeno, cilindro *m* de oxígeno
i cilindro *m* ad ossigeno
n zuurstofcilinder
d Sauerstoffzylinder *m*

3753 OXYGEN LANCE bur
f lance *f* d'oxygène
e lanza *f* de oxígeno
i lancia *f* d'ossigeno
n zuurstoflans
d Sauerstofflanze *f*

3754 OXYGEN LIQUEFIER gas
f installation *f* de liquéfaction d'oxygène
e equipo *m* de licuación de oxígeno
i impianto *m* di liquefazione d'ossigeno
n installatie voor het vloeibaar maken van zuurstof
d Sauerstoffverflüssigungsanlage *f*

3755 OXYGEN REMOVAL PLANT, iwt
DEOXYGENATION PLANT
f installation *f* de désoxygénation
e equipo *m* de desoxigenación
i impianto *m* di disossigenazione
n zuurstofverwijderingsinstallatie
d Sauerstoffbeseitigungsanlage *f*, Sauerstoffentzuganlage *f*

3756 OXYGEN RESPIRATOR hyg,sap
f respirateur *m* à oxygène
e aparato *m* de respiración artificial
i respiratore *m* ad ossigeno
n zuurstofademhalingsapparaat *n*, zuurstofapparaat *n*
d Sauerstoffapparat *m*

3757 OXYGENATOR imp
f appareil *m* d'oxygénation
e aparato *m* de oxigenación
i apparecchio *m* da ossigenare
n oxygenatieapparaat *n*
d Oxygenierapparat *m*

3758 OXYHYDROGEN BURNER, bur,lab
BLOWTORCH BURNER
f chalumeau *m* oxhydrique
e soplete *m* oxídrico
i cannello *m* ossidrico
n knalgasbrander
d Knallgasgebläse *n*

3759 OZONATOR, gas
OZONIZER
f ozonateur *m*, ozoniseur *m*
e ozonador *m*
i ozonizzatore *m*
n ozonisator
d Ozongerät *n*, Ozoniergerät *n*, Ozonisator *m*

3760 OZONE BATTERY gas
f batterie *f* d'ozoniseurs
e batería *f* de ozonadores
i batteria *f* d'ozonizzatori
n ozonbatterij
d Ozonbatterie *f*

3761 OZONE BLEACHING PLANT bl
f installation *f* de blanchiment à l'ozone
e equipo *m* de blanqueo con ozono
i impianto *m* di candeggio all'ozono
n ozonblekerij
d Ozonbleichanlage *f*

3762 OZONE GENERATOR ga
f générateur *m* d'ozone
e generador *m* de ozono
i generatore *m* d'ozono
n ozongenerator
d Ozonentwickler *m*

3763 OZONE PLANT, gas,hy
OZONIZATION PLANT
f installation *f* d'ozonisation
e equipo *m* de ozonización
i impianto *m* per ozonizzazione
n ozoninstallatie
d Ozonanlage *f*

P

3764 PACK CARBURIZING OVEN, htt,met
BOX CARBURIZING OVEN
f four *m* de cémentation en caisse
e horno *m* de cementación en caja
i forno *m* di cementazione in cassetta
n kistcementeeroven
d Kastenzementierofen *m*

3765 PACK DYEING TROUGH, tex
BALE DYEING TROUGH
f cuve *f* de teinture en balles
e cuba *f* de teñido en balas, cuba *f* de teñido en líos
i tino *m* di tintura in balle
n kuip voor het verven van balen stof
d Ballenfärbeküpe *f*

3766 PACKAGE IRRADIATION PLANT bac,nue
f installation de radioexposition d'objets emballés
e equipo *m* de irradiación de objetos emballados
i impianto *m* d'irradiazione d'oggetti imballati
n bestralingsinstallatie voor verpakte voorwerpen
d Bestrahlungsanlage *f* für verpackte Gegenstände

3767 PACKED COLUMN, dis
PACKED TOWER
f colonne *f* garnie
e columna *f* de relleno, torre *f* de rectificación de relleno
i colonna *f* a riempimento
n kolom met vullichamen
d Füllkörpersäule *f*

3768 PACKED SCRUBBER cle
f tour *f* de ruissellement garnie
e torre *f* lavadera de relleno
i torre *f* di depurazione a riempimento
n sproeikolom met vullichamen
d Füllkörperrieselturm *m*

3769 PACKING, cpt
JOINT
f bourrage *m*, garniture *f*
e empaquetadura *f*
i guarnizione *f*
n pakking
d Packung *f*

3770 PACKING APPARATUS gen
f appareil *m* de compression
e aparato *m* de compresión
i apparecchio *m* di compressione
n comprimeerapparaat *n*
d Pressgerät *n*

3771 PACKING MACHINE gen
f machine *f* à emballer
e máquina *f* empaquetadora
i macchina *f* da imballare
n verpakkingsmachine
d Verpackmaschine *f*

3772 PACKING RING, cpt
GASKET RING, JOINT RING
f bague *f* de garniture
e arandela *f* de guarnición
i anillo *m* di guarnizione
n pakkingring
d Dichtungsring *m*, Manschette *f*

3773 PACKING SUPPORTS dis,lab
f spirales *pl* de support
e espiras *pl* de soporte
i spirali *pl* di sopporto
n steunspiralen *pl*
d Stützspiralen *pl*, Wilson-Wendeln *pl*

3774 PADDER, tex
PADDING MACHINE
f foulard *m*
e fulard *m*, máquina *f* de impregnar telas
i macchina *f* da impregnare
n foulardmachine
d Flatzmaschine *f*, Foulardiermaschine *f*, Klotzmaschine *f*

3775 PADDING MANGLE imp,tex
f calandre *f* à imprégner
e calandra *f* de impregnación
i calandra *f* da impregnare
n impregneermangel
d Imprägniermangel *f*

3776 PADDING TROUGH imp,tex
f cuve *f* d'imprégnation
e cuba *f* de impregnación
i tino *m* da impregnare
n impregneertrog
d Imprägniertrog *m*

3777 PADDLE mix
f palette *f*
e paleta *f*
i pala *f*, paletta *f*, spatola *f*
n roerschoep
d Paddel *f*, Schaufel *f*

3778 PADDLE AERATOR hyg
f appareil *m* d'aération à palettes
e aireador *m* de paletas
i aeratore *m* a pale
n schoepenontluchter
d Schaufelentlüfter *m*

3779 PADDLE AGITATOR mix
f agitateur *m* à palettes
e agitador *m* de paletas
i agitatore *m* a pale
n schoepenroerwerk *n*
d Paddelrührwerk *n*, Schaufelrührwerk *n*

3780 PADDLE MIXER, WING MIXER mix
f mélangeur *m* à palettes
e mezclador *m* de paletas
i mescolatrice *f* a pale
n schoepenmenger
d Paddelmischer *m*, Schaufelmischer *m*

3781 PADDLE VAT, PADDLING TANK tan
f tambour *m* agitateur
e tambor *m* agitador
i tamburo *m* agitatore
n roertrommel
d Rührtrommel *f*

3782 PADDLE WHEEL FAN, CENTRIFUGAL FAN cen,clo
f ventilateur *m* centrifuge
e ventilador *m* centrífugo, ventilador *m* de rueda de palas
i aspiratore *m* centrifugo, ventilatore *m* centrifugo
n centrifugaalventilator
d Zentrifugalventilator *m*

3783 PAINT AGITATOR dec
f agitateur *m* de couleurs
e agitador *m* de pinturas
i agitatore *m* di pitture
n verfroerwerk *n*
d Farbenrührwerk *n*

3784 PAINT BATH, PAINT DIP dec
f bain *m* de couleur
e baño *m* de pintura
i bagno *m* di pittura
n verfbad *n*
d Farbbad *n*

3785 PAINT CURING OVEN dec
f four *m* à moufle de vernis
e horno *m* de mufla de barniz
i forno *m* a muffulo di vernice
n lakmoffeloven
d Lackmuffelofen *m*

3786 PAINT MILL, COLOR MILL (US), COLOUR MILL (GB) dec
f moulin *m* à couleurs, moulin *m* à molette
e molino *m* para colores
i mulino *m* a colori
n verfmolen
d Farbmühle *f*

3787 PAINT SHOP dec
f ligne *f* de vernissage
e línea *f* de barnizado
i linea *f* di verniciatura
n lakstraat
d Lackierstrasse *f*

3788 PAINT-SPRAYING BULB dec
f bombe *f* pulvérisatrice de couleur
e bomba *f* atomizadora de pintura
i bomba *f* polverizzatrice di pittura
n verfspuitbol
d Farbbombe *f*

3789 PAINT-SPRAYING GUN dec
f pistolet *m* pulvérisateur de couleurs
e pistola *f* para pintar
i aerografo *m*, pistola *f* per verniciatura a spruzzo
n verfpistool *n*
d Farbspritze *f*

3790 PALLADINIZING BATH el
f bain *m* à revêtement en palladium
e baño *m* para paladinar
i bagno *m* da rivestire con pallad
n palladiumbad *n*
d Palladiumbad *n*

3791 PALLET tr
f plate-forme *f* de chargement
e bandeja *f*, paleta *f*
i piattaforma *f* d'appoggio
n laadbord *n*
d Ladepritsche *f*, Palette *f*

3792 PALLETTE, BATTLEDORE gla
f batte *f*
e rafrentador *m*
i paletta *f*
n palet *n*
d Handbalett *n*

3793 PAN BALANCE lab
f balance *f* à bassins, balance *f* à plateaux
e balanza *f* de platillos
i bilancia *f* a piattelli
n schaalbalans
d Schalenwaage *f*

3794 PAN DRIER, DRYING PAN dry
f bac *m* sécheur
e cuba *f* secadora
i tino *m* essiccatore
n droogpan
d Trockenpfanne *f*

3795 PAN MILL, BLOCK MILL brk
f moulin *m* à blocs de pierre
e molino *m* con bloques de piedra
i mulino *m* a blocchi di pietra
n stootblokmolen
d Stossblockmühle *f*

3796 PAN MIXER bui,mix
f bétonnière *f* malaxeuse
e hormigonera *f*
i betoniera *f* per malta, molazza *f*
n betonmenger
d Mörtelmischmaschine *f*

3797 PAN ROOM sin
f salle *f* de saturation
e cuarto *m* de los tachos, sala *f* de carbonatación
i sala *f* di carbonatazione
n carbonatatieruimte, kookstation *n*
d Scheidepfannenhaus *n*

3798 PAN ROOM soa
f salle *f* des bassins
e cuarto *m* de los tachos
i sala *f* dei bollitori
n ziederij
d Siederei *f*

3799 PANEL MOLD (US), PANEL MOULD (GB) pla
f moule *m* pour panneaux
e molde *m* para paneles
i stampo *m* per pannelli
n paneelvorm
d Paneelform *f*

3800 PAPER CHROMATOGRAPHY JAR chr
f chambre *f* de chromatographie de Hellmann
e cámara *f* de cromatografía de Hellmann
i camera *f* di cromatografia di Hellmann
n papierchromatografiekamer
d Papierchromatographiekammer *f*

3801 PAPER DRIER CYLINDER dry,pap
f cylindre *m* sécheur de papier
e cilindro *m* secador de papel
i cilindro *m* essiccatore di carta
n papierdroogwals
d Papiertrockenwalze *f*

3802 PAPER DRUM DRIER dry,pap
f tambour *m* sécheur de papier
e secador *m* de tambor para papel
i tamburo *m* essiccatore di carta
n papierdroogtrommel
d Papiertrockentrommel *f*

3803 PAPER FACTORY, PAPER MILL pap
f fabrique *f* de papier, papeterie *f*
e fábrica *f* de papel, papelera *f*
i cartiera *f*, fabbrica *f* di carta
n papierfabriek
d Papierfabrik *f*

3804 PAPER FILTER fil
f filtre *m* à papier
e filtro *m* de papel
i filtro *m* in carta
n papieren filter *n*
d Papierfilter *m*

3805 PAPER-LINED CONSTRUCTION elc
f montage *m* au papier
e montaje *m* recubierto de papel
i montaggio *m* con carta impregnata
n constructie met voering
d Papierfutterbauweise *f*, Papierscheiderbauweise *f*

3806 PAPER MACHINE pap
f machine *f* à fabriquer le papier
e máquina *f* para fabricar papel
i macchina *f* per la carta
n papiermachine
d Papiermaschine *f*

3807 PAPER MAT FILTER, PAPER ROLL FILTER cle,fil
f filtre *m* en rouleaux de papier

e filtro *m* en rolletes de papel
i filtro *m* con rulli di carta
n papiermatfilter *n*
d Papierlufttrommelfilter *m*, Rollmattenluftfilter *m*

3808 PAPER PULP THICKENER pap
f épaississeur *m* de pâte papetière
e espesador *m* de pasta papelera, espesador *m* de pulpa de papel
i concentratore *m* di pasta, concentratore *m* di polpa
n papierpulpindikker
d Papierbreieindicker *m*

3809 PAPER SIZING APPARATUS pap
f appareil *m* d'encollage
e aparato *m* de encolaje
i apparecchio *m* di collatura
n papierlijmapparaat *n*
d Papierleimgerät *n*

3810 PARABOLIC NOZZLE, GERMAN NOZZLE ins
f buse *f* parabolique
e tobera *f* parabólica
i ugello *m* parabolico
n parabolisch mondstuk *n*
d Paraboldüse *f*

3811 PARACHUTE bre
f écumeur *m* automatique, parachute *m*
e espumadera *f*
i separatore *m* del lievito
n afschuimer
d Abheber *m*, Hebeabheber *m*

3812 PARAFFIN STOVE lab
f étuve *f* à paraffine
e estufa *f* de parafina
i stufa *f* a paraffina
n paraffinestoof
d Paraffinofen *m*

3813 PARAFFINING TANK coa
f récipient *m* de paraffination
e tanque *m* de parafinado
i serbatoio *m* di paraffinazione
n paraffineertank
d Paraffiniergefäss *n*

3814 PARALLEL (US), SPACER (GB) pla,pre
f cale-témoin *m*, chandelle *f*, témoin *m* de mise en place
e calce *m*, paralelas *pl*
i ponte *m*, spessore *m*
n opvulstuk *n*, vulstuk *n*
d Distanzblock *m*

3815 PARALLEL-CURRENT CONDENSER pre
f condensateur *m* à flux parallèle
e condensador *m* de flujo paralelo
i condensatore *m* a flusso parallelo
n gelijkstroomcondensor
d Gleichstromkondensator *m*, Gleichstromverdichter *m*

3816 PARCEL PLATING BATH, PARTIAL PLATING BATH elc
f bain *m* à dépôt limité
e baño *m* de depósito limitado
i bagno *m* a deposito limitato
n bad *n* voor gedeeltelijke bedekking
d Teilgalvanisierbad *n*

3817 PARCHMENTIZING TANK pap
f récipient *m* à parcheminer
e tanque *m* para apergaminar
i serbatoio *m* a pergamenizzazione
n perkamenteertank
d Pergamentiergefäss *n*

3818 PARISON MOLD (US), BLANK MOLD (US), BLANK MOULD (GB) gla
f moule *m* de paraison
e molde *m* de parisón
i stampo *m* di sbozzo greggio
n glasklompvorm
d Külbelform *f*

3819 PARKERIZING TANK coa
f récipient *m* de parkérisation
e tanque *m* de parkerización
i serbatoio *m* di parkerizzazione
n parkeriseertank
d Parkerisiergefäss *n*

3820 PARTIAL ROASTING OVEN htt,met
f four *m* de grillage partial
e horno *m* de tostación parcial
i forno *m* di desolforazione parzia
n oven voor het gedeeltelijk roosteren
d Teiröstofen *m*

3821 PARTICLE SIZE APPARATUS ins
f appareil *m* granulométrique
e aparato *m* de granulometría
i apparecchio *m* granulometrico
n korrelgroottebepalingsapparaat
d Korngrössebestimmungsapparat

3822 PARTING PLANT met
f installation *f* d'affinage

e equipo *m* de refinación
i impianto *m* di raffinazione
n scheidingsinstallatie
d Scheideanlage *f*

3823 PARTING VESSEL lab,met
f récipient *m* d'affinage, récipient *m* de départ
e cuba *f* de refinación, cuba *f* de separación
i tino *m* di raffinazione, tino *m* di separazione
n scheidingsvat *n*
d Scheidegefäss *n*

3824 PARTITION WALL gen
f cloison *m* de séparation, mur *m* mitoyen
e pantalla *f* divisoria, pared *f* divisoria, tabique *m*
i muro *m* divisorio, parete *f* divisoria, separatore *m*
n tussenschot *n*, tussenwand
d Trennwand *f*, Zwischenwand *f*

3825 PASSAGE VALVE, RUNNING VALVE cpt
f soupape *f* de passage
e válvula *f* de pasaje
i valvola *f* di passaggio
n doorlaatklep
d Durchlaufventil *n*

3826 PASSIVATING DIP elc
f bain *m* de passivation
e baño *m* de pasivación
i bagno *m* di passivazione
n passiveringsbad *n*
d Passivierbad *n*

3827 PASSIVE ELECTRODE, COLLECTING ELECTRODE coa
f électrode *f* passive
e electrodo *m* inerte, electrodo *m* pasivo
i elettrodo *m* passivo
n passieve elektrode
d passive Elektrode *f*

3828 PASTE MIXER mix
f mélangeur *m* de pâte
e mezclador *m* de pasta
i mescolatore *m* di pasta
n deegmenger, pastamenger
d Pastenmischer *m*

3829 PASTE MOLD (US), PASTE MOULD (GB) gla
f moule *m* à pâte de charbon
e molde *m* de pasta de carbón
i stampo *m* a pasta di carbone
n met kool beklede vorm
d mit Kohle gefuttertes Werkzeug *n*

3830 PASTED PLATE, FAURE PLATE elc
f plaque *f* à oxyde rapporté, plaque *f* Faure
e placa *f* de óxido empastado
i piastra *f* ad ossido riportato
n roosterplaat
d pastierte Platte *f*

3831 PASTEURIZER bac
f pasteurisateur *m*
e pasteurizador *m*
i pastorizzatore *m*
n pasteurisatie-apparaat *n*
d Pasteurisator *m*

3832 PATCHING GUN, AIR-BORNE SEALING GUN, HOT PATCHING GUN cer,htt
f pistole *f* à fermer des pores
e pistola *f* para cerrar poros
i pistola *f* da chiudere pori
n poriëndichtend pistool *n*
d porendichtende Pistole *f*

3833 PATENTING BATH met
f bain *m* de trempe au plomb
e baño *m* de temple al plomo, baño *m* isotérmico
i bagno *m* di patentamento, bagno *m* di tempera al piombo
n patenteerbad *n*
d Patentierbad *n*

3834 PATINATION APPARATUS coa
f appareil *m* d'application d'une patine
e aparato *m* para aplicar una pátina
i apparecchio *m* da applicare una patina
n patineerapparaat *n*
d Patinierapparat *m*

3835 PAY-OFF DRUM coa,pla,tra
f tambour *m* alimentateur
e tambor *m* desenrollador
i tamburo *m* alimentatore
n toevoertrommel
d Zufuhrtrommel *f*

3836 PEAR-SHAPED DISTILLING FLASK dis,lab
f ampoule *f* de distillation piriforme
e matraz *m* de destilación piriforme

i matraccio *m* di distillazione a pera
n peervormige destilleerkolf, puntkolf
d birnenförmiger Destillierkolben *m*, Spitzkolben *m*

3837 PEBBLE BED REACTOR, PEBBLE REACTOR — nue
f réacteur *m* à zone active en forme de boules
e reactor *m* de zona activa de bolas
i reattore *m* a letto di sfere
n bolletjesreactor
d Kugelhaufenreaktor *m*

3838 PEBBLE FILTER — fil
f filtre *m* à galets
e filtro *m* de guijarros
i filtro *m* a ciottoli
n kiezelfilter *n*
d Kieselfilter *m*

3839 PEBBLE HEATER, PEBBLE RECUPERATOR — htt
f échangeur *m* de chaleur à galets
e termoconjeador *m* de guijarros
i scambiatore *m* di calore a ciottoli
n warmteuitwisselaar met kiezelvulling
d Wärmeaustauscher *m* mit Kieselfüllung

3840 PEBBLE MILL, BALL GRINDER, BALL MILL — brk
f broyeur *m* à boules, broyeur *m* à galets, jarre *f* de broyage
e molino *m* de bolas
i mulino *m* a palle, mulino *m* a sassi
n kogelmolen
d Kugelmühle *f*

3841 PEEP-HOLE, SPY GLASS — gen
f regard *m*, trou *m* de tête
e mirilla *f*, ventanillo *m*
i foro *m* di spia
n kijkgaatje *n*
d Schauloch *n*

3842 PEEP-WINDOW — gen
f judas *m*
e ventana *f* de observación
i finestra *f* di spia
n kijkvenster *n*
d Schaufenster *n*

3843 PELLET MILL — pre
f presse *f* de boulettes
e prensa *f* de granulos
i pressa *f* di pallottoline
n korrelpers
d Körnerpresse *f*

3844 PELLET SINTERING PLANT — htt,met
f installation *f* de frittage de nodules
e equipo *m* de sinterizar nódulos
i impianto *m* da sinterizzare noduli
n knobbelsinterinstallatie
d Knötchensinteranlage *f*

3845 PELLETING MACHINE (GB), PREFORMING PRESS (GB), TABLETS PRESS (US) — mcy,pre
f pastilleuse *f*
e prensa *f* para empastillar
i pastigliatrice *f*
n tabletteerpers
d Tablettenpresse *f*

3846 PELLETIZING DISK — mcy
f disque *m* pour former des comprimés
e disco *m* para formación de comprimidos
i disco *m* da formare compresse
n tabletteerschijf
d Tablettierscheibe *f*

3847 PELLETIZING DRUM — mcy
f tambour *m* pour former des comprimés
e tambor *m* para formación de comprimidos
i tamburo *m* da formare compresse
n tabletteertrommel
d Tablettiertrommel *f*

3848 PENETRATION TESTER — tes
f appareil *m* à déterminer la mesure de pénétration
e comprobador *m* penetrante
i collaudatore *m* penetrante
n indringdieptemeter
d Eindringtiefenmesser *m*

3849 PENETROMETER — ins,pla
f pénétromètre *m*
e penetrómetro *m*
i penetrometro *m*
n penetrometer
d Penetrometer *n*

3850 PENTANE CYLINDER gas
f cylindre *m* à pentane
e cilindro *m* de pentano
i cilindro *m* a pentano
n pentaancilinder
d Pentanflasche *f*, Pentanzylinder *m*

3851 PENTANE TANK gas
f récipient *m* à pentane
e tanque *m* de pentano
i serbatoio *m* a pentano
n pentaantank
d Pentanbehälter *m*

3852 PEPTIZATION VESSEL oic
f cuve *f* de peptisation
e cuba *f* de peptización
i tino *m* di peptizzazione
n peptisatievat *n*
d Peptisationsgefäss *n*

3853 PERCHING STAND lea
f chevalet *m* d'étirage
e caballete *m* de suavizar
i cavalletto *m* da stirare
n rekraam *n*
d Reckrahmen *m*

3854 PERCOLATING FILTER, CONTINUOUS FILTER fil,iwt
f filtre *m* percolateur
e filtro *m* percolador
i filtro *m* percolatore
n doorzijgfilter *n*, percolator
d Seiher *m*, Sickerfilter *m*

3855 PERCOLATION VESSEL fil
f cuve *f* de filtration, cuve *f* de percolation
e cuba *f* de filtración, cuba *f* de percolación
i tino *m* di filtrazione, tino *m* di percolazione
n filtreervat *n*, percolator, percoleervat *n*
d Filtriergefäss *n*, Perkolator *m*

3856 PERCOLATOR, IRRIGATOR clo
f asperseur *m*
e regadera *f*
i irrigatore *m*
n sproeier
d Rieseler *m*

3857 PERDISTILLATION PLANT dis
f installation *f* de distillation à membrane semiperméable
e equipo *m* de destilación con membrana semipermeable
i impianto *m* di distillazione a membrana semipermeabile, impianto *m* di perdistillazione
n destilleerinstallatie met semipermeabele membraan
d Destillieranlage *f* mit halbdurchlässiger Membrane

3858 PERFECTING ENGINE, REFINER pap
f épurateur *m*
e depurador *m*
i depuratore *m*
n reinigingsmachine
d Reinigungsmaschine *f*

3859 PERFORATED BASKET cen
f tambour *m* perforé
e cesta *f* perforada
i tamburo *m* perforato
n geperforeerde trommel
d gelochte Trommel *f*

3860 PERFORATED BASKET CENTRIFUGE cen
f centrifugeur *m* à tambour perforé
e centrífuga *f* de cesta perforada
i centrifuga *f* a tamburo perforato
n centrifuge met geperforeerde trommel
d Zentrifuge *f* mit gelochter Trommel

3861 PERFORATED BOTTOM fil
f fond *m* perforé
e fondo *m* perforado
i fondo *m* filtrante
n zeefbodem
d Siebboden *m*

3862 PERFORATED PICKLING BASKET elc
f tamis *m* de décapage
e tamiz *m* de decapaje
i vaglio *m* di decapaggio
n beitszeef
d Beizsieb *n*

3863 PERFORATED PLATE, SLOTTED PLATE sep
f tôle *f* perforée
e chapa *f* perforada
i lamiera *f* perforata
n zeefplaat
d Siebblech *n*

3864 PERFORATED PLATE COLUMN dis
f colonne *f* de distillation à plateaux perforés

e columna *f* de destilación de platos perforados
i colonna *f* di distillazione a piatti perforati
n destillatiekolom met zeefschotels
d Destillationskolonne *f* mit Siebboden

3865 PERFORATED TRAY, SIEVE TRAY dis
f plateau *m* perforé
e plato *m* perforado
i piatto *m* perforato
n zeefschotel
d Siebboden *m*

3866 PERFORATED PORCELAIN PLATE lab
f plaque *f* poreuse en porcelaine
e plato *m* poroso de porcelana
i piatto *m* poroso in porcellana
n poreuse porseleinschijf
d Wittsche Scheibe *f*

3867 PERFORATED-WALL CENTRIFUGE cen
f centrifugeur *m* à paroi perforée
e centrífuga *f* de pared perforada
i centrifuga *f* a parete perforata
n centrifuge met geperforeerde wand
d Lochwandzentrifuge *f*

3868 PERIODIC ANNEALING OVEN htt,met
f four *m* à recuire périodique
e horno *m* de recocer cíclico
i forno *m* di ricottura periodico
n kringloopgloeioven
d Kreislaufglühofen *m*

3869 PERMEABILITY FINESESS TESTER ins,tes
f perméabilimètre *m* de Blaine
e permeabilímetro *m* de Blaine
i permeabilimetro *m* di Blaine
n doorlatendheidsmeter
d Durchlässigkeitsmesser *m*

3870 PERMEABLE MEMBRANE bac,sep
f membrane *f* perméable
e membrana *f* permeable
i membrana *f* permeabile
n permeabele membraan
d durchlässige Membrane *f*, permeabele Membrane *f*

3871 PERMUTITE FILTER, ZEOLITE FILTER fil
f filtre *m* à permutite
e filtro *m* de permutita
i filtro *m* a permutite, filtro *m* a zeolite
n zeolietfilter *n*
d Permutitfilter *m*

3872 PEROXIDE BLEACHING BATH tex
f bain *m* de blanchiment à peroxyd
e baño *m* de blanqueo de peróxido
i bagno *m* di candeggio a perossid
n peroxyde-bleekbad *n*
d Peroxydbleichbad *n*

3873 PERVAPORATION PLANT eva
f installation *f* d'évaporation à membrane semi-perméable
e equipo *m* de evaporación con membrana semipermeable, equipo *m* de pervaporación
i impianto *m* d'evaporazione a membrana semipermeabile
n verdampingsinstallatie met semipermeabele membraan
d Verdampfungsanlage *f* mit halbdurchlässiger Membrane

3874 PESTLE lab
f pilon *m*
e majadéro *m*
i pestello *m*
n stamper
d Keule *f*, Pistill *n*

3875 PET COCK (GB), GAGE COCK (US), GAUGE COC (GB), TEST COCK (GB) boi
f robinet *m* d'essai, robinet *m* de jauge, robinet *m* de niveau d'eau
e grifo *m* de prueba, grifo *m* indicador del nivel
i rubinetto *m* di prova
n proefkraan, waterpeilkraan
d Probierhahn *m*, Wasserstandshahn *m*

3876 PETRI DISH, CULTURE DISH, CULTURE PLATE bac
f boîte *f* de Pétri
e fuente *f* de cultivo
i capsula *f* di cultura, capsula *f* di Petri
n cultuurschotel
d Kulturschale *f*, Petri-Schale *f*

3877 PETROL SEPARATOR (GB), PETROL TRAP (GB), GASOLINE SEPARATOR (US), GASOLINE TRAP (US) fue
f séparateur *m* d'essence
e separador *m* de gasolina

i separatore *m* di benzina
n benzineafscheider
d Benzinabscheider *m*

3878 pH-METER ins,lab
f pH-mètre *m*
e medidor *m* de pH
i misuratore *m* di pH
n pH-meter
d pH-Messer *m*

3879 pH-RECORDER ins,lab
f enregistreur *m* de pH
e registrador *m* de pH
i registratore *m* di pH
n pH-indicator
d pH-Geber *m*

3880 PHARMACEUTICAL GRADUATE, GRADUATE lab
f éprouvette *f* graduée de pharmacie
e frasco *m* graduado, vaso *m* graduado
i bicchiere *m* graduato
n apothekersmaatglas *n*
d Apothekermensur *f*

3881 PHASE-CONTRAST MICROSCOPE ins
f microscope *m* à contraste de phase
e microscopio *m* de contraste de fase
i microscopio *m* a contrasto di fase
n fazecontrastmicroscoop
d Phasenkontrastmikroskop *n*

3882 PHENYLATION PLANT oic
f installation *f* de phénylation
e equipo *m* de fenilación
i impianto *m* di fenilazione
n fenyleerinstallatie
d Phenylieranlage *f*

3883 PHOSGENE GENERATOR gas
f générateur *m* de phosgène
e generador *m* de fosgeno
i generatore *m* di fosgene
n fosgeengenerator
d Karbonylchloridgenerator *m*, Phosgengenerator *m*

3884 PHOSPHATE BATH cle,coa
f bain *m* de phosphate
e baño *m* de fosfato
i bagno *m* di fosfato
n fosfaatbad *n*
d Phosphatbad *n*

3885 PHOTOFLUOROMETER ins
f photofluoromètre *m*
e fotofluorómetro *m*
i fotofluorometro *m*
n fotofluorometer
d Photofluorometer *n*

3886 PICKING TOOL, BURLING IRON, TWEEZER tex
f outil *m* d'épluchage
e desmotadora *f*, despinzadora *f*
i utensile *m* di spuntatura
n nopijzer *n*
d Noppeisen *n*

3887 PICKLING BASKET, PICKLING CRATE elc
f panier *m* de décapage
e cesta *f* de decapaje
i paniere *m* di decapaggio
n beitskorf, dompelmand
d Beizkorb *m*

3888 PICKLING BATH elc
f bain *m* de décapage
e baño *m* de decapaje
i bagno *m* di decapaggio
n beitsbad *n*
d Beizbad *n*

3889 PICKLING BATH lea
f bain *m* de piclage
e baño *m* de piclaje, baño *m* de salmuera ácida
i bagno *m* di salamoia acida
n pikkelbad *n*
d Pickelbad *n*

3890 PICKLING VAT elc,met
f cuve *f* de décapage
e cuba *f* de decapaje
i tino *m* di decapaggio
n beitsvat *n*
d Beizgefäss *n*

3891 PIECE KNOCKOUT (US), EJECTING DEVICE (GB) pla,rub
f dispositif *m* de démoulage, dispositif *m* éjecteur
e dispositivo *m* de eyaculación, expulsor *m*, eyector *m*
i dispositivo *m* d'eiezione, dispositivo *m* d'espulsione
n uitstootinrichting
d Auswerfer *m*, Auswurfvorrichtung *f*

3892 PILL MACHINE (GB), BALLING DEVICE (US) mcy
f pilulier *m*
e máquina *f* de hacer píldoras, pildorero *m*

i dispositivo *m* da fare pillole
n pillenmachine, pillenvormer
d Pillenmaschine *f*

3893 PILLAR PRESS, COLUMN PRESS — pre
f presse *f* à colonnes
e prensa *f* de columnas
i pressa *f* a colonne
n kolompers, zuilenpers
d Säulenpresse *f*

3894 PILOT CELL, REFERENCE CELL, SAMPLE CELL — elc
f accumulateur *m* de référence
e acumulador *m* de referencia
i accumulatore *m* di riferimento
n controleaccu
d Kontrollsammler *m*, Prüfakkumulator *m*

3895 PILOT PLANT, EXPERIMENTAL PLANT — tes
f installation *f* d'essai, installation *f* expérimentale
e equipo *m* experimental
i impianto *m* pilota, impianto *m* sperimentale
n proefinstallatie
d Versuchsanlage *f*

3896 PILOT VALVE, RELAY VALVE — con,cpt
f soupape *f* de commande
e válvula *f* de mando
i valvola *f* di comando
n stuurklep
d Steuerventil *n*

3897 PINCH ROLL — pre,rub
f cylindre *m* d'entraînement
e cilindro *m* de arrastre
i cilindro *m* di trascinamento
n meeneemwals
d Förderwalze *f*

3898 PINCH ROLLER, BRIDLE ROLLER — pap,tex
f rouleau *m* de tension
e rodillo *m* cogedor, rodillo *m* de tensión
i rullo *m* di tensione
n trekrol
d Treibrolle *f*

3899 PIOSCOPE — dai,ins
f pioscope *m*
e pioscopio *m*
i pioscopio *m*
n pioscoop
d Pioskop *n*

3900 PIPE — cpt,tra
f conduite *f*
e conducto *m*
i condotta *f*
n leiding
d Leitung *f*

3901 PIPE BELL (US), PIPE SOCKET (GB) — cp
f collet *m*, manchon *m*, mouffle *m*
e tubo *m* abocardado
i bicchiere *m*
n buismof
d Muffenkelch *m*

3902 PIPE COIL HEATING ELEMENT, PLAIN-TUBE HEATING COIL, PRIME-SURFACE HEATING CC — cle,hy
f réchauffeur *m* d'air en tubes ordinaires
e calentador *m* de aire con tubos ordinarios
i riscaldatore *m* d'aria con tubi ordinari
n gladde-buisluchtverhitter
d Glattrohrheizregister *n*, Lufterhitzer *m* mit Stahlglattrö ren

3903 PIPE COOLER — c
f réfrigérant *m* tubulaire
e enfriador *m* tubular
i raffreddatore *m* tubolare
n buiskoeler
d Röhrenkühler *m*

3904 PIPE EXTRUDER — pla,pr
f presse *f* d'extrusion pour tube
e extrusora *f* para tubos
i estrusore *m* per tubi
n buizenpers
d Röhrenpresse *f*

3905 PIPE EXTRUSION DIE — pla,p
f filière *f* pour tubes
e cabezal *m* de extrusión para tu
i filiera *f* per tubi
n matrijs voor buisextrusie
d Rohrziehform *f*

3906 PIPE JOINT — c
f raccord *m* de conduite
e junta *f* de conducto
i giunzione *f* di condotta
n verbindingsmof
d Leitungsverbindung *f*, Rohrverbindung *f*

3907 PIPE STILL, dis
TUBE STILL
f chaudière *f* tubulaire
e retorta *f* tubular
i forno *m* tubolare
n buisverdamper
d Rohrverdampfer *m*

3908 PIPE-TYPE PRECIPITATOR fil
f filtre *m* électrostatique tubulaire
e filtro *m* electrostático tubular
i filtro *m* elettrostatico tubolare
n buisvormig elektrofilter *n*
d röhrenförmiger Elektrofilter *m*

3909 PIPELINE, tra
CLOSED CONDUIT
f tuyautage *m*, tuyauterie *f*
e conducto *m* cerrado, tubería *f*
i tubazione *f*
n pijpleiding *f*
d geschlossene Leitung *f*

3910 PIPELINE FILTER fil
f filtre *m* de pipe-line
e filtro *m* de oleoducto, filtro *m* de tubería
i filtro *m* d'oleodotto
n olieleidingfilter *n*
d Ölleitungsfilter *m*

3911 PIPETTE lab
f pipette *f*
e pipeta *f*
i pipetta *f*
n pipet
d Pipette *f*

3912 PIPETTE CONTROL lab
f appareil *m* à pipetter
e aparato *m* de pipetar
i apparecchio *m* da pipettare
n pipetteerapparaat *n*
d Pipettierhilfe *f*

3913 PISTON COMPRESSOR pre
f compresseur *m* à piston
e compresor *m* de pistón
i compressore *m* a pistone
n zuigercompressor
d Kolbenverdichter *m*

3914 PISTON MANOMETER, ins
DEAD-WEIGHT PRESSURE GAGE (US), DEAD-WEIGHT PRESSURE GAUGE (GB)
f manomètre *m* à piston
e manómetro *m* de pistón
i manometro *m* a pistone
n zuigermanometer
d Kolbendruckmesser *m*, Kolbenmanometer *n*

3915 PISTON PACKING cpt
f garniture *f* de piston
e empaquetadura *f* de pistón
i guarnizione *f* di pistone
n zuigerpakking
d Kolbenpackung *f*

3916 PISTON PUMP pre
f pompe *f* à piston
e bomba *f* de pistón
i pompa *f* a pistone, pompa *f* a stantuffo
n zuigerpomp
d Kolbenpumpe *f*

3917 PISTON RING cpt
f anneau *m* de piston
e aro *m* de pistón
i anello *m* di pistone
n zuigerring
d Kolbenring *m*

3918 PISTON ROD pre
f tige *f* de piston
e barra *f* de pistón, vástago *m* de pistón
i asta *f* di stantuffo, braccio *m* di stantuffo
n zuigerstang
d Kolbenstange *f*

3919 PISTON-TYPE PIPETTE lab
f pipette *f* à piston
e pipeta *f* de pistón
i pipetta *f* a pistone
n zuigerpipet
d Saugkolbenpipette *f*

3920 PISTON VALVE pre
f piston *m* distributeur, tiroir *m* cylindrique
e distribuidor *m* cilíndrico
i saracinesca *f* a pistone
n zuigerklep
d Kolbenschieber *m*

3921 PIT tan
f fosse *f*
e foso *m*
i fossa *f*
n kuil
d Grube *f*

3922 PIT, iwt
REFUSE PIT
f fosse *f* à ordures

e foso *m* de basuras
i fossa *f* d'immondizie
n vuilnisput
d Müllbunker *m*, Müllgrube *f*

3923 PITCH REMOVER bre,cle
f dépoisseur *m*
e desembreadora *f*
i spazzola *f* d'acciaio per levare la pece dal salvadoghe
n teermes
d Pechausreiber *m*

3924 PITCHED BLADE PROPELLER mix
f roue *f* motrice à palettes inclinées
e rueda *f* móvil de paletas inclinadas
i girante *m* a pale inclinate
n propeller met schuin gestelde bladen
d Propeller *m* mit schrägen Flügeln

3925 PITCHING MACHINE coa
f goudronneuse *f*
e embreadora *f*
i impeciatrice *f*
n pikmachine
d Pichapparat *m*

3926 PITCHING TUB, STARTING VESSEL bre
f cuve *f* guilloire
e cuba *f* de reposo, cuba *f* de siembra
i tino *m* per semina
n aanzetkuip
d Anstellbottich *m*, Sammelbottich *m*

3927 PITS, ETCHING PITS met
f piqûres *pl*
e hoyos *pl* de ataque químico, picaduras *pl*
i figure *pl* d'attacco
n putjes *pl*
d Narben *pl*, Poren *pl*

3928 PLAIN EXTRACTOR BODY ext
f corps *m* d'extracteur simple
e cuerpo *m* de extracción simple
i corpo *m* d'estrazione semplice
n enkelvoudig extractielichaam *n*
d einfacher Extraktionskörper *m*

3929 PLAIN ROLLER, STARCH ROLLER tex
f rouleau *m* à amidon
e rodillo *m* de almidón
i rullo *m* ad amido
n verfstofafneemrol
d Farbstoffabstreichrolle *f*

3930 PLAIN-TUBE HEATING COIL, PIPE COIL HEATING ELEMENT, PRIME-SURFACE HEATING COIL cle,hy
f réchauffeur *m* d'air en tubes ordinaires
e calentador *m* de aire con tubos ordinarios
i riscaldatore *m* d'aria con tubi ordinari
n gladde-buisluchtverhitter
d Glattrohrheizregister *n*, Lufterhitzer *m* mit Stahlglattröhren

3931 PLAINING OVEN gl
f four *m* d'affinage
e horno *m* de afino
i forno *m* d'affinazione
n fijnsmeltoven
d Blankschmelzofen *m*, Feinschmelzofen *m*

3932 PLAN SIFTING MACHINE, PLANSIFTER nut,sc
f caisse *f* à tamis plats
e caja *f* de tamices planos
i cassetta *f* a stacci piani
n vlakzeef
d Flachsieb *n*, Plansichter *m*

3933 PLANET WHEEL MIXER, EPICYCLIC MIXER m
f mélangeur *m* épicycloïdal
e mezclador *m* de engranajes planetarios
i mescolatore *m* a ruota epicicloidale
n planeetwielmenger
d Planetenrührwerk *n*

3934 PLANETARY-MOTION AGITATOR m
f agitateur *m* planétaire
e agitador *m* planetario
i agitatore *m* planetario
n roerwerk *n* met planeetaandrijving
d Planetenrührwerk *n*

3935 PLANT, INSTALLATION g
f établissement *m*, installation *f*
e equipo *m*, instalación *f*

i impianto *m*, installazione *f*
n inrichting, installatie
d Anlage *f*

3936 PLANTÉ PLATE, elc
PLATE WITH A LARGE AREA
f plaque *f* à grande surface, plaque *f* Planté
e placa *f* de gran superficie, placa *f* Planté
i piastra *f* a grande superficie, piastra *f* Planté
n groot-oppervlakplaat, plaat met groot oppervlak
d Grossoberflächenplatte *f*

3937 PLASTER OF PARIS MOLD (US), fon
PLASTER OF PARIS MOULD (GB)
f moule *m* de plâtre
e molde *m* de yeso
i forma *f* di gesso
n gipsvorm
d Gipsform *f*

3938 PLASTIC PIPE, cpt
PLASTIC TUBE
f tuyau *m* de résine synthétique, tuyau *m* plastique
e tubo *m* plástico
i tubo *m* di resina plastica, tubo *m* plastico
n kunstharsbuis, plastiekbuis
d Kunstharzrohr *n*, Kunststoffrohr *n*

3939 PLASTIC PIPING tra
f tuyauterie *f* de résine artificielle
e tubería *f* de resina sintética
i tubatura *f* di resina artificiale
n plastiekbuisleiding
d Kunststoffrohrleitung *f*

3940 PLASTIC YIELD TESTER, pla,tes
YIELD POINT TESTER, YIELD VALUE TESTER
f appareil *m* d'essai de contrainte critique
e comprobador *m* de límite de fluidez
i collaudatore *m* di limite di fluidità
n vloeigrensbepalingsapparaat *n*
d Fliessgrenzeprüfgerät *n*

3941 PLASTICATOR, pla,rub
PLASTIFICATOR, MASTICATOR
f masticateur *m*, plastificateur *m*
e masticador *m*, plastificador *m*
i masticatore *m*, plastificatore *m*
n plasticeermachine
d Mastikator *m*, Plastikator *m*, Plastifiziermaschine *f*

3942 PLASTICIZER TANK coa
f récipient *m* de plastification
e tanque *m* de plastificación
i serbatoio *m* di plastificazione
n plastificeertank
d Plastifizierbehälter *m*

3943 PLASTIMETER, ins
PLASTOMETER, CONSISTOMETER, RHETOMETER
f consistomètre *m*, plastomètre *m*
e consistómetro *m*, plastómetro *m*
i consistometro *m*, plastometro *m*
n consistometer, plastometer
d Konsistenzmesser *m*, Plastometer *n*

3944 PLATE elc
f plaque *f*
e placa *f*
i piastra *f*
n plaat
d Platte *f*

3945 PLATE, dis
TRAY
f plateau *m*
e plato *m*
i piatto *m*
n schotel
d Boden *m*

3946 PLATE ANODE elc
f anode *f* en planche, anode *f* en plaque, anode *f* plate
e ánodo *m* de placa
i anodo *m* in piastra
n plaatanode
d Kantenanode *f*, Plattenanode *f*

3947 PLATE COLUMN des
f colonne *f* à plateaux
e columna *f* de platos, columna *f* de platillos
i colonna *f* a piatti
n schotelkolom
d Bodenkolonne *f*, Bodensäule *f*

3948 PLATE COOLER, bre,clo
SURFACE COOLER
f refroidisseur *m* à plaques
e enfriador *m* de placas
i raffreddatore *m* a piastre
n platenkoeler
d Plattenkühler *m*

3949 PLATE DISTANCE dis
f écartement *m* des plateaux
e distancia *f* de las placas, distancia *f* de los platillos
i distanza *f* tra piatti
n schotelafstand
d Bodenabstand *m*

3950 PLATE FEEDER tra
f alimentateur *m* à palettes
e alimentador *m* de paletas
i alimentatore *m* a pale
n vulinstallatie met platen-transporteur
d Füllvorrichtung *f* mit Platten-förderband

3951 PLATE MAGNET sep
f plaque *f* magnétique
e placa *f* magnética
i piastra *f* magnetica
n magnetische plaat, plaatmagneet
d magnetische Platte *f*, Plattenmagnet *m*

3952 PLATE SUPPORT elc
f support *m* de plaque
e soporte *m* de placa
i sopporte *m* di piastra
n plaathouder
d Plattenhalter *m*

3953 PLATE-TYPE PRECIPITATOR fil
f filtre *m* électrostatique à plaques
e filtro *m* electrostático de placas
i filtro *m* elettrostatico a lamine
n plaatelektrofilter *n*
d Plattenelektrofilter *m*

3954 PLATED FRAME PLATE, FRAME PLATE, MASS-TYPE PLATE elc
f plaque *f* à cadre, plaque *f* à masse
e placa *f* de cuadro, placa *f* de masa
i piastra *f* a massa, piastra *f* a telaio
n massaplaat
d Rahmenplatte *f*

3955 PLATEN pla,pre
f plateau *m*
e plato *m*
i piano *m*
n etage
d Etage *f*

3956 PLATEN AREA pla,pre,rub
f surface *f* de montage, surface *f* de plateau
e superficie *f* del plato
i superficie *f* di piano
n montageplaat, opspanplaat
d Aufspannplatte *f*, Montageplatte

3957 PLATFORMING PLANT fue
f installation *f* de réformation avec catalysateur de platine
e equipo *m* de reformación con platina
i impianto *m* di riformazione con platina
n platformeerinstallatie
d Platformieranlage *f*

3958 PLATING BARREL elc
f tonneau *m* de galvanoplastie
e tonel *m* para galvanoplastia
i tamburo *m* di galvanoplastica
n galvaniseertrommel
d Galvanisiertrommel *f*

3959 PLATING GENERATOR el
f générateur *m* de galvanoplastie
e generador *m* de galvanoplastia
i generatore *m* per galvanoplastic
n galvaniseerdynamo
d Galvanisiergenerator *m*

3960 PLATING RACK el
f support *m* d'accrochage
e gancho *m* de asimiento
i sospensione *f*
n galvaniseerrek *n*
d Einhängegestell *n*

3961 PLATINIZING BATH el
f bain *m* de platinage
e baño *m* de platinado
i bagno *m* da platinare
n platinabad *n*
d Platinbad *n*

3962 PLODDER so
f peloteuse-boudineuse *f*
e estruidor *m*
i trafila *f*
n stangenpers
d Strangpresse *f*

3963 PLUNGER cp
f plongeur *m*
e émbolo *m* buzo
i stantuffo *m* tuffante
n plunjer
d Mönchkolben *m*, Plunger *m*, Tauchkolben *m*

3964 PLUNGER (GB), fou
BELL (US)
f cloche *f*
e campana *f*
i campana *f*
n dompelklok
d Tauchglocke *f*

3965 PLUNGER FEEDER tra
f alimentateur *m* à plongeur
e alimentador *m* de émbolo buzo
i alimentatore *m* a stantuffo tuffante
n plunjervoeder
d Tauchkolbenspeiser *m*

3966 PLUNGER MOLDING PRESS (US), pre,rub
PLUNGER MOULDING PRESS (GB)
f presse *f* à plongeur d'injection
e prensa *f* con émbolo buzo de inyección
i pressa *f* con stantuffo tuffante d'iniezione
n injectieplunjerpers, spuitgietpers
d Spritzgusspresse *f*, Spritzkolbenpresse *f*

3967 PLUNGER RETAINER PLATE (GB), pla
FORCE PLATE (US)
f plaque *f* porte-poinçon
e placa *f* de alojamiento de los machos
i piastra *f* portapunzone
n bevestigingsplaat
d Einsatzfutter *n*

3968 PLUNGING SIPHON, nut
SAMPLING TUBE
f tâte-vin *m*
e catavinos *m*
i alzavino *m*
n steekhevel
d Stechheber *m*

3969 PLUTONIUM REACTOR nue
f réacteur *m* à plutonium
e reactor *m* de plutonio
i reattore *m* a plutonio
n plutoniumreactor
d Plutoniumreaktor *m*

3970 PNEUMATIC CLASSIFIER sor
f trieuse *f* pneumatique
e clasificador *m* neumático
i classificatore *m* pneumatico
n persluchtsorteerinrichting
d Luftstromsichter *m*, Pressluftsichter *m*

3971 PNEUMATIC CONCENTRATOR ana
f concentrateur *m* pneumatique
e concentrador *m* neumático
i concentratore *m* pneumatico
n concentreerinrichting met perslucht
d Pressluftkonzentriervorrichtung *f*

3972 PNEUMATIC CONVEYOR tra
f transporteur *m* pneumatique
e transportador *m* neumático
i trasportatore *m* pneumatico
n transportinrichting met perslucht
d Pressluftförderanlage *f*

3973 PNEUMATIC DIPPING DEVICE rub
f appareil *m* pneumatique à tremper
e aparato *m* neumático de inmersión
i apparecchio *m* pneumatico ad immersione
n persluchtdompelinstallatie
d Presslufttauchvorrichtung *f*

3974 PNEUMATIC DRIER dry
f sécheur *m* pneumatique
e secadero *m* neumático
i essiccatoio *m* pneumatico
n drooginrichting met perslucht
d Presslufttrockenanlage *f*

3975 PNEUMATIC FLOTATION MACHINE min
f machine *f* de flottation pneumatique
e máquina *f* de flotación neumática
i macchina *f* di flottazione pneumatica
n met perslucht werkende flotatiemachine
d Pressluftflotationsmaschine *f*

3976 PNEUMATIC LOADER tra
f chargeur *m* pneumatique
e cargador *m* neumático
i caricatore *m* pneumatico
n pneumatische laadinrichting, pneumatische lader
d pneumatische Ladeeinrichtung *f*

3977 POACHER, pap
BLEACHING HOLLANDER,
POTCHER
f hollandais *m* à blanchiment
e batidora *f* de blanqueo
i olandese *m* ad imbiancatura

n bleekhollander
d Bleichholländer *m*

3978 POCKET FEEDER tra
f alimentateur *m* à palettes, alimentateur *m* à tambour
e alimentador *m* de paletas, alimentador *m* de tambor
i alimentatore *m* a pale, alimentatore *m* a tamburo
n schoepentoevoerinrichting, trommeltoevoerinrichting
d Schaufelspeiser *m*, Trommelspeiser *m*

3979 POCKET PLATE elc
f plaque *f* à pochettes
e placa *f* de alvéolos
i piastra *f* a taschini
n pakketjesplaat
d Taschenplatte *f*

3980 POISON GAS GENERATOR, WAR GAS GENERATOR gas,mit
f générateur *m* de gaz asphyxiants, générateur *m* de gaz toxiques
e generador *m* de gases tóxicos
i generatore *m* d'aggressivi chimici
n gifgasgenerator
d Giftgasentwickler *m*

3981 POISONING COMPUTER ins,nue
f calculatrice *f* de l'empoisonnement
e calculadora *f* de envenenamiento
i calcolatrice *f* d'avvelenamento
n vergiftigingscomputer
d Vergiftungsrechenmaschine *f*

3982 POLARIMETER ins
f polarimètre *m*
e polarímetro *m*
i polarimetro *m*
n polarimeter
d Polarimeter *n*

3983 POLARIMETER TUBE ins,lab
f tube *m* de polarimètre
e tubo *m* de polarímetro
i tubo *m* di polarimetro
n polarimeterbuis
d Polarimeterrohr *n*

3984 POLARIZATION FILTER, QUARTER-WAVE FILTER ins
f lamelle *f* quart d'onde
e laminilla *f* de cuarto de onda
i lamina *f* quarto d'onda
n kwartgolflamel, polarisatiefilter *n*
d Filterpolarisator *m*, Polarisationsfilter *m*

3985 POLARIZING MICROSCOPE ins
f microscope *m* à polarisation
e microscopio *m* de polarización
i microscopio *m* a polarizzazione
n polarisatiemicroscoop
d Polarisationsmikroskop *n*

3986 POLING PLANT gla,htt,met
f installation *f* de perchage
e equipo *m* de pertigado
i impianto *m* di trattamento con legno verde
n poolinstallatie
d Polungsanlage *f*

3987 POLISHED CYLINDER ROLL pla,pre,rub
f cylindre *m* rectifié à la meule
e cilindro *m* rectificado
i cilindro *m* passato a rettifica
n nageslepen walsrol
d nachgeschliffene Walze *f*

3988 POLISHING BARREL (US), POLISHING DRUM (GB) po
f tonneau *m* de polissage
e tambor *m* de pulido
i tamburo *m* di pulitura
n polijsttrommel
d Poliertrommel *f*

3989 POLISHING BOB, BOB po
f touffe *m* polisseur
e bruñidor *m*
i brunitoio *m*
n polijstpop
d Schwabbel *f*

3990 POLISHING HEAD, POLISHING LATHE po
f ensemble *m* de polissage
e cabezal *m* pulidor
i gruppo *m* da lucidare
n polijstaggregaat *n*
d Poliergerüst *n*, Polierständer *m*

3991 POLISHING LAP, DOLLY, MOP po
f disque *m* en lisière de tissu, meule *f* flexible en chiffon
e bruñidor *m* de trapos
i brunitoio *m* a pezzi di tessuto
n lappenschijf
d Schwabbelscheibe *f*

3992 POLISHING MACHINE, GRINDING MACHINE — pol
f machine *f* à polir, polissoir *m*
e máquina *f* de pulir
i lucidatrice *f*
n polijstmachine
d Poliermaschine *f*

3993 POLISHING STICK — pol
f astic *m*
e costa *f*
i bisegolo *m*, bussetto *m*
n likbeen *n*, likhout *n*
d Glättholz *n*

3994 POLISHING WHEEL, BUFFING WHEEL — pol
f disque *m* à polir
e disco *m* pulidor
i ruota *f* lucidatrice
n polijstschijf
d Polierscheibe *f*

3995 POLISSOIR — gla,pol
f disque *m* à polir revêtu de feutre
e disco *m* pulidor con revestimiento de fieltre
i disco *m* da lucidare con rivestimento di feltro
n met vilt beklede polijstschijf
d filzbekleideter Holzpuffer *m*

3996 POLYELECTRODE, MULTIPLE ELECTRODE — elc
f électrode *f* multiple, polyélectrode *f*
e electrodo *m* múltiple, polielectrodo *m*
i elettrodo *m* multiplo
n meervoudige elektrode
d Mehrfachelektrode *f*

3997 POLYMERIZATION PLANT — oic
f installation *f* de polymérisation
e equipo *m* de polimerización
i impianto *m* di polimerizzazione
n polymerisatie-installatie
d Polymerisationsanlage *f*

3998 POLYMERIZATION REGULATOR — con,oic
f régulateur *m* de la polymérisation
e regulador *m* de la polimerización
i regolatore *m* della polimerizzazione
n polymerisatieregelaar
d Polymerisationsregler *m*

3999 POLYMERIZING STOVE, BAKING STOVE — htt,tex
f étuve *f* à polymériser
e estufa *f* de polimerización
i stufa *f* di polimerizzazione
n polymeriseerstoof
d Polymerisierofen *m*

4000 PONTIL, PUNTY — gla
f pontil *m*
e pontil *m*, puntel *m*, puntero *m*
i puntello *m*
n hechtijzer *n*, punteerijzer *n*
d Hefteisen *n*

4001 PONYMASHER, CONVERTER, FOREMASHER, PREMASHER — bre
f hydrateur *m*
e premezclador *m*
i premescolatore *m*
n voorbeslagkuip
d Vormaischer *m*

4002 PONY MIXER (GB), CHANGE CAN MIXER (US) — mix
f mélangeur *m* change-bidons
e mezclador *m* con cubas intercambiables
i mescolatore *m* a bidoni intercambiabili
n wisselvatmenger
d Mischer *m* mit auswechselbaren Kannen

4003 POP VALVE — cpt,sap
f soupape *f* de sûreté
e válvula *f* de seguridad de descarga rápida
i valvola *f* di sicurezza
n veiligheidsklep
d Sicherheitsventil *n*

4004 POPETTE (GB), BOOT (US), COVER STONE — gla
f brique *f* de couverture
e ladrillo *m* de recubrimiento
i mattone *m* di copertura
n afdeksteen
d Abdeckstein *m*

4005 POPPET VALVE, AIR INTAKE VALVE, AIR SNIFFLER, MUSHROOM VALVE — cpt
f reniflard *m*
e válvula *f* de entrada de aire, válvula *f* de vástago
i valvola *f* a fungo, valvola *f* d'insufflazione
n snuifklep
d Schnarchventil *n*, Schnüffelventil *n*

4006 PORCELAIN BARREL MILL, PORCELAIN POT MILL brk
f broyeur *m* à boulets de porcelaine
e molino *m* de bolas de porcelana
i frantoio *m* a palle di porcellana
n porseleinpotmolen
d Porzellankugelmühle *f*

4007 PORCELAIN CRUCIBLE lab
f creuset *m* en porcelaine
e crisol *m* de porcelana
i crogiolo *m* di porcellana
n porseleinen kroes
d Porzellantiegel *m*

4008 PORCELAINE EVAPORATING DISH lab
f capsule *f* en porcelaine
e cápsula *f* de porcelana
i capsula *f* essiccativa di porcellana
n porseleinen afdampschaaltje *n*
d Porzellanabdampfschale *f*

4009 PORCELAIN FILTER fil
f filtre *m* en porcelaine
e filtro *m* de porcelana
i filtro *m* di porcellana
n porseleinen filter *n*
d Porzellanfilter *m*

4010 PORCELAIN MORTAR lab
f mortier *m* en porcelaine
e mortero *m* de porcelana
i mortaio *m* di porcellana
n porseleinen mortier, porseleinen vijzel
d Porzellanmörser *m*

4011 PORCELAIN PAINT GRINDER brk,dec
f broyeur *m* en porcelaine de peintures
e molino *m* de porcelana de pinturas
i macinatore *m* di porcellana per pitture
n verfmolen met porseleinen pot
d Porzellanfarbmühle *f*

4012 PORCELAIN PIPING tra
f tuyauterie *f* de porcelaine
e tubería *f* de porcelana
i tubatura *f* di porcellana
n porseleinen buisleiding
d Porzellanrohrleitung *f*

4013 PORCUPINE, GO-DEVIL cle
f hérisson *m*
e raspatubos *m*
i raschiatore *m* per tubi
n buiskrabber, buisschraper
d Molch *m*, Rohrwischer *m*

4014 POROMETER ins
f poromètre *m*
e porímetro *m*
i porometro *m*
n porometer
d Porometer *n*

4015 POROSIMETER ins
f porosimètre *m*
e porosímetro *m*
i porosimetro *m*
n porosimeter
d Porosimeter *n*

4016 POROUS CARBON AERATOR hyg
f aérateur *m* à charbon poreux
e aerador *m* de carbón poroso
i aeratore *m* a carbone poroso
n ontluchter met poreuze kool
d Entlüfter *m* mit poröser Kohle

4017 POROUS CASE, POROUS JACKET elc
f chemise *f* poreuse
e camisa *f* porosa
i camicia *f* porosa
n poreuze huls
d poröse Gehäuse *n*

4018 POROUS FILTER PLATE fi
f plaque *f* filtrante poreuse
e placa *f* filtrante porosa
i piastra *f* filtrante porosa
n poreuze filterplaat
d poröse Filterplatte *f*

4019 POROUS POT el
f pot *m* poreux
e olla *f* porosa
i vaso *m* poroso
n poreuze pot
d poröser Topf *m*

4020 PORT cpt,fu
f orifice *m*
e orificio *m*
i orificio *m*
n opening
d Öffnung *f*

4021 PORTABLE BATTERY el
f batterie *f* transportable
e batería *f* transportable
i batteria *f* portatile
n verplaatsbare batterij
d ortsbewegliche Batterie *f*

4022 PORTABLE MOULD (GB), HAND MOLD (US), HAND MOULD (GB), LOOSE MOLD (US) pla
f moule *m* à main
e molde *m* a mano
i stampo *m* a mano
n handmatrijs
d Handform *f*

4023 POSITIVE, POSITIVE MATRIX elc
f matrice *f* positive, positif *m*
e matriz *f* positiva, positiva *f*
i matrice *f* positiva
n positieve matrijs
d Positiv *n*, positive Matrize *f*

4024 POSITIVE DIE, MALE DIE pla,rub
f moule *m* mâle
e molde *m* macho, molde *m* positivo
i stampo *m* maschio
n positieve mal
d Patrize *f*, Positivform *f*

4025 POSITIVE ELECTRODE elc
f électrode *f* positive
e electrodo *m* positivo
i elettrodo *m* positivo
n positieve elektrode
d positive Elektrode *f*

4026 POSITIVE MOLD (US), POSITIVE MOULD (GB) pre
f moule *m* positif
e molde *m* positivo
i stampo *m* positivo
n matrijs met vulkamer, positieve matrijs
d Füllform *f*, Füllraumform *f*

4027 POSITIVE PLATE elc
f plaque *f* positive
e placa *f* positiva
i piastra *f* positiva
n positieve plaat
d positive Platte *f*

4028 POSITIVE TERMINAL elc
f borne *f* positive, pôle *m* positif
e borna *f* positiva, polo *m* positivo
i morsetto *m* positivo
n positief aansluitcontact *n*, positieve pool
d positiver Pol *m*

4029 POSTCOLUMN PYROLISIS APPARATUS chr
f appareil *m* de pyrolyse à colonne postérieure
e aparato *m* de pirólisis con columna posterior
i apparecchio *m* di pirolisi con colonna posteriore
n pyrolyseapparaat *n* met achterkolom
d Pyrolyseapparat *m* mit Hinterkolonne

4030 POT (US), TRANSFER CHAMBER (GB) pla,pre
f pot *m* de transfert
e cámara *f* de carga, cámara de plastificación
i camera *f* di compressione
n tabletkamer, toevoermatrijs
d Spritztopf *m*

4031 POT ANNEALING OVEN, BOX ANNEALING OVEN, CLOSE ANNEALING OVEN htt,met
f four *m* de recuit en caisse
e horno *m* de recocido en caja
i forno *m* di ricottura in cassetta
n kistgloeioven
d Kastenglühofen *m*

4032 POT ARCH gla
f four *m* de préchauffage
e horno *m* de precalentamiento
i forno *m* di preriscaldamento
n voorverhittingsoven
d Vorheizofen *m*

4033 POT FURNACE gla
f four *m* à creusets
e horno *m* de crisoles
i forno *m* a crogioli
n havenoven, pottenoven
d Hafenofen *m*

4034 POT FURNACE, CRUCIBLE FURNACE, CRUCIBLE MELTING FURNACE htt
f four *m* à creuset
e horno *m* de crisol
i forno *m* a crogiolo
n kroesoven, smeltkroesoven
d Tiegelofen *m*, Tiegelschmelzofen *m*

4035 POT FURNACE EYE, EYE gla
f trou *m* d'admission de gaz combustibles
e mirilla *f*
i orificio *m* per gas combustibili
n ovengat *n*
d Ofenloch *n*

4036 POT PLUNGER (US), TRANSFER PLUNGER (GB) pla,pre
f piston *m* de transfert
e émbolo *m* de transferencia
i pistone *m* d'iniezione
n injectieplunjer, spuitgietplunjer
d Spritzgusskolben *m*, Spritzkolben *m*

4037 POT PRESS fat,pre
f pressoir *m* à huile
e prensa *f* de aceite, trugal *m*
i pressa *f* ad olio, torchio *m*
n oliepers
d Kastenpresse *f*, Ölpresse *f*

4038 POT SPOUT gla
f bec *m* de creuset
e pico *m* de crisol
i beccuccio *m* di crogiolo
n pottuit
d Hafenausguss *m*

4039 POT STILL dis
f alambic *m* chauffé directement par la flamme
e alambique *m* caldeado directamente por la llama
i alambicco *m* riscaldato direttamente per la fiamma, lambicco *m* riscaldato direttamente per la fiamma
n direct door de vlam verhitte destilleerkolf
d Brennkolben *m*

4040 POTASH BULB, ALKALI HYDROMETER, ALKALIMETER ins
f alcalimètre *m*
e alcalímetro *m*, pesaálcalis *m*
i alcalimetro *m*
n alkalimeter, loogzoutmeter
d Alkalimeter *n*, Kaliapparat *m*, Laugenmesser *m*

4041 POTATO FLOUR FACTORY nut
f féculerie *f*, usine *f* de fécule de pommes de terre
e fábrica *f* de fécula de patatas
i fabbrica *f* di fecola di patate
n aardappelmeelfabriek
d Kartoffelmehlfabrik *f*

4042 POTCHER, BLEACHING HOLLANDER, POACHER pap
f hollandais *m* à blanchiment
e batidora *f* de blanqueo
i olandese *m* ad imbiancatura
n bleekhollander
d Bleichholländer *m*

4043 POTHEATER, AUTOCLAVE pre
f autoclave *m*
e autoclave *f*
i autoclave *f*
n autoclaaf
d Autoklav *m*, Dampfkochtopf *m*, Druckkessel *m*

4044 POTTER'S WHEEL, JOLLEY cer
f tour *m* de potier
e torno *m* para moldear loza hueca
i tornio *m* da vasaio
n pottenbakkersschijf
d Töpferscheibe *f*

4045 POUR POINT INDICATOR ins
f signaleur *m* du point de congélation
e indicador *m* del punto de descongelación, indicador *m* del punto de fluidez
i indicatore *m* del punto di congelamento, indicatore *m* del punto di scorrimento
n stolpuntindicator, vloeipunt-indicator
d Stockpunktanzeiger *m*

4046 POWDER FUNNEL tra
f entonnoir *m* à poudre
e embudo *m* de polvo
i imbuto *m* a polvere
n poedertrechter
d Pulvertrichter *m*

4047 PRECHLORINATION BATH, RAW WATER CHLORINATION BATH iw
f bain *m* de préchloration
e baño *m* de precloración
i bagno *m* di clorazione prelimina[re], bagno *m* di preclorazione
n voorchloreerbad *n*
d Vorchlorungsbad *n*

4048 PRECIPITATION BASIN iw
f bassin *m* de précipitation
e artesa *f* de precipitación
i bacino *m* di precipitazione
n bezinkingsbekken *n*
d Fällbecken *n*, Fällungsbecken *n*

4049 PRECIPITATION HARDENING PLANT, AGE-HARDENING PLANT, DISPERSION HARDENING PLANT me[t]

f installation *f* de durcissement structural
e equipo *m* de endurecimiento estructural
i impianto *m* d'incrudimento strutturale
n structurele-hardingsinstallatie
d strukturelle Härtungsanlage *f*

4050 PRECIPITATION TANK sep
f récipient *m* de précipitation
e tanque *m* de precipitación
i serbatoio *m* di precipitazione
n bezinkingstank, neerslagtank, precipitatietank
d Fällgefäss *n*

4051 PRECISION BALANCE lab
f balance *f* de précision
e balanza *f* de precisión
i bilancia *f* di precisione
n precisiebalans
d Präzisionswaage *f*

4052 PRECISION-BORE PIPETTE, PRECISION PIPETTE lab
f pipette *f* à diamètre intérieure de précision
e pipeta *f* de mandrinado de precisión
i pipetta *f* a diametro interiore di precisione
n precisiepipet
d Präzisionspipette *f*

4053 PRECOAT FILTER fil
f filtre *m* à prérevêtement
e filtro *m* de prerrevestimiento
i filtro *m* prerivestito
n voorbedekt filter *n*
d vorbeschichteter Filter *m*

4054 PRECOLUMN PYROLISIS APPARATUS chr
f appareil *m* de pyrolyse à colonne antérieure
e aparato *m* de pirólisis de columna anterior
i apparecchio *m* di pirolisi a colonna anteriore
n pyrolyseapparaat *n* met voorkolom
d Pyrolyseapparat *m* mit Vorderkolonne

4055 PRECOMBUSTION CHAMBER, ANTECHAMBER fue
f antichambre *f* de combustion
e cámara *f* de precombustión
i camera *f* di precombustione
n voorkamer
d Vorkammer *f*

4056 PRECOOLER clo
f préréfrigérateur *m*
e prerrefrigerador *m*
i preraffreddatore *m*
n voorkoeler
d Vorkühler *m*

4057 PREDEFECATION PLANT, PRELIMING PLANT sin
f installation *f* de prédéfécation
e equipo *m* de predefecación
i impianto *m* di predefecazione
n voorsatureerinstallatie
d Vorsaturationsanlage *f*

4058 PREDRIER dry
f présécheur *m*
e presecadero *m*
i preessiccatoio *m*
n voordroger
d Vortrockner *m*

4059 PRE-EVAPORATOR sin
f pré-évaporateur *m*
e preevaporador *m*
i preevaporatore *m*
n voorverdamper
d Vorverdampfer *m*

4060 PREFILL VALVE pla,rub
f soupape *f* de préremplissage
e válvula *f* de preabastecimiento, válvula *f* de prellenado
i valvola *f* di preriempimento
n voorvulklep
d Vorfüllventil *n*

4061 PREFORM pla
f ebauche *f*, préforme *f*
e preforma *f*
i preforma *f*
n voorvorm
d Vorformling *m*

4062 PREFORM SCREEN pla
f écran *m* de préforme
e rejilla *f* de preforma
i griglia *f* di preforma
n voorvormmal
d Saugform *f*

4063 PREFORMING PRESS (GB), PELLETING MACHINE (GB), TABLETS PRESS (US) mcy,pre
f pastilleuse *f*
e prensa *f* para empastillar

i pastigliatrice *f*
n tabletteerpers
d Tablettenpresse *f*

4064 PREHEATER htt
f préchauffeur *m*
e precalentador *m*
i preriscaldatore *m*
n voorverwarmer
d Vorwärmer *m*

4065 PREHEATER BY DIELECTRIC LOSSES, HIGH-FREQUENCY PREHEATER htt,rub
f appareil *m* de préchauffage à haute fréquence, préchauffeur *m* pour radiovulcanisation
e precalentador *m* de alta frecuencia
i preriscaldatore *m* ad alta frequenza
n hoogfrequente voorverwarmingsinstallatie
d Hochfrequenzvorwärmer *m*

4066 PREHEATER BY INFRARED RADIATION, INFRARED PREHEATER pla,rub
f préchauffeur *m* par rayons infrarouges
e precalentador *m* por rayos infrarrojos
i preriscaldatore *m* per raggi infrarossi
n infraroodvoorverwarmer
d Infrarotvorwärmer *m*

4067 PREHEATING CHAMBER tex
f chambre *f* de préchauffage
e cámara *f* de precalentamiento
i camera *f* di preriscaldamento
n voorverwarmingsruimte
d Vorwärmungsraum *m*

4068 PREHEATING MILL, WARMING MILL, WARMING UP MILL pla,rub
f laminoir *m* de réchauffage
e laminador *m* de precalentamiento
i laminatoio *m* di preriscaldamento
n voorwarmwalserij
d Vorwärmwalzwerk *n*

4069 PREHEATING ROLL pla
f rouleau *m* de préchauffage
e rollo *m* de precalentamiento
i rullo *m* di preriscaldamento
n voorwarmrol
d Vorwärmrolle *f*

4070 PREHEATING ZONE de
f zone *f* de préchauffage
e zona *f* de precalefacción
i zona *f* di preriscaldamento
n voorverwarmingszone
d Vorwärmungszone *f*

4071 PRELIMINARY BATH ph
f bain *m* préalable
e baño *m* preliminar
i bagno *m* preliminare
n voorbad *n*
d Vorbad *n*

4072 PRELIMINARY REFINER ru
f mélangeur *m* déchiqueteur
e mezclador *m* prehomogenizado
i raffinatore-mescolatore *m*
n voorhomogeniseerwals
d Vorhomogenisierwalze *f*

4073 PRELIMINARY SEDIMENTATION TANK, PRIMARY CLARIFICATION TANK cle,f
f bassin *m* de clarification préalable
e artesa *f* de clarificación preliminar
i bacino *m* di chiarificazione primaria, vasca *f* di chiarificazione preliminare
n voorklaringsbekken *n*
d Vorbecken *n*, Vorklärbecken *n*

4074 PREMASHER, CONVERTER, FOREMASHER, PONYMASHER br
f hydrateur *m*
e premezclador *m*
i premescolatore *m*
n voorbeslagkuip
d Vormaischer *m*

4075 PRESS gen,pr
f presse *f*
e prensa *f*
i pressa *f*
n pers
d Presse *f*

4076 PRESS CLOTH fa
f étendelle *f*
e tejido *m* para filtrar aceites
i tessuto *m* da filtrare olio
n oliedoek
d Filtriertuch *n*

4077 PRESS CROWN, PRESS HEAD pr
f sommier *m* supérieur, tête *f* de presse

e cabezal *m* de prensa, testa *f* de prensa
i testa *f* di pressa, traversa *f* superiore
n hemel, kop
d Pressbalken *m*, Pressholm *m*

4078 PRESS PLATE pre
f plaque *f* de pression
e placa *f* de presión
i piastra *f* di pressa
n persplaat
d Druckplatte *f*, Pressplatte *f*

4079 PRESS ROLL pap
f rouleau *m* compresseur de machine à papier
e rodillo *m* compresor de máquina de papel
i rullo *m* di pressione per cartiera
n perswals voor papiermachine
d Presswalze *f* für Papiermaschine

4080 PRESS WITH HEATED PLATENS pre
f presse *f* à plateaux chauffants
e prensa *f* con platos de calefacción
i pressa *f* con piani riscaldati
n verwarmde-platenpers
d Heizplattenpresse *f*

4081 PRESSING MACHINE cer
f presse *f* à mouler
e prensa *f* de moldear
i pressa *f* di stampo
n vormpers
d Formpresse *f*

4082 PRESSING MOULD (GB), BLOCK MOLD (US) gla
f moule *m* à presser
e molde *m* a prensar
i stampo *m* di pressa
n persvorm
d Pressform *f*

4083 PRESSION ROLL, PRESSURE ROLL pre
f cylindre *m* presseur
e cilindro *m* de presión
i cilindro *m* di pressione
n perswals
d Presswalze *f*

4084 PRESSION ROLLER, PRESSURE ROLLER pre
f rouleau *m* presseur
e rollo *m* de presión
i rullo *m* di pressione
n aandrukrol
d Andruckrolle *f*

4085 PRESSURE ACCUMULATOR, AIR ACCUMULATOR pre
f accumulateur *m* d'air comprimé
e acumulador *m* de aire comprimido
i accumulatore *m* d'aria compressa
n persluchtaccumulator
d Drucksammler *m*

4086 PRESSURE ADAPTER, PRESSURE CONTROL VALVE, PRESSURE GOVERNOR, PRESSURE REGULATOR pre,sap
f régulateur *m* de pression, soupape *f* de régulation
e regulador *m* de presión, válvula *f* de regulación
i regolatore *m* di pressione, valvola *f* di regolazione
n drukregelaar, drukregelklep
d Druckregler *m*, Druckregulierungsventil *n*

4087 PRESSURE ALARM INDICATOR sap
f avertisseur *m* de surpression, signaleur *m* de surpression
e indicador *m* de sobrepresión
i indicatore *m* di sovrappressione
n overdrukindicator
d Überdruckalarmvorrichtung *f*

4088 PRESSURE BOILER pre
f chaudière *f* à pression
e caldera *f* de presión
i caldaia *f* a pressione
n drukketel
d Druckkessel *m*

4089 PRESSURE BOOSTER pre
f piston-plongeur *m* additionnel
e compresor *m* auxiliar
i elevatore *m* di pressione
n drukaanjager, hulpcompressor
d Zusatztauchkolben *m*

4090 PRESSURE BOTTLE lab,tra
f flacon *m* à pression
e frasco *m* de presión
i boccetta *f* a pressione
n drukfles
d Druckflasche *f*

4091 PRESSURE CONDUIT pre,tra
f conduite *f* de pression
e conducto *m* de presión
i condotta *f* di pressione
n drukleiding
d Druckleitung *f*

4092 PRESSURE CONVERTER nit
f convertisseur *m* de pression
e convertidor *m* de presión
i convertitore *m* di pressione
n drukomzetter
d Druckwandler *m*

4093 PRESSURE COOKER bre
f chaudière *f* à pression
e caldera *f* de presión
i caldaia *f* a pressione
n drukpan
d Druckpfanne *f*

4094 PRESSURE EQUALIZING PLATE, pla
CAUL (US)
f coussin *m* de pression
e almohada *f* compensadora de presión
i cuscino *m* di pressione
n drukegalisator, drukkussen *n*, drukverdeler
d Druckausgleichplatte *f*, Druckkissen *n*

4095 PRESSURE FILTER fil
f filtre *m* à pression
e filtro *m* de presión
i filtro *m* a pressione
n drukfilter *n*
d Druckfilter *m*

4096 PRESSURE GAUGE (GB), ins
BOURDON GAUGE (GB),
CHUCK GAGE (US)
f manomètre *m* de Bourdon
e manómetro *m* de Bourdon
i manometro *m* di Bourdon
n bourdonmanometer
d Bourdonfedermanometer *n*

4097 PRESSURE GOVERNOR, gas,sap
GAS REGULATOR
f régulateur *m* de la pression du gaz
e regulador *m* de la presión del gas
i regolatore *m* della pressione del gas
n gasdrukregelaar
d Gasdruckregler *m*

4098 PRESSURE NOZZLE cpt,pla,pre
f buse *f* de projection
e tobera *f* de presión
i ugello *m* polverizzatore
n drukklep
d Druckdüse *f*

4099 PRESSURE PAD pla
f cales *pl* de renfort, surface *f* d'appui, tampon *m* de pression
e placa *f* base del molde, tampón *m* de presión
i cuscinetto *m* di pressione, listello *m* d'appoggio
n tegendrukplaat
d Druckaufnahmefläche *f*, Gegendruckpolster *n*

4100 PRESSURE PUMP, pr
FORCE PUMP
f pompe *f* élévatoire, pompe *f* foulante
e bomba *f* impelente
i pompa *f* a pressione, pompa *f* premente
n perspomp
d Druckpumpe *f*

4101 PRESSURE REDUCING VALVE, cpt,pr
REDUCING VALVE
f réducteur *m* de pression, soupape *f* de réduction
e válvula *f* reductora
i riduttore *m* di pressione, valvola *f* di riduzione
n reduceerventiel *n*
d Druckminderventil *n*, Reduzierventil *n*

4102 PRESSURE RELIEF VALVE sa
f soupape *f* de réaction, soupape *f* de sûreté
e válvula *f* de desahogo, válvula *f* de seguridad
i valvola *f* di sicurezza
n veiligheidsklep
d Überdruckventil *n*

4103 PRESSURE TANK pre,st
f récipient *m* à pression
e tanque *m* de presión
i serbatoio *m* a pressione
n druktank
d Druckbehälter *m*

4104 PRESSURE TUBE REACTOR nu
f réacteur *m* à tubes sous pressi
e reactor *m* de tubos de presión
i reattore *m* a tubi in pressione
n reactor met buizen onder druk
d Druckröhrenreaktor *m*

4105 PRESSURE VESSEL REACTOR nu
f réacteur *m* à récipient sous pression
e reactor *m* de recipiente de pres

i reattore *m* a recipiente in pressione
n reactor *m* met vat onder druk
d Druckgefässreaktor *m*

4106 PRESSURIZED CASING nue
f gaine *f* résistante à la pression
e envoltura *f* resistente a la presión
i involucro *m* resistente alla pressione
n drukvaste omhulling
d druckfeste Umhüllung *f*

4107 PRESSURIZED HEAVY WATER REACTOR nue
f réacteur *m* à eau lourde sous pression
e reactor *m* de agua pesada de presión
i reattore *m* ad acqua pesante in pressione
n reactor met zwaar water onder druk
d Schwerwasserdruckreaktor *m*

4108 PRESSURIZED WATER REACTOR nue
f réacteur *m* à eau sous pression
e reactor *m* de agua de presión
i reattore *m* ad acqua in pressione
n reactor met water onder druk
d Druckwasserreaktor *m*

4109 PRETANNING TANK tan
f cuve *f* de prétannage
e cuba *f* de precurtido
i tino *m* da preconciare
n voorlooikuip
d Vorgerbebottich *m*

4110 PRIMARY BATTERY elc
f batterie *f* de piles, batterie *f* primaire
e batería *f* de pilas, batería *f* primaria
i batteria *f* di pile, batteria *f* primaria
n batterij van primaire cellen, batterij van primaire elementen, primaire batterij
d Primärbatterie *f*

4111 PRIMARY CELL, BATTERY elc
f élément *m* primaire, pile *f*
e elemento *m* primario, pila *f*
i elemento *m* primario, pila *f*
n batterij, primair element *n*, primaire cel
d Batterie *f*, Primärelement *n*

4112 PRIMARY CLARIFICATION TANK, PRELIMINARY SEDIMENTATION TANK cle,fil
f bassin *m* préalable de clarification
e artesa *f* de clarificación preliminar
i bacino *m* di chiarificazione primaria, vasca *f* di chiarificazione preliminare
n voorklaringsbekken *n*
d Vorbecken *n*, Vorklärbecken *n*

4113 PRIMARY CRUSHER brk
f broyeur *m* primaire
e machacadora *f* primaria
i rompitore *m* primario
n grofbreker
d Grobbrecher *m*, Vorbrecher *m*

4114 PRIME-SURFACE HEATING COIL, PIPE COIL HEATING ELEMENT, PLAIN-TUBE HEATING COIL cle,hyg
f réchauffeur *m* d'air en tubes ordinaires
e calentador *m* de aire con tubos ordinarios
i riscaldatore *m* d'aria con tubi ordinari
n gladde-buisluchtverhitter
d Glattrohrheizregister *n*, Lufterhitzer *m* mit Stahlglattröhren

4115 PRINCIPAL CONDUIT, MAIN CONDUIT, MAIN PIPE LINE cpt,tra
f conduite *f* maîtresse, conduite *f* principale
e conducto *m* principal, tubería *f* principal
i condotta *f* maestra, condotta *f* principale
n hoofdleiding
d Hauptleitung *f*, Stammleitung *f*

4116 PRINTING BLANKET, PRINTING CLOTH tex
f blanchet *m*
e blanqueta *f*
i tessuto *m* per stamperia
n drukdoek, legger
d Drucktuch *n*, Farbtuch *n*

4117 PRINTING MACHINE tex
f machine *f* imprimeuse
e máquina *f* impresora
i macchina *f* stampatrice
n drukmachine
d Druckmaschine *f*

4118 PRINTING ROLLER tex
f cylindre *m* imprimeur
e cilindro *m* impresor
i cilindro *m* stampatore
n drukwals
d Druckwalze *f*

4119 PRISM SPECTROSCOPE ins,spe
f spectroscope *m* à prisme
e espectroscopio *m* de prisma
i spettroscopio *m* a prisma
n prismaspectroscoop
d Prismenspektroskop *n*

4120 PRIVY VAULT, HYDROLYZING TANK, SEPTIC TANK iwt
f récipient *m* septique
e tanque *m* séptico
i serbatoio *m* settico, vasca *f* a flusso di liquame
n vuilnistank, vuilwatertank
d Abwasserfaulraum *m*, Faulgrube *f*, Faulraum *m*

4121 PROCESSING PLANT gen
f installation *f* de traitement
e equipo *m* de elaboración
i impianto *m* di lavorazione
n bewerkingsinstallatie
d Bearbeitungsanlage *f*

4122 PRODUCTION REACTOR, REGENERATIVE REACTOR nue
f réacteur *m* de production
e reactor *m* de producción
i reattore *m* di produzione
n aanmaakreactor
d Produktionsreaktor *m*

4123 PROGRESSIVE HEATING FURNACE, SCANNING HEATING FURNACE htt
f four *m* à induction et chauffage progressif
e horno *m* de inducción con calentamiento progresivo
i forno *m* ad induzione con riscaldamento progressivo
n inductieoven met voortschrijdend werkstuk
d Induktionsofen *m* mit fortschreitendem Werkstück

4124 PROOF BOX, STEAM BOX, STEAM CHEST nut
f boîte *f* à tiroir, boîte *f* de distribution de vapeur, cage *f* de vanne
e caja *f* de distribución de vapor
i camera *f* di vapore
n stoomkast
d Dampfkasten *m*, Schieberkasten

4125 PROPANE-AIR BURNER bu
f brûleur *m* à air et propane
e quemador *m* de aire con propan
i bruciatore *m* ad aria e propano
n propaan-luchtbrander
d Propan-Luftbrenner *m*

4126 PROPANE CYLINDER ga
f cylindre *m* à propane
e cilindro *m* de propano
i cilindro *m* a propano
n propaancilinder
d Propanflasche *f*, Propanzylind

4127 PROPANE DEASPHALTING UNIT ga
f installation *f* de désasphaltage du propane
e equipo *m* de desasfaltado del propano
i impianto *m* di rimozione dell'asfalto del propano
n apparaat *n* voor het onttrekken van asfalt aan propaan
d Propanentasphaltierungsanlage

4128 PROPANE DECARBONIZING TOWER cle,ga
f tour *f* à décarbonisation du propane
e torre *f* de descarbonizar el propano
i torre *f* da decarburazione del propano
n propaanontkolingstoren
d Propanentkohlungsturm *m*

4129 PROPANE TANK ga
f récipient *m* à propane
e tanque *m* de propano
i serbatoio *m* a propano
n propaantank
d Propanbehälter *m*

4130 PROPELLER cle,mix,p
f hélice *f*
e hélice *f*
i elica *f*
n propeller, schroef
d Propeller *m*, Schraube *f*

4131 PROPELLER AGITATED CLEANER, PROPELLER AGITATED CLEANING APPARATUS c
f appareil *m* de dégraissage à agitateur à hélice

e aparato *m* de desengrase con agitador de hélice
i apparecchio *m* sgrassatore con agitatore ad elica
n ontvettingsapparaat *n* met schroefroerwerk
d Reinigungsgerät *n* mit Schraubenrührwerk

4132 PROPELLER AGITATOR, PROPELLER STIRRER mix
f agitateur *m* à hélice
e agitador *m* de hélice
i agitatore *m* ad elica
n propellerroerwerk *n*, schroefroerwerk *n*
d Propellerrührwerk *n*, Schraubenrührwerk *n*

4133 PROPELLER MIXER mix
f mélangeur *m* à hélice
e mezclador *m* de hélice
i mescolatore *m* ad elica
n propellermenger, schroefmenger
d Propellermischer *m*, Schraubenmischer *m*

4134 PROPELLER PUMP, ARCHIMEDEAN SCREW tra
f vis *f* d'Archimède
e tornillo *m* de Arquímedes
i vite *f* d'Archimede
n schroef van Archimedes, schroefpomp, tonmolen
d Transportschnecke *f*

4135 PROPORTIONER gen
f doseur *m*
e dosificador *m*
i apparecchio *m* di dosatura, dosatore *m*
n doseerapparaat *n*
d Dosiervorrichtung *f*

4136 PROPORTIONING PISTON pre
f piston *m* doseur, sabot *m* de dosage
e émbolo *m* dosificador
i pistone *m* dosatore
n doseerplunjer
d Dosierkolben *m*

4137 PROPORTIONING PUMP pre
f pompe *f* de dosage
e bomba *f* dosificadora
i pompa *f* di dosatura
n doseerpomp
d Dosierpumpe *f*

4138 PROTECTIVE ANODE, SACRIFIED ANODE cor
f anode *f* de protection, anode *f* sacrifiée
e ánodo *m* de protección
i anodo *m* di protezione
n beschermende anode
d Schutzanode *f*

4139 PROTECTIVE ATMOSPHERE FURNACE htt
f four *m* à atmosphère protectrice
e horno *m* de atmósfera de protección
i forno *m* ad atmosfera di protezione
n oven met beschermende atmosfeer
d Ofen *m* mit Schutzatmosphäre

4140 PROTECTIVE CLOTHING hyg,sap
f vêtement *m* protecteur
e traje *m* protector
i vestito *m* di protezione
n beschermende kleding
d Schutzanzug *m*, Schutzkleidung *f*

4141 PROTECTIVE COATING coa,sap
f enduit *m* protecteur, revêtement *m* de protection
e revestimiento *m* protector
i rivestimento *m* protettivo
n beschermende laag
d Schutzbelag *m*, Schutzüberzug *m*

4142 PROTECTIVE EQUIPMENT sap
f équipement *m* de protection
e equipo *m* de protección
i equipaggiamento *m* di protezione
n beschermingsapparatuur
d Schutzausrüstung *f*

4143 PROTECTIVE GLOVES sap
f gants *pl* protectifs
e guantes *pl* protectores
i guanti *pl* protettivi
n beschermingshandschoenen *pl*
d Schutzhandschuhe *pl*

4144 PROTECTIVE HELMET sap
f casque *m* protecteur
e casquete *m* protector
i elmetto *m* protettivo
n veiligheidshelm
d Schutzhelm *m*

4145 PSEUDOCARBURIZING OVEN, BLANK CARBURIZING OVEN htt,met
f four *m* de pseudocémentation
e horno *m* de seudocementación
i forno *m* di pseudocementazione
n pseudocementeeroven
d Pseudozementierofen *m*

4146 PSEUDONITRIDING OVEN, BLANK NITRIDING OVEN htt,met
f four *m* de pseudonitruration
e horno *m* de seudonitruración
i forno *m* di pseudonitrurazione
n pseudonitreeroven
d Pseudonitrierofen *m*

4147 PSYCHROMETER ins
f psychromètre *m*
e psicrómetro *m*
i psicrometro *m*
n psychrometer
d Psychrometer *n*

4148 PUERING VAT, BATING VAT lea
f cuve *f* de chipage
e cuba *f* de maceración
i tino *m* di macerazione
n zweldervat *n*, zwelkuip
d Beizbrühegefäss *n*, Schwödbottich *m*

4149 PUG MILL cer,mix
f pétrin *m*
e amasadora *f* de arcilla
i impastatrice *f*
n kleikneder
d Tonkneter *m*

4150 PULL-BACK RAM pre
f piston *m* de retour
e émbolo *m* de retroceso
i pistone *m* di ritorno
n uitstoterplunjer
d Rückdruckkolben *m*

4151 PULL ROLL, TENSION ROLLER pla,pre
f rouleau *m* tendeur
e rodillo *m* tensor
i rullo *m* tenditore
n spanrol, tegendrukrol
d Spannrolle *f*

4152 PULP CATCHER sin
f puiseur *m* de pulpe
e colector *m* de pulpa
i ricuperatore *m* di polpa
n pulpvanger
d Pulpefänger *m*

4153 PULP FILTER bre,fil
f filtre *m* à coquilles, filtre *m* de masse
e filtro *m* de conchas, filtro *m* de masa
i filtro *m* a piastre, filtro *m* di massa
n kiezelgoerfilter *n*
d Massefilter *m*, Schalenfilter *m*

4154 PULP FLUME sin
f caniveau *m* de pulpe
e canal *m* de pulpa
i canale *m* di polpa
n pulpgoot
d Schnitzelschwemme *f*

4155 PULP LIFT sin
f élévateur *m* de pulpe
e elevador *m* de pulpa
i elevatore *m* di polpa
n pulpophaler
d Schnitzelbagger *m*

4156 PULP MOLD (US), PULP MOULD (GB), FELTING SCREEN pap,pla
f moule *m* à grille d'aspiration
e molde *m* de succión
i telaio *m* d'aspirazione
n pulppersmal
d Saugform *f*

4157 PULP SAVER, CATCH-ALL, RECLAIMER, SAVE-ALL pap
f ramasse-pâte *m*
e recogedor *m* de pasta
i ricuperatore *m* di polpa
n vezelstofvanger
d Faserstoffänger *m*, Stoffänger *m*

4158 PULP STRAINER PLATE pap
f plaque *f* de filtre de pâte
e placa *f* de filtro de pasta
i piastra *f* di filtro di polpa
n pulpfilterplaat
d Ganzzeugfilterplatte *f*

4159 PULP WASHER bre
f laveuse *f* de masse filtrante
e lavadora *f* de masa filtrante
i lavatrice *f* per massa filtrante
n kiezelgoerwasser
d Filtermassewaschmaschine *f*

4160 PULP WATER PUMP sin
f pompe *f* à eau de pulpe
e bomba *f* de agua de pulpa
i pompa *f* ad acqua di polpa
n pulpwaterpomp
d Schnitzelwasserpumpe *f*

4161 PULSATION DAMPER (GB), gas,pre
SNUBBER (US)
f amortisseur *m* de pulsations
e amortiguador *m* de pulsaciones
i smorzatore *m* di pulsazioni
n schokdemper
d Stossdämpfer *m*

4162 PULSATOR JIG sor
f crible *m* à grille mobile
e criba *f* de rejilla móvil
i vaglio *m* a griglia mobile
n schudzeefmachine
d Schüttelsiebmaschine *f*

4163 PULSOMETER pre
f pulsomètre *m*
e pulsómetro *m*
i pulsometro *m*
n pulsometer
d Druckautomat *m*, Pulsometer *n*

4164 PULSOMETER PUMP pre
f pompe *f* à pulsomètre
e bomba *f* de pulsómetro
i pompa *f* a pulsometro
n pulsometerpomp
d Dampfdruckpumpe *f*

4165 PULVERIZING MILL brk
f moulin *m* pulvérisateur
e molino *m* pulverizador
i mulino *m* polverizzatore
n fijnmolen, poedermolen
d Feinstmühle *f*

4166 PUMP gen,pre
f pompe *f*
e bomba *f*
i pompa *f*
n pomp
d Pumpe *f*

4167 PUMP VALVE cpt
f soupape *f* de pompe
e válvula *f* de bomba
i valvola *f* di pompa
n pompklep
d Pumpenventil *n*, Pumpventil *n*

4168 PUNCH (GB), pla,pre
FORCE (US), TOP FORCE (GB)
f poinçon *m*
e molde *m* macho, punzón *m*
i punzone *m*
n positief matrijsdeel *n*, stempel
d Stempel *m*

4169 PUNCH PRESS, pla
PUNCHING DIE
f poinçonneuse *f*, presse *f* à poinçonner
e punzonadora *f*, troquel *m* de punzonar
i punzone *m*, punzonatrice *f*
n perforator
d Lochstanze *f*

4170 PUNTY, gla
PONTIL
f pontil *m*
e pontil *m*, puntel *m*, puntero *m*
i puntello *m*
n hechtijzer *n*, punteerijzer *n*
d Hefteisen *n*

4171 PURGE COCK, cpt
DISCHARGE COCK, DRAIN COCK
f robinet *m* de purge, robinet *m* de vidange
e grifo *m* de descarga, grifo *m* de purga
i rubinetto *m* di scarico
n aftapkraan
d Ablasshahn *m*, Entleerungshahn *m*

4172 PURIFIER cle
f épurateur *m*
e depurador *m*, purificador *m*
i depuratore *m*
n zuiveringsapparaat *n*
d Reinigungsgerät *n*

4173 PURIFYING BULB cle,lab
f ballon *m* laveur
e cubeta *f* depuradora
i pallone *m* di depurazione
n absorptieballon
d Absorptionsrohr *n*

4174 PURIFYING JAR cle,lab
f flacon *m* laveur, tube *m* absorbeur
e cilindro *m* de absorción
i cilindro *m* d'assorbimento
n absorptieglas *n*
d Absorptionszylinder *m*, Waschflasche *f*

4175 PUSHBACK cpt,pre
f éjecteur *m* hydraulique
e expulsor *m* hidráulico
i dispositivo *m* eiettore idraulico
n hydraulische uitwerper
d hydraulischer Auswerfer *m*

4176 PUSHER-TYPE FURNACE, htt,met
CONTINUOUS REHEATING FURNACE
f four *m* poussant
e horno *m* empujador
i forno *m* a spinta
n doorschuifoven
d Durchstossofen *m*, Stossofen *m*

4177 PYCNOMETER, PYKNOMETER, DENSITY BOTTLE ins
f pycnomètre *m*
e picnómetro *m*
i picnometro *m*
n pyknometer
d Pyknometer *n*

4178 PYROLYTIC GRAPHITE TUBE nue
f tube *m* à graphite pyrolytique
e tubo *m* de grafito pirolítico
i tubo *m* a grafite pirolitico
n pyrolytische grafietbuis
d pyrolytisches Graphitrohr *n*

4179 PYROMETER ins
f pyromètre *m*
e pirómetro *m*
i pirometro *m*
n pyrometer
d Pyrometer *n*

4180 PYROMETRIC CONES, FUSION CONES, SEGER CONES tes
f cônes *pl* de Seger, cônes *pl* pyrométriques
e conos *pl* de Seger, conos *pl* pirométricos
i coni *pl* di Seger, coni *pl* pirometrici
n segerkegels *pl*, smeltkegels *pl*
d Schmelzkegel *pl*, Segerkegel *pl*

4181 PYROMETRIC CYLINDERS, SENTINEL PYROMETERS ins
f cylindres *pl* pyrométriques
e cilindros *pl* pirométricos
i cilindri *pl* pirometrici
n smeltcilinders *pl*
d Schmelzzylinder *pl*

Q

4182 QUARL BLOCK, bur,gla
BURNER FIRING BLOCK
f brique *f* réfractaire de brûleur
e ladrillo *m* refractario de quemador
i mattone *m* refrattario di bruciatore
n vuurvaste brandersteen
d feuerfester Brennerstein *m*

4183 QUARTER-WAVE FILTER, ins
POLARIZATION FILTER
f lamelle *f* quart d'onde
e laminilla *f* de cuarto de onda
i lamina *f* quarto d'onda
n kwartgolflamel, polarisatiefilter *n*
d Filterpolarisator *m*, Polarisationsfilter *m*

4184 QUARTERING HEAP ana,tes
f tas *m* d'inquartation
e montón *m* de encuartación
i mucchio *m* di quartazione
n kwarteerhoop
d Quartierhaufen *m*

4185 QUARTZ DIFFUSION PUMP vac
f pompe *f* de diffusion à quartz
e bomba *f* de difusión de cuarzo
i pompa *f* di diffusione a quarzo
n kwartsdiffusiepomp
d Quarzdiffusionspumpe *f*

4186 QUARTZ JACKET clo
f chemise *f* de quartz
e camisa *f* de cuarzo
i camicia *f* di quarzo
n kwartsmantel
d Quarzmantel *m*

4187 QUARTZ SPECTROGRAPH ins,spe
f spectrographe *m* à quartz
e espectrógrafo *m* de cuarzo
i spettrografo *m* a quarzo
n kwartsspectrograaf
d Quarzspektrograph *m*

4188 QUASI-ARC WELDING PLANT wel
f installation *f* de soudure à l'arc avec électrodes revêtues d'amiante
e equipo *m* de soldadura al arco con electrodos recubiertos de amianto
i impianto *m* di saldatura all'arco con elettrodi rivestiti d'amianto
n booglasinstallatie met met asbest omhulde elektroden
d Lichtbogenschweissanlage *f* mit asbestummantelten Elektroden

4189 QUENCH AGEING BATH (GB), met
QUENCH AGING BATH (US)
f bain *m* de vieillissement par refroidissement rapide
e baño *m* de envejecimiento por enfriamiento rápido
i bagno *m* d'invecchiamento per raffreddamento rapido
n afschrikverouderingsbad *n*
d Abschreckalterungsbad *n*

4190 QUENCH TANK ctt,met
f récipient *m* de refroidissement rapide
e tanque *m* de enfriamiento rápido
i serbatoio *m* di raffreddamento rapido
n afschriktank
d Abschreckbehälter *m*

4191 QUENCHING BATH met
f bain *m* de refroidissement rapide
e baño *m* de enfriamiento rápido
i bagno *m* di raffreddamento rapido
n afschrikbad *n*
d Abschreckbad *n*

4192 QUENCHING TOWER ctt
f tour *f* d'extinction
e torre *f* de extinción
i torre *f* d'estinzione
n blustoren
d Löschturm *m*

4193 QUICK-FREEZING PLANT clo,nut
f installation *f* de congélation rapide
e equipo *m* de congelación rápida
i impianto *m* di congelazione rapida
n snelvriesinstallatie
d Schnellgefrieranlage *f*

4194 QUICKING BATH coa
f bain *m* à mercure pour placage
e baño *m* de mercurio para platear
i bagno *m* a mercurio per placcatura
n kwikbad *n*

d Quecksilberbad *n*,
Verquickungsbad *n*

4195 QUICKSILVERING BATH coa,elc
f bain *m* d'amalgamation à l'étain
e baño *m* de amalgamación con estaño
i bagno *m* d'amalgamazione allo stagno
n amalgaamtinbad *n*
d Amalgamzinnbad *n*

4196 QUINHYDRONE ELECTRODE, elc
QUINHYDRONE HALF CELL
f demi-cellule *f* à la quinhydrone, électrode *f* à la quinhydrone
e electrodo *m* de quinhidrona, semicelda *f* de quinhidrona
i elettrodo *m* al chinidrone
n chinhydronelekrode
d Chinhydronhalbzelle *f*

R

4197 RABBIT, nue
SHUTTLE
f tube *m* pneumatique
e barquilla *f*, lanzadera *f*
i spola *f*, impianto *m* pneumatico
n buisposthouder
d Rohrpostbüchse *f*

4198 RABBIT HOLE nue
f tube *m* pneumatique
e tubo *m* pneumático
i passaggio *m* per impianto pneumatico
n buizenpostkanaal *n*
d Rohrpostkanal *m*

4199 RABBLER met
f ringard *m*
e hurgón *m*, rable *m*
i asta *f* mandriale
n porstang
d Feuerhaken *m*, Schüreisen *n*

4200 RACE TRACK iso
f piste *f*
e pista *f*
i traccia *f*
n renbaan
d Rennbahn *f*

4201 RACEMIZATION VESSEL oic
f récipient *m* de racémisation
e cuba *f* de racemización
i serbatoio *m* di racemizzazione
n racemisatievat *n*
d Razemisiergefäss *n*

4202 RACKER, brk
RACKING MACHINE, BARREL FILLING MACHINE
f soutireuse *f* à fûts
e llenadora *f* de barriles
i riempifusti *m*
n vatenvulmachine
d Fassabfüllmaschine *f*, Fassfüller *m*

4203 RACKER, bre
STILLION
f bock *m* de soutirage
e estante *m*
i riempifusti *m* barometrico
n tapbok
d Abfüllbock *m*

4204 RACKING COCK bre
f canule *f*
e grifo *m* de llenado
i spinone *m* del riempifusti
n vulkraan
d Abfüllhahn *m*

4205 RACKING PIPE, bre
SPOUT
f tuyau *m* de remplissage
e canula *f*, caña *f*
i cannuccia *f*, pistone *m*
n vulbuis
d Füllrohr *n*

4206 RACKING ROOM bre
f local *m* de soutirage
e sala *f* de envasado, sala *f* de llenado
i locale *m* d'imbottigliamento, locale *m* d'infustamento
n vulafdeling
d Abfüllraum *m*

4207 RACKING SQUARE, bre
RACKING TANK
f cuve *f* de soutirage
e cuba *f* de envasado
i recipiente *m* polmone per infustamento, tino *m* per infustamento
n tapkuip
d Abfüllbütte *f*

4208 RADIAL FLOW AGITATOR mix
f agitateur *m* à flux radial
e agitador *m* de flujo radial
i agitatore *m* a flusso radiale
n roerwerk *n* met radiale stroming
d Rührwerk *n* mit Radialströmung

4209 RADIAL FLOW TRAY dis
f plateau *m* à écoulement radial
e plato *m* de flujo radial
i piatto *m* a flusso radiale
n schotel met radiale stroming
d Radialstromboden *m*

4210 RADIANT TUBE FURNACE htt
f four *m* à tube rayonnant
e horno *m* de tubo de radiación
i forno *m* a tubo di radiazione
n straalbuisoven
d Strahlrohrofen *m*

4211 RADIANT-TYPE BOILER, RADIATION BOILER boi
f chaudière *f* à chauffage rayonnant
e caldera *f* de calefacción radiante
i caldaia *f* a radiazione
n ketel met stralingsverwarming
d Strahlungskessel *m*

4212 RADIATING SURFACE htt
f surface *f* de rayonnement
e superficie *f* de radiación
i superficie *f* di radiazione
n stralingsoppervlak *n*
d Strahlungsfläche *f*

4213 RADIATION CURING OVEN dec
f four *m* à moufle à rayons cathodiques accélérés
e horno *m* de mufla con rayos catódicos acelerados
i forno *m* a muffola a raggi catodici accelerati
n moffeloven met versnelde elektronenstralen
d Muffelofen *m* mit beschleunigten Elektronenstrahlen

4214 RADIATION DETECTOR nue,sap
f signaleur *m* de rayonnement
e detector *m* de radiación
i rivelatore *m* di radiazione
n stralingsdetector
d Strahlungsdetektor *m*

4215 RADIATION FURNACE htt
f four *m* à rayonnement
e horno *m* de radiación
i forno *m* a radiazione
n stralingsoven
d Strahlungsofen *m*

4216 RADIATION PREFINISHING OVEN dec
f four *m* de séchage à rayons cathodiques accélérés
e horno *m* de secado con rayos catódicos acelerados
i forno *m* essiccatore a raggi catodici accelerati
n droogoven met versnelde elektronenstralen
d Trockenofen *m* mit beschleunigten Elektronenstrahlen

4217 RADIATION PYROMETER ins
f pyromètre *m* à rayonnement
e pirómetro *m* de radiación
i pirometro *m* a radiazione
n stralingspyrometer
d Strahlungspyrometer *n*

4218 RADIATOR clo
f radiateur *m*
e radiador *m*
i radiatore *m*
n radiator
d Kühler *m*

4219 RADIATOR htt
f radiateur *m*
e calorífero *m*, radiador *m*
i radiatore *m*
n radiator
d Radiator *m*

4220 RADIOACTIVE EFFLUENT DRAIN PIPE iwt,nue
f tuyau *m* d'écoulement d'eau usée radioactive
e alcantarilla *f* de agua efluente radiactiva
i condotto *m* scaricatoio d'acqua di scarico radioattiva
n afvoerleiding voor radioactief afvalwater
d Abflussleitung *f* für radioaktives Abwasser

4221 RADON CONTAINER nue,sto
f récipient *m* au radon
e recipiente *m* de radón
i recipiente *m* al radon
n radonhouder
d Radonbehälter *m*

4222 RAG BOILER pap
f chaudière *f* de chiffons
e caldera *f* de trapos
i caldaia *f* di stracci
n lompenketel, lompenkoker
d Hadernkocher *m*, Lumpenkocher

4223 RAG CHOPPER, RAG DEVIL, DISINTEGRATOR pap
f effileuse *f*, effilocheuse *f*
e desguince *m*, deshilachadora *f*, lobo *m* de trapos
i lupo *m*, tagliastracci *m*
n lompensnijder, lompenwolf
d Hadernschneider *m*, Lumpenschneider *m*

4224 RAG FERMENTATION TANK, FERMENTATION TANK, ROTTING TANK, STUFF VAT pap
f récipient *m* de pourrissage
e pudridor *m*
i serbatoio *m* di putrefazione

n ontsluitingstank, rottingsvat *n*
d Faulbütte *f*

4225 RAISING BATH lea
f bain *m* de dégraissage de chamois
e baño *m* de desengrase de gamuza
i bagno *m* di sgrassaggio di camoscio
n ontvettingsbad *n* voor gemsleer
d Entfettungsbad *n* für Sämischleder

4226 RAISING PLATFORM tra
f plate-forme *f* élévatrice
e plataforma *f* elevadora
i piattaforma *f* per elevare
n heftafel
d Hebebühne *f*

4227 RAKE, bre
GRAINS STIRRER
f piocheur *m*
e removedora *f*
i mescolatore *m* delle trebbie
n bostelhark
d Aufhackmaschine *f*

4228 RAKE, bre
ROUSER
f balai *m*
e mezclador *m*
i gruccia *f* per mescolare
n roerploeg
d Aufziehkrücke *f*

4229 RAKE CLASSIFIER mis,sor
f classeur *m* à râteaux
e clasificador *m* de rasquetas, clasificador *m* de rastrillos
i classificatore *m* a rastrelli
n harkclassificeerder
d Rechenklassierer *m*

4230 RAKE MIXER mix
f mélangeur *m* à râteaux
e mezclador *m* de rasquetas, mezclador *m* de rastrillos
i mescolatore *m* a rastrelli
n harkmenger
d Rechenmischer *m*

4231 RAKER, cle,rub
CLEANSER, CUP SCRAPER
f curette *f* de tasse à latex
e limpiador *m* para recipiente de látex
i pulitore *m* per tazza di lattice
n cupwisser
d Latexbecherwischer *m*

4232 RAKING AND GRAINS bre
REMOVING MACHINE
f couteaux *pl* de la cuve de filtration
e cuchillas *pl* de la tina de filtración
i rivoltatore *m* delle trebbie
n bostelploeg
d Treberaufhackmaschine *f*

4233 RAM pre
f piston *m*, piston-plongeur *m*
e émbolo *m*, pistón *m*
i pistone *m*
n plunjer
d Kolben *m*, Ramme *f*

4234 RAM AREA pre
f surface *f* du piston-plongeur
e superficie *f* del pistón
i superficie *f* del pistone
n plunjeroppervlak *n*
d Kolbenoberfläche *f*

4235 RAM BOLSTER pre
f tête *f* du piston-plongeur
e cabeza *f* del pistón
i testa *f* del pistone
n plunjerkop
d Kolbenkopf *m*, Stempelkopf *m*

4236 RAM EXTRUDER (GB), pla,rub
HYDRAULIC EXTRUDER,
STUFFER (US)
f expulseur *m* à pression hydraulique, presse *f* à filer
e prensa *f* de extrusión con émbolo hidráulica, prensa *f* hidráulica
i espulsatore *m* a pressione idraulica, estrusore *m* a pistone
n hydraulische spuitpers, plunjerspuitmachine
d hydraulische Strangpresse *f*, Kolbenstrangpresse *f*

4237 RAM-TYPE VULCANIZER rub
f appareil *m* de vulcanisation à piston-plongeur
e prensa *f* vulcanizadora con pistón
i vulcanizzatore *m* a pistone
n vulcanisatiepers met plunjer
d Vulkanisierpresse *f* mit Kolben

4238 RAMMER, pre
TAMPER
f dame *f*, pilonneuse *f*
e pisonador *m*
i mazzapicchio *m*, pilonatrice *f*
n stamper
d Stampfer *m*

4239 RANDOM PACKING dis
f garnissage *m* désordonné, garnissage *m* en vrac
e relleno *m* no distribuído
i riempimento *m* alla rinfusa
n ongeordende vulling
d regellose Füllung *f*

4240 RAPID COOLER sin
f refroidisseur *m* rapide
e enfriador *m* rápido
i raffreddatore *m* rapido
n snelkoeler
d Schnellkühler *m*

4241 RAPID FILTER fil
f filtre *m* rapide
e filtro *m* rápido
i filtro *m* rapido
n snelfilter *n*
d Schnellfilter *m*

4242 RAPID SAND FILTER fil,iwt
f filtre *m* rapide à sable
e filtro *m* rápido de arena
i filtro *m* rapido a sabbia
n snel zandfilter *n*
d Schnellsandfilter *m*

4243 RAREFACTION PLANT vac
f installation *f* de raréfaction
e equipo *m* de rarefacción
i impianto *m* di rarefazione
n luchtverdunningsinstallatie
d Luftverdünnungsanlage *f*

4244 RASCHIG RINGS, FILLING RINGS dis
f anneaux *pl* de remplissage, anneaux *pl* de Raschig
e anillos *pl* llenadores, anillos *pl* de Raschig
i anelli *pl* di Raschig, anelli *pl* di riempimento
n raschigringen *pl*, vulringen *pl*
d Füllringe *pl*, Raschigringe *pl*

4245 RASPING MACHINE nut
f machine *f* à râper
e ralladora *f*
i macchina *f* da grattare
n aardappelraspmachine
d Reibmaschine *f*

4246 RATTLER (US), RUMBLER (GB), TUMBLING BARREL (GB) cle,fou
f tonneau *m* de dessablage
e tambor *m* desarenador
i buratto *m* sterratore
n poetstrommel
d Putztrommel *f*, Scheuertromm[el]

4247 RAW WATER CHLORINATION BATH, PRECHLORINATION BATH iw[t]
f bain *m* de préchloration
e baño *m* de precloración
i bagno *m* di clorazione preliminare, bagno *m* di preclorazione
n voorchloreerbad *n*
d Vorchlorungsbad *n*

4248 REACTION BASIN iw[t]
f bassin *m* de réaction
e artesa *f* de reacción
i bacino *m* di reazione
n reactiebekken *n*
d Reaktionsbecken *n*

4249 REACTION CHAMBER ge[n]
f chambre *f* de réaction
e cámara *f* de reacción
i camera *f* di reazione
n reactieruimte
d Reaktionsraum *m*

4250 REACTION COLUMN ge[n]
f colonne *f* de réaction
e columna *f* de reacción
i colonna *f* di reazione
n reactiekolom
d Reaktionssäule *f*

4251 REACTION-SINTERING PLANT htt,m[et]
f installation *f* de frittage à réaction simultanée
e equipo *m* de sinterización con reacción simultánea
i impianto *m* di sinterizzazione a reazione simultanea
n sinterinstallatie met gelijktijdig plaatsvindende reactie
d Sinteranlage *f* mit gleichzeitig stattfindender Reaktion

4252 REACTION VESSEL, REACTOR ge[n]
f récipient *m* de réaction
e recipiente *m* de reacción
i recipiente *m* di reazione
n reactievat *n*
d Reaktionsgefäss *n*

4253 REACTOR CORE n[uc]
f coeur *m* du réacteur
e núcleo *m* del reactor
i nocciolo *m* del reattore
n actieve zone, reactorhart *n*, splijtzone
d aktive Zone *f*, Spaltzone *f*

54 REAGENT BOTTLE lab
f flacon *m* à réactifs
e frasco *m* para reactivos
i boccetta *f* a reagenti
n reagensfles
d Reagenzflasche *f*

55 REAR SHOE (GB), cpt
BOTTOM CLAMP PLATE (US)
f plaque *f* inférieure
e placa *f* trasera de sujeción del molde
i piastra *f* di base mobile
n onderste opspanplaat
d Aufspannkörper *m* hinten

56 REBOILER, boi, fmi, sin
JUICE BOILER
f récipient *m* de rébullition
e rehervidor *m*
i serbatoio *m* a far bollire di nuovo
n opkoker, opkookvat *n*
d Aufkocher *m*, Aufkochgefäss *n*

57 RECARBURIZATION PLANT, met
CARBON RESTORATION PLANT
f installation *f* de récarburation, installation *f* de régénération de pièces décarburisées
e equipo *m* de recarburación, equipo *m* de regeneración de piezas descarburadas
i impianto *m* di ricarburazione, impianto *m* di rigenerazione di pezzi decarburati
n heropkolingsinstallatie, installatie voor het regenereren van ontkoolde delen
d Aufkohlungsanlage *f*, Rückkohlungsanlage *f*

258 RECEIVER dis
f récepteur
e recipiente *m*
i raccoglitore *m*
n ontvanger
d Auffanggefäss *n*, Vorlage *f*

259 RECEIVER, htt
FOREHEARTH
f avant-creuset *m*
e antecrisol *m*
i avancrogiolo *m*
n voorhaard
d Vorherd *m*

260 RECEIVING TANK, sto
CISTERN, COLLECTING TANK
f récipient *m* collecteur, réservoir *m* à eau
e recipiente *m* de colección, recipiente *m* de agua
i serbatoio *m* collettore, serbatoio *m* d'acqua, serbatoio *m* di raccolta
n verzameltank, watertank
d Sammelbehälter *m*, Wasserbehälter *m*

4261 RECEIVING VESSEL, sto
STORAGE TANK
f cuve *f* de stockage
e tanque *m* de depósito
i serbatoio *m* collettore
n voorraadstank
d Sammelbehälter *m*

4262 RECESSION-TYPE gas
ACETYLENE GENERATOR
f générateur *m* d'acétylène à poussée verticale
e generador *m* de acetileno por desalojamiento del agua
i generatore *m* d'acetilene per sloggiamento dell'acqua
n acetyleengenerator met waterverdringing
d Azetylenentwickler *m* mit Wasserverdrängung

4263 RECHARGE BASIN, iwt
REPLENISHING BASIN
f bassin *m* d'infiltration, bassin *m* de réalimentation, bassin *m* de recharge
e artesa *f* de infiltración
i vasca *f* d'infiltrazione
n infiltratiebekken *n*
d Anreicherungsbecken *n*, Versickerungsbecken *n*

4264 RECIPROCAL LATTICE spe
f réseau *m* réciproque
e retículo *m* recíproco
i reticolo *m* reciproco
n reciprook rooster *n*
d reziprokes Gitter *n*

4265 RECIPROCATING IMPELLER mix
f roue *f* motrice à mouvement alternatif
e rueda *f* móvil de vaivén
i girante *m* a movimento alterno
n heen-en weergaande roerder
d hin-und hergehende Rührstange *f*

4266 RECIPROCATING RAKE sor
CLASSIFIER
f classeur *m* à râteaux à mouvement de va-et-vient
e clasificador *m* de rastrillos de vaivén

i classificatore *m* a rastrelli d'andata e ritorno
n heen-en weergaande hark-classificeerder
d hin-und hergehender Rechenklassierer *m*

4267 RECIPROCATING SHAKER lab
f secoueuse *f* oscillante
e sacudidor *m* oscilante
i apparecchio *m* a scosse oscillanti
n heen-en weergaand schud-apparaat *n*
d Schüttelapparat *m*

4268 RECIPROCATING TABLE sor
f table *f* à secousses
e mesa *f* de sacudidas
i tavola *f* a scosse
n schudtafel
d Schütteltisch *m*

4269 RECIRCULATING PUMP pre
f pompe *f* à circulation continue
e bomba *f* de recirculación
i pompa *f* a circolazione continua
n continue-circulatiepomp
d kontinuierliche Umlaufpumpe *f*

4270 RECIRCULATION HEATING SYSTEM htt
f système *m* de chauffage à circulation continue
e sistema *m* de calentamiento de recirculación
i sistema *m* di riscaldamento a circolazione continua
n kringloopverwarmingssysteem *n*
d Kreislaufheizanlage *f*

4271 RECLAIM MACHINERY rub
f machines *pl* pour régénérés
e maquinaria *f* para caucho regenerado
i macchinario *m* per rigenerati
n machines *pl* voor regenereer-installatie
d Regeneratmaschinen *pl*

4272 RECLAIM MILL brk,rub
f concasseur *m* de régénérés
e machacador *m* para regenerados
i frantoio *m* per rigenerati
n regeneraatwalserij
d Regeneratwalzwerk *n*

4273 RECLAIMER, CATCH-ALL, PULP SAVER, SAVE-ALL pap
f ramasse-pâte *m*
e recogedor *m* de pasta
i ricuperatore *m* di polpa
n vezelstofvanger
d Faserstoffänger *m*, Stoffänger *n*

4274 RECOMPACTING APPARATUS, REPRESSING APPARATUS pr
f appareil *m* de recompression
e aparato *m* de recompresión
i apparecchio *m* di ricompressio[n]
n napersapparaat *n*
d Nachpressgerät *n*

4275 RECORDER in
f enregistreur *m*
e registrador *m*
i registratore *m*
n registreertoestel *n*
d Diagrammschreiber *m*

4276 RECORDING INSTRUMENT in
f instrument *m* enregistreur
e aparato *m* registrador
i strumento *m* registratore
n registrerend instrument *n*
d Registrierapparat *m*

4277 RECOVERY PLANT gen,iw
f installation *f* de récupération
e equipo *m* de recuperación
i impianto *m* di ricupero
n terugwinningsinstallatie
d Rückgewinnungsanlage *f*

4278 RECOVERY TANK iw
f récipient *m* de récupération
e tanque *m* de recuperación
i serbatoio *m* di ricupero
n terugwinningstank
d Rückgewinnungsbehälter *m*

4279 RECTANGULAR ANODE el
f anode *f* rectangulaire
e ánodo *m* rectangular
i anodo *m* rettangolare
n vierkante anode
d rechteckige Anode *f*

4280 RECTANGULAR KILN ce
f four *m* de grillage rectangulair[e]
e horno *m* de tostación rectangula[r]
i forno *m* d'arrostimento rettangolare
n rechthoekige roostoven
d rechteckiger Röstofen *m*

4281 RECTIFICATION COLUMN, RECTIFYING COLUMN di
f colonne *f* de rectification
e columna *f* de rectificación

i colonna *f* di rettificazione
n rectificatiekolom
d Rektifiziersäule *f*

282 RECTIFICATION PLANT dis
f installation *f* de rectification
e equipo *m* de rectificación
i impianto *m* di rettificazione
n rectificeerinstallatie
d Rektifizieranlage *f*

283 RECTIFYING SECTION dis
f section *f* de rectification
e sección *f* rectificadora
i sezione *f* di rettificazione
n rectificeersectie
d Rektifizierteil *m*

284 RECTILINEAR MANIPULATOR, CRANE MANIPULATOR ins,nue
f manipulateur *m* rectiligne
e manipulador *m* rectilineo
i manipolatore *m* rettilineo
n coördinatenmanipulator
d Koordinatenmanipulator *m*

285 RECUPERATIVE AIR HEATER, REGENERATIVE AIR HEATER htt
f chauffeur *m* à air récupératif
e calentador *m* de aire recuperativo
i riscaldatore *m* ad aria ricuperativo
n verwarmingselement *n* met door regeneratie verhitte lucht
d Heizelement *n* mit durch Regeneration geheizter Luft

286 RECUPERATIVE FURNACE glo
f four *m* de récupération
e horno *m* de recuperación
i forno *m* di ricupero
n recuperatoroven
d Rekuperierofen *m*

287 RECUPERATOR htt
f récupérateur *m*
e recuperador *m*
i ricuperatore *m*
n recuperator
d Rekuperator *m*

288 REDISTILLATION PLANT dis
f installation *f* de rédistillation
e equipo *m* de redestilación
i impianto *m* di ridistillazione
n installatie voor herhaalde destillatie
d Anlage *f* zur wiederholten Destillation

4289 REDUCER, REDUCING SOCKET cpt
f manchon *m* de réduction
e manguito *m* de reducción
i manicotto *m* di riduzione
n verloopmof
d Reduzierhülse *f*, Reduzierstück *n*, Übergangsstück *n*

4290 REDUCER BATH pho
f bain *m* affaiblisseur
e baño *m* debilitador
i bagno *m* d'indebilomento
n verzwakkingsbad *n*
d Abschwächbad *n*

4291 REDUCING CROSS cpt
f raccord *m* à quatre voies
e pieza *f* de reducción en cruz
i raccordo *m* a croce
n vierwegskoppeling
d R-Kreuzstück *n*

4292 REDUCING VALVE, PRESSURE REDUCING VALVE cpt,pre
f réducteur *m* de pression, soupape *f* de réduction
e válvula *f* reductora
i riduttore *m* di pressione, valvola *f* di riduzione
n reduceerventiel *n*
d Druckminderventil *n*, Reduzierventil *n*

4293 REDUCTION BATH red
f bain *m* de réduction
e baño *m* de reducción
i bagno *m* di riduzione
n reductiebad *n*
d Reduktionsbad *n*

4294 REDUCTION CELL met,red
f cellule *f* de réduction
e célula *f* de reducción
i cellula *f* di riduzione
n reductiecel
d Reduktionszelle *f*

4295 REDUCTION CLEARING BATH tex
f bain *m* à éliminer les colorants
e baño *m* de eliminación de colorantes
i bagno *m* da eliminare coloranti
n kleurstofverwijderingsbad
d Farbstoffbeseitigungsbad *n*

4296 REDUCTION FURNACE htt,red
f four *m* à réduction
e horno *m* de reducción
i forno *m* di riduzione
n reductieoven
d Reduktionsofen *m*

4297 REDUCTION ROASTING FURNACE met,red
f four *m* de grillage réducteur
e horno *m* de tostación reductora
i forno *m* d'arrostimento riduttore
n reducerende roostoven, reductieroostoven
d Reduktionsröstofen *m*, reduzierender Röstofen *m*

4298 REDUCTION ZONE met,red
f zone *f* de réduction
e zona *f* de reducción
i zona *f* di riduzione
n reductiegebied *n*
d Reduktionsgebiet *n*

4299 REDUCTOR lab,red
f ballon *m* de réduction
e matraz *m* de reducción
i pallone *m* di riduzione
n reduceerkolf
d Reduktor *m*

4300 REEL DYEING BECK, WINCH DYEING BECK tex
f cuve *f* de teinture à moulinet
e barca *f* de devanadora
i tino *m* di tintura a mulinello
n haspelkuip
d Haspelkufe *f*

4301 REENRICHMENT PLANT nue
f installation *f* de réenrichissement
e equipo *m* de reenriquecimiento
i impianto *m* di riarricchimento
n herverrijkingsinstallatie
d Neuanreicherungsanlage *f*

4302 REFERENCE CELL, PILOT CELL, SAMPLE CELL elc
f accumulateur *m* de référence
e acumulador *m* de referencia
i accumulatore *m* di riferimento
n controleaccu
d Kontrollsammler *m*, Prüfakkumulator *m*

4303 REFERENCE ELECTRODE elc,ins
f électrode *f* de référence
e electrodo *m* de referencia
i elettrodo *m* di riferimento
n referentie-elektrode
d Bezugselektrode *f*, Vergleichselektrode *f*

4304 REFERENCE GAGE (US), MASTER GAUGE (GB) ins
f rapporteur *m*
e calibre *m* de referencia, galga *f* patrón
i calibro *m* campione, calibro *m* di riscontro
n standaardkaliber *n*
d Prüflehre *f*

4305 REFINER, NOSE gl
f zone *f* d'affinage
e zona *f* de afino
i zona *f* d'affinazione
n raffineerruimte
d Raffinierraum *m*

4306 REFINER, PERFECTING ENGINE pa
f épurateur *m*
e depurador *m*
i depuratore *m*
n reinigingsmachine
d Reinigungsmaschine *f*

4307 REFINER MILL ru
f laminoir-raffineur *m*, malaxeu de raffinage
e cilindro *m* refinador
i raffinatore *m*
n homogeniseerwals
d Homogenisierwalze *f*, Raffinierungswalze *f*

4308 REFINERY fu
f raffinerie *f*
e refinería *f*
i raffineria *f*
n raffinaderij
d Raffinerie *f*

4309 REFINERY htt,m
f affinerie *f*
e taller *m* de refusión
i affineria *f*
n omsmelthal
d Metallschmelzwerkstatt *f*, Umschmelzwerkstatt *f*

4310 REFINERY GAS GENERATOR g
f générateur *m* de gaz de raffine
e generador *m* de gas de refiner
i generatore *m* di gas di raffineria
n raffinaderijgasgenerator
d Raffineriegasgenerator *m*

4311 REFINING AND DRYING CYLINDER r
f cylindre *m* pour tordre et essorer
e cilindro *m* para refinación y secamiento

i cilindro *m* raffinatore ed asciugatore
n plet-en droogwals
d Homogenisier-und Trocken-walze *f*

4312 REFINING COPPER, SECOND BOILER sin
f claire *f*
e cuba *f* refinadora
i tino *m* raffinatore
n klaarketel, louterpan
d Klärkessel *m*, Läuterkessel *m*

4313 REFINING KETTLE fat
f chaudière *f* de raffinage alcalin
e caldera *f* de desacidificación
i caldaia *f* d'affinazione alcalina
n ontzuringsketel
d Entsäurungskessel *m*

4314 REFLECTED REACTOR nue
f réacteur *m* à réflecteur
e reactor *m* de reflector
i reattore *m* con riflettore
n reflectorreactor
d Reaktor *m* mit Reflektor, reflektierter Reaktor *m*

4315 REFLUX BOILING VESSEL boi
f chaudière *f* à reflux
e caldera *f* de reflujo
i caldaia *f* a riflusso
n terugstroomkookketel
d Rückflusskochkessel *m*

4316 REFLUX CONDENSER pre
f condensateur *m* à reflux
e condensador *m* de reflujo
i condensatore *m* a riflusso
n terugstroomcondensor
d Rückflussverdichter *m*

4317 REFLUX COOLER clo
f réfrigérateur *m* à reflux
e refrigerador *m* de reflujo
i refrigeratore *m* a riflusso
n terugstroomkoeler
d Rückflusskühler *m*

4318 REFLUX DIVIDER, SPLITTER dis
f répartiteur *m* de reflux
e distribuidor *m* de reflujo
i ripartitore *m* di riflusso
n terugstroomverdeler
d Rücklaufverteiler *m*

4319 REFLUX VALVE, BACKFLOW VALVE, CHECK VALVE, STOP VALVE cpt
f soupape *f* d'arrêt
e válvula *f* de retención
i valvola *f* d'arresto, valvola *f* intercettante
n afsluitklep, sperklep
d Absperrventil *n*

4320 REFORMING PLANT fue
f installation *f* de réformation
e equipo *m* de reformación
i impianto *m* di riformazione
n reformeerinstallatie
d Reformierungsanlage *f*

4321 REFRACTORY COATING coa
f revêtement *m* réfractaire
e revestimiento *m* refractario
i rivestimento *m* refrattario
n vuurvaste bekleding
d feuerfeste Auskleidung *f*

4322 REFRACTORY LINING coa,htt
f garnissage *m* réfractaire
e guarnición *f* refractaria
i guarnizione *f* refrattaria
n vuurvaste voering
d feuerfestes Futter *n*

4323 REFRIGERATING COIL, CONDENSING COIL, COOLING COIL clo
f serpentin *m* de réfrigération
e serpentín *m* refrigerador
i serpentina *f* refrigerante
n koelslang
d Kühlschlange *f*, Kühlschlauch *m*

4324 REFRIGERATION CABINET clo
f boîte *f* de réfrigération
e armario *m* de refrigeración
i armadio *m* di refrigerazione
n koelkast
d Kühlkasten *m*

4325 REFRIGERATION PLANT clo
f installation *f* de réfrigération
e equipo *m* de refrigeración
i impianto *m* di refrigerazione
n koelinstallatie
d Kühlanlage *f*

4326 REFRIGERATOR clo
f réfrigérateur *m*
e refrigerador *m*
i refrigeratore *m*
n koelkast
d Kühlschrank *m*

4327 REFUELLING APPARATUS nue
f appareil *m* d'échange du combustible,

appareil *m* de recharge
e aparato *m* de reaprovisionamiento, aparato *m* de reemplazo del combustible
i apparecchio *m* di ricarica
n herladingsapparaat *n*
d Brennstoffwechselgerät *n*, Neubeschickungsgerät *n*, Umladungsgerät *n*

4328 REFUSE INCINERATION PLANT hyg,iwt
f installation *f* d'incinération des ordures
e instalación *f* para incinerar basuras
i impianto *m* d'incenerazione delle immondizie
n vuilverbrandingsinstallatie
d Müllverbrennungsanlage *f*

4329 REFUSE PIT, PIT iwt
f fosse *f* à ordures
e foso *m* de basuras
i fossa *f* d'immondizie
n vuilnisput
d Müllbunker *m*, Müllgrube *f*

4330 REGENERATION AUTOCLAVE, REGENERATION BOILER, DEVULCANIZER, DIGESTER FOR RECLAIMING rub
f autoclave *m* de régénération, chaudière *f* à régénération, devulcaniseur *m*
e caldera *f* de regeneración, desvulcanizador *m*, digestor *m* para regeneración
i caldaia *f* di rigenerazione, devulcanizzatore *m*, digestore *m* per rigenerare
n regenereerautoclaaf, regenereerketel
d Digestor *m*, Entvulkanisierungsgefäss *n*, Regenerierkessel *m*

4331 REGENERATION PLANT rub
f installation *f* de régénération
e equipo *m* de regeneración
i impianto *m* di rigenerazione
n regenereerinstallatie
d Regenerieranlage *f*

4332 REGENERATIVE AIR HEATER, RECUPERATIVE AIR HEATER htt
f chauffeur *m* à air récupératif
e calentador *m* de aire recuperativo
i riscaldatore *m* ad aria ricuperativo
n verwarmingselement *n* met door regeneratie verhitte lucht
d Heizelement *n* mit durch Regeneration geheizter Luft

4333 REGENERATIVE CHAMBER ht
f chambre *f* régénératrice
e cámara *f* regeneradora
i camera *f* rigeneratrice
n regenereerruimte
d Regenerierraum *m*

4334 REGENERATIVE FURNACE gl
f four *m* à régénération
e horno *m* de regeneración
i forno *m* con rigeneratori
n regeneratoroven
d Regenerativofen *m*

4335 REGENERATIVE REACTOR, PRODUCTION REACTOR nu
f réacteur *m* de production
e reactor *m* de producción
i reattore *m* di produzione
n aanmaakreactor
d Produktionsreaktor *m*

4336 REGENERATIVE STOVE fu
f étuve *f* à régénération
e estufa *f* de regeneración
i stufa *f* a rigenerazione
n regeneratordroogstoof
d Regenerativtrockenofen *m*

4337 REGENERATIVE TANK gl
f bassin *m* à régénération
e cuba *f* de regeneración
i tino *m* a rigenerazione
n regeneratorwan
d Regenerativwanne *f*

4338 REGENERATOR ge
f régénérateur *m*
e regenerador *m*
i rigeneratore *m*
n regenerator
d Regenerator *m*

4339 REGISTER, DAMPER, SMOKE SLIDE VAL con,f
f registre *m*
e registro *m*
i paratoia *f*, registro *m*, serran
n regelschuif
d Rauchschieber *m*, Regelschieb Schieberplatte *f*

4340 REGULATING BOX, REGULATOR BOX p
f caisse *f* d'alimentation à régl

e caja *f* reguladora
i cassa *f* regolatrice
n toevoerregelkast
d Regulierkasten *m*

4341 REGULATING TAP con
f robinet *m* de réglage
e grifo *m* de regulación
i rubinetto *m* di regolazione
n regelkraan
d Regulierhahn *m*

4342 REGULATOR con
f régulateur *m*
e regulador *m*
i regolatore *m*
n regulateur
d Regulator *m*

4343 REHEATING FURNACE met
f four *m* à réchauffage
e horno *m* de recalentamiento
i forno *m* a riscaldamento
n herverhittingsoven, opwarmoven
d Wärmeofen *m*, Wiedererhitzungs-ofen *m*

4344 RELAY VALVE, con,cpt
PILOT VALVE
f soupape *f* de commande
e válvula *f* de mando
i valvola *f* di comando
n stuurklep
d Steuerventil *n*

4345 RELIEF SEWER, iwt
DISCHARGE CHANNEL
f canal *m* de décharge, émissaire *m* de dérivation
e canal *m* de descarga
i canale *m* di scarico
n ontlastingskanaal *n*
d Entlastungskanal *m*

4346 RELIEF VALVE cpt
f soupape *f* de décompression
e válvula *f* de alivio
i valvola *f* di decompressione, valvola *f* di riduzione della pressione
n drukverminderingsklep
d Druckreduktionsventil *n*, Entlastungsventil *n*

4347 RELINING APPARATUS (US), fou,htt
GLAZING APPARATUS (GB)
f appareil *m* de brasquage
e aparato *m* de enlucido de refractario
i apparecchio *m* da rivestire
n bekledingsapparaat *n*
d Glasurbrandgerät *n*

4348 REMELTING FURNACE htt,met
f four *m* de refonte
e horno *m* de segunda fusión
i forno *m* di rifusione
n omsmeltoven
d Umschmelzofen *m*

4349 REMILLING AND WASHING MACHINE rub
f déchiqueteur-laveur *m*
e machacadora-lavadora *f*
i macchina *f* per ripassare e lavare
n maal-en wasmachine
d Zerkleinerungs-und Waschmaschine *f*

4350 REMOTE-HANDLING EQUIPMENT nue
f manipulateur *m* à distance
e manipulador *m* de distancia
i manipolatore *m* a distanza
n afstandsbedieningsapparatuur
d Fernbedienungsapparat *m*

4351 REMOTE-READING INSTRUMENT ins
f instrument *m* de lecture à distance
e aparato *m* teleindicador
i strumento *m* da lettura a distanza
n instrument *n* met afstands-aflezing
d Fernablesegerät *n*

4352 REMOTELY OPERATED AUTOCLAVE pre
f autoclave *m* de commande à distance
e autoclave *f* accionada a distancia, autoclave *f* teleaccionada
i autoclave *f* a telecomando
n op afstand bediende autoclaaf
d fernbetätigter Autoklav *m*

4353 RENDERING TANK fat
f récipient *m* de clarification de l'huile
e tanque *m* de clarificación del aceite
i serbatoio *m* da chiarificare l'olio
n olieklaringstank
d Ölklärungsgefäss *n*

4354 RENDERING VESSEL fat
f cuve *f* d'extraction de la graisse, cuve *f* de fonte de la graisse
e paila *f* para extraer la manteca
i tino *m* da estrarre il grasso
n vetuitsmeltvat *n*
d Fettausschmelzgefäss *n*

4355 RENNET SUPPLY TUBE dai,nut
f tube *m* alimentateur de présure
e tubo *m* alimentador de cuajo
i tubo *m* alimentatore di presame
n lebtoevoerbuis
d Labzufuhrrohr *n*

4356 REPLENISHING BASIN, RECHARGE BASIN iwt
f bassin *m* d'infiltration, bassin *m* de réalimentation, bassin *m* de recharge
e artesa *f* de infiltración
i vasca *f* d'infiltrazione
n infiltratiebekken *n*
d Anreicherungsbecken *n*, Versickerungsbecken *n*

4357 REPOLYMERIZATION PLANT oic
f installation *f* de répolymérisation
e equipo *m* de repolimerización
i impianto *m* di ripolimerizzazione
n repolymerisatie-installatie
d Repolymerisationsanlage *f*

4358 REPRECIPITATION TANK cle,sep
f récipient *m* de réprécipitation
e tanque *m* de reprecipitación
i serbatoio *m* di riprecipitazione
n nabezinkingstank
d Nachfällungsgefäss *n*

4359 REPRESSING APPARATUS, RECOMPACTING APPARATUS pre
f appareil *m* de recompression
e aparato *m* de recompresión
i apparecchio *m* di ricompressione
n napersapparaat *n*
d Nachpressgerät *n*

4360 RESEARCH REACTOR nue
f réacteur *m* pour recherches scientifiques
e reactor *m* de investigación
i reattore *m* per ricerca
n researchreactor, speurwerk-reactor
d Forschungsreaktor *m*

4361 RESERVE TANK fue,sto
f récipient *m* de réserve
e tanque *m* de reserva
i serbatoio *m* di riserva
n reservetank
d Reservebehälter *m*

4362 RESERVOIR CHAMBER, CHAMBER OF A RESERVOIR cpt
f compartiment *m* de réservoir
e cámara *f* de recipiente
i vasca *f* di serbatoio
n reservoirkamer
d Behälterkammer *f*

4363 RESIDUES PROCESSING PLANT pho
f installation *f* de régénération de résidus de bains
e equipo *m* de regeneración de residuos de baños
i impianto *m* da rigenerare residui di bagni
n installatie voor het opwerken van badresten
d Anlage *f* zur Aufarbeitung von Badrückständen

4364 RESILIENCE METER, RESILIOMETER ins
f élasticomètre *m*, rebondimètre *m*
e medidor *m* de rebote elástico, resiliómetro *m*
i misuratore *m* d'elasticità, resiliometro
n elasticiteitsmeter
d Elastikometer *n*, Elastizitätsmesser

4365 RESINIFICATION PLANT oic
f installation *f* de résinification
e equipo *m* de resinificación
i impianto *m* di resinificazione
n verharsingsinstallatie
d Verharzungsanlage *f*

4366 RESINTERING FURNACE htt,met
f installation *f* de refrittage
e equipo *m* de resinterización
i impianto *m* di risinterizzazione
n nasinterinstallatie
d Nachsinteranlage *f*

4367 RESIST coa,ele
f couche *f* de protection
e capa *f* protectora
i rivestimento *m* isolante
n afdeklaag
d Abdeckschicht *f*

4368 RESISTANCE FURNACE, RESISTANCE OVEN htt
f four *m* à résistance

e horno *m* de resistencia
i forno *m* a resistenza
n weerstandsoven
d Widerstandsofen *m*

4369 RESISTANCE PYROMETER ins
f pyromètre *m* à résistance
e pirómetro *m* de resistencia
i pirometro *m* a resistenza
n weerstandspyrometer
d Widerstandspyrometer *n*

4370 RESISTOR FURNACE (US), INDIRECT RESISTANCE FURNACE (GB) htt
f four *m* à chauffage indirect par résistance
e horno *m* de calentamiento indirecto por resistencia
i forno *m* a riscaldamento indiretto a resistenza
n weerstandsoven met indirecte verhitting
d indirekter Widerstandsofen *m*

4371 RESPIRATOR sap
f respirateur *m*
e mascarilla *f* de respiración, respirador *m*
i respiratore *m*
n ademhalingsapparaat *n*, respirator
d Atemapparat *m*, Respirator *m*

4372 RESTRAINER BATH pho
f bain *m* retardateur
e baño *m* retardador
i bagno *m* rallentatore
n vertragingsbad *n*
d Verzögerungsbad *n*

4373 RESTRICTED GATE, RESTRICTED GATING pla,pre
f injection *f* capillaire
e bebedero *m* capilar
i iniezione *f* capillare
n puntaanspuiting
d Punktanguss *m*

4374 RESTRICTOR pre
f réducteur *m* de la pression
e reductor *m* de la presión
i riduttore *m* della pressione
n doorstroomregelaar
d Durchflussregler *m*

4375 RESUBLIMATION VESSEL gen
f récipient *m* de résublimation
e tanque *m* de resublimación
i serbatoio *m* di risublimazione
n resublimatievat
d Resublimationsgefäss *n*

4376 RETAINER PLATE pla,pre,rub
f mandrin *m* rapporté
e placa *f* de retención
i piastra *f* di bloccaggio
n bevestigingsplaat
d Einsatzfutter *n*

4377 RETAINER PLATE (GB), LOADING PLATE (US), TRANSFER CHAMBER pla
f plaque *f* contenant la chambre de transfert
e placa *f* intermedia de alojamiento de la cámara de plastificación
i piastra *f* di carica
n tussenplaat, vulplaat
d Füllplatte *f*

4378 RETAINER RING (GB), ADAPTER RING (US) opt
f bague *f* pour douille
e anillo *m* adaptador
i anello *m* di riduzione, riduttore *m*
n pasring
d Haltering *m*, Passring *m*

4379 RETAINING PIN, CARRIER PIN (GB), INSERT PIN (US) pla,pre
f broche *f* à prisonnier, ergot *m* de centrage, goupille *f* de fixation
e clavija *f* de retenida, clavija *f* de sujeción
i spina *f* portainserto
n fixatiepen
d Haltestift *m*

4380 RETAINING RING, ORIFICE PLATE pla,pre
f diaphragme *m* de mesure
e anillo *m* de retención
i diaframma *m* di misura
n orifice-meter, stuwring voor drukverhoging
d Messblende *f*, Stauring *m*

4381 RETANNAGE TANK tan
f récipient *m* de rétannage
e tanque *m* de recurtido
i serbatoio *m* di riconcia
n nalooitank
d Nachgerbegefäss *n*

4382 RETENTION BASIN nue
f bassin *m* de retenue
e estanque *m* de retención
i bacino *m* di ritenuto
n bewaarbekken *n*
d Aufbewahrungsbecken *n*

4383 RETORT htt,lab
f cornue *f*
e retorta *f*
i storta *f*
n retort
d Retorte *f*

4384 RETORT FURNACE, htt
CHAMBER FURNACE
f four *m* à cornues
e horno *m* de retortas
i forno *m* a storte
n retortenoven
d Retortenofen *m*

4385 RETTING PLANT, tex
ROTTING PLANT
f rouisseur *m*, routoir *m*
e enriadero *m*
i impianto *m* di macerazione
n roterij, vlasroterij
d Flachsrösterei *f*, Rösterei *f*

4386 RETTING TANK, tex
ROTTING TANK
f récipient *m* de rouissage
e tanque *m* de enriamiento
i serbatoio *m* di macerazione
n rootkuip, vlasrootkuip
d Röstbottich *m*

4387 RETURN-BEND PASS OVEN rub
f tunnel-séchoir *m* à circulation d'aller et de retour
e horno *m* de túnel con circulación en dos direcciones
i stufa *f* a circolazione d'andata e ritorno
n tunneloven met heen-en weergaand transport
d Ofen *m* mit hin-und hergehendem Durchlauf

4388 RETURN PIN pla,pre
f butée *f* de renvoi, goujon *m* de retour, poussoir *m* de rappel
e espiga *f* de retroceso
i asta *f* respingente
n tegendrukpen, uitstootpen
d Rückdruckstift *m*

4389 RETURN SPRING pla,pre
f ressort *m* de renvoi
e muelle *m* de retroceso
i molla *f* respingente
n tegendrukveer
d Rückdruckfeder *f*

4390 REVERBERATORY FURNACE htt
(GB),
AIR FURNACE (US)
f four *m* réverbère
e horno *m* de reverbero
i forno *m* a riverbero
n vlamoven
d Flammofen *m*

4391 REVERSE-CURRENT elc
CLEANING BATH
f bain *m* de dégraissage anodique
e baño *m* de desengrase anódico
i bagno *m* di sgrassaggio anodico
n anodisch ontvettingsbad *n*
d anodisches Reinigungsbad *n*

4392 REVERSE CYLINDER rub
f cylindre *m* inverseur
e cilindro *m* inversor
i cilindro *m* d'inversione
n omkeerwals
d Umkehrwalze *f*

4393 REVERSE MOLD (US), pla,pre
REVERSE MOULD (GB)
f frette *f*, moule *m* inversé
e molde *m* invertido
i stampo *m* rovescio
n omgekeerde matrijs
d Mantelform *f*

4394 REVERSIBLE CELL elc
f pile *f* réversible
e pila *f* reversible
i pila *f* reversibile
n reversibel element *n*, reversibele cel
d reversibeles Element *n*, umkehrbares Element *n*

4395 REVERSIBLE ELECTRODE elc
f electrode *f* réversible
e electrodo *m* reversible
i elettrodo *m* reversibile
n reversibele elektrode
d reversibele Elektrode *f*

4396 REVERSING EXCHANGER ge
f échangeur *m* de chaleur à inversion
e cambiador *m* de calor de inversión
i scambiatore *m* di calore ad inversione

n zelfreinigende warmte-uitwisselaar
d selbstreinigender Wärmeaustauscher *m*

4397 REVOLVING SCREEN, TROMMEL sor
f trieur *m* rotatif, trommel *m*
e criba *f* giratoria de tambor
i vaglio *m* rotativo a tamburo
n zifttrommel
d Sichtetrommel *f*, Sortiertrommel *f*

4398 RHENISCH RETORT, ELLIPTICAL RETORT htt
f cornue *f* elliptique
e retorta *f* elíptica
i storta *f* ellittica
n ovale retort
d ovale Retorte *f*

4399 RHEOMETER, CONSISTOMETER, PLASTIMETER, PLASTOMETER ins
f consistomètre *m*, plastomètre *m*
e consistómetro *m*, plastómetro *m*
i consistometro *m*, plastometro *m*
n consistometer, plastometer
d Konsistenzmesser *m*, Plastometer *n*

4400 RHODANIZING BATH, RHODIUM PLATING BATH elc
f bain *m* de placage au rhodium
e baño *m* de rodiado
i bagno *m* di placcatura con rodio
n rhodiumbad *n*
d Rhodiumbad *n*

4401 RIBBED ANODE, FLUTED ANODE elc
f anode *f* cannelée
e ánodo *m* acanalado
i anodo *m* scanalato
n geribbelde strookanode
d Riffelanode *f*

4402 RIBBED FUNNEL, FLUTED FUNNEL lab,tra
f entonnoir *m* cannelé
e embudo *m* acanalado
i imbuto *m* scanalato
n geribbeld filter *n*
d Rippentricnter *m*

4403 RIBBON BLENDER mix
f malaxeur *m* à ruban
e mezclador *m* de cinta
i mescolatore *m* a nastro
n lintmenger
d Bandmischer *m*

4404 RIBBON-FLAME BURNER bur
f brûleur *m* à flamme plane
e quemador *m* de llama plana
i bruciatore *m* a fiamma piana
n bandbrander
d Bandbrenner *m*

4405 RICE-HULLING MILL nut
f décortiqueuse *f* de riz
e descascaradora *f* de arroz, molino *m* arrocero
i mulino *m* di pilatura di riso
n rijstmolen, rijstpelmolen
d Reismühle *f*, Reisschälmühle *f*

4406 RICE-HULLING ROLL nut
f cylindre *m* décortiqueur de riz
e cilindro *m* descascarador de arroz
i cilindro *m* di pilatura di riso
n rijstpelwals
d Reisschälwalze *f*

4407 RICE MILL nut
f rizerie *f*
e fábrica *f* de descascarillar arroz
i risaia *f*
n rijstpellerij
d Reismühle *f*

4408 RIDDLE iwt,sor
f tamis *m* de sable
e tamiz *m* de arena
i vaglio *m* di sabbia
n zandzeef
d Sandsieb *n*

4409 RIDER, BALANCE RIDER lab
f cavalier *m*
e caballero *m*, jinetillo *m*
i cavaliere *m*
n ruiter
d Reiter *m*

4410 RIDER CASK bre
f foudre *f* à rangée supérieure
e tanque *m* superior
i botte *f* superiore
n topvat *n*, zadelvat *n*
d Sattelfass *n*

4411 RIFFLE CONCENTRATOR, SLUICE CONCENTRATOR min
f appareil *m* de concentration de minerais à râteau

e concentrador *m* de minerales con rastrillo
i concentratore *m* di minerali a rastrello
n ertsconcentreerinrichting met doorlaatgoot
d Erzanreicherungsanlage *f* mit Durchlassrinne

4412 RIFFLER min,sor
f classeur *m* de minerais
e clasificador *m* de minerales
i classificatore *m* di minerali
n ertsclassificeerder
d Erzklassierer *m*

4413 RIGID DIE pla,pre
f moule *m* rigide
e molde *m* rígido
i forma *f* rigida, stampo *m* rigido
n stijve mal
d starre Form *f*

4414 RING BURNER bur,fue
f brûleur *m* annulaire
e quemador *m* anular
i bruciatore *m* anulare
n ringbrander
d Ringbrenner *m*

4415 RING FOLLOWER pre
f bague *f* filetée
e anillo *m* roscado interiormente
i anello *m* filettato
n schroefdraadmatrijs
d Gewindering *m*, Gewindestift *m*

4416 RING GATE, RING PENCIL GATE, FILTER GATE pla
f attaque *m* annulaire
e entrada *f* anular
i entrata *f* anulare
n ringaansnijding
d Ringanschnitt *m*

4417 RING KILN, ANNULAR KILN, HOFFMANN KILN cer
f four *m* annulaire
e horno *m* anular
i forno *m* anulare
n ringoven
d Ringofen *m*

4418 RING NOZZLE pre
f porte-filière *m* en forme d'anneau
e tobera *f* anular
i ugello *m* anulare
n ringvormig spuitmondstuk *n*
d Ringdüse *f*

4419 RING-ROLLER MILL brk
f broyeur *m* à anneau et rouleaux
e molino *m* de anillo y rodillos
i mulino *m* ad anello e rulli
n ringmolen met rollen
d Ringmühle *f* mit Rollen

4420 RINSER bre
f rinceuse *f*
e enjuagadora *f*
i spruzzatrice *f*
n sproeier
d Ausspritzapparat *m*

4421 RINSING BATH tex
f bain *m* de rinçage
e baño *m* de enjuague
i bagno *m* da risciacquare
n spoelbad *n*
d Spülbad *n*

4422 RINSING TANK tex
f cuve *f* de rinçage
e cuba *f* de enjuague
i tino *m* da risciacquare
n spoelkuip
d Spülbottich *m*

4423 RIPENED FILTER iwt
f filtre *m* préparé
e filtro *m* preparado
i filtro *m* approntato
n geprepareerd filter *n*
d eingearbeiteter Filter *m*

4424 RIPENING BATH pho
f bain *m* de maturation
e baño *m* de maduración
i bagno *m* di maturazione
n rijpingsbad *n*
d Reifungsbad *n*

4425 RIPENING TANK nut
f récipient *m* de maturation
e tanque *m* de maduración
i serbatoio *m* di maturazione
n rijpingstank
d Reifungsgefäss *n*

4426 RIPPLE tex
f drège *f*
e rastrillo *m*
i scapecchiatoio *m*
n repel
d Flachsriffel *m*, Riffelkamm *m*

4427 RIPPLE-ROUND ANODE, elc
GEAR-SHAPED ANODE
f anode *f* crénelée, anode *f* dentelée
e ánodo *m* dentado
i anodo *m* dentato
n tandanode
d Zahnanode *f*

4428 RIPPLE TRAY dis
f plateau *m* perforé à ondulations
e plato *m* perforado ondulado
i piatto *m* forato ondulato
n gegolfde zeefschotel
d Wellsiebboden *m*

4429 RISE, dis
SLOPE
f chute *f*, pente *f*
e pendiente *f*
i pendenza *f*
n afschot *n*
d Steigung *f*

4430 ROAST-SINTERING OVEN, htt,met
BLAST ROASTING OVEN
f four *m* à grillage de frittage, four *m* de frittage sur grille
e horno *m* de sinterización por tostación
i forno *m* ad arrostimento di sinterizzazione
n sinterroostoven
d Sinterröstofen *m*

4431 ROASTING FURNACE, cer
CALCINING FURNACE, KILN
f four *m* de grillage
e horno *m* de tostación
i forno *m* d'arrostimento
n roostoven
d Röstofen *m*

4432 ROASTING PLANT htt,met
f installation *f* de grillage
e equipo *m* de tostación
i impianto *m* d'arrostimento
n roostinstallatie
d Röstanlage *f*

4433 ROCKER min
f berceau *m*, sas *m* mobile
e criba *f* lavadora
i vibrovaglio *m* lavatore
n waszeef
d Schwingtrog *m*

4434 ROCKER CONVEYOR min,tra
f transporteur *m* à secousses
e transportador *m* de sacudidas
i trasportatore *m* a scosse
n schudgoot
d Schüttelrutsche *f*

4435 ROCKING ARC FURNACE htt
f four *m* oscillant à arc
e horno *m* basculante de arco
i forno *m* oscillante ad arco
n schommelende vlamboogoven
d Lichtbogenschaukelofen *m*, schauklender Lichtbogenofen *m*

4436 ROCKING GRATE iwt
f grille *f* oscillante
e rejilla *f* oscilante
i griglia *f* oscillante
n schudrooster *n*
d Schwingrost *m*

4437 ROCKING MILL, brk,min
ARRASTRE
f bocard *m* à pilons
e arrastre *m*
i mulino *m* a pilone, mulino *m* da pestare
n stampmolen
d Arrastra *f*, Pochmühle *f*

4438 ROCKING RESISTOR htt,met
FURNACE
f four *m* oscillant à résistance
e horno *m* basculante de resistencia
i forno *m* oscillante a resistenza
n schommelende weerstandsoven
d Widerstandsschaukelofen *m*, schauklender Widerstandsofen *m*

4439 ROD GRINDER, brk
ROD MILL, BAR MILL
f broyeur *m* à barres
e machacadora *f* de barras
i frantoio *m* a barre
n staafmolen
d Stabmühle *f*

4440 ROLL, gen
BOWL
f cylindre *m*
e cilindro *m*
i cilindro *m*
n wals
d Walze *f*

4441 ROLL BODY, cpt
BODY OF THE ROLL
f corps *m* du cylindre
e cuerpo *m* del cilindro
i corpo *m* del cilindro
n walslichaam *n*
d Walzenkörper *m*

4442 ROLL BOILING TANK tex
DECATIZING TANK
f récipient *m* de décatissage
e cuba *f* de decatizar
i serbatoio *m* di decatissaggio
n afglanstank, decateertank
d Dekatiergefäss *n*

4443 ROLL CALENDER (US), pap
BOWL CALENDER (GB)
f calandre *f* à cylindres
e calandra *f* de cilindros
i calandra *f* a cilindri
n walskalander
d Walzenkalander *m*

4444 ROLL CLEARANCE, pla,rub
BITE, MILL OPENING
f écartement *m*, intervalle *m*, jeu *m*
e espacio *m* perjudicial
i giuoco *m*, interdistanza *f*
n walsafstand, walsspleet
d Spaltweite *f*, Walzenspalt *m*

4445 ROLL COATER pla
f machine *f* à enduire à cylindres
e barnizadora *f* de cilindros, máquina *f* de recubrimiento por cilindros
i spalmatrice *f* a cilindri
n laagopbrenger met walsen
d Walzenbeschichter *m*

4446 ROLL CRUSHER, bre
ROLL MILL
f broyeur *m* à cylindres
e machacadora *f* de cilindros
i frantoio *m* a cilindri
n walsbreker, walsmolen
d Walzenmühle *f*

4447 ROLL FEEDER tra
f alimentateur *m* à cylindres
e alimentador *m* de cilindros
i alimentatore *m* a cilindri
n walsvoedingsapparaat *n*
d Walzenspeisungsapparat *m*, Walzenzuführungsgerät *n*

4448 ROLL MIXER WITH mix
HYDRAULIC ADJUSTMENT
f mélangeur *m* à cylindres à dispositif de serrage hydraulique
e mezclador *m* de cilindros con dispositivo de graduación hidráulico
i mescolatore *m* a cilindri con dispositivo di chiusura idraulico
n mengwals met hydraulische spleetregeling
d Mischwalze *f* mit hydraulischer Einstellung

4449 ROLLED ANODE elc
f anode *f* laminée
e ánodo *m* laminado
i anodo *m* laminato
n gewalste anode
d Walzanode *f*

4450 ROLLER gen
f rouleau *m*
e rodillo *m*
i rullo *m*
n rol
d Rolle *f*

4451 ROLLER CONVEYOR tra
f transporteur *m* à rouleaux
e transportador *m* con rodillos
i trasportatore *m* a rulli
n rollenbaan
d Rollenbahn *f*

4452 ROLLER MILL, pla,rub
MIXING MILL, MIXING
ROLLERS, OPEN MILL
f malaxeur *m* à cylindres, mélangeur *m* ouvert
e amasadora *f* de cilindros, mezclador *m* abierto, mezclador *m* de cilindros
i mescolatore *m* a cilindri, mescolatore *m* aperto
n mengwals, open mengwals
d offene Mischwalze *f*, Walzenmischer *m*

4453 ROLLER PRESS so
f broyeuse *f* à rouleaux
e desmenuzadora *f* de rodillos
i truciolatrice *f* a rulli
n rollenpers
d Rollenpresse *f*

4454 ROLLER PRINTING MACHINE te
f machine *f* d'impression à rouleaux
e estampadora *f* de rodillos
i stampatore *m* a rulli
n roldrukmachine
d Rollendruckmaschine *f*

4455 ROLLING-DOOR hyg,la
STERILIZER
f stérilisateur *m* à volet
e esterilizador *m* de puerta arrollable

i sterilizzatore *m* a saracinesca
n sterilisator met rolluik
d Sterilisator *m* mit Rolladen

4456 ROSE BURNER htt
f brûleur *m* à couronne
e quemador *m* de corona
i bruciatore *m* a corona
n kroonbrander
d Kronenbrenner *m*

4457 ROT-STEEPING BATH tex
f bain *m* alcalin pour éliminer l'apprêt
e baño *m* alcalino para eliminar el apresto
i bagno *m* alcalino per eliminare l'appretto
n alkalisch bad *n* voor het verwijderen van het appreteer-middel
d Alkalibad *n* zur Beseitigung des Appreturmittels

4458 ROTAGRATE (US), ROTARY GRATE MAGNET (GB) sor
f grille *f* magnétique rotative
e pantalla *f* magnética giratoria
i griglia *f* magnetica rotativa
n draaiende roostermagneet
d magnetischer Drehrost *m*

4459 ROTARY AUTOCLAVE pre
f autoclave *m* rotatif
e autoclave *f* giratoria
i autoclave *f* rotativa
n roterende autoclaaf
d Rotationsautoklav *m*

4460 ROTARY BURNER bur
f brûleur *m* rotatif
e quemador *m* giratorio
i bruciatore *m* rotativo
n draaibrander
d Drehbrenner *m*

4461 ROTARY-CATHODE CELL elc
f pile *f* à cathode rotative
e pila *f* de cátodo giratorio
i pila *f* a catodo rotativo
n element *n* met draaiende katode
d Element *n* mit Drehkatode

4462 ROTARY COMPRESSOR pre
f compresseur *m* rotatif
e compresor *m* giratorio
i compressore *m* rotativo
n roterende compressor
d Umlaufverdichter *m*

4463 ROTARY CONE INCINERATOR iwt
f installation *f* d'incinération à cône rotatif
e equipo *m* de incineración de cono giratorio
i impianto *m* d'incenerazione a cono rotativo
n verbrandingsinstallatie met draaiende kegel
d Brennkegelanlage *f*

4464 ROTARY CRUSHER bre
f broyeur *m* rotatif
e machacadora *f* giratoria
i frantoio *m* rotativo
n draaiende breker
d Umlaufbrecher *m*

4465 ROTARY DISK (US), ROTARY SHELF (GB) pla
f plateau *m* rotatif
e plato *m* giratorio
i piatto *m* rotativo
n draaiplateau *n*
d Drehteller *m*

4466 ROTARY DISK COLUMN dis
f colonne *f* à disques rotatifs
e columna *f* de discos giratorios
i colonna *f* a dischi rotativi
n draaischijvenzuil
d Drehscheibenkolonne *f*

4467 ROTARY DISK SPRAYER coa
f atomiseur *m* à disque rotatif
e atomizador *m* de disco giratorio
i atomizzatore *m* a disco rotativo
n draaischijfsproeier
d Drehscheibenspritze *f*

4468 ROTARY DISTILLATION COLUMN dis
f colonne *f* de distillation rotative
e columna *f* de destilación giratoria
i colonna *f* di distillazione rotativa
n draaiende destillatiekolom
d drehende Destillationssäule *f*

4469 ROTARY DRIER dry
f sécheur *m* rotatif
e secadero *m* giratorio
i essiccatoio *m* rotativo
n droogcentrifuge
d Trockenzentrifuge *f*

4470 ROTARY DRUM sep
f tambour *m* rotatif
e tambor *m* giratorio
i tamburo *m* rotativo

n draaitrommel
d Schleudertrommel *f*

4471 ROTARY FILM EVAPORATOR eva
f évaporateur *m* rotatif à couche mince
e evaporador *m* giratorio de capa delgada
i evaporatore *m* rotativo a strato sottile
n roterende dunne-laagverdamper
d Rotationsdünnschichtverdampfer *m*

4472 ROTARY FILTER, DRUM FILTER fil
f filtre *m* à tambour
e filtro *m* giratorio, tambor *m* filtrante
i filtro *m* a tamburo
n trommelfilter *n*
d Trommelfilter *m*

4473 ROTARY FURNACE htt
f four *m* rotatif
e horno *m* giratorio
i forno *m* rotativo
n draaioven
d Drehofen *m*

4474 ROTARY GRATE fue
f grille *f* tournante
e rejilla *f* giratoria
i griglia *f* rotativa
n draaiend rooster *n*
d Drehrost *m*

4475 ROTARY GRATE PRODUCER gas
f générateur *m* de gaz à grille rotative
e generador *m* de gas de rejilla giratoria
i generatore *m* di gas a griglia rotativa
n gasgenerator met draaiend rooster
d Drehrostgasgenerator *m*

4476 ROTARY HEARTH KILN cer
f four *m* à solé rotative
e horno *m* de solera giratoria
i forno *m* a suola rotativa
n draaiende-haardoven
d Drehherdofen *m*

4477 ROTARY HOPPER VACUUM DEWATERER cle
f appareil *m* de déshydratation à vide à trémie rotative
e desaguador *m* de vacío con tolva giratoria
i apparecchio *m* di disidratazione con tramoggia rotativa
n ontwateringsapparaat *n* met draaiende trechter
d Entwässerungsgerät *n* mit Drehtrichter

4478 ROTARY KILN htt
f four *m* à tambour
e horno *m* de tambor
i forno *m* a tamburo
n trommeloven
d Trommelofen *m*

4479 ROTARY MERCERIZER tex
f merceriseuse *f* rotative
e mercerizador *m* giratorio
i mercerizzatore *m* rotativo
n draaiend merceriseerapparaat *n*
d Umlaufmercerisiermaschine *f*

4480 ROTARY PADDLE mix
f palette *f* rotative
e paleta *f* giratoria
i paletta *f* rotativa
n roterende schoep
d Drehschaufel *f*

4481 ROTARY PAN TRAY, SWIVEL TABLE, TURN TABLE rub
f table *f* rotative
e mesa *f* giratoria
i tavola *f* rotativa
n draaibaar spuitblad *n*
d Schwenktablett *n*

4482 ROTARY PELETING MACHINE mcy,pre
f boudineuse *f* de tablettes à couteau rotatif
e budinadora *f* de tabletas con cuchillo giratorio
i trafila *f* pasticcatrice con coltello rotativo
n spuittabletteermachine met draaiend mes
d Tablettenspritzmaschine *f* mit Drehmesser

4483 ROTARY PLATE FEEDER pla,tra
f doseur *m* à plateau rotatif
e alimentador *m* de plato giratorio
i alimentatore *m* a piatto rotativo
n toevoerinrichting met draaiplateau
d Drehplattenspeiser *m*

4484 ROTARY PISTON BLOWER pre
f souffleuse *f* à piston rotatif
e soplante *m* de pistón giratorio

i soffieria *f* a stantuffo rotativo
n blower met draaiende zuiger
d Drehkolbengebläse *n*

4485 ROTARY POT gla
f creuset *m* rotatif
e crisol *m* rotativo
i crogiolo *m* rotante
n draaiende pot
d Drehhafen *m*

4486 ROTARY PRESS (US), DIAL FEED PRESS (GB) pla,rub
f presse *f* à barillet, presse *f* à tourelle revolver
e prensa *f* revólver
i pressa *f* a giostra, pressa *f* rotativa
n carousselpers, roterende pers
d Karussellpresse *f*, Revolverpresse *f*

4487 ROTARY PUMP pre,tra
f pompe *f* rotative
e bomba *f* giratoria
i pompa *f* rotativa
n rotatiepomp
d Umlaufpumpe *f*

4488 ROTARY REGENERATIVE HEATER boi
f réchauffeur *m* régénérateur rotatif
e calentador *m* regenerador giratorio
i riscaldatore *m* rigeneratore rotativo
n draaiende regenerator
d Drehregenerator *m*

4489 ROTARY SCREEN, DRUM SCREEN, SCREENING DRUM sor
f tambour *m* cribleur, tambour *m* tamiseur
e cuba *f* cilíndrica, cuba *f* de tambor
i staccio *m* a tamburo, vaglio *m* rotativo
n zeeftrommel
d Siebtrommel *f*, Trommelrechen *m*, Trommelsieb *n*

4490 ROTARY SHAFT SEAL cpt
f fermeture *f* d'axe rotatif
e cierre *m* de eje giratorio
i chiusura *f* d'asse rotativo
n afsluiting voor roterende as
d Drehachsenverschluss *m*

4491 ROTARY SHAKER lab
f secoueuse *f* rotative
e sacudidor *m* giratorio
i apparecchio *m* a scosse rotativo
n draaiend schudapparaat *n*
d Umlaufschüttelmaschine *f*

4492 ROTARY SHELF DRIER, DISK DRIER dry
f sécheur *m* à plateaux
e secador *m* de platos
i essiccatore *m* a piatti
n draaitafeldroogapparaat *n*
d Tellertrockner *m*

4493 ROTARY SLIDE VALVE cpt
f vanne *f* à plateaux tournants
e válvula *f* de placa giratoria
i cassetta *f* di distribuzione rotativa
n draaiklep, plunjerklep
d Drehschieber *m*

4494 ROTARY SMELTER htt
f four *m* de fusion rotatif
e horno *m* de fusión giratorio
i forno *m* di fusione rotativo
n draaiende smeltoven
d Umlaufschmelzofen *m*

4495 ROTARY SOLDERING TABLE, GENTLEMAN wel
f table *f* rotative pour soudure
e mesa *f* giratoria para soldadura
i tavola *f* rotativa per saldatura
n roterende soldeertafel
d drehender Löttisch *m*

4496 ROTARY STRAINER fil,pap
f épurateur *m* de pâte à papier rotatif
e filtro *m* papel giratorio
i filtro *m* di carta rotativo
n draaiend pulpfilter *n*
d Umlaufstoffänger *m*

4497 ROTARY TABLE PRESS rub
f presse *f* à tableau rotatif
e prensa *f* de mesa giratoria
i pressa *f* con tavola rotativa
n draaitafelpers
d Drehtischpresse *f*

4498 ROTARY TANK gla
f bassin *m* tournant
e cuba *f* rotativa
i tino *m* rotante
n draaiende wan
d Drehwanne *f*

4499 ROTARY TROMMEL CRYSTALLIZER ins,lab
f cristallisoir *m* à tambour rotatif
e cristalizador *m* de tambor giratorio
i cristallizzatore *m* a tamburo rotativo
n kristalliseerapparaat *n* met draaitrommel
d Kristallisierapparat *m* mit Drehtrommel

4500 ROTARY VACUUM DRYING DRUM dry
f séchoir *m* à tambour rotatif à vide
e secadero *m* con tambor giratorio al vacío
i essiccatore *m* con tamburo rotativo in vuoto
n roterende vacuümdroogtrommel
d rotierende Vakuumtrockentrommel *f*

4501 ROTARY VACUUM FILTER fil
f filtre *m* à vide rotatif
e filtro *m* de vacío giratorio
i filtro *m* a vuoto rotativo
n draaiend vacuümfilter *n*
d Vakuumdrehfilter *m*

4502 ROTARY VACUUM PUMP vac
f pompe *f* à vide rotative
e bomba *f* de vacío giratoria
i pompa *f* a vuoto rotativa
n roterende vacuümpomp
d Umlaufvakuumpumpe *f*

4503 ROTARY VALVE cpt
f soupape *f* rotative
e válvula *f* giratoria
i valvola *f* rotativa
n draaiklep
d Drehventil *n*

4504 ROTARY VANE FEEDER, VANE FEEDER cer
f doseur *m* à secteur
e alimentador *m* giratorio de aletas
i alimentatore *m* a settore
n draaiende-schoepenvoeder
d Drehschaufelspeiser *m*

4505 ROTARY VANE SPRAYER coa
f atomiseur *m* à palette rotative
e atomizador *m* de paleta giratoria
i atomizzatore *m* a pala rotativa
n roterende schoepensproeier
d Drehschaufelspritze *f*

4506 ROTARY VISCOMETER, ROTARY VISCOSIMETER in
f viscosimètre *m* rotatif
e viscosímetro *m* giratorio
i viscosimetro *m* rotativo
n rotatieviscosimeter
d Rotationsviskosimeter *n*

4507 ROTARY VULCANIZING DRUM AUTOCLAVE pr
f chaudière *f* autoclave à tambou rotatif
e autoclave *f* de vulcanización con tambor giratorio
i autoclave *f* di meccanizzazione con tamburo rotativo
n roterende-trommelvulcaniseer autoclaaf
d Vulkanisationsdrehtrommel-autoklav *m*

4508 ROTOR, BLADE c
f palette *f*, rotor *m*
e paleta *f*, rotor *m*
i paletta *f*, rotore *m*
n kneedas, rotor
d Knetkörper *m*, Rotor *m*

4509 ROTTING PLANT, RETTING PLANT t
f rouisseur *m*, routoir *m*
e enriadero *m*
i impianto *m* di macerazione
n roterij, vlasroterij
d Flachsrösterei *f*, Rösterei *f*

4510 ROTTING TANK, FERMENTATION TANK, RAG FERMENTATION TANK, STU VAT p
f récipient *m* de pourrissage
e pudridor *m*
i serbatoio *m* di putrefazione
n ontsluitingstank, rottingsvat *n*
d Faulbütte *f*

4511 ROTTING TANK, RETTING TANK t
f récipient *m* de rouissage
e tanque *m* de enriamiento
i serbatoio *m* di macerazione
n rootkuip, vlasrootkuip
d Röstbottich *m*

4512 ROUGHING BASIN iwt
f bassin *m* dégrossisseur
e artesa *f* desbastadora, artesa *f* preliminar
i bacino *m* di sgrossamento
n grofzuiveringsbekken *n*
d Grobklärbecken *n*

4513 ROUGHING FILTER, COARSE FILTER fil
f préfiltre *m*
e filtro *m* grueso
i filtro *m* grosso, filtro *m* sgrossatore
n grof filter *n*
d Grobfilter *m*

4514 ROUND ANODE, STICK ANODE elc
f anode *f* ronde
e ánodo *m* rondo
i anodo *m* rotondo
n ronde anode
d Rundanode *f*

4515 ROUND HOLE SIEVE sor
f tamis *m* à perforations rondes
e criba *f* de agujeros redondos
i staccio *m* a fori circolari
n ronde-gatenzeef
d Rundlochsieb *n*

4516 ROUND KILN, BEEHIVE KILN htt
f étuve *f* à ruche
e estufa *f* de colmena
i stufa *f* ad alveare
n stolpvormige droogstoof
d Bienenkorbtrockenofen *m*

4517 ROUSER, RAKE bre
f balai *m*
e mezclador *m*
i gruccia *f* per mescolare
n roerploeg
d Aufziehkrücke *f*

4518 ROUSING APPARATUS bre
f appareil *m* à mélanger la levure
e mezclador *m* de levadura
i arieggiatore *m* del lievito
n wortluchtapparaat *n*
d Aufziehapparat *m*

4519 RUBBER BLANKET tex
f blanchet *m* de caoutchouc
e tela *f* de caucho
i placca *f* di gomma
n rubberdrukdoek
d Gummituch *n*

4520 RUBBER LINING coa
f garnissage *m* en caoutchouc
e guarnición *f* de goma
i guarnizione *f* in gomma
n rubbervoering
d Gummifutter *n*

4521 RUBBER MIXER rub
f mélangeur *m* de caoutchouc
e mezclador *m* de goma
i mescolatore *m* di gomma
n rubbermengwals
d Gummimischwalze *f*

4522 RUBBER PACKING cpt
f garniture *f* de caoutchouc
e empaquetadura *f* de goma
i guarnizione *f* di gomma
n rubberpakking
d Gummipackung *f*

4523 RUBBER SOLUTION MIXER, CEMENT MIXER rub
f malaxeur *m* à dissolution, mélangeur *m* à dissolution
e mezcladora *f* para soluciones
i agitatore *m* per soluzioni, mescolatore *m* per soluzioni
n solutiemolen
d Gummilöser *m*, Lösungskneter *m*, Lösungsmastikator *m*

4524 RUBBER SPREADER (GB), KNIFE COATER (US) rub
f machine *f* à enduire avec râcle sur rouleau
e máquina *f* engomadora
i spalmatrice *f* a lama libera
n strijkmachine
d Rakelstreichmaschine *f*, Walzenstreichmaschine *f*

4525 RUBBER WASH MILL rub
f crepeuse *f*
e cilindro *m* lavador
i depuratore *m* per lavaggio
n waswals
d Waschwalze *f*

4526 RUBBING CONE, RUBBING LEATHERS, TRUMPET GUIDE, TUNNEL GUIDE gla
f frotteur *m*
e cono *m* de guía, rotafrotador *m*
i imbuto *m*
n wrijfconussen *pl*
d Nitschelwerk *n*

4527 RUBBING LEATHER tex
f cuir *m* frotteur

e cuero *m* frotador
i cuoio *m* da fregare
n frotteerleder *n*
d Frottierleder *n*, Würgelleder *n*

4528 RUGGED-TIP GLASS ELECTRODE elc
f électrode *f* de verre blindé
e electrodo *m* de vidrio blindado
i elettrodo *m* di vetro schermato
n stootvaste glaselektrode
d stossfeste Glasselektrode *f*

4529 RUMBLER (GB), RATTLER (US), TUMBLING BARREL (GB) cle
f tonneau *m* de dessablage
e tambor *m* desarenador
i buratto *m* sterratore
n poetstrommel
d Putztrommel *f*, Scheuertrommel *f*

4530 RUN-DOWN TANK dis
f réservoir *m* intermédiaire
e tanque *m* de descarga
i serbatoio *m* di reparto
n opvangvat *n*
d Auffanggefäss *n*

4531 RUNNER pla,pre
f canal *m* principal
e canal *m* principal
i canale *m* principale
n aanspuitingskanaal *n*
d Angusskanal *m*

4532 RUNNER CLOTH, BACK CLOTH, UNDERCLOTH mat,pla
f bande *f* ensouple, feuille *f* ensouple
e correa *f* acompañante
i nastro *m* d'accompagnamento
n meelopende baan, meelopende foelie
d Mitläuferbad *n*, Mitläuferfolie *f*

4533 RUNNER THROUGH (US), FOUNTAIN, LAUNDER, SPOUT (GB), TAPPING SPOUT (GB) fo
f canal *m* de coulée, chenal *m* de coulée
e canal *m* de colada
i canale *m* di colata
n gietgoot, schenk, tapgoot
d Abstichrinne *f*, Förderrinne *f*, Rutsche *f*

4534 RUNNING-OUT FIRE (GB), DANDY (US) htt,me
f four *m* de premier affinage
e horno *m* de primer afino
i forno *m* di prima affinazione
n eerste raffineeroven
d Frischereiofen *m*

4535 RUNNING VALVE, PASSAGE VALVE cp
f soupape *f* de passage
e válvula *f* de pasaje
i valvola *f* di passaggio
n doorlaatklep
d Durchlaufventil *n*

4536 RUNOFF GUTTER iwt,tra
f gouttière *f*
e vertedor *m*
i canaletto *m* di scolo, condotto *m* di scolo
n afvoergoot
d Ablaufrinne *f*

4537 RUST REMOVAL PLANT cl
f installation *f* de dérouillage
e equipo *m* de remoción del herrumbre
i impianto *m* di rimozione della ruggine
n ontroestingsinstallatie
d Entrostungsanlage *f*

S

4538 SACCHARIFICATION VESSEL bre
f cuve *f* de conversion en sucre, cuve *f* de saccharification
e cuba *f* de sacarificación
i serbatoio *m* di saccarificazione
n sacharificatievat *n*, versuikeringsvat *n*
d Saccharifikationsgefäss *n*, Verzuckerungsgefäss *n*

4539 SACCHARIMETER, SACCHAROMETER, SUGAR HYDROMETER bre,ins
f hydromètre *m* à solution sucrée, saccharimètre *m*
e sacarómetro *m*
i saccarimetro *m*
n sacharimeter, sacharometer
d Saccharimeter *n*, Zuckergehaltsmesser *m*

4540 SACK-FILLING MACHINE tra
f machine *f* à remplir sacs
e máquina *f* rellenasacos
i macchina *f* da riempire sacchi
n zakkenvulmachine
d Absackmaschine *f*

4541 SACK FILTER (US), BAG FILTER (GB) cle,met
f filtre *m* à sac
e filtro *m* de bolsa
i filtro *m* a sacco
n zakfilter *n*
d Beutelfilter *m*, Taschenfilter *m*

4542 SACK FILTRATION PLANT (US), BAG FILTRATION PLANT (GB) fil
f installation *f* de filtrage en sacs
e equipo *m* de filtración en bolsas
i impianto *m* di filtrazione in sacchi
n zakfilterinstallatie
d Taschenfilteranlage *f*

4543 SACRIFIED ANODE, PROTECTIVE ANODE cor
f anode *f* de protection, anode *f* sacrifiée
e ánodo *m* de protección
i anodo *m* di protezione
n beschermende anode
d Schutzanode *f*

4544 SADDLE cer
f support *m* de boîte réfractaire
e soporte *m* de caja refractaria
i sopporto *m* di cassetta refrattaria
n baksteun
d Brennkapselstütze *f*

4545 SADDLE PACKING dis
f garnissage *m* en forme de selle, selle *f* de Berl
e relleno *m* de monturas
i riempimento *m* a sella
n zadelvulling
d Sattelkörper *m*

4546 SAFETY CONTAINER, CONTAINER, CONTAINMENT VESSEL nue
f récipient *m* de sûreté
e recipiente *m* de seguridad
i involucro *m* del reattore, involucro *m* di contenimento
n veiligheidsvat *n*
d Sicherheitsbehälter *m*

4547 SAFETY FUNNEL lab
f tube *m* de sûreté
e tubo *m* de seguridad
i tubo *m* di sicurezza
n veiligheidstrechter
d Sicherheitstrichter *m*

4548 SAFETY FUSE expl
f cordon *m* de sûreté, mèche *f* de sûreté
e mecha *f* de seguridad
i miccia *f* a combustione lenta
n veiligheidslont
d Sicherheitszündschnur *f*

4549 SAFETY INJECTION SYSTEM clo,nue
f système *m* d'inondation de sûreté
e equipo *m* de inundación de seguridad
i impianto *m* d'allagamento di sicurezza
n veiligheidsoverstromingsinstallatie
d Sicherheitsüberflutungsanlage *f*

4550 SAFETY VALVE cpt,sap
f soupape *f* de sûreté
e válvula *f* de seguridad
i valvola *f* di sicurezza
n veiligheidsklep
d Sicherheitsventil *n*

4551 SAGGAR, cer
SAGGER
f boîte *f* réfractaire
e caja *f* refractaria, gaceta *f* refractaria
i casella *f* refrattaria, cassetta *f* refrattaria
n vuurvaste bak
d Brennkapsel *f*

4552 SALIFICATION BATH, sal
SALIFYING BATH
f bain *m* de salification
e baño *m* de salificación
i bagno *m* di salificazione
n zoutvormingsbad *n*
d Salzbildungsbad *n*

4553 SALIMETER (US), ins
SALINIMETER (GB),
SALINOMETER (GB),
SALOMETER (US)
f aréomètre *m* pour saumures, salinomètre *m*
e pesasales *m*, salinómetro *m*
i salinometro *m*
n salinometer, zoutgehaltemeter
d Lakeprober *m*, Salzgehaltmesser *m*, Salzwaage *f*

4554 SALINE, sal
SALT PANS, EVAPORATION PONDS
f marais *pl* salants, saline *f*
e salina *f*
i salina *f*
n zouttuin
d Meersaline *f*, Salzgarten *m*

4555 SALT BATH gen
f bain *m* de sel
e baño *m* de sal
i bagno *m* di sale
n zoutbad *n*
d Salzbad *n*

4556 SALT-BATH FURNACE htt,met
f four *m* à bain de sel
e horno *m* de baño de sal
i forno *m* a bagno di sale
n zoutbadoven
d Salzbadofen *m*

4557 SALT BRIDGE ins,soa
f pont *m* salin
e puente *m* salino
i ponte *m* salino
n zoutbrug
d Salzbrücke *f*

4558 SALT GRAINER soa
f évaporateur *m* de sel
e evaporador *m* de sal
i concentratore *m* di sale a serpentina
n open kookpan, zoutindikker
d Salzeindicker *m*

4559 SALT PIT mir
f mine *f* de sel, saline *f*
e mina *f* de sal, salina *f*
i miniera *f* di sale, miniera *f* di salgemma, salina *f*
n zoutgroeve, zoutmijn
d Salzgrube *f*

4560 SALT SCREEN sp
f écran *m* à tungstate de calcium
e pantalla *f* de tungstato cálcico
i schermo *m* a tungstato di calcio
n calciumwolframaatscherm *n*
d Kalziumwolframatschirm *m*

4561 SALTING BATH ph
f bain *m* de salage, bain *m* de salaison
e baño *m* regulador del contraste
i bagno *m* regolatore del contrast
n zout contrastregelbad *n*
d Salzkontrastreglerbad *n*

4562 SALTING BOX, te
SALTING-ON BOX
f boîte *f* correctrice de couleur
e caja *f* correctora de color
i cassetta *f* correttrice di colore
n kleurcorrectiebak
d Farbenverbesserungskasten *m*

4563 SALTING-OUT PLANT, sc
GRAIN-OUT PLANT,
GRAINING-OUT PLANT
f installation *f* de relargage
e equipo *m* de salado
i impianto *m* di salatura
n uitzoutinstallatie
d Aussalzanlage *f*

4564 SALTING-OUT VESSEL, sc
GRAIN-OUT VESSEL,
GRAINING-OUT VESSEL
f cuve *f* de relargage
e cuba *f* de salado
i tino *m* di salatura
n uitzoutvat *n*
d Aussalzgefäss *n*

4565 SAMPLE CELL, e
PILOT CELL, REFERENCE C
f accumulateur *m* de référence

e acumulador *m* de referencia
i accumulatore *m* di riferimento
n controleaccu
d Kontrollsammler *m*,
Prüfakkumulator *m*

4566 SAMPLE THIEF tes
f outil *m* d'échantillonnage
e captador *m* de muestras
i utensile *m* di campionatura
n monstersteker
d Probenheber *m*

4567 SAMPLING COCK, tes
SAMPLING TAP
f robinet *m* de prise d'échantillons
e grifo *m* de toma de muestras
i rubinetto *m* per la presa di campioni
n monstertapkraan
d Probehahn *m*

4568 SAMPLING DEVICE tes
f appareil *m* de prise d'échantillons
e dispositivo *m* de toma de muestras
i dispositivo *m* per la presa di campioni
n monstersteker
d Probeentnahmegerät *n*

4569 SAMPLING PIPETTE tes
f pipette *f* à prélever des échantillons
e pipeta *f* muestreadora
i pipetta *f* per la presa di campioni
n steekpipet
d Stechpipette *f*

4570 SAMPLING POINT tes
f point *m* de prise des échantillons
e punto *m* de toma de muestras
i punto *m* da presa di campioni
n monstersteekpunt *n*
d Probenahmestelle *f*

4571 SAMPLING TUBE, nut
PLUNGING SIPHON
f tâte-vin *m*
e catavinos *m*
i alzavino *m*
n steekhevel
d Stechheber *m*

4572 SAND BATH htt,lab
f bain *m* de sable
e baño *m* de arena
i bagno *m* di sabbia
n zandbad *n*
d Sandbad *n*

4573 SAND CUTTER, bre
SAND DISINTEGRATOR
f diviseur *m* du sable
e desintegrador *m* de la arena
i disintegratore *m* della sabbia
n zandwolf
d Sandschleuder *f*

4574 SAND FILTER fil,iwt
f filtre *m* à sable
e filtro *m* de arena
i filtro *m* a sabbia
n zandfilter *n*
d Sandfilter *m*

4575 SAND MILL, brk
MULLER, MULLER MIXER,
MULLING MACHINE
f frotteur *m*
e amasador *m*
i molazza-mescolatore *m*
n koldermolen, kollergang
d Kollergang *m*

4576 SAND SEAL cer
f fermeture *f* à sable
e cierre *m* de arena
i chiusura *f* a sabbia
n zandafsluiting
d Sandverschluss *m*

4577 SAND SLUICE fil,iwt
f appareil *m* de dessablage à labyrinthe
e desarenador *m* laberíntico
i filtro *m* di sabbia labirintico
n labyrintzandfilter *n*
d Labyrinthsandfilter *m*

4578 SAND TRAP iwt
f dessableur *m*
e desarenador *m*, filtro *m* de arena
i dissabbiatore *m*
n zandvanger
d Sandfang *m*

4579 SAND TRAP pap
f piège *m* de boue
e caja *f* de fango
i cassetta *f* di fango
n pulpvuilvanger
d Sandfang *m*

4580 SANDBLAST NOZZLE cle
f lance *f* à jet de sable
e tobera *f* lanza-arena
i ugello *m* per getto di sabbia
n zandstraalmondstuk *n*
d Sandstrahldüse *f*

4581 SANDBLASTER, cle
SANDBLASTING MACHINE
f sableuse *f*
e chorreadora *f* de arena
i sabbiatrice *f*
n zandstraalmachine
d Sandstrahlgebläse *n*

4582 SANDING DISK pol
f disque *m* abrasif à sable
e disco *m* abrasivo de arena
i disco *m* abrasivo a sabbia
n fijnslijpschijf
d Feinschleifscheibe *f*

4583 SANFORIZING TANK tex
f récipient *m* de sanforisation
e tanque *m* de sanforización
i serbatoio *m* di sanforizzazione
n sanforiseertank
d Sanforisiergefäss *n*

4584 SANITARY LAND FILL PLOT hyg
(US),
CONTROLLED TIPPING GROUND
(GB)
f terrain *m* de décharge contrôlée
e terreno *m* de descarga
controlada
i terreno *m* di scarico
controllato
n aangewezen vuilbegraafplaats,
aangewezen vuilnisstortplaats
d Grundstück *n* für geordnete
Ablage

4585 SAPONIFICATION FLASK lab,soa
f matras *m* à saponification
e matraz *m* de saponificación
i matraccio *m* a saponificazione
n verzepingskolf
d Verseifungskolben *m*

4586 SAPONIFICATION TANK soa
f récipient *m* de saponification
e recipiente *m* de saponificación
i serbatoio *m* di saponificazione
n verzepingsvat *n*
d Verseifungsgefäss *n*

4587 SATIN-FROSTING PLANT gla
f installation *f* de glaçage,
installation *f* de satinage
e equipo *m* de satinar
i impianto *m* di rasatura
n satineerinstallatie
d Satinieranlage *f*

4588 SATURATED CALOMEL elc
ELECTRODE
f électrode *f* au calomel saturée
e electrodo *m* de calomelanos
saturado
i elettrodo *m* al calomelano
saturato
n verzadigde calomelelektrode
d gesättigte Kalomelhalbzelle *f*

4589 SATURATION PAN, sin
CARBONATION PAN
f cuve *f* de saturation
e cuba *f* de carbonatación
i tino *m* di carbonatazione
n carbonatatieketel
d Entkalkungsgefäss *n*, Scheide-
pfanne *f*

4590 SATURATION TOWER, sin
CARBONATATION TOWER,
CARBONATION TOWER
f colonne *f* de saturation
e columna *f* de carbonatación,
torre *f* de saturación
i colonna *f* di carbonatazione,
colonna *f* di saturazione
n carbonatatiekolom
d Entkalkungskolonne *f*,
Sättigungskolonne *f*,
Saturationssäule *f*

4591 SATURATOR, bre
INJECTION PUMP
f saturateur *m*
e saturador *m*
i pompa *f* di saturazione,
saturatore *m*
n injecteerpomp
d Imprägnierungspumpe *f*

4592 SAUCER cpt
f cavité *f* pour la garniture
e cavidad *f* para la empaquetadur
i alloggiamento *m* per guarnizione
n pakkingkamer
d Packungsraum *m*

4593 SAUCER WHEEL po
f disque *m* à polir en soucoupe
e disco *m* para pulido en platillo
i disco *m* di pulitura in piatto
n schaalvormige polijstschijf
d schalenförmige Polierscheibe *f*

4594 SAVE-ALL, pa
CATCH-ALL, PULP SAVER,
RECLAIMER
f ramasse-pâte *m*
e recogedor *m* de pasta
i ricuperatore *m* di polpa
n vezelstofvanger
d Faserstoffänger *m*, Stoffänger *m*

4595 SAWDUST DRIER dry
f séchoir *m* à sciure
e secadero *m* con serrín de madera
i essiccatoio *m* a segatura
n met zaagsel gevulde droger
d Sägemehltrockner *m*

4596 SAWDUST FLOOR fil
f sol *m* à couche de sciure
e piso *m* con capa de serrín de madera
i piano *m* con strato di segatura
n vloer met zaagsellaag
d Boden *m* mit Sägemehlschicht

4597 SCALDING VESSEL, SEETHING VESSEL, SMOTHERING VESSEL tex
f échaudoir *m*
e cuba *f* de escaldar, escaldadera *f*
i serbatoio *m* da scottare
n broeiketel
d Abbrühgefäss *n*

4598 SCALE BREAKER cle
f laminoir *m* casse-oxyde, laminoir *m* de décalaminage
e laminador *m* descascarador
i laminatoio *m* di rimozione della scagliatura
n oxydelaagbreker
d Zunderbrechwalze *f*

4599 SCALE REMOVAL TOOL boi,cle
f dispositif *m* de désincrustation
e dispositivo *m* de desincrustación, quitacrosta *f*
i martellina *f* di disincrostazione
n ketelsteenafbikhamer
d Kesselsteinausklopfer *m*

4600 SCALING STOVE cer,dec
f étuve *f* à formation d'écailles
e estufa *f* de escamación
i stufa *f* d'ossidazione
n oxydeerstoof
d Verzunderungsofen *m*

4601 SCALPER sor
f tamis *m* de protection
e criba *f* para remover la ganga
i vaglio *m* di protezione
n grove beschermingszeef, voorzeef
d Grobschutzsieb *n*

4602 SCALPER, SLEDGER brk,min
f massette *f*, têtu *m*
e quebrantadora *f* preliminar
i mazza *f*
n grofbreker
d Grobbrecher *m*, Steinbrecher *m*

4603 SCANNING HEATING FURNACE, PROGRESSIVE HEATING FURNACE htt
f four *m* à induction et chauffage progressif
e horno *m* de inducción con calentamiento progresivo
i forno *m* ad induzione con riscaldamento progressivo
n inductieoven met voortschrijdend werkstuk
d Induktionsofen *m* mit fortschreitendem Werkstück

4604 SCARFING BURNER cle
f chalumeau *m* de décalaminage
e soplete *m* escarpador
i cannello *m* di scagliatura
n afbrander
d Abbrenner *m*

4605 SCAVENGER cle
f dispositif *m* de rinçage
e dispositivo *m* de enjuagar
i risciacquatoio *m*
n spoelapparaat *n*
d Spülgerät *n*

4606 SCAVENGING PUMP cle,pre
f pompe *f* de refoulement
e bomba *f* de barrido, bomba *f* de expulsión de gases quemados
i pompa *f* da ricuperare olio
n spoelpomp
d Spülpumpe *f*

4607 SCLEROMETER, DUROMETER, HARDNESS TESTER ins,tes
f duromètre *m*, scléromètre *m*
e durómetro *m*
i durometro *m*
n hardheidsbeproevingsapparaat *n*, hardheidsmeter
d Härteprüfer *m*

4608 SCLEROSCOPE, HARDNESS DROP TESTER ins,tes
f scléroscope *m*
e escleroscopio *m*
i durometro *m* per caduta, scleroscopio *m*
n hardheidsbepalingsapparaat *n* met vallende kogel, scleroscoop
d Fallhärteprüfer *m*

4609 SCOOP, lab
LADLE
f cuiller *f* à pot
e cangilón *m*
i cucchiaia *f*
n scheplepel
d Löffel *m*, Schöpfkelle *f*

4610 SCOOP WHEEL tra
f roue *f* à godets
e rueda *f* de paletas
i ruota *f* a palette
n schoepenrad *n*
d Schaufelrad *n*

4611 SCOPOMETER ins
f scopomètre *m*
e escopómetro *m*
i scopometro *m*
n scopometer
d Skopometer *n*

4612 SCORCHING PLANT rub
f installation *f* de grillage
e equipo *m* de prevulcanización
i impianto *m* di scottatura
n aanvulcaniseerinstallatie
d Anbrennanlage *f*, Anvulkanisier-
anlage *f*

4613 SCORCHING TOOL bur
f dispositif *m* à roussir
e dispositivo *m* para chamuscar
i dispositivo *m* da scottare
n schroeiijzer *n*, zengijzer *n*
d Versengeisen *n*

4614 SCORIFICATION FURNACE htt,min
f four *m* d'affinage, four *m* de
scorification
e horno *m* de afino, horno *m* de
escorificación
i forno *m* di raffinazione, forno *m*
di scorificazione
n louteroven
d Läuterofen *m*

4615 SCORIFIER, htt,min
SCORIFYING CRUCIBLE
f creuset *m* à scorifier,
scorificateur *m*
e crisol *m* de escorificación,
escorificador *m*
i crogiolo *m* da scorificare,
scorificatore
n louterkroes
d Läutertiegel *m*

4616 SCOTCH HEARTH, met,min
ORE HEARTH
f four *m* de fusion des minerais de
plomb
e horno *m* de fundición para
minerales de plomo
i forno *m* per fusione dei minerali
di piombo
n loodertssmeltoven
d Bleierzschmelzofen *m*

4617 SCOTCH KILN, cer,htt
SCOVE, ARCHLESS KILN
f four *m* à briques enfermées en
pâte
e horno *m* de ladrillos encerrados
con barro
i forno *m* a mattoni inchiusi in pa
n oven met in brij opgesloten
vuurvaste stenen
d Ofen *n* mit in Schlicker einge-
schlossenen feuerfesten Steinen

4618 SCOURER nu
f tarare *m* ventilateur
e máquina *f* de mondar
i mulino *m* da vagliare,
scortecciatrice *f*
n graanzuiveringsmachine,
machinale wan, wanmolen
d Getreidereinigungsmaschine *f*,
Wannmaschine *f*

4619 SCOURING BLOCK pc
f corps *m* abrasif à polir les
cylindres
e cuerpo *m* abrasivo para pulir
cilindros
i corpo *m* abrasivo da pulire
cilindri
n walsschuurblok *n*
d Walzenscheuerblock *m*

4620 SCOURING MACHINE m
f machine *f* à décaper
e máquina *f* de decapaje
i macchina *f* di decapaggio
n machine voor het verwijderen
van de oxydelaag
d Entzunderungsmaschine *f*

4621 SCOURING MACHINE, te
WOOL SCOURING MACHINE
f machine *f* de dégraissage,
machine *f* de dessuintage
e desgrasadora *f*, máquina *f* de
desuardado
i macchina *f* di primo lavaggio,
macchina *f* di sgrassaggio
n wolontvettingsmachine
d Entfettungsmaschine *f*,
Entschweissungsmaschine *f*

4622 SCOURING TOOL c
f dispositif *m* de récurage

e dispositivo *m* de fregado
i dispositivo *m* da strofinare
n schuurwerktuig *n*
d Scheuergerät *n*

4623 SCRAPER cle
f benne *f* racleuse, raclette *f*, racleur *m*
e rascador *m*
i raschiatore *m*
n schraper
d Kratzer *m*, Schrapper *m*

4624 SCRAPER BLADE cle
f tôle *f* de racleur
e paleta *f* de rascador
i pala *f* raschiatrice
n schraperblad *n*
d Kratzerblech *n*

4625 SCRAPER CONVEYOR tra
f transporteur *m* à raclettes
e transportador *m* de rasquetas, transportador *m* rascador
i trasportatore *m* a raschiatori
n transportband met meenemers
d Kratzbandförderer *m*

4626 SCRAPER RING, lub
OIL RING
f anneau *m* de lubrification
e anillo *m* de lubricación, anillo *m* engrasador
i anello *m* di lubrificazione
n smeergroef
d Schmierrille *f*

4627 SCRAPING ARM cle
f bras *m* racleur
e brazo *m* rascador
i braccio *m* raschiatore
n schraperarm
d Kratzerarm *m*

4628 SCRAPING TOOL, cer
FETTLING TOOL
f couteau *m* racleur
e cuchillo *m* rascador
i coltello *m* da raschiare
n afschrapmes *n*
d Abschrotmesser *n*

4629 SCRATCH BRUSH cle
f brosse *f* métallique
e cepillo *m* de alambre
i spazzola *f* metallica
n metaalborstel, staaldraadborstel
d Drahtbürste *f*, Metallbürste *f*

4630 SCREEN, sor
SIEVING SCREEN
f grille-tamis *m*
e rejilla-tamiz *m*, zaranda *f*
i griglia *f* da vagliare
n zeefrooster *n*
d Siebrost *m*

4631 SCREEN CENTRIFUGAL- nut,sep
FILTER SEPARATOR
f séparateur *m* à tamis à filtre centrifuge
e separador *m* de cedazo con filtro centrífugo
i separatore *m* a staccio con filtro centrifugo
n scheidingszeef met centrifugaal filter
d Scheidesieb *n* mit Zentrifugalfilter

4632 SCREEN CLEANER, cle
SCREEN RAKE
f nettoyeur *m* de grille
e limpiador *m* de zaranda
i sgrigliatore *m*
n roosterhark
d Rostrechen *m*, Rostreiniger *m*

4633 SCREEN HEAD pla,pre,rub
f tête *f* filtrante
e cabezal *m* tamizador
i testa *f* filtrante
n zeefspuitkop
d Siebkopf *m*

4634 SCREEN HEAD pre,rub
EXTRUDER,
STOCK STRAINER, STRAINER,
STRAINING MACHINE
f boudineuse *f* à tamiser
e budinadora *f* para tamizar
i trafila *f* con testa a filtro
n spuitmachine met zeefspuitkop, zeefpers
d Siebkopfspritzmaschine *f*, Siebpresse *f*

4635 SCREEN PACK, pla
WIRE SCREEN
f toile *f* métallique
e tamiz *m* de tela metálica
i tela *f* metallica
n draadzeef, metaalgaas *n*
d Drahtsieb *n*, Metallgaze *f*

4636 SCREEN PRINTING PLANT, tex
SERIGRAPHY PLANT
f installation *f* d'impression à la pellicule

e equipo *m* de estampar con estarcido
i impianto *m* di stampaggio alla pellicola
n zeefdrukinstallatie
d Netzdruckanlage *f*, Siebdruckanlage *f*

4637 SCREEN SHREDDER, COMMINUTOR, DISINTEGRATOR, TRITURATOR — pap
f appareil *m* de dilacération, dilacérateur *m*
e trituradora *f*
i frantumatrice *f*, trituratore *m*
n fijnmolen
d Rechengutzerkleinerer *m*, Rechenwolf *m*

4638 SCREEN-TYPE CENTRIFUGE — cen
f essoreuse *f* à panier perforé
e centrífuga *f* con tamiz
i centrifuga *f* con paniere forato
n zeeftrommelcentrifuge
d Siebmantelzentrifuge *f*

4639 SCREENING DRUM — min,sor
f tambour-classeur *m*
e tambor *m* clasificador
i tamburo *m* classificatore
n klasseertrommel
d Klassiertrommel *f*

4640 SCREENING DRUM, DRUM SCREEN, ROTARY SCREEN — sor
f tambour *m* cribleur, tambour *m* tamiseur
e cuba *f* cilíndrica, cuba *f* de tambor
i staccio *m* a tamburo, vaglio *m* rotativo
n zeeftrommel
d Siebtrommel *f*, Trommelrechen *m*, Trommelsieb *n*

4641 SCREENING PLANT, SIEVE PLANT — sor
f installation *f* de tamisage
e equipo *m* de tamizado
i impianto *m* di stacciatura, impianto *m* da vagliare
n zeefinstallatie
d Absiebanlage *f*, Siebanlage *f*

4642 SCREENING TABLE — sor
f table *f* de tamisage
e mesa *f* de tamizado
i tavola *f* di stacciatura
n zeeftafel
d Plansichter *m*, Siebtisch *m*

4643 SCREEVE, SEALING RING, JOINT SEALING — cpt
f anneau *m* de fermeture, cordon *m* d'étanchéité
e anillo *m* de cierre, anillo *m* de estanqueidad, tapa-juntas *m*
i anello *m* di chiusura, cordone *m* di sigillatura
n dichtingsgroef, dichtingsring
d Dichtungsrille *f*, Dichtungsring

4644 SCREW CONVEYOR, HELICAL CONVEYOR, WORM CONVEYOR — tra
f transporteur *m* hélicoïdal
e transportador *m* helicoidal
i trasportatore *m* elicoidale
n transportworm
d Förderschnecke *f*

4645 SCREW CONVEYOR DRIER — dry
f sécheur *m* avec transporteur à vis sans fin
e secadero *m* con transportador de tornillo sin fin
i essiccatoio *m* con trasportatore a vite senza fine
n droogapparaat *n* met wormaandrijving
d Trockengerät *n* mit Schneckenantrieb

4646 SCREW CONVEYOR FEEDER — ger
f alimentateur *m* à vis sans fin
e alimentador *m* de tornillo sin fin
i alimentatore *m* a vite senza fine
n toevoerinrichting met wormaandrijving
d Schneckenaufgeber *m*, Schneckenzuführgerät *n*

4647 SCREW CONVEYOR-TYPE EXTRACTOR — ger
f extracteur *m* à vis sans fin
e extractor *m* de tornillo sin fin
i estrattore *m* a vite senza fine
n extraheerapparaat *n* met wormaandrijving
d Extrahiergerät *n* mit Schneckenantrieb

4648 SCREW CORE PIN — pla,pr
f broche *f* à trou fileté
e macho *m* roscado, pitón *m* rosca
i anima *f* filettata
n schroefdraaddoorn
d Gewindestift *m*

4649 SCREW CRYSTALLIZER TANK — cr
f cristallisoir *m* à vis sans fin

e cristalizador *m* de tornillo sin fin
i cristallizzatore *m* a vite senza fine
n kristallisator met worm
d Schneckenkristallisator *m*

4650 SCREW EXTRUDER pla,pre
f boudineuse *f* à vis
e budinadora *f* a tornillo, prensa *f* extruidora de husillo
i estrusore *m* a vite, trafila *f* a vite
n schroefextrusiepers, spuitmachine met worm, wormextrusiepers
d Schneckenspritzmaschine *f*, Schneckenstrangpresse *f*

4651 SCREW GAGE (US), SCREW GAUGE (GB), MICROMETER ins
f micromètre *m* à vis
e micrómetro *m* de tornillo
i micrometro *m* a vite
n schroefmicrometer
d Schraublehre *f*, Schraubmikrometer *n*

4652 SCREW PRESS, FRICTION PRESS, FRICTION SCREW PRESS fat
f balancier *m*, presse *f* à vis
e prensa *f* de husillo, prensa *f* de tornillo
i pressa *f* a vite
n frictiepers, schroefpers, spindelpers
d Spindelpresse *f*

4653 SCRUBBER, SCRUBBER TOWER cle,gas
f tour *f* de lavage à ruissellement
e torre *f* depuradora
i torre *f* di depurazione
n sproeitoren
d Sprühturm *m*

4654 SCUM BREAKER iwt
f malaxeur *m* pour la destruction du chapeau
e rompedor *m* de espuma
i frangicrosta *f*, rompitore *m* della copertura di schiuma
n schuimdekbreker
d Schwimmdeckenzerstörer *m*

4655 SCUM COLLECTOR, SCUM SKIMMER, SKIMMER iwt
f racleur *m* pour l'enlèvement de l'ecume
e colector *m* de espuma
i scolmatore *m* di schiuma
n schuimafstrijker
d Schwimmdeckenabstreifer *m*

4656 SCUTCHING MACHINE, HACKLING MACHINE, HECKLING MACHINE tex
f peigneuse *f*
e máquina *f* peinadora, máquina *f* rastrilladora
i macchina *f* di pettinatura
n hekelmachine
d Hechelmaschine *f*

4657 SCUTCHING MILL tex
f écang *m*
e batán *m*
i gramola *f*
n zwingelmolen
d Schwingmühle *f*

4658 SEA CELL elc
f pile *f* actionnée par eau de mer
e pila *f* accionada por agua marina
i pila *f* azionata per acqua di mare
n door zeewater geactiveerd element *n*
d durch Seewasser betätigtes Element *n*

4659 SEA-WATER DESALINATION PLANT cle
f installation *f* de dessalage d'eau de mer
e equipo *m* de desalación de agua de mar
i impianto *m* di desalificazione d'acqua di mare
n zeewaterontziltingsinstallatie
d Meereswasserentsalzungsanlage *f*

4660 SEAL, SEALING cpt
f étanchement *m*, joint *m* étanche
e cierre *m* hermético, junta *f* hermética
i chiusura *f* ermetica, sigillatura *f*
n afdichting
d Abdichtung *f*, Dichtung *f*

4661 SEALED-IN FUEL ELEMENT (GB), ENCAPSULATED FUEL UNIT (US) nue
f élément *m* combustible scellé
e elemento *m* combustible encamisado
i elemento *m* combustibile incamiciato
n ingekapseld splijtstofelement *n*
d eingehülstes Brennstoffelement *n*

4662 SEALING MACHINE, CAPSULING MACHINE — dai
f machine *f* de capsulage
e máquina *f* tapadora
i capsulatrice *f*
n capsuleermachine
d Verschlussmaschine *f*

4663 SEALING RING — cpt
f bague *f* d'étoupage, rondelle *f* d'étanchéité
e anillo *m* de estanqueidad, aro *m* de cierre hermético
i anello *m* per chiusura ermetica
n afdichtingsring
d Dichtungsring *m*

4664 SEASONING PLANT — lea
f installation *f* d'apprêtage
e equipo *m* de apresto, equipo *m* de lustre
i impianto *m* d'appretto
n glansinstallatie
d Glänzanlage *f*

4665 SEASONING PLANT — pla
f installation *f* de maturation
e equipo *m* de maduración
i impianto *m* di maturazione
n rijpingsinstallatie
d Reifungsanlage *f*

4666 SEASONING PLANT, WOOD SEASONING PLANT — bui
f installation *f* de dessèchement de bois
e equipo *m* de cura de madera, equipo *m* de desecación de madera
i impianto *m* di stagionatura di legno
n houtdrooginstallatie
d Holzauswitterungsanlage *f*

4667 SEAT, SEATING, VALVE SEATING — cpt
f siège *m* de soupape
e asiento *m* de válvula
i seda *f* di valvola
n klepzitting
d Ventilsitz *m*

4668 SECOND BATH — tan
f bain *m* de réduction de l'acide chromique
e baño *m* de reducción del ácido crómico
i bagno *m* di riduzione dell'acido cromico
n chroomzuurreductiebad *n*, reductiebad *n*
d Chromsäurereduktionsbad *n*, Reduktionsbad *n*

4669 SECOND BOILER, REFINING COPPER — sin
f claire *f*
e cuba *f* refinadora
i tino *m* raffinatore
n klaarketel, louterpan
d Klärkessel *m*, Läuterkessel *m*

4670 SECOND RIPENING BATH, AFTERRIPENING BATH — pho
f bain *m* de maturation à refonte
e baño *m* de maduración química
i bagno *m* di maturazione chimica
n chemisch rijpingsbad *n*
d chemisches Reifungsbad *n*, Nachreifungsbad *n*

4671 SECONDARY BATTERY, STORAGE BATTERY — elc
f batterie *f* d'accumulateurs, batterie *f* secondaire
e batería *f* de acumuladores, batería *f* secundaria
i batteria *f* d'accumulatori, batteria *f* secondaria
n accubatterij, secundaire batterij
d Akkumulatorenbatterie *f*, Sekundärbatterie *f*

4672 SECONDARY CELL, ACCUMULATOR, STORAGE CELL — elc
f accumulateur *m*
e acumulador *m*
i accumulatore *m*
n accu, accumulator
d Akkumulator *m*, Sammler *m*

4673 SECONDARY DIGESTION TANK — iwt
f digesteur *m* final
e cuba *f* de digestión final
i vasca *f* di digestione finale
n nadigereertank
d Nachfaulbecken *n*

4674 SECONDARY ELECTRODE, BIPOLAR ELECTRODE — elc
f électrode *f* bipolaire
e electrodo *m* bipolar
i elettrodo *m* bipolare
n bipolaire elektrode
d Bipolarelektrode *f*

4675 SECONDARY ENRICHMENT TANK — min
f récipient *m* d'enrichissement secondaire
e tanque *m* de enriquecimiento secundario
i serbatoio *m* d'arricchimento secondario

n secondaire-concentratietank
d Sekundäranreicherungsgefäss *n*

4676 SECONDARY FERMENTATION VESSEL fmi
f cuve *f* de fermentation secondaire
e cuba *f* de fermentación secundaria
i tino *m* di fermentazione secondaria
n nafermentatievat *n*
d Nachgärungsbehälter *m*

4677 SECONDARY SEDIMENTATION TANK, iwt
FINAL SETTLING TANK
f décanteur *m* secondaire, récipient *m* de clarification finale
e tanque *m* de clarificación final
i serbatoio *m* di sedimentazione finale
n laatste klaringskuip
d Nachklärgefäss *n*

4678 SECONDARY STILL dis
f appareil *m* de distillation secondaire
e aparato *m* de destilación secundaria
i apparecchio *m* di distillazione secondario
n nadestilleerapparaat *n*
d Nachdestillationsapparat *m*

4679 SECTION, elc
NET
f batterie *f*, système *m* multiple
e batería *f*, sistema *m* multiple
i batteria *f*
n serieschakeling van cellen
d Badgruppe *f*

4680 SECTROMETER ins
f potentiomètre *m* avec tube à rayons cathodiques
e potenciómetro *m* con tubo de rayos catódicos
i potenziometro *m* a raggi catodici
n potentiometer met katode-straalbuis
d Potentiometer *n* mit Elektronen-strahlröhre

4681 SEDIMENT BAG bre
f trub-sac *m*
e manga *f*, saco *m* para turbio
i sacco *m* per torbido
n drabzak
d Trubsack *m*

4682 SEDIMENT CONE, lab
SETTLING CONE
f cône *m* à sédimentation, cône *m* pour dépôt
e cono *m* de sédimentación
i bicchiere *m* conico di sedimentazione
n sedimenteerglas *n*
d Sedimentiergefäss *n*

4683 SEDIMENT PRESS bre
f presse *f* de dépôt
e prensa *f* de depósito
i pressa *f* di feccia
n drabpers
d Gelägerpresse *f*

4684 SEDIMENTATION APPARATUS ana,lab
f appareil *m* à lévigation
e aparato *m* de sedimentación
i apparecchio *m* a sedimentazione
n neerslagapparaat *n*
d Schlämmapparat *m*

4685 SEDIMENTATION BASIN, iwt
SETTLING BASIN, SUBSIDENCE BASIN
f bassin *m* de sédimentation
e artesa *f* de sedimentación
i bacino *m* di sedimentazione
n bezinkingsbekken *n*
d Absetzbecken *n*

4686 SEDIMENTATION CHAMBER, iwt
SETTLING COMPARTMENT,
FLOW COMPARTMENT
f chambre *f* de sédimentation
e cámara *f* de sedimentación
i camera *f* di sedimentazione
n bezinkruimte
d Absetzraum *m*, Klärraum *m*

4687 SEDIMENTATION PLANT, iwt
SETTLING PLANT
f installation *f* de sédimentation
e equipo *m* de sedimentación
i impianto *m* di sedimentazione
n bezinkingsinstallatie
d Absetzanlage *f*

4688 SEDIMENTATION TANK, iwt,sto
SETTLING TANK,
CLARIFICATION TANK
f bac *m* de décantation, récipient *m* de clarification
e tanque *m* de clarificación, tanque *m* de sedimentación
i serbatoio *m* di chiarificazione, serbatoio *m* di sedimentazione
n bezinkvat *n*, klaringstank
d Absetzbehälter *m*, Klärbehälter *m*

4689 SEDIMENTATION TUB, SETTLING TUB bre
f bac *m* de sédimentation
e cuba *f* de sedimentación
i tino *m* di sedimentazione
n drabvat *n*
d Absitzbütte *f*

4690 SEEPAGE PIT nue
f puits *m* de percolation
e pozo *m* de percolación
i pozzo *m* di percolazione
n doorsijpelput
d Durchsickergrube *f*

4691 SEETHING VESSEL, SCALDING VESSEL, SMOTHERING VESSEL tex
f échaudoir *m*
e cuba *f* de escaldar, escaldadera *f*
i serbatoio *m* da scottare
n broeiketel
d Abbrühgefäss *n*

4692 SEGER CONES, FUSION CONES, PYROMETRIC CONES tes
f cônes *pl* de Seger, cônes *pl* pyrométriques
e conos *pl* de Seger, conos *pl* pirométricos
i coni *pl* di Seger, coni *pl* pirometrici
n segerkegels *pl*, smeltkegels *pl*
d Schmelzkegel *pl*, Segerkegel *pl*

4693 SEGMENTAL WHEEL pol
f roue *f* à segments
e rueda *f* de segmentos
i ruota *f* a segmenti
n segmentenslijpschijf
d Segmentenschleifscheibe *f*

4694 SEGREGATION VAT, LIQUATION VAT met
f cuve *f* de liquation, cuve *f* de ségrégation
e cuba *f* de licuación, cuba *f* de segregación
i tino *m* di liquazione
n segregatievat *n*, zijgervat *n*
d Ausschwitzgefäss *n*, Seigerungsgefäss *n*

4695 SELECTIVE CRUSHING MILL brk
f broyeur *m* sélectif
e machacadora *f* selectiva
i rompitore *m* selettivo
n selectieve breker
d Selektivbrecher *m*

4696 SELECTIVE FLOTATION PLANT, DIFFERENTIAL FLOTATION PLANT met
f installation *f* de flotation sélective
e equipo *m* de flotación selectiva
i impianto *m* di flottazione selettiva
n selectieve flotatie-installatie
d selektive Flotationsanlage *f*

4697 SELECTIVE FREEZING VESSEL met
f récipient *m* de solidification sélective
e recipiente *m* de solidificación selectiva
i serbatoio *m* di solidificazione selettiva
n vat *n* voor selectieve stolling
d Gefäss *n* für selektive Erstarrung

4698 SELENIUM CELL ele
f cellule *f* au sélénium
e célula *f* de selenio
i cella *f* al selenio
n seleencel
d Selenzelle *f*

4699 SELENIUM RECTIFIER ele
f redresseur *m* au sélénium
e rectificador *m* de selenio
i raddrizzatore *m* al selenio
n seleengelijkrichter
d Selengleichrichter *m*

4700 SELENIUM TONING BATH pho
f bain *m* de virage au sélénium
e baño *m* de viraje al selenio
i bagno *m* di viraggio al selenio
n seleniumkleurbad *n*
d Selentonungsbad *n*

4701 SELF-BAKING ELECTRODE ht
f électrode *f* à auto-cuisson
e electrodo *m* de autococción, electrodo *m* de autocochura
i elettrodo *m* autoformante
n zelfbakkende elektrode
d Dauerelektrode *f*, selbstbacken Elektrode *f*, Sinterelektrode *f*

4702 SELF-CHARGING CENTRIFUGE ce
f centrifugeur *m* à alimentation automatique
e centrífuga *f* de alimentación automática

i centrifuga *f* ad alimentazione automatica
n centrifuge met automatische toevoer
d Zentrifuge *f* mit selbsttätiger Zufuhr

4703 SELF-CLEANING FILTER cle,fil,tex
f filtre *m* à auto-épuration
e filtro *m* autolimpiador
i filtro *m* a pulitura automatica
n zelfreinigend filter *n*
d selbstreinigendes Filter *n*

4704 SELF-DISCHARGING CENTRIFUGE cen
f centrifugeur *m* à décharge automatique
e centrífuga *f* de descarga automática
i centrifuga *f* a scarica automatica
n centrifuge met zelfstandige ontlading
d Zentrifuge *f* mit selbsttätiger Entladung

4705 SELF ELECTRODE spe
f électrode *f* à auto-émission
e electrodo *m* de autoemisión
i elettrodo *m* ad autoemissione
n zelfemitterende elektrode
d Eigenemissionselektrode *f*

4706 SELF-HARDENING PLANT, AIR HARDENING PLANT met
f installation *f* de trempe à l'air
e equipo *m* de temple al aire
i impianto *m* di tempera all'aria
n installatie voor harding aan de lucht
d Lufthärtungsanlage *f*

4707 SELF-PURIFYING AND FRACTIONATING OIL DIFFUSION PUMP vac
f pompe *f* à diffusion d'huile à auto-épuration et à fractionnement
e bomba *f* de difusión de aceite a autodepuración fraccionadora
i pompa *f* a diffusione d'olio depuratrice a frazionamento
n zelfreinigende en fractionerende oliediffusiepomp
d selbstreinigende Fraktionnier-öldiffusionspumpe *f*

4708 SELF-SEALING INJECTION NOZZLE pla,pre,rub
f tuyère *f* d'injection à encliquetage automatique
e tobera *f* de inyector de cierre automático
i ugello *m* d'iniezione con bloccaggio automatico
n zelfsluitend spuitmondstuk *n*
d Einspritzdüse *f* mit selbsttätiger Sperre

4709 SEMICONDUCTOR elc
f semiconducteur *m*
e semiconductor *m*
i semiconduttore *m*
n halfgeleider
d Halbleiter *m*

4710 SEMI-CONTINUOUS KILN cer
f étuve *f* semi-continue
e estufa *f* semicontinua
i stufa *f* semicontinua
n semicontinue droogoven
d halbkontinuierlicher Trockenofen *m*

4711 SEMI-MUFFLE-TYPE FURNACE, OVEN-TYPE FURNACE htt
f four *m* à chambres
e horno *m* de cámaras
i forno *m* a camere
n kameroven
d Kammerofen *m*

4712 SEMI-PERMEABLE MEMBRANE sep
f membrane *f* semi-perméable
e membrana *f* semipermeable
i membrana *f* semipermeabile
n semipermeabele membraan
d halbdurchlässige Membrane *f*

4713 SEMI-POSITIVE MOLD (US), SEMI-POSITIVE MOULD (GB) pla
f moule *m* semi-positif
e molde *m* semipositivo
i stampo *m* semipositivo
n semipositieve matrijs
d Form *f* mit hohem weitem Füllraum und stehendem Grat, halb-positive Form *f*

4714 SEMI-ROTARY PUMP, VANE PUMP pre
f pompe *f* rotative à ailettes
e bomba *f* giratoria de aletas
i pompa *f* rotativa ad alette
n vleugelpomp
d Allweilerpumpe *f*, Flügelpumpe *f*

4715 SENSITIZATION BATH, SENSITIZING BATH pho
f bain *m* sensibilisateur
e baño *m* sensibilizador
i bagno *m* sensibilizzatore
n sensibiliseerbad *n*
d Sensibilisierungsbad *n*

4716 SENTINEL PYROMETERS, PYROMETRIC CYLINDERS ins
f cylindres *pl* pyrométriques
e cilindros *pl* pirométricos
i cilindri *pl* pirometrici
n smeltcilinders *pl*
d Schmelzzylinder *pl*

4717 SEPARATE POT MOLD (US), SEPARATE POT MOULD (GB) pla
f moule *m* à pot de transfert séparé
e molde *m* de cámara de carga separada
i stampo *m* con camera di compressione separata
n vorm met gescheiden toevoer-matrijs
d Form *f* mit Separatspritztopf

4718 SEPARATING BURETTE lab
f burette *f* de séparation
e bureta *f* de separación
i buretta *f* di separazione
n scheidingsburet
d Scheidebürette *f*

4719 SEPARATING CALORIMETER ins
f calorimètre *m* séparateur
e calorímetro *m* separador
i calorimetro *m* separatore
n scheidingscalorimeter
d Trennkalorimeter *n*

4720 SEPARATING CHAMBER chr
f cuvette *f* de séparation
e cubeta *f* de separación
i vaschetta *f* di separazione
n scheidingskamer
d Trennkammer *f*, Trennküvette *f*

4721 SEPARATING COLUMN, SEPARATING TOWER sep
f colonne *f* de séparation
e columna *f* de separación
i colonna *f* di separazione
n scheidingskolom
d Trennsäule *f*

4722 SEPARATING DRUM sep
f tambour *m* séparateur
e tambor *m* separador
i tamburo *m* separatore
n scheidingstrommel
d Scheidetrommel *f*

4723 SEPARATING FUNNEL, EXTRACTION FUNNEL lab
f ampoule *f* à décanter, entonnoir *m* à séparation
e embudo *m* de extracción
i imbuto *m* separatore
n scheitrechter
d Scheidetrichter *m*

4724 SEPARATING SIEVE fil,sep
f tamis *m* de poudre
e tamiz *m* de polvo
i staccio *m* di polvere
n poederzeef
d Pulversieb *n*

4725 SEPARATION PLANT sor
f installation *f* de triage
e equipo *m* de clasificación
i impianto *m* di classificazione
n sorteerinstallatie
d Sichtungsanlage *f*, Sortieranlag

4726 SEPARATION TUBE, THERMAL DIFFUSION PIPE nu
f colonne *f* de diffusion thermiqu
e columna *f* de difusión térmica
i colonna *f* di diffusione termica
n thermische diffusiekolom
d Trennrohr *n*

4727 SEPARATOR mi
f séparateur *m*
e separador *m*
i separatore *m*
n scheider, separator
d Scheider *m*, Separator *m*

4728 SEPARATOR, BATTERY CASE SEPARATOR el
f plaque *f* de séparation
e placa *f* de separación
i separatore *m*
n separator
d Scheider *m*, Separator *m*

4729 SEPARATOR DISK se
f disque *m* de séparation
e disco *m* de separación
i disco *m* di separazione
n separatieschijf
d Abtrennscheibe *f*, Trennscheib

4730 SEPARATOR HEAD pla,pr
f torpille *f*
e cabeza *f* repartidora
i testa *f* ad ogiva

n torpedo
d Verteilerkopf *m*

4731 SEPARATOR MEMBRANE sep
f membrane *f* de séparation
e membrana *f* de separación
i membrana *f* di separazione
n scheidingsmembraan
d Scheidemembrane *f*

4732 SEPARATOR VANE sep
f palette *f* de séparation
e paleta *f* de separación
i pala *f* di separazione
n separatievleugel
d Akkumulatorseparator *m*

4733 SEPTIC TANK, HYDROLYZING TANK, PRIVY VAULT iwt
f récipient *m* septique
e tanque *m* séptico
i serbatoio *m* settico, vasca *f* a flusso di liquame
n vuilwatertank, vuilnistank
d Abwasserfaulraum *m*, Faulgrube *f*, Faulraum *m*

4734 SEQUENCE ETCHING PLANT elc
f bain *m* d'attaque à l'acide composé
e baño *m* de ataque al ácido compuesto
i bagno *m* d'attacco all'acido composto
n bad *n* voor samengestelde etsing
d Folgeätzungsbad *n*

4735 SERIGRAPHY PLANT, SCREEN PRINTING PLANT tex
f installation *f* d'impression à la pellicule
e equipo *m* de estampar con estarcido
i impianto *m* di stampaggio alla pellicola
n zeefdrukinstallatie
d Netzdruckanlage *f*, Siebdruckanlage *f*

4736 SERPENTINE PIPE clo
f tube *m* de serpentin
e tubo *m* de serpentín
i tubo *m* di serpentina
n slangenbuis
d Rohrschlange *f*, Schlangenrohr *n*, Spiralrohr *n*

4737 SERUM CAP bac,mcy
f couvercle *m* à sérum
e tapa *f* para suero
i coperchio *m* per siero
n serumdop
d Serumkappe *f*

4738 SERUM STORAGE BOTTLE bac,lab,mcy
f flacon *m* à sérum, flacon *m* conserve à sérum
e frasco *m* de suero
i boccetta *f* a siero
n serumfles
d Serumflasche *f*

4739 SERVICE PIPE, SUPPLY PIPE tra
f conduite *f* d'alimentation
e conducta *f* de alimentación
i condotta *f* d'alimentazione
n toevoerleiding
d Zufuhrleitung *f*

4740 SERVICE RESERVOIR, CLEAR WATER TANK hyg
f réservoir *m* d'eau potable
e tanque *m* de agua dulce, tanque *m* de agua potable
i serbatoio *m* d'acqua filtrata, serbatoio *m* d'acqua potabile
n drinkwatertank
d Reinwasserbehälter *m*, Trinkwasserbehälter *m*

4741 SETTING, BED cer
f groupe *m* de cornues
e grupo *m* de retortas
i gruppo *m* di storte
n retortengroep
d Retortengruppe *f*

4742 SETTING-OUT BENCH MACHINE lea
f machine *f* à étirer
e máquina *f* de estriado
i macchina *f* da stirare
n strekmachine
d Streckmaschine *f*

4743 SETTING POINT TESTER lub
f appareil *m* d'essai du point de solidification
e probador *m* de punto de solidificación
i collaudatore *m* del punto di ripresa dell'olio
n vloeipuntbepalingsapparaat *n*
d Stockpunktprüfgerät *n*

4744 SETTLER cle,met
f appareil *m* de décantation,

cuve *f* de sédimentation
e decantador *m*, sedimentador *m*
i tino *m* di sedimentazione
n sedimenteervat *n*
d Absetzgefäss *n*

4745 SETTLING BASIN, iwt
SEDIMENTATION BASIN,
SUBSIDENCE BASIN
f bassin *m* de sédimentation
e artesa *f* de sedimentación
i bacino *m* di sedimentazione
n bezinkingsbekken *n*
d Absetzbecken *n*

4746 SETTLING CHAMBER, cle,gas
DUST SETTLING CHAMBER,
GRAVITY SETTLING CHAMBER
f chambre *f* à poussière
e cámara *f* de polvo
i camera *f* a polvere
n stofkamer
d Staubkammer *f*

4747 SETTLING CONE,
SEDIMENT CONE
f cône *m* à sédimentation,
cône *m* pour dépôt
e cono *m* de sedimentación
i bicchiere *m* conico di
sedimentazione
n sedimenteerglas *n*
d Sedimentiergefäss *n*

4748 SETTLING LAGOON, iwt
SLUDGE LAGOON
f étang *m* à boue, étang *m* de
remblaiement
e estanque *m* de deposición
i stagno *m* a fango, stagno *m* di terra
n slibvijver
d Absetzteich *m*, Auflandungs-
teich *m*, Schlammteich *m*

4749 SETTLING PLANT, iwt
SEDIMENTATION PLANT
f installation *f* de sédimentation
e equipo *m* de sedimentación
i impianto *m* di sedimentazione
n bezinkingsinstallatie
d Absetzanlage *f*

4750 SETTLING TANK iwt
f récipient *m* de sédimentation
e tanque *m* de sedimentación
i serbatoio *m* di sedimentazione
n afzettank, sedimentatietank
d Absetzbehälter *m*

4751 SETTLING TANK, bre
HOT WORT RECEIVER
f bac *m* fermé
e tina *f* de sedimentación
i tino *m* di sedimentazione
n gesloten koelbak
d Setzbottich *m*

4752 SETTLING TUB, bre
SEDIMENTATION TUB
f bac *m* de sédimentation
e cuba *f* de sedimentación
i tino *m* di sedimentazione
n drabvat *n*
d Absitzbütte *f*

4753 SEWAGE DISPOSAL PLANT, iwt
SEWAGE PLANT,
CLARIFICATION PLANT
f installation *f* de clarification,
station *f* d'épuration des eaux
d'égout, usine *f* de purification
des eaux d'égout
e equipo *m* de clarificación,
equipo *m* de utilización de
aguas negras
i impianto *m* di chiarificazione,
impianto *m* d'epurazione
delle acque di fogna
n afvalwaterreinigingsinstallatie
d Abwasserreinigungsanlage *f*,
Kläranlage *f*

4754 SEWAGE PUMP iw
f pompe *f* à eau d'égout
e bomba *f* de las aguas cloacales
i pompa *f* per liquame
n rioolwaterpomp
d Abwasserpumpe *f*

4755 SEWER hyg,iw
f canal *m* des eaux d'égouts,
égout *m*
e alcantarilla *f*
i chiarica *f*, fogna *f*
n riool *n*
d Abwasserkanal *m*

4756 SEWER PIPE hyg,iw
f conduit *m* d'égout
e conducto *m* de alcantarilla
i fognatura *f*
n rioolbuis
d Abwasserleitung *f*

4757 SHAFT FURNACE htt,m
f four *m* à cuve
e horno *m* de cuba
i forno *m* a tino
n schachtoven
d Schachtofen *m*

4758 SHAFT KILN htt,met
f four *m* de grillage à cuve
e horno *m* de tostación de cuba
i forno *m* d'arrostimento a tino
n schachtroostoven
d Schachtröstofen *m*

4759 SHAFT PACKING, cpt
f SHAFT SEAL
f étoupe *f* d'arbre
e empaquetadura *f* de eje
i guarnizione *f* d'albero
n asafdichting
d Wellenabdichtung *f*

4760 SHAKER mix
f secoueuse *f*
e sacudidor *m*
i macchina *f* a scosse
n schudmachine
d Schüttelmaschine *f*

4761 SHAKER CONVEYOR tra
f transporteur *m* à secousses
e transportador *m* de sacudidas
i trasportatore *m* a scosse
n schudgoot
d Schüttelrutsche *f*

4762 SHAKING CRYSTALLIZER cry
f cristallisoir *m* à secousses
e cristalizador *m* de sacudidas
i cristallizzatore *m* a scosse
n vibrerend kristallisatieapparaat *n*
d Kristallisierwiege *f*

4763 SHAKING CYLINDER, lab,mix
MIXING CYLINDER
f cylindre *m* mélangeur, éprouvette *f* mélangeuse
e cilindro *m* mezclador
i cilindro *m* mescolatore
n mengmaatcilinder, mengmaatglas *n*
d Mischzylinder *m*

4764 SHAKING FEEDER tra
f alimentateur *m* à va-et-vient
e alimentador *m* con movimiento de vaivén
i alimentatore *m* a viavai
n toevoerinrichting met heen-en weerbeweging
d Pendelzufuhreinrichtung *f*

4765 SHAKING GRATE boi
f grille *f* à secousses
e parrilla *f* de sacudidas
i griglia *f* a scosse
n schudrooster *n*
d Schüttelrost *m*

4766 SHAKING KILN bre
f touraille *f* à secousses
e tostadero *m* de sacudidas
i essiccatoio *m* a scosse
n schokeest
d Rütteldarre *f*

4767 SHAKING SCREEN (GB), sor
SWING SIEVE (US)
f crible *m* oscillant, tamis *m* à secousses
e criba *f* vibrante, tamiz *m* de sacudidas
i vaglio *m* a scosse, vibrovaglio *m*
n schudzeef
d Schüttelsieb *n*, Schwingsieb *n*

4768 SHAKING SHOOT, tra
JIGGING CONVEYOR,
OSCILLATING CHUTE
f couloir *m* à secousses, gouttière *f* à secousses
e canal *m* de sacudidas
i scivolo *m* crivellatore
n schudgoot
d Schüttelrinne *f*

4769 SHALLOW COOLER, bre,clo
COOLSHIP
f bac *m* refroidisseur
e bandeja *f* de enfriamiento
i rinfrescatoio *m*
n koelkuip
d Kühlschiff *n*

4770 SHANK, gla
BEWEL, BOW, LADLE SHANK
f brancard *m*
e maneral *m*
i portante *m*
n draagbeugel, handbeugel
d Tiegelschere *f*, Tragschere *f*

4771 SHAPING BLOCK gla
f mandrin *m* en bois
e molde *m* del pie
i stampo *m* da formare
n vormhout *n*
d Holzlöffel *m*

4772 SHARP-EDGED ORIFICE gas,ins
f orifice *m* percé en mince paroi
e orificio *m* de aristas vivas, orificio *m* en pared delgada
i orificio *m* a boccaglio
n meetflens
d Drosselscheibe *f*, Messblende *f*, Staurand *m*

4773 SHARPENED LIME BATH tan
f pelain *m* renforcé
e pelambre *m* intenso
i bagno *m* rinforzato di macerazione
n geconcentreerd kalkbad *n*
d angeschärfter Äscher *m*

4774 SHARPENING PLANT tex
f installation *f* pour éliminer l'oxygène
e equipo *m* de eliminar el oxígeno
i impianto *m* da eliminare l'ossigeno
n zuurstofverwijderingsinstallatie
d Sauerstoffbeseitigungsanlage *f*

4775 SHAVING MACHINE lea
f machine *f* de dolage, machine *f* à planer
e máquina *f* de adelgazar
i macchina *f* da raschiare
n schaafbank
d Dolliermaschine *f*

4776 SHEATHED PYROMETER ins
f pyromètre *m* à gaine
e pirómetro *m* envainado
i pirometro *m* a guaina
n omhulde pyrometer
d Pyrometer *n* mit Schutzhaube

4777 SHEATHING, TOWER CASING clo
f revêtement *m* de tour
e revestimiento *m* de torre
i rivestimento *m* di torre
n torenbekleding
d Turmauskleidung *f*

4778 SHEET CALENDER pap
f calandre *f* pour papier en feuilles
e calandra *f* de pliegos
i calandra *f* di foglie
n papiervelkalander
d Bogenglättwerk *n*, Bogenkalander *m*

4779 SHEET EXTRUDER pla
f boudineuse *f* pour feuilles
e budinadora *f* de hojas y láminas
i estrusore *m* per lastre, trafila *f* di foglie
n foliespuitmachine
d Folienmaschine *f*

4780 SHEETING MACHINE, SHEETING MILL pla,rub
f laminoir *m* de feuilles
e laminador *m* de hojas
i laminatoio *m* di foglie
n foeliewalswerk *n*
d Folienwalzwerk *n*

4781 SHELL (GB), CASING (US) htt,sap
f carcasse *f*
e carcasa *f*
i carcassa *f*, mantello *m* di forno
n ovenmantel
d Blechmantel *m*

4782 SHELL, FUSED ELECTROLYTE CONTAINER el
f cuve *f* à électrolyte fondu
e cuba *f* de electrólito fundido
i tino *m* ad elettrolito fuso
n gesmolten-elektrolytbak
d Wanne *f* für geschmolzenen Elektrolyt

4783 SHELL-AND-TUBE COOLER cl
f refroidisseur *m* à gaine et tubes
e enfriador *m* de casco y tubos
i raffreddatore *m* ad involucro e tubi
n buizenkoeler
d Röhrenkühler *m*

4784 SHELL-AND-TUBE EXCHANGER ht
f échangeur *m* de chaleur à gaine et tubes
e intercambiador *m* de casco y tu
i scambiatore *m* di calore ad involucro e tubi
n buizenwarmte-uitwisselaar
d Röhrenwärmeaustauscher *m*

4785 SHELL CORE, BLOW STICK pla,pr
f noyau-carapace *m*
e macha *f* en cáscara
i anima *f* a guscio
n holle kern, schaalkern
d Hohlkern *m*

4786 SHELL MOLD (US), SHELL MOULD (GB) p
f moule-carapace *m*
e molde *m* en cáscara
i forma *f* a guscio
n vormschaal
d Maskenform *f*

4787 SHERARDIZING PLANT c
f installation *f* de shérardisatio
e equipo *m* de sherardización

i impianto *m* di sherardizzazione
n sherardiseerinstallatie
d Sherardisieranlage *f*

4788 SHIELD elc
f écran *m*
e pantalla *f*
i schermo *m*
n scherm *n*
d Abschirmung *f*, Schirm *m*

4789 SHIELDED-ARC WELDING PLANT wel
f installation *f* de soudure à arc protégé
e equipo *f* de soldadura por arco protegido
i impianto *m* di saldatura ad arco schermato
n installatie voor afgeschermd booglassen
d Schutzgasschweissanlage *f*

4790 SHIELDING POND, nue,sap
SHIELDING POOL
f réservoir *m* à blindage
e tanque *m* de blindaje
i serbatoio *m* a schermatura
n afschermende tank
d abschirmender Behälter *m*

4791 SHIM, gen
LINER
f chemise *f*
e chapa *f*
i spessore *m*
n blikvoering, busje *n*
d Blecheinlage *f*

4792 SHIP-BORNE DESALINATION PLANT hyg
f installation *f* de dessalage à bord d'un vaisseau
e equipo *m* de desalación embarcado
i impianto *m* di desalinizzazione navotrasportato
n varende ontziltingsinstallatie
d Bordentsalzungsanlage *f*

4793 SHOCK PENDULUM, rub,tes
IMPACT PENDULUM
f appareil *m* à marteau-pendule, mouton *m* pendulaire, mouton-pendule *m*
e péndulo *m* para prueba de choque
i maglio *m* a pendolo, pendolo *m* per prova d'urto
n slagarm, slagpendel, slinger-slagapparaat *n*
d Pendelschlagwerk *n*, Schlagpendel *m*

4794 SHOCK TESTING APPARATUS rub
f appareil *m* d'essai à secousses
e aparato *m* de prueba de sacudidas
i collaudatore *m* a scosse
n schokapparaat *n*
d Stossprüfgerät *n*

4795 SHORT-STROKE PRESS pla,pre
f presse *f* à faible course
e prensa *f* de carrera corte
i pressa *f* a corsa corta
n pers met korte plunjerslag
d Presse *f* mit geringem Hub, Presse *f* mit niedrigem Hub

4796 SHORTERIZING PLANT, htt,met
FLAME HARDENING PLANT.,
TORCH HARDENING PLANT
f installation *f* de trempe au chalumeau
e equipo *m* de temple superficial a la llama
i impianto *m* di tempera alla fiamma
n vlamhardingsinstallatie
d autogene Flammenhärtungs-anlage *f*

4797 SHOT BLASTING MACHINE, cle
GRIT BLASTING MACHINE
f machine *f* de grenaillage
e máquina *f* de chorreado con granalla
i macchina *f* di granigliatura
n gritstraalmachine
d Abstrahlmaschine *f*, Schleuderstrahlmaschine *f*

4798 SHOT PEENING APPARATUS cle
f appareil *m* de grenaillage à écrouissage
e aparato *m* de golpo con granalla
i apparecchio *m* di pallinatura
n staalstraalapparaat *n*
d Granaliengebläse *n*, Strahlhämmergerät *n*

4799 SHREDDER oic
f machine *f* à déchiqueter
e desmenuzadora *f*, trituradora *f*
i trinciatrice *f*
n scheurmachine
d Zerkleinerungsmaschine *f*

4800 SHREDDER pap
f délisseuse *f*, machine *f* à dilacérer
e desfibrador *m*
i macchina *f* da dilacerare
n lompensnijder
d Reisswolf *m*, Zerfaserer *m*

4801 SHRINKAGE BLOCK, COOLING FIXTURE, COOLING JIG clo,pla
f conformateur *m*, jauge *f* conformatrice
e platilla *f* de contracción
i conformatore *m*, controforma *f* di raffreddamento
n krimpblok *n*
d Abkühlvorrichtung *f*, Schrumpfblock *n*

4802 SHRINKING PLANT, SHRINKPROOFING PLANT tex
f installation *f* de traitement irrétrécissable
e equipo *m* de tratamiento antiencogimiento
i impianto *m* di trattamento antiritiro
n installatie voor het krimpvrijmaken
d Anlage *f* zum Krumpfechtmachen

4803 SHRINKING ROLL tex
f cylindre *m* de rétrécissement
e cilindro *m* de encogimiento
i cilindro *m* di ritiro
n krimpwals
d Krumpfwalze *f*

4804 SHRINKING TANK, SHRINKPROOFING TANK tex
f récipient *m* de traitement irrétrécissable
e tanque *m* de tratamiento antiencogimiento
i serbatoio *m* di trattamento antiritiro
n tank voor het krimpvrijmaken
d Gefäss *n* zum Krumpfechtmachen

4805 SHROUDED IMPELLER mix
f roue *f* motrice à palettes renforcées
e rueda *f* móvil de paletas amortajadas
i girante *m* a pale rinforzate
n roerder met versterkte schoepen
d Rührstange *f* mit verstärkten Paddeln

4806 SHUT-OFF VALVE cpt
f soupape *f* d'arrêt, soupape *f* de fermeture
e válvula *f* de cierre
i valvola *f* di chiusura
n afsluitklep
d Absperrventil *n*, Verschlussvent

4807 SHUTTLE, RABBIT nue
f tube *m* pneumatique
e barquilla *f*, lanzadera *f*
i spola *f*, impianto *m* pneumatico
n buisposthouder
d Rohrpostbüchse *f*

4808 SHUTTLE KILN, ENVELOPE KILN, TOP-HAT KILN cer
f four *m* à recuire avec calotte protectrice
e horno *m* cerámico con caperuza de protección
i forno *m* ceramico con cappa di protezione
n kapselbakoven, kokerbakoven
d Kapselbrennofen *m*

4809 SHUTTLE-TYPE FEEDER pre,rub
f alimentateur *m* par conduit vibrant
e alimentador *m* por conducto de vibraciones
i alimentatore *m* per condotto vibrante
n schudgoottoevoerinrichting
d Schüttelrinnenspeiser *m*

4810 SIDE-FIRED FURNACE, CROSS-FIRED FURNACE gla
f four *m* à feu de coupage
e horno *m* de fuego cruzado
i forno *m* a fuoco incrociato
n kruisvuuroven
d Kreuzfeuerofen *m*

4811 SIDE-PLATE PRESS, SLAB SIDE PRESS pre,rub
f presse *f* à parois latérales
e prensa *f* de paredes laterales
i pressa *f* lateralmente chiusa
n pers met plaatvormige zijwanden, trekplatenpers
d Presse *f* mit plattenförmigen Seitenwänden

4812 SIDE POCKET, SLAG POCKET gla,met
f chambre *f* à laitier
e depósito *m* de escorias

i camera *f* delle scorie
n slakkenkamer
d Schlackenfang *m*, Schlacken-
kammer *f*

4813 SIDE VALVE cpt
f soupape *f* latérale
e válvula *f* lateral
i valvola *f* laterale
n zijklep
d seitengesteuertes Ventil *n*

4814 SIEGE, gla
BENCH
f banc *m*
e plaza *f*, solera *f*
i suola *f*
n bank
d Ofenbank *f*, Ofenherd *m*

4815 SIEVE gen
f crible *m*, sas *m*, tamis *m*
e criba *f*, tamiz *m*
i crivello *m*, setaccio *m*, staccio *m*,
vaglio *m*
n zeef
d Sieb *n*

4816 SIEVE DRUM fil
f tambour *m* perforé
e tambor *m* cribador
i tamburo *m* da vagliare
n zeeftrommel
d Siebtrommel *f*

4817 SIEVE GATE fil
f grille *f* à tamis
e parrilla *f* de tamiz
i griglia *f* a vaglio
n zeefrooster *n*
d Siebrost *m*

4818 SIEVE PLANT, sor
SCREENING PLANT
f installation *f* de tamisage
e equipo *m* de tamizado
i impianto *m* di stacciatura,
impianto *m* da vagliare
n zeefinstallatie
d Absiebanlage *f*, Siebanlage *f*

4819 SIEVE-PLANT COLUMN dis
f colonne *f* à plateaux perforés
e columna *f* de platos perforados
i colonna *f* a piatti perforati
n kolom met zeefschotels
d Siebbodenkolonne *f*

4820 SIEVE SHAKER sor
f vibrateur *m* avec une série de
tamis
e vibrador *m* con juego de tamices
i vibratore *m* con una seria di vagli
n zeefschudmachine
d Siebsatzschüttelmaschine *f*

4821 SIEVE TRAY, dis
PERFORATED TRAY
f plateau *m* perforé
e plato *m* perforado
i piatto *m* perforato
n zeefschotel
d Siebboden *m*

4822 SIEVING SCREEN, sor
SCREEN
f grille-tamis *m*
e rejilla-tamiz *m*, zaranda *f*
i griglia *f* da vagliare
n zeefrooster *n*
d Siebrost *m*

4823 SIFTER, cle,sor
SIFTING MACHINE
f tamiseuse *f*
e máquina *f* para tamizar
i macchina *f* per vagliare
n zeefmachine
d Siebmaschine *f*

4824 SIGHT GLASS, cpt,sap
SIGHT HOLE
f regard *m*, trou *m* de tête
e atisbadero *m*, mirilla *f*
i finestra *f* di spia, spia *f*,
vetrino *m* spia
n kijkglas *n*
d Schauglas *n*, Schauloch *n*

4825 SILENCER (GB), cpt
EXHAUST MUFFLER (US),
EXHAUST SILENCER (GB),
MUFFLER (US)
f pot *m* de détente, silencieux *m*
d'échappement
e silenciador *m* del escape
i marmitta *f* di scarico,
silenziatore *m*
n knalpot
d Auspufftopf *m*, Schalldämpfer *m*

4826 SILICA REMOVAL TANK, iwt
DESILICONIZATION TANK
f récipient *m* de désiliciage,
récipient *m* de désiliciation
e tanque *m* de desiliconización
i serbatoio *m* da rimozione della
silice
n zandvanger
d Entkieselungsgefäss *n*,
Sandfänger *m*

4827 SILICONIZING PLANT met
f installation *f* de formation de siliciure
e equipo *m* de siliconización
i impianto *m* di formazione di siliciuro
n siliconiseerinstallatie
d Silizieranlage *f*

4828 SILK BLEACHING VESSEL tex
f cuve *f* à blanchir la soie
e cuba *f* de blanqueo de la seda
i tino *m* di candeggio della seta
n zijdebleekvat *n*
d Seidenbleichgefäss *n*

4829 SILK DEGUMMING VESSEL, tex
SILK SCOURING VESSEL
f cuve *f* de dégommage de la soie
e cuba *f* de desgomación de la seda
i tino *m* di sgommatura della seta
n zijdeontgommingsvat *n*
d Seidendegummiergefäss *n*, Seidenentbastungsgefäss *n*, Seidenentschälungsgefäss *n*

4830 SILK LOADING VESSEL, tex
SILK WEIGHTING VESSEL
f cuve *f* à charger la soie
e cuba *f* de cargado de la seda
i tino *m* di carica della seta
n zijdeverzwaringsvat *n*
d Seidenbeschwerungsgefäss *n*

4831 SILO (GB), nut,tra
ELEVATOR (US)
f silo *m*
e silero *m*, silo *m*
i silo *m*
n silo
d Schachtspeicher *m*, Silo *m*, Zellenspeicher *m*

4832 SILO, bre
GRIST BOX, STORAGE BIN
f trémie *f* à malt, wagon *m* à farine
e tolva *f* de harina
i cassone *m* per macinato
n schrootkaar
d Schrotrumpf *m*

4833 SILT BOX, hyg,iwt
SILT CHAMBER, SILT TRAP
f piège *m* à boue
e depósito *m* de fango
i collettore *m* del fango, pozzetto *m* per la fanghiglia
n slikvanger
d Schlammfang *m*, Schlammkasten *m*

4834 SILVER BRAZING PLANT wel
f installation *f* de soudure forte à l'argent
e equipo *m* de soldadura fuerte co plata
i impianto *m* di brasatura all'argento
n zilverhardsoldeerinstallatie
d Silberhartlötanlage *f*

4835 SILVER CHLORIDE CELL el
f pile *f* à chlorure d'argent
e pila *f* de cloruro de plata
i pila *f* a cloruro d'argento
n chloorzilverelement *n*
d Chlorsilberelement *n*

4836 SILVER DIP, coa
SILVER PLATING BATH
f bain *m* d'argenture
e baño *m* de plateado
i bagno *m* d'argentatura
n zilverbad *n*
d Silberbad *n*

4837 SILVER OXIDE CELL elc
f pile *f* à oxyde d'argent
e pila *f* de óxido de plata
i pila *f* ad ossido d'argento
n zilveroxyde-element *n*
d Silberoxydelement *n*

4838 SILVER-SILVER CHLORIDE elc
ELECTRODE
f électrode *f* argent-chlorure d'argent
e electrodo *m* plata-cloruro de plata
i elettrodo *m* argento-cloruro d'argento
n zilver-zilverchloride-elektrode
d Silber-Chlorsilberelektrode *f*, Silber-Silberchloridelektrode *f*

4839 SILVER STORAGE BATTERY elc
f accumulateur *m* à argent
e acumulador *m* de plata
i accumulatore *m* ad argento
n zilverzinkaccu
d Silber-Zinkakkumulator *m*

4840 SIMPLE BREWHOUSE OUTFIT, bre
TWO VESSEL BREWHOUSE
f salle *f* de brassage simple
e sala *f* de cocimiento simple
i sala *f* di cottura semplice
n enkel brouwhuis *n*
d einfaches Sudwerk *n*, Einfachsudhaus *n*

4841 SIMPLE DISTILLATION PLANT dis
f installation *f* de distillation simple
e equipo *m* de destilación simple
i impianto *m* di distillazione semplice
n eenvoudige destilleerinstallatie
d einfache Destillieranlage *f*

4842 SIMPLE ELECTRODE elc
f électrode *f* simple
e electrodo *m* sencillo
i elettrodo *m* semplice
n enkelvoudige elektrode, gewone elektrode
d Einfachelektrode *f*

4843 SIMPLE-GEAR EXTRUDER pla,pre,rub
f boudineuse *f* à simple harnais d'engrenage
e budinadora *f* de accionamiento no regulable
i trafila *f* ad ingrannaggio semplice
n spuitmachine met niet-regelbare aandrijving
d Strangpresse *f* ohne regelbaren Antrieb

4844 SINGEING APPARATUS, GASSING APPLIANCE tex
f dispositif *m* de flambage, dispositif *m* de gazage
e dispositivo *m* de chamusquina, dispositivo *m* de gaseado
i dispositivo *m* di bruciatura, dispositivo *m* di gasatura
n afzengapparaat *n*, flambeer-apparaat *n*, zengapparaat *n*
d Flambierapparat *m*, Gasiergerät *n*, Senggerät *n*, Verseng-apparat *m*

4845 SINGLE-ACTING PRESS pre
f presse *f* à effet simple
e prensa *f* de efecto simple
i pressa *f* ad effetto semplice
n pers met enkelvoudige werking
d einfachwirkende Presse *f*

4846 SINGLE-ACTION PRESS pre
f presse *f* à single action
e prensa *f* de acción simple
i pressa *f* ad azione semplice
n eenzijdig werkende pers
d einseitig wirkende Presse *f*

4847 SINGLE CAVITY MOLD (US), SINGLE IMPRESSION MOULD (GB) cer
f moule *m* à simple cavité
e molde *m* de simple cavidad
i stampo *m* singolo
n enkelvoudige vorm
d Einfachform *f*

4848 SINGLE-DAYLIGHT PRESS, ONE-DAYLIGHT PRESS pla,pre,rub
f presse *f* à un étage
e prensa *f* de un solo hueco
i pressa *f* ad una apertura
n één-etagepers
d Einentagenpresse *f*

4849 SINGLE DRUM DRIER, DRUM DRIER dry
f séchoir *m* à tambour
e secador *m* de cilindro, secador *m* de tambor
i essiccatore *m* a cilindro, essiccatore *m* a tamburo
n trommeldroger
d Röhrentrockner *m*, Trommeltrockner *m*, Walzentrockner *m*

4850 SINGLE-EFFECT EVAPORATOR pre
f évaporateur *m* à effet simple
e evaporador *m* de efecto simple
i evaporatore *m* ad effetto semplice
n eenvoudige verdamper
d Einkörperverdampfer *m*, Einstufenverdampfer *m*

4851 SINGLE ELECTRODE SYSTEM, HALF CELL, HALF ELEMENT elc
f demi-cellule *f*, hémi-cellule *f*
e semicelda *f*
i semicellula *f*
n halve cel
d Halbzelle *f*

4852 SINGLE-FLOOR KILN, ONE-FLOOR KILN bre
f touraille *f* à un plateau
e tostadero *m* de un plato
i essiccatoio *m* ad un graticcio
n eest met één vloer
d Einhordendarre *f*

4853 SINGLE-PASS TUNNEL OVEN htt
f tunnel-séchoir *m* à circulation en sens unique
e horno *m* de túnel de dirección única

i stufa *f* a galleria a passaggio semplice
n tunneloven met doorgaand transport
d Tunnelofen *m* mit Einrichtungsdurchlauf

4854 SINGLE ROD LATTICE nue
f réseau *m* à barre combustible unique
e red *f* de una barra combustible
i reticolo *m* a barra combustibile semplice
n enkelstaafrooster *n*
d Einstabgitter *n*

4855 SINGLE SCREW EXTRUDER pla
f boudineuse *f* à vis unique
e budinadora *f* de husillo único
i trafila *f* a vite unica
n enkelvoudige spuitpers
d einfache Strangpresse *f*

4856 SINGLE SHELL AUTOCLAVE pre
f autoclave *m* à enceinte unique
e autoclave *f* de una sola pared
i autoclave *f* di parete singola
n enkelwandige autoclaaf
d Einzelwandautoklav *m*

4857 SINGLE SPLIT MOLD (US), pla,rub
SINGLE SPLIT MOULD (GB)
f moule *m* de presse en deux pièces
e molde *m* de dos partes
i stampo *m* bipartito
n tweedelige vorm
d zweiteilige Pressform *f*

4858 SINGLE STAGE COMPRESSOR pre
f compresseur *m* à un étage
e compresor *m* de una sola etapa
i compressore *m* ad uno stadio
n eentrapscompressor
d Einstufenverdichter *m*

4859 SINGLE TOGGLE JAW CRUSHER brk
f concasseur *m* à mâchoires à genouillère unique
e machacadora *f* de mordazas de palanca acodada única
i frantoio *m* a ganasce a ginocchiera unica
n bekkenbreker met één kniehefboom
d Backenbrecher *m* mit einem Kniehebel

4860 SINK hyg,iwt
f évier *m*
e fregadero *m*
i acquaio *m*
n gootsteen
d Ausgussbecken *n*, Gussstein *m*

4861 SINK TRAP, sew
SIPHON, INTERCEPTOR
f siphon *m* inodore
e sifón *m* contra mal olor
i intercettore *m*, sifone *m* contro mal odore
n sifon, stankafsluiter, zwanehals
d Geruchverschluss *m*

4862 SINTERED FILTER fil
f filtre *m* fritté
e filtro *m* sinterizado
i filtro *m* sinterizzato
n sinterfilter *n*
d Sinterfilter *m*

4863 SINTERED GLASS CRUCIBLE lab
f creuset *m* en verre fritté
e crisol *m* de vidrio sinterizado
i crogiolo *m* in vetro sinterizzato
n gesinterde glazen kroes
d Sinterglastiegel *m*

4864 SINTERED GLASS FUNNEL lab
f entonnoir *m* en verre fritté
e embudo *m* de vidrio sinterizado
i imbuto *m* in vetro sinterizzato
n gesinterde glazen trechter
d Sinterglastrichter *m*

4865 SINTERED PLATE elc
f plaque *f* frittée
e placa *f* sinterizada
i piastra *f* sinterizzata
n gesinterde plaat
d Sinterplatte *f*

4866 SINTERING LINE htt
f ligne *f* de frittage
e línea *f* de sinterización
i linea *f* di sinterizzazione
n sinterbaan, sinterstraat
d Sinterstrasse *f*

4867 SINTERING OVEN, htt,met
FRITTING OVEN
f four *m* de frittage
e horno *m* de sinterización
i forno *m* di sinterizzazione
n sinteroven
d Sinterofen *m*

4868 SIPHON gen
f siphon *m*
e sifón *m*
i sifone *m*
n hevel, sifon
d Saugheber *m*, Siphon *m*

4869 SIPHON CONDUIT, SIPHON PIPING cpt
f conduite *f* en siphon
e conducto *m* de sifón
i condotta-sifone *m*
n hevelleiding
d Heberleitung *f*

4870 SIPHON HEAD cpt
f tête *f* de siphon
e cabezal *m* de sifón
i testa *f* di sifone
n sifonkop
d Dückerkopf *m*

4871 SIPHON TRAP hyg,iwt
f fermeture *f* hydraulique
e desagüe *m* de sifón
i pozzetto *m* a sifone
n stankafsluiter
d Geruchverschluss *m*

4872 SIRUP PUMP (US), SYRUP PUMP (GB) sin
f pompe *f* à sirop
e bomba *f* de jarabe
i pompa *f* a sciroppo
n strooppomp
d Siruppumpe *f*

4873 SIX-ROLLER MILL bre,brk
f concasseur *m* à six cylindres
e triturador *m* de seis cilindros
i frantoio *m* a sei cilindri
n schrootmolen met zes walsen
d Sechswalzenmühle *f*

4874 SIZE KETTLE tex
f chaudière *f* à colle
e caldera *f* para apresto
i caldaia *f* di bozzima
n slichtketel
d Schlichtkessel *m*

4875 SIZING PLANT pap
f installation *f* d'engommage
e equipo *m* de engomado
i impianto *m* da collare
n lijmkamer
d Leimkammer *f*

4876 SIZING PLANT, DRESSING PLANT, FINISHING PLANT tex
f installation *f* d'apprêtage
e equipo *m* de apresto
i impianto *m* d'apprettatura
n appreteerinrichting
d Appreturanlage *f*

4877 SIZING PLANT, ORE SIZING PLANT min,sor
f installation *f* de séparation de minerais
e equipo *m* clasificador de minerales
i impianto *m* classificatore di minerali
n ertssorteerinrichting
d Erzscheideanlage *f*

4878 SKIM MILK TEST BOTTLE dai,ins,lab
f butyromètre *m* à lait écrémé
e butirómetro *m* de leche desnatada
i butirometro *m* a latte scremato
n taptemelkbutyrometer
d Magermilchprober *m*

4879 SKIMMER, SCUM COLLECTOR, SCUM SKIMMER iwt
f racleur *m* pour l'enlèvement de l'écume
e colector *m* de espuma
i scolmatore *m* di schiuma
n schuimafstrijker
d Schwimmdeckenabstreifer *m*

4880 SKIMMER (US), SKIMMING TOOL (GB) met
f écumoire *f*
e espumadera *f*
i schiumatoio *m*
n schuimspaan, slakspaan
d Schaumlöffel *m*, Schlackenlöffel *m*

4881 SKIMMING BASIN iwt
f bassin *m* d'écumage
e artesa *f* de espumadura
i bacino *m* di schiumaggio
n schuimbekken *n*
d Schaumbecken *n*

4882 SLAB SIDE PRESS, SIDE-PLATE PRESS pre,rub
f presse *f* à parois latérales
e prensa *f* de paredes laterales
i pressa *f* lateralmente chiusa
n pers met plaatvormige zijwanden, trekplatenpers
d Presse *f* mit plattenförmigen Seitenwänden

4883 SLAG POCKET, gla,met
SIDE POCKET
f chambre *f* à laitier
e depósito *m* de escorias
i camera *f* delle scorie
n slakkenkamer
d Schlackenfang *m*,
Schlackenkammer *f*

4884 SLASHER SIZING MACHINE tex
f machine *f* à écharner,
machine *f* à encoller
e máquina *f* encoladora
i macchina *f* da imbozzimare,
macchina *f* da incollare
n slichtmachine
d Schlichtmaschine *f*

4885 SLAT CONVEYOR tra
f transporteur *m* à palettes
e transportador *m* de listones
i trasportatore *m* a piastre
n lattentransportband
d Plattenbandförderer *m*

4886 SLEDGER, brk,min
SCALPER
f massette *f*, têtu *m*
e quebrantadora *f* preliminar
i mazza *f*
n grofbreker
d Grobbrecher *m*, Steinbrecher *m*

4887 SLEEVE VALVE, cpt
SOCKET VALVE
f pale *f*, soupape *f* à manchon
e válvula *f* de camisa
i valvola *f* a manicotta
n mantelklep
d Muffenventil *n*, Ventilschieber *m*

4888 SLICER cut,sin
f machine *f* à découper les
betteraves
e rebañador *m*
i affettatrice *f*
n bietensnijmachine *f*
d Schnitzelmaschine *f*

4889 SLICKER lea
f étire *m*
e alisador *m*
i scarnatoio *m*
n schaafijzer *n*
d Schabeisen *n*

4890 SLIDE ins
f verre *m* à objet
e platina *f*, portobjeto *m*
i vetrino *m* per microscopio
n objectglaasje *n*
d Objektglas *n*, Objektträger *m*

4891 SLIDE, iwt
CHUTE
f couloir *m*
e canalón *m*
i canaletto *m*
n goot
d Rutsche *f*

4892 SLIDE FOLLOWER pla,pre
f coquille *f* à glissière
e cojinete *f* deslizable
i guancia *f*
n schuifstuk *n*
d Schieberbacke *f*

4893 SLIDE GATE, cpt
SLIDE VALVE
f soupape *f* à glissière
e distribuidor *m* de conchas,
válvula *f* de compuerta
i valvola *f* a tiretto
n schuifklep
d Schieberventil *n*

4894 SLIDE MOLD (US), pla,pre
SLIDE MOULD (GB), SLIDING
BED MOLD (US), SLIDING
CARRIAGE MOULD (GB)
f moule *m* sur glissière
e molde *m* de corredera
i stampo *m* a slitta
n sledematrijs
d Schlittenform *f*

4895 SLIDING BOTTOM pla,pre
f fond *m* à glissière
e fondo *m* corredizo
i fondo *m* a slitta
n doseerschuif
d Schiebeboden *m*

4896 SLIDING CUTOFF cpt
f fermeture *f* à guillotine
e cierre *m* corredizo
i chiusura *f* a ghigliottina
n schuifafsluiter
d Fallschieber *m*

4897 SLIDING JOINT pla,pre
f joint *m* coulissant, joint *m* par
glissement
e junta *f* deslizante
i giunto *m* a scorrimento
n matrijsgeleiding
d gleitende Werkzeugführung *f*

4898 SLIDING PUNCH pla
f poinçon *m* coulissant, poinçon *m* sur glissière
e macho *m* corredizo, punzón *m* deslizante
i punzone *m* scorrevole
n wisselmatrijs
d Schiebestempel *m*

4899 SLIME CHUTE, SLIMES CLASSIFIER min
f classeur *m* de fines, classeur *m* de limon
e clasificador *m* de finos
i classificatore *m* di melma
n slibgoot
d Schlammherd *m*, Schlammrinne *f*, Schlickrinne *f*

4900 SLING PSYCHROMETER, WHIRLING PSYCHROMETER ins
f hygromètre *m* à torsion, psychromètre *m* fronde
e higrómetro *m* de torsión, psicrómetro *m* de torsión
i igrometro *m* a torsione, psicrometro *m* a torsione
n slingerpsychrometer
d Schleuderpsychrometer *n*

4901 SLINGER cer
f machine *f* à lancer
e eslingador *m*
i fionda *f*, frombola *f*
n slingermolen
d Schleudermühle *f*

4902 SLINGING WIRE elc
f fil *m* de suspension
e hilo *m* de suspensión
i filo *m* di sospensione
n ophangdraad, strop
d Einhängedraht *m*

4903 SLIP HOUSE cer
f chambre *f* de pâte d'argile
e cámara *f* de barro
i camera *f* di pasta d'argilla
n kleibrijkamer, paatkamer
d Schlickerkammer *f*, Tonerdebreikammer *f*, Tonerdebreiraum *m*

4904 SLIP KILN cer
f étuve *f* évaporatrice
e estufa *f* evaporadora
i stufa *f* evaporatrice
n afdampoven
d Abdampfofen *m*

4905 SLIT ORIFICE pre
f filière *f* en fente
e orificio *m* de budinadora en forma de ranura
i filiera *f* in forma di fessura
n spuitmatrijs met spleetvormige opening
d Schlitzdüse *f*

4906 SLITTING WHEEL, CUT-OFF DISK, CUTTING WHEEL cpt
f disque *m* de tronçonnage
e disco *m* de cortar
i disco *m* per tranciare
n doorslijpschijf
d Trennscheibe *f*

4907 SLIVER SCREEN pap
f tamis *m* épurateur
e criba *f* depuradora
i vaglio *m* di depuramento
n zuiveringszeef
d Astfänger *m*

4908 SLOPE, RISE dis
f chute *f*, pente *f*
e pendiente *f*
i pendenza *f*
n afschot *n*
d Steigung *f*

4909 SLOPING GAS FLOWMETER ins
f débitmètre *m* à gaz incliné
e flujómetro *m* de gas inclinado
i flussometro *m* a gas inclinato
n gasstromingsmeter met hellende buis
d Schrägrohrgasströmungsmesser *m*

4910 SLOPING GRATE htt
f grille *f* inclinée
e parrilla *f* inclinada
i griglia *f* inclinata
n hellend rooster *n*
d Schrägrost *m*

4911 SLOPING SCREEN sor
f claie *f*
e criba *f* inclinada
i crivello *m* inclinato
n stortzeef
d Schrägsieb *n*, Wurfsieb *n*

4912 SLOT DIE, FLAT SHEETING DIE pla
f filière *f* plate
e boquilla *f* de ranura ancha, tobera *f* de ranura ancha

i filiera *f* piana, testa *f* piana
n spleetmatrijs
d Breitschlitzdüse *f*

4913 SLOTTED PLATE, dis,sep
PERFORATED PLATE
f tôle *f* perforée
e chapa *f* perforada
i lamiera *f* perforata
n zeefplaat
d Siebblech *n*

4914 SLOUGHING-OFF BOX, sor
CLASSIFIER WITH CONICAL TROUGHS
f classeur *m* à bacs coniques
e clasificador *m* de tolvas cónicas
i classificatore *m* a vasche coniche
n classificeerder met conische bakken
d Spitzkasten *m*

4915 SLOW-BURNING PIPETTE ana
f cloche *f* à combustion complète, pipette *f* pour feu lent
e pipeta *f* de combustión lenta
i pipetta *f* a combustione lenta
n verbrandingspipet
d Verbrennungspipette *f*

4916 SLOW FILTER, fil
SLOW SAND FILTER
f filtre *m* lent
e filtro *m* lento
i filtro *m* lento
n langzaam filter *n*
d Langsamfilter *m*

4917 SLUDGE BLANKET FILTER, iwt
FLOATING BLANKET
f filtre *m* à voile de boue
e colchón *m* de fango
i filtro *m* con sospensione di fango
n drijvend filter *n*
d schwebender Filter *m*

4918 SLUDGE-BLANKET REACTOR, iwt
SUSPENDED-SOLIDS CONTACT REACTOR
f bassin *m* de floculation à lit de boue
e artesa *f* de floculación de contacto de fango
i bacino *m* di flocculazione a contatto del fango
n snel uitvlokkingsbekken *n*
d Schlammkontaktflockungsbecken *n*, Schnellflockungsbecken *n*

4919 SLUDGE BOX iwt
f bac *m* à boue
e caja *f* de fango
i cassetta *f* a fango
n slibbak
d Schlammkasten *m*

4920 SLUDGE CAKE iwt
f gâteau *m* de boue, tourteau *m* de résidus de filtrage
e torta *f* de fango
i focaccia *f* di fango, pressati *pl* del filtro
n slibkoek
d Schlammkuchen *m*

4921 SLUDGE CENTRIFUGE iwt
f centrifugeur *m* de boue
e centrífuga *f* de fango
i centrifuga *f* di fango
n slibcentrifuge
d Schlammschleuder *f*

4922 SLUDGE CONCENTRATOR, iwt
SLUDGE THICKENER,
DENSIFIER
f récipient *m* épaississeur de boue
e tanque *m* concentrador de fango
i serbatoio *m* concentratore di fango
n slibindikker
d Schlammeindickbehälter *m*, Schlammeindicker *m*

4923 SLUDGE DEWATERING TANK iwt
f récipient *m* à déhydrater la boue
e tanque *m* de deshidratación de fango
i serbatoio *m* da disidratare il fango
n slibontwateringstank
d Schlammentwasserungsgefäss *n*

4924 SLUDGE DIGESTION iwt
COMPARTMENT,
DIGESTION CHAMBER
f chambre *f* de digestion, chambre *f* de putréfaction de bou
e cámara *f* de digestión
i camera *f* di digestione di fango
n digereerkamer
d Faulraum *m*, Schlammfaulraum *m*

4925 SLUDGE DIGESTION TANK, iwt
DIGESTER, DIGESTER TANK
f digesteur *m*, vase *m* de digestion
e tanque *m* digestor
i vasca *f* di digestione
n digereertank
d Faulbehälter *m*, Schlammfaulbehälter *m*

4926 SLUDGE DISTRIBUTOR iwt
f disque *m* distributeur de boue
e disco *m* distribuidor de fango
i disco *m* distributore di fango
n slibverdeelschijf
d Schlammverteilerscheibe *f*

4927 SLUDGE FILTER PRESS bre
f filtre-presse *m* à trouble
e filtro-prensa *m* para turbio
i pressa *f* per torbido
n drabpers
d Trubfilterpresse *f*, Trubpresse *f*

4928 SLUDGE GATE iwt
f vanne *f* à boue
e esclusa *f* de fango
i saracinesca *f* per il fango
n slibsluis
d Schlammschieber *m*, Schlammschleuse *f*

4929 SLUDGE INCINERATION PLANT iwt
f incinérateur *m* de boue
e incinerador *m* de fango
i inceneratore *m* di fango
n slibverbrandingsinstallatie
d Schlammofen *m*, Schlammverbrenner *m*

4930 SLUDGE LAGOON, SETTLING LAGOON iwt
f étang *m* de boue, étang *m* de remblaiement
e estanque *m* de deposición
i stagno *m* a fango, stagno *m* di terra
n slibvijver
d Absetzteich *m*, Auflandungsteich *m*, Schlammteich *m*

4931 SLUDGE LEVEL INDICATOR iwt
f signaleur *m* de niveau du schlamm
e indicador *m* de nivel del fango
i indicatore *m* di livello del fango
n slibniveau-indicator
d Schlammniveauindikator *m*

4932 SLUDGE PAN, TEMPERING PAN cer
f cuve *f* de gâchage
e cuba *f* de amasado
i tino *m* di preparazione
n aanmaakvat *n*
d Anmachgefäss *n*

4933 SLUDGE PUMP iwt
f pompe *f* de boue
e bomba *f* de fango
i pompa *f* di fango
n slibpomp
d Schlammpumpe *f*

4934 SLUDGE REMOVAL PLANT, DESLUDGING PLANT iwt
f installation *f* d'enlèvement de boue, installation *f* d'évacuation de boue
e equipo *m* de desenlodamiento
i impianto *m* di scarico di fango
n ontslibbingsinstallatie
d Entschlammungsanlage *f*

4935 SLUDGE SCRAPER iwt
f racleur *m* de boue
e rascador *m* de fango
i raschiatore *m* di fango
n slibschraper
d Schlammkratzer *m*

4936 SLUDGE TROUGH iwt
f goulotte *f* de boue
e canal *m* de fango
i canaletto *m* di fango
n slibgoot
d Schlammrinne *f*

4937 SLUDGE VALVE iwt
f soupape *f* de boue
e válvula *f* para barro
i valvola *f* per fango
n slijkklep, slikklep
d Schlammventil *n*

4938 SLUDGER iwt
f débourbeur *m*
e limpiabarros *m*
i separatore *m* di fango
n slibafscheider
d Schlammabscheider *m*

4939 SLUICE CONCENTRATOR, RIFFLE CONCENTRATOR min
f appareil *m* de concentration de minerais à râteau
e concentrador *m* de minerales con rastrillo
i concentratore *m* di minerali a rastrello
n ertsconcentreerinrichting met doorlaatgoot
d Erzanreicherungsanlage *f* mit Durchlassrinne

4940 SLUICE-TYPE REFLUX RATIO HEAD dis,lab
f tête *f* de reflux type écluse
e cabezal *m* de reflujo tipo esclusa

i testa *f* di riflusso tipo chiusa
n opzet voor terugstroom met sluis
d Destillieraufsatz *m* für Rückfluss mit Schleuse

4941 SLUICE VALVE, GATE, GATE VALVE — cpt
f obturateur *m*, robinet-vanne *m*
e válvula *f* de compuerta, válvula *f* de paso directo
i saracinesca *f*
n afsluitschuif
d Absperrschieber *m*, Abzugsschieber *m*

4942 SLURRY DISCHARGE CENTRIFUGE — cen
f centrifugeur *m* séparateur de pâtes
e centrífuga *f* separadora de pastas
i centrifuga *f* da separare paste
n centrifuge voor het afscheiden van dikvloeibare massa's
d dickstoffabscheidende Zentrifuge *f*

4943 SLURRY PUMP — cle
f pompe *f* de substances consistantes
e bomba *f* de substancias consistentes
i pompa *f* di sostanze consistenti
n brijpomp, pomp voor dikvloeibare massa's
d Dickstoffpumpe *f*

4944 SLURRY REACTOR — nue
f réacteur *m* à schlamm
e reactor *m* de combustible en suspensión
i reattore *m* a sospensione
n papreactor, reactor met natte splijtstofsuspensie
d Reaktor *m* mit nasser Suspension, Suspensionsreaktor *m*

4945 SLUSH MOLD (US), SLUSH MOULD (GB) — pla,pre
f moule *m* à plastisols
e molde *m* de embadurnado
i forma *f* per paste
n gietvorm voor pasta's
d Pastengiessform *f*

4946 SMALTHOUSE, ZAFFERBLUE WORKS — cer,dye
f atelier *m* d'azurage, teinturerie *f* en bleu
e tintorería *f* en azul
i officina *f* di vagellaio
n blauwververij
d Blaufarbenwerk *n*

4947 SMELTER, SMELTERY — htt
f fonderie *f*
e hacienda *f* de fundición
i fonderia *f*
n smelterij *f*
d Schmelzhütte *f*

4948 SMELTING FURNACE — red
f four *m* de réduction
e horno *m* de reducción
i forno *m* di riduzione
n reductieoven
d Reduktionsofen *m*

4949 SMELTING FURNACE, MELTING FURNACE — gla,met
f four *m* de fusion
e horno *m* de fusión
i forno *m* di fusione
n smeltoven
d Schmelzofen *m*

4950 SMELTING HEARTH — htt
f fourneau *m* de fonderie
e hogar *m* de fusión
i focolare *m* di fusione
n smelthaard
d Schmelzherd *m*

4951 SMOKE BALL — mit
f bombe *f* fumigène
e bomba *f* fumígena
i bomba *f* fumogena
n rookbom
d Nebelbombe *f*, Rauchbombe *f*

4952 SMOKE CARTRIDGE — exp,mit
f cartouche *f* de fumée
e cartucho *m* de humo
i cartuccia *f* di fumo
n rookpatroon
d Rauchpatrone *f*

4953 SMOKE FILTER — fil
f filtre *m* de gaz de fumée
e filtro *m* de gas de humo
i filtro *m* di gas di fumo
n rookgasfilter *n*
d Rauchgasfilter *m*

4954 SMOKE FLUE — fue
f carneau *m*
e conducto *m* del humo
i condotto *m* del fumo

n rookgang, rookkanaal *n*
d Feuerzug *m*, Fuchs *m*, Rauchkanal *m*

4955 SMOKE FUNNEL fue
f éventouse *f*, sortie *f* de fumée
e salida *f* de humo
i fumaiuolo *m*
n rookgat *n*
d Rauchabzug *m*

4956 SMOKE GAS PREHEATER, ECONOMIZER pre
f économiseur *m*
e calentador *m* del agua de alimentación, economizador *m*
i economizzatore *m*
n economiser, rookgasvoorverwarmer, voedingswatervoorverwarmer
d Ekonomiser *m*, Rauchgasvorwärmer *m*, Sparanlage *f*, Speisewasservorwärmer *m*

4957 SMOKE POINT TESTER fue,tes
f appareil *m* de vérification du point de formation de fumée
e probador *m* del punto de formación de humo
i collaudatore *m* del punto di formazione di fumo
n walmpuntbepalingsapparaat *n*
d Qualmpunktbestimmungsgerät *n*

4958 SMOKE SCREEN mit
f brouillard *m* artificiel, nuage *f* artificielle, rideau *m* de fumée
e cortina *f* de humo
i cortina *f* di fumo, sbarramento *m* di fumo
n nevelgordijn *n*, rookgordijn *n*
d Rauchschleier *m*, Rauchvorhang *m*

4959 SMOKE SHELL mit
f obus *m* fumigène
e proyectil *m* fumígeno
i granata *f* fumogena
n rookgranaat
d Rauchgranate *f*

4960 SMOKE SLIDE VALVE, DAMPER, REGISTER con,fue
f registre *m*
e registro *m*
i paratoia *f*, registro *m*, serranda *f*
n regelschuif
d Rauchschieber *m*, Regelschieber *m*, Schieberplatte *f*

4961 SMOKE TEST APPARATUS hyg,tes
f appareil *m* d'essai des carneaux
e probador *m* de los conductos de humo
i collaudatore *m* dei condotti di fumo
n rookgangcontroleapparaat *n*
d Rauchkanalprüfgerät *n*

4962 SMOTHERED ARC FURNACE, DIRECT ARC FURNACE htt
f four *m* à arc direct
e horno *m* de arco directo
i forno *m* ad arco diretto
n directe vlamboogoven, vlamboogoven met directe verhitting
d direkter Lichtbogenofen *m*

4963 SMOTHERING VESSEL, SCALDING VESSEL, SEETHING VESSEL tex
f échaudoir *m*
e cuba *f* de escaldar, escaldadera *f*
i serbatoio *m* da scottare
n broeiketel
d Abbrühgefäss *n*

4964 SMUT MILL, GRAIN CLEANING MACHINE nut
f tarare *m*
e máquina *f* para limpiar cereales
i macchina *f* per eliminare il carbonchio dei cereali
n graanreinigingsmachine
d Getreidereinigungsmaschine *f*

4965 SNAGGING WHEEL pol
f disque *m* d'ébarbage
e muela *f* desbastadora
i sbavatrice *f*
n poetsschijf
d Fertigputzscheibe *f*

4966 SNEEZE GAS GENERATOR gas,mit
f générateur *m* de diphénylchlorarsine
e generador *m* de difenilclorarsina, generador *m* de gas estornudatorio
i generatore *m* di gas starnutatorio
n niesgasgenerator
d Reizgasgenerator *m*

4967 SNIFTING VALVE, BLOW-THROUGH VALVE cpt
f reniflard *m*
e válvula *f* de purga
i valvola *f* di scarico
n snuifklep

d Schnarchventil *n*, Schnarrventil *n*, Schnüffelventil *n*

4968 SNORE PIECE, STRAINER HEAD — pre
f corps *m* d'aspiration, crépine *f*
e alcachofa *f* de aspiración, parte *f* inferior del tubo de aspiración
i cuffia *f* d'aspirazione, succhieruola *f*
n zuigkorf
d Saugstück *n*, Senkkorb *m*, Siebkopf *m*

4969 SNUBBER (US), PULSATION DAMPER (GB) — gas,pre
f amortisseur *m* de pulsations
e amortiguador *m* de pulsaciones
i smorzatore *m* di pulsazioni
n schokdemper
d Stossdämpfer *m*

4970 SOAK CLEAN, SOAK CLEAN BATH — elc
f bain *m* de nettoyage
e baño *m* de limpieza
i bagno *m* di pulitura
n reinigingsbad *n*
d Reinigungsbad *n*

4971 SOAKING FURNACE, SOAKING PIT — met
f four *m* d'égalisation
e horno *m* de resudar, horno *m* de termodifusión
i forno *m* a pozzo, forno *m* a termodiffusione
n egaliseeroven, voorwarmoven
d Ausgleichofen *m*

4972 SOAKING TANK, INHIBITION TANK — lea
f récipient *m* d'imbibition
e tanque *m* de imbibición
i serbatoio *m* d'imbibizione
n drenkvat *n*
d Tränkgefäss *n*

4973 SOAKING WATER TANK — tan
f récipient *m* d'eau de reverdissage
e tanque *m* de agua de maceración
i serbatoio *m* d'acqua di macerazione
n inweekvat *n*
d Einweichwasserküpe *f*

4974 SOAKING WHEEL, BOTTLE SOAKING MACHINE — bre,nut
f roue *f* trempeuse
e remojadora *f* de botellas
i maceratrice *f*
n flessenreiniger
d Flaschenweiche *f*

4975 SOAP BOILER (GB), SOAP KETTLE (US), SOAP PAN (GB) — soa
f chaudron *m* à savon
e caldera *f* de jabón
i caldaia *f* a sapone
n zeepketel, zeeppan
d Siedekessel *m*, Siedepfanne *f*

4976 SOAP FACTORY — soa
f savonnerie *f*
e jabonería *f*
i fabbrica *f* di sapone, saponeria *f* saponificio *m*
n zeepfabriek
d Seifenfabrik *f*

4977 SOAP MILL — soa
f boudineuse *f*
e molino *m* para jabón
i raffinatrice *f* a cilindri
n pileermachine
d Piliermaschine *f*

4978 SOAPING AFTERTREATMENT BATH — tex
f bain *m* de savonnage
e baño *m* enjabonaduro
i bagno *m* insaponatore
n zeepbad *n*
d Seifenbad *n*

4979 SOCKET (GB), BELL (US) — cpt
f emboîtement *m*, manchon *m* élargi, manchon *m* mandriné
e enchufe *m*, tubo *m* abocarbado
i bicchiere *m*
n buismof, buisverwijding
d Muffenkelch *m*, Rohrmuffe *f*

4980 SOCKET VALVE, SLEEVE VALVE — cpt
f pale *f*, soupape *f* à manchon
e válvula *f* de camisa
i valvola *f* a manicotta
n mantelklep
d Muffenventil *n*, Ventilschieber *m*

4981 SODA EXTRACTION PLANT — sod
f installation *f* d'extraction de soude
e equipo *m* de extracción de sosa
i impianto *m* da estrarre soda
n sodaextractie-installatie
d Sodaausziehanlage *f*

4982 SODA FACTORY sod
f usine *f* de soude
e fábrica *f* de sosa
i fabbrica *f* di soda
n sodafabriek
d Sodafabrik *f*

4983 SODA FURNACE, aic
BALLING FURNACE, BLACK ASH FURNACE
f four *m* de soude
e horno *m* de carbonatar, horno *m* de sosa
i forno *m* di soda
n sodaoven
d Sodaofen *m*

4984 SODA VAT sod
f cuve *f* à soude
e cuba *f* de sosa
i tino *m* a soda
n potaskuip, sodakuip
d Pottaschküpe *f*, Sodaküpe *f*

4985 SODIUM-COOLED FAST REACTOR nue
f réacteur *m* à neutrons rapides à refroidissement par le sodium
e reactor *m* de neutrones rápidos de enfriamiento por sodio
i reattore *m* a neutroni rapidi a raffreddamento al sodio
n met natrium gekoelde snelle reactor
d natriumgekühlter schneller Reaktor *m*

4986 SODIUM GRAPHITE REACTOR nue
f réacteur *m* à sodium-graphite
e reactor *m* sodio-grafito
i reattore *m* a sodio-grafite
n natrium-grafietreactor
d Natrium-Graphitreaktor *m*

4987 SODIUM PEROXIDE CALORIMETER ins
f calorimètre *m* à peroxyde de sodium
e calorímetro *m* de peróxido de sodio
i calorimetro *m* a perossido di sodio
n natriumperoxydecalorimeter
d Natriumperoxyd-Kalorimeter *n*

4988 SODIUM SPOON lab
f cuiller *m* pour sodium
e cuchara *f* para sodio
i cucchiaio *m* per sodio
n natriumlepel
d Natriumlöffel *m*

4989 SODIUM THIOSULPHATE BATH, pho
HYPO BATH
f bain *m* à hyposulfite de sodium
e baño *m* de hiposulfito sódico
i bagno *m* ad iposolfito di sodio
n hypobad *n*
d Fixiernatronbad *n*, Hypobad *n*

4990 SOFT-FACING PLANT coa,met
f installation *f* de rechargement doux
e equipo *m* de recargue suave, equipo *m* de revestimiento suave
i impianto *m* di rivestimento dolce
n zachtmetaalbekledingsinstallatie
d Weichmetallauftraganlage *f*

4991 SOFTENING APPARATUS, hyg,sew
WATER SOFTENER
f appareil *m* pour l'adoucissement de l'eau
e aparato *m* de ablandamiento del agua
i apparecchio *m* d'addolcimento dell'acqua
n wateronthardingsapparaat *n*
d Wasserenthärtungsapparat *m*

4992 SOFTENING FURNACE, met
LEAD IMPROVING FURNACE
f four *m* d'affinage du plomb
e horno *m* de afino del plomo
i forno *m* da affinare il piombo
n loodveredelingsoven
d Bleivergütungsofen *m*, Seigerofen *m*

4993 SOFTENING PLANT lea
f installation *f* d'adoucissement
e equipo *m* de ablandamiento
i impianto *m* d'addolcimento
n installatie voor het soepel maken van leer
d Weichmacheanlage *f*

4994 SOFTENING POINT TESTING APPARATUS, tes
MELTING POINT TESTING APPARATUS
f appareil *m* de détermination du point de ramollissement du bitumen
e comprobador *m* del punto de reblandecimiento del betún
i collaudatore *m* del punto di rammollimento del bitumen
n apparaat *n* voor het bepalen van het verwekingspunt van bitumen
d Erweichungspunktbestimmungsgerät *n* für Bitumen

4995 SOIL FILTER, fil,iwt
INTERMITTENT SOIL FILTER
f filtre *m* par le sol, sol *m* filtrant
e filtro *m* natural
i filtro *m* naturale
n natuurlijk filter *n*
d Bodenfilter *m*

4996 SOIL PIPE, sew
DOWN-TAKE PIPE (GB),
DRAFT TUBE (US)
f tube *m* de chute, tube *m* de descente
e tubo *m* de bajada, tubería *f* descendente
i tubo *m* di caduta, tubo *m* discendente
n valbuis
d Abfallrohr *n*, Fallrohr *n*

4997 SOLAR CELL elc
f pile *f* solaire
e pila *f* solar
i pila *f* solare
n door zonnestralen geactiveerd element *n*
d durch Sonnenstrahlen betätigtes Element *n*

4998 SOLDERING IRON, wel
COPPER BIT
f fer *m* à souder, soudoir *m*
e soldador *m*
i arnese *m* per saldare, saldatoio *m*
n soldeerbout
d Lötkolben *m*

4999 SOLDERING LAMP wel
f lampe *f* à souder
e lámpara *f* de soldadura
i lampada *f* per saldare
n soldeerlamp
d Lötlampe *f*, Lötrohrlampe *f*

5000 SOLID BED, gen
STATIONARY BED
f porteur *m* solide de catalysateur
e portador *m* macizo de catalizador
i portatore *m* solido di catalizzatore
n vaste katalysatordrager
d Festbett *n*

5001 SOLID BOWL-TYPE CENTRIFUGE cen
f centrifugeur *m* à panier plain
e centrífuga *f* de tambor macizo
i centrifuga *f* a paniere senza fori
n trommelcentrifuge
d Trommelzentrifuge *f*

5002 SOLID-LIQUID ROCKET mit
f fusée *f* solide-liquide
e cohete *m* sólido-líquido
i razzo *m* solido-liquido
n vaste stof-vloeistofraket
d Feststoff-Flüssigkeitsrakete *f*

5003 SOLID-PROPELLANT ROCKET mit
f fusée *f* à propulseur solide
e cohete *m* de propulsor macizo
i razzo *m* a propellente solido
n raket met vast voortstuwingsmiddel
d Feststoffrakete *f*

5004 SOLID SAMPLER chr
f dispositif *m* d'introduction d'échantillons solides
e dispositivo *m* de introducción de muestras macizas
i dispositivo *m* da introdurre campioni solidi
n invoerapparaat *n* voor vaste monsters
d Feststoffprobengeber *m*

5005 SOLID-SOLID MIXER mix
f mélangeur *m* de plusieurs matières solides
e mezclador *m* sólido-sólido
i mescolatore *m* solido-solido
n menger voor verschillende vaste stoffen
d Feststoff-Feststoffmischvorrichtung *f*

5006 SOLID WALL CENTRIFUGE cen
f centrifugeur *m* à paroi solide
e centrífuga *f* de pared sólida
i centrifuga *f* a parete solida
n centrifuge met gesloten wand
d Zentrifuge *f* mit geschlossener Wand

5007 SOLIDIFICATION POINT RECORDER, ins
FREEZING POINT RECORDER
f enregistreur *m* du point de solidification
e registrador *m* del punto de solidificación
i registratore *m* del punto di solidificazione
n stolpuntschrijver
d Erstarrungspunktschreiber *m*

5008 SOLUTION HEAT TREATMENT FURNACE htt,me
f four *m* de traitement thermique à dissolution améliorée

e horno *m* de termotratamiento de disolución mejorada
i forno *m* di trattamento termico a dissoluzione migliorata
n oven voor warmtebehandeling met verhoogde oplossing
d Lösungsglühofen *m*, Vergütungsofen *m*

5009 SOLUTION STRIPPING BATH, CHEMICAL STRIPPING BATH cle
f bain *m* de dépouillage chimique
e baño *m* de eliminación química
i bagno *m* di degalvanizzazione chimica
n chemisch laagverwijderingsbad *n*
d chemisches Entplattierungsbad *n*

5010 SOLUTION-TYPE REACTOR nue
f réacteur *m* à combustible en forme de solution
e reactor *m* de combustible en solución
i reattore *m* a soluzione
n oplossingsreactor
d Lösungsreaktor *m*

5011 SOLVATION VESSEL gen
f cuve *f* de solvatation, cuve *f* de solvatisation
e cuba *f* de solvatación
i tino *m* di solvatazione
n solvatatievat *n*, solvatievat *n*
d Solvatationsgefäss *n*

5012 SOLVENT CLEANING BATH elc
f bain *m* de dégraissage au solvant
e baño *m* de desengrase por solvente
i bagno *m* di sgrassaggio con solvente
n ontvettingsbad *n* met oplosmiddel
d Reinigungsbad *n* mit Lösungsmittel

5013 SOLVENT DYEING BATH tex
f bain *m* à solvant de matières colorantes
e baño *m* de solvente para colorantes
i bagno *m* a solvente di coloranti
n oplossingsbad *n* voor kleurstoffen
d Farbstofflösungsbad *n*

5014 SOLVENT EXTRACTION PLANT fue
f installation *f* à extraction par solvant
e equipo *m* de extracción por solvente
i impianto *m* da estrarre per mezzo di solvente
n extraheerinstallatie met oplosmiddel
d Lösungsmittelextraktionsanlage *f*

5015 SOLVENT RECOVERY PLANT dis
f installation *f* pour la récupération du solvant
e equipo *m* de recuperación del solvente
i impianto *m* per ricupero del solvente
n installatie voor terugwinning van het oplosmiddel
d Lösungsmittelwiedergewinnungsanlage *f*, Rückgewinnungsanlage *f* für Lösungsmittel

5016 SOLVENT SCOURING VESSEL tex
f récipient *m* de dégraissage à solvant
e recipiente *m* desgrasador con solvente
i serbatoio *m* sgrassatore con solvente
n wolontvettingskuip met oplosmiddel
d Entfettungsküpe *f* mit Lösungsmittel

5017 SOOT BLOWER, SOOT SWEEPER cle,hyg
f souffleur *m* de suie
e soplahollín *m*
i soffiatore *m* per fuliggine
n roetblazer
d Russgebläse *n*, Russventilator *m*

5018 SOOT CATCHER cle,hyg
f collecteur *m* de suie
e recogehollín *m*
i collettore *m* di fuliggine
n roetvanger
d Russfänger *m*

5019 SOOT CHAMBER boi
f chambre *f* de suie
e cámara *f* de hollín
i camera *f* di fuliggine
n roetkamer
d Russkammer *f*

5020 SOOT PIT boi
f bac *m* de suie
e recipiente *m* de hollín
i recipiente *m* di fuliggine
n roetbak
d Russammelkasten *m*

5021 SORTING MACHINE sor
f trieur *m*, trieuse *f*
e máquina *f* seleccionadora
i selezionatrice *f*
n sorteermachine
d Sortiermaschine *f*

5022 SORTING PLANT sor
f installation *f* de triage
e equipo *m* de tría
i impianto *m* di selezione
n sorteerinstallatie
d Sortieranlage *f*

5023 SOUPLING VESSEL tex
f cuve *f* de dégommage partiel
e cuba *f* de desengomado parcial
i tino *m* di sgommatura parziale
n vat *n* voor gedeeltelijke ontgomming
d Teildegummierungsgefäss *n*, Teilentbastungsgefäss *n*

5024 SOURING BATH ble,tex
f bain *m* de vitriolage
e baño *m* de acidulado
i bagno *m* da acidulare
n zuurbad *n*
d Sauerbad *n*

5025 SOURING TANK, cer
AGEING TANK (GB), AGING TANK (US)
f récipient *m* de mûrissage
e tanque *m* de curar
i serbatoio *m* di maturazione, serbatoio *m* d'omogenizzazione
n homogenisatietank
d Homogeniergefäss *n*

5026 SPACE DEHUMIDIFIER dry
f sécheur *m* d'espace
e deshumidificador *m* de espacio
i deumidificatore *m* di spazio, essiccatore *m* di spazio
n ruimtedroger
d Raumentfeuchter *m*, Raumtrockner *m*

5027 SPACER gen
f entretoise *f*
e espaciador *m*, pieza *f* de distancia
i distanziatore *m*
n afstandhouder
d Abstandsstück *n*, Distanzstück *n*

5028 SPACER pla,pre,rub
f plaque *f* de calage, plaque *f* d'écartement, plaque *f* d'espacement
e placa *f* de distanciamiento placa *f* de separación
i piastra *f* distanziatrice
n opvulplaat, opvulstuk *n*
d Abstandsplatte *f*, Zwischenstück *n*

5029 SPACER (GB), pla,pre
PARALLEL (US)
f cale-témoin *f*, chandelle *f*, témoin *m* de mise en place
e calce *m*, paralelas *pl*
i ponte *m*, spessore *m*
n opvulstuk *n*, vulstuk *n*
d Distanzblock *m*

5030 SPACER (GB), pla,pre
SUPPORT PILLAR (US),
f cale *f*, chandelle *f*, colonnette *f*, pilier *m*
e columna *f*, pilar *m*, soporte *m*
i distanziatore *m*, colonna *f*
n opvulstuk *n*, rond vulstuk *n*, steunkolom
d Abstandsblock *m*, Abstandssäule *f* Stütze *f*, Unterlegstück *n*

5031 SPACER BLOCK pla,pre
f bloc *m* d'écartement
e bloque *m* separador
i distanziatore *m*
n opvulstuk *n*, vulstuk *n*
d Distanzblock *m*

5032 SPACER RING cpt
f anneau *m* d'écartement, anneau *m* intermédiaire
e anillo *m* distanciador, anillo *m* espaciador
i anello *m* distanziatore
n tussenring, vulring
d Abstandsring *m*, Zwischenring *m*

5033 SPARE TANK sto
f récipient *m* de réserve
e tanque *m* de repuesto
i serbatoio *m* di riserva
n reservetank
d Reservebehälter *m*

5034 SPARGER bre
f croix *f* écossaise, installation *f* d'arrosage
e regador *m*
i apparecchio *m* per lavaggio
n sproeier
d Anschwänzapparat *m*, Anschwänzer *m*, Drehkreuz *n*, Spritzkranz *m*

5035 SPARGER, fea
FIRE EXTINGUISHING ROSE
f arroseur *m* extincteur d'incendies
e rociadera *f* extinguidora de incendios
i spruzzatore *m* antincendio
n brandblussproeier
d Feuerlöschbrause *f*

5036 SPARK METAL WORKING PLANT, met
ELECTRO-EROSION METAL-WORKING PLANT
f installation *f* d'usinage par électro-érosion, installation *f* d'usinage par étincelage
e equipo *m* de trabajo por electro-erosión
i impianto *m* di lavorazione a scintillamento
n inrichting *f* voor elektro-erosieve behandeling
d Elektroerosionsbearbeitungs-anlage *f*

5037 SPATTLE, lab
SPATULA
f spatule *f*
e espátula *f*
i spatola *f*
n spatel
d Spachtel *m*, Spatel *m*

5038 SPECIFIC GRAVITY BOTTLE, lab
GRAVITOMETER
f gravitomètre *m*
e gravitómetro *m*
i gravitometro *m*
n gravitometer
d Gravitometer *n*

5039 SPECTROGRAPH ins,spe
f spectrographe *m*
e espectrógrafo *m*
i spettrografo *m*
n spectrograaf
d Spektrograph *m*

5040 SPECTROMETER ins,spe
f spectromètre *m*
e espectrómetro *m*
i spettrometro *m*
n spectrometer
d Spektrometer *n*

5041 SPECTROPHOTOMETER ins,spe
f spectrophotomètre *m*
e espectrofotómetro *m*
i spettrofotometro *m*
n spectrofotometer
d Spektralphotometer *n*

5042 SPECTROSCOPE ins,spe
f spectroscope *m*
e espectroscopio *m*
i spettroscopio *m*
n spectroscoop
d Spektroskop *n*

5043 SPECTRUM BURNER lab
f brûleur *m* pour analyse spectrale
e quemador *m* para análisis espectral
i bruciatore *m* per analisi spettrale
n brander voor spectraalanalyse
d Brenner *m* für Spektralanalyse

5044 SPENT FUEL REPROCESSING PLANT nue
f installation *f* de traitement du combustible usé
e equipo *m* de reprocesamiento del combustible agotado
i impianto *m* di trattamento del combustibile irradiato
n opwerkinstallatie van bestraalde splijtstof
d Wiederaufbereitungsanlage *f* von ausgebranntem Spaltmaterial

5045 SPEW, rub
EXCESS RUBBER OVERFLOW, SPUE
f trop-plein *m*
e rebosamiento *m* del molde
i troppo pieno *m*
n overloop
d Austrieb *m*, Formaustrieb *m*

5046 SPEW CHANNEL, pla,pre
SPEW GROOVE, SPEW WAY
f couloir *m* de trop-plein
e canal *m* de rebose
i canale *m* per troppo pieno
n vormoverloop
d Überlaufrinne *f*

5047 SPHERICAL BOILER pap
f chaudron *m* sphérique
e caldera *f* esférica
i caldaia *f* sferica
n bolvormig kookvat *n*
d Kugelkocher *m*

5048 SPIDER, pre,rub
BREAKER PLATE
f croisillon *m*, étoile *f*
e cruceta *f* esparcidora
i crocetta *f*, crociera *f* spalmatrice
n centerstuk *n* voor doorn
d Armkreuz *n*

5049 SPIDER-LIKE SPRUE pla,pre
f grappe *f* de moulage
e pedúncolo *m*
i ragno *m*
n stervormige aanspuiting
d Angussspinne *f*, Spinne *f*

5050 SPIKE ROLL, tex
EVENER ROLL
f cylindre *m* égaliseur
e cilindro *m* igualador, cilindro *m* de puntas
i cilindro *m* a denti
n getande rol
d Zahnrolle *f*

5051 SPILL VALVE cpt,sap
f soupape *f* de débordement
e válvula *f* de rebaba
i valvola *f* di ribocco
n overloopklep
d Überstromventil *n*

5052 SPILLWAY, sew
OVERFALL, OVERFLOW
f déversoir *m*
e vertedero *m*
i sfioratore *m*, stramazzo *m*
n overloop
d Überfall *m*, Überlauf *m*

5053 SPIN HARDENING PLANT htt,met
f installation *f* d'endurcissement par rotation
e equipo *m* de endurecimiento por rotación
i impianto *m* d'indurimento per rotazione
n rotatiehardingsinstallatie
d Drehungshärtungsanlage *f*

5054 SPIN WELDING APPARATUS pla,wel
f appareil *m* à souder à chaud par rotation
e aparato *m* de sellado por rotación
i apparecchio *m* di saldatura per rotazione
n roterend kunststoflasapparaat *n*
d Drehsiegelapparat *m*

5055 SPINDLE (GB), cpt,gen
ARBOR (US)
f arbre *m*, axe *m*
e árbol *m*, eje *m*
i albero *m*, asse *m*
n as, spil
d Achse *f*, Spille *f*, Spindel *f*, Welle *f*

5056 SPINNERET, tex
SPINNING NOZZLE, DIE-SPINNING NOZZLE
f filière *f*
e tobera *f* de hilatura
i ugello *m* di filiera
n spindop, spinkop
d Spinndüse *f*

5057 SPINNING DISK ATOMIZER, soa
CENTRIFUGAL DISK ATOMIZER
f atomiseur *m* à disque rotatif, disque *m* atomiseur
e atomizador *m* de disco giratorio, disco *m* atomizador
i atomizzatore *m* a disco rotativo, disco *m* atomizzatore
n draaischijfverstuiver
d Fliehkraftzerstäuber *m*, Zerstäuberscheibe *f*

5058 SPINNING FUNNEL tex
f entonnoir *m* de filature
e cono *m* de hilado
i imbuto *m* di filatura
n spintrechter
d Fälltrichter *m*, Spinntrichter *m*

5059 SPINNING MACHINE tex
f machine *f* à filer
e hiladora *f* mecánica, máquina *f* de hilar
i filatoio *m*, macchina *f* per filare
n spinmachine
d Spinnmaschine *f*

5060 SPINNING MILL tex
f filature *f*
e hilandería *f*
i filanda *f*
n spinnerij
d Spinnerei *f*

5061 SPINNING-OFF PLANT sin
f installation *f* de centrifugation
e equipo *m* de centrifugación
i impianto *m* di centrifugazione
n centrifugeerinstallatie
d Zentrifugieranlage *f*

5062 SPINNING PUMP tex
f pompe *f* de filature
e bomba *f* para hilar
i pompa *f* da filare
n spinpomp
d Spinnpumpe *f*

5063 SPINNING VESSEL te
f récipient *m* de liquide filable

e recipiente *m* de liquido para hilar
i recipiente *m* di liquido da filare
n spinvloeistofvat *n*
d Spinnflüssigkeitsbehälter *m*

5064 SPIRAL CONDENSER, clo
SPIRAL RADIATOR
f serpentin *m* de refroidissement
e serpentín *m* de enfriamiento
i serpentina *f* di raffreddamento
n spiraalkoeler
d Schlangenkühler *m*

5065 SPIRAL FIN-TUBE clo,hyg
HEATING COIL,
EXTENDED-SURFACE COIL,
FIN-TYPE HEATING COIL,
FINNED-TUBE HEATING COIL
f réchauffeur *m* d'air en tubes ailetés
e calentador *m* de aire con tubos de aletas
i riscaldatore *m* d'aria con tubo ad alette
n ribbenbuisluchtverhitter
d Bandrippenrohrheizregister *n*

5066 SPIRAL-FLOW TANK iwt
f récipient *m* à circulation hélicoïdale
e tanque *m* de circulación helicoidal
i serbatoio *m* a moto elicoidale
n tank met helicoïdale circulatie
d Umwälzungsbehälter *m*

5067 SPIRAL-FLOW TESTING pla,tes
APPARATUS
f appareil *m* d'essai de flux hélicoïdal
e probador *m* de flujo helicoidal
i collaudatore *m* di flusso elicoidale
n beproevingsapparaat *n* voor helicoïdale stroming
d Prüfgerät *n* für Helikoidalströmung

5068 SPIRAL HEAT EXCHANGER clo,htt
f échangeur *m* de chaleur hélicoïdal
e cambiador *m* de calor helicoidal
i scambiatore *m* di calore elicoidale
n spiraalwarmte-uitwisselaar
d Spiralwärmeaustauscher *m*

5069 SPIRAL RAKE CLASSIFIER sor
f classeur *m* à mouvement hélicoïdal des râteaux
e clasificador *m* con movimiento helicoidal de las rasquetas
i classificatore *m* a moto elicoidale dei rastrelli
n harkclassificeerder met helicoïdale beweging
d Umwälzungsrechenklassierer *m*

5070 SPIRAL RIBBON MIXER mix
f mélangeur *m* à hélice
e mezcladora *f* de hélice
i mescolatore *m* ad elica
n mengmachine met spiraalroerwerk
d Rührschnecke *f*

5071 SPLASH ARM pre
f bras *m* agitateur d'autoclave
e brazo *m* agitador de autoclave
i braccio *m* agitatore d'autoclave
n autoclaafroerarm
d Autoklavenrührarm *m*

5072 SPLASH CONDENSER clo,pre
f condensateur *m* à barbotage, condensateur *m* à ruissellement
e condensador *m* de inyección, condensador *m* de salpicadura
i condensatore *m* a sbattimento
n injectiecondensor, mengcondensor
d Berieselungskondensator *m*, Einspritzkondensator *m*

5073 SPLASH HEAD dis
f ampoule *f* de garde, gardecrotte *m*, tôle *f* d'éclaboussage
e deflector *m* de salpicaduras
i diflettore *m* di spruzzi
n spatvanger
d Tropfenfänger *m*

5074 SPLASH HEAD ADAPTER dis,lab
f adapteur *m* avec ampoule de garde
e racor *m* con ampolla de seguridad
i raccordo *m* con ampolla di sicurezza
n druppelvangerverloopstuk *n*
d Tropfenfängerstück *n*

5075 SPLASH PLATE hyg,iwt
f déflecteur *m* d'eau
e deflector *m* de salpicaduras
i lamiera *f* parafango, paraspruzzi *m*
n spatbord *n*
d Spritzblech *n*

5076 SPLASH-TYPE COOLING TOWER clo
f tour *f* de refroidissement à ruissellement
e torre *f* de enfriamiento por salpicadura
i torre *f* di raffreddamento a sbattimento
n sproeikoeltoren
d Rieselkühlturm *m*

5077 SPLAT COOLING DISK clo,coa
f disque *m* rotatif d'installation de réfrigération de particules métalliques pulvérisées
e disco *m* giratorio de equipo de enfriamiento de partículas pulverizadas
i disco *m* rotativo d'impianto di raffreddamento di particelle polverizzate
n roterende schijf voor koel-installatie van verstoven metaaldeeltjes
d Drehscheibe *f* einer Kühlanlage von verspritzten Metallteilchen

5078 SPLIT, pla,pre
SPLIT FOLLOWER
f coquille *f* de moule, empreinte *f* mobile
e cojinete *m*, coquilla *f*
i guancia *f*, tassello *m*
n matrijsinzetstuk *n*
d Backeneinsatz *m*

5079 SPLIT CAVITY pla,pre
f empreinte *f* en plusieurs pièces
e matriz *f* partida
i impronta *f* a tasselli
n samengestelde matrijs
d mehrteilige Matrize *f*, mehrteiliges Gesenk *n*

5080 SPLIT CAVITY MOLD (US), pla,pre
SPLIT CAVITY MOULD (GB)
f moule *m* à coins, moule *m* à plusieurs pièces d'empreinte
e molde *m* de matriz partida
i stampo *m* a tasselli
n samengestelde vorm
d mehrteilige Form *f*

5081 SPLIT-FOLLOWER MOLD (US), pla,pre
SPLIT-FOLLOWER MOULD (GB)
f moule *m* à coins
e molde *m* de cojinete partido
i stampo *m* a guance
n samengestelde vorm
d Backenform *f*

5082 SPLIT FOLLOWERS, pla,pre
SPLITS
f coins *pl*, coquilles *pl*
e cojinetes *pl*
i tasselli *pl*
n matrijsonderdelen *pl*
d mehrteilige Einsätze *pl*

5083 SPLIT MOLD (US), pla,pre
SPLIT MOULD (GB)
f moule *m* à coins
e molde *m* compuesto
i forma *f* a tasselli
n samengestelde vorm
d zusammengesetztes Werkzeug *n*

5084 SPLIT-FLOW TRAY dis
f plateau *m* à deux passes
e plato *m* de varios flujos
i piatto *m* a doppio flusso
n meerstromenschotel
d mehrflutiger Boden *m*

5085 SPLITTER rub
f machine *f* à fendage, machine *f* à refendre
e máquina *f* para hendir
i affettatrice *f* di gommapiuma
n vellenspuitmachine
d Plattenspalter *m*

5086 SPLITTER, dis
REFLUX DIVIDER
f répartiteur *m* de reflux
e distribuidor *m* de reflujo
i ripartitore *m* di riflusso
n terugstroomverdeler
d Rücklaufverteiler *m*

5087 SPOKE-TYPE DIFFUSER gen
f diffuseur *m* à rais
e difusor *m* de rayos
i diffusore *m* a raggi
n spakendiffuseur
d Speichendiffusor *m*

5088 SPONGE ELECTRODE, elc
DOCTOR
f électrode *f* à éponge
e electrodo *m* macrocelular
i elettrodo *m* macrocellulare
n tampon
d Schwammelektrode *f*

5089 SPONGE PLATING BATH, elc
DOCTOR BATH
f bain *m* de revêtement électrolytique avec électrode à éponge
e baño *m* de revestimiento galvánico con electrodo macrocelular

i bagno *m* di rivestimento galvanico con elettrodo macrocellulare
n tamponeerbad *n*
d Schwammelektrodenbad *n*

5090 SPOT PLATE lab,tes
f plaque *f* pour analyse à la goutte
e placa *f* de gotas
i piastra *f* di gocce
n druppelplaat
d Tüpfelplatte *f*

5091 SPOT TEST APPARATUS lab,tes
f appareil *m* d'essai à la touche
e probador *m* a la gota
i collaudatore *m* alla goccia
n druppelproefapparaat *n*
d Tropfenreaktionsgerät *n*, Tüpfelanalysegerät *n*

5092 SPOUT (GB), FOUNTAIN, LAUNDER, RUNNER THROUGH (US), TAPPING SPOUT (GB) fou
f canal *m* de coulée, chenal *m* de coulée
e canal *m* de colada
i canale *m* di colata
n gietgoot, schenk, tapgoot
d Abstichrinne *f*, Förderrinne *f*, Rutsche *f*

5093 SPOUT, LIP fou,met
f bec *m* de coulée
e pico *m* de colada
i becco *m* di colata
n schenktuit
d Ausguss *m*, Giessschnauze *f*

5094 SPOUT, RACKING PIPE bre
f tuyau *m* de remplissage
e canula *f*, caña *f*
i cannuccia *f*, pistone *m*
n vulbuis
d Füllrohr *n*

5095 SPRAY BANK, SPRAY MANIFOLD cle,hyg
f rideau *m* de pulvérisateurs
e cortina *f* de pulverizadores
i cortina *f* di polverizzatori
n sproeibank
d Düsengitter *n*

5096 SPRAY BURNER bur
f brûleur *m* à ejecteur
e quemador *m* atomizador
i bruciatore *m* atomizzatore
n verstuivingsbrander
d Düsenbrenner *m*

5097 SPRAY CARBURETTER, JET CARBURETTER fue
f carburateur *m* à giclage
e carburador *m* de inyección
i carburatore *m* ad iniezione
n injectiecarburator
d Düsenvergaser *m*

5098 SPRAY CHAMBER gen
f chambre *f* à diffusion
e cámara *f* de difusión
i camera *f* a diffusione
n diffusiekamer
d Diffusionskammer *f*

5099 SPRAY COLUMN dis
f colonne *f* à ruissellement
e columna *f* rociadora
i colonna *f* spruzzatrice
n sproeikolom
d Sprühsäule *f*

5100 SPRAY DRIER dry
f sécheur *m* pulvérisateur
e secador *m* de pulverización
i essiccatoio *m* di polverizzazione
n sproeidroger, verstuivingsdroger
d Zerstäubungstrockner *m*

5101 SPRAY DRYING TOWER soa
f tour *f* d'atomisation sécheuse, tour *f* de séchage par atomisation
e torre *f* secadora de pulverización
i torre *f* essiccatrice per polverizzazione
n verstuivingsdroogtoren
d Zerstäubungstrockenturm *m*

5102 SPRAY DYEING TANK tex
f récipient *m* de teinture par atomisation
e tanque *m* de teñido por atomización
i serbatoio *m* di tintura per atomizzazione
n verfkuip met verstuiver
d Zerstäubungsfarbküpe *f*

5103 SPRAY ELECTRODE elc
f électrode *f* d'atomisation
e electrodo *m* de atomización
i elettrodo *m* a spruzzo
n sproeielektrode
d Sprühelektrode *f*

5104 SPRAY GUN, coa
SPRAYING PISTOL, COATING PISTOL
f pistolet *m* de pulvérisation
e pistola *f* pulverizadora
i pistola *f* a spruzzo
n spuitpistool *n*
d Spritzpistole *f*

5105 SPRAY INJECTION APPARATUS sin
f arroseuse *f* de nettoyage
e aparato *m* de limpieza por aspersión
i spruzzatore *m* da pulire
n sproeier voor panwasser
d Wasserdeckeeinrichtung *f*

5106 SPRAY NOZZLE, cle,pre
SPRINKLER NOZZLE
f tuyère *f* de pulvérisation
e tobera *f* pulverizadora
i ugello *m* polverizzatore
n sproeikop
d Spritzdüse *f*, Sprühdüse *f*

5107 SPRAY QUENCHING PLANT ctt,met
f installation *f* de refroidissement rapide par eau atomisée
e equipo *m* de enfriamiento rápido por agua atomizada
i impianto *m* di raffreddamento rapido per spruzzo d'acqua
n afschrikinstallatie met waterverstuiving
d Abschreckanlage *f* mit Wasserzerstäubung

5108 SPRAY RINSE (GB), elc
SPRAY RINSE BATH (GB), SPRAY SWILL (US), SPRAY SWILL BATH (US)
f bain *m* de rinçage au jet
e baño *m* de enjuague al chorro
i bagno *m* di sciacquatura a spruzzo
n spuitspoelbad *n*
d Spritzspülbad *n*

5109 SPRAY-TYPE HUMIDIFIER hyg
f humecteur *m* à arrosage, humidificateur *m* à arrosage
e humectador *m* por rociado, humidificador *m* por rociado
i umidificatore *m* a spruzzo
n sproeibevochtigingstoestel *n*, vochtsproeier
d Sprühanfeuchtapparat *m*, Sprühbefeuchter *m*

5110 SPRAY WASHER, elc
ALKALINE SPRAY CLEANING MACHINE
f installation *f* de dégraissage alcalin au jet
e equipo *m* de desengrase alcalino al chorro
i impianto *m* di sgrassaggio allo spruzzo
n alkalische spuitreinigings-machine
d alkalische Spritzreinigungs-maschine *f*

5111 SPRAYED METAL COATING coa
f revêtement *m* à métal pulvérisé
e revestimiento *m* de metal pulverizado
i rivestimento *m* a metallo polverizzato
n gespoten metaalbedekking
d Spritzmetallbekleidung *f*

5112 SPRAYER, bre
SPRAYING APPARATUS
f pulvérisateur *m*
e pulverizador *m*
i polverizzatore *m*
n nevelapparaat *n*, verstuiver
d Nebelapparat *m*, Zerstäuber *m*

5113 SPREADER, pla,pre,rub
TORPEDO
f distributeur *m*, torpille *f*
e extendedor *m*, torpedo *m*
i ogiva *f*, punzone *m* distributore
n torpedo, verdeelkop
d Verteiler *m*

5114 SPREADING CALENDER coa
f calandre *f* de gommage
e calandra *f* para engomar
i calandra *f* per spalmatura
n strijkkalander
d Streichkalander *m*

5115 SPREADING CHEST, pre,rub
SPREADING MACHINE CHEST
f table *f* chaude de la machine à engommer
e cámara *f* caliente de la máquina de engomar
i camera *f* calda della spalmatrice
n verwarmingstafel voor strijk-machine
d Dampfbank *f* für Streichmaschine

5116 SPREADING MACHINE coa,rub
f machine *f* à éducation,

machine *f* à enduire, machine *f* à engommer
e máquina *f* de esparcimiento
i macchina *f* di spalmatura, spalmatrice *f*
n machine voor het nat berubberen, strijkmachine
d Streichmaschine *f*

5117 SPREADING SHOP coa
f atelier *m* d'engommage
e taller *m* para engomar
i reparto *m* di spalmatura
n berubberingswerkplaats, strijkafdeling
d Stoffstreicherei *f*, Streichabteilung *f*

5118 SPREADING TABLE coa
f table *f* à engommer
e mesa *f* de engomar
i tavola *f* di spalmatura
n strijktafel
d Streichbank *f*

5119 SPRING-LOADED PRESSURE RELIEF VALVE cpt
f soupape *f* de sûreté tarée par ressort hélicoïdal
e válvula *f* de seguridad de muelle helicoidal
i valvola *f* di sicurezza caricata a molla
n veiligheidsklep met veerbelasting
d federbelastetes Sicherheitsventil *n*

5120 SPRINKLER, AUTOMATIC FIRE SPRINKLER fea
f extincteur *m* automatique d'incendies
e rociador *m* extinguidor automático
i spruzzatore *m* automatico
n automatische sproeier
d automatische Wasserspritze *f*

5121 SPROUT CHAMBER, HEAT CHAMBER, HOT AIR CHAMBER bre
f chambre *f* de chaleur, cochon *m*
e cámara *f* de aire caliente, cámara *f* de calefacción
i camera *f* di riscaldamento
n kleine hel
d Darrsau *f*, Sau *f*, Wärmekammer *f*

5122 SPRUE pre,rub
f canal *m* de carotte, canal *m* d'injection
e bebedero *m*, canal *m* de mazarota
i canale *m* d'iniezione
n aanspuitingskanaal *n*
d Angusskanal *m*

5123 SPRUE BUSH, FEED BUSH pla
f buse *f* du pot d'alimentation, busette *f*, cheminée *f* du moule, douille *f* de carotte
e manguito *m* del bebedero
i bocchello *m*
n aanspuitbus
d Angussbuchse *f*

5124 SPRUE CONE pla,pre
f carotte *f*
e mazarota *f*, muñón *m* de alimentación
i matterozza *f*
n aanspuiting
d Anguss *m*

5125 SPRUE EJECTOR, SPRUE LOCK PIN pla,pre
f éjecteur *m* de la carotte
e espiga *f* de cierre del bebedero, expulsor *m* de la mazarota
i espulsore *m* della matterozza
n aanspuitinguitstoterpen
d Angussauswerfer *m*, Angussdruckstift *m*

5126 SPRUE EJECTOR BAR pla,pre
f barre *f* d'éjection
e tirante *m* de expulsión
i asta *f* d'estrazione
n uitwerperstang
d Ausdrückstange *f*

5127 SPRUE EJECTOR PIN pla,pre
f extracteur-éjecteur *m*
e expulsor *m* de la mazarota
i estrattore *m* della matterozza
n aanspuitinguitstoter
d Angussausstosser *m*

5128 SPUE, EXCESS RUBBER OVERFLOW, SPEW rub
f trop-plein *m*
e rebosamiento *m* del molde
i troppo pieno *m*
n overloop
d Austrieb *m*, Formaustrieb *m*

5129 SPUR WHEEL dye,tex
f roue *f* dentée cylindrique
e rueda *f* dentada cilíndrica
i ruota *f* dentata cilindrica
n getand wiel *n*
d Stirnrad *n*

5130 SPURT PIPE tex
f tube *m* à ruisseler
e tubo *m* rociador
i tubo *m* a spruzzi
n sproeibuis
d Spritzrohr *n*

5131 SPUTTERING APPARATUS coa
f appareil *m* de pulvérisation cathodique
e aparato *m* de pulverización catódica
i apparecchio *m* di polverizzazione catodica
n katodeverstuivingsapparaat *n*
d Katodenzerstäuberapparat *m*

5132 SPY GLASS, PEEP-HOLE gen
f regard *m*, trou *m* de tête
e mirilla *f*, ventanillo *m*
i foro *m* di spia
n kijkgaatje *n*
d Schauloch *n*

5133 SQUEEZE MOLDING PLANT (US), SQUEEZE MOULDING PLANT (GB) met,pre
f installation *f* de serrage par pression
e equipo *m* de apisonado por presión
i impianto *m* di formatura a pressione, impianto *m* di stiratura a pressione
n drukverdichtingsinstallatie
d Druckverdichtungsanlage *f*

5134 SQUEEZE ROLL pre
f cylindre *m* à pressurer
e cilindro *m* aplastador
i cilindro *m* spremitore
n uitperswals
d Ausquetschwalze *f*, Quetschwalze *f*

5135 SQUEEZE ROLLER cle,fat
f rouleau *m* à exprimer l'huile
e rodillo *m* exprimidor de aceite
i rullo *m* per spremitura d'olio
n olie-uitdrukrol
d Ölauspressrolle *f*

5136 STABILIZER, STABILIZING COLUMN fue
f colonne *f* de stabilisation
e columna *f* de estabilización
i colonna *f* di stabilizzazione
n stabilisatiekolom
d Stabilisationssäule *f*

5137 STABILIZING ANNEALING FURNACE, STRESS-RELIEF ANNEALING FURNACE htt,met
f four *m* de recuit de détente
e horno *m* de recocido de atenuación de tensiones
i forno *m* di ricottura di distensione
n herstelgloeioven
d Ofen *m* für spannungsfreies Glühen

5138 STABILIZING BATH pho
f bain *m* de stabilisation
e baño *m* estabilizador
i bagno *m* stabilizzatore
n stabiliseerbad *n*
d Stabilisierungsbad *n*

5139 STACKED PACKING dis
f garnissage *m* ordonné
e relleno *m* ordenado
i riempimento *m* ordinato
n geordende vulling
d geordnete Füllung *f*

5140 STAGE COMPRESSOR pre
f compresseur *m* à plusieurs étages
e compresor *m* de fases escalonadas
i compressore *m* a stadi
n trapsgewijs werkende compress[illegible]
d Stufenkompressor *m*

5141 STAGE VULCANIZING PLANT, STEP-UP CURING PLANT rub
f installation *f* de vulcanisation échelonnée
e equipo *m* de vulcanización escalonada
i impianto *m* di vulcanizzazione a gradini
n trapsgewijze vulcanisatie-inrichting
d Stufenvulkanisationsanlage *f*

5142 STAINING MACHINE, GROUNDING MACHINE pap
f machine *f* à teinter
e máquina *f* de tinción
i macchina *f* di tintura
n grondeermachine
d Fonciermaschine *f*, Grundiermaschine *f*

5143 STAKING MACHINE le[illegible]
f machine *f* à palissonner
e máquina *f* para estirar y ablandar cuero, máquina *f* zurradora

i macchina *f* da ragguagliare
n stoltsmachine *f*
d Stollmaschine *f*

5144 STAKING TOOL lea
f palisson *m*
e dispositivo *m* zurrador
i dispositivo *m* da ragguagliare
n stoltsijzer *n*
d Stolleisen *n*

5145 STALAGMOMETER ins
f stalagmomètre *m*
e estalagmómetro *m*
i stalagmometro *m*
n oppervlaktespanningsmeter, stalagmometer
d Oberflächenspannungsmesser *m*, Stalagmometer *n*

5146 STAMP BATTERY, STAMP MILL brk,pap
f bocard *m*
e batería *f* de bocartes, bocarte *m*, molino *m* de bocartes
i macchina *f* a magli
n stampmachine
d Pochbatterie *f*, Stampfwerk *n*

5147 STAMPING MACHINE soa
f estampeuse *f*, presse-frappeuse *f*
e estampadora *f*, troqueladora *f*
i stampatrice *f*
n stempelmachine
d Stempelmaschine *f*

5148 STAMPING PRESS (US), BLOCKING MACHINE (GB) pla
f presse *f* à marquer
e prensa *f* de embutir, prensa *f* de estampar
i pressa *f* per stampigliare
n stanspers
d Prägepresse *f*

5149 STAND-BY UNIT sap
f installation *f* de réserve
e instalación *f* de urgencia
i impianto *m* ausiliario
n noodaggregaat *n*
d Notaggregat *n*

5150 STANDARD CALOMEL ELECTRODE, NORMAL CALOMEL ELECTRODE elc
f demicullule *f* au calomel normale, électrode *f* au calomel normale
e electrodo *m* de calomelano normal, semicelda *f* de calomelano normal
i elettrodo *m* al calomelano normale
n normale calomelelektrode
d Normalkalomelhalbzelle *f*

5151 STANDARD CELL, NORMAL ELEMENT elc
f pile *f* étalon
e pila *f* patrón
i pila *f* campione
n normaalelement *n*
d Normalelement *n*

5152 STANDARD FRAME pla,pre
f moule *m* normalisé
e molde *m* normalizado
i stampo *m* a matrici intercambiabili, stampo *m* normalizzato
n moedermatrijs
d Einheitsstammform *f*, Normform *f*, Stammform *f*

5153 STAR SHELL, FLARE, LIGHT BOMB mit
f obus *m* éclairant
e cohete *m* luminoso
i razzo *m* luminoso
n lichtgranaat, lichtkogel
d Leuchtbombe *f*

5154 STAR STEAMER tex
f machine *f* de traitement à la vapeur à tourniquet
e máquina *f* de tratamiento al vapor de torniquete
i macchina *f* di trattamento al vapore con arganello
n stoombehandelingsmachine met kruisraam
d Kreuzrahmendampfbehandlungsmaschine *f*

5155 STARCH FACTORY oic
f amidonnerie *f*, fabrique *f* d'amidon
e almidonería *f*, fábrica *f* de almidón
i fabbrica *f* d'amido
n stijfselfabriek
d Stärkefabrik *f*

5156 STARCH ROLLER, PLAIN ROLLER tex
f rouleau *m* à amidon
e rodillo *m* de almidón
i rullo *m* ad amido
n verfstofafneemrol
d Farbstoffabstreichrolle *f*

5157 STARCHING VESSEL tex
f récipient *m* d'amidonnage

e recipiente *m* para almidonar
i recipiente *m* per inamidare
n stijfselvat *n*
d Stärkegefäss *n*

5158 STARTING BATH tex
f bain *m* initial
e baño *m* inicial
i bagno *m* iniziale
n uitgangsbad *n*
d Ausgangsbad *n*

5159 STARTING SHEET elc
f feuille *f* de départ
e hoja *f* de arranque, hoja *f* de partida
i catodo *m* padre, lamierino *m* catodico
n beginplaat
d Mutterblech *n*

5160 STARTING-SHEET BLANK, MOTHER BLANK elc
f plaque-support *m* de feuille de départ
e placa-soporte *m* de hoja de arranque
i piastra *f* di sopporto di lamierino catodico
n beginmatrijs
d Mutterblechkatodenblech *n*

5161 STARTING VESSEL, PITCHING TUB bre
f cuve *f* guilloire
e cuba *f* de reposo, cuba *f* de tiembra
i tino *m* per semina
n aanzetkuip
d Anstellbottich *m*, Sammelbottich *m*

5162 STATIONARY BED, SOLID BED gen
f porteur *m* solide de catalysateur
e portador *m* macizo de catalizador
i portatore *m* solido di catalizzatore
n vaste katalysatordrager
d Festbett *n*

5163 STATIONARY FURNACE htt,met
f four *m* fixe
e horno *m* fijo
i forno *m* fisso
n vaste oven
d ortsfester Ofen *m*

5164 STATIONARY PART OF THE MOLD (US), STATIONARY PART OF THE MOULD (GB) pla,pre
f partie *f* fixe du moule
e semimolde *m* fijo
i parte *f* fissa dello stampo
n vaststaand matrijsgedeelte *n*
d feststehender Formteil *m*

5165 STATIONARY PLATEN (US), FIXED-DIE PLATE (GB) pla
f plaque *f* fixe du moule
e placa *f* fija del molde
i piastra *f* d'impronta fissa
n vaststaande matrijsplaat
d feststehende Formplatte *f*

5166 STEAM ACCUMULATOR boi
f accumulateur *m* de vapeur
e vapoacumulador *m*
i accumulatore *m* di vapore
n stoomaccumulator, stoomreservoir *n*
d Dampfspeicher *m*

5167 STEAM AIR PUMP, STEAM PUMP pre
f pompe *f* actionnée par vapeur
e bomba *f* accionada por vapor
i pompa *f* premente a vapore
n stoompomp
d Dampfluftpumpe *f*

5168 STEAM AUTOCLAVE pre
f autoclave *m* à vapeur
e autoclave *f* de vapor
i autoclave *f* a vapore
n stoomautoclaaf
d Dampfautoklav *m*

5169 STEAM BATH (GB), VAPOR BATH (US) htt,tex
f bain *m* à vapeur
e baño *m* de vapor
i bagno *m* a vapore
n stoombad *n*
d Dampfbad *n*

5170 STEAM BELT, CALANDRIA sin
f chambre *f* à vapeur
e calandria *f*
i camera *f* a vapore
n stoomkamer
d Heizkammer *f*

5171 STEAM BOILER boi
f chaudière *f* à vapeur
e caldera *f* de vapor
i caldaia *f* a vapore
n stoomketel
d Dampfkessel *m*

5172 STEAM BOX, STEAM CHEST, PROOF BOX nut

f boîte *f* à tiroir, boîte *f* de distribution de vapeur, cage *f* de vanne
e caja *f* de distribución de vapor
i camera *f* di vapore
n stoomkast
d Dampfkasten *m*, Schieberkasten *m*

5173 STEAM CHANNEL, STEAM WAY pla
f canal *m* de vapeur, conduit *m* de vapeur
e canal *m* de vapor, conducto *m* de vapor
i canale *m* di vapore, condotto *m* di vapore
n stoomkanaal *n*
d Dampfkanal *m*

5174 STEAM CLEANER cle
f épurateur *m* à vapeur
e limpiador *m* de vapor
i pulitrice *f* a vapore
n stoomreiniger
d Dampfreiniger *m*

5175 STEAM COIL, HEATING COIL pre
f serpentin *m* de chauffage
e serpentín *m* de calentamiento
i serpentina *f* di riscaldamento
n verwarmingsspiraal
d Heizschlange *f*

5176 STEAM-COOLED REACTOR, SUPERHEATED STEAM REACTOR nue
f réacteur *m* à vapeur surchauffée
e reactor *m* de vapor sobrecalentado
i reattore *m* a vapore surriscaldato
n met stoom gekoelde reactor, reactor met oververhitte stoom
d dampfgekühlter Reaktor *m*, Reaktor *m* mit überhitztem Dampf

5177 STEAM COPPER bre
f chaudière *f* pour cuisson à vapeur
e caldera *f* con calentamiento de vapor
i caldaia *f* a riscaldamento a vapore
n stoomkookketel
d Dampfpfanne *f*

5178 STEAM-CORED MOLD (US), STEAM-CORED MOULD (GB) pla
f moule *m* à canaux à vapeur
e molde *m* con canales de vapor
i stampo *m* a canali di vapore
n vorm met stoomkanalen
d Form *f* mit Dampfkanälen

5179 STEAM CURING CHAMBER cer
f chambre *f* de traitement à la vapeur
e cámara *f* de tratamiento al vapor
i camera *f* di trattamento al vapore
n stoombehandelingskamer
d Dampfbehandlungsraum *m*

5180 STEAM CURING PLANT, STEAM VULCANIZING PLANT rub
f installation *f* de vulcanisation en vapeur
e equipo *m* de vulcanización en vapor
i impianto *m* di vulcanizzazione in vapore
n stoomvulcaniseerinstallatie
d Dampfvulkanisationsanlage *f*

5181 STEAM DISTILLATION PLANT dis
f installation *f* de distillation à vapeur
e equipo *m* de destilación por vapor
i impianto *m* di distillazione con vapore
n stoomdestilleerinstallatie
d Dampfdestillationsanlage *f*, Wasserdampfdestillationsanlage *f*

5182 STEAM DOME boi
f dôme *m* de vapeur
e domo *m* de vapor
i duomo *m* di vapore
n stoomdom
d Dampfhaube *f*

5183 STEAM DOME PRESS pre
f presse autoclave *m* avec dôme à vapeur
e prensa-autoclave *f* con domo de vapor
i pressa autoclave *f* a duomo di vapore
n autoclaafpers met stoomkap
d Autoklavenpresse *f* mit Dampfhaube

5184 STEAM DRIER dry
f sécheur *m* à vapeur
e secadero *m* de vapor
i essiccatoio *m* a vapore
n stoomdroogapparaat *n*
d Dampftrockenapparat *m*

5185 STEAM EDUCTOR, STEAM EJECTOR, STEAM JET AIR EJECTOR — vac
- f éjecteur *m* à vapeur
- e eyector *m* accionado por vapor
- i eiettore *m* a vapore
- n stoomejecteur
- d Dampfstrahlluftsauger *m*

5186 STEAM FINISHING SECTION — pap
- f section *f* d'humectation
- e sección *f* de humectación
- i sezione *f* d'umettazione
- n bevochtigingsafdeling
- d Benetzungsabteilung *f*

5187 STEAM FUNNEL — lab
- f entonnoir *m* à vapeur
- e embudo *m* de vapor
- i imbuto *m* a vapore
- n stoomtrechter
- d Dampftrichter *m*

5188 STEAM-GENERATING HEAVY WATER REACTOR — nue
- f réacteur *m* modéré par eau lourde générateur de vapeur
- e reactor *m* moderado con agua pesada generador de vapor
- i reattore *m* moderato per acqua pesante generatore di vapore
- n stoom producerende met zwaar water gemodereerde reactor
- d dampfproduzierender schwer-wassermoderierter Reaktor *m*

5189 STEAM-GENERATING REACTOR — nue
- f réacteur *m* générateur de vapeur
- e reactor *m* generador de vapor
- i reattore *m* generatore di vapore
- n stoom producerende reactor
- d dampferzeugender Reaktor *m*

5190 STEAM-HEATED CYLINDER, HOT ROLLER — pap
- f tambour *m* sécheur
- e tambor *m* secador
- i tamburo *m* essiccatore
- n droogtrommel
- d Dampftrommel *f*, Trocken-trommel *f*

5191 STEAM INJECTOR — boi
- f injecteur *m* de vapeur
- e vapoinyector *m*
- i iniettore *m* di vapore
- n stoominjecteur
- d Dampfinjektor *m*

5192 STEAM JACKET — cpt
- f chemise *f* de vapeur
- e camisa *f* de vapor
- i camicia *f* di vapore
- n stoommantel
- d Dampfmantel *m*

5193 STEAM-JACKETED VULCANIZING PAN — rub
- f chaudière *f* à vulcaniser à chemise de vapeur
- e caldera *f* de vulcanizar con camisa de vapor
- i caldaia *f* di vulcanizzazione con camicia di vapore
- n vulcaniseerketel met stoom-mantel
- d Vulkanisierkessel *m* mit Dampfmantel

5194 STEAM-JET BLOWER — pre
- f soufflante *f* à jet de vapeur
- e soplador *m* de chorro de vapor
- i soffiante *m* a getto di vapore
- n stoomstraalventilator
- d Dampfstrahlgebläse *n*

5195 STEAM-JET CLEANING MACHINE — cle
- f machine *f* de nettoyage à jet de vapeur
- e máquina *f* de limpieza con chorro de vapor
- i pulitrice *f* a getto di vapore
- n stoomstraalreinigingsmachine
- d Dampfstrahlreinigungsmaschine *f*

5196 STEAM-JET EJECTOR — pre
- f éjecteur *m* à jet de vapeur
- e eyector *m* de chorro de vapor
- i eiettore *m* a getto di vapore
- n stoomstraalejecteur
- d Dampfstrahlluftsauger *m*

5197 STEAM KILN — bre
- f touraille *f* à vapeur
- e tostadero *m* de vapor
- i essiccatoio *m* a vapore
- n stoomeest
- d Dampfdarre *f*

5198 STEAM NOZZLE — pre
- f tuyère *f* de vapeur
- e tobera *f* de vapor
- i ugello *m* di vapore
- n stoomstraalpijp
- d Dampfstrahldüse *f*

5199 STEAM PAN — rub
- f vulcanisateur *m* à vapeur

e vulcanizador *m* de vapor
i vulcanizzatore *m* a vapore
n stoomvulcaniseerketel
d Dampfvulkanisierkessel *m*

5200 STEAM PLATEN PRESS pre,rub
f presse *f* à plateaux de vapeur
e prensa *f* con platos de vapor
i pressa *f* a piani di vapore
n met stoom verwarmde etagepers
d dampfgeheizte Etagenpresse *f*

5201 STEAM PRESSURE REGULATOR htt,ins
f régulateur *m* de pression de vapeur
e regulador *m* de presión de vapor
i regolatore *m* di pressione di vapore
n dampdrukregelaar
d Dampfdruckregler *m*

5202 STEAM REFINING PLANT fat
f installation *f* d'épuration d'huile par vapeur
e equipo *m* depurador de aceite con vapor
i impianto *m* di depurazione d'olio con vapore
n stoomolieraffinaderij
d Dampfölraffinerie *f*

5203 STEAM ROLL pap
f cylindre *m* sécheur à vapeur
e cilindro *m* secador de vapor
i cilindro *m* essiccatore a vapore
n stoomdroogwals
d Dampftrockenwalze *f*

5204 STEAM SEAL cpt
f fermeture *f* à vapeur
e cierre *m* de vapor
i chiusura *f* a vapore
n stoomafsluiting
d Dampfabdichtung *f*

5205 STEAM STERILIZER bac
f stérilisateur *m* à vapeur
e esterilizador *m* de vapor
i sterilizzatore *m* a vapore
n stoomsteriliseerapparaat *n*
d Dampfsterilisierapparat *m*

5206 STEAM SUPERHEATER, SUPERHEATER boi
f surchauffeur *m* de vapeur
e recalentador *m* de vapor
i surriscaldatore *m* di vapore
n stoomoververhitter
d Dampfüberhitzer *m*

5207 STEAM SURGE DRUM nue
f récipient *m* à vapeur de compensation
e recipiente *m* de vapor de compensación
i corpo *m* cilindrico di vapore per compensazione
n compenserend stoomvat *n*
d Kompensationsdampfgefäss *n*

5208 STEAM TEMPERING PLANT, HOT PREPARATION PLANT cer
f installation *f* de traitement de l'argile à la vapeur
e equipo *m* de tratamiento de la arcilla al vapor
i impianto *m* di trattamento dell'argilla al vapore
n stoombehandelingsinstallatie voor klei
d Dampfbehandlungsanlage *f* für Ton

5209 STEAM TRAP boi
f séparateur *m* d'eau de condensation
e purgador *m* de agua del vapor
i scaricatore *m* di condensa, separatore *m* di vapore
n condenspot
d Kondenstopf *m*

5210 STEAM WASHER sin
f centrifugeur *m* pour lavage à vapeur
e centrífuga *f* lavadora de vapor
i centrifuga *f* lavatrice a vapore
n stoomwascentrifuge
d Dampfwäschezentrifuge *f*

5211 STEAMING PLANT tex
f installation *f* de traitement à vapeur
e equipo *m* de tratamiento con vapor
i impianto *m* di trattamento con vapore
n stoombehandelingsinstallatie
d Dampfbehandlungsanlage *f*

5212 STEAMING TOWER (GB), TOWER STEAMER (US) tex
f tour *f* de traitement à la vapeur
e torre *f* de tratamiento al vapor
i torre *f* di trattamento al vapore
n stoombehandelingstoren
d Dampfbehandlungsturm *m*

5213 STEEL BALL PEENING PLANT (GB), CLOUDBURSTING PLANT (US) cle

f installation *f* de traitement par jet de balles en acier
e equipo *m* de tratamiento por chorro con perdigonos de acero
i impianto *m* di trattamento con pallottole d'acciaio
n kogelstraalinstallatie
d Kugelstrahlanlage *f*

5214 STEEL MOLD (US), STEEL MOULD (GB) fou
f moule *m* en acier
e molde *m* de acero
i stampo *m* in acciaio
n stalen vorm
d Stahlform *f*

5215 STEELING BATH, ACIERAGE BATH, IRON PLATING BATH coa
f bain *m* d'aciérage
e baño *m* de aceración, baño *m* de hierrado
i bagno *m* d'acciaiatura
n verijzeringsbad *n*, verstalingsbad *n*
d Stählungsbad *n*, Verstählungsbad *n*

5216 STEEP, STEEPING TANK, STEEPING VAT bre
f cuve *f* à tremper, cuve *f* de trempage
e cuba *f* de remojo, remojador *m*
i serbatoio *m* di macerazione
n gerstzwelder
d Gerstenweiche *f*, Quellbottich *m*, Quellstock *m*, Weichbottich *m*, Weiche *f*

5217 STEEPING PLANT, MERCERIZATION PLANT tex
f installation *f* de mercerisage
e equipo *m* de mercerización
i impianto *m* di mercerizzazione
n merceriseerinstallatie
d Mercerieranlage *f*, Mercerisieranlage *f*

5218 STEEPING VAT, MERCERIZER, MERCERIZING TANK tex
f récipient *m* de mercerisage
e tanque *m* de mercerización
i serbatoio *m* di mercerizzazione
n merceriseertank
d Merceriergefäss *n*, Mercerisiergefäss *n*

5219 STENTER, TENTER tex
f élargisseur *m*
e estricador *m*
i apparecchio *m* tenditore
n spanraam *n*
d Spannrahmen *m*

5220 STEP-UP CURING PLANT, STAGE VULCANIZING PLANT rub
f installation *f* de vulcanisation échelonnée
e equipo *m* de vulcanización escalonada
i impianto di vulcanizzazione a gradini
n trapsgewijze vulcanisatie-inrichting
d Stufenvulkanisationsanlage *f*

5221 STEPPED TRAY dis
f plateau *m* à gradins
e plato *m* escalonado
i piatto *m* a gradini
n trapvormige schotel
d Stufenboden *m*

5222 STERILIZATION TOWER bac,iwt
f tour *f* de contact, tour *f* de stérilisation
e torre *f* de contacto, torre *f* de esterilización
i torre *f* di contattto, torre *f* di sterilizzazione
n contacttoren, steriliseertoren
d Kontaktturm *m*, Sterilisierturm *m*

5223 STERILIZER, STERILIZING APPARATUS bac,hyg,iwt
f stérilisateur *m*
e esterilizador *m*
i sterilizzatore *m*
n sterilisator, steriliseer-apparaat *n*
d Sterilisator *m*, Sterilisier-apparat *m*

5224 STERILIZER CAP bac,hyg
f chapeau *m* de stérilisateur
e tapadera *f* de esterilizador
i cappa *f* di sterilizzatore
n sterilisatorkap
d Sterilisatorkappe *f*

5225 STERILIZING FILTER bac,fil
f filtre *m* stérilisateur
e filtro *m* esterilizador
i filtro *m* di sterilizzazione
n steriliserend filter *n*
d Entkeimungsfilter *m*

5226 STICK ANODE, ROUND ANODE ele

f anode *f* ronde
e ánodo *m* rondo
i anodo *m* rotondo
n ronde anode
d Rundanode *f*

5227 STIFFENING RIB elc
f nervure *f* de renforcement
e nervio *m* de refuerzo
i nervatura *f* di rinforzo
n versterkingsrib
d Verstärkungsrippe *f*

5228 STILL, dis
DISTILLER, DISTILLING APPARATUS
f alambic *m*
e alambique *m*
i alambicco *m*
n alambiek, destilleertoestel *n*
d Brennapparat *m*, Destillierapparat *m*

5229 STILL BODY, dis
STILL POT
f cucurbite *f*
e cucúrbita *f*
i storta *f*
n destilleerkolf
d Destillierkolben *m*

5230 STILL CAP dis
f chapiteau *m*
e capitel *m*
i cappelletto *m*
n helm
d Deckel *m*

5231 STILL HEAD, dis
DISTILLATION COLUMN, DISTILLATION TOWER, DISTILLING COLUMN
f colonne *f* à fractionner, colonne *f* de distillation
e columna *f* de destilación
i colonna *f* di distillazione
n destileerkolom
d Destillierkolonne *f*, Destilliersäule *f*

5232 STILLION, bre
RACKER
f bock *m* de soutirage
e estante *m*
i riempifusti *m* barometrico
n tapbok
d Abfüllbock *m*

5233 STIPPLE ROLLER tex
f rouleau *m* à pointillage
e rodillo *m* de punteado
i rullo *m* di punteggiamento
n stippelrol
d Tüpfelrolle *f*

5234 STIRRER, mix
STIRRING MILL, AGITATOR
f agitateur *m*, cuve *f* à brasser, mélangeur *m*, pétrisseur *m*
e agitador *m*
i agitatore *m*
n roerapparaat *n*, roerder, roermachine, roerwerk *n*
d Agitator *m*, Mischer *m*, Rührer *m*, Rührwerk *n*

5235 STITCH WELDING MACHINE, pla
TACK WELDING MACHINE
f machine *f* de soudure à la molette
e máquina *f* de soldadura discontinua
i macchina *f* di saldatura a tratti
n puntlasmachine
d Heftschweissmaschine *f*

5236 STOCK GUIDE, cpt,mix
CHECK, GUIDE PLATE
f flasque *f* de guidage latérale
e carillo *m* de un mezclador
i guancia *f* d'un mescolatore
n keerbord *n*, materiaalgeleider van een mengwals
d Backe *f* einer Mischmaschine

5237 STOCK PAN (GB), mix,rub
BATCH PAN (US)
f baquet *m* à mélange, cuve *f* à mélange
e cuba *f* para mezcla, tina *f* para mezcla
i marnetta *f*
n mengbak
d Mischwanne *f*

5238 STOCK STRAINER, pre,rub
SCREEN HEAD EXTRUDER, STRAINER, STRAINING MACHINE
f boudineuse *f* à tamiser
e budinadora *f* para tamizar
i trafila *f* con testa a filtro
n spuitmachine met zeefspuitkop, zeefpers
d Siebkopfspritzmaschine *f*, Siebpresse *f*

5239 STOCK VAT tex
f cuve-mère *f*
e tina-madre *f*

i tino-madre *m*
n moederkuip
d Stammküpe *f*

5240 STOCKING-TYPE FILTER, fil
CLOTH TUBE FILTER
f filtre *m* à manches
e filtro *m* de mangas
i filtro *m* a maniche
n slangenfilter *n*, zakkenfilter *n*
d Schlauchfilter *m*

5241 STONE CATCHER sin
f épierreur *m*
e colector *m* de piedras
i cacciapietre *m*
n stenenvanger
d Steinfänger *m*

5242 STONE CRUSHER, brk
STONE MILL
f concasseur *m* de pierres
e picapedrero *m*
i rompipietre *m*
n steenbreker
d Steinbrecher *m*

5243 STONEWARE, cer
EARTHENWARE
f argile *f* cuite, terre *f* cuite
e gres *m* cocido, loza *f* de barro
i terracotta *f*, terraglia *f*
n aardewerk *n*
d Steingut *n*

5244 STOP BATH pho
f bain *m* d'arrêt
e baño *m* de parada, baño *m* de paro
i bagno *m* d'arresto, bagno *m* interruttore
n onderbreekbad *n*
d Stopfbad *n*, Unterbrechungsbad *n*

5245 STOP COCK, cpt
STOP TAP
f robinet *m* boisseau, robinet *m* d'arrêt
e grifo *m* de cierre
i rubinetto *m* d'arresto, rubinetto *m* di chiusura
n afsluitkraan
d Absperrhahn *m*

5246 STOP VALVE, cpt
BACKFLOW VALVE, CHECK VALVE, REFLUX VALVE
f soupape *f* d'arrêt
e válvula *f* de retención
i valvola *f* d'arresto, valvola *f* intercettante
n afsluitklep, sperklep
d Absperrventil *n*

5247 STORAGE BATTERY, elc
SECONDARY BATTERY
f batterie *f* d'accumulateurs, batterie *f* secondaire
e batería *f* de acumuladores, batería *f* secundaria
i batteria *f* d'accumulatori, batteria *f* secondaria
n accubatterij, secundaire batterij
d Akkumulatorenbatterie *f*, Sekundärbatterie *f*

5248 STORAGE BIN, bre
GRIST BOX, SILO
f trémie *f* à malt, wagon *m* à farine
e tolva *f* de harina
i cassone *m* per macinato
n schrootkaar
d Schrotrumpf *m*

5249 STORAGE CELL, elc
ACCUMULATOR, SECONDARY CELL
f accumulateur *m*
e acumulador *m*
i accumulatore *m*
n accu, accumulator
d Akkumulator *m*, Sammler *m*

5250 STORAGE CELLAR bre,sto
f cave *f* de garde
e bodega *f* de guarda
i cantina *f* di deposito
n lagerkelder
d Lagerkeller *m*

5251 STORAGE TANK, sto
ACCUMULATOR TANK
f cuve *f* collectrice, récipient *m* d'emmagasinage
e tanque *m* colector
i serbatoio *m* polmone
n verzameltank
d Sammelbehälter *m*

5252 STORAGE TANK, sto
RECEIVING VESSEL
f cuve *f* de stockage
e tanque *m* de depósito
i serbatoio *m* collettore
n voorraadstank
d Sammelbehälter *m*

5253 STORAGE VAT bre,sto
f cuve *f* de repos, tanque *m* de gard
e cuba *f* de reposo
i recipiente *m* di sedimentazione

n lagertank
d Lagerbottich *m*

5254 STORING BASIN, HOLDING BASIN iwt
f bassin *m* d'emmagasinement
e estanque *m* de almacenamiento
i bacino *m* d'immagazzinamento
n vergaarbekken *n*, voorraadsbekken *n*
d Speicherbecken *n*

5255 STOVING BATH, WET STOVING VESSEL tex
f bain *m* de blanchiment réducteur
e baño *m* de blanqueo reductor
i bagno *m* di candeggio ridottore
n reducerend bleekbad *n*
d reduzierendes Bleichbad *n*

5256 STOVING PLANT fou
f installation *f* d'étuvage
e equipo *m* de estufado
i impianto *m* d'essiccazione in stufa
n droogstoofinstallatie
d Trockenanlage *f*

5257 STRAIGHT EXTRUDING NOZZLE pre,rub
f tête *f* de boudineuse droite
e cabezal *m* de budinadora recto
i filiera *f* d'estrusione lineare
n rechte spuitkop
d gerader Spritzkopf *m*

5258 STRAIGHT HEAD (GB), AXIAL HEAD (US) pla
f tête *f* droite
e cabezal *m* de estrusión recto
i testa *f* assiale
n axiale spuitkop
d Längsspritzkopf *m*

5259 STRAIGHT-WAY VALVE, GATE VALVE cpt
f soupape *f* droite
e válvula *f* de corredera, válvula *f* de paso directo
i valvola *f* a saracinesca
n doorlaatklep, in-en uitlaatklep
d Durchgangsventil *n*

5260 STRAIN DISK, STRIATION PLATE tes
f disque *m* à stries
e disco *m* de estrías
i disco *m* a strie
n slierenschijf
d Schlierenblende *f*

5261 STRAIN GAGE (US), STRAIN GAUGE (GB) ins
f extensomètre *m*, jauge *f* de contrainte
e banda *f* extensimétrica, extensímetro
i estensimetro *m*, nastro *m* estensimetrico
n rekstrook
d Dehnungsmessstreifen *m*

5262 STRAIN-HARDENING PLANT, WORK-HARDENING PLANT met
f installation *f* d'écrouissage
e equipo *m* de endurecimiento por deformación en frío
i impianto *m* d'incrudimento per deformazione a freddo
n koudverstevigingsinstallatie
d Kaltverfestigungsanlage *f*

5263 STRAIN PLATE PRESS, FRAME PRESS pla,rub
f presse *f* à cadre
e prensa *f* de cuadro
i pressa *f* a telaio
n raampers, trekplatenpers
d Rahmenpresse *f*

5264 STRAINER iwt
f crépine *f*
e colador *m*
i colatoio *m*
n ziftmachine
d Seiher *m*

5265 STRAINER (GB), BREAKER PLATE (US) cpt,sor
f grille *f*
e criba *f*
i disco *m* forato
n geperforeerde plaat
d Lochscheibe *f*

5266 STRAINER, FALSE BOTTOM bre
f faux fond *m*
e falso fondo *m*
i fondo *m* falso, fondo *m* filtrante
n bodem van de klaringskuip
d Läuterboden *m*, Seihboden *m*, Senkboden *m*

5267 STRAINER, STRAINING MACHINE, SCREEN HEAD EXTRUDER, STOCK STRAINER pre,rub
f boudineuse *f* à tamiser
e budinadora *f* para tamizar
i trafila *f* con testa a filtro

n spuitmachine met zeefspuitkop, zeefpers
d Siebkopfspritzmaschine *f* Siebpresse *f*

5268 STRAINER, SUCTION STRAINER fil
f crépine *f* filtrante, sucette *f*
e colador *m* filtrante
i colatoio *m* filtrante, succhieruola *f*
n filtreermand
d Filterkorb *m*

5269 STRAINER HEAD, SNORE PIECE pre
f corps *m* d'aspiration, crépine *f*
e alcachofa *f* de aspiración, parte *f* inferior del tubo de aspiración
i cuffia *f* d'aspirazione, succhieruola *f*
n zuigkorf
d Saugstück *n*, Senkkorb *m*, Siebkorb *m*

5270 STRAINING FRAME lea
f cadre *m* tendeur
e bastidor *m* de estirado
i telaio *m* per stringere
n spanraam *n*
d Spannrahmen *m*

5271 STRAINING HEAD pre,rub
f tête *f* de boudineuse filtrante
e cabezal *m* de embutidora filtrador
i testa *f* di trafila a filtro
n zeefspuitkop
d Siebspritzkopf *m*

5272 STRAINING VAT, CLARIFYING TUB, FILTER VAT, LAUTER TUB bre
f cuve *f* de filtration, cuve-filtre *m*
e cuba *f* de filtración
i tino *m* di filtrazione
n klaringskuip
d Klärbottich *m*, Läuterbottich *m*

5273 STRAW PULP FACTORY nut,pap
f fabrique *f* de pâte de paille
e fábrica *f* de pasta de paja
i fabbrica *f* di polpa di paglia
n strostoffabriek
d Strohstoffabrik *f*

5274 STRAWBOARD MILL pap
f fabrique *f* de carton de paille
e fábrica *f* de cartón de paja
i fabbrica *f* di cartone di paglia
n strokartonfabriek
d Strohpappenfabrik *f*

5275 STREAK PLATE bac
f plaque *f* de biscuit, plaque *f* de frottis
e placa *f* de porcelana
i piastra *f* di porcellana, vetrina *f* di striscio
n porseleinplaat
d Abstrichplatte *f*, Strichplatte *f*

5276 STRESS-RELIEF ANNEALING FURNACE, STABILIZING ANNEALING FURNACE htt,met
f four *m* de recuit de détente
e horno *m* de recocido de atenuación de tensiones
i forno *m* di ricottura di distensione
n herstelgloeioven
d Ofen *m* für spannungsfreies Glühen

5277 STRETCHER ROLLER (GB), EXPANDING ROLLER (US) rub
f galet *m* tendeur, rouleau *m* déplisseur
e ensanchador *m*, rodillo *m* extensor, rodillo *m* tensor
i rullo *m* di tensione, rullo *m* estensore, stenditore *m*
n spanrol, strekrol
d Breithalter *m*, Spannrolle *f*

5278 STRIKE, STRIKING BATH elc
f bain *m* d'amorçage
e baño *m* de iniciación primario
i bagno *m* d'attivazione
n basislaagbad *n*
d Vorgalvanisierbad *n*

5279 STRINGER, BUNDLE OF FUEL ELEMENTS, CLUSTER OF FUEL ELEMENTS nue
f faisceau *m* d'éléments combustibles
e haz *m* de elementos combustibles
i fascio *m* d'elementi combustibil
n brandstofelementenbundel, splijtstofelementenbundel
d Brennstoffelementenbündel *n*

5280 STRIP ANODE el
f anode *f* en forme de bande
e ánodo *m* de cinta

i anodo *m* a nastro
n stripanode, stookanode
d Streifenanode *f*

5281 STRIP FUSE, sap
THERMAL FUSE
f fusible *m* à lame
e fusible *m* de cinta
i fusibile *m* a nastro
n temperatuurstrip
d Temperaturstreifen *m*

5282 STRIP PAPER chr
CHROMATOGRAPH
f appareil *m* pour la chromatographie à papier
e aparato *m* para la cromatografía de papel
i apparecchio *m* per la cromatografia a carta
n apparaat *n* voor papierchromatografie
d Apparat *m* zur Papierchromatographie

5283 STRIPPER PLATE pla,pre
f plaque *f* de dévêtissage
e placa *f* extractora
i piastra *f* di strappamento
n afdrukplaat, uitwerpplaat
d Abstreiferplatte *f*

5284 STRIPPER PLATE MOLD pla,pre
(US),
STRIPPER PLATE MOULD (GB)
f moule *m* à plaque de dévêtissage
e molde *m* con placa extractora
i stampo *m* con piastra di strappamento
n matrijs met afdrukplaat
d Abstreiferform *f*

5285 STRIPPER TANK elc
f cuve *f* de dépouillage
e tanque *m* de despojamiento
i cella *f* di strappamento
n beginplaatcel
d Mutterblechbad *n*

5286 STRIPPING BATH elc
f bain *m* de dépouillage
e baño *m* de eliminación
i bagno *m* di degalvanizzazione
n verwijderingsbad *n*
d Entplattierungsbad *n*

5287 STRIPPING COLUMN fue
f colonne *f* de dégazolinage, colonne *f* de strippage
e columna *f* de extracción de fracciones livianas
i colonna *f* d'estrazione di frazioni leggeri
n stripkolom
d Stripsäule *f*

5288 STRIPPING SECTION gen
f section *f* d'extraction
e sección *f* agotadora
i sezione *f* d'estrazione
n uitdrijfsectie
d Abtriebsäule *f*

5289 STRIPPING VESSEL tex
f récipient *m* à éliminer les colorants
e recipiente *m* de eliminar colorantes
i recipiente *m* per eliminare coloranti
n kleurstofverwijderingsvat *n*
d Farbstoffbeseitigungsgefäss *n*

5290 STUFF CHEST pap
f cuvier *m* à pâte
e tanque *m* almacenador de pulpa disgregada
i vasca *f* della polpa
n pulpkuip
d Ganzzeugbütte *f*, Stoffbütte *f*

5291 STUFF GRINDER pap
f défibreuse *f* mécanique
e desfibradora *f* mecánica
i sfibratrice *f* meccanica
n houtslijpmachine
d Holzschleifmaschine *f*

5292 STUFF VAT, pap
FERMENTATION TANK,
RAG FERMENTATION TANK,
ROTTING TANK
f récipient *m* de pourrissage
e pudridor *m*
i serbatoio *m* di putrefazione
n ontsluitingstank, rottingsvat *n*
d Faulbütte *f*

5293 STUFFER (US), pla,rub
HYDRAULIC EXTRUDER,
RAM EXTRUDER (GB)
f expulseur *m* à pression hydraulique, presse *f* à filer
e prensa *f* de extrusión con émbolo hidráulica, prensa *f* hidráulica
i espulsatore *m* a pressione idraulica, estrusore *m* a pistone
n hydraulische spuitpers, plunjerspuitmachine
d hydraulische Strangpresse *f*, Kolbenstrangpresse *f*

5294 STUFFING BOX cpt
f presse-étoupe *m*
e prensa-estopa *f*
i premistoppa *m*
n pakkingbus
d Stopfbüchse *f*

5295 STUFFING DRUM lea
f tambour *m* de graissage
e tambor *m* de engrasación
i bottale *m* per ingrasso con aria calda
n invetvat *n*
d Fettlickerfass *n*, Lederschmieregefäss *n*

5296 SUBLIMATION DRYING PLANT dry,sub
f installation *f* de séchage par sublimation
e equipo *m* secador por sublimación
i impianto *m* essiccatore per sublimazione
n drooginstallatie door sublimatie
d Trockenanlage *f* mit Sublimation

5297 SUBLIMING APPARATUS sub
f appareil *m* à sublimer
e aparato *m* de sublimar
i apparecchio *m* per sublimare
n sublimeertoestel *n*
d Sublimiergefäss *n*

5298 SUBLIMING FURNACE htt,sub
f four *m* à sublimer
e horno *m* de sublimar
i forno *m* per sublimare
n sublimeeroven
d Sublimierofen *m*

5299 SUBMERGED ARC-WELDING PLANT wel
f installation *f* de soudure sous flux conducteur
e equipo *m* de soldadura sobre flujo electroconductor
i impianto *m* di saldatura ad arco sommerso
n booglasinstallatie onder poederdak
d Unterpulverschweissanlage *f*

5300 SUBMERGED-COMBUSTION BURNER bur
f brûleur *m* à combustion noyée, brûleur *m* submergé
e quemador *m* de combustión sumergida, quemador *m* de llama sumergida
i bruciatore *m* a combustione sommersa, bruciatore *m* a fiamma sommersa
n brander voor dompelverbranding, onderwaterbrander
d Tauchbrenner *m*, Unterwasserbrenner *m*

5301 SUBMERGED CONDENSER, COLDFINGER (US), IMMERSION CONDENSER pre
f condensateur *m* plongeur, doigt *m* réfrigérant, réfrigérant *m* à immersion
e condensador *m* de inmersión
i condensatore *m* ad immersione
n dompelcondensor
d Einhängekühler *m*, Tauchkondensator *m*

5302 SUBMERGED CONTACT AERATOR, CONTACT AERATOR iwt
f lit *m* submergé
e aireador *m* de contacto
i letto *m* sommerso
n contactaërator
d Emscherfilter *m*, Tauchkörper *m*

5303 SUBMERGED FILTER fil
f filtre *m* noyé
e filtro *m* sumergido
i filtro *m* immerso, filtro *m* sommerso
n geïnundeerd filter *n*
d überstauter Filter *m*

5304 SUBMERGED GAS COVER iwt
f cloche *f* à gaz submergée
e cubierta *f* de gas sumergida
i copertura *f* a gas sommersa
n verzonken gasdek *n*
d versenkte Gasdecke *f*

5305 SUBMERGED GATE pla
f entrée *f* submergée
e entrada *f* sumergida
i entrata *f* sommersa
n verzonken aanspuitopening
d versenkter Anschnitt *m*

5306 SUBSEQUENT FILTER, FINAL FILTER fil
f filtre *m* finisseur
e filtro *m* final
i filtro *m* finale
n eindfilter *n*
d Nachfilter *m*

5307 SUBSIDENCE BASIN, SEDIMENTATION BASIN, SETTLING BASIN iwt
f bassin *m* de sédimentation
e artesa *f* de sedimentación
i bacino *m* di sedimentazione
n bezinkingsbekken *n*
d Absetzbecken *n*

5308 SUCK-OFF APPARATUS gen
f appareil *m* d'aspirazione
e aparato *m* de aspiración
i apparecchio *m* d'aspirazione
n afzuigapparaat *r*
d Absaugapparat *m*

5309 SUCTION BOTTLE, SUCTION FLASK lab
f fiole *f* à vide
e frasco *m* de aspiración, frasco *m* de vacío
i bottiglia *f* per aspirazione
n zuigfles
d Absaugflasche *f*, Saugflasche *f*

5310 SUCTION BOX pap
f caisse *f* d'aspiration
e caja *f* de succión
i cassa *f* d'aspirazione
n zuigkast
d Saugkasten *m*

5311 SUCTION CONDUIT cpt
f conduite *f* d'aspiration
e conducto *m* de aspiración
i condotta *f* d'aspirazione
n zuigleiding
d Saugleitung *f*

5312 SUCTION FILTER fil
f filtre *m* à air aspiré, filtre *m* à aspiration
e filtro *m* de aspiración
i filtro *m* ad aspirazione
n zuigfilter *n*
d Nutschenfilter *m*, Saugfilter *m*

5313 SUCTION FILTER, VACUUM FILTER fil
f filtre *m* à vide
e filtro *m* de vacío
i filtro *m* a vuoto
n vacuümfilter *n*
d Vakuumfilter *m*

5314 SUCTION FUNNEL sep
f entonnoir *m* à aspiration
e embudo *m* de aspiración
i imbuto *m* ad aspirazione
n afzuigtrechter
d Absaugtrichter *m*

5315 SUCTION PIPE cpt
f tube *m* d'aspiration
e tubo *m* de aspiración
i tubo *m* d'aspirazione
n zuigbuis
d Saugrohr *n*

5316 SUCTION PUMP pre
f pompe *f* à aspiration
e bomba *f* de aspiración
i pompa *f* ad aspirazione
n zuigpomp
d Saugpumpe *f*

5317 SUCTION PYROMETER, HIGH-VELOCITY THERMOCOUPLE gla,ins
f pyromètre *m* pour mesurer la température de gaz circulants à vitesse élevée
e pirómetro *m* para medida de la temperatura de gases circulantes a alta velocidad
i pirometro *m* per misurare la temperatura di gas circolanti ad alta velocità
n pyrometer voor temperatuurmeting aan snel stromende gassen
d Pyrometer *n* für Temperaturmessung an schnellströmende Gase

5318 SUCTION STRAINER, STRAINER fil
f crépine *f* filtrante, sucette *f*
e colador *m* filtrante
i colatoio *m* filtrante, succhieruola *f*
n filtreermand
d Filterkorb *m*

5319 SUCTION VALVE iwt
f soupape *f* de retenue
e válvula *f* de aspiración
i valvola *f* d'aspirazione
n aanzuigklep
d Ansaugventil *n*

5320 SUFFOCATING GAS GENERATOR gas,mit
f générateur *m* de gaz suffocant
e generador *m* de gas sofocante
i generatore *m* di gas soffocante
n generator van verstikkende gassen
d Stickgasgenerator *m*

5321 SUGAR CANE CRUSHER sin
f broyeur *m* de canne à sucre
e machacadora *f* de caña de azúcar

i frantoio *m* di canna da zucchero
n suikerrietbreker
d Zuckerrohrbrecher *m*

5322 SUGAR HYDROMETER, SACCHARIMETER, SACCHAROMETER bre,ins
f hydromètre *m* à solution sucrée, saccharimètre *m*
e sacarómetro *m*
i saccarimetro *m*
n sacharimeter, sacharometer
d Saccharimeter *n*, Zuckergehaltsmesser *m*

5323 SUGAR REFINERY sin
f raffinerie *f* de sucre
e refinería *f* de azúcar
i raffineria *f* di zucchero
n suikerraffinaderij
d Zuckerraffinerie *f*

5324 SUGARBEET JUICE PUMP, BEETROOT JUICE PUMP sin
f pompe *f* à jus de betteraves
e bomba *f* de zumo de remolachas
i pompa *f* a succo di barbabietole
n ruwsappomp
d Rübensaftpumpe *f*

5325 SUGARBEET LIFT, BEETROOT LIFT sin,tra
f élévateur *m* de betteraves
e elevador *m* de remolachas
i elevatore *m* di barbabietole
n bietenophaler
d Rübenelevator *m*

5326 SUGARBEET PUMP, BEETROOT PUMP sin,tra
f pompe *f* à betteraves
e bomba *f* de remolachas
i pompa *f* a barbabietole
n bietenpomp
d Rübenpumpe *f*

5327 SUGARBEET SILO, BEETROOT SILO sin,tra
f silo *m* à betteraves
e silo *m* de remolachas
i silo *m* a barbabietole
n bietensilo
d Rübensilo *n*

5328 SUGARBEET WASHER, BEETROOT WASHING MILL sin
f laveuse *f* de betteraves
e lavadora *f* de remolachas
i lavatrice *f* di barbatietole
n bietenwasmolen
d Rübenwäsche *f*

5329 SULFATE COOKING VESSEL (US), SULPHATE COOKING PAN (GB) pap
f chaudière *f* à sulfate
e caldera *f* de sulfato
i caldaia *f* a sulfato
n sulfaatkookketel
d Sulfatkochkessel *m*

5330 SULFATING ROASTING FURNACE (US), SULPHATING ROASTING FURNACE (GB) htt,met
f four *m* de grillage sulfatant
e horno *m* de tostación sulfatizante
i forno *m* di torrefazione a solfatazione
n sulfaterende roostoven
d sulfatisierender Röstofen *m*

5331 SULFIDE TONING BATH (US), SULPHIDE TONING BATH (GB) pho
f bain *m* de virage par sulfuration
e baño *m* de viraje por sulfuración
i bagno *m* di viraggio per solfatura
n zwavelkleurbad *n*
d Schwefeltonbad *n*

5332 SULFITATION TANK (US), SULPHITATION VESSEL (GB) sin
f récipient *m* de sulfitage
e cuba *f* de sulfitación
i serbatoio *m* di solfatura
n zwavelvat *n*
d Schwefelungsgefäss *n*

5333 SULFITE BLACK LIQUOR TANK (US), SULPHITE BLACK LIQUOR TANK (GB) pap
f récipient *m* de liqueur noire
e tanque *m* de líquido residual
i serbatoio *m* di lisciva nera
n zwarte-loogvat *n*
d Schwarzlaugegefäss *n*

5334 SULFITE COOKING VESSEL (US), SULPHITE COOKING PAN (GB) pap
f chaudière *f* à sulfite
e caldera *f* de sulfito
i caldaia *f* a solfito
n sulfietkookketel
d Sulfitkochkessel *m*

5335 SULFITE PULP FACTORY (US), SULPHITE PULP FACTORY (GB) pap
f fabrique *f* de pâte au sulfite
e fábrica *f* de pulpa de sulfito
i fabbrica *f* di polpa al solfito

n sulfietcellulosefabriek
d Sulfitzellstoffabrik *f*

5336 SULFITE WASTE LIQUOR VESSEL (US), pap
SULPHITE WASTE LIQUOR VESSEL (GB)
f récipient *m* de lessive de sulfite résiduelle
e recipiente *m* de solución residual de sulfito
i recipiente *m* di lisciva di solfito residua
n sulfietloogresiduvat *n*
d Sulfitablaugegefäss *n*

5337 SULFONATION FLASK (US), oic
SULPHONATION FLASK (GB)
f ballon *m* mélangeur de Keller, matras *m* de sulfonation
e matraz *m* de sulfonación
i matraccio *m* di solfonazione
n sulfoneerkolf
d Sulfierkolben *m*, Sulfonierkolben *m*

5338 SULFONATOR (US), fue,oic
SULPHONATOR (GB)
f sulfonateur *m*
e sulfonador *m*
i solfonatore *m*
n sulfoneerketel
d Sulfonierkessel *m*

5339 SULFUR BRIDGE (US), lab
SULPHUR BRIDGE (GB)
f pont *m* de soufre
e puente *m* de azufre
i ponte *m* di zolfo
n zwavelbrug
d Schwefelbrücke *f*

5340 SULFUR CHLORIDE VULCANIZATION BATH (US), rub
SULPHUR CHLORIDE VULCANIZATION BATH (GB)
f bain *m* de vulcanisation au chlorure de soufre
e baño *m* de vulcanización al cloruro de azufre
i bagno *m* di vulcanizzazione al cloruro di zolfo
n zwavelchloridevulcanisatiebad *n*
d Schwefelchloridvulkanisationsbad *n*

5341 SULFUR KILN (US), tex
SULPHUR KILN (GB)
f soufroir *m*
e estufa *f* de azufre
i stufa *f* a zolfo
n zwavelstoof
d Schwefelkammer *f*

5342 SULFUR LAMP (US), ana
SULPHUR LAMP (GB)
f appareil *m* de dosage du soufre à la lampe
e aparato *m* dosificador del azufre a la lámpara
i apparecchio *m* di dosaggio dello zolfo alla lampada
n zwavellamp
d Schwefellampe *f*

5343 SULFUR OVEN (US), sin
SULPHUR OVEN (GB)
f four *m* à soufre
e horno *m* de azufre
i forno *m* a zolfo
n zwaveloven
d Schwefelofen *m*

5344 SULFUR TONING BATH (US), pho
SULPHUR TONING BATH (GB)
f bain *m* de virage en tons sépia par sulfuration
e baño *m* de viraje al azufre
i bagno *m* di viraggio allo zolfo
n zwavelkleurbad *n*
d Schwefeltonbad *n*

5345 SULFUR VULCANIZATION BATH (US), rub
SULPHUR VULCANIZATION BATH (GB)
f bain *m* de vulcanisation en soufre
e baño *m* de vulcanización en azufre
i bagno *m* di vulcanizzazione in zolfo
n zwavelvulcanisatiebad *n*
d Schwefelvulkanisationsbad *n*

5346 SULFURATOR (US), bre
SULPHURATCR (GB)
f soufroir *m*
e cámara *f* de azufrado
i camera *f* a zolfo
n zwavelkamer
d Schwefelkammer *f*

5347 SULFURIC ACID WORKS (US), sua
SULPHURIC ACID WORKS (GB)
f fabrique *f* d'acide sulfurique
e fábrica *f* de ácido sulfúrico
i fabbrica *f* d'acido solforico
n zwavelzuurfabriek
d Schwefelsäurefabrik *f*

5348 SUN-BLEACHING PLANT tex
f installation *f* de blanchiment par la lumière solaire
e equipo *m* de blanqueo per la luz solar

i impianto *m* di candeggio per la luce solare
n zonlichtbleekinstallatie
d Sonnenbleichanlage *f*

5349 SUPERCALENDERING ROLL pap
f cylindre *m* de supercalandrage
e cilindro *m* de supercalandrado
i impianto *m* di supercalandratura, impianto *m* di superlucidatura
n superkalanderwals
d Superkalanderwalze *f*

5350 SUPERCHARGER pre
f surcompresseur *m*
e sobrealimentador *m*
i sopracompressore *m*
n overdrukcompressor
d Vorverdichter *m*

5351 SUPERCOOLING VESSEL, UNDERCOOLING VESSEL clo
f récipient *m* de surfusion
e recipiente *m* de sobrefusión
i recipiente *m* di surraffreddamento
n onderkoelingsvat *n*
d Unterkühlungsgefäss *n*

5352 SUPERFINISHING PLANT pol
f installation *f* de superfinissage
e equipo *m* de superpulido
i impianto *m* di soprafinitura
n superfijninstallatie
d Superfeinanlage *f*

5353 SUPERHEATED STEAM REACTOR, STEAM COOLED REACTOR nue
f réacteur *m* à vapeur surchauffée
e reactor *m* de vapor sobrecalentado
i reattore *m* a vapore surriscaldato
n met stoom gekoelde reactor, reactor met oververhitte stoom
d dampfgekühlter Reaktor *m*, Reaktor *m* mit überhitztem Dampf

5354 SUPERHEATER, STEAM SUPERHEATER boi
f surchauffeur *m* de vapeur
e recalentador *m* de vapor
i surriscaldatore *m* di vapore
n stoomoververhitter
d Dampfüberhitzer *m*

5355 SUPERHEATER COIL boi
f serpentin *m* surchauffeur
e serpentín *m* sobrecalentador
i serpentina *f* surriscaldatrice
n oververhittingsspiraal
d Überheizungsschlange *f*

5356 SUPERSATURATION APPARATUS cle,hyg,tex
f appareil *m* de sursaturation
e aparato *m* de supersaturación
i apparecchio *m* di soprassaturazio[ne]
n oververzadigingsapparaat *n*
d Übersättigungsapparat *m*

5357 SUPPLY CHUTE cpt,sin
f conduit *m* d'amenée
e conducto *m* de alimentación
i condotto *m* d'alimentazione
n aanvoergoot
d Zufuhrrinne *f*, Zulaufrinne *f*

5358 SUPPLY PIPE, SERVICE PIPE tra
f conduite *f* d'alimentation
e conducto *m* de alimentación
i condotta *f* d'alimentazione
n toevoerleiding
d Zufuhrleitung *f*

5359 SUPPORT PILLAR (US), SPACER (GB) pla,pre
f cale *f*, chandelle *f*, colonnette *f*, pilier *m*
e columna *f*, pilar *m*, soporte *m*
i distanziatore *m*, colonna *f*
n opvulstuk *n*, rond vulstuk *n*, steunkolom
d Abstandsblock *m*, Abstandssäule *f* Stütze *f*, Unterlegstück *n*

5360 SUPPORTING ELECTRODE spe
f électrode *f* porteuse
e electrodo *m* portador
i elettrodo *m* portatore
n dragerelektrode
d Trägerelektrode *f*

5361 SUPPORTING ROLLER, CARRIER ROLLER, TABLE ROLLER pap
f rouleau *m* de support, rouleau *m* porteur
e rodillo *m* de soporte, rodillo *m* portador
i rullo *m* di sopporto, rullo *m* portatore
n draagrol, steunrol
d Stützrolle *f*, Tragrolle *f*

5362 SURFACE COMBUSTION OVEN htt
f four *m* à combustion superficielle

e horno *m* de combustión superficial
i forno *m* a combustione superficiale
n oppervlakverbrandingsoven
d Oberflächenverbrennungsofen *m*

5363 SURFACE CONDENSER pre
f condensateur *m* à surface
e condensador *m* de superficie
i condensatore *m* a superficie
n oppervlakcondensor
d Oberflächenkondensator *m*

5364 SURFACE CONVERTING FURNACE, htt,met
CASE HARDENING FURNACE
f four *m* de trempe de surface
e horno *m* de temple de superficie
i forno *m* di tempera di superficie
n inzethardingsoven, oppervlakhardingsoven
d Einsatzhärtungsofen *m*, Oberflächenhärtungsofen *m*

5365 SURFACE COOLER, clo
INDIRECT COOLER
f refroidisseur *m* superficiel
e enfriador *m* superficial
i raffreddatore *m* superficiale, refrigerante *m* superficiale
n oppervlakkoeler
d Oberflächenkühler *m*

5366 SURFACE COOLER, bre,clo
PLATE COOLER
f refroidisseur *m* à plaques
e enfriador *m* de placas
i raffreddatore *m* a piastre
n platenkoeler
d Plattenkühler *m*

5367 SURFACE FINISHING APPARATUS pol
f appareil *m* de finissage superficiel
e aparato *m* de acabado superficial
i apparecchio *m* di finitura superficiale
n oppervlakpolijstapparaat *n*
d Oberflächenpoliergerät *n*

5368 SURFACE GRINDING MACHINE pol
f rectifieuse *f* de surface
e máquina *f* de rectificado superficial
i rettificatrice *f* per superficie piane
n oppervlakslijpmachine
d Oberflächenschleifmaschine *f*

5369 SURFACE HARDENING FURNACE htt,met
f four *m* de durcissement superficiel
e horno *m* de endurecimiento superficial
i forno *m* d'indurimento superficiale
n oppervlakhardingsoven
d Oberflächenhärtungsofen *m*

5370 SURFACE LAYER, fil
TOP LAYER
f couche *f* supérieure filtrante
e capa *f* superior filtradora
i strato *m* superiore filtrante
n filterdeklaag
d Filterdeckschicht *f*, Filterschicht *f*

5371 SURFACE TEMPERING FURNACE htt,met
f four *m* de trempe superficielle
e horno *m* de temple superficial
i forno *m* di tempera superficiale
n oppervlakonthardingsoven
d Oberflächenenthärtungsofen *m*

5372 SURFACE TENSION TUBES, lab
TRAUBE STALAGMOMETERS
f trousse *f* de stalagmomètres
e grupo *m* de estalagmómetros
i gruppo *m* di stalagmometri
n stalagmometers *pl* volgens Traube
d Traube-Stalagmometer *pl*

5373 SURFACE TREATMENT PLANT cle,coa,met
f installation *f* de traitement superficial
e equipo *m* de tratamiento superficiel
i impianto *m* di trattamento superficiale
n oppervlakbehandelingsinstallatie
d Oberflächenbehandlungsanlage *f*

5374 SUSPENDED-SOLIDS CONTACT REACTOR, iwt
SLUDGE-BLANKET REACTOR
f bassin *m* de floculation à lit de boue
e artesa *f* de floculación de contacto de fango
i bacino *m* di flocculazione a contatto del fango
n snel uitvlokkingsbekken *n*
d Schlammkontaktflockungsbecken *n*, Schnellflockungsbecken *n*

5375 SWANNECK PRESS, rub
GAP TYPE PRESS, GOOSENECK PRESS, JAW TYPE PRESS, OPEN SIDE PRESS
f presse *f* à col de cygne, presse *f* ouverte sur un côté
e prensa *f* abierta por un lado, prensa *f* acodillada, prensa *f* en cuello de cisne
i pressa *f* a collo di cigno
n muilpers, zwanehalspers
d einseitig offene Presse *f*, Maulpresse *f*, Schwanenhalspresse *f*

5376 SWEATER fue
f four *m* de ressuage
e horno *m* de sudar
i forno *m* di trasudazione
n zweetoven
d Ausschwitzofen *m*, Schwitzofen *m*

5377 SWEATING TANK fue
f récipient *m* de ressuage
e tanque *m* de exudación
i serbatoio *m* di sudata
n zweettank
d Schwitzbehälter *m*

5378 SWEETENING-OFF TANK, sin
SWEETENING-OUT TANK
f récipient *m* de désucrage
e tanque *m* de desazucaramiento
i serbatoio *m* per ricuperare lo zucchero dalla melassa
n melasseontsuikeringstank
d Absüssgefäss *n*

5379 SWEETENING PLANT fue
f installation *f* d'adoucissement, installation *f* de traitement au plombite
e equipo *m* de desmercaptanización
i impianto *m* di trattamento al plombite
n plumbiteerinstallatie
d Süssungsanlage *f*

5380 SWIMMING PARTICLES STRAINER cle,sor
f tamis *m* pour particules flottantes
e tamiz *m* para partículas flotantes
i staccio *m* per particelle flottanti
n zeef voor zwevende deeltjes
d Schwebeteilchenfänger *m*

5381 SWIMMING POOL REACTOR nue
f réacteur *m* piscine
e reactor *m* de piscina
i reattore *m* a piscina
n bassinreactor
d Bassinreaktor *m*, Schwimmbadreaktor *m*

5382 SWING HAMMER CRUSHER (GB), brk
MUD HOG (US)
f broyeur *m* à marteau oscillant rotatif
e machacador *m* de martillo oscilante giratorio
i frantoio *m* a martello oscillante rotativo
n breker met draaiende zwaaihamer
d Brecher *m* mit rotierendem Schwinghammer

5383 SWING PIPE cpt,hyg
f tube *m* articulé
e tubo *m* virable
i tubo *m* girevole
n zwaaipijp
d Schwenkrohr *n*

5384 SWING SIEVE (US), sor
SHAKING SCREEN (GB)
f crible *m* oscillant, tamis *m* à secousses
e criba *f* vibrante, tamiz *m* de sacudidas
i vaglio *m* a scosse, vibrovaglio *m*
n schudzeef
d Schüttelsieb *n*, Schwingsieb *n*

5385 SWIVEL JOINT cpt
f joint *m* à rotule
e junta *f* de nuez, junta *f* de rótula
i giunto *m* girevole
n wartelmoer
d Drehgelenk *n*, Universalgelenk *n*

5386 SWIVEL TABLE, rub
ROTARY PAN TRAY, TURN TABLE
f table *f* rotative
e mesa *f* giratoria
i tavola *f* rotativa
n draaibaar spuitblad *n*
d Schwenktablett *n*

5387 SYRINGE MICROPIPETTE lab
f micropipette *f* à piston
e micropipeta *f* de pistón
i micropipetta *f* a pistone
n microspuitpipet
d Mikrospritzenpipette *f*

5388 SYRUP PUMP (GB), SIRUP PUMP (US) sin

- f pompe *f* à sirop
- e bomba *f* de jarabe
- i pompa *f* a sciroppo
- n strooppomp
- d Siruppumpe *f*

T

5389 TABLE ROLLER, CARRIER ROLLER, SUPPORTING ROLLER — pap
f rouleau *m* de support, rouleau *m* porteur
e rodillo *m* de soporte, rodillo *m* portador
i rullo *m* di sopporto, rullo *m* portatore
n draagrol, steunrol
d Stützrolle *f*, Tragrolle *f*

5390 TABLETS PRESS (US), PELLETING MACHINE (GB), PREFORMING PRESS (GB) — mcy,pre
f pastilleuse *f*
e prensa *f* para empastillar
i pastigliatrice *f*
n tabletteerpers
d Tablettenpresse *f*

5391 TACCIOMETER (GB), TACKMETER (US) — dec,ins
f mesureur *m* de ténacité, mesureur *m* du pouvoir adhésif
e medidor *m* de adhesividad, medidor *m* de pegajosidad
i misuratore *m* d'adesività, misuratore *m* d'appiccicosità
n kleefkrachtmeter
d Klebkraftmesser *m*

5392 TACHOMETER — ins
f tachymètre *m*
e tacómetro *m*
i tachimetro *m*
n tachometer, toerenteller
d Drehzahlmesser *m*, Tachometer *n*, Tourenzähler *m*

5393 TACK WELDING MACHINE, STITCH WELDING MACHINE — pla
f machine *f* de soudure à la molette
e máquina *f* de soldadura discontinua
i macchina *f* di saldatura a tratti
n puntlasmachine
d Heftschweissmaschine *f*

5394 TAKE-OFF ROLLER — pla,pre
f dérouleur *m*, rouleau *m* dévideur
e rodillo *m* de desbobinado
i svolgitore *m*
n afneemrol, afwikkelrol
d Abwickelrolle *f*

5395 TAKE-UP ROLL (US), DRAW ROLL, HAUL-OFF ROLL (GB) — pla
f cylindre *m* de tirage, cylindre récepteur
e cilindro *m* de tracción
i cilindro *m* di tirata
n afneemwals, aftrekwals
d Abzugswalze *f*

5396 TAMPER, RAMMER — pre
f dame *f*, pilonneuse *f*
e pisonador *m*
i mazzapicchio *m*, pilonatrice *f*
n stamper
d Stampfer *m*

5397 TAMPING TOOL — cer
f outil *m* de serrage
e utensilio *m* de apisonado
i utensile *m* di pigiatura
n aanstamper *m*
d Ramme *f*, Stampfer *m*

5398 TAN VAT, TANNING PIT, HANDLER — lea
f fosse *f* à tan
e foso *m* para curtido, noque *m*
i fossa *f* da conciare le pelli
n looikuil
d Lohgrube *f*

5399 TANDEM CALENDER — pap,pla,rub
f calandre *f* en tandem
e calandra *f* en tándem
i calandra *f* in tandem
n tandemkalander
d Doppelkalander *m*

5400 TANGENTIAL-FLOW AGITATOR — mix
f agitateur *m* de flux tangentiel
e agitador *m* de flujo tangencial
i agitatore *m* a flusso tangenziale
n roerwerk *n* met tangentiële stroming
d Rührwerk *n* mit Tangentialströmung

5401 TANGENTIAL-SPIRAL AGITATOR — mix

f agitateur *m* à hélices tangentielles
e agitador *m* de hélices tangenciales
i agitatore *m* ad eliche tangenziali
n roerwerk *n* met tangentieel werkende spiralen
d Rührwerk *n* mit tangentialer Schraubenschaufel

5402 TANK gen
f citerne *f*, récipient *m*, réservoir *m*
e cisterna *f*, cuba *f*, tanque *m*
i cisterna *f*, serbatoio *m*, tank *m*
n tank
d Behälter *m*, Gefäss *n*, Tank *m*

5403 TANK BATH, VERTICAL TANK pho
f bain *m* mort, cuve *f* verticale
e tanque *m* de revelado vertical
i bacinella *f* verticale, vasca *f* per sviluppo lento
n standontwikkelingstank
d Gefässbad *n*, Standentwicklungstank *m*

5404 TANK BLOCK gla
f brique *f* de bassin
e ladrillo *m* de cuba
i mattone *m* di tino
n wansteen
d Wannenstein *m*

5405 TANK BOTTOM, TANK FLOOR sto
f radier *m* de récipient, sole *f* de récipient
e fondo *m* de recipiente, fondo *m* de tanque
i fondo *m* di serbatoio
n tankbodem
d Behältersohle *f*

5406 TANK CRYSTALLIZER cry
f cristallisoir *m* en auge
e tanque *m* de cristalización
i serbatoio *m* di cristallizzazione
n kristallisatiebak
d Kristallisationstrog *m*

5407 TANK DIALYZER oic
f dialyseur *m* en auge
e tanque *m* dializador
i serbatoio *m* dializzatore
n dialysevat *n*
d Trogdialysator *m*

5408 TANK FURNACE, GLASS FURNACE gla
f four *m* à bassins
e horno *m* de cubas
i forno *m* a tini
n wanoven
d Glaswannenofen *m*, Wannenofen *m*

5409 TANK INLET CHAMBER sto
f compartiment *m* d'entrée d'un récipient
e compartimiento *m* de entrada de un tanque
i camera *f* d'arrivo nel serbatoio, camera *f* d'immissione nel serbatoio
n tankinlaat
d Behältereinlaufkammer *f*

5410 TANK LINING coa,gen
f revêtement *m* de récipient
e revestimiento *m* aislante de tanque
i rivestimento *m* di serbatoio
n inwendige tankbekleding, tankvoering
d Gefässauskleidung *f*

5411 TANK PARTITION rub
f partition *f* de récipient
e tabique *m* de tanque
i tramezzo *m* di serbatoio
n tanktussenschot *n*
d Behältertrennwand *f*

5412 TANK REACTOR nue
f réacteur-chaudière *m*
e reactor *m* de tanque
i reattore *m* a cisterna
n tankreactor
d Tankreaktor *m*

5413 TANNER PLATE, TANNER'S BOLSTER tan
f plaque *f* de tanneur, traversin *m* de tanneur
e plancha *f* de curtidor, travesero *m* de curtidor
i piastra *f* di conciatore, sottotrave *f* di conciatore
n looiersplaat, onderlegplaat
d Gerberplatte *f*, Kuschen *f*

5414 TANNERY tan
f tannerie *f*
e curtiduría *f*, tenería *f*
i conceria *f*
n leerlooierij, looierij
d Gerberei *f*

5415 TANNING BATH pho
f bain *m* de tannage
e baño *m* de curtido
i bagno *m* di concia
n looibad *n*
d Gerbbad *n*

5416 TANNING DRUM tan
f tambour *m* de tannage
e tambor *m* de curtido
i tamburo *m* di concia
n looitrommel
d Gerbtrommel *f*

5417 TAP,
COCK cpt
f robinet *m*
e grifo *m*
i rubinetto *m*
n kraan
d Hahn *m*

5418 TAP HOLE fou,pla
f trou *m* de prise
e orificio *m* de toma
i apertura *f* di presa
n afsteekgat *n*, tapgat *n*, werkopening
d Ablassloch *n*, Stichloch *n*

5419 TAP MANIFOLD,
COCK MANIFOLD lab
f rampe *f* à robinets
e rampa *f* de grifos
i rampa *f* di rubinetti
n kranenstraat
d Hahnbrücke *f*

5420 TAP PLUG cpt
f noix *f* de robinet
e macho *m* de grifo
i anima *f* di rubinetto
n kraanplug
d Hahnküken *n*, Küken *n*

5421 TAPERED WHEEL,
CONE WHEEL pol
f disque *m* conique
e disco *m* ahusado
i disco *m* conico
n conische schijf
d Kegelscheibe *f*

5422 TAPPING SPOUT (GB),
FOUNTAIN, LAUNDER,
RUNNER THROUGH (US),
SPOUT (GB) fou
f canal *m* de coulée, chenal *m* de coulée
e canal *m* de colada
i canale *m* di colata
n gietgoot, schenk, tapgoot
d Abstichrinne *f*, Förderrinne *f*, Rutsche *f*

5423 TAR-ACID SEPARATION FUNNEL fue,oic,lab
f ampoule *f* à décanter pour le dosage du phénol
e embudo *m* separador para dosificación del fenol
i imbuto *m* separatore per il dosaggio del fenole
n scheitrechter voor fenolbepaling
d Scheidetrichter *m* für Teersäurebestimmung

5424 TAR SEPARATOR hyg,iwt,pap
f séparateur *m* de goudron
e desalquitranador *m*, separador *m* de alquitrán
i separatore *m* di catrame
n teerafscheider
d Teerabscheider *m*

5425 TARING ROOM mea,sin
f salle *f* de tarage
e sala *f* de tarado
i sala *f* da tarare
n tarreerlokaal *n*
d Tarrierraum *m*

5426 TASIMETER ins
f tasimètre *m*
e tasímetro *m*, termómetro *m* de presión
i tasimetro *m*
n elektrische drukschommelingsmeter, tasimeter
d elektrischer Druckschwankungsmesser *m*, Tasimeter *n*

5427 TAWERY lea
f mégisserie *f*
e curtiduría *f* en blanco
i conceria *f* all'allume
n witlooierij
d Alaungerberei *f*, Weissgerberei

5428 TAWING TANK lea
f récipient *m* de mégisserie
e tanque *m* de curtido en blanco
i serbatoio *m* di concia all'allume
n witlooikuip
d Weissgerbküpe *f*

5429 TEAR GAS BOMB,
LACHRYMATORY BOMB (GB),
LACRIMATORY BOMB (US) gas

f bombe *f* lacrymogène
e bomba *f* de gas lacrimógeno
i spezzone *m* lacrimogeno
n traangasbom
d Tränengasbombe *f*

5430 TEAR GAS GENERATOR, LACHRYMATORY GAS GENERATOR (GB), LACRIMATORY GAS GENERATOR (US) mit
f générateur *m* de gaz lacrymogène
e generador *m* de gas lacrimógeno
i generatore *m* di gas lacrimogeno
n traangasgenerator
d Tränengasgenerator *m*

5431 TEAR TESTING APPARATUS pap,tes
f appareil *m* d'essai de la résistance à la déchirure
e probador *m* de la resistencia al desgarramiento
i collaudatore *m* della resistenza alla lacerazione
n scheursterktebeproevingsapparaat *n*
d Einreissfestigkeitsprüfer *m*

5432 TEDGE tra
f entonnoir *m* d'alimentation
e embudo *m* alimentador
i imbuto *m* alimentatore
n aanvoertrechter
d Speisetrichter *m*

5433 TEE PIECE, THREE-WAY PIPE tra
f branchement *m* en T, raccord *m* en T
e pieza *f* en T
i raccordo *m* a T
n driewegkoppeling, T-stuk *n*
d T-Stück *n*

5434 TELOMERIZATION PLANT oic
f installation *f* de télomérisation
e equipo *m* de telomerización
i impianto *m* di telomerizzazione
n telomerisatie-installatie
d Telomerisationsanlage *f*

5435 TEMPERATURE REGULATOR, THERMOREGULATOR ins
f régulateur *m* thermique, thermorégulateur *m*
e regulador *m* de temperatura, termorregulador *m*
i regolatore *m* di temperatura, termoregolatore *m*
n temperatuurregelaar, thermoregulateur
d Temperaturregler *m*, Wärmeregler *m*

5436 TEMPERATURE SCALE ins
f échelle *f* thermométrique
e escala *f* termométrica
i scala *f* termometrica
n temperatuurschaal, thermometerschaal
d Temperaturskala *f*, Thermometerskala *f*

5437 TEMPERING BATH htt,met
f bain *m* de revenu
e baño *m* de revenido
i bagno *m* di rinvenimento
n ontlaatbad *n*
d Anlassbad *n*

5438 TEMPERING MOLD (US), TEMPERING MOULD (GB) pla,pre
f moule *m* à recuire
e molde *m* para recocer
i stampo *m* per rinvenire
n temperpatrijs
d Temperform *f*

5439 TEMPERING OVEN gla
f four *m* à recuire
e horno *m* para recocer
i forno *m* per tempera
n langzame koeloven, temperoven
d langsamer Kühlofen *m*

5440 TEMPERING PAN, SLUDGE PAN cer
f cuve *f* de gâchage
e cuba *f* de amasado
i tino *m* di preparazione
n aanmaakvat *n*
d Anmachgefäss *n*

5441 TENSIMETER ins
f tensimètre *m*
e tensímetro *m*, tensómetro *m*
i tensimetro *m*
n gasspanningsmeter, tensimeter
d Tensimeter *n*

5442 TENSIOMETER, TENSION METER ins
f tensiomètre *m*
e tensiómetro *m*
i tensiometro *m*
n trekmeter
d Spannungsmesser *m*, Tensiometer *n*

5443 TENSION ROLLER, PULL ROLL pla,pre
f rouleau *m* tendeur

e rodillo *m* tensor
i rullo *m* tenditore
n spanrol, tegendrukrol
d Spannrolle *f*

5444 TENTER, STENTER — tex
f élargisseur *m*
e estricador *m*
i apparecchio *m* tenditore
n spanraam *n*
d Spannrahmen *m*

5445 TERMINAL — elc
f borne *f*, pôle *m*
e borna *f*, polo *m*
i morsetto *m*
n aansluitcontact *n*, pool
d Anschlussklemme *f*, Pol *m*

5446 TEST COCK (GB), GAGE COCK (US), GAUGE COCK (GB), PET COCK (US) — boi
f robinet *m* d'essai, robinet *m* de jauge, robinet *m* de niveau d'eau
e grifo *m* de prueba, grifo *m* indicador del nivel
i rubinetto *m* di prova
n proefkraan, waterpeilkraan
d Probierhahn *m*, Wasserstandshahn *m*

5447 TEST CONE — cer,tes
f cône *m* réfractaire d'essai
e cono *m* refractario de prueba
i cono *m* refrattario campione
n vuurbestendige proefkegel
d feuerfester Probierkegel *m*

5448 TEST CRUCIBLE — tes
f creuset *m* d'essai
e crisol *m* de prueba
i crogiolo *m* di saggio
n cupelleerkroes
d Probiertiegel *m*

5449 TEST FURNACE — tes
f four *m* d'essai
e horno *m* de ensayo
i forno *m* sperimentale
n proefoven
d Probierofen *m*

5450 TEST TUBE — lab,tes
f éprouvette *f*, tube *m* à essai
e probeta *f*, tubo *m* de ensayo
i provetta *f*, tubo *m* da saggio, vetro *m* di prova
n reageerbuisje *n*
d Reagenzglas *n*

5451 TESTING EQUIPMENT — tes
f appareillage *m* d'essai
e aparatos *pl* de prueba
i apparecchiatura *f* di collaudo
n beproevingsapparatuur
d Probeapparatur *f*, Prüfapparatur

5452 TESTING PUMP — tes
f pompe *f* d'essai
e bomba *f* de prueba
i pompa *f* di collaudo alla condotta
n controlepomp
d Probierpumpe *f*

5453 THEORETICAL PLATE — dis
f plateau *m* théorique
e plato *m* teórico
i piatto *m* teorico
n theoretische schotel
d theoretischer Boden *m*

5454 THERMAL ALARM INDICATOR, HEAT ALARM INDICATOR — sap
f dispositif *m* d'alarme thermique
e termoavisador *m*
i dispositivo *m* d'allarme termico
n thermisch alarmapparaat *n*
d wärmebetätigte Alarmvorrichtung

5455 THERMAL BARRIER — gla
f barrage *m* thermique
e barrera *f* térmica
i barriera *f* termica
n warmtekeerwand
d Wärmesperre *f*

5456 THERMAL CRACKING PLANT — fue
f installation *f* de craquage thermique
e equipo *m* de cracking térmico equipo *m* de piezopirólisis térmica
i impianto *m* di cracking termico, impianto *m* di piroscissione termica
n thermische kraakinstallatie
d Thermokrackanlage *f*

5457 THERMAL DIFFUSION PIPE, SEPARATION TUBE — nue
f colonne *f* de diffusion thermique
e columna *f* de difusión térmica
i colonna *f* di diffusione termica
n thermische diffusiekolom
d Trennrohr *n*

5458 THERMAL DIFFUSION SEPARATION PLANT — nue
f installation *f* de séparation par diffusion thermique

e equipo *m* de separación por difusión térmica
i impianto *m* di separazione per diffusione termica
n thermische diffusiescheidingsinstallatie
d thermische Diffusionsscheidungsanlage *f*

5459 THERMAL DISSOCIATION VESSEL gen
f récipient *m* de dissociation thermique
e recipiente *m* de disociación térmica
i recipiente *m* di dissociazione termica
n thermische-dissociatievat *n*
d Thermodissoziationsgefäss *n*

5460 THERMAL EVAPORATION PLANT eva
f installation *f* d'évaporation thermique
e equipo *m* de evaporación térmica
i impianto *m* d'evaporazione termica
n thermische verdampingsinstallatie
d thermische Verdampfungsanlage *f*

5461 THERMAL FUSE, STRIP FUSE sap
f fusible *m* à lame
e fusible *m* de cinta
i fusibile *m* a nastro
n temperatuurstrip
d Temperaturstreifen *m*

5462 THERMAL HIGH-VACUUM GAGE (US), THERMAL HIGH-VACUUM GAUGE (GB) ins
f manomètre *m* thermique à vide élevé
e manómetro *m* térmico de alto vacío
i manometro *m* termico ad alto vuoto
n thermische hoogvacuümmanometer
d Thermohochvakuummanometer *n*

5463 THERMAL REFORMING PLANT fue
f installation *f* de réformation thermique
e equipo *m* de reformación térmica
i impianto *m* di riformazione termica
n thermische reformatie-installatie
d Thermoreformationsanlage *f*

5464 THERMAL SEPARATING COLUMN, THERMAL SEPARATING TOWER sep
f colonne *f* de séparation thermique
e columna *f* de separación térmica
i colonna *f* di separazione termica
n thermische scheidingskolom
d thermische Trennsäule *f*

5465 THERMAL SHIELD nue,sap
f écran *m* thermique
e pantalla *f* térmica
i schermo *m* termico
n warmtescherm *n*
d Wärmeschirm *m*

5466 THERMAL TREATMENT PLANT htt,met
f installation *f* de traitement thermique
e equipo *m* de tratamiento térmico
i impianto *m* di trattamento termico
n warmtebehandelingsinstallatie
d Warmbehandlungsanlage *f*

5467 THERMAL WATER SOFTENING PLANT hyg
f installation *f* thermique d'adoucissement de l'eau
e equipo *m* térmico de ablandamiento del agua
i impianto *m* termico d'addolcimento dell'acqua
n thermische waterontthardingsinstallatie
d thermische Wasserenthärtungsanlage *f*

5468 THERMITE MOLD (US), THERMITE MOULD (GB) wel
f moule *m* pour soudure aluminothermique
e molde *m* para soldadura aluminotérmica
i forma *f* per saldatura alla termite
n vorm voor thermietlassen
d Form *f* für Thermitschweissung

5469 THERMITE WELDING PLANT, ALUMINOTHERMIC WELDING PLANT wel
f installation *f* de soudure aluminothermique
e equipo *m* de soldadura aluminotérmica

i impianto *m* di saldatura alla termite
n termietlasinstallatie
d Thermitschweissanlage *f*

5470 THERMOCOUPLE, THERMOELEMENT elc
f couple *m* thermoélectrique
e par *m* termoeléctrico
i coppia *f* termoelettrica, termocoppia *f*
n thermo-element *n*
d Thermoelement *n*

5471 THERMOELECTRIC THERMOMETER ins
f thermomètre *m* thermoélectrique
e termómetro *m* termoeléctrico
i termometro *m* termoelettrico
n thermo-elektrische thermometer
d thermoelektrisches Thermometer *n*

5472 THERMOHYDROMETER ins
f aréomètre *m* avec thermomètre
e termohidrómetro *m*
i idrometro *m* con termometro
n thermohydrometer
d Thermoaräometer *n*

5473 THERMOMETER ins
f thermomètre *m*
e termómetro *m*
i termometro *m*
n thermometer
d Thermometer *n*

5474 THERMOPILE elc
f pile *f* thermoélectrique, thermopile *f*
e pila *f* termoeléctrica, termopila *f*
i termopila *f*
n thermozuil
d Thermosäule *f*

5475 THERMOREGULATOR, TEMPERATURE REGULATOR ins
f régulateur *m* thermique, thermorégulateur *m*
e regulador *m* de temperatura, thermorregulador *m*
i regolatore *m* di temperatura, termoregolatore *m*
n temperatuurregelaar, thermoregulateur
d Temperaturregler *m*, Wärmeregler *m*

5476 THERMOSTAT con
f thermostat *m*
e termostato *m*
i termostato *m*
n thermostaat
d Thermostat *m*

5477 THICKENER elc
f épaississeur *m*
e espesador *m*, separador *m* sedimentario
i ispessitore *m*
n indikker
d Absetzgefäss *n*, Eindicker *m*

5478 THICKENER met,min
f concentrateur *m*
e concentrador *m*
i concentratore *m*
n ertsconcentrator
d Anreicherungsgerät *n*

5479 THIN-FILM EVAPORATOR eva,lab
f évaporateur *m* à couche mince
e evaporador *m* de capa delgada
i evaporatore *m* a strato sottile
n dunne-laagverdamper
d Dünnschichtverdampfer *m*

5480 THIN-LAYER SPREADER, APPLICATOR ch
f dispositif *m* d'étalement
e extendedor *m* de la capa
i distensore *m* della pellicola
n uitstrijker van de dunne laag
d Dünnschichtstreicher *m*

5481 THIRTING TOOL ce
f couteau *m* à découpage de blocs d'argile
e cuchillo *m* de cortar bloques de arcilla
i coltello *m* per tagliare blocchi d'argilla
n kleiblokkensnijder
d Tonblockenmesser *n*

5482 THISTLE TOP FUNNEL TUBE lab
f tube *m* de sûreté droite
e embudo *m* de campana
i imbuto *m* a campana
n bolkoptrechter
d Glockentrichter *m*

5483 THORIUM REACTOR nu
f réacteur *m* à thorium
e reactor *m* de torio
i reattore *m* a torio
n thoriumreactor
d Thoriumreaktor *m*

5484 THREAD FORMING INSERT pla,pre (GB),
THREAD PLUG (US)
f broche *f* filetée
e inserción *f* para moldeo de roscas
i spina *f* filettata
n schroefdraaddoorn, stift voor het inpersen van schroefdraad
d Einpressgewindestift *m*, Gewindestift *m*

5485 THREE-FLOOR KILN, bre
TREBLE-FLOOR KILN
f touraille *f* à trois plateaux
e tostadera *f* de tres platos
i essiccatoio *m* a tre graticci
n eest met drie vloeren
d Dreihordendarre *f*

5486 THREE-NECKED FLASK, lab
THREE NECKS FLASK
f ballon *m* à trois cols
e matraz *m* de tres cuellos
i matraccio *m* a tre colli
n driehalzige kolf
d Dreihalskolben *m*

5487 THREE-ROLLER MILL bre
f concasseur *m* à trois cylindres
e molino *m* de tres cilindros
i mulino *m* a tre cilindri
n schrootmolen met drie walsen
d Dreiwalzenmühle *f*

5488 THREE-STAGE COMPRESSOR pre
f compresseur *m* à trois étages
e compresor *m* de tres etapas
i compressore *m* a tre stadi
n drietrapscompressor
d Dreistufenkompressor *m*

5489 THREE-WAY COCK cpt,lab
f robinet *m* à trois voies
e grifo *m* de tres vías
i rubinetto *m* a tre vie
n driewegkraan
d Dreiwegehahn *m*

5490 TRHEE-WAY PIPE, tra
TEE PIECE
f branchement *m* en T, raccord m en T
e pieza *f* en T
i raccordo m a T
n driewegkoppeling, T-stuk *n*
d T-Stück *n*

5491 THROTTLE CALORIMETER, ins
CHOKING CALORIMETER
f calorimètre *m* à étranglement
e calorímetro *m* de estrangulación
i calorimetro *m* a strozzatura
n smoorcalorimeter
d Drosselkalorimeter *n*

5492 THROTTLE VALVE cpt
f soupape *f* de détente de vapeur, soupape *f* d'étranglement
e válvula *f* de estrangulación, válvula *f* de retención de vapor
i valvola *f* di ritegno di vapore
n smoorklep
d Dampfabsperrventil *n*, Drosselventil *n*

5493 THROUGH-CARBURIZING htt,met FURNACE (US),
TOTAL CARBURIZING FURNACE (GB)
f four *m* de cémentation totale
e horno *m* de cementación total
i forno *m* di cementazione totale
n door-en-doorcementeeroven
d Durchzementierungsofen *m*

5494 THROUGH-HARDENING htt,met BATH (US),
TOTAL HARDENING BATH (GB)
f bain *m* d'endurcissement total
e baño *m* de endurecimiento total
i bagno *m* d'indurimento totale
n door-en-doorhardingsbad *n*
d Durchhärtungsbad *n*

5495 THRUST ROLL pre,rub
f cylindre *m* de serrage
e cilindro *m* de empuje
i cilindro *m* di spinta
n aanperswals
d Andruckwalze *f*

5496 TILTING DISK CHECK VALVE, cpt
FLAP VALVE
f clapet *m* de fermeture, soupape *f* à clapet
e válvula *f* de charnela, válvula *f* de mariposa
i valvola *f* a cerniera, valvola *f* di chiusura
n deurklep, scharnierende klep
d Klappenventil *n*, Verschlussklappe *f*

5497 TILTING FURNACE, htt
TILTING OPEN-HEARTH FURNACE
f four *m* culbutant
e horno *m* basculante
i forno *m* girevole, forno *m* oscillante

n kanteloven, kipoven
d Kippofen *m*

5498 TILTING-HEAD PRESS pla,pre
f presse *f* à tête renversible
e prensa *f* con cabezal de volquete
i pressa *f* a testa ribaltibile
n kantelkoppers
d Schwenkkopfpresse *f*

5499 TIN, nut
CAN
f boîte *f*
e bote *m*, lata *f*
i barattolo *m*, scatola *f*
n blik *n*
d Dose *f*

5500 TIN SEALING MACHINE (GB), bre
CAN SEALING MACHINE (US)
f sertisseuse *f*
e máquina *f* para tapar latas
i macchina *f* per chiudere barattoli, macchina *f* per chiudere scatole
n blikmachine
d Dosenverschlussmaschine *f*

5501 TINNING MACHINE (GB), bre
CANNING MACHINE (US)
f emboîteuse *f*
e llenadora *f* de latas
i riempitrice *f* per barattoli
n blikvulmachine
d Dosenfüllmaschine *f*

5502 TINTING BATH tex
f bain *m* de nuançage
e baño *m* de tintado
i bagno *m* di tingitura
n bijkleurbad *n*
d Abtönbad *n*

5503 TINTOMETER ins
f tintomètre *m*
e tintómetro *m*
i tintometro *m*
n tintometer
d Tintometer *n*

5504 TISSUE GRINDER brk,lab
f broyeur *m* à tissus
e machacadora *f* de tejidos
i mulino *m* a tessuti
n weefselmaler
d Gewebezerkleinerer *m*

5505 TITRATION APPARATUS ana
f appareil *m* à titrer, titrimètre *m*
e aparato *m* de valoración
i apparecchio *m* di titolazione
n titreerapparaat *n*
d Titrierapparat *m*

5506 TITRATION ASSEMBLY lab
f appareillage *m* de titrage
e apparatos *pl* de valoración
i apparecchiatura *f* di titolazione
n titreerapparatuur
d Titrierapparatur *f*

5507 TITRATION FLASK lab
f matras *m* de titrage
e matraz *m* de valoración
i matraccio *m* di titolazione
n titreerkolf
d Titrierkolben *m*

5508 TITREMETER ins
f titrimètre *m*
e valorímetro *m*
i titrimetro *m*
n titrimeter
d Titrimeter *n*

5509 TOGGLE LEVER PRESS, pre
TOGGLE PRESS
f presse *f* à genouillère
e prensa *f* de palanca acodada
i pressa *f* a ginocchiera
n kniehefboompers
d Kniehebelpresse *f*

5510 TONING AND FIXING BATH, pho
COMBINED BATH
f bain *m* viro-fixateur
e baño *m* de fijado virador
i bagno *m* di viraggio e di fissaggi
n combinatiebad *n*, kleurend fixeerbad *n*
d Tonfixierbad *n*

5511 TONING BATH pho
f bain *m* de virage
e baño *m* virador
i bagno *m* di viraggio
n kleurbad *n*
d Tonbad *n*

5512 TOOTHED ATTRITION MILL (US), brk
TOOTHED DISK MILL (GB)
f broyeur *m* à disques dentés
e molino *m* de discos dentados
i mulino *m* a dischi dentati
n tandschiifmaalmolen
d Exzelsiormühle *f*

5513 TOP BACKING PLATE cpt,pla

f appui *m* supérieur, plaquette *f* d'appui
e contraplaca *f* superior
i piastra *f* di taglio superiore
n bovenste stansplaat
d Stanzauflage *f*

5514 TOP BOARD (US), pla,pre
TOP PLATE (GB)
f plateau *m* de dessus
e plato *m* superior
i piastra *f* superiore
n bovenplaat
d obere Pressplatte *f*

5515 TOP CHROME BATH, tex
AFTERCHROME DYEING BATH
f bain *m* de teinture postérieur au chrome
e baño *m* de postmordentado crómico
i bagno *m* di tintura posteriore al cromo
n nachromeerbad *n*
d Nachchromierungsbad *n*

5516 TOP CLAMP PLATE (US), pla
FRONT SHOE (GB)
f plaque *f* d'entrée, plaque *f* supérieure
e placa *f* delantal de sujeción del molde
i fondello *m*, piastra *f* di base fissa
n bovenste opspanplaat
d Aufspannkörper *m* hinten

5517 TOP-FIRED KILN cer
f four *m* de grillage chauffé par la partie supérieure
e horno *m* de tostación caldeado por la parte alta
i forno *m* di torrefazione con fuoco alla bocca
n bovenvuurroostoven
d Oberfeuerröstofen *m*

5518 TOP FORCE (GB), pla,pre
FORCE (US), PUNCH (GB)
f poinçon *m*
e molde *m* macho, punzón *m*
i punzone *m*
n positief matrijsdeel *n*, stempel
d Stempel *m*

5519 TOP-HAT KILN, cer
ENVELOPE KILN, SHUTTLE KILN
f four *m* à recuire avec calotte protectrice
e horno *m* cerámico con caperuza de protección
i forno *m* ceramico con cappa di protezione
n kapselbakoven, kokerbakoven
d Kapselbrennofen *m*

5520 TOP LAYER, fil
SURFACE LAYER
f couche *f* supérieure filtrante
e capa *f* superior filtradora
i strato *m* superiore filtrante
n filterdeklaag
d Filterdeckschicht *f*, Filterschicht *f*

5521 TOP ROLL sin
f cylindre *m* supérieur
e cilindro *m* superior
i cilindro *m* superiore
n topwals
d Oberwalze *f*

5522 TOPPING COLUMN fue
f colonne *f* de première distillation, installation *f* d'étêtage, installation *f* de séparation des produits de tête
e columna *f* de destilación primaria
i colonna *f* di distillazione primaria
n aftopkolom, primaire destillatiekolom
d Primärdestillationssäule *f*, Topsäule *f*

5523 TORCH, lab
BLOWPIPE
f chalumeau *m*
e soplete *m*
i cannello *m* ferruminatorio
n blaaspijp
d Gebläselampe *f*, Lötrohr *n*

5524 TORCH HARDENING PLANT, htt,met
FLAME HARDENING PLANT, SHORTERIZING PLANT
f installation *f* de trempe au chalumeau
e equipo *m* de temple superficial a la llama
i impianto *m* di tempera alla fiamma
n vlamhardingsinstallatie
d autogene Flammenhärtungsanlage *f*

5525 TORPEDO, pla,pre,rub
SPREADER

f distributeur *m*, torpille *f*
e extendedor *m*, torpedo *m*
i ogiva *f*, punzone *m* distributore
n torpedo, verdeelkop
d Verteiler *m*

5526 TORPEDO MANDREL pla,pre
f mandrin *m* de torpille
e mandril *m* de torpedo
i maschio *m* d'ogiva
n torpedoverlengstuk *n*
d Torpedodorn *m*

5527 TORSION BALANCE lab
f balance *f* à torsion
e balanza *f* de torsión
i bilancia *f* a torsione
n torsiebalans
d Drehwaage *f*, Torsionswaage *f*

5528 TORSION VISCOMETER (US), ins
TORSION VISCOSIMETER (GB)
f viscosimètre *m* à torsion
e viscosímetro *m* de torsión
i viscosimetro *m* a torsione
n torsieviscosimeter
d Torsionsviskosimeter *n*

5529 TOSSING DEVICE met
f écrémeuse *f*
e epurador *m*
i separatore *m* della melma
n vuilvanger
d Schlammfänger *m*

5530 TOTAL CARBURIZING FURNACE (GB), htt,met
THROUGH-CARBURIZING FURNACE (US)
f four *m* de cémentation totale
e horno *m* de cementación total
i forno *m* di cementazione totale
n door-en-doorcementeeroven
d Durchzementierungsofen *m*

5531 TOTAL HARDENING BATH (GB), htt,met
THROUGH-HARDENING BATH (US)
f bain *m* d'endurcissement total
e baño *m* de endurecimiento total
i bagno *m* d'indurimento totale
n door-en-doorhardingsbad *n*
d Durchhärtungsbad *n*

5532 TOWER, gen
COLUMN
f colonne *f*, tour *f*
e columna *f*, torre *f*
i colonna *f*, torre *f*
n kolom, toren, zuil
d Kolonne *f*, Säule *f*, Turm *m*

5533 TOWER CASING, clo
SHEATHING
f revêtement *m* de tour
e revestimiento *m* de torre
i rivestimento *m* di torre
n torenbekleding
d Turmauskleidung *f*

5534 TOWER DRIER, dry,gas
DRYING TOWER
f tour *f* de séchage
e torre *f* de secado
i torre *f* d'essiccamento
n droogtoren
d Trockenturm *m*

5535 TOWER STEAMER (US), tex
STEAMING TOWER (GB)
f tour *f* de traitement à la vapeur
e torre *f* de tratamiento al vapor
i torre *f* di trattamento al vapore
n stoombehandelingstoren
d Dampfbehandlungsturm *m*

5536 TRACE HOLE cer
f couloir *m* de flamme
e paso *m* de llama
i passo *m* di fiamma
n vlamgat *n*
d Flammloch *n*

5537 TRACER AMMUNITION mit
f munitions *pl* traceuses
e munición *f* trazadora
i munizioni *pl* traccianti
n lichtspoormunitie
d Leuchtspurmunition *f*

5538 TRACER CONTROL, con
FEELER CONTROL
f commande *f* à palpeur
e accionamiento *m* por palpador
i azionamento *m* con tastatore
n aftastbesturing
d Fühlersteuerung *f*

5539 TRANSFER CHAMBER, pla
LOADING PLATE (US),
RETAINER PLATE (GB)
f plaque *f* contenant la chambre de transfert
e placa *f* intermedia de alojamiento de la cámara de plastificación
i piastra *f* di carica
n tussenplaat, vulplaat
d Füllplatte *f*

5540 TRANSFER CHAMBER (GB), pla,pre
POT (US)
f pot *m* de transfert
e cámara *f* de carga, cámara *f* de plastificación
i camera *f* di compressione
n tabletkamer, toevoermatrijs
d Spritztopf *m*

5541 TRANSFER MOLD (US), pla,pre
TRANSFER MOULD (GB)
f moule *m* à transfert
e molde *m* de transferencia
i stampo *m* di trasferimento
n spuitpersmatrijs, transfermatrijs
d Pressspritzform *f*, Pressspritzwerkzeug *n*

5542 TRANSFER MOLDING pla,pre
PRESS (US),
TRANSFER MOULDING PRESS (GB)
f presse *f* à moulage par transfert
e prensa *f* para moldear por transferencia
i pressa *f* per stampatura mediante trasferimento
n spuitgietmachine
d Pressspritzmaschine *f*

5543 TRANSFER PLUNGER pla,pre
(GB),
POT PLUNGER (US)
f piston *m* de transfert
e émbolo *m* de transferencia
i pistone *m* d'iniezione
n injectieplunjer, spuitgietplunjer
d Spritzgusskolben *m*, Spritzkolben *m*

5544 TRANSFER PLUNGER pla,pre
RETAINER PLATE
f contreplaque *f* de piston de transfert
e placa *f* de alojamiento del émbolo de transferencia
i piastra *f* portapunzone
n spuitpersplunjerplaat, transferplunjerplaat
d Spritzkolbenplatte *f*

5545 TRANSPARENCY TESTING pap,tes
APPARATUS
f appareil *m* d'examen de la transparence
e aparato *m* de verificación de la transparencia
i collaudatore *m* della trasparenza
n lichtdoorlatendheidsmeter
d Lichtdurchlässigkeitsmesser *m*

5546 TRANSVERSAL pla,rub
EXTRUDER HEAD,
CROSS-EXTRUDER HEAD
f tête *f* d'équerre de boudineuse
e cabezal *m* de budinadora transversal
i testa *f* di trafila ad angolo
n dwarse spuitkop
d Querspritzkopf *m*

5547 TRANSVERSE-ARCH KILN cer
f four *m* annulaire à parois transversales
e horno *m* anular de paredes transversales
i forno *m* anulare con pareti trasversali
n ringoven met radiale tussenwanden
d Ringofen *m* mit Querwänden

5548 TRANSVERSE-FINNED FUEL nue
ELEMENT
f élément *m* combustible à ailettes transversales
e elemento *m* combustible con aletas transversales
i elemento *m* combustibile ad alette trasversali
n van dwarsribben voorzien splijtelement *n*
d Brennstoffelement *n* mit Querberippung

5549 TRAPPING CONDENSER dis,lab
f piège *m* réfrigérant à serpentin
e trampa *f* enfriadora con serpentín
i trappola *f* raffreddatrice a serpentina
n fractievanger met koelslang
d Fraktionssammler *m* mit Kühlschlange

5550 TRASH CATCHER pap
f décrasseur *m*
e recogedor *m* de bagazo
i collettore *m* di ramaglia
n vuilvanger
d Schmutzfänger *m*

5551 TRASH RACK, sor
COARSE BAR SCREEN, COARSE RACK, COARSE SCREEN
f grille *f* grossière
e criba *f* gruesa
i griglia *f* grossa, griglia *f* grossolana
n grove zeef
d Grobrechen *m*

5552 TRAUBE STALAGMOMETERS, lab
SURFACE TENSION TUBES
f trousse *f* de stalagmomètres
e grupo *m* de estalagmómetros
i gruppo *m* di stalagmometri
n stalagmometers *pl* volgens Traube
d Traube-Stalagmometer *pl*

5553 TRAVELLING BAND FILTER, cle,fil
BAND FILTER
f filtre *m* à bande transporteuse
e filtro *m* de cinta transportadora
i filtro *m* a nastro trasportatore
n lopende-bandfilter *n*
d Bandfilter *m*, Wandernutsche *f*

5554 TRAVELLING GRATE tra
f grille *f* mobile
e parrilla *f* móvil
i griglia *f* mobile
n beweegbaar rooster *n*
d Wanderrost *m*

5555 TRAVELLING-GRATE FURNACE htt
f four *m* à grille mobile
e horno *m* de parilla móvil
i forno *m* a griglia mobile
n oven met beweegbaar rooster
d Wanderrostofen *m*

5556 TRAVELLING SPRAY BOOTH dec
f cabine *f* mobile de pistolage
e cubículo *m* móvil para pintar con pistola
i cabina *f* mobile di spruzzatura
n verplaatsbare spuitcabine
d fahrbare Spritzkabine *f*

5557 TRAY elc
f caisse *f* de groupement
e caja *f* de agrupamiento
i cassetta *f* per batterie
n kist
d Batterietrog *m*

5558 TRAY pho
f cuvette *f*
e cubeta *f*
i bacinella *f*
n schaal
d Schale *f*

5559 TRAY, dis
PLATE
f plateau *m*
e plato *m*
i piatto *m*
n schotel
d Boden *m*

5560 TRAY CONVEYOR, tra
TROUGH CONVEYOR
f transporteur *m* à auges
e transportador *m* de artesas
i nastro *m* trasportatore a scompartimento, trasportatore *m* a nastro articolato con cassette
n trogtransportband
d Kastenbandförderer *m*, Kastenförderband *n*, Rinnenförderband *n*, Trogförderband *n*

5561 TRAY DRIER, dry
DRYING TRAY
f bac *m* de séchage
e cubeta *f* secadora
i recipiente *m* essiccatore
n droogbak
d Trockenschale *f*

5562 TREATING PLANT fue
f installation *f* de traitement
e equipo *m* de tratamiento
i impianto *m* di trattamento
n raffineerinstallatie
d Raffinationsanlage *f*, Raffinieranlage *f*

5563 TREBLE-FLOOR KILN, bre
THREE-FLOOR KILN
f touraille *f* à trois plateaux
e tostadera *f* de tres platos
i essiccatoio *m* a tre graticci
n eest met drie vloeren
d Dreihordendarre *f*

5564 TREES AND NODULES elc
f arborescences *pl* et nodules *pl*
e arborescencias *pl* y nódulos *pl*
i arborescenze *pl* e noduli *pl*
n bomen *pl* en wratten *pl*
d Bäumchen *pl* und Knopfen *pl*

5565 TRIANGULAR ANODE elc
f anode *f* triangulaire
e ánodo *m* triangular
i anodo *m* triangolare
n driehoekige anode
d dreieckige Anode *f*

5566 TRIANGULAR CRUCIBLE, htt
HESSIAN CRUCIBLE
f creuset *m* triangulaire
e crisol *m* triangular
i crogiolo *m* triangolare
n driehoekige smeltkroes
d dreieckiger Schmelztiegel *m*

5567 TRICKLING FILTER fil
f filtre *m* à écoulement lent, filtre *m* goutte à goutte
e filtro *m* percolador
i filtro *m* a sgocciolamento
n doorzijgfilter *n*, percolator
d Abtropffilter *m*, Tropffilter *m*

5568 TRICKLING FILTER sew
f lit *m* bactérien
e biofiltro *m*, filtro *m* de goteo
i letto *m* di batteri
n bacteriefilter *n*
d Tropfkörper *m*

5569 TRICKLING TOWER cle,dis
f tour *f* à ruissellement
e torre *f* de escurrimiento
i torre *f* ad irrigazione
n sproeitoren
d Rieselturm *m*

5570 TRIEUR bre
f trieur *m* à alvéoles
e máquina *f* seleccionadora, máquina *f* triadora
i macchina *f* selettrice
n trieur
d Trieur *m*

5571 TRITURATOR brk
f triturateur *m*
e trituradora *f*
i mulino *m* di trituramento
n fijnwrijfmachine
d Zerreibmaschine *f*

5572 TRITURATOR, pap
COMMINUTOR, DISINTEGRATOR, SCREEN SHREDDER
f appareil *m* de dilacération, dilacérateur *m*
e trituradora *f*
i frantumatrice *f*, trituratore *m*
n fijnmaalmolen
d Rechengutzerkleinerer *m*, Rechenwolf *m*

5573 TROMMEL, sor
REVOLVING SCREEN
f trieur *m* rotatif, trommel *m*
e criba *f* giratoria de tambor
i vaglio *m* rotativo a tamburo
n zifttrommel
d Sichtetrommel *f*, Sortiertrommel *f*

5574 TROUGH gla
f auge *f*
e artesa *f*
i cassetta *f*
n trog
d Mulde *f*, Trog *m*

5575 TROUGH, min,sep
FLOAT
f auge *f* de sédimentation
e cuba *f* de sedimentación
i truogolo *m* di sedimentazione
n bezinktrog
d Ablagerungsmulde *f*

5576 TROUGH CONVEYOR, tra
TRAY CONVEYOR
f transporteur *m* à auges
e transportador *m* de artesa
i nastro *m* trasportatore a scompartimento, trasportatore *m* a nastro articolato con cassette
n trogtransportband
d Kastenbandförderer *m*, Kastenförderband *n*, Rinnenförderband *n*, Trogförderband *n*

5577 TRUCK CHAMBER KILN, cer,gla,htt
BOGIE KILN (GB), CAR TUNNEL KILN (US), FIXED-FIRE ZONE KILN
f étuve *f* à chariot
e estufa *f* de carrillo
i stufa *f* a carrello
n wagendroogstoof
d Wagentrockenofen *m*

5578 TRUMPET GUIDE, gla
TUNNEL GUIDE, RUBBING CONE, RUBBING LEATHERS
f frotteur *m*
e cono *m* de guía, rotafrotador *m*
i imbuto *m*
n wrijfconussen *pl*
d Nitschelwerk *n*

5579 TUB bre
f bac *m*, cuve *f*
e cuba *f*, tina *f*
i mastello *m*
n kuip
d Bütte *f*

5580 TUBE BANK cpt
f nid *m* de tubes
e nido *m* de tubos
i nido *m* di tubi
n buizenstel *n*
d Röhrensatz *m*

5581 TUBE CLUSTER, boi
TUBE NEST

f faisceau *m* de tubes
e haz *m* de tubos
i fascio *m* di tubi
n pijpenbundel
d Rohrbündel *n*

5582 TUBE FURNACE htt
f four *m* à tubes fermés
e horno *m* de tubos cerrados
i forno *m* a tubi chiusi
n buisoven
d Röhrenofen *m*

5583 TUBE GRINDER, TUBE MILL sin
f broyeur *m* tubulaire
e machacadora *f* tubular
i mulino *m* a tamburo, mulino *m* a tubo, mulino *m* orizzontale
n buismolen
d Rohrmühle *f*

5584 TUBE PASTEURIZER dai
f pasteur *m* à tubes
e pasteurizador *m* de tubos
i pastorizzatore *m* a tubi
n buizenpasteur
d Röhrenpasteurisator *m*

5585 TUBE STILL, PIPE STILL dis
f chaudière *f* tubulaire
e retorta *f* tubular
i forno *m* tubolare
n buisverdamper
d Rohrverdampfer *m*

5586 TUBES-TYPE PLATE elc
f plaques *f* à tubes
e placa *f* de tubos
i piastra *f* a tubi
n buisjesplaat
d Röhrchenplatte *f*

5587 TUBULAR BOILER, FLAME TUBE BOILER, FLUE BOILER boi
f chaudière *f* tubulaire
e caldera *f* de tubo de llama
i caldaia *f* tubolare
n vlampijpketel
d Flammrohrkessel *m*

5588 TUBULAR CONDENSER, CONTACT CONDENSER pre
f condensateur *m* tubulaire
e condensador *m* tubular
i condensatore *m* tubolare
n buiscondensor
d Röhrenverdichter *m*

5589 TUBULAR DIE pla,pre
f filière *f* annulaire
e boquilla *f* anular
i filiera *f* anulare
n cirkelvormige extrusieopening, ringvormige spuitmatrijs
d Ringdüse *f*

5590 TUBULATED RETORT lab
f cornue *f* tubulée
e retorta *f* tubulada
i storta *f* tubolata
n buizenretort
d Röhrenretorte *f*

5591 TUBULATURE BOTTLE, LEVELLING BOTTLE lab
f flacon *m* d'équilibre de niveau
e frasco *m* de equilibrio de nivel
i bottiglia *f* d'equilibrio di livello
n niveauflesje *n*
d Niveauflasche *f*

5592 TUMBLING BARREL (GB), RATTLER (US), RUMBLER (GB) cle
f tonneau *m* de dessablage
e tambor *m* desarenador
i buratto *m* sterratore
n poetstrommel
d Putztrommel *f*, Scheuertrommel

5593 TUMBLING DRUM, BARREL ROLLING DRUM pol
f tonneau *m* de polissage
e tambor *m* de pulido
i tamburo *m* di pulitura
n polijsttrommel
d Poliertrommel *f*

5594 TUMBLING MIXER mix
f tambour *m* mélangeur rotatif
e tambor *m* mezclador giratorio
i tamburo *m* mescolatore rotativo
n roterende mengtrommel
d drehende Mischtrommel *f*

5595 TUN, VAT, VESSEL bre
f cuve *f*
e cuba *f*, tina *f*
i mastello *m*, tino *m*
n kuip, tobbe
d Bottich *m*

5596 TUNGSTEN ELECTRODE elc
f électrode *f* en tungstène
e electrodo *m* de tungsteno
i elettrodo *m* di tungsteno
n wolfraamelektrode
d Wolframelektrode *f*

5597 TUNNEL CAP dis
f calotte-tunnel *m*
e platillo *m* de túnel
i campanella *f* rettangolare
n tunnelklok
d Tunnelglocke *f*

5598 TUNNEL DRIER cer,dry
f sécheur *m* à tunnel
e secador *m* de túnel
i essiccatoio *m* a tunnel
n droogtunnel
d Trockentunnel *m*

5599 TUNNEL KILN cer,htt
f four *m* à tunnel
e horno *m* de túnel
i forno *m* a tunnel
n tunneloven
d Tunnelofen *m*

5600 TURBIDICATOR, ins
TURBIDIMETER, TURBIDITY INDICATOR
f turbidimètre *m*
e turbidímetro *m*
i torbidimetro *m*
n troebelheidsmeter, turbidimeter
d Trübungsmesser *m*, Turbidimeter *n*

5601 TURBIDIMETRIC ANALYZER, ana
NEPHOLOMETRIC ANALYZER
f analyseur *m* néphélométrique
e analizador *m* nefelométrico
i analizzatore *m* nefelometrico
n nefelometrisch analyseerapparaat *n*
d nephelometrisches Analysiergerät *n*

5602 TURBINE AGITATOR, mix
TURBO-AGITATOR
f agitateur *m* à turbine
e turboagitador *m*
i agitatore *m* a turbina
n turboroerwerk *n*
d Schaufelradmischer *m*, Turborührwerk *n*

5603 TURBINE DRIER, dry
TURBO-DRIER
f sécheur *m* à plateaux tournants, turbo-sécheur *m*
e turbosecador *m*, secadero *m* de turbina
i essiccatore *m* a turbina, turboessiccatore *m*
n turbinedroger
d Büttnertrockner *m*, Ringetagentrockner *m*

5604 TURBULENCE MIXER mix
f mélangeur *m* à turbulence
e mezclador *m* de turbulencia
i mescolatore *m* a turbolenza
n wervelmenger
d Wirbelmischer *m*

5605 TURN TABLE, rub
ROTARY PAN TRAY, SWIVEL TABLE
f table *f* rotative
e mesa *f* giratoria
i tavola *f* rotativa
n draaibaar spuitblad *n*
d Schwenktablett *n*

5606 TURNER, bre
TURNING APPARATUS
f baudet *m* mécanique, retourneur *m*
e descargador *m*, volteador *m*
i mescolatore *m*, scaricatore *m* meccanico
n omzetter
d Wender *m*

5607 TURNING BAR, pla
ANGLE BAR
f barre *f* de retournement, tringle *m* de retournement
e barra *f* de rotación
i barra *f* di rotazione, barra *f* per volgere
n omkeerrol
d Wendestange *f*

5608 TURNING BLADE mix
f pale *f* déflectrice
e paleta *f* para removar
i deflettore *m*
n keerschoep
d Wendeschaufel *f*

5609 TWEEZER, tex
BURLING IRON, PICKING TOOL
f outil *m* d'épluchage
e desmotadora *f*, despinzadora *f*
i utensile *m* di spuntatura
n nopijzer *n*
d Noppeisen *n*

5610 TWEEZERS lab
f brucelles *pl*, précelles *pl*
e bruselas *pl*, tenacillas *pl*
i pinzetta *f*
n pincet
d Pinzette *f*

5611 TWIG TOWER ana
f bâtiment *m* de graduation saline
e torre *f* de graduación salina
i stabilimento *m* graduatorio
n gradeerhuis *n*
d Gradierwerk *n*

5612 TWIN PUMP pre
f pompe *f* à plongeurs jumelés, pompe *f* jumelle
e bomba *f* de dos pistones
i pompa *f* a due pistoni
n dubbelzuigerpomp
d Zwillingspumpe *f*

5613 TWIN-SCREW EXTRUDER MACHINE pre
f boudineuse *f* à vis jumelée
e budinadora *f* de dos tornillos
i trafila *f* a doppia vite
n tweewormige spuitmachine
d Doppelschneckenspritzmaschine *f*

5614 TWIN-SHELL DRY BLENDER dry,mix
f mélangeur-sécheur à double chemise
e mezclador-secador *m* de doble camisa
i mescolatore *m* essiccatore a doppia camicia
n droogmenger met dubbele mantel
d Doppeltrommeltrockenmischer *m*

5615 TWO-ARM KNEADER mix
f pétrisseur *m* à deux bras
e amasadora *f* de dos brazos
i impastatrice *f* a due braccia
n dubbelarmige kneder
d zweiarmiger Kneter *m*

5616 TWO-CYLINDER DRIER bre,brk
f séchoir *m* à deux cylindres
e secador *m* de dos cilindros
i essiccatoio *m* a due cilindri
n droger met twee walsen
d Zweiwalzentrockner *m*

5617 TWO-DAYLIGHT PRESS pre
f presse *f* à deux plateaux
e prensa *f* de dos platos
i pressa *f* a due aperture
n twee-etagepers
d Zweietagenpresse *f*

5618 TWO-FLOOR KILN, DOUBLE-FLOOR KILN bre
f touraille *f* à deux plateaux
e tostadero *m* de dos platos
i essiccatoio *m* a due griglie
n tweevloereneest
d Zweihordendarre *f*

5619 TWO-FLUID CELL elc
f pile *f* à deux liquides
e pila *f* de dos líquidos
i pila *f* a due elettroliti
n element *n* met verschillende elektrolyten
d Volta-Element *n*

5620 TWO-REGION REACTOR nue
f réacteur *m* à deux zones
e reactor *m* de dos zonas
i reattore *m* a due regioni
n tweezonenreactor
d Zweizonenreaktor *m*

5621 TWO-ROLL MILL mix
f malaxeur *m* à deux cylindres
e mezclador *m* de dos cilindros
i mescolatore *m* a due cilindri
n tweewalsenmenger
d Zweiwalzenmischer *m*

5622 TWO-STAGE COMPRESSOR pre
f compresseur *m* à deux étages
e compresor *m* de dos etapas
i compressore *m* a due stadi
n tweetrapscompressor
d Zweistufenkompressor *m*

5623 TWO-STAGE FILTER bre,fil
DOUBLE FILTER
f double-filtre *m*
e filtro *m* doble
i filtro *m* doppio
n dubbelfilter *n*
d Doppelfilter *m*

5624 TWO-VESSEL BREWHOUSE bre
SIMPLE BREWHOUSE OUTFIT
f salle *f* de brassage simple
e sala *f* de cocimiento simple
i sala *f* di cottura semplice
n enkel brouwhuis *n*
d einfaches Sudwerk *n*, Einfachsudhaus *n*

5625 TWO-WAY COCK cpt
f robinet *m* à deux voies
e grifo *m* de dos vias
i rubinetto *m* a due vie
n tweewegkraan
d Zweiwegehahn *m*

5626 TWO-WAY VALVE cpt
f soupape *f* à deux voies
e válvula *f* de doble paso
i valvola *f* commutatrice
n wisselklep
d Wechselventil *n*

U

5627 U-TYPE FURNACE, HAIR-PIN FURNACE cer
f four *m* à épingle à cheveux
e horno *m* de herradura
i forno *m* a forcella
n omloopoven
d Umkehrofen *m*

5628 ULTRACENTRIFUGE, HIGH-SPEED CENTRIFUGE cen
f centrifugeur *m* à vitesse élevée, ultracentrifugeur *m*
e centrífuga *f* de alta velocidad, supercentrífuga *f*, ultracentrífuga *f*
i centrifuga *f* ad alta velocità, ultracentrifuga *f*
n snelle centrifuge, ultracentrifuge
d Schnellschleuder *f*, Schnellzentrifuge *f*, Ultrazentrifuge *f*

5629 ULTRAFILTER fil
f ultrafiltre *m*
e ultrafiltro *m*
i ultrafiltro *m*
n ultrafilter *n*
d Ultrafilter *m*

5630 ULTRAMICROPIPETTE (GB), LAMBDA PIPETTE (US) lab
f ultramicropipette *f*
e ultramicropipeta *f*
i ultramicropipetta *f*
n microlitertransferpipet
d Mikrolitertransferpipette *f*

5631 ULTRAMICROSCOPE ins
f ultramicroscope *m*
e ultramicroscopio *m*
i ultramicroscopio *m*
n ultramicroscoop
d Ultramikroskop *n*

5632 ULTRASONIC INTERFEROMETER ins
f interféromètre *m* ultrasonore
e interferómetro *m* supersónico, interferómetro *m* ultrasónico
i interferometro *m* ultrasonico
n ultrasone interferometer
d Ultraschallinterferometer *n*

5633 ULTRASONIC PRECIPITATOR fil
f dépoussiéreur *m* ultrasonore, filtre *m* ultrasonore
e filtro *m* supersónico, precipitador *m* supersónico
i filtro *m* ultrasonico
n ultrasoon filter *n*
d Ultraschallfilter *m*

5634 ULTRASONIC WELDING APPARATUS pla,wel
f appareil *m* de soudure ultrasonore
e aparato *m* de soldadura supersónica, aparato *m* de soldadura ultrasónica
i apparecchio *m* di saldatura ultrasonica
n ultrasoon lasapparaat *n*
d Ultraschallschweissgerät *n*

5635 ULTRAVIOLET MICROSCOPE ins
f microscope *m* à rayonnement ultraviolet
e microscopio *m* de radiación ultravioleta
i microscopio *m* a radiazione ultravioletta
n ultravioletmicroscoop
d Ultraviolettmikroskop *n*

5636 UNDERBACK, GRANT bre
f bac *m* reverdoir, reverdoir *m*
e recipiente *m* colector
i recipiente *m* di raccolta, vasca *f* di raccolta
n lekgoot
d Läutergrand *m*, Läuterrinne *f*

5637 UNDERBACK, UNDERLET, GRANT bre
f batterie *f* de filtration
e batería *f* de filtración
i batteria *f* di chiarificazione
n serie klaringskuipen
d Läuterbatterie *f*, Würzegrand *m*

5638 UNDERCLOTH, BACK CLOTH, RUNNER CLOTH mat,pla
f bande *f* ensouple, feuille *f* ensouple
e correa *f* acompañante
i nastro *m* d'accompagnamento
n meelopende baan, meelopende foelie
d Mitläuferband *n*, Mitläuferfolie *f*

5639 UNDERCOOLING VESSEL, SUPERCOOLING VESSEL — clo
f récipient *m* de surfusion
e recipiente *m* de sobrefusión
i recipiente *m* di surraffreddamento
n onderkoelingsvat *n*
d Unterkühlungsgefäss *n*

5640 UNDERDRAIN — sew
f canal *m* d'écoulement souterrain
e canal *m* de desagüe subterráneo
i condotto *m* di spurgo sotterraneo
n ondergronds afvloeikanaal *n*
d Dohle *f*, unterirdischer Abzugskanal *m*

5641 UNDERDRAINED SETTLING BASIN — iwt
f bassin *m* d'égouttage
e artesa *f* de avenamiento subterráneo
i bacino *m* drenato
n sijpelbekken *n*
d Sickerbecken *n*

5642 UNDERFEED FUEL BED — htt
f lit *m* de combustion à alimentation d'en bas
e lecho *m* de combustión alimentado por la parte baja
i letto *m* di combustione con alimentazione dal basso
n verbrandingslaag met ondervuur
d Verbrennungsschicht *f* mit Unterfeuerung

5643 UNHAIRING BATH — tan
f bain *m* de dépilage, bain *m* d'épilage
e baño *m* de apelambrado
i bagno *m* di depilazione
n ontharingsbad *n*
d Enthaarungsbad *n*

5644 UNHAIRING MACHINE — tan
f machine *f* de dépilage, machine *f* d'épilage
e máquina *f* de apelambrado
i macchina *f* di depilazione
n ontharingsmachine
d Enthaarungsmaschine *f*

5645 UNHAIRING TANK, DEPILATION TANK — lea
f récipient *m* de dépilage, récipient *m* d'épilage
e tanque *m* de depilación
i serbatoio *m* di depilazione
n blootvat *n*, ontharingskuip
d Enthaarungsgefäss *n*

5646 UNIMOLECULAR LAYER, MOLECULAR LAYER, MONOMOLECULAR LAYER — coa
f couche *f* monomoléculaire
e capa *f* monomolecular
i strato *m* monomolecolare
n monomoleculaire laag
d Monomolekularschicht *f*

5647 UNIT MOLD (US), UNIT MOULD (GB) — pla,pre
f élément *m* rapporté, pièce *f* complémentaire
e molde *m* de un juego normalizado, molde *m* intercambiable
i matrice *f* intercambiabile
n onderdeel *n* van moedermatrijs
d Einheitsstammformeinsatz *m*, Formeinheit *f*

5648 UNIVERSAL BURNER, UNIVERSAL COMBUSTION BURNER — htt,lab
f brûleur *m* universel
e quemador *m* universal
i bruciatore *m* universale
n universele brander
d Allzweckbrenner *m*, Universalbrenner *m*

5649 UNIVERSAL ELECTROPLATING BATH — elc
f bain *m* de galvanoplastie universel
e baño *m* de galvanoplastia universal
i bagno *m* di galvanoplastica universale
n universeel galvanostegiebad *n*
d Universalelektroplattierungsbad *n*, Universalgalvanostegiebad *n*

5650 UNIVERSAL TESTING MACHINE — tes
f machine *f* d'essais universelle
e máquina *f* de pruebas universal
i macchina *f* di prove universale
n universele beproevingsmachine
d Universalprüfmaschine *f*

5651 UNLOADER, UNLOADING DEVICE — tra
f déchargeur *m*
e descargador *m*
i scaricatore *m*
n ontlader
d Entladevorrichtung *f*

5652 UNLOADING VALVE, DISCHARGE VALVE, OUTLET VALVE — cpt

f soupape *f* de décharge
e válvula *f* de descarga
i valvola *f* di scarico
n uitlaatklep
d Ablaufventil *n*, Auslassventil *n*

5653 UNSHIELDED METAL ARC WELDING PLANT wel
f installation *f* de soudure à arc non-protégé
e equipo *m* de soldadura con arco descubierto
i impianto *m* di saldatura ad arco nudo
n installatie voor onbeschermd booglassen
d Anlage *f* für ungeschütztes Lichtbogenschweissen

5654 UNWRAPPED CONSTRUCTION, elc
NON-LINED CONSTRUCTION
f montage *m* sans habillage
e montaje *m* sin recubrimiento
i montaggio *m* senza rivestimento
n constructie zonder voering
d wickellose Bauweise *f*

5655 UP-STROKE MULTI-DAYLIGHT PRESS pre,rub
f presse *f* à intervalles multiples à piston inférieur
e prensa *f* de platos múltiples de carrera ascendente
i pressa *f* a molti piani di corsa ascendente
n meervoudige etagepers met opgaande slag
d Mehretagenpresse *f* mit Aufwärtshub

5656 UP-STROKE PRESS, pla
BOTTOM RAM PRESS
f presse *f* ascendante, presse *f* à piston ascendant
e prensa *f* de plato inferior móvil
i pressa *f* ascendente
n pers met onderplunjer
d Unterdruckpresse *f*, Unterkolbenpresse *f*

5657 UPPER MOLD SECTION (US), pre,rub
UPPER MOULD SECTION (GB)
f coquille *f* supérieure, demi-moule *m* supérieur
e parte *f* superior del molde
i parte *f* superiore dello stampo
n bovenmatrijs
d obere Formhälfte *f*

5658 UPPER PLATEN pla,pre
f plateau *m* supérieur
e mesa *f* superior, plato *m* superior
i piano *m* superiore
n bovenste persplaat
d oberer Pressentisch *m*

5659 URANIUM FUELLED REACTOR, nue
URANIUM REACTOR
f réacteur *m* à uranium
e reactor *m* de uranio
i reattore *m* ad uranio
n uraniumreactor
d Uranreaktor *m*

5660 URANIUM-THORIUM REACTOR nue
f réacteur *m* à uranium-thorium
e reactor *m* de uranio-torio
i reattore *m* ad uranio-torio
n uranium-thoriumreactor
d Uran-Thoriumreaktor *m*

5661 UREAMETER, ins,mcy
UREOMETER
f uréomètre *m*
e ureómetro *m*
i ureometro *m*
n ureometer
d Ureometer *n*

5662 URINOMETER ins,mcy
f pèse-urine *f*, urinomètre *m*
e urinómetro *m*
i orinometro *m*
n urinemeter
d Urinometer *n*, Urinsenkwaage *f*

V

5663 VACUUM AUGMENTER, pre
VACUUM BOOSTER
f éjecteur *m* d'air renforçateur du vide
e vacuomultiplicador *m* de fuerza, vacuorreforzador *m*
i eiettore *m* d'aria ausiliario
n ejecteur ter verhoging van het vacuüm
d Vakuumzusatzpumpe *f*

5664 VACUUM BREAKER sap
f soupape *f* de retenue sous vide
e válvula *f* de alivio, válvula *f* vacuorreguladora
i valvola *f* di ritegno sotto vuoto
n terugslagklep tegen vacuüm
d Rückschlagventil *n* gegen Vakuum

5665 VACUUM CARBON TRAIN tes
f appareil *m* de détermination de la teneur en carbone sous vide
e aparato *m* de determinación del contenido de carbono en el vacío
i apparecchio *m* di determinazione del tenore di carbonio sotto vuoto
n koolstofgehaltebepalings-apparaat *n* in vacuüm
d Kohlenstoffgehaltsbestimmungs-gerät *n* in Vakuum

5666 VACUUM CASTING APPARATUS fou,pla
f appareil *m* de coulée sous vide
e aparato *m* de colada en vacío
i apparecchio *m* di colata sotto vuoto
n apparaat *n* voor het gieten in vacuüm
d Vakuumgiessapparat *m*

5667 VACUUM COATING APPARATUS, coa
VACUUM METALLIZING APPARATUS
f appareil *m* de métallisation sous vide
e aparato *m* de vacuometalización
i apparecchio *m* di metallizzazione sotto vuoto
n vacuümmetalliseerapparaat *n*
d Vakuummetallisiergerät *n*

5668 VACUUM CONVEYOR tra
f transporteur *m* pneumatique
e transportador *m* neumático
i trasportatore *m* pneumatico
n door zuiglucht voortbewogen transportband
d Saugluftförderer *m*, Unterdruck-förderer *m*

5669 VACUUM CRYSTALLIZER cry
f cristalloir *m* à vide
e cristalizador *m* de vacío
i cristallizzatore *m* a vuoto
n vacuümkristallisator
d Vakuumkristallisator *m*

5670 VACUUM DE-ICER tra
f dégivreur *m* pneumatique
e descongelador *m* neumático
i antighiaccio *m* pneumatico
n pneumatische ijsbestrijdings-installatie
d Saugluftenteiser *m*

5671 VACUUM DESICCATOR, dry,lab
VACUUM ESSICCATOR
f dessiccateur *m* sous vide
e vacuodesecador *m*
i essiccatore *m* sotto vuoto
n vacuümexsiccator
d Vakuumexsikkator *m*

5672 VACUUM DIFFUSION PUMP vac
f pompe *f* à diffusion sous vide
e bomba *f* difusora por vacío
i pompa *f* a diffusione sotto vuoto
n vacuümdiffusiepomp
d Vakuumdiffusionspumpe *f*

5673 VACUUM DISTILLATION APPARATUS, dis
VACUUM STILL
f appareil *m* de distillation sous vide
e aparato *m* de destilación en vacío
i apparecchio *m* di distillazione sotto vuoto
n vacuümdestileerapparaat *n*
d Vakuumdestillationsgerät *n*

5674 VACUUM DRIER dry
f séchoir *m* à vide
e secador *m* al vacío, vacuosecador *m*
i essiccatore *m* sotto vuoto
n vacuümdroger
d Vakuumtrockner *m*

5675 VACUUM DRYING OVEN dry
f four *m* sécheur à vide
e horno *m* secadero de vacío
i forno *m* essiccatore a vuoto
n vacuümdroogoven
d Vakuumtrockenofen *m*

5676 VACUUM EVAPORATOR eva
f évaporateur *m* sous vide
e evaporador *m* de vacío
i evaporatore *m* sotto vuoto
n vacuümverdamper
d Unterdruckverdampfer *m*, Vakuumverdampfer *m*

5677 VACUUM FERMENTATION VAT fmi
f récipient *m* de fermentation sous vide
e recipiente *m* de fermentación en vacío
i recipiente *m* di fermentazione sotto vuoto
n vacuümfermenteervat *n*
d Unterdruckgärungsgefäss *n*

5678 VACUUM FILTER, SUCTION FILTER fil
f filtre *m* à vide
e filtro *m* de vacío
i filtro *m* a vuoto
n vacuümfilter *n*
d Vakuumfilter *m*

5679 VACUUM FIRING OVEN cer
f four *m* de cuisson à vide
e horno *m* de cocción de vacío
i forno *m* di cottura a vuoto
n vacuümbakoven
d Vakuumbrennofen *m*

5680 VACUUM FORMING MOLD (US), VACUUM FORMING MOULD (GB) pla,pre
f moule *m* pour la formation sous vide
e molde *m* para la formación en vacío
i stampo *m* di formazione sotto vuoto
n vacuümvorm
d Saugform *f*

5681 VACUUM FUSING FURNACE, VACUUM MELTING FURNACE htt,met
f four *m* de fusion sous vide
e horno *m* de vacuofusión
i forno *m* di fusione sotto vuoto
n vacuümsmeltoven
d Vakuumschmelzofen *m*

5682 VACUUM GAGE (US), VACUUM GAUGE (GB), VACUUM MANOMETER ins
f vacuomètre *m*
e aforo *m* de vacío, vacuómetro *m*
i vuotometro *m*
n vacuümdrukmeter
d Unterdruckmesser *m*, Vakuummesser *m*

5683 VACUUM GRATING SPECTROGRAPH ins,spe
f spectrographe *m* à réseau sous vide
e espectrógrafo *m* de rejilla en vacío
i spettrografo *m* a reticolo sotto vuoto
n vacuümtraliespectrograaf
d Vakuumgitterspektrograph *m*

5684 VACUUM HEAT TREATMENT OVEN htt,met
f four *m* de traitement thermique dans le vide
e horno *m* de vacuotermotratamiento
i forno *m* di trattamento termico nel vuoto
n vacuümoven voor warmtebehandeling
d Vakuumofen *m* zur Warmbehandlung

5685 VACUUM HEATING PLANT htt,met
f installation *f* de chauffage sous vide
e equipo *m* de calentamiento en vacío
i impianto *m* di trattamento sotto vuoto
n vacuümverhittingsinstallatie
d Vakuumheizungsanlage *f*

5686 VACUUM IMPREGNATION APPARATUS imp
f appareil *m* d'imprégnation sous vide
e aparato *m* de vacuoimpregnación
i apparecchio *m* d'impregnazione sotto vuoto
n vacuümimpregneerapparaat *n*
d Vakuumimprägniergerät *n*

5687 VACUUM JACKET cl
f chemise *f* à vide

e camisa *f* de vacío
i camicia *f* a vuoto
n vacuümmantel
d Vakuummantel *m*

5688 VACUUM MIXER cer,mix
f mélangeur *m* sous vide
e mezclador *m* en vacío
i mescolatore *m* sotto vuoto
n vacuümmenger
d Vakuummischer *m*

5689 VACUUM OIL PUMP vac
f pompe *f* à vide à l'huile
e bomba *f* de vacío al aceite
i pompa *f* a vuoto all'olio
n olievacuümpomp
d Ölvakuumpumpe *f*

5690 VACUUM OIL REFINING PLANT fat
f installation *f* d'épuration d'huile dans le vide
e equipo *m* de depuración de aceite en el vacío
i impianto *m* di depurazione dell'olio nel vuoto
n vacuümolieraffinaderij
d Vakuumölraffinerie *f*

5691 VACUUM OVEN htt
f four *m* à vide
e horno *m* al vacío, vacuohorno *m*
i forno *m* a vuoto
n vacuümoven
d Vakuumofen *m*

5692 VACUUM PAN nut,sin
f vaisseau *m* de jus concentré sous vide
e tacho *m* de vacío
i caldaia *f* di succo concentrato sotto vuoto
n diksapvacuümpan
d Dicksaftvakuumpfanne *f*, Vakuumkörper *m* für Erstprodukt

5693 VACUUM-PAN HOUSE nut,sin
f salle *f* de vaisseaux de jus concentré
e casa *f* de tachos de vacío
i sala *f* di caldaie di succo concentrato
n diksappankamer
d Kochstation *f*

5694 VACUUM PUG cer,pre
f pétrin *m* à vide
e amasadora *f* de vacío
i impastatrice *f* a vuoto
n vacuümkneder
d Vakuumkneter *m*

5695 VACUUM PUMP vac
f pompe *f* à vide
e bomba *f* de vacío
i pompa *f* a vuoto
n vacuümpomp
d Vakuumpumpe *f*

5696 VACUUM RECEIVER gen
f récipient *m* à vide
e recipiente *m* de vacío
i recipiente *m* a vuoto
n vacuümketel
d Vakuumkessel *m*

5697 VACUUM REFRIGERATOR clo
f réfrigérateur *m* à vide
e refrigerador *m* de vacío
i frigorifero *m* a vuoto, refrigeratore *m* a vuoto
n vacuümkoelkast
d Vakuumkühlschrank *m*

5698 VACUUM SEALING DOOR hyg
f porte *f* à fermeture étanche
e puerta *f* de cierre hermético
i porta *f* a chiusura ermetica
n hermetisch sluitende deur
d vakuumdichte Tür *f* einer Klimakammer

5699 VACUUM SHELF DRIER dry
f sécheur *m* à plateaux à vide
e secador *m* de platos de vacío
i essiccatore *m* a piani a vuoto
n vacuümetagedroger
d Vakuumetagentrockner *m*

5700 VACUUM SINTERING OVEN htt,met
f four *m* de frittage sous vide
e horno *m* de vacuosinterización
i forno *m* di sinterizzazione sotto vuoto
n vacuümsinteroven
d Vakuumsinterofen *n*

5701 VACUUM SPECTROGRAPH ins,spe
f spectrographe *m* à vide
e espectrógrafo *m* de vacío
i spettrografo *m* a vuoto
n vacuümspectrograaf
d Vakuumspektrograph *m*

5702 VACUUM SPUTTERING DEVICE coa
f appareil *m* de pulvérisation cathodique à vide

e aparato *m* de pulverización catódica de vacío
i apparecchio *m* di polverizzazione catodica a vuoto
n vacuümkatodeverstuivingsapparaat *n*
d Vakuumkatodenzerstäubungsgerät *n*

5703 VACUUM SUBLIMATION APPARATUS dis,lab
f appareil *m* à sublimation sous vide
e aparato *m* de sublimación en vacío
i apparecchio *m* di sublimazione sotto vuoto
n vacuümsublimeerapparaat *n*
d Vakuumsublimierapparat *m*

5704 VACUUM TAP vac
f robinet *m* à vide
e grifo *m* de vacío
i rubinetto *m* a vuoto
n vacuümkraan
d Vakuumhahn *m*

5705 VACUUM TRAP lab,vac
f piège *m* à vide
e purgador *m* de vacío, trampa *f* al vacío
i trappola *f* a vuoto
n vacuümsluis
d Kühlfalle *f*, Vakuumschleuse *f*

5706 VALVE cpt,gen
f clapet *m*, soupape *f*
e válvula *f*
i valvola *f*
n klep
d Ventil *n*

5707 VALVE BALL cpt
f boulet *m* de soupape
e bola *f* de válvula
i sfera *f* di valvola
n klepkogel
d Ventilkugel *f*

5708 VALVE BODY, VALVE HOUSING cpt
f corps *m* de soupape
e cuerpo *m* de válvula
i corpo *m* di valvola
n klephuis *n*
d Ventilgehäuse *n*

5709 VALVE BOX, VALVE CHEST boi,cpt
f boîte *f* de soupape
e caja *f* de válvula, cámara *f* de válvula
i camera *f* di valvola, scatola *f* di valvola
n klepkast
d Ventilkasten *m*

5710 VALVE DISK, VALVE HEAD cpt
f tête *f* de soupape
e disco *m* de válvula
i disco *m* di valvola
n klepkop
d Ventilteller *m*

5711 VALVE FACE cpt
f plaque *f* de soupape
e placa *f* de válvula
i piastra *f* di valvola
n klepvlak *n*
d Ventilplatte *f*

5712 VALVE NEEDLE cpt
f pointeau *m* obturateur
e aguja *f* de válvula
i spillo *m* di valvola
n klepnaald
d Ventilnadel *f*

5713 VALVE ROCKER cpt
f levier *m* de soupape
e balancín *m* de válvula
i alzavalvole *m*
n klephefboom
d Ventilhebel *m*

5714 VALVE SEATING, SEAT, SEATING cpt
f siège *m* de soupape
e asiento *m* de válvula
i seda *f* di valvola
n klepzitting
d Ventilsitz *m*

5715 VALVE SPRING cpt
f ressort *m* de soupape
e muelle *m* de válvula
i molla *f* di valvola
n klepveer
d Ventilfeder *f*

5716 VALVE STEM cpt
f tige *f* de soupape
e vástago *m* de válvula
i stelo *m* di valvola
n klepstang, klepsteel
d Ventilspindel *m*

5717 VALVE TRAY dis
f plateau *m* à clapets

e plato *m* de válvulas
i piatto *m* a valvole
n kleppenschotel
d Ventilboden *m*

5718 VANE FEEDER, cer
ROTARY VANE FEEDER
f doseur *m* à secteur
e alimentador *m* giratorio de aletas
i alimentatore *m* a settore
n draaiende-schoepenvoeder
d Drehschaufelspeiser *m*

5719 VANE PUMP, pre
SEMI-ROTARY PUMP
f pompe *f* rotative à ailettes
e bomba *f* giratoria de aletas
i pompa *f* rotativa ad alette
n vleugelpomp
d Allweilerpumpe *f*, Flügelpumpe *f*

5720 VANE-TYPE DRAFT GAGE (US), ins
VANE-TYPE DRAUGHT GAUGE (GB)
f indicateur *m* de tirage à palette
e indicador *m* del tiro de paleta
i indicatore *m* di tiraggio a paletta
n vaantrekmeter
d Flügelzugmesser *m*

5721 VANNER min,sor
f vanneur *m*
e vibroseparador *m*
i vagliatore *m*
n zeeftrommelbak
d Schwingschaufel *f*

5722 VAPOR BATH (US), htt,tex
STEAM BATH (GB)
f bain *m* à vapeur
e baño *m* de vapor
i bagno *m* a vapore
n stoombad *n*
d Dampfbad *n*

5723 VAPOR BLASTING PLANT (US), pol
VAPOUR BLASTING PLANT (GB), LIQUID HONING PLANT
f installation *f* de honage au jet de vapeur
e equipo *m* de rectificación al chorro de vapor
i impianto *m* di rettificatura al getto di vapore
n hooninstallatie met dampstraal
d Honanlage *f* mit Dampfstrahl

5724 VAPOR CONDENSER (US), bre
VAPOUR CONDENSER (GB)
f condensateur *m* de buées
e condensador *m* de vahos
i condensatore *m* di fumana
n dampencondensor
d Schwadendunstkondensator *m*, Wrasendunstkondensator *m*

5725 VAPOR DEGREASING TANK (US), cle
VAPOUR DEGREASING TANK (GB)
f cuve *f* de dégraissage à vapeur de solvant
e cuba *f* desengrasadora por vapor de solvente
i tino *m* di sgrassaggio mediante vapore di solvente
n ontvettingstank met verdampt oplosmiddel
d Entfettungsgefäss *n* mit verflüchtigtem Lösungsmittel

5726 VAPOR DIVIDING FRACTIONING HEAD (US), dis,lab
VAPOUR DIVIDING FRACTIONING HEAD (GB)
f tête *f* à fractionner avec réglage en phase vapeur
e cabezal *m* de fracciones de regulación en fase de vapor
i testa *f* da frazione a regolazione in fase di vapore
n kolomkop met dampverdeler
d Kolonnenkopf *m* mit Dampfverteiler

5727 VAPOR JET PUMP (US), vac
VAPOUR JET PUMP (GB)
f pompe *f* à jet de vapeur
e bomba *f* de chorro de vapor
i pompa *f* a getto di vapore
n dampstraalpomp
d Dampfstrahlpumpe *f*

5728 VAPOR-LIQUID EQUILIBRIUM STILL (US), dis
VAPOUR-LIQUID EQUILIBRIUM STILL (GB)
f appareil *m* de distillation à équilibre liquide-vapeur
e aparato *m* de destilación con equilibrio líquido-vapor
i apparecchio *m* di distillazione con equilibrio liquido-vapore
n destilleerapparaat *n* met vloeistof-dampevenwicht
d Destillationsapparat *m* mit Dampf-Flüssigkeitgleichgewicht

5729 VAPOR LOCK (US), pre
VAPOUR LOCK (GB)

f barrage *m* en bulles de vapeur
e bolsa *f* de vapor, obstrucción *f* por burbujas
i tampone *m* di vapore
n stoombellenversperring
d Blasensperre *f*

5730 VAPOR-PHASE CHLORINATION TANK (US), VAPOUR-PHASE CHLORINATION TANK (GB) fue
f récipient *m* de chloruration en phase vapeur
e tanque *m* de cloruración en fase vapor
i serbatoio *m* di clorurazione in fase vapore
n dampfazechloreertank
d Dampfphasenchloriergefäss *n*

5731 VAPOR-PHASE CRACKING PLANT (US), VAPOUR-PHASE CRACKING PLANT (GB) fue
f installation *f* de craquage en phase vapeur
e equipo *m* de craqueo en fase vapor, equipo *m* de pirodesintegración en fase vapor
i impianto *m* di piroscissione in fase vapore
n dampfazekraakinstallatie
d Dampfphasenkrackanlage *f*

5732 VAPOR-PHASE DEPOSITION APPARATUS (US), VAPOUR-PHASE DEPOSITION APPARATUS (GB) coa
f appareil *m* de déposition de la phase de vapeur
e aparato *m* de deposición de la fase de vapor
i apparecchio *m* di deposizione della fase di vapore
n apparaat *n* voor neerslaan uit de dampfaze
d Apparat *m* zum Niederschlagen aus der Dampfphase

5733 VAPOR-PHASE HYDROGENATION PLANT (US), VAPOUR-PHASE HYDROGENATION PLANT (GB) fue
f installation *f* d'hydrogénation en phase vapeur
e equipo *m* de hidrogenación en fase vapor
i impianto *m* d'idrogenazione in fase vapore
n dampfazehydreerinstallatie
d Dampfphasenhydrieranlage *f*

5734 VAPORIMETER ins
f vaporimètre *m*
e vaporímetro *m*
i vaporimetro *m*
n dampspanningsmeter, vaporimeter
d Dampfspannungsmesser *m*, Vaporimeter *n*

5735 VAPORIZER eva,pre
f vaporisateur *m*
e vaporizador *m*
i spruzzatore *m*, vaporizzatore *m*
n vaporisator, verstuiver
d Evaporator *m*, Verdampfer *m*, Zerstäuber *m*

5736 VAPORIZER, EVAPORATOR dry,eva
f évaporateur *m*
e desecador *m*, evaporador *m*, vaporizador *m*
i evaporatore *m*
n verdamper, verdamplichaam *n*
d Verdampfer *m*

5737 VARIABLE DELIVERING PUMP pre
f pompe *f* à débit variable
e bomba *f* de gasto variable
i pompa *f* a portata variabile
n perspomp met variabele opbrengst
d Pumpe *f* mit variabler Förderleistung

5738 VARNISH FACTORY, LACQUER FACTORY dec
f fabrique *f* de laque, fabrique *f* de vernis
e fábrica *f* de barniz, fábrica *f* de laca
i fabbrica *f* di lacca, fabbrica *f* di vernice
n lakfabriek, vernisfabriek
d Firnisfabrik *f*, Lackfabrik *f*

5739 VARNISHING MACHINE dec
f machine *f* à laquer, machine *f* à vernir
e máquina *f* para barnizar
i macchina *f* per verniciare
n lakmachine, vernismachine
d Lackiermaschine *f*

5740 VARNISHING OVEN dec
f four *m* de laquage, four *m* de vernissage
e horno *m* de barnizado
i forno *m* da verniciare
n lakoven
d Lackierofen *m*

5741 VARNISHING TABLE dec
f table *f* de vernissage
e mesa *f* de barnizar
i tavola *f* per verniciare
n laktafel
d Lackiertisch *m*

5742 VAT tex
f cuve *f*
e cuba *f*
i tino *m*
n vat *n*
d Küpe *f*

5743 VAT, TUN, VESSEL bre
f cuve *f*
e cuba *f*, tina *f*
i mastello *m*, tino *m*
n kuip, tobbe
d Bottich *m*

5744 VEGETABLE TANNING VESSEL tan
f cuve *f* de tannage végétal
e cuba *f* de curtido vegetal
i tino *m* di concia vegetale
n vat *n* voor plantaardige looistoffen
d Gefäss *n* für pflänzliche Gerbstoffe

5745 VELVETING PLANT rub
f installation *f* à velouter, installation *f* de suédage
e equipo *m* para aterciopelar
i impianto *m* per vellutare
n velveteerinstallatie
d Velourisierungsanlage *f*

5746 VENT fon,pla
f trou *m* d'évent
e agujero *m* de aire
i apertura *f* di sfogo, sfogatoio *m*, sfogo *m* per l'aria
n ontluchtingskanaal *n*
d Entlüftungsbohrung *f*, Entlüftungskanal *m*

5747 VENT COCK, AIR RELIEF COCK sap
f robinet *m* d'évacuation d'air
e grifo *m* de escape de aire
i rubinetto *m* di sfogo
n ontluchtingskraan
d Entlüftungshahn *m*

5748 VENT PIPE, EXHAUST PIPE, OUTLET VENT cpt
f tuyau *m* d'évent
e tubo *m* de escape
i sfiatoio *m*, tubo *m* di sfiato
n afblaaspijp
d Abblasrohr *n*, Dunstrohr *n*

5749 VENTILATOR, BLOWER, FAN clo
f ventilateur *m*
e ventilador *m*
i ventilatore *m*
n ventilator
d Lüfter *m*, Ventilator *m*

5750 VERTICAL BOILER boi
f chaudière *f* verticale
e caldera *f* vertical
i caldaia *f* verticale
n staande ketel
d stehender Kessel *m*

5751 VERTICAL CROSS-TUBE BOILER boi
f chaudière *f* verticale à bouilleurs transversaux
e caldera *f* vertical de hervidores transversales
i caldaia *f* verticale con bollitori trasversali
n staande ketel met dwarse vlampijpen
d stehender Flammrohrofen *m* mit Quersiedern

5752 VERTICAL FLASH RING, FLASH RING pla
f gorge *f* circulaire
e anillo *m* de rebaba, ranura *f* de rebaba
i bava *f* verticale
n afknijpring
d Abquetschring *m*

5753 VERTICAL GAS FLOWMETER ins
f débitmètre *m* à gaz vertical
e flujómetro *m* de gas vertical
i flussometro *m* a gas verticale
n verticale gasstromingsmeter
d aufrechter Gasströmungsmesser *m*

5754 VERTICAL IMMERSION PUMP, VERTICAL WET-PIT PUMP pre
f pompe *f* plongeante verticale
e bomba *f* de inmersión vertical
i pompa *f* d'immersione verticale
n staande onderwaterpomp
d stehende Unterwasserpumpe *f*, Tauchpumpe *f*

5755 VERTICAL KILN bre
f touraille *f* verticale

e tostadero *m* vertical
i essiccatoio *m* verticale
n verticale eest
d Vertikaldarre *f*

5756 VERTICAL RECOVERY BEND, dis,lab
DISTILLING CONNECTING TUBE
f coude *m* à prises parallèles
e codo *m* de tomas paralelas
i gomito *m* a prese parallele
n destillatieoverloop
d Destillierbrücke *f*

5757 VERTICAL RETORT met
f cornue *f* verticale
e retorta *f* vertical
i storta *f* verticale
n verticale retort
d Vertikalretorte *f*

5758 VERTICAL RETORT OVEN met
f four *m* à cornues verticales
e horno *m* de retortas verticales
i forno *m* a storte verticali
n verticale retortenoven
d Vertikalretortenofen *m*

5759 VERTICAL SCREW MIXER mix
f mélangeur *m* à hélice verticale
e mezclador *m* de hélice vertical
i mescolatore *m* ad elica verticale
n verticale-schroefmenger
d senkrechter Schraubenmischer *m*

5760 VERTICAL TANK, pho
TANK BATH
f bain *m* mort, cuve *f* verticale
e tanque *m* de revelado vertical
i bacinella *f* verticale, vasca *f* per sviluppo lento
n standontwikkelingstank
d Gefässbad *n*, Standentwicklungstank *m*

5761 VERTICAL TUBE BOILER boi
f chaudière *f* à tubes verticaux
e caldera *f* de tubos verticales
i caldaia *f* a tubi verticali
n staande pijpketel
d stehender Röhrenkessel *m*, Steilrohrkessel *m*

5762 VERTICAL TUBE EVAPORATOR eva,pre
f évaporateur *m* à tubes verticaux
e evaporador *m* de tubos verticales
i evaporatore *m* a tubi verticali
n staande-pijpenverdamper
d stehender Röhrenverdampfer *m*, Steilrohrverdampfer *m*

5763 VERTICAL TWO-ROLL CALENDER pap,rub
f calandre *f* verticale à deux cylindres
e calandra *f* vertical de dos cilindros
i calandra *f* verticale a due cilindri
n verticale tweewalsenkalander
d stehender Zweiwalzenkalander *m*

5764 VESSEL, bre
TUN, VAT
f cuve *f*
e cuba *f*, tina *f*
i mastello *m*, tino *m*
n kuip, tobbe
d Bottich *m*

5765 VIBRATING BALL MILL brk
f broyeur *m* à boules à secousses
e molino *m* de bolas de sacudidas
i mulino *m* a palle a scosse
n schuddende kogelmolen
d schwingende Kugelmühle *f*

5766 VIBRATING CONVEYOR (GB), tra
VIBRATING DRAIN (US)
f transporteur *m* par vibration
e transportador *m* vibrante
i trasportatore *m* vibrante
n triltransporteur
d Schwingförderer *m*

5767 VIBRATING FEEDER tra
f alimentateur *m* vibrant
e alimentador *m* vibratorio
i alimentatore *m* a vibrazioni
n trillende toevoerinrichting, trillende voedingsinrichting
d vibrierende Speisevorrichtung *f*

5768 VIBRATING MILL brk
f broyeur *m* vibrant
e molino *m* vibratorio
i mulino *m* a vibrazioni
n trilmolen
d Schwingmühle *f*

5769 VIBRATING SCREEN, sor
VIBRATING SIEVE
f tamis *m* vibrant
e tamiz *m* vibratorio
i staccio *m* vibrante, vaglio *m* vibratore
n trilzeef
d Rüttelsieb *n*, Schüttelsieb *n*, Schwingsieb *n*

5770 VIBRATION DAMPER gen
f amortisseur *m* de vibrations
e amortiguador *m* de vibraciones
i smorzatore *m* di vibrazioni
n trillingsdemper
d Schwingungsdämpfer *m*

5771 VIBRATION TESTING APPARATUS tes
f appareil *m* d'essais à vibrations
e aparato *m* de prueba de vibraciones
i collaudatore *m* a vibrazioni
n trillingsproefapparaat *n*
d Schwingungsprobegerät *n*

5772 VIBRATOR gen
f vibrateur *m*, vibreur *m*
e vibrador *m*
i vibratore *m*
n schudder, triller
d Rüttler *m*

5773 VIBRATORY HOPPER tra
f distributeur *m* à vibrations, trémie *f* vibrante
e tolva *f* vibradora
i tramoggia *f* vibratrice
n trilvuller
d Vibrierfüller *m*

5774 VIGREUX COLUMN dis,lab
f colonne *f* de Vigreux
e columna *f* de Vigreux
i colonna *f* di Vigreux
n destillatieopzet volgens Vigreux, spijkeropzet
d Vigreux-Kolonne *f*

5775 VINEGAR FACTORY aam
f vinaigrerie *f*
e vinagrería *f*
i fabbrica *f* d'aceto
n azijnfabriek
d Essigfabrik *f*

5776 VINEGAR GENERATOR aam
f cuve *f* génératrice à vinaigre
e cuba *f* generadora de vinagre
i tino *m* generatore d'aceto
n snelazijnkuip, standkuip
d Gradierfass *n*, Schnellessigbildner *m*, Ständer *m*

5777 VINOMETER, OENOMETER ins
f oenomètre *m*, pèse-vin *m*, vinomètre *m*
e enómetro *m*
i enometro *m*
n vinometer, wijnmeter
d Weinmesser *m*, Weinwaage *f*

5778 VISCOSITY BREAKING COLUMN fue
f colonne *f* de réduction de la viscosité
e columna *f* de reducción de la viscosidad
i colonna *f* di riduzione della viscosità
n viscositeitsverlagingskolom
d Viskositätsbrechsäule *f*

5779 VISCOUS BRIGHT DIP elc
f bain *m* de brillantage à viscosité élevée
e baño *m* de abrillantado de viscosidad elevada
i bagno *m* di brillantaggio a viscosità elevata
n dikvloeibaar glansbad *n*
d dickflüssiges Glänzbad *n*, hochviskoses Glänzbad *n*

5780 VITRIFIED GRINDING WHEEL, VITRIFIED WHEEL pol
f roue *f* à meuler vitrifiée
e muela *f* vitrificada
i mola *f* con impasto vetrificato
n verglaasde slijpschijf
d verglaste Schleifscheibe *f*

5781 VOLATILIZATION APPARATUS eva,pre
f appareil *m* de volatilisation
e aparato *m* de volatilización
i apparecchio *m* di volatilizzazione
n vervluchtigingsapparaat *n*
d Verdunstungsapparat *m*, Verflüchtigungsapparat *m*

5782 VOLTOLIZATION APPARATUS fat
f appareil *m* de voltolyse
e aparato *m* de voltólisis
i apparecchio *m* di voltolisi
n voltolyseapparaat *n*
d Voltolisierungsapparat *m*

5783 VOLUME RECORDER, FLOW RECORDER ins
f appareil *m* enregistreur de volume
e registrador *m* de volumen
i registratore *m* di portata, registratore *m* di quantità, registratore *m* di volume
n volumeopnemer
d Mengenschreiber *m*

5784 VOLUMENOMETER, VOLUMETER ins
f voluménomètre *m*, volumomètre *m*
e balanza *f* volumétrica, volumenómetro *m*, volúmetro *m*
i strumento *m* per la misura diretta dei volumi, volumenometro *m*, volumetro *m*
n volumemeter, volumeter
d Mengenmesser *m*, Volumenometer *n*

5785 VOLUMETRIC FLASK lab
f matras *m* jaugé
e matraz *m* aforado
i matraccio *m* graduato
n maatkolf
d Messkolben *m*

5786 VOLUMETRIC WATER DETERMINATION APPARATUS ana
f appareil *m* de détermination du teneur en eau
e aparato *m* de determinación del contenido de agua
i apparecchio *m* di determinazione del contenuto d'acqua
n waterbepalingsapparaat *n*
d Wasserbestimmungsapparat *m*

5787 VULCANIZATION AUTOCLAVE rub
f autoclave *m* de vulcanisation
e autoclave *f* de vulcanización
i autoclave *f* di vulcanizzazione
n vulcaniseerautoclaaf
d Vulkanisierautoklav *m*

5788 VULCANIZATION TUB rub
f cuve *f* de vulcanisation
e cuba *f* de vulcanización
i tino *m* di vulcanizzazione
n vulcaniseerbak
d Vulkanisationswanne *f*

5789 VULCANIZER, VULCANIZING APPARATUS rub
f vulcanisateur *m*
e vulcanizador *m*
i vulcanizzatore *m*
n vulcaniseerapparaat *n*
d Vulkanisierapparat *m*, Vulkanisator *m*

5790 VULCANIZING BOILER, CURING PAN rub
f chaudière *f* de vulcanisation
e caldera *f* de vulcanización
i caldaia *f* di vulcanizzazione
n vulcaniseerketel
d Vulkanisierkessel *m*

5791 VULCANIZING MOLD (US), VULCANIZING MOULD (GB) rub
f moule *m* de vulcanisation
e molde *m* de vulcanización
i stampo *m* di vulcanizzazione
n vulcaniseervorm
d Vulkanisierform *f*

5792 VULCANIZING OVEN, CURING OVEN rub
f étuve *f* de vulcanisation
e estufa *f* de vulcanización
i riscaldatore *m* di vulcanizzazione
n vulcaniseerstoof
d Vulkanisierofen *m*

5793 VULCANIZING PRESS, CURING PRESS rub
f presse *f* de vulcanisation
e prensa *f* de vulcanización
i pressa *f* di vulcanizzazione
n vulcaniseerpers
d Vulkanisierpresse *f*

5794 VULCANIZING SHOP rub
f atelier *m* de vulcanisation
e taller *m* de vulcanización
i officina *f* di vulcanizzazione
n vulcaniseerruimte
d Vulkanisierwerkstatt *f*

W

5795 WALKING-BEAM KILN cer
f four *m* à poutre mobile
e horno *m* de viga móvil
i forno *m* a tavola di sollevamento
n hefbalkoven
d Hebebalkenofen *m*

5796 WAR GAS GENERATOR, POISON GAS GENERATOR gas,mit
f générateur *m* de gaz asphyxiants, générateur *m* de gaz toxiques
e generador *m* de gases tóxicos
i generatore *m* d'aggressivi chimici
n gifgasgenerator
d Giftgasentwickler *m*

5797 WARMING MILL, WARMING UP MILL, PREHEATING MILL pla,rub
f laminoir *m* de réchauffage
e laminador *m* de precalentamiento
i laminatoio *m* di preriscaldamento
n voorwarmwalserij
d Vorwärmwalzwerk *n*

5798 WARMING PLATE, HEATING PLATE lab
f plaque *f* chauffante, plateau *m* chauffant
e chapa *f* caliente, placa *f* calentadora
i piastra *f* riscaldante
n verwarmingsplaat
d Heizplatte *f*

5799 WASH GATE, WASH MILL cer
f cuve *f* à laver l'argile
e cuba *f* de lavar arcilla
i tino *m* da lavare argilla
n kleiwasbak
d Tonwäsche *f*

5800 WASH TROUGH min
f auge *f* à laver
e artesa *f* de lavar
i vasca *f* da lavare
n wastrog
d Setzfass *n*

5801 WASHER cpt
f rondelle *f*
e arandela *f*
i ranella *f*
n dichtingsring, onderlegplaatje *n*
d Federring *m*, Unterlegscheibe *f*

5802 WASHING BATH pho
f bain *m* de lavage
e baño *m* de lavado
i bagno *m* di lavaggio
n spoelbak
d Spülbad *n*, Wässerungsbad *n*

5803 WASHING BOTTLE lab
f pissette *f*
e redoma *f* con sifón
i spruzzetta *f*
n spuitfles
d Spritzflasche *f*

5804 WASHING CLASSIFIER, WET CLASSIFIER sor
f classeur *m* à eau courante
e clasificador *m* por corriente de agua
i classificatore *m* a corrente d'acqua
n slibgoot
d Stromklassierer *m*

5805 WASHING HOLLANDER pap
f hollandais *m* de lavage
e holander *m* lavador
i olandese *m* lavatore
n washollander
d Waschholländer *m*

5806 WASTE DE-INKING PLANT pap
f installation *f* de désencrage
e equipo *m* de extracción de la tinta
i impianto *m* di disinchiostratura
n inktextractie-installatie
d Entschwärzungsanlage *f*

5807 WASTE GRINDER iwt
f broyeur *m* à déchets
e machacadora *f* de desechos
i macinatore *m* per cascami
n afvalmolen
d Abfallzerkleinerungsmaschine *f*

5808 WASTE HEAT BOILER boi
f chaudière *f* à récupération
e caldera *f* de recuperación
i caldaia *f* a ricupero
n recupereerketel
d Abgaskessel *m*, Abhitzekessel *m*

5809 WASTE HEAT DRIER cer,dry
f sécheur *m* à récupération
e secador *m* de recuperación
i essiccatore *m* a ricupero
n recupereerdroger
d Abhitzetrockner *m*

5810 WASTE NEUTRALIZATION PLANT iwt
f installation *f* à neutralisation de déchets
e equipo *m* de neutralización de borras
i impianto *m* a neutralizzazione di cascami
n afvalneutraliseerinstallatie
d Abfallneutralisierungsanlage *f*

5811 WASTE PIPE iwt
f tuyau *m* de décharge
e caño *m* de desagüe
i tubo *m* di scarico
n afvoerpijp
d Ablassrohr *n*

5812 WASTE WATER PURIFYING PLANT iwt,sew
f installation *f* de purification d'eaux d'égout
e equipo *m* de depuración de aguas residuales
i impianto *m* d'epurazione dell'acqua di fogna
n afvalwaterreinigingsinstallatie
d Abwasserreinigungsanlage *f*

5813 WATCH GLASS, BEAKER COVER GLASS lab
f verre *m* de montre
e vidrio *m* de reloj
i vetro *m* da orologio
n horlogeglaasje *n*
d Uhrgläschen *n*, Uhrglasschale *f*

5814 WATER BAG pla,pre
f sac *m* à eau chaude
e bolsa *f* de agua caliente
i sacco *m* ad acqua calda
n heetwaterzak
d Heisswasserbeutel *m*

5815 WATER BAG, CURING BAG rub
f sac *m* à cuisson, sac *m* à eau
e cámara *f* vulcanizadora de agua
i camera *f* per vulcanizzazione ad acqua
n hulpband, vulcanisatiekern, waterzak
d Wasserheizschlauch *m*

5816 WATER BATH lab
f bain-marie *m*
e baño-maria *m*
i bagno-maria *m*
n bain-marie, waterbad *n*
d Wasserbad *n*

5817 WATER BOILER REACTOR, HOMOGENEOUS SOLUTION-TYPE REACTOR nue
f réacteur *m* à solution homogène
e reactor *m* de solución homogénea
i reattore *m* omogeneo acquoso
n homogene oplossingsreactor
d homogener Lösungsreaktor *m*, Wasserkocherreaktor *m*

5818 WATER CHANNEL (GB), WATER LINE (US) clo,pla,pre
f canal *m* d'eau, passage *m* pour l'eau
e canal *m* de agua, tubería *f* de agua
i canale *m* per l'acqua
n koelwaterkanaal *n*
d Wasserkanal *m*

5819 WATER CIRCULATOR clo,lab
f appareil *m* à faire circuler l'eau, chemise *f* de circulation
e camisa *f* de circulación, circulador *m* de agua
i apparecchio *m* per la circolazione dell'acqua, camicia *f* di circolazione
n mantel met watercirculatie
d Umlaufkühler *m*

5820 WATER COLLAR pla,pre
f colle *f* à eau
e reborde *m* de agua
i collare *m* ad acqua
n waterkraag
d Wasserkragen *m*

5821 WATER-COOLED HEAVY WATER REACTOR nue
f réacteur *m* à eau lourde à refroidissement à l'eau
e reactor *m* de agua pesada de enfriamiento por agua
i reattore *m* ad acqua pesante raffreddato per acqua
n met water gekoelde zwaar-waterreactor
d wassergekühlter Schwerwasser-reaktor *m*

5822 WATER-COOLED MOLD (US), WATER-COOLED MOULD (GB) fou
f moule *m* à refroidissement par eau

e molde *m* de enfriamiento por agua
i forma *f* a raffreddamento per acqua
n met water gekoelde vorm
d wassergekühlte Form *f*

5823 WATER DEVELOPMENT BATH pho
f bain *m* développeur à action retardée
e baño *m* de revelado de acción lenta
i bagno *m* rivelatore ad azione lenta
n verdund ontwikkelbad *n*
d langsam arbeitendes Entwicklungsbad *n*

5824 WATER DISTILLING APPARATUS, dis
WATER STILL
f appareil *m* à eau distillée
e aparato *m* de destilación de agua
i apparecchio *m* a distillazione d'acqua
n waterdestilleerapparaat *n*
d Wasserdestillierapparat *m*

5825 WATER-FILM PIPE PRECIPITATOR cle,fil
f dépoussiéreur *m* à couche d'eau
e electrofiltro *m* húmedo
i elettrofiltro *m* umido
n nat elektrofilter *n*
d Nasselektrofilter *m*

5826 WATER GAGE (US), boi,ins
WATER GAUGE (GB), WATER-LEVEL INDICATOR
f indicateur *m* de niveau d'eau
e indicador *m* de nivel de agua
i indicatore *m* del livello d'acqua
n waterpeilindicator
d Wasserstandsanzeiger *m*

5827 WATER GAS GENERATOR gas
f générateur *m* de gaz bleu, générateur *m* de gaz d'eau
e generador *m* de gas azul, generador *m* de gas de agua
i generatore *m* di gas d'acqua
n blauwwatergasgenerator
d Blauwassergasgenerator *m*

5828 WATER HAMMER ARRESTER pre
f anti-bélier *m*
e regulador *m* de golpo de ariete
i regolatore *m* di colpo d'ariete
n drukstootdemper
d Widderstossregler *m*

5829 WATER HARDENING BATH met
f bain *m* de durcissement dans l'eau
e baño *m* de endurecimiento en agua
i bagno *m* d'indurimento in acqua
n waterhardingsbad *n*
d Wasserhärtungsbad *n*

5830 WATER JACKET clo
f chemise *f* à eau de refroidissement
e camisa *f* de agua de enfriamiento
i camicia *f* ad acqua refrigerante
n koelwatermantel
d Kühlwassermantel *m*

5831 WATER JET ELECTRIC HEATER, boi,htt
WATER JET HEATER
f chaudière *f* à jet d'eau
e caldera *f* de chorro de agua
i caldaia *f* a getto d'acqua
n waterstraalelektrodenketel
d Wasserstrahlelektrodenkessel *m*

5832 WATER JET PUMP, pre,vac
WATER PRESSURE PUMP
f pompe *f* à eau aspirante et soufflante, pompe *f* à jet d'eau
e bomba *f* de chorro de agua
i pompa *f* a getto d'acqua
n waterstraalpomp
d Wasserstrahlpumpe *f*

5833 WATER LATTICE nue
f réseau *m* d'éléments combustibles modéré à l'eau
e red *f* de elementos combustibles moderada con agua
i reticolo *m* d'elementi combustibili moderato ad acqua
n met water gemodereerd splijtstofrooster *n*
d wassermoderiertes Brennstoff-raster *n*

5834 WATER-MODERATED REACTOR, nue
WATER REACTOR, LIGHT WATER REACTOR
f réacteur *m* à eau légère
e reactor *m* de agua liviana
i reattore *m* ad acqua leggera
n licht-waterreactor, waterreactor
d Leichtwasserreaktor *m*, Wasserreaktor *m*

5835 WATER-REPELLENT FINISHING PLANT tex
f installation *f* d'apprêt hydrofuge
e equipo *m* de apresto hidrófugo

i impianto *m* d'apprettatura idrofuga
n installatie voor het water-afstotend maken
d Anlage *f* zum Wasserabstossend-machen

5836 WATER RESERVOIR, WATER TANK sto
f réservoir *m* d'eau
e tanque *m* de agua
i serbatoio *m* d'acqua
n waterreservoir, watertank
d Wasserbehälter *m*, Wasserreservoir *n*

5837 WATER ROLLING TANK cle
f récipient *m* de nettoyage rotatif à l'eau de savonnage
e tanque *m* de limpieza giratoria con agua jabonosa
i serbatoio *m* di pulitura rotativa con acqua di sapone
n tank voor het roterend spoelen in zeepsop
d Reinigungsdrehtrommel *f* mit Seifenwasser

5838 WATER SEAL cpt
f fermeture *f* à l'eau
e cierre *m* de agua
i chiusura *f* all'acqua
n waterafdichting, watersluiting
d Wasserabdichtung *f*, Wasserverschluss *m*

5839 WATER-SEAL AGITATOR mix
f agitateur *m* à fermeture hydraulique
e agitador *m* de cierre hidráulico
i agitatore *m* a chiusura idraulica
n roerwerk *n* met watersluiting
d Rührwerk *n* mit Wasserverschluss

5840 WATER SEPARATOR bre,cle
f séparateur *m* d'eau
e separador *m* de agua
i separatore *m* d'acqua
n waterafscheider *m*
d Wasserabscheider *m*

5841 WATER SOFTENER, SOFTENING APPARATUS hyg,sew
f appareil *m* pour l'adoucissement de l'eau
e aparato *m* de ablandamiento del agua
i apparecchio *m* d'addolcimento dell'acqua
n waterontthardingsapparaat *n*
d Wasserenthärtungsapparat *m*

5842 WATER TOUGHENING BATH met
f bain *m* d'augmentation de la ténacité
e baño *m* de aumento de la tenacidad
i bagno *m* d'aumento della ténacité
n taaiheidsverhogend bad *n*
d Zähigkeitsverbesserungsbad *n*

5843 WATER TREATING PLANT bre
f installation *f* pour la correction de l'eau
e instalación *f* de tratamiento del agua
i impianto *m* per il trattamento dell'acqua
n waterzuiveringsinstallatie
d Wasseraufbereitungsanlage *f*

5844 WATER-TUBE BOILER boi
f chaudière *f* à tubes d'eau
e caldera *f* acuotubular, caldera *f* de tubos de agua
i caldaia *f* a tubi d'acqua
n waterpijpketel
d Wasserrohrkessel *m*

5845 WATERMARK SCREEN pap
f grille *f* de filigrane
e rejilla *f* de filigrana
i griglia *f* di filigrana
n watermerkraster
d Wasserzeichenraster *n*

5846 WATERPROOFING VESSEL tex
f récipient *m* d'imperméabilisation
e recipiente *m* de impermeabili-zación
i recipiente *m* d'impermeabilizzazi[one]
n vat *n* voor waterdichtmaken
d Gefäss *n* zum Wasserdichtmache[n]

5847 WAVER, DISTRIBUTING ROLL pap
f cylindre *m* broyeur
e cilindro *m* distribuidor
i cilindro *m* distributore, cilindro *m* mascinatore
n distributiewals
d Reibwalze *f*, Verreibwalze *f*, Verteilerwalze *f*

5848 WAX FINISHING APPARATUS lea
f appareil *m* d'apprêt de cire
e aparato *m* de apresto con cera
i apparecchio *m* d'apprettatura con cera
n wasappreteerapparaat *n*
d Wachsappreturgerät *n*

5849 WAX VENT fou
f rat *m* de cave
e mecha *f* para vientos
i cerino *m*
n waskoord *n*
d Wachsschnur *f*

5850 WAXING ROLLER pap
f cylindre *m* de cirage
e cilindro *m* de encerado
i cilindro *m* d'inceratura
n paraffineerwals
d Wachswalze *f*

5851 WEAR AND TEAR TESTING MACHINE tes
f machine *f* d'essai d'usure
e máquina *f* de prueba de desgaste
i macchina *f* di prova ad usura
n levensduurbepalingsmachine
d Lebensdauerprüfmaschine *f*

5852 WEAR TESTING MACHINE tes
f machine *f* à mesurer l'usure
e máquina *f* para medir el desgaste
i macchina *f* per prova di resistenza all'usura
n slijtagemachine
d Abnutzungsmaschine *f*, Verschleissmaschine *f*

5853 WEARING LINING coa
f garnissage *m* à l'usure
e guarnición *f* al uso
i guarnizione *f* all'usura
n slijtvoering
d Verschleissfutter *n*

5854 WEATHERING TEST TUBE lab
f tube *m* de centrifugeur pour gaz naturel
e tubo *m* de centrífuga para gas natural
i tubo *m* di centrifuga per gas naturale
n aardgascentrifugebuis
d Erdgaszentrifugenglas *n*

5855 WEATHERING TESTING SHOP dec,pla
f atelier *m* d'essai du vieillissement aux intempéries
e taller *m* de prueba de envejecimiento por intemperie
i officina *f* collaudatrice d'invecchiamento per intemperie
n weerbestendigheidsonderzoek-ruimte
d Bewitterungsraum *m*

5856 WEATHERING TROUGH cer
f cuve *f* de désagrégation
e cuba *f* de disgregación
i tino *m* di stagionamento
n opslagkuip
d Auswitterungsküpe *f*

5857 WEATHEROMETER dec,ins
f mesureur *m* de désintégration
e medidor *m* de disgregación
i veterometro *m*
n verweringsmeter
d Verwitterungsmesser *m*

5858 WEDGE POLARIMETER ins
f polarimètre *m* à coin
e polarímetro *m* de cuña
i polarimetro *m* a cuneo
n wigpolarimeter
d Keilpolarimeter *n*

5859 WEDGE PYROMETER ins
f pyromètre *m* à coin
e pirómetro *m* de cuña
i pirometro *m* a cuneo
n wigpyrometer
d Keilpyrometer *n*

5860 WEDGE ROASTER met
f four *m* de grillage cylindrique multiple
e horno *m* de tostación cilíndrico múltiple
i forno *m* d'arrostimento cilindrico multiplo
n meervoudige ronde roostoven
d runder Mehrfachröstofen *m*

5861 WEDGE-WIRE SCREEN fil
f tamis *m* à fils metalliques profilé
e tamiz *m* de hilos metálicos perfilado
i vaglio *m* di fili metallici profilato
n profieldraadzeef
d Profildrahtsieb *n*

5862 WEEDKILLER FACTORY, HERBICIDE FACTORY agr,cle
f usine *f* d'herbicides
e fábrica *f* de herbicidas
i fabbrica *f* d'erbacide
n fabriek van onkruidverdelgende middelen
d Fabrik *f* von Unkrautvertilgungs-mitteln

5863 WEEP HOLE fou
f chantepleure *f*
e agujero *m* de drenaje

i foro *m* di drenaggio
n draineergat *n*
d Tropfloch *n*

5864 WEIGHING BOAT lab
f main *f* pour poudres
e cápsula *f* de pesado
i navicella *f* di pesatura
n weegschuitje *n*
d Wägeschiffchen *n*

5865 WEIGHING BOTTLE lab
f pese-filtre *m*, vase *m* à peser
e jarro *m* de pesar
i pesafiltro *m*
n weegflesje *n*
d Wägeglas *n*

5866 WEIGHING BURETTE lab
f burette *f* à peser
e bureta *f* de pesar
i buretta *f* di pesatura
n weegburet
d Wägebürette *f*

5867 WEIGHING PIG ana
f cochonnet *m* à peser
e —
i —
n weegvarkentje *n*
d Wägeschweinchen *n*

5868 WEIGHING PIPETTE lab
f pipette *f* à peser
e pipeta *f* de pesar
i pipetta *f* di pesatura
n weegpipet
d Wägepipette *f*

5869 WEIGHING ROOM lab
f salle *f* des balances
e sala *f* de las balanzas
i reparto *m* delle bilancie
n weegkamer
d Wägezimmer *n*

5870 WEIGHING TUBE ana
f pèse-substance *m*
e tubo *m* de pesadas
i tubo *m* di pesatura
n weegbuisje *n*
d Wägeröhrchen *n*

5871 WEIGHT FEEDER (US), DOSING FEEDER (GB) cpt,pla
f dispositif *m* de dosage
e alimentador *m* ponderal, dispositivo *m* de dosificación
i dosatore *m*
n gewichtsdoseerinrichting
d Dosiervorrichting *f* nach Gewicht

5872 WEIGHT-LOADED ACCUMULATOR pla,pre
f hydroaccumulateur à pression par poids
e hidroacumulador *m* de presión por pesos
i idroaccumulatore *m* a pressione per pesi
n gewichtsaccumulator
d Gewichtsakkumulator *m*

5873 WEIGHTING VESSEL, LOADING VESSEL tex
f cuve *f* à charger
e cuba *f* de cargado
i tino *m* di carica
n verzwaringskuip
d Beschwerungsküpe *f*

5874 WELDING GROUND (US), BACKING ELECTRODE (GB) wel
f contre-électrode *f*
e contraelectrodo *m*
i controelettrodo *m*
n onderlegelektrode
d Gegenelektrode *f*

5875 WELDING GUN pla,wel
f pistole *f* à souder
e pistola *f* de soldar
i cannello *m* per saldare
n laspistool *n*
d Schweisspistole *f*

5876 WELDING PLANT htt,pla,wel
f installation *f* de soudure
e equipo *m* de soldadura
i impianto *m* di saldatura
n lasinstallatie
d Schweissanlage *f*

5877 WELDING ROD wel
f baguette *f* de soudure
e varilla *f* de soldadura
i bacchetta *f* di saldatura
n lasstaaf
d Schweissstab *m*

5878 WELDING ROD, FILLER ROD pla,wel
f baguette *f* d'apport
e varilla *f* de aportación
i bacchetta *f* di rapporto
n oplasstaaf
d Aufschweissstab *m*

5879 WELDING ROLLER htt,pla,wel
f rouleau *m* de soudure
e rodillo *m* de soldadura
i rullo *m* di saldatura
n lasrol
d Schweissrolle *f*

5880 WELDING TORCH htt,pla,wel
f chalumeau *m* de soudure
e soplete *m* de soldadura
i cannello *m* per saldatura
n lasbrander
d Lötrohr *n*

5881 WET CELL elc
f pile *f* liquide
e pila *f* líquida
i pila *f* a liquido
n nat element *n*, natte cel
d Füllelement *n*, Nasselement *n*

5882 WET CLASSIFIER, WASHING CLASSIFIER sor
f classeur *m* à eau courante
e clasificador *m* por corriente de agua
i classificatore *m* a corrente d'acqua
n slibgoot
d Stromklassierer *m*

5883 WET ENAMELLING APPARATUS, DIP ENAMELLING APPARATUS coa
f appareil *m* d'émaillage au trempé
e aparato *m* de esmaltado por inmersión
i apparecchio *m* di smaltatura ad immersione
n dompelemailleerapparaat *n*
d Nassemaillierungsgerät *n*

5884 WET END pap
f groupe *m* liquide
e grupo *m* líquido
i gruppo *m* liquido
n natte sectie
d Nasspartie *f*

5885 WET FASTNESS TESTING APPARATUS tex
f appareil *m* d'essai de la résistance à l'humidité
e probador *m* de la resistencia a la humedad
i collaudatore *m* della resistenza all'umidità
n vochtbestendigheidsbeproevingsapparaat *n*
d Feuchtbeständigkeitsprüfgerät *n*

5886 WET FINISHING APPARATUS tex
f appareil *m* d'apprêtage liquide
e aparato *m* de aprestado líquido
i apparecchio *m* d'apprettatura liquida
n nat appreteertoestel *n*
d Nassappreturgerät *n*

5887 WET MILLING MACHINE brk
f machine *f* de broyage sous liquide
e máquina *f* de molimiento al líquido
i macchina *f* di molitura per via liquida
n natmaalmachine
d Nassmahlmaschine *f*

5888 WET SOLUTION MIXER mix
f malaxeur *m* à dissolution sous liquide
e mezclador *m* de soluciones al líquido
i mescolatore *m* per soluzione per via liquida
n natte solutiemolen
d nasser Gummilösungsmischer *m*

5889 WET SPINNING PLANT tex
f installation *f* de filage dans un liquide
e equipo *m* de hilatura en un líquido
i impianto *m* di filatura in un liquido
n natspininstallatie
d Nassspinnanlage *f*

5890 WET-STEAM COOLED REACTOR nue
f réacteur *m* à refroidissement par vapeur humide
e reactor *m* enfriado por vapor húmedo
i reattore *m* refrigerato per vapore umido
n met natte stoom gekoelde reactor
d nassdampfgekühlter Reaktor *m*

5891 WET STOVING VESSEL, STOVING BATH tex
f bain *m* de blanchiment réducteur
e baño *m* de blanqueo reductor
i bagno *m* di candeggio ridottore
n reducerend bleekbad *n*
d reduzierendes Bleichbad *n*

5892 WET TUMBLING BARREL cle,pla
f tambour *m* de nettoyage sous liquide
e tambor *m* de limpieza al líquido
i tamburo *m* di pulitura per via liquida
n natte poetstrommel
d nasse Putztrommel *f*

5893 WET VACUUM PUMP vac
f pompe *f* à vide avec liquide

e bomba *f* de vacío con líquido
i pompa *f* a vuoto con liquido
n vloeistofvacuümpomp
d Flüssigkeitsvakuumpumpe *f*

5894 WETTED-WALL COLUMN dis
f colonne *f* à ruissellement
e columna *f* de pared mojada
i colonna *f* a parete bagnata
n kolom met natte wand
d Dünnschichtkolonne *f*, Nasswandkolonne *f*

5895 WHETSTONE, HONE pol
f affiloir *m*, queue *f*, queux *m*
e piedra *f* aguzadera, piedra *f* de afilar
i cote *f*, pietra *f* per affilare
n aanzetsteen, wetsteen
d Absetzstein *m*, Wetzstein *m*

5896 WHEY VESSEL dai,nut
f récipient *m* à petit lait
e recipiente *m* de suero de lecho
i recipiente *m* a siero di latte
n huivat *n*, weivat *n*
d Käsewassergefäss *n*

5897 WHIPPER, WHISK mix,nut
f batteur *m*, fouet *m*
e batidor *m*
i frullino *m*
n klopper, klutser
d Quirl *m*

5898 WHIRL GATE, WHIRL GATE DIRT TRAP cle
f piège *f* à crasser à canal tangentiel
e trampa *f* para escorias en canal tangencial
i fermascoria *f* a pozzetto tangenziale
n cycloonslakvanger
d Kreisel *m*, Schaumtrichter *m*

5899 WHIRLING PSYCHROMETER, SLING PSYCHROMETER ins
f hygromètre *m* à torsion, psychromètre *m* fronde
e higrómetro *m* de torsión, psicrómetro *m* de torsión
i igrometro *m* a torsione, psicrometro *m* a torsione
n slingerpsychrometer
d Schleuderpsychrometer *n*

5900 WHISKING MACHINE, BEATING MACHINE, FOAM BEATER, FOAMING MACHINE rub
f batteuse *f* à mousse, fouetteuse *f*
e batidora *f* mecánica
i frullatore *m*, produttore *m* di schiuma, sbattitore *m*
n klutsmachine, opklopmachine, schuimklopper
d Schaumschlagmaschine *f*, Schlagmaschine *f*

5901 WHISTLER fou
f évent *m*
e viento *m*
i respiro *m*, sfiatatoio *m*
n ontluchtingskanaal *n*
d Luftpfeife *f*

5902 WHITE PICKLING BATH cle
f bain *m* de décapage supplémenta
e baño *m* de decapaje suplementar
i bagno *m* di decapaggio supplementare
n tweede beitsbad *n*
d Nachbeizbad *n*

5903 WHITE SOURING BATH te
f bain *m* de vitriolage blanc
e baño *m* de acidulado blanco
i bagno *m* da acidulare bianco
n wit zuurbad *n*
d weisses Säurebad *n*

5904 WHIZZER, HYDROEXTRACTOR dry
f essoreuse *f* centrifuge, hydro-extracteur *m*, toupie *f* mécanique
e extractor *m* centrífugo, hidroextractor *m*
i estrattore *m* centrifugo, idroestrattore *m*
n droogcentrifuge
d Zentrifugaltrockenmaschine *f*

5905 WICKET cer,ht
f paroi *f* de fermeture
e pared *f* de cierre
i parete *f* di chiusura
n sluitwand
d Abschliesswand *f*

5906 WIDE-NECKED FLASK la
f matras *m* à col large
e matraz *m* de boca ancha
i matraccio *m* a collo largo
n kolf met wijde hals
d Weithalskolben *m*

5907 WILLEY, WILLOW, DISINTEGRATING MILL, DISINTEGRATOR, OPENER — tex
- f broyeur *m* à couronnes dentées, loup *m* batteur, ouvreuse *f*
- e abridora *f*, batidora *f*, batuar *m*, diablo *m*
- i apritoio *m*, lupo *m*
- n duivel, snar, wolf
- d Klopfwolf *m*, Reisswolf *m*

5908 WINCH DYEING BECK, REEL DYEING BECK — tex
- f cuve *f* de teinture à moulinet
- e barca *f* de devanadora
- i tino *m* di tintura a mulinello
- n haspelkuip
- d Haspelkufe *f*

5909 WINE CASK — tra
- f pièce *f* de vin
- e pipa *f*
- i botte *f* da vino
- n wijnvat *n*
- d Weinfass *n*

5910 WINE PRESS — nut
- f pressoir *m*
- e prensa *f* de vino
- i ammostatoio *m* di vino
- n wijnpers
- d Kelter *f*

5911 WING MIXER, PADDLE MIXER — mix
- f mélangeur *m* à palettes
- e mezclador *m* de paletas
- i mescolatrice *f* a pale
- n schoepenmenger
- d Paddelmischer *m*, Schaufelmischer *m*

5912 WINTERING VESSEL — fat
- f cuve *f* d'extraction de la stéarine
- e cuba *f* de extracción de la estearina
- i tino *m* d'estrazione della stearina
- n stearine-extractievat *n*
- d Stearineextraktionsgefäss *n*

5913 WIRE CLOTH — pap
- f tamis *m* à fils métalliques, toile *f* métallique
- e tamiz *m* de hilos metálicos, tela *f* metálica
- i staccio *m* a fili metallici, tela *f* metallica
- n draadgaaszeef
- d Metalldrahtsieb *n*

5914 WIRE COATING EXTRUSION MACHINE — coa,pla
- f boudineuse *f* à revêtir des fils métalliques
- e budinadora *f* para revestir hilos metálicos
- i filiera *f* da rivestire fili metallici
- n spuitgietmachine voor draadbekleding
- d Spritzpresse *f* für Drahtbekleidung

5915 WIRE END — pap
- f section *f* de tamis
- e sección *f* de tamices
- i sezione *f* di stacci
- n zeefafdeling
- d Siebabteilung *f*

5916 WIRE FRAME — pap
- f table *f* de tamisage
- e mesa *f* de los tamices
- i tavola *f* degli stacci
- n zeeftafel
- d Siebtisch *m*

5917 WIRE GAUZE (GB), WIRE GAZE (US) — lab
- f toile *f* métallique
- e tela *f* metálica
- i tela *f* metallica
- n draadgaasje *n*
- d Metallgaze *f*

5918 WIRE KILN FLOOR — bre
- f plateau *m* en fil métallique
- e plato *m* de tela metálica
- i griglia *f* a filo metallico
- n draadvloer
- d Drahthorde *f*

5919 WIRE SCREEN, SCREEN PACK — pla
- f toile *f* métallique
- e tamiz *m* de tela metálica
- i tela *f* metallica
- n draadzeef, metaalgaas *n*
- d Drahtsieb *n*, Metallgaze *f*

5920 WITNESS — lab
- f piège *m* de liquides
- e trampa *f* de líquidos
- i trappola *f* di liquidi
- n vloeistofvanger
- d Flüssigkeitsfänger *m*

5921 WOOD DISTILLATION PLANT dis
f installation *f* de distillation du bois
e equipo *m* de destilación de la madera
i impianto *m* di distillazione del legno
n houtdestilleerderij
d Holzdestillieranlage *f*

5922 WOOD PULP FACTORY pap
f fabrique *f* de pâte de bois
e fábrica *f* de pulpa de madera
i fabbrica *f* di polpa di legno
n houtpulpfabriek, houtslijpfabriek
d Holzschleiferei *f*, Holzstoffabrik *f*

5923 WOOD PULP HYDROLYSIS PLANT aam
f installation *f* d'hydrolyse du bois
e equipo *m* de hidrólisis de la madera
i impianto *m* d'idrolizzazione del legno
n houthydroliseerinstallatie
d Holzverzuckerungsanlage *f*

5924 WOOD PULP THICKENING DRUM pap
f tambour *m* épaississeur de pâte de bois
e tambor *m* espesador de pulpa de madera
i ispessitore *m* di polpa di legno
n houtpulpindikker, houtslijpindikker
d Holzschliffeindicker *m*, Holzstoffeindicker *m*

5925 WOOD SEASONING PLANT, SEASONING PLANT bui
f installation *f* de dessèchement de bois
e equipo *m* de cura de madera, equipo *m* de desecación de madera
i impianto *m* di stagionatura di legno
n houtdrooginstallatie
d Holzauswitterungsanlage *f*

5926 WOOD STEAMER pap
f chaudière *f* à étuver le bois
e caldera *f* para la cocción de madera
i caldaia *f* per cucinare il legno
n houtkookketel
d Holzkochkessel *m*

5927 WOODEN TANK sua
f cuve *f* en bois
e cuba *f* de madera
i tino *m* di legno
n houten kuip
d Holzküpe *f*

5928 WOOL SCOURING MACHINE, SCOURING MACHINE tex
f machine *f* de dégraissage, machine *f* de dessuintage
e desgrasadora *f*, máquina *f* de desuardado
i macchina *f* di primo lavaggio, macchina *f* di sgrassaggio
n wolontvettingsmachine
d Entfettungsmaschine *f*, Entschweissungsmaschine *f*

5929 WORK COIL, APPLICATOR, HEATING INDUCTOR htt
f inducteur *m* de chauffage
e inductor *m* de equipo de caldeo
i induttore *m* riscaldante
n verhittingsinductor, verwarmingsinductor, werkspoel
d Heizinduktor *m*, Heizschleife *f*

5930 WORK HARDENING PLANT, STRAIN HARDENING PLANT met
f installation *f* d'écrouissage
e equipo *m* de endurecimiento por deformación en frío
i impianto *m* d'incrudimento per deformazione a freddo
n koudverstevigingsinstallatie
d Kaltverfestigungsanlage *f*

5931 WORKING PLATFORM, OPERATING PLATFORM gen
f plate-forme *f* de manoeuvre
e plataforma *f* de servicio
i piattaforma *f* di servizio
n bedieningsbordes *n*
d Arbeitsbühne *f*

5932 WORKS, FACTORY, MILL gen
f établissement *m*, fabrique *f*, manufacture *f*, usine *f*
e fábrica *f*
i fabbrica *f*, stabilimento *m*
n fabriek
d Fabrik *f*, Werk *n*

5933 WORKSHOP gen
f atelier *m*
e taller *m*

i bottega *f*, officina *f*, reparto *m*
n werkplaats
d Werkstatt *f*

5934 WORM, dis
COIL
f serpentin *m*
e serpentín *m*
i serpentina *f*
n spiraalkoeler
d Rohrschlange *f*

5935 WORM, cpt
ENDLESS SCREW
f vis *f* sans fin
e tornillo *m* sin fin
i vite *f* senza fine
n schroef zonder eind, worm
d endlose Schraube *f*, Schnecke *f*

5936 WORM CONVEYOR, tra
HELICAL CONVEYOR,
SCREW CONVEYOR
f transporteur *m* hélicoïdal
e transportador *m* helicoidal
i trasportatore *m* elicoidale
n transportworm
d Förderschnecke *f*

5937 WORT CIRCULATING PUMP bre
(GB),
CIRCULATOR (US)
f circulateur
e bomba *f* de prefiltración
i pompa *f* per prefiltro
n voorlooppomp
d Vorlaufpumpe *f*

5938 WORT COOLER bre
f refroidisseur *m* de moût
e enfriador *m* de mosto
i refrigerante *m* per mosto
n wortkoeler
d Würzekühler *m*

5939 WORT COPPER bre
f chaudière *f* à moût
e caldera *f* de mosto
i caldaia *f* di mosto
n wortketel
d Würzepfanne *f*

5940 WORT PIPE bre
f tuyau *m* de vidange du moût
e tubo *m* de filtración
i tubo *m* di filtrazione
n wortfilterbuis
d Läuterrohr *n*

5941 WOVEN FABRIC DUST cle,gas
COLLECTOR
f collecteur *m* de poussière à
tissu
e colector *m* de polvo de tejido
plano
i collettore *m* di polvere a
tessuto
n stofverzamelaar uit weefsel
d Webstoffstaubsammler *m*

5942 WOVEN WIRE sor
f toile *f* métallique
e tela *f* metálica
i tela *f* metallica
n draadgaas *n*
d Drahtgewebe *n*

X

5943 XENON LIQUEFIER gas
f installation *f* de liquéfaction de xénon
e equipo *m* de licuación de xenón
i impianto *m* di liquefazione di xenon
n installatie voor het vloeibaar maken van xenon
d Xenonverflüssigungsanlage *f*

Y

5944 YARN DYEING VAT, BUNDLE DYEING VAT, HANK DYEING VAT — tex
f cuve *f* de teinture en écheveaux
e cuba *f* de teñido en suspensión de las madejas
i tino *m* di tintura delle matasse
n strengenverfkuip
d Strähnenfärbküpe *f*

5945 YARYAN VESSEL — pap
f cuve *f* de récupération de la soude
e cuba *f* de recuperación de la sosa
i tino *m* di ricupero della soda
n sodaterugwinningsvat *n*
d Sodarückgewinnungsgefäss *n*

5946 YEAST COUNTING APPARATUS, HEMATIMETER, HEMATOMETER — bre,ins
f hématimètre *m*
e hematímetro *m*
i ematimetro *m*
n gisttelraam *n*
d Hämatimeter *n*, Hefezählvorrichtung *f*

5947 YEAST COVER, YEAST HEAD — bre
f couvercle *m*
e cubierta *f*
i coperta *f*
n schuimdek *n*
d Gärdecke *f*

5948 YEAST FACTORY, YEAST WORKS — fmi
f fabrique *f* de levure, levurerie *f*
e fábrica *f* de levadura
i fabbrica *f* di lievito
n gistfabriek
d Hefefabrik *f*

5949 YEAST PRESS — bre
f presse *f* de levure
e prensa *f* de levadura
i pressa *f* per lievito
n gistpers
d Hefepresse *f*

5950 YEAST PROPAGATOR — bre
f appareil *m* à culture pure de levure, appareil *m* à propagation pure de levure
e aparato *m* de cultivo puro de levadura, aparato *m* para producción de levadura pura
i apparecchio *m* per cultura pura di lievito, apparecchio *m* di produzione di lievito puro
n cultuurvat *n* voor zuivere gist
d Hefekulturapparat *m*, Reinzuchtapparat *m*

5951 YEAST ROUSER — bre
f râteau *m* à levure
e mezclador *m* de levadura
i mescolatore *m* di lievito
n gistroerder
d Satzkrücke *f*

5952 YEAST ROUSING APPARATUS — bre
f oeuf *m* à levure
e agitador *m* de levadura
i areatore *m* per lievito
n beluchtapparaat *n*
d Hefeaufziehapparat *m*

5953 YEAST SPOON — bre
f cuiller *f* de levure
e cucharón *m* de levadura
i cucchiaio *m* di lievito
n gistlepel
d Satzlöffel *m*

5954 YEAST TANK — bre
f récipient *m* à levure
e tanque *m* de levadura
i serbatoio *m* per lievito
n gisttank
d Hefebehälter *m*

5955 YEAST TUB — bre
f bac *m* à levure, bacquet *m* de levure, cuvier *m*
e tina *f* de levadura
i tino *m* di lievito, vasca *f* di lievito
n gistbak
d Bränke *f*, Hefewanne *f*, Zeugwanne *f*

5956 YIELD POINT TESTER, YIELD VALUE TESTER, PLASTIC YIELD TESTER — pla,tes
f appareil *m* d'essai de contrainte critique

e comprobador *m* de límite de fluidez
i collaudatore *m* di limite di fluidità
n vloeigrensbepalingsapparaat *n*
d Fliessgrenzeprüfgerät *n*

5957 YPERITE DETECTOR, ins,mit,sap
MUSTARD GAS DETECTOR
f détecteur *m* de gaz moutarde, détecteur *m* d'ypérite
e detector *m* de gas mostaza, detector *m* de iperita
i rivelatore *m* d'iprite
n mosterdgasdetector, mosterdgasverklikker, yperietverklikker
d Senfgaswarngerät *n*, Yperitanzeigegerät *n*

5958 YPERITE GENERATOR, gas,mit
MUSTARD GAS GENERATOR
f générateur *m* d'ypérite
e generador *m* de iperita
i generatore *m* d'iprite
n mosterdgasgenerator
d Senfgasgenerator *m*

Z

5959 ZAFFERBLUE WORKS, SMALTHOUSE cer,dye
f atelier *m* d'azurage, teinturerie *f* en bleu
e tintorería *f* en azul
i officina *f* di vagellaio
n blauwververij
d Blaufarbenwerk *n*

5960 ZEOLITE FILTER, PERMUTITE FILTER fil
f filtre *m* à permutite
e filtro *m* de permutita
i filtro *m* a permutite, filtro *m* a zeolite
n zeolietfilter *n*
d Permutitfilter *m*

5961 ZEOLITE WATER SOFTENER hyg
f appareil *m* pour l'adoucissement de l'eau par permutite
e aparato *m* para el tratamiento con permutita
i apparecchio *m* per il trattamento alla permutite
n permutietwateronthardingsapparaat *n*
d Permutitenthärter *m*

5962 ZIGZAG FILTER cle,fil,tex
f filtre *m* zigzag
e filtro *m* zigzag
i filtro *m* zigzag
n zigzagfilter *n*
d Zickzackfilter *m*

5963 ZIGZAG KILN cer
f four *m* à conduite en zig-zag du feu
e horno *m* de conducto en zigzag del fuego
i forno *m* a conduzione in zigzag del fuoco
n oven met zigzaggeleiding van het vuur
d Ofen *m* mit Zickzackführung des Feuers

5964 ZINC ABSORPTION JAR, ZINC ABSORPTION TUBE lab
f cylindre *m* absorbeur à grenaille de zinc
e cilindro *m* absorbador de granalla de cinc
i cilindro *m* d'assorbimento a pallini di zinco
n absorptiecilinder met zinkvulling
d Absorptionszylinder *m* für Zinkfüllung

5965 ZINC COATING BATH, ZINC PLATING BATH coa
f bain *m* de zingage
e baño *m* de cincado
i bagno *m* di zincatura
n zinkbad *n*
d Zinkbad *n*

5966 ZINC CONDENSOR met
f condensateur *m* de zinc
e condensador *m* de cinc
i condensatore *m* di zinco
n zinkcondensor
d Vorlage *f*, Zinkverdichter *m*

5967 ZINC FURNACE met
f four *m* de zinc
e horno *m* de cinc
i forno *m* di zinco
n zinkoven
d Zinkdestillierofen *m*

5968 ZINC PHOSPHATIZING BATH coa
f bain *m* de revêtement en phosphate de zinc
e baño *m* de revestimiento de fosfato de cinc
i bagno *m* di rivestimento a fosfato di zinco
n zinkfosfaatbad *n*
d Zinkphosphatbad *n*

5969 ZINC REDISTILLATION PLANT met
f installation *f* de distillation sélective de zinc
e equipo *m* de destilación selectiva de cinc
i impianto *m* di distillazione selettiva di zinco
n selectieve zinkdestilleerinstallatie
d selektive Zinkdestillationsanlage *f*

5970 ZINC RETORT dis,met
f cornue *f* à zinc
e retorta *f* de cinc

i storta *f* a zinco
n zinkretort
d Zinkmuffel *f*, Zinkretorte *f*

5971 ZINC SMELTER htt,met
f four *m* de fusion du zinc
e horno *m* de fusión del cinc
i forno *m* di fusione dello zinco
n zinksmeltoven
d Zinkschmelzofen *m*

5972 ZINCATE TREATMENT BATH coa
f bain *m* de traitement au zincate
e baño *m* de tratamiento al cincato
i bagno *m* trattamento allo zincato
n zinkaatbehandelingsbad *n*
d Zinkatbehandlungsbad *n*

5973 ZYMOMETER fer,ins
f zymomètre *m*
e zimómetro *m*
i zimometro *m*
n zymometer
d Zymometer *n*

FRANÇAIS

F

ESPAÑOL

E

ITALIANO

I

impianto d'evaporazione termica 5460
- di bioaerazione 584
- di brasatura 721
- di brasatura ad induzione 2955
- di brasatura all'argento 4834
- di brasatura per immersione 1609
- di brillantaggio 744
- di cadmiatura 810
- di calorizzazione 837
- di candeggio 611
- di candeggio all'ozono 3761
- di candeggio per la luce solare 5348
- di carbonizzazione 896
- di carbonizzazione incompleta 3282
- di cementazione al cianuro 872
- di centrifugazione 5061
- di chetonizzazione 3092
- di chiarificazione 1109
- di cianurizzazione 1422
- di classificazione 4725
- di clorinazione intermittente 3017
- di comando del carburatore 905
- di congelazione rapida 4193
- di cracking 1371
- di cracking termico 5456
- di cromatura 1089
- di decapaggio al gas 2466
- di decapaggio catodico 951
- di decapaggio continuo 1297, 1313
- di decapaggio per soda caustica 965
- di decarbonizzazione 1461
- di declorazione 1464
- di defecazione 1476
- di defenolizzazione 1525
- di deflagrazione 1481
- di defosforazione 1527
- di deidroalogenizzazione 1493
- di deidrogenazione 1492
- di deionizzazione a letto misto 3463
- di demanganizzazione 1504
- di demineralizzazione 1506
- di denaturazione 1509
- di deparaffinazione 1523
- di deposito per contatto 1278
- di depurazione biologica 585
- di depurazione dell'olio nel vuoto 5690
- di depurazione di gas inerti 2963
- di depurazione d'olio con vapore 5202
- di desalificazione d'acqua di mare 4659
- di desalinizzazione navotrasportato 4792
- di desolforazione con calce 3215
- di devitrificazione 1562

impianto di disacidificazione a calce 3223
- di disinchiostratura 5806
- di disincrostazione 1536
- di disossidazione 1521
- di disossigenazione 1522
- di distillazione a bassa temperatura 3282
- di distillazione a controcorrente 1356
- di distillazione a lotti 501
- di distillazione a membrana semipermeabile 3857
- di distillazione a secco 1546
- di distillazione a vuoto spinto 2769
- di distillazione continua 1286
- di distillazione con vapore 5181
- di distillazione del legno 5921
- di distillazione d'ammoniaca 258
- di distillazione discontinua 501
- di distillazione distruttiva 1546
- di distillazione equilibrata in corrente di vapore 2244
- di distillazione estrattiva 2052
- di distillazione frazionata 2367
- di distillazione molecolare 3501
- di distillazione selettiva di zinco 5969
- di distillazione semplice 4841
- di doratura al fuoco 2204
- d'idratazione 2854
- di drenaggio 1491
- d'idrogenazione 2874
- d'idrogenazione a bassa temperatura 3284
- d'idrogenazione ad alta pressione 2760
- d'idrogenazione di grasso 2080
- d'idrogenazione in fase liquida 3248
- d'idrogenazione in fase vapore 5733
- d'idrogeno pesante 2723
- d'idrolisi 2876
- d'idrolizzazione del legno 5923
- d'idrossilazione 2883
- di fabbricazione di gas ad alta pressione 2757
- di fenilazione 3882
- di fermentazione effervescente 1855
- di filatura in un liquido 5889
- di filtrazione 2161
- di filtrazione in sacchi 421
- di finissaggio 2193
- di finitura 2194
- di fissazione dell'azoto 3607
- di flottazione 2297
- di flottazione mediante schiuma 2389
- di flottazione selettiva 1591
- di fluorizzazione 2326

impianto di formatura a pressione 5133
- di formazione del granello 2589
- di formazione di siliciuro 4827
- di galvanizzazione a caldo 2835
- di galvanoplastica in tamburo 483
- di gas naturale 3563
- di gelatinizzazione 2513
- di generazione di ammoniaca 255
- di laccare 3126
- di lavaggio di gas 2482
- di lavorazione 4121
- di lavorazione a scintillamento 1897
- di liquefazione 3236
- di liquefazione d'argon 327
- di liquefazione d'aria 194
- di liquefazione d'azoto 3608
- di liquefazione di cripton 3113
- di liquefazione di neon 3575
- di liquefazione di xenon 5943
- di liquefazione d'ossigeno 3754
- di lisciviazione 3161
- di lisciviazione in mucchi 2684
- di lucidatura 2555, 3290
- di macerazione 4385
- di maturazione 4665
- di mercerizzazione 3394
- di metallizzazione 3428
- di metallizzazione elettrolitica 1921
- di metanizzazione 3432
- di metilazione 3434
- d'impregnazione 2926
- d'impregnazione del legname con sublimato corrosivo 3114
- di naftalina 3558
- d'incenerazione a cono rotativo 4463
- d'incenerazione delle immondizie 4328
- d'incrudimento per deformazione a freddo 5262
- d'incrudimento strutturale 126
- d'indurimento al cianuro 872
- d'indurimento per rotazione 5053
- di nitrazione 3598
- di nitrificazione 3604
- di perdistillazione 3857
- di piroscissione 1371
- di piroscissione a letto mobile 3522
- di piroscissione con catalizzatore fluido 2316
- di piroscissione in fase liquida 3247
- di piroscissione in fase mista 3466
- di piroscissione in fase vapore 5731

NEDERLANDS

NL

DEUTSCH

D

ELSEVIER'S

DICTIONARY OF CHEMICAL ENGINEERING

VOLUME 2

CHEMICAL ENGINEERING; PROCESSES AND PRODUCTS

ELSEVIER PUBLISHING COMPANY
335 JAN VAN GALENSTRAAT
P.O. BOX 211, AMSTERDAM, THE NETHERLANDS

ELSEVIER PUBLISHING CO. LTD.
BARKING, ESSEX, ENGLAND

AMERICAN ELSEVIER PUBLISHING COMPANY, INC.
52 VANDERBILT AVENUE
NEW YORK, NEW YORK 10017

LIBRARY OF CONGRESS CARD NUMBER: 68–54865
STANDARD BOOK NUMBER 444–40715–4

PRINTED IN THE NETHERLANDS

ELSEVIER'S

DICTIONARY OF CHEMICAL ENGINEERING

VOLUME 2

CHEMICAL ENGINEERING; PROCESSES AND PRODUCTS

IN SIX LANGUAGES

ENGLISH/AMERICAN - FRENCH - SPANISH
ITALIAN - DUTCH - GERMAN

COMPILED AND ARRANGED ON
AN ENGLISH ALPHABETICAL BASIS BY

W. E. CLASON

FORMERLY HEAD OF THE TRANSLATION DEPARTMENT
N.V. PHILIPS, GLOEILAMPENFABRIEKEN
EINDHOVEN (THE NETHERLANDS)

ELSEVIER PUBLISHING COMPANY
AMSTERDAM/LONDON/NEW YORK
1968

ELSEVIER'S

DICTIONARY OF CHEMICAL ENGINEERING

VOLUME 1
CHEMICAL ENGINEERING AND LABORATORY EQUIPMENT

VOLUME 2
CHEMICAL ENGINEERING; PROCESSES AND PRODUCTS

A complete catalogue listing all Elsevier dictionaries will be sent upon request

PUBLISHER'S NOTE

By the most conservative estimates, there are at least two thousand basic languages in current use today. By some definitions, this figure could reach more than five thousand languages, not including a vast number of dialects. Added to this, the incredible current advances in engineering and science have rendered most existing technical dictionaries obsolete. The accumulated result has been a fast growing need for specialized multilingual dictionaries, particularly in the face of international scientific cooperation and exchange of knowledge.

To meet this need, Elsevier publishes a number of multilingual technical dictionaries relating to special fields of science and industry. Edited under authoritative auspices, they draw upon rich sources of knowledge.

In planning this dictionary, the author and publishers have been guided by the principles proposed by the United Nations Educational, Scientific and Cultural Organization (UNESCO). The aim is to ensure that each dictionary will fit into place in a pattern which may progressively extend over all interrelated fields of science and technology and cover all necessary languages.

For each language, there is an alphabetical list of words referring to corresponding numbers in the basic table. The system of thumb-indexing enables one to find any language at one. The binding, smooth paper and convenient size result in an enjoyable and valuable reference book.

PREFACE

In compiling this dictionary the author has taken the view that, in general, it is unnecessary to include purely chemical terms, such as natrium-chloride or installations and processes where the name is derived from a proper noun. This basic principle was, however, not adhered to in the case of:

1. a chemical substance with more than one name
2. an installation or process also known under a name not derived from a proper noun.

In view of the great diversity of terms, the author has had to limit the subjects under abbreviations to the most important ones, whereby the selection was not always an easy matter.

Finally, for some items an attempt has been made to normalize the terminology in current use.

The author wishes to thank those both at home and abroad who have assisted him in his work. A special word of thanks is due to his wife, who took care of the entire manuscript, and checked the spelling and terminology with great accuracy.

September, 1968 W.E. Clason

ABBREVIATIONS

aam	acetic acid manufacture
aap	auxiliary apparatus
abs	absorption, adsorption etc.
act	acid treatment
agr	agricultural chemistry
aic	inorganic industrial chemistry
ana	analysis
app	general apparatus
bac	bacteriological chemistry
ble	bleaching
bre	brewing
brk	breaking and milling
boi	boilers and boiling
bui	building materials
bur	burners and burning
cen	centrifuges and centrifuging
cer	ceramics
chr	chromatography
cle	cleaning and degreasing
clo	cooling
coa	coating(s)
col	colloids
con	control and regulation
cor	corrosion
cov	conserving industry
cpt	component parts
cry	crystallography and crystallization
ctt	cold treatment
dai	dairy industry
dec	decorating and painting
dis	distillation
dry	drying
dye	dyeing and printing
elc	electrochemistry
eva	evaporation
exp	explosives
ext	extraction
fat	fats and oils
fea	fire extinguishers and extinguishing
fer	fertilizers
fil	filters and filtration
fmi	fermentation
fou	foundry
fue	fuels, petroleum industry
gen	general subjects
gla	glass
glu	glues
htt	high temperature treatment
hyg	hygienics
imp	impregnation
ins	instruments
iso	isotopes
iwt	industrial waste treatment
lab	laboratory apparatus
lea	leather industry
lub	lubrication
mat	sundry materials
mcy	medical and pharmaceutical chemistry
met	metals and alloys

min	mining and minerals	sap	safety appliances
mit	military chemistry	sep	separators and separating
mix	mixers and mixing	sin	sugar industry
		soa	soap industry
nue	nuclear chemistry	sor	sorting
nit	nutrition including beverages	spe	spectroscopy and spectrography
		sto	storage
oic	organic industrial chemistry	sua	sulfuric acid manufacture
pap	paper industry	tan	tanning industry
per	perfumery	tes	testing apparatus and methods
pho	photography	tex	textile chemistry
pla	plastics	tra	transport
pol	polishing		
pre	pressurized apparatus and processes	vac	vacuum processes and apparatus
rad	radiation chemistry	wel	welding and soldering
rub	rubber industry		

---

adj	adjective	f	Français
f	feminine	e	Español
GB	English, British usage	i	Italiano
m	masculine	n	Nederlands
n	neuter	d	Deutsch
pl	plural		
US	English, American usage		
v	verb		

BASIC TABLE

A

1 ABAISER (US), dec
IVORY BLACK (GB)
f noir *m* d'ivoire
e negro *m* de marfil
i nero *m* d'avorio
n ivoorzwart *n*
d Elfenbeinschwarz *n*

2 ABELITE exp
f abélite *f*
e abelita *f*
i abelite *f*
n abeliet *n*
d Abelit *m*

3 ABLATION coa,htt
f ablation *f*
e ablación *f*
i ablazione *f*
n afsmelting
d Abbrennen *n*, Abschmelzung *f*

4 ABLATIVE RESIN coa,mat,sap
f résine *f* ablative
e resina *f* ablativa
i resina *f* ablativa
n afsmeltende kunsthars
d Abschmelzkunstharz *n*

5 ABLUTION, iwt
WASHING OUT
f ablution *f*
e ablución *f*
i abluzione *f*
n uitwassing
d Abspülung *f*, Abwaschung *f*, Auswaschung *f*

6 ABRADANT, mat
ABRASIVE
f abrasif *m*, agent *m* abrasif
e abrasivo *m*
i abradente *m*, abrasivo *m*
n schuurmiddel *n*, slijpmiddel *n*
d Abreibungsmittel *n*, Putzkörper *m*, Schleifmittel *n*

7 ABRASIAN OIL fat
f huile *f* d'abrasin, huile *f* de bois d'Indochine, huile *f* de faux baucoulier
e aceite *m* de abrasina
i olio *m* d'abrasine
n abrasineolie, montanahoutolie
d Abrasinöl *n*

8 ABRASION, wea
ABRASIVE WEAR, WEAR BY FRICTION
f abrasion *f*, usure *f*
e abrasión *f*, desgaste *m* por abrasión
i abrasione *f*, usura *f*
n slijtage
d Abnutzung *f* durch Reibung, Verschleiss *m*

9 ABRASION TEST tes
f essai *m* d'usure
e prueba *f* de desgaste por abrasión
i prova *f* d'usura
n slijtageproef
d Abnutzungsprüfung *f*, Verschleissprüfung *f*

10 ABRASIVE FINISHING pol
f finissage *m* par abrasifs
e acabado *m* con abrasivos
i finitura *f* con abrasivi
n schurend polijsten *n*
d Feinen *n* mit Schleifmittel

11 ABRIDGED MASHING bre
SYSTEM
f procédé *m* à trempe courte
e procedimiento *m* de mezcla de corta duración y a alta temperatura
i processo *m* a tempera rapida
n snelbeslagprocédé *n*
d Hochkurzmaischverfahren *n*

12 ABSINTHIUM OIL, fat
ABSINTHOLE, WORMWOOD OIL
f essence *f* d'absinthe
e esencia *f* de ajenjo
i olio *m* d'assenzio
n absintol *n*, absintolie
d Wermutöl *n*

13 ABSORBATE abs
f substance *f* absorbée
e absorbato *m*
i sostanza *f* assorbita
n absorptief *n*
d Absorptiv *n*

14 ABSORBENT abs
f absorbant *m*
e absorbente *m*
i assorbente *m*

n absorbens *n*
d Absorbens *n*, absorbierendes Mittel *n*

15 ABSORPTION abs
f absorption *f*
e absorción *f*
i assorbimento *m*
n absorptie
d Absorption *f*

16 ABSORPTION OIL, WASHING OIL cle,gas
f huile *f* lourde pour débenzolage du gaz
e aceite *m* de absorción
i olio *m* d'assorbimento
n wasolie
d Waschöl *n*

17 ABSORPTIVENESS, ABSORPTIVITY abs
f absorptivité *f*
e absorbencia *f*
i assorbilità *f*
n absorptievermogen *n*
d Absorptionsfähigkeit *f*, Aufnahmefähigkeit *f*, Saugfähigkeit *f*

18 ACACIA GUM, ARABINE, GUM ARABIC, SENEGAL GUM glu
f gomme *f* arabique, gomme *f* d'acacia
e goma *f* arábiga, goma *f* Senegal
i gomma *f* arabica, gomma *f* d'acacia
n acaciagom, Arabische gom, kordofangom, senegalgom
d Akaziengummi *m*, Arabin *n*, Gummiarabicum *n*, Senegalgummi *m*

19 ACCELERATED AGEING TEST (GB), ACCELERATED AGING TEST (US) tes
f essai *m* accéléré de vieillissement
e prueba *f* acelerada de envejecimiento
i collaudo *m* accelerato d'invecchiamento
n versnelde verouderingsproef
d beschleunigte Alterungsprüfung *f*, Schnellalterungsversuch *m*

20 ACCELERATED FILTRATION fil
f filtrage *m* accéléré, filtration *f* accélérée
e filtración *f* acelerada
i filtrazione *f* accelerata
n versneld filtreren *n*
d Saugfilterung *f*, Schnellfilterung

21 ACCELERATED WEATHERING TEST tes
f essai *m* accéléré de vieillissemen par intempéries
e prueba *f* acelerada de intemperismo
i collaudo *m* d'invecchiamento per intemperie
n versnelde verweringsproef
d beschleunigte Bewitterungsprüfung *f*

22 ACCELERATION gen,rub
f accélération *f*
e aceleración *f*
i accelerazione *f*
n versnelling
d Beschleunigung *f*

23 ACCELERATOR gen,rub
f accélérateur *m*
e acelerador *m*, subtancia *f* aceleradora
i accelerante *m*
n versneller
d Beschleuniger *m*, Vulkanisationsbeschleuniger *m*

24 ACCELERATOR, HARDENER pla
f agent *m* durcisseur
e acelerador *m*, agente *m* endurecedor
i accelerante *m*
n hardingsmiddel *n*
d Härter *m*, Härtungsmittel *n*

25 ACCELERATOR ACTIVATOR rub
f activateur *m* de l'accélérateur
e activador *m* del acelerador
i attivatore *m* dell'accelerante
n versnelleractivator
d Beschleunigeraktivator *m*

26 ACCELERATOR RETARDER rub
f retardateur *m* de l'accélérateur
e retardador *m* del acelerador
i ritardante *m* dell'accelerante
n versnellervertrager
d Beschleunigerverzögerer *m*

27 ACCEPTOR gen
f accepteur *m*
e aceptor *m*
i accettore *m*
n acceptor
d Akzeptor *m*

28 ACCLIMATATION (US), ACCLIMATISATION (GB) col
f acclimatation *f*
e aclimatación *f*
i acclimatazione *f*
n acclimatisering
d Akklimatisation *f*, Akklimatisierung *f*

29 ACCRETION cry
f accroissement *m*
e acrecimiento *m*, acreción *f*
i accrescimento *m*
n aangroeiing
d Anwachsen *n*

30 ACCUMULATION gen
f accumulation *f*
e acumulación *f*
i accumulazione *f*, addensamento *m*
n accumulatie, ophoping, verzameling
d Akkumulation *f*, Anhäufung *f*

31 ACETALDEHYDE, ACETYL HYDRIDE, ETHANAL oic
f acétaldéhyde *m*, aldéhyde *m* acétique
e acetaldehido *m*
i acetaldeide *f*, aldeide *f* acetica, aldeide *f* etilica
n acetaldehyde *n*, ethanal *n*
d Azetaldehyd *n*

32 ACETANILIDE, ANTIFEBRIN, PHENYL-ACETAMIDE mcy
f acétanilide *f*, antifébrile *m*
e acetanilida *f*, antifebrina *f*, fenilacetamida *f*
i acetanilide *f*
n acetanilide *n*, antifebrine *n*
d Antifebrin *n*, Azetanilid *n*

33 ACETATE FILM, SAFETY FILM pho
f film *m* de sécurité, pellicule *f* à l'acétate de cellulose
e película *f* de acetato de celulosa
i pellicola *f* all'acetato di cellulosa
n acetaatfilm
d Azetatfilm *m*, Sicherheitsfilm *m*

34 ACETATE PROCESS tex
f procédé *m* acétate
e procedimiento *m* acetato
i processo *m* acetato
n acetaatprocédé *n*
d Azetatverfahren *n*

35 ACETATE RAYON, ACETATE SILK tex
f rayonne *f* d'acétate
e rayón *m* de acetato de celulosa
i raion *m* all'acetato
n acetaatzijde
d Azetatseide *f*

36 ACETATE STAPLE FIBRE tex
f fil *m* coupé d'acétate
e fibra *f* cortada de acetato
i fibra *f* tagliata d'acetato
n acetaatstapelvezel
d Azetatstapelfaser *f*

37 ACETIC ACID BACTERIA, ACETOBACTERIA aam
f bactéries *pl* acétiques, mycodermes *pl* du vinaigre
e acetobacterias *pl* bacterias *pl* acéticas
i acetobacteri *pl*, batteri *pl* acetici
n azijnzuurbacteriën *pl*
d Essigbakterien *pl*, Essigsäurebakterien *n*

38 ACETIC ESTER, ACETIC ETHER, ETHYL ACETATE mcy,oic
f acétate *m* d'éthyle, ester *m* acétique
e acetato *m* etílico, éster *m* acético
i acetato *m* d'etile, etere *m* acetico
n azijnzure ester, ethylacetaat *n*
d Äthylazetat *n*, Essigäther *m*, Essigester *m*, Essigsäure-äthylester *m*

39 ACETIC FERMENTATION, ACETIFICATION, ACETOUS FERMENTATION aam,fmi
f fermentation *f* acétique
e fermentación *f* acética
i fermentazione *f* acetica
n azijnzure gisting
d Essiggärung *f*, Essigsäure-gärung *f*

40 ACETIFIER aam
f acétifiant *m*
e acetificador *m*
i acetificante *m*
n snelverzuuragens *n*
d Schnellsäurer *m*

41 ACETIMETRY, ACETOMETRY ana
f acétimétrie *f*

e acetimetría *f*
i acetimetria *f*
n acetometrie, azijnzuurmeting
d Azetometrie *f*, Essigsäure-
messung *f*

42 ACETOACETIC ACID, oic
ACETYL ACETIC ACID,
BUTANONE ACID
f acide *m* acétylacétique
e ácido *m* acetoacético
i acido *m* acetoacetico
n acetylazijnzuur *n*
d Azetessigsäure *f*

43 ACETOLYSIS oic
f acétolyse *f*
e acetólisis *f*
i acetolisi *f*
n acetolyse
d Azetolyse *f*

44 ACETONE, oic
DIMETHYL KETONE,
KETOPROPANE, PROPANONE
f acétone *f*, éther *m* pyroacétique
e acetona *f*, dimetil-cetona *f*
i acetone *m*, dimetilchetone *m*
n aceton *n*, dimethylketon *n*,
propanon *n*
d Azeton *n*, Dimethylketon *n*

45 ACETYLATION, oic,tex
ACETYLIZATION
f acétylation *f*
e acetilación *f*
i acetilazione *f*
n acetyleren *n*
d Azetylierung *f*

46 ACETYLATOR tex
f agent *m* d'acétylation
e acetilador *m*
i agente *m* d'acetilazione
n acetyleermiddel *n*
d Azetylator *m*, Azetylierungs-
mittel *n*

47 ACETYLCELLULOSE, oic
CELLULOSE ACETATE
f acétate *m* de cellulose,
acétocellulose *f*
e acetato *m* celulósico,
acetocelulosa *f*
i acetato *m* di cellulosa
n acetylcellulose, cellulose-
acetaat *n*
d Azetylzellulose *f*, Zellulose-
azetat *n*

48 ACETYLCHLORIDE oic
f chlorure *f* d'acétyle
e cloruro *m* de acetilo
i cloruro *m* d'acetile
n acetylchloride *n*,
methaancarbochloride *n*
d Azetylchlorid *n*,
Chlorazetyl *n*

49 ACETYLENE, oic
ALDEHYDENE, ETHINE
f acétylène *m*, éthyne *m*
e acetileno *m*
i acetilene *m*
n acetyleen *n*, ethyn *n*
d Äthyn *n*, Azetylen *n*

50 ACETYLENE BLACK oic
f noir *m* d'acétylène
e hollín *m* de acetileno,
negro *m* de humo de acetileno
i nero *m* d'acetilene
n acetyleenzwart *n*
d Azetylenruss *m*

51 ACHROMIC PERIOD fmi,gl
f période *f* achromique
e período *m* acrómico
i periodo *m* acromico
n kleurloze tijd
d farblose Zeit *f*

52 ACID ge
f acide *m*
e ácido *m*
i acido *m*
n zuur *n*
d Säure *f*

53 ACID AGEING (GB), te
ACID AGING (US)
f traitement *m* par vapeur acide
e tratamiento *m* por húmedo ácid
i trattamento *m* per vapore acido
n zure stoombehandeling
d Säuredampfbehandlung *f*

54 ACID ANNEALING me
f recuit *m* acide
e recocido *m* ácido
i ricottura *f* acida
n zuur uitgloeien *n*
d saures Ausglühen *n*

55 ACID COOKING LIQUOR pa
f liquide *m* acide de cuisson
e líquido *m* ácido de cocción
i liquido *m* acido di cottura
n kookzuur *n*
d Kochsäure *f*

56 ACID COOKING PROCESS pap
ACID PULPING
f préparation *f* de la pâte chimique à l'acide, procédé *m* de cuisson acide
e procedimiento *m* de cocción ácida, pulpado *m* ácido
i processo *m* di cottura acida, riduzione *f* della polpa con acido
n zuur kookproces *n*, zuur ontsluitingsproces *n*
d saurer Aufschluss *m*, saures Aufschliessungsverfahren *n*, saures Kochverfahren *n*

57 ACID EMBOSSING, elc
ACID ETCHING
f gravure *f* à l'acide
e grabado *m* al ácido
i incisione *f* all'acido
n zuuretsen *n*
d Säureätzen *n*

58 ACID FIXING pho
f fixage *m* acide
e fijación *f* ácida
i fissazione *f* acida
n zuurfixeren *n*
d Säurefixieren *n*

59 ACID FORMATION bre
f formation *f* d'acide
e formación *f* de ácido
i formazione *f* d'acido
n zuurvorming
d Säurebildung *f*

60 ACID FROSTING coa,gla
f matage *m* à l'acide
e esmerilado *m* al ácido
i smerigliatura *f* all'acido
n zuurmatteren *n*
d Säuremattieren *n*

61 ACID FULLING, tex
ACID MILLING
f foulage *m* en liquide acide
e batanado *m* en líquido ácido
i follatura *f* in liquido acido
n vollen *n* in zuur, zuurvollen *n*
d Walken *n* in Säure

62 ACID INDEX, bre,gen
ACID NUMBER, ACID VALUE
f indice *m* d'acide, valeur *f* d'acide, valeur *f* d'acidité
e índice *m* de acidez
i indice *m* d'acido, valore *m* d'acidità
n zuurgetal *n*
d Säuregrad *m*, Säurezahl *f*

63 ACID PICKLING elc
f décapage *m* à l'acide
e decapaje *m* al ácido
i decapaggio *m* con acido
n zuurbeitsen *n*
d Säurebeizen *n*

64 ACID POLISHING gla
f polissage *m* à l'acide
e pulimento *m* al ácido
i pulitura *f* all'acido
n zuurpolijsten *n*
d Säurepolieren *n*

65 ACID PROCESS met
f procédé *m* acide
e procedimiento *m* ácido
i processo *m* acido
n zuur proces *n*
d saures Verfahren *n*

66 ACID PROCESS rub
REGENERATION,
ACID RECLAIMING PROCESS
f procédé *m* de régénération à l'acide
e procedimiento *m* de regeneración al ácido
i processo *m* di rigenerazione all'acido
n zuur-regenereerproces *n*
d Regenierierungsverfahren *n* mit Säure, Säure-Regenerierverfahren *n*

67 ACID-RESISTING PAINT dec
f peinture *f* anti-acide
e pintura *f* antiácida
i pittura *f* antiacida
n zuurvaste verf
d säurefeste Anstrichfarbe *f*

68 ACID SALT, gen
HYDROGEN SALT
f sel *m* acide
e sal *f* ácida
i sale *m* acido
n zuur zout *n*
d Sauersalz *n*, saures Salz *n*

69 ACID SEAL PAINT coa,dec,rub
f peinture *f* résistante aux acides pour revêtements
e pintura *f* ácidorresistente para revestimientos
i vernice *f* acidoresistente per rivestimenti

n zuurbestendige bekledingslak
d säurefester Anstrich *m* für Auskleidung

70 ACID SPOTTING tes,tex
f essai *m* de résistance à l'acide
e prueba *f* de resistencia al ácido
i collaudo *m* di resistenza all'acido
n zuurbestendigheidsproef
d Säurefestigkeitsprüfung *f*

71 ACID STEEL, BESSEMER STEEL met
f acier *m* acide
e acero *m* ácido
i acciaio *m* acido
n bessemerstaal *n*
d Bessemerstahl *m*

72 ACID TINNING elc
f étamage *m* à l'acide
e estañado *m* al ácido
i stagnatura *f* all'acido
n zuurvertinnen *n*
d Säureverzinnen *n*

73 ACIDIFICATION, ACIDULATION gen
f acidification *f*
e acidificación *f*
i acidificazione *f*
n aanzuren *n*
d Ansäuern *n*, Sauermachen *n*, Säuern *n*, Säuerung *f*

74 ACIDIMETRY ana
f acidimétrie *f*
e acidimetría *f*
i acidimetria *f*
n acidimetrie, zuurgraadmeting
d Azidimetrie *f*, Säurebestimmung *f*

75 ACIDPROOF MATERIAL mat
f matière *f* résistante aux acides
e materia *f* acidorresistente
i materia *f* resistente agli acidi
n zuurbestendig materiaal *n*
d säurefestes Material *n*

76 ACIDPROOF TEST tes
f essai *m* de résistance à l'acide
e ensayo *m* de resistencia al ácido
i collaudo *m* di resistenza all'acido
n zuurvastheidsproef
d Säurebeständigkeitsprüfung *f*

77 ACIERAGE, IRON PLATING, STEELING co
f aciérage *m*
e acieración *f*, hierrado *m*
i acciaiatura *f*
n verstaling, verijzering
d Stählung *f*, Verstählung *f*

78 ACROLEIC ACID, ACRYLIC ACID, PROPENE ACID oi
f acide *f* acrylique
e ácido *m* acrílico
i acido *m* acrilico
n acrylzuur *n*, etheencarbonzuur
d Akrylsäure *f*, Propensäure *f*

79 ACROLEINE, ACRYLALDEHYDE, PROPENAL oi
f acroléine *f*, aldéhyde *f* acrylique
e acraldehido *m*, acroleína *f* aldehido *m* acrílico
i acroleina *f*, aldeide *m* acrilico
n acroleïne *n*, acrylaldehyde *n*
d Akrolein *n*, Akrylaldehyd *n*

80 ACTIVATED ADSORPTION, CHEMISORPTION ger
f adsorption *f* activée
e adsorción *f* activada
i adsorbimento *m* attivato
n chemisorptie, geactiveerde adsorptie
d aktivierte Adsorption *f*, Chemisorption *f*

81 ACTIVATED CARBON, ACTIVATED CHARCOAL, ACTIVE CARBON abs
f charbon *m* actif
e carbón *m* activado, carbón *m* activo
i carbone *m* attivo
n actieve kool
d Aktivkohle *f*

82 ACTIVATED CHALK rub
f craie *f* activée
e creta *f* activada
i gesso *m* attivato
n geactiveerd krijt *n*
d aktivierte Kreide *f*

83 ACTIVATED SINTERING elc
f frittage *m* activé
e sinterización *f* activada
i sinterizzazione *f* attivata
n geactiveerd sinteren *n*
d aktiviertes Sintern *n*

84 ACTIVATED SLUDGE iwt
f boue *f* activée
e fango *m* activado, fango *m* biológicamente activo
i fanghiglia *f* attivata
n geactiveerd slib *n*
d belebter Schlamm *m*, Belebtschlamm *m*

85 ACTIVATED SLUDGE PROCESS iwt
f procédé *m* d'épuration pour boue activée
e procedimiento *m* con fango biológicamente activo
i processo *m* con fanghiglia attivata
n geactiveerd-slibproces *n*
d Belebtschlammverfahren *n*

86 ACTIVATED WATER rad
f eau *f* activée
e agua *f* activada
i acqua *f* attivata
n radioactief water *n*
d radioaktives Wasser *n*

87 ACTIVATION gen,htt,sew
f activation *f*
e activación *f*
i attivazione *f*
n activering
d Aktivierung *f*, Belebung *f*

88 ACTIVATION nue
f activation *f*
e activación *f*
i attivazione *f*
n activering
d Aktivierung *f*

89 ACTIVATION ANALYSIS ana
f analyse *f* par activation
e análisis *f* por activación
i analisi *f* per attivazione
n activeringsanalyse
d Aktivierungsanalyse *f*

90 ACTIVATION ENERGY gen
f énergie *f* d'activation
e energía *f* de activación
i energia *f* d'attivazione
n activeringsenergie
d Aktivierungsenergie *f*

91 ACTIVE COLOR (US), ACTIVE COLOUR (GB), ACTIVE PAINT dec
f pigment *m* chimiquement actif
e pigmento *m* químicamente activo
i pigmento *m* chimicamente attivo
n actief pigment *n*
d aktives Pigment *n*, seifenbildendes Pigment *n*

92 ACTIVE FILLER, REINFORCING FILLER rub
f charge *f* active, charge *f* renforçante
e carga *f* activa, carga *f* reforzadora
i carica *f* attiva, carica *f* rinforzante
n actieve vulstof, versterkende vulstof
d aktiver Füllstoff *m*, aktives Füllmittel *n*, Verstärkerfüllmittel *n*

93 ACTIVE HYDROGEN, ATOMIC HYDROGEN gas
f hydrogène *m* actif
e hidrógeno *m* activo
i idrogeno *m* attivo
n actieve waterstof, atomaire waterstof
d aktiver Wasserstoff *m*, atomarer Wasserstoff *m*

94 ACTIVE INGREDIENT gen
f ingrédient *m* actif
e ingrediente *m* activo
i ingrediente *m* attivo
n werkzaam bestanddeel *n*, werkzaam ingrediënt *n*
d Wirkstoff *m*

95 ACTIVE NITROGEN gas
f azote *m* actif
e nitrógeno *m* activo
i nitrogeno *m* attivo
n actieve stikstof, atomaire stikstof
d aktiver Stickstoff *m*, atomarer Stickstoff *m*

96 ACTIVE OXYGEN gas
f oxygène *m* actif
e oxígeno *m* activo
i ossigeno *m* attivo
n actieve zuurstof, atomaire zuurstof
d aktiver Sauerstoff *m*, atomarer Sauerstoff *m*

97 ACTIVE SOLVENT dec
f solvant *m* actif
e solvente *m* activo
i solvente *m* attivo

n actief oplosmiddel *n*
d echtes Lösungsmittel *n*

98 ACYLATION oic
f acylation *f*
e acilación *f*
i acilazione *f*
n acyleren *n*
d Azylierung *f*

99 ADAMANTINE CLINKER bui
f brique *f* très dure
e ladrillo *m* muy duro
i mattone *m* molto duro
n lichtrode baksteen
d hellroter Klinker *m*

100 ADDITION, gen
ADDITION REACTION,
ADDITIVE REACTION
f réaction *f* d'addition
e reacción *f* por adición
i reazione *f* d'addizione
n additie
d Addition *f*, Additionsreaktion *f*,
Anlagerungsreaktion *f*

101 ADDITION AGENT elc
f agent *m* d'addition
e agente *m* de adición
i agente *m* d'addizione
n badtoevoeging
d Badzusatz *m*

102 ADDITIVE COMPOUND gen
f composé *m* d'addition
e compuesto *m* de adición
i composto *m* derivato per addizione
n additieverbinding
d Additionsverbindung *f*

103 ADERMIN, oic
PYRIDOXYN
f adermine *f*, pyridoxyne *f*
e piridoxina *f*
i adermina *f*, piridoxina *f*
n adermine *n*, vitamine *n* B_6
d Adermin *n*, Pyridoxin *n*,
Vitamin *n* B_6

104 ADHESION gen
f adhésion *f*
e adhesión *f*
i adesione *f*
n adhesie
d Adhäsion *f*

105 ADHESIVE glu
f adhésif *m*, agglutinant *m*
e adhesivo *m*
i adesivo *m*
n kleefmiddel *n*, kleefstof,
plakmiddel *n*
d Bindemittel *n*, Klebemittel *n*,
Klebstoff *m*

106 ADHESIVE FABRIC glu
f tissu *m* adhésif, tissu *m* autocollant
e tela *f* adhesiva
i tessuto *m* adesivo
n zelfklevend weefsel *n*
d selbstklebendes Gewebe *n*

107 ADHESIVE FOIL, pho
ADHESIVE TISSUE,
MOUNTING TISSUE
f feuille *f* adhésive pour montage sec
e tisú *m* para montar en seco
i foglia *f* adesiva per montare le copie
n adhesieve foelie, droogopzet-foelie
d Klebefolie *f*, Trockenklebefolie *f*

108 ADHESIVE INSULATING TAPE elc
f ruban *m* adhésif isolant
e cinta *f* adhesiva aisladora
i nastro *m* adesivo isolante
n klevend isolatieband *n*
d Klebeisolierband *n*

109 ADHESIVE PLASTER (GB), mcy
STICKING PLASTER (US)
f emplâtre *m* adhésif,
emplâtre *m* agglutinatif
e emplaste *m* adhesivo,
esparadrapo *m*, parche *m* adhesivo
i cerotto *m*, impiastro *m* adesivo
n hechtpleister, plakpleister
d Heftpflaster *n*

110 ADHESIVE TAPE (GB), cpt
SELF-ADHESIVE TAPE (GB),
STICKING TAPE (US)
f ruban *m* adhésif, ruban *m* collant
e cinta *f* adhesiva
i nastro *m* adesivo
n plakband *n*
d Klebeband *n*, Klebstreifen *m*

111 ADHESIVE VARNISH coa,dec
f vernis *m* adhésif
e barniz *m* adhesivo
i vernice *f* adesiva
n kleeflak
d Klebelack *m*

112 ADJECTIVE DYE dye
f colorant *m* adjectif
e colorante *m* adjetivo
i colorante *m* adiettivo
n adjectieve kleurstof
d adjektiver Farbstoff *m*

113 ADSORBATE abs
f substance *f* adsorbée
e adsorbato *m*
i sostanza *f* adsorbita
n adsorptief *n*
d Adsorptiv *n*

114 ADSORBENT abs
f adsorbant *m*
e adsorbante *m*
i adsorbente *m*
n adsorbens *n*
d Adsorbens *n*

115 ADSORPTION abs
f adsorption *f*
e adsorción *f*
i adsorbimento *m*
n adsorptie
d Adsorption *f*

116 ADSORPTION ANALYSIS ana
f analyse *f* par adsorption
e análisis *f* por adsorción
i analisi *f* per adsorbimento
n adsorptieanalyse
d Adsorptionsanalyse *f*

117 ADSORPTION CATALYSIS gen
f catalyse *f* par adsorption
e catálisis *f* por adsorción
i catalisi *f* per adsorbimento
n adsorptiekatalyse
d Adsorptionskatalyse *f*

118 ADSORPTION RECOVERY rub
f récupération *f* par adsorption
e recuperación *f* por adsorción
i ricupero *m* per adsorbimento
n terugwinning door adsorptie
d Adsorptionsrückgewinnung *f*

119 ADULTERATION gen
f adultération *f*
e adulteración *f*
i adulterazione *f*
n vervalsing
d Fälschung *f*, Verfälschung *f*

120 ADULTERING AGENT gen
f agent *m* d'adultération
e agente *m* adulterante
i agente *m* adulterante
n vervalsingsmiddel *n*
d Fälschungsmittel *n*, Verfälschungsmittel *n*

121 AERATED CONCRETE, CELLULAR CONCRETE, FOAMED CONCRETE, GAS CONCRETE bui
f béton *m* cellulaire
e hormigón *m* aerado
i calcestruzzo *m* poroso
n cellenbeton *n*
d Zellenbeton *m*

122 AERATED PLASTIC, FOAMED PLASTIC pla
f matière *f* plastique aérée, matière *f* plastique mousse
e materia *f* plástica celular
i materia *f* plastica cellulare, materia *f* plastica schiuma
n schuimkunststof, schuimplastiek *n*
d Schaumkunststoff *m*, Schaumstoff *m* mit nichtporösen Zellen

123 AERATED WATER, SODA WATER bre
f eau *f* de Seltz, eau *f* gazeuse
e agua *f* de Seltz, agua *f* gaseosa, gaseosa *f*
i acqua *f* di Seltz, acqua *f* gassosa
n sodawater *n*, spuitwater *n*
d kohlensaures Wasser *n*, Sodawasser *n*

124 AERATION iwt
f aérage *m*, aération *f*
e aeración *f*, aireación *f*
i aerazione *f*
n aëratie
d Belüftung *f*

125 AERATION CELL CORROSION, DIFFERENTIAL AERATION CORROSION cor
f corrosion *f* par aération différentielle
e corrosión *f* por aeración diferencial
i corrosione *f* per aerazione differenziale
n corrosie door differentiële aëratie
d Korrosion *f* durch unterschiedliche Belüftung

126 AERIAL OXIDATION, AIR OXIDATION cor
f oxydation *f* à l'air

e oxidación *f* al aire
i ossidazione *f* all'aria
n oxydatie aan de lucht
d Luftoxydation *f*, Oxydation *f* an der Luft, Oxydierung *f* an der Luft

127 AEROSOL col
f aérosol *m*
e aerosol *m*
i aerosole *m*
n aërosol *n*
d Aerosol *n*

128 AEROSOL LACQUER, PUSH-BUTTON PAINT dec
f peinture *f* aérosol
e pintura *f* de aerosol
i pittura *f* d'aerosole
n aërosollak
d Aerosollack *m*, Sprühdosenlack *m*

129 AERUGO, VERDIGRIS dec
f verdet *m*, vert-de-gris *m*
e cardenillo *m*, verdín *m*
i erugine *f*, verderame *m*
n kopergroen *n*
d Edelrost *m*, Grünspan *m*

130 AFFINATION sin
f affination *f*
e afinación *f*
i affinazione *f*
n raffineren *n*
d Affination *f*, Raffinieren *n*

131 AFTERBODYING, AFTERTHICKENING dec
f post-épaississement *m*
e post espesamiento *m*
i post addensamento *m*
n nadikken *n*
d Nachdicken *n*

132 AFTERCHROME DYEING, BACK-CHROME DYEING, TOP CHROMING tex
f chromatage *m*, teinture *f* postérieure au chrome
e cromado *m* secundario, postmordentado *m* crómico
i tintura *f* posteriore al cromo
n nachromeren *n*
d Nachchromieren *n*

133 AFTERCONTRACTION bui
f contraction *f* secondaire
e contracción *f* secundaria, postcontracción *f*
i contrazione *f* secondaria
n nakrimpen *n*
d Nachschwinden *n*

134 AFTERCOOLING clo
f réfrigération *f* secondaire
e postrefrigeración *f*
i postrefrigerazione *f*
n nakoelen *n*
d Nachkühlung *f*

135 AFTERCOPPERING tex
f traitement *m* postérieur au sel de cuivre
e postratamiento *m* con sal de cobre
i trattamento *m* posteriore con sale di rame
n nabehandeling met koperzout
d Nachbehandlung *f* mit Kupfersalz

136 AFTERCURE, AFTERVULCANISATION rub
f post-vulcanisation *f*
e postvulcanización *f*
i postvulcanizzazione *f*
n navulcanisatie
d Nachvulkanisation *f*

137 AFTEREXPANSION bui,cer
f expansion *f* secondaire
e expansión *f* secundaria
i espansione *f* secondaria
n na-uitzetting
d Nachdehnung *f*

138 AFTERFERMENTATION fmi
f fermentation *f* secondaire
e fermentación *f* secundaria
i fermentazione *f* secondaria
n nagisten *n*
d Nachgärung *f*

139 AFTERFRACTIONATION dis
f fractionnement *m* postérieur
e fraccionamiento *m* posterior
i frazionamento *m* posteriore
n nafractioneren *n*
d Nachfraktionierung *f*

140 AFTERRIPENING, CHEMICAL RIPENING, DIGESTION, SECOND RIPENING pho
f maturation *f* chimique, maturation *f* en refonte
e maduración *f* química, segunda maduración *f*
i maturazione *f* chimica
n chemische rijping
d chemische Reifung *f*, Nachreifun

141 AFTERRUN, LAST RUNNING, SECOND RUNNING dis
f queue *f* de distillation
e cola *f* de destilación
i coda *f* di distillazione
n naloop
d Nachlauf *m*

142 AFTERTACK, STICKINESS dec
f tendance *f* au collant
e pegajosidad *f*
i appiccicatura *f*, collosità *f*
n kleverigheid, pikkigheid
d Klebrigkeit *f*

143 AFTERTREATMENT tex
f traitement *m* postérieur, traitement *m* secondaire
e postratamiento *m*, tratamiento *m* secundario
i trattamento *m* posteriore, trattamento *m* secondario
n nabehandeling
d Nachbehandlung *f*

144 AFTERWORT, SECOND RUN OF WORT bre
f trempe *f* de lavage
e segundo mosto *m*
i secondo mosto *m*
n onderste wort, tweede wort
d Glatterwasser *n*, Nachwürze *f*

145 AGATE min
f agate *f*
e ágata *f*
i agata *f*
n agaat *n*
d Achat *m*

146 AGE HARDENING, DISPERSION HARDENING, PRECIPITATION HARDENING met
f durcissement *m* structural
e endurecimiento *m* estructural
i incrudimento *m* strutturale
n structurele harding
d strukturelle Härtung *f*

147 AGE RESISTOR, ANTI-AGER, ANTIOXIDANT rub
f antioxydant *m*, antioxygène *m*, antivieillisseur *m*
e antioxidante *m*
i agente *m* contro invecchiamento, anti-invecchiante *m*
n middel *n* tegen veroudering
d Alterungsschutzmittel *n*

148 AGEING (GB), AGING (US) met,rub
f vieillissement *m*
e envejecimiento *m*
i invecchiamento *m*
n veroudering
d Alterung *f*

149 AGEING (GB), AGING (US) tex
f traitement *m* par vapeur
e vaporizado *m* húmedo
i trattamento *m* al vapore
n stoombehandeling
d Dampfbehandlung *f*

150 AGEING (GB), SOURING (US) cer
f mûrissage *m*
e curación *f*
i maturazione *f*, omogenizzazione *f*
n homogenisatie
d Homogenisierung *f*

151 AGEING TEST (GB), AGING TEST (US) rub
f essai *m* de vieillissement
e ensayo *m* de envejecimiento
i collaudo *m* d'invecchiamento
n verouderingsproef
d Alterungsprüfung *f*

152 AGGLOMERATE pre
f aggloméraț *m*
e aglomeración *f*, amontonamiento *m*
i agglomerato *m*
n agglomeraat *n*
d Agglomerat *n*

153 AGGLOMERATE (TO) pre
f agglomérer v
e aglomerar v, amontonar v
i agglomerarsi v
n agglomereren v, opeenhopen v, samenklonteren v
d agglomerieren v, zusammenballen v

154 AGGLOMERATION gen
f agglomération *f*
e aglomeración *f*
i agglomerazione *f*
n agglomeratie, opeenhoping, samenklontering
d Agglomeration *f*, Zusammenballung *f*

155 AGGLUTINATION bac,gen
f agglutination *f*
e aglutinación *f*

i agglutinazione *f*
n agglutinatie, samenkleving
d Agglutination *f*, Verklebung *f*

156 AGITATED CLEANING elc
f dégraissage *m* à agitation
e desengrase *m* con agitación
i sgrassaggio *m* con agitazione
n ontvetting door roeren
d Flutreinigung *f*

157 AGITATION, STIRRING mix
f agitation *f*
e agitación *f*
i agitazione *f*
n roeren *n*
d Rühren *n*

158 AGRICULTURAL CHEMISTRY agr
f chimie *f* agricole
e agroquímica *f*, química *f* agrícola
i chimica *f* agraria
n landbouwchemie
d Ackerbauchemie *f*, Agrikulturchemie *f*

159 AGRICULTURAL LIME, FERTILIZING LIME fer
f engrais *m* calcique
e cal *f* de abonar
i calce *f* da fertilizzare
n kalkmeststof
d Düngekalk *m*

160 AIR BELL, AIR BUBBLE, BUBBLE gen
f bulle *f* d'air
e burbuja *f* de aire
i bolla *f* d'aria
n luchtbel
d Luftblase *f*

161 AIR BLAST PROCESS rub
f procédé *m* par soufflage d'air
e procedimiento *m* por chorro de aire
i soffiatura *f* con aria a pressione
n blaasluchtproces *n*
d Blasluftverfahren *n*

162 AIR BLOWING fat
f affinage *m* par courant d'air
e refino *m* con corriente de aire
i affino *m* con corrente d'aria
n raffineren *n* door luchtdoorblazen
d Raffinieren *n* mit Luftstrom

163 AIR CONDITIONING hyg
f climatisation *f*, conditionnement *m* de l'air
e acondicionamiento *m* del aire, aeroacondicionamiento *m*, climatización *f*
i condizionamento *m* dell'aria
n luchtconditionering, luchtverzorging
d Klimatisierung *f*

164 AIR COOLED clo
f refroidi par l'air
e enfriado al aire
i raffreddato ad aria
n met lucht gekoeld
d lufgekühlt adj

165 AIR COOLING clo
f refroidissement *m* de l'air
e enfriamiento *m* del aire
i raffreddamento *m* dell'aria
n koeling van lucht
d Luftabkühlung *f*

166 AIR COOLING clo
f refroidissement *m* par air
e enfriamiento *m* al aire
i raffreddamento *m* ad aria
n luchtkoeling
d Luftkühlung *f*

167 AIR CURING rub
f vulcanisation *f* par l'air chaud
e vulcanización *f* por aire caliente
i vulcanizzazione *f* ad aria calda
n vulcaniseren *n* met hete lucht
d Heissluftvulkanisation *f*

168 AIR DRIED MALT bre
f malt *m* au vent
e malta *f* secada al aire
i malto *m* essiccato ad aria
n luchtmout
d Luftmalz *n*

169 AIR DRYING dry,pla
f séchage *m* à l'air
e secado *m* al aire
i essiccazione *f* all'aria
n drogen *n* aan de lucht
d Trocknung *f* an der Luft

170 AIR DRYING dry,pap
f séchage *m* de l'air
e secado *m* del aire
i essiccazione *f* dell'aria
n luchtdroging
d Freilufttrocknung *f*, Lufttrocknung *f*

171 AIR ELUTRIATION, cle
AIR SEPARATION
f élutriation *f* par l'air
e elutriación *f* por el aire, levigación *f* por el aire
i elutriazione *f* per l'aria
n elueren *n* door lucht
d Lufteluierung *f*, Luftschlämmung *f*

172 AIR-ENTRAINED CONCRETE bui
f béton *m* avec occlusion d'air
e hormigón *m* con aire ocluído
i calcestruzzo *m* con occlusione d'aria
n luchthoudend beton *n*
d Luftsporenbeton *m*

173 AIR ENTRAINING bui
f occlusion *f* d'air
e oclusión *f* de aire
i occlusione *f* d'aria
n luchtbelvorming
d Plastifizierung *f*

174 AIR FLOW DRYING dry
f séchage *m* par courant d'air
e secado *m* por corriente de aire
i essiccazione *f* per corrente d'aria
n droging met luchtstroom
d Luftstromtrocknung *f*

175 AIR GAS, gas
DRY GAS
f gaz *m* à l'air
e gas *m* de aire
i gas *m* d'aria
n luchtgas *n*
d Aerogengas *n*, Luftgas *n*

176 AIR HARDENING, met
SELF-HARDENING
f trempe *f* à l'air
e temple *m* al aire
i tempera *f* all'aria
n harding aan de lucht
d Lufthärtung *f*

177 AIR HUMIDIFICATION dry
f humectation *f* de l'air, mouillage *m* de l'air
e humectación *f* del aire
i umidificazione *f* dell'aria
n luchtbevochtiging
d Luftanfeuchtung *f*

178 AIR-KNIFE COATING coa,pla
f enduit *m* avec lame d'air
e recubrimiento *m* por chorro de aire dosificador
i spalmatura *f* a lama d'aria
n koten *n* met luchtmes
d Luftbürstenauftrag *m*

179 AIR LIQUEFYING ctt,gas
f liquéfaction *f* d'air
e licuación *f* de aire
i liquefazione *f* d'aria
n vloeibaar maken *n* van lucht
d Luftverflüssigung *f*

180 AIR OVEN AGEING (GB), rub,gas
AIR OVEN AGING (US)
f essai *m* de vieillissement à l'air chaud
e ensayo *m* de envejecimiento con aire caliente
i prova *f* d'invecchiamento ad aria calda
n verouderingsproef met hete lucht
d Heissluftalterungsprüfung *f*

181 AIR OXIDATION, cor
AERIAL OXIDATION
f oxydation *f* à l'air
e oxidación *f* al aire
i ossidazione *f* all'aria
n oxydatie aan de lucht
d Luftoxydation *f*, Oxydation *f* an der Luft, Oxydierung *f* an der Luft

182 AIR POLLUTION hyg
f pollution *f* de l'air
e contaminación *f* del aire, polución *f* del aire
i inquinamento *m* dell'aria
n luchtverontreiniging
d Luftverunreinigung *f*

183 AIR PRESSURE HEAT AGEING (GB), rub,tes
AIR PRESSURE HEAT AGING (US)
f essai *m* de vieillissement à l'air chaud sous pression
e ensayo *m* de envejecimiento con aire caliente bajo presión
i prova *f* d'invecchiamento ad aria calda sotto pressione
n verouderingsproef met hete lucht onder druk
d Überdruckheissluftalterungsprüfung *f*

184 AIR PURIFICATION cle,hyg
f épuration *f* de l'air
e depuración *f* del aire
i depurazione *f* dell'aria

n luchtreiniging
d Luftreinigung *f*

185 AIR REFINING, CONVERTING — met
f affinage *m* par le vent
e refinación *f* por aire
i affinazione *f* mediante aria
n raffineren *n* in thomasconvector
d Windfrischen *n*

186 AIRBORNE SEALING, HOT PATCHING — cer
f fermeture *f* de pores
e cerrado *m* de poros
i chiusura *f* di pori
n poriëndichting
d Porendichtung *f*

187 AIRFOAM CUSHIONING, LATEX FOAM CUSHIONING — rub
f rembourrage *m* de mousse de latex
e almohadillado *m* de espuma de látex
i imbottitura *f* di gommapiuma
n schuimrubbervulling
d Schaumgummipolsterung *f*

188 AIRLESS SPRAY PAINTING, AIRLESS SPRAYING, HYDRAULIC ATOMIZATION — dec
f atomisation *f* sans air comprimé, pistolage *m*
e atomización *f* sin aire
i atomizzazione *f* senza aria
n luchtloos spuiten *n*, spuiten *n* zonder perslucht
d Hochdruckfarbspritzen *n*

189 AIRPROOF CLOTH — tex
f tissu *m* imperméable à l'air
e tejido *m* impermeable al aire
i tessuto *m* impermeabile all'aria
n luchtdichte stof
d luftdichter Stoff *m*

190 AJOWAN OIL — fat
f essence *f* d'ajowan
e esencia *f* de ajowan
i olio *m* d'aiowan
n ajowanolie
d Ajowanöl *n*

191 ALABASTER, COMPACT GYPSUM — min
f albâtre *m*
e alabastro *m*
i alabastro *m*
n albast *n*, gipsalbast *n*
d Alabaster *m*, dichter Gips *m*

192 ALABASTER GLASS, OPALESCENT GLASS — gla
f verre *m* opalescent
e vidrio *m* opalescente
i vetro *m* opalescente
n albastglas *n*, melkglas *n*
d Achatglas *n*, Alabasterglas *n*, Reisglas *n*

193 ALBUMEN — nut
f albumen *m*, blanc *m* d'oeuf
e albumen *m*, clara *f* del huevo
i albume *m*, bianco *m* dell'uovo, chiara *f* dell'uovo
n eiwit *n*
d Eiweiss *n*

194 ALBUMIN — oic
f albumine *f*
e albúmina *f*
i albumina *f*
n albumine *n*, eiwitstof
d Albumin *n*, Eiweissstoff *m*

195 ALBUMINIZED PAPER — pap,pho
f papier *m* albuminé
e papel *m* albuminado
i carta *f* albuminata
n albuminepapier *n*
d Albuminpapier *n*, Eiweisspapier

196 ALBUMINOID, SCLEROPROTEIN — oic
f albuminoïde *m*, scléroprotéine *f*
e albuminoide *m*, escleroproteína
i albuminoide *m*, scleroproteina *f*
n albuminoïde *n*, scleroproteïne *n*
d Skleroprotein *n*

197 ALCHLOR, ALUMINIUM CHLORIDE PROCESS (GB), ALUMINUM CHLORIDE PROCESS (US) — lub
f dépuration *f* d'un lubrificant par chlorure d'aluminium
e depuración *f* de un lubricante por cloruro de aluminio
i depurazione *f* di un lubrificante per cloruro d'alluminio
n smeermiddelreiniging met aluminiumchloride
d Schmierölreinigung *f* mittels Aluminiumchlorid

198 ALCOHOLIC FERMENTATION, VINOUS FERMENTATION — fmi
f fermentation *f* alcoolique
e fermentación *f* alcohólica

i fermentazione *f* alcoolica
n alkoholische gisting
d alkoholische Gärung *f*

199 ALCOHOLIC LYE oic
f lessive *f* d'alcool
e lejía *f* alcohólica
i lisciva *f* d'alcool
n alkoholische loog
d alkoholische Lauge *f*

200 ALCOHOLIC POTASH EXTRACT oic
f extrait *m* à la potasse alcoolique
e extracto *m* de potasa alcohólica
i estratto *m* di potassa alcoolica
n alkoholisch loogextract *n*
d alkoholischer Potaschextrakt *m*

201 ALCOHOLOMETRY ana
f alcoolimétrie *f*
e alcoholimetría *f*
i alcoolimetria *f*
n alkoholmeting
d Alkoholmessung *f*

202 ALDEHYDENE, ACETYLENE, ETHINE oic
f acétylène *m*, éthyne *m*
e acetileno *m*
i acetilene *m*
n acetyleen *n*, ethyn *n*
d Äthyn *n*, Azetylen *n*

203 ALE bre
f ale *m*
e ale *m*, cerveza *f* inglesa de alta fermentación
i ale *f*, birra *f* inglese d'alta fermentazione
n aal *n*
d obergäriges englisches Bier *n*

204 ALITE bui
f alite *f*
e alita *f*
i alite *f*
n gemalen klinker
d gemahlener Portlandzement-klinker *m*

205 ALITIZING, CALORIZING met
f calorisation *f*
e calorización *f*
i calorizzazione *f*, termitizzazione *f*
n caloriseren *n*
d Alitieren *n*, Kalorisieren *n*

206 ALKALI LYE TREATMENT ble
f traitement *m* par lessive alcaline
e tratamiento *m* por lejía alcalina
i trattamento *m* per lisciva alcalina
n loogbehandeling
d Laugebehandlung *f*

207 ALKALI PROCESS OF REGENERATION, ALKALI RECLAIMING PROCESS rub
f procédé *m* de régénération à la soude
e regeneración *f* por procedimiento alcalino
i processo *m* di rigenerazione all'alcali
n alkaliregenereerproces *n*
d Alkaliregenerierverfahren *n*

208 ALKALI RECLAIM rub
f régénéré *m* à l'alcali
e regenerado *m* al alcali
i rigenerato *m* all'alcali
n alkaliregeneraat *n*
d Alkaliregenerat *n*

209 ALKALI SPOTTING tes,tex
f essai *m* de résistance à l'alcali
e prueba *f* de resistencia al alcali
i collaudo *m* di resistenza all'alcali
n alkalibestendigheidsproef
d Alkalifestigkeitsprüfung *f*

210 ALKALIMETRY ana
f alcalimétrie *f*
e alcalimetría *f*
i alcalimetria *f*
n alkalimetrie
d Alkalimessung *f*, Alkalimetrie *f*

211 ALKALINE CLEANING, ALKALINE DEGREASING cle,elc
f dégraissage *m* alcalin
e desengrase *m* alcalino
i sgrassaggio *m* alcalino
n alkalische ontvetting
d Reinigung *f* mit alkalischen Lösungsmitteln

212 ALKALINE COOKING LIQUOR pap
f liquide *m* alcalin de cuisson
e líquido *m* alcalino de cocción
i liquido *m* alcalino di cottura
n kookloog
d Kochlauge *f*

213 ALKALINE COOKING PROCESS, ALKALINE PULPING pap
f préparation *f* de la pâte

chimique à l'alcali,
procédé *m* de cuisson
alcaline
e procedimiento *m* de cocción
alcalina, pulpado *m* alcalino
i processo *m* di cottura alcalina,
riduzione *f* della polpa con
alcali
n alkalisch kookproces *n*,
alkalisch ontsluitingsproces *n*
d alkalischer Aufschluss *m*,
alkalisches Aufschliessungs-
verfahren *n*, alkalisches Koch-
verfahren *n*

214 ALKALINE IMMERSION CLEANING cle,elc
f dégraissage *m* alcalin par immersion
e desengrase *m* alcalino por inmersión
i sgrassaggio *m* alcalino per immersione
n alkalische dompelontvetting
d alkalische Tauchreinigung *f*

215 ALKALINE MEDIUM gen
f milieu *m* alcalin
e medio *m* alcalino
i ambiente *m* alcalino
n alkalisch milieu *n*
d alkalische Umgebung *f*, alkalisches Milieu *n*

216 ALKALINE PAINT STRIPPING dec
f décapage *m* alcalin
e decapaje *m* alcalino
i rimozione *f* alcalina della vernice
n alkalisch afbijten *n*
d alkalisches Entlacken *n*

217 ALKALINE REACTION gen
f réaction *f* alcaline
e reacción *f* alcalina
i reazione *f* alcalina
n alkalische reactie
d alkalische Reaktion *f*

218 ALKALINE SOAK-CLEANING cle
f dégraissage *m* alcalin par imbibition
e desengrase *m* alcalino por imbibición
i sgrassaggio *m* alcalino per imbibizione
n alkalische ontvetting door drenken
d alkalische Reinigung *f* durch Tränkung

219 ALKALINE SOAKER (US), ALKALINE STEEPING AGENT tex
f agent *m* d'imbibition alcalin
e agente *m* de imbibición alcalino
i agente *m* d'imbibizione alcalino
n alkalisch drenkmiddel *n*
d alkalisches Tränkmittel *n*

220 ALKALINE SPRAY CLEANING cle,elc
f dégraissage *m* alcalin au jet
e desengrase *m* alcalino al chorro
i sgrassaggio *m* alcalino allo spruzzo
n alkalische spuitreiniging
d alkalische Spritzreinigung *f*

221 ALKALINITY DECREASING gen
f réduction *f* de l'alcalinité
e reducción *f* de la alcalinidad
i riduzione *f* dell'alcalinità
n alkaliniteitsverlaging
d Alkalinitätsherabsetzung *f*

222 ALKALIZATION gen
f alcalisation *f*
e alcalización *f*
i alcalizzazione *f*
n alkaliseren *n*
d Alkalisierung *f*

223 ALKANIZATION fue
f alcanisation *f*
e alcanización *f*
i alcanizzazione *f*
n alkaniseren *n*
d Alkanisierung *f*

224 ALKYLATION, ALKYLIZATION oic
f alcoylation *f*, alkylation *f*, alkylisation *f*
e alkilización *f*, alquilación *f*
i alchilazione *f*
n alkyleren *n*
d Alkylation *f*, Alkylierung *f*

225 ALL-SLIMING PROCESS min
f procédé *m* de réduction à boue très fine
e procedimiento *m* de reducción a la finura de fangos, procedimiento *m* de trituración fina
i processo *m* di riduzione in melma finissima
n fijnslibproces *n*
d Feinschlammverfahren *n*

226 ALLIGATOR SKIN, met
ORANGE PEEL EFFECT
f effet *m* de pelure d'orange
e efecto *m* de corteza de naranja
i effetto *m* di buccia d'arancio
n sinaasappelschileffect *n*
d Orangenschaleneffekt *m*

227 ALLOTROPIC CHANGE, gen
ALLOTROPIC TRANSFORMATION
f modification *f* allotropique
e transformación *f* alotrópica
i trasformazione *f* allotropica
n allotrope modificatie
d allotrope Modifikation *f*

228 ALLOYING met
f alliage *m*
e aleación *f*
i legamento *m*
n legeren *n*
d Legieren *n*

229 ALLYLENE, oic
PROPINE
f allylène *m*, méthylacétylène *m*
e metil-acetileno *m*
i allilene *m*
n allyleen *n*, methylacetyleen *n*, propyn *n*
d Allylen *n*, Methylazetylen *n*

230 ALMOND OIL fat,soa
f huile *f* d'amandes
e aceite *m* de almendras
i olio *m* di mandorle
n amandelolie
d Mandelöl *n*

231 ALUM mat
f alun *m*
e alumbre *m*
i allume *m*
n aluin *n*
d Alaun *m*

232 ALUM TANNING lea,tan
f tannage *m* à l'alun
e curtimiento *m* con alumbre
i concia *f* all'allume
n aluinlooiing
d Alaungerbung *f*, Weissgerbung *f*

233 ALUMETIZATION, coa
ALUMINIZATION, ALUMIZING
f aluminisation *f*
e aluminiación *f*, aluminización *f*
i alluminiatura *f*
n alumineren *n*, aluminiseren *n*
d Aluminieren *n*, Veraluminieren *n*

234 ALUMETIZING coa
f atomisation *f* d'aluminium
e atomización *f* de aluminio
i allumetizzazione *f*
n aluminiumverstuiving
d Aluminiumzerstäubung *f*

235 ALUMINA, mat
OXIDE OF ALUMINIUM
f alumine *f*
e alumina *f*
i allumina *f*
n alumina *n*, aluminiumoxyde *n*,
korund *n*
d Aluminiumoxyd *n*, Tonerde *f*

236 ALUMINA PORCELAIN cer
f porcelaine *f* à l'alumine
e porcelana *f* de alúmina
i porcellana *f* all'allumina
n aluminahoudend porselein *n*
d tonerdehaltiges Porzellan *n*

237 ALUMINIUM LUBRICATING lub
GREASE (GB),
ALUMINUM LUBRICATING
GREASE (US)
f graisse *f* à l'aluminium
e grasa *f* de aluminio
i grasso *m* all'alluminio
n aluminiumhoudend smeervet *n*
d aluminiumhaltiges Schmierfett *n*

238 ALUMINOTHERMIC PROCESS, met
THERMITE PROCESS
f procédé *m* aluminothermique,
procédé *m* thermite
e procedimiento *m* aluminotérmico,
procedimiento *m* termita
i processo *m* alluminotermico
n aluminothermisch proces *n*,
thermietproces *n*
d aluminothermisches Verfahren *n*,
Thermitverfahren *n*

239 ALUMINOTHERMIC WELDING, wel
THERMITE WELDING
f soudage *m* aluminothermique
e soldadura *f* aluminotérmica
i saldatura *f* alla termite
n thermietlassen *n*
d Thermitschweissung *f*

240 ALUMINOUS CEMENT (GB), bui
ALUMINUM CEMENT (US)
f ciment *m* alumineux,
ciment *m* fondu
e cemento *m* aluminoso
i cemento *m* alluminoso
n aluminiumcement *n*

d Aluminiumzement *m*, Tonerdezement *m*

241 AMALGAMATION elc,met
f amalgamation *f*
e amalgamación *f*
i amalgamazione *f*
n amalgamering
d Amalgamierung *f*

242 AMBER ACID, ETHYLENE SUCCINIC ACID, SUCCINIC ACID oic
f acide *m* d'ambre, acide *m* d'éthylène succinique, acide *m* succinique
e ácido *m* de ámbar, ácido *m* succínico
i acido *m* d'ambra, acido *m* succinico
n barnsteenzuur *n*, ethyleenbarnsteenzuur *n*
d Äthylenbernsteinsäure *f*, Bernsteinsäure *f*

243 AMBER OIL fat
f huile *f* d'ambre
e esencia *f* de ámbar
i olio *m* d'ambra
n barnsteenolie
d Bernsteinöl *n*

244 AMERICAN CLOTH (GB), AMERICAN LEATHER (GB), ENAMELLED CLOTH (US) coa
f toile *f* cirée
e encerado *m*, hule *m*
i tela *f* cerata, tessuto *m* lucido
n wasdoek *n*
d Wachsleinwand *f*, Wachstuch *n*

245 AMERICAN ZINC OXIDE PROCESS, DIRECT ZINC OXIDE PROCESS met
f procédé *m* direct pour la fabrication d'oxyde de zinc
e procedimiento *m* directo para la fabricación de óxido de cinc
i processo *m* diretto per la fabbricazione d'ossido di zinco
n direct proces *n* voor de vervaardiging van zinkoxyde
d direktes Zinkoxydverfahren *n*

246 AMIDATION oic
f formation *f* d'amides
e amidación *f*
i formazione *f* d'amidi
n amidevorming
d Amidenbildung *f*

247 AMINATION oic
f amination *f*
e aminación *f*
i formazione *f* d'amine
n aminevorming
d Aminierung *f*

248 AMINOBENZENE, AMINOPHEN (US), ANILINE, PHENYLAMINE dye
f aniline *f*, phénylamine *f*
e anilina *f*, fenilamina *f*
i anilina *f*, fenilamina *f*
n aminobenzol *n*, aniline, fenylamine *n*
d Aminobenzol *n*, Anilin *n*, Phenylamin *n*

249 AMMONIA OXIDATION PROCESS aic
f procédé *m* d'oxydation d'ammoniaque
e procedimiento *m* de oxidación de amoníaco
i processo *m* d'ossidazione d'ammoniaco
n ammoniakoxydatieproces *n*
d Ammoniakoxydationsverfahren *n*

250 AMMONIA-SODA PROCESS, SOLVAY PROCESS aic
f procédé *m* de la soude à l'ammoniaque, procédé *m* Solvay
e procedimiento *m* de sosa amoniaca
i processo *m* Solvay
n solvayproces *n*
d Ammoniaksodaverfahren *n*

251 AMMONIA SUPERPHOSPHATE fer
f superphosphate *m* ammoniacal
e superfosfato *m* amoníaco
i superfosfato *m* ammoniaco
n ammoniumsuperfosfaat *n*
d ammoniakhaltiges Superphosphat, Ammoniaksuperphosphat *n*

252 AMMONIA WASTE, SPENT AMMONIA LIQUOR, SPENT GAS LIQUOR iwt
f eaux *pl* résiduaires ammoniacales
e agua *f* agotada amoniacal
i acqua *f* di rifiuto ammoniacale
n ammoniakhoudend afvalwater *n*
d Ammoniakabwasser *n*

253 AMMONIACAL BRINE, AMMONIATED BRINE aic
f saumure *f* ammoniacale
e salmuera *f* amoniacal

i salamoia *f* ammoniacale
n ammoniakhoudende pekel
d ammoniakhaltige Sole *f*

254 AMMONIACAL LIQUOR, GAS LIQUOR, GAS WATER gas
f eau *f* ammoniacale
e agua *f* amoniacal
i acqua *f* ammoniacale
n gaswater *n*
d Gaswasser *n*

255 AMMONIATION rub
f rendre v ammoniacal
e hacer v amoniacal
i aggiunta *f* d'ammoniaca
n ammoniëren *n*
d Ammonieren *n*

256 AMORPHOUS SULFUR (US), AMORPHOUS SULPHUR (GB) mat
f soufre *m* amorphe
e azufre *m* amorfo
i zolfo *m* amorfo
n amorfe zwavel
d amorpher Schwefel *m*

257 AMYLASE, DIASTASE sin
f amylase *f*, diastase *f*
e amilasa *f*, diastasa *f*
i amilasi *f*, diastase *f*
n amylase, diastase
d Amylase *f*, Diastase *f*

258 AMYLODEXTRIN, SOLUBLE STARCH gla
f amidon *m* soluble
e almidón *m* soluble
i amido *m* solubile
n oplosbare stijfsel
d lösliche Stärke *f*

259 AMYLUM, STARCH oic
f amidon *m*
e almidón *m*
i amido *m*
n stijfsel
d Stärke *f*

260 AMYMOLITIC FERMENTATION, STARCH FERMENTATION fmi
f fermentation *f* amymolitique
e fermentación *f* amimolítica
i fermentazione *f* amimolitica
n stijfselfermentatie
d Stärkefermentation *f*

261 ANALYSIS ana
f analyse *f*
e análisis *f*
i analisi *f*
n analyse
d Analyse *f*

262 ANAPHORESIS elc
f anaphorèse *f*
e anaforesis *f*
i anaforesi *f*
n anaforese *f*
d Anaphorese *f*

263 ANGELICA OIL fat
f essence *f* d'angélique
e esencia *f* de angélica
i olio *m* d'angelica
n angelicaolie
d Angelikaöl *n*

264 ANHYDRIC AMMONIA, ANHYDROUS AMMONIA mat
f ammoniaque *m* anhydre
e amoníaco *m* anhidro
i ammoniaco *m* anidro
n watervrije ammoniak
d Ammoniakanhydrid *n*, wasserfreies Ammoniak *n*

265 ANHYDRITE, ANHYDROUS SULFATE OF LIME (US), ANHYDROUS SULPHATE OF LIME (GB) aic
f anhydrite *f*
e anhidrita *f*, sulfato *m* de calcio anhidro
i anidrite *f*, solfato *m* di calcio anidro
n anhydriet *n*, doodgebrand gips *n*, watervrij calciumsulfaat *n*
d Anhydrit *m*, wasserfreier Gips *m*

266 ANILINE BLACK, OXIDATION BLACK, SUSPENSION BLACK dye
f noir *m* d'aniline
e negro *m* de anilina
i nero *m* d'anilina
n anilinezwart *n*
d Anilinschwarz *n*

267 ANILINE DYE dye
f colorant *m* azoique
e colorante *m* azóico, colorante *m* de anilina
i colore *m* d'anilina
n anilinekleurstof
d Anilinfarbe *f*

268 ANILINE HYDROCHLORIDE, ANILINE SALT
f hydrochlorure *f* d'aniline

e clorhidrato *m* de anilina
i cloridrato *m* d'anilina
n anilinehydrochloride *n*
d Anilinchlorhydrat *n*, Anilinsalz *n*

269 ANILINE OIL oic
f huile *f* d'aniline
e aceite *m* de anilina
i olio *m* d'anilina
n anilineolie, blauwolie
d Anilinöl *n*

270 ANIMAL BLACK, ANIMAL CHARCOAL, BONE BLACK mat,oic
f charbon *m* animal, noir *m* animal
e carbón *m* animal, negro *m* animal, negro *m* de huesos
i carbone *m* animale, carbone *m* d'ossa, nero *m* animale, nero *m* d'ossa
n beenderkool, beenzwart *n*
d Knochenkohle *f*, Knochenschwarz *n*, Tierschwarz *n*

271 ANIMAL FAT fat
f graisse *f* animale
e grasa *f* animal
i grasso *m* animale
n dierlijk vet *n*
d Tierfett *n*

272 ANIMAL-SIZED PAPER pap
f papier *m* collé avec gélatine
e papel *m* encolado con gelatina
i carta *f* incollata con gelatina
n met gelatine gelijmd papier *n*
d gelatinegeleimtes Papier *n*

273 ANIMALIZING tex
f animaliser v
e animalizar v
i animalizzare v
n animaliseren *n*
d Animalisieren *n*

274 ANION EXCHANGE, ANIONIC EXCHANGE elc
f échange *m* d'anions
e intercambio *m* de aniones
i scambio *m* d'anioni
n anionenuitwisseling
d Anionenaustausch *m*

275 ANISE OIL fat
f essence *f* d'anis
e esencia *f* de anís
i olio *m* d'anice
n anijsolie
d Anisöl *n*

276 ANISE-STAR OIL, STAR-ANISE OIL fat
f essence *f* d'anis étoilé, essence *f* de badiane
e esencia *f* de anís estrellado
i olio *m* d'anice stellato
n steranijsolie
d Sternanisöl *n*

277 ANISOTROPIC LIQUID, CRYSTALLINE LIQUID, LIQUID CRYSTAL cry,gla
f cristal *m* liquide, liquide *m* anisotropique
e cristal *m* líquido, líquido *m* anisotrópico
i cristallo *m* liquido, liquido *m* anisotropico
n anisotrope vloeistof, vloeibaar kristal *n*
d anisotrope Flüssigkeit *f*, flüssige Kristalle *pl*, kristalline Flüssigkeit *f*

278 ANNEALING met
f recuit *m*
e recocido *m*
i ricottura *f*
n gloeien *n*, uitgloeien *n*
d Ausglühen *n*, Glühen *n*

279 ANNEALING, COOLING gla
f refroidissement *m*
e enfriamiento *m*
i raffreddamento *m*
n koeling, temperen *n*
d Kühlung *f*, Tempern *n*

280 ANNEALING RANGE gla
f champ *m* de refroidissement
e campo *m* de enfriamiento
i campo *m* di raffreddamento
n koelgebied *n*, tempertraject *n*
d Kühlstrecke *f*

281 ANODE MUD, ANODE SLIME elc
f boue *f* d'anode, schlamm *m* anodique
e limo *m* de ánodo, lodo *m* anódico
i fanghi *pl* d'anodo
n anodeslib *n*
d Anodenrückstand *m*, Anodenschlamm *m*

282 ANODIC BRIGHTENING, elc
ELECTROBRIGHTENING
f brillantage *m* anodique
e abrillantado *m* anódico
i brillantaggio *m* anodico
n anodisch glanzen *n*
d anodisches Glänzen *n*

283 ANODIC CLEANING, elc
REVERSE CURRENT CLEANING
f dégraissage *m* anodique
e desengrase *m* anódico
i sgrassaggio *m* anodico
n anodische ontvetting
d anodische Reinigung *f*

284 ANODIC ETCHING elc
f attaque *f* anodique, gravure *f* anodique
e grabado *m* anódico
i incisione *f* anodica
n anodisch etsen *n*
d anodische Ätzung *f*

285 ANODIC OXIDATION coa,elc
ANODIC TREATMENT,
ANODIZING
f oxydation *f* anodique
e anodización *f*, eloxación *f*
i ossidazione *f* anodica
n anodiseren *n*
d Eloxieren *n*

286 ANODIC PICKLING elc
f décapage *m* anodique
e decapaje *m* anódico
i decapaggio *m* anodico
n anodisch beitsen *n*
d anodische Beizung *f*

287 ANONA OIL, fat
CANANGA OIL, YLANG YLANG OIL
f essence *f* d'ilang-ilang
e esencia *f* de cananga
i olio *m* di cananga odorata
n kanangaolie, ylang-ylangolie
d Anonaöl *n*, Kanangaöl *n*, Ylang-Ylangöl *n*

288 ANTI-AGER, rub
ANTIOXIDANT, AGE RESISTOR
f antioxydant *m*, antioxygène *m*, antivieillisseur *m*
e antioxidante *m*
i agente *m* contro invecchiamento, anti-invecchiante *m*
n middel *n* tegen veroudering
d Alterungsschutzmittel *n*

289 ANTICHLORINE cle,tex
GREEN MORDANT, SODA HYPOSULFITE (US), SODIUM THIOSULPHATE (GB)
f antichlore *m*
e anticloro *m*
i anticloro *m*
n antichloor *n*, natriumthiosulfaat *n*
d Antichlor *n*, Natriumthiosulphat *n*

290 ANTICORROSIVE PAINT dec
f peinture *f* anticorrosive
e pintura *f* anticorrosiva
i pittura *f* anticorrosiva
n anticorrosieverf, corrosiewerende verf
d Korrosionsschutzfarbe *f*

291 ANTIDETONANT, fue
ANTIKNOCK ADDITIVE,
BLENDING AGENT
f produit *m* antidétonant
e aditivo *m* antidetonante
i sostanza *f* antidetonante
n antiklopmiddel *n*
d Antiklopfmittel *n*, klopfhindernder Zusatz *m*

292 ANTIFEBRIN, mcy
ACETANILIDE,
PHENYLACETAMIDE
f acétanilide *f*, antifébrile *m*
e acetanilida *f*, antifebrina *f*, fenilacetamida *f*
i acetanilide *m*
n acetanilide *n*, antifebrine *n*
d Antifebrin *n*, Azetanilid *n*

293 ANTIFLOODING AGENT, dec
ANTIFLOTATION AGENT
f agent *m* antiflottage, agent *m* antinuançage
e agente *m* antiflotación
i agente *m* antiflottazione
n opdrijven verhinderend middel *n*
d Aufschwimmverhütungsmittel *n*

294 ANTIFOAMING AGENT, tex
ANTIFROTHING AGENT,
DEFOAMING AGENT
f agent *m* antimousse
e agente *m* antiespuma
i agente *m* antischiuma
n antischuimmiddel *n*, schuimwerend middel *n*
d Antischaummittel *n*, Schaumverhütungsmittel *n*

295 ANTIFOGGING AGENT, pho
ANTISTAIN AGENT
f antivoile *m*

e agente *m* antivelo
i antivelo *m*
n antisluiermiddel *n*
d Antischleiermittel *n*

296 ANTIFOULING COMPOSITION, ANTIFOULING PAINT coa,dec
f enduit *m* préservatif, peinture *f* antisallisures
e pintura *f* antipútrida
i pittura *f* antiincrostazione, pittura *f* sottomarina
n aangroeiwerende verf, antifoulingverf, onderwaterverf
d anwuchsverhindernde Farbe *f*, Schutzanstrich *m*, Unterwasseranstrich *m*

297 ANTIFREEZE clo,sap
f antigel *m*
e agente *m* anticongelante
i anticongelante *m*, antigelo *m*
n antivries
d Frostschutz *m*, Gefrierschutzmittel *n*

298 ANTIFRICTION METAL, BEARING METAL met
f métal *m* antifriction
e metal *m* antifricción
i metallo *m* antifrizione
n antifrictiemetaal *n*, lagermetaal *n*
d Antifriktionsmetall *n*, Lagermetall *n*, Lagerweissmetall *n*

299 ANTI-INCRUSTATOR, DISINCRUSTANT boi
f anti-incrostant *m*, tartrifuge *m*
e antiincrustante *m*, desincrustante *m*
i sostanza *f* che evita le incrostazioni
n ketelsteenverhinderend middel *n*
d Kesselsteinverhütungsmittel *n*

300 ANTIKNOCK VALUE, KNOCK RATING, PRE-IGNITION RESISTANCE fue
f pouvoir *m* antidétonant
e índice *m* de detonación, poder *m* antidetonante
i potere *m* antidetonante
n klopvastheid
d Klopffestigkeit *f*

301 ANTILIVERING AGENT dec
f agent *m* anti-épaississant
e agente *m* antiespesador
i agente *m* anti-addensamento
n middel *n* tegen verdikken
d Eindickungsverhütungsmittel *n*

302 ANTIMONY BLACK dec
f noir *m* d'antimoine, sulfure *m* noir d'antimoine
e negro *m* de antimonio, sulfuro *m* de antimonio
i nero *m* d'antimonio
n antimoontrisulfide *n*
d Antimontrisulfid *n*

303 ANTINOISE PAINT coa,dec
f peinture *f* amortisseuse
e pintura *f* antirruido
i pittura *f* antirombo
n geluiddempende verf
d Antidröhnanstrich *m*

304 ANTIOZIDANT, ANTIOZONANT gen
f agent *m* antiozone
e agente *m* antiozono
i agente *m* antiozono
n ozoninwerkingtegengaand middel *n*
d Antiozonisator *m*, Ozonschutzmittel *n*

305 ANTIRETARDER pho
f antiretardateur *m*
e antiretardador *m*
i antiritardante *m*
n vertragingtegenwerkend middel *n*
d Antihemmkörper *m*

306 ANTIRIVELLING AGENT dec
f agent *m* antirides
e agente *m* antiarrugas
i agente *m* anti-increpature
n antischroeimiddel *n*
d Runzelverhütungsmittel *n*

307 ANTIRUST AGENT coa
f agent *m* antirouille
e agente *m* anticorrosivo
i agente *m* antiruggine
n roestwerend middel *n*
d Rostschutzmittel *n*

308 ANTIRUST PAINT coa,dec
f enduit *m* antirouille
e pintura *f* anticorrosiva
i pittura *f* antiruggine
n roestwerende verf
d Rostschutzfarbe *f*

309 ANTISAG AGENT dec
f agent *m* anticoulures

e agente *m* antiarrufo
i agente *m* antismotta
n middel *n* ter voorkoming van zakkers
d Läuferverhütungsmittel *n*

310 ANTISCORCHER, ANTISCORCHING AGENT, SCORCH RETARDER — rub
f antigrilleur *m*, retardateur *m* de grillage
e retardador *m* de la prevulcanización
i agente *m* antiscottatura
n vertrager van de aanvulcanisatie
d Vorvulkanisationsverzögerer *m*

311 ANTISEPTIC ASPHALT MIXTURE, TREE PUTTY — rub
f mastic *m* antiseptique
e mezcla *f* antiséptica de asfalto
i mastice *m* a base d'asfalto antisettico
n antiseptisch asfaltmengsel *n*
d antiseptische Asphaltmischung *f*

312 ANTISHRINKING AGENT — tex
f agent *m* prévenant le rétrécissement
e agente *m* antiencogimiento
i agente *m* antiritiro
n krimpvrijmakend middel *n*
d krumpffreimachendes Mittel *n*

313 ANTISKID PAINT, ANTISLIP PAINT — dec,sap
f peinture *f* antiglissante
e pintura *f* antideslizante
i pittura *f* antisdrucciolevole
n antislipverf
d rutschfeste Farbe *f*

314 ANTISKIMMING AGENT — mix
f agent *m* anti-émulsionnant
e agente *m* antiemulgente
i agente *m* antiemulsionante
n anti-emulgator
d Antiemulgierungsmittel *n*

315 ANTISKINNING AGENT — dec
f agent *m* antifrisant, agent *m* antipeaux
e agente *m* antidescascarillador
i agente *m* antipelle
n velvormingverhinderend middel *n*
d Hautverhütungsmittel *n*

316 ANTISTATIC AGENT — sap
f agent *m* antistatique
e agente *m* antistático
i agente *m* antistatico
n antistatisch middel *n*
d antistatisches Mittel *n*

317 ANTISTATIC RUBBER — rub
f caoutchouc *m* antistatique
e goma *f* antistática
i gomma *f* antistatica
n antistatische rubber
d antistatischer Gummi *m*

318 ANTISWELLING AGENT — pho
f agent *m* antigonflant
e agente *m* antihinchamiento
i agente *m* antigonfiamento
n zwellingverhinderend middel *n*
d quellungverhütendes Mittel *n*

319 ANTIWEAR ADDITIVE — lub
f additif *m* anti-usure
e aditivo *m* antidesgaste
i additivo *m* antiusura
n slijtageverhinderend toevoegsel *n*
d verschleisshindernde Zugabe *f*

320 APPARENT ATTENUATION — bre
f atténuation *f* apparente
e atenuación *f* aparente
i attenuazione *f* apparente
n schijnbare vergisting
d scheinbare Vergärung *f*

321 APPARENT DEGREE OF ATTENUATION — bre
f degré *m* d'atténuation apparente
e grado *m* de atenuación aparente
i grado *m* d'attenuazione apparente
n schijnbare vergistingsgraad
d scheinbarer Vergärungsgrad *m*

322 APPLE ACID, MALIC ACID — oic
f acide *m* malique
e ácido *m* málico
i acido *m* malico
n appelzuur *n*
d Apfelsäure *f*

323 APPLE BRANDY (GB), APPLE JACK (US) — nut
f calvados *m*, eau *f* de vie de cidre
e aguardiente *m* de manzana
i acquavite *f* di sidro, calvados *m*
n appelbrandewijn, calvados
d Apfelschnaps *m*

324 APPLICATION PRINTING, DIRECT PRINTING — tex
f impression *f* directe
e estampa *f* directa
i stampa *f* diretta
n directe textieldruk
d direkter Textildruck *m*

325 APPRET — tex
f apprêt *m*
e apresto *m*
i appretto *m*
n appreteermiddel *n*
d Appreturmittel *n*

326 AQUA FORTIS, CONCENTRATED NITRIC ACID — dec
f eau *f* forte
e agua *f* fuerte
i acquaforte *f*
n aqua fortis, ets, sterk water *n*
d Ätzwasser *n*, konzentrierte Salpetersäure *f*, Scheidewasser *n*

327 AQUA REGIA, CHLOROAZOTIC ACID (US), CHLORONITROUS ACID, NITROHYDROCHLORIC ACID, NITROMURIATIC ACID — aic,met
f eau *f* régale
e agua *f* regia
i acqua *f* regia
n koningswater *n*
d Königswasser *n*

328 AQUEOUS SOLUTION OF FORMALDEHYDE, FORMALIN — oic
f formaline *f*, formol *m*
e formalina *f*, formol *m*
i formalina *f*, formol *m*
n formalien *n*, formaline *n*, formol *n*
d Formalin *n*, Formol *n*

329 ARABINE, ACACIA GUM, GUM ARABIC, SENEGAL GUM — glu
f gomme *f* arabique, gomme *f* d'acacia
e goma *f* arábiga, goma *f* Senegal
i gomma *f* arabica, gomma *f* d'acacia
n acaciagom, Arabische gom, kordofangom, senegalgom
d Akaziengummi *m*, Arabin *n*, Gummiarabicum *n*, Senegalgummi *m*

330 ARACHIS OIL, EARTHNUT OIL, GROUND OIL, PEANUT OIL — fat
f huile *f* d'arachide
e aceite *m* de cacahuete
i olio *m* d'arachide
n aardnotenolie
d Erdnussöl *n*

331 ARC HEATING — elc
f chauffage *m* par arc électrique
e calentamiento *m* por arco eléctrico
i riscaldamento *m* ad arco elettrico
n vlamboogverhitting
d Lichtbogenheizung *f*

332 ARC PROCESS — gas
f procédé *m* à l'arc électrique
e procedimiento *m* al arco eléctrico
i processo *m* all'arco elettrico
n lichtboogproces *n*
d direkte Stickstoffgewinnung *f* durch Lichtbogen, Schönherrverfahren *n*

333 ARC SPRAYING — coa
f vaporisation *f* par arc électrique
e vaporización *f* por arco eléctrico
i vaporizzazione *f* mediante arco elettrico
n lichtboogverstuiving
d Lichtbogenzerstäubung *f*

334 ARCHITECTURAL PAINT, HOUSE PAINT — dec
f peinture *f* pour bâtiments
e pintura *f* de casas
i pittura *f* per case
n huisschilderverf
d Bautenfarbe *f*

335 ARDENT SPIRITS, SPIRITUOUS LIQUORS — oic
f liqueurs *pl* fortes, spiritueux *pl*
e bebidas *pl* espirituosas
i bevande *pl* alcooliche
n spiritualiën *pl*, sterke drank
d hochprozentige alkoholische Getränke *pl*, Spirituosen *pl*

336 AREA MONITORING — nue
f contrôle *f* de radioactivité de terrain
e vigilancia *f* del área

i controllo *m* di radioattività di zona
n terreinbewaking
d Flächenüberwachung *f*, Raumüberwachung *f*

337 ARGOL, CRUDE CREAM OF TARTAR, WINE LEES cor
f argol *m*, crème *f* de tartre brute
e crémore *m* tártaro crudo
i argol *m*, cremore *m* di tartaro crudo
n ruwe wijnsteen
d Argal *n*, roher Weinstein *m*

338 ARGON LIQUEFYING gas
f liquéfaction *f* d'argon
e licuación *f* de argón
i liquefazione *f* d'argon
n vloeibaar maken *n* van argon
d Argonverflüssigung *f*

339 ARMED CONCRETE, FERROCONCRETE, REINFORCED CONCRETE bui
f béton *m* armé
e hormigón *m* armado
i cemento *m* armato, conglomerato *m* cementizio armato
n gewapend beton *n*
d bewehrter Beton *m*, Eisenbeton *m*, Stahlbeton *m*

340 ARMOURED GLASS, WIRE ROLLED GLASS gla
f verre *m* armé, verre *m* métallifié
e vidrio *m* armado
i vetro *m* retinato
n gewapend glas *n*, draadglas *n*
d armiertes Glas *n*, Drahtglas *n*

341 AROMATIZATION oic
f aromatisation *f*
e aromatización *f*
i aromatizzazione *f*
n aromatiseren *n*
d Aromatisierung *f*

342 ARRESTED CRUSHING brk
f broyage *m* préliminaire
e trituración *f* limitada
i frantumazione *f* preliminare
n voorbreken *n*
d Vorbrechen *n*

343 ARSENIC DETECTION, ARSENIC DETERMINATION lab,mcy
f dosage *m* de l'arsenic
e dosificación *f* de arsénico
i dosaggio *m* d'arsenico
n arseenbepaling, arsenicumbepaling
d Arsenbestimmung *f*

344 ARSENICAL SOLUTION, FOWLER SOLUTION mcy
f liqueur *m* de Fowler
e arsénico *m* en solución de Fowler
i soluzione *f* di Fowler
n liquor Fowleri
d Fowlersche Lösung *f*

345 ARSINE GENERATION gas
f génération *f* d'arsine
e generación *f* de arsino
i generazione *f* d'arsino
n arsinegeneratie, arsine-ontwikkeling
d Arsenwasserstofferzeugung *f*, Arsinerzeugung *f*

346 ARSONATION oic
f traitement *m* par acide arsénique
e tratamiento *m* con ácido arsénico
i trattamento *m* con acido arsenico
n behandeling met arseenzuur
d Behandlung *f* mit Arsensäure

347 ART PAPER pap
f papier *m* couché
e papel *m* couché, papel *m* de arte
i carta *f* patinata
n kunstdrukpapier *n*
d Kunstdruckpapier *n*

348 ARTIFICIAL AGEING (GB), ARTIFICIAL AGING (US) fou,pla
f vieillissement *m* accéléré
e envejecimiento *m* acelerado
i invecchiamento *m* artificiale
n kunstmatige veroudering
d Warmauslagerung *f*

349 ARTIFICIAL FERTILIZER, ARTIFICIAL MANURE fer
f engrais *pl* chimiques
e abonos *pl* químicos
i concime *m* chimico
n kunstmest
d Kunstdünger *m*, künstlicher Dünger *m*

350 ARTIFICIAL FIBRE tex
f fibre *f* artificielle
e fibra *f* artificial
i fibra *f* artificiale

n kunstvezel
d Chemiefaser *f*, Kunstfaser *f*

351 ARTIFICIAL HORN, GALALITE — pla
f corne *f* artificielle, galalithe *f*
e cuerno *m* artificial, galalita *f*
i corno *m* artificiale, galalite *f*
n galaliet *n*, kunsthoorn *n*
d Galalith *m*, Kunsthorn *n*

352 ARTIFICIAL LEATHER, IMITATION LEATHER, LEATHERETTE — lea
f cuir *m* artificiel
e cuero *m* artificial, similcuero *m*
i cuoio *m* artificiale, dermoide *f*
n kunstleer *n*
d Kunstleder *n*

353 ARTIFICIAL RESIN, SYNTHETIC RESIN — pla
f résine *f* synthétique
e resina *f* sintética
i resina *f* sintetica
n kunsthars *n*
d Kunstharz *n*

354 ARTIFICIAL RUBBER, MAN-MADE RUBBER (US), SYNTHETIC RUBBER (GB) — rub
f caoutchouc *m* artificiel, caoutchouc *m* synthétique, élastomère *f* synthétique
e caucho *m* sintético, goma *f* artificial, goma *f* sintética
i gomma *f* sintetica
n kunstrubber, synthetische rubber
d Kunstgummi *m*, synthetischer Gummi *m*, synthetischer Kautschuk *m*

355 ARTIFICIAL SILK, RAYON — tex
f soie *f* artificielle, rayonne *f*
e rayón *m*, seda *f* artificial
i raion *m*, seta *f* artificiale
n kunstzijde
d Kunstseide *f*, Rayon *m*, Reyon *m*

356 ARTIFICIAL STONE, RECONSTRUCTED STONE — bui
f pierre *f* artificielle
e piedra *f* artificial
i pietra *f* artificiale
n kunststeen
d Kunststein *m*

357 ARTIFICIAL WOOL, MAN-MADE WOOL (US), SYNTHETIC WOOL (GB) — tex
f laine *f* artificielle
e lana *f* artificial
i lana *f* artificiale
n kunstwol
d Zellwolle *f*

358 ARYLATION — oic
f arylation *f*
e arilación *f*
i arilazione *f*
n aryleren *n*
d Arylierung *f*

359 ASBESTOS — min
f amiante *m*, asbeste *m*
e amianto *m*, asbesto *m*, corcho *m* fósil
i amianto *m*, asbesto *m*
n asbest *n*
d Asbest *m*

360 ASBESTOS BOARD, ASBESTOS MILLBOARD — bui
f carton *m* d'amiante
e cartón *m* de amianto
i cartone *m* d'amianto
n asbestkarton *n*
d Asbestpappe *f*

361 ASBESTOS CEMENT — bui
f ciment *m* d'amiante
e cemento *m* amiantado, cemento-asbesto *m*, cemento *m* de amianto
i fibrocemento *m*
n asbestcement *n*
d Asbestzement *m*

362 ASBESTOS CEMENT SHEETING, ASBESTOS SHEET — bui
f tôle *f* en amiante-ciment
e hoja *f* de cemento-amianto
i piastra *f* di copertura in cemento-amianto
n asbestcementplaten *pl*
d Asbestzementplatten *pl*

363 ASBESTOS CLOTH, ASBESTOS FABRIC — sap
f tissu *m* d'amiante, toile *f* d'amiante
e tejido *m* de amianto, tela *f* de amianto
i tessuto *m* d'amianto
n asbestweefsel *n*
d Asbestgewebe *n*, Asbesttuch *n*

364 ASBESTOS CORD, ASBESTOS ROPE, ASBESTOS STRING — mat
f tresse *f* d'amiante
e cordón *m* de amianto, cuerda *f* de amianto, trenza *f* de amianto
i cordone *m* d'amianto, treccia *f* d'amianto
n asbestkoord *n*
d Asbestschnur *f*, Asbestseil *n*

365 ASBESTOS CURTAIN, FIREPROOF CURTAIN — sap
f rideau *m* d'amiante
e telón *m* de amianto
i sipario *m* d'amianto
n asbestgordijn *n*
d Asbestvorhang *m*

366 ASBESTOS FELT — mat
f feutre *m* d'amiante
e fieltro *m* de amianto
i feltro *m* d'amianto
n asbestvilt *n*
d Asbestfilz *m*

367 ASBESTOS PAINT — dec
f enduit *m* d'amiante
e pintura *f* de amianto
i pittura *f* d'amianto
n asbestverf
d Asbestanstrich *m*, Asbestfarbe *f*

368 ASBESTOS PAPER — pap
f papier *m* d'amiante
e papel *m* de amianto
i carta *f* d'amianto
n asbestpapier *n*
d Asbestpapier *n*

369 ASBESTOS PLY, INTERMEDIATE ASBESTOS LAYER — bui
f couche *f* d'amiante
e capa *f* intermedia de amianto
i strato *m* intermedio d'amianto
n asbesttussenlaag
d Asbesteinlage *f*

370 ASBESTOS SHIDE, ASBESTOS SHINGLE — bui
f bardeau *m* d'amiante, essente *f* d'amiante, tavaillon *m* d'amiante
e tejamanil *m* de amianto
i assicella *f* d'amianto
n asbestschindel
d Asbestschindel *m*

371 ASBESTOS YARN — mat
f fil *m* d'amiante
e hilaza *f* de amianto
i filo *m* d'amianto
n asbestdraad
d Asbestfaden *m*

372 ASH — iwt
f cendre *f*
e ceniza *f*
i cenere *f*
n as
d Asche *f*

373 ASH HANDLING, ASH REMOVAL — iwt
f enlèvement *m* des cendres
e manejo *m* de cenizas
i asportazione *f* delle ceneri, rimozione *f* delle ceneri
n asafvoer
d Aschenabfuhr *f*, Aschenentfernung *f*

374 ASHING — pla
f polissage *m*
e pulido *m*
i pulitura *f*
n polijsten *n*
d Polieren *n*

375 ASPHALT, BITUMEN, JUDEAN PITCH, MINERAL PITCH — bui
f asphalte *m*, poix *f* de Judée
e asfalto *m*
i asfalto *m*
n aardpek *n*, asfalt *n*, jodenpek *n*
d Asphalt *m*, Bitumen *n*, Erdpech *n*

376 ASPHALT BINDER — glu
f mastic *m* asphaltique
e ligador *m* asfáltico
i mastice *m* d'asfalto
n asfaltkit
d Asphaltkitt *m*

377 ASPHALT BLOCK, ASPHALT CAKE, MASTIC BLOC — bui
f bloc *m* d'asphalte
e pan *m* de asfalto
i asfalto *m* in pani, mattonella *f* d'asfalto
n asfaltblok *n*, asfaltbrood *n*
d Asphaltbrot *n*, Asphaltstein *m*

378 ASPHALT CARDBOARD — bui
f carton *m* asphalté
e cartón *m* asfaltado
i cartone *m* asfaltato
n asfaltpapier *n*
d Asphaltpappe *f*

379 ASPHALT CEMENT (GB), ASPHALTIC CEMENT (US) bui
f ciment *m* asphaltique
e cemento *m* asfáltico, mástique *m* asfáltico
i cemento *m* asfaltico
n asfaltcement *n*
d Asphaltzement *m*

380 ASPHALT CONCRETE (GB) ASPHALTIC CONCRETE (US) bui
f béton *m* asphaltique
e hormigón *m* asfáltico
i calcestruzzo *m* asfaltico
n asfaltbeton *n*
d Asphaltbeton *m*

381 ASPHALT FELT (GB), ASPHALTIC FELT (US) bui
f feutre *m* asphalté
e fieltro *m* asfaltado
i feltro *m* asfaltato
n asfaltvilt *n*
d Asphaltfilz *m*

382 ASPHALT GROUT, ASPHALT MORTAR (GB), ASPHALTIC MORTAR (US) bui
f mortier *m* asphaltique
e mortero *m* asfáltico
i malta *f* asfaltica
n asfaltmortel
d Asphaltmörtel *m*

383 ASPHALT-GROUTED MACADAM bui
f empierrement *m* asphaltique
e macadam *m* construido por penetración
i macadam *m* d'asfalto, massicciata *f* con impregnazione di bitume
n asfaltmacadam *n*, met bitumen gedrenkt macadam *n*
d Asphalttränkmakadam *m*

384 ASPHALT OIL fue,oic
f pétrole *m* riche en asphalte
e pez *m* de petróleo
i petrolio *m* ricco d'asfalto
n asfaltolie
d Destillationsrückstand *m* von Erdöl

385 ASPHALT PAINT (GB), ASPHALTIC PAINT (US) dec
f peinture *f* asphaltique
e pintura *f* asfáltica
i colore *m* a base d'asfalto, mano *m* d'asfalto, pittura *f* asfaltica
n asfaltverf
d Asphaltfarbe *f*

386 ASPHALT PAPER bui
f papier *m* asphalté
e papel *m* asfaltado
i carta *f* asfaltata
n asfaltpapier *n*
d Asphaltpapier *n*

387 ASPHYXIANT GAS mit
f gaz *m* asphyxiant
e gas *m* asfixiante
i gas *m* asfissiante
n verstikkend gas *n*
d erstickendes Gas *n*

388 ASSAYING ana
f coupellation *f*, essai *m*
e copelación *f*, ensayo *m*
i assaggio *m*, coppellazione *f*
n cupelleren *n*, essaaieren *n*
d Kupellieren *n*, Prüfen *n*

389 ASSISTANT tex
f agent *m* auxiliaire de l'industrie textile
e producto *m* auxiliar en la industria textil
i agente *m* ausiliario dell'industria tessile
n textielhulpmiddel *n*
d Textilhilfsmittel *n*

390 ASYMMETRIC SYNTHESIS mcy
f synthèse *f* asymétrique
e síntesis *f* asimétrica
i sintesi *f* asimmetrica
n asymmetrische synthese
d Synthese *f* optisch aktiver Verbindungen

391 ATMOSPHERIC CORROSION cor
f corrosion *f* atmosphérique
e corrosión *f* atmosférica
i corrosione *f* atmosferica
n verwering
d Verwitterung *f*

392 ATOMIC ARC WELDING, ATOMIC HYDROGEN WELDING wel
f soudure *f* à l'hydrogène atomiqu
e soldadura *f* por hidrógeno atómico
i saldatura *f* ad idrogeno atomico
n lassen *n* met atomaire waterstof
d Arkatomschweissen *n*

393 ATOMIC HYDROGEN, ACTIVE HYDROGEN gas
f hydrogène *m* actif

e hidrógeno *m* activo
i idrogeno *m* attivo
n actieve waterstof, atomaire waterstof
d aktiver Wasserstoff *m*, atomarer Wasserstoff *m*

394 ATOMIZATION, ATOMIZING, NEBULIZATION coa,pre
f atomisation *f*, pulvérisation *f*, vaporisation *f*
e atomización *f*, nebulización *f*, pulverización *f*
i atomizzazione *f*, nebulizzazione *f*, polverizzazione *f*
n verneveling, verstuiving
d Zerstäubung *f*

395 ATTAR OIL, OTTO OIL, ROSE OIL fat
f essence *f* de roses, huile *f* rosat
e esencia *f* de rosas
i essenza *f* di rose, olio *m* rosato
n rozenolie
d Rosenessenz *f*, Rosenöl *n*

396 ATTEMPERATION con
f régulation *f* de température
e atemperación *f*, regulación *f* de temperatura
i regolazione *f* di temperatura
n temperatuurregeling
d Temperaturregelung *f*

397 ATTENUATION bre,fmi
f atténuation *f*
e atenuación *f*
i attenuazione *f*
n vergisting
d Vergärung *f*

398 ATTRITION brk
f trituration *f*
e trituración *f*
i triturazione *f*
n fijnwrijven *n*
d Zerreibung *f*

399 ATTRITION TEST tes
f essai *m* d'usure complète
e ensayo *m* de desgaste completo
i prova *f* d'attrito
n uitputtende slijtageproef
d Abriebsdauerversuch *m*

400 AUER METAL, MISCH METAL met
f mischmétal *m*
e mischmetal *m*
i mischmetallo *m*
n mischmetaal *n*
d Mischmetall *n*

401 AUSTENIZING HEAT TREATMENT htt,met
f austénitisation *f*
e austenización *f*
i austenitizzazione *f*
n austenitiseren *n*
d Austenitisieren *n*

402 AUSTRIAN CINNABAR, BASIC LEAD CHROMATE, CHROME RED, PHOENICITE dec
f chromate *m* de plomb basique, rouge *m* de chrome
e cromato *m* de plomo básico, rojo *m* de cromo
i cromato *m* di piombo basico, rosso *m* di cromo
n basisch loodchromaat *n*, chromaatrood *n*
d basisches Bleichromat *n*, Chromrot *n*

403 AUTOCHROME PROCESS, CHROMATE PRINTING PROCESS tex
f impression *f* autochrome
e impresión *f* en fotocromía
i stampa *f* autocroma
n autochroomdruk
d Autochromdruck *m*

404 AUTOGENOUS WELDING wel
f soudure *f* autogène
e soldadura *f* autógena
i saldatura *f* autogena
n autogeenlassen *n*
d Autogenschweissen *n*

405 AUTO-OXIDATION (GB), AUTOXIDATION (US) aic
f auto-oxydation *f*
e autooxidación *f*
i autossidazione *f*
n zelfoxydatie
d Autoxydation *f*, Selbstoxydation *f*

406 AUTOREDUCTION gen
f autoréduction *f*
e autorreducción *f*
i autoriduzione *f*
n zelfreductie
d Autoreduktion *f*, Selbstreduktion *f*

407 AVIATION KEROSENE, AVIATION SPIRIT, WIDE-CUT FUEL fue
f essence *f* d'aviation
e gasolina *f* para aviación
i benzina *f* d'aviazione
n vliegtuigbenzine
d Fliegerbenzin *n*

408 AVIVAGE tex
f avivage *m*
e avivaje *m*
i avivazione *f*
n aviveren *n*
d Avivierung *f*

409 AXITE (US),
CORDITE (GB) exp
f cordite *f*
e cordita *f*
i cordite *f*
n cordiet *n*
d Kordit *m*

410 AZEOTROPIC DISTILLATION dis
f distillation *f* azéotropique
e destilación *f* azeotrópica
i distillazione *f* azeotropica
n azeotrope destillatie
d azeotrope Destillation *f*

411 AZEOTROPIC MIXTURE,
CONSTANT BOILING MIXTURE gen
f mélange *m* azéotropique
e mezcla *f* azeotrópica
i miscela *f* azeotropica
n azeotroop mengsel *n*,
azeotropisch mengsel *n*
d azeotropes Gemisch *n*

412 AZEOTROPIC PROCESS dec
f procédé *m* aux solvants,
procédé *m* azéotropique
e procedimiento *m* con solventes
i processo *m* ai solventi
n azeotroop proces *n*
d azeotropes Verfahren *n*,
Lösungsmittelverfahren *n*

413 AZOTIZING aic,oic
f azotation *f*
e azoación *f*, nitrogenación *f*
i azotazione *f*
n met stikstof verbinden v
d Azotieren *n*, mit Stickstoff
verbinden v

B

414 BACK-CHROME DYEING, dye
AFTERCHROME DYEING,
TOP CHROMING
f chromatage *m*, teinture *f* postérieure au chrome
e cromado *m* secundario, postmordentado *m* crómico
i tintura *f* posteriore al cromo
n nachromeren *n*
d Nachchromieren *n*

415 BACKFLOW, dis, gen
REFLUX
f reflux *m*
e reflujo *m*
i riflusso *m*
n terugloop, terugvloeien *n*
d Rückfluss *m*, Rücklauf *m*, Zurückfliessen *n*

416 BACKING SAND, fou
BLACK SAND
f sable *m* de remplissage
e arena *f* de relleno
i terra *f* di riempimento
n vulzand *n*
d Füllsand *m*, Haufensand *m*

417 BACKPRESSURE pla,pre
f contre-pression *f*
e contrapresión *f*
i contropressione *f*
n tegendruk
d Gegendruck *m*

418 BACKPRESSURE FILLING vac
f remplissage *m* à contre-pression
e relleno *m* de contrapresión
i riempimento *m* a contropressione
n tegendrukvulling
d Gegendruckfüllung *f*

419 BACKREACTION gen
f réaction *f* inverse
e reacción *f* inversa
i reazione *f* inversa
n omkeerreactie
d Rückreaktion *f*

420 BACKTANNING tex
f traitement *m* postérieur à l'acide tannique
e postratamiento *m* con ácido tánico
i trattamento *m* posteriore all'acido tannico
n looizuurnabehandeling
d Gerbsäurenachbehandlung *f*

421 BACKTITRATION ana
f titrage *m* en retour
e revalorización *m*, valorización *f* inversa
i titolazione *f* inversa
n terugtitreren *n*
d Rücktitration *f*, Umkehrtitration *f*

422 BACKWASH fue
f lavage *m* à inversion de courant
e lavado *m* a contracorriente
i disoliatura *f*, lavaggio *m* ad inversione di flusso
n uitwassen *n* in terugloop
d Rückwäsche *f*

423 BACKWASHING tex
f dégraissage *m* de la laine
e alisado *m* de la lana
i sgrassaggio *m* della lana
n wolontvetting
d Wollentfettung *f*

424 BACON RUBBER rub
f gomme-lard *m*
e goma *f* tocino
i gomma-prosciutto *m*
n spekrubber
d Speckgummi *m*

425 BACTERIA CULTURE mcy
f culture *f* de bactéries
e cultivo *m* de bacterias
i cultura *f* di batteri
n bacteriëncultuur, bacteriënkweek
d Bakterienzüchtung *f*

426 BACTERIAL ACTION bac
f action *f* bactérienne
e acción *f* bacteriana
i azione *f* batterica
n bacteriewerking, werking van micro-organismen
d Bakterienwirkung *f*

427 BACTERIAL COAGULATION bac
f coagulation *f* bactérienne
e coagulación *f* bacteriana
i coagulazione *f* batterica
n coagulatie door micro-organismen
d bakterielle Koagulation *f*

428 BACTERIAL FERMENTATION fmi
f fermentation *f* bactérienne
e fermentación *f* bacteriana
i fermentazione *f* batterica
n gisting door micro-organismen
d bakterielle Gärung *f*

429 BACTERICIDE mcy
f bactéricide *f*
e bactericida *m*
i battericida *m*
n bactericide *n*, bacteriëndodend middel *n*
d bakterientötendes Mittel *n*, Bakterizid *n*

430 BAD SPIRITS, FUSEL nut
f alcool *m* amylique, flegmes *pl*, mauvais goûts *pl* de tête
e alcohol *m* amílico
i alcool *m* fatturato, cattiva acquavite *f*, flemma *f*
n foezel
d Fusel *m*

431 BAG FILTRATION (GB), SACK FILTRATION (US) fil
f filtrage *m* en sacs
e filtración *f* en bolsas
i filtrazione *f* in sacchi
n zakfiltreren *n*
d Taschenfiltrierung *f*

432 BAG MOLDING (US), BAG MOULDING (GB) pla
f moulage *m* à sac de caoutchouc
e moldeo *m* por saco de goma
i formatura *f* a sacco di gomma
n rubberzakproces *n*
d Gummisackverfahren *n*

433 BAGASSE sin
f bagasse *f*
e bagazo *m*
i bagassa *f*
n bagasse
d Ampas *m*, Bagasse *f*, gemahlenes Zuckerrohr *n*

434 BAGGING, SACKCLOTH tra
f serpillière *f*, toile *f* à sac
e arpillera *f*, tela *f* de saco
i grossa tela *f*, tela *f* da sacco
n zakkengoed *n*, zakkenlinnen *n*
d Packleinwand *f*, Sackleinwand *f*

435 BAKING, BURNING cer
f cuite *f*
e quemado *m*
i cottura *f*
n bakken *n*
d Brennen *n*

436 BAKING, GAS EXPELLING htt
f expulsion *f* de gaz
e expulsión *f* de gas
i espulsione *f* di gas
n gasuitdrijving
d Gasaustreibung *f*

437 BAKING, POLYMERIZING tex
f polymérisation *f*
e polimerización *f*
i polimerizzazione *f*
n polymerisatie
d Polymerisierung *f*

438 BAKING OUT elc
f dégazage *m* par recuit
e desgasificación *f* por recocido
i degassaggio *m* mediante ricottura
n ontgassen *n* door uitgloeien
d Entgasung *f* durch Ausglühen

439 BAKING POWDER, BAKING SODA (GB), SALERATUS (US) nut
f levure *f* artificielle, poudre *f* levain
e bicarbonato *m* sódico, levadura *f* en polvo
i bicarbonato *m* di soda, lievito *m* in polvere
n bakpoeder *n*
d Backpulver *n*

440 BALANCED REACTION, INCOMPLETE REACTION gen
f réaction *f* incomplète, réaction *f* réversible
e reacción *f* incompleta
i reazione *f* bilanciata
n evenwichtsreactie
d reversible Reaktion *f*, unvollständige Reaktion *f*

441 BALE DYEING, CIRCULATION DYEING, PACK DYEING tex
f teinture *f* en balles
e teñido *m* en balas, teñido *m* en líos
i tintura *f* in balle
n balenverven *n*, verven *n* in balen
d Ballenfärbung *f*

442 BALL BURNISHING met,pol
f polissage *m* à billes
e bruñido *m* con bolas
i brunitura *f* per palle
n kogelpolijsten *n*
d Kugelpolierung *f*

443 BALL CLAY, cer
POTTER'S CLAY
f argile *f* de potier, argile *f* figuline
e arcilla *f* de alfarero, arcilla *f* figulina
i argilla *f* per terraglie
n pottenbakkersklei
d Töpferton *m*

444 BALLING brk,cer
f formation *f* de grumeaux
e formación *f* de grumos
i formazione *f* di grumi
n klonteren *n*
d Klumpenbildung *f*

445 BALLING met
f formation *f* des loupes
e formación *f* de las bolas
i formazione *f* dei masselli
n puddelbalvorming
d Luppenmachen *n*

446 BALLING (US), mcy
PILL MAKING
f formation *f* de pilules
e formación *f* de píldoras
i formazione *f* di pillole
n persen *n* van pillen
d Pillenformung *f*

447 BALM, glw,mcy
BALSAM
f baume *m*
e bálsamo *m*
i balsamo *m*
n balsem
d Balsam *m*

448 BALM GILEAD, mcy,pho
BALSAM OF FIR, CANADA BALSAM
f baume *m* du Canada
e bálsamo *m* del Canadá
i balsamo *m* dell'abete resinoso
n canadabalsem
d Kanadabalsam *m*

449 BALSAM OF PERU, mcy
BLACK BALSAM
f baume *m* du Pérou
e bálsamo *m* del Perú
i balsamo *m* del Perù
n perubalsem
d Perubalsam *m*

450 BALSAM OF TOLU mcy
f baume *m* du Tolu
e bálsamo *m* del Tolú
i balsamo *m* del Tolù
n tolubalsem
d Tolubalsam *m*

451 BAND COAGULATION rub
f coagulation *f* en ruban
e coagulación *f* en cinta
i coagulazione *f* in nastro
n lintcoagulatie
d Bandkoagulation *f*

452 BAND CONVEYING, tra
BELT CONVEYING
f transport *m* à bande
e transporte *m* de cinta
i trasporto *m* a nastro
n bandtransport *n*
d Bandförderung *f*

453 BAND DRYING, dry
BELT DRYING
f séchage *m* à bandes transporteuses
e secado *m* con cintas transportadoras
i essiccazione *f* a nastri trasportatori
n etagedroging
d Bandtrocknung *f*

454 BAND FEEDING, tra
BELT FEEDING
f alimentation *f* à bande
e alimentación *f* de cinta
i alimentazione *f* a nastro
n bandtoevoer, bandvoeding
d Bandzufuhr *f*

455 BAND SINTERING htt,met
f frittage *m* sur bande
e sinterización *f* sobre cinta
i agglomerazione *f* su nastro
n bandsinteren *n*
d Bandsintern *n*

456 BARE SPOT, dec
HOLIDAY, UNCOATED AREA
f absence *f*, blanc *m*, manque *m*, zone *f* non couverte
e superficie *f* pasada por alto
i zona *f* non coperta
n heiligedag
d blanke Stelle *f*, Jute *f*, Jutlochstelle *f*

457 BARIUM CONCRETE bui,nue
f béton *m* au barium
e hormigón *m* al bario
i calcestruzzo *m* al bario
n bariumbeton *n*
d Bariumbeton *m*

458 BARIUM MEAL mcy
f bouillie *f* opaque
e papilla *f* de bario
i pasto *m* opaco
n bariumpap
d Bariumbrei *m*

459 BARIUM PLASTER bui,nue
f plâtre *m* baryté
e yeso *m* de bario
i gesso *m* di bario
n bariumgips *n*
d Bariummörtel *m*, Bariumweisskalk *m*

460 BARK lea,tan
f tannée *f*
e tanino *m* consumido
i tannino *m* consumato
n kif *n*, uitgewerkte run
d Gerbereiabfälle *pl*

461 BARK TANNING tan
f tannage *m* à l'écorce
e curtido *m* con casca
i concia *f* alla scorza
n runlooien *n*
d Lohgerbung *f*

462 BARKING pap
f décorticage *m*
e descortezamiento *m*
i scortecciatura *f*
n ontschorsing, schillen *n*
d Entrindung *f*

463 BARLEY BREE bre
f cervoise *f* forte
e cerveza *f* de cebada fuerte
i birra *f* d'orzo forte
n sterk gerstebier *n*
d Starkbier *n*

464 BARLEY MALT bre
f malt *m* d'orge
e malta *f* de cebada
i malto *m* d'orzo
n gerstemout *n*
d Gerstenmalz *n*

465 BARLEY WINE bre
f cervoise *f*
e cerveza *f* de cebada de alta calidad
i birra *f* d'orzo d'alta qualità
n fijn gerstebier *n*
d feines Gerstenbier *n*

466 BARM BEER bre
f bière *f* de levure
e cerveza *f* de restos
i resti *pl*
n gistbier *n*
d Hefebier *n*

467 BARM YEAST, PITCHING YEAST bre
f levain *m*
e levadura *f* de siembra
i lievito *m* di semenza
n aanzetgist, entgist, moedergist
d Anstellhefe *f*, Zeug *m*

468 BAROPHORESE gen
f barophorèse *f*
e baroforesis *f*
i baroforesi *f*
n baroforese
d Barophorese *f*

469 BARREL ELECTROPLATING, BARREL PLATING elc
f galvanoplastie *f* au tonneau
e galvanoplastia *f* en tambor
i galvanoplastica *f* in tamburo, rotogalvanostegia *f*
n trommelgalvaniseren *n*
d Trommelgalvanisierung *f*

470 BARREL ROLLING, BARRELLING, TUMBLING pol
f polissage *m* au tonneau
e pulido *m* en tambor
i pulitura *f* in tamburo
n trommelpolijsten *n*
d Trommelpolierung *f*

471 BARYTA PAPER pap
f papier *m* baryté
e papel *m* revestido de sulfato de bario
i carta *f* con rivestimento di solfato di bario
n barietpapier *n*
d Barytpapier *n*

472 BARYTA WATER aic
f eau *f* de baryte
e agua *f* de barita
i acqua *f* di barite
n barietwater *n*
d Barytwasser *n*

473 BARYTES, HEAVY SPAR min
f barytine *f*, spath *m* pesant

e baritina *f*, espato *m* pesado
i baritina *f*, spato *m* pesante
n bariet *n*, zwaarspaat *n*
d Baryt *m*, Schwerspat *m*

474 BASAN, BASIL — lea
f basane *f*
e badana *f*
i bazzana *f*
n bezaanleer *n*
d ungefärbtes Schafleder *n*

475 BASE — gen
f base *f*
e base *f*
i base *f*
n base
d Base *f*

476 BASE — pho
f support *m* de l'émulsion
e soporte *m* de la emulsión
i sopporto *m* dell'emulsione
n emulsiedrager
d Emulsionsträger *m*

477 BASE BULLION, WORK LEAD — met
f plomb *m* d'oeuvre
e plomo *m* de obra
i piombo *m* argentifero, piombo *m* impuro
n ongezuiverd lood *n*, werklood *n*
d silberhaltiges Blei *n*, Werkblei *n*

478 BASE EXCHANGE — gen
f échange *m* de base
e cambio *m* de base
i scambio *m* di base
n base-uitwisseling
d Basenaustausch *m*

479 BASE EXCHANGE SOFTENING — hyg
f adoucissement *m* par résine cationique
e ablandamiento *m* por cambiador básico
i addolcimento *m* mediante scambio di base
n ontharding door base-uitwisseling
d Basenaustauschenthärtung *f*

480 BASE MATERIAL, BASIC MATERIAL — gen,mat
f matière *f* première
e materia *f* prima
i materia *f* prima
n grondstof, uitgangsmateriaal *n*
d Ausgangsmaterial *n*, Ausgangsstoff *m*

481 BASE METAL — met
f métal *m* basique
e metal *m* básico
i metallo *m* basico
n basisch metaal *n*
d basisches Metall *n*

482 BASE METAL — met
f métal *m* de base, métal *m* principal
e elemento *m* metálico principal, metal *m* de base
i metallo *m* di base, metallo *m* principale
n basismetaal *n*
d Basismetall *n*

483 BASE METAL — met
f métal *m* non-précieux
e metal *m* no noble
i metallo *m* ignobile
n onedel metaal *n*
d Nicht-Edelmetall *n*

484 BASE MIX (US), BASIC MATERIAL (GB), BLANK MIX (US), BLANK STOCK (GB) — gen
f composé *m* de base, mélange *m* de base
e mezcla *f* básica
i mescolanza *f* base, mescolanza *f* di sfondo
n basismengsel *n*
d Grundmischung *f*

485 BASE OIL, OXIDIZED OIL, THICKENED OIL — fat,lub
f huile *f* épaissie, huile *f* oxydée, huile *f* soufflée
e aceite *m* oxidado
i olio *m* ossidato, olio *m* soffiato
n geblazen olie, geoxydeerde olie
d geblasenes Öl *n*, oxydiertes Öl *n*

486 BASE ORE, LOW GRADE ORE — min
f minerai *m* à faible teneur, minerai *m* de qualité inférieure, minerai *m* pauvre
e mineral *m* de baja ley, mineral *m* pobre
i minerale *m* di qualità cattiva
n arm erts *n*, laagwaardig erts *n*
d armes Erz *n*, geringwertiges Erz *n*, taubes Erz *n*

487 BASE STRENGTH gen
f force *f* de base
e fortaleza *f* básica
i forza *f* di base
n basesterkte
d Basenstärke *f*

488 BASIC BESSEMER PROCESS, THOMAS PROCESS met
f procédé *m* Bessemer basique, procédé *m* Thomas
e procedimiento *m* Thomas
i processo *m* al convertitore basico, processo *m* Thomas
n thomasproces *n*
d Thomas-Verfahren *n*

489 BASIC DYE dye
f colorant *m* basique
e materia *f* colorante básica
i colore *m* basico
n basische kleurstof
d basischer Farbstoff *m*

490 BASIC LEAD CARBONATE, WHITE LEAD dec
f blanc *m* de céruse, céruse *f*, plomb *m* blanc
e albayalde *m*, blanco *m* de plomo, cerusa *f*
i biacca *f* di piombo, cerussa *f*
n basisch loodcarbonaat *n*, loodwit *n*
d Bleiweiss *n*, Deckweiss *n*

491 BASIC LEAD CHROMATE, AUSTRIAN CINNABAR, CHROME RED, PHOENICITE dec
f chromate *m* de plomb basique, rouge *m* de chrome
e cromato *m* de plomo básico, rojo *m* de cromo
i cromato *m* di piombo basico, rosso *m* di cromo
n basisch loodchromaat *n*, chromaatrood *n*
d basisches Bleichromat *n*, Chromrot *n*

492 BASIC LINING coa,htt
f garnissage *m* basique, revêtement *m* basique
e guarnición *f* básica, revestimiento *m* básico
i guarnizione *f* basica, rivestimento *m* basico
n basische bekleding, basische voering
d basische Auskleidung *f*, basisches Futter *n*

493 BASIC OPEN-HEARTH PROCESS htt
f procédé *m* Martin basique
e procedimiento *m* Siemens-Martin básico
i processo *m* Martin a suolo basico
n basisch siemens-martinproces *n*
d basisches Siemens-Martin-Verfahren *n*

494 BASIC OXIDE gen
f oxyde *m* basique
e óxido *m* básico
i ossido *m* basico
n basisch oxyde *n*
d basisches Oxyd *n*

495 BASIC REFRACTORY met
f réfractaire *m* basique
e refractario *m* básico
i refrattario *m* basico
n basisch vuurvast materiaal *n*
d basischer feuerfester Stoff *m*

496 BASIC SALT gen
f sel *m* basique
e sal *f* básica
i sale *m* basico
n basisch zout *n*
d basisches Salz *n*

497 BASIC SLAG fer
f laitier *m* basique, scorie *f* basique
e escoria *f* básica, escoria *f* de desfosforación
i scoria *f* basica
n basische slak, thomasslakkenmeel
d basische Schlacke *f*, Thomasschlacke *f*

498 BASIC SOLUTION gen
f solution *f* basique
e solución *f* básica
i soluzione *f* basica
n basische oplossing
d basische Lösung *f*

499 BASICITY gen
f basicité *f*
e basicidad *f*
i basicità *f*
n basiciteit
d Basenvermögen *n*, basische Eigenschaft *f*, Basizität *f*

500 BASIFICATION, BASIFYING gen
f basification *f*, formation *f* de base

e basificación *f*, formación *f* de base
i basificazione *f*, formazione *f* di base
n basevorming
d Basenbildung *f*

501 BASIFIED VISCOSE RAYON tex
f soie *f* de viscose à addition de base
e rayón *m* de viscosa con adición de base
i raion *m* alla viscosa con addizione di base
n viscosezijde met basetoevoeging
d Viskoseseide *f* mit Basenzugabe

502 BASIFIER gen
f générateur *m* de base
e generador *m* de base
i generatore *m* di base
n basevormer
d Basenbilder *m*

503 BASILICON OIL fat
f essence *f* de basilic
e esencia *f* de albahaca
i olio *m* di basilico
n basilicumolie
d Basilikumöl *n*

504 BAST LIQUOR, BOILED-OFF LIQUOR tex
f liquide *m* de décreusage, liquide *m* de dégommage
e líquido *m* de descrudado, líquido *m* de desengomado
i liquido *m* di sgommatura
n zijdeafkooksel *n*
d Abkochflüssigkeit *f*, Degummierungsflüssigkeit *f*

505 BASTARD FLATTING (US), EGG-SHELL GLOSS (GB) dec
f aspect *m* coquille d'oeuf
e lustre *m* suave
i lucidatura *f* opaca
n eiglans
d Eierschalenglanz *m*

506 BATCH gen
f lot *m*
e carga *f*, lote *m*
i carica *f*, lotto *m*
n afgemeten hoeveelheid, portie
d Beschickung *f*, Charge *f*, Posten *m*

507 BATCH CARBONATATION sin
f saturation *f* discontinue
e carbonatación *f* por etapas
i carbonatazione *f* a lotti
n stapsgewijze carbonatatie
d Einzelsaturation *f*

508 BATCH CHARGING, BATCH FEEDING gla
f alimentation *f* de composition
e alimentación *f* de mezclas
i alimentazione *f* di miscugli
n gemengtoevoer, gemengvoeding
d Gemengespeisung *f*

509 BATCH CRYSTALLIZATION cry
f cristallisation *f* à lots, cristallisation *f* intermittente
e cristalización *f* intermitente
i cristallizzazione *f* intermittente
n niet-continue kristallisatie
d Chargenkristallisation *f*, diskontinuierliche Kristallisation *f*, Satzkristallisation *f*

510 BATCH DISTILLATION dis
f distillation *f* discontinue, distillation *f* intermittente
e destilación *f* discontinua, destilación *f* intermitente, destilación *f* por etapas
i distillazione *f* a lotti, distillazione *f* discontinua, distillazione *f* intermittente
n intermitterende destillatie, niet-continue destillatie
d Blasendestillation *f*, intermittierende Destillation *f*

511 BATCH DYEING tex
f teinture *f* discontinue
e teñido *m* no continuo
i tintura *f* intermittente
n niet-continu verven *n*
d diskontinuierliche Färbung *f*

512 BATCH PRODUCTION gen
f production *f* discontinue
e producción *f* por lotes
i produzione *f* intermittente
n niet-continue produktie
d absatzweiser Betrieb *m*, Chargenbetrieb *m*

513 BATCH REFINING fat
f désacidification *f*
e desacidificación *f*
i disacidificazione *f*
n ontzuring
d Entsäuerung *f*

514 BATCH-THROUGH CIRCULATION DRYING dry
f séchage *m* à circulation intermittente
e secado *m* de circulación intermitente
i essiccazione *f* a circolazione intermittente
n niet-continue doorloopdroging
d diskontinuierliche Durchlauftrocknung *f*

515 BATCHING gen
f dosage *m*
e dosificación
i dosaggio *m*
n dosering
d Dosierung *f*

516 BATCHING tex
f huilage *m*
e engrase *m*
i oliaggio *m*
n smouten *n*
d Schmälzen *n*

517 BATCHING OIL tex
f saindoux *m*
e manteca *f*
i strutto *m*
n smout *n*
d Schmalz *n*

518 BATE tan
f bain *m* de chipage, confit *m*, mordant *m*
e baño *m* de maceración, mordiente *m*
i bagno *m* di macerazione
n etsloog, zwelder
d Ätzlauge *f*, Beizbrühe *f*, Gerbbeize *f*

519 BATH gen
f bain *m*
e baño *m*
i bagno *m*
n bad *n*
d Bad *n*

520 BATH LUBRICATION lub
f lubrification *f* en bain d'huile
e lubricación *f* por baño, lubricación *f* por borboteo
i lubrificazione *f* a bagno d'olio, lubrificazione *f* ad immersione
n dompelsmering
d Tauchschmierung *f*

521 BATH RATIO tex
f rapport *m* de bain de teinture
e relación *f* de baño de teñido
i rapporto *m* di bagno di tintura
n verfbadverhouding
d Flottenverhältnis *n*

522 BATHOTONIC edn
f reduisant la tension de surface
e batotónico adj
i batotonico adj
n met de oppervlaktespanning verminderende werking
d die Oberflächenspannung vermindernd, selbstvernetzend adj

523 BATING, PUERING lea
f chipage *m*
e maceración *f*
i macerazione *f*
n zwellen *n*
d Schwöden *n*

524 BATTERY ACID elc
f acide *f* d'accumulateur
e ácido *m* de acumulador
i acido *m* d'accumulatore
n accuzuur *n*
d Akkumulatorensäure *f*, Batteriesäure *f*

525 BATTERY MANGANESE, BLACK OXIDE OF MANGANESE, GLASSMAKER'S SOAP, MANGANESE DIOXIDE min
f pyrolusite *f*
e pirolusita *f*
i perossido *m* di manganese
n bruinsteen *n*, mangaansuperoxyde *n*
d Braunstein *m*, Mangandioxyd *n*, Pyrolusit *m*

526 BAUXITE, BEAUXITE min
f bauxite *f*
e bauxita *f*
i bauxite *f*
n bauxiet *n*
d Bauxit *m*

527 BAY OIL, BAYBERRY OIL, LAUREL OIL fat
f essence *f* de laurier
e esencia *f* de laurel
i olio *m* d'alloro, olio *m* di lauro, olio *m* laurino
n bayolie, laurierolie
d Bayöl *n*, Lorbeeröl *n*, Loröl *n*

528 BEAD POLYMERIZATION pla
f polymérisation *f* en perles
e polimerización *f* en suspensión
i polimerizzazione *f* in perle
n polymerisatie in korrelvorm
d Perlpolymerisation *f*,
Suspensionspolymerisation *f*

529 BEADING cer
f bordage *m*
e rebordeado *m*
i bordatura *f*
n randen *n*
d Rändern *n*

530 BEADING, dec
CISSING, CRAWLING
f contraction *f*, rampage *m*,
retrait *m*
e avejigamiento *m*
i falta *f* di presa
n kralen *n*, samentrekken *n*,
schiften *n*
d Kriechen *n*, Perlen *n*

531 BEADING, dec
PUCKERING, WRINKLING
f fripe *f*, frisage *m*
e arrugado *m*
i raggrinzamento *m*
n schroeien *n*
d Faltenbildung *f*, Runzelbildung *f*

532 BEADING ENAMEL cer
f émail *m* de bordage
e esmalte *m* de rebordeado
i smalto *m* di bordatura
n randemail *n*
d Ränderemail *n*

533 BEAM DYEING tex
f teinture *m* en ensouple
e teñido *m* en enjullo, teñido *m*
en plegador
i tintura *f* in subbi
n boomverven *n*
d Baumfärbung *f*

534 BEAN OIL, fat
SOY OIL, SOYA BEAN OIL
f huile *f* de soya
e aceite *m* de soja
i olio *m* di soia
n sojaolie
d Sojaöl *n*

535 BEARING METAL, met
ANTIFRICTION METAL
f métal *m* antifriction
e metal *m* antifricción
i metallo *m* antifrizione
n antifrictiemetaal *n*,
lagermetaal *n*
d Antifriktionsmetall *n*,
Lagermetall *n*, Lagerweiss-
metall *n*

536 BEATING pap
f broyage *m*
e molienda *f*
i macinatura *f*, macinazione *f*
n hollanderen *n*
d Zermahlen *n*

537 BEAUMONTAGE bui
f badigeon *m*, mastic *m* à
reboucher
e mástique *m* para tapar defectos
i mastice *m* da riempire fori
n opvulkit
d Kitt *m* zum Ausfüllen von
Löchern

538 BEER bre
f bière *f*
e cerveza *f*
i birra *f*
n bier *n*
d Bier *n*

539 BEER DRAWN OFF FROM bre
BOTTOM,
DRAWINGS
f bière *f* de fond
e cerveza *f* de colador
i resti *pl* dalle botti
n afgefiltreerd bier *n*
d Abseihbier *n*

540 BEER FROM THE WOOD, bre
CASK BEER, DRAFT BEER (US),
DRAUGHT BEER (GB)
f bière *f* en fût
e cerveza *f* de barril
i birra *f* a spina, birra *f* in fusto
n bier *n* van het vat, vatbier *n*
d Bier *n* vom Fass, Fassbier *n*

541 BEER SCALE bre
f tartre *m*
e piedra *f* de cerveza
i tartaro *m*
n biersteen
d Bierstein *m*

542 BEESWAX oic
f cire *f* d'abeilles
e cera *f* de abejas
i cera *f* d'api
n bijenwas
d Bienenwachs *m*

543 BEET JUICE sin
f jus *m* de betteraves, jus *m* sucré
e jugo *m* de remolachas
i succo *m* di barbabietole
n suikerbietensap *n*
d Rübensaft *m*

544 BEET PULP, sin
BEETROOT PULP,
EXHAUSTED CHIPS
f pulpe *f* de betteraves
e pulpa *f* de remolachas
i polpa *f* di barbabietole
n bietenpulp
d Brei *m*, Rübenbrei *m*

545 BEET SLICES, sin
CHIPS, COSSETTES
f cossettes *pl*
e cosetas *pl*, rebanadas *pl*
i ritagli *pl* di barbabietole
n bietensnijdsels *pl*
d Rübenschnitzel *pl*

546 BEET SLICING sin
f coupage *m* de betteraves
e rebanado *m* de remolachas
i affettazione *f* di barbatietole
n bietensnijden *n*
d Rübenzerschneidung *f*

547 BEET SPIRIT, sin
BEETROOT SPIRIT
f alcool *m* de betteraves, alcool *m* de mélasse
e alcohol *m* de melaza, alcohol *m* de remolachas
i alcool *m* di barbabietole, alcool *m* di melassa
n beetwortelspiritus, melasse-spiritus
d Melassesprit *m*

548 BEET SUGAR, sin
BEETROOT SUGAR
f sucre *m* de betteraves
e azúcar *m* de remolachas
i zucchero *m* di barbabietole
n bietsuiker
d Rübenzucker *m*

549 BEET SYRUP, sin
MOLASSES
f mélasse *f*
e melaza *f*
i melassa *f*
n melasse
d Melasse *f*, Rübensirup *m*

550 BEETLING tex
f pilage *m*
e enfurtido *m*, estampado *m*
i pestatura *f*
n stampen *n*
d Stampfen *n*

551 BELL METAL met
f bronze *m* de cloche, métal *m* de cloche
e bronce *m* de campana, metal *m* campanil
i bronzo *m* da campane, metallo *m* per campane
n klokkebrons *n*, klokkespijs
d Glockenmetall *n*, Glockenspeise *f*

552 BELT CONVEYING, tra
BAND CONVEYING
f transport *m* à bande
e transporte *m* de cinta
i trasporto *m* a nastro
n bandtransport *n*
d Bandförderung *f*

553 BELT DRYING, dry
BAND DRYING
f séchage *m* à bandes transporteus
e secado *m* con cintas trans-portadoras
i essiccazione *f* a nastri trasportatori
n etagedroging
d Bandtrocknung *f*

554 BELT FEEDING, tra
BAND FEEDING
f alimentation *f* à bande
e alimentación *f* de cinta
i alimentazione *f* a nastro
n bandtoevoer, bandvoeding
d Bandzufuhr *f*

555 BEND LEATHER, lea
BUTT
f cœur *m*
e cuero *m* crupón
i cuoio *m* per suola di prima qualità
n kernleer *n*
d Kernleder *n*

556 BENEFICIATION (US), min
ORE DRESSING (GB)
f traitement *m* de minerais
e beneficio *m* de minerales
i trattamento *m* di minerali
n ertsbewerking
d Erzaufbereitung *f*

557 BENNE OIL, fat
SESAME OIL
f huile *f* de sésame
e aceite *m* de sésamo
i olio *m* di sesamo
n sesamolie
d Sesamöl *n*

558 BENZALDEHYDE, fat,soa
BITTER ALMOND OIL
f benzaldéhyde *m*, essence *f* d'amandes amères
e aceite *m* de almendras amargas, benzaldehido *m*
i benzaldeide *f*
n benzaldehyde *n*, bittere-amandelolie
d Benzaldehyd *m*, Bittermandelöl *n*

559 BENZENE RECOVERY fue
f récupération *f* du benzol
e recuperación *f* del benzol
i ricupero *m* del benzolo
n benzolterugwinning
d Benzolrückgewinnung *f*

560 BENZOL-ACETONE PROCESS, fue
GOVER PROCESS
f procédé *m* benzol-acétone
e procedimiento *m* benzol-acetona
i processo *m* benzolo-acetone
n benzol-acetonproces *n*
d Benzol-Azetonverfahren *n*

561 BERGAMOT OIL fat
f essence *f* de bergamote
e esencia *f* de bergamot
i olio *m* di bergamotto
n bergamotolie
d Bergamottöl *n*

562 BERGIUS PROCESS, fue
COAL HYDROGENATION PROCESS
f procédé *m* Bergius
e procedimiento *m* de hidrogenación de carbón
i processo *m* di bergiusurazione
n bergiusproces *n*, steenkool-hydrogenatieproces *n*, zweelproces *n*
d Bergius-Verfahren *n*, Kohlehydrierungsverfahren *n*, Schwelverfahren *n*

563 BERLIN BLUE, dec
PRUSSIAN BLUE
f bleu *m* de Prusse
e azul *m* de Prusia
i blu *m* di Prussia
n Berlijns blauw *n*
d Preussischblau *n*

564 BESSEMER STEEL, met
ACID STEEL
f acier *m* acide
e acero *m* ácido
i acciaio *m* acido
n bessemerstaal *n*
d Bessemerstahl *m*

565 BIBLE PAPER, pap
INDIA PAPER
f papier *m* Bible, papier *m* indien
e papel *m* de Biblia
i carta *f* Cina
n bijbelpapier *n*, dundrukpapier *n*
d Bibeldruckpapier *n*, Chinapapier *n*, Dünndruckpapier *n*

566 BICHROMATED GELATINE pho
f gélatine *f* au bichromate, gélatine *f* bichromatée
e gelatina *f* bicromatada
i gelatina *f* al bicromato
n chroomgelatine, dichromaat-gelatine
d Bichromatgelatine *f*

567 BIMOLECULAR REACTION gen
f réaction *f* bimoléculaire
e reacción *f* bimolecular
i reazione *f* bimolecolare
n bimoleculaire reactie
d bimolekulare Reaktion *f*

568 BINARY COMPOUND gen
f composé *m* binaire
e compuesto *m* binario
i composto *m* binario
n binaire verbinding
d binäre Verbindung *f*

569 BINARY ELECTROLYTE elc
f électrolyte *m* binaire
e electrólito *m* binario
i elettrolito *m* binario
n binaire elektrolyt
d binärer Elektrolyt *m*

570 BINARY MIXTURE mix
f mélange *m* binaire
e mezcla *f* binaria
i miscela *f* binaria
n binair mengsel *n*
d Zweistoffgemisch *n*

571 BINDER, cer,mix
BINDING AGENT
f liant *m*
e aglomerante *m*, aglutinante *m*
i agglomerante *m*
n bindmiddel *n*
d Bindemittel *n*

572 BIO-AERATION iwt
f bio-aération *f*
e bioaireación *f*
i bioaerazione *f*
n biologische aëratie, slibactivering
d biologische Aeration *f*, Schlammbelebung *f*

573 BIO-ASSAY, BIOLOGICAL ASSAY mcy
f essai *m* biologique
e bioensayo *m*, ensayo *m* biológico
i collaudo *m* biologico
n biologische proef
d biologische Prüfung *f*

574 BIOCHEMICAL ANALYSIS ana
f analyse *f* biochimique
e análisis *f* bioquímica
i analisi *f* biochimica
n biochemische analyse
d biochemische Analyse *f*

575 BIOCHEMICAL OXYGEN DEMAND, B.O.D. hyg
f demande *f* biochimique d'oxygène
e demanda *f* bioquímica de oxígeno
i esigenza *f* biochimica d'ossigeno
n biochemische zuurstofbehoefte
d biochemischer Sauerstoffbedarf *m*, BSB

576 BIOCHEMICAL RESOLUTION bio
f dédoublement *m* biochimique
e resolución *f* bioquímica
i risoluzione *f* biochimica
n biochemische splitsing
d biochemische Spaltung *f*

577 BIOLOGICAL CLEANING cle,iwt
f épuration *f* biologique, purification *f* biologique
e depuración *f* biológica, purificación *f* biológica
i depurazione *f* biologica
n biologische reiniging
d biologische Reinigung *f*

578 BIRCH TAR fat
f huile *f* de bouleau
e alquitrán *m* de abedul
i catrame *m* di betulla
n berkenteer
d Birkenteer *m*

579 BISCUIT cer
f biscuit *m*
e bizcocho *m* de porcelana
i biscotto *m* di porcellana
n biscuit
d Biskuitporzellan *n*, unglasiertes Porzellan *n*

580 BISCUIT pla
f comprimé *m*, pastille *f*
e pastilla *f*, pieza *f* moldeada
i compressa *f*, pastiglia *f*
n persstuk *n*, tablet *n*
d Pressling *m*, Tablette *f*

581 BISCUIT BAKING, BISCUIT FIRING cer
f cuisson *m* de biscuit
e abizcochura *f*
i cottura *f* di biscotto
n biscuitbrand
d Biskuitbrand *m*

582 BITTER ALMOND OIL, BENZALDEHYDE fat,soa
f benzaldéhyde *m*, essence *f* d'amandes amères
e aceite *m* de almendras amargas, benzaldehido *m*
i benzaldeide *f*
n benzaldehyde *n*, bittere-amandelolie
d Benzaldehyd *m*, Bittermandelöl *n*

583 BITTER ORANGE FLOWER OIL, NEROLI OIL, ORANGE BLOSSOM OIL fat
f essence *f* de fleurs d'oranger, essence *f* de néroli
e esencia *f* de neroli
i olio *m* di fiori d'arancio, olio *m* di neroli
n neroliolie, oranjebloesemolie
d Neroliöl *n*, Orangenblütenöl *n*, Pomeranzblütenöl *n*

584 BITTER SALT, EPSOM SALT mcy
f sel *m* amer, sel *m* d'Epsom
e epsomita *f*, sal *f* amarga
i solfato *m* di magnesio
n bitterzout *n*, Engels zout *n*, magnesiumsulfaat *n*
d Bittersalz *n*, Epsomsalz *n*, Magnesiumsulfat *n*

585 BITTERING POWER bre
f pouvoir *m* d'amertume
e poder *m* de amargor
i valore *m* amaro
n bitterheidsgraad
d Bitterwert *m*

586 BITTERN, aic,gen
LEACH BRINE, MOTHER-LYE
f eau *f* mère
e agua *f* madre
i acqua *f* madre
n moederloog
d Bitterlauge *f*, Mutterlauge *f*, Muttersole *f*

587 BITTINESS, dec
GRITTINESS
f aspect *m* granuleux
e contextura *f* arenosa
i aspetto *m* di granelli
n gruizigheid, zaverigheid
d Punktbildung *f*

588 BITUMEN, bui
ASPHALT, JUDEAN PITCH, MINERAL PITCH
f asphalte *m*, poix *f* de Judée
e asfalto *m*
i asfalto *m*
n aardpek *n*, asfalt *n*, jodenpek *n*
d Asphalt *m*, Bitumen *n*, Erdpech *n*

589 BITUMINOUS CEMENT, rub
BITUMINOUS JOINT FILLING COMPOUND
f mastic *m* bitumineux
e mástique *m* bituminoso
i mastice *m* bituminoso
n asfaltkit
d Asphaltkitt *m*

590 BITUMINOUS LACQUER, dec
BITUMINOUS PAINT
f vernis *m* bitumineux
e pintura *f* bituminosa
i pittura *f* a base bituminosa
n asfaltlak *n*
d Asphaltlack *m*

591 BITUMINOUS PLASTICS pla
f plastiques *pl* bitumineuses
e productos *pl* plásticos bituminosos
i plastici *m* bituminosi
n asfaltplastieken *pl*
d Asphaltpressmassen *pl*

592 BIURET REACTION, ana
BIURET TEST
f réaction *f* du biuret
e reacción *f* del biuret
i reazione *f* biuretica
n biuretreactie
d Biuretreaktion *f*

593 BLACK ANNEALING, met
OPEN ANNEALING
f recuit *m* en noir
e recocido *m* en negro
i ricottura *f* in nero
n zwartgloeien *n*
d Schwarzglühen *n*

594 BLACK ANODIZING coa
f oxydation *f* anodique en noir
e anodización *f* en negro, eloxación *f* en negro
i ossidazione *f* anodica in nero
n zwartanodiseren *n*
d Schwarzeloxieren *n*

595 BLACK BALSAM, mcy
BALSAM OF PERU
f baume *m* du Pérou
e bálsamo *m* del Perú
i balsamo *m* del Perù
n perubalsem
d Perubalsam *m*

596 BLACK BLASTING POWDER, exp
BLACK POWDER, GUNPOWDER
f poudre *f*
e pólvora *f*
i polvere *f*
n buskruit *n*, kruit *n*
d Pulver *n*, Schiesspulver *n*, Schwarzpulver *n*

597 BLACK JAPAN dec
f vernis *m* noir japonais semitransparent
e barniz *m* semitransparente de secado rápido
i vernice *f* nera semitrasparente
n zwarte japanlak
d schwarzer Japanlack *m*

598 BLACK LIQUOR, pap
BLACK MORDANT, IRON ACETATE LIQUOR
f solution *f* d'acétate de fer
e mordiente *m* para negro, solución *f* de acetato de hierro
i soluzione *f* d'acetato di ferro
n ijzeracetaatoplossing
d Eisenazetatlösung *f*

599 BLACK MALT, bre
CARAMEL MALT, ROASTED MALT
f malt *m* noir, malt *m* torréfié
e malta *f* colorante, malta *f* tostada
i malto *m* torrefatto
n licht geroosterde mout
d Farbmalz *n*, Röstmalz *n*

600 BLACK MORTAR bui
f mortier *m* noir
e mortero *m* negro
i malta *f* nera
n zwarte mortel, zwarte specie
d schwarzer Mörtel *m*

601 BLACK OXIDE OF MANGANESE, min
BATTERY MANGANESE,
GLASSMAKER'S SOAP,
MANGANESE DIOXIDE
f pyrolusite *f*
e pirolusita *f*
i perossido *m* di manganese
n bruinsteen *n*, mangaansuperoxyde *n*
d Braunstein *m*, Mangandioxyd *n*, Pyrolusit *m*

602 BLACK SAND, fou
BACKING SAND
f sable *m* de remplissage
e arena *f* de relleno
i terra *f* di riempimento
n vulzand *n*
d Füllsand *m*, Haufensand *m*

603 BLACKENING coa
f noircissement *m*
e ennegrecimiento *m*
i annerimento *m*
n zwarten *n*, zwarting
d Schwärzen *n*, Schwärzung *f*

604 BLACKING, fou
BLACKWASH, COATING,
FACING
f enduit *m*
e unto *m* de revestimiento
i intonaco *m*
n vormdekmiddel *n*
d Schlichte *f*, Schwärze *f*

605 BLANC DE CHINE cer
f blanc *m* de Chine
e vidriado *m* blanco brillante
i bianco *m* di Cina
n Chinees wit *n*
d Chinesischweiss *n*

606 BLANC FIXE, dec
PERMANENT WHITE
f blanc *m* fixe
e blanco *m* fijo
i bianco *m* fisso
n barietwit *n*
d Barytweiss *n*, Blankfix *n*, Permanentweiss *n*

607 BLANK CARBURIZING, htt,met
PSEUDOCARBURIZING
f pseudocémentation *f*
e seudocementación *f*
i pseudocementazione *f*
n pseudocementeren *n*
d Pseudozementierung *f*

608 BLANK DETERMINATION ana
f essai *m* à blanc
e prueba *f* en blanco
i prova *f* in bianco
n blinde proef
d Blindprobe *f*

609 BLANK MIX (US), gen
BLANK STOCK (GB), BASE
MIX (US), BASIC MATERIAL (GI
f composé *m* de base, mélange *m* de base
e mezcla *f* básica
i mescolanza *f* base, mescolanza di sfondo
n basismengsel *n*
d Grundmischung *f*

610 BLANK NITRIDING, htt,met
PSEUDONITRIDING
f pseudonitruration *f*
e seudonitruración *f*
i pseudonitrurazione *f*
n pseudonitreren *n*
d Pseudonitrierung *f*

611 BLAST FURNACE CEMENT bui
f ciment *m* de laitier de haut fourneau
e cemento *m* de escoria de alto horno
i cemento *m* di scorie d'alto forno
n hoogovencement *n*
d Hochofenzement *m*

612 BLAST FURNACE GAS fue,gas
f gaz *m* de gueulard, gaz *m* de haut fourneau
e gas *m* de alto horno
i gas *m* d'alto forno
n hoogovengas *n*
d Gichtgas *n*, Hochofengas *n*

613 BLAST FURNACE SLAG htt
f laitier *m* de haut fourneau
e escoria *f* de alto horno
i scorie *pl* d'alto forno
n hoogovenslak
d Hochofenschlacke *f*

614 BLAST ROASTING, ROAST SINTERING htt,met
f frittage *m* sur grille, grillage *m* de frittage
e sinterización *f* por tostación
i arrostimento *m* di sinterizzazione, sinterizzazione *f* su griglia
n sinterroosten *n*
d Sinterrösten *n*

615 BLEACH-OUT PROCESS pho
f procédé *m* de blanchiment
e procedimiento *m* de blanqueo
i processo *m* d'imbianchimento
n uitbleekproces *n*
d Ausbleichen *n*, Ausbleichverfahren *n*

616 BLEACHED PULP pap
f pâte *f* blanchie
e pulpa *f* blanqueada
i polpa *f* imbiancata
n gebleekte pulp
d gebleichtes Ganzzeug *n*

617 BLEACHING ble
f blanchiment *m*
e blanqueo *m*
i candeggio *m*, imbianchimento *m*, sbianca *f*, sbiancamento *m*
n bleken *n*
d Bleichen *n*

618 BLEACHING AGENT ble
f agent *m* de blanchiment
e agente *m* de blanqueo
i agente *m* di candeggio, agente *m* di sbianca
n bleekmiddel *n*
d Bleichmittel *n*

619 BLEACHING BATH pho
f bain *m* de blanchiment
e baño *m* de blanqueo
i bagno *m* d'imbianchimento
n bleekbad *n*
d Bleichbad *n*

620 BLEACHING CLAY, BLEACHING EARTH cle,fue
f argile *f* décoloratrice
e arcilla *f* descoloradora, tierra *f* de jaboneros
i argilla *f* decoloratrice
n bleekaarde, ontkleuringsklei
d Bleicherde *f*, Entfärbungston *m*

621 BLEACHING FIXING pho
f blanchiment *m* et fixage *m*
e blanqueo *m* y fijado *m*
i imbianchimento *m* e fissaggio *m*
n blekend fixeren *n*
d Bleichfixierung *f*

622 BLEACHING LIQUOR ble
f eau *f* de Javel, eau *f* de Javelle
e agua *f* de cloro
i acqua *f* da imbiancare
n bleekloog, bleekwater *n*
d Bleichflotte *f*, Chlorlauge *f*, Kalziumhypochloritlösung *f*

623 BLEACHING POWDER, CHLORINATED LIME ble
f chlorure *m* de chaux
e cal *f* clorada, cloruro *m* de cal, hipoclorito *m* cálcico
i calce *f* clorurata, cloruro *m* di calce
n bleekpoeder *n*, chloorkalk
d Bleichpulver *n*, Chlorkalk *m*, Kalziumhypochlorit *n*

624 BLEACHING PROCESS ble
f procédé *m* de blanchiment
e procedimiento *m* de blanqueo
i processo *m* di candeggio, processo *m* di sbianca
n bleekproces *n*
d Bleichverfahren *n*

625 BLEED THROUGH pla
f pénétration *f* de l'adhésif
e penetración *f* del adhesivo
i penetrazione *f* dell'adesivo
n uitlopen *n* van het kleefmiddel
d Klebemittelauslauf *m*

626 BLEEDER STEAM boi
f vapeur *f* de soutirage
e vapor *m* de escape, vapor *m* de purga
i vapore *m* di scappata
n ontsnappende stoom
d Anzapfdampf *m*

627 BLEEDING dec
f dégorgement *m*, saignement *m*
e sangrado *m*
i sanguinamento *m*
n bloeden *n*, doorslaan *n*
d Ausbluten *n*, Durchschlagen *n*

628 BLEEDING tex
f déteinte *f*
e corrimiento *m*
i struggimento *m*
n doorlopen *n*
d Verwaschung *f*

629 BLEMISH, bui
EFFLORESCENCE, SCUM
f carie *f*, efflorescence *f*, salpêtre *m*
e eflorescencia *f*
i efflorescenza *f* murale
n muurkanker, muursalpeter, muuruitslag
d Ausblühung *f*, Auswitterung *f*, Mauerfrass *m*, Salpeterfrass *m*

630 BLENDING, mix
MIXING
f mélange *m*, mixtion *f*
e mezcla *f*, mixtión *f*
i mescolamento *m*, mescolanza *f*
n menging
d Mischung *f*

631 BLENDING AGENT, fue
ANTIDETONANT, ANTIKNOCK ADDITIVE
f produit *m* antidétonant
e aditivo *m* antidetonante
i sostanza *f* antidetonante
n antiklopmiddel *n*
d Antiklopfmittel *n*, klopfhindernder Zusatz *m*

632 BLISTERING, dec
PIMPLING
f cloquage *m*
e formación *f* de vejigas
i vescicatura *f*
n blaasjesvorming
d Blasenbildung *f*

633 BLOCKING dis
f blocage *m*, obstruction *f*
e obstrucción *f*
i blocco *m*
n verstopping
d Verstopfung *f*

634 BLOCKING pla
f adhérence *f* entre feuilles, collage *m* de couches
e adherencia *f* entre capas
i aderenza *f* tra foglie, incollatura *f*
n plakken *n* van lagen
d Ankleben *n* von Schichten

635 BLOOM fue
f fluorescence *f*
e fluorescencia *f*
i fluorescenza *f*
n fluorescentie
d Fluoreszenz *f*

636 BLOOM rub
f efflorescence *f*
e eflorescencia *f*
i efflorescenza *f*
n uitbloeming
d Effloreszenz *f*

637 BLOOM-FREE (US), rub
NON-BLOOMING (GB)
f non-efflorescent adj
e no eflorescente adj
i non efflorescente adj
n vrij van uitbloemen
d nicht ausblühend adj, nicht ausschwefelnd adj

638 BLOOM GLOSS, dec
HAZY GLOSS
f brillant *m* terni, brillant *m* voilé
e lustre *m* velado
i lucidatura *f* velata
n spekglans
d Speckglanz *m*

639 BLOOMING, dec
BLUEING, HAZING
f bleuissement *m*, brouillage *m*, voilage *m*
e falta *f* de lustre, ponerse v mate
i divenire v torbido, velamento *m*
n vorming van een blauw waas, wazen *n*
d Blauanlaufen *n*, Hauchbildung *f*, Schleierbildung *f*

640 BLOOMING OF SULFUR (US), aic
BLOOMING OF SULPHUR (GB)
f fleur *f* de soufre
e flor *f* de azufre
i fiori *pl* di zolfo
n bloem van zwavel, zwavelbloem
d Schwefelblume *f*

641 BLOTTING-PAPER pap
f buvard *m*, papier *m* brouillard
e papel *m* secante, secante *m*
i carta *f* assorbente, carta *f* suga
n vloei *n*, vloeipapier *n*
d Löschpapier *n*

642 BLOW-AND-BLOW PROCESS gla
f procédé *m* de soufflage de verre en deux phases
e procedimiento *m* de soplado de vidrio de dos fases
i processo *m* di soffiatura di vetro a due fasi
n tweetrapsglasblaasproces *n*
d Zweistufenglasblaseverfahren *n*

643 BLOW MOLDING (US), BLOW MOULDING (GB) pre
f procédé *m* de soufflage en moule
e procedimiento *m* de soplado en molde
i processo *m* di soffiatura in forma
n blazen *n* in de vorm
d Herstelllung *f* in Blasform

644 BLOW PRESSURE pla,pre
f pression *f* de soufflage
e presión *f* de soplado
i pressione *f* di soffiatura
n blaasdruk
d Blasedruck *m*

645 BLOW STAGE fue
f temps *m* de soufflage
e tiempo *m* de soplado
i tempo *m* di soffiatura
n blaastijd
d Blasezeit *f*

646 BLOWING pla,pre
f soufflage *m*
e soplado *m*
i soffiatura *f*
n blazen *n*
d Blasen *n*

647 BLOWING rub
f gonflement *m*
e hinchamiento *m*, sopleo *m*
i rigonfiamento *m*
n opblazen *n*
d Blähen *n*, Treiben *n*

648 BLOWING tex
f traitement *m* à la vapeur
e tratamiento *m* al vapor
i trattamento *m* al vapore
n stoombehandeling
d Dampfbehandlung *f*

649 BLOWING AGENT, INFLATING AGENT, SPONGING AGENT rub
f agent *m* gonflant
e agente *m* esponjante, agente *m* hinchante
i agente *m* rigonfiante
n blaasmiddel *n*
d Blähmittel *n*, Treibmittel *n*

650 BLOWN BITUMEN fue
f bitume *m* soufflé
e betún *m* soplado
i bitume *m* soffiato
n geblazen asfalt *n*
d geblasener Asphalt *m*

651 BLOWN EXTRUSION PROCESS rub
f procédé *m* de moulage par boudinage et soufflage
e procedimiento *m* combinado de budinado y soplado
i processo *m* di stampatura per trafilatura e soffiatura
n gecombineerd spuit-blaasproces *n*
d kombiniertes Spritz-Ausblasverfahren *n*

652 BLOWN FILM pla,rub
f pellicule *f* soufflée
e película *f* soplada
i pellicola *f* soffiata
n geblazen film
d geblasener Film *m*

653 BLOWN LATEX FOAM rub
f mousse *f* de latex par soufflage
e espuma *f* de látex soplada
i schiuma *f* di lattice soffiata
n geblazen schuimrubber *n*
d geblähter Schaumgummi *m*

654 BLOWN OIL lub
f huile *f* soufflée
e aceite *m* soplado
i olio *m* soffiato
n geblazen olie
d Trockenöl *n*

655 BLOWPIPE ANALYSIS ana
f analyse *f* au chalumeau
e análisis *f* al soplete
i analisi *f* al cannello
n blaaspijpanalyse
d Lötrohranalyse *f*

656 BLUBBERING lea
f dégraissage *m*
e desengrase *m*
i sgrassatura *f*
n ontvetting
d Entfettung *f*

657 BLUE ANNEALING htt
f bleuissage *m*
e pavonado *m*
i azzurrimento *m*, scaldare v al calore blu
n blauwgloeien *n*
d Blauglühen *n*

658 BLUE DUST, ZINC DUST met
f poudre *f* de zinc
e polvo *m* de cinc
i polvere *f* di zinco
n zinkstof *n*
d Zinkstaub *m*

659 BLUE GAS, gas
BLUE WATER GAS, WATER GAS
f gaz *m* à l'eau, gaz *m* bleu
e gas *m* azul, gas *m* de agua
i gas *m* blu, gas *m* d'acqua
n blauw watergas *n*, watergas *n*
d Blauwassergas *n*, Koksgas *n*, Wassergas *n*

660 BLUE GAS GENERATION, gas
BLUE WATER GAS GENERATION
f génération *f* de gaz à l'eau
e generación *f* de gas de agua
i generazione *f* di gas d'acqua
n watergasontwikkeling
d Wassergaserzeugung *f*

661 BLUE LEAD met
f plomb *m* métallique
e plomo *m* metálico
i piombo *m* metallico
n metalliek lood *n*
d metallisches Blei *n*

662 BLUE LEAD, met
GALENA, LEAD GLANCE
f galène *f*
e alquifol *m*, galena *f*
i galena *f*
n galeniet *n*, loodglans *n*
d Bleiglanz *m*, Galenit *m*

663 BLUE OIL fat
f huile *f* bleue
e aceite *m* azul
i olio *m* azzurro
n blauwolie
d Blauöl *n*

664 BLUE-PRINT PAPER pap
f papier *m* à tirage en bleu
e papel *m* al ferroprusiato
i carta *f* cianografica
n blauwdrukpapier *n*
d Blaupauspapier *n*

665 BLUE PRINTING pho
f tirage *m* en bleu
e copia *f* en papel al ferroprusiato
i cianografica *f*
n blauwdrukken *n*
d eine Blaupause anfertigen v

666 BLUE STONE, min
BLUE VITRIOL, CHALCANTHITE, CYANOSITE
f chalcanthite *f*, vitriol *m* bleu
e calcantita *f*, vitriolo *m* azul
i calcantite *f*, vitriolo *m* azzurro
n blauw vitriool *n*, blauwsteen *n*, chalcantiet *n*
d Blaustein *n*, Blauvitriol *n*, Chalkantit *m*, Kupfervitriol *n*

667 BLUEING (GB), met
BLUING (US)
f bleuissage *m*
e azulado *m*
i rinvenire v al blu
n blauw aanlopen *n*
d blau anlassen *n*

668 BLUEING (GB), tex
BLUING (US)
f azurage *m*, bleutage *m*
e azulado *m*
i processo *m* da azzurrare, processo *m* da azzurreggiare
n blauwen *n*
d Blauen *n*

669 BLUEING BATH (GB), tex
BLUING BATH (US)
f bain *m* d'azurage, bain *m* de bleutage
e baño *m* de azulado
i bagno *m* d'azzurraggio
n blauwbad *n*
d Bläuungsbad *n*

670 BLUSHING dec
f formation *f* d'une toile laiteuse, louchissement *m*
e opacidad *f* superficial
i imbianchimento *m*, lattiginosità
n wit uitslaan *n*
d Weissanlaufen *n*

671 BOARD, mat,pap
CARDBOARD, PASTEBOARD
f carton *m*
e cartón *m*
i cartone *m*
n karton *n*
d Pappe *f*

672 BOARDING, lea
GRAINING
f grenure *f*
e granulado *m*
i granitura *f*
n krispelen *n*
d Krispeln *n*

673 B.O.D., hyg
BIOCHEMICAL OXYGEN DEMA
f demande *f* biochimique d'oxygèr
e demanda *f* bioquímica de oxígen
i esigenza *f* biochimica d'ossigen

n biochemische zuurstofbehoefte
d biochemischer Sauerstoff-
bedarf *m*, BSB

674 BODIED OIL, dec
STAND OIL
f standolie *f*
e aceite *m* secante espesado
i olio *m* di lino cotto
n standolie
d Standöl *n*

675 BODY bre
f corps *m*
e cuerpo *m*
i corpo *m*
n gehalte *n*
d Gehalt *m*

676 BODY COLOR (US), dec
BODY COLOUR (GB)
f couleur *f* dominante
e color *m* dominante
i colore *m* dominante
n overheersende kleur
d vorherrschende Farbe *f*

677 BODY COLOR (US), dec
BODY COLOUR (GB)
f peinture *f* couvrante
e pintura *f* opaca
i pittura *f* opaca
n dekverf
d Deckfarbe *f*

678 BODY OF A PAINT dec
f consistence *f* d'une peinture
e consistencia *f* de una pintura
i consistenza *f* d'una pittura
n lijvigheid
d Dickflüssigkeit *f*

679 BODY OF A VARNISH dec
f corps *m* d'un vernis, pouvoir *m*
garnissant
e cuerpo *m* de un barniz
i corpo *m* d'una vernice
n vulkracht
d Füllkraft *f*

680 BOIL (TO) gen
f bouillir v
e bullir v
i bollire v
n koken v
d kochen v, sieden v

681 BOIL-UP RATE dis
f taux *m* de rebouillage
e reflujo *m* interno de evaporación
i velocità *f* d'evaporazione
n verkokingssnelheid
d Dampfbelastung *f*

682 BOILED-OFF LIQUOR, tex
BAST LIQUOR
f liquide *m* de décreusage,
liquide *m* de dégommage
e líquido *m* de descrudado,
líquido *m* de desengomado
i liquido *m* di sgommatura
n zijdeafkooksel *n*
d Abkochflüssigkeit *f*,
Degummierungsflüssigkeit *f*

683 BOILED-OFF SILK, tex
DEGUMMED SILK
f soie *f* dégommée
e seda *f* descrudada, seda *f*
desgomada
i seta *f* cotta
n afgekookte zijde, ontgomde zijde
d entschälte Seide *f*, linde Seide *f*

684 BOILED OIL, dec
BOILED LINSEED OIL,
DRYING OIL
f huile *f* de lin cuite
e aceite *m* de linaza cocido,
aceite *m* secante
i olio *m* di lino cotto
n lijnolievernis *n*
d Leinölfirnis *m*, Ölfirnis *m*

685 BOILER COMPOSITION (GB), boi
BOILER COMPOUND (US)
f anti-incrostant *m*, tartrifuge *m*
e desincrustante *m* para calderas
i composizione *f* antiincrostazione
n antiketelsteenmiddel *n*
d Kesselsteinlösemittel *n*,
Kesselsteinverhütungsmittel *n*

686 BOILER FEED WATER boi
f eau *f* d'alimentation de chaudière
e agua *f* de alimentación de caldera
i acqua *f* d'alimentazione di
caldaio
n ketelvoedingswater *n*, ketelwater *n*
d Kesselspeisewasser *n*

687 BOILER SCALE boi
f tartre *m*
e incrustación *f*
i incrostazione *f*
n ketelsteen *n*
d Kesselstein *m*, Wasserstein *m*

688 BOILING gen
f ébullition *f*

e ebullición *f*
i ebollizione *f*
n koken *n*
d Kochen *n*

689 BOILING DOWN, bre,gen
CONCENTRATION
f épaississement *m*, réduction *f*
e concentración *f*
i concentrazione *f*
n indikking, inkoken *n*
d Eindickung *f*, Einkochen *n*

690 BOILING FERMENTATION bre,fmi
f fermentation *f* bouleuse
e fermentación *f* tumultuosa
i fermentazione *f* in ebollizione
n sterke gisting
d kochende Gärung *f*

691 BOILING-OFF tex
f décreusage *m*, dégommage *m*
e descrudado *m*
i cottura *f*
n afkoken *n*
d Abkochen *n*

692 BOILING-OVER gen
f débordement *m*
e rebosamiento *m*
i traboccamento *m*
n overkoken *n*
d Überkochen *n*

693 BOILING RANGE gen
f intervalle *m* d'ébullition
e intervalo *m* de ebullición
i intervallo *m* d'ebollizione
n kookgebied *n*, kooktraject *n*
d Kochgebiet *n*

694 BOILING TO GRAIN sin
f cuite *f* en grains
e cocción *f* hasta formación del grano
i cottura *f* in grano
n op korrel koken *n*
d Kornkochen *n*

695 BOLTING nut,sor
f blutage *m*
e cernido *m*, tamizado *m*
i burattatura *f*, stacciatura *f*
n zeven *n*
d Sieben *n*

696 BOMB mit
f bombe *f*
e bomba *f*
i bomba *f*
n bom
d Bombe *f*

697 BOND TESTING tes
f essai *m* de collage
e ensayo *m* de adhesión
i prova *f* d'adesione
n hechtkrachtproef
d Haftfestigkeitsprüfung *f*

698 BONDING glu,pla
f encollage *m*
e encoladura *f*, encolamiento *m*
i incollatura *f*
n lijmen *n*, plakken *n*
d Leimung *f*

699 BONDING RESIN glu,pla
f résine *f* adhésive
e resina *f* adhesiva
i resina *f* adesiva
n klevende hars
d klebendes Harz *n*

700 BONE ASH, cer
BONE EARTH
f cendre *f* d'os, terre *f* d'os
e ceniza *f* de huesos, tierra *f* de huesos
i cenere *pl* d'ossa, terra *f* d'ossa
n beenderaarde, beenderas
d Knochenerde *f*

701 BONE BLACK, mat,oic
ANIMAL BLACK, ANIMAL CHARCOAL
f charbon *m* animal, noir *m* animal
e carbón *m* animal, negro *m* anima[l], negro *m* de huesos
i carbone *m* animale, carbone *m* d'ossa, nero *m* animale, nero *m* d'ossa
n beenderkool, beenzwart *n*
d Knochenkohle *f*, Knochenschwar[z], Tierschwarz *n*

702 BONE CHINA, cer
BONE PORCELAIN
f porcelaine *f* mince
e porcelana *f* blanda inglesa
i porcellana *f* sottile
n beenderporselein *n*
d dünnes Porzellan *n*

703 BONE DUST, fer
BONE MANURE, BONE MEAL
f engrais *m* d'os
e abono *m* de huesos, harina *f* de huesos
i farina *f* d'ossa

n beendermeel *n*
d Knochenmehl *n*

704 BONE FAT, BONE TALLOW — fat
f graisse *f* d'os
e sebo *m* de huesos
i grasso *m* d'ossa
n beendervet *n*
d Knochenfett *n*

705 BONE GLUE — glu
f colle *f* d'os
e cola *f* de huesos, osteocola *f*
i colla *f* d'ossa
n beenderlijm
d Knochenleim *m*

706 BONE OIL, DIPPEL'S OIL — fat
f huile *f* d'os
e aceite *m* de huesos
i olio *m* d'ossa
n beenderolie
d Knochenöl *n*

707 BOOST MELTING — gla
f fusion *f* à chauffage additionnel
e fusión *f* de calentamiento adicional
i fusione *f* a riscaldamento addizionale
n smelten *n* met bijverhitting
d Schmelzen *n* mit zusätzlicher Heizung

708 BORAX BEAD — ana
f perle *f* de borax
e perla *f* de borax
i perla *f* di borace
n boraxparel
d Boraxperle *f*

709 BORAX GLASS, FUSED SODIUM BORATE — gla
f verre *m* au borate de soude
e vidrio *m* al borato de sodio
i vetro *m* al borato di sodio
n boraxglas *n*
d Boraxglas *n*

710 BOROSILICATE GLASS — gla
f verre *m* au silicate de bore
e vidrio *m* al borosilicato
i vetro *m* al borosilicato
n boriumsilicaatglas *n*
d Borosilikatglas *n*

711 BORT — min
f bort *m*
e bort *m*, carbonado *m*
i diamante *m* bort
n boort *n*
d Bort *m*

712 BOTTING CLAY — htt
f argile *f* de bouchage
e arcilla *f* de taponamiento
i argilla *f* di tamponatura
n afsluitklei, propklei
d Abschlusstonerde *f*, Stopfentonerde *f*

713 BOTTLE GLASS — gla
f verre *m* à bouteilles, verre *m* vert
e vidrio *m* de botellas
i vetro *m* di bottiglie
n flessenglas *n*, groen glas *n*
d Flaschenglas *n*, grünes Glas *n*

714 BOTTLED BEER — bre
f bière *f* en bouteille
e cerveza *f* de botella
i birra *f* in bottiglia
n flessebier *n*
d Flaschenbier *n*

715 BOTTLING — bre
f embouteillage *m*, mise *f* en bouteilles
e embotellado *m*
i imbottigliatura *f*
n bottelen *n*
d Abziehen *n* auf Flaschen, Flaschenfüllung *f*

716 BOTTOM EJECTION — pla,pre
f expulsion *f* inférieure
e expulsión *f* inferior
i espulsione *f* inferiore
n onderlossing
d Bodenauswurf *m*

717 BOTTOM FERMENTATION — bre,fmi
f fermentation *f* basse
e fermentación *f* baja
i fermentazione *f* bassa
n ondergisting
d Untergärung *f*

718 BOTTOM FERMENTATION YEAST, BOTTOM YEAST — bre,fmi
f levure *f* basse
e levadura *f* de fermentación baja
i lievito *m* per fermentazione bassa
n ondergist
d untergärige Hefe *f*, Unterzeug *m*

719 BOTTOM LAYER OF YEAST bre
f couche *f* inférieure de levure
e capa *f* inferior de levadura
i strato *m* inferiore di lievito
n ondergistlaag
d Unterzeugschicht *f*

720 BOTTOM PRODUCT dis
f produit *m* de fond
e producto *m* de cola
i prodotto *m* di fondo
n bodemprodukt *n*
d Sumpfprodukt *n*

721 BOTTOMING tex
f lavage *m* intensif
e lavado *m* intensivo
i lavaggio *m* intensivo
n grondige reiniging, intensieve wassing
d gründliche Reinigung *f*

722 BOTTOMING, GROUNDING tex
f teinture *f* préliminaire
e teñido *m* preliminar
i tintura *f* preliminare
n voorverven *n*
d Vorfärbung *f*

723 BOTTOMS des,fue
f résidus *pl* de distillation
e residuos *pl* de destilación
i fondi *pl* di distillazione
n destillatieresidu's *pl*
d hochsiedender Rückstand *m*

724 BOTTOMS, CASK DEPOSIT bre
f dépôt *m*, fond *m* de foudre
e fondo *m* de barril
i feccia *f*
n vatbezinksel *n*
d Fassgeläger *n*

725 BOUND OXYGEN gen
f oxygène *m* lié
e oxígeno *m* combinado
i ossigeno *m* legato
n gebonden zuurstof *n*
d gebundener Sauerstoff *m*

726 BOUND RUBBER rub
f caoutchouc *m* lié
e caucho *m* combinado
i gomma *f* legata
n gebonden rubber
d gebundener Gummi *m*

727 BOUND SULFUR (US), COMBINED SULPHUR (GB) rub
f soufre *m* combiné
e azufre *m* combinado
i zolfo *m* combinato
n gebonden zwavel
d gebundener Schwefel *m*

728 BOUND WATER gen
f eau *f* liée
e agua *f* combinada
i acqua *f* legata
n gebonden water *n*
d gebundenes Wasser *n*

729 BOUNDARY LUBRICATION, MARGINAL LUBRICATION lub
f lubrification *f* dans la couche limite
e lubricación *f* en la capa límite
i lubrificazione *f* nello strato limite
n grenssmering
d Grenzschmierung *f*

730 BOX ANNEALING, CLOSE ANNEALING, POT ANNEALING htt,met
f recuit *m* en caisse
e recocido *m* en caja
i ricottura *f* in cassetta
n kistgloeien *n*
d Kastenglühen *n*

731 BOX CALF lea
f box-calf *m*, veau *m* chromé
e boxcalf *m*, piel *f* de becerro al cro
i pelle *f* di vitello al cromo
n bokskalf *n*, boxcalf *n*
d Boxcalf *n*, Boxkalb *n*

732 BOX CARBURIZING, PACK CARBURIZING htt,met
f cémentation *f* en caisse
e cementación *f* en caja
i cementazione *f* in cassetta
n kistcementeren *n*
d Kastenzementierung *f*

733 BOX GRAIN lea
f box-calf *m* grenelé
e boxcalf *m* graneado
i pelle *f* di vitello al cromo a granitura
n gekrispeld boxcalf *n*
d Krispelboxcalf *n*

734 BOX MALTING bre
f malterie *f* à cases
e maltería *f* en cajas
i malteria *f* a casse
n kastmouten *n*
d Kastenmälzen *n*

735 BRAIN SUGAR (US), sin
CEREBROSE (GB),
GALACTOSE
f galactose *f*
e galactosa *f*
i galattosio *m*
n galactose
n Cerebrose *f*, Galaktose *f*

736 BRAKE FLUID cpt
f liquide *m* pour freins hydrauliques
e líquido *m* para frenos hidráulicos
i liquido *m* per freni idraulici
n remvloeistof
d Bremsflüssigkeit *f*

737 BRAKE LINING cpt
f garniture *f* de frein
e forro *m* de freno, guarnición *f* de freno
i guarnizione *f* di freno
n remvoering
d Bremsbelag *m*

738 BRAN DECOCTION BATH dye
f bain *m* de son
e baño *m* de salvado
i bagno *m* di crusca
n zemelbad *n*
d Kleienabsudbad *n*

739 BRAN OIL (US), oic
FURAN
f furanne *m*
e furano *m*
i furano *m*
n furaan *n*, furfuraan *n*
d Furan *n*, Furfuran *n*

740 BRASS, met
YELLOW BRASS
f laiton *m*
e latón *m*
i ottone *m*
n messing *n*
d Gelbguss *m*, Messing *m*

741 BRASS PLATING elc
f placage *m* en laiton
e latonado *m*
i ottonatura *f*
n vermessingen *n*
d Vermessingung *f*

742 BRASS POLISH, cle
METAL POLISH
f moyen *m* de polissage, produit *m* à fourbir
e lustre *m* para metales, pasta *f* limpiadora
i lucidatrice *f* per metalli
n koperpoets, metaalpoets
d Putzmittel *n*

743 BRAZIL WAX, fat
CARNAUBA WAX
f cire *f* de carnauba
e cera *f* carnáubica
i cera *f* carnauba
n carnaubavet *n*
d Karnaubawachs *m*

744 BRAZING, wel
HARD SOLDERING
f soudure *f* forte
e soldadura *f* fuerte
i brasatura *f*
n hardsolderen *n*
d Hartlöten *n*

745 BRAZING SOLDER wel
f soudure *f* forte
e cobresoldadura *f*, soldadura *f* fuerte
i lega *f* per brasatura
n hardsoldeer *n*
d Hartlot *n*

746 BREAK bre
f cassure *f*
e coagulación *f*
i coagulo *m*
n breuk
d Bruch *m*

747 BREAK FORMATION bre
f formation *f* du tranché
e formación *f* del coágulo
i formazione *f* del coagulo
n breukvorming
d Bruchbildung *f*

748 BREAKDOWN, gen
DEGRADATION
f décomposition *f*
e degradación *f*, descomposición *f*
i disgregazione *f*
n afbraak, ontbinding
d Abbau *m*

749 BREAKING, brk
SIZE REDUCTION
f broyage *m*
e quebrantamiento *m*
i frantumazione *f*
n vermaling
d Zerkleinerung *f*

750 BREAKING OF AN EMULSION rub
f décomposition *f* d'émulsions
e descomposición *f* de emulsiones
i decomposizione *f* d'emulsioni
n emulsieafbraak
d Emulsionsabbau *m*

751 BREATHING CYCLE, DEGASSING CYCLE gen
f cycle *m* de désaération
e ciclo *m* de desaireación
i ciclo *m* di disareazione
n ontluchtingscyclus
d Entlüftungszyklus *m*

752 BREEZE BRICK, SLAG BRICK bui
f brique *f* en ciment et braise de houille
e ladrillo *m* de cemento y cenizas de coque
i mattone *m* di cemento e cenere di coke
n slakkensteen
d Schlackenstein *m*

753 BREEZING, CULM LAYER cer
f couche *f* en poussier d'anthracite
e capa *f* de cisco
i strato *m* di polvere d'antracite
n laag van antracietpoeder
d Kohlenlöscheschicht *f*

754 BREW, GYLE bre
f brassin *m*
e cocción *f*, cocimiento *m*
i cotta *f*
n brouwsel *n*
d Gebräu *n*, Sud *m*

755 BREWER'S GRAINS bre
f drêches *pl*
e orujo *m*
i trebbie *pl*
n bostel
d Treber *pl*, Trester *pl*

756 BREWER'S PITCH oic
f poix *f* de fûts
e pez *f* para barriles
i pece *f* per fusti
n brouwerspek *n*
d Brauerpech *n*, Fasspech *n*

757 BREWER'S YEAST bre
f levure *f* à bière
e levadura *f* de cerveza
i lievito *m* di birra
n biergist
d Bierhefe *f*

758 BREWING bre
f brassage *m*
e fabricación *f* de cerveza
i fabbricazione *f* di birra
n brouwen *n*
d Brauen *n*, Sud *m*

759 BREWING BARLEY bre
f orge *f* de brasserie
e cebada *f* cervecera
i orzo *m* da birra
n brouwgerst
d Braugerste *f*

760 BREWING LIQUOR bre
f eau *f* de brassage
e agua *f* de cocimiento
i acqua *f* per birra
n brouwwater *n*
d Brauwasser *n*

761 BREWING VALUE bre
f valeur *f* brassicole
e valor *m* cervecero
i valore *m* in rapporto alla fabbricazione della birra
n brouwwaarde
d Brauwert *m*

762 BRICK bui
f brique *f*
e ladrillo *m*
i mattone *m*
n baksteen
d Backstein *m*, Ziegelstein *m*

763 BRICK CLAY, BRICK EARTH bui
f terre *f* glaise
e arcilla *f* para ladrillos, tierra *f* de ladrillos
i argilla *f* per mattoni, terra *f* da mattoni
n klei, leem *n*
d Ziegelerde *f*, Ziegelton *m*

764 BRIDGING ATOM gen
f atome *m* de pont
e átomo *m* de puente
i atomo *m* di ponte
n brugatoom *n*
d Brückenatom *n*

765 BRIGHT ANNEALING htt
f recuit *m* blanc
e recocido *m* brillante
i ricottura *f* antiossidante

n blankgloeien *n*
d Blankglühen *n*

766 BRIGHT DIPPING elc
f brillantage *m*
e abrillantamiento *m*
i brillantaggio *m*
n glansbeitsen *n*
d Glanzbrennen *n*

767 BRIGHT FINISH pol
f affinage *m* brillant
e afino *m* brillante
i finitura *f* brillante
n glansfijnen *n*
d Glanzfeinen *n*

768 BRIGHT STOCK lub
f lubrificant *m* de pétrole résiduel
e lubricante *m* de petróleo residual
i lubrificante *m* di petrolio residuale
n geëxtraheerde smeerolie
d extrahiertes Schmieröl *n*

769 BRIGHTENER elc
f agent *m* de brillantage
e agente *m* de abrillantado
i agente *m* per brillantaggio
n glansmiddel *n*
d Glanzzusatz *m*

770 BRIGHTENING pol
f polissage *m*
e abrillantado *m*, pulimentado *m*
i brillantatura *f*
n polijsten *n*
d Polieren *n*

771 BRIGHTENING tex
f catissage *m*, lissage *m*
e lustración *f*
i lustramento *m*
n glanzen *n*
d Glänzen *n*

772 BRIGHTENING AGENT, OPTICAL BRIGHTENER tex
f agent *m* de blanchiment optique, azurant *m* optique
e agente *m* blanqueante óptico, blanqueador *m* óptico
i azzurrante *m* ottico, sbiancante *m* ottico
n optisch bleekmiddel *n*, optisch witmiddel *n*
d optischer Aufheller *m*, Weisstöner *m*

773 BRILLANT GREEN dye,hyg
f vert *m* brillant
e verde *m* brillante
i verde *m* brillante
n briljantgroen *n*
d Brillantgrün *n*

774 BRINDLED BRICK bui
f brique *f* tachetée
e ladrillo *m* listeado
i mattone *m* striato
n gestreepte baksteen, gevlekte baksteen
d gefleckter Backstein *m*, scheckiger Backstein *m*

775 BRINE, PICKLE gen
f saumure *f*
e salmuera *f*
i salamoia *f*
n pekel
d Pökel *m*, Pökellauge *f*

776 BRINE, SALT-WATER gen
f solution *f* de sel
e solución *f* de sal
i soluzione *f* salina
n zoutoplossing
d Kochsalzlösung *f*, Salzlake *f*, Sole *f*, Sulze *f*

777 BRINE BATH gen
f bain *m* de sel
e baño *m* de sal
i bagno *m* di sale
n zoutbad *n*
d Kochsalzbad *n*

778 BRINE CURING, BRINING lea
f salage *m*, saumurage *m*
e curado *m* en salmuera, salazón *m*
i stagionatura *f* in salamoia
n pikkelen *n*
d Salzlakenbehandlung *f*, Solbadbehandlung *f*

779 BRINE GRADUATION aic
f graduation *f* de saumure
e graduación *f* de salmuera
i graduazione *f* di salamoia
n pekelgraduatie
d Solengradierung *f*

780 BRIQUETTE fue
f briquette *f*
e briqueta *f*
i brichetta *f*
n briket
d Brikett *n*

781 BRIQUETTING fue
f briquetage *m*
e briqueteado *m*
i brichettaggio *m*
n briketteren *n*
d Brikettieren *n*

782 BRISANCE, exp
DETONATING VIOLENCE,
SHATTERING POWER
f pouvoir *m* brisant
e brisancia *f*, efecto *m* rompedor
i effetto *m* dirompente, effetto *m* esplosivo
n brisante werking, brisantie
d Brisanz *f*

783 BRISANCE TEST, exp
SHATTERING TEST,
VIOLENCE TEST
f essai *m* du pouvoir brisant
e prueba *f* de la brisancia
i collaudo *m* dell'effetto dirompente
n brisantieproef
d Brisanzprüfung *f*

784 BRISTOL BOARD, pap
BRISTOL PAPER
f carton *m* Bristol
e cartón *m* Bristol, cartulina *f*
i cartone *m* Bristol
n bristol *n*, bristolpapier *n*
d kartonstarkes Zeichenpapier *n*, Wichmannpapier *n*

785 BRITISH GUM (US), glu
DEXTRINE (GB), STARCH GUM (GB)
f dextrine *f*
e dextrina *f*
i destrina *f*
n dextrine *n*
d Dextrin *n*, Stärkegummi *m*

786 BROAD IRRIGATION, sew
OVERHEAD IRRIGATION,
SURFACE IRRIGATION
f irrigation *f* superficielle
e riego *m* con aguas cloacales, riego *m* con aguas negras
i disperdimento *m* per spandimento su terreni naturali permeabili, irrigazione *f* superficiale
n sproeibevloeiing
d Oberflächenberieselung *f*, Rieselfeldreinigung *f*

787 BROKEN WHITE, dec
OFF-WHITE
f blanc *m* cassé, blanchâtre adj
e blanco *m* mate, blancuzco *m*, blanquecino adj
i biancastro adj
n gebroken wit *n*
d gebrochenes Weiss *n*

788 BROMIDE PAPER, pho
SILVER BROMIDE PAPER
f papier *m* au bromure d'argent
e papel *m* al bromuro de plata
i carta *f* al bromuro d'argento
n broomzilverpapier *n*
d Bromsilberpapier *n*

789 BROMOIL PROCESS pho
f bromoil *m*, oléobromie *f*
e bromóleo *m*
i bromoleotipia *f*, bromolio *m*, oleobromia *f*
n broomoliedruk, broomverfdruk
d Bromölverfahren *n*

790 BRONZE PIGMENTED dec
LACQUER,
GOLD PAINT
f laque *f* de bronze
e laca *f* de bronce
i lacca *f* di bronzo
n bronslak
d Bronzelack *m*

791 BRONZE POWDER, coa
PURPURINA
f poudre *f* à bronzer
e polvo *m* para broncear
i polvere *f* da bronzare
n bronspoeder *n*
d Bronzefarbe *f*, Bronzepulver *n*

792 BRONZING coa
f bronzage *m*
e bronzeado *m*
i bronzatura *f*
n bronsen *n*
d Bronzieren *n*

793 BROWN SOURING BATH tex
f bain *m* de vitriolage brun
e baño *m* de acidulado marrón
i bagno *m* da acidulare bruno
n bruin zuurbad *n*
d braunes Sauerbad *n*

794 BROWNING coa
f brunissage *m*
e pavonado *m* pardo
i brunitura *f*
n bruineren *n*
d Braunieren *n*

795 BROWNING (GB), BURNT SUGAR (US), CARAMEL (GB) — sin
f caramel *m*
e caramelo *m*
i caramello *m*
n gebrande suiker, karamel
d gebrannter Zucker *m*, Gerstenzucker *m*, Karamel *m*, Zuckercouleur *f*

796 BRUSH DYEING, STAINING — lea
f coloration *f* du cuir à la brosse
e coloración *f* del cuero con la brocha
i colorazione *f* del cuoio al pennello
n leder verven *n* met borstel
d Lederfärben *n* mittels Einbürsten

797 BRUSH SOLUTION — dec,rub
f dissolution *f* pour brosse
e disolución *f* para brocha
i soluzione *f* per pennello
n tamponneersolutie
d Bürstenlösung *f*

798 BRUSHING LACQUER (GB), BRUSHING VARNISH (US) — dec
f laque *f* au pinceau
e laca *f* para pincel
i vernice *f* al pennello
n strijklak
d Streichlack *m*

799 BUBBLE, AIR BELL, AIR BUBBLE — gen
f bulle *f* d'air
e burbuja *f* de aire
i bolla *f* d'aria
n luchtbel
d Luftblase *f*

800 BUBBLE FERMENTATION — bre,fmi
f fermentation *f* bulbeuse
e fermentación *f* con burbujas
i fermentazione *f* a bolle
n borrelende gisting
d Blasengärung *f*

801 BUBBLE POINT — dis
f point *m* de bulle
e punto *m* de burbujeo, punto *m* de ebullición
i punto di bolla
n borrelpunt *n*
d Sprudelpunkt *m*

802 BUBBLING — gen
f barbotage *m*
e borboteo *m*, burbujeo *m*
i formazione *f* di bolle
n borrelen *n*, luchtbelvorming
d Blasenbildung *f*

803 BUBBLY CLAY — cer
f argile *f* à inclusions organiques
e arcilla *f* de inclusiones orgánicas
i argilla *f* ad inclusioni organiche
n klei met organische verontreinigingen
d Tonerde *f* mit organischen Verunreinigungen

804 BUCKING — tex
f lessivage *m*
e colada *f*
i liscivazione *f*
n beuken *n*
d Beuchen *n*

805 BUFF LEATHER — lea
f buffle *m*
e cuero *m* de búfalo
i cuoio *m* di buffalo
n buffelleer *n*
d Buffelleder *n*, Ochsenleder *n*

806 BUFFER, BUFFER SALT — elc
f tampon *m*, sel *m* tampon
e sal *f* tampón, tampón *m*
i tampone *m*
n buffer
d Puffersubstanz *f*

807 BUFFER ACTION — gen
f action *f* tampon
e acción *f* tampón
i azione *f* tampone
n bufferwerking
d Pufferung *f*, Pufferwirkung *f*

808 BUFFER BATTERY — elc
f batterie *f* tampon, batterie *f* volant
e batería *f* compensadora, batería *f* volante
i batteria *f* tampone
n bufferbatterij
d Pufferbatterie *f*

809 BUFFER CAPACITY — gen
f pouvoir *m* tampon
e poder *m* tampón
i potere *m* tampone
n buffervermogen *n*
d Pufferkapazität *f*

810 BUFFER INDEX, BUFFER VALUE — gen
f valeur *f* du tampon
e valor *m* del tampón
i valore *m* del tampone
n bufferwaarde
d Pufferwert *m*

811 BUFFER SOLUTION, REGULATOR MIXTURE — gen
f solution *f* tampon
e solución *f* reguladora, solución *f* tampón
i soluzione *f* tampone
n bufferoplossing
d Pufferlösung *f*

812 BULK CONCENTRATION — gen
f concentration *f* volumétrique
e concentración *f* volumétrica
i concentrazione *f* volumetrica
n volumeconcentratie
d Volumenkonzentration *f*

813 BULK DENSITY — gen
f densité *f* volumétrique, masse *f* volumique apparente
e densidad *f* volumétrica
i densità *f* in mucchio, densità *f* volumetrica
n bulkdichtheid, effectieve dichtheid, volumedichtheid
d Schüttdichte *f*

814 BULK FACTOR — pla,pre
f rapport *m* densité d'une pièce moulée/densité poudre utilisée
e relación *f* de la densidad de una pieza moldeada a la densidad del polvo constituyente
i rapporto *m* della densità d'una pezza stampata alla densità della polvere usata
n verhouding dichtheid geperst produkt tot dichtheid gebruikt poeder
d Füllkonstante *f*, Schüttvolumen *n*

815 BULK POLYMERIZATION — oic
f polymérisation *f* en masse
e polimerización *f* en masa
i polimerizzazione *f* in massa
n blokpolymerisatie
d Blockpolymerisation *f*, trockene Polymerisation *f*

816 BULKING, MOISTURE EXPANSION — fon
f gonflement *m* dû à l'humidité
e hinchazón *f* por humedad
i gonfiamento *m* dovuto all'umidità
n zwelling door vocht
d Feuchtigkeitsausdehnung *f*, Quellung *f*

817 BUMPING — dis,gen
f ébullition *f* à soubresauts
e ebullición *f* a golpes
i ebollizione *f* a scosse
n stoten *n*, stotend koken *n*
d Stossen *n*, stossweises Sieden *n*

818 BUNDLE DYEING, HANK DYEING, YARN DYEING — tex
f teinture *f* en écheveaux
e teñido *m* en suspensión de madejas
i tintura *f* delle matasse
n strengenverving
d Strähnenfärbung *f*

819 BURL DYEING — tex
f teinture *f* additionnelle de nopes
e teñido *m* adicional de motas
i tintura *f* addizionale di bruscoli
n bijkleuren *n* van noppen
d Farbengleichung *f*, Noppenfärbu[ng]

820 BURN (TO) — gen
f brûler v
e quemar v
i bruciare v
n branden v, verbranden v
d brennen v, verbrennen v

821 BURNING, BAKING — cer
f cuite *f*
e quemado *m*
i cottura *f*
n bakken *n*
d Brennen *n*

822 BURNING, COMBUSTION — gen
f combustion *f*
e combustión *f*
i combustione *f*
n verbranding
d Verbrennung *f*

823 BURNING-OFF — cer,dec
f décapage *m* à la lampe à brûler
e decapaje *m* con el soplete
i decapaggio *m* col cannello
n afbranden *n*
d Abbrennen *n*

824 BURNISHING — cer
f brunissage *m*

e bruñido *m*
i brunitura *f*
n bruineren *n*
d Bräunung *f*, Brünierung *f*

825 BURNT LIMESTONE, aic
QUICKLIME, UNSLAKED LIME
f chaux *f* vive
e cal *f* viva
i calce *f* viva
n ongebluste kalk
d Ätzkalk *m*, gebrannter Kalk *m*, ungelöschter Kalk *m*

826 BURNT OCHRE, dec
RED OCHRE
f ocre *f* rouge
e ocre *m* rojo
i ocra *f* rossa
n rode oker
d roter Ocker *m*

827 BURNT RUBBER, rub
SCORCHED RUBBER
f caoutchouc *m* brûlé
e caucho *m* quemado
i gomma *f* bruciata
n aangebrande rubber
d angebrannter Gummi *m*

828 BURNT SIENNA dec
f terre *f* de Sienne brûlée
e siena *f* tostada
i terra *f* di Siena bruciata
n terrasiëna
d hellgelber Ocker *m*, Sienaerde *f*

829 BURNT SUGAR (US), sin
BROWNING (GB), CARAMEL (GB)
f caramel *m*
e caramelo *m*
i caramello *m*
n gebrande suiker, karamel
d gebrannter Zucker *m*, Gerstenzucker *m*, Karamel *m*, Zuckercouleur *f*

830 BURNT UMBER dec
f terre *f* d'ombre brûlée
e tierra *f* de ombra tostada
i terra *f* d'ombra bruciata
n gebrande omber
d gebrannter Umber *m*, Umbraun *n*

831 BUTANONE ACID, oic
ACETOACETIC ACID, ACETYL ACETIC ACID
f acide *m* acétylacétique
e ácido *m* acetoacético
i acido *m* acetoacetico
n acetylazijnzuur *n*
d Azetessigsäure *f*

832 BUTT, lea
BEND LEATHER
f cœur *m*
e cuero *m* crupón
i cuoio *m* per suola di prima qualità
n kernleer *n*
d Kernleder *n*

833 BUTTER dai
f beurre *m*
e mantequilla *f*
i burro *m*
n boter
d Butter *f*

834 BUTTERMILK dai
f babeurre *m*, petit-lait *m*
e suero *m* de manteca
i latticello *m*, siero *m* del burro
n karnemelk
d Buttermilch *f*

835 BUTYRIC FERMENTATION fmi
f fermentation *f* butyrique
e fermentación *f* butírica
i fermentazione *f* butirrica
n boterzuurgisting
d Buttersäuregärung *f*

836 BY-PRODUCT gen
f sous-produit *m*
e subproducto *m*
i sottoprodotto *m*
n bijprodukt *n*
d Nebenprodukt *n*

837 BYPASS FILTERING fil
f filtration *f* en dérivation auxiliaire
e filtración *f* en derivación
i filtraggio *m* in derivazione
n omloopfiltreren *n*
d Umgangsfiltrierung *f*

C

838 CABINET VARNISH dec
f vernis *m* d'ébéniste
e barniz *m* de muñeca
i vernice *f* da ebanista
n meubelvernis *n*
d Möbellack *m*, Politur *f*

839 CABLE COMPOUND elc
f brai *m* de câble
e composición *f* aislante para cables, compuesto *m* de obturación
i massa *f* d'otturazione, massa *f* isolante per sigillatura dei cavi
n kabelmassa, vulmassa
d Ausgussmasse *f*, Kabelvergussmasse *f*

840 CABLE WAX elc
f cire *f* isolante pour câbles formée par bombardement ionique
e cera *f* formada por bombardeo iónico del aceite de cable
i cera *f* isolante per cavi formata per bombardamento ionico
n door ionenbombardement van de kabelolie gevormde was
d Kabelwachs *n*

841 CACAO BUTTER, CACAONUT OIL, THEOBROMA OIL fat
f beurre *m* de cacao
e manteca *f* de cacao
i burro *m* di cacao
n cacaoboter
d Kakaobutter *f*

842 CADE OIL, JUNIPER-BERRY OIL, JUNIPER OIL fat
f essence *f* de baies de genièvre, essence *f* de genièvre
e esencia *f* de enebrina
i olio *m* di ginepro
n jeneverbesolie
d Wacholderöl *n*

843 CADMIFEROUS met
f cadmifère adj
e cadmífero adj
i cadmifero adj
n cadmiumhoudend adj
d kadmiumhaltig adj

844 CADMIUM ELECTRODE elc
f électrode *f* de cadmium
e electrodo *m* de cadmio
i elettrodo *m* di cadmio
n cadmiumelektrode
d Kadmiumelektrode *f*

845 CADMIUM ORANGE, CADMIUM SELENIDE, CADMIUM YELLOW dec
f jaune *m* de cadmium
e amarillo *m* de cadmio
i giallo *m* di cadmio
n cadmiumgeel *n*
d Kadmiumgelb *n*

846 CADMIUM PLATING elc
f cadmiage *m*
e cadmiado *m*
i cadmiatura *f*
n vercadmiummen *n*
d Kadmierung *f*, Verkadmung *f*

847 CADMIUM RED dec
f rouge *m* de cadmium
e rojo *m* de cadmio
i rosso *m* di cadmio
n cadmiumrood *n*
d Kadmiumrot *n*

848 CADMIUM TEST elc,tes
f essai *m* au cadmium
e ensayo *m* al cadmio
i collaudo *m* al cadmio
n cadmiumproef
d Kadmiummessung *f*

849 CAJUPUT OIL fat
f essence *f* de cajeput
e esencia *f* de cayeputi
i olio *m* di caieput
n kajapoetolie
d Kajeputöl *n*

850 CAKE tex
f gâteau *m* de filature
e corona *f* de rayón, rodete *m* de rayón
i torta *f* di filatura
n garenkoek
d Spinnkuchen *m*

851 CAKING gen
f agglomération *f*, agglutination *f*
e aglutinación *f*

i agglutinazione *f*, impaccamento *m*
n samenkoeken *n*
d Balligwerden *n*, Zusammenbacken *n*

852 CALAMUS OIL fat
f essence *f* de calamus
e esencia *f* de cálamo
i olio *m* di calamo, olio *m* di fusto, olio *m* di stelo
n kalmoesolie
d Kalmusöl *n*

853 CALCAREOUS gen
f calcaire adj
e calcáreo adj , calizo adj
i calcareo adj
n kalkachtig adj
d kalkartig adj

854 CALCIFEROUS gen
f calcifère adj
e calcífero adj
i calcifero adj
n kalkhoudend adj
d kalkhaltig adj

855 CALCIFICATION mcy
f calcification *f*
e calcificación *f*
i calcificazione *f*
n verkalking
d Verkalkung *f*

856 CALCINATION, CALCINING htt
f incinération *f*
e incineración *f*
i incenerazione *f*
n verassing
d Veraschung *f*

857 CALCINATION, CALCINING, ROASTING met
f calcination *f*, grillage *m*
e calcinación *f*, tostación *f*
i arrostimento *m*, calcinazione *f*
n calcineren *n*, roosten *n*
d Kalzinieren *n*, Rösten *n*

858 CALCIUM CHELATING POWER, CALCIUM SEQUESTERING POWER soa
f pouvoir *m* anticalcaire
e poder *m* anticalcáreo, poder *m* secuestrante cálcico
i potere *m* sequestrante per il calcio
n kalkbindend vermogen *n*
d Kalzium-Komplexbindungsvermögen *n*, Kalkbindevermögen *n*

859 CALCIUM FERTILIZER, LIME FERTILIZER fer
f engrais *m* calcique
e abono *m* a base de cal
i fertilizzante *m* a base di calce
n kalkmeststof
d Kalkdünger *m*

860 CALCIUM HARDNESS soa
f dureté *f* calcaire, dureté *f* calcique
e dureza *f* cálcica, dureza *f* de calcio
i durezza *f* calcarea, durezza *f* calcica
n kalkhardheid
d Kalkhärte *f*

861 CALCIUM SOAP, LIME SOAP cle,tex
f savon *m* calcique, savon *m* de chaux
e jabón *m* de cal
i sapone *m* di calce
n kalkzeep
d Kalkseife *f*

862 CALCIUM SOAP LUBRICANT lub
f graisse *f* à base de savon calcique
e lubricante *m* jabonoso cálcico
i lubrificante *m* a base di sapone di calce
n kalkzeepsmeervet *n*
d Kalkseifenschmierfett *n*

863 CALCIUM SODA ELECTROLYSIS elc
f électrolyse *f* à sels de calcium et sodium
e electrólisis *f* con sales de calcio y sodio
i elettrolisi *f* a sali di calcio e sodio
n calcium-natriumelektrolyse
d Kalziumnatriumelektrolyse *f*

864 CALCIUM SUCRATE SUGAR LIME mcy
f sucrate *m* de calcium
e sacarato *m* de calcio
i saccarato *m* di calcio
n calciumsacharaat *n*, suikerkalk
d Zuckerkalk *m*

865 CALENDER SHRINKAGE tex
f effet *m* de calandrage
e contracción *f* de calandrado
i effetto *m* di calandra
n kalandereffect *n*
d Kalandereffekt *m*

866 CALENDERED COATING coa
f enduction *f* à la calandre, revêtement *m* par calandrage
e revestimiento *m* por calandrado
i rivestimento *m* per calandratura
n gekalandreerde bedekking
d Kalanderauftrag *m*

867 CALENDERED PAPER pap
f papier *m* satiné
e papel *m* satinado
i carta *f* calandrata
n gesatineerd papier *n*
d geglättetes Papier *n*, satiniertes Papier *n*

868 CALENDERED SHEET pla,rub
f feuille *f* calandrée
e hoja *f* calandrada
i foglia *f* calandrata
n gekalandreerde foelie
n Kalanderfolie *f*

869 CALENDERING tex
f calandrage *m*
e calandrado *m*
i calandratura *f*, spianatura *f*
n glanzen *n*, kalanderen *n*
d Glätten *n*, Kalandern *n*

870 CALENDERING, SATINING pap
f satinage *m*
e satinado *m*
i calandratura *f*, spianatura *f*
n satineren *n*
d Satinieren *n*

871 CALIBRATION, GAGING (US), GAUGING (GB) mea
f calibrage *m*, étalonnage *m*
e calibración *f*
i calibratura *f*, taratura *f*
n ijken *n*, kalibreren *n*
d Eichen *n*, Kalibrierung *f*

872 CALKING (US), CAULKING (GB) shi
f calfatage *m*
e calafateo *m*
i calafatura *f*, cianfrinatura *f*
n breeuwen *n*, kalfateren *n*
d Kalfatern *n*, Stemmen *n*

873 CALKING TOW (US), CAULKING FELT (GB) shi
f ploc *m*
e estopa *f* para calafatear
i capecchio *m* da calafatare, stoppa *f* da calafatare
n breeuwwerk *n*
d Kalfaterwerg *n*

874 CALOMEL, MERCUROUS CHLORIDE elc
f calomel *m*, chlorure *m* de mercure
e calomelano *m*, cloruro *m* mercurioso
i calomelano *m*, cloruro *m* mercuroso
n kalomel *n*, mercurochloride *n*
d Kalomel *n*, Mercurochlorid *n*, Quecksilberchlorur *n*

875 CALOMEL PASTE mcy
f pâte *f* au calomel
e pasta *f* de calomelano
i pasta *f* al calomelano
n kalomelpasta
d Kalomelpaste *f*

876 CALORIFYING clo
f réchauffage *m* au serpentin
e calentamiento *m* con serpentín
i riscaldamento *m* alla serpentina
n labyrintverwarming
d Labyrintherwärmung *f*

877 CALORIMETRY mea
f calorimétrie *f*
e calorimetría *f*
i calorimetria *f*
n calorimetrie, warmtemeting
d Kalorimetrie *f*, Wärmemessung

878 CALORIZING, ALITIZING met
f calorisation *f*
e calorización *f*
i calorizzazione *f*, termitizzazion
n caloriseren *n*
d Alitieren *n*, Kalorisieren *n*

879 CAMBOGIA, GAMBOGE dec,mcy
f gomme-gutte *f*
e gutagamba *f*
i gommaguta *f*
n gittegom, guttegom
d Gummigutt *n*

880 CAMBRIC GRASS, CHINA GRASS, RAMIE tex
f ramie *f*
e ramio *m*
i ramia *f*, ramie *f*
n ramee, ramie
d Ramie *f*

881 CAMOMILE OIL fat
f essence *f* de camomille romaine
e esencia *f* de manzanilla
i olio *m* di camomilla
n kamilleolie
d Kamillenöl *n*

882 CAMOUFLAGE PAINTING dec,mit
f peinture *f* de camouflage
e pintura *f* de enmascaramiento
i pittura *f* di mascheramento, pittura *f* di mimetizzazione
n camouflageverf
d Tarnanstrich *m*

883 CAMPHOR oic
f camphre *m*
e alcanfor *m*
i canfora *f*
n kamfer
d Kampfer *m*

884 CAMPHOR LINIMENT, CAMPHORATED OIL fat,mcy
f onguent *m* camphré
e ungüento *m* alcanforado
i unguento *m* canforato
n kamferzalf
d Kampfereinreibung *f*, Kampferliniment *n*

885 CAMPHOR OIL fat
f essence *f* de camphre
e esencia *f* de alcanfor
i olio *m* di canfora
n kamferolie
d Kampferöl *n*

886 CAMPHORATED SPIRIT mcy
f alcool *m* camphré
e alcohol *m* alcanforado
i spirito *m* canforato
n kamferspiritus
d Kampfergeist *m*, Kampferspiritus *m*

887 CANADA BALSAM, BALM GILEAD, BALSAM OF FIR mcy,pho
f baume *m* du Canada
e bálsamo *m* del Canadá
i balsamo *m* dell'abete resinoso
n canadabalsem
d Kanadabalsam *m*

888 CANANGA OIL, ANONA OIL, YLANG YLANG OIL fat
f essence *f* d'ilang-ilang
e esencia *f* de cananga
i olio *m* di cananga odorata
n kanangaolie, ylang-ylangolie
d Anonaöl *n*, Kanangaöl *n*, Ylang-Ylangöl *n*

889 CANARY LITHARGE dec
f litharge *f* jaune
e litargirio *m* amarillo
i litargirio *m* giallo
n geel loodglit *n*
d gelbe Bleiglätte *f*

890 CANE, SUGAR CANE sin
f canamelle *f*, cannamelle *f*, canne *f* à sucre
e caña *f* de azúcar, caña *f* dulce, caña *f* melar
i canna *f* da zucchero, cannamele *f*
n suikerriet *n*
d Zuckerrohr *n*

891 CANE JUICE sin
f vesou *m*
e guarapo *m*
i succo *m* della canna da zucchero
n rietsuikersap *n*
d Rohrzuckersaft *m*

892 CANE SUGAR, SUCROSE sin
f sucre *m* de canne
e azúcar *m* de caña
i zucchero *m* di canna
n rietsuiker
d Rohrzucker *m*

893 CANNING (US), TINNING (GB) nut
f mise *f* en boîtes, mise *f* en conserves
e procedimiento *m* de conservar en latas
i conservazione *f* in scatola
n in blik conserveren *n*
d in Blechdosen Konservieren *n*

894 CAOUTCHOUC, ELASTIC GUM, INDIA RUBBER rub
f caoutchouc *m*
e caucho *m*
i caucciù *m*
n caoutchouc *n*
d Gummielasticum *n*, Kautschuk *m*, Rohgummi *m*

895 CAPACITY TO TAKE THE FINISH tex
f capacité *f* d'apprêtage
e capacidad *f* de aprestar
i capacità *f* d'apprettatura

n appreteerbaarheid
d Appreturfähigkeit *f*

896 CAPILLARITY, CAPILLARY ATTRACTION gen
f capillarité *f*
e capilaridad *f*
i capillarità *f*
n capillariteit
d Kapillarität *f*, Kapillarwirkung *f*

897 CAPILLARY CONSTANT fue
f constante *f* capillaire
e constante *f* capilar
i costante *f* capillare
n capillariteitsconstante
d Kapillaritätskonstante *f*

898 CAPILLARY PRESSURE gen
f pression *f* capillaire
e presión *f* capilar
i pressione *f* capillare
n capillaire druk
d Kapillardruck *m*

899 CAPITAL LEES aci
f lessive *f* de potasse concentrée
e lejía *f* de potasa concentrada
i lisciva *f* di potassa concentrata
n sterke potasloog
d starke Potaschenlauge *f*

900 CARAMEL (GB), BROWNING (GB), BURNT SUGAR (US) sin
f caramel *m*
e caramelo *m*
i caramello *m*
n gebrande suiker, karamel
d gebrannter Zucker *m*, Gerstenzucker *m*, Karamel *m*, Zuckercouleur *f*

901 CARAMEL MALT, BLACK MALT, ROASTED MALT bre
f malt *m* noir, malt *m* torréfié
e malta *f* colorante, malta *f* tostada
i malto *m* torrefatto
n lichtgeroosterde mout
d Farbmalz *n*, Röstmalz *n*

902 CARAWAY OIL fat
f essence *f* de carvi
e esencia *f* de alcaravea
i olio *m* di carvi
n karwijolie
d Kümmelöl *n*

903 CARBIDE ETCHING elc
f attaque *f* à l'acide du carbure
e ataque *m* ácido del carburo
i incisione *f* all'acquaforte del carburo
n carbide-etsing
d Karbidätzung *f*

904 CARBIDE NITRIFICATION fer
f fixation *f* d'azote sur le carbure de calcium
e nitrogenación *f* de carburo
i nitrificazione *f* di carburo
n stikstofbinding aan calciumcarbid
d Azotieren *n*

905 CARBIDE PRECIPITATION aic
f précipitation *f* de carbure
e precipitación *f* de carburo
i precipitazione *f* di carburo
n carbideafscheiding
d Karbidausscheidung *f*

906 CARBO-HYDROGEN, OILGAS fue,gas
f gaz *m* de pétrole
e gas *m* de petróleo
i gas *m* di petrolio
n oliegas *n*, vetgas *n*
d Fettgas *n*, Ölgas *n*

907 CARBO-NITRIDING, CYANIDE HARDENING met
f cémentation *f* au cyanure, endurcissement *m* au cyanure
e cementación *f* al cianuro, endurecimiento *m* al cianuro
i cementazione *f* al cianuro, indurimento *m* al cianuro
n cementering met cyanide, cyaanbadharding
d Zyansalzbadhärtung

908 CARBOHYDRASES bio
f carbohydrases *pl*
e carbohidrasas *pl*
i carboidrasi *pl*
n carbohydrasen *pl*
d Karbohydrasen *pl*

909 CARBOHYDRATES oic
f glucides *pl*, hydrates *pl* de carbon
e carbohidratos *pl*, hidratos *pl* de carbono
i carboidrati *pl*, idrati *pl* di carboni
n koolhydraten *pl*
d Kohlehydrate *pl*, Kohlenhydraten *pl*

910 CARBOLIC ACID, PHENOL oic
f acide *m* carbolique, phénol *m*
e ácido *m* fénico
i acido *m* carbolico

n carbolzuur *n*, fenol *n*, hydroxybenzeen *n*
d Benzophenol *n*, Karbolsäure *f*, Phenol *n*

911 CARBOLIC OIL, MIDDLE OIL fue
f huile *f* carbolique
e aceite *m* carbólico
i olio *m* carbolico
n carbololie, middelolie
d Karbolöl *n*, Mittelöl *n*, Phenolöl *n*

912 CARBOLINEUM, CARBONIL bui,dec
f carboléum *m*, carbolinéum *m*
e carbolíneo *m*
i carbolineum *m*
n carbolineum
d Karbolineum *n*

913 CARBON COMPOUNDS oic
f composés *pl* de carbone
e compuestos *pl* de carbono
i composti *pl* di carbonio
n koolstofverbindingen *pl*
d Kohlenstoffverbindungen *pl*

914 CARBON DIOXIDE SNOW, DRY ICE mcy
f neige *f* carbonique
e hielo *m* seco, nieve *f* carbónica
i ghiaccio *m* secco
n droog ijs *n*
d Kohlensäureschnee *m*, Trockeneis *n*

915 CARBON MONOXIDE (GB), WHITE DAMP (US) gas
f oxyde *m* de carbone
e óxido *m* de carbone
i monossido *m* di carbonio, ossido *m* di carbonio
n koolmonoxyde *n*
d giftige Wetter *pl*, Kohlenoxyd *n*

916 CARBON PAPER pap
f papier *m* carbone, papier *m* carboné
e papel *m* carbón
i carta *f* carbone
n carbonpapier *n*
d Durchschreibepapier *n*

917 CARBON PRINTING, CARBON PROCESS pho
f procédé *m* au charbon
e impresión *f* sobre el papel pigmentado
i stampo *m* al pigmento
n kooldruk, pigmentdruk
d Kohlendruck *m*, Pigmentdruck *m*

918 CARBON RESTORATION, RECARBURIZATION met
f récarburation *f*, régénération *f* de pièces décarburisées
e recarburación *f*, regeneración *f* de piezas descarburadas
i ricarburazione *f*, rigenerazione *f* di pezzi decarburati
n regenereren *n* van ontkoolde delen
d Aufkohlung *f*, Rückkohlung *f*

919 CARBON TETRACHLORIDE (GB), PHENOXIN (US) oic
f tétrachlorure *m* de carbone
e tetracloruro *m* de carbono
i tetracloruro *m* di carbonio
n tetra, tetrachloorkoolstof
d Tetra *n*, Tetrachlorkohlenstoff *m*

920 CARBON TISSUE pho
f papier *m* au charbon, papier *m* pigmentaire
e papel *m* al carbón
i carta *f* al carbone
n pigmentpapier *n*
d Pigmentpapier *n*

921 CARBONACEOUS gen
f carboné adj, charbonneux adj
e carbonado adj, carbonoso adj
i carbonaceo adj
n koolstofachtig adj, koolstofhoudend adj
d kohlenstoffartig adj, kohlenstoffhaltig adj

922 CARBONATATION, CARBONATION sin
f saturation *f*
e carbonatación *f*
i carbonatazione *f*
n carbonatatie
d Entkalkung *f*, Sättigung *f*

923 CARBONATATION MUD, CARBONATION MUD sin
f boue *f* de saturation
e fango *m* de carbonatación
i fango *m* di carbonatazione
n carbonatatieslib *n*
d Sättigungsschlamm *m*

924 CARBONATE BUFFER elc
f tampon *m* à carbonate
e tampón *m* de carbonato

i tampone *m* a carbonato
n carbonaatbuffer
d Karbonatpuffer *m*

925 CARBONATE HARDNESS bre
f dureté *f* de carbonates
e dureza *f* de carbonatos
i durezza *f* di carbonati
n carbonatenhardheid
d Karbonatenhärte *f*

926 CARBONATE WATER, CHALK WATER bre
f eau *f* calcifère
e agua *f* caliza, agua *f* selenitosa
i acqua *f* calcifera
n kalkhoudend water *n*
d kalkhaltiges Wasser *n*

927 CARBONATED LIQUID gas
f liquide *m* carbonaté
e líquido *m* carbonatado
i liquido *m* carbonato
n koolzuurhoudende vloeistof
d kohlensäurehaltige Flüssigkeit *f*

928 CARBONIC ACID aic
f acide *m* carbonique
e ácido *m* carbónico
i acido *m* carbonico
n koolzuur *n*
d Kohlensäure *f*

929 CARBONIZATION htt
f carbonisation *f*
e carbonización *f*
i carbonizzazione *f*
n carbonisatie, verkoling
d Karbonisierung *f*, Verkohlung *f*

930 CARBONIZATION, CARBONIZING tex
f carbonisage *m*
e carbonización *f*
i carbonizzazione *f*
n carboniseren *n*
d Karbonisieren *n*

931 CARBONIZATION, CEMENTATION met
f cémentation *f*
e cementación *f*
i cementazione *f*
n cementeren *n*
d Aufkohlung *f*, Zementieren *n*

932 CARBRO PROCESS, OZOBROME PROCESS pho
f carbro *m*, ozobromie *f*
e procedimiento *m* al bromocarbón
i ozobromia *f*, processo *m* al bromocarbone
n carbroproces *n*
d Karbrodruck *m*, Ozobromdruck *n*

933 CARBURATION, CARBURETION fue
f carburation *f*
e carburación *f*
i carburazione *f*
n carburatie
d Karburierung *f*, Vergasung *f*

934 CARBURETTED WATER GAS, ENRICHED WATER GAS fue
f gaz *m* à l'eau carburée
e gas *m* de agua carburada
i gas *m* d'acqua arricchita
n gecarbureerd watergas *n*
d angereichertes Wassergas *n*, karburiertes Wassergas *n*

935 CARBURIZATION met
f carburation *f*, cémentation *f* au carbone
e carburación *f*, cementación *f* al carbono
i carburazione *f*, cementazione *f* al carbonio
n cementeren *n* met koolstof
d Zementieren *n* mit Kohlenstoff

936 CARBURIZER, CARBURIZING AGENT (GB), PIGGING BACK (US) met
f agent *m* carburant
e agente *m* cementador
i agente *m* cementatore
n verkolend middel *n*
d Aufkohlungsmittel *n*, Kohlungsstoff *m*

937 CARD-BOARD, BOARD, PASTE-BOARD mat,pap
f carton *m*
e cartón *m*
i cartone *m*
n karton *n*
d Pappe *f*

938 CARDAMOM OIL fat
f essence *f* de cardamome
e esencia *f* de cardamomo
i olio *m* d'amomo, olio *m* di cardamomo
n kardemomolie
d Kardamomenöl *n*

939 CARMINE dec
f carmin *m*

e carmín *m*
i carminio *m*
n karmijn *n*
d Karmin *n*, Karminrot *n*

940 CARNAUBA WAX, BRAZIL WAX — fat
f cire *f* de carnauba
e cera *f* carnáubica
i cera *f* carnauba
n carnaubavet *n*
d Karnaubawachs *m*

941 CARPENTER'S GLUE, HIDE GLUE, JOINER'S GLUE — glu
f colle *f* forte
e cola *f* de carnaza, cola *f* de piel
i colla *f* forte
n houtlijm
d Hautleim *m*

942 CARRIER FILM — pho
f support *m* pelliculaire
e película *f* de soporte
i pellicola *f* di sopporto
n draagfilm
d Trägerfolie *f*

943 CARRIER VAPOR DISTILLATION (US), CARRIER VAPOUR DISTILLATION (GB) — dis
f entraînement *m* à la vapeur
e destilación *f* por arrastre de vapor de agua
i distillazione *f* in corrente di vapore
n destillatie in waterdampstroom
d Trägerdampfdestillation *f*

944 CARTRIDGE PAPER, MANILLA PAPER — pap
f papier *m* bulle
e papel *m* manila
i carta *f* manilla
n manillapapier *n*
d Manillapapier *n*

945 CASE HARDENING, SURFACE CONVERTING — met
f trempe *f* de surface
e temple *m* de superficie
i tempera *f* di superficie
n inzetharding, oppervlakharding
d Einsatzhärtung *f*, Oberflächenhärtung *f*

946 CASE HARDENING COMPOSITION, CEMENTING POWDER — met
f agent *m* de trempe de surface
e agente *m* de temple de superficie
i agente *m* di tempera di superficie
n oppervlakhardingsmiddel *n*
d Einsatzpulver *n*, Härtepulver *n*

947 CASED GLASS, FLASHED GLASS, PLY GLASS — gla
f verre *m* à deux couches, verre *m* doublé
e vidrio *m* contrachapado
i vetro *m* a due strati
n tweelaagsglas *n*
d Überfangglas *n*

948 CASEIN — dai,pla
f caséine *f*
e caseína *f*
i caseina *f*
n caseïne *n*, kaasstof
d Kasein *n*

949 CASEIN FIBRE — tex
f fibre *f* de caséine
e fibra *f* de caseína
i fibra *f* di caseina
n caseïnevezel
d Kaseinfaser *f*

950 CASEIN GLUE — glu
f colle *f* à la caséine
e cola *f* de caseína
i colla *f* alla caseina
n caseïnelijm
d Kaseinleim *m*, wasserfester Kaltleim *m*

951 CASEIN MORDANT — oic
f mordant *m* à la caséine
e mordiente *m* de caseína
i mordente *m* alla caseina
n caseïnebijtmiddel *n*
d Kaseinbeize *f*

952 CASEIN PAINT — dec
f couleur *f* à base de caséine
e pintura *f* a base de caseína
i pittura *f* a base di caseina
n caseïneverf
d Kaseinfarbe *f*

953 CASEIN PLASTICS — pla
f plastiques *pl* à base de caséine
e plásticos *pl* a base de caseína
i plastici *pl* a base di caseina
n caseïneplastica *pl*, kunststoffen *pl* op caseïnebasis
d Kaseinkunststoffe *pl*

954 CASINGHEAD GASOLINE — fue
f gaz *m* de tête de puits

e gas *m* de boca de pozo, gas *m* natural
i prodotto *m* di testa gassoso
n putgasbenzine
d Rohrkopfbenzin *n*

955 CASK BEER, BEER FROM THE WOOD, DRAFT BEER (US), DRAUGHT BEER (GB) bre
f bière *f* en fût
e cerveza *f* de barril
i birra *f* a spina, birra *f* in fusto
n bier *n* van het vat, vatbier *n*
d Bier *n* vom Fass, Fassbier *n*

956 CASK DEPOSIT, BOTTOMS bre
f dépôt *m*, fond *m* de foudre
e fondo *m* de barril
i feccia *f*
n vatbezinksel *n*
d Fassgeläger *n*

957 CASSIA OIL, CHINESE OIL fat
f essence *f* de cassie
e esencia *f* de cañafistula
i olio *m* di cassia
n kassiaolie
d Kassiaöl *n*

958 CAST CONCRETE, IN SITE CONCRETE, ON SITE CONCRETE bui
f béton *m* coulé
e hormigón *m* colado
i calcestruzzo *m* colato
n gestort beton *n*, gietbeton *n*
d Gussbeton *m*, Schüttbeton *m*

959 CAST FILM pla
f film *m* coulé, pellicule *f* coulée
e película *f* colada
i pellicola *f* colata
n gegoten film
d gegossener Film *m*, Gussfolie *f*

960 CASTING cer
f coulée *f* en moule
e colada *f* en molde
i colata *f* in forma
n vormgieten *n*
d Giessen *n* in Gipsform

961 CASTING, COATING pho
f étendage *m* d'une émulsion
e emulsionamiento *m*
i emulsionatura *f*
n opbrengen *n* van een emulsie
d Auftragen *n* einer Emulsion

962 CASTING, POURING, TEEMING fou,pla
f coulée *f*
e colada *f*
i colata *f*
n gieten *n*
d Giessen *n*

963 CASTING COMPOUND pla
f mélange *m* de coulage
e compuesto *m* de colada
i mescolanza *f* di colata
n gietmengsel *n*
d Vergiessmasse *f*, Vergussmasse *f*

964 CASTING RESIN pla
f résine *f* à couler
e resina *f* de colada
i resina *f* da colare
n gietbare hars
d Giessharz *n*, Vergussharz *n*

965 CASTOR OIL, RICINUS OIL fat
f huile *f* de ricin
e aceite *m* de ricino
i olio *m* di ricino
n castorolie, ricinusolie, wonderolie
d Rizinusöl *n*

966 CAT CRACKING, CATALYTIC CRACKING fue
f craquage *m* catalytique
e pirodesintegración *f* catalítica, termodesintegración catalítica
i piroscissione *f* catalitica
n katalytisch kraken *n*
d katalytisches Kracken *n*

967 CATALYSIS gen
f catalyse *f*
e catálisis *f*
i catalisi *f*
n katalyse
d Katalyse *f*

968 CATALYST gen
f catalyseur *m*
e catalizador *m*
i catalizzatore *m*
n katalysator
d Katalysator *m*

969 CATALYST CARRIER gen
f porte-catalyseur *m*
e portacatalizador *m*
i portacatalizzatore *m*
n katalysatordrager
d Katalysatorträger *m*

970 CATALYST DRY REGENERATION gen
f régénération *f* du catalyseur par l'hydrogène
e regeneración *f* del catalizador con hidrógeno
i rigenerazione *f* del catalizzatore con idrogeno
n katalysatorregeneratie met waterstof
d Katalysatorregenierung *f* mit Wasserstoff

971 CATALYST POISON, PARALYZER gen
f poison *m* de catalyseur
e veneno *m* de catalizador
i veleno *m* catalitico
n katalysatorvergif *n*
d Katalysatorgift *n*, Katalytgift *n*, Kontaktgift *n*

972 CATALYST POISONING gen
f empoisonnement *m* du catalyseur
e envenenamiento *m* del catalizador
i avvelenamento *m* del catalizzatore
n katalysatorvergiftiging
d Katalysatorvergiftung *f*, Kontaktvergiftung *f*

973 CATALYST REACTIVATION gen
f réactivation *f* du catalyseur
e reactivación *f* del catalizador
i riattivazione *f* del catalizzatore
n reactivering van de katalysator
d Reaktivierung *f* des Katalysators

974 CATALYST REGENERATION gen
f régénération *f* du catalyseur
e regeneración *f* del catalizador
i rigenerazione *f* del catalizzatore
n regeneratie van de katalysator
d Regenerierung *f* des Katalysators

975 CATALYTIC DECOMPOSITION gen
f décomposition *f* catalytique
e descomposición *f* catalítica
i decomposizione *f* catalitica
n katalytische ontleding
d katalytische Zersetzung *f*

976 CATALYTIC FADING tex
f décoloration *f* catalytique
e descoloración *f* catalítica
i scolorimento *m* catalitico
n katalytisch verschieten *n*
d katalytisches Verschiessen *n*

977 CATALYTIC HYDROGENATION gen
f hydrogénation *f* catalytique
e hidrogenación *f* catalítica
i idrogenazione *f* catalitica
n katalytisch hydreren *n*
d katalytische Hydrierung *f*, katalytische Hydrogenierung *f*

978 CATALYTIC POLYMERIZATION oic
f polymérisation *f* catalytique
e polimerización *f* catalítica
i polimerizzazione *f* catalitica
n katalytische polymerisatie
d katalytische Polymerisation *f*

979 CATALYTIC REDUCTION gen
f réduction *f* catalytique
e reducción *f* catalítica
i riduzione *f* catalitica
n katalytische reductie
d katalytische Reduktion *f*

980 CATALYTIC REFORMING fue
f réformation *f* catalytique
e reformación *f* catalítica
i riformazione *f* catalitica
n katalytisch reformeren *n*, katalytische omvorming
d katalytisches Reformieren *n*

981 CATALYZED LACQUER, CHEMICALLY CURING LACQUER, REACTION LACQUER dec
f laque *f* à durcissement chimique, vernis *m* réactionnel
e laca *f* reaccional
i lacca *f* ad indurimento chimico, lacca *f* di reazione
n reactielak
d Kalthärtelack *m*, Reaktionslack *m*

982 CATAPHORESIS, ELECTROPHORESIS gen
f cataphorèse *f*
e cataforesis *f*
i cataforesi *f*
n kataforese
d Kataphorese *f*

983 CATHODE COPPER, ELECTROLYTIC COPPER elc,met
f cuivre *m* électrolytique
e cobre *m* de cátodo, cobre *m* electrolítico

i rame *m* elettrolitico
n elektrolytisch koper *n*
d Elektrolytkupfer *n*,
Katodenkupfer *n*

984 CATHODE DEPOSIT elc
f dépôt *m* cathodique
e depósito *m* catódico
i deposito *m* catodico
n katodeneerslag
d Katodenniederschlag *m*

985 CATHODE SPUTTERING, DISINTEGRATION elc
f pulvérisation *f* cathodique
e pulverización *f* catódica
i polverizzazione *f* catodica
n katodeverstuiving
d Katodenzerstäubung *f*

986 CATHODIC PROTECTION coa
f protection *f* cathodique
e protección *f* catódica
i protezione *f* catodica
n katodische bescherming
d galvanischer Korrosionsschutz *m*,
katodischer Schutz *m*

987 CATHODIC REDUCTION elc
f réduction *f* cathodique
e reducción *f* catódica
i riduzione *f* catodica
n katodische reductie
d katodische Reduktion *f*

988 CATHOLYTE, CATOLYTE elc
f catholyte *m*
e católito *m*
i catolito *m*
n katoliet
d Katolyt *m*

989 CATION elc
f cation *m*
e catión *m*
i catione *m*
n kation *n*
d Kation *n*

990 CATION EXCHANGE SOFTENING hyg,sew
f adoucissement *m* par échangeur cationique
e ablandamiento *m* por cambiador catiónico
i addolcimento *m* mediante scambiatore di cationi
n waterontharding door kationenuitwisseling
d Basenaustauschenthärtung *f*,
Wasserenthärtung *f* durch Kationenaustausch

991 CATIONIC DYE dye
f colorant *m* cationique
e colorante *m* catiónico
i colorante *m* cationico
n kationische kleurstof
d kationischer Farbstoff *m*

992 CATIONIC SOAP soa
f savon *m* cationique
e jabón *m* catiónico
i sapone *m* cationico
n kationische zeep
d kationische Seife *f*

993 CAULKING (GB), CALKING (US) shi
f calfatage *m*
e calafateo *m*
i calafatura *f*, cianfrinatura *f*
n breeuwen *n*, kalfaten *n*
d Kalfatern *n*, Stemmen *n*

994 CAULKING FELT (GB), CALKING TOW (US) shi
f ploc *m*
e estopa *f* para calafatear
i capecchio *m* da calafatare,
stoppa *f* da calafatare
n breeuwwerk *n*
d Kalfaterwerg *n*

995 CAUSTIC ALCOHOL, SODIUM ETHYLATE oic
f éthylate *m* de sodium
e etilato *m* de sodio
i etilato *m* di sodio
n natriumethylaat *n*
d Natriumäthylat *n*

996 CAUSTIC DIP elc
f solution *f* de soude caustique
e solución *f* de hidróxido sódico
i soluzione *f* d'idrossido sodico
n bijtende-sodaoplossing
d Ätznatronlösung *f*

997 CAUSTIC LIME aic
f chaux *f* caustique
e cal *f* cáustica, hidrato *m* cálcico
i calce *f* caustica
n gebrande kalk
d gebrannter Kalk *m*, Kalziumoxyd

998 CAUSTIC LYE aic
f lessive *f* de soude caustique
e lejía *f* de hidróxido sódico

i lisciva *f* d'idrossido sodico
n bijtende-sodaloog
d Ätznatronlauge *f*

999 CAUSTIC POTASH, POTASSIUM HYDROXIDE aic
f hydroxyde *m* de potasse
e hidróxido *m* potásico
i potassa *f* caustica
n bijtende potas
d Ätzkali *n*, Kalziumhydroxyd *n*

1000 CAUSTIC SODA, SODIUM HYDROXIDE aic
f hydroxyde *m* de sodium
e sosa *f* cáustica
i soda *f* caustica
n bijtende soda
d Ätznatron *n*, Natriumhydroxyd *n*

1001 CEDER WOOD OIL fat
f essence *f* de bois de cèdre, essence *f* de cèdre
e aceite *m* de madera de cedro
i olio *m* di legno di cedro
n cederhoutolie
d Zedernholzöl *n*

1002 CELL DUST REVERSING PRECIPITATION iwt
f dépoussiérage *m* à chicanes
e pulviprecipitación *f* reversible
i precipitazione *f* di polvere a celle ribaltabili
n stofverwijdering met omkeerbare cellen
d Umkehrzellenentstäubung *f*

1003 CELLOPHANE pla
f cellophane *m*
e celofán *m*
i cellofane *m*
n cellofaan *n*
d Cellophan *n*, Zellglas *n*

1004 CELLULAR CONCRETE, AERATED CONCRETE, FOAMED CONCRETE, GAS CONCRETE bui
f béton *m* cellulaire
e hormigón *m* aerado
i calcestruzzo *m* poroso
n cellenbeton *n*
d Zellenbeton *m*

1005 CELLULAR POLYETHYLENE pla
f polyéthylène *m* cellulaire
e polietileno *m* celular
i polietilene *m* cellulare
n cellenpolyethyleen *n*
d Zellenpolyäthylen *n*

1006 CELLULAR RUBBER rub
f caoutchouc *m* alvéolaire
e goma *f* alveolar
i gomma *f* cellulare
n cellenrubber, celrubber
d Zellengummi *m*, Zellgummi *m*

1007 CELLULOID pla
f celluloïd *m*, celluloïde *m*
e celuloide *m*
i celluloide *m*
n celluloid *n*, celluloïde *n*
d Zellhorn *n*, Zelluloid *n*

1008 CELLULOID FILM pho
f pellicule *f* à celluloïde
e película *f* de celuloide
i pellicola *f* a celluloide
n celluloid film
d Zelluloidfilm *m*

1009 CELLULOSE oic
f cellulose *f*
e celulosa *f*
i cellulosa *f*
n cellulose, celstof
d Zellstoff *m*

1010 CELLULOSE ACETATE, ACETYLCELLULOSE oic
f acétate *m* de cellulose, acétocellulose *f*
e acetato *m* celulósico, acetocelulosa *f*
i acetato *m* di cellulosa
n acetylcellulose, celluloseacetaat *n*
d Azetylzellulose *f*, Zelluloseazetat *n*

1011 CELLULOSE ACETATE SILK tex
f rayonne *f* d'acétate de cellulose
e rayón *m* de acetato celulósico
i raion *m* d'acetato di cellulosa
n celluloseacetaatzijde
d Zelluloseazetatseide *f*

1012 CELLULOSE LACQUER dec
f vernis *m* cellulosique
e barniz *m* celulósico
i vernice *f* cellulosica
n celluloselak
d Zelluloselack *m*

1013 CELLULOSE NITRATE, NITROCELLULOSE exp
f nitrocellulose *f*
e nitrocelulosa *f*, piroxilina *f*
i nitrocellulosa *f*
n nitrocellulose
d Nitrozellulose *f*, Zellulosenitrat *n*

1014 CELLULOSE PAPER, WOODPULP PAPER — pap
f papier *m* de cellulose
e papel *m* de celulosa
i carta *f* di cellulosa
n cellulosepapier *n*
d Zellulosepapier *n*

1015 CELLULOSE PLASTICS — pla
f plastiques *pl* cellulosiques
e plásticos *pl* celulósicos
i plastici *pl* cellulosici
n cellulosekunststoffen *pl*
d Zellulosekunststoffe *pl*

1016 CELLULOSE WADDING — mcy
f ouate *m* de cellulose
e algodón *m* artificial, guata *f* de celulosa
i ovatta *f* di cellulosa
n celstofwatten *pl*
d Zellstoffwatte *f*

1017 CELLULOSE XANTHATE — pla
f xanthate *m* de cellulose
e xantato *m* de celulosa
i xantato *m* di cellulosa, xantogenato *m* di cellulosa
n cellulosexanthaat *n*
d Zellulosexanthogenat *n*

1018 CELLULOSE YARN — pla
f fil *m* de cellulose
e hilo *m* de celulosa
i filo *m* di cellulosa
n cellulosegaren *n*
d Zellulosegarn *n*

1019 CEMENT — bui
f ciment *m*
e cemento *m*
i cemento *m*
n cement *n*
d Zement *m*

1020 CEMENT, GLUE — glu
f adhésif *m*, colle *f*
e adhesivo *m*, cola *f*
i adesivo *m*, colla *f*
n kleefmiddel *n*, lijm, plakmiddel *n*
d Klebemittel *n*, Klebestoff *m*, Leim *m*

1021 CEMENT, PUTTY — bre,glu
f mastic *m*
e masilla *f*, mástique *m*
i mastice *m*
n kit
d Kitt *m*

1022 CEMENT, RUBBER SOLUTION — rub
f dissolution *f*
e disolución *f*
i soluzione *f*
n solutie
d Gummilösung *f*

1023 CEMENT BLENDING — bui,mix
f préparation *f* de mortier de ciment
e preparación *f* de mortero de cemento
i preparazione *f* di malta di cemento
n aanmaken *n* van cementspecie
d Anfertigung *f* von Zementmörtel

1024 CEMENT CAKE TEST (GB), PAT TEST (US) — bui
f essai *m* à galette de ciment
e ensayo *m* de pastilla de cemento
i collaudo *m* a pastilla di cemento
n cementkoekproef
d Zementkuchenprobe *f*

1025 CEMENT CLAY — bui
f argile *f* pour ciment
e arcilla *f* para cemento
i argilla *f* per cemento
n cementklei
d Zementton *m*

1026 CEMENT COPPER — met
f cuivre *m* brut
e cáscara *f* de cobre, cobre *m* de cementación
i rame *m* impuro
n onzuiver koper *n*
d Rohkupfer *n*, Zementkupfer *n*

1027 CEMENT DOSAGE — bui
f dosage *m* de ciment
e dosificación *f* de cemento
i dosaggio *m* di cemento
n cementverdeling
d Zementbeschickung *f*

1028 CEMENT GROUT — bui
f bouillie *f* de ciment, mortier *m* liquide de ciment
e lechada *f* de cemento
i cemento *m* liquido
n cementbrij, dunvloeibare cement
d dünnflüssiger Zement *m*, Zementbrei *m*

1029 CEMENT GROUTING bui
f injection *f* de ciment liquide
e inyección *f* de lechada de cemento
i iniezione *f* di cemento liquido
n cementbrijinspuiting
d Zementbreieinspritzung *f*

1030 CEMENT MORTAR bui
f mortier *m* de ciment
e mortero *m* de cemento
i malta *f* di cemento
n cementmortel, cementspecie
d Zementmörtel *m*

1031 CEMENT QUICK SETTING bui
f prise *f* rapide de ciment
e fraguado *m* rápido de cemento
i indurimento *m* rapido di cemento
n snel binden *n* van cement
d Schnellbinden *n* von Zement

1032 CEMENT SETTING bui
f prise *f* de ciment
e fraguado *m* de cemento
i indurimento *m* di cemento
n binden *n* van cement
d Abbinden *n* von Zement

1033 CEMENTATION pla
f prise *f*
e fraguado *m*
i indurimento *m*
n harden *n*
d Aushärten *n*, Härten *n*

1034 CEMENTATION, CARBONIZATION met
f cémentation *f*
e cementación *f*
i cementazione *f*
n cementeren *n*
d Aufkohlung *f*, Zementieren *n*

1035 CEMENTED CARBIDE, SINTERED CARBIDE met
f carbure *m* fritté
e carburo *m* sinterizado
i carburo *m* sinterizzato
n sintercarbide *n*
d Metallkarbid *n*, Sinterhartmetall *n*, Sinterkarbid *n*

1036 CEMENTING POWDER, CASE HARDENING COMPOSITION met
f agent *m* de trempe de surface
e agente *m* de temple de superficie
i agente *m* di tempera di superficie
n oppervlakhardingsmiddel *n*
d Einsatzpulver *n*, Härtepulver *n*

1037 CEMENTITE, CEMENTITE CARBIDE met
f cémentite *f*
e cementita *f*
i cementite *f*
n cementiet *n*
d Eisenkarbid *n*, Zementit *m*

1038 CENTRAL DISK ATOMIZATION, SPINNING DISK ATOMIZATION soa
f atomisation *f* à disque tournant
e atomización *f* por disco giratorio
i atomizzazione *f* mediante disco rotante
n verstuiving met draaiende schijf
d Zerstäubung *f* mit rotierendem Teller

1039 CENTRALIZED LUBRICATION, CHASSIS LUBRICATION, ONE SHOT LUBRICATION lub
f graissage *m* central, graissage *m* centralisé
e lubricación *f* centralizada
i lubrificazione *f* centrale, lubrificazione *f* centralizzata
n centrale smering
d Eindruckschmierung *f*, Zentralschmierung *f*

1040 CERAMIC MATERIAL cer
f matière *f* céramique
e materia *f* cerámica
i materiale *m* ceramico
n keramisch materiaal *n*
d keramisches Material *n*

1041 CERAMICS cer
f céramique *f*
e cerámica *f*
i ceramica *f*
n keramiek
d Keramik *f*

1042 CERATE fat
f cérat *m*
e cerato *m*
i cerotto *m*
n ceratum *n*, waszalf
d Wachssalbe *f*, Zerat *n*

1043 CEREBROSE (GB), BRAIN SUGAR (US), GALACTOSE sin
f galactose *f*

e galactosa *f*
i galattosio *m*
n galactose
d Cerebrose *f*, Galaktose *f*

1044 CERESINE, fat
CERINE, FOSSIL WAX,
OZOKERITE
f cire *f* minérale, ozocérite *f*,
suif *m* de montagne
e cera *f* fósil, ceresina *f*
i cera *f* minerale, ozocherite *f*
n aardwas, ozokeriet *n*
d Erdwachs *n*, mineralisches
Wachs *n*, Ozokerit *m*

1045 CERIC SULFATE ana
TITRATION (US),
CERIC SULPHATE TITRATION
(GB)
f cérimétrie *f*
e cerimetría *f*
i cerimetria *f*
n cerimetrie
d Zerimetrie *f*

1046 CEROTE PASTE, pho
ENCAUSTIC PASTE
f cérotine *f*, encaustique *f*
e pasta *f* encáustica
i pasta *f* encaustica
n glanswas
d Glanzwachs *n*

1047 CETACEUM, fat
SPERMACETI
f blanc *m* de baleine, spermaceti *m*,
sperme *m* de baleine
e blanco *m* de ballena, esperma *f*
de ballena
i spermaceti *m*
n cetaceum *n*, spermaceti,
walschot *n*
d Walrat *n*

1048 CETANE NUMBER, fue
CETANE RATING
f indice *m* de cétane
e número *m* de cetano
i numero *m* di cetano
n cetaangetal *n*
d CaZ, Cetanzahl *f*

1049 CETIN (US), soa
PALMITIC ACID (GB)
f acide *m* palmitique
e ácido *m* palmítico, cetina *f*
i acido *m* palmitico
n palmitinezuur *n*
d Palmitinsäure *f*

1050 CHAIN gen
f chaîne *f*
e cadena *f*
i catena *f*
n keten
d Kette *f*

1051 CHAIN REACTION gen
f réaction *f* en chaîne
e reacción *f* en cadena
i reazione *f* a catena
n kettingreactie
d Kettenreaktion *f*

1052 CHALAT (US), oic
ETHYLCHLORIDE (GB),
MONOCHLORETHANE
f chlorure *m* d'éthyl
e cloruro *m* etílico
i cloruro *m* d'etile
n ethylchloride *n*, monochloor-
ethaan *n*
d Äthylchlorid *n*, Chelen *n*,
Chloräthyl *n*

1053 CHALCANTHITE, min
BLUE STONE, BLUE VITRIOL,
CYANOSITE
f chalcanthite *f*, vitriol *m* bleu
e calcantita *f*, vitriolo *m* azul
i calcantite *f*, vitriolo *m* azzurro
n blauw vitriool *n*, blauwsteen *n*,
chalcantiet *n*
d Blaustein *n*, Blauvitriol *n*,
Chalkantit *m*, Kupfervitriol *n*

1054 CHALK min
f craie *f*
e creta *f*, tiza *f*
i creta *f*, gesso *m*
n krijt *n*
d Kreide *f*

1055 CHALK LIME, aic
SLACK LIME, SLAKED LIME
f chaux *f* éteinte
e cal *f* apagada
i calce *f* spenta
n gebluste kalk
d Äscherkalk *m*, Gerberkalk *m*,
Löschkalk *m*

1056 CHALK TEST, tes
LIME BAG TEST
f essai *m* à la craie
e prueba *f* a la creta
i collaudo *m* alla creta
n krijtproef
d Kreideprobe *f*

1057 CHALK WATER, bre
CARBONATE WATER
f eau *f* calcifère
e agua *f* caliza, agua *f* selenitosa
i acqua *f* calcifera
n kalkhoudend water *n*
d kalkhaltiges Wasser *n*

1058 CHALKING coa,rub
f talquage *m*
e entalcado *m*
i intalcatura *f*
n talkeren *n*
d Pudern *n*

1059 CHALKING dec,pla
f farinage *m*, poudrage *m*
e formación *f* de una capa de tiza
i difetto *m* superficiale con apparenza di macchia di gesso
n krijten *n*
d Abkreiden *n*, Kreiden *n*

1060 CHALKING met
f pulvérisation *f*
e pulverización *f*
i polverizzamento *m*
n verpoederen *n*
d Verpulverung *f*

1061 CHALYBEATE WATER, hyg
FERRIFERROUS WATER,
FERRIGINOUS WATER
f eau *f* ferrigineuse
e agua *f* ferruginosa
i acqua *f* ferrigenosa
n ijzerhoudend water *n*
d eisenhaltiges Wasser *n*

1062 CHAMBER ACID sua
f acide *m* sulfurique des chambres de plomb
e ácido *m* sulfúrico de las cámaras de plomo
i acido *m* solforico delle camere di piombo
n kamerzuur *n*
d Kammersäure *f*

1063 CHAMBER CRYSTALS sua
f cristaux *pl* d'acide nitroso-sulfurique
e cristales *pl* de ácido nitroso-sulfúrico
i cristalli *pl* d'acido nitroso-solforico
n nitrosylzwavelzuurkristallen *pl*
d Bleikammerkristalle *pl*, Nitrosylschwefelsäurekristalle *pl*

1064 CHAMBER PROCESS sua
f procédé *m* des chambres de plomb
e procedimiento *m* en las cámaras de plomo
i processo *m* delle camere del piombo
n loden-kamerproces *n*
d Bleikammerverfahren *n*, Kammerverfahren *n*

1065 CHAMOIS DRESSING, tan
CHAMOISING
f chamoisage *m*
e agamuzado *m*
i concia *f* a mezzo d'olio di fegato di merluzzo
n zeemtouwen *n*
d Sämischgerben *n*

1066 CHAMOIS LEATHER lea
f chamois *m*, peau *f* de chamois
e gamuza *f*, piel *f* de gamuza
i pelle *f* di camoscio
n gemsleer *n*, gemzeleder *n*, zeemleer *n*
d Chamoisleder *n*, Fensterleder *n*, Sämischleder *n*

1067 CHAMOTTE, cer
FIRECLAY, SPAVIN,
WARRANT
f apyre *f*, argile *f* réfractaire, chamotte *f*
e arcilla *f* refractaria, chamota *f*
i argilla *f* refrattaria, terra *f* refrattaria
n chamotte, vuurklei
d feuerfester Ton *m*, Schamotte *f*

1068 CHAMOTTE BRICK bui,cer
f brique *f* de chamotte
e ladrillo *m* de chamota
i mattone *m* di terra refrattaria
n chamottesteen
d Schamottestein *m*

1069 CHAMOTTE CEMENT bui
f ciment *m* de chamotte
e cemento *m* de chamota
i cemento *m* di terra refrattaria
n chamottecement *n*
d Schamottemörtel *m*

1070 CHAMPAC OIL fat
f essence *f* de champac
e esencia *f* de champaca
i olio *m* di champaca
n champacaolie, tjempakaolie
d Tjampaköl *n*

1071 CHANGE OF STATE gen
f changement *m* d'état
e cambio *m* de estado
i cambiamento *m* di stato
n toestandsverandering, verandering van de aggregatietoestand
d Aggregatszustandsänderung *f*, Zustandsänderung *f*

1072 CHANGE OF STEEP WATER bre
f renouvellement *m* d'eau
e cambio *m* de agua
i cambio *m* d'acqua
n waterverversing
d Wasserwechsel *m*

1073 CHANNEL BLACK, GAS BLACK dec
f noir *m* au canal, noir *m* de gaz
e negro *m* de gas
i nero *m* di gas
n gasroet *n*
d Gasruss *m*

1074 CHANNELLING dis
f formation *f* de chemins préférentiels, renardage *m*
e formación *f* de canales
i formazione *f* di canalizzazione
n kanaalvorming
d Kanalbildung *f*

1075 CHARCOAL fue
f charbon *m* de bois
e carbón *m* de leña, carbón *m* vegetal
i carbone *m* di legna
n houtskool
d Holzkohle *f*

1076 CHARCOAL BLACKING, CHARCOAL DUST, FOUNDRY BLACKING fou
f noir *m* de fonderie
e negro *m* de fundición, polvo *m* de carbón
i nero *m* per fonderia, polvere *f* di carbone
n koolpoeder *n*
d Kohlenstaub *m*

1077 CHARCOAL GAS gas
f gaz *m* de charbon de bois
e gas *m* de carbón de leña
i gas *m* di carbone di legna
n houtskoolgas *n*
d Holzkohlengas *n*

1078 CHARGE METERING pla,rub
f dosage *m* de l'alimentation
e dosificación *f* de la alimentación
i dosaggio *m* dell'alimentazione
n voedingsdosering
d Speisemessung *f*

1079 CHARRING htt
f carbonisation *f*
e carbonización *f*
i carbonizzazione *f*
n verkolen *n*
d Verkohlen *n*

1080 CHASSIS LUBRICATION, CENTRALIZED LUBRICATION, ONE SHOT LUBRICATION lub
f graissage *m* central, graissage *m* centralisé
e lubricación *f* centralizada
i lubrificazione *f* centrale, lubrificazione *f* centralizzata
n centrale smering
d Eindruckschmierung *f*, Zentralschmierung *f*

1081 CHECKERBERRY OIL, WINTERGREEN OIL fat
f essence *f* de gaulthérie, essence *f* de wintergreen
e esencia *f* de gaulteria
i olio *m* di gaulteria
n wintergroenolie
d Wintergrünöl *n*

1082 CHECKING, CRACKING, CRAZING cer,dec
f craquelure *f*, formation *f* de criques, formation *f* de gerces
e agrietamiento *m*
i incrinatura *f*, screpolamento *m*
n barsten *n*, haarscheurvorming
d Rissbildung *f*

1083 CHEDDITE exp
f cheddite *f*
e chedita *f*
i cheddite *f*
n cheddiet *n*
d Cheddit *m*

1084 CHEESE nut
f fromage *m*
e queso *m*
i formaggio *m*
n kaas
d Käse *m*

1085 CHEESE, CROSS-WOUND BOBBIN tex
f bobine *f* croisée
e bobina *f* cruzada
i filo *m* avvolto nella bobina

n kruisspoel
d Kreuzspule *f*

1086 CHEESE DYEING tex
f teinture *f* de bobines croisées
e teñido *m* de bobinas cruzadas
i tintura *f* di fili avvolti nella bobina
n kruisspoelverven *n*
d Kreuzspulenfärbung *f*

1087 CHELATE FORMATION, CHELATION, COMPLEXING ACTION, SEQUESTRATION oic
f action *f* complexante
e acción *f* quelacante, facultad *f* de formar complejos
i azione *f* complessante, azione *f* sequestrante, chelazione *f*
n chelateren *n*, complexvorming, klauwkoppeling, klauwverbinding
d Chelatbildung *f*, Chelierung *f*, Komplexbildung *f*

1088 CHELATING AGENT, COMPLEXING AGENT, SEQUESTERING AGENT, SEQUESTRANT gen
f agent *m* séquestrant complexant, chélatant *m*
e agente *m* secuestrador reactivo, agente *m* secuestrante, formador *m* de complejos
i agente *m* chelante, agente *m* sequestrante
n complexvormer
d chelatbildendes Reagenz *n*, Komplexbildner *m*

1089 CHEMIC tex
f chlorure *m* de chaux
e cloruro *m* de cal
i cloruro *m* di calce
n calciumchloride *n*
d Kalziumchlorid *n*

1090 CHEMICAL AFFINITY gen
f affinité *f* chimique
e afinidad *f* química
i affinità *f* chimica
n chemische affiniteit
d chemische Affinität *f*

1091 CHEMICAL ANALYSIS ana
f analyse *f* chimique
e análisis *f* química
i analisi *f* chimica
n chemische analyse
d chemische Analyse *f*

1092 CHEMICAL BOND gen
f liaison *f* chimique
e enlace *m* químico
i legame *m* chimico
n chemische binding
d chemische Bindung *f*

1093 CHEMICAL CHANGE gen
f transformation *f* chimique
e transformación *f* química
i trasformazione *f* chimica
n chemische omzetting
d chemische Umwandlung *f*

1094 CHEMICAL COLORING (US), CHEMICAL COLOURING (GB) dec
f colorage *m* chimique
e coloreado *m* químico
i colorazione *f* chimica
n chemisch kleuren *n*
d chemisches Färben *n*

1095 CHEMICAL COMPOSITION gen
f composition *f* chimique
e composición *f* química
i composizione *f* chimica
n chemische samenstelling
d chemische Zusammensetzung *f*

1096 CHEMICAL COMPOUND gen
f composé *m* chimique
e compuesto *m* químico
i composto *m* chimico
n chemische verbinding
d chemische Verbindung *f*

1097 CHEMICAL CONSTITUTION gen
f constitution *f* chimique
e constitución *f* química
i costituzione *f* chimica
n atoomrangschikking, chemische constitutie
d Atomanordnung *f*, chemische Konstitution *f*

1098 CHEMICAL FADE pho
f décoloration *f* chimique
e descoloramiento *m* químico
i scolorimento *m* chimico
n chemisch verbleken *n*
d chemische Schwindung *f*, chemische Verblassung *f*

1099 CHEMICAL FOG pho
f voile *f* chimique
e velo *m* químico
i velatura *f* chimica, velo *m* chimico
n chemische sluier
d chemischer Schleier *m*

1100 CHEMICAL LEAD   met
f plomb *m* chimiquement pur
e plomo *m* químicamente puro
i piombo *m* chimicamente puro
n chemisch zuiver lood *n*
d chemisch reines Blei *n*

1101 CHEMICAL PICKLING   elc
f décapage *m* chimique
e decapaje *m* químico
i decapaggio *m* chimico
n chemisch beitsen *n*
d chemisches Beizen *n*

1102 CHEMICAL PLATING, ELECTROLESS PLATING   soa
f placage *m* chimique
e quimioplastia *f*
i placcatura *f* chimica
n chemisch pleten *n*
d chemische Plattierung *f*

1103 CHEMICAL PRECIPITATION   sew
f précipitation *f* chimique
e precipitación *f* química
i precipitazione *f* chimica
n chemische afzetting
d chemisches Absetzverfahren *n*

1104 CHEMICAL PULP, CHEMICAL WOODPULP, CHEMIGROUND   pap
f pâte *f* chimique
e pulpa *f* química
i polpa *f* chimica
n chemische pulp
d Chemipulpe *f*, chemischer Holzstoff *m*

1105 CHEMICAL REACTION   gen
f réaction *f* chimique
e reacción *f* química
i reazione *f* chimica
n chemische reactie
d chemische Reaktion *f*

1106 CHEMICAL RIPENING, AFTERRIPENING, DIGESTION, SECOND RIPENING   pho
f maturation *f* chimique, maturation *f* en refonte
e maduración *f* química, segunda maduración *f*
i maturazione *f* chimica
n chemische rijping
d chemische Reifung *f*, Nachreifung *f*

1107 CHEMICAL SEWAGE TREATMENT   sew
f traitement *m* chimique d'eaux d'égout
e tratamiento *m* químico de aguas cloacales
i trattamento *m* chimico di fognatura
n chemische afvalwaterverwerking
d chemische Abwasserverarbeitung

1108 CHEMICAL STRIPPING, SOLUTION STRIPPING   coa
f dépouillage *m* chimique
e eliminación *f* química
i degalvanizzazione *f* chimica
n chemische laagverwijdering
d chemische Entplattierung *f*

1109 CHEMICAL TONING   pho
f virage *m* chimique
e viraje *m* químico
i viraggio *m* chimico
n chemisch kleuren *n*
d chemische Tonung *f*

1110 CHEMICAL WARFARE, GAS WARFARE   mit
f guerre *f* chimique
e guerra *f* química
i guerra *f* chimica
n gasoorlogvoering
d Gaskrieg *m*

1111 CHEMICALLY CURING LACQUER, CATALYZED LACQUER, REACTION LACQUER   dec
f laque *f* à durcissement chimique, vernis *m* réactionnel
e laca *f* reaccional
i lacca *f* ad indurimento chimico, lacca *f* di reazione
n reactielak
d Kalthärtelack *m*, Reaktionslack *m*

1112 CHEMICKING   tex
f traitement *m* au chlorure de chaux
e tratamiento *m* al cloruro de cal
i trattamento *m* al cloruro di calce
n calciumchloridebehandeling
d Kalziumchloridbehandlung

1113 CHEMICO CONTACT PROCESS, CONTACT PROCESS   sua
f procédé *m* de contact
e procedimiento *m* de contacto
i processo *m* di contatto
n contactproces *n*
d Kontaktverfahren *n*

1114 CHEMISORPTION, gen
ACTIVATED ADSORPTION
f adsorption *f* activée
e adsorción *f* activada
i adsorbimento *m* attivato
n chemisorptie, geactiveerde adsorptie
d aktivierte Adsorption *f*, Chemisorption *f*

1115 CHENOPODIUM OIL, fat
SANTONICA OIL,
WORMSEED OIL
f essence *f* de chénopodium
e esencia *f* de quenopodio
i olio *m* di chenopodio
n wormzaadolie
d Chenopodiumöl *n*, Wurmsamenöl *n*

1116 CHEQUE PAPER pap
f papier *m* pour chèques
e papel *m* de cheques
i carta *f* per assegni
n chequepapier *n*
d Scheckpapier *n*

1117 CHEVERIL, lea
KID LEATHER
f chevrotin *m*, peau *f* de chèvre
e cabritilla *f*
i pelle *f* di capra, pelle *f* di capretto
n geiteleder *n*
d Bockleder *n*, Ziegenleder *n*

1118 CHILE NITRE, fer
CHILE SALTPETRE,
CUBIC SALTPETRE
f nitre *m* cubique, nitre *m* du Chile
e nitrato *m* de Chile, salitre *m* de Chile
i salnitro *m* del Cile
n chilisalpeter
d Chilisalpeter *m*, Natronsalpeter *m*

1119 CHILL CRYSTALS fon
f cristaux *pl* dus au refroidissement rapide
e cristales *pl* debidos al enfriamiento rápido
i cristalli *pl* dovuti al raffreddamento rapido
n door snelle afkoeling ontstane kristallen
d durch Schnellkühlung entstandene Kristalle *pl*

1120 CHILL HAZE bre,clo
f trouble *m* au froid
e enturbiamiento *m* por frío
i intorbidamento *m* per freddo
n vertroebeling bij afkoeling
d Kältetrübung *f*, Kühltrübung *f*

1121 CHILLING cer,clo
f refroidissement *m* brusque
e enfriamiento *m* brusco
i raffreddamento *m* brusco
n snel afkoelen *n*
d Abschrecken *n*

1122 CHILLING dec
f ternissement *m*
e oscurecimiento *m*
i perdita *f* di lustro
n glansverlies *n*
d Glanzverlust *m*

1123 CHINA CLAY, cer
KAOLIN
f kaolin *m*, terre *f* à porcelaine
e caolín *m*
i caolino *m*
n kaoline *n*, porseleinaarde
d Kaolin *n*, Porzellanerde *f*

1124 CHINA GRASS, tex
CAMBRIC GRASS, RAMIE
f ramie *f*
e ramio *m*
i ramia *f*, ramie *f*
n ramee, ramie
d Ramie *f*

1125 CHINA WOOD OIL, fat
TUNG OIL
f huile *f* de bois de Chine, huile *f* de Tung
e aceite *m* de madera, aceite *m* de palo, aceite *m* de tung
i olio *m* di legno, olio *m* essiccante cinese
n houtolie, tungolie
d Holzöl *n*, Tungöl *n*

1126 CHINESE OIL, fat
CASSIA OIL
f essence *f* de cassie
e esencia *f* de cañafistula
i olio *m* di cassia
n kassiaolie
d Kassiaöl *n*

1127 CHINESE RED, dec
DERBY RED
f rouge *m* de Chine
e sulfuro *m* mercúrico rojo
i rosso *m* di Cina
n Chinees rood *n*
d rotes Quecksilbersulfid *n*

1128 CHINESE WHITE, dec
FLOWERS OF ZINC
f blanc *m* de Chine
e blanco *m* al óxido de cinc
i biacca *f* all'ossido di zinco
n Chinees wit *n*
d Chinaweiss *n*, Zinkoxydweiss *n*

1129 CHIPS, sin
BEET SLICES, COSSETTES
f cossettes *pl*
e cosetas *pl*, rebanadas *pl*
i ritagli *pl* di barbabietole
n bietensnijdsels *pl*
d Rübenschnitzel *pl*

1130 CHIT MALT bre
f malt *m* à peine germé
e malta *f* poco germinada
i malto *m* poco germogliato
n nauwelijks ontkiemde mout
d Spitzmalz *n*

1131 CHITTERING cer
f rupture *f* de bord
e ruptura *f* de borde
i rottura *f* di bordo
n randbreuk
d Randbruch *m*

1132 CHITTING, bre
SPROUTING
f piquage *m*
e apuntado *m*
i spuntatura *f*
n kiemverwijdering
d Spitzen *n*

1133 CHLORATE EXPLOSIVE exp
f explosif *m* chloraté
e explosivo *m* de clorato
i esplosivo *m* a clorato
n chloraatspringstof
d Chloratsprengstoff *m*

1134 CHLORIDE PAPER pho
f papier *m* au chlorure d'argent
e papel *m* de cloruro de plata
i carta *f* al cloruro d'argento
n zilverchloridepapier *n*
d Chlorsilberpapier *n*

1135 CHLORIDIZING ROASTING met
f grillage *m* chlorurant
e tostación *f* clorurante
i torrefazione *f* al cloruro
n chlorerend roosten *n*
d chlorierende Röstung *f*

1136 CHLORINATED LIME, ble
BLEACHING POWDER
f chlorure *m* de chaux
e cal *f* clorada, cloruro *m* de cal, hipoclorito *m* cálcico
i calce *f* clorurata, cloruro *m* di calce
n bleekpoeder *n*, chloorkalk
d Bleichpulver *n*, Chlorkalk *m*, Kalziumhypochlorit *n*

1137 CHLORINATED RUBBER rub
f caoutchouc *m* chloré
e goma *f* clorada
i gomma *f* clorurata
n chloorrubber
d chlorierter Gummi *m*, Chlorkautschuck *m*

1138 CHLORINATED RUBBER dec,rub
PAINT
f peinture *f* à base de caoutchouc chloré
e pintura *f* a base de goma clorada
i pittura *f* a base di gomma clorurata
n chloorrubberverf
d Chlorgummifarbe *f*

1139 CHLORINATED WATER hyg,tex
f eau *f* chlorée, eau *f* javellisée, eau *f* verdunisée
e agua *f* clorada
i acqua *f* clorata
n gechloreerd water *n*
d chloriertes Wasser *n*, gechlortes Wasser *n*

1140 CHLORINATED WOOL tex
f laine *f* chlorée
e lana *f* clorada
i lana *f* clorata
n gechloreerde wol, met chloor behandelde wol
d chlorierte Wolle *f*

1141 CHLORINATION gen
f chloruration *f*
e cloriración *f*
i clorazione *f*
n chloreren *n*
d Chlorierung *f*

1142 CHLORINE ADDITION gen
f addition *f* de chlore
e adición *f* de cloro
i addizione *f* di cloro
n chloortoevoeging
d Chlorgabe *f*, Chlorierung *f*

1143 CHLORINE BLEACH LIQUID tex
f liquide *m* de blanchiment en chlore
e líquido *m* de blanqueo por cloro
i liquido *m* di candeggio con cloro
n chloorbleekmiddel *n*
d Chlorbleichmittel *n*

1144 CHLORINE BLEACHING tex
f blanchiment *m* en chlore, chlorage *m*
e blanqueo *m* por cloro
i candeggio *m* con cloro
n chloorbleking
d Chlorbleichung *f*

1145 CHLORINE DOSING cle
f dosage *m* de chlore
e dosificación *f* de cloro
i dosaggio *m* di cloro
n chloordosering
d Chlordosierung *f*

1146 CHLORINE RETENTION tex
f absorption *f* de chlore
e absorción *f* de cloro
i assorbimento *m* di cloro
n chloorabsorptie
d Chlorabsorption *f*

1147 CHLORINE WATER, cle
CHLORURETTED WATER
f eau *f* de chlore
e agua *f* de cloro
i acqua *f* di cloro
n chloorwater *n*
d Chlorwasser *n*

1148 CHLORITE BLEACHING tex
f blanchiment *m* à chlorite
e blanqueo *m* de clorito
i candeggio *m* a clorite
n chlorietbleking
d Chloritbleichung *f*

1149 CHLOROAZOTIC ACID (US), aic,met
CHLORONITROUS ACID, AQUA REGIA, NITROHYDROCHLORIC ACID, NITROMURIATIC ACID
f eau *f* régale
e agua *f* regia
i acqua *f* regia
n koningswater *n*
d Königswasser *n*

1150 CHOCOLATE nut
f chocolat *m*
e chocolate *m*
i cioccolata *f*
n chocolade
d Schokolade *f*

1151 CHROMATE DYEING, dye,tex
METACHROME PROCESS, MONOCHROME PROCESS
f teinture *f* à chromate
e teñido *m* de cromato
i tintura *f* a cromato
n chromaatverven *n*
d Chromatfärbung *f*

1152 CHROMATE MORDANT, dye
METACHROME MORDANT, MONOCHROME MORDANT
f mordant *m* au chromate
e mordiente *m* metacromático
i mordente *m* al cromato
n chromaatbijtmiddel *n*
d Chromatbeize *f*

1153 CHROMATE PRINTING PROCESS, tex
AUTOCHROME PROCESS
f impression *f* autochrome
e impresión *f* en fotocromía
i stampa *f* autocroma
n autochroomdruk
d Autochromdruck *m*

1154 CHROMATING cle,met
f mordançage *m*
e decapaje *m* al bicromato
i mordenzatura *f*
n beitsbehandeling, bichromaatbeitsen *n*
d Bichromatbeizung *f*

1155 CHROMATOGRAPHIC ANALYSIS, ana
CHROMATOGRAPHY
f analyse *f* chromatographique, chromatographie *f*
e análisis *f* cromatográfica, cromatografía *f*
i analisi *f* cromatografica, cromatografia *f*
n chromatografie, chromatografische analyse
d Chromatographie *f*, chromatographische Analyse *f*

1156 CHROMATOGRAPHIC RESOLUTION ana
f pouvoir *m* résolvant chromatographique
e poder *m* resolvente cromatográfico
i potere *m* risolvente cromatografico
n chromatografisch scheidend vermogen *n*

d chromatographisches Auflösungsvermögen *n*

1157 CHROME ALUM dye,tan
f alun *m* chromo-potassique
e alumbre *m* crómico, alumbre *m* de cromo
i allume *m* di cromo
n chroomaluin
d Chromalaun *m*

1158 CHROME BATH elc
f bain *m* de chromage
e baño *m* de cromato
i bagno *m* di cromatura
n chroombad *n*, verchromingsbad *n*
d Chrombad *n*, Verchromungsbad *n*

1159 CHROME BRICK bui
f brique *f* à chromite
e ladrillo *m* de cromita
i mattone *m* a cromite
n chromietsteen
d Chromitstein *m*

1160 CHROME DYES dye
f colorants *pl* au chrome
e colorantes *pl* al cromo
i coloranti *pl* al cromo
n chroomkleurstoffen *pl*
d Chromfarbstoffe *pl*

1161 CHROME GREEN (GB), EMERAUDE GREEN (US) dec
f vert *m* Casalis, vert *m* de chrome
e verde *m* de cromo
i verde *m* di cromo
n chromaatgroen *n*
d Guignetgrün *n*, Milorigrün *n*

1162 CHROME LEATHER lea
f cuir *m* chromé
e cuero *m* al cromo, cuero *m* cromado
i cuoio *m* al cromo
n chroomleder *n*
d Chromleder *n*

1163 CHROME PRINTING tex
f impression *f* chromique
e impresión *f* crómica
i stampatura *f* cromica
n chroomdruk
d Chromdruck *m*

1164 CHROME PRINTING DYES dye
f colorants *pl* pour l'impression chromique
e colorantes *pl* para la impresión crómica
i coloranti *pl* per la stampatura cromica
n chroomdrukkleurstoffen *pl*
d Chromdruckfarbstoffe *pl*

1165 CHROME RED, AUSTRIAN CINNABAR, BASIC LEAD CHROMATE, PHOENICITE dec
f chromate *m* de plomb basique, rouge *m* de chrome
e cromato *m* de plomo básico, rojo *m* de cromo
i cromato *m* di piombo basico, rosso *m* di cromo
n basisch loodchromaat *n*, chromaatrood *n*
d basisches Bleichromat *n*, Chromrot *n*

1166 CHROME TANNING tan
f tannage *m* au chrome
e curtimiento *m* al cromo
i concia *f* con acido cromico
n chroomlooien *n*
d Chromgerbung *f*

1167 CHROME TANNING LIQUOR tan
f liqueur *f* de chrome
e líquido *m* para curtimiento al cromo
i acido *m* crómico
n chroomlooivloeistof
d Chromgerbbrühe *f*

1168 CHROME-TIN PINK, DUTCH PINK, ENGLISH PINK, ITALIAN PINK dec
f pigment *m* jaune transparent
e pigmento *m* amarillo transparent
i pigmento *m* giallo trasparente
n gele lakverf
d gelbe Lackfarbe *f*, Schüttgelb *n*

1169 CHROME YELLOW, LEAD CHROMATE, PARIS YELLOW dec
f jaune *m* de chrome
e amarillo *m* de cromo, cromato *m* de plomo
i giallo *m* di cromo
n chroomgeel *n*, Parijs geel *n*
d Bleichromat *n*, Chromgelb *n*, Parisergelb *n*

1170 CHROMIUM PLATING elc
f chromage *m*
e chromado *m*
i cromatura *f*
n verchroming
d Verchromung *f*

1171 CHROMIZING coa
f cémentation *f* par le chrome
e cromización *f*
i cromizzazione *f*
n chromeren *n*
d Inkromieren *n*

1172 CHROMO PAPER pho
f papier *m* pour chromolithographie
e papel *m* para cromolitografía
i carta *f* per cromolitografia
n chromolithografiepapier *n*
d Chromolithographiepapier *n*

1173 CHROMOLITHOGRAPHY pri
f chromolithographie *f*
e cromolitografía *f*
i cromolitografia *f*
n kleurensteendruk
d Farbensteindruck *m*

1174 CHROMOUS SALTS aic
f sels *pl* chromeux
e sales *pl* cromosos
i sali *pl* cromosi
n chromozouten *pl*
d Chrom-II-Salze *pl*

1175 CHURNING dai,nut
f barattage *m*
e batido *m*
i sbattatura *f* del latte nella zangola
n karnen *n*
d Buttern *n*

1176 CIDER nut
f cidre *m*
e sidra *f*
i sidro *m*
n appelwijn, cider
d Apfelwein *m*

1177 CINCHONA, mcy
PERUVIAN BARK
f écorce *f* de quinquina, quinquina *f*
e corteza *f* de quino, quina *f*
i china *f*, corteccia *f* di china
n kina, kinabast
d Chinarinde *f*, Cinchonarinde *f*, Fieberrinde *f*

1178 CINDER, gen
CLINKER, SLAG
f laitier *m*, mâchefer *m*, scorie *f*
e escoria *f*
i scoria *f*
n sintel, slak
d Schlacke *f*

1179 CINNABAR, dec
VERMILION, VERMILLION
f cinabre *m*, vermillon *m*
e bermellón *m*, cinabrio *m*
i cinabro *m*, vermiglione *m*
n cinnaber *n*, vermiljoen *n*
d Zinnoberrot *n*

1180 CINNAMALDEHYDE (US), per
CINNAMIC ALDEHYDE (GB),
PHENYLACROLEIN
f aldéhyde *m* cinnamique
e aldehido *m* cinámico
i aldeide *f* cinnamica
n kaneelaldehyde *n*
d Zimtaldehyd *n*

1181 CINNAMIC ALCOHOL (GB), per
PERUVIN (US)
f alcool *m* cinnamique
e alcohol *m* cinámico
i alcool *m* cinnamico
n kaneelalcohol
d Styron *n*, Zimtalkohol *m*

1182 CINNAMON OIL fat
f essence *f* de cannelle
e esencia *f* de canela
i olio *m* di cannella
n kaneelolie
d Zimtöl *n*

1183 CINNAMYL ACETATE per
f acétate *m* cinnamique
e acetato *m* cinámico
i acetato *m* cinnamico
n kaneelacetaat *n*
d Zimtazetat *n*

1184 CIRCUIT, gen
CIRCULATION
f circuit *m*, cycle *m*
e ciclo *m*, circuito *m*
i ciclo *m*, circuito *m*
n kringloop
d Kreislauf *m*, Umlauf *m*

1185 CIRCULATING REFLUX dis
f reflux *m* de récyclage
e reflujo *m* circulante
i riflusso *m* circolante
n circulerende terugstroom
d umlaufender Rücklauf *m*

1186 CIRCULATING SYSTEM lub
LUBRICATION
f lubrification *f* par circulation
e lubricación *f* por circulación
i lubrificazione *f* per circolazione
n spoelsmering
d Umlaufschmierung *f*

1187 CIRCULATION COOLING clo
f refroidissement *m* par circulation
e enfriamiento *m* por circulación
i raffreddamento *m* per circolazione
n omloopkoeling
d Umlaufkühlung *f*

1188 CIRCULATION DYEING, BALE DYEING, PACK DYEING tex
f teinture *f* en balles
e teñido *m* en balas, teñido *m* en líos
i tintura *f* in balle
n balenverven *n*, verven *n* in balen
d Ballenfärbung *f*

1189 CIRCULATION OF ELECTROLYTE elc
f agitation *f* du bain
e agitación *f* del baño
i agitazione *f* del bagno
n badbeweging
d Badbewegung *f*

1190 CISSING, BEADING, CRAWLING dec
f contraction *f*, rampage *m*, retrait *m*
e avejigamiento *m*
i falta *f* di presa
n kralen *n*, samentrekken *n*, schiften *n*
d Kriechen *n*, Perlen *n*

1191 CITRON YELLOW, ZINC YELLOW dec
f chromate *m* de zinc, jaune *f* de zinc
e amarillo *m* de cinc, amarillo *m* limón
i giallo *m* di zinco
n zinkchromaat *n*, zinkgeel *n*
d Zinkgelb *n*, Zitronengelb *n*

1192 CITRONELLA OIL fat
f essence *f* de citronnelle
e esencia *f* de citronela
i olio *m* di citronella
n citronella, grasolie
d Citronelöl *n*

1193 CITY WATER, MUNICIPAL WATER, TAP WATER hyg
f eau *f* du robinet, eau *f* potable
e agua *f* corriente, agua *f* potable
i acqua *f* di rubinetto, acqua *f* potabile
n drinkwater *n*, leidingwater *n*
d Leitungswasser *n*, Trinkwasser *n*

1194 CIVET per
f civette *m*, viverréum *m*
e algalia *f*, civeto *m*
i zibetto *m*
n civet
d Zibet *n*

1195 CLAIRECOLLE, CLEARCOLE, CLERECOLE dec
f claircolle *f*, détrempe *f*
e temple *m*
i tempera *f*
n voorlijm
d Vorleim *m*

1196 CLARIFICATION iwt
f clarification *f*
e clarificación *f*
i chiarificazione *f*
n klaring
d Klärung *f*

1197 CLARIFIER, CLARIFYING AGENT gen
f agent *m* clarifiant
e agente *m* clarificador
i agente *m* chiarificante
n klaringsmiddel *n*
d Kläre *f*, Klärmittel *n*

1198 CLASSIFICATION sor
f classification *f*, triage *m*
e clasificación *f*
i classificazione *f*
n classificatie
d Klassierung *f*

1199 CLATHRATE COMPOUNDS gen
f clathrates *pl*, composés *pl* en cag
e compuestos *pl* clatrados
i composti *pl* clatrati
n clathraatverbindingen *pl*, kooiverbindingen *pl*
d Einschlussverbindungen *pl*, Klathrate *pl*

1200 CLAY min
f argile *f*
e arcilla *f*
i argilla *f*
n klei
d Ton *m*

1201 CLAY FILLING bui
f remplissage *m* en argile
e relleno *m* en arcilla
i riempimento *m* in argilla
n kleivulling
d Tonfüllung *f*

1202 CLAY GLAZE cer
f émail *m* d'argile
e vidriado *m* de arcilla
i smalto *m* d'argilla
n leemglazuur *n*
d Erdglasur *f*, Lehmglasur *f*

1203 CLAY WASH (TO) fou
f glaiser v
e enlucir v
i dare v la creta
n met klei insmeren v
d mit Lehmbrühe anstreichen v

1204 CLAY WASHING cer
f débourbage *m* de l'argile
e lavado *m* de la arcilla
i lavaggio *m* dell'argilla
n slibben *n*
d Schlämmen *n*

1205 CLAYEY cer
f argilleux adj
e arcilloso adj
i argilloso adj
n kleiachtig adj
d lehmig adj, tonhaltig adj, tonig adj

1206 CLEAN GAS (US), DUST-FREE GAS (GB) gas
f gaz *m* dépoussiéré
e gas *m* desempolvado
i gas *m* depurato della polvere
n fijngas *n*, stofvrij gas *n*
d entstaubtes Gas *n*, staubfreies Gas *n*

1207 CLEANER cle
f agent *m* dégraisseur
e desengrasador *m*, desgrasador *m*
i sgrassatore *m*
n ontvettingsmiddel *n*
d Entfettungsmittel *n*

1208 CLEANING elc
f dégraissage *m*
e desengrase *m*
i sgrassaggio *m*
n ontvetting
d Reinigung *f*

1209 CLEANING, DRESSING bre
f nettoyage *m*
e limpieza *f*
i pulitura *f*
n poetsen *n*
d Putzen *n*

1210 CLEANING EFFECT cle
f effet *m* nettoyant
e efecto *m* limpiador
i risultato *m* di lavaggio
n reinigende werking
d Reinigungswirkung *f*

1211 CLEANING EFFICIENCY, DETERGENT ACTION, WASHING PERFORMANCE cle
f action *f* détergente
e acción *f* detergente
i azione *f* detergente
n waswerking
d Waschwirkung *f*

1212 CLEANSING PROCESS, UNION PROCESS bre,fmi
f procédé *m* de fermentation en tonnes
e procedimiento *m* de fermentación en barriles
i processo *m* di fermentazione in botte
n vatgistingsproces *n*
d Fassgärungsverfahren *n*

1213 CLEAR LACQUER dec
f vernis *m* transparent
e barniz *m* transparente
i vernice *f* trasparente
n transparantlak
d Klarlack *m*

1214 CLEARING pho
f clarification *f*
e clarificación *f*
i chiarificazione *f*
n klaring
d Klärung *f*

1215 CLEARING tex
f traitement *m* à l'hypochlorite de sodium
e tratamiento *m* al hipoclorito sódico
i trattamento *m* all'ipoclorito di sodio
n behandeling met natrium-hypochloriet
d Behandlung *f* mit Natrium-hypochlorit

1216 CLEARING AGENT (GB), CLEARING FLUID (US), IMMERSION OIL ana
f huile *f* d'immersion
e aceite *m* de inmersión
i olio *m* d'immersione
n immersieolie
d Immersionsöl *n*

1217 CLEARING BATH pho
f bain *m* de clarification
e baño *m* de clarificación
i bagno *m* di chiarificazione
n klaringsbad *n*
d Klärbad *n*

1218 CLEAVING cry
f clivage *m*
e clivaje *m*, exfoliación *f*
i clivaggio *m*, sfaldatura *f*
n kloven *n*, splijten *n*
d Spalten *n*

1219 CLEAVING gen
f clivage *m*
e escisión *f*
i scissione *f*
n splijting
d Spaltung *f*

1220 CLIMBING-FILM EVAPORATION eva
f évaporation *f* à couche montante
e evaporación *f* de capa ascendente
i evaporazione *f* con strato di salita
n verdamping met stijgende laag
d Aufsteigschichtverdampfung *f*, Kletterverdampfung *f*

1221 CLINKER bui
f clinker *m*
e clinquer *m*
i clinker *m*
n klinker
d Klinker *m*

1222 CLINKER, gen
CINDER, SLAG
f laitier *m*, mâchefer *m*, scorie *f*
e escoria *f*
i scoria *f*
n sintel, slak
d Schlacke *f*

1223 CLINKER ASPHALT bui
f clinker-asphalte *m*
e clinker *m* asfáltico
i asfalto *m* di scorie
n klinkerasfalt *n*
d Klinkerasphalt *m*

1224 CLINKERING cer
f frittage *m*, vitrification *f*
e aglutinación *f*, sinterización *f*
i sinterizzazione *f*, vetrificazione *f*
n klinkeren *n*
d Sintern *n*

1225 CLINKERING htt
f décrassage *m*
e remoción *f* de escorias
i scorificazione *f*
n ontslakking
d Entschlackung *f*

1226 CLOD, cer
CLOT
f caillot *m*, grumeau *m*
e grumo *m*
i grumo *m*
n klomp, klont
d Klumpen *m*

1227 CLOGGING, iwt
PONDING
f encrassement *m*
e atascamiento *m*, obstrucción *f*
i intasatura *f*
n verstopping
d Verschlammung *f*

1228 CLOSE ANNEALING, htt,met
BOX ANNEALING, POT ANNEALING
f recuit *m* en caisse
e recocido *m* en caja
i ricottura *f* in cassetta
n kistgloeien *n*
d Kastenglühen *n*

1229 CLOSE-BOILING MIXTURE dis
f mélange *m* à points d'ébullition rapprochés
e mezcla *f* de puntos de ebullición próximos
i miscela *f* a stretto intervallo d'ebollizione
n mengsel *n* met klein kooktraject
d engsiedendes Gemisch *n*

1230 CLOSE PLATING coa
f application *f* d'une couche mince métallique par soudure
e aplicación *f* de una capa delgada de metal por soldadura
i aggiustaggio *m* d'uno strato sottile metallico mediante saldatura
n opbrengen *n* van een dunne metaallaag door solderen
d Auflötung *f* einer dünnen Metallplatte

1231 CLOSED-CUP FLASH-POINT TEST tes
f vérification *f* du point d'inflammation en vase fermé

e prueba *f* del punto de deflagración en copa cerrada
i collaudo *m* del punto d'infiammazione in coppa chiusa
n vlampuntproef in gesloten vat
d Flammpunktprüfung *f* in geschlossenem Behälter

1232 CLOTTING, CURDLING dec
f caillage *m*, grumelage *m*
e agrumación *f*
i formazione *f* di grumi
n klonteren *n*
d Käsen *n*, Klümperigwerden *n*

1233 CLOUD POINT, TURBIDITY POINT fue
f point *m* de trouble
e punto *m* de enturbiamiento
i punto *m* d'intorbidimento
n troebelingspunt *n*
d Trübungspunkt *m*

1234 CLOUDBURSTING (US), STEEL BALL PEENING (GB) cle
f traitement *m* par jet de balles en acier
e tratamiento *m* por chorro con perdigones de acero
i trattamento *m* con pallottole d'acciaio
n kogelstralen *n*
d Kugelstrahlen *n*

1235 CLOUDING OF GLASS gla
f trouble *m* du verre
e empañamiento *m* del vidrio, enturbiamiento *m* del vidrio
i torbidezza *f* del vetro
n glastroebeling
d Trübung *f* des Glases

1236 CLOVE OIL fat
f essence *f* de girofle
e esencia *f* de clavo
i olio *m* di chiodo di garofano
n kruidnagelolie
d Gewürznelkenöl *n*, Nelkenöl *n*

1237 COACERVATION col
f coacervation *f*
e coacervación *f*
i coacervazione *f*
n coacervatie
d Koazervierung *f*

1238 COAGULANT col
f agent *m* coagulant, coagulant *m*
e agente *m* coagulante, coagulante *m*
i agente *m* coagulante, coagulante *m*
n coagulant, coaguleermiddel *n*
d Koagulant *n*, Koaguliermittel *n*

1239 COAGULANT CASTING PROCESS pla,rub
f procédé *m* de moulage à coagulant
e procedimiento *m* de moldeo con coagulante
i processo *m* di stampatura con coagulante
n coagulantgietproces *n*
d Koagulierungsverfahren *n*

1240 COAGULATION col
f coagulation *f*
e coagulación *f*
i coagulazione *f*
n coagulatie, samenvlokken *n*
d Ausflockung *f*, Koagulierung *f*

1241 COAGULATION, CURDLING col,dai,nut
f caillage *m*, coagulation *f*
e coagulación *f*, cuajamiento *m*
i accagliamento *m*, coagulazione *f*
n coagulatie, stremmen *n*
d Gerinnung *f*, Koagulierung *f*

1242 COAL GAS gas
f gaz *m* d'éclairage, gaz *m* de houille
e gas *m* de hulla, gas *m* del alumbra
i gas *m* di carbone, gas *m* illuminante
n lichtgas *n*
d Leuchtgas *n*

1243 COAL GAS GENERATION gas
f génération *f* de gaz d'éclairage, génération *f* de gaz de houille
e generación *f* de gas de hulla, generación *f* de gas del alumbrado
i generazione *f* di gas di carbone, generazione *f* di gas illuminante
n lichtgasontwikkeling
d Leuchtgasentwicklung *f*

1244 COAL GASIFICATION gas
f gazéification *f* du charbon
e gasificación *f* del carbón
i gassificazione *f* del carbone
n steenkoolvergassing
d Steinkohlenvergasung *f*

1245 COAL HYDROGENATION PROCESS, BERGIUS PROCESS fue
f procédé *m* Bergius

e procedimiento *m* de hidrogenación de carbón
i processo *m* di bergiusurazione
n bergiusproces *n*, steenkool-hydrogenatieproces *n*, zweelproces *n*
d Bergius-Verfahren *n*, Kohlehydrierungsverfahren *n*, Schwelverfahren *n*

1246 COAL TAR oic
f goudron *m* de houille
e alquitrán *m* de hulla
i catrame *m* di carbone
n koolteer
d Steinkohlenteer *m*

1247 COAL TAR DYES dye
f colorants *pl* dérivés du goudron
e colorantes *pl* de alquitrán
i coloranti *pl* a base di catrame
n teerkleurstoffen *pl*
d Teerfarbstoffe *pl*

1248 COAL WATER GAS gas
f gaz *m* double
e gas *m* doble
i gas *m* doppio
n gas *n* ontstaan bij totale steenkoolvergassing
d Kohlenwassergas *n*

1249 COARSE GRINDING brk
f broyage *m* grossier, concassage *m* préalable
e molienda *f* grossa
i frantumazione *f* in pezzi grossi, prima frantumazione *f*
n grofmalen *n*
d Grobzerkleinerung *f*, Vorzerkleinerung *f*

1250 COARSE SLUDGE bre
f cassure *f* à l'ébullition
e coágulo *m* de ebullición
i coagulo *m* a caldo
n hete draf
d Heisstrub *m*

1251 COARSE STUFF, MORTAR FOR FIRST LAYER OF RENDERING bui
f mortier *m* pour la couche de fond de l'enduit
e enfoscado *m* primo, mezcla *f* de mortero de cal con pelos
i impasto *m* di calce e peli per intonaco
n grondspecie
d Grundmörtel *m*

1252 COARSELY GROUND MALT bre
f mouture *f* grossière
e molienda *f* gruesa
i macinato *m* grosso
n grof maalsel *n*
d Grobschrot *n*

1253 COAT, COATING dec
f couche *f* de peinture, main *f*
e capa *f* de pintura, mano *f*
i mano *f*, strato *m* di pittura
n verflaag
d Anstrich *m*, Farbschicht *f*

1254 COATED PAPER pap
f papier *m* couvert de kaolin
e papel *m* revestido de caolín
i carta *f* ricoperta con caolino
n met kaolin bedekt papier *n*
d kaolinbeschichtetes Papier *n*

1255 COATED PAPER pap,pla
f papier *m* couvert de résine synthétique
e papel *m* revestido de resina sintética
i carta *f* ricoperta con resina sintetica
n met kunstharslaag bedekt papier
d kunstharzbeschichtetes Papier *n*

1256 COATED PIGMENT, TREATED PIGMENT dec
f pigment *m* enrobé
e pigmento *m* revestido
i pigmento *m* rivestito
n omhuld pigment *n*, nabehandeld pigment *n*
d Hüllenpigment *n*, Kernpigment *n*

1257 COATING coa
f recouvrement *m*, revêtement *m*
e recubrimiento *m*, revestimiento *m*
i ricoprimento *m*, rivestimento *m*
n bedekken *n*
d Beschichten *n*

1258 COATING, BLACKING, BLACKWASH, FACING fou
f enduit *m*
e unto *m* de revestimiento
i intonaco *m*
n vormdekmiddel *n*
d Schlichte *f*, Schwärze *f*

1259 COATING, CASTING pho
f étendage *m* d'une émulsion

e emulsionamiento *m*
i emulsionatura *f*
n opbrengen *n* van een emulsie
d Auftragen *n* einer Emulsion

1260 COATING COMPOUND, SPREADING MIXTURE coa
f mélange *m* d'enduisage
e mezcla *f* de recubrimiento
i mescolanza *f* per ricoperto
n strijkmengsel *n*
d Streichmischung *f*, Überzugsmischung *f*

1261 COATING PAPER pap
f papier *m* non rogné
e papel *m* no talado
i carta *f* non tagliata
n onbesnoeid papier *n*
d unbeschnittenes Papier *n*

1262 COATING VARNISH dec
f vernis *m* de revêtement
e barniz *m* de revestimiento
i vernice *f* di rivestimento
n blanke deklak
d farbloser Überzugslack *m*

1263 COBALT BLUE, HUNGARY BLUE, PARIS BLUE dec
f bleu *m* de cobalt, bleu *m* de Paris
e azul *m* de cobalto, azul *m* de París
i blu *m* di cobalto, blu *m* di Parigi
n Berlijns blauw *n*, kobaltblauw *n*, Parijs blauw *n*
d Berliner Blau *n*, Kobaltblau *n*, Königsblau *n*, Wiener Blau *n*

1264 COBALT DRIER dec
f siccatif *m* au cobalt
e desecante *m* al cobalto
i essiccante *m* al cobalto
n kobaltsiccatief *n*
d Kobalttrockenmittel *n*

1265 COBALT GREEN, GREEN SMALT dec
f vert *m* de cobalt
e verde *m* de cobalto
i verde *m* di cobalto
n kobaltgroen *n*
d Kobaltgrün *n*, Zinkgrün *n*

1266 COBALT YELLOW dec
f jaune *m* de cobalt
e amarillo *m* de cobalto
i giallo *m* di cobalto
n kobaltgeel *n*
d Kobaltgelb *n*

1267 COBBLING tex
f reteinture *f*
e reteñido *m*
i ritintura *f*
n naverven *n*
d Nachfärben *n*

1268 COBWEBBING dec
f rayonnage *m*
e descolgamiento *m*
i filamentosità *f*
n draden trekken *n*, spinnewebvorming
d Fadenziehen *n*, Netzbildung *f*

1269 COCONDENSATION pre
f condensation *f* mixte
e condensación *f* mezclada
i condensazione *f* per mescolanza
n mengcondensatie
d Mischkondensation *f*

1270 COCONUT FIBRE (GB), COIR (US) tex
f bastin *m*, coir *m*, fibre *f* de coco
e fibra *f* de coco
i fibra *f* di cocco
n kokosvezel
d Kokosbast *m*, Kokosfaser *f*

1271 COCONUT OIL fat
f huile *f* de coco
e aceite *m* de coco
i olio *m* di cocco
n kokosolie
d Kokosfett *n*, Kokosnussöl *n*

1272 COCONUT OIL SOAP, SALTWATER SOAP sao
f savon *m* à l'huile de coco
e jabón *m* de coco, jabón *m* para agua de mar
i sapone *m* di cocco
n kokoszeep, zeewaterzeep
d Salzwasserseife *f*

1273 COD LIVER OIL fat,mcy
f huile *f* de foie de morue
e aceite *m* de hígado de bacalao
i olio *m* di fegato di merluzzo
n levertraan
d Lebertran *m*

1274 COHESION gen
f cohésion *f*
e cohesión *f*
i coesione *f*
n cohesie
d Kohäsion *f*

1275 COHESION PRESSURE gen
f pression *f* de cohésion
e presión *f* de cohesión
i pressione *f* di coesione
n cohesiedruk
d Kohäsionsdruck *m*

1276 COHESIVE FORCE gen
f force *f* de cohésion
e fuerza *f* de cohesión
i forza *f* di coesione
n cohesiekracht
d Kohäsionskraft *f*

1277 COHOBATION dis
f cohobation *f*
e destilación *f* repetida
i coobazione *f*
n herhaalde destillatie
d mehrmalige Destillation *f*

1278 COIL COATING dec
f laquage *m* de rubans métalliques
e laqueado *m* de hojas metálicas
i laccatura *f* di nastri metallici
n lakken *n* van bandmetaal
d Metallbandlackierung *f*

1279 COKE OVEN GAS gas
f gaz *m* de cokerie, gaz *m* de four à coke
e gas *m* de coquería, gas *m* de horno de coque
i gas *m* di cokeria, gas *m* di forno a coke
n cokesovengas *n*
d Kokereigas *n*, Koksofengas *n*

1280 COKE OVEN GAS GENERATION gas
f génération *f* de gaz de cokerie, génération *f* de gaz de four à coke
e generación *f* de gas de coquería, generación *f* de gas de horno de coque
i generazione *f* di gas di cokeria, generazione *f* di gas di forno a coke
n cokesovengasontwikkeling
d Kokereigasentwicklung *f*, Koksofengasentwicklung *f*

1281 COKEY pla
f trempé incomplèment adj
e insuficiamente templado adj
i non completamente temprato adj
n niet-volledig gehard adj
d nichtdurchgehärtet adj

1282 COLATURE sep
f liquide *m* filtré
e líquido *m* filtrado
i liquido *m* filtrato
n gefiltreerde vloeistof
d durchgeseihte Flüssigkeit *f*

1283 COLCOTHAR, RED IRON OXIDE dec
f colcotar *m*
e colcótar *m*
i colcotar *m*
n colcothar, dodekop *n*
d Colcothar *m*, Eisenoxydrot *n*, Totenkopf *m*

1284 COLD BLEACHING ble
f blanchiment *m* à froid
e blanqueo *m* en frío
i candeggio *m* a freddo
n koudbleken *n*
d Kaltbleichen *n*

1285 COLD BOND cer
f liant *m* à base d'argile colloïdale
e aglutinante *m* a base de arcilla coloidal
i agente *m* legante a base d'argilla colloidale
n colloïdaal bindmiddel *n* op kleibasis
d kolloidales tonhaltiges Bindemittel *n*

1286 COLD BREAK bre
f cassure *f* à froid
e coágulo *m* en frío
i coagulo *m* a freddo
n koude breuk
d Kaltbruch *m*

1287 COLD CURE, COLD VULCANIZATION rub
f vulcanisation *f* à froid
e vulcanización *f* en frío
i vulcanizzazione *f* a freddo
n koudvulcanisatie
d Kaltvulkanisation *f*

1288 COLD CUT VARNISH dec
f vernis *m* détrempé à froid
e barniz *m* desleido en frío
i vernice *f* preparata a freddo
n koud aangemaakt vernis *n*
d kalt vermischter Lack *m*

1289 COLD ENAMELLING coa
f émaillage *m* à froid
e esmaltado *m* en frío
i smaltatura *f* a freddo

n koudemailleren *n*
d Kaltemaillierung *f*, Tauchemaillierung *f*

1290 COLD EXTRACTION ext
f extraction *f* avec des solvants refroidis
e extracción *f* con solventes enfriados
i estrazione *f* con solventi raffreddati
n koudextraheren *n*
d Kaltextrahierung *f*

1291 COLD FLOW gen
f fluage *m* à froid
e escurrimiento *m* en frío
i scorrimento *m* a freddo
n koudvloeien *n*
d kalter Fluss *m*

1292 COLD FLOW TESTING tes
f essai *m* d'écoulement à froid
e ensayo *m* de escurrimiento en frío
i prova *f* di scorrimento a freddo
n koudevloeiproef
d Prüfung *f* für kalten Fluss

1293 COLD GALVANIZING elc
f zingage *m* à froid
e cincado *m* en frío
i zincatura *f* a freddo
n koudverzinken *n*
d Kaltverzinkung *f*

1294 COLD MASHING bre
f empâtage *m* à froid
e infusión *f* de malta en frío
i estrazione *f* a freddo
n koud beslag *n*
d kalter Satz *m*

1295 COLD MOLDING (US), COLD MOULDING (GB) pla,pre
f moulage *m* à froid
e moldeo *m* en frío
i modellatura *f* a freddo, stampatura *f* a freddo
n koudpersen *n*
d Kaltpressen *n*

1296 COLD PROCESS soa
f saponification *f* à froid
e saponificación *f* en frío
i saponificazione *f* a freddo
n koude verzeping
d Verseifung *f* auf kaltem Wege

1297 COLD SETTING pla
f durcissant adj à froid
e endureciendo adj en frío
i indurente adj a freddo
n koudhardend adj
d kalthärtend adj

1298 COLD SETTING ADHESIVE, COLD SETTING GLUE glu
f colle *f* à froid
e cola *f* fría
i colla *f* con indurimento a freddo
n koudlijm
d kalthärtender Leim *m*

1299 COLD SETTING LACQUER dec
f vernis *m* durcissant à l'air
e barniz *m* endureciendo al aire
i vernice *f* indurente all'aria
n aan de lucht hardende lak
d an der Luft härtender Lack *m*

1300 COLD SETTING RESIN dec
f résine *f* durcissante à froid
e resina *f* endurecienda en frío
i resina *f* indurente a freddo
n koudhardende hars
d kalthärtender Harz *m*

1301 COLD SETTLING PROCESS fue
f déparaffinage *m* par sédimentation à froid
e desparafinaje *m* per sedimentación en frío
i rimozione *f* di paraffina mediante sedimentazione a freddo
n ontparaffinering door middel van koud neerslaan
d Entparaffinierung *f* mittels kalter Sedimentation

1302 COLD SLUDGE bre
f trouble *m* fin
e turbio *m* por frío
i torbido *m* per freddo
n koude draf
d Kältetrub *m*, Kühltrub *m*

1303 COLD WELDING wel
f soudage *m* à froid
e soldadura *f* en frío
i saldatura *f* a freddo
n koudwellen *n*
d Kaltschweissung *f*

1304 COLIC ROOT OIL fat
f essence *f* de galanga
e esencia *f* de galanga
i olio *m* di galanga
n galangaolie, galantolie
d Galgantöl *n*

1305 COLLARGOL col,met
f argent *m* colloïdal
e plata *f* coloidal
i argento *m* colloidale
n colloïdaal zilver *n*
d kolloidales Silber *n*

1306 COLLINGITE (US), PHOSGENE (GB) gas
f phosgène *m*
e fosgeno *m*
i fosgeno *m*
n fosgeen *n*
d Karbonylchlorid *n*, Phosgen *n*

1307 COLLODION mcy
f coton *m* de collodion
e algodón *m* de colodión
i cotone *m* di collodio
n collodiumkatoen *n*
d Kollodiumwolle *f*, Lackwolle *f*

1308 COLLOID col
f colloïde *m*
e coloide *m*
i colloide *m*
n colloïde *n*
d Kolloid *n*

1309 COLLOID EQUIVALENT col
f équivalent *m* colloïdal
e equivalente *m* coloidal
i equivalente *m* colloidale
n colloïde-equivalent *n*
d Äquivalentaggregat *n*, Elektroäquivalent *n* der Mizellionen, Kolloidäquivalent *n*

1310 COLLOIDAL DYE dye
f colorant *m* colloïdal
e colorante *m* coloidal
i colorante *m* colloidale
n colloïdale kleurstof
d kolloidaler Farbstoff *m*

1311 COLLOIDAL ELECTROLYTE elc
f électrolyte *f* colloïdale
e electrólito *m* coloidal
i elettrolito *m* colloidale
n colloïdale elektrolyt
d Kolloidalelektrolyt *m*

1312 COLLOIDAL FUEL col,fue
f combustible *m* colloïdal
e combustible *m* coloidal
i combustibile *m* colloidale
n stookolie-steenkoolsuspensie
d Fliesskohle *f*, Kolloidkohle *f*

1313 COLLOIDAL SOLUTION, COLLOIDAL SUSPENSION, PSEUDOSOLUTION col
f solution *f* colloïdale, suspension *f* colloïdale
e solución *f* coloidal, suspensión *f* coloidal
i soluzione *f* colloidale, sospensione *f* colloidale
n colloïdale oplossing, colloidale suspensie
d kolloidale Lösung *f*, Kolloidsuspension *f*

1314 COLLOIDAL STATE col
f état *m* colloïdal
e estado *m* coloidal
i stato *m* colloidale
n colloïdale toestand
d Kolloidzustand *m*

1315 COLOR AGGLUTINANT (US), COLOUR AGGLUTINANT (GB) dec
f liant *m* de peinture
e aglutinante *m* de pintura
i legamento *m* di pittura
n verfbindmiddel *n*
d Farbenbindemittel *n*

1316 COLOR DEVELOPER (US), COLOUR DEVELOPER (GB) pho
f révélateur *m* chromogène
e revelador *m* cromógeno
i sviluppatore *m* cromogeno
n kleurenontwikkelaar
d Farbentwickler *m*

1317 COLOR FASTNESS (US), COLOUR FASTNESS (GB) tex
f grand teint *m*, solidité *f* de la couleur
e permanencia *f* del color, solidez *f* del color
i stabilità *f* del colore
n kleurbestendigheid
d Farbbeständigkeit *f*, Farbechtheit *f*

1318 COLOR FILM (US), COLOUR FILM (GB) pho
f film *m* en couleurs, pellicule *f* en couleurs
e film *m* en colores, película *f* en colores
i film *m* a colori, pellicola *f* a colori
n kleurenfilm
d Farbfilm *m*

1319 COLOR LAKE (US), dec
COLOUR LAKE (GB)
f laque *f* colorée
e laca *f* coloreada
i lacca *f* colorata
n kleurlak
d Farblack *m*

1320 COLOR PRINTING (US), pho
COLOUR PRINTING (GB)
f chromotypographie *f*, impression *f* en couleurs
e cromotipia *f*, impresión *f* en colores
i stampa *f* a colori
n chromotypie, kleurendruk
d Buntdruck *m*, Chromotypie *f*, Farbendruck *m*

1321 COLOR SEPARATION (US), pho
COLOUR SEPARATION (GB)
f sélection *f* des couleurs
e selección *f* de los colores
i selezione *f* dei colori
n kleurselectie
d Farbauszug *m*

1322 COLORANT dye
f colorant *m*, matière *f* colorante
e colorante *m*, materia *f* colorante
i colorante *m*, materia *f* colorante
n kleurstof
d Farbstoff *m*

1323 COLORIMETRY ana
f colorimétrie *f*
e colorimetría *f*
i colorimetria *f*
n colorimetrie
d Farbmessung *f*, Kolorimetrie *f*

1324 COLORING PIGMENT (US), dec
COLOURING PIGMENT (GB), STAINER
f pigment *m* colorant
e pigmento *m* de color
i pigmento *m* colorante
n pigmentkleurstof
d Farbpigment *n*

1325 COLUMN CHROMATOGRAPHY ana
f chromatographie *f* sur colonne
e cromatografía *f* sobre columna
i cromatografia *f* a colonna
n zuilenchromatographie
d Säulenchromatographie *f*

1326 COLZA OIL, fat
RAPE SEED OIL
f huile *f* de colza
e aceite *m* de colza
i olio *m* di colza
n raapolie
d Kohlsaatöl *n*, Rapöl *n*, Rüböl *n*

1327 COMBINATION TANNING, tan
COMPOUND TANNING
f tannage *m* mixte
e curtimiento *m* mixto
i concia *f* mista
n gemengde looiing
d gemischte Gerbung *f*

1328 COMBINED BATH, pho
TONING AND FIXING BATH
f bain *m* viro-fixateur
e baño *m* de viraje y fijación
i bagno *m* di virofissaggio
n kleur-fixeerbad *n*
d Tonfixierbad *n*

1329 COMBINED CARBON gen
f carbone *m* combiné
e carbono *m* combinado
i carbonio *m* combinato
n gebonden koolstof
d gebundener Kohlenstoff *m*

1330 COMBINED SULPHUR (GB), rub
BOUND SULFUR (US)
f soufre *m* combiné
e azufre *m* combinado
i zolfo *m* combinato
n gebonden zwavel
d gebundener Schwefel *m*

1331 COMBUSTION, gen
BURNING
f combustion *f*
e combustión *f*
i combustione *f*
n verbranding
d Verbrennung *f*

1332 COMMINUTION, pap
FINE CRUSHING, FINE TRITURATION
f broyage *m* fin, dilacération *f*
e trituración *f* fina
i frantumazione *f* fina, triturazione *f* fina
n fijnmalen *n*
d Feinzerkleinerung *f*

1333 COMPACT GYPSUM, min
ALABASTER
f albâtre *m*
e alabastro *m*
i alabastro *m*
n albast *n*, gipsalbast *n*
d Alabaster *m*, dichter Gips *m*

1334 COMPACT SLUDGE iwt,sew
f boue *f* épaisse
e fango *m* compacto
i fango *m* compatto
n compact slib *n*
d stichfester Schlamm *m*

1335 COMPACTING pre
f compression *f*
e compactación *f*, compresión *f*
i compressione *f*
n comprimeren *n*
d Komprimieren *n*

1336 COMPENSATING DEVELOPER pho
f révélateur *m* compensateur
e revelador *m* compensador
i sviluppatore *m* compensatore
n compenserende ontwikkelaar
d Ausgleichsentwickler *m*

1337 COMPLETE ARTIFICIAL MANURE fer
COMPOUND ARTIFICIAL MANURE
f engrais *m* minéral complet
e abono *m* mineral completo
i concime *m* chimico completo
n complete kunstmest, mengmeststof
d künstlicher Volldünger *m*

1338 COMPLETE COMBUSTION, PERFECT COMBUSTION fue
f combustion *f* complète
e combustión *f* completa
i combustione *f* completa
n volledige verbranding
d vollständige Verbrennung *f*

1339 COMPLETE GELATION col
f gélification *f* intégrale
e gelificación *f* completa
i gelizzazione *f* completa
n volledige gelering
d Durchgelierung *f*

1340 COMPLETE REACTION, IRREVERSIBLE REACTION gen
f réaction *f* complète, réaction *f* irréversible
e reacción *f* completa, reacción *f* irreversible
i reazione *f* completa, reazione *f* irreversibile
n niet-omkeerbare reactie, volledige reactie
d nicht umkehrbare Reaktion *f*, vollständige Reaktion *f*

1341 COMPLEXING ACTION, CHELATE FORMATION, CHELATION, SEQUESTRATION oic
f action *f* complexante
e acción *f* quelacante, facultad *f* de formar complejos
i azione *f* complessante, azione *f* sequestrante, chelazione *f*
n chelateren *n*, complexvorming, klauwkoppeling, klauwverbinding
d Chelatbildung *f*, Chelierung *f*, Komplexbildung *f*

1342 COMPLEXING AGENT, CHELATING AGENT, SEQUESTERING AGENT, SEQUESTRANT gen
f agent *m* séquestrant complexant, chélatant *m*
e agente *m* secuestrador reactivo, agente *m* secuestrante, formador *m* de complejos
i agente *m* chelatante, agente *m* sequestrante
n complexvormer
d chelatbildendes Reagenz *n*, Komplexbildner *m*

1343 COMPONENTS gen
f composants *pl*
e componentes *pl*
i componenti *pl*
n bestanddelen *pl*, componenten *pl*
d Bestandteile *pl*

1344 COMPOSITION gen
f composition *f*, constitution *f*
e composición *f*
i composizione *f*
n samenstelling
d Zusammensetzung *f*, Zusammenstellung *f*

1345 COMPOUND ETHER, ESTER, ETHERAL SALT oic
f ester *m*, éther *m* composé
e éster *m*
i estere *m*
n ester
d Ester *m*

1346 COMPRESSED YEAST bre,fer
f levure *f* pressée
e levadura *f* prensada
i lievito *m* pressato
n persgist
d Presshefe *f*

1347 COMPRESSION IGNITION fue
f allumage *m* par compression
e ignición *f* por compresión
i accensione *f* per combustione
n zelfontsteking
d Eigenzündung *f*, Selbstzündung *f*

1348 COMPRESSION MOLDING (US), pla,pre,rub
COMPRESSION MOULDING (GB)
f moulage *m* sous compression
e moldeo *m* baja compresión
i stampa *f* sotto pressione
n vormpressen *n*
d Formpressen *n*

1349 COMPRESSION RATIO, fue
RATIO OF COMPRESSION
f rapport *m* de compression
e relación *f* de compresión
i rapporto *m* di compressione
n compressieverhouding
d Verdichtungsverhältnis *n*

1350 COMPRESSION REFRIGERATION cle
f réfrigération *f* par compression
e refrigeración *f* por compresión
i refrigerazione *f* a compressione
n compressiekoeling
d Kompressionskühlung *f*

1351 CONCENTRATE min
f substance *f* concentrée
e concentrado *m*
i concentrato *m*, sostanza *f* concentrata
n concentraat *n*
d Konzentrat *n*

1352 CONCENTRATED DYE LIQUOR dye
f liquide *m* colorant concentré
e líquido *m* colorante concentrado
i liquido *m* colorante concentrato
n geconcentreerde verfvloeistof
d kurze Flotte *f*

1353 CONCENTRATED NITRIC ACID, dec
AQUA FORTIS
f eau *f* forte
e agua *f* fuerte
i acquaforte *f*
n aqua fortis, ets, sterk water *n*
d Ätzwasser *n*, konzentrierte Salpetersäure *f*, Scheidewasser *n*

1354 CONCENTRATION gen
f concentration *f*
e concentración *f*
i concentrazione *f*
n concentratie
d Entwasserung *f*, Konzentration *f*

1355 CONCENTRATION, bre,gen
BOILING DOWN
f épaississement *m*, réduction *f*
e concentración *f*
i concentrazione *f*
n indikking, inkoken *n*
d Eindickung *f*, Einkochen *n*

1356 CONCENTRATION BY FLOTATION met,min
f concentration *f* par flottation
e concentración *f* por flotación
i concentrazione *f* per flottazione
n natte bewerking
d Nassaufbereitung *f*, Schwimmaufbereitung *f*

1357 CONCRETE bui
f béton *m*
e hormigón *m*
i calcestruzzo *m*
n beton *n*
d Beton *m*

1358 CONDENSATE dis,gen
f liquide *m* condensé
e líquido *m* condensado
i liquido *m* condensato
n condensaat *n*
d Kondensat *n*

1359 CONDENSATION gen
f condensation *f*
e condensación *f*
i condensazione *f*
n condensatie
d Kondensation *f*, Kondensierung *f*, Verdichtung *f*

1360 CONDENSATION POLYMERIZATION, pla
POLYCONDENSATION
f polymérisation *f* par condensation
e policondensación *f*
i polimerizzazione *f* per condensazione
n polycondensatie
d Polykondensation *f*

1361 CONDENSATION RESINS pla
f résines *pl* de condensation
e resinas *pl* de condensación
i resine *pl* di condensazione
n condensatieharsen *pl*
d Kondensationsharze *pl*

1362 CONDENSATION WATER, CONDENSED WATER — boi
f eau *f* condensée, eau *f* de condensation
e agua *f* condensada, agua *f* de condensación
i acqua *f* condensata, acqua *f* di condensazione
n condenswater *n*
d Kondenswasser *n*

1363 CONDENSED MILK — dai
f lait *m* condensé
e leche *f* condensada
i latte *m* condensato
n gecondenseerde melk
d kondensierte Milch *f*, Kondensmilch *f*

1364 CONDENSED SURFACE FILM — eva
f pellicule *f* superficielle condensée
e película *f* superficial condensada
i pellicola *f* superficiale condensata
n gecondenseerde oppervlaklaag
d kondensierte Oberflächenschicht *f*

1365 CONDENSING AGENT — gen
f agent *m* de condensation
e agente *m* condensador
i agente *m* condensatore
n condensatiemiddel *n*
d Kondensationsmittel *n*, Kondensiermittel *n*

1366 CONDITIONING — gla
f réglage *m* de la température
e regulación *f* de la temperatura
i regolazione *f* della temperatura
n temperatuurregeling
d Temperaturregelung *f*

1367 CONDITIONING — tex
f conditionnement *m*
e acondicionamiento *m*
i condizionatura *f*
n conditioneren *n*
d Konditionieren *n*, Prüfen *n*

1368 CONDUCTIMETRIC ANALYSIS, CONDUCTOMETRIC ANALYSIS — cond
f analyse *f* conductométrique, conductométrie *f*
e análisis *f* conductimétrica, conductometría *f*
i analisi *f* conduttimetrica, conduttometria *f*
n conductometrie, conductometrische analyse
d Konduktometrie *f*, konduktometrische Massanalyse

1369 CONDUCTIVITY WATER — mea
f eau *f* très pure pour mesures de conductivité
e agua *f* muy pura para medidas de conductividad
i acqua *f* purissima per misurazione di conduttività
n zeer zuiver water *n* voor geleidendheidsmetingen
d sehr reines Wasser *n* für Leitfähigkeitsmessungen

1370 CONFECTIONER'S SUGAR, ICING SUGAR — nut
f sucre *m* en poudre
e azúcar *m* en polvo
i zucchero *m* in polvere
n poedersuiker
d Puderzucker *m*, Staubzucker *m*

1371 CONFIGURATION, CONFORMATION — gen
f configuration *f*
e configuración *f*
i configurazione *f*
n chemische structuur, configuratie
d chemische Struktur *f*, Konfiguration *f*

1372 CONFINED WATER — sew
f eau *f* captive, nappe *f* captive
e agua *f* artesiana
i acqua *f* sotterranea artesiana
n artesisch water *n*
d gespanntes Grundwasser *n*

1373 CONGO PAPER, CONGO RED PAPER — ana
f papier *m* impregné avec une solution de rouge Congo
e papel *m* impregnado con solución de rojo Congo
i carta *f* impregnata con una soluzione di rosso Congo
n kongopapier *n*
d Herzberg's Papier *n*, Kongopapier *n*, Riegel's Papier *n*

1374 CONGO RED — dec
f rouge *m* Congo
e rojo *m* Congo
i rosso *m* Congo

n kongorood *n*
d Kongorot *n*

1375 CONGRUENT MELTING gen
f fusion *f* congruente
e fusión *f* congruente
i fusione *f* congruente
n congruent smelten *n*
d kongruentes Schmelzen *n*

1376 CONING AND QUARTERING TESTING tes
f inquartation *f*, quartation *f*
e cuarteamiento *m*, cuarteo *m*
i riduzione *f* del campione col metodo dei quarti
n kwartering
d Quartierung *f*

1377 CONJUGATE SOLUTION gen
f solution *f* conjugée
e solución *f* conjugada
i soluzione *f* coniugata
n geconjugeerde oplossing
d konjugierte Lösung *f*

1378 CONSECUTIVE REACTIONS gen
f réactions *pl* consécutives
e reacciones *pl* consecutivas
i reazioni *pl* consecutive
n opeenvolgende reacties *pl*
d aufeinander folgende Reaktionen *pl*

1379 CONSERVANCY, CONSERVATION hyg
f conservation *f*
e conservación *f*
i protezione *f* dall'inquinamento
n vermijding van waterverontreiniging
d Reinhaltung *f* der Gewässer

1380 CONSISTENCE, CONSISTENCY fat
f consistance *f*
e consistencia *f*
i consistenza *f*
n consistentie, lijvigheid
d Dichte *f*, Dichtigkeit *f*, Dicke *f*, Konsistenz *f*

1381 CONSISTENCY TEST tes
f essai *m* de consistance
e prueba *f* de consistencia
i collaudo *m* di consistenza
n consistentieproef
d Konsistenzprobe *f*

1382 CONSOLUTE mix
f complètement miscible adj
e consoluto adj
i completamente mescolabile adj
n volledig mengbaar adj
d vollständig mischbar ajd

1383 CONSTANT BOILING MIXTURE, AZEOTROPIC MIXTURE gen
f mélange *m* azéotropique
e mezcla *f* azeotrópica
i miscela *f* azeotropica
n azeotroop mengsel *n*, azeotropisch mengsel *n*
d azeotropes Gemisch *n*

1384 CONSTANT MOLAL REFLUX dis
f reflux *m* molaire constant
e reflujo *m* molar constante
i riflusso *m* molare costante
n constante molaire terugstroom
d konstanter Rücklauf *m* in Molen

1385 CONSTITUTIONAL FORMULA, GRAPHIC FORMULA, STRUCTURAL FORMULA gen
f formule *f* de constitution
e fórmula *f* de constitución, fórmula *f* estructural
i formula *f* di struttura
n structuurformule
d Konstitutionsformel *f*, Strukturformel *f*

1386 CONSTITUTIONAL WATER, WATER OF CRYSTALLIZATION cry
f eau *f* de cristallisation
e agua *f* de cristalización
i acqua *f* di cristallizzazione
n kristalwater *n*
d Konstitutionswasser *n*, Kristallwasser *n*

1387 CONSUMED OXYGEN, OXYGEN CONSUMED FROM PERMANGANATE iwt
f demande *f* en permanganate de potassium, oxydabilité *f* au permanganate
e consumo *m* de permanganato
i consumo *m* di permanganato
n permanganaatverbruik *n*
d Kaliumpermanganatverbrauch *m*, Permanganatverbrauch *m*

1388 CONTACT ADHESIVE, IMPACT ADHESIVE glu
f adhésif *m* à action directe
e adhesivo *m* de acción directa
i adesivo *m* ad azione diretta
n direct hechtende kleefstof
d direkt haftendes Klebemittel *n*

1389 CONTACT AERATION iwt
f aération *f* par contact
e aireación *f* por contacto
i aerazione *f* per contatto
n contactaëratie
d Kontaktbelüftung *f*

1390 CONTACT COOLING clo
f refroidissement *m* direct
e enfriamiento *m* directo
i raffreddamento *m* diretto
n directe koeling
d direkte Kühlung *f*

1391 CONTACT INSECTICIDE agr
f insecticide *m* par contact
e insecticida *m* por contacto
i insetticida *m* per contatto
n contactgif *n*, contactinsekticide *n*
d Berührungsinsektizid *n*, Kontaktgift *n*

1392 CONTACT LAMINATING, LOW PRESSURE LAMINATING pla
f laminage *m* à basse pression
e formación *f* de laminados a baja presión
i laminazione *f* a bassa pressione
n lamineren *n* bij lage druk
d Niederdruckschichtstoffherstel-lung *f*

1393 CONTACT MATERIAL OF A CONTACT BED sew
f matière *f* du lit de contact
e materia *f* del lecho de contacto
i materiale *m* del letto di contatto
n contactbedmateriaal *n*
d Füllkörpermaterial *n*

1394 CONTACT PLATING coa
f dépôt *m* par contact
e depósito *m* por contacto
i deposito *m* per contatto
n contactbedekking
d Kontaktplattierung *f*

1395 CONTACT PRINT pho
f copie *f* par contact
e prueba *f* por contacto
i stampa *f* ottenuta per contatto con negativa
n kontaktafdruk
d Kontaktkopie *f*

1396 CONTACT PROCESS sew
f procédé *m* par contact
e procedimiento *m* de contacto
i processo *m* per contatto
n contactproces *n*
d Füllverfahren *n*

1397 CONTACT PROCESS, CHEMICO CONTACT PROCESS sua
f procédé *m* de contact
e procedimiento *m* de contacto
i processo *m* di contatto
n contactproces *n*
d Kontaktverfahren *n*

1398 CONTACT PROCESS, REACTION GROUND COAT PROCESS dec
f procédé *m* de contact
e procedimiento *m* de mano primera reactiva
i processo *m* di mano prima reattiva
n contactgrondproces *n*, reactieve-grondlaagmethode
d Aktivgrundverfahren *n*, Reaktionsgrundverfahren *n*, Zweistufenlackierverfahren *n*

1399 CONTACT RESINS pla
f résines *pl* pour le procédé à basse pression
e resinas *pl* para el procedimiento a baja presión
i resine *pl* per il processo a bassa pressione
n lage-drukharsen *pl*
d Kontaktpressharze *pl*, Niederdruckpressharze *pl*

1400 CONTACT SERIES, ELECTROCHEMICAL SERIES, ELECTROMOTIVE SERIES elc
f série *f* électrochimique
e serie *f* electroquímica
i serie *f* elettrochimica
n spanningsreeks
d elektrochemische Spannungs-reihe *f*

1401 CONTAMINATION hyg,iwt
f contamination *f*, infection *f*
e contaminación *f*, enviciamiento *m*
i contaminazione *f*, infezione *f*
n contaminatie, vervuiling
d Verseuchung *f*, Verunreinigung *f*

1402 CONTAMINATION nue
f contamination *f*
e contaminación *f*
i contaminazione *f*
n contaminatie
d Kontamination *f*

1403 CONTENT gen
f teneur
e contenido *m*

i contenuto *m*
n gehalte *n*
d Gehalt *m*

1404 CONTINUOUS BELT MIXING mix
f mélange *m* à bande continue
e mezclado *m* de cinta continua
i mescolanza *f* a nastro continuo
n eindloze-bandmenging
d kontinuierliche Bandmischung *f*

1405 CONTINUOUS DEVELOPMENT pho
f développement *m* en continu
e revelado *m* continuo
i sviluppo *m* continuo
n continue ontwikkeling
d Durchlauf *m*, Durchlaufentwicklung *f*

1406 CONTINUOUS DIFFUSION sin
f diffusion *f* continue
e difusión *f* continua
i diffusione *f* continua
n continue diffusie
d kontinuierliche Diffusion *f*

1407 CONTINUOUS DISTILLING dis
f distillation *f* continue
e destilación *f* continua
i distillazione *f* continua
n continue destillatie
d kontinuierliche Destillation *f*

1408 CONTINUOUS DYEING dye
f teinture *f* continue
e teñido *m* continuo
i tintura *f* continua
n continu verven *n*
d kontinuierliches Färben *n*

1409 CONTINUOUS EXTRACTING ext
f extraction *f* continue
e extracción *f* continua
i estrazione *f* continua
n continue extractie
d kontinuierliche Extraktion *f*

1410 CONTINUOUS FILAMENT tex
f fil *m* de filature continu
e hilo *m* de hilado continuo
i filo *m* di filatura continuo
n eindloze gesponnen draad
d endloses Gespinst *n*

1411 CONTINUOUS FILTERING, PERCOLATION iwt
f percolation *f*
e percolación *f*
i filtrazione *f* continua
n druppelfiltratie, percoleren *n*
d Perkolation *f*, Tropffiltrierung *f*

1412 CONTINUOUS PRINTING pho
f tirage *m* continu
e tiraje *m* de copias continuo
i stampa *f* continua
n continu kopiëren *n*
d Durchlaufkopieren *n*

1413 CONTINUOUS PROCESS gen
f procédé *m* continu
e procedimiento *m* continuo
i processo *m* continuo
n continu proces *n*
d kontinuierliches Verfahren *n*

1414 CONTINUOUS SHEETING COAGULATION rub
f coagulation *f* en ruban continu
e coagulación *f* en cinta continua
i coagulazione *f* in nastro continuo
n lintcoagulatie
d Bandkoagulation *f*

1415 CONTRACTION, SHRINKAGE pla
f retrait *m*
e contracción *f*, retracción *f*
i ritiro *m*
n krimpen *n*
d Schrumpfen *n*, Schwinden *n*, Schwund *m*

1416 CONTRACTION ALLOWANCE, MOLDING ALLOWANCE (US), MOULDING ALLOWANCE (GB) pla
f double retrait *m*
e doble contracción *f*
i doppio ritiro *m*
n kruiptoegift
d doppeltes Schwindmass *n*

1417 CONTRAST DEVELOPER, CONTRAST INTENSIFIER pho
f renforçateur *m* de contraste
e reforzador *m* de contraste
i rinforzatore *m* di contrasto
n contrastversterker
d Kontrastverstärker *m*

1418 CONTROLLED-BURNING DUMPING iwt
f décharge *f* de boue à combustion contrôlée
e descarga *f* de fango de combustión controlada
i scarico *m* di fango a combustione controllata

n vuilstorting met gecontroleerde verbranding
d Ablagerung *f* mit geordneter Verbrennung

1419 CONTROLLED-SUDSING DETERGENT, LOW FOAMING DETERGENT, RESTRICTED-SUDSING DETERGENT — cle
f détergent *m* peu moussant
e detergente *m* poco espumante
i detergente *m* poco schiumogeno
n weinig schuimend wasmiddel *n*
d wenig schäumendes Waschmittel *n*

1420 CONVERSION — gen
f conversion *f*
e conversión *f*
i conversione *f*
n omzetting
d Umformung *f*, Umsetzung *f*, Umwandlung *f*

1421 CONVERSION COATING, PASSIVATION — elc
f passivation *f*
e pasivación *f*
i passivazione *f*
n passiveren *n*
d Passivierung *f*

1422 CONVERTING, AIR REFINING — met
f affinage *m* par le vent
e refinación *f* por aire
i affinazione *f* mediante aria
n raffineren *n* in thomasconvector
d Windfrischen *n*

1423 COOKER LIQUOR, COOKING LIQUOR, DIGESTER LIQUOR — pap
f lessive *f* de cuisson
e lejía *f* de cocción
i lisciva *f* di cottura
n kookvloeistof
d Kocherlauge *f*

1424 COOLANT, COOLING AGENT — clo
f réfrigérant *m*
e refrigerante *m*
i refrigerante *m*
n koelmiddel *n*
d Kühlmittel *n*

1425 COOLER SLUDGE — bre
f dépôt *m* de bac
e precipitado *m*, turbio *m*
i torbido *m*
n draf
d Trub *m*

1426 COOLING — clo,gen
f réfrigération *f*, refroidissement *m*
e enfriamiento *m*, refrigeración *f*
i raffreddamento *m*, refrigerazione
n afkoeling, koeling
d Abkühlung *f*, Kühlung *f*

1427 COOLING, ANNEALING — gla
f refroidissement *m*
e enfriamiento *m*
i raffreddamento *m*
n koeling, temperen *n*
d Kühlung *f*, Tempern *n*

1428 COP DYEING — dye
f teinture *f* en cannette
e teñido *m* en canilla, teñido *m* en husada
i tintura *f* in cannetta
n knotverven *n*, spoelverven *n*
d Garnwickelfärben *n*, Knäuelfärben *n*

1429 COPAIBA OIL — dec,mcy,pap
f baume *m* de copahu
e bálsamo *m* de copaiba
i balsamo *m* di copaive
n copaïvabalsem
d Kopaivabalsam *m*

1430 COPOLYMERIZATION — oic
f copolymérisation *f*
e copolimerización *f*, polimerización *f* mixta
i copolimerizzazione *f*
n copolymerisatie
d Mischpolymerisation *f*

1431 COPPER-CLAD LAMINATES — elc,pla
f matières *pl* stratifiées cuivrées
e laminados *pl* chapados de cobre
i materie *pl* laminate rivestite di rame
n met koper beklede gelaagde kunststoffen *pl*
d verkupferte Schichtstoffe *pl*

1432 COPPER INDEX, COPPER NUMBER — gen
f indice *m* de cuivre
e índice *m* de cobre
i indice *m* di rame
n kopergetal *n*
d Kupferzahl *f*

1433 COPPER PLATING coa
f cuivrage *m*
e cobreado *m*
i ramatura *f*
n verkoperen *n*
d Verkupfern *n*

1434 COPPER SPRAY agr
f bouillie *f* bordelaise
e caldo *m* bordelés
i miscela *f* bordolese
n Bordeauxse pap
d Bordelaiser Brühe *f*, Kupferkalkbrühe *f*

1435 COPPER YIELD bre
f rendement *m* de brassage, rendement *m* de chaudière
e rendimiento *m* en sala de cocimiento
i rendimento *m* in sala di cottura
n brouwrendement *n*
d Sudhausausbeute *f*

1436 COPYING pho
f tirage *m*
e tiraje *m*
i stampatura *f*
n afdrukken *n*
d Abziehen *n*

1437 CORAL RED cer,dec
f corallin *m*
e rojo *m* de coralina
i rosso *m* di corallo
n koraalrood *n*
d Korallenrot *n*

1438 CORDITE (GB), AXITE (US) exp
f cordite *f*
e cordita *f*
i cordite *f*
n cordiet *n*
d Kordit *m*

1439 CORE pla
f couche *f* intermédiaire
e capa *f* intermedia
i strato *m* intermedio
n tussenlaag
d Zwischenschicht *f*

1440 CORE SAND fou
f sable *m* à noyaux
e arena *f* para machos
i sabbia *f* per anime, terra *f* per anime
n kernzand *n*
d Kernsand *m*

1441 CORED SOLID SOLUTION met
CORING
f microségrégation *f*
e microsegregación *f*
i microsegregazione *f*
n microsegregatie, microzijging, ontmenging
d Entmischung *f*, Mikroseigerung *f*

1442 CORIANDER OIL fat
f essence *f* de coriandre
e esencia *f* de cilantro
i olio *m* di coriandolo, olio *m* di coriandro
n korianderolie
d Korianderöl *n*

1443 CORK mat
f liège *m*
e corcho *m*
i sughero *m*
n kurk *n*
d Kork *m*

1444 CORK PLATE, CORK SHEET, CORK SLAB mat
f plaque *f* de liège
e plancha *f* de corcho
i lastra *f* di sughero
n kurkplaat
d Korkplatte *f*

1445 CORKRUBBER SHEET rub
f feuille *f* en caoutchouc-liège
e hoja *f* de goma y corcho
i foglia *f* di gomma-sughero
n kurkrubberplaat
d Korkgummiplatte *f*

1446 CORN FLOUR (US), MAIZE FLOUR (GB) nut
f farine *f* de maïs
e harina *f* de maíz
i farina *f* di granturco, farina *f* di mais
n maïsmeel *n*, maizena
d Maismehl *n*, Maizena *n*

1447 CORN OIL (US), MAIZE OIL (GB) fat
f huile *f* de maïs
e aceite *m* de maíz
i olio *m* di granturco, olio *m* di mais
n maïsolie
d Maisöl *n*

1448 CORN STARCH (US), MAIZE STARCH (GB) nut
f amidon *m* de maïs, fécule *f* de maïs

e almidón *m* de maíz, fécula *f* de maíz
i amido *m* di granturco, amido *m* di mais, fecola *f* di granturco, fecola *f* di mais
n maïsstijfsel
d Maisstärke *f*

1449 CORRODING AGENT, CORROSIVE — cor
f agent *m* corrosif
e agente *m* corroedor, corrosivo *m*
i agente *m* corrosivo
n corroderende stof, corrosieve stof
d Korrosionsmittel *n*

1450 CORROSION — cor
f corrosion *f*
e corrosión *f*
i corrosione *f*
n corrosie
d Korrosion *f*

1451 CORROSION FATIGUE — cor,met
f fatigue *f* par corrosion
e fatiga *f* por corrosión
i fatica-corrosione *f*
n corrosievermoeiing
d Korrosionsermüdung *f*

1452 CORROSION INHIBITOR — cor
f anticorrosif *m*
e corrosiófugo *m*, sustancia *f* anticorrosiva
i anticorrosivo *m*
n corrosiewerende stof
d Korrosionsschutzmittel *n*

1453 CORROSIVE SUBLIMATE, MERCURIC CHLORIDE — cor
f sublimé *m* corrosif
e sublimado *m* corrosivo
i sublimato *m* corrosivo
n bijtend sublimaat *n*
d Ätzquecksilber *n*, Ätzsublimat *n*

1454 CORUNDUM — mat
f corindon *m*
e corindón *m*
i corindone *m*
n korund *n*
d Korund *m*

1455 COSMETICS — soa
f cosmétiques *pl*
e cosméticos *pl*
i cosmetici *pl*
n cosmetica *pl*, kosmetische middelen *pl*
d Hautpflegemittel *pl*, kosmetische Mittel *pl*, Schönheitsmittel *pl*

1456 COSSETTES, BEET SLICES, CHIPS — sin
f cossettes *pl*
e cosetas *pl*, rebanadas *pl*
i ritagli *pl* di barbabietole
n bietensnijdsels *pl*
d Rübenschnitzel *pl*

1457 COTTON — mat,tex
f coton *m*
e algodón *m*
i cotone *m*
n katoen *n*
d Baumwolle *f*

1458 COTTON BLEACHING — tex
f blanchiment *m* de coton
e blanqueo *m* del algodón
i decolorazione *f* del cotone
n katoenbleken *n*
d Baumwollbleichung *f*

1459 COTTON FLOCK FILLER — pla
f charge *f* de bourres de coton
e carga *f* de copos de algodón
i carica *f* di fiocchi di cotone
n katoenvlokkenvulsel *n*
d Baumwollflockungfüllstoff *m*

1460 COTTON LINTERS — pap,pla
f duvets *pl* de coton
e borra *f* de algodón, linters *pl* de algodón
i cascami *pl* di cotone
n katoenlinters *pl*
d Linters *pl*

1461 COTTON OIL, COTTON SEED OIL — tex
f huile *f* de coton
e aceite *m* de algodón
i olio *m* di semi di cotone
n katoenolie, katoenzaadolie
d Baumwollsamenöl *n*, Baumwollöl *n*, Kottonöl *n*

1462 COTTON WASTE — cle,tex
f bourre *f* de coton, étoupe *f*
e estopa *f*
i stoppa *f*
n poetskatoen *n*
d Putzwolle *f*

1463 COTTON WOOL — mcy
f ouate *f*
e algodón *m* en rama, algodón *m* hidrófilo, guata *f*

i cotone *m* idrofilo, ovatta *f*
n watten *pl*
d Watte *f*

1464 COTTONIZATION tex
f cotonisation *f*
e algodonización *f*, cotonización *f*
i cotonizzazione *f*
n cotoniseren *n*
d Kotonisieren *n*

1465 COUCH, GROUND COAT, PRIMER COAT dec
f main *f* d'impression, première couche *f*
e capa *f* base, mano *f* de imprimación
i prima mano *f* di colore
n grondlaag
d Grund *m*

1466 COUCHING bre
f mise *f* du malt en couche
e extensión *f* de una capa de malta
i distendimento *m* del malto in strato
n uitspreiden *n* van mout
d Aufschütten *n* des Malzes

1467 COUCHING pap
f couchage *m*
e apilamiento *m*
i impilatura *f*
n koetsen *n*
d Gautschen *n*

1468 COUNTERCURRENT gen
f contre-courant *m*
e contracorriente *f*
i controcorrente *f*
n tegenstroom
d Gegenstrom *m*, Gegenströmung *f*

1469 COUNTERCURRENT JUICE HEATING sin
f préchauffage *m* en contre-courant
e precalefacción *f* en contracorriente
i preriscaldamento *m* in controcorrente
n tegenstroomvoorverwarming
d Gegenstromvorwärmung *f*

1470 COUNTERCURRENT PROCESS gen
f procédé *m* à contre-courant
e procedimiento *m* de contracorriente
i processo *m* a controcorrente
n tegenstroomproces *n*
d Gegenstromverfahren *n*

1471 COURSE OF REACTION gen
f déroulement *m* de la réaction
e curso *m* de la reacción, marcha *f* de la reacción
i progresso *m* della reazione
n reactieverloop *n*
d Reaktionsverlauf *m*

1472 COVER COAT, FINISHING COAT cer,dec
f main *f* de finissage, couche *f* de teinte
e capa *f* de cubierta
i mano *f* a finire
n deklaag
d Deckanstrich *m*

1473 COVERAGE, OBSCURATION, SPREADING CAPACITY dec
f pouvoir *m* couvrant en surface
e extensión *f* específica
i distendimento *m* specifico
n uitstrijkrendement *n*
d Ausgiebigkeit *f*

1474 COVERING POWER, HIDING POWER dec
f opacité *f*, pouvoir *m* couvrant
e capacidad *f* encubridora, opacidad *f*, poder *m* cubridor
i opacità *f*, potere *m* coprente, potere *m* di ricoprimento
n dekkracht, dekvermogen *n*
d Deckfähigkeit *f*, Deckkraft *f*

1475 CRABBING tex
f fixage *m* du tissu
e fijación *f* del tejido
i fissaggio *m* della lucentezza del tessuto
n branden *n*
d Einbrennen *n*, Krabben *n*

1476 CRACKED GAS fue
f gaz *m* craqué
e gas *m* craqueado, gas *m* piezopirolizado
i gas *m* di piroscissione
n kraakgas *n*
d Krackgas *n*

1477 CRACKED GASOLINE fue
f essence *f* craquée
e gasolina *f* craqueada, gasolina *f* piezopirolizada
i gasolina *f* di piroscissione

n kraakbenzine
d Krackbenzin *n*

1478 CRACKING fue
f cracking *m*, craquage *m*
e cracking *m*, craqueo *m*, piezopirólisis *f*, pirodesintegración *f*
i cracking *m*, piroscissione *f*
n kraken *n*
d Kracken *n*

1479 CRACKING, CRAZING, CHECKING cer,dec
f craquelure *f*, formation *f* de criques, formation *f* de gerces
e agrietamiento *m*
i incrinatura *f*, screpolamento *m*
n barsten *n*, haarscheurvorming
d Rissbildung *f*

1480 CRAWLING, BEADING, CISSING dec
f contraction *f*, rampage *m*, retrait *m*
e avejigamiento *m*
i falta *f* di presa
n kralen *n*, samentrekken *n*, schiften *n*
d Kriechen *n*, Perlen *n*

1481 CRAZE, CRAZING, CRIZZLING gla,met,pla
f craquelure *f* superficielle
e agrietamiento *m* superficial
i incrinatura *f* superficiale, screpolatura *f* superficiale
n oppervlaktehaarscheurtjes *pl*
d Oberflächenhaarrisse *pl*

1482 CREAM dai,nut
f crème *f*
e crema *f*, nata *f*
i crema *f*, panna *f*
n room
d Rahm *m*, Sahne *f*

1483 CREAM-LAID PAPER, CREAM-WOVE PAPER pap
f papier *m* vergé blanc
e papel *m* acanillado, papel *m* blanco con filigrana, papel *m* verjurado blanco
i carta *f* da lettere filigranata
n wit schrijfpapier *n* met watermerk
d weisses Briefpapier *n* mit Wasserzeichen

1484 CREAM OF TARTAR, POTASSIUM HYDROGEN TARTRATE, PURIFIED TARTAR dye
f crème *m* de tartre, tartrate *m* de potassium
e crémore *m* tártaro, tartrato *m* de potasio
i cremore *m* di tartaro, tartrato *m* di potassio
n gezuiverde wijnsteen, kaliumtartraat *n*
d gereinigter Weinstein *m*, Kaliumtartrat *n*

1485 CREAMED LATEX rub
f latex *m* écrémé
e látex *m* desnatado
i lattice *m* scremato
n opgeroomde latex, oproomlatex
d aufgerahmter Latex *m*, Rahmlatex *m*

1486 CREAMERY WASTE, DAIRY WASTE dai,nut
f eau *f* résiduaire de laiterie
e agua *f* cloacal de lechería
i acqua *f* di rifiuto di latteria
n afvalwater *n* van melkfabriek
d Molkereiabwasser *n*

1487 CREAMING pla,rub
f écrémage *m*
e desnate *m*
i scrematura *f*
n opromen *n*
d Aufrahmen *n*

1488 CREASE-RESIST FINISHING tex
f finissage *m* résistant au froissement
e acabado *m* inarrugable, acabado *m* resistente al arrugamiento
i finissaggio *m* resistente alla gualcitura
n kreukvrij maken *n*
d Knitterfestmachen *n*

1489 CREEP cry,eva
f formation *f* de cristaux à la paroi du récipient
e formación *f* de cristales reptantes encima de la superficie de un líquido
i formazione *f* di cristalli sulle pareti d'un recipiente
n uitkristalliseren *n* aan de vatwan
d Auskristallisation *f* an der Gefässwand

1490 CREEP met,pla
f fluage *m*
e arrastre *m*, fluencia *f*, flujo *m*
i scorrimento *m*
n kruip, kruipen *n*
d Kriechen *n*

1491 CREEP-PATH FORMATION, TRACKING pla
f formation *f* d'un passage conducteur
e formación *f* de un paso conductor
i formazione *f* d'un passaggio conduttore
n kruipwegvorming
d Kriechwegbildung *f*

1492 CREEP TESTING met
f essai *m* de fluage
e prueba *f* de fluencia, prueba *f* de flujo
i collaudo *m* di scorrimento
n kruipproef
d Dauerstandsversuch *m*, Kriechprobe *f*

1493 CREEPAGE, CREEPING OF SALT elc
f sels *pl* grimpants
e ascensión *f* capilar, sales *pl* trepadoras
i azione *f* di bagnamento dell'elettrolito
n opkruipen *n* van elektrolyt, opkruipen *n* van zout
d Hochkriechen *n* von Salz, Kriechstrom *m*

1494 CREEPING OF CONCRETE bui
f fluage *m* de béton
e escurrimiento *m* plástico de hormigón
i scorrimento *m* di calcestruzzo
n kruipen *n* van beton
d Kriechen *n* von Beton, Wandern *n* von Beton

1495 CREEPING OF PAINT, SAGGING OF PAINT dec
f formation *f* de rideaux
e derramamiento *m*
i scorrimento *m*
n uitzakken *n*
d Ausbauchen *n*, Ausweichen *n*

1496 CREMNITZ WHITE (GB), CREMUS WHITE (US) dec
f blanc *m* de Cremnitz
e blanco *m* de Cremnitz
i bianco *m* di Cremnitz
n Berlijns wit *n*
d Berlinerweiss *n*, Kremserweiss *n*, Venezianerweiss *n*, Vitriolbleiweiss *n*

1497 CREOSOTE OIL, DEAD OIL fat
f huile *f* de créosote
e aceite *m* de creosota
i olio *m* di creosoto
n creosootolie
d Kreosotöl *n*

1498 CREOSOTING OF WOOD bui
f créosotage *m*
e creosotación *f* de madera
i creosotizzazione *f* di legno
n creosoteren *n* van hout
d Holzimprägnierung *f*

1499 CREPE RUBBER rub
f caoutchouc *m* crêpe
e goma *f* crepe
i gomma *f* crepe
n crêperubber
d Kreppgummi *m*

1500 CRIMSON LAKE dec
f vernis *m* cramoisi
e barniz *m* carmesí
i vernice *f* cremisi
n karmozijnlak
d Karmesinlack *m*

1501 CRITICAL SOLUTION TEMPERATURE mix
f température *f* critique de solution
e temperatura *f* crítica de una solución
i temperatura *f* critica d'una soluzione
n kritieke oplossingstemperatuur
d kritische Lösungstemperatur *f*

1502 CRIZZLING, CRAZE, CRAZING gla,met,pla
f craquelure *f* superficielle
e agrietamiento *m* superficial
i incrinatura *f* superficiale, screpolatura *f* superficiale
n oppervlaktehaarscheurtjes *pl*
d Oberflächenhaarrisse *pl*

1503 CROCKING FASTNESS (US), RUBBING FASTNESS tex
f résistance *f* au frottement
e resistencia *f* al frotamiento, solidez *f* al borroneo

i resistenza *f* allo strofinamento
n wrijvingsbestendigheid
d Reibfestigkeit *f*

1504 CROSS DYEING dye
f teinture *f* croisée
e teñido *m* cruzado
i tintura *f* incrociata
n gekruisd verven *n*
d Überfärben *n*

1505 CROSS FLOW dis
f courants *pl* croisés
e corrientes *pl* cruzadas
i flusso *m* incrociato
n kruisstroming
d Kreuzstrom *m*

1506 CROSS-LAMINATED SHEET pla
f plaque *f* à lamination croisée
e hoja *f* de laminación cruzada
i lamiera *f* a stratificazione incrociata
n kruisgewijs gelamineerde plaat
d kreuzweise geschichtete Platte *f*

1507 CROSS-WOUND BOBBIN, CHEESE tex
f bobine *f* croisée
e bobina *f* cruzada
i filo *m* avvolto nella bobina
n kruisspoel
d Kreuzspule *f*

1508 CROWN GLASS gla
f crownglass *m*
e crownglass *m*
i vetro *m* corona, vetro *m* crown
n kroonglas *n*
d Kronglas *n*

1509 CROWN LEATHER, HELVETIA LEATHER lea
f cuir *m* helvétique
e cuero *m* helvético
i cuoio *m* elvetico
n kroonleder *n*
d Crownleder *n*, Kronleder *n*

1510 CRUDE CREAM OF TARTAR, ARGOL, WINE LEES cor
f argol *m*, crème *f* de tartre brute
e crémore *m* tártaro crudo
i argol *m*, cremore *m* di tartaro crudo
n ruwe wijnsteen
d Argal *n*, roher Weinstein *m*

1511 CRUDE OIL, KEROSENE (US), PETROLEUM (GB) fue
f pétrole *m* brut
e petróleo *m* bruto
i petrolio *m* greggio
n ruwe petroleum
d Erdöl *n*, Petroleum *n*, Steinöl *n*

1512 CRUDE OIL, RESIDUAL OIL fue
f huile *f* non-raffinée
e aceite *m* no rafinado
i olio *m* greggio
n ruwe olie
d Roherdöl *n*, Rohöl *n*

1513 CRUDE SEWAGE, CRUDE WASTE, RAW SEWAGE sew
f eau *f* brute usée
e agua *f* cruda de alcantarillo
i acqua *f* bruta di rifiuto
n ongezuiverd afvalwater *n*
d Rohabwasser *n*, rohes Abwasser *n*

1514 CRUDE SLUDGE, FRESH SLUDGE, RAW SLUDGE iwt
f boues *pl* fraîches, bourbe *f*, limon *m*
e fango *m* crudo
i fango *m* fresco
n onbewerkt slib *n*
d Frischschlamm *n*, Rohschlamm *m*

1515 CRUDE WAX, PETROLEUM WAX, SLACK WAX fat
f paraffine *f* brute, paraffine *f* non-déshuilée
e cera *f* parafínica muy aceitosa, mezcla *f* de cera parafínica y de petróleo
i paraffina *f* greggia, paraffina *f* molle
n ruwe paraffine, uit aardolie verkregen paraffine
d Rohparaffin *n*

1516 CRUSHED MALT, GRIST bre
f mouture *f*
e malta *f* molida, molienda *f* de malta
i macinato *m* di malto
n moutmeel *n*, moutschroot *n*
d Malzschrot *n*, Schrot *n*

1517 CRUSHING brk
f broyage *m*, concassage *m*
e machaqueo *m*, quebrantadura *f*
i frantumazione *f*

n breken *n*
d Brechen *n*, Zerkleinern *n*

1518 CRUSHING tex
f égalisation *f*
e igualación *f*
i uguagliamento *m*
n egaliseren *n*
d Egalisieren *n*

1519 CRUSTIZATION cor
f incrustation *f*
e incrustación *f*
i incrostazione *f*
n aankorsting
d Verkrustung *f*

1520 CRYOSCOPIC METHOD, ana
FREEZING-POINT METHOD
f méthode *f* cryoscopique
e método *m* crioscópico
i metodo *m* crioscopico
n kryoscopie
d Kryoskopie *f*

1521 CRYSTAL GLASS gla
f cristal *m*
e cristal *m*, vidrio *m* ornamental, vidrio *m* sonoro
i cristallo *m*
n kristal *n*
d Kristallglas *n*

1522 CRYSTAL MALT bre
f malt *m* caramel
e malta *f* caramelo
i malto *m* caramello
n karamelmout
d Karamelmalz *n*

1523 CRYSTALLINE LIQUID, cry,gla
ANISOTROPIC LIQUID,
LIQUID CRYSTAL
f cristal *m* liquide, liquide *m* anisotropique
e cristal *m* líquido, líquido *m* *m* anisotrópico
i cristallo *m* liquido, liquido *m* anisotropico
n anisotrope vloeistof, vloeibaar kristal *n*
d anisotrope Flüssigkeit *f*, flüssige Kristalle *pl*, kristalline Flüssigkeit *f*

1524 CRYSTALLIZATION cry
f cristallisation *f*
e cristalización *f*
i cristallizzazione *f*
n kristalliseren *n*
d Kristallisieren *n*

1525 CUBIC SALTPETRE, fer
CHILE NITRE, CHILE
SALTPETRE
f nitre *m* cubique, nitre *m* du Chile
e nitrato *m* de Chile, salitre *m* de Chile
i salnitro *m* del Cile
n chilisalpeter
d Chilisalpeter *m*, Natronsalpeter *m*

1526 CUBING cer
f cubage *m*
e cubicaje *m*
i cubatura *f*
n kubusvorming
d Würfelbildung *f*

1527 CUDBEAR, dye
ORCHIL
f orseille *f*
e orcina *f*
i orcina *f*
n orseille *n*
d Cudbear *m*, Orseillenfarbe *f*

1528 CULM LAYER, cer
BREEZING
f couche *f* en poussier d'anthracite
e capa *f* de cisco
i strato *m* di polvere di antracite
n laag van antracietpoeder
d Kohlenlöscheschicht *f*

1529 CULTURE bac
f culture *f*
e cultivo *m*
i cultura *f*
n cultuur
d Kultur *f*, Zucht *f*

1530 CULTURE MEDIUM bac
f milieu *m* de culture, terrain *m* nutritif
e medio *m* de cultivo
i terreno *m* di cultura, terreno *m* nutritivo
n voedingsbodem
d Nährboden *m*

1531 CULTURE YEAST bre
f levure *f* de culture
e levadura *f* artificial, levadura *f* de cultivo
i lievito *m* di cultura
n cultuurgist, gekweekte gist
d Kulturhefe *f*

1532 CUMIN OIL fat
f essence *f* de cumin
e esencia *f* de comino

i olio *m* di cumino
n komijnolie, komijnzaadolie
d Cuminöl *n*

1533 CUPEL ASSAY, met,tes
CUPELLING, FIRE ASSAYING
f coupellation *f*
e copelación *f*
i coppellazione *f*
n copelleren *n*
d Kapellenprobe *f*, Kupellation *f*

1534 CUPRIFEROUS met
f cuprifère adj
e cuprífero adj
i cuprifero adj
n koperhoudend adj
d kupferhaltig adj

1535 CUPROUS ION PROCESS dye
f teinture *f* à production d'ions cuivreux
e teñido *m* con producción de iones cuprosos
i tintura *f* a produzione d'ioni rameosi
n verven *n* met gelijktijdige ontwikkeling van cupro-ionen
d Färbung *f* mit gleichzeitiger Kupro-Ionenentwicklung

1536 CURCUMIN PAPER dis
f papier *m* curcuma
e papel *m* de cúrcuma
i carta *f* di curcuma
n kurkumapapier *n*
d Curcumapapier *n*

1537 CURD dai,mit
f caillebotte *f*
e naterón *m*, requesón *m*
i ricotta *f*
n kwark, wrongel
d Quark *m*

1538 CURD SOAP (GB), soa
LAUNDRY SOAP (US),
SODIUM SOAP (US)
f savon *m* dur
e jabón *m* duro
i sapone *m* duro
n harde zeep, kernzeep
d harte Seife *f*, Kernseife *f*

1539 CURD SUSPENSION dai,nut
f suspension *f* de caillebotte
e suspensión *f* de naterón, suspensión *f* de requesón
i sospensione *f* di ricotta
n kwarksuspensie, wrongelsuspensie
d Quarksuspension *f*

1540 CURDLING dec
f épaississement *m*
e posamiento *m*
i raggrumamento *m*
n indikken *n*, indrogen *n*
d Gerinnen *n*

1541 CURDLING, dec
CLOTTING
f caillage *m*, grumelage *m*
e agrumación *f*
i formazione *f* di grumi
n klonteren *n*
d Käsen *n*, Klümperigwerden *n*

1542 CURDLING, col,dai,nut
COAGULATION
f caillage *m*, coagulation *f*
e coagulación *f*, cuajamiento *m*
i accagliamento *m*, coagulazione *f*
n coagulatie, stremmen *n*
d Gerinnung *f*, Koagulierung *f*

1543 CURDLING, dai
TURNING INTO CHEESE
f procédé *m* de manufacture du fromage
e procedimiento *m* de arrequesonar
i processo *m* di fabbricazione del formaggio
n kazen *n*
d Käsen *n*, Verkäsen *n*

1544 CURED MALT, bre
KILN-DRIED MALT
f malt *m* touraillé
e malta *f* tostada
i malto *m* essiccato
n geroost mout *n*
d Darrmalz *n*

1545 CURING lea
f salage *m*
e salazón *f*
i salata *f*, salatura *f*
n inzouten *n*
d Einsalzen *n*

1546 CURING nut
f conservation *f*
e curación *f*
i conservazione *f*
n conserveren *n*
d Haltbarmachung *f*

1547 CURING pla
f moulage *m*
e moldeo *m*
i indurimento *m*
n uitharden *n*
d Härten *n*

1548 CURING, ROASTING, TORRIFYING — bre
f torréfaction *f*
e tostación *f*
i torrefazione *f*
n roosting
d Röstung *f*

1549 CURING, VULCANIZING — rub
f vulcanisation *f*
e vulcanización *f*
i vulcanizzazione *f*
n vulcaniseren *n*
d Vulkanisieren *n*

1550 CURING TIME — pla
f temps *m* de moulage
e tiempo *m* de moldeo
i tempo *m* d'indurimento
n uithardingstijd
d Härtungszeit *f*

1551 CURRYING, TOWING — lea
f corroyage *m*
e tratamiento *m* con mezclas grasas
i trattamento *m* con mescolanze grasse
n touwen *n*
d Zurichtung *f*

1552 CURTAIN COATING — dec
f vernissage *m* à rideau
e barnizado *m* de cortina
i verniciatura *f* a tendina
n aanbrengen *n* met een lakgordijn
d Druck-Giess-Lackierverfahren *n*

1553 CUT, FRACTION — dis,fue
f fraction *f*
e corriente *f*, fracción *f*
i frazione *f*
n fractie
d Fraktion *f*

1554 CUTBACK (US), CUTBACK BITUMEN (GB) — fue
f cutback *m*
e betún *m* reblandecido
i bitume *m* ricostituito
n vloeibitumen *n*
d Kaltasphalt *m*, Verschnittasphalt *m*

1555 CUTTING, GRAINING — soa
f dessalage *m*
e desalación *f*, separación *f* del jabón de la glicerina
i dissalatura *f*
n uitzouten *n*
d Aussalzen *n*

1556 CUTTING OIL — lub
f huile *f* pour couper les métaux
e aceite *m* cortante
i olio *m* per il taglio dei metalli
n snijolie
d Schneidöl *n*

1557 CYANAMIDE PROCESS — gas
f procédé *m* de fabricage de l'azote au cyanamide
e procedimiento *m* de fabricación del nitrógeno a la cianamida cálcica
i processo *m* di fabbricazione dell'azoto alla cianamide
n stikstoffabricage volgens het kalkstikstofproces
d Stickstoffgewinnung *f* nach dem Kalkstickstoffverfahren

1558 CYANIDATION, CYANIDING — met
f extraction *f* de l'or par cyanuration
e beneficio *m* del oro por cianuración
i cianurizzazione *f*
n cyanide-uitloogproces *n*
d Zyanidlaugerei *f*, Zyanidlaugungsverfahren *n*

1559 CYANIDE COPPER PLATING — elc
f cuivrage *m* au cyanure de potassium
e cobreado *m* con cianuro de potasio
i ramatura *f* al cianuro di potassio
n cyaankaliverkopering
d Zyankaliverkupferung *f*

1560 CYANIDE HARDENING, CARBO-NITRIDING — met
f cémentation *f* au cyanure, endurcissement *m* au cyanure
e cementación *f* al cianuro, endurecimiento *m* al cianuro
i cementazione *f* al cianuro, indurimento *m* al cianuro
n cementering met cyanide, cyaanbadharding
d Zyansalzbadhärtung *f*

1561 CYANOSITE, BLUE STONE, BLUE VITRIOL, CHALCANTHITE — min
f chalcanthite *f*, vitriol *m* bleu

e calcantita *f*, vitriolo *m* azul
i calcantite *f*, vitriolo *m* azzurro
n blauw vitriool *n*, blauwsteen *n*, chalcantiet *n*
d Blaustein *n*, Blauvitriol *n*, Chalkantit *m*, Kupfervitriol *n*

1562 CYLINDRICAL SCREEN FEEDING cer

f alimentation *f* à écran cylindrique
e alimentación *f* con pantalla cilíndrica
i alimentazione *f* a griglia cilindrica
n rolschermtoevoer
d Rollschirmspeisung *f*

D

1563 DAIRY WASTE, dai,nut
CREAMERY WASTE
f eau *f* résiduaire de laiterie
e agua *f* cloacal de lechería
i acqua *f* di rifiuto di latteria
n afvalwater *n* van melkfabriek
d Molkereiabwasser *n*

1564 DAMP, fue,gas
FIREDAMP, PIT GAS (GB),
WILD GAS (US)
f grisou *m*, mofette *f*
e grisú *m*, mofeta *f*
i grisu *m*
n mijngas *n*
d Grubengas *n*, Schlagwetter *n*

1565 DAMP DETECTION, sap
FIREDAMP DETECTION,
PIT GAS DETECTION (GB),
WILD GAS DETECTION (US)
f détection *f* de grisou,
détection *f* de mofette
e detección *f* de grisú, detección *f*
de mofeta
i rivelazione *f* di grisu
n mijngasdetectie
d Grubengasanzeige *f*,
Schlagwetteranzeige *f*

1566 DAMP-PROOF bui
f étanche adj , hydrofuge ajd
e resistente adj a la humedad
i resistente adj all'umidità
n vochtbestendig adj
d feuchtigkeitsfest adj

1567 DAMP-PROOFING bui
f rendre v étanche à l'eau
e hidrofugación *f*
i far v resistente all'umidità
n vochtbestendigmaken *n*
d Feuchtigkeitsfestmachen *n*

1568 DAMPING pap
f humidification *f*, mouillage *m*
e humectación *f*
i umidificazione *f*
n bevochtiging
d Anfeuchtung *f*

1569 DARK GLOWING met
f recuit *m* bleu
e recocido *m* azul
i ricottura *f* azzurra,
scaldata *f* al colore blu
n blauwgloeien *n*
d Blauglühen *n*

1570 DAUBING coa
f application *f* d'un revêtement
en matière réfractaire
e aplicación *f* de recubrimiento
de materia refractaria
i applicazione *f* d'un rivestimento
di materiale refrattario
n aanbrengen *n* van een vuurvaste
laag
d feuerbeständige Beschichtung *f*

1571 DEACIDIFICATION, cle
DISACIDIFICATION
f désacidification *f*
e desacidificación *f*
i disacidificazione *f*
n ontzuring
d Entsäuerung *f*

1572 DEACTIVATION gen
f désactivation *f*
e desactivación *f*
i deattivazione *f*
n inactiveren *n*, ontactiveren *n*
d Entaktivierung *f*

1573 DEAD, gen
DEAD MATERIAL
f matière *f* détérioriée
e materia *f* deteriorada
i materia *f* deteriorata
n uitgeput materiaal *n*
d ausgenütztes Material *n*

1574 DEAD-BURNT LIME mat
f chaux *f* morte
e cal *f* morte
i calce *f* morta
n doodgebrande kalk
d totgebrannter Kalk *m*

1575 DEAD-BURNT LIMESTONE mat
f pierre *f* à chaux morte,
pierre *f* calcaire morte
e piedra *f* de cal morte
i pietra *f* calcarea morta
n doodgebrand kalksteen *n*
d totgebrannter Kalkstein *m*

1576 DEAD-BURNT MAGNESITE mat
f magnésite *f* frittée
e magnesita *f* sinterizada
i magnesite *f* sinterizzata

n sintermagnesiet *n*
d Sintermagnesit *m*

1577 DEAD COLORING (US), DEAD COLOURING (GB) dec
f première main *f* de couleur
e primera mano *f* de color
i prima mano *f* di colore
n eerste kleurlaag
d erster Farbanstrich *m*

1578 DEAD FINISH, DULL FINISH, FLAT FINISH dec
f finissage *m* mat
e acabado *m* mate, terminación *f* mate
i finissaggio *m* matto, rifinitura *f* opaca
n mat afstrijken *n*, matte deklaag
d Mattglanz *m*, matte Deckschicht *f*

1579 DEAD MILLING, KILLING OF RUBBER rub
f mastication *f* à mort
e masticación *f* a muerte
i masticazione *f* a morte
n doodmaken *n*, doodwalsen *n*
d Totmastizieren *n*, Totwälzen *n*

1580 DEAD OIL, CREOSOTE OIL fat
f huile *f* de créosote
e aceite *m* de creosota
i olio *m* di creosoto
n creosootolie
d Kreosotöl *n*

1581 DEAD ROASTING met
f grillage *m* total
e tostación *f* total
i torrefazione *f* totale
n totale roosting
d Totröstung *f*

1582 DEAD STEAM, EXHAUST STEAM, WASTE STEAM boi
f vapeur *f* d'échappement
e vapor *m* de escape
i vapore *m* di ritorno
n afgewerkte stoom
d Abdampf *m*

1583 DE-AERATION hyg
f désaération *f*
e desaireación *f*
i disaerazione *f*
n ontluchting
d Entlüftung *f*

1584 DEALKYLATION oic
f désalkylation *f*
e desalkilación *f*
i desalchilazione *f*
n ontalkylering
d Dealkylierung *f*

1585 DEALUMINIZING cle
f dégagement *m* de l'aluminium
e desaluminiación *f*
i separazione *f* dell'alluminio
n aluminiumafscheiding
d Aluminiumabscheidung *f*

1586 DEAMMONIATION aic
f déammoniation *f*
e desamonización *f*
i eliminazione *f* d'ammoniaca
n ammoniakafscheiding
d Ammoniakabtrennung *f*

1587 DEARSENICATION sua
f extraction *f* de l'arsenic
e extracción *f* del arsénico
i estrazione *f* dell'arsenico
n arseenextractie
d Arsenextraktion *f*

1588 DEASPHALTATION fue
f désasphaltage *m*
e desasfaltación *f*
i estrazione *f* dell'asfalto
n asfaltextractie
d Entasphaltierung *f*

1589 DEBENZOLIZATION fue
f débenzolage *m*
e desbenzolación *f*
i estrazione *f* del benzolo
n benzolextractie
d Entbenzolung *f*

1590 DEBISMUTHIZING met
f élimination *f* du bismuth
e eliminación *f* del bismuto
i eliminazione *f* del bismuto
n bismutonttrekking
d Wismuthbeseitigung *f*

1591 DEBITTERIZING bre
f désamérisation *f*
e desamargado *m*
i deamarizzazione *f*
n ontbitteren *n*
d Entbittern *n*

1592 DEBUTANIZING fue
f élimination *f* du butane
e desbutanización *f*
i estrazione *f* del butano

n butaanextractie
d Entbutanisierung *f*

1593 DECALCINATION aic
f déchaulage *m*
e descalcinación *f*
i decalcinazione *f*
n ontkalken *n*
d Entkalkung *f*

1954 DECANNING, DECLADDING, DEJACKETING nue
f dégainage *m*
e descamiciado *m*
i spogliatura *f*
n ontmanteling
d Enthülsung *f*

1995 DECANTING gen,lab
f décantage *m*, décantation *f*
e decantación *f*
i decantazione *f*
n afschenken *n*, decanteren
d Abgiessen *n*, Dekantieren *n*, Umgiessen *n*

1596 DECARBONIZATION iwt
f décarbonisation *f*
e descarbonización *f*
i decarbonizzazione *f*, eliminazione *f* dell'acido carbonico
n koolzuurverwijdering
d Entkarbonisierung *f*

1597 DECARBONIZING, DECARBURIZING met
f décarburation *f*
e descarburación *f*
i decarburazione *f*
n ontkolen *n*
d Entkohlung *f*

1598 DECARBOXYLATION oic
f décarboxylation *f*
e descarboxilación *f*
i decarbossilazione *f*
n decarboxyleren *n*
d Dekarboxylieren *n*

1599 DECATIZING, ROLL BOILING, STEAMING tex
f décatissage *m*
e decatizado *m*, vaporización *f*
i decatissaggio *m*
n afglanzen *n*, decateren *n*
d Dekatieren *n*, Dekatur *f*, Glanzkrumpe *f*

1600 DECAY, DISINTEGRATION nue
f désintégration *f*
e desintegración *f*
i decadimento *m*, disintegrazione *f*
n desintegratie, verval *n*
d Umwandlung *f*, Zerfall *m*

1601 DECHLORINATION hyg,iwt
f déchloration *f*
e descloración *f*, descloridación *f*
i declorazione *f*
n ontchloring
d Entchlorung *f*

1602 DECOBALTIZING met
f élimination *f* du cobalt
e remoción *f* del cobalto
i separazione *f* del cobalto
n kobaltafscheiding
d Entkobaltierung *f*

1603 DECOCTION, EXTRACT ext
f décocté *m*
e decocción *f*
i decotto *m*
n afkooksel *n*
d Abkochung *f*, Absud *m*, Dekokt *m*

1604 DECOCTION, EXTRACTION BY BOILING ext
f décoction *f*
e decocción *f*
i decozione *f*
n afkoken *n*
d Abkochen *n*, Absieden *n*

1605 DECOCTION METHOD, THICK MASH METHOD bre
f méthode *f* par décoction
e método *m* por decocción
i metodo *m* per decozione
n afkookmethode, decoctiemethode
d Dekoktionsverfahren *n*

1606 DECOLORIZATION fue,gla,tex
f décoloration *f*
e descoloración *f*
i decolorazione *f*
n ontkleuring
d Entfärbung *f*

1607 DECOLORIZER, DECOLORIZING AGENT gla
f décolorant *m*
e descolorante *m*
i decolorante *m*
n ontkleuringsmiddel *n*
d Entfärbungsmittel *n*

1608 DECOLORIZING CHARCOAL cle
f charbon *m* décolorant
e carbón *m* descolorador
i carbone *m* decolorante
n ontkleuringskool
d Entfärbungskohle *f*

1609 DECOMPOSITION gen
f décomposition *f*
e descomposición *f*
i decomposizione *f*
n afbraak, ontleding
d Abbau *m*, Abbrúch *m*, Zersetzung *f*

1610 DECOMPOSITION REACTION gen
f réaction *f* de décomposition
e reacción *f* de descomposición
i reazione *f* di decomposizione
n afbraakreactie
d Abbaureaktion *f*, Abbruchreaktion *f*

1611 DECOMPOSITION VOLTAGE (GB), DISCHARGE VOLTAGE (US) elc
f tension *f* de décomposition
e tensión *f* de descomposición
i tensione *f* di decomposizione, tensione *f* di scarica
n ontledingsspanning
d Zersetzungsspannung *f*

1612 DECONTAMINATION, DETOXICATION hyg,nue
f décontamination *f*, désintoxication *f*
e descontaminación *f*, desintoxicación *f*
i decontaminazione *f*, disintossicazione *f*
n ontsmetting
d Dekontaminierung *f*, Entgiftung *f*

1613 DECORATION FIRING, ENAMELLING cer,htt
f émaillage *m*
e esmaltado *m*
i smaltatura *f*
n emailleren *n*
d Emaillieren *n*

1614 DECORTICATION pap,tan
f décorticage *m*, décortication *f*
e descortezamiento *m*
i scorticatura *f*
n ontschorsen *n*, schillen *n*
d Entrinden *n*, Schälen *n*

1615 DECREPITATION cry
f décrépitation *f*
e decrepitación *f*
i crepitazione *f*, decrepitazione *f*
n decrepiteren *n*
d Dekrepitieren *n*

1616 DE-EMULSIFICATION, DEMULSIFICATION lub
f démulsification *f*
e desmulsificación *f*
i demulsificazione *f*
n demulgeren *n*
d Demulgierung *f*

1617 DE-EMULSIFICATION NUMBER, DEMULSIFICATION NUMBER lub
f indice *m* de démulsification
e índice *m* de desmulsificación
i numero *m* di demulsificazione
n demulgeergetal *n*
d Demulgierungszahl *f*

1618 DE-EMULSIFIER, DEMULSIFIER lub
f agent *m* démulsificateur
e agente *m* desmulsificador
i agente *m* demulsificatore
n demulgeermiddel *n*
d Demulgiermittel *n*

1619 DE-ENAMELLING cer,met
f élimination *f* de l'émail
e desesmaltación *f*, desesmaltado
i rimozione *f* dello smalto
n emailverwijdering
d Entemaillierung *f*

1620 DEEP-ETCH TESTING, MACRO-ETCH TESTING tes
f essai *m* de macro-attaque
e prueba *f* de ataque profundo, prueba *f* de macroataque
i collaudo *m* a macroincisione all'acquaforte, collaudo *m* ad attacco chimico profondo
n diepetsproef
d Tiefätzprobe *f*

1621 DEEP FREEZING, LOW TEMPERATURE REFRIGERATION clo
f congélation *f*, réfrigération *f* à basse température
e congelación *f*, refrigeración *f* de baja temperatura
i congelamento *m*, congelazione *f*, refrigerazione *f* a bassa temperatura

n diepvriezen *n*
d Tiefkühlung *f*

1622 DEFECATING sin
f défécation *f*
e defecación *f*
i defecazione *f*
n defecatie, nasatureren *n*
d Defekation *f*, Scheide-saturation *f*, Schlammsaftsaturation *f*

1623 DEFECATION SLIME sin
f boue *f* de défécation
e fango *m* de defecación
i fango *m* di defecazione
n defecatieslib *n*
d Klärschlamm *m*, Scheideschlamm *m*

1624 DEFERRIFICATION, DEFERRIZATION, DE-IRONING hyg,iwt
f déferrisation *f*
e desferrización *f*
i eliminazione *f* del ferro
n ontijzering
d Enteisenung *f*

1625 DEFERVESCENCE gen
f fin d'ébullition
e fin *m* de ebullición
i fine *m* d'ebollizione
n ophouden *n* met koken
d Ende *n* des Siedens

1626 DEFIBERIZING, DEFIBRATION pap
f défibrage *m*
e desfibración *f*
i sfibramento *m*
n defibratie, vezelen *n*
d Zerfaserung *f*

1627 DEFINING, DEFLASHING cle,pol
f débarbage *m*
e rebabado *m*
i sbavatura *f*
n ontgraten *n*
d Entgratung *f*

1628 DEFLAGRATION lab
f déflagration *f*
e deflagración *f*
i deflagrazione *f*
n deflagratie
d Deflagration *f*, schnelle Verbrennung *f*

1629 DEFLOCCULATION sin
f défloculation *f*
e desfloculación *f*
i separazione *f* sotta formazione di fiocchi
n ontvlokking, redispergeren *n* van een flocculaat
d Entflockung *f*

1630 DEFOAMING, DEFROTHING cle
f démoussage *m*
e despumación *f*
i destruzione *f* della schiuma
n schuimverdrijving
d Schaumbeseitigung *f*

1631 DEFOAMING AGENT, ANTIFOAMING AGENT, ANTIFROTHING AGENT tex
f agent *m* antimousse
e agente *m* antiespuma
i agente *m* antischiuma
n antischuimmiddel *m*, schuimwerend middel *n*
d Antischaummittel *n*, Schaumverhütungsmittel *n*

1632 DEFROSTING clo
f décongélation *f*
e descongelación *f*
i decongelazione *f*
n ontdooiing, ontvriezing
d Auftauung *f*, Enteisen *n*, Entfrostung *f*

1633 DEGASIFICATION (US), DEGASIFYING (GB), DEGASSING (GB) gas
f dégazification *f*, dégazage *m*
e desgaseado *m*, desgasificación *f*
i degassaggio *m*, degassamento *m*
n ontgassing
d Entgasung *f*

1634 DEGASSING CYCLE, BREATHING CYCLE gen
f cycle *m* de désaération
e ciclo *m* de desaireación
i ciclo *m* di disareazione
n ontluchtingscyclus
d Entlüftungszyklus *m*

1635 DEGENERATED YEAST fer
f levure *f* dégénérée
e levadura *f* degenerada
i lievito *m* degenerato
n gedegenereerde gist
d verkümmerte Hefe *f*

1636 DEGRADATION, BREAKDOWN gen
f décomposition *f*
e degradación *f*, descomposición *f*
i disgregazione *f*
n afbraak, ontbinding
d Abbau *m*

1637 DEGRAS, MOELLEN tan
f dégras *m*
e churre *f*
i grasso *m* ricavato
n afvalvet *n*, dégras *n*, traan-wateremulsie
d Abfett *n*, Degras *n*, Gerberfett *n*, Lederschmiere *f*, Moellon *n*

1638 DEGREASING cle
f dégraissage *m*
e desengrase *m*
i sgrassaggio *m*, sgrassatura *f*
n ontvetting
d Entfettung *f*, Reinigung *f*

1639 DEGREE OF OXIDATION, OXIDATION RATIO gen
f degré *m* d'oxydation
e grado *m* de oxidación *f*
i grado *m* d'ossidazione
n oxydatiegraad
d Oxydationsgrad *m*, Oxydationsstufe *f*

1640 DEGREE OF REDUCTION, REDUCTION RATIO gen
f degré *m* de réduction
e grado *m* de reducción
i grado *m* di riduzione
n reductiegraad
d Reduktionsgrad *m*, Reduktionsstufe *f*

1641 DEGRESSIVE BURNING POWDER exp
f explosif *m* dégressif
e explosivo *m* degresivo
i esplosivo *m* digressivo
n degressieve springstof
d degressiver Sprengstoff *m*

1642 DEGUMMED SILK, BOILED-OFF SILK tex
f soie *f* dégommée
e seda *f* descrudada, seda *f* desgomada
i seta *f* cotta
n afgekookte zijde, ontgomde zijde
d entschälte Seide *f*, linde Seide *f*

1643 DEGUMMING tex
f dégommage *m*
e desengomado *m*
i sgommatura *f*
n ontgomming
d Degummierung *f*, Entbastung *f*

1644 DEHALOGENATION aic
f déshalogénation *f*
e deshalogenación *f*
i desalogenazione *f*
n halogeenafscheiding
d Halogenabtrennung *f*

1645 DEHUMIDIFYING hyg
f élimination *f* de l'humidité
e deshumectación *f*, deshumidificación *f*
i deumidificazione *f*
n ontvochting
d Entfeuchtung *f*

1646 DEHYDRATION, DEWATERING dry,hyg
f déshydratation *f*
e deshidratación *f*
i asciugamento *m*, disidratazione *f*, prosciugament[o]
n dehydratatie, wateronttrekking
d Dehydratisierung *f*, Entwässerung *f*, Wasserentziehung *f*

1647 DEHYDROFREEZING hyg
f déshydrocongélation *f*
e deshidrocongelación *f*
i deidrocongelazione *f*
n diepvriezen *n* na voorafgaande droging
d Tiefkühlung *f* in vorgetrocknete[m] Zustand

1648 DEHYDROGENATION oic
f déshydrogénation *f*
e deshidrogenación *f*
i deidrogenazione *f*
n dehydreren *n*, waterstofextracti[e]
d Dehydrierung *f*, Wasserstoffextrahierung *f*

1649 DEHYDROHALOGENATION fue,oic
f déshydrohalogénation *f*
e deshidrohalogenación *f*
i deidroalogenazione *f*
n waterstofhalogeenextractie
d Halogenwasserstoffabspaltung *f*, Wasserstoffhalogenextraktion *f*

1650 DE-IONIZATION elc
f désionisation *f*
e desionización *f*
i deionizzazione *f*
n desionisatie
d Entionisierung *f*

1651 DE-IRONING, DEFERRIFICATION, DEFERRIZATION hyg,iwt
f déferrisation *f*
e desferrización *f*
i eliminazione *f* del ferro
n ontijzering
d Enteisenung *f*

1652 DEJACKETING, DECANNING, DECLADDING nue
f dégainage *m*
e descamiciado *m*
i spogliatura *f*
n ontmanteling
d Enthülsung *f*

1653 DELIMING tan
f déchaulage *m*
e descalcificación *f*
i decalcificazione *f*
n ontkalking
d Entkalkung *f*

1654 DELIQUESCENCE gen
f déliquescence *f*
e delicuescencia *f*
i deliquescenza *f*
n vervloeien *n*
d Zerfliessen *n*

1655 DELUSTERING tex
f délustrage *m*
e deslustre *m*
i delucidazione *f*
n ontglanzen *n*
d Entglänzung *f*

1656 DELUSTRANT tex
f agent *m* délustrant
e deslustrante *m*
i agente *m* di delucidazione
n ontglanzingsmiddel *n*
d Entglänzungsmittel *n*

1657 DEMANGANIZING hyg
f démanganisation *f*
e desmanganización *f*
i demanganizzazione *f*
n ontmanganen *n*
d Entmanganung *f*

1658 DEMINERALIZATION cle
f déminéralisation *f*
e desmineralización *f*
i demineralizzazione *f*
n verwijderen *n* van minerale zouten
d Entmineralisieren *n*

1659 DEMULSIFICATION, DE-EMULSIFICATION lab
f démulsification *f*
e desmulsificación *f*
i demulsificazione *f*
n demulgeren *n*
d Demulgierung *f*

1660 DEMULSIFICATION NUMBER, DE-EMULSIFICATION NUMBER lub
f indice *m* de démulsification
e índice *m* de desmulsificación
i numero *m* di demulsificazione
n demulgeergetal *n*
d Demulgierungszahl *f*

1661 DEMULSIFIER, DE-EMULSIFIER lub
f agent *m* démulsificateur
e agente *m* desmulsificador
i agente *m* demulsificatore
n demulgeermiddel *n*
d Demulgiermittel *n*

1662 DENATURANT gen
f dénaturant *m*
e desnaturalizante *m*
i denaturante *m*
n denatureermiddel *n*
d Denaturierungsmittel *n*, Vergällungsmittel *n*

1663 DENATURATED ALCOHOL oic
f alcool *m* dénaturé
e alcohol *m* de quemar, alcohol *m* desnaturalizado
i alcool *m* denaturato
n gedenatureerde alcohol
d denaturierter Alkohol *m*

1664 DENATURATION oic
f dénaturation *f*
e desnaturalización *f*
i denaturazione *f*
n denatureren *n*
d Denaturierung *f*

1665 DENICKLEFICATION cle
f dénickelage *m*
e desniquelificación *f*
i eliminazione *f* del nichel
n ontnikkeling
d Entnickelung *f*

1666 DENITRATION, DENITRIFICATION oic
f dénitrification *f*
e desnitrificación *f*
i denitrificazione *f*
n denitrificeren *n*
d Denitrifikation *f*

1667 DENITRIDING sua
f dénitruration *f*
e desnitruración *f*
i denitrurazione *f*
n denitreren *n*
d Denitrierung *f*

1668 DENUDING eva
f élimination *f* de fractions volatiles
e destrozo *m* de fracciones volátiles
i eliminazione *f* di frazioni volatili
n verwijdering van vluchtige componenten
d Beseitigung *f* von flüchtigen Bestandteilen

1669 DEODORANT, DEODORIZING AGENT hyg
f désodorisant *m*
e desodorante *m*, desodorizante *m*
i deodorante *m*, deodorizante *m*
n deodorans *n*, desodoriseermiddel *n*
d Desodorisierungsmittel *n*, Geruchbeseitigungsmittel *n*

1670 DEODORIZING iwt
f désodorisation *f*
e desodorización *f*
i deodorizzazione *f*
n desodorisatie, stankverdrijving
d Desodorisieren *n*, Geruchsbeseitigung *f*

1671 DEOXIDIZER met
f désoxydant *m*
e desoxidante *m*
i disossidante *m*
n desoxydatiemiddel *n*
d Desoxydationsmittel *n*, Entoxydierungsmittel *n*

1672 DEOXIDIZING met
f désoxydation *f*
e desoxidación *f*
i disossidazione *f*
n desoxydatie
d Desoxydierung *f*, Sauerstoffentziehung *f*

1673 DEOXYGENATION, OXYGEN REMOVAL iwt
f désoxygénation *f*
e desoxigenación *f*
i disossigenazione *f*
n zuurstofverwijdering
d Entziehung *f* des Sauerstoffs, Sauerstoffbeseitigung *f*

1674 DEPARAFFINIZATION, DEWAXING fue
f déparaffinage *m*
e desparafinación *f*
i deparaffinazione *f*
n ontparaffineren *n*
d Entparaffinieren *n*

1675 DEPASSIVATION cle
f dépassivation *f*
e despasivación *f*
i depassivazione *f*
n ontpassiveren *n*
d Entpassivieren *n*

1676 DEPHENOLIZING iwt
f déphénolage *m*
e desfenolización *f*
i defenolaggio *m*, defenolizzazione *f*
n fenolafscheiding
d Entphenolung *f*

1677 DEPHLEGMATION dis
f déflegmation *f*
e desflegmación *f*
i deflegmazione *f*
n deflegmeren *n*
d Dephlegmieren *n*

1678 DEPHOSPHORIZING met
f déphosphoration *f*
e desfosforación *f*
i defosforazione *f*
n fosforverdrijving
d Entphosphorung *f*

1679 DEPICKLING tan
f ramollissement *m*
e remoción *f* de la salazón
i rammollimento *m*
n weekmaken *n*
d Weichmachen *n*

1680 DEPILATION, UNHAIRING lea
f dépilage *m*, épilage *m*
e depilación *f*
i depilazione *f*
n bloten *n*, ontharen *n*
d Enthaarung *f*

1681 DEPITCHING, tex
DETARRING
f dégoudronnage *m*
e desalquitranaje *m*
i rimozione *f* del catrame
n ontteren *n*
d Entpechung *f*

1682 DEPLETION nue
f appauvrissement *m*
e agotamiento *m*
i impoverimento *m*
n verarming
d Abreicherung *f*

1683 DEPOLARIZATION elc
f dépolarisation *f*
e despolarización *f*
i depolarizzazione *f*
n depolarisatie
d Depolarisation *f*

1684 DEPOLARIZER elc
f dépolarisant *m*
e despolarizador *m*
i depolarizzatore *m*
n depolarisator
d Depolarisator *m*

1685 DEPOLARIZING MIX elc
f mélange *m* dépolarisant
e mezcla *f* despolarizadora
i miscela *f* depolarizzante
n depolarisatormengsel *n*
d Depolarisationsgemisch *n*, Depolarisationsmasse *f*

1686 DEPOLYMERIZATION oic
f dépolymérisation *f*
e despolimerización *f*
i depolimerizzazione *f*
n depolymerisatie
d Depolymerisation *f*

1687 DEPOSITING-OUT, elc
LIBERATION
f dépôt *m* total, libération *f*
e depósito *m* total, liberación *f*
i esaurimento *m*
n afwerken *n*
d Entkupferung *f*, Entmetallisierung *f*

1688 DEPOSITION EFFICIENCY, elc
METAL DISTRIBUTION RATIO
f rapport *m* de distribution du métal
e relación *f* de distribución del metal
i rapporto *m* di distribuzione del metallo
n dikteverhouding van het neerslag
d Niederschlagsverteilungsverhältnis *n*

1689 DEPRESSION OF FREEZING POINT gen
f abaissement *m* du point de congélation
e descenso *m* del punto de congelación
i abbassamento *m* del punto di congelamento
n vriespuntverlaging
d Gefrierpunktserniedrigung *f*

1690 DEPROTEINIZATION oic
f déprotéination *f*
e desproteinización *f*
i deproteinizzazione *f*
n eiwitafscheiding
d Eiweissabtrennung *f*

1691 DEPTH CHARGE exp,mit
f grenade *f* sous-marine
e carga *f* antisubmarina, carga *f* de profundidad
i bomba *f* di profondità
n dieptebom
d Unterwasserbombe *f*

1692 DERBY RED, dec
CHINESE RED
f rouge *m* de Chine
e sulfuro *m* mercúrico rojo
i rosso *m* di Cina
n Chinees rood *n*
d rotes Quecksilbersulfid *n*

1693 DERIVATIVE gen
f dérivé *m*
e derivado *m*, producto *m* derivado
i prodotto *m* derivato
n derivaat *n*
d Abkömmling *m*, Derivat *n*

1694 DERUSTING cle
f dérouillage *m*
e remoción *f* de la herrumbre
i rimozione *f* della ruggine
n ontroesten *n*
d Entrosten *n*

1695 DESACCHARIFICATION sin
f désucrage *m*
e desacarificación *f*
i dissaccarificazione *f*
n ontsuikeren *n*
d Entzuckern *n*

1696 DESALINATION, cle
DESALTING
f dessalage *m*

e desalación *f*
i dissalinizzazione *f*
n ontzilting
d Entsalzung *f*

1697 DESATURATION oic
f désaturation *f*
e desaturación *f*
i dissaturazione *f*
n onverzadigd maken *n*
d Desaturation *f*

1698 DESCALING met
f décalaminage *m*, détartrage *m*
e descascarillado *m*
i disincrostazione *f*, scagliatura *f*
n oxydehuidverwijdering
d Entzunderung *f*

1699 DESENSITIZATION pho
f désensibilisation *f*
e desensibilización *f*
i desensibilizzazione *f*
n desensibilisatie
d Desensibilisierung *f*

1700 DESENSITIZATION, STABILIZATION exp
f flegmatisation *f*, phlegmatisation *f*
e flematización *f*, insensibilización *f*
i insensibilizzazione *f*
n flegmatiseren *n*, stabiliseren *n*
d Phlegmatisierung *f*

1701 DESENSITIZER pho
f désensibilisateur *m*
e desensibilizador *m*
i desensibilizzatore *m*
n desensibilisator
d Desensibilisator *m*

1702 DESICCANT dry
f agent *m* desséchant
e desecante *m*
i disseccante *m*
n droogmiddel *n*
d Trocknungsmittel *n*

1703 DESICCATION gen
f dessication *f*
e desecación *f*
i essiccamento *m*, essiccazione *f*
n uitdroging
d Austrocknung *f*

1704 DESILICONIZATION iwt
f désiliciage *m*, désiliciation *f*
e desiliconización *f*
i rimozione *f* della silice
n zandverwijdering
d Entkieselung *f*, Siliziumabbau *m*

1705 DESILTING, DESLUDGING, SLUDGE REMOVAL iwt
f débourbage *m*, enlèvement *m* de boue, évacuation *f* de boue
e desenlodamiento *m*
i scarico *m* di fango
n ontslibbing
d Entschlammung *f*, Schlammabscheidung *f*

1706 DESILVERING pho
f désargentation *f*
e desplatación *f*, desplateado *m*
i rimozione *f* dell'argento
n ontzilveren *n*
d Entsilberung *f*

1707 DESIZING tex
f déparementage *m*, désencollage *m*
e desapresto *m*, desencolado *m*
i sbozzimatura *f*
n ontslichting
d Entschlichtung *f*

1708 DESORPTION abs
f désorption *f*
e desabsorción *f*, desorción *f*
i desorbimento *m*
n desorptie
d Desorption *f*

1709 DESTRUCTION, INCINERATION hyg,iwt
f destruction *f*, incinération *f*
e destrucción *f*, incineración *f*
i incenerizzazione *f*
n destructie, verbranding
d Verbrennung *f*

1710 DESTRUCTIVE DISTILLATION, DRY DISTILLATION dis
f distillation *f* destructive, distillation *f* sèche
e destilación *f* destructiva, destilación *f* seca
i distillazione *f* a secco, distillazione *f* distruttiva
n droge destillatie
d Trockendestillation *f*, Zersetzungsdestillation *f*

1711 DESTRUCTIVE HYDROGENATION fue
f hydrogénolyse *f*

e hidrogenación *f* con descomposición
i idrogenolisi *f*
n hydrogenolyse
d spaltende Hydrierung *f*

1712 DESTRUCTIVE TESTING tes
f essai *m* destructif
e prueba *f* destructiva
i collaudo *m* distruttivo, prova *f* di rottura
n destructieve beproeving
d zerstörende Prüfung *f*

1713 DESULFURIZING (US), DESULPHURIZATION (GB) met
f dessoufrage *m*, désulfuration *f*
e desazufración *f*, desulfuración *f*
i desolforazione *f*
n ontzwaveling
d Entschwefelung *f*

1714 DETARRING, DEPITCHING tex
f dégoudronnage *m*
e desalquitranaje *m*
i rimozione *f* del catrame
n ontteren *n*
d Entpechung *f*

1715 DETERGENT, WASHING AGENT cle
f détergent *m*, détersif *m*
e detergente *m*, detersivo *m*
i detergente *m*, detersivo *m*
n detergens *n*, wasmiddel *n*
d Detergens *n*, Waschmittel *n*

1716 DETERGENT ACTION, CLEANING EFFICIENCY, WASHING PERFORMANCE cle
f action *f* détergente
e acción *f* detergente
i azione *f* detergente
n waswerking
d Waschwirkung *f*

1717 DETERGENT BATH cle
f bain *m* détergent, bain *m* détersif
e baño *m* detergente, baño *m* detersivo
i bagno *m* detergente, bagno *m* detersivo
n wasvloeistof
d Waschflotte *f*

1718 DETERRENT exp
f agent *m* de flegmatisation, agent *m* de phlegmatisation
e agente *m* de flematización, agente *m* de insensibilización
i agente *m* d'insensibilizzazione
n flegmatiseermiddel *n*, stabiliseermiddel *n*
d Phlegmatisator *m*

1719 DETINNING elc
f désétamage *m*
e desestañado *m*
i ricupero *m* dello stagno
n tinterugwinning
d Entzinnung *f*, Zinnrückgewinnung *f*

1720 DETONATING GAS, OXYHYDROGEN GAS gas
f gaz *m* tonnant
e gas *m* detonante
i gas *m* ossidrico, gas *m* tonante
n knalgas *n*
d Knallgas *n*

1721 DETONATING VIOLENCE, BRISANCE, SHATTERING POWER exp
f pouvoir *m* brisant
e brisancia *f*, efecto *m* rompedor
i effetto *m* dirompente, effetto *m* esplosivo
n brisante werking, brisantie
d Brisanz *f*

1722 DETONATION exp
f détonation *f*
e detonación *f*
i detonazione *f*
n detonatie, ontploffing
d Detonation *f*, Explosion *f*

1723 DETONATOR exp
f capsule *f* fulminante
e cápsula *f* detonadora
i detonatore *m*
n ontstekingskap, slaghoedje *n*
d Sprengkapsel *f*, Sprengzünder *m*

1724 DETONATOR exp
f fulminant *m*
e detonador *m*, fulminante *m*
i detonante *m*, detonatore *m*
n slagsas
d Zündstoff *m*

1725 DETOXICATION cle
f désintoxication *f*
e desintoxicación *f*
i disintossicazione *f*
n ontgiften *n*
d Entgiftung *f*

1726 DETOXICATION, DECONTAMINATION hyg,nue
f décontamination *f*, désintoxication *f*
e descontaminación *f*, desintoxicación *f*
i decontaminazione *f*, disintossicazione *f*
n ontsmetting
d Dekontaminierung *f*, Entgiftung *f*

1727 DEUTERIUM, HEAVY HYDROGEN nue
f hydrogène *m* lourd
e hidrógeno *m* pesado
i idrogeno *m* pesante
n zware waterstof
d schwerer Wasserstoff *m*

1728 DEUTERIUM OXIDE, HEAVY WATER nue
f eau *f* lourde
e agua *f* pesada
i acqua *f* pesante
n zwaar water *n*
d schweres Wasser *n*

1729 DEVELOPED DYE, DEVELOPING DYE dye
f colorant *m* à développement
e colorante *m* de revelado
i colorante *m* a sviluppo
n ontwikkelingskleurstof
d Entwicklungsfarbstoff *m*

1730 DEVELOPER pho
f révélateur *m*
e revelador *m*
i sviluppatore *m*
n ontwikkelaar
d Entwickler *m*

1731 DEVELOPING pho
f développement *m*
e revelado *m*
i sviluppo *m*
n ontwikkelen *n*
d Entwicklung *f*

1732 DEVELOPMENT OF CARBONIC ACID gas
f dégagement *m* d'acide carbonique
e desprendimiento *m* de ácido carbónico
i generazione *f* d'acido carbonico
n koolzuurontwikkeling
d Kohlensäureentwicklung *f*

1733 DEVILLING, SCORING dec,pla
f rayage *m*
e raspado *m*
i rigatura *f*
n krassen *n*
d Kerben *n*

1734 DEVITRIFICATION cle
f dévitrification *f*
e desvitrificación *f*
i devitrificazione *f*
n ontglazing
d Entglasung *f*

1735 DEVULCANIZING, RECLAIMING, REGENERATION rub
f dévulcanisation *f*, régénération *f*
e regeneración *f*
i rigenerazione *f*
n regeneratie
d Regeneration *f*

1736 DEWATERING hyg,tra
f vidange *m*
e escurrimiento *m*
i prosciugamento *m*, vuotamento *m*
n afvoeren *n*, lozen *n*
d Entleerung *f*

1737 DEWATERING, DEHYDRATION dry,hyg
f déshydratation *f*
e deshidratación *f*
i asciugamento *m*, disidratazione *f*, prosciugamento *m*
n dehydratatie, wateronttrekking
d Dehydratisierung *f*, Entwässerung *f*, Wasserentzieh[ung *f*]

1738 DEWAXING, DEPARAFFINIZATION fue
f déparaffinage *m*
e desparafinación *f*
i deparaffinazione *f*
n ontparaffineren *n*
d Entparaffinieren *n*

1739 DEXTRINE (GB), BRITISH GUM (US), STARCH GUM (GB) glu
f dextrine *f*
e dextrina *f*
i destrina *f*
n dextrine *n*
d Dextrin *n*, Stärkegummi *m*

1740 DEXTROSE, GLUCOSE, GRAPE SUGAR sir
f dextrose *f*, glucose *m*
e dextrosa *f*, glucosa *f*
i destrosio *m*, glucosio *m*, zucchero *m* d'uva

n dextrose, druivesuiker, glucose
d Dextrose *f*, Glukose *f*, Glykose *f*, Traubenzucker *m*

1741 DEZINCIFICATION, DEZINKING cer,met
f dégagement *m* du zinc, dézincification *f*
e descincado *m*
i dezincatura *f*, rimozione *f* dello zinco
n ontzinken *n*
d Entzinkung *f*

1742 DIALYSATE, DIFFUSATE ana
f solution *f* d'une substance dialysée
e dializado *m*
i soluzione *f* d'una sostanza dializzata
n dialysaat *n*
d Dialysat *n*

1743 DIALYSIS gen
f dialyse *f*
e diálisis *f*
i dialisi *f*
n dialyse
d Dialyse *f*

1744 DIAMOND DUST pol
f poudre *f* de diamant
e brujido *m*, polvo *m* de diamante
i polvere *f* di diamante
n diamantpoeder *n*
d Diamantpulver *n*

1745 DIASTASE, AMYLASE sin
f amylase *f*, diastase *f*
e amilasa *f*, diastase *f*
i amilasi *f*, diastase *f*
n amylase, diastase
d Amylase *f*, Diastase *f*

1746 DIATOM EARTH, DIATOMITE, INFUSORIAL EARTH, KIESELGUHR mat
f farine *f* fossile, tripoli *m*
e harina *f* fósil, trípoli *m*
i diatomite *m*, farina *f* fossile, tripolo *m*
n diatomeeënaarde, diatomiet *n*, infusoriënaarde, kiezelgoer *n*
d Diatomeenerde *f*, Infusorienerde *f*, Kieselgur *m*, Tripel *m*

1747 DIAZO COMPOUND dye,pho
f composé *m* diazoïque
e diazo-compuesto *m*
i diazo-composto *m*
n diazoverbinding
d Diazoverbindung *f*

1748 DIAZOTIZATION pho
f diazotation *f*
e diazotización *f*
i diazotizzazione *f*
n diazoteren *n*
d Diazotieren *n*

1749 DIAZOTYPE pho
f diazotypie *f*
e diazotipia *f*
i diazotipia *f*
n diazotypie
d Diazotypie *f*

1750 DICHROIC FOG pho
f voile *f* dichroïque
e velo *m* dicroico
i nebbia *f* dicroica
n dichroïtische sluier, tweekleurige sluier
d dichroitischer Schleier *m*

1751 DIELECTRIC DRYING, HIGH-FREQUENCY DRIER dry
f séchage *m* par pertes diélectriques
e secado *m* por pérdidas dieléctricas
i essiccazione *f* mediante perdite dielettriche
n capacitief drogen *n*
d dielektrische Trocknung *f*, Hochfrequenztrocknung *f*

1752 DIELS-ALDER REACTION, DIENE SYNTHESIS oic
f synthèse *f* de diène
e síntesis *f* de dieno
i sintesi *f* di dieno
n dieensynthese
d Diensynthese *f*

1753 DIENE oic
f diène *m*
e dieno *m*
i dieno *m*
n dieen *n*
d Dien *n*

1754 DIENE NUMBER, DIENE VALUE oic
f indice *m* de diène
e índice *m* de dieno

i indice *m* di dieno
n dieengetal *n*
d Dienzahl *f*

1755 DIESEL OIL, GAS OIL, SOLAR OIL — fat
f gas-oil *m*, huile *f* pour moteurs diesel
e gasoil *m*
i gasolio *m*, nafta *f* per motori diesel
n dieselolie
d Dieselöl *n*

1756 DIFFERENTIAL AERATION, SELECTIVE AERATION — cor
f aération *f* sélective
e aireación *f* selectiva
i aerazione *f* selettiva
n selectieve aëratie
d unterschiedliche Belüftung *f*

1757 DIFFERENTIAL AERATION CORROSION, AERATION CELL CORROSION — cor
f corrosion *f* par aération différentielle
e corrosión *f* por aeración diferencial
i corrosione *f* per aerazione differenziale
n corrosie door differentiële aëratie
d Korrosion *f* durch unterschiedliche Belüftung

1758 DIFFERENTIAL COOLING, SELECTIVE COOLING — clo
f refroidissement *m* sélectif
e enfriamiento *m* selectivo
i raffreddamento *m* selettivo
n selectieve koeling
d selektive Kühlung *f*

1759 DIFFERENTIAL DISTILLATION — dis
f distillation *f* différentielle, distillation *f* élémentairement équilibrée
e destilación *f* diferencial
i distillazione *f* differenziale, distillazione *f* progressiva
n open destillatie
d offene Destillation *f*

1760 DIFFERENTIAL FLOTATION, SELECTIVE FLOTATION — met
f flottation *f* sélective
e flotación *f* selectiva
i flottazione *f* selettiva
n selectieve flotatie
d selektive Flotation *f*

1761 DIFFERENTIAL HARDENING, SELECTIVE HARDENING — pho
f durcissement *m* sélectif
e endurecimiento *m* selectivo
i indurimento *m* selettivo
n selectieve harding
d selektive Härtung *f*, Teilhärtung *f*

1762 DIFFERENTIATING SOLVENT — gen
f solvant *m* sélectif
e disolvente *m* selectivo
i solvente *m* selettivo
n selectief oplosmiddel *n*
d selektives Lösungsmittel *n*

1763 DIFFUSION — gen
f diffusion *f*
e difusión *f*
i diffusione *f*
n diffusie
d Diffusion *f*

1764 DIFFUSION ANALYSIS — ana
f analyse *f* par diffusion
e análisis *f* por difusión
i analisi *f* per diffusione
n diffusieanalyse
d Diffusionsanalyse *f*

1765 DIFFUSION JUICE — sin
f jus *m* de diffusion
e jugo *m* de difusión
i succo *m* di diffusione
n diffusiesap *n*
d Diffusionssaft *m*

1766 DIFFUSION SINTERING — met
f frittage *m* à diffusion
e sinterización *f* de difusión
i sinterizzazione *f* a diffusione
n diffusiesinteren *n*
d Diffusionssintern *n*

1767 DIGESTED SLUDGE — iwt
f boue *f* digérée
e lodo *m* digerido
i fango *m* digerito
n gewassen slib *n*
d ausgefaulter Schlamm *m*

1768 DIGESTER LIQUOR, COOKER LIQUOR, COOKING LIQUOR — pap
f lessive *f* de cuisson
e lejía *f* de cocción
i lisciva *f* di cottura
n kookvloeistof
d Kocherlauge *f*

1769 DIGESTER RECLAIM — rub
f régénéré *m* à l'alcali

e regenerado *m* alcalino
i rigenerato *m* all'alcali
n alkaliregeneraat *n*
d Alkaliregenerat *n*

1770 DIGESTION, AFTERRIPENING, CHEMICAL RIPENING, SECOND RIPENING pho
f maturation *f* chimique, maturation *f* en refonte
e maduración *f* química segunda maduración *f*
i maturazione *f* chimica
n chemische rijping
d chemische Reifung *f*, Nachreifung *f*

1771 DILUENT, DILUTING AGENT, THINNER dec
f diluant *m*
e diluyente *m*
i diluente *m*
n verdunner, verdunningsmiddel *n*
d Streckmittel *n*, Verdünnungsmittel *n*

1772 DILUENT OIL (GB), THINNING OIL (US) dec
f huile *f* diluante
e aceite *m* diluyente
i olio *m* diluente
n verdunnende olie
d Verdünnungsöl *n*

1773 DILUTE SOLUTION gen
f solution *f* diluée
e solución *f* diluída
i soluzione *f* diluita
n verdunde oplossing
d verdünnte Lösung *f*

1774 DILUTION gen
f dilution *f*
e desleimiento *m*, dilución *f*
i diluizione *f*
n aanlengen *n*, verdunnen *n*
d Konzentrationsverringerung *f*, Verdünnung *f*

1775 DIMETHYL KETONE, ACETONE, KETOPROPANE, PROPANONE oic
f acétone *f*, éther *m* pyroacétique
e acetona *f*, dimetil-cetona *f*
i acetone *m*, dimetilchetone *m*
n aceton *n*, dimethylketon *n*, propanon *n*
d Azeton *n*, Dimethylketon *n*

1776 DIP BRAZING wel
f soudure *f* forte par immersion
e soldadura *f* fuerte por inmersión
i brasatura *f* per immersione
n hardsolderen *n* met dompelbad
d Immersionshartlötung *f*

1777 DIP COAT coa
f potée *f*
e polvo *m* en emulsión
i luto *m* liquido
n vormsuspensie
d Formschlicker *m*, Überzugsmasse *f*

1778 DIP COATING, IMMERSION COATING cer,pla
f couche *f* au trempé
e capa *f* por inmersión
i rivestimento *m* per immersione
n deklaag verkregen door indompelen
d Tauchüberzug *m*

1779 DIP COATING, IMMERSION COATING dec
f vernissage *m* au trempé
e barnizado *m* por inmersión
i verniciatura *f* per immersione
n dompellakken *n*
d Tauchlackierung *f*

1780 DIP ENAMELLING, WET ENAMELLING coa
f émaillage *m* au trempé
e esmaltado *m* por inmersión
i smaltatura *f* ad immersione
n dompelemailleren *n*
d Nassemaillierung *f*

1781 DIP PLATING, IMMERSION PLATING coa
f dépôt *m* au trempé
e depósito *m* por inmersión
i deposito *m* per immersione
n chemisch neerslaan *n*
d Ansieden *n*, Eintauchplattierung *f*

1782 DIPPEL'S OIL, BONE OIL fat
f huile *f* d'os
e aceite *m* de huesos
i olio *m* d'ossa
n beenderolie
d Knochenöl *n*

1783 DIPPING COMPOUND coa
f mélange *m* de trempage
e mezcla *f* para inmersión
i mescolanza *f* per immersione
n dompelmengsel *n*
d Tauchmischung *f*

1784 DIPPING LACQUER coa,dec
f vernis *m* à tremper
e barniz *m* para inmersión
i vernice *f* per immersione
n dompellak
d Tauchlack *m*

1785 DIRECT COTTON DYE, DIRECT DYE dye
f colorant *m* direct
e colorante *m* directo
i colorante *m* diretto
n directe kleurstof
d Direktfarbstoff *m*

1786 DIRECT DYEING dye
f teinture *f* directe
e teñido *m* directo
i tintura *f* diretta
n directe verving
d direkte Färbung *f*

1787 DIRECT FIRING cer
f cuite *f* à feu direct
e cocción *f* de fuego directo
i cottura *f* a fuoco diretto
n bakken *n* in open vuur
d Brennen *n* in offenem Feuer

1788 DIRECT OXIDATION met
f oxydation *f* superficielle
e oxidación *f* superficial
i ossidazione *f* superficiale
n oppervlakkige oxydatie
d oberflächlige Oxydierung *f*, oberflächliche Oxydation *f*

1789 DIRECT PRINTING, APPLICATION PRINTING tex
f impression *f* directe
e estampa *f* directa
i stampa *f* diretta
n directe textieldruk
d direkter Textildruck *m*

1790 DIRECT PROCESS, KASIKA PROCESS, TWIN HEAD PROCESS dec
f procédé *m* de pistolage à deux composants
e procedimiento *m* de chorreado con dos componentes
i processo *m* di spruzzo con due componenti
n tweecomponentenspuitproces *n*
d Zweikomponentenspritzverfahren *n*

1791 DIRECT ZINC OXIDE PROCESS, AMERICAN ZINC OXIDE PROCESS met
f procédé *m* direct pour la fabrication d'oxyde de zinc
e procedimiento *m* directo para la fabricación de óxido de cinc
i processo *m* diretto per la fabbricazione d'ossido di zinco
n direct proces *n* voor de vervaardiging van zinkoxyde
d direktes Zinkoxydverfahren *n*

1792 DISACIDIFICATION, DESACIDIFICATION cle
f désacidification *f*
e desacidificación *f*
i disacidificazione *f*
n ontzuring
d Entsäuerung *f*

1793 DISCHARGE iwt
f décharge *f*
e descarga *f*
i scarica *f*
n lozing
d Entlastung *f*

1794 DISCHARGE AGENT tex
f rongeant *m*
e corrosivo *m*
i agente *m* corrosivo
n beitsmiddel *n*, bijtmiddel *n*
d Ätzmittel *n*

1795 DISCHARGE PRINTING tex
f impression *f* au rongeant
e estampado *m* por corrosión
i stampatura *f* al corrosivo
n beitsdruk
d Ätzdruck *m*

1796 DISCHARGE VOLTAGE (US), DECOMPOSITION VOLTAGE (GB) elc
f tension *f* de décomposition
e tensión *f* de descomposición
i tensione *f* di decomposizione, tensione *f* di scarica
n ontledingsspanning
d Zersetzungsspannung *f*

1797 DISCHARGING tex
f rongeage *m*
e corrosión *f*
i corrosione *f*
n beitsen *n*, bijten *n*
d Ätzen *n*

1798 DISCOLORATION OF PAPER pap
f jaunissement *m* de papier
e descoloración *f* de papel
i ingiallimento *m* di carta
n vergelen *n* van papier
d Vergilbung *f* von Papier

1799 DISCOLORING (US), dec
DISCOLOURING (GB)
f déteinte *f*
e desteñidura *f*
i cambio *m* di colore
n verkleuren *n*
d Entfärben *n*, Farbumschlag *m*

1800 DISINCRUSTANT, boi
ANTI-INCRUSTATOR
f anti-incrostant *m*, tartrifuge *m*
e antiincrustante *m*, desincrustante *m*
i sostanza *f* che evita le incrostazioni
n ketelsteenverhinderend middel *n*
d Kesselsteinverhütungsmittel *n*

1801 DISINFECTANT hyg,mcy
f désinfectant *m*
e desinfectante *m*
i disinfettante *m*
n desinfecteermiddel *n*, ontsmettingsmiddel *n*
d Desinfektionsmittel *n*

1802 DISINFECTION hyg,mcy
f désinfection *f*
e desinfección *f*
i disinfezione *f*
n desinfectie, ontsmetting
d Desinfektion *f*, Entseuchung *f*

1803 DISINFESTANT hyg
f agent *m* d'épouillage
e agente *m* de espulgo
i agente *m* da spidocchiare
n ontluizingsmiddel *n*
d Entläusungsmittel *n*

1804 DISINTEGRATION gen
f désintégration *f*
e desintegración *f*
i disintegrazione *f*
n uiteenvallen *n*
d Zersetzung *f*

1805 DISINTEGRATION, elc
CATHODE SPUTTERING
f pulvérisation *f* cathodique
e pulverización *f* catódica
i polverizzazione *f* catodica
n katodeverstuiving
d Katodenzerstäubung *f*

1806 DISINTEGRATION, nue
DECAY
f désintégration *f*
e desintegración *f*
i decadimento *m*, disintegrazione *f*
n desintegratie, verval *n*
d Umwandlung *f*, Zerfall *m*

1807 DISINTEGRATION, gen
PULVERIZATION
f pulvérisation *f*
e pulverización *f*
i polverizzazione *f*
n verpoedering, verpulvering
d Verpulverung *f*

1808 DISPERGATOR, col
DISPERSING AGENT
f agent *m* dispersif
e agente *m* dispersante
i agente *m* disperdente
n dispergator, dispergeermiddel *n*
d Dispergator *m*, Dispergierungsmittel *n*

1809 DISPERSE DYE dye
f colorant *m* à dispersion
e colorante *m* de dispersión
i colorante *m* a dispersione
n dispersiekleurstof
d Dispersionsfarbstoff *m*

1810 DISPERSE PHASE RULE, gas
SOLID PHASE RULE
f règle *f* de la phase dispersée
e regla *f* de la fase dispersa
i regola *f* della fase dispersa
n disperse-fazeregel
d disperse Phasenregel *f*

1811 DISPERSION gen
f dispersion *f*
e dispersión *f*
i dispersione *f*
n dispersie
d Dispersion *f*

1812 DISPERSION HARDENING, met
AGE HARDENING,
PRECIPITATION HARDENING
f durcissement *m* structural
e endurecimiento *m* estructural
i incrudimento *m* strutturale
n structurele harding
d strukturelle Härtung *f*

1813 DISPERSION MEDIUM gen
f milieu *m* de dispersion
e medio *m* de dispersión
i mezzo *m* di dispersione
n dispersiemedium *n*, dispersiemiddel *n*
d Dispergens *n*, Dispersionsmedium *n*, Dispersionsmittel *n*

1814 DISPERSION RESINS pla
f résines *pl* à dispersion
e resinas *pl* de dispersión
i resine *pl* a dispersione
n dispersieharsen *pl*
d Dispersionsharze *pl*

1815 DISSOCIATION gen
f dissociation *f*
e disociación *f*
i dissociazione *f*
n dissociatie
d Dissoziation *f*

1816 DISSOLUTION, SOLUTION gen
f dissolution *f*, solution *f*
e disolución *f*, solución *f*
i dissoluzione *f*, soluzione *f*
n oplossing
d Auflösung *f*, Lösung *f*

1817 DISSOLVED ACETYLENE gas,wel
f acétylène-dissous *m*
e acetileno *m* disuelto
i acetilene *m* disciolto
n acétylène-dissous *n*, dissous-gas *n*
d Dissousgas *n*, gelöstes Azetylen *n*

1818 DISSOLVED CARBON gen
f carbone *m* dissous
e carbono *m* disuelto
i carbonio *m* disciolto
n opgeloste koolstof
d gelöster Kohlenstoff *m*

1819 DISSOLVING gen
f dissolution *f*
e disolución *f*
i dissoluzione *f*
n oplossen *n*
d Auflösen *n*

1820 DISTEMPER, TEMPER dec
f détrempe *f*
e templa *f*
i tempera *f*
n lijmverf, temperaverf
d Leimfarbe *f*, Temperafarbe *f*

1821 DISTEMPER, WATER PAINT dec
f peinture *f* à l'eau
e aguada *f*
i colore *m* ad acqua
n waterverf
d Wasserfarbe *f*

1822 DISTILLATE dis
f distillat *m*
e destilado *m*
i distillato *m*
n destillaat *n*
d Destillat *n*

1823 DISTILLATION dis
f distillation *f*
e destilación *f*
i distillazione *f*
n destilleren *n*
d Destillieren *n*

1824 DISTILLED VINEGAR aam
f acide *f* acétique
e ácido *m* acético
i acido *m* acetico
n azijnzuur *n*
d Essigsäure *f*

1825 DISTILLED WATER dis
f eau *f* distillée
e agua *f* destilada
i acqua *f* distillata
n gedestilleerd water *n*
d destilliertes Wasser *n*

1826 DISTILLER'S MALT, LONG MALT, OVERGROWN MALT bre
f hussards *pl*
e malta *f* de germinación larga
i malto *m* a germogliazione spinta, ussari *pl*
n huzaren *pl*, langmout *n*
d Langmalz *n*

1827 DISTILLER'S WASH, SLOP, SPENT MASH, STILLAGE, VINASSE nut
f drêches *pl*, vinasses *pl* de distillerie
e orujo *m*, vinazas *pl* de destilación
i burlanda *f*, trebbia *f*, vinacce *pl* distillate
n drab
d abgebrannte Würze *f*, Schlempe *f*, Spülicht *n*

1828 DISTILLERY WASTE dis,iwt
f eaux *pl* résiduaires de distillerie
e aguas *pl* cloacales de destilería
i acque *pl* di rifiuto di distilleria
n afvalwater *n* van een destilleerderi
d Brennereiabwässer *pl*

1829 DOCTOR SOLUTION fue
f solution *f* de plombite de sodium

e solución *f* de sodio plúmbico
i soluzione *f* di piombite di sodio
n natriumplumbietoplossing
d Natriumplumbitlösung *f*

1830 DOCTOR TEST ana,fue
f détermination *f* du soufre
e determinación *f* del azufre
i determinazione *f* dello zolfo
n zwavelbepaling
d Doktortest *m*, Schwefelbestimmung *f*

1831 DOCTOR TREATMENT fue
f raffination *f* pour éliminer le mercaptan
e desodorización *f* de petróleos, tratamiento *m* doctor
i raffinazione *f* per eliminare cattivi odori
n doctorraffinage, mercaptaanvrij maken *n*
d Merkaptanbeseitigung *f*

1832 DOCTORING pap
f raclement *m*
e rascadura *f*
i spalmatura *f*
n afschrapen *n*, afstrijken *n*
d Abstreichen *n*, Rakeln *n*

1833 DOCTORING, SPONGE PLATING elc
f revêtement *m* électrolytique avec électrode à éponge
e revestimiento *m* galvánico con electrodo macrocelular
i rivestimento *m* galvanico con elettrodo macrocellulare
n tamponneren *n*
d Galvanisieren *n* mit Schwammelektrode

1834 DOMESTIC PAINT, HOUSEHOLD PAINT dec
f peinture *f* ménagère
e pintura *f* casera
i pittura *f* domestica
n bussenverf, huishoudverf
d Farbe *f* für Haushaltzwecke

1835 DOMESTIC REFUSE (GB), HOUSEHOLD GARBAGE (US) hyg
f ordures *pl* ménagères
e basuras *pl* domésticas
i immondizie *pl* domestiche, rifiuti *pl* domestici
n huisvuil *n*
d Hausmüll *m*

1836 DOPE coa
f couche *f* protectrice
e capa *f* de protección
i strato *m* di protezione
n beschermende laag
d Schutzüberzug *m*

1837 DOSING gen
f dosage *m*
e dosificación *f*
i dosaggio *m*
n doseren *n*
d Dosieren *n*

1838 DOT WELDING wel
f soudage *m* d'apport
e soldadura *f* de aporto
i saldatura *f* d'apporto
n oplassen *n*
d Aufschweissen *n*, Auftragschweissen *n*

1839 DOUBLE-BATH DYEING dye
f teinture *f* en deux bains
e teñido *m* de doble baño
i tintura *f* in due bagni
n tweebadsverving
d zweibadige Färbung *f*, Zweibadverfahren *n*

1840 DOUBLE-COATED FILM pho
f pellicule *f* à deux couches sensibles
e película *f* de dos capas sensibles
i pellicola *f* a doppio strato sensibile
n tweelaagsfilm
d Doppelschichtfilm *m*, Zweischichtfilm *m*

1841 DOUBLE TRANSFER PROCESS pho
f procédé *m* à double report
e procedimiento *m* de doble transporte
i processo *m* a doppio trasporto
n dubbeloverdrukproces *n*
d Doppelübertragungsverfahren *n*

1842 DOWN-FLOW, SAGGING dec
f affaisement *m*, formation *f* de festons
e arrugamiento *m*
i insaccamento *m*
n vorming van zakkers
d Gardinenbildung *f*, Vorhängebildung *f*

1843 DRAFT BEER (US), bre
DRAUGHT BEER (GB),
BEER FROM THE WOOD,
CASK BEER
f bière *f* en fût
e cerveza *f* de barril
i birra *f* a spina, birra *f* in fusto
n bier *n* van het vat, vatbier *n*
d Bier *n* vom Fass, Fassbier *n*

1844 DRAFT IN DIFFUSION (US), sin
DRAUGHT IN DIFFUSION (GB),
JUICE DRAW-OFF
f décocté *m* de jus
e decocción *f* de jugo
i decotto *m* di succo
n sapaftreksel *n*
d Saftabzug *m*

1845 DRAG-IN elc
f solution *f* adhérente
e solución *f* adherida
i soluzione *f* aderente
n inbrengverliezen *pl*
d eingeschleppte Lösung *f*, Eintrag *m*

1846 DRAG-OUT elc
f solution *f* entraînée
e solución *f* arrastrada
i soluzione *f* asportata, soluzione *f* estratta
n uithaalverliezen *pl*
d Austrag *m*, herausgeschleppte Lösung *f*

1847 DRAGON'S BLOOD dec,pla
f sang *m* de dragon, sang-dragon *m*
e sangre *f* de drago
i sangue *m* di drago
n drakebloed *n*
d Drachenblut *n*

1848 DRAWINGS, bre
BEER DRAWN OFF FROM BOTTOM
f bière *f* de fond
e cerveza *f* de colador
i resti *pl* dalle botti
n afgefiltreerd bier *n*
d Abseihbier *n*

1849 DRAWN-IN SCALE met
f encroûtement *m*
e cascarilla *f* incrustada
i scaglia *f* incrostata
n ingevreten oxydelaag
d eingefressener Zunder *m*

1850 DREDGING cer,coa
f poudrage *m*, poudrement *m*
e espolvereo *m*
i spargimento *m* di polvere
n bestrooiing
d Aufpudern *n*, Aufstreuen *n*

1851 DREGS, bre
SPENT BARLEY
f drêches *pl* de malt
e bagazo *m* de malta, orujos *pl* de malta
i trebbie *pl* di malto
n moutdraf
d Malztreber *m*

1852 DRENCHING lea
f trempage *m*
e remojado *m*
i rammollimento *m*
n weken *n*
d Weichen *n*

1853 DRESSING tex
f encollage *m*
e adobado *m*, encolado *m*, zurrado *m*
i imbozzimatura *f*
n slichten *n*
d Schlichten *n*

1854 DRESSING, bre
CLEANING
f nettoyage *m*
e limpieza *f*
i pulitura *f*
n poetsen *n*
d Putzen *n*

1855 DRESSING, tex
FINISHING, SIZING
f apprêtage *m*
e apresto *m*
i apprettatura *f*
n appreteren *n*
d Appretur *f*

1856 DRESSING, min
ORE CONCENTRATION
f enrichissement *m* des minerais
e enriquecimiento *m* de los minerales
i arricchimento *m* dei minerali
n ertsconcentratie
d Erzanreicherung *f*

1857 DRIER, dec
SICCATIVE
f siccatif *m*
e secante *m*, secativo *m*

i essiccativo *m*
n droogmiddel *n*, siccatief *n*
d Trockenmittel *n*, Sikkativ *n*

1858 DRIP COOLING clo
f refroidissement *m* à gouttes
e enfriamiento *m* de gotas
i raffreddamento *m* a gocce
n druppelkoeling
d Tropfkühlung *f*

1859 DROP COLORIMETRY ana
f colorimétrie *f* à la goutte
e colorimetría *f* de gota
i colorimetria *f* a goccia
n droppelcolorimetrie
d Tüpfelkolorimetrie *f*

1860 DROSS (US), SCUM (GB) met
f écume *f*
e espuma *f*
i feccia *f*, scoria *f* metallifera
n droes, schuim *n*
d Krätze *f*, Schaum *m*

1861 DROSSING (US), SKIMMING (GB) met
f décrassage *m*, écrémage *m*
e despumación *f*
i rimozione *f* della feccia,
n afschuimen *n*
d Schlickern *n*

1862 DRUM DYEING tex
f teinture *f* à tambour
e teñido *m* de tambor
i tintura *f* a tamburo
n trommelverven *n*
d Trommelfärben *n*

1863 DRUMMING lea
f battement *m* du cuir
e batanadura *f*
i bottalatura *f*
n lederkloppen *n*
d Lederklopfen *n*

1864 DRY AIR CURING, DRY HEAT VULCANIZATION, HOT AIR CURING rub
f vulcanisation *f* à l'air chaud
e vulcanización *f* al aire caliente
i vulcanizzazione *f* in aria calda
n hete-luchtvulcanisatie
d Heissluftvulkanisation *f*

1865 DRY ASSAYING, FIRE ASSAYING ana
f essai *m* à sec
e ensayo *m* por vía seca
i assaggio *m* per via secca
n essaaieren *n* langs de droge weg
d Prüfen *n* auf trocknem Wege

1866 DRY CLEANING cle
f nettoyage *m* à sec
e limpieza *f* en seco
i lavaggio *m* a secco, pulitura *f* a secco
n chemisch wassen *n*
d chemische Reinigung *f*

1867 DRY CLEANING NAPHTA cle
f essence *f* de pétrole pour nettoyage à sec
e nafta *f* para limpieza en seco
i nafta *f* per lavaggio a secco, nafta *f* per pulitura a secco
n wasbenzine
d Waschbenzin *n*

1868 DRY COLOR (US), DRY COLOUR (GB), PIGMENT dec
f couleur *f* en poudre, couleur *f* sèche, pigment *m*
e materia *f* colorante, pigmento *m*
i materia *f* colorante, pigmento *m*
n pigment *n*, verfstof
d Farbkörper *m*, Körperfarbe *f*, Pigment *n*, Trockenfarbe *f*

1869 DRY CYANIDING, GAS CYANIDING met
f cyanuration *f* à l'ammoniaque
e cianuración *f* al amoníaco
i cianurazione *f* all'ammoniaca
n cyaancementeren *n* in ammoniak
d Zyanierung *f* in Ammoniak

1870 DRY DISTILLATION, DESTRUCITIVE DISTILLATION dis
f distillation *f* destructive, distillation *f* sèche
e destilación *f* destructiva, destilación *f* seca
i distillazione *f* a secco, distillazione *f* distruttiva
n droge destillatie
d Trockendestillation *f*, Zersetzungsdestillation *f*

1871 DRY ENAMELLING coa,met
f émaillage *m* au poudré
e esmaltado *m* con polvo
i smaltatura *f* a spolvere
n poederemailleren *n*
d Puderemaillieren *n*

1872 DRY FILLER pla,rub
f charge *f* sèche
e carga *f* seca
i carica *f* secca
n droge vulstof
d Trockenfüllstoff *m*

1873 DRY FLUE GAS fue,gas
f fumées *pl* sèches
e humos *pl* secos
i fumi *pl* secchi
n droge rookgassen *pl*
d Trockenabgase *pl*, trockene Rauchgase *pl*

1874 DRY GAS gas
f gaz *m* naturel sans hydrocarbures
e gas *m* natural sin hidrocarburos
i gas *m* naturale senza idrocarburi
n aardgas *n* vrij van koolwaterstoffen
d Erdgas *n* frei von Kohlenwasserstoffen

1875 DRY GAS, AIR GAS gas
f gaz *m* à l'air
e gas *m* de aire
i gas *m* d'aria
n luchtgas *n*
d Aerogengas *n*, Luftgas *n*

1876 DRY GAS CLEANING dry,gas
f dépoussiérage *m* à sec du gaz
e depuración *f* seca del gas
i epurazione *f* a secco del gas
n droge gasreiniging
d trockene Gasreinigung *f*

1877 DRY HOPPING bre
f houblonnage *m* en fût, houblonnage *m* en garbe
e lupulado *m* en crudo, lupulado *m* en seco
i luppolatura *f* in cantina, luppolatura *f* in crudo
n drooghoppen *n*
d Hopfen *n* vor dem Abfüllen, Hopfenstopfen *n*

1878 DRY ICE, CARBON DIOXIDE SNOW mcy
f neige *f* carbonique
e hielo *m* seco, nieve *f* carbónica
i ghiaccio *m* secco
n droog ijs *n*
d Kohlensäureschnee *m*, Trockeneis *n*

1879 DRY LUBRICANT, LUBRICATING LACQUER dec
f peinture *f* lubrifiante
e pintura *f* lubricante
i pittura *f* lubrificante
n glijlak
d Gleitlack *m*

1880 DRY MATTER, TOTAL SOLIDS rub
f matières *pl* solides
e materia *f* seca
i materia *f* secca
n droge stof, indampresidu *n*
d Trockensubstanz *f*

1881 DRY MILLING dec
f broyage *m* à sec
e bocarteado *m* en seco
i frantumazione *f* a secco
n droog malen *n*
d Trockenvermahlung *f*

1882 DRY PROCESS bui
f procédé *m* par voie sèche
e procedimiento *m* por vía seca
i processo *m* a secco
n droog proces *n*
d Trockenverfahren *n*

1883 DRY PROCESS pla
f procédé *m* de moulage à quantité d'eau insuffisante
e procedimiento *m* de moldeo con cuantidad de agua insuficiente
i processo *m* di formazione a quantità d'acqua insufficiente
n persproces *n* met ondermaat water
d Pressverfahren *n* mit Untermass an Wasser

1884 DRY RESIDUE, RESIDUE ON EVAPORATION eva
f résidu *m* sec, résidu *m* fixe
e residuo *m* seco
i deposito *m* secco, residuo *m* fisso
n droog residu *n*
d Trockenrückstand *m*

1885 DRY-SALTED HIDE tan
f peau *f* salée à sec
e piel *f* salada en seco
i pelle *f* a salatura a secco
n drooggezouten huid
d trockengesalzene Haut *f*

1886 DRY SATURATED STEAM, DRY STEAM boi
f vapeur *f* sèche

e vapor *m* seco
i vapore *m* secco, vapore *m* surriscaldato
n droge stoom, oververhitte stoom
d Trockendampf *m*

1887 DRY SPINNING tex
f filature *f* à sec
e hilado *m* en seco, hilatura *f* en seco
i filatura *f* a secco
n droogspinnen *n*
d Trockenspinnverfahren *n*

1888 DRY SPRAYING coa,dec
f pistolage *m* à sec
e chorreado *m* en seco
i spruzzo *m* a secco
n droogspuiten *n*
d Trockenspritzen *n*

1889 DRYING dry
f séchage *m*
e secado *m*
i essiccazione *f*
n drogen *n*
d Trocknen *n*

1890 DRYING AGENT pla
f accélérateur *m* de durcissement
e acelerador *m* de endurimiento
i acceleratore *m* d'indurimento
n hardingsversneller
d Härtungsbeschleuniger *m*

1891 DRYING BY EVAPORATION dry,eva
f séchage *m* par évaporation
e secado *m* por evaporación
i essiccazione *f* per evaporazione
n verdampingsdroging
d Trocknen *n* durch Verdampfung

1892 DRYING OIL, dec
BOILED OIL, BOILED LINSEED OIL
f huile *f* de lin cuite
e aceite *m* de linaza cocido, aceite *m* secante
i olio *m* di lino cotto
n lijnolievernis *n*
d Leinölfirnis *m*, Ölfirnis *m*

1893 DRYING OUT dry
f desséchement *m*
e desecación *f*, desecamento *m*
i prosciugamento *m*
n uitdrogen *n*
d Austrocknen *n*

1894 DUFF, sor
THROUGH, UNDERSIZE
f passé *m* de crible
e materia *f* cribada, materia *f* que pasa por la criba
i materia *f* passante, materia *f* vagliata
n zeefsel *n*, ziftsel *n*
d Siebdurchfall *m*, Unterkorn *n*

1895 DULL DRESSING, tex
DULL FINISHING, DULL SIZING
f apprêtage *m* mat
e apresto *m* mate
i apprettatura *f* matta
n mat appreteren *n*
d Mattappretur *f*

1896 DULL FINISH, dec
DEAD FINISH, FLAT FINISH
f finissage *m* mat
e acabado *m* mate, terminación *f* mate
i finissaggio *m* matto, rifinitura *f* opaca
n mat afstrijken *n*, matte deklaag
d matte Deckschicht *f*, Mattglanz *m*

1897 DULL FINISH, cle
MAT ETCHING
f décapage *m* au mat
e decapaje *m* mate
i decapaggio *m* matto
n matbeitsen *n*
d Mattbeizung *f*

1898 DUO-SOL PROCESS lub
f extraction *f* à deux solvants
e extracción *f* con dos solventes
i estrazione *f* a due solventi
n extractie met twee oplosmiddelen
d Extraktion *f* mit zwei Lösungsmitteln

1899 DURABLE DRESSING, tex
DURABLE FINISHING, DURABLE SIZING
f apprêt *m* permanent
e apresto *m* permanente
i apprettatura *f* permanente
n duurzaam appreteren *n*
d dauerhaftes Appretieren *n*

1900 DUST EXHAUSTION cle
f aspiration *f* de poussières, dépoussiérage *m*
e aspiración *f* de polvos, despolverización *f*
i aspirazione *f* di polveri, depolverazione *f*

n stofafzuiging
d Staubabsaugung *f*

1901 DUST-FREE GAS (GB), CLEAN GAS (US) gas
f gaz *m* dépoussiéré
e gas *m* desempolvado
i gas *m* depurato della povere
n fijngas *n*, stofvrij gas *n*
d entstaubtes Gas *n*, staubfreies Gas *n*

1902 DUSTING met
f poudrage *m* au soufre
e espolvoreo *m* con azufre
i impolverizzazione *f* con zolfo
n zwavelbepoedering
d Einstauben *n* mit Schwefel, Schwefelbestäubung *f*

1903 DUSTING (GB), SOAP STONING (US) coa
f talquage *m*
e entalcado *m*
i impolverizzazione *f* con talco
n met talk bepoederen *n*, talken *n*
d Talkumieren *n*

1904 DUTCH ENAMEL, STANDOIL PAINT dec
f peinture *f* à la standolie
e pintura *f* al aceite secante
i pittura *f* all'olio di lino cotto
n standolieverf
d Standölfarbe *f*

1905 DUTCH PINK, CHROME-TIN PINK, ENGLISH PINK, ITALIAN PINK dec
f pigment *m* jaune transparent
e pigmento *m* amarillo transparente
i pigmento *m* giallo trasparente
n gele lakverf
d gelbe Lackfarbe *f*, Schüttgelb *n*

1906 DYE BATH dye
f bain *m* de teinture
e baño *m* de tinte
i bagno *m* di tintura
n verfbad *n*
d Färbebad *n*

1907 DYE-FIXING AGENT dye
f moyen *m* fixateur de colorants
e medio *m* de fixación de los colorantes
i mezzo *m* da fissare i coloranti
n kleurstoffixeermiddel *n*
d Farbstoffixiermittel *n*

1908 DYE TONING pho
f virage *m* par teinture sur mordantage
e viraje *m*
i viraggio *m*
n kleuren *n*
d Bunttonung *f*, Farbtonung *f*

1909 DYEING dye
f teinture *f*
e teñido *m*
i tintura *f*
n verven *n*
d Färben *n*

1910 DYESTUFFS dye
f colorants *pl*
e colorantes *pl*
i coloranti *pl*
n kleurstoffen *pl*
d Farbstoffe *pl*

E

1911 EARLY SETTING CEMENT(US),bui
QUICK SETTING CEMENT (GB),
RAPID HARDENING CEMENT
f ciment *m* à prise rapide,
ciment *m* prompt
e cemento *m* de fraguado rápido
i cemento *m* a presa rapida,
cemento *m* pronto
n snelbindend cement *n*
d schnellbindender Zement *m*,
Schnellbinder *m*

1912 EARTH COLORS (US), dec
EARTH COLOURS (GB),
NATURAL EARTH PIGMENTS
f couleurs *pl* minérales,
pigments *pl* terreux
e pigmentos *pl* de tierras naturales
i colori *pl* a base di terra
n aardverfstoffen *pl*
d Erdfarben *pl*, natürliche
Mineralfarben *pl*

1913 EARTHNUT OIL, fat
ARACHIS OIL, GROUND OIL,
PEANUT OIL
f huile *f* d'arachide
e aceite *m* de cacahuete
i olio *m* d'arachide
n aardnotenolie
d Erdnussöl *n*

1914 EASY FIRING cer
f chauffage *m* insuffisant
e caldeo *m* insuficiente
i riscaldamento *m* insufficiente
n te lage verhitting
d zu niedrige Heizung *f*

1915 EASY FLOW pla
f fluidité *f* élevée
e fluidez *f* fácil
i fluidità *f* facile
n goede vloeibaarheid
d gute Fliessfähigkeit *f*

1916 EBONITE (GB), pla
HARD RUBBER (US),
VULCANITE
f ébonite *f*
e ebonita *f*
i ebanite *f*
n eboniet *n*
d Ebonit *m*, Hartgummi *m*

1917 EBULLITION gen
f ébullition *f*
e ebullición *f*
i ebollizione *f*
n koken *n*
d Sieden *n*

1918 EDIBLE FAT nut
f graisse *f* alimentaire
e grasa *f* comestible
i grasso *m* comestibile
n eetbaar vet *n*, spijsvet *n*
d Speisefett *n*

1919 EDIBLE OIL nut
f huile *f* alimentaire
e aceite *m* comestible
i olio *m* comestibile
n eetbare olie, spijsolie
d Speiseöl *n*

1920 EDUCT gen
f produit *m* de décomposition
e educto *m*, producto *m* de
descomposición
i prodotto *m* di decomposizione
n edukt *n*, ontledingsprodukt *n*
d Zersetzungsprodukt *n*

1921 EDUCTION hyg,iwt
f écoulement *m*
e educción *f*, escape *m*
i efflusso *m*
n afvoer
d Abführung *f*, Abzug *m*, Auslass *m*,
Auspuff *m*

1922 EDULCORATION cle
f édulcoration *f*
e purificación *f* por medio de
lavado
i processo *m* edulcorante
n uitwassen *n*
d Auswaschen *n*

1923 EFFERVESCENCE gen
f effervescence *f*
e efervescencia *f*
i effervescenza *f*
n bruisen *n*
d Aufbrausen *n*, Brausen *n*

1924 EFFERVESCENT FERMEN- fmi
TATION
f fermentation *f* effervescente,
fermentation *f* tumultueuse

e fermentación *f* efervescente
i fermentazione *f* effervescente
n gisting met schuimvorming, stormachtige gisting
d Schaumgärung *f*, stürmische Gärung *f*

1925 EFFLORESCENCE cer,fue,gen
f efflorescence *f*
e eflorescencia *f*
i efflorescenza *f*
n efflorescentie, uitbloeiing
d Ausblühung *f*, Effloreszenz *f*

1926 EFFLORESCENCE, BLEMISH, SCUM bui
f carie *f*, efflorescence *f*, salpêtre *m*
e eflorescencia *f*
i efflorescenza *f* murale
n muurkanker, muursalpeter, muuruitslag
d Ausblühung *f*, Auswitterung *f*, Mauerfrass *m*, Salpeterfrass *m*

1927 EFFLUENT MONITORING nue
f contrôle *m* de radioactivité d'effluent
e control *m* de radiactividad de efluente
i controllo *m* di radioattività d'effluente
n afvoerbewaking
d Abflussüberwachung *f*

1928 EFFLUENTS hyg,iwt
f eaux *pl* d'égout
e efluentes *pl*
i effluenti *pl*
n afvalwater *n*
d Abwässer *pl*

1929 EFFUSION gen
f effusion *f*
e efusión *f*
i effusione *f*
n effusie, uitstroming
d Ausströmung *f*, Effusion *f*

1930 EGG-SHELL FINISH pap
f non-calandré adj
e no calandrado adj
i non calandrato adj
n niet-gekalanderd adj
d unkalandert adj

1931 EGG-SHELL FINISH tex
f apprêt *m* mat
e acabado *m* mate
i apprettatura *f* matta
n mat appreteren *n*
d Mattappretur *f*

1932 EGG-SHELL GLOSS (GB), BASTARD FLATTING (US) dec
f aspect *m* coquille d'oeuf
e lustre *m* suave
i lucidatura *f* opaca
n eiglans
d Eierschalenglanz *m*

1933 EGG-SHELL PORCELAIN cer
f coquille *f* d'oeuf
e porcelana *f* muy fina
i porcellana *f* guscio d'uovo
n eierschaalporselein *n*
d Eierschalenporzellan *n*

1934 EJECTION pla,rub
f éjection *f*
e expulsión *f*, eyección *f*
i eiezione *f*, espulsione *f*
n uitstoten *n*
d Ausstossen *n*, Entformung *f*, Herauswerfen *n*

1935 ELASTIC AFTEREFFECT pla
f élasticité *f* de suite, récupération *f* élastique retardée
e elasticidad *f* subsequiente, recuperación *f* elástica retardada
i elasticità *f* susseguente, ricupero *m* elastico ritardato
n elastische nawerking
d elastische Nachwirkung *f*

1936 ELASTIC BITUMEN, ELATERITE, MINERAL CAOUTCHOUC min
f bitume *m* élastique, élatérite *f*
e caucho *m* mineral, elaterita *f*
i elaterite *f*
n elateriet *n*, mineraal caoutchouc veerkrachtig aardpek *n*
d elastisches Erdpech *n*, Elaterit *m*

1937 ELASTIC GUM, CAOUTCHOUC, INDIA RUBBER rub
f caoutchouc *m*
e caucho *m*
i caucciù *m*
n caoutchouc *n*
d Gummielasticum *n*, Kautschuk *m*, Rohgummi *m*

1938 ELASTOMER pla,rub
f élastomère *m*

e elastómero *m*
i elastomero *m*
n elastisch polymeer *n*, elastomeer *n*
d elastisches Polymer *n*,
Elastomer *n*

1939 ELASTOMERIC PLASTICS pla
f plastiques *pl* élastomériques
e plásticos *pl* elastoméricos
i plastici *pl* elastomerici
n elastomere kunststoffen *pl*
d elastomere Kunststoffe *pl*

1940 ELECTRIC SOLDERING wel
f soudure *f* électrique
e soldadura *f* eléctrica
i saldatura *f* elettrica
n elektrisch solderen *n*
d Elektrolötung *f*

1941 ELECTRICAL PRECIPITATION, coa
ELECTROSTATIC PRECIPITATION
f précipitation *f* électrique
e precipitación *f* eléctrica
i precipitazione *f* elettrica
n elektrisch neerslaan *n*
d elektrische Abscheidung *f*

1942 ELECTROANALYSIS ana
f électroanalyse *f*
e electroanálisis *f*
i elettroanalisi *f*
n elektroanalyse
d Elektroanalyse *f*

1943 ELECTROBRIGHTENING, elc
ANODIC BRIGHTENING
f brillantage *m* anodique
e abrillantado *m* anódico
i brillantaggio *m* anodico
n anodisch glanzen *n*
d anodisches Glänzen *n*

1944 ELECTROCHEMICAL OXIDATION, gen
ELECTROLYTIC OXIDATION
f oxydation *f* électrochimique,
oxydation *f* électrolytique
e oxidación *f* electroquímica,
oxidación *f* electrolítica
i ossidazione *f* elettrochimica,
ossidazione *f* elettrolitica
n elektrochemische oxydatie,
elektrolytische oxydatie
d elektrochemische Oxydation *f*,
elektrolytische Oxydation *f*

1945 ELECTROCHEMICAL elc
PASSIVITY
f passivité *f* électrochimique
e pasividad *f* electroquímica
i passività *f* elettrochimica
n elektrochemische passiviteit
d elektrochemische Passivität *f*

1946 ELECTROCHEMICAL SERIES, elc
ELECTROMOTIVE SERIES,
CONTACT SERIES
f série *f* électrochimique
e serie *f* electroquímica
i serie *f* elettrochimica
n spanningsreeks
d elektrochemische Spannungsreihe *f*

1947 ELECTRODECANTATION cle,sep
f décantation *f* électrique
e electrodecantación *f*
i decantazione *f* elettrica
n elektrisch decanteren *n*
d elektrisches Dekantieren *n*

1948 ELECTRODEPOSITION coa,elc
f dépôts *pl* électrolytiques
e depósitos *pl* electrolíticos
i deposizione *f* elettrolitica
n galvanotechniek
d elektrolytische Abscheidung *f*

1949 ELECTRODIALYSIS, elc
ELECTRO-ENDOSMOSIS
f électro-endosmose *f*,
électro-osmose *f*
e electroendósmosis *f*,
electroósmosis *f*
i elettroosmosi *f*
n elektro-osmose
d Elektrodialyse *f*,
Elektroendosmose *f*,
Elektroosmose *f*

1950 ELECTRODIPPING dec
f vernissage *m* électrique au
trempé
e barnizado *m* eléctrico por
inmersión
i verniciatura *f* elettrica per
immersione
n elektrisch dompellakken *n*
d Elektrotauchlackierung *f*

1951 ELECTRODISPERSION col,met
f électrodispersion *f*
e electrodispersión *f*
i elettrodispersione *f*
n elektrodispersie
d Elektrodispersion *f*

1952 ELECTRODISSOLUTION elc
f électrodissolution *f*
e electrodisolución *f*

i elettrodissoluzione *f*
n elektrolytisch oplossen *n*
d elektrolytische Auflösung *f*

1953 ELECTRO-EROSION METALWORKING, SPARK METALWORKING met
f usinage *m* par électro-érosion, usinage *m* par étincelage
e trabajo *m* por electro-erosión
i erosione *f* mediante scintillio, lavorazione *f* a scintillamento
n elektro-erosieve behandeling
d Elektroerosionsbehandlung *f*

1954 ELECTRO-EXTRACTION, ELECTROWINNING elc,met
f extraction *f* par voie électrolytique
e electroextracción *f*, extracción *f* por vía electrolítica
i estrazione *f* per via elettrolitica
n elektrolytische winning
d elektrolytische Metallgewinnung *f*

1955 ELECTROFACING coa
f recouvrement *m* électrolytique en métal dur
e recargue *m* galvánico
i rivestimento *m* elettrolitico con metallo duro
n elektrolytische hardmetaalbekleding
d elektrolytische Hartmetallbekleidung *f*

1956 ELECTROFORMING, GALVANOPLASTY elc
f électroformage *m*, galvanoplastie *f*
e electroformación *f*
i galvanoplastica *f*
n galvanoplastiek
d Galvanoplastik *f*

1957 ELECTROGALVANIZING coa,elc
f galvanisation *f* électrolytique, zingage *m* électrolytique
e galvanización *f* electrolítica
i galvanizzazione *f* elettrolitica
n elektrolytisch galvaniseren *n*
d galvanische Verzinkung *f*

1958 ELECTROGILDING elc
f dorure *f* électrolytique
e doradura *f* electrolítica
i doratura *f* elettrolitica
n elektrolytisch vergulden *n*
d elektrolytische Vergoldung *f*

1959 ELECTROGRANODIZING elc
f phosphatage *m* électrolytique préventif
e fosfatización *f* electrolítica preventiva
i fosfatizzazione *f* elettrolitica preventiva
n elektrolytisch fosfateren *n*
d elektrolytische Schutzphosphatierung *f*

1960 ELECTROLESS PLATING, CHEMICAL PLATING coa
f placage *m* chimique
e quimioplastia *f*
i placcatura *f* chimica
n chemisch pleten *n*
d chemische Plattierung *f*

1961 ELECTROLYSIS elc
f électrolyse *f*
e electrólisis *f*
i elettrolisi *f*
n elektrolyse
d Elektrolyse *f*

1962 ELECTROLYSIS OF FUSED SALTS elc
f électrolyse *f* en bain fondu, électrolyse *f* ignée
e electrólisis *f* en sales fundidas
i elettrolisi *f* in sali fusi
n smeltbadelektrolyse
d Schmelzflusselektrolyse *f*

1963 ELECTROLYSIS OF SOLUTION elc
f électrolyse *f* aqueuse
e electrólisis *f* de solución
i elettrolisi *f* di soluzione
n natte elektrolyse
d Nasselektrolyse *f*

1964 ELECTROLYTE elc
f électrolyte *m*
e electrólito *m*
i elettrolito *m*
n elektrolyt
d Elektrolyt *m*

1965 ELECTROLYTIC CLEANING cle,elc
f dégraissage *m* électrolytique
e desengrase *m* electrolítico
i sgrassaggio *m* elettrolitico
n elektrolytisch ontvetten *n*
d elektrolytische Reinigung *f*

1966 ELECTROLYTIC COPPER, CATHODE COPPER elc,met
f cuivre *m* électrolytique

e cobre *m* de cátodo, cobre *m* electrolítico
i rame *m* elettrolitico
n elektrolytisch koper *n*
d Elektrolytkupfer *n*, Katodenkupfer *n*

1967 ELECTROLYTIC CORROSION cor,elc
f corrosion *f* électrolytique
e corrosión *f* electrolítica
i corrosione *f* elettrolitica
n elektrolytische corrosie
d elektrolytische Korrosion *f*

1968 ELECTROLYTIC DEPOLARIZATION elc
f dépolarisation *f* électrolytique
e despolarización *f* electrolítica
i depolarizzazione *f* elettrolitica
n elektrolytische depolarisatie
d elektrolytische Depolarisation *f*

1969 ELECTROLYTIC DISSOCIATION elc
f dissociation *f* électrolytique
e disociazión *f* electrolítica
i dissociazione *f* elettrolitica
n elektrolytische dissociatie
d elektrolytische Dissoziation *f*

1970 ELECTROLYTIC PARTING, ELECTROLYTIC SEPARATION, ELECTROPARTING elc
f séparation *f* électrolytique
e separación *f* electrolítica
i separazione *f* elettrolitica
n elektrolytische metaalscheiding
d elektrolytische Metallscheidung *f*

1971 ELECTROLYTIC PICKLING elc
f décapage *m* électrolytique
e decapaje *m* electrolítico
i decapaggio *m* elettrolitico
n elektrolytisch beitsen *n*
d elektrolytisches Beizen *n*

1972 ELECTROLYTIC REDUCTION elc,met
f réduction *f* électrolytique
e reducción *f* electrolítica
i riduzione *f* elettrolitica
n elektrolytische reductie
d elektrolytische Reduktion *f*

1973 ELECTROLYTIC REFINING, ELECTROREFINING elc,met
f électroraffinage *m*, raffination *f* électrolytique
e afinado *m* electrolítico, electroafino *m*
i elettroraffinazione *f*, raffinazione *f* elettrolitica
n elektrolytisch raffineren *n*, elektrolytische zuivering
d elektrolytische Vergütung *f*

1974 ELECTROLYTIC SOLUTION elc
f solution *f* électrolytique
e solución *f* electrolítica
i soluzione *f* elettrolitica
n elektrolytoplossing
d Elektrolytlösung *f*

1975 ELECTROLYTIC SOLUTION PRESSURE elc
f pression *f* de solution électrolytique
e presión *f* de solución electrolítica
i pressione *f* di soluzione elettrolitica
n elektrolytoplossingsdruk
d Elektrolytlösungsdruck *m*

1976 ELECTROLYTIC STRIPPING elc
f dépouillage *m* électrolytique
e eliminación *f* electrolítica
i degalvanizzazione *f* elettrolitica
n elektrolytische laagverwijdering
d elektrolytische Entplattierung *f*

1977 ELECTROLYTIC TINNING coa,elc
f étamage *m* électrolytique
e estañado *m* electrolítico
i stagnatura *f* elettrolitica
n elektrolytisch vertinnen *n*
d elektrolytische Verzinnung *f*

1978 ELECTROGMAGNETIC SEPARATION sor
f séparation *f* électromagnétique, triage *m* électromagnétique
e separación *f* electromagnética
i separazione *f* elettromagnetica
n elektromagnetische afscheiding
d elektromagnetisches Scheiden *n*, Magnetausscheidung *f*

1979 ELECTROMETALLIZATION elc
f métallisation *f* électrolytique
e metalización *f* electrolítica
i metallizzazione *f* elettrolitica
n elektrolytische metallisatie
d elektrolytische Metallisierung *f*

1980 ELECTROMETRIC TITRATION ana
f titrage *m* potentiométrique
e valoración *f* electrométrica

i titolazione *f* elettrometrica
n elektrometrisch titreren *n*
d elektrometrische Massanalyse *f*

1981 ELECTROPHORESIS, CATAPHORESIS gen
f cataphorèse *f*
e cataforesis *f*
i cataforesi *f*
n kataforese
d Kataphorese *f*

1982 ELECTROPLATING, GALVANOSTEGY elc
f galvanoplastie *f*
e galvanoplastia *f*, galvanostegia *f*
i galvanoplastica *f*, galvanostegia *f*
n galvanostegie
d Elektroplattierung *f*, Galvanostegie *f*

1983 ELECTROPOLISHING pol
f polissage *m* électrolytique
e pulido *m* electrolítico
i pulitura *f* elettrolitica
n elektrolytisch glanzen *n*, elektrolytisch polijsten *n*
d Elektropolieren *n*

1984 ELECTROSTATIC GAS CONDITIONING gas
f dépoussiérage *m* électrostatique de gaz
e purificación *f* electrostática de gas
i purificazione *f* elettrostatica di gas
n elektrostatische gasreiniging
d elektrostatische Gasreinigung *f*

1985 ELECTROSTATIC PRECIPITATION, ELECTRICAL PRECIPITATION coa
f précipitation *f* électrique
e precipitación *f* eléctrica
i precipitazione *f* elettrica
n elektrisch neerslaan *n*
d elektrische Abscheidung *f*

1986 ELECTROSTATIC SPRAYING coa
f pistolage *m* électrostatique
e chorreado *m* electrostático
i spruzzatura *f* elettrostatica
n elektrostatisch spuiten *n*
d elektrostatisches Spritzen *n*

1987 ELECTROTYPING, GALVANOTYPING pri
f électrotypie *f*
e electrotipia *f*, galvanotipia *f*
i elettrotipia *f*
n galvanotypie
d Elektrotypie *f*

1988 ELECTROWINNING, ELECTRO-EXTRACTION elc,met
f extraction *f* par voie électrolytique
e electroextracción *f*, extracción *f* por vía electrolítica
i estrazione *f* per via elettrolitica
n elektrolytische winning
d elektrolytische Metallgewinnung *f*

1989 ELEMENT gen
f élément *m*
e elemento *m*
i elemento *m*
n element *n*
d Element *n*

1990 ELEMENTARY ANALYSIS, ORGANIC ANALYSIS, ULTIMATE ANALYSIS ana
f analyse *f* élémentaire
e análisis *f* elemental
i analisi *f* elementare
n elementaire analyse
d Elementaranalyse *f*, Verbrennungsanalyse *f*

1991 ELEVATION OF BOILING POINT boi
f relèvement *m* du point d'ébullition
e aumento *m* ebulliscópico, elevación *f* del punto de ebullición
i elevazione *f* del punto d'ebollizione
n kookpuntverhoging
d Siedepunktserhöhung *f*

1992 ELIXIR mcy
f élixir *m*
e elixir *m*
i elisir *m*
n elixer *n*, elixir *n*
d Elixir *n*

1993 ELUTION sin
f élution *f*
e elución *f*
i eluzione *f*
n uitwassen *n*
d Auswaschen *n*

1994 ELUTRIATION min,sor
f élutriation *f*

e elutriación *f*, levigación *f*
i dilavamento *m*, lavaggio *m* a controcorrente
n elueren *n*, opslibben *n*
d Abschlämmen *n*, Schlämmen *n*

1995 EMANATIONS gas,nue
f émanations *pl*
e emanaciones *pl*
i emanazioni *pl*
n emanaties *pl*
d Emanationen *pl*

1996 EMBOSSING lea,pap,tex
f gaufrage *m*, grainage *m*
e gofrado *m*
i goffratura *f*
n gaufreren *n*, pregen *n*
d Gaufrieren *n*, Prägen *n*

1997 EMERAUDE GREEN (US), CHROME GREEN (GB) dec
f vert *m* Casalis, vert *m* de chrome
e verde *m* de cromo
i verde *m* di cromo
n chromaatgroen *n*
d Guignetgrün *n*, Milorigrün *n*

1998 EMERY pol
f émeri *m*
e esmeril *m*
i smeriglio *m*
n amaril *n*, smergel
d Schmirgel *m*

1999 EMERY CLOTH pol
f toile *f* émeri, toile *f* émerisée
e tela *f* esmeril
i tela *f* smeriglio
n schuurlinnen *n*
d Schmirgelleinen *n*

2000 EMERY PAPER pol
f papier *m* émeri
e papel *m* esmeril
i carta *f* smeriglio
n schuurpapier *n*
d Schmirgelpapier *n*

2001 EMPEROR GREEN (US), PARIS GREEN (GB), SCHWEINFURT GREEN dec
f vert *m* de Scheele
e cardenillo *m* arsenioso, verde *m* de París
i verde *m* di Parigi
n Parijs groen *n*
d Parisergrün *n*, Schweinfurter Grün *n*

2002 EMPIRICAL FORMULA gen
f formule *f* empyrique
e fórmula *f* empírica
i formula *f* empirica
n empirische formule, stoechiometrische formule
d empirische Formel *f*, Erfahrungsformel *f*

2003 EMULSIBILITY mix
f emulsibilité *f*
e emulsibilidad *f*
i emulsibilità *f*
n emulgeerbaarheid
d Emulgierbarkeit *f*

2004 EMULSIFICATION, EMULSIFYING mix
f émulsionnement *m*
e emulsionamiento *m*
i emulsionamento *m*
n emulgeren *n*
d Emulgieren *n*

2005 EMULSIFYING AGENT mix
f agent *m* émulsionnant
e agente *m* emulsionante
i agente *m* emulsionante
n emulgator, emulgeermiddel *n*
d Emulgator *m*, Emulgiermittel *n*

2006 EMULSION mix
f émulsion *f*
e emulsión *f*
i emulsione *f*
n emulsie
d Emulsion *f*

2007 EMULSION BREAKING mix
f désintégration *f* de l'émulsion, rupture *f* de l'émulsion
e rotura *f* de la emulsión
i rottura *f* dell'emulsione
n emulsieafbraak
d Emulsionszerstörung *f*

2008 EMULSION CARRIER pho
f porteur *m* de l'émulsion
e portador *m* de la emulsión
i portatore *m* dell'emulsione
n emulsiedrager
d Emulsionsträger *m*

2009 EMULSION CASTING, EMULSION COATING pho
f coulée *f* de l'émulsion
e colada *f* de la emulsión
i colata *f* dell'emulsione
n emulsiegieten *n*
d Emulsionsguss *m*

2010 EMULSION FORMING fat
f séparation *f* de l'eau
e separación *f* del agua
i separazione *f* dell'acqua
n uitzakken *n*
d Zuwassergehen *n*

2011 EMULSION HARDENING pho
f durcissement *m* de l'émulsion
e endurecimiento *m* de la emulsión
i indurimento *m* dell'emulsione
n emulsieharding
d Emulsionshärtung *f*

2012 EMULSION LAYER pho
f couche *f* d'émulsion
e capa *f* de emulsión
i strato *m* d'emulsione
n emulsielaag
d Emulsionsschicht *f*

2013 EMULSION PAINT dec
f couleur *f* en émulsion
e pintura *f* de emulsión
i colore *m* in emulsione
n emulsieverf
d Emulsionsfarbe *f*

2014 EMULSION POLYMERIZATION rub
f polymérisation *f* par émulsion
e polimerización *f* por emulsión
i polimerizzazione *f* mediante emulsione
n emulsiepolymerisatie
d Emulsionspolymerisierung *f*

2015 EMULSION THICKENING dye
f épaississement *m* de l'émulsion
e espesamiento *m* de la emulsión
i addensamento *m* dell'emulsione
n emulsie-indikking
d Emulsionseindickung *f*

2016 EMULSOID, col
LYOPHILIC COLLOID
f colloïde *m* liophyle
e coloide *m* liófilo
i colloide *m* liofilo
n emulsoïde *n*, reversibel colloïde *n*
d Emulsoid *n*, Emulsionskolloid *n*

2017 ENAMEL coa,dec
f émail *m*
e esmalte *m*
i smalto *m*
n email *n*
d Email *n*

2018 ENAMEL COATING coa
f revêtement *m* émaillé
e revestimiento *m* de esmalte
i rivestimento *m* di smalto
n emaillaag
d Schmelzüberzug *m*

2019 ENAMEL PAINT, dec
ENAMEL VARNISH
f vernis *m* à émail
e barniz *m* de esmalte
i vernice *f* a smalto
n emaillak
d Emaillack *m*

2020 ENAMELLED CLOTH (US), coa
AMERICAN CLOTH (GB),
AMERICAN LEATHER (GB)
f toile *f* cirée
e encerado *m*, hule *m*
i tela *f* cerata, tessuto *m* lucido
n wasdoek *n*
d Wachsleinwand *f*, Wachstuch *n*

2021 ENAMELLED LEATHER, coa,lea
JAPANNED LEATHER,
PATENT LEATHER
f cuir *m* verni
e charol *m*
i pelle *f* di lustrino, pelle *f* verniciata
n lakleder *n*
d Glanzleder *n*, Lackleder *n*

2022 ENAMELLED PAPER, pap
GLAZED PAPER
f papier *m* glacé
e papel *m* glazeado
i carta *f* lucida
n glanspapier *n*
d Glacépapier *n*, Glanzpapier *n*, Kreidepapier *n*

2023 ENAMELLING pho
f glaçage *m*
e satinado *m*
i lucidamento *m*
n hoogglanzen *n*
d Hochglänzen *n*

2024 ENAMELLING, cer,htt
DECORATION FIRING
f émaillage *m*
e esmaltado *m*
i smaltatura *f*
n emailleren *n*
d Emaillieren *n*

2025 ENCAUSTIC PASTE, pho
CEROTE PASTE
f cérotine *f*, encaustique *f*

e pasta *f* encáustica
i pasta *f* encausta
n glanswas
d Glanzwachs *n*

2026 END LIQUOR sin
f eau *f* mère
e lejía *f* final
i lisciva *f* finale
n eindloog
d Endlauge *f*

2027 END POINT ana
f point *m* final
e punto *m* final
i punto *m* d'equilibrio
n eindpunt *n*, omslagpunt *n*
d Endpunkt *m*, Umschlagspunkt *m*

2028 END-POINT COLOR (US), END-POINT COLOUR (GB) dec,dye
f teinte *f* de virage
e tinta *f* de viraje
i tinta *f* di viraggio
n omslagkleursoort
d Umschlagsfarbton *m*

2029 END PRODUCT, FINAL PRODUCT gen
f produit *m* final
e producto *m* final
i prodotto *m* finale
n eindprodukt *n*
d Endprodukt *n*

2030 ENDING dye
f teinture *f* inégale
e teñido *m* desigual
i tintura *f* ineguale
n ongelijkmatige verving
d ungleichmässige Färbung *f*

2031 ENDOERGIC REACTION, ENDOTHERMIC REACTION gen
f réaction *f* endothermique
e reacción *f* endotérmica
i reazione *f* endotermica
n endotherme reactie
d endotherme Reaktion *f*

2032 ENDURANCE TEST, FATIGUE TEST tes
f essai *m* de résistance à la fatigue
e prueba *f* de fatiga
i collaudo *m* di fatica
n vermoeiingsproef
d Dauerfestigkeitsprüfung *f*

2033 ENFLEURAGE per
f enfleurage *m*
e enfleuraje *m*
i estrazione *f* di profumo
n enfleurage
d Parfümextraktion *f*

2034 ENGLISH PINK, CHROME-TIN PINK, DUTCH PINK, ITALIAN PINK dec
f pigment *m* jaune transparent
e pigmento *m* amarillo transparente
i pigmento *m* giallo trasparente
n gele lakverf
d gelbe Lackfarbe *f*, Schüttgelb *n*

2035 ENGOBE cer
f engobe *m*
e enlucido *m* cerámico
i strato *m* sottile d'argilla
n dunne kleilaag, engobe
d Beguss *m*, Engobe *f*

2036 ENGRAVING dye
f gravure *f*
e grabado *m*
i incisione *f*
n graveren *n*
d Gravieren *n*

2037 ENRICHED WATER GAS, CARBURETTED WATER GAS fue
f gaz *m* à l'eau carburée
e gas *m* de agua carburada
i gas *m* d'acqua arricchito
n gecarbureerd watergas *n*
d angereichertes Wassergas *n*, karburiertes Wassergas *n*

2038 ENRICHING dis,nue
f enrichissement *m*
e enriquecimiento *m*
i arricchimento *m*
n aanrijking, verrijking
d Anreicherung *f*

2039 ENSILAGE agr,sto
f mise *f* en silo
e ensilaje *m*
i insilazione *f*
n in silo opslaan, inzouten *n*
d Einsäuern *n*, Silieren *n*

2040 ENTRAINER dis
f entraîneur *m*, tiers-corps *m*
e agente *m* adicional
i agente *m* trasportatore
n meesleepmiddel *n*
d Schleppmittel *n*

2041 ENTRAINMENT dis,fil
f entraînement *m*
e arrastre *m*
i trascinamento *m*
n meeslepen *n*
d Hochreissen *n*, Mitreissen *n*

2042 ENUMERATION OF BACTERIA bac
f numération *f* des germes
e contaje *m* de bacterias
i conteggio *m* batterico, numerazione *f* dei germi
n kiemtelling
d Keimzählung *f*

2043 ENZYMATIC BATE tan
f bain *m* de chipage enzymatique, confit *m* enzymatique, mordant *m* enzymatique
e baño *m* de maceración enzimático, mordiente *m* enzimático
i bagno *m* di macerazione enzimatico, purga *f* enzimatica
n enzymatische etsloog, enzymatische zwelder
d enzymatische Ätzlauge *f*, enzymatische Beizbrühe *f*, enzymatische Gerbbeize *f*

2044 ENZYMATIC DEPILATION tan
f dépilation *f* enzymatique
e depilación *f* enzimática
i depilazione *f* enzimatica
n enzymatisch bloten *n*, enzymatische ontharing
d enzymatische Enthaarung *f*

2045 ENZYMATIC FAT SPLITTING fat
f dédoublement *m* enzymatique de graisse
e escisión *f* enzimática de grasa
i scissione *f* enzimatica di grasso
n enzymatische vetsplitsing
d enzymatische Fettspaltung *f*

2046 ENZYME, FERMENT fmi
f diastase *m*, ferment *m*
e fermento *m*
i fermento *m*
n enzym *n*, ferment *n*, giststof
d Enzym *n*, Ferment *n*, Gärungsmittel *n*

2047 ENZYME ACTION fmi
f action *f* diastasique
e acción *f* fermentativa
i azione *f* fermentativa
n fermentwerking
d Fermentwirkung *f*

2048 ENZYME PREPARATION fmi
f préparation *f* enzymatique
e preparado *m* enzimático
i preparato *m* enzimatico
n enzympreparaat *n*
d Enzympräparat *n*

2049 ENZYMIC ACTIVITY fmi
f activité *f* enzymatique
e actividad *f* enzimática
i attività *f* enzimatica
n enzymwerkzaamheid
d Enzymtätigkeit *f*

2050 E.P.-LUBRICANT, EXTREME PRESSURE LUBRICANT lub
f lubrifiant *m* pour pressions très élevées
e lubricante *m* para hiperpresiones
i lubrificante *m* per sovrapressioni
n superhoge-druksmeermiddel *n*
d E.P.-Schmiermittel *n*

2051 EPICOTE RESINS, EPOXIDE RESINS pla
f résines *pl* époxydes
e resinas *pl* epoxídicas
i resine *pl* epossidriche
n epoxyharsen *pl*
d Äthoxylinharze *pl*, Epoxydharze *pl* Epoxyharze *pl*

2052 EPSOM SALT, BITTER SALT mcy
f sel *m* amer, sel *m* d'Epsom
e epsomita *f*, sal *f* amarga
i solfato *m* di magnesio
n bitterzout *n*, Engels zout *n*, magnesiumsulfaat *n*
d Bittersalz *n*, Epsomsalz *n*, Magnesiumsulfat *n*

2053 EQUILIBRIUM FLASH EVAPORATION dis,fue
f évaporation *f* par détente en équilibre
e evaporación *f* instantánea en equilibrio
i evaporazione *f* istantanea in equilibrio
n evenwichtsdestillatie
d Gleichgewichtsdestillation *f*

2054 EQUIVALENT CONDUCTANCE elc
f conductance *f* équivalente

e conductancia *f* equivalente
i conduttanza *f* equivalente
n equivalentgeleiding
d Äquivalentleitwert *m*

2055 EQUIVALENT CONDUCTIVITY elc
f conductivité *f* équivalente
e conductividad *f* equivalente
i conduttività *f* equivalente
n equivalentgeleidendheid
d Äquivalentleitfähigkeit *f*

2056 ESSENCE fat,per
f essence *f*, huile *f*
e aceite *m*, esencia *f*
i essenza *f*, olio *m*
n essence, olie
d Essenz *f*, Öl *n*

2057 ESSENTIAL OILS, fat,per
ETHEREAL OILS, VOLATILE OILS
f huiles *pl* essentielles
e aceites *pl* esenciales
i oli *pl* essenziali
n etherische oliën *pl*
d ätherische Öle *pl*

2058 ESTER, oic
ETHERAL SALT, COMPOUND ETHER
f ester *m*, éther *m* composé
e éster *m*
i estere *m*
n ester
d Ester *m*

2059 ESTER GUM, pla
ROSIN ESTER
f résine *f* estérifiée
e resina *f* esterificada
i resina *f* esterificata
n esterhars
d Estergummi *m*, Esterharz *n*

2060 ESTER NUMBER, fat,oic
ESTER VALUE
f indice *m* d'ester
e índice *m* de éster
i indice *m* d'estere
n estergetal *n*
d Esterzahl *f*

2061 ESTERIFICATION fat
f estérification *f*
e esterificación *f*
i esterificazione *f*
n verestering
d Veresterung *f*

2062 ETCHANT, elc,met
ETCHING AGENT
f caustique *m*, corrosif *m*, réactif *m* d'attaque à l'acide
e reactivo *m* para ataque al ácido
i reattivo *m* per attacco all'acido
n etsmiddel *n*, etsvloeistof
d Ätzflüssigkeit *f*, Ätzmittel *n*

2063 ETCHING elc,met
f attaque *f* chimique
e ataque *m* químico
i attacco *m* chimico
n etsen *n*
d Ätzen *n*

2064 ETCHING PRIMER (GB), dec
METAL CONDITIONS (US), WASH PRIMER (GB)
f peinture *f* pour couche primaire phosphante
e pintura *f* de mano primera fostatizante
i pittura *f* di mano prima a base di fosfato
n hechtgrondverf voor metalen
d Haftgrundmittel *n*, reaktionsfähiger Haftgrund *m*, Metallwandler *m*

2065 ETCHING TEST tes
f essai *m* pour l'attaque chimique
e prueba *f* por el ataque químico
i prova *f* per l'attacco chimico
n beitsproef
d Ätzprobe *f*

2066 ETHANAL, oic
ACETALDEHYDE, ACETYL HYDRIDE
f acétaldéhyde *m*, aldéhyde *m* acétique
e acetaldehido *m*
i acetaldeide *f*, aldeide *f* acetica, aldeide *f* etilica
n acetaldehyde *n*, ethanal *n*
d Azetaldehyd *n*

2067 ETHENOID RESINS pla
f résines *pl* éthénoïdes
e resinas *pl* etenoides
i resine *pl* etenoidi
n etheenharsen *pl*
d Äthenharze *pl*

2068 ETHER oic
f éther *m*
e éter *m*
i etere *m*
n ether
d Äther *m*

2069 ETHER EXTRACT oic
f extrait *m* à l'éther
e extracto *m* al éter
i estratto *m* etereo
n etherextract *n*
d Ätherextrakt *m*

2070 ETHER EXTRACTION oic
f extraction *f* à l'éther
e extracción *f* al éter
i estrazione *f* all'etere
n etherextractie
d Ausäthern *n*

2071 ETHER INDEX ana
f indice *m* d'éther
e índice *m* de éter
i indice *m* d'etere
n ethergetal *n*
d Ätherzahl *f*

2072 ETHERIFICATION oic,tex
f éthérification *f*
e eterificación *f*
i eterificazione *f*
n ethervorming
d Ätherbildung *f*

2073 ETHINE, oic
ACETYLENE, ALDEHYDENE
f acétylène *m*, éthyne *m*
e acetileno *m*
i acetilene *m*
n acetyleen *n*, ethyn *n*
d Äthyn *n*, Azetylen *n*

2074 ETHYL ACETATE, mcy,oic
ACETIC ESTER, ACETIC ETHER
f acétate *m* d'éthyle, ester *m* acétique
e acetato *m* etílico, éster *m* acético
i acetato *m* d'etile, estere *m* acetico
n azijnzure ester, ethylacetaat *n*
d Athylazetat *n*, Essigäther *m*, Essigester *m*, Essigsäure-äthylester *m*

2075 ETHYL ALCOHOL, nut
GRAIN ALCOHOL
f alcool *m* éthylique
e alcohol *m* etílico
i alcool *m* etilico
n ethylalcohol
d Äthylalkohol *m*, Gärungsalkohol *m*

2076 ETHYLATION oic
f éthylation *f*
e etilación *f*
i etilázione *f*
n ethyleren *n*
d Äthylieren *n*

2077 ETHYLCHLORIDE (GB), oic
CHALAT (US),
MONOCHLORETHANE
f chlorure *m* d'éthyl
e cloruro *m* etílico
i cloruro *m* d'etile
n ethylchloride *n*, monochloorethaan
d Äthylchlorid *n*, Chelen *n*, Chloräthyl *n*

2078 ETHYLENE, gas,oic
OLEFIANT GAS
f éthylène *m*
e etileno *m*
i etilene *m*
n ethyleen *n*
d Äthylen *n*, Elyal *n*

2079 ETHYLENE SUCCINIC ACID, oic
AMBER ACID, SUCCINIC ACID
f acide *m* d'ambre, acide *m* d'éthylène succinique, acide *m* succinique
e ácido *m* de ámbra, ácido *m* succínico
i acido *m* d'ambra, acido *m* succinico
n barnsteenzuur *n*, ethyleenbarnsteenzuur *n*
d Äthylenbernsteinsäure *f*, Bernsteinsäure *f*

2080 EUCALYPTUS OIL fat
f essence *f* d'eucalyptus
e esencia *f* de eucalipto
i olio *m* d'eucalitto
n eucalyptusolie
d Eukalyptusöl *n*

2081 EUCOLLOIDS col
f eucolloïdes *pl*
e eucoloides *pl*
i eucolloidi *pl*
n eucolloïden *pl*
d Eukolloide *pl*

2082 EUTECTIC MIXTURE, mix
EUTECTOID
f eutectoïde *m*, mélange *m* eutectique
e eutectoide *m*, mezcla *f* eutéctica
i eutettoide *m*, mescolanza *f* eutettica
n eutectisch mengsel *n*, eutectoïde *n*
d eutektische Mischung *f*, Eutektoid *n*

2083 EVAPORATED LATEX rub
f latex *m* concentré par évaporation
e látex *m* evaporado
i lattice *m* evaporato
n ingedampte latex
d eingedampfter Latex *m*, eingedickter Latex *m*

2084 EVAPORATION eva
f évaporation *f*
e evaporación *f*
i evaporazione *f*
n verdamping
d Verdampfung *f*

2085 EVAPORATION CONCENTRATE eva
f concentré *m* d'évaporation
e concentrado *m* de evaporación
i concentrato *m* d'evaporazione
n indampconcentraat *n*
d Eindampfkonzentrat *n*

2086 EVAPORATION LOSSES eva
f pertes *pl* d'évaporation
e pérdidas *pl* de evaporación
i perdite *pl* d'evaporazione
n verdampingsverliezen *pl*
d Verdampfungsverluste *pl*

2087 EVAPORATIVE COOLING clo,eva
f refroidissement *m* par évaporation
e enfriamiento *m* por evaporación
i raffreddamento *m* mediante evaporazione
n verdampingskoeling
d Verdampfungskühlung *f*, Verdunstungskühlung *f*

2088 EVEN DYEING, LEVEL DYEING dye
f teinture *f* égale
e teñido *m* igual
i tintura *f* uguale
n gelijkmatige verving
d gleichmässige Färbung *f*

2089 EXCESS CHLORINATION, HIGH CHLORINATION, SUPERCHLORINATION aic
f chloration *f* à dose élevée, surchloration *f*
e sobrecloración *f*
i sovraclorazione *f*, superclorazione *f*
n overchloreren *n*
d Hochchlorung *f*, Überchlorung *f*, Überschusschlorung *f*

2090 EXCESS LIME PROCESS iwt
f désacidification *f* par chaux en excès
e procedimiento *m* de sobredosaje de cal
i processo *m* di sovradosaggio di calce
n kalkovermaatproces *n*
d Kalküberschussverfahren *n*, Überschusskalkung *f*

2091 EXCESS MATERIAL gen
f matière *f* en excès
e exceso *m* de material
i materiale *m* eccedente
n overtollig materiaal *n*
d überschüssiges Material *n*

2092 EXCHANGE REACTION gen
f réaction *f* d'échange
e reacción *f* de intercambio
i reazione *f* di scambio
n uitwisselingsreactie
d Austauschreaktion *f*

2093 EXCIPIENT, MENSTRUM mcy
f excipient *m*
e excipiente *m*
i eccipiente *m*
n menstruüm *n*, vloeibaar oplossingsmiddel *n*
d Arzneimittelträger *m*

2094 EXFOLIATING, PEELING OFF, SCALING dec
f écaillage *m*
e despellegamiento *m*
i screpolatura *f*
n afbladderen *n*
d Abblättern *n*, Abschälen *n*

2095 EXFOLIATION, FLAKING, SPALLING met
f écaillage *m*
e desconchado *m*
i scagliatura *f*, screpolatura *f*
n afbladderen *n*
d Abblättern *n*, Häutung *f*

2096 EXHAUST GASES, WASTE GASES fue,gas,iwt
f gaz *pl* d'échappement
e gases *pl* de escape
i gas *pl* di scappamento
n afvoergassen *pl*, uitlaatgassen *pl*
d Abgase *pl*

2097 EXHAUST STEAM, DEAD STEAM, WASTE STEAM boi
f vapeur *f* d'échappement
e vapor *m* de escape
i vapore *m* di ritorno
n afgewerkte stoom
d Abdampf *m*

2098 EXHAUSTED PULP, BEET PULP, BEETROOT PULP sin
f pulpe *f* de betteraves
e pulpa *f* de remolachas
i polpa *f* di barbabietole
n bietenpulp
d Brei *m*, Rübenbrei *m*

2099 EXHAUSTING AGENT dye
f agent *m* augmentant l'absorption
e agente *m* de aumento de la absorción
i agente *m* d'aumento dell'assorbimento
n absorptieverhogend middel *n*
d Mittel *n* zur Steigerung der Absorption

2100 EXHAUSTIVE BROMINATION gen
f bromuration *f* totale
e bromación *f* a fondo
i bromurazione *f* totale
n uitputtende bromering
d erschöpfende Bromierung *f*

2101 EXHAUSTIVE CHLORINATION gen
f chloruration *f* totale
e cloración *f* a fondo
i clorurazione *f* totale
n uitputtende chlorering
d erschöpfende Chlorung *f*

2102 EXHAUSTIVE METHYLATION, HOFMANN DEGRADATION oic
f méthylation *f* totale
e metilación *f* a fondo
i metilazione *f* totale
n uitputtende methylering
d erschöpfende Methylierung *f*

2103 EXOTHERMIC REACTION gen
f réaction *f* exothermique
e reacción *f* exotérmica
i reazione *f* esotermica
n exotherme reactie
d exotherme Reaktion *f*

2104 EXPANDING CEMENT bui
f ciment *m* expansif
e cemento *m* expansivo
i cemento *m* espansivo
n zwellend cement *n*
d Schwellzement *m*

2105 EXPELLOR WASTE bre,iwt
f eau *f* de presse
e agua *f* de prensa
i acqua *f* di spremitura
n perswater *n*
d Biertreberpresswasser *n*

2106 EXPLOSION exp
f explosion *f*
e explosión *f*
i esplosione *f*
n explosie
d Explosion *f*

2107 EXPLOSIVE exp
f explosif *m*
e explosivo *m*
i esplosivo *m*
n explosieve stof, springmiddel *n*
d Explosivstoff *m*, Sprengmittel *n*

2108 EXPLOSIVE GELATINE exp
f gélatine *f* à 93
e gelatina *f* explosiva
i gelatina *f* esplosiva
n springgelatine
d Sprenggallerte *f*, Sprenggelatine *f*

2109 EXPLOSIVE MIXTURE exp
f mélange *m* explosif
e mezcla *f* explosiva
i mescolanza *f* esplosiva
n explosief mengsel *n*
d Explosionsgemisch *n*

2110 EXPLOSIVE OIL (US), NITROGLYCERINE (GB) exp
f nitroglycérine *f*, trinitrate *m* de glycérine
e nitroglicerina *f*, trinitrato *m* de glicerina
i nitroglicerina *f*, trinitrato *m* di glicerina
n glyceroltrinitraat *n*, nitroglycerine, nitroglycerol *n*
d Glyzerintrinitrat *n*, Nitroglyzerin *n*, Trinitrin *n*

2111 EXPLOSIVE PRESSING pre
f moulage *m* à choc
e moldeo *m* de choque
i formazione *f* ad urto
n schokpersen *n*, slagpersen *n*
d Schlagpressen *n*

2112 EXPLOSIVE TEST tes
f essai *m* d'explosion
e prueba *f* de explosión
i collaudo *m* d'esplosione
n explosieproef
d Explosionsprobe *f*

2113 EXPRESSED BEET PULP sin
f pâte *f* de betteraves exprimée
e pulpa *f* de remoladas exprimida
i polpa *f* di barbabietole spremuta

n uitgeperste bietenpulp
d ausgepresster Rübenbrei *m*

2114 EXTENDER dec,pla
f charge *f* inerte
e blanco *m* de carga, extensor *m*
i riempitivo *m*, sostanza *f* inerte
n strekmiddel *n*, vulstof
d Füllstoff *m*, Streckmittel *n*

2115 EXTENDER PLASTICIZER, pla
SECONDARY PLASTICIZER
f plastifiant *m* inerte
e plastificante *m* diluente
i plastificante *m* inerte
n secondair plastificeermiddel *n*
d Beschwermittel *n*

2116 EXTERIOR FINISH dec
f couleur *f* extérieur
e pintura *f* exterior
i colore *m* esteriore
n buitenverf
d Aussenanstrich *m*

2117 EXTINCTION pla
f extinction *f*
e extinción *f*
i estinzione *f*, spegnimento *m*
n blussing
d Löschung *f*

2118 EXTRACT, ext
DECOCTION
f décocté *m*
e decocción *f*
i decotto *m*
n afkooksel *n*
d Abkochung *f*, Absud *m*, Dekokt *m*

2119 EXTRACTABLE SULPHUR (GB), rub
FREE SULFUR (US)
f soufre *m* libre
e azufre *m* libre
i zolfo *m* libero
n vrije zwavel
d freier Schwefel *m*

2120 EXTRACTANT, ext
EXTRACTION SOLVENT,
EXTRACTIVE AGENT
f agent *m* d'extraction, agent *m* extractif
e extractante *m*, extraente *m*, solvente *m* de extracción
i agente *m* estrattivo, estrattore *m*
n extractiemiddel *n*, extraheermiddel *n*
d Extraktionsmittel *n*, selektives Lösungsmittel *n*

2121 EXTRACTED MEAL, fat
LINSEED MEAL
f farine *f* de grain de lin
e harina *f* de linaza
i farina *f* di seme di lino
n lijnzaadmeel *n*
d Leinsamenmehl *n*

2122 EXTRACTION ext
f extraction *f*
e extracción *f*
i estrazione *f*
n extractie, extraheren *n*
d Extrahieren *n*, Extraktion *f*

2123 EXTRACTION BY BOILING, ext
DECOCTION
f décoction *f*
e decocción *f*
i decozione *f*
n afkoken *n*
d Abkochen *n*, Absieden *n*

2124 EXTRACTION LIQUID, ext,met
EXTRACTION LIQUOR
f liquide *m* d'extraction
e líquido *m* de extracción
i liquido *m* d'estrazione
n uitloogvloeistof
d Auslaugeflüssigkeit *f*

2125 EXTRACTIVE DISTILLATION dis
f distillation *f* extractive
e destilación *f* extractiva
i distillazione *f* estrattiva
n extractieve destillatie
d extraktive Destillation *f*

2126 EXTREME PRESSURE LUBRICANT, lub
E.P.-LUBRICANT
f lubrifiant *m* pour pressions très élevées
e lubricante *m* para hiperpresiones
i lubrificante *m* per sovrapressioni
n superhoge-druksmeermiddel *n*
d E.P. Schmiermittel *n*

2127 EXTRUSION pla
f boudinage *m*, extrusion *f*
e budinado *m*, extrusión *f*
i estrusione *f*, trafilatura *f*
n extrusie, spuitgieten *n*
d Fliesspressen *n*, Kaltspritzen *n*, Strangpressen *n*

2128 EXTRUSION AID pla,pre
f agent *m* de boudinage, agent *m* d'extrusion
e agente *m* de budinado, agente *m* de extrusión

i agente *m* d'estruzione,
agente *m* di trafilatura
n spuitvloeimiddel *n*
d Spritzbarmacher *m*

2129 EXTRUSION COATING, pla
EXTRUSION LAMINATION
f boudinage-laminage *m*,
doublage *m* par extrusion-laminage
e extrusión-laminación *f*
i rivestimento *m* con estrusore
n extrusiekoten *n*
d Beschichten *n* über die Strangpresse

2130 EXTRUSION MOLDING (US), pla
EXTRUSION MOULDING (GB)
f moulage *m* par extrusion
e moldeo *m* por extrusión
i formazione *f* per estrusione
n spuitgieten *n* in vorm
d Strangpressen *n*

2131 EXUDATION, rub
SWEATING OUT
f exsudation *f*
e exudación *f*
i essudazione *f*
n exsudatie, uitzweten *n*
d Ausschwitzen *n*, Exsudation *f*

F

2132 FABRIC tex
f étoffe *f*, tissu *m*
e tejido *m*, tela *f*
i stoffa *f*, tessuto *m*
n stof, weefsel *n*
d Gewebe *n*, Stoff *m*

2133 FACING, fon
BLACKING, BLACKWASH,
COATING
f enduit *m*
e unto *m* de revestimiento
i intonaco *m*
n vormdekmiddel *n*
d Schlichte *f*, Schwärze *f*

2134 FACING SAND fou
f sable *m* de contact
e arena *f* de contacto
i terra *f* da modello, terra *f* fina
n modelzand *n*
d Modellsand *m*

2135 FACTICE rub
f factice *m*
e facticio *m*
i factice *m*, fatturato *m*
n factis
d Faktis *m*, Ölkautschuk *m*

2136 FACTITIOUS gen
f artificiel adj, factice adj
e artificial adj, facticio adj
i artificiale adj, fattizio adj
n kunstmatig adj
d künstlich adj

2137 FACTORIAL DEVELOPMENT, pho
WATKINS DEVELOPMENT
f développement *m* semi-automatique basé sur le coefficient de Watkins
e revelado *m* precalculado
i sviluppo *m* precalcolato
n van te voren berekende ontwikkeling
d Faktorenentwicklung *f*, vorberechnete Entwicklung *f*

2138 FADING tex
f déteinte *f*
e desteñidura *f*
i cambio *m* di colore
n verkleuring
d Verfärbung *f*

2139 FADING TEST dec
f essai *m* de déteinte
e ensayo *m* de desteñidura
i collaudo *m* di cambio di colore
n lichtechtheidsproef
d Lichtechtheitsprobe *f*

2140 FALLING-FILM DISTILLATION dis
f distillation *f* à couche descendante
e destilación *f* de película bajante
i distillazione *f* con pellicola discendente
n destillatie met dalende vloeistoffilm
d Destillation *f* mit herabsinkender Flüssigkeitsschicht, Fallfilmdestillation *f*

2141 FALSE BODY, dec
THIXOTROPY
f faux corps *m*, thixotropie *f*
e tixotropía *f*
i tixotropia *f*
n thixotropie
d falscher Körper *m*, Thixotropie *f*

2142 FARINA, tex
STARCH PASTE
f apprêt *m* à base d'amidon
e apresto *m* de base de almidón
i appretto *m* a base d'amido
n pap
d Stärkekleister *m*

2143 FAST ACCELERATOR, rub
FAST CURING ACCELERATOR
f accélérateur *m* rapide
e acelerante *m* rápido
i accelerante *m* ad azione rapida
n actieve versneller, ultraversneller
d schnellwirkender Beschleuniger *m*

2144 FAST BURNING FUSE exp
f étoupille *f*, mèche *f* à combustion rapide
e mecha *f* de combustión rápida
i miccia *f* a combustione rapida
n gezwinde lont
d Schnellzündschnur *f*

2145 FAST COLOR (US), dec,pla
FAST COLOUR (GB)
f bon teint *m*, grand teint *m*, couleur *f* solide

e color *m* estable, color *m* fijo, color *m* sólido
i colore *m* solido
n bestendige kleur
d dauerhafte Farbe *f*, echte Farbe *f*, stehende Farbe *f*

2146 FAST FERMENTATION bre,fmi
f fermentation *f* rapide
e fermentación *f* rápida
i fermentazione *f* rapida
n snelle gisting
d schnelle Gärung *f*

2147 FAST-MILLING PIGMENT dec
f pigment *m* à dispersion rapide
e pigmento *m* de dispersión rápida
i pigmento *m* a dispersione rapida
n snelmengend pigment *n*
d leichtmischendes Pigment *n*

2148 FASTNESS gen
f solidité *f*, résistance *f*
e estabilidad *f*
i solidità *f*
n bestendigheid, echtheid
d Dauerhaftigkeit *f*, Echtheit *f*

2149 FASTNESS TO WASHING, WASHING FASTNESS tex
f solidité *f* au lavage
e resistencia *f* al lavado
i solidità *f* al lavaggio
n wasbestendigheid
d Waschfestheit *f*

2150 FASTNESS TO WATER, WATER FASTNESS tex
f solidité *f* à l'eau
e resistencia *f* al agua
i solidità *f* all'acqua
n waterbestendigheid
d Wasserbeständigkeit *f*

2151 FAT fat
f graisse *f*
e grasa *f*
i grasso *m*
n vet *n*
d Fett *n*

2152 FAT COLOR (US), FAT COLOUR (GB) dec
f peinture *f* collante
e pintura *f* pegante
i pittura *f* appiccicaticcia
n kleverig geworden verf
d klebrige Farbe *f*

2153 FAT EDGES dec
f excès *m* de peinture
e exceso *m* de pintura
i eccesso *m* di pittura
n zakker
d Tropfnasenbildung *f*

2154 FAT HARDENING, FAT HYDROGENATION fat,oic
f hydrogénation *f* d'huile
e hidrogenación *f* de grasa
i idrogenazione *f* di grasso
n vetharding
d Fetthärtung *f*, Ölhärtung *f*

2155 FAT LIME, PURE LIME, RICH LIME bui
f chaux *f* grasse
e cal *f* grasa
i calce *f* grassa
n kluitkalk, Luikse kalk, steenkalk
d Fettkalk *m*, Speckkalk *m*

2156 FAT LIME, WHITE LIME, WHITEWASH dec
f chaux *f* à badigeonner, lait *m* de chaux
e jalbegue *m*, lechada *f* de cal
i calce *f* da imbianchino, latte *m* di calce
n witkalk
d Tünche *f*, Weisskalk *m*

2157 FAT LIQUID, FAT LIQUOR tan
f graisse *f* d'imprégnation
e grasa *f* de impregnación
i grasso *m* da impregnare
n looivet *n*
d Fettschmiere *f*

2158 FAT LIQUORING tan
f graissage *m* du cuir, huilage *m* du cuir
e impregnación *f* del cuero con aceite
i impregnazione *f* del cuoio con olio
n vetlooien *n*
d Fettgerbung *f*

2159 FAT RECOVERY, GREASE RECOVERY fat,oic
f récupération *f* de graisse
e recuperación *f* de grasa
i ricupero *m* di grasso
n vetterugwinning
d Fettrückgewinnung *f*

2160 FAT RENDERING fat
f extraction *f* de la graisse

e extracción *f* de la manteca
i estrazione *f* del grasso
n uitsmelten *n* van vet
d Fettausschmelzung *f*

2161 FAT SAND fou
f sable *m* fort
e arena *f* grasa
i terra *f* grassa
n vet vormzand *n*
d fetter Sand *m*

2162 FAT SEPARATING, cle,fat
GREASE SEPARATING
f séparation *f* de graisse
e separación *f* de grasa
i separazione *f* di grasso
n vetafscheiding
d Fettabscheidung *f*

2163 FAT SPLITTING fat,oic
f dédoublement *m* de graisse
e disgregación *f* de grasa
i disgregazione *f* di grasso
n vetsplijting
d Fettspaltung *f*

2164 FATIGUE TEST, tes
ENDURANCE TEST
f essai *m* de résistance à la fatigue
e prueba *f* de fatiga
i collaudo *m* di fatica
n vermoeiingsproef
d Dauerfestigkeitsprüfung *f*

2165 FATTY ACIDS fat
f acides *pl* gras
e ácidos *pl* grasos
i acidi *pl* grassi
n vetzuren *pl*
d Fettsäuren *pl*

2166 FATTY OILS, fat
FIXED OILS
f huiles *pl* fixes
e aceites *pl* fijos, aceites *pl* grasos
i oli *pl* fissi
n oliën *pl* van dierlijke en plantaardige oorsprong
d animalische und vegetabilische Öle *pl*

2167 FEATHERWEIGHT PAPER pap
f papier *m* bouffant
e papel *m* pluma
i carta *f* sgonfiata
n opdikkend papier *n*
d aufbauschendes Papier *n*

2168 FEEBLY HYDRAULIC LIME bui
f chaux *f* légèrement hydraulique
e cal *f* débilmente hidráulica
i calce *f* leggermente idraulica
n licht-hydraulische kalk
d Kalk *m* aus Kalkstein mit 5-12% Tongehalt, niedrighydraulischer Kalk *m*

2169 FEED REGULATING, con,gen
FEED TIMING
f réglage *m* de l'alimentation
e regulación *f* de la alimentación
i regolazione *f* dell'alimentazione
n toevoerregeling, voedingsregeling
d Speiseregelung *f*, Zufuhrregelung *f*

2170 FEED WATER boi,pre
f eau *f* d'alimentation
e agua *f* de alimentación
i acqua *f* d'alimentazione
n voedingswater *n*
d Speisewasser *n*

2171 FEEDING gen
f alimentation *f*
e alimentación *f*
i alimentazione *f*
n aanvoer, voeding
d Speisung *f*, Zufuhr *f*

2172 FEEDING LIQUOR dye
f liquide *m* supplémentaire
e líquido *m* suplementario
i liquido *m* di supplemento
n aanvullende vloeistof
d Ergänzungsflüssigkeit *f*

2173 FELDSPAR, min
FELSPAR
f feldspath *m*
e feldespato *m*
i feldspato *m*
n veldspaat *n*
d Feldspat *m*

2174 FELDSPAR GLAZE cer
f glaçure *f* de feldspath
e vidriado *m* feldespático
i invetriatura *f* feldspatica
n veldspaatglazuur *n*
d Feldspatglasur *f*

2175 FELT tex
f feutre *m*
e fieltro *m*
i feltro *m*
n vilt *n*
d Filz *m*

2176 FELT BLANKET pap
f manchon *m*
e correa *f* de fieltro
i strato *m* di feltro
n viltlaag
d Papiermacherfilzschicht *f*

2177 FELT DRYING, pap
FELTING
f séchage *m* à manchon
e secado *m* por correa de fieltro
i essiccazione *f* per strato di feltro
n drogen *n* op viltlaag
d Trocknen *n* auf Filzschicht

2178 FELT PAPER bui
f papier-feutre *m*
e papel *m* fieltro
i cartone *m* feltro
n viltpapier *n*
d Filzpapier *n*

2179 FELTING tex
f feutrage *m*
e fieltramiento *m*
i feltratura *f*
n vervilten *n*
d Filzen *n*

2180 FENNEL OIL fat
f essence *f* de fenouil
e esencia *f* de inojo
i olio *m* di finocchio
n venkelolie
d Fenchelöl *n*

2181 FERMENT, fmi
ENZYME
f diastase *m*, ferment *m*
e fermento *m*
i fermento *m*
n enzym *n*, ferment *n*, giststof
d Enzym *n*, Ferment *n*, Gärungsmittel *n*

2182 FERMENT OIL fat
f huile *f* essentielle obtenue par un ferment végétal
e aceite *m* esencial obtenido por un fermento vegetal
i olio *m* essenziale ottenuto per un fermento vegetale
n door plantaardig ferment ontwikkelde etherische olie
d durch Pflanzenferment erzeugtes ätherisches Öl *n*

2183 FERMENTATION fmi
f fermentation *f*
e fermentación *f*
i fermentazione *f*
n fermentatie, fermenteren *n*, gisting
d Fermentierung *f*, Gärung *f*

2184 FERMENTATION oic
AMYLALCOHOL,
ISOAMYLALCOHOL
f alcool *m* isoamylique
e alcohol *m* isoamílico
i alcool *m* isoamilico
n gistingsamylalcohol
d Gärungsamylalkohol *m*, Isoamylalkohol *m*

2185 FERMENTING cer
f pourrissage *m*
e pudrición *f*, putrefacción *f*
i putrefazione *f*
n laten uitwerken *n*, rottingsproces
d Mauken *n*

2186 FERRIFERROUS WATER, hyg
FERRIGINOUS WATER,
CHALYBEATE WATER
f eau *f* ferrigineuse
e agua *f* ferrugenosa
i acqua *f* ferrigenosa
n ijzerhoudend water *n*
d eisenhaltiges Wasser *n*

2187 FERRITE YELLOW, dec
YELLOW IRON OXIDE
f jaune *m* d'oxyde de fer
e amarillo *m* de óxido de hierro
i giallo *m* d'ossido di ferro
n oxydegeel *n*, ijzeroxydegeel *n*
d Eisenoxydgelb *n*, gelbes Eisenoxyd *n*

2188 FERROCONCRETE, bui
ARMED CONCRETE,
REINFORCED CONCRETE
f béton *m* armé
e hormigón *m* armado
i cemento *m* armato, conglomerat *m* cementizio armato
n gewapend beton *n*
d bewehrter Beton *m*, Eisenbeton *n*, Stahlbeton *m*

2189 FERROPRUSSIATE PAPER pap
f papier *m* au ferroprussiate
e papel *m* ferroprusiato
i carta *f* al ferroprussiato
n ferrocyanidepapier *n*
d Ferrozyanidpapier *n*

2190 FERROSILICON HYDROGEN gas
PROCESS

f procédé *m* de fabrication d'hydrogène par moyen de ferrosilicium
e procedimiento *m* de fabricación de hidrógeno con ferrosilicio
i processo *m* di fabbricazione d'idrogeno con ferrosilicio
n waterstoffabricageproces *n* met ferrosilicium
d Wasserstofferzeugungsverfahren *n* mit Ferrosilizium

2191 FERROTYPE pho
f ferrotypie *f*
e ferrotipia *f*
i ferrotipia *f*
n ferrotypie
d Ferrotypie *f*

2192 FERTILE MATERIAL nue
f matière *f* fertile
e material *m* fértil
i materiale *m* fertile
n kweekmateriaal *n*
d Brutmaterial *n*, Brutstoff *m*

2193 FERTILIZER agr,fer
f engrais *m*
e fertilizador *m*
i fertilizzante *m*
n meststof
d Düngemittel *n*, Dünger *m*

2194 FERTILIZING LIME, AGRICULTURAL LIME fer
f engrais *m* calcique
e cal *f* de abonar
i calce *f* da fertilizzare
n kalkmeststof
d Düngekalk *m*

2195 FESTOON DRYING, LOOP DRYING dry,tex
f séchage *m* à guirlandes
e secado *m* de festones
i essiccazione *f* a festoni
n guirlandedroging
d Guirlandentrocknung *f*

2196 FESTOON STEAMING tex
f traitement *m* de guirlandes supportées à vapeur
e tratamiento *m* de festones con vapor
i trattamento *m* di festoni con vapore
n guirlandestoombehandeling
d Schleifendampfbehandlung *f*

2197 FESTOONED CREPE rub
f crêpe *f* festonnée
e crepe *m* afestonado
i foglia *f* crepe a festone
n in slingers opgehangen crepe
d in Schleifen aufgehängte Kreppfelle

2198 FETTLING, SCRAPING cer
f raclage *m*
e rascadura *f*
i raschiamento *m*, raschiatura *f*
n afschrappen *n*
d Abschroten *n*

2199 FIBER (US), FIBRE (GB) gen
f fibre *f*
e fibra *f*
i fibra *f*
n vezel
d Faser *f*

2200 FIBER DUSTING (US), FIBRE DUSTING (GB) coa
f flocage *m*
e revestimiento *m* con fibras
i rivestimento *m* con fibre
n vezelverstuiving
d Beflockung *f*

2201 FIBER RECOVERING (US), FIBRE RECOVERING (GB) pap
f récupération *f* de fibres
e recuperación *f* de fibras
i ricupero *m* di fibre
n opwerken *n* van vezels
d Faseraufarbeitung *f*

2202 FIBERBOARD (US), FIBREBOARD (GB) bui
f board *m*, panneaux *pl* de fibres
e cartón *m* de fibra, fibra-cartón *m*
i cartone *m* di fibra
n board *n*, vezelplaten *pl*
d Faserplatten *pl*, Holzfaserplatten *pl*

2203 FIBROUS CONCRETE bui
f béton *m* à fibres
e hormigón *m* de fibras
i calcestruzzo *m* a fibre
n vezelhoudend beton *n*
d Faserbeton *n*

2204 FIELD TEST, NATURAL EXPOSURE TEST, NATURAL WEATHERING TEST tes
f essai *m* de tenue aux intempéries
e ensayo *m* de exposición a las intemperies
i collaudo *m* d'esposizione all'intemperie

n buitenproef, dakproef, weerrekproef
d Bewitterungsprobe *f*, Freilagerversuch *m*

2205 FILLED CLOTH, WEIGHTED CLOTH tex
f tissu *m* chargé
e tela *f* cargada
i tessuto *m* caricato
n verzwaard weefsel *n*
d beschwertes Gewebe *n*

2206 FILLER dec
f blanc *m* de charge
e sellaporos *m*
i mestica *f*
n plamuur
d Grundierlack *m*, Streichkitt *m*

2207 FILLER gen
f charge *f*, substance *f* de remplissage
e substancia *f* de relleno
i materiale *m* di carica, materiale *m* di riporto, riempitivo *m*
n vulstof
d Füllstoff *m*

2208 FILLER METAL wel
f métal *m* de remplissage
e metal *m* de aportación, metal *m* de relleno
i metallo *m* d'apporto, metallo *m* di riempimento
n vulmetaal *n*
d Füllmetall *n*

2209 FILLING gen
f remplissage *m*
e relleno *m*
i riempimento *m*
n vullen *n*
d Abfüllen *n*, Füllen *n*

2210 FILLING-UP dec
f bouchage *m*
e relleno *m*
i riempimento *m*
n stoppen *n*
d Stopfen *n*

2211 FILM gen
f couche *f*
e capa *f*
i strato *m*
n laag
d Schicht *f*

2212 FILM pho
f film *m*, pellicule *f*
e film *m*, película *f*
i film *m*, pellicola *f*
n film
d Film *m*

2213 FILM BASE, FILM SUPPORT pho
f support *m* pelliculaire
e soporte *m* de película
i portapellicola *m*
n emulsiedrager, filmdrager
d Filmträger *m*, Filmunterlage *f*, Schichtträger *m*

2214 FILM BOILING boi
f ébullition *f* pelliculaire
e ebullición *f* pelicular
i ebollizione *f* pellicolare
n filmkoken *n*
d Filmsieden *n*

2215 FILM CASTING pho,pla
f coulée *f* de pellicules
e colada *f* de películas
i colata *f* di pellicole
n filmgieten *n*
d Filmgiessen *n*

2216 FILM FORMING YEAST, MYCODERMA bre
f mycoderme *m*
e micoderma *m*
i micoderma *m*
n kaamgist, mycoderma *n*
d Kahmhefe *f*

2217 FILM GLUE pla
f feuille *f* adhésive
e hoja *f* adhesiva
i foglia *f* adesiva
n lijmfoelie
d Leimfolie *f*

2218 FILM PROCESSING pho
f traitement *m* du film
e elaboración *f* de la película
i lavorazione *f* del film
n filmbewerking
d Filmbearbeitung *f*

2219 FILTER pla
f filtre *m*
e filtro *m*
i filtro *m*
n filter *n*
d Filter *m*

2220 FILTER ALUM, SULFATE OF ALUMINUM (US), SULPHATE OF ALUMINIUM(GB) mat
f sulfate *m* d'aluminium
e sulfato *m* de alumina
i solfato *m* d'alluminio
n aluminiumsulfaat *n*
d Aluminiumsulfat *n*, schwefelsauere Tonerde *f*, Tonerdesulfat *n*

2221 FILTER BED PROCESS sew
f procédé *m* à couche filtrante
e procedimiento *m* de lecho filtrante
i processo *m* a letto filtrante
n filterbedproces *n*
d Filterbettverfahren *n*

2222 FILTER PAPER fil,pap
f papier-filtre *m*
e papel *m* de filtro
i carta *f* da filtro
n filtreerpapier *n*
d Filtrierpapier *n*

2223 FILTER PULP pap
f pâte *m* filtrée
e pulpa *f* filtrada
i pannello *m* di fibre di tela
n filterpulp
d Filterbrei *m*

2224 FILTERING fil
f filtrage *m*, filtration *f*
e filtración *f*
i filtrazione *f*
n filteren *n*, filtreren *n*
d Filtrieren *n*

2225 FILTRATE fil
f liquide *m* filtré
e filtrado *m*
i filtrato *m*
n filtraat *n*
d Filtrat *n*

2226 FILTRATION RESIDUE fil
f résidu *m* de filtration
e residuo *m* de filtración
i residuo *m* di filtrazione
n filterresidu *n*
d Filterrückstand *m*

2227 FINAL BLEACHING tex
f blanchiment *m* final
e blanqueo *m* final
i candeggio *m* finale
n aanbleken *n*, nableken *n*
d Nachbleichen *n*

2228 FINAL BOILING POINT boi
f point *m* d'ébullition final
e punto *m* de ebullición final
i punto *m* d'ebollizione finale
n eindkookpunt *n*
d Endsiedepunkt *m*

2229 FINAL CLEANING tex
f nettoyage *m* final
e limpieza *f* final
i pulitura *f* finale
n nareiniging
d Nachreinigung *f*

2230 FINAL HARDING met
f durcissement *m* final
e endurecimiento *m* final
i indurimento *m* finale
n naharding
d Nachhärtung *f*

2231 FINAL PRODUCT, END PRODUCT gen
f produit *m* final
e producto *m* final
i prodotto *m* finale
n eindprodukt *n*
d Endprodukt *n*

2232 FINAL SETTLING, SECONDARY SEDIMENTATION iwt
f clarification *f* finale, décantation *f* secondaire
e clarificación *f* final
i sedimentazione *f* finale
n laatste klaring
d Nachklärung *f*

2233 FINAL SINTERING htt,met
f frittage *m* final
e sinterización *f* final
i sinterizzazione *f* finale
n afsinteren *n*
d Fertigsintern *n*

2234 FINE CRUSHING FINE TRITURATION, COMMINUTION pap
f broyage *m* fin, dilacération *f*
e trituración *f* fina
i frantumazione *f* fina, triturazione *f* fina
n fijnmalen *n*
d Feinzerkleinerung *f*

2235 FINE-GRAIN DEVELOPMENT pho
f développement *m* à grain fin
e revelado *m* de grano fino
i sviluppo *m* finegranulante
n fijnkorrelontwikkeling
d Feinkornentwicklung *f*

2236 FINE-GRAIN EMULSION pho
f émulsion *f* à grain fin
e emulsión *f* de grano fino
i emulsione *f* a grano fine
n fijnkorrelemulsie
d Feinkornemulsion *f*

2237 FINENESS gen
f finesse *f*
e finura *f*
i finezza *f*
n fijnheid
d Feinheit *f*

2238 FINENESS TEST tes
f essai *m* de finesse de broyage
e ensayo *m* del grado de finura
i collaudo *m* del grado di finezza
n fijnheidsproef, maalgraadproef
d Mahlgradprobe *f*

2239 FINES sep,sor
f particules *pl* très fines
e partículas *pl* menores de 44 micrómetros
i stacciatura *f* fina
n fijnste deeltjes *pl*
d Abrieb *m*, die feinsten Kornfraktionen *pl*, Feinabsiebung *f*

2240 FINING gla
f affinage *m*
e afinación *f*, eliminación *f* de burbujas de gas
i chiarificazione *f*
n louteren *n*
d Läutern *n*

2241 FINING met
f affinage *m*
e afino *m*
i affinazione *f*
n fijnen *n*
d Feinen *n*

2242 FINING AGENT gla
f agent *m* d'affinage
e agente *m* de afinación
i agente *m* di chiarificazione
n louteringsmiddel *n*
d Läuterungsmittel *n*

2243 FINING OFF, PLASTERING, RENDERING, RUNNING OFF bui
f finissage *m* par du plâtre
e colocación *f* de un emplasto
i intonacatura *f*
n afpleisteren *n*
d Verputzen *n*

2244 FININGS bre
f clarifiant *m*
e clarificante *m*
i chiarificante *m*
n klaringsmiddel *n*
d Klärungsmittel *n*

2245 FINISHING gen
f finition *f*
e acabado *m*
i finitura *f*
n nabewerken *n*
d Endbearbeitung *f*, Nachbearbeitung *f*, Zurichten *n*

2246 FINISHING lea
f finissage *m*
e acabado *m*
i finissaggio *m*
n afwerken *n*
d Aufbereitung *f*

2247 FINISHING, DRESSING, SIZING tex
f apprêtage *m*
e apresto *m*
i apprettatura *f*
n appreteren *n*
d Appretur *f*

2248 FINISHING AGENT gen
f agent *m* de finition
e agente *m* de acabado
i agente *m* di finitura
n nabewerkingsmiddel *n*
d Nachbearbeitungsmittel *n*

2249 FINISHING COAT, COVER COAT cer,dec
f main *f* de finissage, couche *f* de teinte
e capa *f* de cubierta
i mano *f* a finire
n deklaag
d Deckanstrich *m*

2250 FIR LEAF OIL, FIR NEEDLE OIL, PINE NEEDL OIL fat
f huile *f* de sapin
e esencia *f* de pino
i olio *m* di pino
n dennenaaldolie
d Fichtennadelöl *n*, Kiefernnadelöl *n*

2251 FIRE ASSAYING, DRY ASSAYING ana
f essai *m* à sec
e ensayo *m* por vía seca

i assaggio *m* per via secca
n essaaieren *n* langs de droge weg
d Prüfen *n* auf trocknem Wege

2252 FIRE ASSAYING, CUPEL ASSAY, CUPELLING met,tes
f coupellation *f*
e copelación *f*
i coppellazione *f*
n copelleren *n*
d Kapellenprobe *f*, Kupellation *f*

2253 FIRE CRACK, HEAT CHECK cer
f crique *f* due à la chaleur
e grieta *f* debida al calor
i incrinatura *f* dovuta al calore
n warmtescheur
d Brandriss *m*, Wärmeriss *m*

2254 FIRE DISTILLATION dis
f distillation *f* à chauffage direct
e destilación *f* de calentamiento directo
i distillazione *f* a riscaldamento diretto
n destillatie met directe verhitting
d Destillation *f* mit direkter Heizung

2255 FIRE GILDING coa
f dorure *f* au feu
e doradura *f* al fuego
i doratura *f* al fuoco
n thermisch vergulden *n*
d Feuervergoldung *f*

2256 FIRE-OFF DIPPING elc
f pré-décapage *m*
e decapaje *m* preliminar
i decapaggio *m* preliminare
n voorbeitsen *n*
d Gelbbrennen *n*

2257 FIRE POLISHING gla
f polissage *m* au feu
e pulimento *m* al fuego
i lucidatura *f* al fuoco
n thermisch polijsten *n*
d Feuerpolitur *f*

2258 FIRE REFINING htt,met
f raffinage *m* au feu
e afino *m* al fuego
i affinazione *f* al fuoco
n thermisch veredelen *n*
d Feuervergütung *f*, thermische Vergütung *f*

2259 FIRE SCALE met
f couche *f* d'oxyde de cuivre
e capa *f* de óxido de cobre
i strato *m* d'ossido di rame
n koperoxydelaag
d Kupferoxydschicht *f*

2260 FIRECLAY, CHAMOTTE, SPAVIN, WARRANT cer
f apyre *f*, argile *f* réfractaire, chamotte *f*
e arcilla *f* refractaria, chamota *f*
i argilla *f* refrattaria, terra *f* refrattaria
n chamotte, vuurklei
d feuerfester Ton *m*, Schamotte *f*

2261 FIREDAMP, DAMP, PIT GAS (GB), WILD GAS (US) fue,gas
f grisou *m*, mofette *f*
e grisú *m*, mofeta *f*
i grisu *m*
n mijngas *n*
d Grubengas *n*, Schlagwetter *n*

2262 FIREDAMP DETECTION, DAMP DETECTION, PIT GAS DETECTION (GB), WILD GAS DETECTION (US) sap
f détection *f* de grisou, détection *f* de mofette
e detección *f* de grisú, detección *f* de mofeta
i rivelazione *f* di grisu
n mijngasdetectie
d Grubengasanzeige *f*, Schlagwetteranzeige *f*

2263 FIREPROOF sap
f incombustible adj, réfractaire adj
e incombustible adj, refractario adj
i incombustibile adj, refrattario adj
n vuurbestendig adj
d feuerfest adj, feuersicher adj

2264 FIREPROOF CEMENT bui
f ciment *m* réfractaire
e cemento *m* refractario
i cemento *m* refrattario
n vuurbestendig cement *n*
d feuerfester Zement *m*

2265 FIREPROOF CURTAIN, ASBESTOS CURTAIN sap
f rideau *m* d'amiante

e telón *m* de amianto
i sipario *m* d'amianto
n asbestgordijn *n*
d Asbestvorhang *m*

2266 FIREPROOF PAINT dec
f couleur *f* ignifuge
e pintura *f* ignífuga
i pittura *f* ignifuga
n tegen vuur beschermende verf
d feuerhemmender Anstrich *m*, Feuerschutzanstrich *m*

2267 FIRERESISTANT sap
f résistant adj au feu
e resistente adj al fuego
i resistente adj al fuoco
n vuurwerend adj
d feuerhemmend adj

2268 FIRERETARDANT sap
f résistant ajd au feu
e reistente adj al fuego
i resistente adj al fuoco
n vuurvertragend adj
d feuerhemmend adj

2269 FIREWORKS exp
f feu *m* d'artifice
e fuegos *pl* artificiales
i fuochi *pl* artificiali
n vuurwerk *n*
d Feuerwerk *n*

2270 FIRING boi
f chargement *m*, chauffage *m*
e encendido *m*
i alimentazione *f*
n stoken *n*
d Feuerung *f*

2271 FIRING cer
f cuisson *m*, cuite *f*
e cocción *f*
i cottura *f*
n bakken *n*
d Brennen *n*

2272 FIRING fue
f amorçage *m*
e encendido *m*
i accensione *f*
n ontsteken *n*
d Zünden *n*

2273 FIRST BATH tan
f bain *m* premier
e primer baño *m*
i primo bagno *m*
n impregneerbad *n*
d Imprägnierbad *n*

2274 FIRST ENAMEL COAT dec
f couche *f* de fond pour peinture-émail
e primera mano *f* para esmaltar
i prima mano *f* per smaltare
n grondlaklaag
d Lackgrundierung *f*

2275 FIRST RUNNING, FORE-RUNNING dis
f première fraction *f*, produit *m* de tête
e cabeza *f* de destilación, primera fracción *f*
i prima frazione *f*, prodotto *m* di testa
n eerste fractie, voorloop
d Vorlauf *m*, Vorprodukt *n*

2276 FISH GLUE, ISINGLASS glu
f colle *f* de poisson, ichtyocolle *f*
e cola *f* de pescado
i colla *f* di pesce
n vislijm
d Fischleim *m*, Hausenblase *f*

2277 FISH OIL, TRAIN OIL fat
f huile *f* de baleine, huile *f* de poisson
e aceite *m* de ballena, aceite *m* de pescado
i olio *m* di balena, olio *m* di pesce
n traan, visolie
d Fischtran *m*, Tran *m*

2278 FISSION nue
f fission *f*
e fisión *f*
i fissione *f*
n splijting
d Spaltung *f*

2279 FISSIONABLE MATERIAL nue
f matière *f* fissible
e materia *f* físil
i materiale *m* fissile
n splijtbaar materiaal *n*, splijtmateriaal *n*
d Spaltstoff *m*

2280 FIXATION OF NITROGEN gas
f fixation *f* de l'azote
e fijación *f* del nitrógeno
i fissazione *f* dell'azoto
n stikstofbinding
d Stickstoffbindung *f*

2281 FIXATIVE per,pri
f fixatif *m*

e fijador *m*
i fissativo *m*
n fixatief *n*
d Fixativ *n*

2282 FIXED-BED CATALYST fue
f catalysateur *m* de lit fixe
e catalizador *m* de lecho fijo
i catalizzatore *m* di letto fisso
n vast opgestelde katalysator
d Festbettkatalysator *m*

2283 FIXED-BED CRACKING fue
f craquage *m* à lit fixe
e pirodesintegración *f* en lecho fijo
i piroscissione *f* in letto fisso
n kraken *n* met vast opgestelde katalysator
d Festbettkracken *n*

2284 FIXED-BED ION EXCHANGE elc
f échange *m* d'ions à lit fixe
e cambio *m* de iones de lecho fijo
i scambio *m* d'ioni a letto fisso
n ionenuitwisseling met vast harsbed
d Festbettionenaustausch *m*

2285 FIXED EXPANSION pre
f expansion *f* constante
e expansión *f* constante
i espansione *f* costante
n constante expansie
d konstante Expansion *f*

2286 FIXED OILS, FATTY OILS fat
f huiles *pl* fixes
e aceites *pl* fijos, aceites *pl* grasos
i oli *pl* fissi
n oliën *pl* van dierlijke en plantaardige oorsprong
d animalische und vegetabilische Öle *pl*

2287 FIXED-PAN MILLING brk
f broyage *m* à cuve fixe
e molienda *f* de cuba fija
i molitura *f* a tino fisso
n malen *n* met stilstaande bodem
d Kolleren *n* mit feststehendem Teller

2288 FIXED RESIDUE, IGNITION RESIDUE fue
f résidu *m* calciné, résidu *m* fixe après calcination
e residuo *m* de calcinación
i residuo *m* calcinato, residuo *m* di calcinazione, residuo *m* fisso
n calcinatieresidu *n*, gloeirest
d Glührückstand *m*

2289 FIXING pho,tex
f fixage *m*
e fijado *m*
i fissaggio *m*
n fixeren *n*
d Fixieren *n*

2290 FLAKE gen
f flocon *m*
e copo *m*, escama *f*
i scaglia *f*
n vlok
d Flocke *f*

2291 FLAKE GRAPHITE mat
f graphite *m* lamellaire
e grafito *m* laminar
i grafite *m* a lamelle, grafite *m* lamellare
n bladgrafiet *n*, lamellair grafiet *n*
d A-Graphit *m*, Lamellengraphit *m*

2292 FLAKING dec
f écaillement *m*
e desconchado *m*
i scagliamento *m*
n afbladderen *n*
d Abblättern *n*

2293 FLAKING, EXFOLIATION, SPALLING met
f écaillage *m*
e desconchado *m*
i scagliatura *f*, screpolatura *f*
n afbladderen *n*
d Abblättern *n*, Häutung *f*

2294 FLAME bur
f flamme *f*
e llama *f*
i fiamma *f*
n vlam
d Flamme *f*

2295 FLAME ANNEALING htt,met
f recuit *m* au chalumeau
e recocido *m* con soplete
i ricottura *f* alla fiamma
n vlamgloeien *n*
d Flammglühung *f*, Oberflächenglühung *f*

2296 FLAME CLEANING cle,met
f décalaminage *m* au chalumeau
e limpia *f* con soplete
i pulitura *f* alla fiamma
n hamerslagverwijdering aan de vlam
d Flammentzunderung *f*

2297 FLAME COLORING (US), FLAME COLOURING (GB) ana
f coloration *f* à la flamme
e colorización *f* a la llama
i colorazione *f* alla fiamma
n vlamkleuring
d Flammenfärbung *f*

2298 FLAME COLORING TEST (US), FLAME COLOURING TEST (GB) tes
f essai *m* à la flamme
e ensayo *m* a la llama
i collaudo *m* alla fiamma
n vlamkleuringsanalyse
d Flammenfärbungsprobe *f*

2299 FLAME CUTTING, GAS CUTTING, TORCH CUTTING met
f coupage *m* à l'autogène, oxycoupage *m*
e corte *m* por llama oxiacetilénica, oxicorte *m*
i taglio *m* per fiamma ossiacetilene
n autogeen snijden *n*
d Brennschneiden *n*

2300 FLAME HARDENING, SHORTERIZING, TORCH HARDENING htt,met
f trempe *f* au chalumeau
e temple *m* superficial a la llama
i tempera *f* alla fiamma
n vlamharding
d Flammenhärtung *f*, Oberflächenhärtung *f*

2301 FLAME PLATING coa
f placage *m* à la flamme
e chaplado *m* a la llama
i placcatura *f* alla fiamma
n vlampleten *n*
d Flammplattierung *f*

2302 FLAME SCALING cle,met
f décalaminage *m* au chalumeau
e descascarillado *m* con soplete
i disincrostazione *f* alla fiamma
n oxydehuidverbranding
d thermische Entzunderung *f*

2303 FLAME SPRAYING bui,coa
f métallisation *f* par jet
e metalización *f* por chorro
i metallizzazione *f* con getto
n metalliseren *n* met metaalspuit
d Metallspritzverfahren *n*

2304 FLAME TREATING pla
f traitement *m* superficiel à la flamme
e tratamiento *m* superficial a la llama
i trattamento *m* superficiale alla fiamma
n oppervlakbehandeling
d Oberflächenbehandlung *f*

2305 FLAMELESS COMBUSTION, SURFACE COMBUSTION fue
f combustion *f* sans flamme
e combustión *f* sin llama
i combustione *f* senza fiamma
n vlamloze verbranding
d flammenlose Verbrennung *f*

2306 FLAMELESS EXPLOSIVE exp
f explosif *m* de sûreté
e explosivo *m* de seguridad
i esplosivo *m* di sicurezza
n mijngasveilige springstof
d Kohlendynamit *m*, Sicherheitssprengstoff *m*

2307 FLAMEPROOF gen
f ininflammable adj
e antideflagante adj, ininflamable adj
i non infiammabile adj
n onontvlambaar adj
d feuersicher adj, flammensicher adj

2308 FLASH AGEING (GB), FLASH AGING (US), tex
f traitement *m* rapide par vapeur
e tratamiento *m* húmedo rápido
i trattamento *m* rapido al vapore
n snelstoombehandeling
d Schnelldämpfung *f*

2309 FLASH DISTILLATION dis
f distillation *f* à évaporation instantanée
e destilación *f* con evaporación instantánea
i distillazione *f* equilibrata in corrente di vapore
n afdampen *n*
d Kurzwegdestillation *f*

2310 FLASH DRYING dry
f séchage *m* rapide
e secado *m* rápido
i essiccazione *f* rapida
n snelle droging
d Schnelltrocknung *f*, Stromtrocknung *f*

2311 FLASH EVAPORATION eva
f évaporation *f* à basse pression

e evaporación *f* de baja presión
i evaporazione *f* a bassa pressione
n lage-drukverdamping
d Niederdruckverdampfung *f*

2312 FLASH PLATING coa
f voile *f*
e velo *m*
i vela *f*
n kleurlaag
d Schnellüberzug *m*

2313 FLASH POINT fue
f point *m* d'éclair
e punto *m* inflamador
i punto *m* d'infiammabilità
n vlampunt *n*
d Flammpunkt *m*

2314 FLASH POINT TEST tes
f vérification *f* du point d'éclair
e prueba *f* del punto inflamador
i collaudo *m* del punto d'infiammabilità
n vlampuntproef
d Flammpunktprobe *f*

2315 FLASH ROASTING htt,min
f grillage *m* de concentrés pulvérulents
e tostación *f* de concentrados pulverizados
i arrostimento *m* di concentrati polverizzati
n vlamroosten *n*
d Flammrösten *n*

2316 FLASHBACK bui
f retour *m* de flamme
e retorno *m* de llama
i ritorno *m* di fiamma
n inslaan *n* van een brander
d Rückschlag *m* der Flamme

2317 FLASHED GLASS, CASED GLASS, PLY GLASS gla
f verre *m* à deux couches, verre *m* doublé
e vidrio *m* contrachapado
i vetro *m* a due strati
n tweelaagsglas *n*
d Überfangglas *n*

2318 FLAT COAT dec
f couche *f* d'apprêt
e fondo *m* de aparejo
i fondo *m* di mestica
n plamuurlaag
d Spachtelauftrag *m*

2319 FLAT CURING, PLATEAU EFFECT rub
f effet *m* de plateau
e efecto *m* de meseta
i effetto *m* di piano
n plateaueffect *n*
d Plateau-Effekt *m*

2320 FLAT FINISH, DEAD FINISH, DULL FINISH dec
f finissage *m* mat
e acabado *m* mate, terminación *f* mate
i finissaggio *m* matto, rifinitura *f* opaca
n mat afstrijken *n*, matte deklaag
d Mattglanz *m*, matte Deckschicht *f*

2321 FLAT SHEET EXTRUSION, SLOT EXTRUSION pla
f extrusion *f* à filière plate
e extrusión *f* de laminas
i estrusione *f* di foglie piane
n foeliespuitmethode
d Breitbandspritzverfahren *n*

2322 FLAT VARNISH dec
f vernis *m* mat
e barniz *m* mate
i vernice *f* opaca
n matlak
d Mattlack *m*

2323 FLATTING, PRIMER dec
f apprêt *m*
e aparejo *m*
i mestica *f*
n plamuur
d Spachtel *m*

2324 FLATTING AGENT, GLOSS-REDUCING AGENT dec
f agent *m* de matage
e producto *m* para opacificar
i agente *m* opacificante
n matteermiddel *n*
d Mattierungsmittel *n*

2325 FLATTING LACQUER dec
f laque *f* à poncer
e laca *f* de aparejo
i lacca *f* a mestica
n slijplak
d Schleiflack *m*, Spachtellack *m*

2326 FLAYING lea
f écorchement *m*, équarrissage *m*, équarrissement *m*

e desolladura *f*
i scorticamento *m*
n villen *n*
d Abdecken *n*

2327 FLESH SIDE lea
f côte *f* chair, croûte *f*
e carnaza *f*
i carne *f*
n vleeszijde
d Aasseite *f*, Fleischseite *f*

2328 FLESHING lea
f écharnage *m*, écharnement *m*
e descarnadura *f*
i scarnatura *f*
n ontvlezen *n*
d Entfleischen *n*

2329 FLINT min
f silex *m*
e pedernal *m*, silex *m*
i flint *m*, selce *f*
n flint, silex *n*, vuursteen *n*
d Feuerstein *m*, Flintstein *m*

2330 FLINT, pap
FLINT COATED PAPER,
FLINT PAPER
f carton *m* brillanté
e papel *m* de lija
i carta *f* brillante
n glansbordkarton *n*, glanskarton *n*
d Flintpapier *n*, Glanzpappe *f*

2331 FLINT GLASS gla
f flint-glass *m*, verre *m* flint
e flint-glass *m*
i vetro *m* flint
n flintglas *n*
d Flintglas *n*

2332 FLOATING LAYER, iwt,sew
SURFACE SCUM
f boue *f* flottante, chapeau *m* d'écume
e fango *m* flotante, nata *f* espumosa
i cappellaccio *m*, fango *m* galleggiante
n schuimlaag
d Schwimmdecke *f*, Schwimmschlamm *m*

2333 FLOCCULATED LATEX rub
f latex *m* floculé
e látex *m* floculado
i lattice *m* flocculato
n uitgevlokte latex
d ausgeflockter Latex *m*, flockulierter Latex *m*

2334 FLOCCULATING YEAST bre
f levure *f* caséeuse, levure *f* floculante
e levadura *f* floculante
i lievito *m* agglutinante
n uitvlokkende gist
d Bruchhefe *f*

2335 FLOCCULATION col
f floculation *f*
e floculación *f*
i flocculazione *f*
n floculeren *n*, uitvlokken *n*
d Ausflockung *f*, Flockung *f*

2336 FLOCCULATOR, gen
FLOCCULENT
f agent *m* floculant
e agente *m* floculante
i agente *m* flocculante
n uitvlokmiddel *n*
d Flockungsmittel *n*

2337 FLOCK tex
f flocon *m*
e borra *f*
i fiocco *m*
n vlok
d Flocke *f*

2338 FLOCK FINISHING dec
f laque *f* à flocons
e laca *f* de boras
i lacca *f* a fiocchi
n vlokkenlak
d Flockenlack *m*

2339 FLOCK SPRAYING coa,dec
f pulvérisation *f* de flocons
e pulverización *f* de borras
i polverizzazione *f* di fiocchi
n vlokkenverstuiving
d Flockenzerstäubung *f*

2340 FLOOD LUBRICATION lub
f lubrification *f* à bain d'huile
e lubricación *f* con baño de aceite
i lubrificazione *f* in bagno d'olio
n oliebadsmering
d Ölbadschmierung *f*, Ölspülung *f*

2341 FLOODING dis,gen
f engorgement *m*, noyage *m*
e anegación *f*, inundación *f*
i ingolfamento *m*, ingorgo *m*, inondazione *f*
n overstromen *n*
d Fluten *n*, Überfluten *n*

2342 FLOODING POINT dis
f point *m* d'engorgement

e punto *m* de anegación, punto *m* de inundación
i punto *m* d'ingolfamento
n overstromingspunt *n*
d Flutpunkt *m*, Überflutungspunkt *m*

2343 FLOORING bre
f conduite *f* de la germination
e conducción *f* de la germinación
i condotto *m* della germinazione
n kiemingsbeïnvloeding
d Haufenführung *f*

2344 FLOTATION met,min
f flottation *f*
e flotación *f*
i flottazione *f*
n flotatie
d Flotation *f*, Schwimmaufbereitung *f*

2345 FLOUR nut
f farine *f*
e harina *f*
i farina *f*
n bloem, meel *n*
d Kernmehl *n*, Mehl *n*

2346 FLOW gen
f courant *m*, écoulement *m*, flux *m*
e corriente *f*, flujo *m*, gasto *m*
i corrente *f*, flusso *m*
n stroming, stroom
d Strom *m*, Strömung *f*

2347 FLOW COATING coa
f vernissage *m* par arrosage
e barnizadura *f* por rociado
i verniciatura *f* per annaffiamento
n lakgieten *n*
d Lackguss *m*

2348 FLOW CONTROL con,dis
f réglage *m* de débit
e regulación *f* de caudal
i regolazione *f* di portata
n stromingsregeling
d Strömungsregelung *f*

2349 FLOW TEST tes
f essai *m* d'écoulement
e ensayo *m* de fluidez
i collaudo *m* di fluidità
n vloeiproef
d Fliessprobe *f*

2350 FLOWABILITY (GB), FLUIDITY (US) tes
f pouvoir *m* d'écoulement
e poder *m* de fluidez
i potere *m* scorrente
n vloeivermogen *n*
d Fliessfähigkeit *f*

2351 FLOWERS, SPANGLES elc
f fleurs *pl* de zinc
e floreado *m*
i fiori *pl* di zinco
n zinkbloemen *pl*
d Zinkblumen *pl*

2352 FLOWERS OF SULFUR (US), FLOWERS OF SULPHUR (GB) aic
f fleurs *pl* de soufre
e flores *pl* de azufre
i fiori *pl* di zolfo
n bloem van zwavel
d Schwefelblume *f*

2353 FLOWERS OF ZINC, CHINESE WHITE dec
f blanc *m* de Chine
e blanco *m* al óxido de cinc
i biacca *f* all'ossido di zinco
n Chinees wit *n*
d Chinaweiss *n*, Zinkoxydweiss *n*

2354 FLUE GAS fue,gas
f fumées *pl*
e gases *pl* de combustión
i fumi *pl*, gas *pl* di combustione
n rookgas *n*
d Rauchgas *n*

2355 FLUE GAS PURIFICATION cle,gas
f dépoussiérage *m* des fumées
e depuración *f* de los gases de combustión
i depurazione *f* dei fumi
n rookgasreiniging
d Rauchgasreinigung *f*

2356 FLUE SWEAT, PERSPIRATION WATER fue
f eau *f* de condensation de cheminée
e agua *f* de condensación de chimenea
i acqua *f* di condensazione di camino
n zweetwater *n*
d Schwitzwasser *n*

2357 FLUFFING lea
f ponçage *m*
e tratamiento *m* con la piedra pómez
i trattamento *m* con pomice
n puimen *n*
d Bimssteinbehandlung *f*

2358 FLUID, LIQUID gen
f liquide *m*
e líquido *m*
i liquido *m*
n vloeistof
d Flüssigkeit *f*

2359 FLUID CATALYTIC CRACKING fue
f craquage *m* catalytique fluide
e pirodesintegración *f* con catalizador flúido
i piroscissione *f* con catalizzatore fluido
n kraken *n* met gefluïdiseerde katalysator
d Fliessbettkrackung *f*, Wirbelschichtkrackung *f*

2360 FLUID DYNAMIC PRESSURE mea
f pression *f* dynamique d'un liquide
e presión *f* dinámica de un líquido
i pressione *f* dinamica d'un liquido
n dynamische vloeistofdruk
d dynamischer Flüssigkeitsdruck *m*

2361 FLUID EXTRACT mcy
f extract *m* fluide
e extracto *m* flúido
i estratto *m* fluido
n vloeibaar extract *n*
d Fluidextrakt *m*

2362 FLUID INTERFACE nue
f couche *f* intermédiaire entre fluides
e zona *f* interfacial entre flúidos
i strato *m* di separazione tra fluidi
n vloeistofgrensvlak *n*
d Flüssigkeitsgrenzfläche *f*

2363 FLUID LUBRICATION lub
f lubrification *f* idéale
e lubricación *f* ideal
i lubrificazione *f* fluida
n vloeistofwrijving, volledige smering
d Idealschmierung *f*

2364 FLUID SCALE cor,met
f battitures *pl* fluides
e cascarilla *f* flúida
i scaglia *f* fluida
n vloeiende oxydelaag
d flüssiger Zunder *m*

2365 FLUIDIFICATION, LIQUEFACTION, LIQUEFYING gas
f liquéfaction *f*
e licuación *f*, licuefacción *f*
i liquefazione *f*
n vloeibaar maken *n*
d Verflüssigung *f*

2366 FLUIDITY gen
f fluidité *f*
e fluidez *f*
i fluidità *f*
n vloeibaarheid
d Flüssigkeitsgrad *m*

2367 FLUIDIZED BED DRYING dry
f séchage *m* à lit fluidisé
e secado *m* de lecho fluidificado
i essiccazione *f* a letto fluidizzato
n werveldrogen *n*
d Wirbeltrocknung *f*

2368 FLUIDIZED GASIFICATION fue
f gazéification *f* du charbon par fluidisation
e gasificación *f* del carbón en lecho flúido
i gassificazione *f* del carbone in letto fluido
n gefluïdiseerd vergassen *n*
d Wirbelschichtverfahren *n*

2369 FLUORIDATION hyg
f fluorisation *f*
e fluorización *f*
i fluorizzazione *f*
n fluorideren *n*, fluortoevoeging
d Fluorierung *f*, Fluorzusatz *m*

2370 FLUORINATION aic
f fluoration *f*
e fluoración *f*
i fluorazione *f*
n fluoreren *n*
d Fluorieren *n*, Fluorination *f*

2371 FLUORSPAR dec,min
f chaux *f* fluatée, spath *m* fluor, fluorine *f*
e espato *m* fluor, fluorita *f*
i fluorite *f*
n vloeispaat *n*
d Fluorit *m*, Flussspat *m*

2372 FLUSHING sew
f curage *m*
e limpieza *f* por descarga
i cacciata *f*
n doorspoelen *n*
d Ausschwemmen *n*, Ausspülen *n*

2373 FLUX gen
f fondant *m*
e fundente *m*
i fondente *m*
n toeslag, vloeimiddel *n*
d Flüssmittel *n*, Zuschlag *m*

2374 FLUX OIL fat
f huile *f* fluidifiante
e aceite *m* fluidificante
i olio *m* fluidificante
n fluxolie
d Verflüssigungsöl *n*

2375 FOAM, gen
FROTH
f écume *f*
e espuma *f*
i schiuma *f*
n schuim *n*
d Schaum *m*

2376 FOAM GLASS gla
f verre *m* spongieux
e vidrio *m* alveolar, vidrio *m* celular
i vetro *m* spugnoso
n schuimglas *n*
d Schaumglas *n*

2377 FOAM GLUE glu
f colle *f* écumante
e cola *f* de espuma
i colla *f* schiumante
n schuimlijm
d Schaumleim *m*

2378 FOAM LATEX, rub
FOAM RUBBER, LATEX FOAM
f caoutchouc *m* spongieux, mousse *f* de latex
e espuma *f* de látex
i gommapiuma *f*, lattice *m* schiuma, schiuma *f* di lattice
n schuimrubber
d Latexschaum *m*, Schaumgummi *m*

2379 FOAMED CONCRETE, bui
AERATED CONCRETE,
CELLULAR CONCRETE,
GAS CONCRETE
f béton *m* cellulaire
e hormigón *m* aerado
i calcestruzzo *m* poroso
n cellenbeton *n*
d Zellenbeton *m*

2380 FOAMED PLASTIC, pla
AERATED PLASTIC
f matière *f* plastique aérée, matière *f* plastique mousse
e materia *f* plástica celular
i materia *f* plastica cellulare, materia *f* plastica schiuma
n schuimkunststof, schuimplastiek *n*
d Schaumkunststoff *m*, Schaumstoff *m* mit nichtporösen Zellen

2381 FOAMING AGENT gen
f agent *m* moussant, générateur *m* d'écume
e agente *m* espumante, generador *m* de espuma
i agente *m* schiumante, generatore *m* di schiuma
n schuimmiddel *n*
d Schaummittel *n*

2382 FODDER YEAST, fmi
FOOD YEAST
f levure *f* alimentaire
e levadura *f* para alimentación
i lievito *m* per alimentazione
n voedingsgist
d Nährhefe *f*

2383 FOG dec
f nuage *m*
e niebla *f*
i spruzzo *m*
n verfnevel
d Farbnebel *m*

2384 FOG pho
f voile *m*
e velo *m*
i velatura *f*, velo *m*
n sluier
d Schleier *m*

2385 FOG QUENCHING met
f refroidissement *m* rapide en vapeur d'eau
e enfriamiento *m* rápido por neblina de agua
i raffreddamento *m* rapido per nebbia d'acqua
n afschrikken *n* in waterdamp
d Abschrecken *n* in Wasserdampf

2386 FOIL met
f feuille *f* de métal
e hoja *f* delgada
i foglia *f*
n bladmetaal *n*, foelie
d Folie *f*

2387 FOLIATED, met,pla
LAMINATED
f laminé adj

e laminado adj
i laminato adj
n gelaagd adj , gelamelleerd adj
d geschichtet adj , lamelliert adj

2388 FOOD nut
f matière *f* nutritive
e substancia *f* nutritiva
i sostanza *f* nutritiva
n voedsel *n*
d Nährstoff *m*

2389 FOOD PRESERVATIVES nut
f conservants *pl*
e substancias *pl* conservantes
i sostanze *pl* per la conservazione dei cibi
n conserveringsmiddelen *pl*
d Konservierungsmittel *pl*

2390 FOOL-PROOF gen
f à toute épreuve
e a prueba de fallas
i sicuro adj
n feilzeker adj
d misbrauchbetriebssicher adj, narrensicher adj

2391 FOOTS fat
f crasses *pl*, effondrilles *pl*
e heces *pl*, poso *m*
i sedimento *m* da olio
n neerslag uit olie
d Niederschlag *m* aus Öl

2392 FORCE FEED, FORCED LUBRICATION, FULL FORCE FEED lub
f lubrification *f* forcée
e lubricación *f* a presión
i lubrificazione *f* forzata
n druksmering
d Öldruckschmierung *f*

2393 FORCED CIRCULATION, FORCED FLOW boi,gen
f circulation *f* forcée
e circulación *f* forzada
i circolazione *f* forzata
n gedwongen circulatie
d erzwungener Umlauf *m*, Zwangsumlauf *m*

2394 FORCED-CIRCULATION EVAPORATION eva
f évaporation *f* à circulation forcée
e evaporación de circulación forzada
i evaporazione *f* a circolazione forzata
n verdamping met gedwongen circulatie
d Umwalzverdampfung *f*, Verdampfung *f* mit erzwungenem Umlauf, Zwangsumlaufverdampfu

2395 FORCED DRAFT (US), INDUCED DRAUGHT (GB) fue
f tirage *m* forcé
e tiro *m* forzado
i tiraggio *m* forzato
n gedwongen trek
d künstlicher Zug *m*, Unterwind *m*

2396 FORCED DRAFT COOLING (US), INDUCED DRAUGHT COOLING (GB) clo
f refroidissement *m* à tirage forcé
e enfriamiento *m* de tiro forzado
i raffreddamento *m* a tiraggio forzato
n koeling met kunstmatige trek
d Kühlung *f* mit künstlichem Zug

2397 FORE-RUNNING, FIRST RUNNING dis
f produit *m* de tête
e cabeza *f* de destilación
i prodotto *m* di testa
n voorloop
d Vorlauf *m*

2398 FORE VACUUM vac
f vide *m* préliminaire élevé
e vacío *m* preliminar elevado
i vuoto *m* preliminare elevato
n voorvacuüm *n*
d Vorvakuum *n*

2399 FORECOOLING clo
f pré-refroidissement *m*, refroidissement *m* préliminaire
e enfriamiento *m* preliminar, preenfriamiento *m*
i preraffreddamento *m*, raffreddamento *m* preliminare
n voorkoeling
d Vorkühlung *f*

2400 FOREIGN MATTER gen
f matière *f* étrangère
e substancia *f* extraña
i impurità *pl*, sostanza *f* estranea
n verontreinigingen *pl*, vreemde bestanddelen *pl*
d Fremdkörper *pl*, Verunreinigung

2401 FORENSIC ANALYSIS ana
f analyse *f* légale
e análisis *f* judicial
i analisi *f* legale
n gerechtelijke analyse
d Gerichtsanalyse *f*

2402 FORGE SCALE met
f battitures *pl* de forge, mâchefer *m*
e batiduras *pl* a forja
i scaglia *f* di fucinatura
n hamerslag *n*
d Hammerschlag *m*

2403 FORMALDEHYDE RESINS pla
f résines *pl* de formaldéhyde
e resinas *pl* de formaldehide
i resine *pl* di formaldeide
n formaldehydeharsen *pl*
d Formaldehydharze *pl*

2404 FORMALIN, AQUEOUS SOLUTION OF FORMALDEHYDE oic
f formaline *f*, formol *m*
e formalina *f*, formol *m*
i formalina *f*, formol *m*
n formalien *n*, formaline *n*, formol *n*
d Formalin *n*, Formol *n*

2405 FORMED PLATE elc
f plaque *f* formée
e placa *f* formada
i piastra *f* formata
n gevormde plaat
d formierte Platte *f*

2406 FORMIC ACID oic
f acide *m* formique
e ácido *m* fórmico
i acido *m* formico
n mierezuur *n*
d Ameisensäure *f*

2407 FORMIC SPIRIT mcy
f esprit *m* formique
e espíritu *m* fórmico
i spirito *m* formico
n mierespiritus
d Ameisengeist *m*, Ameisenspiritus *m*

2408 FORMULA gen
f formule *f*
e fórmula *f*
i formula *f*
n formule
d Formel *f*

2409 FORMYLATION oic
f formylation *f*
e formilación *f*
i formilazione *f*
n formylatie
d Formylierung *f*

2410 FOSSIL WAX, CERESINE, CERINE, OZOKERITE fat
f cire *f* minérale, ozocérite *f*, suif *m* de montagne
e cera *f* fósil, ceresina *f*
i cera *f* minerale, ozocherite *f*
n aardwas, ozokeriet *n*
d Erdwachs *n*, mineralisches Wachs *n*, Ozokerit *m*

2411 FOUL CLAY, PLASTIC CLAY, PURE CLAY, STRONG CLAY bui
f argile *f* plastique
e arcilla *f* plástica
i galestro *m*
n zuivere klei
d reiner Ton *m*, reines Aluminiumsilikat *n*

2412 FOUL ELECTROLYTE elc
f électrolyte *m* impropre
e electrólito *m* impuro
i elettrolito *m* impuro, elettrolito *m* inquinato
n onzuivere elektrolyt
d verbrauchter Elektrolyt *m*

2413 FOULING dec
f encrassement *m*
e incrustación *f*
i incrostazione *f*
n aangroeiing, aanvreten *n*
d Anwuchs *m*, Bewuchs *m*

2414 FOULING FACTOR dis
f facteur *m* d'encrassement
e factor *m* de ensuciamiento
i fattore *m* di sporcamento
n vervuilingsfactor
d Verkrustungsfaktor *m*, Verschmutzungsfaktor *m*

2415 FOUND gla
f fusion *f*
e fusión *f*
i fusione *f*
n smelt
d Schmelze *f*

2416 FOUNDRY BLACKING, CHARCOAL BLACKING, CHARCOAL DUST fou
f noir *m* de fonderie

e negro *m* de fundición, polvo *m* de carbón
i nero *m* per fonderia, polvere *f* di carbone
n koolpoeder *n*
d Kohlenstaub *m*

2417 FOWLER SOLUTION, ARSENICAL SOLUTION mcy
f liqueur *f* de Fowler
e arsénico *m* en solución de Fowler
i soluzione *f* di Fowler
n liquor Fowleri
d Fowlersche Lösung *f*

2418 FRACTION, CUT dis,fue
f fraction *f*
e corriente *f*, fracción *f*
i frazione *f*
n fractie
d Fraktion *f*

2419 FRACTIONAL COAGULATION col
f coagulation *f* fractionnée
e coagulación *f* fraccionada
i coagulazione *f* frazionata
n gefractioneerde coagulatie
d fraktionierte Koagulierung *f*

2420 FRACTIONAL COMBUSTION fue
f combustion *f* fractionnée
e combustión *f* fraccionada
i combustione *f* frazionata
n gefractioneerde verbranding
d fraktionierte Verbrennung *f*

2421 FRACTIONAL CONDENSATION clo
f condensation *f* fractionnée
e condensación *f* fraccionada
i condensazione *f* frazionata
n gefractioneerde condensatie
d fraktionierte Kondensation *f*

2422 FRACTIONAL CRYSTALLIZATION cry
f cristallisation *f* fractionnée
e cristalización *f* fraccionada
i cristallizzazione *f* frazionata
n gefractioneerde kristallisatie
d fraktionierte Kristallisation *f*, Kristallisationsdifferentiation *f*

2423 FRACTIONAL DISTILLATION, FRACTIONATING dis
f distillation *f* fractionnée
e destilación *f* fraccionada
i distillazione *f* frazionata
n gefractioneerde destillatie
d fraktionierte Destillation *f*

2424 FRACTIONAL PRECIPITATION, FRACTIONAL SEPARATION sep
f précipitation *f* fractionnée
e precipitación *f* fraccionada
i precipitazione *f* frazionata
n gefractioneerde precipitatie
d fraktionierte Fällung *f*

2425 FRACTIONATION dis
f fractionnement *m*
e fraccionamiento *m*
i frazionamento *m*
n fractioneren *n*
d Fraktionieren *n*

2426 FRAGRANCE per
f odeur *f* agréable, odeur *f* suave
e fragrancia *f*
i fragranza *f*
n aangename geur
d Duft *m*, Wohlgeruch *m*

2427 FRANKFORT BLACK, GERMAN BLACK pri
f noir *m* d'Allemagne
e negro *m* de Alemania
i nero *m* di Germania
n koperdrukzwart *n*
d Kupferdruckschwarz *n*

2428 FRANKFORT BLACK, VINE BLACK dec
f noir *m* de fumée de pampres
e negro *m* de humo de sarmientes de vid
i nero *m* di fumo di viticci della vite
n rankenzwart *n*, wijngaardzwart *n*
d Frankfurter Schwarz *n*, Rebenschwarz *n*, Weinrebenschwarz *n*

2429 FREE ENERGY gen
f énergie *f* libre
e energía *f* libre
i energia *f* libera
n vrije energie
d freie Energie *f*

2430 FREE MOISTURE gen
f teneur *f* libre d'humidité
e contenido *m* libre de humedad
i tenore *m* libero d'umidità
n vrij vochtgehalte *n*
d freier Feuchtigkeitsgehalt *m*

2431 FREE RADICAL oic
f radical *m* libre

e radical *m* libre
i radicale *m* libero
n vrij radicaal *n*
d freies Radikal *n*

2432 FREE RADICAL POLYMERIZATION oic
f polymérisation *f* radiculaire
e polimerización *f* por radical libre
i polimerizzazione *f* per radicale libero
n polymerisatie met vrije radicaal
d Freiradikalpolymerierung *f*

2433 FREE SULFUR (US), EXTRACTABLE SULPHUR (GB) rub
f soufre *m* libre
e azufre *m* libre
i zolfo *m* libero
n vrije zwavel
d freier Schwefel *m*

2434 FREEZE CASTING cer
f coulée *f* à congélation
e colada *f* de congelación
i colata *f* a congelazione
n vriesgieten *n*
d Gefrierguss *m*

2435 FREEZE DRYING clo,dry
f séchage *m* par congélation
e secado *m* por congelación
i deidrazione *f* a freddo
n vriesdrogen *n*
d Gefriertrocknung *f*

2436 FREEZE RESISTANCE clo
f résistance *f* à la gelée
e resistencia *f* a la helada
i resistenza *f* al gelo
n vorstvrijheid
d Frostbeständigkeit *f*

2437 FREEZING clo
f congélation *f*, gelée *f*
e congelación *f*, helada *f*
i congelazione *f*, gelo *m*
n bevriezen *n*
d Gefrieren *n*

2438 FREEZING, SOLIDIFICATION gen
f solidification *f*
e solidificación *f*
i solidificazione *f*
n stolling
d Erstarrung *f*

2439 FREEZING MIXTURE, FRIGORIC MIXTURE clo
f mélange *m* frigorifique
e mezcla *f* frigorífica
i mescolanza *f* frigorifera
n vriesmengsel *n*
d Gefriermischung *f*, Kältemischung *f*

2440 FREEZING POINT gen
f point *m* de congélation
e punto *m* de congelación
i punto *m* di congelazione
n vriespunt *n*
d Eispunkt *m*, Gefrierpunkt *m*, Nullpunkt *m*

2441 FREEZING POINT, SOLIDIFICATION POINT gen
f point *m* de solidification
e punto *m* de solidificación
i punto *m* di solidificazione
n stolpunt *n*
d Erstarrungspunkt *m*

2442 FREEZING-POINT METHOD, CRYOSCOPIC METHOD ana
f méthode *f* cryoscopique
e método *m* crioscópico
i metodo *m* crioscopico
n kryoscopie
d Kryoskopie *f*

2443 FRENCH CHALK (US), SOAPSTONE, TALC (GB), TALCUM (GB) mat
f talc *m*
e piedra *f* de jabón, talco *m*
i talco *m*
n talk
d Talkum *n*

2444 FRENCH CHALK, VIENNA CHALK hyg,lub
f chaux *f* de Vienne
e cal *f* de Viena
i calce *f* di Vienna
n Weense kalk
d Wiener Kalk *m*

2445 FRENCH POLISH dec
f vernis *m* à l'alcool, vernis *m* au tampon
e barniz *m* de muñeca
i vernice *f* a spirito
n schellakpolitoer *n*
d Schellackpolitur *f*

2446 FRENCH WHITE, SILVER WHITE dec
f blanc *m* d'argent

e blanco *m* de plata
i bianco *m* d'argento
n zuiver loodwit *n*
d reinstes Bleiweiss *n*

2447 FRENCH ZINC OXIDE PROCESS, INDIRECT ZINC OXIDE PROCESS — met
f procédé *m* indirect d'oxyde de zinc
e procedimiento *m* indirecto de óxido de cinc
i processo *m* indiretto d'ossido di zinco
n indirect zinkoxydeproces *n*
d indirektes Zinkoxydverfahren *n*

2448 FRESH SLUDGE, CRUDE SLUDGE, RAW SLUDGE — iwt
f boues *pl* fraîches, bourbe *f*, limon *m*
e fango *m* crudo
i fango *m* fresco
n onbewerkt slib *n*
d Frischschlamm *m*, Rohschlamm *m*

2449 FRESH WATER — hyg
f eau *f* douce
e agua *f* dulce
i acqua *f* dolce
n zoet water *n*
d Frischwasser *n*, Süsswasser *n*

2450 FRESHENER — rub
f refraîchissant *m*
e reactivante *m*
i reattivante *m*
n opfrisser
d Erfrischungsmittel *n*

2451 FRETTING CORROSION — cor
f corrosion *f* par frottage
e corrosión *f* por rozamiento
i corrosione *f* per attrito
n wrijvingscorrosie
d Reibungskorrosion *f*

2452 FRICTION — gen
f friction *f*, frottement *m*
e fricción *f*, rozamiento *m*
i attrito *m*, frizione *f*
n frictie, wrijving
d Reibung *f*

2453 FRICTION GLAZING — pap
f satinage *m* par friction
e satinado *m* por rozamiento
i lucidazione *f* per attrito
n wrijvingssatineren *n*
d Reibungssatinierung *f*

2454 FRICTIONING — tex
f frictionnage *m*
e friccionado *m*
i frizionatura *f*
n frictioneren *n*
d Friktionierung *f*

2455 FRIGORIFIC AGENT — clo
f agent *m* frigorifique
e agente *m* frigorífico
i agente *m* frigorifico
n vriesmiddel *n*
d Gefriermittel *n*

2456 FRIT, GLASS COMPOSITION — gla
f fritte *f*
e frita *f*
i fritta *f*
n frit, halfgesmolten glasmassa
d Fritte *f*, Glasmasse *f*, Glassatz *m*, Schmelze *f*

2457 FRITTED GLASS, SINTERED GLASS — gla
f verre *m* fritté
e vidrio *m* fritado
i vetro *m* sinterizzato
n gesinterd glas *n*
d Sinterglas *n*

2458 FRITTED GLAZE — cer,coa
f émail *m* fritté
e esmalte *m* fritado
i fritta *f* di smalto, vetrina *f*
n sinteremail *n*
d Frittenemail *n*

2459 FRITTING, SINTERING — htt,met
f frittage *m*
e sinterización *f*
i sinterizzazione *f*
n sinteren *n*
d Sintern *n*

2460 FROSTED GLASS, OBSCURED GLASS — bui,gla
f verre *m* dépoli
e vidrio *m* esmerilado
i vetro *m* smerigliato
n matglas *n*
d Mattglas *n*

2461 FROSTING — gla
f matage *m*
e esmerilado *m*

i smerigliatura *f*
n matteren *n*
d Mattieren *n*

2462 FROTH, FOAM — gen
f écume *f*
e espuma *f*
i schiuma *f*
n schuim *n*
d Schaum *m*

2463 FROTH FLOTATION — met
f flottation *f* par écume
e flotación *f* por espuma
i flottazione *f* mediante schiuma
n schuimflotatie
d Schaumflotation *f*

2464 FROTHER — min
f générateur *m* d'écume
e espumante *m*
i schiumatrice *f*
n schuimmiddel *n*, schuimontwikkelaar
d Schaumbildner *m*, Schaummittel *n*

2465 FUEL — fue
f combustible *m*
e combustible *m*
i combustibile *m*
n brandstof
d Brennstoff *m*

2466 FUEL OIL — fue
f mazout *m*
e mazut *m*
i nafta *f*
n stookolie
d Heizöl *n*

2467 FUGACITY — gas
f fugacité *f*
e fugacidad *f*
i fugacità *f*
n vluchtigheid
d Flüchtigkeit *f*

2468 FUGITIVE TINTING COLORS (US), SIGHTING COLOURS (GB) — dec
f couleurs *pl* fugitives
e colores *pl* fugaces
i colori *pl* che sbiadiscono facilmente
n onbestendige kleurstoffen *pl*
d unbeständige Farbstoffe *pl*

2469 FULL FORCE FEED, FORCE FEED, FORCED LUBRICATION — lub
f lubrification *f* forcée
e lubricación *f* a presión
i lubrificazione *f* forzata
n druksmering
d Öldruckschmierung *f*

2470 FULL GLOSS, HIGH GLOSS — dec
f brillant *m* élevé
e gran brillo *m*
i gran lucidezza *f*
n hoge glans, hoogglans
d Hochglanz *m*

2471 FULLER'S EARTH, FULLING CLAY — mat
f terre *f* à foulon
e arcilla *f* esmética, tierra *f* de batán
i argilla *f* saponifera, argilla *f* smettica
n bleekaarde, vette klei, vollersaarde
d Bleicherde *f*, Fetton *m*, Seifenerde *f*, Walkerde *f*

2472 FULLING, MILLING — tex
f foulage *m*
e batanadura *f*
i follatura *f*
n vollen *n*
d Walken *n*

2473 FULMINATING GOLD — exp
f or *m* fulminant
e fulminato *m* de oro
i fulminato *m* d'oro
n goudfulminaat *n*, knalgoud *n*
d Goldfulminat *n*, Knallgold *n*

2474 FULMINATING MERCURY — exp
f fulminante *f* mercurique, mercure *f* fulminant
e fulminato *m* de mercurio
i fulminato *m* di mercurio
n knalkwik *n*, kwikfulminaat *n*
d Knallquecksilber *n*, Quecksilberfulminat *n*

2475 FULMINATING PLATINUM — exp
f platine *m* fulminant
e fulminato *m* de platina
i fulminato *m* di platino
n platinafulminaat *n*
d Platinfulminat *n*

2476 FULMINATING SILVER — exp
f argent *m* fulminant
e fulminato *m* de plata
i fulminato *m* d'argento

n knalzilver *n*, zilverfulminaat *n*
d knallsaures Silber *n*, Knallsilber *n*

2477 FUME gen
f fumée *f*
e humo *m*
i esalazione *f*, fumo *m*
n rook
d Rauch *m*

2478 FUMIGANT mit
f produit *m* fumigène
e producto *m* fumigante
i fumigante *m*
n rook ontwikkelend middel *n*
d Räuchermittel *n*

2479 FUMIGATION, SMOKING agr
f fumigation *f*
e fumigación *f*
i fumigazione *f*
n roken *n*, uitroken *n*
d Ausräuchern *n*, Räuchern *n*

2480 FUMING pho
f traitement *m* à l'ammoniaque
e tratamiento *m* al amoníaco
i trattamento *m* all'ammoniaco
n ammoniakbehandeling
d Ammoniakbehandlung *f*

2481 FUMING LIQUIDS aic
f liquides *pl* fumants
e líquidos *pl* fumantes
i liquidi *pl* fumanti
n rokende vloeistoffen *pl*
d rauchende Flüssigkeiten *pl*

2482 FUMING NITRIC ACID aic
f acide *m* nitrique fumant
e ácido *m* nítrico fumante
i acido *m* nitrico fumante
n rokend salpeterzuur *n*
d rauchende Salpetersäure *f*

2483 FUMING SULFURIC ACID (US), FUMING SULPHURIC ACID (GB), NORDHAUSEN SULPHURIC ACID aic
f acide *m* sulfurique fumant
e ácido *m* sulfúrico fumante
i acido *m* solforico fumante
n oleum *n*, rokend zwavelzuur *n*
d Oleum *n*, rauchende Schwefelsäure *f*

2484 FUNGICIDAL PAINT dec
f couleur *f* fungicide
e pintura *f* fungicida
i pittura *f* antricrittogamica
n zwamverhinderende verf
d schwammverhindernde Farbe *f*

2485 FUNGICIDE agr
f fungicide *m*
e fungicida *m*
i sostanza *f* anticrittogamica
n fungicide *n*, zwamverhinderend middel *n*
d Fungizid *n*

2486 FURAN, BRAN OIL (US) oic
f furanne *m*
e furano *m*
i furano *m*
n furaan *n*, furfuraan *n*
d Furan *n*, Furfuran *n*

2487 FURAN RESINS pla
f résines *pl* à furanne
e resinas *pl* furánicas
i resine *pl* a furano
n furaanharsen *pl*
d Furanharze *pl*

2488 FURFURAL RESINS pla
f résines *pl* à furfural
e resinas *pl* furfurólicas
i resine *pl* a furfurolo
n furfuralharsen *pl*, furfurolharsen *pl*
d Furfurolharze *pl*

2489 FURNACE BLACK rub
f noir *m* de fourneau
e negro *m* de horno
i nero *m* di forno
n ovenroet *n*
d Ofenruss *m*

2490 FUSE exp
f mèche *f*
e mecha *f*
i miccia *f*
n lont
d Zündschnur *f*

2491 FUSE POWDER exp
f poudre *m* explosif de mèche
e polvo *m* explosivo de mecha
i polvere *f* esplosiva di miccia
n lontpoeder *n*
d Zündschnurpulver *n*

2492 FUSED ELECTROLYTE, FUSED SALT elc
f électrolyte *m* fondu, sel *m* fondu
e electrólito *m* fundido, sal *f* fundida

i elettrolito *m* fuso, sale *m* fuso
n smelt
d Schmelze *f*

2493 FUSED SILICA, QUARTZ GLASS — gla
f verre *m* quarzeux
e cristal *m* de cuarzo
i vetro *m* di quarzo
n kwartsglas *n*
d Quarzglas *n*

2494 FUSED SODIUM BORATE, BORAX GLASS — gla
f verre *m* au borate de soude
e vidrio *m* al borato de sodio
i vetro *m* al borato di sodio
n boraxglas *n*
d Boraxglas *n*

2495 FUSEL, BAD SPIRITS — nut
f alcool *m* amylique, flegmes *pl*, mauvais goûts *pl* de tête
e alcohol *m* amílico
i alcool *m* fatturato, cattiva acquavite *f*, flemma *f*
n foezel
d Fusel *m*

2496 FUSEL OIL — fat
f huile *f* de fusel
e aceite *m* de fusel
i olio *m* di flemma
n foezelolie
d Fuselöl *n*

2497 FUSION — gen
f fusion *f*
e fusión *f*
i fusione *f*
n smelten *n*
d Schmelzen *n*

2498 FUSION CASTING — cer
f coulée *f* à fusion
e colada *f* de fusión
i colata *f* a fusione
n smeltgieten *n*
d Schmelzguss *m*

2499 FUSION POINT — gen
f point *m* de fusion
e punto *m* de fusión
i punto *m* di fusione
n smeltpunt *n*
d Schmelzpunkt *m*

2500 FUSION WELDING — wel
f soudure *f* par fusion
e soldadura *f* por fusión
i saldatura *f* per fusione
n smeltlassen *n*
d Schmelzschweissung *f*

2501 FUSION ZONE — wel
f zone *f* de fusion
e zona *f* de fusión
i zona *f* di fusione
n smeltzone
d Schmelzzone *f*

G

2502 GAGING (US), GAUGING (GB), CALIBRATION — mea
f calibrage *m*, étalonnage *m*
e calibración *f*
i calibratura *f*, taratura *f*
n ijken *n*, kalibreren *n*
d Eichen *n*, Kalibrierung *f*

2503 GAIZE — bui
f pouzzolane *f* artificielle
e arsenisca *f* arcillosa friable
i pozzolana *f* artificiale
n pozzolanocement *n*
d künstliche Puzzolanerde *f*

2504 GALACTOSE, BRAIN SUGAR (US), CEREBROSE (GB) — sin
f galactose *f*
e galactosa *f*
i galattosio *m*
n galactose
d Cerebrose *f*, Galaktose *f*

2505 GALALITE, ARTIFICIAL HORN — pla
f corne *f* artificielle, galalithe *f*
e cuerno *m* artificial, galalita *f*
i corno *m* artificiale, galalite *f*
n galaliet *n*, kunsthoorn *n*
d Galalith *m*, Kunsthorn *n*

2506 GALENA, BLUE LEAD, LEAD GLANCE — met
f galène
e alquifol *m*, galena *f*
i galena *f*
n galeniet *n*, loodglans *n*
d Bleiglanz *m*, Galenit *m*

2507 GALL OF GLASS, SANDIVER — gla
f suin *m* de verre, suint *m* de verre
e sulfato *m* sódico no reducido
i solfato *m* sodico non ridotto
n glasgal
d Glasgalle *f*

2508 GALLIC ACID — oic
f acide *m* gallique
e ácido *m* gálico
i acido *m* gallico
n galluszuur *n*
d Gallussäure *f*, Trioxybenzoesäure *f*

2509 GALLING, SEIZING — met
f écorchure *f*
e excoriación *f*
i scorticatura *f*
n aanvreten *n* door wrijving
d Fressen *n*

2510 GALVANIC CORROSION — cor
f corrosion *f* galvanique
e corrosión *f* galvánica
i corrosione *f* galvanica
n galvanische corrosie
d galvanische Korrosion *f*

2511 GALVANIZATION, GALVANIZING — coa,elc
f galvanisation *f*, zingage *m*
e cincado *m*, galvanización *f*
i galvanizzazione *f*, zincatura *f*
n galvaniseren *n*, verzinken *n*
d Galvanisieren *n*, Verzinken *n*

2512 GALVANIZED IRON — coa,met
f fer *m* galvanisé
e hierro *m* galvanizado
i ferro *m* galvanizzato
n gegalvaniseerd ijzer *n*
d galvanisiertes Blech *n*, galvanisiertes Eisen *n*

2513 GALVANOPLASTY, ELECTROFORMING — elc
f électroformage *m*, galvanoplastie *f*
e electroformación *f*
i galvanoplastica *f*
n galvanoplastiek
d Galvanoplastik *f*

2514 GALVANOSTEGY, ELECTROPLATING — elc
f galvanoplastie *f*
e galvanoplastia *f*, galvanostegia *f*
i galvanoplastica *f*, galvanostegia
n galvanostegie
d Elektroplattierung *f*, Galvanostegie *f*

2515 GALVANOTYPING, ELECTROTYPING — pri
f électrotypie *f*
e electrotipia *f*, galvanotipia *f*
i elettrotipia *f*
n galvanotypie
d Elektrotypie *f*

2516 GAMBOGE, CAMBOGIA dec,mcy
f gomme-gutte *f*
e gutagamba *f*
i gommaguta *f*
n gittegom, guttegom
d Gummigutt *n*

2517 GARBAGE GRINDING iwt
f broyage *m* d'ordures, délacération *f*
e trituración *f* de basuras
i lacerazione *f* d'immondizie
n malen *n* van afval, malen *n* van vuilnis
d Müllzerkleinerung *f*

2518 GAS gas
f gaz *m*
e gas *m*
i gas *m*
n gas *n*
d Gas *n*

2519 GAS ANALYSIS ana
f analyse *f* de gaz
e análisis *f* de gas
i analisi *f* di gas
n gasanalyse
d Gasanalyse *f*

2520 GAS BLACK, CHANNEL BLACK dec
f noir *m* au canal, noir *m* de gaz
e negro *m* de gas
i nero *m* di gas
n gasroet *n*
d Gasruss *m*

2521 GAS BUBBLE gen
f bulle *f* de gaz
e burbuja *f* de gas
i bolla *f* di gas
n gasbel, gasblaasje *n*
d Gasblase *f*

2522 GAS CARBON htt
f charbon *m* de cornue
e carbón *m* de retorta
i carbone *m* di storta
n retortekool
d Gaskohle *f*, Retortenkohle *f*

2523 GAS CARBURIZING met
f cémentation *f* par gaz
e cementación *f* gaseosa
i cementazione *f* a mezzo di gas
n verkolen *n* met gas
d Gasaufkohlen *n*

2524 GAS CHROMATOGRAPHY ana
f chromatographie *f* en phase vapeur
e cromatografía *f* en fase vapor
i cromatografia *f* in fase vapore
n gaschromatografie
d Gaschromatographie *f*

2525 GAS CLEANING, GAS CONDITIONING, GAS PURIFICATION gas
f épuration *f* de gaz
e depuración *f* de gas
i depurazione *f* di gas
n gasreiniging
d Gasreinigung *f*

2526 GAS COMPRESSION gas,pre
f compression *f* de gaz
e compresión *f* de gas
i compressione *f* di gas
n gascompressie
d Gaskompression *f*, Gasverdichtung *f*

2527 GAS CONCRETE, AERATED CONCRETE, CELLULAR CONCRETE FOAMED CONCRETE bui
f béton *m* cellulaire
e hormigón *m* aerado
i calcestruzzo *m* poroso
n cellenbeton *n*
d Zellenbeton *m*

2528 GAS COOLING clo,gas
f refroidissement *m* au gaz
e enfriamiento *m* con gas
i raffreddamento *m* a gas
n koeling met gas
d Gaskühlung *f*

2529 GAS COOLING clo,gas
f refroidissement *m* de gaz
e enfriamiento *m* de gas
i raffreddamento *m* di gas
n gaskoeling
d Gaskühlung *f*

2530 GAS CUTTING, FLAME CUTTING, TORCH CUTTING met
f coupage *m* à l'autogène, oxycoupage *m*
e corte *m* por llama oxiacetilénica, oxicorte *m*
i taglio *m* per fiamma ossiacetilena
n autogeen snijden *n*
d Brennschneiden *n*

2531 GAS CYANIDING, DRY CYANIDING met
f cyanuration *f* à l'ammoniaque
e cianuración *f* al amoníaco
i cianurazione *f* all'ammoniaca
n cyaancementeren *n* in ammoniak
d Zyanierung *f* in Ammoniak

2532 GAS DETECTION min,sap
f indication *f* de grisou, signalisation *f* de grisou
e detección *f* de grisú, indicación *f* de grisú
i indicazione *f* di grisu, rivelazione *f* di grisu
n mijngasdetectie
d Grubengasanzeige *f*, Schlagwetteranzeige *f*

2533 GAS DRYING dry,gas
f séchage *m* de gaz
e secado *m* de gas
i essiccazione *f* di gas
n gasdroging
d Gastrocknung *f*

2534 GAS EVOLUTION, GAS LIBERATION htt,met
f dégagement *m* de gaz
e desprendimiento *m* de gas
i liberazione *f* di gas, sviluppo *m* di gas
n gasafgifte, gasontwikkeling
d Ausgasen *n*, Gasen *n*, Gasentwicklung *f*

2535 GAS EXHAUSTION gas
f extraction *f* de gaz
e expulsión *f* de gas
i estrazione *f* di gas
n gasafzuiging
d Absaugen *n* von Gas

2536 GAS EXPELLING, BAKING htt
f expulsion *f* de gaz
e expulsión *f* de gas
i espulsione *f* di gas
n gasuitdrijving
d Gasaustreibung *f*

2537 GAS FILLING elc
f remplissage *m* à gaz
e relleno *m* de gas
i riempimento *m* di gas
n gasvulling
d Gasfüllung *f*

2538 GAS FIRING fue,htt
f chauffage *m* au gaz
e calentamiento *m* al gas
i riscaldamento *m* al gas
n gasverhitting
d Gasheizung *f*

2539 GAS GENERATION gas
f génération *f* de gaz
e generación *f* de gas
i generazione *f* di gas
n gasgeneratie, gasontwikkeling
d Gasentwicklung *f*, Gasgeneration *f*

2540 GAS IONIZATION, GASEOUS IONIZATION elc,gas
f ionisation *f* de gaz
e ionización *f* de gas
i ionizzazione *f* di gas
n gasionisatie
d Gasionisation *f*, Gasionisierung

2541 GAS-LIGHT PAPER pho
f papier *m* au gélatine-bromure
e papel *m* de imagen latente, papel *m* fotográfico al bromuro de plata
i carta *f* al bromuro
n gaslichtpapier *n*
d Gaslichtpapier *n*

2542 GAS LIME gas
f chaux *f* de gaz
e cal *f* de gas
i calce *f* di gas
n gaskalk
d Gaskalk *m*

2543 GAS-LIQUID CHROMATOGRAPHY ana
f chromatographie *f* gaz-liquide
e cromatografía *f* gas-líquido
i cromatografia *f* gas-liquido
n gas-vloeistofchromatografie
d Gas-Flüssigchromatographie *f*

2544 GAS LIQUOR, GAS WATER, AMMONIACAL LIQUOR gas
f eau *f* ammoniacale
e agua *f* amoniacal
i acqua *f* ammoniacale
n gaswater *n*
d Gaswasser *n*

2545 GAS MIXTURE gas,mix
f mélange *m* gazeux
e mezcla *f* gaseosa
i miscela *f* gassosa
n gasmengsel *n*
d Gasgemenge *n*, Gasgemisch *n*

2546 GAS OIL, DIESEL OIL, SOLAR OIL — fat
f gas-oil *m*, huile *f* pour moteurs diesel
e gasoil *m*
i gasolio *m*, nafta *f* per moteurs diesel
n dieselolie
d Dieselöl *n*

2547 GAS PERMEABILITY — gas
f perméabilité *f* au gaz
e permeabilidad *f* para gas
i permeabilità *f* al gas
n gasdoorlaatbaarheid
d Gasdurchlässigkeit *f*

2548 GAS PICKLING — cle
f décapage *m* au gaz
e decapaje *m* con gas
i decapaggio *m* al gas
n beitsen *n* met gas
d Beizen *n* mit Gas, Gasbeizung *f*

2549 GAS POCKET — pla
f inclusion *f* de gaz
e bolsa *f* de gas, inclusión *f* de gas
i sacca *f* di gas
n gasgal, gasinsluitsel *n*
d Gaseinschluss *m*

2550 GAS POLYMERIZATION, GASEOUS POLYMERIZATION — oic
f polymérisation *f* en phase de gaz
e polimerización *f* en fase de gas
i polimerizzazione *f* in fase di gas
n polymerisatie in de gasfaze
d Polymerisation *f* in der Gasphase

2551 GAS REGENERATION — gas
f régénération *f* de gaz
e regeneración *f* de gas
i rigenerazione *f* di gas
n gasregeneratie
d Gasregenierierung *f*

2552 GAS SAMPLING — tes
f échantillonnage *m* de gaz
e toma *f* de muestras de gas
i campionatura *f* di gas
n gasmonsterneming
d Gasprobenahme *f*

2553 GAS SCRUBBING — cle
f lavage *m* de gaz
e lavado *m* de gas
i lavaggio *m* di gas
n gaswassen *n*
d Gaswaschen *n*

2554 GAS SEPARATION — fue
f séparation *f* de gaz
e separación *f* de gas
i separazione *f* di gas
n gasafscheiding
d Gasabscheidung *f*

2555 GAS-SOLID CHROMATOGRAPHY — ana
f chromatographie *f* gaz-solide
e cromatografía *f* gas-sólido
i cromatografia *f* gas-solido
n gas-vaste-stofchromatografie
d Gas-Feststoffchromatographie *f*

2556 GAS TAR — oic
f goudron *m* de gaz
e alquitrán *m* de gas
i catrame *m* di gas
n gasteer
d Gasteer *m*

2557 GAS-TIGHT — vac
f étanche adj au gaz
e hermético adj al gas
i a tenuta di gas
n gasdicht adj
d gasdicht adj

2558 GAS WARFARE, CHEMICAL WARFARE — mit
f guerre *f* chimique
e guerra *f* química
i guerra *f* chimica
n gasoorlogvoering
d Gaskrieg *m*

2559 GAS WASHING — cle
f lavage *m* de gaz
e lavado *m* de gas
i lavaggio *m* di gas
n gaswassen *n*
d Gaswaschen *n*

2560 GAS WELDING — wel
f soudure *f* à flamme de gaz
e soldadura *f* por llama de gas
i saldatura *f* per fiamma di gas
n gasvlamlassen *n*, lassen *n* met gasvlam
d Gasflammschweissung *f*, Schweissen *n* mit Gasflamme

2561 GASEOUS — gas
f en état gazeux, gazéiforme adj

e gaseoso adj
i gassoso adj
n gasachtig adj , gasvormig adj
d gasförmig adj

2562 GASEOUS FUEL fue
f combustible *m* gazeux
e combustible *m* gaseoso
i combustibile *m* gassoso
n gasvormige brandstof
d gasförmiger Brennstoff *m*

2563 GASIFYING fue
f gazéification *f*
e gasificación *f*
i gassificazione *f*
n vergassing
d Vergasung *f*

2564 GASOLENE, GASOLINE, PETROLEUM ETHER fue
f gazoline *f*
e éter *m* de petróleo
i gasolina *f*
n gasoline
d Gasäther *m*, Gasolin *n*

2565 GASOLINE (US), MOTOR SPIRIT, PETROL (GB) fue
f essence *f*
e gasolina *f*
i benzina *f*
n benzine
d Benzin *n*

2566 GASOLINE SEPARATION (US), PETROL SEPARATION (GB) fue
f séparation *f* d'essence
e separación *f* de gasolina
i separazione *f* di benzina
n benzineafscheiding
d Benzinabscheidung *f*

2567 GASSING elc
f bouillonnement *m*, dégagement *m* de gaz
e burbujeo *m*, desprendimiento *m* de gases
i ebollizione *f*
n koken *n*
d Gasen *n*, Kochen *n*

2568 GASSING SINGEING tex
f flambage *m*, gazage *m*
e chamusquina *f*, gaseado *m*
i bruciatura *f*, gasatura *f*
n afzengen *n*, zengen *n*
d Flambieren *n*, Gasieren *n*, Sengen *n*, Versengen *n*

2569 GATHERING dec
f moucheture *f*
e moteado *m*
i difetto *m* di verniciatura dovuto all'assorbimento
n vlekkigheid
d Fleckigkeit *f*

2570 GATHERING gla
f cueillage *m*
e colecta *f*
i raccoglimento *m*
n keien *n*
d Entnahme *f*

2571 GAUZE (GB), GAZE (US) tex
f gaze *f*
e gasa *f*
i garza *f*
n gaas *n*
d Gaze *f*

2572 GEL col
f gel *m*
e gel *m*
i gel *m*
n gel *n*
d Gel *n*

2573 GEL PAINT, JELLY PAINT, THIXOTROPIC PAINT dec
f peinture *f* thixotrope
e pintura *f* tixotrópica
i pittura *f* tixotropica
n thixotrope verf
d thixotrope Farbe *f*

2574 GELATIN, GELATINE oi
f gélatine *f*
e gelatina *f*
i gelatina *f*
n gelatine
d Gallert *n*, Gelatine *f*

2575 GELATINIZATION oi
f gélatinisation *f*, gélification *f*
e gelatinización *f*
i gelatinizzazione *f*
n gelatinieren *n*
d Gelatinieren *n*, Gelbildung *f*

2576 GELATINIZER, SOLVENT PLASTICIZER oi
f plastifiant *m* gélatinisant
e plastificante *m* gelatinizador
i plastificante *m* da gelatinizzare
n gelatinerende weekmaker
d gelatinierender Weichmacher *m*

2577 GELATINIZING AGENT oic
f agent *m* gélifiant
e agente *m* gelatinizador
i agente *m* da gelatinizzare
n gelatineringsmiddel *n*
d Gelatinierungsmittel *n*

2578 GELATINO-CHLORIDE PROCESS pho
f procédé *m* gélatine-chlorure
e procedimiento *m* gelatina-cloruro
i processo *m* gelatina-cloruro
n chloorzilverproces *n*
d Chlorsilberverfahren *n*

2579 GELATINOUS oic
f gélatineux adj
e gelatinoso adj
i gelatinoso adj
n gelatineus adj
d gallertartig adj , gelatinartig adj

2580 GELATION, GELLING, PECTIZATION col
f formation *f* de gel, gélification *f*
e formación *f* de gel, gelificación *f*
i gelizzazione *f*
n geleren *n*, gelvorming, stijf worden *n*
d Gelbildung *f*, Gelieren *n*, Steifwerden *n*

2581 GELIGNITE exp
f gélignite *f*
e gelignita *f*
i gelignite *f*
n gelatinedynamiet *n*, geligniet *n*
d Gelatinedynamit *n*

2582 GELLING AGENT col
f agent *m* de prise
e agente *m* gelificante
i agente *m* gelizzante
n geleermiddel *n*
d Geliermittel *n*

2583 GELLING POINT col
f point *m* de gélification
e punto *m* de gelificación
i punto *m* di gelizzazione
n geleringspunt *n*
d Gelierungspunkt *m*

2584 GENTIAN VIOLET, PYOCTANIN BLUE mcy
f violet *m* de gentiane
e violeta *f* de genciana
i violetto *m* di genziana
n gentiaanviolet *n*
d Gentianaviolett *n*

2585 GERANIUM OIL fat
f essence *f* de géranium
e esencia *f* de geranio
i olio *m* di geranio
n geraniumolie
d Geranienöl *n*

2586 GERM-REPELLING PAINT, GERMICIDAL PAINT, SELF-SANITIZING PAINT dec
f peinture *f* biocide
e pintura *f* germicida
i pittura *f* germicida
n biocide verf, kiemdodende verf
d biologische Selbstschutzfarbe *f*, biozide Farbe *f*, keimtötende Farbe *f*

2587 GERMAN BLACK, FRANKFORT BLACK pri
f noir *m* d'Allemagne
e negro *m* de Alemania
i nero *m* di Germania
n koperdrukzwart *n*
d Kupferdruckschwarz *n*

2588 GERMICIDE bac
f biocide *m*, germicide *m*
e germicida *m*
i germicida *m*
n kiemdodend middel *n*
d keimtötendes Mittel *n*

2589 GERMINATION bre
f germination *f*
e germinación *f*
i germinazione *f*
n kiemen *n*
d Keimen *n*

2590 GETTER cle,elc,gas
f dégazeur *m*, getter *m*
e afinador *m* del vacío, getter *m*
i assorbitore *m*, getter *m*
n getter *m*, vangstof
d Fangstoff *m*, Getter *m*

2591 GILDING coa
f dorure *f*
e doradura *f*
i doratura *f*
n vergulden *n*
d Vergolden *n*

2592 GILDING WAX, GOLD SIZE, GOLD WAX dec
f colle *f* d'or,

vernis *m* d'apprêt
e cera *f* de dorador
i cera *f* da indoratore
n vergulderslijm
d Anlegeöl *n*, Blattgoldgrundöl *n*, Glühwachs *n*, Vergolderwachs *n*

2593 GIN nut
f genièvre *m*
e ginebra *f*
i acquavite *f* di ginepro, gin *m*
n genever, jenever
d Wacholderbranntwein *m*

2594 GINGER BEER nut
f boisson *f* gazeuse au gingembre
e agua *f* gaseosa al jengibre
i acqua *f* gassosa al zenzero
n gemberbier *n*
d Ingwerbier *n*

2595 GINGER OIL (GB), GINGEROL (US) fat
f essence *f* de gingembre, gingérol *m*
e esencia *f* de jengibre
i olio *m* di zenzero
n gemberolie
d Ingweröl *n*

2596 GLACÉ KID, GLACÉ LEATHER lea
f cuir *m* glacé
e cabritilla *f*
i pelle *f* lucida
n glacé *n*
d Glacéleder *n*

2597 GLACIAL ACETIC ACID aam,oic
f acide *m* acétique glacial
e ácido *m* acético glacial
i acido *m* acetico glaciale
n ijsazijn
d Eisessig *m*

2598 GLASS gla
f verre *m*
e vidrio *m*
i vetro *m*
n glas *n*
d Glas *n*

2599 GLASS, MADE OF GLASS gla
f de verre, en verre
e de vidrio, vítreo adj
i di vetro, vitreo adj
n glazen adj
d gläsern adj

2600 GLASS COMPOSITION, FRIT gla
f fritte *f*
e frita *f*
i fritta *f*
n frit, halfgesmolten glasmassa
d Fritte *f*, Glasmasse *f*, Glassatz *m*, Schmelze *f*

2601 GLASS DEVITRIFICATION gla
f dévitrification *f* du verre
e devitrificación *f* del vidrio
i devitrificazione *f* del vetro
n devitrificatie, ontglazing
d Entglasung *f*

2602 GLASS ETCHING dec,gla
f gravure *f* chimique du verre
e grabado *m* químico del vidrio
i incisione *f* all'acquaforte del vetro
n etsen *n* van glas
d Glasätzung *f*

2603 GLASS FIBER (US), GLASS FIBRE (GB) gla
f fibre *f* de verre
e fibra *f* de vidrio
i fibra *f* di vetro
n glasvezel
d Glasfaser *f*

2604 GLASS LAMINATE, LAMINATED GLASS, SAFETY GLASS gla,sap
f verre *m* de sûreté
e vidrio *m* laminado
i vetro *m* di sicurezza
n veiligheidsglas *n*
d Sicherheitsglas *n*

2605 GLASS-TO-METAL-SEAL ele
f scellement *m* verre-métal
e cierre *m* vidrio-metal, unión *f* vidrio-metal
i chiusura *f* vetro-metallo
n glas-metaalversmelting
d Glas-Metallverschmelzung *f*

2606 GLASS WOOL, SPUN GLASS gla
f coton *m* de verre, laine *f* de verre
e guata *f* de vidrio
i lana *f* di vetro
n glaswol
d Glaswolle *f*

2607 GLASSINE (US), PERGAMYNE (GB), TRANSPARENT PARCHMENT pa
f papier *m* cristal,

papier *m* serpente, serpente *f*
e papel *m* cristal
i carta *f* glassene, carta *f* trasparente
n glaspapier *n*, kristalpapier *n*
d Glashaut *f*, Glassin *n*, Pergamin *n*

2608 GLASSMAKER'S SOAP, BATTERY MANGANESE, BLACK OXIDE OF MANGANESE, MANGANESE DIOXIDE — min
f pyrolusite *f*
e pirolusita *f*
i perossido *m* di manganese
n bruinsteen *n*, mangaansuperoxyde *n*
d Braunstein *m*, Mangandioxyd *n*, Pyrolusit *m*

2609 GLASSPAPER — pap
f papier *m* à verre
e papel *m* de lija
i carta *f* vetrata
n glaspapier *n*
d Glaspapier *n*

2610 GLASSY, VITREOUS — gla
f vitreux adj
e vítreo adj
i vetroso adj
n glasachtig adj
d glasartig adj , glasig adj

2611 GLAZED BOARD — pap
f carton *m* satiné
e cartulina *f* satinada
i cartone *m* lucido
n glanskarton *n*
d Glanzkarton *m*, Glanzpappe *f*

2612 GLAZED PAPER, ENAMELLED PAPER — pap
f papier *m* glacé
e papel *m* glaseado
i carta *f* lucida
n glanspapier *n*
d Glacépapier *n*, Glanzpapier *n*, Kreidepapier *n*

2613 GLAZING — gla
f vitrification *f*
e vidriado *m*
i vetrificazione *f*
n verglazen *n*
d Verglasen *n*

2614 GLAZING — lea
f criage *m*, parage *m*
e abrillantado *m*
i lucidatura *f*
n glanzen *n*
d Glänzen *n*

2615 GLAZING — pap,pla
f satinage *m*
e satinado *m*
i rasamento *m*
n satineren *n*
d Glänzen *n*, Satinieren *n*

2616 GLAZING — pho
f glaçage *m*
e glaseado *m*
i lucidatura *f*
n glanzen *n*
d Glänzen *n*

2617 GLAZING — pol
f émoussage *m*
e embotamiento *m*
i smussatura *f*
n ontruwen *n*
d Schmirgeln *n*

2618 GLAZING — pol
f polissage *m* préliminaire
e pulido *m* preliminar
i lucidatura *f* preliminare
n voorpolijsten *n*
d Vorpolieren *n*

2619 GLAZING, GLAZING COMPOSITION — dec
f glacis *m*
e barniz *m* vítreo
i vernice *f* da lucidare
n glaceermiddel *n*, glaceervernis *n*
d Glasur *f*, Lasurfarbe *f*

2620 GLAZING, RELINING — fou
f brasquage *m*
e enlucido *m*
i rivestimento *m*
n bekleding
d Glasurbrand *m*

2621 GLAZING COLOR (US), GLAZING COLOUR (GB) — dec
f vernis *m* transparent
e barniz *m* transparente
i vernice *f* trasparente
n transparantvernis *n*
d durchscheinende Farbe *f*

2622 GLOSS, LUSTRE, SHEEN — dec,tex
f brillant *m*, éclat *m*, luisance *f* lustre *m*

e brillo *m*, lustre *m*
i lucentezza *f*, lustro *m*, splendore *m*
n glans
d Glanz *m*

2623 GLOSS PAINT dec
f peinture *f* brillante
e pintura *f* brillante
i pittura *f* rilucente
n glansverf
d Glanzlackfarbe *f*

2624 GLOSS-REDUCING AGENT, FLATTING AGENT dec
f agent *m* de matage
e producto *m* para opacificar
i agente *m* opacificante
n matteermiddel *n*
d Mattierungsmittel *n*

2625 GLUCOSE, DEXTROSE, GRAPE SUGAR sin
f dextrose *f*, glucose *m*
e dextrosa *f*, glucosa *f*
i destrosio *m*, glucosio *m*, zucchero *m* d'uva
n dextrose, druivesuiker, glucose
d Dextrose *f*, Glukose *f*, Glykose *f*, Traubenzucker *m*

2626 GLUE, CEMENT glu
f adhésif *m*, colle *f*
e adhesivo *m*, cola *f*
i adesivo *m*, colla *f*
n kleefmiddel *n*, lijm, plakmiddel *n*
d Klebemittel *n*, Klebestoff *m*, Leim *m*

2627 GLUE LAYER glu
f couche *f* de colle
e capa *f* de cola
i strato *m* di colla
n lijmlaag
d Klebeschicht *f*

2628 GLUE STOCK glu
f carnasses *pl*, débris *pl* cornés de peau
e carnaza *f*, recortadura *f* de piel, recortes *pl*
i carniccio *m*, ritagli *pl* di pelle per colla
n afvalleder *n*
d Leimleder *n*

2629 GLUEING (GB), GLUING (US) glu
f encollage *m*
e encolado *m*
i incollatura *f*
n lijmen *n*
d Leimen *n*

2630 GLUTEN, OSMAZOME, VEGETABLE GLUTEN glu
f gluten *m*, phytocolle *f*
e gluten *m*
i gluten *m*
n gluten *n*
d Gluten *n*, Tritizin *n*

2631 GLYCERIN, GLYCERINE, GLYCEROL oic
f glycérine *f*
e glicerina *f*
i glicerina *f*, glicerolo *m*
n glycerine, glycerol
d Glyzerin *n*

2632 GLYCERINE LITHARGO CEMENT glu
f colle *f* à base de glycérine et litharge
e cola *f* a base de glicerina y litargirio
i collante *m* a base di glicerina e litargirio
n loodglycerine
d Bleiglyzerin *n*, Glyzerinkitt *m*

2633 GLYCEROL-PHTALIC RESINS, GLYPTAL RESINS, PHTALIC GLYCERIDE RESINS pla
f résines *pl* glycérides phtaliques
e resinas *pl* sintéticas alquídicas
i resine *pl* gliceroftaliche
n glyptaalharsen *pl*
d Glyptalharze *pl*

2634 GOLD DIP coa,elc
f dorure *f* par immersion
e doradura *f* por inmersión
i doratura *f* per immersione
n dompelvergulden *n*
d Tauchvergoldung *f*

2635 GOLD-FILM GLASS gla
f verre *m* stratifié à couche intermédiaire d'or pulvérisé
e vidrio *m* laminado con capa intermedia de oro pulverizado
i vetro *m* laminato con strato intermedio di polvere d'oro
n gelaagd glas *n* met goudpoedertussenlaag
d geschichtetes Glas *n* mit Goldpulverzwischenschicht

2636 GOLD FOIL, dec
GOLD LEAF
f feuille *f* d'or, or *m* en feuilles
e pan *m* de oro
i foglia *f* d'oro
n bladgoud *n*
d Blättergold *n*, Blattgold *n*, Goldfolie *f*

2637 GOLD NUMBER met
f nombre *m* d'or
e número *m* de oro
i numero *m* d'oro
n goudgetal *n*
d Goldzahl *f*, Schutzzahl *f*

2638 GOLD PAINT, dec
BRONZE PIGMENTED LACQUER
f laque *f* de bronze
e laca *f* de bronce
i lacca *f* di bronzo
n bronslak
d Bronzelack *m*

2639 GOLD PLATING coa
f placage *m* à l'or
e plaqueado *m* al oro
i placcatura *f* all'oro
n goudpleten *n*
d Goldplattierung *f*

2640 GOLD REDUCTION min
f extraction *f* d'or
e extracción *f* de oro
i estrazione *f* d'oro
n goud uitsmelten *n*, goudextractie
d Goldausschmelzen *n*

2641 GOLD SIZE, dec
GOLD WAX, GILDING WAX
f colle *f* d'or, vernis *m* d'apprêt
e cera *f* de dorador
i cera *f* da indoratore
n vergulderslijm
d Anlegeöl *n*, Blattgoldgrundöl *n*, Glühwachs *n*, Vergolderwachs *n*

2642 GOLD SIZE, dec
PAINT HARDENER
f vernis *m* colle d'or
e endurecedor *m* de pintura
i induritore *m* di pittura
n verfverharder, verhardingslak
d Härtelack *m*, Harttrocken-glanzöl *n*

2643 GOLD STOVING VARNISH dec
f vernis *m* d'or
e barniz *m* de oro
i vernice *f* d'oro
n goudlak
d Goldlack *m*

2644 GOLDBEATER'S SKIN dec
f baudruche *f*
e película *f* de batihoja
i budello *m* di battiloro
n goudslagersvlies *n*
d Goldschlägerhaut *f*

2645 GOLDBEATING coa
f battement *m* d'or
e batidura *f* de oro
i battitura *f* d'oro
n goudslagerij
d Blattgoldschlagerei *f*

2646 GOLDNATURED GLASS gla
f verre *m* trempé à grande plage de températures
e vidrio *m* templado con gran campo de temperaturas
i vetro *m* temprato con gamma di temperature estesa
n over groot temperatuurgebied spanningsvrij glas *n*
d langes Glas *n*

2647 GOUGING TEST tes
f essai *m* d'émail
e ensayo *m* de esmalte
i collaudo *m* di smalto
n emailbeproeving
d Emailprüfung *f*

2648 GOULARD'S POWDER, dec
LEAD ACETATE, SALT OF SATURN
f acétate *m* de plomb, sel *m* de Saturne
e acetato *m* de plomo
i acetato *m* di piombo, sale *m* di Saturno
n loodsuiker
d Bleizucker *m*, Goulardsches Pulver *n*

2649 GOVER PROCESS, fue
BENZOL-ACETONE PROCESS
f procédé *m* benzol-acétone
e procedimiento *m* benzol-acetona
i processo *m* benzolo-acetone
n benzol-acetonproces *n*
d Benzol-Azetonverfahren *n*

2650 GRADING sor
f triage *m*
e clasificación *f*
i cernitura *f*
n sorteren *n*
d Sortieren *n*

2651 GRADING, SAND SIEVE GRADING — sor
f granulométrie *f*
e granulometría *f*
i granulometria *f*
n korrelverdeling
d Korngrössenverteilung *f*

2652 GRADUAL PURIFICATION — cle
f dépuration *f* échelonnée
e depuración *f* escalonada
i depurazione *f* graduale
n trapsgewijze reiniging
d schrittweise Reinigung *f*

2653 GRAFFITO, SGRAFFITO — cer,dec
f sgraffite *m*
e esgrafiado *m*, obra *f* esgrafiada
i graffito *m*
n graffito *n*
d Sgraffitoarbeit *f*

2654 GRAFT POLYMERIZATION — oic
f polymérisation *f* en greffage
e polimerización *f* de injerta
i polimerizzazione *f* ad innestamento
n polymerisatie door enten
d Pfropfpolymerisation *f*

2655 GRAIN — lea
f grain *m*, grenure *f*
e flor *f*, grano *m*
i grano *m*
n nerf
d Narbe *f*

2656 GRAIN — pho
f grain *m*
e grano *m*
i grana *f*
n korrel
d Korn *n*

2657 GRAIN ALCOHOL, ETHYL ALCOHOL — nut
f alcool *m* éthylique
e alcohol *m* etílico
i alcool *m* etilico
n ethylalcohol
d Äthylalkohol *m*, Gärungsalkohol *m*

2658 GRAIN IRRADIATION — nue,nut
f irradiation *f* de céréales, irradiation *f* du blé
e irradiación *f* de cereales
i irradiazione *f* del grano
n graanbestraling
d Getreidebestrahlung *f*

2659 GRAIN LEATHER, SHAGREEN — lea
f chagrin *m*
e cuero *m* granulado, cuero *m* chagrinado
i pelle *f* di pescecane
n chagrijnleder *n*, segrijnleder *n*, Turks leder *n*
d Körnerleder *n*, Narbenleder *n*

2660 GRAIN SIDE, HAIR SIDE — tan
f côté *m* poil du cuir, fleur *f* de la peau
e cara *f* del grano, cara *f* del pelo
i lato *m* del pelo, lato *m* granitico del cuoio
n haarzijde, nerfkant
d Haarseite *f*, Narbenseite *f*

2661 GRAIN SPLIT — lea
f fleur *f* de cuir fendu
e flor *f* curtida
i pelle *f* spaccata
n splitleder *n* van de nerfkant
d genarbtes Spaltleder *n*

2662 GRAININESS — pho
f granulation *f*
e granosidad *f*, granularidad *f*
i granitura *f*, granulosità *f*
n korreligheid
d Körnigkeit *f*, Körnung *f*

2663 GRAINING — cer,dec
f faire v du faux bois, marbrage *m*
e jaspeado
i venatura *f*
n aderen *n*, houten *n*, marmeren *n*
d Maserieren *n*

2664 GRAINING, BOARDING — lea
f grenure *f*
e granulado *m*
i granitura *f*
n krispelen *n*
d Krispeln *n*

2665 GRAINING, CUTTING — soa
f désalage *m*
e graneado *m*, separación *f* de la glicerina del jabón
i granitura *f*
n uitzouten *n*
d Aussalzen *n*

2666 GRAINING, SHAGREENING — lea
f crépissage *m*

e chargrinado *m*,
granelación *f*
i spaccatura *f*
n chagrineren *n*
d Narben *n*

2667 GRAINING OUT, SALTING OUT soa
f relargage *m*
e saladura *f*
i salatura *f*
n uitzouten *n*
d Aussalzen *n*

2668 GRANULATED SUGAR sin
f sucre *m* cristallisé
e azúcar *m* cristalizado,
azúcar *m* granulado
i zucchero *m* semolino
n kristalsuiker
d Kastorzucker *m*, Kornzucker *m*,
Kristallzucker *m*

2669 GRANULATING brk
f granulation *f*
e granulación *f*
i granulazione *f*
n granulatie, korreling
d Granulierung *f*, Kornbildung *f*

2670 GRAPE SUGAR, DEXTROSE, GLUCOSE sin
f dextrose *f*, glucose *m*
e dextrosa *f*, glucosa *f*
i destrosio *m*, glucosio *m*,
zucchero *m* d'uva
n dextrose, druivesuiker, glucose
d Dextrose *f*, Glukose *f*, Glykose *f*,
Traubenzucker *m*

2671 GRAPHIC FORMULA, CONSTITUTIONAL FORMULA, STRUCTURAL FORMULA gen
f formule *f* de constitution
e fórmula *f* de constitución,
fórmula *f* estructural
i formula *f* di struttura
n structuurformule
d Konstitutionsformel *f*,
Strukturformel *f*

2672 GRAPHITE aic
f graphite *m*
e grafito *m*
i grafite *f*
n grafiet *n*
d Graphit *m*

2673 GRAPHITE GREASE, GRAPHITIC LUBRICANT lub
f graisse *f* graphitisée
e grasa *f* grafitada
i grasso *m* grafitizzato
n grafietsmeermiddel *n*
d Graphitschmiere *f*,
Graphitschmiermittel *n*

2674 GRAPHITE PAINT dec
f peinture *f* au graphite
e pintura *f* al grafito, pintura *f*
grafitada
i pittura *f* alla grafite
n grafietverf
d Graphitfarbe *f*

2675 GRAPHITE SOFTENING, GRAPHITIC ACTION, GRAPHITIC CORROSION cor
f corrosion *f* graphitique
e corrosión *f* grafítica
i corrosione *f* grafitica
n grafietcorrosie
d faulende Rostanfressung *f*,
graphitische Aufweichung *f*,
Graphitkorrosion *f*

2676 GRAPHITIZATION coa
f graphitisation *f*
e grafitación *f*
i grafitizzazione *f*
n grafiteren *n*
d Graphitierung *f*

2677 GRAVIMETRIC ANALYSIS, QUANTITIVE ANALYSIS ana
f analyse *f* gravimétrique
e análisis *f* gravimétrica
i analisi *f* gravimetrica
n kwantitatieve analyse
d Gewichtsanalyse *f*,
gravimetrische Bestimmung *f*

2678 GRAVITATIONAL DUST COLLECTOR cle,gas
f collecteur *m* de poussière à
gravitation
e colector *m* de polvo por
gravitación
i collettore *m* di polvere a
gravitazione
n stofverzamelaar volgens het
zwaartekrachtprincipe
d Staubsammler *m* mit
Schwerewirkung

2679 GREASE fat,lub
f graisse *f*
e grasa *f* consistente
i grasso *m*, olio *m* denso
n smeermiddel *n*
d Schmiere *f*, Schmiermittel *n*

2680 GREASE PAINT fat
f fard *m*, maquillage *m*
e afeite *m*
i cerone *m*
n grimeersel *n*, schmink
d Schminke *f*

2681 GREASE RECOVERY, FAT RECOVERY fat,oic
f récupération *f* de graisse
e recuperación *f* de grasa
i ricupero *m* di grasso
n vetterugwinning
d Fettrückgewinnung *f*

2682 GREASE SEPARATING, FAT SEPARATING cle,fat
f séparation *f* de graisse
e separación *f* de grasa
i separazione *f* di grasso
n vetafscheiding
d Fettabscheidung *f*

2683 GREASE WOOL tex
f laine *f* en suint
e lana *f* suarda
i lana *f* sucida
n vette wol
d Schmierwolle *f*, Schmutzwolle *f*

2684 GREASEPROOF pap
f imperméable adj à la graisse
e impermeable adj a la grasa
i impermeabile adj al grasso
n vetvast adj , vetvrij adj
d fettdicht adj , fettfrei adj

2685 GREASEPROOF PAINT dec
f peinture *f* résistante à la graisse
e pintura *f* resistente a la grasa
i pittura *f* resistente al grasso
n tegen vet resistente verf
d fettfeste Farbe *f*

2686 GREASEPROOF PAPER pap
f papier *m* imperméable à la graisse, papier *m* parcheminé
e papel *m* impermeable por la grasa
i carta *f* pergamena vegetale
n vetwerend papier *n*
d fettdichtes Papier *n*

2687 GREASING fat
f graissage *m*
e engrasación *f*
i oliatura *f*
n invetten *n*
d Einfetten *n*

2688 GREEN COMPOUND (US), UNCURED COMPOUND (GB) rub
f mélange *m* non-vulcanisé
e mezcla *f* no vulcanizada
i mescolanza *f* non vulcanizzata
n ongevulcaniseerd mengsel *n*
d unvulkanisiertes Gemisch *n*

2689 GREEN HIDE tan
f peau *f* crue, peau *f* verte
e cuero *m* verde, piel *f* sin curtir
i pelle *f* non conciata
n ongelooide huid
d grüne Haut *f*, Rohhaut *f*

2690 GREEN MALT bre
f malt *m* vert
e malta *f* verde
i malto *m* verde
n groenmout *n*
d Grünmalz *n*

2691 GREEN MORDANT, ANTICHLORINE, SODA HYPOSULFITE (US), SODIUM THIOSULPHATE (GB) cle,tex
f antichlore *m*
e anticloro *m*
i anticloro *m*
n antichloor *n*, natriumthiosulfaat *n*
d Antichlor *n*, Natriumthiosulphat *n*

2692 GREEN ROT cor
f corrosion *f* par oxydation intercristalline
e corrosión *f* por oxidación intercristalina
i corrosione *f* dovuta all'ossidazione intercristallina
n corrosie door interkristallijne oxydatie
d Grünsäule *f*

2693 GREEN SAND fou
f sable *m* à vert
e arena *f* en verde
i terra *f* a verde
n groen vormzand *n*
d Nassgussand *m*

2694 GREEN SMALT, COBALT GREEN dec
f vert *m* de cobalt
e verde *m* de cobalto
i verde *m* di cobalto
n kobaltgroen *n*
d Kobaltgrün *n*, Zinkgrün *n*

2695 GREEN SOAP soa
f savon *m* vert

e jabón *m* de potasa, jabón *m* verde
i sapone *m* di potassa
n groene zeep
d grüne Schmierseife *f*

2696 GREEN SYRUP (GB), WASH SIRUP (US), WASH SYRUP (GB) sin
f eaux *pl* mères de deuxième et troisième jet
e jarabe *m* verde
i sciroppo *m* verde
n groene stroop
d Deckablauf *m*, Waschsirup *m*

2697 GREY SOUR tex
f procédé *m* d'élimination de substances alcalines
e procedimiento *m* de eliminación de substancias alcalinas
i processo *m* da eliminare sostanze alcaline
n proces *n* voor het verwijderen van alkaliresten
d Verfahren *n* zur Beseitigung alkalischer Rückstände

2698 GRINDING brk
f broyage *m*
e trituración *f*
i frantumazione *f*
n breken *n*
d Brechen *n*

2699 GRINDING pap
f défibration *f*
e desfibración *f*
i sfibratura *f*
n houtslijpen *n*
d Holzschleifen *n*, Zerfasern *n*

2700 GRINDING pol
f meulage *m*
e amoladura *f*
i molatura *f*
n slijpen *n*
d Schleifen *n*

2701 GRINDING tex
f aiguisage *m*
e esmerilado *m*
i affilazione *f*
n slijpen *n*
d Schleifen *n*

2702 GRINDING, POLISHING pol
f polissage *m*
e pulido *m*
i lucidatura *f*, pulitura *f*
n polijsten *n*
d Polieren *n*

2703 GRINDING AGENT, GRINDING MEDIUM, MILL-BASE dec
f base *f* de broyage, liant *m* de broyage
e aglutinante *m* de molido
i legante *m* di molitura
n maalbindmiddel *n*
d Bindemittel *n* zum Anreiben

2704 GRINDING VARNISH dec
f vernis *m* de base
e barniz *m* de base
i vernice *f* di base
n maalvernis *n*
d Anreibelack *m*

2705 GRIST, CRUSHED MALT bre
f mouture *f*
e malta *f* molida, molienda *f* de malta
i macinato *m* di malto
n moutmeel *n*, moutschroot *n*
d Malzschrot *n*, Schrot *n*

2706 GRIT BLASTING, SHOT BLASTING cle
f grenaillage *m*
e rebarbado *m* por chorro de arenilla
i granigliatura *f*
n gritstralen *n*, gruisstralen *n*
d Abstrahlen *n*, Kiesstrahlen *n*, Schleuderstrahlen *n*

2707 GRITTINESS, BITTINESS dec
f aspect *m* granuleux
e contextura *f* arenosa
i aspetto *m* di granelli
n gruizigheid, zaverigheid
d Punktbildung *f*

2708 GROUND BASIC SLAG fer
f farine *f* de scories basiques
e harina *f* de escorias Thomas
i farina *f* di scorie Thomas, scorie *pl* Thomas macerate
n thomasslakkenmeel *n*
d Thomasschlackenmehl *n*

2709 GROUND COAT, COUCH, PRIMER COAT dec
f main *f* d'impression, première couche *f*

e capa *f* base, mano *f* de imprimación
i prima mano *f* di colore
n grondlaag
d Grund *m*

2710 GROUND OIL, ARACHIS OIL, EARTHNUT OIL, PEANUT OIL — fat
f huile *f* d'arachide
e aceite *m* de cacahuete
i olio *m* d'arachide
n aardnotenolie
d Erdnussöl *n*

2711 GROUNDING, BOTTOMING — tex
f teinture *f* préliminaire
e teñido *m* preliminar
i tintura *f* preliminare
n voorverven *n*
d Vorfärbung *f*

2712 GROUNDING, STAINING — pap
f teinture *f*
e tinción *f*
i tintura *f*
n gronderen *n*
d Foncieren *n*, Grundieren *n*

2713 GROUP REACTION — ana
f réaction *f* de groupe
e reacción *f* de grupo
i reazione *f* di gruppo
n groepreactie
d Gruppenreaktion *f*

2714 GUM — glu,pla
f gomme *f*
e goma *f*
i gomma *f*
n gom
d Gummi *n*

2715 GUM ARABIC, ACACIA GUM, ARABINE, SENEGAL GUM — glu
f gomme *f* arabique, gomme *f* d'acacia
e goma *f* arábiga, goma *f* Senegal
i gomma *f* arabica, gomma *f* d'acacia
n acaciagom, Arabische gom, kordofangom, senegalgom
d Akaziengummi *m*, Arabin *n*, Gummiarabicum *n*, Senegalgummi *m*

2716 GUM BICHROMATE PROCESS — pho
f procédé *m* à la gomme bichromatée
e procedimiento *m* a la goma bicromatada
i processo *m* alla gomma bicromatata
n gomdrukproces *n*
d Gummidrucklackverfahren *n*

2717 GUM GUAIAC — oic
f résine *f* de gaïac
e resina *f* de guayación
i resina *f* di guaiaco
n guajakhars *n*
d Guajakharz *n*, Guajakum *n*

2718 GUMMED PAPER — glu,pap
f papier *m* gommé
e papel *m* engomado
i carta *f* gommata
n gegomd papier *n*, gompapier *n*
d gummiertes Papier *n*

2719 GUMMING — dec
f engommage *m*
e engomado *m*
i rivestimento *m* di gomma
n verharsen *n*
d Verharzen *n*

2720 GUNCOTTON — exp
f coton-poudre *m*, fulmicoton *m*
e fulmicotón *m*, pólvora *f* de algodón
i cotone *m* fulminante, fulmicotone *m*
n schietkatoen *n*
d Schiessbaumwolle *f*

2721 GUNK — iwt
f crasse *f*
e mugre *f*
i lordura *f*
n smurrie
d Schmutz *m*

2722 GUNPOWDER, BLACK BLASTING POWDER, BLACK POWDER — exp
f poudre *f*
e pólvora *f*
i polvere *f*
n buskruit *n*, kruit *n*
d Pulver *n*, Schiesspulver *n*, Schwarzpulver *n*

2723 GYLE, BREW — bre
f brassin *m*
e cocción *f*, cocimiento *m*
i cotta *f*
n brouwsel *n*
d Gebräu *n*, Sud *m*

2724 GYPSUM, bui
PLASTER OF PARIS
f plâtre *m* de moulage,
plâtre *m* de Paris
e yeso *m*
i gesso *m*
n gebrand gips *n*
d gebrannter Gips *m*, Kalkgips *m*

H

2725 HAARLEM OIL, SULFURATED TURPENTINE OIL (US), SULPHURATED TURPENTINE OIL (GB) fat
f huile *f* de Haarlem
e aceite *m* de Harlem
i olio *m* di Haarlem
n haarlemmerolie
d Haarlemer Öl *n*

2726 HACKLING, HECKLING, SCUTCHING tex
f peignage *m*
e peinado *m*, rastrillado *m*
i pettinatura *f*
n hekelen *n*
d Hecheln *n*

2727 HAIR SIDE, GRAIN SIDE tan
f côté *m* poil du cuir, fleur *f* de la peau
e cara *f* del grano, cara *f* del pelo
i lato *m* del pelo, lato *m* granitico del cuoio
n haarzijde, nerfkant
d Haarseite *f*, Narbenseite *f*

2728 HAIRLINES cer,dec
f craquelure *f*, gerçure *f*, tressaillure *f*
e agrietamiento *m*
i screpolatura *f*
n haarscheurtjes *pl*, oppervlakbarstjes *pl*
d Haarrisse *pl*

2729 HALATION pho
f halation *f*
e halación *f*
i formazione *f* dell'alone
n halatie
d Lichthofbildung *f*

2730 HALF AND HALF, PLUMBER'S SOLDER wel
f soudure *f* de plombier
e soldadura *f* de plomero
i stagno *m* da idraulico
n loodgieterssoldeer *n*
d Lötmörtel *m*

2731 HALF LIFE, HALF-LIFE PERIOD iso,nue
f période *f* de demi-valeur, période *f* de demi-vie
e período *m* de vida media
i tempo *m* di dimezzamento
n halveringstijd
d Halbwertszeit *f*

2732 HALF STUFF pap
f défilé *m*, pâte *f* effilochée
e pulpa *f* semiacabada
i polpa *f* semiallestita
n halfgoed *n*, halfstof
d Halbzeug *n*

2733 HALF THICKNESS, HALF-VALUE LAYER nue
f couche *f* de demi-absorption
e espesor *m* mediador
i spessore *m* di dimezzamento
n halveringsdikte
d Halbwertsschicht *f*

2734 HALF-TONE PROCESS pho
f autotypie *f*, similigravure *f*
e autotipia *f*, procedimiento *m* de grabado a media tinta
i autotipia *f*
n autotypie
d Autotypie *f*, Halbtonätzung *f*

2735 HALIBUT LIVER OIL fat,mcy
f huile *f* de foie de flétan
e aceite *m* de hígado de mero
i olio *m* di fegato di passera di mare
n heilbotlevertraan
d Heilbuttlebertran *m*

2736 HALIDE, HALOGENIDE gen
f halogénure *m*
e halogenuro *m*, haluro *m*
i alogenuro *m*
n halogenide *n*
d Halogenid *n*

2737 HALOGENATION oic
f halogénation *f*
e halogenación *f*
i alogenazione *f*
n halogeneren *n*
d Halogenieren *n*, Halogenisierung

2738 HAMMER SCALE met
f battitures *pl* de martelage
e batiduras *pl* de forja
i scaglia *f* di fucinatura
n hamerslag *n*
d Glühspäne *pl*, Hammerschlag *m*

2739 HAND MOLDING (US), pla
HAND MOULDING (GB)
f moulage *m* à main
e moldeo *m* a mano
i formazione *f* a mano
n met de hand vormen
d Handformung *f*

2740 HANDLERS tan
f fosses *pl* à tan
e noques *pl*
i fosse *pl* da conciar le pelle
n looikuilen *pl*
d Lohgruben *pl*

2741 HANDMADE PAPER pap
f papier *m* à la main, papier *m* cuve
e papel *m* hecho a mano, papel *m* de tina
i carta *f* fatta a mano
n geschept papier *n*, met de hand vervaardigd papier *n*
d Büttenpapier *n*, mit der Hand geschöpftes Hadernpapier *n*

2742 HANK, tex
SKEIN
f écheveau *m*
e madeja *f*
i matassa *f*
n streng
d Strähne *f*, Strang *m*

2743 HANK DYEING, tex
BUNDLE DYEING, YARN DYEING
f teinture *f* en écheveaux
e teñido *m* en suspensión de madejas
i tintura *f* delle matasse
n strengenverving
d Strähnenfärbung *f*

2744 HARD CHROMIUM PLATING, coa,elc
HARD PLATING
f chromage *m* dur
e cromado *m* duro
i cromatura *f* dura
n hardverchroming
d Hartverchromung *f*

2745 HARD FACING coa,met
f rechargement *m* dur
e recargue *m* duro, revestimiento *m* duro
i placchetta *f* di metallo duro, rivestimento *m* duro
n hardmetaalbekleding
d Hartmetallauftrag *m*

2746 HARD FIBER (US), tex
HARD FIBRE (GB)
f fibre *f* de feuille
e fibra *f* dura
i fibra *f* di foglia
n bladvezel
d Blattfaser *f*

2747 HARD FLOW, pla
STIFF FLOW
f basse fluidité *f*
e mala fluidez *f*
i bassa fluidità *f*
n slecht vloeivermogen *n*
d harter Fluss *m*

2748 HARD GLASS gla
f verre *m* incassable, verre *m* trempé
e vidrio *m* poco fusible
i vetro *m* temprato
n hardglas *n*
d Hartglas *n*

2749 HARD GLOSS PAINT dec
f peinture *f* dure brillante
e pintura *f* dura brillante
i pittura *f* dura lustra
n harde glansverf
d Hartglanzfarbe *f*

2750 HARD LEAD met
f plomb *m* aigre, plomb *m* dur
e plomo *m* endurecido
i piombo *m* duro
n hardlood *n*
d Hartblei *n*

2751 HARD PARAFFIN, fat
PARAFFIN WAX
f cire *f* de paraffine
e cera *f* parafínica
i cera *f* di paraffina
n paraffinewas, vaste paraffine
d Paraffinwachs *n*

2752 HARD RUBBER (US), rub
EBONITE (GB), VULCANITE
f ébonite *f*
e ebonita *f*
i ebanite *f*
n eboniet *n*
d Ebonit *m*, Hartgummi *m*

2753 HARD SOAP soa
f savon *m* dur
e jabón *m* de sosa, jabón *m* duro
i sapone *m* duro
n harde zeep, kernzeep
d harte Seife *f*, Kernseife *f*

2754 HARD SOLDER wel
f brasure *f*
e soldadura *f* fuerte
i brasatura *f*, saldatura *f* forte
n hardsoldeer *n*
d Hartlot *n*

2755 HARD SOLDERING, BRAZING wel
f soudure *f* forte
e soldadura *f* fuerte
i brasatura *f*
n hardsolderen *n*
d Hartlöten *n*

2756 HARD WATER hyg
f eau *f* inconstante
e agua *f* dura, agua *f* gorda
i acqua *f* dura
n hard water *n*
d hartes Wasser *n*

2757 HARD WATER SALTS, HARDNESS SALTS hyg
f sels *pl* de dureté de l'eau
e sales *pl* de dureza del agua
i sali *pl* di durezza dell'acqua
n hardheidvormers *pl*
d Härtebildner *pl*

2758 HARD ZINC sep
f sédiment *m* zincifère
e sedimento *m* cincífero
i sedimento *m* zincoso
n zinkhoudend neerslag *n*
d zinkhaltiger Schlamm *m*

2759 HARDBOARD bui
f carton *m* pressé, panneau *m* dur en fibres de bois
e tabla *f* de fibras prensadas
i piastra *f* dura in fibre di legno
n hardboard *n*
d Hartfaserplatte *f*, Pressplatte *f*

2760 HARDENER, ACCELERATOR pla
f agent *m* durcisseur
e acelerador *m*, agente *m* endurecedor
i accelerante *m*
n hardingsmiddel *n*
d Härter *m*, Härtungsmittel *n*

2761 HARDENER, STIFFENER rub
f durcisseur *m*, raidisseur *m*
e endurecedor *m*
i indurente *m*
n verhardingsmiddel *n*, verstijvingsmiddel *n*
d Härtungsmittel *n*, Versteifungsmittel *n*

2762 HARDENING htt,met
f durcissement *m*, trempe *f*
e endurecimiento *m*, temple *m*
i indurimento *m*, tempera *f*
n harding
d Härtung *f*

2763 HARDENING pho
f durcissement *m*
e refuerzo *m*
i indurimento *m*
n harding
d Härtung *f*

2764 HARDENING MEDIUM met
f bain *m* de durcissement, bain *m* de trempe
e medio *m* de endurecimiento, medio *m* de temple
i mezzo *m* d'indurimento, mezzo *m* temprante
n hardingsbad *n*
d Härtemittel *n*

2765 HARDENING OF OIL, OIL HARDENING fat
f hydrogénation *f* d'huile
e hidrogenación *f* de aceite
i idrogenazione *f* d'olio
n harding van vloeibare vetten, olieharding
d Fetthärtung *f*

2766 HARDNESS gen
f dureté *f*
e dureza *f*
i durezza *f*
n hardheid, hardte
d Härte *f*

2767 HARSHNESS tex
f rudesse *f*
e aspereza *f*, asperidad *f*
i rigidezza *f*
n stugheid
d Sprödigkeit *f*

2768 HARTSHORN SALTS, SAL VOLATILE mcy
f sel *m* volatil
e sal *f* volátil
i sale *m* volatile
n hertshoornzout *n*, vlugzout *n*
d flüchtiges Laugensalz *n*, Hirschhornsalz *n*

2769 HAZE dec
f voile *f*

e brumazón *m*
i velo *m*
n waas *n*
d Glanzschleier *m*

2770 HAZE pla
f état *m* trouble
e turbiedad *f*
i torbidezza *f*
n troebelheid
d Schleier *m*, Trübung *f*

2771 HAZING, BLOOMING, BLUEING dec
f bleuissement *m*, brouillage *m*, voilage *m*
e falta *f* de lustre, ponerse v mate
i divenire v torbido, velamento *m*
n vorming van een blauw waas, wazen *n*
d Blauanlaufen *n*, Hauchbildung *f*, Schleierbildung *f*

2772 HAZY GLOSS, BLOOM GLOSS dec
f brillant *m* terni, brillant *m* voilé
e lustre *m* velado
i lucidatura *f* velata
n spekglans
d Speckglanz *m*

2773 HEAD bre
f chapeau *m*
e sombrero *m*
i cappello *m*
n schuimkop
d Schaumhaube *f*

2774 HEAD RETENTION bre
f tenue *f* de la mousse
e conservación *f* de la espuma
i resistenza *f* della schiuma
n schuimbestendigheid
d Schaumbeständigkeit *f*, Schaumhaltigkeit *f*

2775 HEAP LEACHING sep
f lessivation *f* en amas, lixiviation *f* en amas
e lixiviación *f* en montones
i lisciviazione *f* in mucchi
n massa-uitloging
d Haufenauslaugung *f*

2776 HEART CUT dis
f fraction *f* de coeur
e corte *m* principal
i taglio *m* centrale
n kernaftap
d Kernschnitt *m*

2777 HEAT gen
f chaleur *f*
e calor *m*
i calore *m*
n hitte, warmte
d Hitze *f*, Wärme *f*

2778 HEAT ABSORPTION gen
f absorption *f* de la chaleur
e absorción *f* del calor
i assorbimento *m* del calor
n warmteabsorptie
d Wärmeabsorption *f*, Wärmeaufnahme *f*

2779 HEAT ACCUMULATION htt
f accumulation *f* de chaleur
e acumulación *f* de calor
i accumulazione *f* di calore
n warmteaccumulatie
d Wärmespeicherung *f*, Wärmestauung *f*

2780 HEAT CHECK, FIRE CRACK cer
f crique *f* due à la chaleur
e grieta *f* debida al calor
i incrinatura *f* dovuta al calore
n warmtescheur
d Brandriss *m*, Wärmeriss *m*

2781 HEAT COMPENSATION gen
f compensation *f* de la chaleur
e compensación *f* del calor
i compensazione *f* del calore
n warmtecompensatie
d Wärmeausgleich *m*

2782 HEAT CONSUMPTION gen
f consumption *f* de chaleur
e consumo *m* de calor
i consumo *m* di calore
n warmteverbruik *n*
d Wärmeaufwand *m*, Wärmeverbrauch *m*

2783 HEAT DISPLACEMENT gen
f déplacement *m* de chaleur
e desplazamiento *m* de calor
i spostamento *m* di calore
n warmteverplaatsing
d Wärmeverschiebung *f*

2784 HEAT DISSIPATION gen
f dispersion *f* de chaleur
e dispersión *f* de calor
i dissipazione *f* di calore
n warmteverspreiding
d Wärmeableitung *f*

2785 HEAT DISTORTION pla
f déformation *f* thermique,
distorsion *f* thermique
e deformación *f* por calor,
distorsión *f* térmica
i deformazione *f* termica,
distorsione *f* termica
n thermische vervorming
d thermische Verzerrung *f*

2786 HEAT DISTORTION POINT tes
f point *m* de distorsion thermique,
température *f* de déformation
thermique
e punto *m* de distorsión térmica,
temperatura *f* de deformación
por calor
i punto *m* di distorsione termica,
temperatura *f* di deformazione
termica
n vormvastheidstemperatuur
d Formbeständigkeitstemperatur *f*

2787 HEAT DISTORTION TESTING tes
f essai *m* de déformation thermique,
essai *m* de distorsion thermique
e prueba *f* de deformación
térmica, prueba *f* de distorsión
térmica
i collaudo *m* di deformazione
termica, collaudo *m* di distorsione
termica
n vormvastheidsbeproeving
d Formbeständigkeitsprüfung *f*

2788 HEAT EXCHANGE, gen
HEAT EXCHANGING
f échange *m* de chaleur
e cambio *m* de calor,
intercambio *m* de calor
i scambio *m* di calore
n warmte-uitwisseling
d Wärmeaustausch *m*

2789 HEAT FLUSH gas
f séparation *f* d'hélium^3 et hélium^4
par choc de chaleur
e separación *f* de helio^3 y helio^4
por choque de calor
i separazione *f* d'elio^3 ed elio^4
per urto di calore
n scheiden *n* van helium^3 en helium^4
door warmtestoot
d Helium^3-Helium^4-Trennung *f*
mittels Wärmestoss

2790 HEAT FLUX DENSITY boi
f flux *m* de chaleur
e flujo *m* de calor
i flusso *m* di calore
n warmtestroomdichtheid
d Heizflächenbelastung *f*,
Wärmestromdichte *f*

2791 HEAT-HARDENABLE RESIN, pla
HOT SETTING RESIN,
THERMOSETTING RESIN
f résine *f* durcissable par
chaleur
e resina *f* termoendurecente
i resina *f* termoindurente
n met warmte uithardbare hars,
thermohardende hars
d in Wärme aushärtendes
Kunstharz *n*

2792 HEAT INDUCTION htt
f induction *f* de chaleur
e inducción *f* de calor
i induzione *f* di calore
n warmte-inductie
d Wärmeinduktion *f*

2793 HEAT INSULATING clo
MATERIAL
f matière *f* calorifuge
e materia *f* calorífuga,
materia *f* termoaisladora
i materiale *m* da isolare
termicamente
n warmte-isolatiemateriaal *n*
d Wärmeisoliermittel *n*,
Wärmeschutzmittel *n*

2794 HEAT INSULATION gen
f isolation *f* thermique
e aislamiento *m* térmico
i isolamento *m* termico
n warmte-isolatie
d Wärmeisolierung *f*

2795 HEAT LOSSES gen
f pertes *pl* thermiques
e pérdidas *pl* térmicas
i perdite *pl* termiche
n warmteverliezen *pl*
d Wärmeverluste *pl*

2796 HEAT OF ADSORPTION abs
f chaleur *f* d'adsorption
e calor *m* de adsorción
i calore *m* d'adsorbimento
n adsorptiewarmte
d Adsorptionswärme *f*

2797 HEAT OF COAGULATION gen
f chaleur *f* de coagulation
e calor *m* de coagulación
i calore *m* di coagulazione
n coagulatiewarmte
d Koagulierungswärme *f*

2798 HEAT OF COMBINATION gen
f chaleur *f* de combinaison
e calor *m* de combinación
i calore *m* di combinazione
n verbindingswarmte
d Verbindungswärme *f*

2799 HEAT OF COMBUSTION fue
f chaleur *f* de combustion
e calor *m* de combustión
i calore *m* di combustione
n verbrandingswarmte
d Verbrennungwärme *f*

2800 HEAT OF COMPRESSION pre
f chaleur *f* de compression
e calor *m* de compresión
i calore *m* di compressione
n compressiewarmte
d Kompressionswärme *f*, Verdichtungswärme *f*

2801 HEAT OF CONDENSATION clo
f chaleur *f* de condensation
e calor *m* de condensación
i calore *m* di condensazione
n condensatiewarmte
d Kondensationswärme *f*

2802 HEAT OF CRYSTALLIZATION, LATENT HEAT OF SOLIDIFICATION cry,gen
f chaleur *f* de cristallisation
e calor *m* de cristalización
i calore *m* di cristallizzazione
n kristallisatiewarmte
d Kristallisationswärme *f*

2803 HEAT OF DECOMPOSITION gen
f chaleur *f* de décomposition
e calor *m* de descomposición
i calore *m* di decomposizione
n ontledingswarmte
d Zersetzungswärme *f*

2804 HEAT OF DILUTION gen
f chaleur *f* de dilution
e calor *m* de dilución
i calore *m* di diluzione
n verdunningswarmte
d Verdünnungswärme *f*

2805 HEAT OF DISSOCIATION gen
f chaleur *f* de dissociation
e calor *m* de disociación
i calore *m* di dissociazione
n dissociatiewarmte
d Dissoziationswärme *f*

2806 HEAT OF EVAPORATION, HEAT OF VAPORIZATION dis
f chaleur *f* d'évaporation, chaleur *f* de vaporisation
e calor *m* de evaporación, calor *m* de vaporización
i calore *m* d'evaporazione, calore *m* di vaporizzazione
n verdampingswarmte
d Verdampfungswärme *f*

2807 HEAT OF FORMATION gen
f chaleur *f* de formation
e calor *m* de formación
i calore *m* di formazione
n vormingswarmte
d Bildungswärme *f*, Entstehungswärme *f*

2808 HEAT OF FUSION, LATENT HEAT OF FUSION gen,htt
f chaleur *f* de fusion
e calor *m* de fusión
i calore *m* di fusione
n smeltwarmte
d Schmelzwärme *f*

2809 HEAT OF HYDRATION gen
f chaleur *f* de hydratation
e calor *m* de hidratación
i calore *m* d'idratazione
n hydratatiewarmte
d Hydratationswärme *f*, Hydrationwärme *f*

2810 HEAT OF MIXING mix
f chaleur *f* de mixtion
e calor *m* de mezclado
i calore *m* di miscelazione
n mengingswarmte, mengwarmte
d Mischungswärme *f*

2811 HEAT OF NEUTRALIZATION gen
f chaleur *f* de neutralisation
e calor *m* de neutralización
i calore *m* di neutralizzazione
n neutralisatiewarmte
d Neutralisierungswärme *f*

2812 HEAT OF RACEMIZATION gen
f chaleur *f* de racémisation
e calor *m* de racemización
i calore *m* di racemizzazione
n racemisatiewarmte
d Razemisierungswärme *f*

2813 HEAT OF REACTION gen
f chaleur *f* de réaction
e calor *m* de reacción
i calore *m* di reazione

n reactiewarmte
d Reaktionswärme *f*,
Wärmetönung *f*

2814 HEAT OF SOLIDIFICATION gen
f chaleur *f* de solidification
e calor *m* de solidificación
i calore *m* di solidificazione
n stollingswarmte
d Erstarrungswärme *f*

2815 HEAT OF SOLUTION gen
f chaleur *f* de solution
e calor *m* de solución
i calore *m* di soluzione
n oploswarmte
d Lösungswärme *f*

2816 HEAT OF SWELLING, gen
HEAT OF TUMEFACTION
f chaleur *f* de gonflage
e calor *m* de hinchamiento
i calore *m* di rigonfiamento
n zwellingswarmte
d Quellungswärme *f*

2817 HEAT OF TRANSFORMATION gen
f chaleur *f* de transformation
e calor *m* de transformación
i calore *m* di trasformazione
n omzettingswarmte
d Umwandlungswärme *f*

2818 HEAT OF TRANSITION gen
f chaleur *f* de transition
e calor *m* de transición
i calore *m* di transizione
n overgangswarmte
d Übergangswärme *f*

2819 HEAT OF WETTING gen
f chaleur *f* de mouillage
e calor *m* de humectación
i calore *m* d'umidificazione
n bevochtigingswarmte
d Benetzungswärme *f*

2820 HEAT PERMEABILITY gen
f perméabilité *f* pour la chaleur
e permeabilidad *f* para el calor
i permeabilità *f* per il calore
n warmtedoorlatendheid
d Wärmedurchlässigkeit *f*

2821 HEAT REGENERATION gen
f régénération *f* de chaleur
e regeneración *f* de calor
i rigenerazione *f* di calore
n warmteregeneratie
d Wärmeregenerierung *f*

2822 HEAT RESISTANCE gen
f résistance *f* à la chaleur
e resistencia *f* al calor
i resistenza *f* al calore
n hittevastheid
d Hitzebeständigkeit *f*

2823 HEAT RESISTING, gen
THERMODURIC (US)
f résistant adj à la chaleur,
résistant adj au feu
e resistente adj al calor,
resistente adj al fuego
i resistente adj al calore,
resistente adj al fuoco
n hittevast adj , vuurvast adj
d feuerfest adj , hitzebeständig
adj

2824 HEAT SEALING, pla
HIGH-FREQUENCY WELDING
f soudure *f* à chaud
e sellado *m* en caliente
i sigillatura *f* a caldo,
termosaldatura *f*
n warmlassen *n*
d Heissiegeln *n*

2825 HEAT STERILIZATION bac,hyg
f thermostérilisation *f*
e termoesterilización *f*
i sterilizzazione *f* a caldo
n thermisch steriliseren *n*
d Thermosterilisation *f*,
Thermosterilisierung *f*

2826 HEAT TINTING dec,ht
f coloration *f* thermique
e termocoloración *f*
i colorazione *f* termica
n thermisch kleuren *n*
d Wärmetönung *f*

2827 HEAT TRANSFER clo,nu
f transfert *m* de chaleur
e termotransferencia *f*
i trasmissione *f* di calore
n warmteoverdracht
d Wärmeübertragung *f*

2828 HEAT TRANSFER SALT ge
f sel *m* réfrigérant
e sal *f* refrigerante
i sale *m* refrigerante
n koelzout *n*
d Kühlsalz *n*, Warmbadsalz *n*

2829 HEAT TREATMENT htt,m
f traitement *m* thermique
e termotratamiento *m*

i trattamento *m* termico
n warmtebehandeling
d Vergütung *f*, Wärmebehandlung *f*

2830 HEATER RECLAIMING PROCESS, PAN RECLAIMING PROCESS rub
f procédé *m* de régénération par la vapeur et l'huile
e procedimiento *m* de regeneración con vapor y aceite
i processo *m* di rigenerazione in vapore ed olio
n stoom-olie-regenereerproces
d Dampf-Ölregenerierungs-verfahren *n*

2831 HEATING gen
f chauffage *m*
e calefacción *f*, calentamiento *m*
i riscaldamento *m*
n verhitting, verwarming
d Erhitzung *f*, Erwärmung *f*, Heizung *f*

2832 HEATING STEAM boi
f vapeur *f* de chauffage
e vapor *m* calentador
i vapore *m* riscaldatore
n verhittingsstoom
d Heizdampf *m*

2833 HEAVY AGGREGATE CONCRETE nue,sap
f béton *m* chargé
e hormigón *m* cargado
i calcestruzzo *m* caricato
n verzwaard beton *n*
d Schwerbeton *m*

2834 HEAVY HYDROGEN, DEUTERIUM nue
f hydrogène *m* lourd
e hidrógeno *m* pesado
i idrogeno *m* pesante
n zware waterstof
d schwerer Wasserstoff *m*

2835 HEAVY KEY dis
f constituant-clef *m* lourd
e componente *m* clave pesado
i componente *m* chiave pesante
n zware sleutelcomponent
d schwere Schlüsselkomponente *f*

2836 HEAVY OIL fue
f huile *f* lourde
e aceite *m* pesado
i olio *m* pesante
n zware olie
d Schweröl *n*

2837 HEAVY SPAR, BARYTES min
f barytine *f*, spath *m* pesant
e baritina *f*, espato *m* pesado
i baritina *f*, spato *m* pesante
n bariet *n*, zwaarspaat *n*
d Baryt *m*, Schwerspat *m*

2838 HEAVY WATER, DEUTERIUM OXIDE nue
f eau *f* lourde
e agua *f* pesada
i acqua *f* pesante
n zwaar water *n*
d schweres Wasser *n*

2839 HECKLING, HACKLING, SCUTCHING tex
f peignage *m*
e peinado *m*, rastrillado *m*
i pettinatura *f*
n hekelen *n*
d Hecheln *n*

2840 HEEL, RESIDUE dis
f résidu *m*
e residuo *m*
i residuo *m*
n residu *n*
d Rückstand *m*

2841 HELIUM LEAK DETECTION nue
f détection *f* de pertes d'hélium
e detección *f* de pérdidas de helio
i rivelazione *f* di perdite d'olio
n heliumdetectie
d Heliumlecksuche *f*

2842 HELIUM LIQUEFACTION gas
f liquéfaction *f* d'hélium
e licuación *f* de helio
i liquefazione *f* d'elio
n vloeibaar maken *n* van helium
d Heliumverflüssigung *f*

2843 HELVETIA LEATHER, CROWN LEATHER lea
f cuir *m* helvétique
e cuero *m* helvético
i cuoio *m* elvetico
n kroonleder *n*
d Crownleder *n*, Kronleder *n*

2844 HEMICOLLOID col
f hémicolloïde *m*
e hemicoloide *m*
i emicolloide *m*
n grof-dispers colloïde *n*
d Halbkolloid *n*, Hemikolloid *n*, Übergangskolloid *n*

2845 HEMP tex
f chanvre *m*
e cáñamo *m*
i canapa *f*
n hennep
d Hanf *m*

2846 HEMP OIL, fat
HEMP SEED OIL
f huile *f* de chanvre
e aceite *m* de cáñamo
i olio *m* di canapa
n hennepzaadolie
d Hanföl *n*

2847 HERBICIDE, agr
WEEDKILLER
f herbicide *m*
e herbicida *m*
i erbicida *m*
n onkruidverdelgend middel *n*
d Unkrautvertilgungsmittel *n*

2848 HETEROGENEOUS gen
f hétérogène ajd
e heterogéneo adj
i eterogeneo adj
n heterogeen adj
d heterogen adj

2849 HIDE GLUE, glu
CARPENTER'S GLUE,
JOINER'S GLUE
f colle *f* forte
e cola *f* de carnaza, cola *f* de piel
i colla *f* forte
n houtlijm
d Hautleim *m*

2850 HIDING PIGMENT, dec
PRIME PIGMENT
f pigment *m* opacifiant
e pigmento *m* cubridor
i pigmente *m* coprente
n dekkend pigment *n*
d deckendes Pigment *n*

2851 HIDING POWER, dec
COVERING POWER
f opacité *f*, pouvoir *m* couvrant
e capacidad *f* encubridora,
opacidad *f*, poder *m* cubridor
i opacità *f*, potere *m* coprente,
potere *m* di ricoprimento
n dekkracht, dekvermogen *n*
d Deckfähigkeit *f*, Deckkraft *f*

2852 HIGH-ACTIVITY WASTE nue
f déchets *pl* de haute activité
e desechos *pl* fuertemente
radiactivos
i rifiuti *pl* altamente radioattivi
n zeer radioactief afval *n*
d Abfall *m* mit hoher Aktivität

2853 HIGH AROMATIC WHITE fue
SPIRIT
f essence *f* de pétrole pour vernis
à haute solubilité
e gasolina *f* con gran proporción
de hidrocarburos aromáticos
i benzina *f* per vernice ad alta
solubilità
n aromaatrijke lakbenzine
d hocharomatisches Lackbenzin *n*

2854 HIGH ASH CONTENT gen
f teneur *f* élevée en cendres
e alto contenido *m* de cenizas
i alto contenuto *m* in ceneri
n hoog asgehalte *n*
d hoher Aschengehalt *m*

2855 HIGH-ATTENUATING bre,fmi
YEAST
f levure *f* à forte atténuation
e levadura *f* de alta atenuación
i lievito *m* ad alto grado
d'attenuazione
n snelwerkende gist
d hochvergärende Hefe *f*

2856 HIGH BACKING-PRESSURE vac
f contre-pression *f* élevée
e alta contrapresión *f*
i alta contropressione *f*
n hoge tegendruk
d hoher Gegendruck *m*

2857 HIGH-BOILING FRACTION dis
f fraction *f* à point d'ébullition
élevé
e fracción *f* de alto punto de
ebullición
i frazione *f* altobollente
n fractie met hoog kookpunt
d hochsiedende Fraktion *f*

2858 HIGH CHLORINATION, aic
EXCESS CHLORINATION,
SUPERCHLORINATION
f chloration *f* à dose élevée,
surchloration *f*
e sobrecloración *f*
i sovraclorazione *f*,
superclorazione *f*
n overchloreren *n*
d Hochchlorung *f*, Überchlorung *f*
Überschusschlorung *f*

2859 HIGH EXPLOSIVE exp
f explosif *m* puissant, grand explosif *m*
e explosivo *m* rompedor
i alto esplosivo *m*
n explosieve stof met hoge brisantie, hoogbrisante springstof
d hochbrisanter Explosivstoff *m*, hochbrisantes Sprengmittel *n*

2860 HIGH-FREQUENCY DRIER, DIELECTRIC DRYING dry
f séchage *m* par pertes diélectriques
e secado *m* por pérdidas dieléctricas
i essiccazione *f* mediante perdite dielettriche
n capacitief drogen *n*
d dielektrische Trocknung *f*, Hochfrequenztrocknung *f*

2861 HIGH GLOSS, FULL GLOSS dec
f brillant *m* élevé
e gran brillo *m*
i gran lucidezza *f*
n hoge glans, hoogglans
d Hochglanz *m*

2862 HIGH GREEN (US), HIGH GREEN SYRUP (GB) sin
f sirop *m* vert
e jarabe *m* verde
i sciroppo *m*
n groene stroop, ongedekte stroop
d Feinsirup *m*, Grünsirup *m*, Weissablauf *m*

2863 HIGH LIGHTS elc
f points *pl* brillants
e puntos *pl* brillantes
i punti *pl* brillanti
n hoogglanspunten *pl*
d Glanzlichter *pl*, Glanzpunkte *pl*

2864 HIGH-OCTANE FUEL fue
f combustible *m* à indice d'octane élevé
e combustible *m* de índice de octano elevado
i combustibile *m* a numero d'ottano elevato
n brandstof met hoog octaangehalte
d Brennstoff *m* mit hoher Oktanzahl

2865 HIGH PRESSURE acc
f haute pression *f*, pression *f* élevée
e alta presión *f*
i alta pressione *f*
n hoge druk
d Hochdruck *m*, hoher Druck *m*

2866 HIGH-PRESSURE MOLDING (US), HIGH-PRESSURE MOULDING (GB) pla
f moulage *m* à haute pression
e moldeo *m* de alta presión
i stampatura *f* ad alta pressione
n vormen *n* onder hoge druk
d Hochdruckpressung *f*

2867 HIGH-PRESSURE STEAM PROCESS REGENERATION rub
f régénération *f* à la vapeur sous haute pression
e procedimiento *m* de regeneración con vapor de alta presión
i processo *m* di rigenerazione con vapore sotto alta pressione
n hoge-druk-stoomregeneratie-proces *n*
d Hochdampfdruck-Regenerier-verfahren *n*

2868 HIGH-RATE ACTIVATED SLUDGE PROCESS, MODIFIED AERATION iwt
f procédé *m* des boues activées à haut rendement
e procedimiento *m* de fango activado de alto rendimiento
i processo *m* a fango attivato ad alta dosatura
n geactiveerd-slibproces *n* met hoog rendement
d hochbelastetes Belebungs-verfahren *n*, Hochleistungs-belebtschlammverfahren *n*

2869 HIGH RAW BOILING sin
f cuite *f* au filet, deuxième cuisson *f*
e segunda cocción *f*
i seconda cottura *f*
n koken *n* op middelprodukt
d Blankkochen *n*

2870 HIGH VACUUM vac
f vide *m* élevé, vide *m* poussé
e alto vacío *m*
i alto vuoto *m*, vuoto *m* spinto
n hoogvacuüm *n*
d Hochvakuum *n*

2871 HIGH WASH (US), MOTHER SYRUP (GB) sin
f eau-mère *f* jaunâtre

e jarabe *m* blanco
i sciroppo *m* bianco
n moederstroop
d Muttersirup *m*

2872 HINDERED SETTLING sep
f sédimentation *f* troublée
e sedimentación *f* retrasada
i sedimentazione *f* non libera
n gestoord neerslaan *n*
d gestörtes Absetzen *n*

2873 HOFMANN DEGRADATION, EXHAUSTIVE METHYLATION oic
f méthylation *f* totale
e metilación *f* a fondo
i metilazione *f* totale
n uitputtende methylering
d erschöpfende Methylierung *f*

2874 HOLD-UP nue
f charge *f* ouvrable
e carga *f* práctica
i contenuto *m*
n materiaalinbrengst
d Materialeinsatz *m*

2875 HOLDING, STORING sto
f emmagasinage *m*, emmagasinement *m*
e almacenamiento *m*
i immagazzinamento *m*
n vergaren *n*, verzamelen *n*
d Speichern *n*

2876 HOLIDAY, BARE SPOT, UNCOATED AREA dec
f absence *f*, blanc *m*, manque *m*, zone *f* non couverte
e superficie *f* pasada por alto
i zona *f* non coperta
n heiligedag
d blanke Stelle *f*, Jute *f*, Jutlochstelle *f*

2877 HOLLOW GLASS gla
f verrerie *f* creuse
e artículos *pl* de vidrio hueco
i vetro *m* concavo
n holglas *n*
d Hohlglas *n*

2878 HOMOGENEOUS gen
f homogène adj
e homogéneo adj
i omogeneo adj
n homogeen adj
d homogen adj

2879 HOMOGENEOUS DYE dye
f colorant *m* unicolore
e colorante *m* unicolor
i colorante *m* unicolore
n eènkleurige kleurstof
d einfarbiger Farbstoff *m*

2880 HOMOGENIZING tex
f homogénéisation *f*
e homogenización *f*
i omogenizzazione *f*
n homogenisatie
d Homogenisierung *f*

2881 HONING pol
f affilage *m*
e afiladura *f*, aguzadura *f*
i affilatura *f*
n aanzetten *n*, wetten *n*
d Abziehen *n*, Wetzen *n*

2882 HOP BITTER, LUPULIN bre
f lupuline *f*
e lupolina *f*
i luppolina *f*
n hopbitter *n*, lupuline
d Hopfenbitter *n*, Lupulin *n*

2883 HOP EXTRACTION bre
f extraction *f* de houblon, lavage *m* de houblon
e extracción *f* de lúpulo
i estrazione *f* di luppolo
n hopuitloging
d Hopfenentlaugung *f*

2884 HOP OIL bre,fat
f huile *f* de houblon
e aceite *m* de lúpulo
i olio *m* di luppolo
n hopolie
d Hopfenöl *n*

2885 HOP PRESS LIQUOR bre,iwt
f eau *f* de presse de houblon
e agua *f* de presión de lúpulo
i acqua *f* di rifiuto della spremitura del luppolo
n hoppersafvalwater *n*
d Hopfenpresswasser *n*

2886 HOPPED WORT bre
f moût *m* houblonné
e mosto *m* lupulado
i mosto *m* luppolato
n met hop behandelde most
d gehopfte Würze *f*

2887 HOT AIR AGEING (GB), HOT AIR AGING (US) rub
f vieillissement *m* à air chaud

e envejecimiento *m* en aire caliente
i invecchiamento *m* in aria calda
n hete-luchtveroudering
d Heissluftalterung *f*

2888 HOT AIR CURING, rub
DRY AIR CURING,
DRY HEAT VULCANIZATION
f vulcanisation *f* à l'air chaud
e vulcanización *f* al aire caliente
i vulcanizzazione *f* in aria calda
n hete-luchtvulcanisatie
d Heissluftvulkanisation *f*

2889 HOT AIR DRYING dry
f séchage *m* à air chaud
e secado *m* en aire caliente
i essiccazione *f* ad aria calda
n hete-luchtdroging
d Heisslufttrocknung *f*

2890 HOT AIR STERILIZATION hyg,lab
f stérilisation *f* à air chaud
e sterilización *f* en aire caliente
i sterilizzazione *f* ad aria calda
n hete-luchtsterilisatie
d Heissluftsterilisierung *f*

2891 HOT COMPACTING, met,pre
HOT PRESSING
f compression *f* à chaud
e compresión *f* en caliente
i compressione *f* a caldo
n warmpersen *n*
d Warmpressen *n*

2892 HOT CURING rub
f vulcanisation *f* à chaud
e vulcanización *f* en caliente
i vulcanizzazione *f* a caldo
n vulcanisatie bij hoge temperatuur
d Heissvulkanisation *f*

2893 HOT-DIP COATING coa
f revêtement *m* par immersion à bain chaud
e revestimiento *m* por inmersión con baño caliente
i rivestimento *m* per immersione a bagno caldo
n dompelmetalliseren *n* in warm bad
d Tauchmetallisierung *f* in heissem Bad

2894 HOT ENAMELLING coa,htt
f émaillage *m* à chaud
e esmaltado *m* en caliente
i smaltatura *f* a caldo
n warmemailleren *n*
d Puderemaillierung *f*

2895 HOT FERMENTATION, fmi,iwt
THERMOPHILIC AEROBIC DIGESTION
f fermentation *f* à chaud
e fermentación *f* en caliente
i fermentazione *f* a caldo
n fermenteren *n* door broeien
d Heissvergärung *f*

2896 HOT FINISHING htt,pol
f finissage *m* à chaud
e acabado *m* en caliente
i finitura *f* a caldo
n warmveredeling
d Warmvergütung *f*

2897 HOT FORMING pla
f formation *f* à chaud
e formación *f* en caliente
i formazione *f* a caldo
n warmvormen *n*
d Warmformgebung *f*

2898 HOT GALVANIZING coa
f galvanisation *f* à chaud
e galvanización *f* en caliente
i galvanizzazione *f* a caldo
n thermische verzinking
d Feuerverzinkung *f*

2899 HOT GAS WELDING pla,wel
f soudure *f* à gaz chaud
e soldadura *f* con gas caliente
i saldatura *f* a gas caldo
n heet-gaslassen *n*
d Heissglasschweissung *f*

2900 HOT-GROUND PULP, pap
MECHANICAL WOODPULP
f pâte *f* de bois
e pulpa *f* de madera
i polpa *f* di legno
n houtslijp *n*, houtstof *n*
d Holzschliff *m*, Holzstoff *m*

2901 HOT LACQUER, dec
HOT SPRAY LACQUER
f laque *f* à atomiser à chaud
e laca *f* de atomización en caliente
i lacca *f* da atomizzazione a caldo
n warmspuitlak
d Heissspritzlack *m*

2902 HOT MOLDING (US), pla
HOT MOULDING (GB)
f moulage *m* à chaud

e moldeo *m* en caliente
i stampatura *f* a caldo
n warmpersen *n*
d Warmpressung *f*

2903 HOT OIL PROCESSING tex
f teinture *f* à huile chaude
e teñido *m* con aceite caliente
i tintura *f* ad olio caldo
n hete-olieverving
d Heissölfärbung *f*

2904 HOT PATCHING, AIRBORNE SEALING cer
f fermeture *f* de pores
e cerrado *m* de poros
i chiusura *f* di pori
n poriëndichting
d Porendichtung *f*

2905 HOT-PLATE SPALLING TEST cer,tes
f essai *m* d'écaillage à plaque chauffée
e prueba *f* de fisuración con placa calentada
i collaudo *m* di sfaldamento con piastra riscaldata
n afspringproef met verhitte plaat
d Ausplatzprobe *f* mit geheizter Platte

2906 HOT PREPARATION, STEAM TEMPERING cer
f traitement *m* de l'argile à la vapeur
e tratamiento *m* de la arcilla al vapor
i trattamento *m* dell'argilla con vapore
n stoombehandeling voor klei
d Dampfbehandlung *f* für Ton

2907 HOT-PRESSED PAPER pap
f papier *m* satiné par plaques chauffées
e papel *m* satinado con placas calentadas
i carta *f* rasata con piastre riscaldate
n tussen verwarmde platen gesatineerd papier *n*
d zwischen Heizplatten satiniertes Papier *n*

2908 HOT RINSING (GB), HOT SWILLING (US), elc
f rinçage *m* à chaud
e enjuague *m* en caliente
i sciacquatura *f* a caldo
n warm spoelen *n*
d Heissspülen *n*, Warmspülen *n*

2909 HOT-ROLLED PAPER pap
f papier *m* satiné par cylindres chauffés
e papel *m* satinado con cilindros calentados
i carta *f* rasata con cilindri riscaldati
n tussen verhitte walsen gesatineerd papier *n*
d zwischen erhitzten Walzen satiniertes Papier *n*

2910 HOT SETTING GLUE glu
f colle *f* durcissante à chaud
e cola *f* endurecente en caliente
i colla *f* indurente a caldo
n thermohardende lijm
d heisshärtender Leim *m*, warmhärtender Leim *m*

2911 HOT SETTING RESIN, HEAT-HARDENABLE RESIN, THERMOSETTING RESIN pla
f résine *f* durcissable par chaleur
e resina *f* termoendurecente
i resina *f* termoindurente
n met warmte uithardbare hars, thermohardende hars
d in Wärme aushärtendes Kunstharz *n*

2912 HOT SPRAY PAINTING, HOT SPRAYING dec
f pistolage *m* à chaud
e atomización *f* de pintura en caliente
i atomizzazione *f* di puntura a caldo
n warmspuiten *n*
d Warmspritzen *n*

2913 HOT TINNING coa
f étamage *m* à chaud
e estañado *m* en caliente
i stagnatura *f* a caldo
n thermische vertinning
d Feuerverzinnung *f*

2914 HOT WATER MERCERIZATION, HOT WATER STEEPING tex
f mercerisage *m* à l'eau chaude
e mercerización *f* con agua caliente
i mercerizzazione *f* ad acqua cal[illegible]
n merceriseren *n* met heet water
d Mercerisieren *n* mit heissem Wa[illegible]

2915 HOT WORKING met
f travail *m* à chaud
e trabajo *m* en caliente
i lavorazione *f* a caldo
n warmbewerking
d Warmbearbeitung *f*

2916 HOUSE PAINT, dec
ARCHITECTURAL PAINT
f peinture *f* pour bâtiments
e pintura *f* de casas
i pittura *f* per uso interno
n huisschilderverf
d Bautenfarbe *f*

2917 HOUSEHOLD GARBAGE (US), hyg
DOMESTIC REFUSE (GB)
f ordures *pl* ménagères
e basuras *pl* domésticas
i immondizie *pl* domestiche, rifiuti *pl* domestici
n huisvuil *n*
d Hausmüll *m*

2918 HOUSEHOLD PAINT, dec
DOMESTIC PAINT
f peinture *f* ménagère
e pintura *f* casera
i pittura *f* domestica
n bussenverf, huishoudverf
d Farbe *f* für Haushaltzwecke

2919 HUE, dec
SHADE
f teinte *f*, tonalité *f* chromatique
e tinte *m*, tonalidad *f* cromática
i tinta *f*, tonalità *f* cromatica
n kleurtoon
d Farbton *m*, Ton *m*

2920 HUMECTANT gen
f agent *m* de humectation
e substancia *f* humectante
i mezzo *m* umidificatore
n bevochtigingsmiddel *n*
d Befeuchtungsmittel *n*

2921 HUMIC ACID, fer
ULMIC ACID
f acide *m* humique
e ácido *m* húmico
i acido *m* umico
n humuszuur *n*
d Huminsäure *f*, Humussäure *f*

2922 HUMIDIFYING, hyg
MOISTENING
f humectation *f*, humidification *f*
e humectación *f*, humidificación *f*
i umidificazione *f*
n bevochtiging
d Anfeuchtung *f*, Befeuchtung *f*

2923 HUMIDITY CONTROL hyg
f réglage *m* de l'humidité
e higrorregulación *f*, regulación *f* de la humedad
i regolazione *f* dell'umidità
n vochtigheidsregeling
d Feuchtigkeitsregelung *f*

2924 HUMIDITY DRYING cer,dry
f séchage *m* à air chaud humide
e secado *m* en aire caliente húmedo
i essiccazione *f* in aria calda umida
n drogen *n* met vochtige lucht
d Trocknen *n* mit feuchter Luft

2925 HUMIDITY RESISTANCE gen
f résistance *f* à l'humidité
e resistencia *f* a la humedad
i resistenza *f* all'umidità
n vochtwerendheid
d Feuchtbeständigkeit *f*

2926 HUNGARY BLUE, dec
COBALT BLUE, PARIS BLUE
f bleu *m* de cobalt, bleu *m* de Paris
e azul *m* de cobalto, azul *m* de París
i blu *m* di cobalto, blu *m* di Parigi
n Berlijns blauw *n*, kobaltblauw *n*, Parijs blauw *n*
d Berliner Blau *n*, Kobaltblau *n*, Königsblau *n*, Wiener Blau *n*

2927 HYDRATION gen
f hydratation *f*
e hidratación *f*
i idratazione *f*
n hydratatie
d Hydratation *f*, Hydration *f*

2928 HYDRATION WATER sin
f eau *f* d'hydratation
e agua *f* de hidratación
i acqua *f* d'idratazione
n hydraatwater *n*, hydratiewater *n*
d Hydratationswasser *n*, Hydratwasser *n*

2929 HYDRAULIC ATOMIZATION, dec
AIRLESS SPRAY PAINTING,
AIRLESS SPRAYING
f atomisation *f* sans air comprimé, pistolage *m*
e atomización *f* sin aire
i atomizzazione *f* senza aria
n luchtloos spuiten *n*, spuiten *n* zonder perslucht
d Hochdruckfarbspritzen *n*

2930 HYDRAULIC CEMENT bui
f ciment *m* hydraulique
e cemento *m* hidráulico
i cemento *m* idraulico
n hydraulisch cement *n*, watercement *n*
d hydraulischer Zement *m*, unterwasserhärtender Zement *m*, Wasserzement *m*

2931 HYDRAULIC CLASSIFICATION sor
f classification *f* hydraulique, hydrotriage *m*
e hidraclasificación *f*
i classificazione *f* idraulica
n hydraulisch classificeren *n*
d hydraulisches Klassieren *n*

2932 HYDRAULIC DESCALING cle,fou
f décalaminage *m* par jet d'eau
e descascarillado *m* hidráulico
i scagliatura *f* idraulica
n afspuiten *n*
d Abspritzen *n*

2933 HYDRAULIC GLUE glu
f colle *f* hydrofuge
e cola *f* hidrófuga
i colla *f* idrofuga, colla *f* resistente all'umidità
n vochtwerende lijm
d feuchthemmender Leim *m*

2934 HYDRAULIC GRADIENT, LIQUID GRADIENT dis
f gradient *m* hydraulique
e gradiente *m* hidráulico
i gradiente *m* idraulico
n vloeistofgradiënt
d Flüssigkeitsgefälle *n*

2935 HYDRAULIC LIME, LEAN LIME, MEAGRE LIME bui
f chaux *f* hydraulique
e cal *f* hidráulica, cal *f* magra
i calce *f* idraulica
n hydraulische kalk, magere kalk
d Wasserkalk *m*

2936 HYDRAULIC MORTAR bui
f mortier *m* à chaux hydraulique
e mortero *m* hidráulico
i malta *f* idraulica
n hydraulische mortel, watermortel
d hydraulischer Kalkmörtel *m*, hydraulischer Mörtel *m*, Wasserkalkmörtel *m*

2937 HYDROBLAST cle
f nettoyage *m* par jet d'eau
e hidrochorro *m* abrasivo
i pulitura *f* per getto d'acqua
n waterstralen *n*
d Wasserstrahlen *n*

2938 HYDROCHLORIC ACID (GB), MURIATIC ACID, SALT ACID(US) aic
f acide *m* chlorhydrique
e ácido *m* clorhídrico, ácido *m* muriático
i acido *m* cloridrico, acido *m* muriatico
n chloorwaterstofzuur *n*, zoutzuur *n*
d Chlorwasserstoffsäure *f*, Salzsäure *f*

2939 HYDROCHLORINATED RUBBER rub
f caoutchouc *m* hydrochloré
e goma *f* hidroclorada
i gomma *f* idroclorurato
n zoutzure rubber
d Kautschukhydrochlorid *m*

2940 HYDROCRACKING fue
f craquage *m* catalytique à l'hydrogène
e hidropirólisis *f*
i piroscissione *f* catalitica all'idrogeno
n katalytisch kraken *n* met waterstof
d katalytisches Kracken *n* mit Wasserstoff

2941 HYDROCYANIC ACID, PRUSSIC ACID oic
f acide *m* cyanhydrique, acide *m* prussique
e ácido *m* cianhídrico
i acido *m* cianidrico, acido *m* prussico
n blauwzuur *n*, cyaanwaterstof, cyaanwaterstofzuur *n*
d Blausäure *f*, Hydrozyansäure *f*, Zyanwasserstoffsäure *f*

2942 HYDRODESULFURATION(US), HYDRODESULPHURATION (GB) fue
f hydrodésulfuration *f*
e hidrodesulfuración *f*
i idrodesolforazione *f*
n hydrogenerend ontzwavelen *n*
d Wasserstoffentschwefelung *f*

2943 HYDROFORMING fue
f hydroréformation *f*
e hidroreformación *f*
i idroriformazione *f*, riformazione *f* con un catalizzat[ore] d'ossido di molibdeno

n dehydrerend aromatiseren *n*, hydroformeren *v*
d Reformieren *n* unter Wasserstoffdruck

2944 HYDROFUGE gen
f hydrofuge adj
e hidrófugo ajd
i idrofugo adj
n vochtwerend adj
d feuchtabstossend adj

2945 HYDROGEN aic
f hydrogène *m*
e hidrógeno *m*
i idrogeno *m*
n waterstof
d Wasserstoff *m*

2946 HYDROGEN ADDITION gen
f addition *f* d'hydrogène
e adición *f* de hidrógeno
i addizione *f* d'idrogeno
n waterstofadditie
d Wasserstoffanlagerung *f*

2947 HYDROGEN ANNEALING htt
f recuit *m* à l'hydrogène
e recocido *m* al hidrógeno
i ricottura *f* all'idrogeno
n gloeifrissen *n* in waterstof
d Glühfrischen *n* in Wasserstoff

2948 HYDROGEN EMBRITTLEMENT, cor
PICKLING EMBRITTLEMENT
f fragilité *f* par décapage
e fragilidad *f* por decapaje
i fragilità *f* per decapaggio
n beitsingsverbrozing
d Beizsprödigkeit *f*

2949 HYDROGEN ION CONCENTRATION elc
f concentration *f* ionique d'hydrogène
e concentración *f* iónica de hidrógeno
i concentrazione *f* ionica d'idrogeno
n waterstofionenconcentratie
d Wasserstoffionenkonzentration *f*

2950 HYDROGEN SALT, gen
ACID SALT
f sel *m* acide
e sal *f* ácida
i sale *m* acido
n zuur zout *n*
d Sauersalz *n*, saures Salz *n*

2951 HYDROGEN SULPHIDE (GB), gas
STINK DAMP (US),
SULFURETTED HYDROGEN (US),
SULPHURETTED HYDROGEN (GB)
f hydrogène *m* sulfuré
e hidrógeno *m* sulfurado
i idrogeno *m* sulforato
n zwavelwaterstof
d Schwefelwasserstoff *m*

2952 HYDROGENATION fat,fue
f hydrogénation *f*
e hidrogenación *f*
i idrogenazione *f*
n hydreren *n*, hydrogeneren *n*
d Hydrieren *n*, Hydrogenieren *n*, Wasserstoffanlagerung *f*

2953 HYDROLYSIS iwt
f hydrolyse *f*
e hidrólisis *f*
i idrolisi *f*
n hydrolyse
d Hydrolyse *f*

2954 HYDROPHILIC COLLOID col
f colloïde *m* hydrophile
e coloide *m* hidrófilo
i colloide *m* idrofilo
n hydrofiel colloïde
d hydrophiles Kolloid *n*

2955 HYDROPHILIC FIBER (US), tex
HYDROPHILIC FIBRE (GB)
f fibre *f* hydrophile
e fibra *f* hidrófila
i fibra *f* idrofila
n wateropnemende vezel
d wasseranziehende Faser *f*

2956 HYDROPHOBIC COLLOID col
f colloïde *m* hydrophobe
e coloide *m* hidrófobo
i colloide *m* idrofobo
n hydrofoob colloïde *n*
d hydrophobes Kolloid *n*

2957 HYDROPHOBIC FIBER (US), tex
HYDROPHOBIC FIBRE (GB)
f fibre *f* hydrophobe
e fibra *f* hidrófoba
i fibra *f* idrofoba
n waterafstotende vezel
d wasserabstossende Faser *f*

2958 HYDROTOPE, soa
SOLUBILIZER
f agent *m* de solubilisation, tiers *m* solvant
e agente *m* solubilizante, fracción *f* disolvente

i agente *m* di solubilizzazione
n solubiliserend middel *n*
d Lösungsvermittler *m*

2959 HYDROXYLATION oic
f hydroxylation *f*
e hidroxilación *f*
i idrossilazione *f*
n hydroxyleren *n*
d Hydroxylieren *n*

2960 HYDROZINCITE, met
ZINC BLOOM
f fleurs *pl* de zinc,
hydrozincite *f*, zincosine *f*
e hidrocincita *f*, zinconita *f*
i zinconite *f*
n hydrozinkiet *n*, zinkbloem
d Zinkblüte *f*

2961 HYGROSCOPIC gen
f hygroscopique adj
e higroscópico adj
i igroscopico adj
n hygroscopisch adj ,
vochtaantrekkend adj
d feuchtigkeitanziehend adj ,
hygroskopisch adj

2962 HYPERSENSITIZATION pho
f hypersensibilisation *f*
e hipersensibilación *f*
i ipersensibilizzazione *f*
n hypersensibilisatie
d Hypersensibilisierung *f*

2963 HYPOCHLORITE BLEACHING tex
f blanchiment *m* à l'hypochlorite
e blanqueo *m* con hipoclorito
i candeggio *m* con ipoclorito
n bleken *n* met hypochloriet
d Hypochloritbleichung *f*

I

2964 ICE gen
f glace *f*
e hielo *m*
i ghiaccio *m*
n ijs *n*
d Eis *n*

2965 ICECREAM nut
f crême *f* à la glace, glace *f* à la crême
e mantecado *m* helado
i gelato *m* alla crema
n roomijs *n*
d Eiskrem *m*, Rahmeis *n*, Speiseeis *n*

2966 ICICLING dec
f coulage *m*
e corrimiento *m*
i scorrimento *m*
n vorming van zakkers
d starke Tropfnasenbildung *f*

2967 ICING SUGAR, CONFECTIONER'S SUGAR nut
f sucre *m* en poudre
e azúcar *m* en polvo
i zucchero *m* in polvere
n poedersuiker
d Puderzucker *m*, Staubzucker *m*

2968 IDEAL GAS, PERFECT GAS gas
f gaz *m* idéal, gaz *m* parfait
e gas *m* ideal, gas *m* perfecto
i gas *m* ideale, gas *m* perfetto
n ideaal gas *n*
d ideales Gas *n*

2969 IGNITING AGENT exp
f agent *m* allumeur
e agente *m* explosor, medio *m* detonante
i accenditore *m*
n ontstekingsmiddel *n*
d Zündmittel *n*

2970 IGNITION gen
f allumage *m*, amorçage *m*
e encendido *m*, ignición *f*
i accensione *f*
n ontsteking
d Zündung *f*

2971 IGNITION MIXTURE exp
f mélange *m* allumeur
e detonador *m*, mezcla *f* detonante
i mescolanza *f* accenditrice
n ontstekingsmengsel *n*
d Zündmasse *f*

2972 IGNITION POINT, KINDLING POINT fue,gas
f point *m* d'inflammation
e punto *m* de inflamación
i punto *m* d'infiammazione
n ontvlammingspunt *n*
d Entflammungspunkt *m*

2973 IGNITION RESIDUE, FIXED RESIDUE fue
f résidu *m* calciné, résidu *m* fixe après calcination
e residuo *m* de calcinación
i residuo *m* calcinato, residuo *m* di calcinazione, residuo *m* fisso
n calcinatieresidu *n*, gloeirest
d Glührückstand *m*

2974 IGNITION TEMPERATURE exp
f température *f* d'explosion
e temperatura *f* de explosión
i temperatura *f* d'esplosione
n explosietemperatuur
d Explosionstemperatur *f*

2975 IGNITION TEMPERATURE, KINDLING TEMPERATURE fue,gas
f température *f* d'inflammation
e temperatura *f* de inflamación
i temperatura *f* d'infiammazione
n ontvlammingstemperatuur
d Entflammungstemperatur *f*

2976 IHRIGIZING met
f imprégnation *f* au silicium
e silicionación *f*
i impregnazione *f* con silicio
n impregneren *n* met silicium
d Imprägnieren *n* mit Silizium

2977 ILLUMINATING GAS, MANUFACTURED GAS, TOWN GAS gas
f gaz *m* d'éclairage, gaz *m* de ville
e gas *m* de ciudad, gas *m* del alumbrado
i gas *m* di città, gas *m* illuminante
n lichtgas *n*, stadsgas *n*
d Leuchtgas *n*, Stadtgas *n*

2978 IMBIBITION, SOAKING — pho
f imbibition *f*
e imbibición *f*
i imbibizione *f*
n doordrenken *n*, drenken *n*
d Durchtränken *n*, Imbibierung *f*, Tränken *n*

2979 IMBIBITION PRINTING — pho
f tirage *m* par imbibition
e positivado *m* por imbibición
i tiraggio *m* per imbibizione
n imbibitiedruk
d Einsaugekopieren *n*

2980 IMITATION LEATHER, ARTIFICIAL LEATHER, LEATHERETTE — lea
f cuir *m* artificiel
e cuero *m* artificial, similcuero *m*
i cuoio *m* artificiale, dermoide *f*
n kunstleer *n*
d Kunstleder *n*

2981 IMITATION PARCHMENT — pap
f papier *m* parcheminé, papier *m* pergamine
e papel *m* pergamino
i carta *f* pergamena d'imitazione
n perkamentpapier *n*
d Pergamentersatz *m*, Pergamin *n*

2982 IMMERSION BATH — gen
f bain *m* à immersion
e baño *m* de inmersión
i bagno *m* ad immersione
n dompelbad *n*
d Tauchbad *n*

2983 IMMERSION COATING, DIP COATING — cer,pla
f couche *f* au trempé
e capa *f* por inmersión
i rivestimento *m* per immersione
n deklaag verkregen door indompelen
d Tauchüberzug *m*

2984 IMMERSION COATING, DIP COATING — dec
f vernissage *m* au trempé
e barnizado *m* por inmersión
i verniciatura *f* per immersione
n dompellakken *n*
d Tauchlackierung *f*

2985 IMMERSION OIL, CLEARING AGENT (GB), CLEARING FLUID (US) — ana
f huile *f* d'immersion
e aceite *m* de inmersión
i olio *m* d'immersione
n immersieolie
d Immersionsöl *n*

2986 IMMERSION PLATING, DIP PLATING — coa
f dépôt *m* au trempé
e depósito *m* por inmersión
i deposito *m* per immersione
n chemisch neerslaan *n*
d Ansieden *n*, Eintauchplattierung *f*

2987 IMMISCIBILITY — mix
f immiscibilité *f*
e inmiscibilidad *f*
i immiscibilità *f*
n onvermengbaarheid
d Nichtmischbarkeit *f*, Unvermischbarkeit *f*

2988 IMMISCIBLE — mix
f immiscible adj
e inmiscible adj
i immiscibile adj
n onvermengbaar adj
d nicht mischbar adj, unvermischbar adj

2989 IMPACT ADHESIVE, CONTACT ADHESIVE — glu
f adhésif *m* à action directe
e adhesivo *m* de acción directa
i adesivo *m* ad azione diretta
n direct hechtende kleefstof
d direkt haftendes Klebemittel *n*

2990 IMPACT CLEANING — cle
f nettoyage *m* à jet de sable
e limpieza *f* por chorro de arena
i pulitura *f* a getto di sabbia
n reinigen *n* met zandstraal
d Putzen *n* mit Sandstrahl

2991 IMPACT DRYING — dry
f séchage *m* à chocs d'air chaud
e secado *m* por choques de aire caliente
i essiccazione *f* per urti d'aria calda
n drogen *n* met warme-luchtstoten
d Trocknen *n* mittels Heissluftanpr

2992 IMPACT MOLDING (US), IMPACT MOULDING (GB) — pla
f moulage *m* à chocs, moulage *m* à frappes
e moldeo *m* por impacto
i stampatura *f* a percussione

n slagpersen *n*
d Schlagpressen *n*

2993 IMPACT TEST tes
f essai *m* au choc
e ensayo *m* de resistencia al choque
i collaudo *m* d'urto
n slagproef
d Schlagprobe *f*

2994 IMPASTATION, IMPASTO cer,dec
f empâtement *m*
e empaste *m*
i impasto *m*
n dikke verflaag
d dicker Farbauftrag *m*, Impastierung *f*

2995 IMPERMEABILITY gen
f imperméabilité
e impermeabilidad *f*
i impermeabilità *f*
n ondoorlatendheid
d Undurchdringlichkeit *f*, Undurchlässigkeit *f*

2996 IMPERMEABLE gen
f imperméable adj
e impermeable adj
i impermeabile adj, stagno adj
n ondoorlatend adj
d undurchlässig adj

2997 IMPERVEOUS gen
f impénétrable adj
e impenetrable adj
i impenetrabile adj
n ondoordringbaar adj
d undurchdringlich adj

2998 IMPERVEOUSNESS gen
f impénétrabilité *f*
e impenetrabilidad *f*
i impenetrabilità *f*
n ondoordringbaarheid
d Undurchdringlichkeit *f*

2999 IMPINGEMENT BLACK, SOOT BLACK dec
f noir *m* de combustion, pigment *m* noir déposé
e negro *m* de combustión, tizna *f*
i nero *m* di combustione, nerofumo *m*
n roetzwart *n*, zwartsel *n*
d Farbruss *m*, Russschwarz *n*

3000 IMPOVERISHMENT min,nue
f appauvrissement *m*
e emprobecimiento *m*
i impoverimento *m*
n verarming
d Verarmung *f*

3001 IMPREGNATED CARBON elc
f charbon *m* panchromatique
e carbón *m* mineralizado
i carbone *m* a fiamma
n effectkool
d Effektkohle *f*, Flammenkohle *f*

3002 IMPREGNATING imp
f imprégnation *f*
e impregnación *f*
i impregnazione *f*
n drenken *n*, impregneren *n*
d Imprägnieren *n*, Tränken *n*

3003 IMPREGNATING VARNISH, PENETRATING VARNISH dec
f vernis *m* d'imprégnation
e barniz *m* de impregnación
i vernice *f* d'impregnazione
n impregneerlak
d Imprägnierlack *m*

3004 IMPRESSION pri
f empreinte *f*
e impresión *f*
i impressione *f*
n afdruk
d Abdruck *m*, Prägung *f*

3005 IMPROVER, WHITENING AGENT nut
f agent *m* blanchissant de farine
e blanqueador *m* de harina
i imbiancatore *m* di farina
n meelbleekmiddel *n*
d Mehlbleichmittel *n*

3006 IMPURE gen
f impur adj
e impuro adj
i impuro adj
n onzuiver adj, verontreinigd adj
d unrein adj, verunreinigt adj

3007 IMPURITY gen
f impureté *f*
e impuredad *f*
i impurità *f*
n onzuiverheid, verontreiniging
d Fremdkörper *m*, Unreinheit *f*, Verunreinigung *f*

3008 IN BATCHES gen
f en lots
e en lotes
i in lotti

n in porties
d satzweise

3009 IN SITE CONCRETE, CAST CONCRETE, ON SITE CONCRETE — bui
f béton *m* coulé
e hormigón *m* colado
i calcestruzzo *m* colato
n gestort beton *n*, gietbeton *n*
d Gussbeton *m*, Schüttbeton *m*

3010 INACTIVATION — gen
f rendre v inactif
e inactivación *f*
i inattivazione *f*
n inactiveren *n*
d Inaktivierung *f*

3011 INACTIVE — gen
f inactif adj
e inactivo adj
i inattivo adj
n onwerkzaam adj
d unwirksam adj

3012 INACTIVE FILLER, INERT FILLER — dec
f charge *f* inactive, substance *f* de remplissage inactive
e carga *f* inactiva, substancia *f* de relleno inactiva
i carica *f* inattiva, materiale *m* di carica inattiva, riempitivo *m* inattivo
n inactieve vulstof
d inaktiver Füllstoff *m*

3013 INCINERATION — lab
f incinération *f*
e incineración *f*
i incenerazione *f*
n verassing
d Veräschung *f*

3014 INCINERATION, DESTRUCTION — hyg,iwt
f destruction *f*, incinération *f*
e destrucción *f*, incineración *f*
i incenerazione *f*
n destructie, verbranding
d Verbrennung *f*

3015 INCLUSION — fou,met,pla
f inclusion *f*
e inclusión *f*
i inclusione *f*
n insluitsel *n*
d Einschluss *m*

3016 INCOMBUSTIBLE — fue
f incombustible adj
e incombustible adj
i incombustibile adj
n onbrandbaar adj, onverbrandbaar adj
d unbrennbar adj, unverbrennbar adj

3017 INCOMPLETE REACTION, BALANCED REACTION — gen
f réaction *f* incomplète, réaction *f* réversible
e reacción *f* incompleta
i reazione *f* bilanciata
n evenwichtsreactie
d reversible Reaktion *f*, unvollständige Reaktion *f*

3018 INCRUSTATION, SCALE FORMATION — cor
f entartrage *m*, incrustation *f*
e incrustación *f*
i incrostazione *f*
n ketelsteenvorming
d Kesselsteinbildung *f*, Wassersteinansatz *m*

3019 INCUBATION — bac
f incubation *f*
e incubación *f*
i incubazione *f*
n incubatie
d Inkubation *f*

3020 INDIA PAPER, BIBLE PAPER — pap
f papier *m* Bible, papier *m* indien
e papel *m* de Biblia
i carta *f* Cina
n bijbelpapier *n*, dundrukpapier *n*
d Bibeldruckpapier *n*, Chinapapier *n*, Dünndruckpapier

3021 INDIA RUBBER, CAOUTCHOUC, ELASTIC GUM — rub
f caoutchouc *m*
e caucho *m*
i caucciù *m*
n caoutchouc *n*
d Gummielasticum *n*, Kautschuk *m* Rohgummi *m*

3022 INDIAN INK — dec
f encre *f* de Chine, noir *m* de Chi
e tinta *f* china
i inchiostro *m* di Cina
n Oostindische inkt
d Ausziehtusche *f*, Chinesische Tusche *f*

3023 INDICATOR — lab
f indicateur *m*

e indicador *m*
i indicatore *m*
n indicator
d Indikator *m*

3024 INDICATOR, iso
TRACER
f indicateur *m*, traceur *m*
e indicador *m*, trazador *m*
i tracciante *m*
n indicator, tracer
d Indikator *m*, Leitisotop *n*, Tracer *m*

3025 INDICATOR EXPONENT lab
f exponent *m* indicateur
e exponente *m* indicador
i esponente *m* indicatore
n omslagpunt *n*
d Umschlagspunkt *m*

3026 INDICATOR PAPER lab
f papier *m* d'indicateur
e papel *m* de indicador
i carta *f* d'indicatore
n indicatorpapier *n*
d Indikatorpapier *n*

3027 INDICATOR RANGE lab
f gamme *f* de l'indicateur
e gama *f* del indicador
i gamma *f* dell'indicatore
n indicatorgebied *n*
d Indikatorbereich *m*

3028 INDIFFERENT GAS gas
f gaz *m* indifférent
e gas *m* indiferente
i gas *m* indifferente
n indifferent gas *n*
d indifferentes Gas *n*

3029 INDIRECT COOLING, clo
SURFACE COOLING
f refroidissement *m* superficiel
e enfriamiento *m* superficial
i raffreddamento *m* superficiale
n oppervlakkoeling
d Oberflächenkühlung *f*

3030 INDIRECT ZINC OXIDE PROCESS, met
FRENCH ZINC OXIDE PROCESS
f procédé *m* indirect d'oxyde de zinc
e procedimiento *m* indirecto de óxido de cinc
i processo *m* indiretto d'ossido di zinco
n indirect zinkoxydeproces *n*
d indirektes Zinkoxydverfahren *n*

3031 INDISSOLUBLE, gen
INSOLUBLE
f insoluble adj
e insoluble adj
i insolubile adj
n onoplosbaar adj
d unlöslich adj

3032 INDOOR PAINT, dec
INTERIOR PAINT
f peinture *f* intérieure
e pintura *f* interior
i pittura *f* per interno
n binnenverf
d Innenfarbe *f*

3033 INDUCED DRAUGHT (GB), fue
FORCED DRAFT (US)
f tirage *m* forcé
e tiro *m* forzado
i tiraggio *m* forzato
n gedwongen trek
d künstlicher Zug *m*, Unterwind *m*

3034 INDUCED DRAUGHT COOLING (GB), clo
FORCED DRAFT COOLING (US)
f refroidissement *m* à tirage forcé
e enfriamiento *m* de tiro forzado
i raffreddamento *m* a tiraggio forzato
n koeling met kunstmatige trek
d Kühlung *f* mit künstlichem Zug

3035 INDUCED REACTION gen
f réaction *f* induite
e reacción *f* inducida
i reazione *f* indotta
n geïnduceerde reactie
d induzierte Reaktion *f*

3036 INDUCTION BRAZING wel
f soudure *f* forte à induction
e soldadura *f* fuerte por corriente inducida
i brasatura *f* ad induzione
n inductief hardsolderen *n*
d induktives Hartlöten *n*

3037 INDURATED gla,met
f durci adj
e endurecido adj, indurado adj
i indurito adj
n gehard adj
d gehärtet adj

3038 INERT COLOR (US), dec
INERT COLOUR (GB), INERT PIGMENT
f pigment *m* chimiquement inerte

e pigmento *m* inactivo
i pigmento *m* chimicamente inerte
n inert pigment *r*
d inaktives Pigment *n*, indifferentes Pigment *n*, inertes Pigment *n*

3039 INERT FILLER, INACTIVE FILLER — dec
f charge *f* inactive, substance *f* de remplissage inactive
e carga *f* inactiva, substancia *f* de relleno inactiva
i carica *f* inattiva, materiale *m* di carica inattiva, riempitivo *m* inattivo
n inactieve vulstof
d inaktiver Füllstoff *m*

3040 INERT GAS, NOBLE GAS, RARE GAS — gas
f gaz *m* inerte, gaz *m* noble
e gas *m* inerte, gas *m* noble
i gas *m* inerte, gas *m* nobile
n edelgas *n*, inert gas *n*
d Edelgas *n*, inertes Gas *n*

3041 INFILTRATION — sew
f infiltration *f*
e infiltración *f*
i infiltrazione *f*
n infiltratie, insijpelen *n*
d Einsickern *n*, Infiltration *f*

3042 INFINITE REFLUX — dis
f reflux *m* infini
e reflujo *m* infinito
i riflusso *m* infinito
n totale terugloop
d totaler Rücklauf *m*

3043 INFLAMMABILITY — fue,gas
f inflammabilité *f*
e inflamabilidad *f*
i infiammabilità *f*
n ontvlambaarheid
d Entflammbarkeit *f*, Feuergefährlichkeit *f*

3044 INFLAMMABLE — fue,gas
f inflammable adj
e inflamable adj
i infiammabile adj
n ontvlambaar adj, vuurgevaarlijk adj
d entflammbar adj, feuergefährlich adj

3045 INFLATING AGENT, BLOWING AGENT, SPONGING AGENT — rub
f agent *m* gonflant
e agente *m* esponjante, agente *m* hinchante
i agente *m* rigonfiante
n blaasmiddel *n*
d Blähmittel *n*, Treibmittel *n*

3046 INFRARED DRYING — dry
f séchage *m* à l'infrarouge
e secado *m* por rayos infrarrojos
i essiccazione *f* per raggi infrarossi
n drogen *n* met infrarode stralen
d Infrarottrocknung *f*

3047 INFRARED EMULSION — pho
f émulsion *f* sensible à l'infraroug[e]
e emulsión *f* sensible al infrarroj[o]
i emulsione *f* sensibile all'infrar[osso]
n voor infrarood gevoelige emulsie
d infrarotempfindliche Emulsion *f*

3048 INFUSE (TO) — sin
f infuser v
e infundir v
i infondere v, fare v una infusione
n aftrekken v, opgieten v
d aufgiessen v, einweichen v, ziehen lassen v

3049 INFUSIBLE — gen
f infusible adj
e infusible adj
i infusibile adj
n niet-smeltbaar adj, onsmeltbaar adj
d nicht schmelzbar adj, unschmelzbar adj

3050 INFUSION — mcy
f infusion *f*
e infusión *f*
i infusione *f*
n aftreksel *n*, infusie, opgietsel *n*
d Aufguss *m*, Infusion *f*

3051 INFUSION METHOD — mcy
f méthode *f* d'infusion
e método *m* por infusión
i metodo *m* d'infusione
n infusiemethode
d Aufgussverfahren *n*, Infusionsverfahren *n*

3052 INFUSORIAL EARTH, DIATOM EARTH, DIATOMITE, KIESELGUHR — mat
f farine *f* fossile, tripoli *m*
e harina *f* fósil, trípoli *m*
i diatomite *m*, farina *f* fossile, tripolo *m*

n diatomeeënaarde, diatomiet *n*, infusoriënaarde, kiezelgoer *n*
d Diatomeenerde *f*, Infusorienerde *f*, Kieselgur *m*, Tripel *m*

3053 INGREDIENT gen
f ingrédient *m*
e ingrediente *m*
i ingrediente *m*
n bestanddeel *n*, ingrediënt *n*
d Bestandteil *m*, Ingredienz *n*

3054 INHIBITOR gen
f inhibiteur *m*
e inhibidor *m*
i inibitore *m*
n inhibitor, remstof, vertrager
d Hemmstoff *m*, Inhibitor *m*

3055 INHIBITORY PHASE col
f phase *f* d'inhibition
e fase *f* de inhibición
i fase *f* d'inibizione
n beschuttingsfaze
d Inhibitionsphase *f*

3056 INITIAL BOILING POINT gen
f point *m* d'ébullition initial
e punto *m* de ebullición inicial
i punto *m* d'ebollizione iniziale
n beginkookpunt *n*
d Anfangssiedepunkt *m*, Siedebeginn *m*

3057 INITIAL TEMPERATURE, STRIKING HEAT bre
f degré *m* initial, température *f* d'empâtage
e temperatura *f* inicial de mezcla
i temperatura *f* di miscela
n begintemperatuur van het beslag
d Einmaischtemperatur *f*

3058 INITIATOR gen
f molécule *f* initiatrice, substance *f* initiatrice
e molécula *f* iniciadora, substancia *f* iniciadora
i molecola *f* iniziatrice, sostanza *f* iniciatrice
n uitgangsmolecule, uitgangsstof
d Ausgangsmaterie *f*, Ausgangsmolekül *n*

3059 INJECTION pla
f injection *f*
e inyección *f*
i iniezione *f*
n spuitgieten *n*
d Spritzguss *m*

3060 INJECTION BLOW MOLDING (US), INJECTION BLOW MOULDING (GB) pla,pre
f moulage *m* par injection à soufflage
e moldeo *m* por inyección de soplado
i stampatura *f* per iniezione a soffiatura
n spuitgieten *n* met blaasmethode
d Spritzguss *m* mit Blasverfahren

3061 INJECTION MOLDING (US), INJECTION MOULDING (GB) pla
f moulage *m* à injection
e moldeo *m* por inyección
i stampatura *f* per iniezione
n spuitgieten *n*
d Spritzguss *m*

3062 INK dec
f encre *f*
e tinta *f*
i inchiostro *m*
n inkt
d Tinte *f*

3063 INNOCUOUS, INNOXIOUS gen
f innocent adj, inoffensif adj
e inocente adj, innocuo adj
i innocuo adj
n onschadelijk adj
d harmlos adj, unschädlich adj

3064 INOCULATION cry
f inoculation *f*
e inoculación *f*
i inoculazione *f*
n enten *n*, oculatie
d Impfung *f*, Okulation *f*

3065 INODOROUS gen
f inodore adj
e inodoro adj
i inodoro adj
n reukloos adj
d geruchlos adj

3066 INORGANIC CHEMISTRY aic,gen
f chimie *f* minérale
e química *f* inorgánica
i chimica *f* inorganica
n anorganische chemie
d anorganische Chemie *f*

3067 INSECTICIDAL PAINT dec
f peinture *f* insecticide
e pintura *f* insecticida
i pittura *f* insetticida
n insektendodende verf
d insektentötende Farbe *f*

3068 INSECTICIDE agr,hyg
f insecticide *m*
e insecticida *m*
i insetticida *m*
n insektendodend middel *n*, insekticide *n*
d Insektizid *n*, Schädlingsbekämpfungsmittel *n*

3069 INSOLUBLE, INDISSOLUBLE gen
f insoluble adj
e insoluble adj
i insolubile adj
n onoplosbaar adj
d unlöslich adj

3070 INSPISSATION sin
f épaississement *m*
e condensación *f*
i inspessimento *m*
n indikken *n*
d Eindicken *n*

3071 INSULATING OIL elc
f huile *f* isolante
e aceite *m* aislante
i olio *m* isolante
n isolerende olie
d Isolieröl *n*

3072 INSULATING VARNISH dec,elc
f vernis *m* isolant
e barniz *m* aislante
i vernice *f* isolante
n isolatielak
d Isolierlack *m*

3073 INSULATION gen
f isolation *f*
e aislamiento *m*
i isolamento *m*
n isolatie, isoleren *n*
d Isolation *f*, Isolieren *n*

3074 INTENSIFICATION pho
f renforcement *m*
e refuerzo *m*
i rinforzo *m*
n versterking
d Verstärkung *f*

3075 INTENSIFYING LIQUID pho
f liquide *m* de renforcement
e líquido *m* reforzador
i liquido *m* di rinforzo
n versterkvloeistof
d Verstärkungsflüssigkeit *f*

3076 INTERACTION gen
f interaction *f*
e interacción *f*
i azione *f* di scambio
n wisselwerking
d Wechselwirkung *f*

3077 INTERCRYSTALLINE CORROSION, INTERGRANULAR CORROSION cor
f corrosion *f* fissurante, corrosion *f* intercristalline
e corrosión *f* intercristalina
i corrosione *f* intercristallina
n corrosie aan de kristalgrenzen, interkristallijne corrosie
d interkristalline Korrosion *f*, Korngrenzenbruch *m*

3078 INTERIOR PAINT, INDOOR PAINT dec
f peinture *f* intérieure
e pintura *f* interior
i pittura *f* per interno
n binnenverf
d Innenfarbe *f*

3079 INTERMEDIATE ASBESTOS LAYER, ASBESTOS PLY bui
f couche *f* d'amiante
e capa *f* intermedia de amianto
i strato *m* intermedio d'amianto
n asbesttussenlaag
d Asbesteinlage *f*

3080 INTERMEDIATE COATING, OVERCOATING, UNDERCOATI[NG] dec
f main *f* intermédiaire, peinture de sous-couche
e mano *f* intermedia
i mano *f* intermedia
n overgrondverf, voorlak
d Voranstrich *m*, Zwischenschicht

3081 INTERMEDIATE FRACTION dis
f fraction *f* intermédiaire
e fracción *f* intermedia
i frazione *f* intermedia
n middenfractie
d Zwischenfraktion *f*

3082 INTERMEDIATE PRODUCT ge[n]
f produit *m* intermédiaire
e producto *m* intermedio
i prodotto *m* intermedio

n tussenprodukt *n*
d Zwischenprodukt *n*

3083 INTERMITTENT CHLORINATION hyg
f chloration *f* intermittente
e cloración *f* intermitente
i clorazione *f* intermittente
n discontinue chlorering
d Stosschlorung *f*

3084 INTERMITTENT DOWNWARD FILTRATION, INTERMITTENT SOIL FILTRATION sew
f filtration *f* intermittente par le sol
e filtración *f* intermitente a través del suelo
i filtrazione *f* intermittente attraverso il suolo
n intermitterende bodemfiltratie
d intermittierende Bodenfilterung *f*, Stauberieselung *f*, Staufiltration *f*

3085 INTERNAL COMBUSTION fue
f combustion *f* interne
e combustión *f* interna
i combustione *f* interna
n inwendige verbranding
d Verbrennung *f* im Zylinder

3086 INTERNAL INDICATOR lab
f indicateur *m* incorporé dans une solution
e indicador *m* interno
i indicatore *m* interno
n in een oplossing aanwezige indicator
d in einer Lösung befindlicher Indikator *m*

3087 INTUMESCENT PAINT dec
f peinture *f* ignifuge
e pintura *f* ignífuga
i pittura *f* ignifuga
n brandwerende verf
d aufblähende Farbe *f*

3088 INVASION gas
f solution *f* d'un gaz dans un liquide
e solución *f* de un gas en un líquido
i soluzione *f* d'un gas in un liquido
n gasoplossing in een vloeistof
d Gaslösung *f* in einer Flüssigkeit

3089 INVERSION sin
f inversion *f*
e inversión *f*
i inversione *f*
n inversie
d Inversion *f*

3090 INVERT SUGAR sin
f sucre *m* inverti
e azúcar *m* invertido
i zucchero *m* invertito
n invertsuiker
d Invertzucker *m*

3091 INVERTASE (GB), INVERTIN (US), SACCHARASE, SUCRASE sin
f invertase *f*
e invertasa *f*
i invertasi *f*
n invertase, saccharase
d Invertase *f*, Invertin *n*, Saccharase *f*

3092 INVESTMENT MOULDING (GB), LOST-WAX PROCESS (US), PRECISION MOULDING (GB) fou
f moulage *m* à la cire perdue
e moldeo *m* a la cera perdida
i formatura *f* a cera spersa
n vormen *n* volgens de verloren-wasmethode
d Modellausschmelzverfahren *n*

3093 ION EXCHANGE elc
f échange *m* d'ions
e cambio *m* de iones
i scambio *m* d'ioni
n ionenuitwisseling
d Ionenaustausch *m*

3094 ION EXCHANGE CHROMATOGRAPHY ana
f chromatographie *f* à échange d'ions
e cromatografía *f* de cambio de iones
i cromatografia *f* a scambio d'ioni
n chromatografie met ionen-uitwisseling
d Chromatographie *f* mit Ionenaustausch

3095 ION EXCHANGE RESINS pla
f résines *pl* à échange d'ions
e resinas *pl* de cambio de iones
i resine *pl* a scambio d'ioni
n harsen *pl* met ionenuitwisseling
d Ionenaustauschharze *pl*

3096 IONIC MIGRATION elc
f migration *f* d'ions

e migración *f* de iones
i migrazione *f* d'ioni
n ionenzwerving
d Ionenwanderung *f*

3097 IONIC MOBILITY elc
f mobilité *f* d'ions
e movilidad *f* de iones
i mobilità *f* d'ioni
n ionenbeweeglijkheid
d Ionenbeweglichkeit *f*

3098 IONIC PRODUCT, SOLUBILITY PRODUCT elc
f produit *m* de solubilité, produit *m* ionique
e producto *m* de solubilidad, producto *m* iónico
i prodotto *m* di solubilità, prodotto *m* ionico
n ionisch produkt *n*, oplosbaarheidsprodukt *n*
d Ionenprodukt *n*, Löslichkeitsprodukt *n*

3099 IONIC STRENGTH elc
f force *f* ionique
e fuerza *f* iónica
i forza *f* ionica
n ionensterkte
d Ionenkraft *f*

3100 IONIZATION elc
f ionisation *f*
e ionización *f*
i ionizzazione *f*
n ionisatie, ioniseren *n*
d Ionisation *f*, Ionisieren *n*

3101 IONOGENIC elc
f ionogène adj
e ionogéneo adj
i ionogeneo adj
n ionogeen adj
d ionogen adj

3102 IRIS OIL fat
f essence *f* d'iris
e esencia *f* de íride
i olio *m* di giaggiuolo
n irisolie
d Irisöl *n*

3103 IRON ACETATE LIQUOR, BLACK LIQUOR, BLACK MORDANT pap
f solution *f* d'acétate de fer
e mordiente *m* para negro, solución *f* de acetato de hierro
i soluzione *f* d'acetato di ferro
n ijzeracetaatoplossing
d Eisenazetatlösung *f*

3104 IRON PLATING, ACIERAGE, STEELING coa
f aciérage *m*
e aceración *f*, hierrado *m*
i acciaiatura *f*
n verijzering, verstaling
d Stählung *f*, Verstählung *f*

3105 IRON REMOVAL sew
f déferrisation *f*
e desferrización *f*
i eliminazione *f* di ferro
n ontijzering
d Enteisenung *f*

3106 IRON-STEAM PROCESS gas
f fabrication *f* d'hydrogène par le procédé vapeur-fer
e fabricación *f* de hidrógeno con el procedimiento vapor-hierro
i fabbricazione *f* d'idrogeno con il processo vapore-ferro
n waterstofbereiding volgens het ijzer-stoomproces
d Wasserstofferzeugung *f* nach dem Eisen-Dampf-Verfahren

3107 IRON TURBIDITY bre
f trouble *m* de fer
e enturbiamiento *m* de hierro
i intorbidamento *m* per ferro
n ijzervertroebeling
d Eisentrübung *f*

3108 IRREVERSIBLE COLLOID, LYOPHOBIC COLLOID col
f colloïde *m* irréversible, colloïde *m* lyophobe
e coloide *m* irreversible, coloide *m* liófobo
i colloide *m* irreversibile, colloide *m* liofobo
n irreversibel colloïde *n*, lyofoob colloïde *n*
d irreversibeles Kolloid *n*, lyophobes Kolloid *n*

3109 IRREVERSIBLE REACTION, COMPLETE REACTION ger
f réaction *f* complète, réaction *f* irréversible
e reacción *f* completa, reacción *f* irreversible
i reazione *f* completa, reazione *f* irreversibile
n niet-omkeerbare reactie, volledige reactie

d nicht umkehrbare Reaktion *f*, vollständige Reaktion *f*

3110 IRRIGATION sew
f irrigation *f*
e irrigación *f*
i irrigazione *f*
n bevloeiing, irrigatie
d Bewasserung *f*, Irrigation *f*, Wasserberieselung *f*

3111 IRRIGATION COOLING clo
f refroidissement *m* à ruissellement
e enfriamiento *m* de riego
i raffreddamento *m* a pioggia
n sproeikoeling
d Berieselungskühlung *f*

3112 ISINGLASS, FISH GLUE glu
f colle *f* de poisson, ichtyocolle *f*
e cola *f* de pescado
i colla *f* di pesce
n vislijm
d Fischleim *m*, Hausenblase *f*

3113 ISOAMYLALCOHOL, FERMENTATION AMYLALCOHOL oic
f alcool *m* isoamylique
e alcohol *m* isoamílico
i alcool *m* isoamilico
n gistingsamylalcohol
d Gärungsamylalkohol *m*, Isoamylalkohol *m*

3114 ISOLATION bac
f isolation *f*
e aislamiento *m*
i isolamento *m*
n afzondering, isolatie
d Absonderung *f*

3115 ISOTOPE iso
f isotope *m*
e isótopo *m*
i isotopo *m*
n isotoop *n*
d Isotop *n*

3116 ISOTOPIC DILUTION ANALYSIS ana,iso
f analyse *f* par dilution isotopique
e análisis *f* por dilución isotópica
n analisi *f* per diluzione isotopica
n analyse met isotopenverdunning
d Isotopenverdünnungsanalyse *f*

3117 ITALIAN PINK, CHROME-TIN PINK, DUTCH PINK, ENGLISH PINK dec
f pigment *m* jaune transparent
e pigmento *m* amarillo transparente
i pigmento *m* giallo trasparente
n gele lakverf
d gelbe Lackfarbe *f*, Schüttgelb *n*

3118 IVORY BLACK (GB), ABAISER (US) dec
f noir *m* d'ivoire
e negro *m* de marfil
i nero *m* d'avorio
n ivoorzwart *n*
d Elfenbeinschwarz *n*

3119 IVORY PAPER pap
f papier *m* ivoire
e papel *m* marfil
i carta *f* avorio
n ivoorpapier *n*
d Elfenbeinpapier

J

3120 JACKET COOLING clo
f refroidissement *m* à chemise
e enfriamiento *m* por chaquete
i raffreddamento *m* a camicia
n mantelkoeling
d Mantelkühlung *f*

3121 JACKETED clo
f à double paroi
e de doble pared
i a doppia parete
n dubbelwandig adj
d doppelwandig adj

3122 JAM nut
f confiture *f*
e confitura *f*
i conserva *f* di frutta, marmellata *f*
n jam
d Konfitüre *f*

3123 JAPAN dec
f laque *f* de Chine, laque *f* du Japon
e laca *f* de China, laca *f* del Japón
i lacca *f* del Giappone
n japanlak
d Japanlack *m*

3124 JAPAN TALLOW, JAPAN WAX, SUMACH WAX fat
f cire *f* du Japon
e cera *f* del Japón
i cera *f* del Giappone
n Japanse was
d Japanwachs *n*, Japantalg *m*

3125 JAPANESE PAPER pap
f papier *m* du Japon
e papel *m* del Japón
i carta *f* giapponese
n japanpapier *n*
d Japanpapier *n*

3126 JAPANESE TISSUE pap,pho
f papier *m* d'empreinte japonais
e papel *m* de impresa del Japón
i carta *f* da stampa giapponese
n ongelijmd japanpapier *n*
d ungeleimtes Japanpapier *n*

3127 JAPANESE VELLUM, MULBERRY PAPER pap
f papier *m* d'écorce de mûrier
e papel *m* de corteza de moral
i cartapecora *f* giapponese
n Chinees papier *n*
d chinesisches Papier *n*, Maulbeerbaumpapier *n*

3128 JAPANNED LEATHER, ENAMELLED LEATHER, PATENT LEATHER coa,lea
f cuir *m* verni
e charol *m*
i pelle *f* di lustrino, pelle *f* verniciata
n lakleder *n*
d Glanzleder *n*, Lackleder *n*

3129 JAPANNING dec
f vernissage *m* à la laque de Chine vernissage *m* à la laque du Japon
e barnizado *m* con laca de China, barnizado *m* con laca del Japón
i verniciatura *f* con lacca del Giappone
n met japanlak vernissen *n*
d Überziehen *n* mit Japanlack

3130 JASMINE OIL, JESSAMINE OIL fat
f essence *f* de jasmin
e esencia *f* de jazmín
i olio *m* di gelsomino
n jasmijnolie
d Jasminöl *n*

3131 JASPER cer
f jaspe *m*
e jaspe *m*
i diaspro *m*
n jaspis
d Jasper *m*, Jaspis *m*

3132 JASPER PORCELAIN, JASPERATED PORCELAIN cer
f porcelanite *f*
e porcelana *f* jaspeada
i porcellana *f* a diaspro
n jaspisporselein *n*
d Jasperware *f*, Jaspisporzellan *n*

3133 JAVEL WATER, JAVELLE WATER ble
f eau *f* de Javel, eau *f* de Javelle
e agua *f* de Javel
i acqua *f* di Javel

n eau de Javelle
d Eau de Javelle

3134 JELLY PAINT, dec
GEL PAINT, THIXOTROPIC PAINT
f peinture *f* thixotrope
e pintura *f* tixotrópica
i pittura *f* tixotropica
n thixotrope verf
d thixotrope Farbe *f*

3135 JET CONDENSATION clo
f condensation *f* par injection
e condensación *f* de inyección
i condensazione *f* ad iniezione
n condensatie door waterstraal
d Einspritzkondensation *f*

3136 JET DEVELOPMENT, pho
SPRAY DEVELOPMENT
f développement *m* sous jet
e revelado *m* de chorro
i sviluppo *m* con getto
n ontwikkelen *n* met sproeien
d Spritzentwicklung *f*

3137 JET DRYING dry
f séchage *m* à jet d'air
e secado *m* con chorro de aire
i essiccazione *f* a getto d'aria
n drogen *n* met luchtstraal
d Trocknen *n* mit Luftanprall

3138 JET FLAME htt
f dard *m* de flamme, flamme *f* coupante
e llama *f* acuta
i fiamma *f* a dardo, fiamma *f* acuta
n steekvlam
d Stichflamme *f*

3139 JET MOLDING (US), pla
JET MOULDING (GB)
f boudinage *m*
e budinado *m*
i trafila *f*
n strangpersen *n*
d Strangpressen *n*

3140 JET RINSING pho
f lavage *m* sous jet, rinçage *m* sous jet
e enjuague *m* sobre chorro
i risciacquatura *f* a getto
n uitwassen *n* met sproeier
d Spritzwässerung *f*

3141 JEWELLER'S ROUGE, pol
POLISHING CROCUS, POLISHING ROUGE, ROUGE
f rouge *m* à polir, rouge *m* d'Angleterre, rouge *m* de fer
e óxido *m* férrico, rojo *m* de Inglaterra, rojo *m* de pulir
i ossido *m* ferrico, rosso *m* inglese
n Berlijns rood *n*, Engels rood *n*, gele bolus, polijstrood *n*
d Englischrot *n*, Pariserrot *n*, Polierrot *n*

3142 JIG DYEING dye
f teinture *f* en jigger
e teñido *m* en jigger
i tintura *f* al largo
n verven *n* met de breedverfmachine
d Färbung *f* mit der Breitfärbemaschine

3143 JIG WELDING pla,wel
f soudure *f* sur gabarit
e soldadura *f* sobre plantilla
i saldatura *f* a taglio di contorno
n contourlassen *n*
d Umrissschweissen *n*

3144 JIGGING sor
f triage *m* à crible vibrant
e clasificación *f* con criba vibradora
i classificazione *f* a crivello oscillante
n sorteren *n* met deintoestel
d Sortieren *n* mit Siebsetzmaschine

3145 JOINER'S CEMENT glu
f colle *f* du menuisier
e cola *f* del carpintero, cola *f* fuerte
i colla *f* del falegname
n timmermanslijm
d Leimkitt *m*

3146 JOINER'S GLUE, glu
CARPENTER'S GLUE, HIDE GLUE
f colle *f* forte
e cola *f* de carnaza, cola *f* de piel
i colla *f* forte
n houtlijm
d Hautleim *m*

3147 JOINTER, bui
JOINTING COMPOUND, JOINTING MATERIAL
f matière *f* d'étoupage
e mástique *m* para juntas
i mastice *m* per giunzioni

n dichtmiddel *n*
d Dichtungsmaterial *n*, Dichtungsmittel *n*

3148 JUCTEN, RUSSIAN LEATHER, YUFTEN lea
f cuir *m* de Russie
e cuero *m* de Rusia, vaqueta *f* de Moscovia
i bulgaro *m*, cuoio *m* di Russia
n juchtleder *n*
d Juchtenleder *n*

3149 JUDEAN PITCH, ASPHALT, BITUMEN, MINERAL PITCH bui
f asphalte *m*, poix *f* de Judée
e asfalto *m*
i asfalto *m*
n aardpek *n*, asfalt *n*, jodenpek *n*
d Asphalt *m*, Bitumen *n*, Erdpech *n*

3150 JUICE nut
f jus *m*
e jugo *m*, zumo *m*
i succhio *m*, succo *m*, sugo *m*
n sap *n*
d Saft *m*

3151 JUICE DECANTATION sin
f décantation *f* du jus
e decantación *f* del jugo
i decantazione *f* del succo
n decanteren *n* van het sap
d Saftabsetzung *f*

3152 JUICE DRAW-OFF, DRAFT IN DIFFUSION, DRAUGHT IN DIFFUSION (GB) sin
f décocté *m* de jus
e decocción *f* de jugo
i decotto *m* di succo
n sapaftreksel *n*
d Saftabzug *m*

3153 JUICE STRAINING sin
f filtration *f* du jus
e filtración *f* del jugo
i filtrazione *f* del succo
n sapfiltratie, zeven *n* van het sap
d Saftabsiebung *f*

3154 JUMPING cer
f crevassage *m*
e hendidura *f*
i screpolatura *f*
n openspringen *n*
d Aufspringen *n*

3155 JUNIPER-BERRY OIL, JUNIPER OIL, CADE OIL fat
f essence *f* de baies de genièvre, essence *f* de genièvre
e esencia *f* de enebrina
i olio *m* di ginepro
n jeneverbesolie
d Wacholderöl *n*

3156 JUNK iwt
f déchets *pl*
e desechos *pl*, desperdicios *pl*
i rottami *pl*
n afvalrommel
d Ramschware *pl*

3157 JUTE, PAUT HEMP tex
f chanvre *m* de Calcutta, jute *m*
e cáñamo *m* de Calcuta, yute *m*
i iuta *f*
n jute
d Jute *f*, Pahthanf *m*, ostindischer Hanf *m*

K

3158 KAOLIN, cer
CHINA CLAY
f kaolin *m*, terre *f* à porcelaine
e caolín *m*
i caolino *m*
n kaoline *n*, porseleinaarde
d Kaolin *n*, Porzellanerde *f*

3159 KASIKA PROCESS, dec
DIRECT PROCESS, TWIN HEAD PROCESS
f procédé *m* de pistolage à deux composants
e procedimiento *m* de chorreado con dos componentes
i processo *m* di spruzzo con due componenti
n tweecomponentenspuitproces *n*
d Zweikomponentenspritzverfahren *n*

3160 KASPINE LEATHER lea
f cuir *m* blanc lavable
e cuero *m* blanco lavable
i cuoio *m* bianco lavabile
n wit wasbaar leder *n*
d weisses waschbares Leder *n*

3161 KELP ASHES aic
f cendres *pl* de varech
e cenizas *pl* de algas marinas
i ceneri *pl* di laminaria
n kelpas, zeewieras
d Tangasche *f*

3162 KELP SODA, aic
RAW SODA
f soude *f* brute, soude *f* des varechs
e sosa *f* bruta
i soda *f* greggia
n ruwe soda
n Steinsoda *f*

3163 KERNEL OIL fat
f huile *f* de pépins
e aceite *m* de pepitas
i olio *m* di granelli
n pittenolie
d Kernöl *n*

3164 KEROSENE (US), fue
CRUDE OIL, PETROLEUM (GB)
f pétrole *m* brut
e petróleo *m* bruto
i petrolio *m* greggio
n ruwe petroleum
d Erdöl *n*, Petroleum *n*, Steinöl *n*

3165 KETONIZATION oic
f cétonisation *f*
e cetonización *f*
i cetonizzazione *f*
n ketonisatie
d Ketonisieren *n*

3166 KETOPROPANE, oic
ACETONE, DIMETHYL KETONE, PROPANONE
f acétone *f*, éther *m* pyroacétique
e acetona *f*, dimetil-cetona *f*
i acetone *m*, dimetilchetone *m*
n aceton *n*, dimethylketon *n*, propanon *n*
d Azeton *n*, Dimethylketon *n*

3167 KETTLE OXIDATION, fou,met
TOSSING
f oxydation *f* par cuillerées
e oxidación *f* por cucharadas
i ossidazione *f* per cucchiaiate
n oxydatie door lepelen
d Oxydation *f* durch Löffeln

3168 KEY COMPONENT dis
f constituant-clef *m*
e componente *m* clave
i componente *m* chiave
n sleutelcomponent
d Schlüsselkomponente *f*

3169 KIBBLING brk
f égrugeage *m*
e trituración *f*
i sgrossatura *f*
n grofmalen *n*
d Grobmahlen *n*

3170 KID LEATHER, lea
CHEVERIL
f chevrotin *m*, peau *f* de chèvre
e cabritilla *f*
i pelle *f* di capra, pelle *f* di capretto
n geiteleder *n*
d Bockleder *n*, Ziegenleder *n*

3171 KIESELGUHR, mat
DIATOM EARTH, DIATOMITE, INFUSORIAL EARTH
f farine *f* fossile, tripoli *m*

e harina *f* fósil, trípoli *m*
i diatomite *m*, farina *f* fossile, tripolo *m*
n diatomeeënaarde, diatomiet *n*, infusoriënaarde, kiezelgoer *n*
d Diatomeenerde *f*, Infusorienerde *f*, Kieselgur *m*, Tripel *m*

3172 KILLED RUBBER rub
f gomme *f* malaxée à mort
e goma *f* masticada a muerte
i gomma *f* masticata a morte
n doodgeplasticeerde rubber
d totplastizierter Gummi *m*

3173 KILLED SPIRITS, SOLDERING SOLUTION wel
f eau *f* à souder
e agua *f* de soldar
i acqua *f* per saldare
n soldeerwater *n*
d Lötwasser *n*

3174 KILLED STEEL, SOLID STEEL met
f acier *m* calmé
e acero *m* reposado
i acciaio *m* calmo
n rustig staal *n*
d beruhigter Stahl *m*, ruhiger Stahl *m*

3175 KILLING OF RUBBER, DEAD MILLING rub
f mastication *f* à mort
e masticación *f* a muerte
i masticazione *f* a morte
n doodmalen *n*, doodwalsen *n*
d Totmastizieren *n*, Totwalzen *n*

3176 KILLING PICKLE elc,pol
f bain *m* épuisé
e baño *m* disipado
i bagno *m* spento
n afgewerkt bad *n*
d ausgenütztes Bad *n*

3177 KILN-DRIED MALT, CURED MALT bre
f malt *m* touraillé
e malta *f* tostada
i malto *m* essiccato
n geroost mout *n*
d Darrmalz *n*

3178 KILN HEATING SYSTEM bre
f système *m* à touraille
e tostación *f*
i essiccazione *f*
n eesten *n*
d Darren *n*

3179 KINDLING POINT, IGNITION POINT fue,gas
f point *m* d'inflammation
e punto *m* de inflamación
i punto *m* d'infiammazione
n ontvlammingspunt *n*
d Entflammungspunkt *m*

3180 KINDLING TEMPERATURE, IGNITION TEMPERATURE fue,gas
f température *f* d'inflammation
e temperatura *f* de inflamación
i temperatura *f* d'infiammazione
n ontvlammingstemperatuur
d Entflammungstemperatur *f*

3181 KISH met
f flocons *pl* de graphite
e copos *pl* de grafito
i fiocchi *pl* di grafite, schiuma *f* di grafite
n grafietschuim *n*, grafietvlokken *pl*
d Garschaum *m*, Graphitflocken *pl*

3182 KNEADABLE mix,pla
f pétrissable adj
e amasable adj
i impastabile adj
n kneedbaar adj
d knetbar adj

3183 KNEADING mix,pla
f malaxage *m*, pétrissage *m*
e amasadura *f*
i impastatura *f*
n kneden *n*
d Kneten *n*

3184 KNIFE COATING (US), RUBBER SPREADING (GB) pap,rub
f enduisage *m* à la racle sur rouleau
e recubrimiento *m* con cuchilla sobre rodillo
i spalmatura *f* a lama su cilindro
n strijken *n* met rakel op wals
d Streichen *n* mit Rakel auf Walze

3185 KNIFING FILLER dec
f apprêt *m* pour couteau-palette
e tapaporas *m* para cuchilla
i mestica *f* da coltello
n mesplamuur
d Messerspachtel *m*, Ziehspachtel

3186 KNOCK RATING, ANTI-KNOCK VALUE, PRE-IGNITION RESISTANCE fue
f pouvoir *m* anti-détonant
e índice *m* de detonación, poder *m* antidetonante

i potere *m* antidetonante
n klopvastheid
d Klopffestigkeit *f*

3187 KRAFT PAPER pap
f papier *m* de pâte au sulphate
e papel *m* kraft
i carta *f* fatta di polpa al solfato
n bruin pakpapier *n*, kraftpapier *n*
d Hartpapier *n*, Kraftpapier *n*

3188 KRAFT PULP, SULFATE PULP (US), SULPHATE PULP (GB) pap
f pâte *f* au sulphate
e pulpa *f* de kraft, pulpa *f* de sulfato
i polpa *f* a solfato
n sulfaatpulp
d Natronzellstoff *m*, Sulfatzellstoff *m*

3189 KRAFT TURPENTINE, SULFATE TURPENTINE (US), SULPHATE TURPENTINE (GB) oic
f essence *f* de térébenthine résiduaire, essence *f* de térébenthine suédoise
e esencia *f* de trementina al sulfato
i trementina *f* al solfato
n sulfaatterpentijn
d schwedisches Terpentinöl *n*, Sulfatterpentinöl *n*, Zelluloseterpentinöl *n*

3190 KRYPTON LIQUEFACTION, KRYPTON LIQUEFYING gas
f liquéfaction *f* de krypton
e licuación *f* de criptón
i liquefazione *f* di cripton
n vloeibaar maken *n* van krypton
d Kryptonverflüssigung *f*

3191 KYANIZING imp
f imprégnation *f* de sublimé corrosif
e kianización *f*
i impregnazione *f* con sublimato corrosivo
n kyanisatie
d Kyanisieren *n*

L

3192 LABYRINTH HEATING htt
f chauffage *m* à labyrinthe
e calefacción *f* de laberinto
i riscaldamento *m* a labirinto
n labyrintverwarming
d Labyrinthheizung *f*

3193 LAC, dec
SHELLAC
f gomme *f* lacque
e goma *f* laca
i gomma *f* lacca
n schellak
d Schellack *m*

3194 LAC WAX dec
f cire *f* de gomme laque
e cera *f* de goma laca
i cera *f* di gomma lacca
n schellakwas
d Schellackwachs *m*

3195 LACERATION pap,tex
f déchirure *f*, lacération *f*
e laceración *f*
i lacerazione *f*
n verscheuren *n*
d Zerreissen *n*

3196 LACHRYMATORY GAS (GB), mit
LACRIMATORY GAS (US),
TEAR GAS
f gaz *m* lacrymogène
e gas *m* lacrimógeno
i gas *m* lacrimogeno
n traangas *n*
d Tränengas *n*

3197 LACMUS, ana
LITMUS
f tournesol *m*
e tornasol *m*
i laccamuffa *f*
n lakmoes *n*
d Lackmus *m*

3198 LACMUS PAPER, ana
LITMUS PAPER
f papier *m* tournesol
e papel *m* tornasol
i carta *f* laccamuffa
n lakmoespapier *n*
d Lackmuspapier *n*

3199 LACQUER dec
f laque *f*
e laca *f*
i lacca *f*
n lak
d Lack *m*

3200 LACQUER SPRAYING dec
f atomisation *f* de laque
e atomización *f* de laca
i atomizzazione *f* di lacca
n lakspuiten *n*
d Lackspritzen *n*

3201 LACQUERING dec
f laquage *m*
e laqueado *m*
i laccatura *f*
n lakken *n*
d Lackieren *n*

3202 LACTIC ACID oic
f acide *m* lactique
e ácido *m* láctico
i acido *m* lattico
n melkzuur *n*
d Milchsäure *f*

3203 LACTIC ACID FERMENTA- fmi
TION
f fermentation *f* lactique
e fermentación *f* láctica
i fermentazione *f* lattica
n melkzuurgisting
d Milchsäuregärung *f*

3204 LACTIC ACID REST bre,fmi
f acidification *f* de la trempe par
développement du ferment
lactique
e reposo *m* del ácido láctico
i sosta *f* per acido lattico
n melkzuurpauze
d Milchsäurerast *f*

3205 LACTOBIOSE, dai,sir
LACTOSE, MILK SUGAR
f lactose *f*
e lactosa *f*
i lattosio *m*
n melksuiker
d Milchzucker *m*

3206 LAGER BEER nu
f bière *f* de fermentation basse
e cerveza *f* de fermentación
baja

i birra *f* normale, birra *f* stagionata
n lagerbier *n*
d Lagerbier *n*

3207 LAGGING clo,soa
f recouvrement *m* calorifuge
e chaqueta *f* calorífuga
i rivestimento *m* termoisolante
n warmte-isolerende bedekking
d wärmeisolierende Futterung *f*

3208 LAITANCE, bui
SCUM
f écume *f* de béton
e exudación *f* del hormigón, lechada *f* de cemento
i schiuma *f* di calcestruzzo
n betonschuim *n*
d Betonschaum *m*

3209 LAKE dec
f laque *f*
e laca *f*
i lacca *f*
n substraatpigment *n*, verflak
d Farbstofflack *m*, Lackpigment *n*, Substratfarbe *f*

3210 LAKE fil
f masse *f* de ségrégation
e masa *f* de segregación
i massa *f* di segregazione
n zijgermassa
d Seigermasse *f*

3211 LAKE ASPHALT bui
f asphalte *m* du lac
e asfalto *m* del lago
i asfalto *m* del lago
n meerasfalt *n*
d Seeasphalt *m*

3212 LAMELLAR PIGMENT dec
f pigment *m* lamellaire
e pigmento *m* lamelar
i pigmento *m* lamellare
n schubvormig pigment *n*
d Blätterpigment *n*, schuppenförmiges Pigment *n*

3213 LAMINAR FILLER pla,rub
f charge *f* en lamelles
e carga *f* de laminas
i riempitivo *m* in lamelle
n plaatjesvormige vulstof, schilferachtige vulstof
d lamellenartiger Füllstoff *m*

3214 LAMINAR FLOW fil
f écoulement *m* laminaire
e flujo *m* laminar
i flusso *m* laminare
n laminaire stroming
d Laminarströmung *f*

3215 LAMINATED, met,pla
FOLIATED
f laminé adj
e laminado adj
i laminato adj
n gelaagd adj , gelamelleerd adj
d geschichtet adj , lamelliert adj

3216 LAMINATED CLOTH, tex
LAMINATED FABRIC,
TEXTILE LAMINATE
f tissu *m* laminé
e tela *f* laminada
i tessuto *m* laminato
n hardweefsel *n*
d Hartgewebe *n*

3217 LAMINATED GLASS, gla,sap
GLASS LAMINATE, SAFETY GLASS
f verre *m* de sûreté
e vidrio *m* laminado
i vetro *m* di sicurezza
n veiligheidsglas *n*
d Sicherheitsglas *n*

3218 LAMINATED PAPER pap
f stratifié *m* à base de papier
e papel *m* laminado
i laminato *m* di carta
n hardpapier *n*
d Hartpapier *n*

3219 LAMINATED PLASTIC pla
f matière *f* plastique stratifiée
e plástico *m* laminado
i plastico *m* laminato, plastico *m* stratificato
n gelaagde kunststof
d geschichteter Pressstoff *m*, Schichtpressstoff *m*

3220 LAMINATION pla
f lamination *f*
e laminación *f*
i laminazione *f*, stratificazione *f*
n lamineren *n*, uit lagen opbouwen *n*
d Laminieren *n*

3221 LAMPBLACK dec
f noir *m* de fumée, noir *m* de lampe
e negro *m* de humo, negro *m* de lámpara
i nerofumo *m*
n lamperoet *n*, lampzwart *n*
d Lampenruss *n*, Lampenschwarz *n*

3222 LANCE CUTTING, OXYGEN CUTTING htt,met
f oxycoupage *m*
e corte *m* con lanzo, oxicorte *m*
i ossitaglio *m*, taglio *m* con ossigeno
n snijden *n* met zuurstoflans
d Brennschneiden *n*

3223 LANOLIN, WOOL FAT fat
f graisse *f* de laine, lanoline *f*
e lanolina *f*
i lanolina *f*
n lanoline, wolvet *n*
d Lanolin *n*, Wollfett *n*

3224 LAP WELDING, LAPPED SEAM WELDING pla
f soudure *f* à recouvrement
e soldadura *f* de recubrimiento
i saldatura *f* a ricoprimento, saldatura *f* a sovrapposizione
n overlappend lassen *n*
d Überlappungsschweissen *n*

3225 LAPPING pol
f rodage *m*
e repasado *m*, tribopulimento *m*
i lappatura *f*
n leppen *n*
d Läppen *n*

3226 LARD OIL fat
f huile *f* de saindoux
e aceite *m* de tocino
i olio *m* di grasso di maiale
n reuzelolie
d Schmalzöl *n*, Specköl *n*

3227 LARGE BUBBLE AERATION iwt,sew
f aération *f* par grandes bulles
e aireación *f* de grandes burbujas
i aerazione *f* con grosse bolle
n aëratie met grote luchtbellen
d grobblasige Belüftung *f*

3228 LAST RUNNING, AFTERRUN, SECOND RUNNING dis
f queue *f* de distillation
e cola *f* de destilación
i coda *f* di distillazione
n naloop
d Nachlauf *m*

3229 LAST RUNNINGS bre
f dernier lavage *m*, repasse *f*
e mosto *m* del último lavado
i acqua *f* dell'ultimo lavaggio
n naloop
d Glattwasser, Nachlauf *m*

3230 LATENSIFICATION, LATENT IMAGE INTENSIFICAT pho
f renforcement *m* de l'image latent
e refuerzo *m* de la imagen latente
i rinforzo *m* dell'imagine latente
n versterking van het latente beeld
d Verstärkung *f* des latenten Bildes

3231 LATENT HEAT gen
f chaleur *f* latente
e calor *m* latente
i calore *m* latente
n latente warmte
d latente Wärme *f*

3232 LATENT HEAT OF FUSION, HEAT OF FUSION gen,htt
f chaleur *f* de fusion
e calor *m* de fusión
i calore *m* di fusione
n smeltwarmte
d Schmelzwärme *f*

3233 LATENT HEAT OF SOLIDIFICATION, HEAT OF CRYSTALLIZATION cry,gen
f chaleur *f* de cristallisation
e calor *m* de cristalización
i calore *m* di cristallizzazione
n kristallisatiewarmte
d Kristallisationswärme *f*

3234 LATENT IMAGE pho
f image *f* latente
e imagen *f* latente
i immagine *f* latente
n latent beeld *n*
d latentes Bild *n*

3235 LATEX rub
f latex *m*
e látex *m*
i lattice *m*
n latex
d Kautschukmilch *f*, Latex *m*

3236 LATEX ADHESIVE glu,rub
f adhésif *m* de latex
e cola *f* de látex
i adesivo *m* di lattice
n latexlijm, melklijm
d Kautschukmilchleim *m*, Latexleim *m*

3237 LATEX CLARIFYING rub
f clarification *f* du latex
e clarificación *f* del látex
i chiarificazione *f* del lattice
n latexklaring
d Kautschukmilchklärung *f*

3238 LATEX FOAM, rub
FOAM LATEX, FOAM RUBBER
f caoutchouc *m* spongieux, mousse *f* de latex
e espuma *f* de látex
i gommapiuma *f*, lattice *m* schiuma, schiuma *f* di lattice
n schuimrubber
d Latexschaum *m*, Schaumgummi *m*

3239 LATEX FOAM CUSHIONING, rub
AIRFOAM-CUSHIONING
f rembourrage *m* de mousse de latex
e almohadillado *m* de espuma de látex
i imbottitura *f* di gommapiuma
n schuimrubbervulling
d Schaumgummipolsterung *f*

3240 LATHER (GB), soa
SUDS (US)
f mousse *f* de savon
e espuma *f* de jabón
i schiuma *f* di sapone
n zeepschuim *n*
d Seifenschaum *m*

3241 LATHER BOOSTER soa
f exalteur *m* de mousse, renforçateur *m* de mousse
e aumentador *m* de espuma, favorecedor *m* de espuma, mejorador *m* de espuma
i esaltatore *m* di schiuma
n schuimverbeteraar
d Schaumverbesserer *m*

3242 LATHER COLLAPSE soa
f brisage *m* de la mousse, chute *f* de la mousse
e rompimiento *m* de la espuma
i caduta *f* della schiuma, rottura *f* della schiuma
n breken *n* van het schuim
d Zusammenbruch *m* des Schaumes

3243 LATHER VALUE, soa
LATHERING POWER
f indice *m* de mousse
e índice *m* espumante
i indice *m* di schiuma
n schuimgetal *n*
d Schaumwert *m*, Schaumzahl *f*

3244 LAUGHING GAS, gas
NITROUS OXIDE
f gaz *m* hilarant, protoxyde *m* d'azote
e gas *m* hilarante
i gas *m* esilarante, ossido *m* nitroso
n lachgas *n*
d Lachgas *n*, Stickstoffoxydul *n*

3245 LAUNDRY BLUE, cle
STARCHING BLUE, WASHING BLUE
f bleu *m* d'azur, bleu *m* d'empois
e azulete *m*
i turchinetto *m*
n blauwsel *n*
d Waschblau *n*

3246 LAUNDRY SOAP (US), soa
CURD SOAP (GB), SODIUM SOAP (US)
f savon *m* dur
e jabón *m* duro
i sapone *m* duro
n harde zeep, kernzeep
d harte Seife *f*, Kernseife *f*

3247 LAUREL OIL, fat
BAY OIL, BAYBERRY OIL
f essence *f* de laurier
e esencia *f* de laurel
i olio *m* d'alloro, olio *m* di lauro, olio *m* laurino
n bayolie, laurierolie
d Bayöl *n*, Lorbeeröl *n*, Loröl *n*

3248 LAUTER MASH bre
f trempe *f* claire
e braceado *m* claro
i tempera *f* liquida
n helder beslag *n*
d Lautermaische *f*

3249 LAVENDER OIL, fat
SPIKE OIL
f essence *f* de lavande
e esencia *f* de espliego
i olio *m* di lavanda, olio *m* di spigo
n lavendelolie
d Lavendelöl *n*

3250 LEACH BRINE, aic,gen
BITTERN, MOTHER-LYE
f eau *f* mère
e agua *f* madre
i acqua *f* madre
n moederloog
d Bitterlauge *f*, Mutterlauge *f*, Muttersole *f*

3251 LEACHING, tan
LEECHING
f extraction *f* de l'acide tannique
e extracción *f* del ácido tánnico

i estrazione *f* dell'acido tannico
n uitlogen *n* van het looizuur
d Auslaugen *n* der Gerbsäure

3252 LEACHING, LIXIVIATION gen
f lessivage *m*, lixiviation *f*
e lixiviación *f*
i lisciviazione *f*
n logen *n*, uitlogen *n*
d Auslaugen *n*, Laugen *n*

3253 LEAD ACETATE, GOULARD'S POWDER, SALT OF SATURN dec
f acétate *m* de plomb, sel *m* de Saturne
e acetato *m* de plomo
i acetato *m* di piombo, sale *m* di Saturno
n loodsuiker
d Bleizucker *m*, Goulardsches Pulver *n*

3254 LEAD ANNEALING met
f recuit *m* en plomb fondu
e recocido *m* en plomo fundido
i ricottura *f* in piombo fuso
n uitgloeien *n* in gesmolten lood
d Ausglühen *n* in geschmolzenem Blei

3255 LEAD ATOMIZATION cao
f pulvérisation *f* de plomb
e pulverización *f* de plomo
i polverizzazione *f* di piombo
n loodverstuiving
d Bleizerstäubung *f*

3256 LEAD-BASE PRIMER dec
f primaire *f* plombifère
e pintura *f* de imprimación a base de plomo
i pittura *f* d'imprimatura a base di piombo
n loodhoudende grondverf
d bleihaltige Grundfarbe *f*

3257 LEAD BATH met
f bain *m* de plomb
e baño *m* de plomo
i bagno *m* di piombo
n loodbad *n*
d Bleibad *n*

3258 LEAD BURNING wel
f soudure *f* par fusion du plomb
e soldadura *f* por fusión del plomo
i saldatura *f* per fusione del piombo
n loodsolderen *n* door versmelten
d Bleilöten *n* durch Verschmelzen

3259 LEAD CHROMATE, CHROME YELLOW, PARIS YELLOW dec
f jaune *m* de chrome
e amarillo *m* de cromo, cromato *m* de plomo
i giallo *m* di cromo
n chroomgeel *n*, Parijs geel *n*
d Bleichromat *n*, Chromgelb *n*, Parisergelb *n*

3260 LEAD DESILVERIZATION met
f désargentation *f* du plomb
e desplatación *f* del plomo
i rimozione *f* dell'argento dal piombo
n loodontzilvering
d Bleientsilberung *f*

3261 LEAD GASOLENE (US), LEAD PETROL (GB) fue
f essence *f* éthylée
e gasolina *f* etilada
i benzina *f* etilata
n ethylbenzine
d Äthylbenzin

3262 LEAD GLANCE, BLUE LEAD, GALENA met
f galène *f*
e alquifol *m*, galena *f*
i galena *f*
n galeniet *n*, loodglans *n*
d Bleiglanz *m*, Galenit *m*

3263 LEAD GLASS gla
f verre *m* plombeux
e vidrio *m* plomizo
i vetro *m* piombico
n loodglas *n*
d Bleiglas *n*

3264 LEAD GLAZE, LEAD GLAZING cer
f émail *m* au plomb
e vidriado *m* de plomo
i vetrina *f* di piombo
n loodglazuur *n*
d Bleiglasur *f*

3265 LEAD HARDENING met
f trempe *f* en plomb
e temple *m* en plomo
i tempera *f* in piombo
n harden *n* in lood
d Härten *n* in Blei

3266 LEAD IMPROVING, SOFTENING met
f affinage *m* du plomb
e afino *m* del plomo
i affinatura *f* del piombo
n loodveredeling
d Bleivergütung *f*

3267 LEAD MELTING, LEAD SMELTING met
f fusion *f* du plomb
e fusión *f* del plomo
i fusione *f* del piombo
n loodsmelten *n*
d Bleischmelzen *n*

3268 LEAD MONOXIDE, LITHARGE, MASSICOT dec
f litharge *f*, oxyde *m* de plomb
e litargirio *m*, óxido *m* de plomo
i litargirio *m*, ossido *m* di piombo
n loodglit *n*, massicot *n*
d Bleiglätte *f*, Massikot *m*

3269 LEAD PAINT dec
f peinture *f* à base de plomb
e pintura *f* a base de plomo
i pittura *f* a base di piombo
n loodverf
d Bleifarbe *f*

3270 LEAD PATENTING, LEAD QUENCHING met
f trempe *f* au plomb
e temple *m* al plomo
i patentamento *m* al piombo
n afschrikken *n* van lood
d Bleipatentierung *f*

3271 LEAD SOAP dec
f savon *m* plombifère
e jabón *m* de plomo
i sapone *m* a piombo
n loodzeep
d Bleiseife *f*

3272 LEAD SOLUBILITY TESTING tes
f essai *m* de la solubilité du plomb
e ensayo *m* de la solubilidad del plomo
i collaudo *m* della solubilità del piombo
n loodoplosbaarheidsproef
d Bleilöslichkeitsprüfung *f*

3273 LEAD WOOL hyg,iwt
f filasse *f* de plomb, laine *f* de plomb
e cuerda *f* de hilos de plomo
i lana *f* di piombo
n loodwol
d Bleiwolle *f*

3274 LEADLESS GLAZE cer
f émail *m* sans plomb
e vitriado *m* sin plomo
i vetrina *f* senza piombo
n loodvrij glazuur *n*
d bleifreie Glasur *f*

3275 LEAFING dec
f étalement *m* en surface
e peliculación *f*
i formazione *f* di pellicole
n spiegelvorming
d Schwimmen *n*

3276 LEAKAGE WATER, SEEPAGE WATER hyg,sew
f eau *f* de fuites
e agua *f* de infiltración
i acqua *f* di fughe
n lekwater *n*, sijpelwater *n*
d Leckwasser *n*, Sickerwasser *n*

3277 LEAN CONCRETE, POOR CONCRETE bui
f béton *m* maigre
e hormigón *m* pobre
i calcestruzzo *m* magro
n mager beton *n*
d Magerbeton *m*, Sparbeton *m*

3278 LEAN GAS, PRODUCER GAS gas
f gaz *m* pauvre
e gas *m* pobre
i gas *m* povero
n arm gas *n*
d armes Gas *n*

3279 LEAN LIME, HYDRAULIC LIME, MEAGRE LIME bui
f chaux *f* hydraulique
e cal *f* hidráulica, cal *f* magra
i calce *f* idraulica
n hydraulische kalk, magere kalk
d Wasserkalk *m*

3280 LEATHER lea
f cuir *m*
e cuero *m*, piel *f*
i cuoio *m*, pelle *f*
n leder *n*, leer *n*
d Leder *n*

3281 LEATHERCLOTH lea
f toile-cuir *m*

e tela *f* cuero
i tessuto *m* pelle
n leerdoek *n*
d Ledertuch *n*

3282 LEATHERETTE, ARTIFICIAL LEATHER, IMITATION LEATHER — lea
f cuir *m* artificiel
e cuero *m* artificial, similcuero *m*
i cuoio *m* artificiale, dermoide *f*
n kunstleer *n*
d Kunstleder *n*

3283 LEAVENING, PITCHING — bre
f mise *f* en levain
e sembradura *f* con levadura
i aggiuntatura *f* del lievito
n gisttoevoeging
d Hefegabe *f*

3284 LEECHING, LEACHING — tan
f extraction *f* de l'acide tannique
e extracción *f* del ácido tánnico
i estrazione *f* dell'acido tannico
n uitlogen *n* van het looizuur
d Auslaugen *n* der Gerbsäure

3285 LEGUMINOUS COVER — fer
f couverture *f* de légumineuses
e abono *m* de leguminosas
i coltura *f* a leguminose
n groenbemesting
d Gründüngung *f*

3286 LEMON GRASS OIL — fat
f essence *f* de schénante
e esencia *f* de andropogon citratus
i olio *m* d'andropogon citratus
n lemoengrasolie
d Lemongrasöl *n*

3287 LEMON OIL — fat
f citrine *f*, essence *f* de citron
e esencia *f* de limón
i essenza *f* di limone
n citroenolie
d Limonenöl *n*, Zitronenöl *n*

3288 LENGTH OF BATH — tex
f volume *m* de bain
e relación *f* de baño
i volume *m* di bagno
n badverhouding
d Flottenverhältnis *n*

3289 LESS VOLATILE COMPOUND — dis
f constituant *m* moins volatil
e componente *m* menos volátil
i componente *m* meno volatile
n moeilijker vluchtige component
d schwererflüchtige Komponente

3290 LEVEL DYEING, EVEN DYEING — dye
f teinture *f* égale
e teñido *m* igual
i tintura *f* uguale
n gelijkmatige verving
d gleichmässige Färbung *f*

3291 LEVELLER, LEVELLING AGENT — gen
f agent *m* égalisateur
e agente *m* igualador
i agente *m* uguagliatore
n egaliseermiddel *n*, nivelleermiddel *n*
d Ausgleichmittel *n*, Egalisiermittel *n*

3292 LEVELLING — dye
f égalisation *f*
e igualación *f*
i uguagliamento *m*
n egalisatie
d Egalisierung *f*

3293 LEVELLING EFFECT, LEVELLING POWER — elc
f pouvoir *m* égalisant
e poder *m* de igualación
i potere *m* d'uguagliamento
n egaliseervermogen *n*
d Egalisiervermögen *n*

3294 LEVELLING SOLUTION — met
f solution *f* égalisante
e solución *f* igualadora
i soluzione *f* uguagliatrice
n egaliserende oplossing
d egalisierende Lösung *f*

3295 LEVELLING SOLVENT — gen
f solvant *m* égalisant
e solvente *m* igualador
i solvente *m* uguagliatore
n egaliserend oplosmiddel *n*
d egalisierendes Lösungsmittel *n*

3296 LEVIGATION — cle
f lévigation *f*
e levigación *f*
i levigazione *f*
n selectieve bezinking
d Abschlämmung *f*

3297 LIBERATION, DEPOSITING-OUT — elc
f depôt *m* total, libération *f*

e depósito *m* total, liberación *f*
i esaurimento *m*
n afwerken *n*
d Entkupferung *f*, Entmetallisierung *f*

3298 LICORICE ROOT mcy
f réglisse *f* en bâton
e orozuz *m*, regaliz *m*
i liquirizia *f*
n zoethout *n*
d Lakritze *f*, Süssholz *n*

3299 LIGHT AGEING (GB), LIGHT AGING (US) pla,rub
f vieilissement *m* à la lumière
e envejecimiento *m* por la luz
i invecchiamento *m* alla luce
n veroudering door licht
d Alterung *f* durch Licht

3300 LIGHT GASOLENE (US), LIGHT PETROL (GB) fue
f essence *f* légère
e gasolina *f* ligera
i benzina *f* leggera
n lichte benzine
d Leichtbenzin *n*

3301 LIGHT OIL fue
f huile *f* légère
e aceite *m* flúido, aceite *m* ligero
i olio *m* leggero
n dunvloeibare olie, lichte olie
d dünnflüssiges Öl *n*, Leichtöl *n*

3302 LIGHT RESISTANT, LIGHTFAST tex
f inaltérable adj , résistant adj à la lumière
e fotorresistente adj , sólido adj a la luz
i resistente adj alla luce
n lichtvast adj
d lichtecht adj

3303 LIGHT-SENSITIVE EMULSION pho
f émulsion *f* photosensible
e emulsión *f* fotosensible
i emulsione *f* sensibile alla luce
n lichtgevoelige emulsie
d lichtempfindliche Emulsion *f*

3304 LIGHT SPAR (US), NATURAL CALCIUM SULPHATE (GB), TERRA ALBA min
f blanc *m* brillant, blanc *m* de gypse, sulfate *m* de calcium naturel
e sulfato *m* de calcio natural
i terra *f* alba
n lenzine, lichtspaat *n*
d Federweiss *n*, Kalziumsulfat *n*, Leichtspat *m*, Lenzin *m*

3305 LIGHTPROOF gen
f opaque adj
e opaco adj
i opaco adj
n lichtdicht adj
d lichtdicht adj

3306 LIGNALOE OIL, LINALOE OIL fat
f essence *f* de linaloé
e esencia *f* de lináloe
i olio *m* di linaloe
n linaloëolie
d Linaloeöl *n*, Rosenholzöl *n*

3307 LIGNITE OIL mcy
f huile *f* de lignite
e aceite *m* de lignito
i olio *m* di lignite
n bruinkoololie
d Braunkohlenöl *n*

3308 LIGNITE WAX fat
f cire *f* de lignite
e cera *f* de lignito
i cera *f* di lignite
n bruinkoolwas
d Braunkohlenwachs *n*, Erdwachs *n*

3309 LIME bui
f chaux *f*
e cal *f*
i calce *f*
n kalk
d Kalk *m*

3310 LIME BAG TEST, CHALK TEST tes
f essai *m* à la craie
e prueba *f* a la creta
i collaudo *m* alla creta
n krijtproef
d Kreideprobe *f*

3311 LIME BURNING htt
f cuisson *f* de la chaux
e cochura *f* de la cal
i cottura *f* della calce
n kalkbranden *n*
d Kalkbrennen *n*

3312 LIME CAKE, LIME MUD, LIME SLUDGE sin
f boue *f* de chaux
e lodo *m* de cal
i fango *m* di calce

n defecatieresidu *n*
d Scheideschlamm *m*

3313 LIME-CEMENT MORTAR bui
f mortier *m* à chaux et ciment
e mortero *m* de cal y cemento
i malta *f* di calce e cemento
n cement-kalkmortel
d Zementkalkmörtel *m*

3314 LIME COATING coa
f revêtement *m* en chaux
e revestimiento *m* de cal
i rivestimento *m* di calce
n kalklaag
d Kalkschicht *f*

3315 LIME DEFECATION, sin
LIMING
f défécation *f*, saturation *f*
e alcalización *f*, defecación *f*
i defecazione *f*, saturazione *f*
n defecatie, saturatie
d Scheiden *n*

3316 LIME DESULFURIZING (US), met
LIME DESULPHURIZATION (GB)
f désulfuration *f* par la chaux
e desulfuración *f* con cal
i desolforazione *f* con calce
n ontzwaveling met kalk
d Entschwefelung *f* mit Kalk

3317 LIME FERTILIZER, fer
CALCIUM FERTILIZER
f engrais *m* calcique
e abono *m* a base de cal
i fertilizzante *m* a base di calce
n kalkmeststof
d Kalkdünger *m*

3318 LIME MIXING sin
f chaulage *m*
e mezcladura *f* de la cal
i mescolatura *f* della calce
n kalkmenging
d Kalkmischung *f*

3319 LIME MORTAR bui
f mortier *m* de chaux
e mortero *m* de cal
i malta *f* di calce
n kalkmortel
d Kalkmörtel *m*

3320 LIME POWDER, bui
POWDERED LIME
f chaux *f* vive en poudre
e cal *f* apagada en polvo
i calce *f* viva in polvere
n verweerde kalk
d Kalkstaub *m*, Staubkalk *m*,
verwitterter Kalk *m*

3321 LIME SLURRY, iwt
MILK OF LIME
f lait *m* de chaux
e lechada *f* de cal
i latte *m* di calce
n kalkmelk
d Kalkbrühe *f*, Schlämme *f*

3322 LIME SOAP, cle,tex
CALCIUM SOAP
f savon *m* calcique, savon *m* de
chaux
e jabón *m* de cal
i sapone *m* di calce
n kalkzeep
d Kalkseife *f*

3323 LIME-SOAP DISPERSION soa
f dispersion *f* de savon de chaux
e dispersión *f* de jabón cálcico
i dispersione *f* di sapone di calce
n dispergeren *n* van de kalkzeep
d Kalkseifendispergierung *f*

3324 LIME-SODA PROCESS hyg
f procédé *m* de soude et chaux
e procedimiento *m* ablandador de
sosa y cal
i processo *m* di soda e calce
n kalk-sodaproces *n*
d Kalk-Sodaverfahren *n*

3325 LIME STAIN tan
f tache *f* de plainage
e macho *m* de pelado
i macchia *f* di pelatura
n plootvlek
d Äscherschatten *m*

3326 LIME SULFUR WASH (US), bre
LIME SULPHUR WASH (GB)
f bouillie *f* sulfocalcique
e caldo *m* sulfocálcico
i miscela *f* solfocalcica
n zwavelkalkbrij
d Schwefelkalkbrühe *f*

3327 LIME WATER dec
f eau *f* de chaux
e agua *f* de cal
i acqua *f* di calce
n kalkwater *n*
d Kalkwasser *n*

3328 LIME-WATER SOFTENING hyg,sev
f désacidification *f* par l'eau de
chaux

e desacidificación *f* por agua de cal
i disacidificazione *f* a calce
n ontzuring met kalkwater
d Entsäuerung *f* mit Kalkwasser

3329 LIMESTONE FILLER pla,rub
f charge *f* calcaire
e carga *f* calcárea
i riempitivo *m* calcareo
n vulstof uit kalksteen
d Kalksteinfüllstoff *m*

3330 LIMETTA OIL fat
f essence *f* de limette
e esencia *f* de limeta
i olio *m* di limetta
n limetolie
d Limettaöl *n*

3331 LIMEWASH, WHITENING, WHITEWASH dec
f badigeon *m*
e lechada *f* de cal
i latte *m* di calce
n witkalk
d Kalkmilch *f*

3332 LIMEWASHING dec
f passage *m* au lait de chaux
e blanqueo *m* con cal, encalado *m*
i imbiancatura *f*
n witten *n*
d Tünchen *n*, Weissen *n*

3333 LIMING bre
f chaulage *m*
e incalado *m*
i dare v la calce
n kalktoevoeging
d Kalkern *n*

3334 LIMING tan
f pelainage *m*, plainage *m*, plamage *m*
e pelado *m*, pelambre *m*, pelote *m*
i pelatura *f*
n ploten *n*
d Äschern *n*

3335 LIMING AFTER UNHAIRING tan
f pelainage *m* après la dépilation, plainage *m* postérieur, plamage *m* postérieur
e pelado *m* después de la depilación
i pelatura *f* dopo la depilazione
n ploten *n* na ontharing
d Äschern *n* nach Enthaarung

3336 LIMITING DENSITY gas
f densité *f* limite
e densidad *f* límite
i densità *f* limite
n grensdichtheid, kritieke dichtheid
d Grenzdichte *f*

3337 LINEN tex
f toile *f*
e lienzo *m*, tela *f* de lino
i pannolino *m*, tela *f*
n linnen *n*
d Leinwand *f*

3338 LINGEL lea
f ligneul *m*
e hilo *m* encerado, liñuelo *m*
i spago *m* da calzolaio
n pekdraad, pikdraad
d Pechdraht *m*

3339 LINIMENT, OINTMENT mcy
f liniment *m*, onguent *m*
e linimento *m*, ungüento *m*
i linimento *m*, unguento *m*
n liniment *n*, zalf
d Liniment *n*, Salbe *f*

3340 LINOLEUM bui
f linoléum *m*
e linóleo *m*
i linoleum *m*
n linoleum *n*
d Linoleum *n*

3341 LINSEED MEAL, EXTRACTED MEAL fat
f farine *f* de grain de lin
e harina *f* de linaza
i farina *f* di seme di lino
n lijnzaadmeel *n*
d Leinsamenmehl *n*

3342 LINSEED OIL fat
f huile *f* de lin
e aceite *m* de linaza
i olio *m* di seme di lino
n lijnolie
d Leinöl *n*

3343 LINSEED OIL PAINT dec
f peinture *f* à l'huile de lin
e pintura *f* de aceite de linaza
i pittura *f* ad olio di seme di lino
n lijnolieverf
d Leinölfarbe *f*

3344 LINSEED OIL VARNISH dec
f vernis *m* à l'huile de lin

e barniz *m* de aceite de linaza
i vernice *f* ad olio di seme di lino
n lijnolievernis *n*
d Leinöllack *m*

3345 LINTERS pla,rub
f bourres *pl* de coton
e borras *pl* de algodón, linters *pl*
i cascami *pl* di cotone
n linters *pl*
d Linters *pl*

3346 LIQUATION, met
SEGREGATION
f liquation *f*, ségrégation *f*
e licuación *f*, segregación *f*
i liquazione *f*, segregazione *f*
n segregatie, zijgeren *n*
d Ausschwitzen *n*, Seigerung *f*

3347 LIQUEFACTION, gas
LIQUEFYING, FLUIDIFICATION
f liquéfaction *f*
e licuación *f*, licuefacción *f*
i liquefazione *f*
n vloeibaar maken *n*
d Verflüssigung *f*

3348 LIQUEFIED PETROLEUM GASES, fue
LPG
f gaz *pl* de pétrole liquefiés
e gases *pl* de petróleo licuados
i gas *pl* di petrolio liquefatti
n lpg-gas *n*
d Flüssiggas *n*

3349 LIQUID, gen
FLUID
f liquide *m*
e líquido *m*
i liquido *m*
n vloeistof
d Flüssigkeit *f*

3350 LIQUID AIR gas
f air *m* liquide
e aire *m* líquido
i aria *f* liquida
n vloeibare lucht
d flüssige Luft *f*

3351 LIQUID BATH htt
f bain *m* liquide
e baño *m* líquido
i bagno *m* liquido
n smeltbad *n*
d Schmelzbad *n*

3352 LIQUID CARBURIZING, met
LIQUID HEAT TREATMENT,
SALT BATH TREATMENT
f cémentation *f* par liquide
e cementación *f* por líquido
i cementazione *f* per liquido
n met vloeistof cementeren *n*
d Flüssigzementierung *f*

3353 LIQUID CRYSTAL, cry,gla
ANISOTROPIC LIQUID,
CRYSTALLINE LIQUID
f cristal *m* liquide, liquide *m* anisotropique
e cristal *m* líquido, líquido *m* anisotrópico
i cristallo *m* liquido, liquido *m* anisotropico
n anisotrope vloeistof, vloeibaar kristal *n*
d anisotrope Flüssigkeit *f*, flüssige Kristalle *pl*, kristalline Flüssigkeit *f*

3354 LIQUID DETERGENT soa
f détergent *m* liquide
e detergente *m* líquido
i detergente *m* liquido
n vloeibaar wasmiddel *n*
d flüssiges Waschmittel *n*

3355 LIQUID FUEL fue
f combustible *m* liquide
e combustible *m* líquido
i combustibile *m* liquido
n vloeibare brandstof
d flüssiger Brennstoff *m*

3356 LIQUID GLASS, aic
SILICATE, SOLUBLE GLASS,
WATER GLASS
f silicate *m* de sodium, verre *m* soluble
e silicato *m* sódico líquido, vidrio *m* soluble
i tetrasilicato *m* di sodio
n silicaat *n*, waterglas *n*
d Silikat *n*, Wasserglas *n*

3357 LIQUID GRADIENT, dis
HYDRAULIC GRADIENT
f gradient *m* hydraulique
e gradiente *m* hidráulico
i gradiente *m* idraulico
n vloeistofgradiënt
d Flüssigkeitsgefälle *n*

3358 LIQUID HEAD dis
f charge *f* de liquide
e altura *f* de líquido

i altezza *f* di liquido
n vloeistofhoogte
d Flüssigkeitshöhe *f*

3359 LIQUID HONING, VAPOR BLASTING (US), VAPOUR BLASTING (GB) pol
f honage *m* au jet de vapeur
e rectificación *f* al chorro de vapor
i rettificatura *f* al getto di vapore
n honen *n* met vloeistofstraal
d Honen *n* mit Dampfstrahl

3360 LIQUID-LIQUID EXTRACTION nue
f extraction *f* par partage liquide-liquide
e extracción *f* por repartición líquido-líquido
i estrazione *f* per ripartizione liquido-liquido
n vloeistof-vloeistofextractie
d Flüssig-Flüssigextraktion *f*

3361 LIQUID LOAD dis
f débit *m* massique de liquide
e carga *f* de líquido
i carico *m* di liquido
n vloeistofbelasting
d Flüssigkeitsbelastung *f*

3362 LIQUID METAL COOLING nue
f refroidissement *m* par métal fondu
e enfriamiento *m* por metal líquido
i raffreddamento *m* per metallo liquido
n koeling met vloeibaar metaal
d Flüssigmetallkühlung *f*

3363 LIQUID NITROGEN gas
f azote *m* liquide
e nitrógeno *m* líquido
i azoto *m* liquido, nitrogeno *m* liquido
n vloeibare stikstof
d flüssiger Stickstoff *m*

3364 LIQUID OXYGEN gas
f oxygène *m* liquide
e oxígeno *m* líquido
i ossigeno *m* liquido
n vloeibare zuurstof
d flüssiger Sauerstoff *m*

3365 LIQUID-OXYGEN EXPLOSIVES exp
f oxylignites *pl*, oxyliquites *pl*
e explosivos *pl* de oxígeno líquido
i esplosivi *pl* ad ossigeno liquido
n oxyliquiten *pl*
d Oxyliquite *pl*

3366 LIQUID PARAFFIN, PARAFFIN OIL fat
f huile *f* de paraffine
e aceite *m* de parafina
i olio *m* di paraffina
n paraffineolie
d Paraffinöl *n*, Vaselineöl *n*

3367 LIQUID-PHASE CRACKING fue
f craquage *m* en phase liquide
e pirólisis *f* en fase líquida
i piroscissione *f* in fase liquida
n kraken *n* volgens het vloeistof-fazeproces
d Kracken *n* nach dem Flüssig-phaseverfahren

3368 LIQUID-PHASE HYDROGENATION fat
f hydrogénation *f* en phase liquide
e hidrogenación *f* en fase líquida
i idrogenazione *f* in fase liquida
n hydreren *n* volgens het vloeistoffazeproces
d Hydrieren *n* nach dem Flüssigphaseverfahren

3369 LIQUID-PHASE SINTERING htt
f frittage *m* en présence de la phase liquide
e sinterización *f* con la fase líquida
i sinterizzazione *f* con la fase liquida
n smeltsinteren *n*
d Schmelzsintern *n*

3370 LIQUID RESIN, TALL OIL, TALLOL fat,pap
f huile *f* de Thal, résine *f* liquide
e líquido *m* resinoso aceitoso, resina *f* líquida
i olio *m* di sego
n tallolie
d Tallöl *n*

3371 LIQUID SOAP soa
f savon *m* liquide
e jabón *m* líquido
i sapone *m* liquido
n vloeibare zeep
d flüssige Seife *f*

3372 LIQUOR bre
f eau *f* de brassage
e agua *f* de fabricación

i acqua *f* per fare la birra
n brouwwater *n*
d Brauwasser *n*

3373 LIQUOR FINISHING coa,met
f étamage *m* préliminaire
e estañado *m* preliminar
i stagnatura *f* preliminare
n voorvertinnen *n*
d Vorverzinnung *f*

3374 LITHARGE,
LEAD MONOXIDE,
MASSICOT dec
f litharge *f*, oxyde *m* de plomb
e litargirio *m*, óxido *m* de plomo
i litargirio *m*, ossido *m* di piombo
n loodglit *n*, massicot *n*
d Bleiglätte *f*, Massikot *m*

3375 LITHOGRAPHIC PAPER pap
f papier *m* lithographique
e papel *m* litográfico
i carta *f* litografica
n steendrukpapier *n*
d Lithographenpapier *n*,
Steindruckpapier *n*

3376 LITHOPONE dec
f lithopone *f*
e litopón *m*
i litopone *m*
n lithopoon *n*, patentwit *n*
d Emailweiss *n*, Lithopone *f*,
Zinksulfidweiss *n*

3377 LITMUS,
LACMUS ana
f tournesol *m*
e tornasol *m*
i laccamuffa *f*
n lakmoes *n*
d Lackmus *m*

3378 LITMUS PAPER,
LACMUS PAPER ana
f papier *m* tournesol
e papel *m* tornasol
i carta *f* laccamuffa
n lakmoespapier *n*
d Lackmuspapier *n*

3379 LIVE STEAM boi,bre
f vapeur *f* vive
e vapor *m* directo, vapor *m* vivo
i vapore *m* diretto, vapore *m*
saturo
n open stoom, vrije stoom
d direkter Dampf *m*, Frischdampf *m*

3380 LIVER OF SULFUR (US), mcy,pho
LIVER OF SULPHUR (GB)
f soie *m* de soufre
e hígado *m* de azufre
i fegato *m* di zolfo
n zwavellever
d Schwefelleber *f*

3381 LIVERING dec
f coagulation *f* de peinture
e coagulación *f* de pintura
i coagulazione *f* di pittura
n dik worden *n* van verf
d Dickwerden *n* der Farbe

3382 LIXIVIAL SALTS aic
f sels *pl* de lixiviation
e sales *pl* lixiviales
i sali *pl* di lisciviazione
n uitlogende zouten *pl*
d auslaugende Salze *pl*

3383 LIXIVIATION,
LEACHING gen
f lessivage *m*, lixiviation *f*
e lixiviación *f*
i lisciviazione *f*
n logen *n*, uitlogen *n*
d Auslaugen *n*, Laugen *n*

3384 LOADED MIXTURE pla,rub
f mélange *m* chargé
e mezcla *f* cargada
i mescolanza *f* caricata
n mengsel *n* met vulstof
d gefülltes Gemisch *n*

3385 LOADING pap
f addition *f* à la pâte
e adición *f* de la pulpa
i addizione *f* alla polpa
n pulpbijmenging
d Ganzstoffbeimischung *f*

3386 LOADING,
WEIGHTING tex
f chargement *m*
e cargado *m*
i caricamento *m*
n verzwaren *n*
d Beschweren *n*

3387 LOADING AGENT,
WEIGHTING AGENT tex
f matière *f* de charge
e material *m* de carga
i materiale *m* di carica
n verzwaringsmiddel *n*
d Beschwerungsmittel *n*

3388 LOAF SUGAR sin
f pain *m* de sucre
e pan *m* de azúcar
i pane *m* di zucchero
n suikerbrood *n*
d Hut *m*

3389 LOAM, bui
MILD CLAY, SANDY CLAY
f terre *f* glaise
e tierra *f* arcillosa
i terra *f* argillosa, terra *f* grassa
n leem *n*
d Lehm *m*

3390 LOAN pap
f papier *m* à billets
e papel *m* de trapos
i carta *f* per biglietti di banca
n bankbiljettenpapier *n*
d Banknotenpapier *n*

3391 LOCAL ACTION, cor
LOCAL CORROSION, SELF CORROSION
f action *f* locale
e acción *f* local
i azione *f* locale
n plaatselijke inwerking
d örtliche Einwirkung *f*

3392 LOCAL BOILING, boi
SURFACE BOILING
f ébullition *f* locale
e ebullición *f* local
i ebollizione *f* locale
n lokaal koken *n*
d örtliches Sieden *n*, Verdampfen *n* in unterkühlter Flüssigkeit

3393 LONG DURATION TEST tes
f essai *m* à longue durée
e ensayo *m* de larga duración
i collaudo *m* di lunga durata
n duurproef
d Dauerprobe *f*

3394 LONG MALT, bre
DISTILLER'S MALT,
OVERGROWN MALT
f hussards *pl*
e malta *f* de germinación larga
i malto *m* a germogliazione spinta, ussari *pl*
n huzaren *pl*, langmout *n*
d Langmalz *n*

3395 LONG-OIL VARNISH dec
f vernis *m* à forte teneur en huile
e barniz *m* con exceso de aceite
i vernice *f* a contenuto elevato in olio
n vet vernis *n*
d fetter Klarlack *m*

3396 LONG RESIDUE fue
f résidu *m* long
e residuo *m* largo, residuo *m* surtido
i residuo *m* lungo
n residu *n* met smeeroliefracties
d Toprückstand *m*

3397 LOOP DRYING, dry,tex
FESTOON DRYING
f séchage *m* à guirlandes
e secado *m* de festones
i essiccazione *f* a festoni
n guirlandedroging
d Guirlandentrocknung *f*

3398 LOST-WAX PROCESS (US), fon
INVESTMENT MOULDING (GB),
PRECISION MOULDING (GB)
f moulage *m* à la cire perdue
e moldeo *m* a la cera perdida
i formatura *f* a cera spersa
n vormen *n* volgens de verloren-wasmethode
d Modellausschmelzverfahren *n*

3399 LOW-BOILING FRACTION dis
f fraction *f* à bas point d'ébullition
e fracción *f* de bajo punto de ebullición
i frazione *f* bassobollente
n fractie met laag kookpunt
d niedrigsiedende Fraktion *f*

3400 LOW EXPLOSIVE exp
f explosif *m* à détonation lente
e explosivo *m* lento
i esplosivo *m* a detonazione lenta
n springstof met kleine detonatiesnelheid
d langsam wirkender Sprengstoff *m*

3401 LOW-FOAMING DETERGENT, cle
CONTROLLED SUDSING DETERGENT, RESTRICTED SUDSING DETERGENT
f détergent *m* peu moussant
e detergente *m* poco espumante
i detergente *m* poco schiumogeno
n weinig schuimend wasmiddel *n*
d wenig schäumendes Waschmittel *n*

3402 LOW GRADE ORE, min
BASE ORE
f minerai *m* à faible teneur,

minerai *m* de qualité inférieure, minerai *m* pauvre
e mineral *m* de baja ley, mineral *m* pobre
i minerale *m* di qualità cattiva
n arm erts *n*, laagwaardig erts *n*
d armes Erz *n*, geringwertiges Erz *n*, taubes Erz *n*

3403 LOW GRADE STOCK gen
f mélange *m* de mauvaise qualité
e mezcla *f* de calidad inferior
i mescolanza *f* povera
n mengsel *n* van slechte kwaliteit
d geringwertiges Gemisch *n*

3404 LOW GRADE SUGAR sin
f sucre *m* de sous-produit
e azúcar *m* de segundo producto
i zucchero *m* di secondo prodotto
n naprodukt *n*
d Nachprodukt *n*, Nachzucker *m*

3405 LOW HEAT CEMENT bui
f ciment *m* à basse chaleur de prise
e cemento *m* de bajo calor de fraguado
i cemento *m* a basso calore d'indurimento
n cement *n* met lage bindingswarmte
d Zement *m* mit niedriger Abbindewärme

3406 LOW PRESSURE LAMINATING, CONTACT LAMINATING pla
f laminage *m* à basse pression
e formación *f* de laminados a baja presión
i laminazione *f* a bassa pressione
n lamineren *n* bij lage druk
d Niederdruckschichtstoffherstellung *f*

3407 LOW TEMPERATURE CARBONIZATION fue
f semicarbonisation *f*
e carbonización *f* de baja temperatura
i carbonizzazione *f* a bassa temperatura
n carboniseren *n* bij lage temperatuur, zwelen *n*
d Schwelen *n*, Tieftemperaturverkokung *f*

3408 LOW TEMPERATURE REFRIGERATION, DEEP-FREEZING clo
f congélation *f*, réfrigération *f* à basse température
e congelación *f*, refrigeración *f* de baja temperatura
i congelamento *m*, congelazione *f*, refrigerazione *f* a bassa temperatura
n diepvriezen *n*
d Tiefkühlung *f*

3409 LPG, LIQUEFIED PETROLEUM GASES fue
f gaz *pl* de pétrole liquefiés
e gases *pl* de petróleo licuados
i gas *pl* di petrolio liquefatti
n lpg-gas *n*
d Flüssiggas *n*

3410 LUBRICANT lub
f lubrifiant *m*
e lubricante *m*
i lubrificante *m*
n smeermiddel *n*
d Schmiermittel *n*

3411 LUBRICATING GREASE lub
f graisse *f* lubrifiant
e grasa *f* lubricante
i grasso *m* lubrificante
n smeervet *n*
d Schmiere *f*, Schmierfett *n*

3412 LUBRICATING LACQUER, DRY LACQUER dec
f peinture *f* lubrifiante
e pintura *f* lubricante
i pittura *f* lubrificante
n glijlak
d Gleitlack *m*

3413 LUBRICATING OIL lub
f huile *f* lubrifiante
e aceite *m* lubricante
i olio *m* lubrificante
n smeerolie
d Schmieröl *n*

3414 LUBRICATION lub
f lubrification *f*
e lubricación *f*, lubrificación *f*
i lubrificazione *f*
n smering
d Schmierung *f*

3415 LUBRICITY lub
f pouvoir *m* lubrifiant
e poder *m* lubricante
i potere *m* lubrificante
n smerend vermogen *n*
d Schmierfähigkeit *f*

3416 LUMINOUS GAS FLAME bur,fue
f flamme *f* lumineuse
e llama *f* luminosa
i fiamma *f* luminosa
n lichtgevende vlam
d leuchtende Flamme *f*

3417 LUMINOUS PAINT dec
f peinture *f* luminescente
e pintura *f* luminescente
i pittura *f* luminescente
n lichtgevende verf
d Leuchtfarbe *f*

3418 LUMINOUS PIGMENT dec,pla
f pigment *m* luminescent
e pigmento *m* luminescente
i pigmento *m* luminescente
n lichtgevend pigment *n*
d Leuchtpigment *n*

3419 LUPULIN,
HOP BITTER bre
f lupuline *f*
e lupolina *f*
i luppolina *f*
n hopbitter *n*, lupuline
d Hopfenbitter *n*, Lupulin *n*

3420 LUSTRE,
GLOSS, SHEEN dec,tex
f brillant *m*, éclat *m*, luisance *f*, lustre *m*
e brillo *m*, lustre *m*
i lucentezza *f*, lustro *m*, splendore *m*
n glans
d Glanz *m*

3421 LUSTRE FINISH,
LUSTRING tex
f apprêt *m* à lustre
e acabado *m* lustrado
i lucidatura *f*
n glansappreteren *n*
d Glanzappretur *f*

3422 LUTEOLIN,
WELD dye
f lutéoline *f*
e luteolina *f*
i luteolina *f*
n luteoline
d Luteolin *n*

3423 LYE gen
f lessive *f*
e lejía *f*
i lisciva *f*
n loog
d Lauge *f*

3424 LYOPHILIC COLLOID,
EMULSOID col
f colloïde *m* liophyle
e coloide *m* liófilo
i colloide *m* liofilo
n emulsoïde *n*, reversibel colloïde *n*
d Emulsoid *n*, Emulsionskolloid *n*

3425 LYOPHOBIC COLLOID,
IRREVERSIBLE COLLOID col
f colloïde *m* irréversible
colloïde *m* lyophobe
e coloide *m* irreversible, coloide *m* liófobo
i colloide *m* irreversibile, colloide *m* liofobo
n irreversibel colloïde *n*, lyofoob colloïde *n*
d irreversibeles Kolloid *n*, lyophobes Kolloid *n*

M

3426 MACE OIL fat
f huile *f* de macis
e aceite *m* de macis
i olio *m* di macis
n macisolie
d Macisöl *n*, Muskatöl *n*

3427 MACERATION tan
f macération *f*
e maceración *f*
i macerazione *f*
n maceratie, uittrekken *n*, zweden *n*
d Aufquellen *n*, Mazeration *f*

3428 MACHINABILITY pla
f usinabilité *f*
e maquinabilidad *f*, trabajabilidad *f*
i lavorabilità *f*
n bewerkbaarheid
d Bearbeitbarkeit *f*

3429 MACHINE-FINISHED PAPER pap
f papier *m* à finissage à la machine
e papel *m* sin satinado
i carta *f* finita a macchina
n op de machine geglansd papier *n*
d maschinenglattes Papier *n*

3430 MACHINE PAPER pap
f papier *m* fabriqué à la machine
e papel *m* fabricado a máquina
i carta *f* a macchina
n machinaal vervaardigd papier *n*
d Maschinenpapier *n*

3431 MACHINERY OIL lub
f huile *f* pour machines
e aceite *m* para transmisiones
i olio *m* per macchine
n machineolie
d Maschinenöl *n*

3432 MACROCHEMISTRY gen
f macrochimie *f*
e macroquímica *f*
i macrochimica *f*
n macrochemie
d Makrochemie *f*

3433 MACRO-ETCH TESTING, DEEP-ETCH TESTING tes
f essai *m* de macro-attaque
e prueba *f* de ataque profundo, prueba *f* de macroataque
i collaudo *m* a macroincisione all'acquaforte, collaudo *m* ad attacco chimico profondo
n diepetsproef
d Tiefätzprobe *f*

3434 MADDER LAKE dec
f laque *f* de garance
e laca *f* de rubia
i lacca *f* di robbia
n kraplak, meekraplak
d Krapplack *m*

3435 MADE OF GLASS, GLASS gla
f de verre, en verre
e de vidrio, vítreo adj
i di vetro, vitreo adj
n glazen adj
d gläsern adj

3436 MAGNESIA GLASS gla
f verre *m* à magnésie
e vidrio *m* de magnesia
i vetro *m* a magnesia
n magnesiaglas *n*
d Magnesiaglas *n*

3437 MAGNESIUM SOAP soa
f savon *m* à magnésium
e jabón *m* de magnesio
i sapone *m* di magnesio
n magnesiumzeep
d Magnesiumseife *f*

3438 MAGNETIC SEPARATION sep
f séparation *f* magnétique
e separación *f* magnética
i separazione *f* magnetica
n magnetische scheiding
d magnetische Trennung *f*, Magnetscheidung *f*

3439 MAGNETOCHEMISTRY gen
f magnétochimie *f*
e magnetoquímica
i magnetochimica *f*
n magnetochemie
d Magnetochemie *f*

3440 MAIN FRACTION dsi
f fraction *f* principale
e fracción *f* principal

i frazione *f* principale
n hoofdfractie
d Hauptfraktion *f*

3441 MAIZE FLOUR (GB), CORN FLOUR (US) nut
f farine *f* de maïs
e harina *f* de maíz
i farina *f* di granturco, farina *f* di mais
n maïsmeel *n*, maizena
d Maismehl *n*, Maizena *n*

3442 MAIZE OIL (GB), CORN OIL (US) fat
f huile *f* de maïs
e aceite *m* de maíz
i olio *m* di granturco, olio *m* di mais
n maïsolie
d Maisöl *n*

3443 MAIZE STARCH (GB), CORN STARCH (US) nut
f amidon *m* de maïs, fécule *f* de maïs
e almidón *m* de maíz, fécula *m* de maíz
i amido *m* di granturco, amido *m* di mais, fecola *f* di granturco, fecola *f* di mais
n maïsstijfsel
d Maisstärke *f*

3444 MAJOLICA cer
f majolique *f*
e mayolica *f*
i maiolica *f*
n majolica *n*
d Majolika *f*

3445 MAKE-UP WATER boi
f eau *f* d'appoint, eau *f* de renivellement
e agua *f* de relleno
i acqua *f* di rimpiazzo
n bijvulwater *n*
d Zusatzwasser *n*

3446 MAKE-UP WATER cer
f eau *f* de pétrissage
e agua *f* de desleimiento
i acqua *f* d'impastatura
n aanmaakwater *n*
d Anreibewasser *n*

3447 MALDISTRIBUTION dis
f mauvaise répartition *f*
e mala distribución *f*
i cattiva distribuzione *f*
n slechte verdeling
d schlechte Verteilung *f*

3448 MALIC ACID, APPLE ACID oic
f acide *m* malique
e ácido *m* málico
i acido *m* malico
n appelzuur *n*
d Apfelsäure *f*

3449 MALM, WASHED CLAY bui
f marne *f* artificielle
e marga *f* artificial, mezcla *f* de greda y arcilla
i marna *f* artificiale
n kunstmergel
d künstlicher Mergel *m*

3450 MALT bre
f malt *m*
e malta *f*
i malto *m*
n mout *n*
d Malz *n*

3451 MALT CHARGE bre
f versement *m*
e carga *f* de malta
i carica *f* di malto
n moutlading
d Malzschüttung *f*

3452 MALT CLEANING, MALT DEGERMINATION bre
f dégermination *f*, nettoyage *m*
e desgerminación *f*, limpia *f*
i pulitura *f*
n moutpoetsen *n*
d Malzentkeimung *f*, Malzputzen *n*

3453 MALT COUCH, MALT PIECE bre
f tas *m* de malt
e montón *m* de malta, pila *f* de malta
i mucchio *m* di malto
n mouthoop
d Malzhaufen *m*

3454 MALT CRUSHING, MALT GRINDING bre
f concassage *m*
e trituración *f*
i frantumazione *f* di malto
n schroten *n* van mout
d Malzbrechen *n*

3455 MALT POLISHING bre
f polissage *m* du malt
e pulimento *m* de la malta
i lucidatura *f* del malto
n polijsten *n* van mout

d Abreiben *n* von Malz,
Polieren *n* von Malz

3456 MALT SUGAR, MALTOBIOSE, MALTOSE nut
f maltose *f*, sucre *m* de malt
e maltosa *f*
i maltosio *m*
n maltose, moutsuiker
d Maltose *f*, Malzzucker *m*

3457 MALTING bre
f maltage *m*
e maltaje *m*, malteo *m*
i fare v il malto, maltaggio *m*
n mouten *n*
d Malzen *n*

3458 MALTING LOSS bre
f freinte *f* de maltage
e merma *f* de malteo
i perdita *f* di maltaggio
n moutverlies *n*
d Mälzungsschwand *m*

3459 MAN-MADE FIBER (US), SYNTHETIC FIBRE (GB) tex
f fibre *f* artificielle
e fibra *f* artificial
i fibra *f* artificiale
n kunstvezel
d Chemiefaser *f*

3460 MAN-MADE RUBBER (US), ARTIFICIAL RUBBER, SYNTHETIC RUBBER (GB) rub
f caoutchouc *m* artificiel, caoutchouc *m* synthétique, élastomère *f* synthétique
e caucho *m* sintético, goma *f* artificial, goma *f* sintética
i gomma *f* sintetica
n kunstrubber, synthetische rubber
d Kunstgummi *m*, synthetischer Gummi *n*, synthetischer Kautschuck *m*

3461 MAN-MADE WOOL (US), ARTIFICIAL WOOL, SYNTHETIC WOOL (GB) tex
f laine *f* artificielle
e lana *f* artificial
i lana *f* artificiale
n kunstwol
d Zellwolle *f*

3462 MANGANESE DIOXIDE, BATTERY MANGANESE, BLACK OXIDE OF MANGANESE, BATTERY MANGANESE min
f pyrolusite *f*
e pirolusita *f*
i perossido *m* di manganese
n bruinsteen *n*, mangaansuperoxyde *n*
d Braunstein *m*, Mangandioxyd *n*, Pyrolusit *m*

3463 MANILLA PAPER, CARTRIDGE PAPER pap
f papier *m* bulle
e papel *m* manila
i carta *f* manilla
n manillapapier *n*
d Manillapapier *n*

3464 MANUFACTURED GAS, ILLUMINATING GAS, TOWN GAS gas
f gaz *m* d'éclairage, gaz *m* de ville
e gas *m* de ciudad, gas *m* del alumbrado
i gas *m* di città, gas *m* illuminante
n lichtgas *n*, stadsgas *n*
d Leuchtgas *n*, Stadtgas *n*

3465 MANUFACTURING PROCESS gen
f procès *m* de fabricage
e procedimiento *m* de fabricación
i processo *m* di fabbricazione
n fabricageproces *n*
d Herstellungsverfahren *n*

3466 MANURE fer
f fumier *m*
e abono *m*
i concime *m*
n mest
d Dünger *m*

3467 MANURING agr,fer
f fumage *m*
e abonado *m*
i concimazione *f*
n mesten *n*
d Düngung *f*

3468 MAPLE SUGAR sin
f sucre *m* d'érable
e azúcar *m* de arce
i zucchero *m* d'acero
n ahornsuiker
d Ahornzucker *m*

3469 MAPLE SYRUP sin
f sirop *m* d'érable
e jarabe *m* de arce
i sciroppo *m* d'acero
n ahornstroop
d Ahornsirup *m*

3470 MARBLED CLEAR THROUGH dec
f marbré adj sur toute l'épaisseur
e jaspeado adj hasta el fondo
i marmorizzato adj fino al fondo
n door en door geaderd adj ,
door en door gemarmerd adj
d durchmarmoriert adj

3471 MARBLING dec
f marbrage *m*
e jaspeado *m*, marmolado *m*
i marmorizzazione *f*
n aderen *n*, marmeren *n*
d Marmorieren *n*, Masern *n*

3472 MARGARINE nut
f margarine *f*
e margarina *f*
i margarina *f*
n margarine
d Margarine *f*

3473 MARGINAL LUBRICATION, lub
BOUNDARY LUBRICATION
f lubrification *f* dans la couche limite
e lubricación *f* en la capa límite
i lubrificazione *f* nello strato limite
n grenssmering
d Grenzschmierung *f*

3474 MARINE GLUE bui,glu
f colle *f* marine
e mezcla *f* de goma bruta, goma *f* laca y brea *f*
i colla *f* marina
n watervaste lijm
d wasserbeständiger Leim *m*

3475 MARJORAM OIL fat
f essence *f* de marjolaine
e esencia *f* de mejorana
i olio *m* di maggiorana
n marjoleinolie
d Majoranöl *n*

3476 MARKING OFF dye
f déteinte *f*
e desteñido *m*
i stintura *f*
n afgeven *n*
d Abfärben *n*

3477 MARKING INK tex
f encre *f* à marquer
e tinta *f* para marcar
i inchiostro *m* da marcare
n merkinkt
d Wäschetinte *f*

3478 MARL bui
f marne *f*
e marga *f*
i marna *f*
n mergel
d Mergel *m*

3479 MASH bre
f trempe *f*
e mezcla *f*
i tempera *f*
n beslag *n*
d Maische *f*

3480 MASH GOODS bre
f ensemble *m* de la trempe
e total *m* de la mezcla
i impasto *m* totale
n beslagmassa
d Maischgut *n*

3481 MASH LIQUOR bre
f eau *f* d'empâtage
e agua *f* de mezcla
i acqua *f* per l'impasto
n beslagwater *n*
d Einmaischwasser *n*,
Guss *m*, Maischwasser *n*

3482 MASHING nut
f empâtage *m*
e empastado *m*
i impastatura *f*
n kneden *n*
d Einteigen *n*

3483 MASHING SYSTEM bre
f méthode *f* de brassage
e método *m* de braceado
i metodo *m* d'ammostatura
n beslagsysteem *n*
d Maischverfahren *n*

3484 MASK SPRAYING dec
f pistolage *m* à masque
e pintura *f* con estarcido
i pittura *f* a maschera
n sjabloonspuiten *n*
d Schablonenspritzung *f*

3485 MASS COLORATION dye
f teinture *f* en masse
e teñido *m* de masa
i tintura *f* in massa
n massaverven *n*
d Massenfärbung *f*

3486 MASS CONCRETE bui
f béton *m* non-armé
e hormigón *m* sin armor
i calcestruzzo *m* senza armatura
n ongewapend beton *n*
d unbewehrter Beton *m*

3487 MASS DIFFUSION sep
f diffusion *f* massique
e difusión *f* másica
i diffusione *f* di massa
n massadiffusie
d Massendiffusion *f*

3488 MASS POLYMERIZATION oic
f polymérisation *f* massique
e polimerización *f* másica
i polimerizzazione *f* in massa
n massapolymerisatie
d Massenpolymerisation *f*

3489 MASS SPECTROGRAPHY spe
f spectrographie *f* de masse(s)
e espectrografía *f* de masa
i spettrografia *f* di massa
n massaspectrografie
d Massenspektrographie *f*

3490 MASSECUITE sin
f massecuite *f*
e masacocida *f*
i massacotta *f*
n kookmassa
d Füllmasse *f*

3491 MASSICOT, dec
LEAD MONOXIDE, LITHARGE
f litharge *f*, oxyde *m* de plomb
e litargirio *m*, óxido *m* de plomo
i litargirio *m*, ossido *m* di piombo
n loodglit *n*, massicot *n*
d Bleiglätte *f*, Massikot *m*

3492 MASTER COLORS (US), dec
MASTER COLOURS (GB)
f teintes *pl* de base, teintes *pl* mère
e colores *pl* de base
i colori *pl* di base
n basiskleuren *pl*, moederkleuren *pl*
d Basisfarben *pl*

3493 MASTIC bui
f mastic *m*
e mástique *m*
i mastice *m*
n mastiek
d Mastix *m*

3494 MASTIC dec
f ciment *m*, lut *m*
e masilla *f*
i mastice *m*
n stopverf
d Glaserkitt *m*

3495 MASTIC ASPHALT bui
f asphalte *m* coulé, mastic *m* d'asphalte
e asfalto *m* fundido, mástique *m* asfaltado
i mastice *m* d'asfalto
n asfaltmastiek, gietasfalt *m*
d Gussasphalt *m*, Splittasphalt *m*

3496 MASTIC BLOC, bui
ASPHALT BLOCK, ASPHALT CAKE
f bloc *m* d'asphalte
e pàne *m* de asfalto
i asfalto *m* in pani, mattonella *f* d'asfalto
n asfaltblok *n*, asfaltbrood *n*
d Asphaltbrot *n*, Asphaltstein *m*

3497 MASTIC GUM dec,mcy,nut
f mastic *m*
e almáciga *f*
i gomma *f* di lentischio
n lantisk, mastiek, mastix
d Mastixharz *n*

3498 MASTICATION pla,rub
f mastication *f*, plastification *f*
e amasado *m*, masticación *f*, plastificación *f*
i masticazione *f*, plastificazione
n plastificeren *n*
d Mastikation *f*, Plastizieren *n*

3499 MAT ETCHING, cle
DULL FINISH
f décapage *m* au mat
e decapaje *m* mate
i decapaggio *m* matto
n matbeitsen *n*
d Mattbeizung *f*

3500 MATCHING dye
f appareillement *m*
e armonización *f*
i adattamento *m*
n aanpassing
d Farbgleichheit *f*

3501 MATERIAL TESTING tes
f essai *m* de matériaux
e ensayo *m* de materiales
i collaudo *m* di materiali
n materiaalonderzoek *n*
d Materialprüfung *f*

3502 MATURATION bre,cer
f maturation *f*
e maduración *f*
i maturazione *f*
n rijping
d Reifung *f*

3503 MATURING RANGE cer
f gamme *f* de maturation
e gama *f* de maduración
i gamma *f* di maturazione
n rijpingsgebied *n*
d Reifungsbereich *m*

3504 MEAGRE LIME, HYDRAULIC LIME, LEAN LIME bui
f chaux *f* hydraulique
e cal *f* hidráulica, cal *f* magra
i calce *f* idraulica
n hydraulische kalk, magere kalk
d Wasserkalk *m*

3505 MEAL POWDER, MEALED POWDER exp
f pulvérin *m*
e polverín *m*
i polverino *m*
n fijn kruit *n*, pankruit *n*
d Zündpulver *n*

3506 MEASURE mea
f mesure *f*
e medida *f*
i misura *f*
n maat
d Mass *n*

3507 MEASUREMENT mea
f mesure *f*
e medida *f*
i misura *f*
n meting
d Messung *f*

3508 MEASURING mea
f mesure *f*
e medida *f*
i misurazione *f*
n meten *n*
d Messen *n*

3509 MECHANICAL FROTH FLOTATION min
f flottation *f* par moussage mécanique
e flotación *f* por espuma mécanica
i flottazione *f* per schiuma meccanica
n mechanische schuimflotatie
d mechanische Schwimmaufbereitung *f*, mechanische Schaumflotation *f*

3510 MECHANICAL POLISHING, SANDING pol
f polissage *m* mécanique
e pulimento *m* mecánico
i lucidatura *f* meccanica
n mechanisch polijsten *n*
d maschinelle Polierung *f*

3511 MECHANICAL PROPERTIES gen
f propriétés *pl* mécaniques
e propiedades *pl* mecánicas
i proprietà *pl* meccaniche
n mechanische eigenschappen *pl*
d mechanische Eigenschaften *pl*

3512 MECHANICAL STRIPPING coa
f dépouillage *m* mécanique
e eliminación *f* mecánica
i degalvanizzazione *f* meccanica
n strippen *n*
d Abschleifen *n*, Abziehen *n*

3513 MECHANICAL TEST tes
f essai *m* mécanique
e ensayo *m* mecánico
i collaudo *m* meccanico
n mechanische beproeving
d mechanische Prüfung *f*

3514 MECHANICAL WOODPULP, HOT-GROUND PULP pap
f pâte *f* de bois
e pulpa *f* de madera
i polpa *f* di legno
n houtslijp *n*, houtstof *n*
d Holzschliff *m*, Holzstoff *m*

3515 MEDICATED SOAP mcy,soa
f savon *m* médicinal
e jabón *m* medicinal
i sapone *m* medicinale
n medicinale zeep
d medizinale Seife *f*

3516 MEDICINAL OIL mcy
f huile *f* médicinale
e aceite *m* medicinal
i olio *m* medicinale
n medicinale olie
d Medizinalöl *n*

3517 MEDICINE mcy
f médicament *m*, remède *m*
e medicamento *m*, medicina *f*, remedio *m*
i medicamento *m*, medicina *f*, rimedio *m*
n geneesmiddel *n*, medicijn
d Arznei *f*, Heilmittel *n*, Medikament *n*

3518 MEDIUM-SIZE GRINDING brk
f broyage *m* mi-fin

e trituración *f* de tamaño medio
i frantumazione *f* a dimensione media
n breken *n* op middelgrootte
d Nachbrechen *n*

3519 MEERSCHAUM, SEPIOLITE mat
f écume *f* de mer, sépiolite *f*
e espuma *f* de mar, sepiolita *f*
i schiuma *f* di mare, sepiolite *f*
n meerschuim *n*, sepioliet *n*
d Meerschaum *m*, Sepiolith *m*

3520 MELISSA OIL fat
f essence *f* de mélisse
e esencia *f* de melisa
i olio *m* di melissa
n melisseolie
d Melissenöl *n*

3521 MELLOWING tan
f perte *f* d'astringence
e pérdida *f* de astringencia
i perdita *f* del carattere astringente
n ouderen *n* van de looivloeistof
d Altern *n* der Brühe

3522 MELT gla
f fonte *f*
e masa *f* fundida
i fusione *f*
n smelt
d Schmelze *f*

3523 MELTABLE gen
f fusible adj
e fusible adj
i fusibile adj
n smeltbaar adj
d schmelzbar adj

3524 MELTING, SMELTING gen
f fusion *f*
e fusión *f*
i fusione *f*
n smelten *n*
d Schmelzen *n*

3525 MELTING LOSS htt
f pertes *pl* de fusion
e pérdida *f* de fusión
i perdita *f* di fusione
n smeltverlies *n*
d Abbrand *m*, Verlust *m*

3526 MELTING POINT gen
f point *m* de fusion
e punto *m* de fusión
i punto *m* di fusione
n smeltpunt *n*
d Schmelzpunkt *m*

3527 MELTING RANGE htt
f intervalle *m* de fusion
e intervalo *m* de fusión
i intervallo *m* di fusione
n smeltinterval *n*, smelttraject *n*
d Schmelzbereich *m*, Schmelzintervall *n*

3528 MELTING RATE htt
f vitesse *f* de fusion
e velocidad *f* de fusión
i velocità *f* di fusione
n smeltsnelheid
d Schmelzgeschwindigkeit *f*

3529 MENSTRUM, EXCIPIENT mcy
f excipient *m*
e excipiente *m*
i eccipiente *m*
n menstruum *n*, vloeibaar oplossingsmiddel *n*
d Arzneimittelträger *m*

3530 MERCERIZATION, STEEPING tex
f mercerisage *m*
e mercerización *f*
i mercerizzazione *f*
n merceriseren *n*
d Mercerieren *n*, Mercerisieren *n*

3531 MERCURATION, MERCURIZATION oic
f mercuration *f*
e mercuración *f*
i mercurazione *f*
n mercuriseren *n*
d Mercurieren *n*

3532 MERCURIC CHLORIDE, CORROSIVE SUBLIMATE cor
f sublimé *m* corrosif
e sublimado *m* corrosivo
i sublimato *m* corrosivo
n bijtend sublimaat *n*
d Ätzquecksilber *n*, Ätzsublimat *n*

3533 MERCURIFICATION met
f extraction *f* du mercure
e mercurificación *f*
i estrazione *f* del mercurio
n kwikwinning
d Quecksilbergewinnung *f*

3534 MERCUROUS CHLORIDE, elc
CALOMEL
f calomel *m*, chlorure *m* de mercure
e calomelano *m*, cloruro *m* mercurioso
i calomelano *m*, cloruro *m* mercuroso
n kalomel *n*, mercurochloride *n*
d Kalomel *n*, Mercurochlorid *n*, Quecksilberchlorur *n*

3535 MERCURY-CATHODE METHOD OF ELECTROLYSIS elc
f électrolyse *f* à cathode de mercure
e electrólisis *f* con cátodo de mercurio
i elettrolisi *f* con catodo di mercurio
n elektrolyse met kwikkatode
d Elektrolyse *f* mit Quecksilber-elektrode

3536 MERCURY PENETRATION METHOD cer,tes
f essai *m* de pénétration de mercure
e ensayo *m* de penetración de mercurio
i collaudo *m* di penetrazione di mercurio
n kwikindringproef
d Quecksilberdurchdringungs-verfahren *n*

3537 METACHROME MORDANT, dye
CHROMATE MORDANT,
MONOCHROME MORDANT
f mordant *m* au chromate
e mordiente *m* metacromático
i mordente *m* al cromato
n chromaatbijtmiddel *n*
d Chromatbeize *f*

3538 METACHROME PROCESS, dye,tex
CHROMATE DYEING,
MONOCHROME PROCESS
f teinture *f* à chromate
e teñido *m* de cromato
i tintura *f* a cromato
n chromaatverven *n*
d Chromatfärbung *f*

3539 METAFILTRATION fil
f filtration *f* à bourrelets
e filtración *f* de fajas
i filtrazione *f* a strisce
n stripfiltratie
d Streifenfiltrierung *f*

3540 METAL BATH met
f bain *m* de métal fondu
e baño *m* de metal fundido
i bagno *m* di metallo fuso
n gesmolten-metaalbad *n*, smeltbad *n*
d Metallbad *n*, Schmelzbad *n*

3541 METAL-CERAMICS cer
f céramique *f* métallique
e metalocerámica *f*
i ceramica *f* metallica
n metaalkeramiek
d keramische Metallurgie *f*, Metallkeramik *f*

3542 METAL-COMPLEX DYE dye
f colorant *m* à base de composés métalliques complexes
e colorante *m* a base de compuestos metálicos complejos
i colorante *m* a base di composti metallici complessi
n kleurstof op basis van complexe metaalverbindingen
d Farbstoff *m* auf Basis von komplexen Metallverbindungen

3543 METAL CONDITIONS (US), dec
ETCHING PRIMER (GB),
WASH PRIMER (GB)
f peinture *f* pour couche primaire phosphante
e pintura *f* de mano primera fosfatizante
i pittura *f* di mano prima a base di fosfato
n hechtgrondverf voor metalen
d Haftgrundmittel *n*, reaktionsfähiger Haftgrund *m*, Metallwandler *m*

3544 METAL DISTRIBUTION RATIO, elc
DEPOSITION EFFICIENCY
f rapport *m* de distribution du métal
e relación *f* de distribución del metal
i rapporto *m* di distribuzione del metallo
n dikteverhouding van het neerslag
d Niederschlagsverteilungs-verhältnis *n*

3545 METAL FOG, elc,gen
METAL MIST
f brouillard *m* métallique
e metal *m* en suspensión, neblina *f* metálica
i nebbia *f* catodica, nebbia *f* metallica

n metaalnevel
d Metallnebel *m*

3546 METAL PLATING, METALLIZATION coa
f métallisation *f*
e metalización *f*
i metallizzazione *f*
n metalliseren *n*
d Metallisieren *n*

3547 METAL POLISH, BRASS POLISH cle
f moyen *m* de polissage, produit *m* à fourbir
e lustre *m* para metales, pasta *f* limpiadora
i lucidatrice *f* per metalli
n koperpoets, metaalpoets
d Putzmittel *n*

3548 METAL SPRAYING coa
f métallisation *f* par projection, pulvérisation *f* de métal
e metalización *f* por chorreado, pulverización *f* de metal
i metallizzazione *f* a spruzzo, polverizzazione *f* di metallo
n metaalspuiten *n*, metaalverstuiving
d Metallspritzen *n*, Metallzerstäubung *f*

3549 METAL TONERS pla
f pigments *pl* organiques à base de sels de métaux lourds
e pigmentos *pl* orgánicos a base de sales de metales pesados
i pigmenti *pl* organici a base di sali di metalli pesanti
n organische pigmenten *pl* op basis van zouten van zware metalen
d organische Pigmente *pl* auf Basis von Schwermetallsalzen

3550 METAL TURBIDITY gen
f trouble *m* dû au métal
e enturbiamiento *m* por metal
i intorbidamento *m* per metallo
n metaaltroebeling
d Metalltrübung *f*

3551 METALLIC COATING coa
f revêtement *m* métallique
e revestimiento *m* metálico
i rivestimento *m* metallico
n metaalbedekking
d Metallüberzug *m*

3552 METALLIC PIGMENT dec,dye
f pigment *m* métallique
e pigmento *m* metálico
i pigmento *m* metallico
n metalliek pigment *n*
d metallisches Pigment *n*

3553 METALLIC SOAP dec
f savon *m* métallique
e jabón *m* metálico
i sapone *m* metallico
n metaalzeep
d Metallseife *f*

3554 METALLIC TONING pho
f virage *m* métallique
e viraje *m* metálico
i viraggio *m* metallico
n metalliek kleuren *n*
d Metalltonung *f*

3555 METALLOCHROMY met
f métallochromie *f*
e metalocromía *f*
i metallocromia *f*
n chemische metaalkleuring
d chemische Metallfärbung *f*

3556 METALLURGICAL CEMENT, SULFUR CEMENT (US), SULPHUR CEMENT (GB) bui
f ciment *m* de fer, ciment *m* métallurgique
e cemento *m* de azufre
i cemento *m* di zolfo, mastice *m* di ferro
n ijzercement *n*
d Eisenkitt *m*, Schwefelkitt *m*, Schwefelzement *m*

3557 METHACRYLATE RESINS pla
f résines *pl* méthacryliques
e resinas *pl* metacrílicas
i resine *pl* metacriliche
n methacrylharsen *pl*
d Methakrylharze *pl*

3558 METHANATION, METHANIZATION oic
f méthanisation *f*
e metanización *f*
i metanizzazione *f*
n methaniseren *n*
d Methanisieren *n*

3559 METHANOL, METHYL ALCOHOL, WOOD SPIRIT dis
f alcool *m* méthylique, esprit *m* de bois, esprit *m* pyroxylique

e alcohol *m* metílico, espíritu *m* de madera, metanol *m*
i alcool *m* metilico, spirito *m* di legno
n houtgeest, methylalcohol
d Holzgeist *m*, Methylalkohol *m*

3560 METHYLATION oic
f méthylation *f*
e metilación *f*
i metilazione *f*
n methyleren *n*
d Methylieren *n*

3561 MICROANALYSIS ana
f microanalyse *f*
e microanálisis *f*
i microanalisi *f*
n microanalyse
d Mikroanalyse *f*

3562 MICROBALLOON eva
f microballon *m* en matière plastique
e microbalón *m* en material plástico
i microsfera *f* in materiale plastico
n microluchtballon uit kunststof
d Mikroluftballon *m*

3563 MIDDLE DISTILLATE dis
f moyen distillat *m*
e destilado *m* medio
i distillato *m* medio
n middendestillaat *n*
d Mitteldestillat *n*

3564 MIDDLE OIL, CARBOLIC OIL fue
f huile *f* carbolique
e aceite *m* carbólico
i olio *m* carbolico
n carbololie, middelolie
d Karbolöl *n*, Mittelöl *n*, Phenolöl *n*

3565 MIGRATION dye,elc
f migration *f*
e migración *f*
i migrazione *f*
n migratie, zwerven *n*
d Migration *f*, Wanderung *f*

3566 MIGRATION INHIBITOR dye
f inhibiteur *m* de migration
e inhibidor *m* de migración
i inhibitore *m* di migrazione
n migratie-inhibitor
d Migrationsinhibitor *m*

3567 MIGRATION OF PLASTICIZER pla,rub
f migration *f* d'un plastifiant
e migración *f* de un plastificante
i migrazione *f* d'un plastificante
n migratie van weekmaker
d Weichmacherwanderung *f*

3568 MILD CLAY bui
LOAM, SANDY CLAY
f terre *f* glaise
e tierra *f* arcillosa
i terra *f* argillosa, terra *f* grassa
n leem *n*
d Lehm *m*

3569 MILDEW INHIBITOR, agr
MILDEW PREVENTING AGENT
f agent *m* antimildiou
e inhibidor *m* de mildiu
i inhibitore *m* di muffa
n meeldauwwerende stof
d meltauverhütendes Mittel *n*

3570 MILITARY GAS gas,mit
f gaz *m* de guerre
e gas *m* de guerra
i gas *m* di guerra
n oorlogsgas *n*
d Kriegsgas *n*

3571 MILK dai,nut
f lait *m*
e leche *f*
i latte *m*
n melk
d Milch *f*

3572 MILK GLASS gla
f verre *m* opale
e vidrio *m* opalino
i vetro *m* latteo, vetro *m* opalo
n melkglas *n*
d Milchglas *n*

3573 MILK OF LIME iwt
LIME SLURRY
f lait *m* de chaux
e lechada *f* de cal
i latte *m* di calce
n kalkmelk
d Kalkbrühe *f*, Schlämme *f*

3574 MILK SUGAR, dai,sin
LACTOBIOSE, LACTOSE
f lactose *f*
e lactosa *f*
i lattosio *m*
n melksuiker
d Milchzucker *m*

3575 MILL-BASE, dec
GRINDING AGENT,
GRINDING MEDIUM
f base *f* de broyage, liant *m* de broyage
e aglutinante *m* de molido
i legante *m* di molitura
n maalbindmiddel *n*
d Bindemittel *n* zum Anreiben

3576 MILLBOARD pap
f fort carton *m*, gros carton *m*
e cartón *m* grueso
i cartone *m* grossolano
n grof karton *n*
d Rohpappe *f*

3577 MILLING, tex
FULLING
f foulage *m*
e batanadura *f*
i follatura *f*
n vollen *n*
d Walken *n*

3578 MILLING (GB), brk
MULLING (US)
f frottage *m*
e amasado *m*
i molazzatura *f*
n kolleren *n*
d Kollern *n*

3579 MINERAL min
f minerai *m*
e mineral *m*
i minerale *m*
n delfstof, mineraal *n*
d Mineral *n*

3580 MINERAL CAOUTCHOUC, min
ELASTIC BITUMEN,
ELATERITE
f bitume *m* élastique, élatérite *f*
e caucho *m* mineral, elaterita *f*
i elaterite *f*
n elateriet *n*, mineraal caoutchouc *n*, veerkrachtig aardpek *n*
d elastisches Erdpech *n*, Elaterit *m*

3581 MINERAL COLORS (US), dec
MINERAL PIGMENTS (GB)
f matières *pl* colorantes minerales
e materias *pl* colorantes minerales
i materie *pl* coloranti minerali
n minerale verfstoffen *pl*
d Mineralfarben *pl*

3582 MINERAL OIL fue
f huile *f* minérale
e aceite *m* mineral
i olio *m* minerale
n aardolie
d Erdöl *n*

3583 MINERAL PITCH, bui
ASPHALT, BITUMEN,
JUDEAN PITCH
f asphalte *m*, poix *f* de Judée
e asfalto *m*
i asfalto *m*
n aardpek *n*, asfalt *n*, jodenpek *n*
d Asphalt *m*, Bitumen *n*, Erdpech *n*

3584 MINERAL SPIRIT, dec
MINERAL TURPENTINE,
WHITE SPIRIT
f essence *f* minérale, white-spiri
e aguarrás *m* mineral, trementina mineral
i surrogato *m* d'acqua ragia
n lakbenzine
d Lackbenzin *n*, Schwerbenzin *n*, Terpentinersatz *m*, Testbenzin *n*

3585 MINERAL WOOL, clo
SLAG WOOL
f laine *f* de laitier
e lana *f* de escorias
i lana *f* di scorie
n slakkenwol
d Schlackenwolle *f*

3586 MINING EXPLOSIVE exp
f explosif *m* de sécurité
e explosión *f* de minería
i esplosivo *m* da miniera
n mijngasveilige springstof
d Grubensprengstoff *m*

3587 MINIUM, dec
RED LEAD
f minium *m* de plomb
e minio *m* de plomo
i minio *m* di piombo
n loodmenie, menie
d Bleimennige *f*, Mennige *f*

3588 MISCH METAL, AUER METAL — met
f mischmétal *m*
e mischmetal *m*
i mischmetallo *m*
n mischmetaal *n*
d Mischmetall *n*

3589 MISCIBILITY — mix
f miscibilité *f*
e miscibilidad *f*
i miscibilità *f*
n mengbaarheid
d Mischbarkeit *f*

3590 MISCIBILITY GAP — mix
f intervalle *m* de miscibilité
e intervalo *m* de miscibilidad
i intervallo *m* di miscibilità
n onderbreking van de mengbaarheid
d Mischungslücke *f*

3591 MISCIBLE — mix
f miscible adj
e miscible adj
i miscibile adj
n mengbaar adj
d mischbar adj

3592 MISFIRING — exp
f raté *m*
e fallo *m*
i far v cilecca
n ketsen *n*
d Fehlzündung *f*

3593 MIXED ACID — aic
f acide *m* sulfonitrique
e mezcla *f* sulfonítrica
i miscela *f* solfonitrica
n nitreerzuur *n*
d Nitriersäure *f*

3594 MIXED CATALYST — cat
f catalyseur *m* mixte
e catalizador *m* mixto
i catalizzatore *m* misto
n samengestelde katalysator
d Mehrstoffkatalysator *m*, Mischkatalysator *m*

3595 MIXED CRYSTAL, PRIMARY SOLUTION, SOLID SOLUTION — gen
f cristal *m* mixte, solution *f* solide
e cristal *m* mixto, solución *f* sólida
i cristallo *m* misto, soluzione *f* solida
n mengkristal *n*, vaste oplossing
d feste Lösung *f*, Mischkristall *m*

3596 MIXED FERTILIZER — fer
f engrais *m* composé
e abono *m* compuesto
i concime *m* composto
n samengestelde meststof
d Mischdünger *m*

3597 MIXED PHASE CRACKING — fue
f craquage *m* en phase mixte
e pirólisis *f* en fase mixta
i piroscissione *f* in fase mista
n kraken *n* volgens het gemengdefazenproces
d Kracken *n* nach dem Gemischtphaseverfahren

3598 MIXED YEAST — fmi
f mélange *m* de levures
e mezcla *f* de levaduras
i miscela *f* di lieviti
n gistmengsel *n*
d Mischhefe *f*

3599 MIXING — mix
f mélange *m*
e mezcladura *f*
i mescolatura *f*
n mengen *n*
d Mischen *n*

3600 MIXING, BLENDING — mix
f mélange *m*, mixtion *f*
e mezcla *f*, mixtión *f*
i mescolamento *m*, mescolanza *f*
n menging
d Mischung *f*

3601 MIXTURE — mcy
f mixture *f*
e mixtura *f*, poción *f*
i mistura *f*
n mixtuur
d Mixtur *f*

3602 MIXTURE — mix
f mélange *m*
e mezcla *f*
i mescolanza *f*, miscela *f*, miscuglio *m*
n mengsel *n*
d Gemisch *n*, Mischung *f*

3603 MIXTURE DYE — dye
f colorant *m* mélangé
e colorante *m* mezclado
i colorante *m* misto
n gemengde kleurstof
d Mischfarbstoff *m*

3604 MIXTURE RATIO mix
f proportions *pl* du mélange
e proporciones *pl* de la mezcla
i rapporto *m* della miscela
n mengverhouding
d Mischungsverhältnis *n*

3605 MODERATE ACCELERATOR rub
f accélérateur *m* à action modérée
e acelerador *m* de acción moderada
i accelerante *m* ad azione moderata
n matig sterke versneller
d mittelstarker Beschleuniger *m*

3606 MODERATELY HARD WATER hyg
f eau *f* moyennement dure
e agua *f* moderadamente dura
i acqua *f* di durezza media
n matig hard water *n*
d mässig hartes Wasser *n*

3607 MODERATOR nue
f modérateur *m*
e moderador *m*
i moderatore *m*
n moderator, remstof
d Bremsstoff *m*, Moderator *m*

3608 MODERATOR COOLANT nue
f modérateur-refroidisseur *m*
e moderador *m* refrigerante
i moderatore *m* raffreddatore
n koelende moderator, koelende remstof
d Kühlmoderator, Kühlbremsstoff *m*

3609 MODIFICATION bre
f désagrégation *f*
e desagregación *f*
i disgregazione *f*
n ontbinding
d Auflösung *f*

3610 MODIFIED AERATION, HIGH-RATE ACTIVATED SLUDGE PROCESS iwt
f procédé *m* des boues activées à haut rendement
e procedimiento *m* de fango activado a alto rendimiento
i processo *m* a fango attivato ad alta dosatura
n geactiveerd-slibproces *n* met hoog rendement
d hochbelastetes Belebungsverfahren *n*, Hochleistungsbelebtschlammverfahren *n*

3611 MODIFIED RESIN pla
f résine *f* modifiée
e resina *f* modificada
i resina *f* modificata
n gemodificeerd hars *n*
d modifiziertes Harz *n*

3612 MODIFIER, REGULATOR pla
f agent *m* modificateur
e agente *m* modificador
i agente *m* modificatore
n modificator
d Abänderungsmittel *n*

3613 MOELLEN, DEGRAS tan
f dégras *m*
e churre *f*
i grasso *m* ricavato
n afvalvet *n*, dégras *n*, traan-wateremulsie
d Abfett *n*, Degras *n*, Gerberfett *n*, Lederschmiere *f*, Moellon *n*

3614 MOIL gla
f déchets *pl* de verrier
e desperdicios *pl* de vidrio
i rifiuti *pl* di soffiatore
n glasafval *n*
d Glasabfall *m*

3615 MOISTENING, HUMIDIFYING hyg
f humectation *f*, humidification *f*
e humectación *f*, humidificación *f*
i umidificazione *f*
n bevochtiging
d Anfeuchtung *f*, Befeuchtung *f*

3616 MOISTURE-BARRIER PAINT COAT dec
f couche *f* de peinture imperméable à l'humidité
e mano *f* de pintura hidrófuga
i mano *f* di pittura idrofuga
n vochtafsluitende verflaag
d feuchtigkeitsabschliessender Anstrich *m*

3617 MOISTURE CONTENT ana
f pourcentage *m* d'humidité
e porcentaje *m* de humedad
i percentuale *m* d'umidità
n vochtgehalte *n*
d Feuchtigkeitsgehalt *m*

3618 MOISTURE EXPANSION, BULKING for
f gonflement *m* dû à l'humidité
e hinchazón *f* por humedad
i gonfiamento *m* dovuto all'umidità

n zwelling door vocht
d Feuchtigkeitsausdehnung *f*, Quellung *f*

3619 MOISTUREPROOF tex
f hydrofuge adj
e hidrófugo adj
i idrofugo adj
n tegen vocht beschermd
d mit Feuchtigkeitsschutz

3620 MOLASSES, BEET SYRUP sin
f mélasse *f*
e melaza *f*
i melassa *f*
n melasse
d Melasse *f*, Rübensirup *m*

3621 MOLASSES PULP bre
f pulpe *f* de mélasse
e pulpa *f* de melaza
i polpa *f* di melassa
n melassepulp
d Melasseschnitzel *pl*, melassierte Schnitzel *pl*

3622 MOLASSES SLOP sin
f rinçure *f* de mélasse
e lavazas *pl* de melaza
i risciacquatura *f* di melassa
n spoelwater *n*
d Spülwasser *n*

3623 MOLD LOAD (US), MOULD CHARGE (GB) pla
f charge *f* de moule
e carga *f* de molde
i carica *f* di forma
n vormvulling
d Pressformbeschickung *f*

3624 MOLD PARTING AGENT (US), RELEASE AGENT (GB) pla
f agent *m* de démoulage
e agente *m* de desmoldeado
i agente *m* di sformatura
n losmiddel *n*
d Trennmittel *n*

3625 MOLDING (US), MOULDING (GB) pla,pre
f moulage *m*
e moldeo *m*
i stampatura *f*
n persen *n*
d Pressen *n*

3626 MOLDING ALLOWANCE (US), MOULDING ALLOWANCE (GB), CONTRACTION ALLOWANCE pla
f double retrait *m*
e doble contracción *f*
i doppio ritiro *m*
n kruiptoegift
d doppeltes Schwindmass *n*

3627 MOLDING POWDER (US), MOULDING POWDER (GB) pla
f poudre *m* de moulage
e polvo *m* para moldeo
i polvere *f* per formare
n perspoeder *n*
d Presspulver *n*

3628 MOLDING SHRINKAGE (US), MOULDING SHRINKAGE (GB) pla
f retrait *m* de moule
e contracción *f* de moldeo
i ritiro *m* di forma
n vormkrimp
d Formschwindung *f*

3629 MOLECULAR DISTILLATION dis
f distillation *f* moléculaire
e destilación *f* molecular
i distillazione *f* molecolare
n moleculaire destillatie
d Kurzwegdestillation *f*, Molekulardestillation *f*

3630 MOLTEN METAL DYEING PROCESS dye
f procédé *m* de teinture continue en métal fondu
e procedimiento *m* de teñido continuo en metal fundido
i processo *m* di tintura continua in metallo fuso
n verfproces *n* in gesmolten metaal
d Färbeverfahren *n* in geschmolzenem Metall

3631 MONOCHLORETHANE, CHALAT (US), ETHYLCHLORIDE (GB) oic
f chlorure *m* d'éthyl
e cloruro *m* etílico
i cloruro *m* d'etile
n ethylchloride *n*, monochloorethaan *n*
d Äthylchlorid *n*, Chelen *n*, Chloräthyl *n*

3632 MONOCHROME MORDANT, CHROMATE MORDANT, METACHROME MORDANT dye
f mordant *m* au chromate
e mordiente *m* metacromático
i mordente *m* al cromato
n chromaatbijtmiddel *n*
d Chromatbeize *f*

3633 MONOCHROME PROCESS, dye,tex
CHROMATE DYEING,
METACHROME PROCESS
f teinture *f* à chromate
e teñido *m* de cromato
i tintura *f* a cromato
n chromaatverven *n*
d Chromatfärbung *f*

3634 MONOMOLECULAR LAYER, coa,fat
UNIMOLECULAR LAYER
f couche *f* monomoléculaire
e capa *f* monomolecular
i strato *m* monomolecolare
n monomoleculaire laag
d monomolekulare Schicht *f*

3635 MONTAN WAX, fat
STONE WAX
f cire *f* de lignite
e cera *f* montana
i cera *f* montana
n montaanwas
d Montanwachs *n*

3636 MORDANT, dye
ROTSTEEP
f mordant *m*
e mordiente *m*
i mordente *m*
n beitsmiddel *n*, bijtmiddel *n*
d Ätzwasser *n*, Beize *f*,
Beizmittel *n*, Schutzbeize *f*

3637 MORDANT DYE dye
f colorant *m* mordant
e colorante *m* mordiente
i colorante *m* mordente
n beitskleurstof
d Beizenfarbstoff *m*

3638 MORDANTING dye,pho
f mordançage *m*
e mordentación *f*
i mordenzatura *f*
n beitsen *n*
d Beizen *n*

3639 MORDANTING BATH dye
f bain *m* de mordançage
e baño *m* de mordentación
i bagno *m* di mordenzatura
n beitsbad *n*
d Beizbad *n*

3640 MORE VOLATILE dis
COMPONENT
f constituant *m* plus volatil
e componente *m* más volátil
i componente *m* più volatile
n gemakkelijker vluchtige
component
d leichterflüchtige Komponente *f*

3641 MOROCCO lea
f maroquin *m*
e marroquí *m*, tafilete *m*
i marocchino *m*
n marokijn *n*
d Marokanischleder *n*, Maroquin *n*,
Saffian *n*

3642 MORTAR bui
f mortier *m*
e mortero *m*
i malta *f*
n mortel, specie
d Mörtel *m*

3643 MORTAR FOR FIRST LAYER bui
OF RENDERING,
COARSE STUFF
f mortier *m* pour la couche de
fond de l'enduit
e enfoscado *m* primo, mezcla *f* de
mortero de cal con pelos
i impasto *m* di calce e peli per
intonaco
n grondspecie
d Grundmörtel *m*

3644 MOSAIC GOLD (GB), dec
STANNIC SULFIDE (US)
f or *m* musif
e oro *m* musivo
i oro *m* musivo
n musiefgoud *n*, vals schelpgoud *n*
d Musivgold *n*, unechtes
Muschelgold *n*, Zinn-IV-Sulfid *n*

3645 MOSSY ZINC met
f grenaille *f* de zinc
e cinc *m* metálico desintegrado,
granalla *f* de cinc
i granuli *pl* di zinco
n zinkkorrels *pl*
d Zinkgries *m*

3646 MOTHER LIQUOR cry
f eau *f* de cristallisation
e agua *f* de cristalización
i acqua *f* di cristallizzazione
n kristallisatiewater *n*
d Kristallisierungswasser *n*

3647 MOTHER-LYE, aic,gen
BITTERN, LEACH BRINE
f eau *f* mère
e agua *f* madre
i acqua *f* madre

n moederloog
d Bitterlauge *f*, Mutterlauge *f*, Muttersole *f*

3648 MOTHER SYRUP (GB), HIGH WASH (US) sin
f eau-mère *f* jaunâtre
e jarabe *m* blanco
i sciroppo *m* bianco
n moederstroop
d Muttersirup *m*

3649 MOTOR OIL lub
f huile *f* pour moteurs
e aceite *m* para motores
i olio *m* per motori
n motorolie
d Motorenöl *n*, Treiböl *n*

3650 MOTOR SPIRIT (GB), GASOLINE (US), PETROL fue
f essence *f*
e gasolina *f*
i benzina *f*
n benzine
d Benzin *n*

3651 MOTTLING pla
f marbrage *m*
e jaspeado *m*, moteado *m*
i marmorizzazione *f*
n marmeren *n*
d Marmorieren *n*

3652 MOUNTING TISSUE, ADHESIVE FOIL, ADHESIVE TISSUE pho
f feuille *f* adhesive pour montage sec
e tisú *m* para montar en seco
i foglia *f* adesiva per montare le copie
n adhesieve foelie, droogopzetfoelie
d Klebefolie *f*, Trockenklebefolie *f*

3653 MOVING BED CATALYTIC CRACKING fue
f craquage *m* à lit mobile
e pirólisis *f* sobre leche móvil
i piroscissione *f* a letto mobile
n kraken *n* met bewegend bed
d Wanderbettkracken *n*

3654 MUD, SILT, SLUDGE iwt
f boues *pl*, limon *m* épais
e fangos *pl*
i fanghi *pl*, melma *f*, mota *f*
n modder, slijk *n*
d Schlamm *m*, Schlick *m*

3655 MUFFLING coa
f traitement *m* au moufle
e tratamiento *m* de mufla
i trattamento *m* alla muffola
n moffelen *n*
d Muffeln *n*

3656 MULBERRY PAPER, JAPANESE VELLUM pap
f papier *m* d'écorce de mûrier
e papel *m* de corteza de moral
i cartapecora *f* giapponese
n Chinees papier *n*
d chinesisches Papier *n*, Maulbeerbaumpapier *n*

3657 MULLING (US), MILLING (GB) brk
f frottage *m*
e amasado *m*
i molazzatura *f*
n kolleren *n*
d Kollern *n*

3658 MULTICOLOR FINISH (US), POLYCHROME FINISH (GB) dec
f finition *f* multicolore, finition *f* polychrome
e acabado *m* policromo
i finitura *f* policroma
n meerkleurendeklaag
d Mehrfarbendeckschicht *f*

3659 MULTIPLE-EFFECT EVAPORATION eva
f évaporation *f* à effet multiple
e evaporación *f* de efecto múltiple
i evaporazione *f* ad effetto multiplo
n multipel-effectverdamping
d Mehrfachverdampfung *f*

3660 MUNICIPAL WATER, CITY WATER, TAP WATER hyg
f eau *f* du robinet, eau *f* potable
e agua *f* corriente, agua *f* potable
i acqua *f* di rubinetto, acqua *f* potabile
n drinkwater *n*, leidingwater *n*
d Leitungswasser *n*, Trinkwasser *n*

3661 MURIATIC ACID, HYDROCHLORIC ACID (GB), SALT ACID (US) aic
f acide *m* chlorhydrique
e ácido *m* clorhídrico, ácido *m* muriático
i acido *m* cloridrico, acido *m* muriatico
n chloorwaterstofzuur *n*, zoutzuur *n*

d Chlorwasserstoffsäure *f*, Salzsäure *f*

3662 MUST nut
f moût *m*
e mosto *m*
i mosto *m*
n most
d Most *m*

3663 MUSTARD GAS, YPERITE gas,mit
f gaz *m* moutarde, ypérite *f*
e gas *m* mostaza, iperita *f*
i iprite *f*
n mosterdgas *n*, yperiet *n*
d Senfgas *n*, Yperit *m*

3664 MUSTARD OIL fat
f essence *f* de sénevé
e esencia *f* de mostaza
i olio *m* di senape
n mosterdolie
d Senföl *n*

3665 MYCODERMA, FILM FORMING YEAST bre
f mycoderme *m*
e micoderma *m*
i micoderma *m*
n kaamgist, mycoderma *n*
d Kahmhefe *f*

3666 MYRTLE OIL fat
f essence *f* de myrte
e esencia *f* de arrayán
i olio *m* di mirto
n mirte-olie
d Myrtenöl *n*

N

3667 NAIL TEST exp
f essai *m* au clou
e ensayo *m* al clavo
i collaudo *m* al chiodo
n spijkerproef
d Nagelprobe *f*

3668 NAPHTOLATION tex
f naphtolage *m*
e naftolización *f*
i naftolizzazione *f*
n naftoleren *n*
d Naphtolgrundierung *f*, Naphtolieren *n*

3669 NAPLES GREEN dec
f vert *m* de Naples
e verde *m* de Nápoles
i verde *m* di Napoli
n Napels groen
d Neapelgrün *n*

3670 NAPLES YELLOW dec
f jaune *m* de Naples
e amarillo *m* de Nápoles
i giallo *m* di Napoli
n Napels geel *n*
d Neapelgelb *n*

3671 NARROW CUT dis
f coupe *f* étroite, coupe *f* serrée
e separación *f* exacta
i taglio *m* stretto
n fractie met klein kooktrajekt
d scharfer Schnitt *m*

3672 NASCENT GAS gas
f gaz *m* naissant
e gas *m* naciente
i gas *m* nascente
n gas *n* in status nascendi
d Gas *n* im Entstehen, naszierendes Gas *n*

3673 NASCENT STATE gas
f état *m* naissant
e estado *m* naciente
i stato *m* nascente
n status nascendi
d im Entstehen begriffen

3674 NATIVE ASPHALT, ROCK ASPHALT bui
f asphalte *m* naturel, roche *f* asphaltique
e asfalto *m* natural, roca *f* asfáltica
i roccia *f* asfaltica naturale
n natuurasfalt *n*
d Naturasphalt *m*

3675 NATURAL BASE oic
f base *f* végétale
e base *f* vegetal
i base *f* vegetale
n natuurlijke base, plantaardige base
d Alkaloid *n*, Pflanzenbase *f*

3676 NATURAL CALCIUM SULPHATE (GB), LIGHT SPAR (US), TERRA ALBA min
f blanc *m* brillant, blanc *m* de gypse, sulfate *m* de calcium naturel
e sulfato *m* de calcio natural
i terra *f* alba
n lenzine, lichtspaat *n*
d Federweiss *n*, Kalziumsulfat *n*, Leichtspat *m*, Lenzin *m*

3677 NATURAL CEMENT bui
f ciment *m* naturel
e cemento *m* natural
i cemento *m* naturale
n natuurcement *n*
d Naturzement *m*

3678 NATURAL DRAFT (US), NATURAL DRAUGHT (GB) clo
f tirage *m* naturel
e tiro *m* natural
i tiraggio *m* naturale
n natuurlijke trek
d natürlicher Zug *m*, Selbstzug *m*

3679 NATURAL DYE dye
f colorant *m* naturel
e colorante *m* natural
i colorante *m* naturale
n natuurlijke kleurstof
d Naturfarbstoff *m*

3680 NATURAL EARTH PIGMENTS, EARTH COLORS (US), EARTH COLOURS (GB) dec
f couleurs *pl* minérales, pigments *pl* terreux
e pigmento *m* de tierras naturales
i colori *pl* a base di terra
n aardverfstoffen *pl*
d Erdfarben *pl*, natürliche Mineralfarben *pl*

3681 NATURAL EXPOSURE TEST, tes
NATURAL WEATHERING TEST,
FIELD TEST
f essai *m* de tenue aux intempéries
e ensayo *m* de exposición a las intemperies
i collaudo *m* d'esposizione all'intemperie
n buitenproef, dakproef, weerrekproef
d Bewitterungsprobe *f*, Freilagerversuch *m*

3682 NATURAL FIBER (US), tex
NATURAL FIBRE (GB)
f fibre *f* naturelle
e fibra *f* natural
i fibra *f* naturale
n natuurlijke vezel, plantaardige vezel
d Naturfaser *f*, Pflanzenfaser *f*

3683 NATURAL GAS fue,gas
f gaz *m* naturel
e gas *m* natural
i gas *m* naturale
n aardgas *n*
d Erdgas *n*, Naturgas *n*

3684 NATURAL HARDNESS hyg
f dureté *f* naturelle
e dureza *f* natural
i durezza *f* naturale
n natuurlijke hardheid
d Naturhärte *f*, natürliche Härte *f*

3685 NATURAL PURE CULTURE bac,fmi
f culture *f* pure naturelle
e cultivo *m* puro natural
i cultura *f* pura naturale
n natuurlijke reincultuur, natuurlijke zuivere cultuur
d natürliche Reinzucht *f*

3686 NATURAL RAW POTASSIC min
SALTS
f sels *pl* de potasse bruts
e sales *pl* de potasio brutas
i sali *pl* di potassio greggi
n ruwe kaliumzouten *pl*
d Abraumsalze *pl*, Kalirohsalze *pl*

3687 NATURAL SEASONING bui
f séchage *m* du bois à l'air
e cura *f* de la madera al aire
i stagionamento *m* naturale
n hout drogen *n* aan de lucht
d Lufttrocknung *f*, natürliche Ablagerung *f*

3688 NATURAL WATER, hyg
UNTREATED WATER
f eau *f* naturelle
e agua *f* natural
i acqua *f* naturale
n ruw water *n*
d Rohwasser *n*

3689 NATURAL WEATHERING tes
f exposition *f* aux intempéries
e ensayo *m* de envejecimiento para intemperie
i collaudo *m* d'invecchiamento per intemperie
n blootstellen *n* aan verwering
d Freibewitterung *f*

3690 NATURE PRINT elc,met
f empreinte *f* d'attaque à l'acide
e imprenta *f* de ataque al ácido
i stampa *f* d'attacco all'acido
n etsafdruk, etsprent
d Ätzabdruck *m*

3691 NAUSEATING, gas,gen
OBNOXIOUS
f nauséabond adj, nauséeux adj
e nauseabundo adj, nauseativo adj
i nauseabondo adj, nauseante adj
n misselijk makend adj
d übelkeiterregend adj

3692 NEAT CEMENT bui
f mortier *m* de ciment sans sable
e mezcla *f* de cemento y agua
i impasto *m* di cemento senza sabb
n zandvrije cementspecie
d sandfreier Zementmörtel *m*

3693 NEAT'S FOOT OIL lub
f huile *f* de pied de boeuf
e aceite *m* de pie de buey
i olio *m* di piede di bue
n klauweolie, klauwevet *n*
d Klauenfett *n*, Klauenöl *n*

3694 NEBULIZATION, coa,pre
ATOMIZATION, ATOMIZING
f atomisation *f*, pulvérisation *f*, vaporisation *f*
e atomización *f*, nebulización *f*, pulverización *f*
i atomizzazione *f*, nebulizzazione polverizzazione *f*
n verneveling, verstuiving
d Zerstäubung *f*

3695 NEGATIVE pho
f négatif *f*

e negativo *m*
i negativo *m*
n negatief *n*
d Negativ *n*

3696 NEGATIVE CATALYST, REACTION INHIBITING CATALYST gen
f catalyseur *m* négatif
e catalizador *m* negativo
i catalizzatore *m* negativo
n negatieve katalysator, reactievertragende katalysator
d negativer Katalysator *m*, reaktionshemmender Katalysator *m*

3697 NEON LIQUEFYING gas
f liquéfaction *f* de néon
e licuación *f* de neón
i liquefazione *f* di neon
n vloeibaar maken *n* van neon
d Neonverflüssigung *f*

3698 NEPHELOMETRIC ANALYSIS ana
f analyse *f* néphélométrique
e análisis *f* nefelométrica
i analisi *f* nefelometrica
n nefelometrische analyse
d nephelometrische Analyse *f*

3699 NEROLI OIL, BITTER ORANGE FLOWER OIL, ORANGE BLOSSOM OIL fat
f essence *f* de fleurs d'oranger, essence *f* de néroli
e esencia *f* de neroli
i olio *m* di fiori d'arancio, olio *m* di neroli
n neroliolie, oranjebloesemolie
d Neroliöl *n*, Orangenblütenöl *n*, Pomeranzblütenöl *n*

3700 NERVE GAS gas,mit
f gaz *m* énervant
e gas *m* enervante
i gas *m* da apoplessia nervosa
n zenuwgas *n*
d Nervengas *n*

3701 NETWORK-FORMING ION col,gla
f ion *m* de branchement
e ión *m* de ramificación
i ione *m* di diramazione
n netvormend ion *n*, vertakkingsion *n*
d Vernetzungsion *n*

3702 NEUTRAL POINT, NON-SLIP POINT gen
f ligne *f* neutre, zone *f* neutre
e zona *f* neutra
i zona *f* neutra
n punt *n* van gelijke snelheid
d Fliessscheide *f*

3703 NEUTRAL RECLAIMING PROCESS rub
f procédé *m* de régénération neutre
e procedimiento *m* de regeneración neutro
i processo *m* di rigenerazione neutro
n neutraal regenereringsproces *n*
d neutrales Regenerierverfahren *n*

3704 NEUTRAL SALT EFFECT gen
f effet *m* de sel neutre
e efecto *m* de sal neutra
i effetto *m* di sale neutro
n neutraal-zouteffect *n*
d Neutralsalzeffekt *m*

3705 NEUTRAL SOLUTION gen
f solution *f* neutre
e solución *f* neutra
i soluzione *f* neutra
n neutrale oplossing
d neutrale Lösung *f*

3706 NEUTRALIZATION gen
f neutralisation *f*
e neutralización *f*
i neutralizzazione *f*
n neutralisatie
d Neutralisierung *f*

3707 NEUTRALIZATION OF WASTE sew
f neutralisation *f* d'égouts
e neutralización *f* de aguas residuales
i neutralizzazione *f* d'acqua di scarico
n afvalwaterontzuring
d Entsäuerung *f* von Abwässer

3708 NEUTRALIZER elc
f neutralisant *m*
e neutralizador *m*
i neutralizzatore *m*
n neutralisator
d Neutralysator *m*

3709 NEUTRALIZING bui,dec
f neutralisation *f* de chaux
e neutralización *f* de la cal
i neutralizzazione *f* della calce
n kalkneutralisatie
d Kalkneutralisierung *f*

3710 NEW BEER bre
f bière *f* jeune

e cerveza *f* nueva
i birra *f* giovane
n jong bier *n*
d Jungbier *n*

3711 NEW COMPOUND rub
f mélange *m* non-vulcanisé
e mezcla *f* no vulcanizada
i mescolanza *f* di nuovo impiego
n ongevulcaniseerd mengsel *n*
d unvulkanisiertes Gemisch *n*

3712 NEWSPRINT pap
f papier-journal *m*
e papel *m* para diarios
i carta *f* da giornali
n krantenpapier *n*
d Zeitungspapier *n*

3713 NIB htt,met
f compact *m* terminé
e pastilla *f* terminada
i compatto *m* terminato
n afgewerkte pastille
d fertiger Pressling *m*

3714 NICKEL FLASHING elc
f nickelage *m* rapide
e niquelado *m* rápido
i nichelatura *f* rapida
n snelvernikkelen *n*
d Schnellvernicklung *f*

3715 NICKEL PLATING elc
f nickelage *m*
e niquelado *m*
i nichelatura *f*
n vernikkelen *n*
d Vernickelung *f*

3716 NICKEL STRIKING elc
f amorçage *m* de nickelage
e niquelado *m* preliminar
i nichelatura *f* preliminare
n voorvernikkelen *n*
d Vorvernickelung *f*

3717 NITRALIZING htt,met
f immersion *f* en nitrate de sodium fondu
e inmersión *f* en nitrato de sodio fundido
i immersione *f* in nitrato di sodio fuso
n dompelen *n* in gesmolten natriumnitraat
d Eintauchen *n* in geschmolzenem Natriumnitrat

3718 NITRATE FILM pho
f pellicule *f* de nitrate de cellulose
e película *f* al nitrato de celulosa
i pellicola *f* al nitrato di cellulosa
n cellulosenitraatfilm, nitraatfilm
d Nitratfilm *m*, Zellulosenitratfilm

3719 NITRATE PULPING pap
f décomposition *f* à l'acide nitrique
e descomposición *f* al ácido nítrico
i decomposizione *f* all'acido nitrico
n ontsluiting met salpeterzuur
d salpetersaurer Aufschluss *m*

3720 NITRATION bac
f nitration *f*
e nitración *f*
i nitrazione *f*
n nitratie *n*
d Nitration *f*

3721 NITRE, POTASSIUM NITRATE, SALTPETRE aic
f salpêtre *m*
e salitre *m*
i salnitro *m*
n salpeter
d Salpeter *m*

3722 NITRE BATH gen
f bain *m* de salpêtre
e baño *m* de salitre
i bagno *m* di salnitro
n salpeterbad *n*
d Salpeterbad *n*

3723 NITRIC ACID aic
f acide *m* nitrique
e ácido *m* nítrico
i acido *m* nitrico
n salpeterzuur *n*
d Salpetersäure *f*

3724 NITRIC ACID TEST tes
f essai *m* à l'acide nitrique
e ensayo *m* al ácido nítrico
i collaudo *m* all'acido nitrico
n salpeterzuurproef
d Salpetersäureprobe *f*

3725 NITRIDING me[t]
f nitruration *f*
e nitruración *f*
i nitrurazione *f*
n nitreren *n*
d Nitrieren *n*

3726 NITRIFICATION, NITRIFYING ag[r]
f nitrification *f*

e nitrificación *f*
i nitrificazione *f*
n nitrificeren *n*
d Nitrifizieren *n*

3727 NITROBACTERIA, bac,fer
NITROGEN FIXING BACTERIA
f nitrobactéries *pl*
e bacterias *pl* nitrificadoras
i batterie *pl* da assimilare azoto
n stikstofbindende bacteriën *pl*
d stickstoffbindende Bakterien *pl*

3728 NITROCELLULOSE, exp
CELLULOSE NITRATE
f nitrocellulose *f*
e nitrocelulosa *f*, piroxilina *f*
i nitrocellulosa *f*
n nitrocellulose
d Nitrozellulose *f*,
Zellulosenitrat *n*

3729 NITROCELLULOSE LACQUER, dec
NITROLACQUER, PYROXYLIN
LACQUER
f laque *f* nitrocellulosique
e laca *f* nitrocelulósica
i lacca *f* nitrocellulosica
n nitrocelluloselak
d Nitrolack *m*, Nitrozelluloselack *m*

3730 NITROCELLULOSE POLISHING dec
LACQUER
f laque *f* nitrocellulosique à polir
e laca *f* nitrocelulósica
i lacca *f* nitrocellulosica per
lustrare
n nitrocellulosepolijstlak
d Nitropolierlack *m*

3731 NITROGEN ARC WELDING, wel
NITROGEN SHIELDED ARC
WELDING
f soudure *f* à arc en azote
e soldadura *f* a arco en
nitrógeno
i saldatura *f* ad arco in azoto
n lassen *n* in stikstofatmosfeer
d Lichtbogenschweissen *n* unter
Stickstoff

3732 NITROGEN FIXATION gas
f fixage *m* de l'azote
e fijación *f* del nitrógeno
i fissazione *f* dell'azoto
n stikstofbinding
d Stickstoffbindung *f*

3733 NITROGEN LIQUEFACTION, gas
NITROGEN LIQUEFYING
f liquéfaction *f* d'azote
e licuación *f* de nitrógeno
i liquazione *f* d'azoto
n vloeibaar maken *n* van stikstof
d Stickstoffverflüssigung *f*

3734 NITROGENIZING oic
f nitrogénation *f*
e nitrogenación *f*
i nitrogenazione *f*
n nitrogeneren *n*
d Nitrogenieren *n*

3735 NITROHYDROCHLORIC aic,met
ACID,
NITROMURIATIC ACID, AQUA
REGIA, CHLOROAZOTIC ACID
(US), CHLORONITROUS ACID
f eau *f* régale
e agua *f* regia
i acqua *f* regia
n koningswater *n*
d Königswasser *n*

3736 NITROGLYCERINE (GB), exp
EXPLOSIVE OIL (US)
f nitroglycérine *f*,
trinitrate *m* de glycérine
e nitroglicerina *f*, trinitrato *m*
de glicerina
i nitroglicerina *f*, trinitrato *m*
di glicerina
n glyceroltrinitraat *n*,
nitroglycerine, nitroglycerol *n*
d Glyzerintrinitrat *n*,
Nitroglyzerin *n*, Trinitrin *n*

3737 NITROUS ACID aic
f acide *m* nitreux
e ácido *m* nitroso
i acido *m* nitroso
n salpeterig zuur
d Salpetrigsäure *f*

3738 NITROUS OXIDE, gas
LAUGHING GAS
f gaz *m* hilarant, protoxyde *m*
d'azote
e gas *m* hilarante
i gas *m* esilarante, ossido *m*
nitroso
n lachgas *n*
d Lachgas *n*, Stickstoffoxydul *n*

3739 NO-FINES CONCRETE bui
f béton *m* sans fines
e hormigón *m* con agregado de un
solo tamaño
i calcestruzzo *m* granulare
n korrelbeton *n*
d Einkornbeton *m*, Kornbeton *m*

3740 NOBLE GAS, INERT GAS, RARE GAS — gas
f gaz *m* inerte, gaz *m* noble
e gas *m* inerte, gas *m* noble
i gas *m* inerte, gas *m* nobile
n edelgas *n*, inert gas *n*
d Edelgas *n*, inertes Gas *n*

3741 NOBLE METALS, PRECIOUS METALS — met
f métaux *pl* nobles, métaux *pl* précieux
e metales *pl* nobles
i metalli *pl* nobili
n edele metalen *pl*
d Edelmetalle *pl*

3742 NOBLE POTENTIAL, STANDARD ELECTRODE POTENTIAL — elc
f tension *f* standard d'une électrode
e tensión *f* normal de un electrodo
i potenziale *m* normale d'un elettrodo
n normaalpotentiaal van een elektrode
d Normalpotential *n* einer Elektrode

3743 NODULE — elc
f nodule *m*
e nódulo *m*
i nodulo *m*
n knobbel, noduul *n*
d Knötchen *n*

3744 NODULIZING — htt,met
f formation *f* de nodules
e nodulación *f*
i formazione *f* di noduli
n noduleren *n*
d Knötchenbildung *f*

3745 NON-BLACK REINFORCING FILLER, WHITE REINFORCING FILLER — rub
f charge *f* renforçante blanche
e carga *f* reforzadora blanca
i carica *f* rinforzante bianca
n witte versterkende vulstof
d heller Verstärkerfüllstoff *m*

3746 NON-BLOOMING (GB), BLOOM-FREE (US) — rub
f non-efflorescent adj
e no eflorescente adj
i non efflorescente adj
n vrij van uitbloemen
d nicht ausblühend adj, nicht ausschwefelnd adj

3747 NON-DESTRUCTIVE TESTING — tes
f essai *m* non-destructif
e ensayo *m* indestructivo
i collaudo *m* non distruttivo
n niet-destructieve beproeving
d zerstörungsfreie Prüfung *f*

3748 NON-DRYING OIL — dec,fat
f huile *f* non-séchante
e aceite *m* no secante
i olio *m* non essiccante
n niet-drogende olie
d nichttrocknendes Öl *n*

3749 NON-FLOCULATING YEAST, POWDERY YEAST — bre,fmi
f levure *f* poussiéreuse
e levadura *f* pulverulenta
i lievito *m* polverulente
n gistpoeder *n*
d Staubhefe *f*

3750 NON-IONIC DYE — dye
f colorant *m* non-ionique
e colorante *m* iniónico
i colorante *m* non ionico
n niet-ionische kleurstof
d nichtionischer Farbstoff *m*

3751 NON-METAL — aic
f non-métal *m*
e no-metal *m*
i non-metallo *m*
n niet-metaal *n*
d Nichtmetall *n*

3752 NON-MIGRATORY PLASTICIZER — pla
f plastifiant *m* non-migrant
e plastificante *m* no-migrante
i plastificatore *m* non migrante
n niet-zwervende weekmaker
d nichtwandernder Weichmacher *m*

3753 NON-ODOROUS, ODORLESS (US), ODOURLESS (G — per
f inodore adj
e inodoro adj
i inodoro adj
n reukeloos adj , reukloos adj
d geruchlos adj

3754 NON-SKID PAINT, NON-SLIP PAINT — dec
f peinture *f* antidérapante
e pintura *f* antirresbaladiza
i pittura *f* antisdrucciolevole
n antislipverf
d Antirutschfarbe *f*, Gleitschutzfa

3755 NON-SLIP POINT, NEUTRAL POINT gen
f ligne *f* neutre, zone *f* neutre
e zona *f* neutra
i zona *f* neutra
n punt *n* van gelijke snelheid
d Fliessscheide *f*

3756 NON-STAINING PAINT dec
f peinture *f* non-tachante
e pintura *f* que no mancha
i pittura *f* che non macchia
n niet-afgevende verf
d nichtabfärbende Farbe *f*

3757 NORDHAUSEN SULPHURIC ACID, FUMING SULFURIC ACID (US), FUMING SULPHURIC ACID (GB) aic
f acide *m* sulfurique fumant
e ácido *m* sulfúrico fumante
i acido *m* solforico fumante
n oleum *n*, rokend zwavelzuur *n*
d Oleum *n*, rauchende Schwefelsäure *f*

3758 NORMAL SALT aic
f sel *m* normal
e sal *f* normal
i sale *m* normale
n normaalzout *n*
d Normalsalz *n*

3759 NORMAL SOLUTION gen
f solution *f* normale
e solución *f* normal
i soluzione *f* normale
n normaaloplossing
d Normallösung *f*

3760 NORMALITY (GB), TITER (US) gen
f normalité *f* de la solution
e normalidad *f* de la solución
i normalità *f* della soluzione
n normaliteit van een oplossing
d Normalität *f*, Titer *m*

3761 NOXIOUS gen
f nuisible adj
e nocivo adj
i nocivo adj
n schadelijk adj
d schädlich adj

3762 NUCLEAR BOILING, NUCLEATE BOILING boi
f ébullition *f* nuclée, ébullition *f* par centres
e ebullición *f* nucleada
i ebollizione *f* nucleata
n kernkoken *n*
d Blasensieden *n*, Blasenverdampfung *f*

3763 NUCLEAR CHEMISTRY nue
f chimie *f* nucléaire
e química *f* nuclear
i chimica *f* nucleare
n kernchemie
d Kernchemie *f*

3764 NUCLEATION boi
f formation *f* de bulles
e formación *f* de burbujas
i formazione *f* di bolle
n kernvorming
d Keimbildung *f*

3765 NUCLEATION cry
f nucléation *f*
e nucleación *f*
i nucleazione *f*
n kiemvorming
d Keimbildung *f*

3766 NUTMEG OIL fat
f essence *f* de noix de muscade
e esencia *f* de nuez moscada
i olio *m* di noce moscata
n muskaatolie
d Muskatnussöl *n*

3767 NUTRIENT nut
f substance *f* nutritive
e substancia *f* nutritiva
i elemento *m* nutritivo
n voedingsstof
d Nährstoff *m*

3768 NYLON tex
f nylon *m*
e nilón *m*, nylón *m*
i nailon *m*
n nylon *n*
d Nylon *n*

O

3769 OAK BARK tan
f écorce *f* de chêne
e corteza *f* de roble
i corteccia *f* di quercia
n eikeschors
d Eichenrinde *f*

3770 OAKUM mat
f étoupe *f*
e estopa *f*
i capecchio *m*, stoppa *f*
n werk *n*
d Werg *m*

3771 OBJECTIONABLE ODOR(US), gen
OBJECTIONABLE ODOUR (GB),
OFFENSIVE ODOUR (GB)
f odeur *f* désagréable
e olor *m* desagradable
i odore *m* sgradevole,
odore *m* spiacevole
n onaangename reuk, stank
d ekelhafter Geruch *m*,
unangenehmer Geruch *m*

3772 OBNOXIOUS, gas,gen
NAUSEATING
f nauséabond adj , nauséeux adj
e nauseabundo adj , nauseativo
adj
i nauseabondo adj , nauseante
adj
n misselijk makend adj
d übelkeiterregend adj

3773 OBSCURATION, dec
COVERAGE, SPREADING
CAPACITY
f pouvoir *m* couvrant en surface
e extensión *f* específica
i distendimento *m* specifico
n uitstrijkrendement *n*
d Ausgiebigkeit *f*

3774 OBSCURED GLASS, bui,gla
FROSTED GLASS
f verre *m* dépoli
e vidrio *m* esmerilado
i vetro *m* smerigliato
n matglas *n*
d Mattglas *n*

3775 OBSERVATION gen
f observation *f*
e observación *f*
i osservazione *f*
n waarneming
d Beobachtung *f*

3776 OCCLUDED GAS gas
f gaz *m* absorbé, gaz *m* occlus
e gas *m* absorbido, gas *m* ocluido
i gas *m* assorbito, gas *m* occluso
n geabsorbeerd gas *n*,
geoccludeerd gas *n*
d absorbiertes Gas *n*,
eingeschlossenes Gas *n*

3777 OCCURRENCE min
f venue *f*
e suceso *m*
i manifestazione *f*
n voorkomen *n*
d Vorkommen *n*

3778 OCHER, dec
OCHRE
f ocre *f*
e ocre *m*
i ocra *f*
n oker
d Ocker *m*

3779 OCTANE NUMBER (GB), fue
OCTANE RATING (US)
f indice *m* d'octane
e índice *m* de octano
i indice *m* d'ottano
n octaangetal *n*
d Oktanzahl *f*

3780 ODOR (US), per
ODOUR (GB), SCENT, SMELL
f odeur *f*
e olor *m*
i odore *m*
n geur, reuk
d Duft *m*, Geruch *m*

3781 ODORANT per
f odorant *m*
e odorante *m*
i odorante *m*
n reukstof
d Riechstoff *m*

3782 ODORIFEROUS per
f odoriférant adj
e bienoliente adj , odorífero *m* ac
oloroso adj

i odoroso adj , odorifero adj
n welriekend adj
d wohlriechend adj

3783 ODORIMETRY, OLFACTOMETRY per
f odorimétrie *f*
e odorimetría *f*
i odorimetria *f*
n odorimetrie
d Geruchsmessung *f*

3784 ODORLESS (US), ODOURLESS (GB), NON-ODOROUS per
f inodore adj
e inodoro adj
i inodoro adj
n reukeloos adj , reukloos adj
d geruchlos adj

3785 OFF-WHITE, BROKEN WHITE dec
f blanc *m* cassé, blanchâtre adj
e blanco *m* mate, blancuzco *m*, blanquecino adj
i biancastro adj
n gebroken wit *n*
d gebrochenes Weiss *n*

3786 OIL fat,lub
f huile *f*
e aceite *m*
i olio *m*
n olie
d Öl *n*

3787 OIL ABSORPTION dec
f absorption *f* d'huile
e absorción *f* de aceite
i assorbimento *m* d'olio
n olieopneming
d Ölaufnahme *f*

3788 OIL BATH met
f bain *m* d'huile
e baño *m* de aceite
i bagno *m* d'olio
n oliebad *n*
d Ölbad *n*

3789 OIL CAKE agr,nut
f tourteau *m*
e torta *f* de borujo
i pannello *m* di residui di semi oleosi
n oliekoek
d Ölkuchen *m*

3790 OIL COOLED clo
f refroidi adj par huile
e enfriado adj por aceite
i raffreddato adj per olio
n met olie gekoeld adj
d mit Ölkühlung, ölgekühlt adj

3791 OIL COOLING clo
f refroidissement *m* à l'huile
e enfriamiento *m* de aceite
i raffreddamento *m* all'olio
n oliekoeling
d Ölkühlung *f*

3792 OIL-EXTENDED RUBBER, OIL-SOFTENED RUBBER rub
f caoutchouc *m* étendu à l'huile
e goma *f* enriquecida con aceite
i gomma *f* diluita all'olio
n met olie gestrekte rubber, met olie versneden rubber
d mit Öl gestreckter Kautschuk *m*

3793 OIL FINISHING tan
f apprêt *m* gras
e apresto *m* graso
i apprettatura *f* all'olio, finissaggio *m* all'olio
n vet appreteren *n*
d Fettappretur *f*, Nachbehandlung *f* in Öl

3794 OIL FIRING htt
f chauffage *m* au mazout
e caldeo *m* por aceite combustible
i riscaldamento *m* all'olio combustibile
n oliestook
d Ölheizung *f*

3795 OIL HARDENING htt,met
f trempe *f* en huile
e temple *m* en aceite
i tempera *f* in olio
n olieharding
d Ölhärtung *f*

3796 OIL HARDENING, HARDENING OF OIL fat
f hydrogénation *f* d'huile
e hidrogenación *f* de aceite
i idrogenazione *f* d'olio
n harding van vloeibare vetten, olieharding
d Fetthärtung *f*

3797 OIL IMMERSION ana
f immersion *f* à l'huile
e inmersión *f* de aceite
i immersione *f* ad olio
n olie-immersie
d Ölimmersion *f*

3798 OIL-IMMERSION TEST tes
f essai *m* d'immersion dans l'huile
e ensayo *m* de inmersión en aceite
i collaudo *m* d'immersione in olio
n oliedompelproef, oliezwelproef
d Öleintauchprobe *f*

3799 OIL-IMPREGNATED PAPER elc,pap
f papier *m* imprégné d'huile
e papel *m* impregnado de aceite
i carta *f* impregnata d'olio
n met olie geïmpregneerd papier *n*
d mit Öl imprägniertes Papier *n*

3800 OIL LENGTH dec
f teneur *f* en huile
e relación *f* de aceite
i tenore *m* d'olio
n oliegehalte *n*
d Ölgehalt *m*

3801 OIL MIST lub
f brouillard *m* d'huile
e niebla *f* de aceite
i nebbia *f* d'olio
n olienevel
d Ölnebel *m*

3802 OIL OF PINETAR, PINETAR OIL fat
f huile *f* de goudron de pin
e aceite *m* de alquitrán de pino
i olio *m* di catrame di pino
n pijnboomteerolie
d Kienöl *n*, Nadelholzteeröl *n*

3803 OIL PAINT dec
f peinture *f* à l'huile
e pintura *f* al aceite, pintura *f* al oleo
i pittura *f* ad olio
n olieverf
d Ölfarbe *f*

3804 OIL PASTE, PASTE PAINT dec
f pigment *m* broyé à l'huile
e pasta *f* de aceite
i pasta *f* ad olio
n dik in olie gemalen pasta, oliepasta
d in Öl angeriebene Farbe *f*, Ölpaste *f*

3805 OIL PRESSURE RELIEF con,sap
f réglage *m* de la pression de l'huile
e regulación *f* de la presión del aceite
i regolazione *f* della pressione dell'olio
n oliedrukregeling
d Öldruckregelung *f*

3806 OIL PROOFING bui
f traitement *m* du béton à l'huile
e tratamiento *m* del hormigón con aceite
i trattamento *m* del calcestruzzo all'olio
n oliebehandeling van beton
d Ölbehandlung *f* für Beton

3807 OIL QUENCHING met
f refroidissement *m* rapide à l'huile
e enfriamiento *m* rápido al aceite
i raffreddamento *m* rapido all'olio
n afschrikken *n* in olie
d Ölabschreckung *f*

3808 OIL REFINING fue
f raffination *f* du pétrole
e refinado *m* del petróleo
i raffinazione *f* del petrolio
n raffineren *n* van olie
d Ölraffination *f*, Ölraffinierung *f*

3809 OIL REMOVAL, OIL SEPARATION cle
f déshuilage *m*
e desaceitación *f*
i eliminazione *f* dell'olio
n olieverwijdering
d Entölung *f*

3810 OIL RENDERING fat
f clarification *f* d'huile
e clarificación *f* de aceite
i chiarificazione *f* d'olio
n olieklaring
d Ölklärung *f*

3811 OIL RESISTANT fat
f résistant adj à l'huile
e resistente adj al aceite
i resistente adj all'olio
n oliewerend adj
d ölbeständig adj

3812 OIL TANNING tan
f tannage *m* à l'huile
e curtimiento *m* al aceite
i concia *f* con olio
n looien *n* met olie
d Fettgerbung *f*

3813 OIL VARNISH, OLEO-RESINOUS VARNISH dec
f vernis *m* à l'huile, vernis *m* gras
e barniz *m* al aceite
i vernice *f* all'olio

n olievernis *n*
d Öllack *m*

3814 OILCLOTH tex
f tissu *m* huilé
e tejido *m* aceitado
i tela *f* oliata
n oliedoek *n*
d Öltuch *n*, Ölzeug *n*

3815 OILED COTTON tex
f coton *m* huilé
e algodón *m* aceitado
i cotone *m* oliato
n geolied katoen *n*
d Öltuch *n*

3816 OILED FINISHING met
f finissage *m* à l'huile
e acabado *m* al aceite
i finitura *f* all'olio
n oliefijnen *n*
d Ölvergütung *f*

3817 OILED PAPER pap
f papier *m* huilé
e papel *m* aceitado
i carta *f* oliata
n geolied papier *n*
d geöltes Papier *n*

3818 OILED SILK, tex
OILSILK
f soie *f* huilée
e seda *f* aceitada
i seta *f* oliata
n geoliede zijde
d Ölseide *f*

3819 OILED SILK TAPE tex
f ruban *m* de soie huilée
e cinta *f* de seda aceitada
i nastro *m* di seta oliata
n geolied zijdelint *n*
d Ölseidenband *n*

3820 OILGAS, fue,gas
CARBO-HYDROGEN
f gaz *m* de pétrole
e gas *m* de petróleo
i gas *m* di petrolio
n oliegas *n*, vetgas *n*
d Fettgas *n*, Ölgas *n*

3821 OILINESS fat
f aspect *m* graisseux, état *m* graisseux
e oleaginosidad *f*
i oleosità *f*
n olieachtigheid
d Öligkeit *f*

3822 OILING tex
f ensimage *m*
e ensaimado *m*
i oliatura *f*
n invetten *n*, smouten *n*
d Schmälzen *n*

3823 OILPROOF fat
f inattaquable adj à l'huile
e inatacable adj por el aceite
i inattaccabile adj all'olio
n bestand tegen olie
d ölbeständig adj , ölfest adj

3824 OILSTONE pol
f affiloir *m*, pierre *f* à l'huile
e piedra *f* de aceite
i pietra *f* d'olio
n oliesteen
d Ölstein *m*

3825 OILY fat
f huileux adj
e aceitoso adj , oleoso adj
i oleoso adj
n olieachtig adj
d ölig adj

3826 OINTMENT, mcy
LINIMENT
f liniment *m*, onguent *m*
e linimento *m*, ungüento *m*
i linimento *m*, unguento *m*
n liniment *n*, zalf
d Liniment *n*, Salbe *f*

3827 OLEAGENOUS fat
f huileux adj , oléagineux adj
e aceitoso adj , oleaginoso adj
i oleoso adj
n olieachtig adj , vettig adj
d ölartig adj , ölhaltig adj

3828 OLEFIANT GAS, gas,oic
ETHYLENE
f éthylène *m*
e etileno *m*
i etilene *m*
n ethyleen *n*
d Äthylen *n*, Elyal *n*

3829 OLFACTOMETRY, per
ODORIMETRY
f odorimétrie *f*
e odorimetría *f*
i odorimetria *f*
n odorimetrie
d Geruchsmessung *f*

3830 OLIBANUM OIL fat
f essence *f* d'oliban

e esencia *f* de olíbano
i olio *m* d'olibano
n olibanumolie
d Olibanumöl *n*

3831 OLIVE OIL, fat,nut
SWEET OIL
f huile *f* d'olive
e aceite *m* de oliva
i olio *m* d'oliva
n olijfolie
d Olivenöl *n*

3832 ON SITE CONCRETE, bui
CAST CONCRETE, IN SITE CONCRETE
f béton *m* coulé
e hormigón *m* colado
i calcestruzzo *m* colato
n gestort beton *n*, gietbeton *n*
d Gussbeton *m*, Schüttbeton *m*

3833 ONCE-THROUGH COOLING clo,nue
f refroidissement *m* à passage unique
e enfriamiento *m* de pasaje único
i raffreddamento *m* a passaggio unico
n koeling met enkelvoudige doorloop van het koelmiddel
d Kühlung *f* mit einmaligem Kühlmitteldurchlauf

3834 ONE-BATH DYEING dye
f teinture *f* à bain unique
e teñido *m* de baño único
i tintura *f* a bagno unico
n eenbadverving
d Einbadfärbung *f*

3835 ONE-SHOT LUBRICATION, lub
CENTRALIZED LUBRICATION,
CHASSIS LUBRICATION
f graissage *m* central,
graissage *m* centralisé
e lubricación *f* centralizada
i lubrificazione *f* centrale,
lubrificazione *f* centralizzata
n centrale smering
d Eindruckschmierung *f*,
Zentralschmierung *f*

3836 ONE-SHOT MOLDING (US), pla
ONE-SHOT MOULDING (GB)
f moulage *m* à opération unique
e moldeo *m* de operación única
i stampatura *f* ad operazione unica
n vorming in enkelvoudige operatie
d Formung *f* im Einzelverfahren

3837 OPACIFIER cer,gla
f substance *f* opacifiante
e substancia *f* opacificadora
i sostanza *f* opacificatrice
n troebelmakend middel *n*
d Trübungsmittel *n*

3838 OPAL GLASS gla
f verre *m* opalin
e vidrio *m* opalino
i vetro *m* opalino
n opaalglas *n*
d Opalglas *n*, Trübglas *n*

3839 OPALESCENCE fue
f opalescence *f*
e opalescencia *f*
i opalescenza *f*
n opalescentie
d Opaleszenz *f*, Opalisieren *n*

3840 OPALESCENT GLASS, gla
ALABASTER GLASS
f verre *m* opalescent
e vidrio *m* opalescente
i vetro *m* opalescente
n albastglas *n*, melkglas *n*
d Achatglas *n*, Alabasterglas *n*,
Reisglas *n*

3841 OPAQUE GLASS gla
f verre *m* opaque
e vidrio *m* opaco
i vetro *m* opaco
n ondoorschijnend glas *n*,
opaak glas *n*
d Opakglas *n*

3842 OPEN ANNEALING, met
BLACK ANNEALING
f recuit *m* en noir
e recocido *m* en negro
i ricottura *f* in nero
n zwartgloeien *n*
d Schwarzglühen *n*

3843 OPEN BOILING, tex
OPEN KIER BOILING
f traitement *m* à chaud en récipient ouvert
e tratamiento *m* en caliente en recipiente abierto
i trattamento *m* a caldo in serbatoio aperto
n warmbehandeling in open vat
d Warmbehandlung *f* in offenem Gefäss

3844 OPEN CURING rub
f vulcanisation *f* à nu,
vulcanisation *f* libre

e vulcanización *f* al aire
i vulcanizzazione *f* in aria
n vrije vulcanisatie
d freie Vulkanisation *f*

3845 OPEN STEAM htt,rub
f vapeur *f* libre
e vapor *m* libre
i vapore *m* libero
n vrije stoom
d Freidampf *m*

3846 OPEN STEAM CURE rub
f vulcanisation *f* en vapeur libre
e vulcanización *f* en vapor libre
i vulcanizzazione *f* in vapore libero
n vulcanisatie met vrije stoom
d Freidampfheizung *f*

3847 OPERATION CONTROL cer,gen
f contrôle *m* de régime,
contrôle *m* de service
e control *m* de operación
i controllo *m* d'esercizio
n bedrijfscontrole
d Betriebskontrolle *f*,
Betriebsüberwachung *f*

3848 OPTICAL BRIGHTENER, tex
BRIGHTENING AGENT
f agent *m* de blanchiment optique,
azurant *m* optique
e agente *m* blanqueante óptico,
blanqueador *m* óptico
i azzurrante *m* ottico,
sbiancante *m* ottico
n optisch bleekmiddel *n*,
optisch witmiddel *n*
d optischer Aufheller *m*,
Weisstöner *m*

3849 OPTICAL GLASS gla
f verre *m* optique
e vidrio *m* óptico
i vetro *m* ottico
n optisch glas *n*
d optisches Glas *n*

3850 ORANGE BLOSSOM OIL, fat
BITTER ORANGE FLOWER OIL,
NEROLI OIL
f essence *f* de fleurs d'oranger,
essence *f* de néroli
e esencia *f* de neroli
i olio *m* di fiori d'arancio,
olio *m* di neroli
n neroliolie, oranjebloesemolie
d Neroliöl *n*, Orangenblütenöl *n*,
Pomeranzblütenöl *n*

3851 ORANGE LEAD, dec
ORANGE MINERAL
f rouge-orange *m*
e albayalde *m* anaranjado,
óxido *m* rojo de plomo
i arancio *m* al piombo, tetrossido *m*
di piombo
n oranje menie
d Orangemennige *f*

3852 ORANGE PEEL EFFECT, met
ALLIGATOR SKIN
f effet *m* de pelure d'orange
e efecto *m* de corteza de naranja
i effetto *m* di buccia d'arancio
n sinaasappelschileffect *n*
d Orangenschaleneffekt *m*

3853 ORANGE PEELING dec
f effet *m* de pelure d'orange
e efecto *m* de corteza de naranja
i effetto *m* di buccia d'arancio
n kratervorming
d Apfelsinenschaleneffekt *m*,
Kraterbildung *f*

3854 ORCHIL, dye
CUDBEAR
f orseille *f*
e orcina *f*
i orcina *f*
n orseille *n*
d Cudbear *m*, Orseillenfarbe *f*

3855 ORE CONCENTRATION, min
DRESSING
f enrichissement *m* des minerais
e enriquecimiento *m* de los minerales
i arricchimento *m* dei minerali
n ertsconcentratie
d Erzanreicherung *f*

3856 ORE DRESSING (GB), min
BENEFICIATION (US)
f traitement *m* de minerais
e beneficio *m* de minerales
i trattamento *m* di minerali
n ertsbewerking
d Erzaufbereitung *f*

3857 ORE GRADING min
f classification *f* de minerais,
triage *m* de minerais
e clasificación *f* de minerales
i classificazione *f* di minerali
n ertsclassificatie
d Erzklassierung *f*

3858 ORGANIC ANALYSIS, ana
ELEMENTARY ANALYSIS,
ULTIMATE ANALYSIS
f analyse *f* élémentaire

e análisis *f* elemental
i analisi *f* elementare
n elementaire analyse
d Elementaranalyse *f*, Verbrennungsanalyse *f*

3859 ORGANIC CHEMISTRY oic
f chimie *f* organique
e química *f* orgánica
i chimica *f* organica
n organische chemie
d organische Chemie *f*

3860 ORGANIC PIGMENT dec
f pigment *m* organique
e pigmento *m* orgánico
i pigmento *m* organico
n organisch pigment *n*, organische verfstof
d organischer Farbstoff *m*, organisches Pigment *n*

3861 ORIGANUM OIL fat
f essence *f* d'origan
e esencia *f* de orégano
i olio *m* d'origano
n wilde-marjoleinolie
d wildes Majoranöl *n*, Spanischhopfenöl *n*

3862 ORIGINAL WORT bre
f moût *m* original
e primer mosto *m*
i mosto *m* originale
n eerste most
d Stammwürze *f*

3863 OSMAZOME, GLUTEN, VEGETABLE GLUTEN glu
f gluten *m*, phytocolle *f*
e gluten *m*
i gluten *m*
n gluten *n*
d Gluten *n*, Tritizin *n*

3864 OSMOTIC PRESSURE col
f pression *f* osmotique
e presión *f* osmótica
i pressione *f* osmotica
n osmotische druk
d osmotischer Druck *m*

3865 OSMOTIC SWELLING col
f gonflement *m* osmotique
e hinchamiento *m* osmótico
i ringonfiamento *m* osmotico
n osmotische zwelling
d osmotische Quellung *f*

3866 OSTWALD RIPENING, PHYSICAL RIPENING, RECRYSTALLIZATION pho
f maturation *f* physique
e maduración *f* física
i maturazione *f* fisica
n ostwaldse rijping
d Ostwaldsche Reifung *f*, physikalische Reifung *f*, Umkristallisierung *f*, Vorreifung *f*

3867 OTTO OIL, ATTAR OIL, ROSE OIL fat
f essence *f* de roses, huile *f* rosat
e esencia *f* de rosas
i essenza *f* di rose, olio *m* rosato
n rozenolie
d Rosenessenz *f*, Rosenöl *n*

3868 OUTFLOW sew
f écoulement *m*
e gasto *m*
i efflusso *m*
n uitloop
d Ausfluss *m*, Entleerung *f*

3869 OVERALL PLATE EFFICIENCY dis
f coefficient *m* d'efficacité total par plateau
e coeficiente *m* de rendimiento total por plato
i coefficiente *m* di rendimento totale per piatto
n totaal nuttig effect *n* per schotel, totale schotelverhouding
d Gesamtbodenwirkungsgrad *m*, Gesamtstufenwirkungsgrad *m*

3870 OVERBAKING (US), OVERSTOVING (GB) cer,dec
f surcuisson *m*
e sobrecocción *f*
i sovracottura *f*
n te ver doormoffelen *n*
d Überbrennen *n*

3871 OVERCOATING, INTERMEDIATE COATING, UNDERCOATING dec
f main *f* intermédiaire, peinture *f* de sous-couche
e mano *f* intermedia
i mano *f* intermedia
n overgrondverf, voorlak
d Voranstrich *m*, Zwischenschicht

3872 OVEREXPOSURE pho
f surexposition *f*

e sobreexposición *f*
i sovraesposizione *f*
n overbelichting
d Überbelichtung *f*

3873 OVERFLOW pre,rub
f déversage *m*
e desbordamiento *m*, rebosamiento *m*
i sfioro *m*, troppo pieno *m*
n overloop
d Überlauf *m*

3874 OVERGROWN MALT, DISTILLER'S MALT, LONG MALT bre
f hussards *pl*
e malta *f* de germinación larga
i malto *m* a germogliazione spinta, ussari *pl*
n huzaren *pl*, langmout *n*
d Langmalz *n*

3875 OVERHEAD IRRIGATION, BROAD IRRIGATION, SURFACE IRRIGATION sew
f irrigation *f* superficielle
e riego *m* con aguas cloacales, riego *m* con aguas negras
i disperdimento *m* per spandimento su terreni naturali permeabili, irrigazione *f* superficiale
n sproeibevloeiing
d Oberflächenberieselung *f*, Rieselfeldreinigung *f*

3876 OVERHEAD REFLUX dis
f reflux *m* en tête
e reflujo *m* en cabeza
i riflusso *m* in testa
n terugstroming in de kop
d Rücklauf *m* am Kopf

3877 OVERHEAD TEMPERATURE dis
f température *f* de tête
e temperatura *f* de cabeza
i temperatura *f* di testa
n koptemperatuur
d Kopftemperatur *f*

3878 OVERHEAD VAPOR (US), OVERHEAD VAPOUR (GB) dis
f vapeur *f* de tête
e vapor *m* de cabeza
i vapore *m* di testa
n damp in de kolomkop
d Dampf *m* am Kolonnenkopf

3879 OVERLIMING tan
f surchaulage *m*, surplainage *m*
e exceso *m* de pelambre
i pelatura *f* eccessiva
n overdadig ploten *n*
d Überäschern *n*

3880 OVERMASTICATION, OVERMILLING rub
f mastication *f* excessive
e masticación *f* excesiva
i masticazione *f* eccessiva
n te ver plasticeren *n*
d übermässige Mastizierung *f*

3881 OVERPICKLING cer
f surdécapage *m*
e sobredecapaje *m*
i sovradecapaggio *m*
n te sterk beitsen v
d zu stark beizen v

3882 OVERSPRAY dec
f perte *f* de peinture pulvérisée
e pérdida *f* de pintura pulverizada
i perdita *f* di pittura polverizzata
n spuitverlies *n*
d Locksprühverlust *m*

3883 OVERSTEEPING bre
f excès *m* de trempe
e remojo *m* excesivo
i eccesso *m* di macerazione
n overdadig weken *n*
d Überweiche *f*

3884 OXIDATION gen
f oxydation *f*
e oxidación *f*
i ossidazione *f*
n oxydatie
d Oxydation *f*

3885 OXIDATION BLACK, ANILINE BLACK, SUSPENSION BLACK dye
f noir *m* d'aniline
e negro *m* de anilina
i nero *m* d'anilina
n anilinezwart *n*
d Anilinschwarz *n*

3886 OXIDATION CATALYST, PRO-OXYGENIC AGENT cat,rub
f catalyseur *m* d'oxydation
e catalizador *m* de oxidación
i catalizzatore *m* d'ossidazione
n oxydatiekatalysator
d Oxydationskatalysator *m*

3887 OXIDATION RATIO, DEGREE OF OXIDATION gen
f degré *m* d'oxydation

e grado *m* de oxidación
i grado *m* d'ossidazione
n oxydatiegraad
d Oxydationsgrad *m*, Oxydationsstufe *f*

3888 OXIDATION-REDUCTION PROCESS gen
f procédé *m* oxydation-réduction
e procedimiento *m* oxidación-reducción, procedimiento *m* oxidorreducción
i processo *m* ossidazione-riduzione, processo *m* ossiriduzione
n oxydoreductieproces *n*
d Redoxverfahren *n*

3889 OXIDE OF ALUMINIUM, ALUMINA mat
f alumine *f*
e alumina *f*
i allumina *f*
n alumina *n*, aluminiumoxyde *n*, korund *n*
d Aluminiumoxyd *n*, Tonerde *f*

3890 OXIDIZED OIL, BASE OIL, THICKENED OIL fat,lub
f huile *f* épaissie, huile *f* oxydée, huile *f* soufflée
e aceite *m* oxidado
i olio *m* ossidato, olio *m* soffiato
n geblazen olie, geoxydeerde olie
d geblasenes Öl *n*, oxydiertes Öl *n*

3891 OXYGEN gas
f oxygène *m*
e oxígeno *m*
i ossigeno *m*
n zuurstof
d Sauerstoff *m*

3892 OXYGEN BALANCE, OXYGEN EQUILIBRIUM sew
f équilibre *m* d'oxygène
e equilibrio *m* de oxígeno
i equilibrio *m* d'ossigeno
n zuurstofevenwicht *n*
d Sauerstoffgleichgewicht *n*, Sauerstoffhaushalt *m*

3893 OXYGEN BLEACHING ble
f blanchiment *m* à l'oxygène
e blanqueo *m* al oxígeno
i candeggio *m* all'ossigeno
n zuurstofbleken *n*
d Sauerstoffbleichen *n*

3894 OXYGEN BOMB TEST tes
f essai *m* à la bombe à oxygène
e ensayo *m* en la bomba de oxígeno
i collaudo *m* nella bomba ad ossigeno
n verouderingsproef met zuurstofbom
d Sauerstoffalterung *f* in der Druckbombe

3895 OXYGEN CONSUMED FROM PERMANGANATE, CONSUMED OXYGEN iwt
f demande *f* en permanganate de potassium, oxydabilité *f* au permanganate
e consumo *m* de permanganato
i consumo *m* di permanganato
n permanganaatverbruik *n*
d Kaliumpermanganatverbrauch *m*, Permanganatverbrauch *m*

3896 OXYGEN CUTTING, LANCE CUTTING htt,met
f oxycoupage *m*
e corte *m* con lanzo, oxicorte *m*
i ossitaglio *m*, taglio *m* con ossigeno
n snijden *n* met zuurstoflans
d Brennschneiden *n*

3897 OXYGEN LANCE htt
f lance *f* d'oxygène
e lanza *f* de oxígeno
i lancia *f* d'ossigeno
n zuurstoflans
d Sauerstofflanze *f*

3898 OXYGEN LIQUEFACTION gas
f liquéfaction *f* d'oxygène
e licuación *f* de oxígeno
i liquefazione *f* d'ossigeno
n vloeibaar maken *n* van zuurstof
d Sauerstoffverflüssigung *f*

3899 OXYGEN REMOVAL, DEOXYGENATION iwt
f désoxygénation *f*
e desoxigenación *f*
i disossigenazione *f*
n zuurstofverwijdering
d Entziehung *f* des Sauerstoffs, Sauerstoffbeseitigung *f*

3900 OXYGENATION imp
f oxygénation *f*
e oxigenación *f*
i ossigenazione *f*
n oxygenatie
d Oxygenierung *f*

3901 OXYHYDROGEN GAS, DETONATING GAS gas
f gaz *m* tonnant

e gas *m* detonante
i gas *m* ossidrico,
gas *m* tonante
n knalgas *n*
d Knallgas *n*

3902 OXYHYDROGEN WELDING wel
f soudure *f* oxyhydrique
e soldadura *f* oxídrica
i saldatura *f* ossidrica
n waterstof-zuurstoflassen *n*
d Wasserstoff-Sauerstoff-
schweissung *f*

3903 OZOBROME PROCESS, pho
CARBRO PROCESS
f carbro *m*, ozobromie *f*
e procedimiento *m* al
bromocarbón
i ozobromia *f*, processo *m* al
bromocarbone
n carbroproces *n*
d Karbrodruck *m*, Ozobromdruck *m*

3904 OZOKERITE, fat
CERESINE, CERINE,
FOSSIL WAX
f cire *f* minérale, ozocérite *f*,
suif *m* de montagne
e cera *f* fósil, ceresina *f*
i cera *f* minerale, ozocherite *f*
n aardwas, ozokeriet *n*
d Erdwachs *n*, mineralisches
Wachs *n*, Ozokerit *m*

3905 OZONATION, gas
OZONIZING
f ozonisation *f*
e ozonización *f*
i ozonizzazione *f*
n ozoniseren *n*
d Ozonisieren *n*

3906 OZONE gas
f ozone *m*
e ozono *m*
i ozono *m*
n ozon *n*
d Ozon *n*

3907 OZONE BLEACHING ble
f blanchiment *m* à l'ozone
e blanqueo *m* al ozono
i candeggio *m* all'ozono
n ozonbleken *n*
d Ozonbleichung *f*

3908 OZONE PAPER pap
f papier *m* ozonoscopique
e papel *m* sensible al ozono
i carta *f* sensibile all'ozono
n op ozon reagerend papier *n*
d Ozonpapier *n*, Ozonreagenz-
papier *n*

P

3909 PACK ANNEALING htt,met
f recuit *m* emballé, recuit *m* en paquet
e recocido *m* en paquete
i ricottura *f* a pacchetto
n pakketgloeien *n*
d Paketglühen *n*

3910 PACK CARBURIZING, BOX CARBURIZING htt,met
f cémentation *f* en caisse
e cementación *f* en caja
i cementazione *f* in cassetta
n kistcementeren *n*
d Kastenzementierung *f*

3911 PACK DYEING, BALE DYEING, CIRCULATION DYEING tex
f teinture *f* en balles
e teñido *m* en balas, teñido *m* en líos
i tintura *f* in balle
n balenverven *n*, verven *n* in balen
d Ballenfärbung *f*

3912 PACKAGING (US), PACKING (GB) gen
f emballage *m*
e empaquetamiento *m*
i imballaggio *m*
n verpakking
d Verpackung *f*

3913 PACKING dis
f garnissage *m*
e relleno *m*
i riempimento *m*
n vullichamen *pl*, vulling
d Füllkörper *pl*, Füllung *f*

3914 PACKING mcy,met
f compression *f*
e compresión *f*
i compressione *f*
n comprimeren *n*
d Zusammenpressen *n*

3915 PAD-ROLL DYEING SYSTEM dye
f teinture *f* par imprégnation à rouleau
e teñido *m* por impregnación con rodillo
i tintura *f* per impregnazione con rullo
n verven *n* door impregneren met rol
d Färben *n* durch Imprägnierung mit Rolle

3916 PAD-STEAM PROCESS dye
f teinture *f* par imprégnation à vapeur
e teñido *m* por impregnación con vapor
i tintura *f* per impregnazione con vapore
n verven *n* door impregneren met stoom
d Färben *n* durch Imprägnieren mit Dampf

3917 PADDING tex
f teinture *f* par imprégnation
e teñido *m* por impregnación
i tintura *f* per impregnazione
n verven *n* met foulardmachine
d Foulardieren *n*, Klatschen *n*, Klotzen *n*

3918 PADDLE DYEING dye
f teinture *f* à agitation du bain
e teñido *m* con agitación del baño
i tintura *f* con agitazione del bagno
n verven *n* in in beweging gehouden bad
d Färben *n* in Bad mit Rührwerk

3919 PADDLING tan
f tannage *m* à agitation du bain
e curtimiento *m* con agitación del baño
i concia *f* con agitazione del bagno
n looien *n* in in beweging gehouden looistof
d Gerben *n* in Gerbflüssigkeit mit Rührwerk

3920 PAINT dec
f couleur *f*, peinture *f*
e color *m*, pintura *f*
i colore *m*, pittura *f*
n verf
d Anstrichfarbe *f*, Farbe *f*

3921 PAINT ADHESION dec
f adhésion *f* de peinture
e adhesión *f* de pintura
i adesione *f* di pittura
n kleurhechting
d Farbenhaftkraft *f*

3922 PAINT FILM dec
f film *m* de peinture
e película *f* de pintura
i pellicola *f* di pittura
n verffilm
d Anstrichfilm *m*, Farbfilm *m*

3923 PAINT HARDENER, GOLD SIZE dec
f vernis *m* colle d'or
e endurecedor *m* de pintura
i induritore *m* di pittura
n verfverharder, verhardingslak
d Härtelack *m*, Harttrockenglanzöl *n*

3924 PAINT REMOVER dec
f décapant *m*
e decapante *m* para pinturas, quitapinturas *m*
i agente *m* di decapaggio
n afbijtmiddel *n*
d Farbenbeize *f*, Farbenabbeizmittel *n*

3925 PAINT SPRAYING, SPRAY PAINTING dec
f pistolage *m*
e pintura *f* con pistola
i spruzzatura *f* di pittura
n verfspuiten *n*
d Spritzlackierung *f*

3926 PAINTING dec
f peinture *f*
e pintura *f*
i pittura *f*
n schilderen *n*, verven *n*
d Anstreichen *n*, Streichen *n*

3927 PAINTING MATERIALS, SURFACE COATING MATERIALS dec
f matériaux *pl* pour la peinture, peintures *pl* et vernis *pl*
e materiales *pl* para pintar
i materiali *pl* per pitturare
n verfwaren *pl*
d Anstrichmittel *pl*, Farbwaren *pl*

3928 PALE ALE bre
f bière *f* blanche
e cerveza *f* clara
i birra *f* chiara
n helder aal *n*
d helles obergäriges Bier *n*

3929 PALE LAGER, PILSNER BEER bre
f bière *f* de Pilsen
e cerveza *f* de Pilsen
i birra *f* di Pilsen
n pilsener *n*
d Pilsnerbier *n*

3930 PALE LEAF GOLD coa
f feuille *f* en alliage d'or et argent
e hoja *f* en aleación de oro y plata
i foglia *f* in lega d'oro e argento
n goudzilverfoelie
d Goldsilberfolie *f*

3931 PALLADINIZED ASBESTOS mat
f amiante *m* au palladium
e amianto *m* paladinizado
i amianto *m* al palladio
n palladiumasbest *n*
d Palladiumasbest *m*

3932 PALLADINIZING elc
f revêtement *m* au palladium
e paladinización *f*
i rivestimento *m* al palladio
n palladiumbedekking
d Palladiumbeschichtung *f*

3933 PALLADIUM PLATING coa
f placage *m* au palladium
e paladiado *m*
i placcatura *f* al palladio
n palladiumpleten *n*
d Palladiumplattierung *f*

3934 PALLADIUM SPONGE met
f éponge *f* au palladium, mousse *f* au palladium
e esponja *f* al paladio
i palladio *m* spugnoso
n palladiumspons
d Palladiumschwamm *m*

3935 PALM KERNEL OIL, PALM NUT OIL fat
f huile *f* de palmiste
e aceite *m* de nuez de palma
i olio *m* di noce di palma
n palmkernolie, palmpittenolie
d Palmkernöl *n*

3936 PALM OIL fat
f huile *f* de palme
e aceite *m* de palma
i olio *m* di palma
n palmolie
d Palmöl *n*

3937 PALMAROSA OIL fat
f essence *f* de palmarose

e esencia *f* de palmarosa
i olio *m* di palmarosa
n palmarosaolie
d Palmarosaöl *n*, Rusaöl *n*

3938 PALMITATE SOAP soa
f savon *m* de l'acide palmitique
e jabón *m* de palmitato
i sapone *m* di palmitato
n palmitaatzeep
d Palmitatseife *f*

3939 PALMITIC ACID (GB), CETIN (US) soa
f acide *m* palmitique
e ácido *m* palmítico, cetina *f*
i acido *m* palmitico
n palmitinezuur *n*
d Palmitinsäure *f*

3940 PAN PROCESS rub
f régénération *f* à la vapeur sous haute pression
e procedimiento *m* en caldero
i processo *m* in padella, rigenerazione *f* sotto vapore ad alta pressione
n hoge-druk-stoomregeneratie
d Hochdruckdampfregenerierung *f*

3941 PAN RECLAIMING PROCESS, HEATER RECLAIMING PROCESS rub
f procédé *m* de régénération par la vapeur et l'huile
e procedimiento *m* de regeneración con vapor y aceite
i processo *m* di rigenerazione in vapore ed olio
n stoom-olie-regenereerproces *n*
d Dampf-Ölregenerierungsverfahren *n*

3942 PANCHROMATIC EMULSION pho
f émulsion *f* panchromatique
e emulsión *f* pancromática
i emulsione *f* pancromatica
n panchromatische emulsie
d panchromatische Emulsion *f*

3943 PAPER pap
f papier *m*
e papel *m*
i carta *f*
n papier *n*
d Papier *n*

3944 PAPER FILLER pla
f matière *f* de charge en papier
e materia *f* de carga de papel
i materia *f* di carica in carta
n papieren vulmiddel *n*
d Papierfüllmittel *n*

3945 PAPER LAMINATE pap
f papier *m* stratifié
e papel *m* de capas superpuestas
i carta *f* stratificata
n hardpapier *n*
d Hartpapier *n*

3946 PAPER PULP pap
f pâte *f* de papier
e pulpa *f* de papel
i polpa *f* di carta
n papierpulp
d Papierbrei *f*

3947 PAPER PULP THICKENING pap
f concentration *f* de la pâte de papier
e concentración *f* de la pulpa de papel
i addensamento *m* della polpa di carta
n indikken *n* van de papierpulp
d Eindicken *n* der Papierbrei

3948 PAPER SIZING pap
f encollage *m* de papier
e encolaje *m* de papel
i incollatura *f* di carta
n lijmen *n* van papier
d Leimen *n* von Papier

3949 PAPER TAPE pap
f ruban *m* de papier
e cinta *f* de papel
i nastro *m* di carta
n papierlint *n*, papierstrook
d Papierband *n*, Papierstreifen *n*

3950 PAPER VARNISH dec
f vernis *m* pour papier
e barniz *m* para papel
i vernice *f* per carta
n papiervernis *n*
d Papierlack *m*

3951 PAPIER-MÂCHÉ pap
f carton *m* pâte
e cartón *m* piedra
i cartapesta *f*
n papier-maché *n*
d Papiermaché *n*

3952 PARA RUBBER rub
f caoutchouc *m* de Para, caoutchouc *m* para
e goma *f* de Para
i para *f*
n pararubber
d Paragummi *m*

3953 PARAFFIN HYDROCARBONS, PARAFFINS fat,oic
f hydrocarbures *pl* paraffiniques
e hidrocarburos *pl* parafínicos
i idrocarburi *pl* di paraffina
n paraffinekoolwaterstoffen *pl*
d Paraffinkohlenwasserstoffe *pl*

3954 PARAFFIN OIL, LIQUID PARAFFIN fat
f huile *f* de paraffine
e aceite *m* de parafina
i olio *m* di paraffina
n paraffineolie
d Paraffinöl *n*, Vaselinöl *n*

3955 PARAFFIN SCALE fat
f paraffine *f* en écailles
e parafina *f* en escamas
i paraffina *f* squamosa
n paraffineschubben *pl*
d Schuppenparaffin *n*

3956 PARAFFIN WAX, HARD PARAFFIN fat
f cire *f* de paraffine
e cera *f* parafínica
i cera *f* di paraffina
n paraffinewas, vaste paraffine
d Paraffinwachs *n*

3957 PARAFFINING coa
f paraffinage *m*
e parafinado *m*
i rivestimento *m* alla paraffina
n paraffineren *n*
d Paraffinieren *n*

3958 PARALLEL FLOW cen
f flux *m* parallèle
e flujo *m* paralelo
i flusso *m* parallelo
n gelijkstroming, parallelle stroming
d Gleichstrom *m*

3959 PARALYZER, CATALYST POISON gen
f poison *m* de catalysateur
e veneno *m* de catalizador
i veleno *m* catalitico
n katalysatorvergif *n*
d Katalysatorgift *n*, Katalytgift *n*, Kontaktgift *n*

3960 PARCEL PLATING, PARTIAL PLATING elc
f dépôt *m* limité
e depósito *m* limitado
i deposito *m* limitato
n gedeeltelijke bedekking
d Teilgalvanisierung *f*

3961 PARCHMENT lea
f parchemin *m*
e pergamino *m*
i cartapecora *f*, pergamena *f*
n perkament *n*
d Pergament *n*

3962 PARCHMENT PAPER pap
f papier *m* parcheminé
e papel *m* pergamino
i carta *f* pergamena
n perkamentpapier *n*
d Pergamentpapier *n*

3963 PARCHMENTIZING pap
f parcheminage *m*
e apergaminado *m*
i pergamenizzazione *f*
n perkamentachtig maken *n*
d Pergamentieren *n*, Transparentieren *n*

3964 PARENT METAL wel
f métal *m* base
e metal *m* base
i metallo *m* base
n basismetaal *n*
d Grundmetall *n*, Muttermetall *n*

3965 PARENT SUBSTANCE gen
f substance *f* base
e substancia *f* madre
i sostanza *f* base
n uitgangsstof
d Mutterstoff *m*, Stammstoff *m*

3966 PARIS BLUE, COBALT BLUE, HUNGARY BLUE dec
f bleu *m* de cobalt, bleu *m* de Paris
e azul *m* de cobalto, azul *m* de París
i blu *m* di cobalto, blu *m* di Parigi
n Berlijns blauw *n*, kobaltblauw *n*, Parijs blauw *n*
d Berliner Blau *n*, Kobaltblau *n*, Königsblau *n*, Wiener Blau *n*

3967 PARIS GREEN (GB), EMPEROR GREEN (US), SCHWEINFURT GREEN dec
f vert *m* de Scheele
e cardenillo *m* arsenioso, verde *m* de París
i verde *m* di Parigi
n Parijs groen *n*
d Parisergrün *n*, Schweinfurter Grün *n*

3968 PARIS YELLOW, CHROME YELLOW, LEAD CHROMATE dec
f jaune *m* de chrome
e amarillo *m* de cromo, cromato *m* de plomo
i giallo *m* di cromo
n chroomgeel *n*, Parijs geel *n*
d Bleichromat *n*, Chromgelb *n*, Parisergelb *n*

3969 PARTIAL MISCIBILITY mix
f miscibilité *f* partielle
e miscibilidad *f* parcial
i miscibilità *f* parziale
n beperkte mengbaarheid
d beschränkte Mischbarkeit *f*

3970 PARTIAL PYRITIC SMELTING htt,met
f fusion *f* pyritique partielle
e fusión *f* pirítica parcial
i fusione *f* piritica parziale
n gedeeltelijke pyrietsmelting
d teilweise Pyritschmelzung *f*

3971 PARTIAL REACTION gen
f réaction *f* partielle
e reacción *f* parcial
i reazione *f* parziale
n gedeeltelijke reactie
d Teilreaktion *f*

3972 PARTIAL REFLUX dis
f reflux *m* partiel
e reflujo *m* parcial
i riflusso *m* parziale
n gedeeltelijke terugstroming
d teilweiser Rücklauf *m*

3973 PARTIAL ROASTING htt,met
f grillage *m* partiel
e tostación *f* parcial
i arrostimento *m* parziale, torrefazione *f* parziale
n gedeeltelijke roosting
d Teilröstung *f*

3974 PARTIAL SOLID SOLUBILITY met
f solubilité *f* partielle dans l'état solide
e solubilidad *f* parcial en el estado sólido
i solubilità *f* parziale nello stato solido
n gedeeltelijke oplosbaarheid in de vaste toestand
d partielle Löslichkeit *f* im festen Zustand

3975 PARTICLE MEAN SIZE, PARTICLE SIZE gen
f dimension *f* des grains, dimension *f* des particules
e dimensión *f* de los granos, dimensión *f* de las partículas
i dimensione *f* dei grani, dimensione *f* delle particelle
n grootte van het deeltje
d Teilchengrösse *f*

3976 PARTICLE-SIZE ANALYSIS ana
f analyse *f* granulométrique
e análisis *f* granulométrica
i analisi *f* granulometrica
n granulometrische analyse
d granulometrische Analyse *f*

3977 PARTICLE-SIZE DISTRIBUTION gen
f distribution *f* granulométrique
e distribución *f* granulométrica
i distribuzione *f* granulometrica
n granulometrische verdeling
d granulometrische Verteilung *f*, Teilchengrössenverteilung *f*

3978 PARTICLE-SIZE FRACTION, SIZE FRACTION gen
f fraction *f* granulométrique
e fracción *f* granulométrica
i frazione *f* granulometrica
n deeltjesfractie
d Kornfraktion *f*, Teilchenfraktion

3979 PARTICLE-SIZE RANGE gen
f intervalle *m* de classement granulométrique
e intervalo *m* de clasificación granulométrica
i intervallo *m* di classificazione granulometrica
n deeltjesklasse
d Kornklasse *f*, Teilchengrössenklasse *f*

3980 PARTING AGENT pla,rub
f agent *m* de démoulage
e substancia *f* separadora
i sostanza *f* separatrice
n glijmiddel *n*
d Gleitmittel *n*, Trennmittel *n*

3981 PARTING OF NOBLE METALS met
f départ *m* de métaux précieux
e separación *f* de metales nobles
i separazione *f* di metalli nobili
n scheiding van edele metalen
d Edelmetallscheidung *f*

3982 PARTITION CHROMATOGRAPHY ana
f chromatographie *f* de partage
e chromatografía *f* de división
i cromatografia *f* di divisione
n verdelingschromatografie
d Verteilungschromatographie *f*

3983 PASSIVATING DIP (US), PASSIVATING SOLUTION (GB) elc
f solution *f* de passivation
e solución *f* pasivadora
i soluzione *f* di passivazione
n passiverende oplossing
d passivierende Lösung *f*

3984 PASSIVATION, CONVERSION COATING elc
f passivation *f*
e pasivación *f*
i passivazione *f*
n passiveren *n*
d Passivierung *f*

3985 PASSIVE METAL coa
f métal *m* passif
e metal *m* pasivo
i metallo *m* passivo
n passief metaal *n*
d selbstpassivierendes Metall *n*

3986 PASSIVITY coa,met
f passivité *f*
e pasividad *f*
i passività *f*
n passiviteit
d Passivität *f*

3987 PASTE cer
f pâte *f*
e pasta *f*
i pasta *f*
n brij, pasta
d Brei *f*

3988 PASTE elc
f pâte *f*
e pasta *f*
i pasta *f* gelificante
n gelei
d Paste *f*

3989 PASTEBOARD, BOARD, CARD-BOARD mat,pap
f carton *m*
e cartón *m*
i cartone *m*
n karton *n*
d Pappe *f*

3990 PASTE PAINT, OIL PASTE dec
f pigment *m* broyé à l'huile
e pasta *f* de aceite
i pasta *f* ad olio
n dik in olie gemalen pasta, oliepasta
n in Öl angeriebene Farbe *f*, Ölpaste *f*

3991 PASTE SOLDER wel
f pâte *f* à souder
e pasta *f* para soldar
i pasta *f* per saldare
n soldeerpasta
d Lötpaste *f*

3992 PASTEL SHADE dec,dye
f couleur *f* pastel
e color *m* pastel
i colore *m* pastello
n pastelkleur
d Pastellfarbe *f*

3993 PASTEURIZATION, PASTEURIZING bac,dai
f pasteurisation *f*
e pasteurización *f*
i pastorizzazione *f*
n pasteuriseren *n*
d Pasteurisieren *n*

3994 PASTING gen
f encollage *m*
e empaste *m*
i incollatura *f*
n lijmen *n*
d Verleimung *f*

3995 PAT TEST (US), CEMENT CAKE TEST (GB) bui
f essai *m* à galette de ciment
e ensayo *m* de pastilla de cemento
i collaudo *m* a pastilla di cemento
n cementkoekproef
d Zementkuchenprobe *f*

3996 PATCHOULI OIL fat
f essence *f* de patchouli
e esencia *f* de patchouli
i olio *m* di patchouli
n patschoeliolie
d Patschuliöl *n*

3997 PATENT LEATHER, ENAMELLED LEATHER, JAPANNED LEATHER coa,lea
f cuir *m* verni
e charol *m*

i pelle *f* di lustrino, pelle *f* verniciata
n lakleder *n*
d Glanzleder *n*, Lackleder *n*

3998 PATENT PLATE (US), PLATE GLASS (GB) gla
f glace *f* sans tain
e cristal *m* de espeja, cristal *m* de luna
i cristallo *m* da specchio, luce *f*
n spiegelglas *n*
d Spiegelglas *n*

3999 PATENTING met
f trempe *f* au plomb
e temple *m* al plomo, temple *m* isotérmico
i patentamento *m*
n afschrikken *n* in lood
d Patentieren *n*

4000 PATINA coa
f patine *f*
e pátina *f*
i patina *f*
n patina *n*
d Patina *f*

4001 PATINA FINISH coa
f patine *f* artificielle
e pátina *f* artificial
i patina *f* artificiale
n namaakpatina *n*
d künstliche Patina *f*

4002 PATINATION coa
f application *f* de patine
e aplicación *f* de pátina
i patinatura *f*
n patijnen *n*, patineren *n*
d Patinieren *n*

4003 PAUT HEMP, JUTE tex
f chanvre *m* de Calcutta, jute *m*
e cáñamo *m* de Calcuta, yute *m*
i iuta *f*
n jute
d Jute *f*, Pahthanf *m*, ostindischer Hanf *m*

4004 PEAK FLUX DENSITY boi
f flux *m* de chaleur maximal
e flujo *m* de calor máximo
i flusso *m* di calore massimo
n maximale warmtestroom-dichtheid in kernkoken
d maximale Heizflächenbelastung *f*

4005 PEANUT OIL, ARACHIS OIL, EARTHNUT OIL, GROUND OIL fat
f huile *f* d'arachide
e aceite *m* de cacahuete
i olio *m* d'arachide
n aardnotenolie
d Erdnussöl *n*

4006 PEARL ASH, PURIFIED POTASH aic
f perlasse *f*
e perlasa *f*, potasa *f* purificada
i potassa *f* purificata
n Amerikaanse potas, parelas
d Aschensalz *n*, Perlasche *f*

4007 PEARSHAPED gen
f piriforme adj
e piriforme adj
i piriforme adj
n peervormig adj
d birnenförmig adj

4008 PEAT fue
f tourbe *f*
e turba *f*
i torba *f*
n turf
d Torf *m*

4009 PEAT LOW TEMPERATURE CARBONIZATION fue
f semicarbonisation *f* de tourbe
e carbonización *f* de baja temperatura de turba
i carbonizzazione *f* a bassa temperatura di torba
n carbonisatie van turf bij lage temperatuur, zwelen *n* van turf
d Torfschwelung *f*

4010 PEAT TAR fue
f goudron *m* de tourbe
e alquitrán *m* de turba
i catrame *m* di torba
n turfteer
d Torfteer *m*

4011 PECTIZATION, GELATION, GELLING col
f formation *f* de gel, gélification *f*
e formación *f* de gel, gelificación *f*
i gelizzazione *f*
n geleren *n*, gelvorming, stijf worden *n*
d Gelbildung *f*, Gelieren *n*, Steifwerden *n*

4012 PEEL-OFF COATING STRIPPABLE COATING dec
f vernis *m* pelable

e barniz *m* pelable
i lacca *f* a spelamento
n afpelbare lak, striplak
d Abziehlack *m*

4013 PEELING elc
f écaillage *m*
e escamado *m*
i scagliatura *f*, sfogliatura *f*
n bladderen *n*
d Abblättern *n*

4014 PEELING OF GLAZE, SHIVERING cer
f écaillage *m* de l'émail
e descamación *f* en el vidriado
i scheggiatura *f* dell'invetriatura
n schilferen *n* van de glazuur
d Abrollen *n* der Glasur

4015 PEELING OFF, EXFOLIATING, SCALING dec
f écaillage *m*
e despellegamiento *m*
i screpolatura *f*
n afbladderen *n*
d Abblättern *n*, Abschälen *n*

4016 PEENING SHOT coa
f grenaille *f* d'acier
e granalla *f* de acero
i graniglia *f* d'acciaio
n staalschroot *n*
d Stahlschrot *m*

4017 PELLET pla
f boulette *f*, pastille *f*
e granulo *m*, pastilla *f*
i granulo *m*, pallottolina *f*
n bolletje *n*, korrel
d Korn *n*, Kügelchen *n*, Tablette *f*

4018 PELLET, TABLET pla
f pastille *f* précomprimée
e pastilla *f* precomprimida
i pasticca *f* precompressa
n voorgeperst tablet *n*
d vorgepresste Tablette *f*

4019 PELLETIZING mcy
f formation *f* des comprimés
e formación *f* de comprimidas
i formazione *f* di compresse
n tabletteren *n*
d Tablettieren *n*

4020 PELLETIZING rub
f réduction *f* de caoutchouc en boulettes
e reducción de goma a pastillas
i riduzione *f* di gomma in pallottoline
n korrelen *n* van rubber
d Kautschukkörnung *f*

4021 PELT tan
f peau *f* en tripe
e piel *f* sin curtir
i pelle *f* non conciata
n bloot
d Blösse *f*, Hautblösse *f*

4022 PENETRANT, PENETRATING AGENT dye,mcy,tex
f agent *m* pénétrant
e agente *m* de penetración
i agente *m* penetrante
n indringmiddel *n*
d Durchdringungsmittel *n*

4023 PENETRATING OIL cle
f huile *f* pénétrante
e aceite *m* penetrante
i olio *m* penetrante
n kruipolie, roest oplossende olie
d rostlösendes Öl *n*

4024 PENETRATING VARNISH, IMPREGNATING VARNISH dec
f vernis *m* d'imprégnation
e barniz *m* de impregnación
i vernice *f* d'impregnazione
n impregneerlak
d Imprägnierlack *m*

4025 PENETRATION bui
f pénétration *f*
e penetración *f*
i penetrazione *f*
n penetratie
d Durchdringung *f*

4026 PENETRATION TEST tes
f essai *m* de pénétration
e ensayo *m* de penetración
i collaudo *m* di penetrazione
n penetratieproef
d Eindringtiefeprüfung *f*

4027 PEPPERMINT OIL fat
f essence *f* de menthe poivrée
e esencia *f* de hierbabuena
i olio *m* di menta piperita
n pepermuntolie
d Pfefferminzöl *n*

4028 PEPTIZATION col
f peptisation *f*
e peptización *f*

i peptizzazione *f*
n gel-sol-overgang, peptisatie
d Gel-Sol-Umwandlung *f*, Peptisierung *f*

4029 PEPTIZER, PEPTIZING AGENT — rub
f agent *m* peptisant
e agente *m* peptizante
i agente *m* peptizzante
n chemisch afbraakmiddel *n*, peptiseermiddel *n*
d Abbaumittel *n*, Peptisiermittel *n*

4030 PEPTONIZATION — col
f peptonisation *f*
e peptonización *f*
i peptonizzazione *f*
n afbraak van eiwitten, peptoniseren *n*
d Peptonisierung *f*, Stärkeabbau *m*

4031 PEPTONIZING REST — bre
f relais *m* d'albumine
e intervalo *m* de peptonización
i sosta *f* di peptonizzazione
n peptonisatierust
d Eiweissrast *m*

4032 PERCHING — lea
f étirage *m*
e suavidazación *f*
i stiratura *f*
n rekken *n*
d Recken *n*

4033 PERCOLATION, CONTINUOUS FILTERING — iwt
f percolation *f*
e percolación *f*
i filtrazione *f* continua
n druppelfiltratie, percoleren *n*
d Perkolation *f*, Tropffiltrierung *f*

4034 PERCRYSTALLIZATION — cry
f cristallisation *f* à membrane semiperméable
e cristalización *f* con membrana semipermeable, percristalización *f*
i cristallizzazione *f* a membrana semipermeabile, percristallizzazione *f*
n kristallisatie met semipermeabele membraan
d Kristallisation *f* bei der Dialyse, Kristallisieren *n* mit halbdurchlässiger Membrane

4035 PERDISTILLATION — dis
f distillation *f* à membrane semiperméable
e destilación *f* con membrana semipermeable, perdestilación *f*
i distillazione *f* a membrana semipermeabile, perdistillazione *f*
n destilleren *n* met semipermeabele membraan
d Destillieren *n* mit halbdurchlässiger Membrane

4036 PERFECT COMBUSTION, COMPLETE COMBUSTION — fue
f combustion *f* complète
e combustión *f* completa
i combustione *f* completa
n volledige verbranding
d vollständige Verbrennung *f*

4037 PERFECT GAS, IDEAL GAS — gas
f gaz *m* idéal, gaz *m* parfait
e gas *m* ideal, gas *m* perfecto
i gas *m* ideale, gas *m* perfetto
n ideaal gas *n*
d ideales Gas *n*

4038 PERFUME, SCENT — per
f parfum *m*
e perfume *m*
i profumo *m*
n odeur, parfum *n*
d Parfüm *n*

4039 PERFUME BASE — per
f base *f* de parfum
e base *f* de perfume
i base *f* di profumo
n parfumgrondstof
d Parfümgrundlage *f*

4040 PERFUME FIXATIVE — per
f fixatif *m* de parfum
e fixativo *m* de perfume
i fissativo *m* di profumo
n parfumfixatief *n*
d Parfümfixativ *n*

4041 PERGAMYNE (GB), GLASSINE (US), TRANSPARENT PARCHMENT (GB) — pap
f papier *m* cristal, papier *m* serpente, serpente *f*
e papel *m* cristal
i carta *f* glassene, carta *f* trasparente
n glaspapier *n*, kristalpapier *n*
d Glashaut *f*, Glassin *n*, Pergamin *n*

4042 PERMANENT FINISH tex
f apprêt *m* permanent
e apresto *m* permanente
i appretto *m* permanente
n permanent appreteren *n*
d permanentes Appretieren *n*

4043 PERMANENT HARDNESS hyg
f dureté *f* permanente
e dureza *f* permanente
i durezza *f* permanente
n blijvende hardheid
d bleibende Härte *f*, permanente Härte *f*, Sulfathärte *f*

4044 PERMANENT WHITE, BLANC FIXE dec
f blanc *m* fixe
e blanco *m* fijo
i bianco *m* fisso
n barietwit *n*
d Barytweiss *n*, Blankfix *n*, Permanentweiss *n*

4045 PERMEATION fue
f imprégnation *f*, infiltration *f*
e permeación *f*
i attraversamento *m*
n doordringing
d Durchdringung *f*

4046 PERMISSIBLE EXPLOSIVE exp
f explosif *m* de sécurité
e explosivo *m* autorizado, explosivo *m* de seguridad
i esplosivo *m* di sicurezza
n mijngasveilige springstof
d Sicherheitssprengstoff *m*

4047 PERRY nut
f cidre *m* de poire
e sidra *f* de pera
i sidro *m* di pera
n perencider
d Birnenmost *m*

4048 PERSORPTION abs
f absorption *f* intensive
e absorción *f* intensiva
i assorbimento *m* intensivo
n intensieve absorptie
d intensive Absorption *f*

4049 PERSPIRATION WATER, FLUE SWEAT fue
f eau *f* de condensation de cheminée
e agua *f* de condensación de chimenea
i acqua *f* di condensazione di camino
n zweetwater *n*
d Schwitzwasser *n*

4050 PERUVIAN BARK, CINCHONA mcy
f écorce *f* de quinquina, quinquina *f*
e corteza *f* de quino, quina *f*
i china *f*, corteccia *f* di china
n kina, kinabast
d Chinarinde *f*, Cinchonarinde *f*, Fieberrinde *f*

4051 PERUVIN (US), CINNAMIC ALCOHOL (GB) per
f alcool *m* cinnamique
e alcohol *m* cinámico
i alcool *m* cinnamico
n kaneelalcohol
d Styron *n*, Zimtalkohol *m*

4052 PERVAPORATION eva
f évaporation *f* à membrane semiperméable
e evaporación *f* con membrana semipermeable
i evaporazione *f* a membrana semipermeabile
n verdamping met semipermeabele membraan
d Verdampfung *f* mit halbdurchlässiger Membrane

4053 PETROCHEMICALS fue
f produits *pl* chimiques dérivés du pétrole
e productos *pl* químicos derivados del petróleo
i prodotti *pl* chimici derivati del petrolio
n aardoliederivaten *pl*
d Erdölderivate *pl*

4054 PETROL (GB), GASOLINE (US), MOTOR SPIRIT fue
f essence *f*
e gasolina *f*
i benzina *f*
n benzine
d Benzin *n*

4055 PETROL SEPARATION (GB), GASOLINE SEPARATION (US) fue
f séparation *f* d'essence
e separación *f* de gasolina
i separazione *f* di benzina
n benzineafscheiding
d Benzinabscheidung *f*

4056 PETROLATUM, PETROLEUM JELLY lub
f pétrolatum *m*, vaseline *f*
e petrolato *m*, vaselina *f*
i petrolato *m*, vaselina *f*
n petrolatum *n*, vaseline
d Petrolatum *n*, Vaselin *n*, Vaseline *f*

4057 PETROLEUM (GB), CRUDE OIL, KEROSENE (US) fue
f pétrole *m* brut
e petróleo *m* bruto
i petrolio *m* greggio
n ruwe petroleum
d Erdöl *n*, Petroleum *n*, Steinöl *n*

4058 PETROLEUM ETHER, GASOLENE, GASOLINE fue
f gazoline *f*
e éter *m* de petróleo
i gasolina *f*
n gasoline
d Gasäther *m*, Gasolin *n*

4059 PETROLEUM WAX, CRUDE WAX SLACK WAX fat
f paraffine *f* brute, paraffine *f* non-déshuilée
e cera *f* parafínica muy aceitosa, mezcla *f* de cera parafínica y de petróleo
i paraffina *f* greggia, paraffina *f* molle
n ruwe paraffine, uit aardolie verkregen paraffine
d Rohparaffin *n*

4060 PHASE CONTRAST MICROSCOPY ana
f microscopie *f* à contraste de phase
e microscopia *f* por contraste de fase
i metodo *m* del contrasto di fase
n fazecontrastmicroscopie
d Phasenkontrastverfahren *n*

4061 PHASE INVERSION, PHASE REVERSAL gen
f inversion *f* de phase
e inversión *f* de fase
i cambiamento *m* di fase
n fazeomzetting
d Phasenumkehrung *f*

4062 PHASE REACTION gen
f réaction *f* de phase
e reacción *f* de fase
i reazione *f* di fase
n fazereactie
d Phasenreaktion *f*

4063 PHENOL, CARBOLIC ACID oic
f acide *m* carbolique, phénol *m*
e ácido *m* fénico
i acido *m* carbolico
n carbolzuur *n*, fenol *n*, hydroxybenzeen *n*
d Benzophenol *n*, Karbolsäure *f*, Phenol *n*

4064 PHENOL-FORMALDEHYDE RESINS pla
f résines *pl* phénol-formal
e resinas *pl* de fenol-formaldehidos
i resine *pl* fenoliche aldeidiche
n fenolformaldehydeharsen *pl*
d Phenolformaldehydharze *pl*

4065 PHENOLIC RESINS pla
f résines *pl* phénoliques
e resinas *pl* fenólicas
i resine *pl* fenoliche
n fenolharsen *pl*
d Phenolharze *pl*

4066 PHENOXIN (US), CARBON TETRACHLORIDE (GB) oic
f tétrachlorure *f* de carbone
e tetracloruro *m* de carbono
i tetracloruro *m* di carbonio
n tetra, tetrachloorkoolstof
d Tetra *n*, Tetrachlorkohlenstoff *m*

4067 PHENYLACETAMIDE, ACETANILIDE, ANTIFEBRIN mcy
f acétanilide *f*, antifébrile *m*
e acetanilida *f*, antifebrina *f*, fenilacetamida *f*
i acetanilide *f*
n acetanilide *n*, antifebrine *n*
d Antifebrin *n*, Azetanilid *n*

4068 PHENYLACROLEIN, CINNAMALDEHYDE (US), CINNAMIC ALDEHYDE (GB) per
f aldéhyde *m* cinnamique
e aldehido *m* cinámico
i aldeide *f* cinnamica
n kaneelaldehyde *n*
d Zimtaldehyd *n*

4069 PHENYLAMINE, AMINOBENZENE, AMINOPHEN (US), ANILINE dye
f aniline *f*, phénylamine *f*
e anilina *f*, fenilamina *f*
i anilina *f*, fenilamina *f*

n aminobenzol *n*, aniline, fenylamine *n*
d Aminobenzol *n*, Anilin *n*, Phenylamin *n*

4070 PHENYLATION oic
f phénylation *f*
e fenilación *f*
i fenilazione *f*
n fenyleren *n*
d Phenylieren *n*

4071 PHLEGMATIZING AGENT exp
f flegmatisant *m*
e agente *m* flemador
i agente *m* ritardatore
n flegmatiseermiddel *n*
d Phlegmatisator *m*

4072 PHOENICITE, AUSTRIAN CINNABAR, BASIC LEAD CHROMATE, CHROME RED dec
f chromate *m* de plomb basique, rouge *m* de chrome
e cromato *m* de plomo básico, rojo *m* de cromo
i cromato *m* di piombo basico, rosso *m* di cromo
n basisch loodchromaat *n*, chromaatrood *n*
d basisches Bleichromat *n*, Chromrot *n*

4073 PHONOCHEMISTRY gen
f phonochimie *f*
e fonoquímica *f*
i fonochimica *f*
n ultrasoonchemie
d Ultraschallchemie *f*

4074 PHOSGENE (GB), COLLINGITE (US) gas
f phosgène *m*
e fosgeno *m*
i fosgene *m*
n fosgeen *n*
d Karbonylchlorid *n*, Phosgen *n*

4075 PHOSPHATE COATING coa
f couche *f* de phosphate, revêtement *m* phosphatique
e capo *m* de fosfato, revestimiento *m* fosfático
i strato *m* di fosfato, rivestimento *m* fosfatico
n fosfaatlaag
d Phosphatschicht *f*, Phosphatüberzug *m*

4076 PHOSPHOR elc
f substance *f* luminescente
e substancia *f* luminiscente
i sostanza *f* luminescente
n fluorescerende stof
d Leuchtphosphor *m*, Leuchtstoff *m*

4077 PHOSPHORIZED COPPER met
f cuivre *m* phosphorisé
e cobre *m* fosforado
i rame *m* fosforizzato
n gefosforiseerd koper *n*
d phosphatiertes Kupfer *n*, phosphorisiertes Kupfer *n*

4078 PHOTOCHEMISTRY gen
f photochimie *f*
e fotoquímica *f*
i fotochimica *f*
n fotochemie
d Photochemie *f*

4079 PHOTODECOMPOSITION, PHOTODISSOCIATION gen
f photodissociation *f*
e fotodisociación *f*
i fotodissociazione *f*
n fotodissociatie
d photochemische Zersetzung *f*

4080 PHTALIC GLYCERIDE RESINS, GLYCEROL-PHTALIC RESINS, GLYPTAL RESINS pla
f résines *pl* glycérides phtaliques
e resinas *pl* sintéticas alquídicas
i resine *pl* gliceroftaliche
n glyptaalharsen *pl*
d Glyptalharze *pl*

4081 PHYSICAL RIPENING, OSTWALD RIPENING, RECRYSTALLIZATION pho
f maturation *f* physique
e maduración *f* física
i maturazione *f* fisica
n ostwaldse rijping
d Ostwaldsche Reifung *f*, physikalische Reifung *f*, Umkristallisierung *f*, Vorreifung *f*

4082 PICKLE, BRINE gen
f saumure *f*
e salmuera *f*
i salamoia *f*
n pekel
d Pökel *m*, Pökellauge *f*

4083 PICKLE STAIN elc
f tache *f* de décapage

e mancha *f* de decapaje
i macchia *f* di decapaggio
n beitsvlek
d Beizfleck *m*

4084 PICKLE TEST tes
f essai *m* d'immersion en acide
e ensayo *m* de inmersión en ácido
i collaudo *m* d'immersione in acido
n zuurdompelproef
d Säure-Eintauchprobe *f*

4085 PICKLING elc
f décapage *m*
e decapaje *m*
i decapaggio *m*
n beitsen *n*
d Beizen *n*

4086 PICKLING lea
f piclage *m*
e piclaje *m*
i mettere v in salamoia
n pikkelen *n*
d Pickeln *n*

4087 PICKLING EMBRITTLEMENT, cor
HYDROGEN EMBRITTLEMENT
f fragilité *f* par décapage
e fragilidad *f* por decapaje
i fragilità *f* per decapaggio
n beitsingsverbrozing
d Beizsprödigkeit *f*

4088 PICKLING INHIBITOR elc
f inhibiteur *m* de décapage
e inhibidor *m* de decapaje
i inibitore *m* di decapaggio
n beitsrem, beitsvertrager
d Beizzusatz *m*

4089 PICKUP met
f métal *m* adhérent
e metal *m* adherente
i metallo *m* aderente
n aanzetsel *n*
d Ansatz *m*

4090 PIEZOCHEMISTRY gen
f piëzochimie *f*
e piezoquímica *f*
i piezochimica *f*
n piëzochemie
d Piezochemie *f*

4091 PIGGING BACK (US), met
CARBURIZER, CARBURIZING
AGENT (GB)
f agent *m* carburant
e agente *m* cementador
i agente *m* cementatore
n verkolend middel *n*
d Aufkohlungsmittel *n*,
Kohlungsstoff *m*

4092 PIGMENT, dec
DRY COLOR (US), DRY
COLOUR (GB)
f couleur *f* en poudre, couleur *f*
sèche, pigment *m*
e materia *f* colorante, pigmento *m*
i materia *f* colorante, pigmento *m*
n pigment *n*, verfstof
d Farbkörper *m*, Körperfarbe *f*,
Pigment *n*, Trockenfarbe *f*

4093 PIGMENT/BINDER RATIO, dec
RATIO OF PIGMENT TO BINDER
f rapport *m* pigment/liant
e relación *f* pigmento-aglutinante
i rapporto *m* pigmento-legante
n pigment/bindmiddelverhouding
d Farbe/Bindemittelverhältnis *n*

4094 PIGMENT PASTE dec
f pâte *f* de pigment
e pasta *f* de pigmento
i pasta *f* di pigmento
n aangemaakt pigment *n*, aangemaakt
verfstof
d angeriebener Farbstoff *m*,
angeriebenes Pigment *n*

4095 PIGMENT PROCESS pho
f procédé *m* pigmentaire
e procedimiento *m* pigmentario
i processo *m* pigmentario
n pigmentproces *n*
d Pigmentverfahren *n*

4096 PIGMENTING CAPACITY dec
f pouvoir *m* de pigmentation
e poder *m* de absorción de un
pigmento, potencia *f* pigmentador
i potere *m* di pigmentazione
n pigmentopnemingsvermogen *n*
d Pigmentaufnahmevermögen *n*

4097 PILCHARD OIL fat
f huile *f* de pilchard
e aceite *m* de arenque
i olio *m* di sardinella
n pelserolie
d Sardellenöl *n*

4098 PILL mcy
f pilule *f*
e píldora
i pillola *f*
n pil
d Pille *f*

4099 PILL MAKING, mcy
BALLING (US)
f formation *f* de pilules
e formación *f* de píldoras
i formazione *f* di pillole
n persen *n* van pillen
d Pillenformung *f*

4100 PILSNER BEER, bre
PALE LAGER
f bière *f* de Pilsen
e cerveza *f* de Pilsen
i birra *f* di Pilsen
n pilsener *n*
d Pilsnerbier *n*

4101 PIMPLING, dec
BLISTERING
f cloquage *m*
e formación *f* de vejigas
i vescicatura *f*
n blaasjesvorming
d Blasenbildung *f*

4102 PINCH POINT dis
f point *m* d'invariance
e punto *m* crítico de separación
i punto *m* a composizione costante
n overganspunt *n* naar constante samenstelling
d Übergangspunkt *m* zu konstanter Zusammensetzung

4103 PINE NEEDLE OIL, fat
FIR LEAF OIL, FIR NEEDLE OIL
f huile *f* de sapin
e esencia *f* de pino
i olio *m* di pino
n dennenaaldolie
d Fichtennadelöl *n*, Kiefernnadelöl *n*

4104 PINE RESIN pla
f résine *f* de bois de pin
e galipodio *m*
i resina *f* di pino
n pijnhars
d Fichtenharz *n*, Kiefernharz *n*

4105 PINETAR OIL, fat
OIL OF PINETAR
f huile *f* de goudron de pin
e aceite *m* de alquitrán de pino
i olio *m* di catrame di pino
n pijnboomteerolie
d Kienöl *n*, Nadelholzteeröl *n*

4106 PINHOLES elc
f piqûres *pl*
e porosidades *pl*
i camolatura *f*
n poriën *pl*
d Nadelstichporen *pl*

4107 PIPAGE tra
f transport *m* par tuyautage
e transporte *m* por tuberías
i trasporto *m* per tubatura
n pijpleidingtransport *n*
d Röhrenleitungstransport *m*

4108 PIPECLAY cer
f argile *f* de pipe, terre *f* de pipe
e blanquizal *m*, tierra *f* de pipas
i argilla *f* fina da pipe, terra *f* da pipe
n pijpaarde
d Pfeifenton *m*

4109 PIT GAS (GB), fue,gas
DAMP, FIREDAMP, WILD GAS (US)
f grisou *m*, mofette *f*
e grisú *m*, mofeta *f*
i grisu *m*
n mijngas *n*
d Grubengas *n*, Schlagwetter *n*

4110 PIT GAS DETECTION (GB), sap
DAMP DETECTION, FIREDAMP DETECTION, WILD GAS DETECTION (US)
f détection *f* de grisou, détection *f* de mofette
e detección *f* de grisú, detección *f* de mofeta
i rivelazione *f* di grisu
n mijngasdetectie
d Grubengasanzeige *f*, Schlagwetteranzeige *f*

4111 PITCH bre,fue
f poix *f*
e pez *f*
i pece *f*
n pek *n*, pik *n*
d Pech *n*

4112 PITCH REMOVAL cle
f dépoissage *m*
e desembreado *m*
i eliminazione *f* della pece
n pekafscheiding
d Pechaustreibung *f*

4113 PITCHING bre
f ensemencement *m*
e siembra *f*
i semina *f*
n uitzaaiing
d Anstellen *n*

4114 PITCHING, LEAVENING bre
f mise *f* en levain
e sembradura *f* con levadura
i aggiuntatura *f* del lievito
n gisttoevoeging
d Hefegabe *f*

4115 PITCHING RATE bre
f dose *f* de levure
e dosis *f* de levadura
i aggiunto *m* di lievito
n gistdosis
d Hefegabe *f*

4116 PITCHING YEAST, BARM YEAST bre
f levain *m*
e levadura *f* de siembra
i lievito *m* di semenza
n aanzetgist, entgist, moedergist
d Anstellhefe *f*, Zeug *m*

4117 PITTED SURFACE cor
f surface *f* grippée
e superficie *f* atacada
i superficie *f* a vaciolatura
n aangevreten oppervlak *n*
d angefressene Oberfläche *f*

4118 PITTING CORROSION cor
f corrosion *f* sélective
e corrosión *f* crateriforme
i corrosione *f* selettiva
n aanvretende corrosie
d grübchenartige Rostanfressung *f*, Lochfrass *m*

4119 PLAINING gla
f affinage *m*
e afino *m*
i affinazione *f*
n fijnsmelten *n*
d Blankschmelze *f*, Feinschmelze *f*

4120 PLAIT POINT gen,nue
f point *m* de pliage
e punto *m* de plegado
i punto *m* di piega
n kritiek ontmengingspunt *n*, plooipunt *n*
d Faltpunkt *m*, kritischer Entmischungspunkt *m*

4121 PLASTER OF PARIS, GYPSUM bui
f plâtre *m* de moulage, plâtre *m* de Paris
e yeso *m*
i gesso *m*
n gebrand gips *n*
d gebrannter Gips *m*, Kalkgips *m*

4122 PLASTERING, FINING OFF, RENDERING, RUNNING OFF bui
f finissage *m* par du plâtre
e colocación *f* de un emplasto
i intonacatura *f*
n afpleisteren *n*
d Verputzen *n*

4123 PLASTIC CLAY, FOUL CLAY, PURE CLAY, STRONG CLAY bui
f argile *f* plastique
e arcilla *f* plástica
i galestro *m*
n zuivere klei
d reiner Ton *m*, reines Aluminium-silikat *n*

4124 PLASTIC FLOW pla
f écoulement *m* plastique
e flujo *m* plástico
i scorrimento *m* plastico
n elastisch vloeien *n*
d elastisches Fliessen *n*

4125 PLASTIC PAINT, RELIEF PAINT, TEXTURE FINISH dec
f peinture *f* plastique, peinture *f* relief
e pintura *f* plástica
i pittura *f* plastica
n plastiekverf, reliefverf
d plastische Anstrichfarbe *f*

4126 PLASTIC YIELD pla
f commencement *m* d'écoulement plastique
e principio *m* de la fluencia
i principio *m* dello snervamento
n begin *n* van het plastisch vloeien
d Beginn *m* des Fliessens

4127 PLASTICIZATION, PLASTIFICATION rub
f plastification *f*
e plastificación *f*
i plastificazione *f*
n plasticeren *n*, plastificeren *n*, weekmaken *n*
d Plastifizieren *n*, Plastizieren *n*, Weichmachen *n*

4128 PLASTICIZER gen
f plastifiant *m*
e plastificante *m*

i plastificante *m*
n plasticeermiddel *n*, plastificeermiddel *n*, weekmaker
d Plastifiziermittel *n*, Plastiziermittel *n*, Weichmacher *m*

4129 PLASTICS pla
f substances *pl* plastiques
e substancias *pl* plásticas
i sostanze *pl* plastiche
n kunstharsstoffen *pl*, plastieken *pl*
d Kunstharzstoffe *pl*, Plaste *pl*

4130 PLATE GLASS (GB), PATENT PLATE (US) gla
f glace *f* sans tain
e cristal *m* de espeja, cristal *m* de luna
i cristallo *m* da specchio, luce *f*
n spiegelglas *n*
d Spiegelglas *n*

4131 PLATEAU EFFECT, FLAT CURING rub
f effet *m* de plateau
e efecto *m* de meseta
i effetto *m* di piano
n plateaueffect *n*
d Plateau-Effekt *m*

4132 PLATFORMING, PLATINUM REFORMING fue
f réformation *f* à catalyseur de platine
e reformación *f* con platino
i riformazione *f* con platino
n platformeren *n*
d Platformieren *n*

4133 PLATING bac
f culture *f* sur boîte de Pétri
e cultivo *m* en caja de Petri
i cultura *f* su piastre
n platencultuur
d Plattenkultur *f*

4134 PLATING coa
f placage *m*
e chapeado *m*, plateadura *f*
i placcatura *f*
n plateren *n*, pleten *n*
d Plattieren *n*

4135 PLATINIZED ASBESTOS mat
f amiante *m* platiné
e amianto *m* platinado
i amianto *m* platinato
n platina-asbest *n*
d Platinasbest *m*

4136 PLATINIZING coa
f platinage *m*
e platinado *m*
i platinatura *f*
n platineren *n*
d Platinieren *n*

4137 PLATINUM BLACK coa
f noir *m* de platine
e negro *m* de platino
i nero *m* di platino
n platinazwart *n*
d Platinmohr *n*, Platinschwarz *n*

4138 PLATINUM SPONGE, SPONGY PLATINUM met
f éponge *f* de platine
e esponja *f* de platino
i platino *m* spugnoso
n platinaspons *n*
d Platinschwamm *m*

4139 PLODDING soa
f extrusion *f* de barres de savon
e extrusión *f* de barras de jabón
i estrusione *f* di barre di sapone
n persen *n* van zeepstangen
d Pressen *n* von Seifenstangen

4140 PLUMBER'S SOLDER, HALF AND HALF wel
f soudure *f* de plombier
e soldadura *f* de plomero
i stagno *m* da idraulico
n loodgieterssoldeer *n*
d Lötmörtel *m*

4141 PLUMPING lea
f bassage *m*, gonflement *m*
e hinchamiento *m*
i gonfiamento *m*
n zwellen *n*
d Schwellen *n*

4142 PLY GLASS, CASED GLASS, FLASHED GLASS gla
f verre *m* à deux couches, verre *m* doublé
e vidrio *m* contrachapado
i vetro *m* a due strati
n tweelaagsglas *n*
d Überfangglas *n*

4143 PLYWOOD bui
f contre-plaqué
e madera *f* contrachapada, madera *f* contraplacada
i legno *m* compensato
n multiplex *n*, triplex *n*
d Sperrholz *n*

4144 PNEUMATIC CLASSIFICATION sor
f classification *f* pneumatique, triage *m* pneumatique
e clasificación *f* neumática
i classificazione *f* pneumatica
n sorteren *n* met perslucht
d Luftstromsichtung *f*, Pressluftsichtung *f*

4145 PNEUMATIC CONVEYING tra
f transport *m* pneumatique
e transporte *m* neumático
i trasporto *m* pneumatico
n transport *n* met perslucht
d pneumatischer Transport *m*

4146 PNEUMATIC DRYING dry
f séchage *m* pneumatique
e secado *m* neumático
i essiccazione *f* pneumatica
n drogen *n* met perslucht
d Presslufttrocknen *n*

4147 PNEUMATIC FLOTATION sep
f flottation *f* pneumatique
e flotación *f* neumática
i flottazione *f* pneumatica
n flotatie met perslucht
d Pressluftflotation *f*

4148 POISING elc
f maintenir v constant le potentiel d'oxydation-réduction
e mantener v constante el potencial de oxidación-reducción
i mantenere v costante il potenziale d'ossidazione-riduzione
n bufferen *n*, constant houden *n* van de oxydoreductiepotentiaal
d Konstanthaltung *f* des Redoxpotentials, Puffern *n*

4149 POISON mcy
f poison *m*
e veneno *m*
i veleno *m*
n gif *n*, vergif *n*
d Gift *n*

4150 POISON GAS, mit
WAR GAS
f gaz *m* asphyxiant, gas *m* toxique
e gas *m* tóxico
i gas *m* aggressivo chimico
n gifgas *n*
d Giftgas *n*

4151 POISONING agr,hyg,mcy
f empoisonnement *m*
e envenenamiento *m*
i avvelenamento *m*
n vergiftiging
d Vergiftung *f*

4152 POISONOUS gen
f toxique adj
e venenoso adj
i velenoso adj
n vergiftig adj
d giftig adj

4153 POLING met
f perchage *m*
e pertigado *m*
i trattamento *m* con pali di legno verde
n polen *n*
d Polen *n*

4154 POLISH ETCHING pol
f attaque *f* chimique polisseuse
e ataque *m* para pulimento
i attacco *m* da pulimento
n polijstend etsen *n*
d polierende Ätzung *f*

4155 POLISHING, pol
GRINDING
f polissage *m*
e pulido *m*
i lucidatura *f*, pulitura *f*
n polijsten *n*
d Polieren *n*

4156 POLISHING CROCUS, pol
POLISHING ROUGE,
JEWELLER'S ROUGE, ROUGE
f rouge *m* à polir, rouge *m* d'Angleterre, rouge *m* de fer
e óxido *m* férrico, rojo *m* de Inglaterra, rojo *m* de pulir
i ossido *m* ferrico, rosso *m* inglese
n Berlijns rood *n*, Engels rood *n*, gele bolus, polijstrood *n*
d Englischrot *n*, Pariserrot *n*, Polierrot *n*

4157 POLISHING PASTE pol
f pâte *f* à fourbir
e pasta *f* para pulir
i pasta *f* da pulire
n poetspomade
d Putzpomade *f*

4158 POLLUTION hyg,iwt
f pollution *f*, souillure *f*
e ensuciamiento *m*
i inquinamento *m*
n vervuiling
d Verschmutzung *f*, Verunreinigung

4159 POLYCHROME FINISH (GB), dec
MULTICOLOR FINISH (US)
f finition *f* multicolore, finition *f* polychrome
e acabado *m* policromo
i finitura *f* policroma
n meerkleurendeklaag
d Mehrfarbendeckschicht *f*

4160 POLYCONDENSATION, pla
CONDENSATION POLYMERIZATION
f polymérisation *f* par condensation
e policondensación *f*
i polimerizzazione *f* per condensazione
n polycondensatie
d Polykondensation *f*

4161 POLYMERIZATION oic
f polymérisation *f*
e polimerización *f*
i polimerizzazione *f*
n polymerisatie
d Polymerisation *f*, Polymerisierung *f*

4162 POLYMERIZATION IN SOLUTION oic
f polymérisation *f* en solution
e polimerización *f* en solución
i polimerizzazione *f* in soluzione
n polymerisatie in oplossing
d Lösungspolymerisation *f*

4163 POLYMERIZATION INHIBITOR oic
f inhibiteur *m* de polymérisation
e inhibidor *m* de polimerización
i inibitore *m* di polimerizzazione
n polymerisatie-inhibitor, polymerisatieremstof
d Polymerisationsinhibitor *m*, Polymerisationsremmstoff *m*

4164 POLYMERIZATION MODIFIER oic
f modificateur *m* de polymérisation
e modificador *m* de polimerización
i modificatore *m* di polimerizzazione
n polymerisatiemodificator
d Polymerisationsmodifikator *m*

4165 POLYMERIZING, tex
BAKING
f polymérisation *f*
e polimerización *f*
i polimerizzazione *f*
n polymerisatie
d Polymerisierung *f*

4166 POMACE nut
f marc *m*
e pulpa *f* residual
i rifiuti *pl* di pressione
n uitgeperste massa
d gepresste Masse *f*, Pressrückstände *pl*

4167 PONDING, iwt
CLOGGING
f encrassement *m*
e atascamiento *m*, obstrucción *f*
i intasatura *f*
n verstopping
d Verschlammung *f*

4168 POOR CONCRETE, bui
LEAN CONCRETE
f béton *m* maigre
e hormigón *m* pobre
i calcestruzzo *m* magro
n mager beton *n*
d Magerbeton *m*, Sparbeton *m*

4169 POOR IN OXYGEN gas
f pauvre adj en oxygène
e pobre adj) en oxígeno
i povero adj d'ossigeno
n zuurstofarm adj
d sauerstoffarm adj

4170 POOR SOLVENT gen
f mauvais solvant *m*, solvant *m* pauvre
e solvente *m* pobre
i solvente *m* povero
n zwak oplosmiddel *n*
d schwaches Lösungsmittel *n*

4171 POORLY ATTENUATING YEAST bre,fmi
f levure *f* à faible atténuation
e levadura *f* de baja atenuación
i lievito *m* di bassa attenuazione
n slecht vergistende gist
d niedrigvergärende Hefe *f*

4172 POPPY SEED OIL fat
f huile *f* d'oeillettes, huile *f* de pavot
e aceite *m* de semillas de adormidera
i olio *m* di papavero
n maanolie, maanzaadolie, papaverolie
d Mohnöl *n*

4173 PORCELAIN cer
f porcelaine *f*
e porcelana *f*
i porcellana *f*
n porselein *n*
d Porzellan *n*

4174 POREWATER cer
f eau *f* interstitielle
e agua *f* de los poros
i acqua *f* contenuta nei pori
n poriënwater *n*
d Porenwasser *n*

4175 POROSITY gen
f porosité *f*
e porosidad *f*
i porosità *f*
n poreusheid
d Porosität *f*

4176 POROUS CHROME HARDENING coa
f chromage *m* dur poreux
e cromado *m* duro poroso
i cromatura *f* dura porosa
n poreus hardverchromen *n*
d poröse Hartverchromung *f*

4177 POSITIVE gen,pho
f positif *m*
e positivo *m*
i positivo *m*
n positief *n*
d Positiv *n*

4178 POSTCHLORINATION pla
f chloruration *f* postérieure
e cloruración *f* posterior
i clorurazione *f* posteriore
n nachloren *n*, nachloreren *n*
d Nachchlorieren *n*, Nachchlorung *f*

4179 POSTCURING pla
f traitement *m* postérieur
e tratamiento *m* posterior
i trattamento *m* posteriore
n nabehandeling
d Nachbehandlung *f*

4180 POSTHEATING htt
f réchauffage *m*
e recalentamiento *m*
i reriscaldamento *m*
n naverwarming
d Nachheizung *f*, Nachverwärmung *f*

4181 POT ANNEALING, htt,met
BOX ANNEALING, CLOSE ANNEALING
f recuit *m* en caisse
e recocido *m* en caja
i ricottura *f* in cassetta
n kistgloeien *n*
d Kastenglühen *n*

4182 POT DISTILLATION dis
f distillation *f* à chauffage direct
e destilación *f* de calentamiento directo
i distillazione *f* a riscaldamento diretto
n destillatie met directe verhitting
d Brennkolbendestillation *f*

4183 POT SPINNING tex
f filage *m* centrifuge
e hilatura *f* centrífuga
i filatura *f* centrifuga
n potspinnen *n*
d Topfspinnen *n*

4184 POTASH SOFT SOAP soa
f savon *m* mou à la potasse
e jabón *m* blando de potasa
i sapone *m* molle di potassa
n zachte kalizeep
d Kalischmierseife *f*

4185 POTASSIUM HYDROGEN TARTRATE, dye
CREAM OF TARTAR, PURIFIED TARTAR
f crème *m* de tartre, tartrate *m* de potassium
e crémore *m* tártaro, tartrato *m* de potasio
i cremore *m* di tartaro, tartrato di potasio
n gezuiverde wijnsteen, kaliumtartraat *n*
d gereinigter Weinstein *m*, Kaliumtartrat *n*

4186 POTASSIUM HYDROXIDE, ai
CAUSTIC POTASH
f hydroxyde *m* de potasse
e hidróxido *m* potásico
i potassa *f* caustica
n bijtende potas
d Ätzkali *n*, Kalziumhydroxyd *n*

4187 POTASSIUM NITRATE, ai
NITRE, SALTPETRE
f salpêtre *m*
e salitre *m*
i salnitro *m*
n salpeter
d Salpeter *m*

4188 POTATO FLOUR nu
f fécule *f* de pommes de terre
e fécula *f* de patatas
i fecola *f* di patate
n aardappelmeel *n*
d Kartoffelmehl *n*

4189 POTATO STARCH gl
f amidon *m* de pommes de terre

e almidón *m* de patatas
i amido *m* di patate
n aardappelstijfsel
d Kartoffelstärke *f*

4190 POTENTIAL MEDIATOR elc
f substance *f* accélérante l'ajustage du potentiel
e substancia *f* aceleradora del ajuste del potencial
i sostanza *f* accelerante l'aggiustaggio del potenziale
n potentiaalinsteller
d Potentialvermittler *m*

4191 POTENTIOMETRIC TITRATION ana
f titrage *m* potentiométrique
e valoración *f* potenciométrica
i titolazione *f* potenziometrica
n potentiometrisch titreren *n*
d potentiometrische Titration *f*

4192 POTSHERDS gla
f tessons *pl* de vieux creusets pour verre
e cascos *pl* de crisoles viejos para vidrio
i cocci *pl* di crogioli usati per vetro
n scherven *pl* van gebruikte glaspotten
d Glashafenscherben *pl*

4193 POTTER'S CLAY, BALL CLAY cer
f argile *f* de potier, argile *f* figuline
e arcilla *f* de alfarero, arcilla *f* figulina
i argilla *f* per terraglie
n pottenbakkersklei
d Töpferton *m*

4194 POTTERY cer
f poterie *f*
e alfarería *f*
i terraglia *f*
n aardewerk *n*
d Tonwaren *pl*

4195 POTTERY cer
f poterie *f*
e alfarería *f*, ollería *f*
i fabbrica *f* di terraglie
n pottenbakkerij, potterij
d Töpferei *f*

4196 POTTING pla
f encastrement *m*
e encapsulación *f*
i incastratura *f*
n inkapselen *n*
d Einkapseln *n*

4197 POTTING tex
f décatissage *m* par voie humide
e decatizaje *m* en húmedo
i decatissaggio *m* per via umida
n decatiseren *n* langs de vochtige weg
d Nassdekatieren *n*

4198 POUR POINT lub
f point *m* de congélation
e punto *m* de descongelación, punto *m* de fluidez
i punto *m* di congelazione, punto *m* di scorrimento
n stolpunt *n*, vloeipunt *n*
d Stockpunkt *m*

4199 POUR POINT DEPRESSING lub
f abaissement *m* du point de congélation
e descenso *m* del punto de descongelación
i abbassamento *m* del punto di congelazione, abbassamento *m* del punto di scorrimento
n stolpuntverlaging, vloeipunt-verlaging
d Stockpunkterniedrigung *f*

4200 POURABLE SEALING COMPOUND mat
f produit *m* de jointement coulant
e material *m* líquido para obturaciones
i mescolatura *f* per giunzione a colata
n gietbare vulmassa
n Vergussmasse *f*

4201 POURING, CASTING, TEEMING fou,pla
f coulée *f*
e colada *f*
i colata *f*
n gieten *n*
d Giessen *n*

4202 POWDER exp
f poudre *f*
e pólvora
i polvere *f*
n buskruit *n*
d Pulver *n*

4203 POWDER gen
f poudre *f*

e polvo *m*
i polvere *f*
n poeder *n*
d Pulver *n*

4204 POWDER BURNING, POWDER CUTTING — htt
f oxycoupage *m* à la poudre de fer
e oxicorte *m* con polvo de hierro, pulvicorte *m*
i taglio *m* con polvere di ferro in ossigeno
n snijden *n* met ijzerpoeder in zuurstofvlam
d Brennschneiden *n* mit Eisenpulver

4205 POWDER COMPACTING — pre
f moulage *m* par pression
e compresión *f* del polvo
i compressione *f* della polvere
n poederpersen *n*
d Pulverpressen *n*

4206 POWDER COMPOSITION, POWDER PULP — exp
f galette *f*
e galleta *f*
i barlotto *m*
n kruitkoek
d Pulversatz *m*

4207 POWDERED LIME, LIME POWDER — bui
f chaux *f* vive en poudre
e cal *f* apagada en polvo
i calce *f* viva in polvere
n verweerde kalk
d Kalkstaub *m*, Staubkalk *m*, verwitterter Kalk *m*

4208 POWDERED SOAP — soa
f savon *m* en poudre
e jabón *m* en polvo
i sapone *m* in polvere
n zeeppoeder *n*
d Seifenpulver *n*

4209 POWDERING — dec
f poudrage *m*
e espolvorización *f*
i polverizzazione *f*
n verpulveren *n*
d Abkreiden *n*, Kreiden *n*

4210 POWDERY YEAST, NON-FLOCULATING YEAST — bre,fmi
f levure *f* poussiéreuse
e levadura *f* pulverulenta
i lievito *m* polverulente
n gistpoeder
d Staubhefe *f*

4211 POZZOLANA, POZZUOLANA — bui
f pouzzolane *f*
e pozolana *f*
i pozzolana *f*
n puzzolaan *n*
d Bröckeltuff *m*, Puzzolanerde *f*

4212 POZZOLANA CEMENT, POZZUOLANA CEMENT — bui
f ciment *m* à pouzzolane
e cemento *m* de pozolana
i cemento *m* a pozzolana
n puzzolaancement *n*
d Puzzolanzement *m*

4213 PRECHLORINATION — iwt
f préchloration *f*
e precloración *f*
i clorazione *f* preliminare, preclorazione *f*
n voorchloren *n*
d Vorchlorung *f*

4214 PRECIOUS METALS, NOBLE METALS — met
f métaux *pl* nobles, métaux *pl* précieux
e metales *pl* nobles
i metalli *pl* nobili
n edele metalen *pl*
d Edelmetalle *pl*

4215 PRECIPITANT, PRECIPITATING AGENT, SEED CHARGE — gen
f agent *m* de précipitation, précipitant *m*
e agente *m* de precipitación, precipitante *m*
i agente *m* precipitante
n neerslagmiddel *n*
d Fällmittel *n*, Niederschlagsmittel *n*

4216 PRECIPITATE — gen
f précipité *m*
e precipitado *m*
i precipitato *m*
n neerslag *n*, precipitaat *n*
d Niederschlag *m*, Präzipitat *n*

4217 PRECIPITATION — gen
f précipitation *f*
e precipitación *f*
i precipitazione *f*
n neerslaan *n*
d Fällung *f*

4218 PRECIPITATION HARDENING, AGE HARDENING, DISPERSION HARDENING met
f durcissement *m* structural
e endurecimiento *m* estructural
i incrudimento *m* strutturale
n structurele harding
d strukturelle Härtung *f*

4219 PRECISION MOULDING (GB), INVESTMENT MOULDING (GB), LOST-WAX PROCESS (US) fon
f moulage *m* à la cire perdue
e moldeo *m* a la cera perdida
i formatura *f* a cera spersa
n vormen *n* volgens de verlorenwasmethode
d Modellausschmelzverfahren *n*

4220 PRECOATING fil
f prérevêtement *m*
e prerrevestimiento *m*
i prerivestimento *m*
n voorbedekking
d Vorbeschichtung *f*

4221 PRECONDENSATION clo
f précondensation *f*
e precondensación *f*
i precondensazione *f*
n voorcondensatie
d Vorkondensation *f*

4222 PRECOOLING clo
f préréfrigération *f*
e prerrefrigeración *f*
i preraffreddamento *m*
n voorkoeling
d Vorkühlung *f*

4223 PRECURING, PREVULCANIZATION rub
f prévulcanisation *f*
e prevulcanización *f*
i prevulcanizzazione *f*
n voorvulcanisatie
d Vorvulkanisation *f*

4224 PREDEFECATION, PRELIMING sin
f prédéfécation *f*
e predefecación *f*
i predefecazione *f*
n voorsatureren *n*
d Vorsaturation *f*, Vorscheidung *f*

4225 PREDRYING dry
f préséchage *m*
e presecado *m*
i pre-essiccazione *f*
n voordroging
d Vortrocknung *f*

4226 PRE-EVAPORATION eva
f prévaporisation *f*
e prevaporización *f*
i prevaporizzazione *f*
n voorverdamping
d Vorverdampfung *f*

4227 PREFORMING pla
f préformation *f*
e preformación *f*
i prestampatura *f*
n voorvorming
d Vorformung *f*

4228 PREHEATING htt
f préchauffage *m*
e precalentamiento *m*
i preriscaldamento *m*
n voorverwarming
d Vorwärmung *f*

4229 PRE-IGNITION fue
f allumage *m* prématuré
e preencendido *m*
i accensione *f* prematura
n voorontsteking
d Glühzündung *f*

4230 PRE-IGNITION RESISTANCE, ANTI-KNOCK VALUE, KNOCK RATING fue
f pouvoir *m* anti-détonant
e índice *m* de detonación, poder *m* antidetonante
i potere *m* antidetonante
n klopvastheid
d Klopffestigkeit *f*

4231 PREMASHING bre
f empâtage *m* préalable
e mezcla *f* prematura
i premescolatura *f*
n voormengen *n*
d Vormaischen *n*

4232 PREPARING SALT, STANNATE OF SODIUM tex
f sel *m* d'apprêt, stannate *m* de sodium
e estañato *m* sódico, sal *f* de apresto
i sale *m* d'appretto, stannato *m* di sodio
n natriumstannaat *n*, prepareerzout *n*
d Präpariersalz *n*

4233 PREROASTING met
f prégrillage *m*
e pretostación *f*
i pretorrefazione *f*
n voorroosten *n*
d Vorrösten *n*

4234 PRESERVATION nut
f conservation *f*, préservation *f*
e conservación *f*, preservación *f*
i conservazione *f*, preservazione *f*
n conserveren *n*, verduurzamen *n*
d Konservierung *f*

4235 PRESERVATIVE nut
f agent *m* de conservation, agent *m* préservant
e agente *m* de conservación, agente *m* preservador
i agente *m* preservante, agente *m* preservativo
n conserveermiddel *n*, verduurzamingsmiddel *n*
d Konservierungsmittel *n*, Präservierungsmittel *n*

4236 PRESOFTENING pla
f plastification *f* préalable
e plastificación *f* preliminar
i plastificazione *f* preliminare
n voorplasticeren *n*
d Vorplastizieren *n*

4237 PRESS BONDING pla
f encollage *m* sous pression
e encoladura *f* bajo presión
i incollatura *f* sotto pressione
n onder druk lijmen *n*
d Leimen *n* unter Druck

4238 PRESS CURING rub
f vulcanisation *f* sous presse
e vulcanización *f* bajo prensa
i vulcanizzazione *f* in pressa
n persvulcanisatie
d Pressevulkanisation *f*

4239 PRESS POLISHING pla
f polissage *m* sous pression
e pulido *m* bajo presión
i lucidatura *f* sotto pressione
n polijsten *n* onder druk
d Druckpolierung *f*

4240 PRESSBOARD elc,pap
f carton *m* isolant
e cartón *m* aislante
i cartone *m* isolante
n isolatiekarton *n*, pletbord *n*
d Pressspahn *m*

4241 PRESSED GLASS gla
f verre *m* pressé
e vidrio *m* prensado
i vetro *m* compresso
n persglas *n*
d Pressglas *n*

4242 PRESSED OIL fat
f huile *f* pressée
e aceite *m* prensado
i olio *m* pressato
n geslagen olie
d Pressöl *n*

4243 PRESSURE ATOMIZATION pre
f pulvérisation *f* sous pression
e pulverización *f* bajo presión
i polverizzazione *f* sotto pressione
n verstuiving onder druk
d Düsenzerstäubung *f*

4244 PRESSURE CASTING cer
f coulée *f* sous pression
e colada *f* bajo presión
i colata *f* sotto pressione
n persgieten *n*
d Druckguss *m*

4245 PRESSURE DYEING dye
f teinture *f* sous pression
e teñido *m* bajo presión
i tintura *f* sotto pressione
n verven *n* onder druk
d Druckfärbung *f*

4246 PRESSURE-FED OIL LUBRICATION lub
f lubrification *f* à huile sous pression
e lubricación *f* con aceite bajo presión
i lubrificazione *f* mediante olio sotto pressione
n oliedruksmering
d Öldruckschmierung *f*

4247 PRESSURE GASIFICATION fue
f gazéification *f* sous pression
e gasificación *f* bajo presión
i gassificazione *f* sotto pressione
n vergassing onder druk
d Druckvergasung *f*

4248 PRESSURE LEACHING cle
f lessivage *m* sous pression
e lixiviación *f* bajo presión
i lisciviazione *f* sotto pressione
n uitlogen *n* onder druk
d Auslaugen *n* unter Druck

4249 PRESSURE SENSITIVE ADHESIVE glu
f adhésif *m* auto-collant
e adhesivo *m* piezosensible
i adesivo *m* autocollante
n zelfklevend plakmiddel *n*
d Selbstkleber *m*

4250 PRESSURE SINTERING htt,met
f frittage *m* sous charge
e piezosinterización *f*
i sinterizzazione *f* sotto pressione
n sinteren *n* onder druk
d Drucksintern *n*

4251 PRESSURE STILL TAR fue
f résidu *m* solide de craquage
e residuo *m* sólido de pirólisis
i residuo *m* solido di piroscissione
n kraakresidu *n*
d Krackrückstand *m*

4252 PRESTRESSED CONCRETE bui
f béton *m* précontraint
e hormigón *m* pretensado
i calcestruzzo *m* precompresso
n voorgespannen beton *n*
d Spannbeton *m*, vorgespannter Beton *m*

4253 PRETANNING tan
f prétannage *m*
e precurtido *m*
i preconcia *f*
n voorlooien *n*
d Vorgerbung *f*

4254 PRETREATMENT gen
f traitement *m* antérieur
e tratamiento *m* anterior
i trattamento *m* anteriore
n voorbewerking
d Vorbearbeitung *f*

4255 PREVULCANIZATION, PRECURING rub
f prévulcanisation *f*
e prevulcanización *f*
i prevulcanizzazione *f*
n voorvulcanisatie
d Vorvulkanisation *f*

4256 PRIMARY COLORS (US), PRIMARY COLOURS (GB) dec
f couleurs *pl* fondamentales
e colores *pl* elementales, colores *pl* fundamentales
i colori *pl* fondamentali
n primaire kleuren *pl*
d Grundfarben *pl*, Primärfarben *pl*

4257 PRIMARY GRAPHITE mat
f graphite *m* primaire
e grafito *m* primario
i grafite *f* primaria
n primair grafiet *n*
d C-Graphit *m*, Gärschaum-graphit *m*, Primärgraphit *m*

4258 PRIMARY SOLUTION, MIXED CRYSTAL, SOLID SOLUTION gen
f cristal *m* mixte, solution *f* solide
e cristal *m* mixto, solución *f* sólida
i cristallo *m* misto, soluzione *f* solida
n mengkristal *n*, vaste oplossing
d feste Lösung *f*, Mischkristall *m*

4259 PRIME PIGMENT, HIDING PIGMENT dec
f pigment *m* opacifiant
e pigmento *m* cubridor
i pigmente *m* coprente
n dekkend pigment *n*
d deckendes Pigment *n*

4260 PRIMER, FLATTING dec
f apprêt *m*
e aparejo *m*
i mestica *f*
n plamuur
d Spachtel *m*

4261 PRIMER, PRIMING COMPOSITION exp
f charge *f* amorce
e carga *f* iniciadora
i innesco *m*
n ontstekingsmiddel *n*
d Initialsprengstoff *m*

4262 PRIMER COAT, COUCH, GROUND COAT dec
f main *f* d'impression, première couche *f*
e capa *f* base, mano *f* de imprimación
i prima mano *f* di colore
n grondlaag
d Grund *m*

4263 PRIMER SEALER dec
f apprêt *m* pour bois, impression *f*
e imprimación *f*
i appretto *m* per legno
n voedgrond, voorstrijkmiddel *n*
d Absperrmittel *n*, Einlassmittel *n*, Sperrgrund *m*

4264 PRIMES gen
f produits *pl* de haute qualité
e productos *pl* de alta calidad
i prodotti *pl* d'alta qualità
n hoogwaardige produkten *pl*
d hochwertige Produkte *pl*

4265 PRINCIPAL REACTION gen
f réaction *f* principale
e reacción *f* principal
i reazione *f* principale
n hoofdreactie
d Hauptreaktion *f*

4266 PRINT pho
f copie *f*, épreuve *f*
e prueba *f*
i stampa *f*
n afdruk
d Kopie *f*

4267 PRINTING tex
f impression *f*
e estampación *f*
i stampaggio *m*
n bedrukken *n*
d Druck *m*

4268 PRINTING INK pri
f encre *f* d'imprimerie
e tinta *f* de imprenta
i inchiostro *m* di stampa
n drukinkt
d Druckerfarbe *f*, Druckerschwärze *f*

4269 PRINTING PAPERS pap
f papiers *pl* d'impression
e papeles *pl* de imprenta
i carte *pl* da stampa
n drukpapier *n*
d Druckpapier *n*

4270 PRINTING PASTE tex
f pâte *f* d'impression
e pasta *f* para estampación
i pasta *f* da stampatura
n drukpasta
d Druckerpaste *f*

4271 PROCESS OILS, PROCESSING OILS fat
f huiles *pl* ramollissantes
e aceites *pl* emolientes
i oli *pl* emollienti
n weekmakende oliën *pl*
d Weichmacheröle *pl*

4272 PROCESSING gen
f mise *f* en oeuvre, ouvraison *f*
e elaboración *f*
i lavorazione *f*
n behandeling, bewerking
d Bearbeitung *f*, Behandlung *f*

4273 PRODUCER GAS, LEAN GAS gas
f gaz *m* pauvre
e gas *m* pobre
i gas *m* povero
n arm gas *n*
d armes Gas *n*

4274 PROGRESSIVE BURNING POWDER exp
f poudre *f* progressive
e pólvora *f* progresiva
i polvere *f* progressiva
n progressief verbrandend kruit *n*
d progressiv verbrennendes Pulver

4275 PROGRESSIVE DECOMPOSITION gen
f décomposition *f* progressive
e descomposición *f* progresiva
i decomposizione *f* progressiva
n progressieve ontleding
d progressive Zersetzung *f*

4276 PROMOTER, PROMOTOR cat
f substance *f* activante
e substancia *f* activadora
i sostanza *f* attivatrice
n activator
d Aktivierungsmittel *n*, Beschleuniger *m*, Promotor *m*

4277 PROOF SPIRIT oic
f alcool *m* à 57%
e alcohol *m* de 57%
i alcool *m* a 57%
n normaalalcohol (57 vol.%)
d Normalstärke *f* alkoholischer Getränke (57 Vol.%)

4278 PROOFING, RUBBERIZING, TOPPING WITH GUM coa,rub
f gommage *m*, recouvrement *m* de caoutchouc
e engomadura *m*, recubrimiento *m* con goma
i gommatura *f*, ricopertura *f* con gomma
n berubberen *n*, gommeren *n*
d Gummieren *n*

4279 PRO-OXYGENIC AGENT, OXIDATION CATALYST cat,rub
f catalyseur *m* d'oxydation

e catalizador *m* de oxidación
i catalizzatore *m* d'ossidazione
n oxydatiekatalysator
d Oxydationskatalysator *m*

4280 PROPANE DEASPHALTING gas
f désasphaltage *m* de propane
e desasfaltado *m* de propano
i eliminazione *f* dell'asfalto di propano
n asfaltonttrekking aan propaan
d Propanentasphaltierung *f*

4281 PROPANE DECARBONIZING cle,gas
f décarbonisation *f* de propane
e descarbonización *f* de propano
i decarburazione *f* di propano
n propaanontkoling
d Propanentkohlung *f*

4282 PROPANONE, ACETONE, DIMETHYL KETONE, KETOPROPANE oic
f acétone *f*, éther *m* pyroacétique
e acetona *f*, dimetil-cetona *f*
i acetone *m* dimetilchetone *m*
n aceton *n*, dimethylketon *n*, propanon *n*
d Azeton *n*, Dimethylketon *n*

4283 PROPELLANT exp
f poudre *f* propulsive
e pólvora *f* de propulsión
i polvere *f* propellente
n voortdrijfmiddel *n*
d Treibpulver *n*

4284 PROPENAL, ACROLEINE, ACRYLALDEHYDE oic
f acroléine *f*, aldéhyde *f* acrylique
e acraldehido *m*, acroleína *f*, aldehido *m* acrílico
i acroleina *f*, aldeide *m* acrilico
n acroleïne *n*, acrylaldehyde *n*
d Akrolein *n*, Akrylaldehyd *n*

4285 PROPENE ACID, ACROLEIC ACID, ACRYLIC ACID oic
f acide *f* acrylique
e ácido *m* acrílico
i acido *m* acrilico
n acrylzuur *n*, etheencarbonzuur *n*
d Akrylsäure *f*, Propensäure *f*

4286 PROPINE, ALLYLENE oic
f allylène *m*, méthylacéthylène *m*
e metil-acetileno *m*
i allilene *m*
n allyleen *n*, methylacetyleen *n*, propyn *n*
d Allylen *n*, Methylazetylen *n*

4287 PROPORTIONING gen
f dosage *m*
e dosificación *f*
i dosatura *f*
n doseren *n*
d Dosieren *n*

4288 PROPORTIONING BY VOLUME gen
f dosage *m* volumique
e dosificación *f* de volumen
i dosaggio *m* di volume
n volumedosering
d Volumendosierung *f*

4289 PROPORTIONING BY WEIGHT gen
f dosage *m* pondéral
e dosificación *f* ponderal
i dosaggio *m* ponderale
n gewichtsdosering
d Gewichtsdosierung *f*

4290 PROTECTION AGAINST CORROSION cor
f protection *f* contre la corrosion
e protección *f* contra la corrosión
i protezione *f* contro la corrosione
n bescherming tegen corrosie
d Korrosionsschutz *m*

4291 PROTECTIVE ACTION col
f action *f* protectrice
e acción *f* protectora
i azione *f* protettiva
n beschermende werking
d Schutzwirkung *f*

4292 PROTECTIVE ATMOSPHERE htt,wel
f atmosphère *f* protectrice
e atmósfera *f* de protección
i atmosfera *f* di protezione
n beschermende atmosfeer
d Schutzatmosphäre *f*

4293 PROTECTIVE COLLOID col
f colloïde *m* protecteur
e coloide *m* protector
i colloide *m* protettivo
n beschermingscolloïde *n*
d Schutzkolloid *n*

4294 PROTECTIVE FOOD nut
f denrée *f* alimentaire vitaminisée

e nutritivo *m* vitaminizado
i cibo *m* vitaminizzato
n gevitaminiseerd voedsel *n*
d vitaminisierter Nährstoff *m*

4295 PRUSSIAN BLUE, BERLIN BLUE dec
f bleu *m* de Prusse
e azul *m* de Prusia
i blu *m* di Prussia
n Berlijns blauw *n*
d Preussischblau *m*

4296 PRUSSIC ACID, HYDROCYANIC ACID oic
f acide *m* cyanhydrique, acide *m* prussique
e ácido *m* cianhídrico
i acido *m* cianidrico, acido *m* prussico
n blauwzuur *n*, cyaanwaterstof, cyaanwaterstofzuur *n*
d Blausäure *f*, Hydrozyansäure *f*, Zyanwasserstoffsäure *f*

4297 PSEUDOCARBURIZING, BLANK CARBURIZING htt,met
f pseudocémentation *f*
e seudocementación *f*
i pseudocementazione *f*
n pseudocementeren *n*
d Pseudozementierung *f*

4298 PSEUDOCATALYSIS gen
f pseudocatalyse *f*
e seudocatálisis *f*
i pseudocatalisi *f*
n pseudokatalyse *f*
d Pseudokatalyse *f*

4299 PSEUDONITRIDING, BLANK NITRIDING htt,met
f pseudonitruration *f*
e seudonitruración *f*
i pseudonitrurazione *f*
n pseudonitreren *n*
d Pseudonitrierung *f*

4300 PSEUDOSOLUTION, COLLOIDAL SOLUTION, COLLOIDAL SUSPENSION col
f solution *f* colloïdale, suspension *f* colloïdale
e solución *f* coloidal, suspensión *f* coloidal
i soluzione *f* colloidale, sospensione *f* colloidale
n colloïdale oplossing, colloïdale suspensie
d kolloidale Lösung *f*, Kolloidsuspension *f*

4301 PSEUDOYEAST fmi
f levure *f* fausse
e seudolevadura *f*
i pseudolievito *m*
n onechte gist
d unechte Hefe *f*

4302 PUCKERING, BEADING, WRINKLING dec
f fripe *f*, frisage *m*
e arrugado *m*
i raggrinzamento *m*
n schroeien *n*
d Faltenbildung *f*, Runzelbildung *f*

4303 PUDDLING dye
f préparation *f* humide de mélanges de colorants
e preparación *f* húmeda de mezclas de colorantes
i preparazione *f* umida di mescolanze di coloranti
n nat aanmaken *n* van kleurstofmengsels
d Nassanfertigung *f* von Farbstoffgemische

4304 PUDDLING met
f puddlage *m*
e pudelaje *m*
i puddellaggio *m*
n puddelen *n*
d Puddeln *n*

4305 PUERING, BATING lea
f chipage *m*
e maceración *f*
i macerazione *f*
n zwellen *n*
d Schwöden *n*

4306 PUFFING AGENT, THICKENER dec
f matière *f* d'épaississement
e agente *m* espesante
i agente *m* d'addensamento
n verdikkingsmiddel *n*
d Verdickungsmittel *n*

4307 PUGGING cer
f malaxage *m* d'argile
e amasadura *f* de arcilla
i impastatura *f* d'argilla
n kneden *n* van klei
d Kneten *n* von Ton

4308 PULP pap,sin
f pâte *f*
e pasta *f*, pulpa *f*
i pasta *f*, polpa *f*
n brij, pulp
d Brei *m*

4309 PULP PRESS WATER sin
f eaux *pl* de presse
e agua *f* de prensa de rebanadas
i acqua *f* di scarico dalla spremitura di polpe
n snijdselwater *n*
d Schnitzelpresswasser *n*

4310 PULPING pap
f préparation *f* de la pâte
e pulpado *m*
i preparazione *f* della polpa
n ontsluiten *n*
d Aufschliessen *n*

4311 PULPWOOD bui
f panneau *m* de fibres de bois
e placa *f* de fibras de madera
i piastra *f* di fibre legnose
n board *n*, houtvezelplaat
d Holzfaserplatte *f*

4312 PULVERIZATION, gen
DISINTEGRATION
f pulvérisation *f*
e pulverización *f*
i polverizzazione *f*
n verpoedering, verpulvering
d Verpulverung *f*

4313 PULVERIZED FUEL fue
f charbon *m* pulvérisé
e carbón *m* pulverizado
i carbone *m* in polvere
n poederkool
d Kohlenstaub *m*, Staubkohle *f*

4314 PUMICE dec
f pierre *f* ponce, pumicite *f*
e piedra *f* pómez, pumita *f*
i pietra *f* pomice, pomice *f*
n puimsteen *n*
d Bimsstein *m*

4315 PUMICING dec
f ponçage *m*
e apomazado *m*
i pomiciatura *f*
n afpuimen *n*
d Bimsen *n*

4316 PUMPING gen
f pompage *m*
e bombeo *m*
i pompata *f*
n pompen *n*
d Pumpen *n*

4317 PUNGENT ODOR (US), hyg
PUNGENT ODOUR (GB)
f odeur *f* âcre, odeur *f* piquante
e olor *m* picante
i odore *m* acre, odore *m* piccante
n prikkelende geur
d scharfer Geruch *m*

4318 PURE CLAY, bui
FOUL CLAY, PLASTIC CLAY,
STRONG CLAY
f argile *f* plastique
e arcilla *f* plástica
i galestro *m*
n zuivere klei
d reiner Ton *m*, reines Aluminiumsilikat *n*

4319 PURE CULTURE fmi
f culture *f* pure
e cultivo *m* puro
i cultura *f* pura
n zuivere cultuur
d Reinkultur *f*

4320 PURE-CULTURE YEAST fmi
f levure *f* pure
e levadura *f* pura
i lievito *m* puro
n zuivere gist
d Reinhefe *f*

4321 PURE GUM VULCANIZATE (US), rub
UNLOADED VULCANIZATE (GB)
f vulcanisat *m* sans charge
e vulcanizado *m* sin carga
i vulcanizzato *m* di gomma pura
n A-mengsel *n*, vulcanisaat *n* zonder vulstof
d Reinkautschukvulkanisat *n*

4322 PURE LIME, bui
FAT LIME, RICH LIME
f chaux *f* grasse
e cal *f* grasa
i calce *f* grassa
n kluitkalk, Luikse kalk, steenkalk
d Fettkalk *m*, Speckkalk *m*

4323 PURE YEAST CULTURE fmi
f culture *f* de levure pure
e cultivo *m* de levadura pura
i cultura *f* di lievito puro
n cultuur van zuivere gist
d Hefereinzucht *f*

4324 PURIFICATION, PURIFYING cle
f dépuration *f*, épuration *f*, purification *f*
e depuración *f*, purificación *f*
i depurazione *f*, purificazione *f*
n zuivering
d Reinigung *f*

4325 PURIFIED POTASH, PEARL ASH aic
f perlasse *f*
e perlasa *f*, potasa *f* purificada
i potassa *f* purificata
n Amerikaanse potas, parelas
d Aschensalz *n*, Perlasche *f*

4326 PURIFIED TARTAR, CREAM OF TARTAR, POTASSIUM HYDROGEN TARTRATE dye
f crème *m* de tartre, tartrate *m* de potassium
e crémore *m* tártaro, tartrato *m* de potasio
i cremore *m* di tartaro, tartrato di potassio
n gezuiverde wijnsteen, kaliumtartraat *n*
d gereinigter Weinstein *m*, Kaliumtartrat *n*

4327 PURPURINA, BRONZE POWDER coa
f poudre *f* à bronzer
e polvo *m* para broncear
i polvere *f* da bronzare
n bronspoeder *n*
d Bronzefarbe *f*, Bronzepulver *n*

4328 PUSH-BUTTON PAINT, AEROSOL LACQUER dec
f peinture *f* aérosol
e pintura *f* de aerosol
i pittura *f* d'aerosole
n aërosollak
d Aerosollack *m*, Sprühdosenlack *m*

4329 PUTREFACTION hyg,iwt
f pourriture *f*, putréfaction *f*
e podredura *f*, putrefacción *f*
i putrefazione *f*
n rotting
d Fäulnisprozess *m*, Fäulung *f*

4330 PUTREFACTIVE BACTERIA, TERMOBACTERIA fmi,iwt
f bactéries *pl* putrides
e bacterias *pl* de la putrefacción
i batteri *pl* della putrefazione
n rottingsbacteriën *pl*
d Fäulnisbakterien *pl*

4331 PUTTY, CEMENT bre,glu
f mastic *m*
e masilla *f*, mástique *m*
i mastice *m*
n kit
d Kitt *m*

4332 PUTTY, STOPPER dec
f mastic *m*
e masilla *f*
i mastice *m*
n stopverf
d Kitt *m*

4333 PUTTY POWDER gla,pol
f potée *f* d'étain
e cenizas *pl* de estaño
i polveri *pl* di stagno
n tinas
d Polierasche *f*, Zinnasche *f*

4334 PYOCTANIN BLUE, GENTIAN VIOLET mcy
f violet *m* de gentiane
e violeta *f* de genciana
i violetto *m* di genziana
n gentiaanviolet *n*
d Gentianaviolett *n*

4335 PYRIDOXYN, ADERMIN oic
f adermine *f*, pyridoxyne *f*
e piridoxina *f*
i adermina *f*, piridoxina *f*
n adermine *n*, vitamine *n* B_6
d Adermin *n*, Pyridoxin *n*, Vitamin *n* B_6

4336 PYRITIC ROASTING met
f grillage *m* pyriteux
e tostación *f* pirítica
i torrefazione *f* piritica
n pyritisch roosten *n*
d Kiesröstung *f*

4337 PYRITIC SMELTING htt,met
f fusion *f* pyriteuse
e fusión *f* piritosa
i fusione *f* piritica
n pyritisch smelten *n*
d Kiesschmelzen *n*, Rohschmelzen *n*

4338 PYROLISIS fue
f pyrolyse *f*

e pirólisis *f*
i pirolisi *f*, piroscissione *f*
n pyrolyse
d Pyrolyse *f*, thermische Zersetzung *f*

4339 PYROMETRY mea
f pyrométrie *f*
e pirometría *f*
i pirometria *f*
n pyrometrie
d Pyrometrie *f*

4340 PYROPHORIC exp
f pyrophore adj
e pirofórico adj
i piroforico adj
n pyrofoor adj , zelfontbrandend adj
d pyrophor adj , selbstzündend adj

4341 PYROXYLIN LACQUER, NITROCELLULOSE LACQUER, NITROLACQUER dec
f laque *f* nitrocellulosique
e laca *f* nitrocelulósica
i lacca *f* nitrocellulosica
n nitrocelluloselak
d Nitrolack *m*, Nitrozelluloselack *m*

Q

4342 Q-gas gas
f mélange *m* de gaz à 98,7% d'hélium et 1,3% de butane
e mezcla *f* de gas con 98,7% de helio y 1,3% de butano
i mescolatura *f* di gas a 98,7% d'elio et 1,3% di butano
n gasmengsel *n* van 98,7% helium en 1,3% butaan
d Gasgemisch *n* mit 98,7% Helium und 1,3% Butan

4343 QUALITATIVE ANALYSIS ana
f analyse *f* qualitative
e análisis *f* cualitativa
i analisi *f* qualitativa
n kwalitatieve analyse
d qualitative Analyse *f*

4344 QUANTITIVE ANALYSIS, GRAVIMETRIC ANALYSIS ana
f analyse *f* gravimétrique
e análisis *f* gravimétrica
i analisi *f* gravimetrica
n kwantitative analyse
d Gewichtsanalyse *f*, gravimetrische Bestimmung *f*

4345 QUARTATION tes
f inquartation *f*
e encuartación *f*
i quartazione *f*
n kwarteren *n*
d Quartieren *n*

4346 QUARTERING tes
f réduction *f* de la prise d'essai
e reducción *f* de la muestra
i riduzione *f* del campione
n monsterreductie
d Quartieren *n* des Probegutes

4347 QUARTZ gla,mat
f quartz *m*
e cuarzo *m*
i quarzo *m*
n kwarts *n*
d Quarz *m*

4348 QUARTZ GLASS, FUSED SILICA gla
f verre *m* quarzeux
e cristal *m* de cuarzo
i vetro *m* di quarzo
n kwartsglas *n*
d Quarzglas *n*

4349 QUARTZ POWDER gla,mat
f poudre *f* de quartz
e polvo *m* de cuarzo
i polvere *f* di quarzo
n kwartspoeder *n*
d Quarzmehl *n*

4350 QUASI-ARC WELDING wel
f soudure *f* à arc avec électrode enrobée d'amiante
e soldadura *f* de arco con electrodo forrado de amianto
i saldatura *f* quasiarco
n booglassen *n* met met asbest bedekte elektrode
d Lichtbogenschweissung *f* mit asbestbekleideter Elektrode

4351 QUENCH AGEING (GB), QUENCH AGING (US) met
f vieillissement *m* de trempe
e envejecimiento *m* de temple
i invecchiamento *m* di tempera
n afschrikveroudering
d Abschreckalterung *f*

4352 QUENCH HARDENING met
f durcissement *m* par trempe
e endurecimiento *m* por temple
i indurimento *m* per tempera
n afschrikharding
d Abschreckhärtung *f*

4353 QUENCHING ctt
f extinction *f*
e extinción *f*
i estinzione *f*
n doving
d Löschung *f*

4354 QUENCHING met
f refroidissement *m* rapide, trempe *f*
e enfriamiento *m* rápido, temple *m*
i raffreddamento *m* rapido, tempera *f*
n afschrikken *n*
d Abschrecken *n*

4355 QUENCHING OIL met
f huile *f* de trempe
e aceite *m* de temple
i olio *m* di tempera
n afschrikolie
d Abschrecköl *n*

4356 QUERCITANNIC ACID, TANNIN tan
f tanin *m*, tannin *m*
e tanino *m*
i tannino *m*
n looizuur *n*
d Gerbsäure *f*, Tannin *n*

4357 QUICK AGEING (GB), QUICK AGING (US) met
f trempe *f* rapide
e temple *m* rápido
i tempera *f* rapida
n snelle afschrikking
d Schnellabschreckung *f*

4358 QUICK-BURNING POWDER exp
f poudre *f* vive
e pólvora *f* de quemado rápido, pólvora *f* viva
i polvere *f* viva
n snel ontbrandend kruit *n*
d schnellbrennendes Pulver *n*

4359 QUICK CURING pla
f à durcissement rapide
e de indurecimiento rápido
i ad indurimento rapido
n snelhardend adj
d schnellhärtend adj

4360 QUICK FERMENTATION bre,fmi
f fermentation *f* rapide
e fermentación *f* rápida
i fermentazione *f* rapida
n snelle gisting
d Schnellgärung *f*

4361 QUICK FREEZING clo,nut
f congélation *f* rapide
e congelación *f* rápida
i congelazione *f* rapida
n diepvriezen *n*
d Tiefkühlung *f*

4362 QUICK FUSE exp
f étoupille *f*
e mecha *f* de detonación rápida
i miccia *f* a detonazione rapida
n gezwinde lont
d detonierende Zündschnur *f*, Knallzündschnur *f*

4363 QUICK SETTING CEMENT (GB), EARLY SETTING CEMENT (US), RAPID HARDENING CEMENT bui
f ciment *m* à prise rapide, ciment *m* prompt
e cemento *m* de fraguado rápido
i cemento *m* a presa rapida, cemento *m* pronto
n snelbindend cement *n*
d schnellbindender Zement *m*, Schnellbinder *m*

4364 QUICK VINEGAR PROCESS aam
f procédé *m* d'acétification rapide
e procedimiento *m* rápido de fabricación de vinagre
i processo *m* rapido di fabbricazione d'aceto
n snelazijnproces *n*
d Schnellessigherstellungsverfahren *n*

4365 QUICKING coa
f placage *m* au mercure
e plateado *m* con mercurio
i placcatura *f* con mercurio
n verkwikzilveren *n*
d Verquicken *n*

4366 QUICKING SOLUTION coa
f solution *f* de mercure à plaquer
e solución *f* de mercurio para platear
i soluzione *f* di mercurio per placcare
n kwikoplossing voor pleetwerk
d Quecksilberlösung *f* für Plattierarbeiten

4367 QUICKLIME, BURNT LIMESTONE, UNSLAKED LIME aic
f chaux *f* vive
e cal *f* viva
i calce *f* viva
n ongebluste kalk
d Ätzkalk *m*, gebrannter Kalk *m*, ungelöschter Kalk *m*

4368 QUICKSILVERING coa
f amalgamation *f* à l'étain
e amalgamación *f* con estaño
i amalgamazione *f* allo stagno
n amalgameren *n* met tin
d Amalgamieren *n* mit Zinn

R

4369 RABBLING met
f brassage *m*
e rableado *m*
i rimescolamento *m*
n roeren *n*
d Krählen *n*, Umrühren *n*

4370 RACEMATION (US), RACEMIZATION (GB) oic
f inactivation *f* optique, racémisation *f*
e inactivación *f* óptica, racemización *f*
i inattivazione *f* ottica, racemizzazione *f*
n opheffing van het draaiingsvermogen, racemisatie, racemiseren *n*
d optische Inaktivierung *f*, Razemisierung *f*

4371 RACEMIC ACID oic
f acide *m* tartrique racémique
e ácido *m* tártrico racémico
i acido *m* tartarico racemico
n druivezuur *n*, racemisch wijnsteenzuur *n*
d razemische Weinsäure *f*, Traubensäure *f*

4372 RACEMIC COMPOUND, RACEMIC SUBSTANCE oic
f mélange *m* racémique
e mezcla *f* racémica
i mescolanza *f* racemica
n racemisch mengsel *n*
d Razemkörper *m*

4373 RACKING bre
f décuvage *m*, mise *f* en tonneaux, soutirage *m*
e trasvase *m* a los barriles
i infustamento *m*
n aftappen *n* in fust
d Abfüllen *n* in Fass

4374 RADIAL FLOW mix
f flux *m* radial
e flujo *m* radial
i flusso *m* radiale
n radiale stroming
d Radialströmung *f*

4375 RADIATION nue
f rayonnement *m*
e radiación *f*
i radiazione *f*
n straling
d Strahlung *f*

4376 RADIATION CHEMISTRY nue
f chimie *f* du rayonnement
e química *f* de la radiación
i chimica *f* della radiazione
n stralingschemie
d Strahlungschemie *f*

4377 RADIOCHEMISTRY nue
f chimie *f* des éléments radioactifs
e radioquímica *f*
i radiochimica *f*
n chemie van de radioactieve stoffen
d Chemie *f* radioaktiver Stoffe

4378 RAFFINATE dis,fue
f produit *m* raffiné
e refinado *m*
i prodotto *m* raffinato
n raffinaat *n*
d Raffinat *n*

4379 RAG BOILING pap
f décreusage *m*, lessivage *m* de chiffons
e cocción *f* de trapos
i cottura *f* di stracci
n koken *n* van lompen
d Kochen *n* von Hadern, Kochen *n* von Lumpen

4380 RAG CHOPPING, RAG CUTTING pap
f coupage *m* de chiffons
e corte *m* de trapos
i taglio *m* di stracci
n lompensnijden *n*
d Hadernschneiden *n*, Lumpenschneiden *n*

4381 RAG PAPER pap
f papier *m* de chiffons
e papel *m* de trapos
i carta *f* di stracci
n lompenpapier *n*
d Hadernpapier *n*, Lumpenpapier *n*

4382 RAG STOCK rub,tex
f déchets *pl* de caoutchouc avec tissu

e desechos *pl* de goma con tejido
i cascame *m* di gomma con tessuto
n rubbertextielafval
d Kautschukabfall *m* mit Textil

4383 RAGGING min
f concentration *f* préalable
e concentración *f* previa
i concentrazione *f* grossolana
n grove concentratie
d grobe Anreicherung *f*, Vorsortieren *n*

4384 RAGS pap
f chiffons *pl*
e trapos *pl*
i stracci *pl*
n lompen *pl*
d Hadern *pl*, Lumpen *pl*

4385 RAISING lea
f dégraissage *m*
e desengrase *m*
i sgrassaggio *m*
n ontvetting
d Entfettung *f*

4386 RAMIE, CAMBRIC GRASS, CHINA GRASS tex
f ramie *f*
e ramio *m*
i ramia *f*, ramie *f*
n ramee, ramie
d Ramie *f*

4387 RAMMED CONCRETE bui
f béton *m* banché, béton *m* compacté, béton *m* damé
e hormigón *m* apisonado
i calcestruzzo *m* battuto
n gestampt beton *n*, stampbeton *n*
d Stampfbeton *m*

4388 RANCID nut
f rance adj
e rancio adj
i rancido adj
n ransig adj
d ranzig adj

4389 RANCIDITY nut
f rancidité *f*
e rancidez *f*, ranciedad *f*
i rancidezza *f*, rancidità *f*
n ransigheid
d Ranzigheit *f*

4390 RAPE SEED OIL, COLZA OIL fat
f huile *f* de colza
e aceite *m* de colza
i olio *m* di colza
n raapolie
d Kohlsaatöl *n*, Rapöl *n*, Rüböl *n*

4391 RAPID COOLING clo
f refroidissement *m* rapide
e enfriamiento *m* rápido
i raffreddamento *m* rapido
n snelle koeling
d Schnellkühlung *f*

4392 RAPID FIXER pho
f fixateur *m* rapide
e fijador *m* rápido
i fissatore *m* rapido
n snel fixeerzout *n*
d Schnellfixiersalz *n*

4393 RAPID HARDENING CEMENT, EARLY SETTING CEMENT (US), QUICK SETTING CEMENT (GB) bui
f ciment *m* à prise rapide, ciment *m* prompt
e cemento *m* de fraguado rápido
i cemento *m* a presa rapida, cemento *m* pronto
n snelbindend cement *n*
d schnellbindender Zement *m*, Schnellbinder *m*

4394 RARE GAS, INERT GAS, NOBLE GAS gas
f gaz *m* inerte, gaz *m* noble
e gas *m* inerte, gas *m* noble
i gas *m* inerte, gas *m* nobile
n edelgas *n*, inert gas *n*
d Edelgas *n*, inertes Gas *n*

4395 RAREFACTION vac
f raréfaction *f*
e rarefacción *f*
i rarefazione *f*
n verdunning
d Verdünnung *f*

4396 RAREFIED GAS gas
f gaz *m* raréfié
e gas *m* rarificado
i gas *m* rarefatto
n verdund gas *n*
d verdünntes Gas *n*

4397 RASPING nut
f râpage *m*
e raspadura *f*, raspamiento *m*
i grattatura *f*
n raspen *n*
d Zerreiben *n*

4398 RASPING PROCESS rub
f procédé *m* de râpage
e procedimiento *m* por raspado
i processo *m* da irruvidire, processo *m* di raspatura
n ruwproces *n*
d Aufrauhverfahren *n*, Raspelungsverfahren *n*

4399 RAT POISON, RATSBANE mcy
f mort-aux-rats *m*
e matarratas *m*
i veleno *m* pei topi
n rattenkruit *n*
d Rattengift *n*

4400 RATE elc
f régime *m*
e régimen *m*
i corrente *m* di regime
n ontlaadstroom
d Nennentladestrom *m*

4401 RATE OF DEPOSITION wel
f vitesse *f* de dépôt
e velocidad *f* de deposición
i velocità *f* di deposizione
n neerlassnelheid
d Abbrenngeschwindigkeit *f*

4402 RATE OF FLOW gen
f vitesse *f* d'écoulement
e velocidad *f* de flujo, velocidad *f* de paso
i velocità *f* di flusso
n doorstroomsnelheid
d Durchflussgeschwindigkeit *f*

4403 RATE OF REACTION, REACTION RATE gen
f vitesse *f* de réaction
e velocidad *f* de reacción
i velocità *f* di reazione
n reactiesnelheid
d Reaktionsgeschwindigkeit *f*

4404 RATING OF STORAGE BATTERY elc
f capacité *f* nominale
e capacidad *f* nominal
i capacità *f* nominale
n nominale capaciteit
d Nennkapazität *f*, Nennleistung *f*

4405 RATIO OF COMPRESSION, COMPRESSION RATIO fue
f rapport *m* de compression
e relación *f* de compresión
i rapporto *m* di compressione
n compressieverhouding
d Verdichtungsverhältnis *n*

4406 RATIO OF PIGMENT TO BINDER, PIGMENT/BINDER RATIO dec
f rapport *m* pigment/liant
e relación *f* pigmento-aglutinante
i rapporto *m* pigmento-legante
n pigment/bindmiddelverhouding
d Farbe/Bindemittelverhältnis *n*

4407 RATTLING (US), RUMBLING (GB), TUMBLING (US) cle
f dessablage *m*
e desarenado *m*
i sterratura *f*
n poetsen *n*
d Putzen *n*, Scheuern *n*

4408 RAW COMPOUND rub
f mélange *m* brut, mélange *m* vierge
e composición *f* bruta
i composto *m* grezzo
n ongevulkaniseerd mengsel *n*
d Rohmischung *f*

4409 RAW FLAVOR (US), RAW FLAVOUR (GB) bre
f bouquet *m* de jeune
e aroma *m* de cerveza nueva
i profumo *m* di birra giovane
n aroma *n* van jong bier
d Jungbukettstoff *m*

4410 RAW GRAIN, UNMALTED GRAIN bre
f grain *m* cru
e grano *m* crudo
i grano *m* crudo
n ongemout graan *n*
d Rohfrucht *f*

4411 RAW HIDE tan
f peau *f* non-tannée
e piel *f* cruda, piel *f* verde
i pelle *f* non conciata
n ongelooide huid
d grüne Haut *f*, rohe Haut *f*

4412 RAW JUICE sin
f jus *m* de diffusion, vesou *m*
e jugo *m* bruto
i succo *m* di diffusione
n ruwsap *n*
d Diffusionssaft *m*, Rohsaft *m*

4413 RAW MATERIALS gen
f matières *pl* primaires

e materias *pl* primas
i materie *pl* prime
n onbewerkte grondstoffen *pl*
d Ausgangsstoffe *pl*, Grundstoffe *pl*, Rohstoffe *pl*

4414 RAW SEWAGE, CRUDE SEWAGE, CRUDE WASTE — sew
f eau *f* brute usée
e agua *f* cruda de alcantarillo
i acqua *f* bruta di rifiuti
n ongezuiverd afvalwater *n*
d Rohabwasser *n*, rohes Abwasser *n*

4415 RAW SLUDGE, CRUDE SLUDGE, FRESH SLUDGE — iwt
f boues *pl* fraîches, bourbe *f*, limon *m*
e fango *m* crudo
i fango *m* fresco
n onbewerkt slib *n*
d Frischschlamm *m*, Rohschlamm *m*

4416 RAW SODA, KELP SODA — aic
f soude *f* brute, soude *f* des varechs
e sosa *f* bruta
i soda *f* greggia
n ruwe soda
d Steinsoda *f*

4417 RAW SUGAR — sin
f sucre *m* roux
e azúcar *m* bruto
i zucchero *m* di primo prodotto
n ruwe suiker
d Rohzucker *m*

4418 RAYON, ARTIFICIAL SILK — tex
f soie *f* artificielle, rayonne *f*
e rayón *m*, seda *f* artificial
i raion *m*, seta *f* artificiale
n kunstzijde
d Kunstseide *f*, Rayon *m*, Reyon *m*

4419 REACTANT — gen
f corps *m* réagissant
e cuerpo *m* de reacción
i corpo *m* di reazione
n reactiestof
d Reaktant *m*, Reaktionsteilnehmer *m*

4420 REACTION — gen
f réaction *f*
e reacción *f*
i reazione *f*
n reactie
d Reaktion *f*

4421 REACTION CHAIN — gen
f chaîne *f* de réactions
e cadena *f* de reacciones
i catena *f* di reazioni
n reactieketen
d Reaktionskette *f*

4422 REACTION GROUND COAT PROCESS, CONTACT PROCESS — dec
f procédé *m* de contact
e procedimiento *m* de mano primera reactiva
i processo *m* di mano prima reattiva
n contactgrondproces *n*, reactieve-grondlaagmethode
d Aktivgrundverfahren *n*, Reaktionsgrundverfahren *n*, Zweistufenlackierverfahren *n*

4423 REACTION INHIBITING CATALYST, NEGATIVE CATALYST — gen
f catalyseur *m* négatif
e catalizador *m* negativo
i catalizzatore *m* negativo
n negatieve katalysator, reactievertragende katalysator
d negativer Katalysator *m*, reaktionshemmender Katalysator *m*

4424 REACTION LACQUER, CATALYZED LACQUER, CHEMICALLY CURING LACQUER — dec
f laque *f* à durcissement chimique, vernis *m* réactionnel
e laca *f* reaccional
i lacca *f* ad indurimento chimico, lacca *f* di reazione
n reactielak
d Kalthärtelack *m*, Reaktionslack *m*

4425 REACTION LIMIT — cor
f limite *f* de corrosion
e límite *m* de corrosión
i limite *m* di corrosione
n corrosiegrens
d Korrosionsgrenze *f*

4426 REACTION PRODUCT, RESULTANT — gen
f produit *m* de la réaction
e producto *m* de la reacción
i prodotto *m* della reazione
n reactieprodukt *n*
d Reaktionsprodukt *n*

4427 REACTION-SINTERING — met
f frittage *m* à réaction chimique simultanée

e sinterización *f* de reacción química simultánea
i sinterizzazione *f* a reazione chimica simultanea
n reactiesinteren *n*
d Reaktionssintern *n*

4428 REACTIVATION gen
f réactivation *f*
e reactivación *f*
i riattivazione *f*
n reactiveren *n*
d Reaktivieren *n*

4429 READY-MIXED PAINT dec
f peinture *f* prête à l'usage
e pintura *f* preparada al uso
i pittura *f* preparata all'uso
n gebruiksklare verf, voor gebruik gerede verf
d anstrichfertige Farbe *f*

4430 REAGENT ana
f réactif *m*
e reactivo *m*
i reagente *m*
n reagens *n*
d Reagenz *n*

4431 REAL ATTENUATION bre,fmi
f atténuation *f* réelle
e atenuación *f* real
i attenuazione *f* reale
n werkelijke vergisting
d wirkliche Vergärung *f*

4432 REAL DEGREE OF ATTENUATION, REAL DEGREE OF FERMENTATION bre,fmi
f degré *m* d'atténuation réelle
e grado *m* de atenuación real
i grado *m* d'attenuazione reale
n werkelijke vergistingsgraad
d wirklicher Vergärungsgrad *m*

4433 REBOILING boi,fmi,sin
f rebouillement *m*
e recocción *f*
i ribollimento *m*
n opkoken *n*
d Abkochen *n*, Aufkochen *n*

4434 RECARBURIZATION, CARBON RESTORATION met
f récarburation *f*, régénération *f* de pièces décarburisées
e recarburación *f*, regeneración *f* de piezas descarburadas
i ricarburazione *f*, rigenerazione *f* di pezzi decarburati
n regenereren *n* van ontkoolde delen
d Aufkohlung *f*, Rückkohlung *f*

4435 RECARBURIZER met
f agent *m* récarburateur
e agente *m* de recarburación
i agente *m* di ricarburazione
n regenerator van ontkoolde delen
d Aufkohlungssubstanz *f*

4436 RECIRCULATION pre
f circulation *f* continue
e recirculación *f*
i circolazione *f* continua
n continue circulatie
d kontinuierlicher Umlauf *m*

4437 RECLAIM (GB), RECUPERATED WASTE VULCANIZED RUBBER (US) rub
f caoutchouc *m* régénéré
e regenerado *m* de goma
i rigenerato *m* di gomma
n geregenereerde rubber, regeneraat *n*
d Kautschukregenerat *n*, Regenerat *n*

4438 RECLAIMING, DEVULCANIZING, REGENERATION rub
f dévulcanisation *f*, régénération *f*
e regeneración *f*
i rigenerazione *f*
n regeneratie
d Regeneration *f*

4439 RECOMPACTING, REPRESSING pre
f recompression *f*
e recompresión *f*
i ricompressione *f*
n napersen *n*
d Nachpressen *n*

4440 RECONSTRUCTED STONE, ARTIFICIAL STONE bui
f pierre *f* artificielle
e piedra *f* artificial
i pietra *f* artificiale
n kunststeen
d Kunststein *m*

4441 RECOVERY, RECUPERATION gen
f récupération *f*
e recuperación *f*
i ricupero *m*

n terugwinning
d Rückgewinnung *f*,
Wiedergewinnung *f*

4442 RECRYSTALLIZATION, OSTWALD RIPENING, PHYSICAL RIPENING pho
f maturation *f* physique
e maduración *f* física
i maturazione *f* fisica
n ostwaldse rijping
d Ostwaldsche Reifung *f*, physikalische Reifung *f*, Umkristallisierung *f*, Vorreifung *f*

4443 RECRYSTALLIZATION ANNEALING htt,met
f recuit *m* à récristallisation
e recocido *m* de recristalización
i ricottura *f* a ricristallizzazione
n rekristallisatiegloeien *n*
d Umkristallisationsglühen *n*

4444 RECRYSTALLIZATION TEMPERATURE cry,met
f température *f* de récristallisation
e temperatura *f* de recristalización
i temperatura *f* di ricristallizzazione
n rekristallisatietemperatuur
d Umkristallisationstemperatur *f*

4445 RECTIFICATION dis,gas
f rectification *f*
e rectificación *f*
i rettificazione *f*
n rectificeren *n*
d Rektifizieren *n*

4446 RECUPERATIVE HEATING, REGENERATIVE HEATING boi
f chauffage *m* à récupération
e calentamiento *m* recuperativo
i riscaldamento *m* a ricupero
n recuperatieve verhitting
d Rekuperativheizung *f*

4447 RED IRON OXIDE, COLCOTHAR dec
f colcotar *m*
e colcótar *m*
i colcotar *m*
n colcothar, dodekop *n*
d Colcothar *m*, Eisenoxydrot *n*, Totenkopf *m*

4448 RED LEAD, MINIUM dec
f minium *m* de plomb
e minio *m* de plomo
i minio *m* di piombo
n loodmenie, menie
d Bleimennige *f*, Mennige *f*

4449 RED OCHRE, BURNT OCHRE dec
f ocre *f* rouge
e ocre *m* rojo
i ocra *f* rossa
n rode oker
d roter Ocker *m*

4450 REDISTILLATION dis
f distillation *f* répétée
e destilación *f* repetida
i distillazione *f* ripetuta
n herhaalde destillatie
d wiederholte Destillierung *f*

4451 REDISTILLED ZINC dis,met
f zinc *m* redistillé
e cinc *m* redestilado
i zinco *m* fino
n fijnzink *n*
d Feinzink *n*

4452 REDOX POTENTIAL elc
f potentiel *m* d'oxyréduction
e potencial *m* de óxidorreducción
i redopotenziale *m*
n redoxpotentiaal
d Redoxpotential *n*

4453 REDUCED CHROME MORDANT dye
f mordant *m* à chrome réduit
e mordiente *m* de cromo reducido
i mordente *m* a cromo ridotto
n gereduceerde chroombijtstof
d reduzierte Chrombeize *f*

4454 REDUCER pho
f affaiblisseur *m*
e debilitador *m*
i indebolitore *m*, riduttore *m*
n verzwakker
d Abschwächer *m*

4455 REDUCING AGENT gen
f agent *m* réducteur
e agente *m* de reducción
i agente *m* riducente
n reductiemiddel *n*
d Reduktionsmittel *n*

4456 REDUCING FLAME bur
f flamme *f* réductrice
e llama *f* reductora
i fiamma *f* riducente
n reducerende vlam
d Reduktionsflamme *f*

4457 REDUCTION gen
f réduction *f*
e reducción *f*
i riduzione *f*
n reduceren *n*
d Reduktion *f*

4458 REDUCTION pho
f affaiblissement *m*
e debilitamiento *m*
i indebolimento *m*, riduzione *f*
n verzwakking
d Abschwächung *f*

4459 REDUCTION CLEARING dye
f élimination *f* du colorant non-usé
e eliminación *f* del colorante no consumado
i eliminazione *f* del colorante non consumato
n verwijdering van ongebruikte kleurstof
d Beseitigung *f* des nicht verarbeiteten Farbstoffes

4460 REDUCTION RATIO, DEGREE OF REDUCTION gen
f degré *m* de réduction
e grado *m* de reducción
i grado *m* di riduzione
n reductiegraad
d Reduktionsgrad *m*, Reduktionsstufe *f*

4461 REDUCTION ROASTING met
f grillage *m* réductif
e tostación *f* reductora
i arrostimento *m* riducente, torrefazione *f* riducente
n reducerend roosten *n*
d reduzierendes Rösten *n*

4462 REEL DYEING, WINCH DYEING dye
f teinture *f* à dévidoir
e teñido *m* con aspadera
i tintura *f* ad arcolaio
n verven *n* op haspel
d Haspelfärbung *f*

4463 RE-ENRICHMENT nue
f réenrichissement *m*
e reenriquecimiento *m*
i riarricchimento *m*
n herverrijking
d Wiederanreicherung *f*

4464 REFINED LINSEED OIL dec
f huile *f* de lin blanchie, huile *f* de lin raffinée
e aceite *m* de linaza refinado
i olio *m* di seme di lino imbiancato
n gebleekte lijnolie
d gebleichtes Leinöl *n*, Lackleinöl *n*

4465 REFINED SUGAR sin
f raffinade *f*
e azúcar *m* refinado, refino *m* de azúcar
i zucchero *m* raffinato
n geraffineerde suiker, raffinade
d Raffinade *f*, Raffinadezucker *m*

4466 REFINING fat
f raffinage *m* alcalin
e desacidificación *f*
i disacidificazione *f*
n raffineren *n* van olie
d Entsäuerung *f*

4467 REFINING gen
f épuration *f*
e depuración *f*
i depurazione *f*
n zuivering
d Reinigung *f*

4468 REFINING min
f affinage *m*
e afino *m*, refinación *f*
i affinazione *f*
n affineren *n*
d Raffinieren *n*

4469 REFINING FURNACE htt,met
f four *m* d'affinage
e horno *m* de afino
i forno *m* d'affinazione, forno *m* per bonifica
n raffineeroven, veredelingsoven
d Vergütungsofen *m*

4470 REFINING OF METALS met
f affinement *m* de métaux
e afino *m* de metales
i bonifica *f* di metalli
n metaalveredeling
d Metallvergütung *f*

4471 REFLUX, BACKFLOW dis,gen
f reflux *m*
e reflujo *m*
i riflusso *m*
n terugloop, terugvloeien *n*
d Rückfluss *m*, Rücklauf *m*, Zurückfliessen *n*

4472 REFLUX RATIO dis
f taux *m* de reflux
e relación *f* de reflujo
i rapporto *m* di riflusso
n terugstroomverhouding
d Rückflussverhältnis *n*, Rücklaufverhältnis *n*

4473 REFORMING fue
f réformation *f*
e reformación *f*
i riformazione *f*
n reformeren *n*
d Reformieren *n*

4474 REFRACTORY gen
f réfractaire adj
e refractario adj
i refrattario adj
n hittevast adj, vuurvast adj
d feuerfest adj

4475 REFRACTORY CEMENT bui
f ciment *m* réfractaire
e cemento *m* refractario
i cemento *m* refrattario
n vuurvaste specie
d feuerfester Zement *m*

4476 REFRIGERANT clo
f réfrigérant *m*
e refrigerante *m*
i refrigerante *m*
n koelmiddel *n*
d Kältemittel *n*, Kühlmittel *n*

4477 REFRIGERATION clo
f réfrigération *f*
e refrigeración *f*
i refrigerazione *f*
n koeling
d Kühlung *f*

4478 REFUSE, WASTE iwt
f déchets *pl*, refus *m*
e basuras *pl*, desechos *pl*, detritos *pl*
i cascame *m*, rifiuti *pl*, scarto *m*
n afvalstoffen *pl*
d Abfall *m*

4479 REGENERATED LEACH LIQUOR aic
f lessive *f* régénérée
e lejía *f* regenerada
i lisciva *f* rigenerata
n arme loog, verarmde loog
d regenerierte Lauge *f*

4480 REGENERATION gen
f régénération *f*
e regeneración *f*
i rigenerazione *f*
n regeneratie
d Regenerierung *f*

4481 REGENERATION, DEVULCANIZING, RECLAIMING rub
f dévulcanisation *f*, régénération *f*
e regeneración *f*
i rigenerazione *f*
n regeneratie
d Regeneration *f*

4482 REGENERATIVE HEATING, RECUPERATIVE HEATING boi
f chauffage *m* à récupération
e calentamiento *m* recuperativo
i riscaldamento *m* a ricupero
n recuperatieve verhitting
d Rekuperativheizung *f*

4483 REGULATOR, MODIFIER pla
f agent *m* modificateur
e agente *m* modificador
i agente *m* modificatore
n modificator
d Abänderungsmittel *n*

4484 REGULATOR MIXTURE, BUFFER SOLUTION gen
f solution *f* tampon
e solución *f* reguladora, solución *f* tampón
i soluzione *f* tampone
n bufferoplossing
d Pufferlösung *f*

4485 REGULINE DEPOSIT elc
f dépôt *m* régulin
e depósito *m* regulin
i deposito *m* compatto
n gezond neerslag *n*
d regulinischer Niederschlag *m*

4486 REINFORCED CONCRETE, ARMED CONCRETE, FERROCONCRETE bui
f béton *m* armé
e hormigón *m* armado
i cemento *m* armato, conglomerato *m* cementizio armato
n gewapend beton *n*
d bewehrter Beton *m*, Eisenbeton *m*, Stahlbeton *m*

4487 REINFORCED PLASTICS pla
f matière *f* plastique renforcée

e materia *f* plástica reforzada
i materia *f* plastica rinforzata
n versterkte kunststof
d verstärkter Kunststoff *m*

4488 REINFORCING FILLER, ACTIVE FILLER rub
f charge *f* active, charge *f* renforçante
e carga *f* activa, carga *f* reforzadora
i carica *f* attiva, carica *f* rinforzante
n actieve vulstof, versterkende vulstof
d aktiver Füllstoff *m*, aktives Füllmittel *n*, Verstärkerfüllmittel *n*

4489 RELIEF PAINT, PLASTIC PAINT, TEXTURE FINISH dec
f peinture *f* plastique, peinture *f* relief
e pintura *f* plástica
i pittura *f* plastica
n plastiekverf, reliefverf
d plastische Anstrichfarbe *f*

4490 RELEASE AGENT (GB), MOLD PARTING AGENT (US) pla
f agent *m* de démoulage
e agente *m* de desmoldeado
i agente *m* di sformatura
n losmiddel *n*
d Trennmittel *n*

4491 RELINING, GLAZING fou
f brasquage *m*
e enlucido *m*
i rivestimento *m*
n bekleding
d Glasurbrand *m*

4492 REMELTING htt
f refonte *f*
e refusión *f*
i rifusione *f*
n omsmelten *n*
d Umschmelzen *n*

4493 RENDERING, FINING OFF, PLASTERING, RUNNING OFF bui
f finissage *m* par du plâtre
e colocación *f* de un emplasto
i intonacatura *f*
n afpleisteren *n*
d Verputzen *n*

4494 RENNET dai,nut
f présure *f*
e cuajo *m*
i presame *m*
n leb
d Lab *n*

4495 REPRESSING, RECOMPACTING pre
f recompression *f*
e recompresión *f*
i ricompressione *f*
n napersen *n*
d Nachpressen *n*

4496 RERACKING bre
f transvasement *m*
e transvasado *m*, trasegación *f*
i travasamento *m*
n overschenken *n*
d Umfüllen *n*

4497 RESIDUAL ACID dis
f acide *m* résiduel
e ácido *m* residual
i acido *m* residuo
n zuurrest
d Säurerückstand *m*

4498 RESIDUAL CURRENT elc
f courant *m* résiduel
e corriente *f* residual
i corrente *f* residua
n reststroom
d Reststrom *m*

4499 RESIDUAL FUEL OIL fue
f huile *f* combustible résiduelle
e aceite *m* combustible residual
i olio *m* combustibile residuo
n residuale stookolie
d Rückstandsöl *n*

4500 RESIDUAL GASES elc
f gaz *pl* résiduels
e gases *pl* résiduales
i gas *pl* residui
n restgassen *pl*
d Restgase *pl*

4501 RESIDUAL OIL, CRUDE OIL fue
f huile *f* non-raffinée
e aceite *m* no rafinado
i olio *m* greggio
n ruwe olie
d Roherdöl *n*, Rohöl *n*

4502 RESIDUAL SOIL, SOIL RESIDUE elc

f crasse *f* restante, souillure *f* restante
e residuo *m* de la suciedad, suciedad *f* restante
i residuo *m* di sporco
n resterend vuil *n*
d Restschmutz *m*

4503 RESIDUE, dis
HEEL
f résidu *m*
e residuo *m*
i residuo *m*
n residu *n*
d Rückstand *m*

4504 RESIDUE ON EVAPORATION, eva
DRY RESIDUE
f résidu *m* sec, résidu *m* fixe
e residuo *m* seco
i deposito *m* secco, residuo *m* fisso
n droog residu *n*
d Trockenrückstand *m*

4505 RESIDUES pho
f résidus *pl* de bains
e residuos *pl* de baños
i residui *pl* di bagni
n badresten *pl*
d Badrückstände *pl*

4506 RESIDUES PROCESSING pho
f régénération *f* de résidus de bains
e regeneración *f* de residuos de baños
i rigenerazione *f* di residui di bagni
n opwerken *n* van badresten
d Aufarbeitung *f* von Badrückständen

4507 RESIN GLUE, glu,pap
RESIN SIZE
f colle *f* de résine
e cola *f* resinosa
i colla *f* di resina
n harslijm
d Harzleim *m*

4508 RESIN MILK pap
f lait *m* de colle
e leche *f* resinosa
i latte *m* resinoso
n harsmelk
d Harzmilch *f*

4509 RESIN OIL, fat
ROSIN OIL
f huile *f* de résine
e aceite *m* de resina
i olio *m* di resina
n harsolie
d Harzöl *n*

4510 RESIN SOAP, soa
SODIUM RESINATE
f résinate *m* de soude, savon *m* de résine
e resinato *m* de sodio
i resinato *m* di sodio
n harszeep, natriumresinaat *n*
d Harzseife *f*, Natriumresinat *n*

4511 RESINIFICATION oic
f résinification *f*
e resinificación *f*
i resinificazione *f*
n verharsing
d Verharzung *f*

4512 RESINOUS CEMENT glu
f colle *f* résineuse
e cola *f* resinosa
i colla *f* resinosa
n harskit
d Harzkitt *m*

4513 RESINTERING htt,met
f refrittage *m*
e resinterización *f*
i risinterizzazione *f*
n nasinteren *n*, oversinteren *n*
d Nachsintern *n*, wiederholtes Sintern *n*

4514 RESIST elc
f matière de protection
e materia *f* protectora
i rivestimento *m* isolante
n afdekmateriaal *n*
d Abdecker *m*, Abdeckmittel *n*

4515 RESIST tex
f réserve *f*
e reserva *f*
i riserva *f*
n reserve
d Reserve *f*

4516 RESIST PRINTING tex
f impression *f* par réserves
e estampación *f* por reservas
i stampaggio *m* per riservazione
n reserveringsdruk
d Reservagedruck *m*, Reservedruck *m*, Schutzdruck *m*

4517 RESOLUTION gen
f rédissolution *f*
e redisolución *f*

i ridissoluzione *f*
n heroplossing
d Wiederauflösung *f*

4518 RESORPTION abs
f résorption *f*
e resorción *f*
i risorbimento *m*
n resorptie
d Resorption *f*

4519 REST MALTING bre
f malterie *f* avec relais d'acide carbonique
e maltería *f* con relé de ácido carbónico
i malteria *f* per lavoro con soste
n mouten *n* in de rusttoestand
d Rastmälzerei *f*

4520 RESTRAINER pho
f retardateur *m*
e retardator *m*
i ritardatore *m*
n verzwakker
d Abschwächer *m*

4521 RESTRICTED-SUDSING DETERGENT, CONTROLLED-SUDSING DETERGENT, LOW-FOAMING DETERGENT cle
f détergent *m* peu moussant
e detergente *m* poco espumante
i detergente *m* poco schiumogeno
n weinig schuimend wasmiddel *n*
d wenig schäumendes Waschmittel *n*

4522 RESULTANT, REACTION PRODUCT gen
f produit *m* de la réaction
e producto *m* de la reacción
i prodotto *m* della reazione
n reactieprodukt *n*
d Reaktionsprodukt *n*

4523 RETANNING tan
f rétannage *m*
e recurtido *m*
i riconcia *f*
n nalooien *n*
d Nachgerben *n*

4524 RETARDER cat
f retardateur *m*
e retardador *m*
i ritardatore *m*
n vertrager
d Verzögerer *m*

4525 RETENTION gen,sto
f retenue *f*
e retención *f*
i ritenuto *m*
n bewaren *n*, opslaan *n*
d Aufbewahren *n*

4526 RETROGRADE CONDENSATION dis
f condensation *f* rétrograde
e condensación *f* retrógrada
i condensazione *f* retrograda
n achterwaartse condensatie
d retrograde Kondensation *f*, Rückwärtsverdichtung *f*

4527 RETTING, ROTTING tex
f rouissage *m*
e enriamiento *m*
i macerazione *f*
n roten *n*
d Rösten *n*

4528 RETURN AIR COOLING clo
f refroidissement *m* à air en circuit fermé
e enfriamiento *m* por circuito de aire
i raffreddamento *m* ad aria ritornata
n luchtkoeling met gesloten circuit
d Luftumlaufkühlung *f*

4529 RE-USED WOOL tex
f laine *f* renaissance
e lana *f* reutilizada
i lana *f* riusata
n geregenereerde wol
d aufgearbeitete Wolle *f*

4530 REVERSE CURRENT CLEANING, ANODIC CLEANING elc
f dégraissage *m* anodique
e desengrase *m* anódico
i sgrassaggio *m* anodico
n anodische ontvetting
d anodische Reinigung *f*

4531 REVERSIBLE REACTION gen
f réaction *f* réversible
e reacción *f* reversible
i reazione *f* reversibile
n omkeerbare reactie
d umkehrbare Reaktion *f*

4532 REVIVIFICATION fue
f réactivation *f*
e reactivación *f*
i riattivazione *f*

n gloeien *n*, reactiveren *n*
d Glühen *n*, Reaktivieren *n*

4533 RHODANIZING elc
f placage *m* au rhodium
e rodiado *m*
i placcatura *f* con rodio, rodanizzazione *f*
n pleten *n* met rhodium, verrhodiummen *n*
d Rhodiumplattierung *f*, Rhodonisieren *n*

4534 RICE HULLING nut
f décorticage *m* de riz
e descascarillado *m* de arroz
i pilatura *f* di riso
n pellen *n* van rijst
d Reisenthülsung *f*

4535 RICE PAPER pap
f papier *m* de Chine, papier *m* de riz
e papel *m* de China, papel *m* de paja de arroz
i carta *f* di riso
n rijstpapier *n*
d Reispapier *n*

4536 RICE STARCH glu
f amidon *m* anglais, amidon *m* de riz
e almidón *m* de arroz
i amido *m* di riso
n rijststijfsel
d Reisstärke *f*

4537 RICH GAS gas
f gaz *m* riche
e gas *m* rico
i gas *m* ricco
n rijk gas *n*
d reiches Gas *n*

4538 RICH IN OXYGEN gas
f riche adj en oxygène
e rico adj en oxígeno
i ricco adj d'ossigeno
n zuurstofrijk adj
d sauerstoffreich adj

4539 RICH LIME, bui
FAT LIME, PURE LIME
f chaux *f* grasse
e cal *f* grasa
i calce *f* grassa
n kluitkalk, Luikse kalk, steenkalk
d Fettkalk *m*, Speckkalk *m*

4540 RICINUS OIL, fat
CASTOR OIL
f huile *f* de ricin
e aceite *m* de ricino
i olio *m* di ricino
n kasterolie, ricinusolie, wonderolie
d Rizinusöl *n*

4541 RIM FERMENTATION bre,fmi
f fermentation *f* aux bords de la cuve
e fermentación *f* a los bordes
i fermentazione *f* ai bordi
n randgisting
d Randgärung *f*

4542 RINSING, cle
WASHING OFF
f rinçage *m*
e enjuague *m*
i risciacquamento *m*
n spoelen *n*
d Spülen *n*

4543 RIPENING pho
f maturation *f*
e maduración *f*
i maturazione *f*
n rijping
d Reifung *f*

4544 RIPENING ACCELERATOR,nut,pho
RIPENING AGENT
f adjuvant *m* de la maturation
e acelerador *m* de la maduración
i acceleratore *m* della maturazione
n rijpingsversneller
d Reifkörper *m*

4545 RIPPLE FINISH dec
f finissage *m* à ondulations
e acabado *m* de ondulaciones
i finitura *f* ad ondulazioni
n rimpeleffect *n*
d Kräuselverfahren *n*

4546 RIPPLING tex
f égrenage *m*, égrugage *m*
e desgrane *m*, rastrillado *m*
i scapecchiamento *m*
n repelen *n*
d Reffen *n*, Riffeln *n*

4547 RISING OF THE YEAST bre,fmi
f levée *f* de la levure
e subida *f* de la levadura
i levata *f* del lievito
n rijzen *n* van de gist
d Hefetrieb *m*

4548 ROAST SINTERING, htt,met
BLAST ROASTING
f frittage *m* sur grille, grillage *m* de frittage

e sinterización *f* por tostación
i arrostimento *m* di sinterizzazione, sinterizzazione *f* su griglia
n sinterroosten *n*
d Sinterrösten *n*

4549 ROASTED MALT, BLACK MALT, CARAMEL MALT — bre
f malt *m* noir, malt *m* torréfié
e malta *f* colorante, malta *f* tostada
i malto *m* torrefatto
n licht geroosterde mout
d Farbmalz *n*, Röstmalz *n*

4550 ROASTING, CALCINATION, CALCINING — met
f calcination *f*, grillage *m*
e calcinación *f*, tostación *f*
i arrostimento *m*, calcinazione *f*
n calcineren *n*, roosten *n*
d Kalzinieren *n*, Rösten *n*

4551 ROASTING, CURING, TORRIFYING — bre
f torréfaction *f*
e tostación *f*
i torrefazione *f*
n roosting
d Röstung *f*

4552 ROCK ASPHALT, NATIVE ASPHALT — bui
f asphalte *m* naturel, roche *f* asphaltique
e asfalto *m* natural, roca *f* asfáltica
i roccia *f* asfaltica naturale
n natuurasfalt *n*
d Naturasphalt *m*

4553 ROLL BOILING, DECATIZING, STEAMING — tex
f décatissage *m*
e decatizado *m*, vaporización *f*
i decatissaggio *m*
n afglanzen *n*, decateren *n*
d Dekatieren *n*, Dekatur *f*, Glanzkrumpe *f*

4554 ROLLED GOLD — coa
f doublé *m* d'or
e oro *m* chapado
i oro *m* laminato
n gouddoublé *n*
d Golddublée *n*, Walzgold *n*

4555 ROLLED-IN SCALE — htt
f battitures *pl* enfoncées
e escamas *pl* de laminación embebidas
i scaglia *f* laminata dentro
n ingewalste oxydelaag
d eingewalzter Zunder *m*

4556 ROLLER PRINTING — tex
f impression *f* à rouleaux
e impresión *f* de rodillos
i stampatura *f* a rulli
n roldruk
d Rollendruck *m*

4557 ROLLER SCALE — met
f croûte *f* de laminage
e cascarilla *f* de laminación
i scaglia *f* di laminazione
n walsoxyde *n*
d Walzzunder *m*

4558 ROPY FERMENTATION — bre,fmi
f fermentation *f* visqueuse
e fermentación *f* viscosa
i fermentazione *f* viscosa
n slijmerige gisting
d schleimige Gärung *f*

4559 ROSE OIL, ATTAR OIL, OTTO OIL — fat
f essence *f* de roses, huile *f* rosat
e esencia *f* de rosas
i essenza *f* di rose, olio *m* rosato
n rozenolie
d Rosenessenz *f*, Rosenöl *n*

4560 ROSEMARY OIL — fat
f essence *f* de romarin
e esencia *f* de rosmarino
i olio *m* di rosmarino
n rozemarijnolie
d Rosmarinöl *n*

4561 ROSETTE GRAPHITE — mat
f graphite *m* en rosettes
e grafito *m* en rosetas
i grafite *f* a rosette
n rozetvormig grafiet *n*
d B-Graphit *m*, Rosettengraphit *m*

4562 ROSIN ESTER, ESTER GUM — pla
f résine *f* estérifiée
e resina *f* esterificada
i resina *f* esterificata
n esterhars
d Estergummi *m*, Esterharz *n*

4563 ROSIN OIL, RESIN OIL — fat
f huile *f* de résine
e aceite *m* de resina

i olio *m* di resina
n harsolie
d Harzöl *n*

4564 ROT STEEPING tex
f élimination *f* de l'apprêt
e eliminación *f* del apresto
i eliminazione *f* dell'appretto
n verwijdering van het appreteermiddel
d Beseitigung *f* des Appreturmittels

4565 ROTSTEEP, dye
MORDANT
f mordant *m*
e mordiente *m*
i mordente *m*
n beitsmiddel *n*, bijtmiddel *n*
d Ätzwasser *n*, Beize *f*, Beizmittel *n*, Schutzbeize *f*

4566 ROTTING, tex
RETTING
f rouissage *m*
e enriamiento *m*
i macerazione *f*
n roten *n*
d Rösten *n*

4567 ROUGE, pol
JEWELLER'S ROUGE,
POLISHING CROCUS,
POLISHING ROUGE
f rouge *m* à polir, rouge *m* d'Angleterre, rouge *m* de fer
e óxido *m* férrico, rojo *m* de Inglaterra, rojo *m* de pulir
i ossido *m* ferrico, rosso *m* inglese
n Berlijns rood *n*, Engels rood *n*, gele bolus, polijstrood *n*
d Englischrot *n*, Pariserrot *n*, Polierrot *n*

4568 ROUGH GRINDING pol
f dégrossissage *m* à la meule
e amoladura *f* bruta
i sgrossatura *f*
n grofslijpen *n*
d Zuschleifen *n*

4569 RUBBER ADHESIVE, glu
SOLUTION
f solution *f* de caoutchouc
e solución *f* de goma
i adesivo *m* di gomma
n rubberlijm, solutie
d Gummiklebstoff *m*, Klebkitt *m*

4570 RUBBER SOLUTION, rub
CEMENT
f dissolution *f*
e disolución *f*
i soluzione *f*
n solutie
d Gummilösung *f*

4571 RUBBER SPREADING (GB),pap,rub
KNIFE COATING (US)
f enduisage *m* à la racle sur rouleau
e recubrimiento *m* con cuchilla sobre rodillo
i spalmatura *f* a lama su cilindro
n strijken *n* met rakel op wals
d Streichen *n* mit Rakel auf Walze

4572 RUBBERIZING, coa,rub
PROOFING, TOPPING WITH GUM
f gommage *m*, recouvrement *m* de caoutchouc
e engomadura *f*, recubrimiento *m* con goma
i gommatura *f*, ricopertura *f* con gomma
n berubberen *n*, gommeren *n*
d Gummieren *n*

4573 RUBBING FASTNESS, tex
CROCKING FASTNESS (US)
f résistance *f* au frottement
e resistencia *f* al frotamiento, solidez *f* al borroneo
i resistenza *f* allo strofinamento
n wrijvingsbestendigheid
d Reibfestigkeit *f*

4574 RUMBLING (GB), cle
RATTLING (US), TUMBLING (US)
f dessablage *m*
e desarenado *m*
i sterratura *f*
n poetsen *n*
d Putzen *n*, Scheuern *n*

4575 RUNNING, dec
TEARING
f coulage *m*, formation *f* de coulures
e derramamiento *m*
i scolamento *m*
n vorming van lopers
d Abrinnen *n*, Läuferbildung *f*

4576 RUNNING OFF, bui
FINING OFF, PLASTERING,
RENDERING
f finissage *m* par du plâtre
e colocación *f* de un emplasto
i intonacatura *f*
n afpleisteren *n*
d Verputzen *n*

4577 RUSSIAN LEATHER, JUCTEN, YUFTEN — lea
f cuir *m* de Russie
e cuero *m* de Rusia, vaqueta *f* de Moscovia
i bulgaro *m*, cuoio *m* di Russia
n juchtleder *n*
d Juchtenleder *n*

4578 RUST — cor
f rouille *f*
e herrumbre *f*, orín *m*
i ruggine *f*
n roest
d Rost *m*

4579 RUST INHIBITOR, RUST PREVENTING MEDIUM — cor
f agent *m* antirouille, antirouille *m*
e inhibidor *m* antioxidante, substancia *f* antiherrumbrosa
i antiruggine *f*
n roestwerend middel *n*
d Rostschutzmittel *n*

4580 RUST REMOVAL — cle
f dérouillage *m*
e eliminación *f* de la herrumbre
i rimozione *f* della ruggine
n ontroesten *n*
d Entrostung *f*

4581 RUST REMOVER — cle
f dissolvant *m* de rouille
e disolvente *m* de herrumbre
i dissolvente *m* di ruggine
n ontroestingsmiddel *n*, roestoplosmiddel *n*
d Entrostungsmittel *n*, Rostlösungsmittel *n*

4582 RUSTING — cor
f formation *f* de rouille, rouillure *f*
e formación *f* de herrumbre
i arrugginimento *m*
n roesten *n*
d Rosten *n*

4583 RUSTING, SULLING — elc
f oxydation *f* après décapage
e oxidación *f* después del decapaje
i ossidazione *f* dopo il decapaggio
n oxydatie na beitsen
d Oxydation *f* nach dem Beizen

4584 RUSTLESS STEEL — met
f acier *m* inoxydable
e acero *m* inoxidable
i acciaio *m* inossidabile
n niet-roestend staal *n*, roestvast staal *n*
d rostfreier Stahl *m*

S

4585 SACCHARASE, sin
INVERTASE (GB), INVERTIN (US),
SUCRASE
f invertase *f*
e invertasa *f*
i invertasi *f*
n invertase, saccharase
d Invertase *f*, Invertin *n*,
Saccharase *f*

4586 SACCHARIFICATION bre
f saccharification *f*
e sacarificación *f*
i saccarificazione *f*
n sacharificatie, versuikering
d Saccharifikation *f*,
Verzuckerung *f*

4587 SACCHARIFICATION REST bre
f pause *f* de saccharification
e estacionamiento *m* por
sacarificación
i sosta *f* per saccarificazione
n versuikeringspauze
d Verzuckerungspause *f*

4588 SACCHARIFICATION PERIOD bre
f durée *f* de saccharification
e duración *f* de sacarificación
i durata *f* di saccarificazione
n versuikeringstijd
d Verzuckerungszeit *f*

4589 SACCHARIN (GB), nut
SACCHARINOSE (US)
f saccharine *f*, sulfimide *m*
benzoïque
e sacarina *f*
i saccarina *f*
n sacharine
d Benzoesäuresulfimid *n*,
Saccharin *n*

4590 SACK FILTRATION (US), fil
BAG FILTRATION (GB)
f filtrage *m* en sacs
e filtración *f* en bolsas
i filtrazione *f* in sacchi
n zakfiltreren *n*
d Taschenfiltrierung *f*

4591 SACKCLOTH, tra
BAGGING
f serpillière *f*, toile *f* à sac
e arpillera *f*, tela *f* de saco
i grossa tela *f*, tela *f* da sacco
n zakkengoed *n*, zakkenlinnen *n*
d Packleinwand *f*, Sackleinwand *f*

4592 SADDENING-DOWN dec
f perte *f* de lustre
e pérdida *f* de lustre
i perdita *f* di lucentezza
n glansverlies *n*
d Glanzverlust *m*, Stumpfwerden *n*

4593 SAFETY EXPLOSIVE exp
f explosif *m* de sûreté
e explosivo *m* de seguridad
i esplosivo *m* di sicurezza
n mijngasveilige springstof
d Sicherheitssprengstoff *m*

4594 SAFETY FILM, pho
ACETATE FILM
f film *m* de sécurité, pellicule *f* à
l'acétate de cellulose
e película *f* de acetato de celulosa
i pellicola *f* all'acetato di
cellulosa
n acetaatfilm
d Azetatfilm *m*, Sicherheitsfilm *m*

4595 SAFETY GLASS, gla,sap
GLASS LAMINATE,
LAMINATED GLASS
f verre *m* de sûreté
e vidrio *m* laminado
i vetro *m* di sicurezza
n veiligheidsglas *n*
d Sicherheitsglas *n*

4596 SAFFLOR OIL (US), fat
SAFFLOWER OIL (GB),
SAFFRON OIL (GB)
f huile *f* de carthame
e aceite *m* de cártamo
i olio *m* di cartamo
n saffloerolie
d Karthamusöl *n*, Saffloröl *n*

4597 SAFFLOR RED (US), dec
SAFFLOWER CARMIN (GB),
SAFFLOWER RED (GB)
f rouge *m* de carthame
e rojo *m* de cártamo
i rosso *m* di cartamo
n saffloerrood *n*
d Karthamin *n*, Safflorrot *n*

4598 SAFFLOR YELLOW (US), SAFFLOWER YELLOW (GB) dec
f jaune *m* de carthame
e amarillo *m* de cártamo, azafranina *f*
i giallo *m* di cartamo
n saffloergeel *n*
d Safflorgelb *n*

4599 SAGE OIL fat
f essence *f* de sauge
e esencia *f* de salvia
i olio *m* di salvia
n salieolie
d Salbeiöl *n*

4600 SAGGING, DOWN-FLOW dec
f affaisement *m*, formation *f* de festons
e arrugamiento *m*
i insaccamento *m*
n vorming van zakkers
d Gardinenbildung *f*, Vorhängebildung *f*

4601 SAGGING OF PAINT, CREEPING OF PAINT dec
f formation *f* de rideaux
e derramiento *m*
i scorrimento *m*
n uitzakken *n*
d Ausbauchen *n*, Ausweichen *n*

4602 SAL AMMONIAC min
f chlorure *m* d'ammonium, salmiac *m*, sel *m* ammoniac
e sal *f* amoníaco, salmiac *m*
i cloruro *m* d'ammonio, sale *m* ammoniaco
n salmiak
d Ammoniumchlorid *n*, Chlorammonium *n*, Salmiak *m*

4603 SAL SODA aic
f soude *f* cristallisée
e sosa *f* cristalizada
i soda *f* cristallizzata
n kristalsoda
d Kristallsoda *f*

4604 SAL VOLATILE, HARTSHORN SALTS mcy
f sel *m* volatil
e sal *f* volátil
i sale *m* volatile
n hertshoornzout *n*, vlugzout *n*
d flüchtiges Laugensalz *n*, Hirschhornsalz *n*

4605 SALAD OIL nut
f huile *f* comestible
e aceite *m* de comer
i olio *m* alimentare
n slaolie
d Salatöl *n*

4606 SALERATUS (US), BAKING POWDER, BAKING SODA (GB) nut
f levure *f* artificielle, poudre *f* levain
e bicarbonato *m* sódico, levadura *f* en polvo
i bicarbonato *m* di soda, lievito *m* in polvere
n bakpoeder *n*
d Backpulver *n*

4607 SALIFICATION, SALIFYING sal
f salification *f*
e salificación *f*
i salificazione *f*
n verzouten *n*, zoutvorming
d Salzbildung *f*

4608 SALINE gen
f salin adj
e salino adj
i salino adj
n zout adj , zouthoudend adj
d salzartig adj , salzhaltig adj , salzig adj

4609 SALINE SOLUTION gen
f solution *f* saline
e solución *f* salina
i soluzione *f* salina
n zoutoplossing
d Salzlösung *f*

4610 SALINITY gen
f salinité *f*
e salinidad *f*
i salinità *f*
n zoutheid, zoutte
d Salzhaltigkeit *f*, Salzigkeit *f*

4611 SALT gen
f sel *m*
e sal *f*
i sale *m*
n zout *n*
d Salz *n*

4612 SALT ACID (US), HYDROCHLORIC ACID (GB), MURIATIC ACID aic
f acide *m* chlorhydrique

e ácido *m* clorhídrico, ácido *m* muriático
i acido *m* cloridrico, acido *m* muriatico
n chloorwaterstofzuur *n*, zoutzuur *n*
d Chlorwasserstoffsäure *f*, Salzsäure *f*

4613 SALT BATH QUENCH met
f trempe *f* en bain de sel
e temple *m* en baño de sal
i tempera *f* in bagno di sale
n afschrikken *n* in zoutbad
d Salzbadabschrecken *n*

4614 SALT BATH TREATMENT, LIQUID CARBURIZING, LIQUID HEAT TREATMENT met
f cémentation *f* par liquide
e cementación *f* por líquido
i cementazione *f* per liquido
n met vloeistof cementeren *n*
d Flüssigzementierung *f*

4615 SALT CAKE (US), SODIUM SULPHATE (GB) aic
f sulfate *m* de sodium commercial
e sulfato *m* de sodio comerciale
i solfato *m* di sodio commerciale
n handelsnatriumsulfaat *n*
d technisches Natriumsulfat *n*

4616 SALT LINKAGE soa
f liaison *f* saline
e enlace *m* salino
i legame *m* salino
n zoutband
d Salzbindung *f*

4617 SALT OF SATURN, GOULARD'S POWDER, LEAD ACETATE dec
f acétate *m* de plomb, sel *m* de Saturne
e acetato *m* de plomo
i acetato *m* di piombo, sale *m* di Saturno
n loodsuiker
d Bleizucker *m*, Goulardsches Pulver *n*

4618 SALT SPRAY TEST tes
f essai *m* au brouillard salin
e ensayo *m* a la niebla salina
i collaudo *m* alla nebbia salina
n zoutnevelproef, zoutsproeiproef
d Salznebelversuch *m*

4619 SALT STAIN tan
f tache *f* de sel
e mancha *f* de sal
i macchia *f* di sale
n zoutvlek
d Salzfleck *m*

4620 SALT WATER aic
f eau *f* salée
e agua *f* salada
i acqua *f* salsa
n zout water *n*
d Salzsole *f*, Salzwasser *n*, Sole *f*

4621 SALT WATER, SEA WATER gen
f eau *f* de mer
e agua *f* de mar
i acqua *f* di mare
n zeewater *n*
d Meerwasser *n*, Seewasser *n*

4622 SALTING ON dye
f addition *f* de colorant
e adición *f* de colorante
i addizione *f* di colorante
n bijvoegen *n* van kleurstof
d Zugabe *f* von Farbstoff

4623 SALTING OUT, GRAINING OUT soa
f relargage *m*
e saladura *f*
i salatura *f*
n uitzouten *n*
d Aussalzen *n*

4624 SALTPETRE, NITRE, POTASSIUM NITRATE aic
f salpêtre *m*
e salitre *m*
i salnitro *m*
n salpeter
d Salpeter *m*

4625 SALTWATER, BRINE gen
f solution *f* de sel
e solución *f* de sal
i soluzione *f* salina
n zoutoplossing
d Kochsalzlösung *f*, Salzlake *f*, Sole *f*, Sulze *f*

4626 SALTWATER SOAP, COCONUT OIL SOAP soa
f savon *m* à l'huile de coco
e jabón *m* de coco, jabón *m* para agua de mar
i sapone *m* di cocco
n kokoszeep, zeewaterzeep
d Salzwasserseife *f*

4627 SAMPLE tes
f échantillon *m*
e muestra *f*, prueba *f*
i campione *m*
n monster *n*
d Muster *n*, Probe *f*

4628 SAMPLING tes
f échantillonnage *m*
e muestreo *m*
i prelievo *m* di campione, presa *f* di campione
n monster nemen *n*, monster trekken *n*
d Probeentnahme *f*, Probenahme *f*

4629 SAND fou
f sable *m*
e arena *f*
i sabbia *f*, terra *f*
n zand *n*
d Sand *m*

4630 SAND FILTERING, SAND FILTRATION hyg,iwt
f filtration *f* sur sable
e filtración *f* con arena
i filtrazione *f* a sabbia
n filtreren *n* met zand
d Sandfilterung *f*

4631 SAND LEACHING min
f lessivage *m* de sable aurifère
e lixiviación *f* de arena aurífera
i lisciviazione *f* di sabbia aurifera
n uitlogen *n* van goudhoudend zand
d Auslaugen *n* von goldhaltigem Sand

4632 SAND SIEVE GRADING, GRADING sor
f granulométrie *f*
e granulometría *f*
i granulometria *f*
n korrelverdeling
d Korngrössenverteilung *f*

4633 SAND TEST exp,tes
f essai *m* de sable
e ensayo *m* con proyectil lastrado
i collaudo *m* di sabbia
n zandproef
d Sandversuch *m*

4634 SANDALWOOD OIL, SANTAL OIL fat
f essence *f* de bois de santal, essence *f* de santal
e esencia *f* de sándalo
i olio *m* di sandalo
n sandelhoutolie
d Sandelholzöl *n*, Sandelöl *n*

4635 SANDBLASTING cle
f dessablage *m*, sablage *m*
e chorreo *m* con arena
i sabbiatura *f*
n straalpoetsen *n*, zandstralen *n*
d Sandstrahlen *n*, Sandelung *f*, Strahlputzen *n*

4636 SANDING, MECHANICAL POLISHING pol
f polissage *m* mécanique
e pulimento *m* mecánico
i lucidatura *f* meccanica
n mechanisch polijsten *n*
d maschinelle Polierung *f*

4637 SANDIVER, GALL OF GLASS gla
f suin *m* de verre, suint *m* de verré
e sulfato *m* sódico no reducido
i solfato *m* sodico non ridotto
n glasgal
d Glasgalle *f*

4638 SANDPAPER pap,pol
f papier *m* de verre, papier-verré *m*
e papel *m* de lija
i carta *f* vetrata
n schuurpapier *n*
d Glaspapier *n*, Sandpapier *n*

4639 SANDY CLAY, LOAM, MILD CLAY bui
f terre *f* glaise
e tierra *f* arcillosa
i terra *f* argillosa, terra *f* grassa
n leem *n*
d Lehm *m*

4640 SANFORIZING tex
f sanforisation *f*
e sanforización *f*
i sanforizzazione *f*
n kreukvrij maken *n*, sanforiseren *n*
d Krumpfechtmachen *n*, Sanforisieren *n*

4641 SANITARY PAPER hyg,pap
f papier *m* hygiénique
e papel *m* higiénico
i carta *f* igienica
n closetpapier *n*, toiletpapier *n*
d Klosettpapier *n*, Toilettenpapier *n*

4642 SANITARY WALLPAPER bui,pap
f papier *m* de tenture lavable

e empapelado *m* lavable
i carta *f* da parato lavabile
n afwasbaar behang *n*, afwasbaar behangselpapier *n*
d abwaschbare Tapete *f*

4643 SANTONIA OIL, CHENOPODIUM OIL, WORMSEED OIL — fat
f essence *f* de chénopodium
e esencia *f* de quenopodio
i olio *m* di chenopodio
n wormzaadolie
d Chenopodiumöl *n*, Wurmsamenöl *n*

4644 SAPONACEOUS, SOAPY — soa
f saponacé adj , savonneux adj
e jabonoso adj , saponáceo adj
i saponaceo adj , saponoso adj
n zeepachtig adj , zepig adj
d Seifenartig adj , seifig adj

4645 SAPONIFIABLE — soa
f saponifiable adj
e saponificable adj
i saponificabile adj
n verzeepbaar adj
d verseifbar adj

4646 SAPONIFICATION — soa
f saponification *f*
e saponificación *f*
i saponificazione *f*
n verzeping
d Verseifung *f*

4647 SAPONIFICATION EQUIVALENT — soa
f équivalent *m* de saponification
e equivalente *m* de saponificación
i equivalente *m* di saponificazione
n verzepingsequivalent *n*
d Verseifungsäquivalent *n*

4648 SAPONIFICATION NUMBER, SAPONIFICATION VALUE — soa
f indice *m* de saponification
e índice *m* de saponificación
i indice *m* di saponificazione
n verzepingsgetal *n*
d Verseifungszahl *f*

4649 SASSAFRAS OIL — fat
f essence *f* de sassafras
e esencia *f* de sasafraso
i olio *m* di sassafrasso
n sassafrasolie
d Sassafrasöl *n*

4650 SATIN-FROSTING — gla
f glaçage *m*, satinage *m*
e satinado *m*
i smerigliatura *f* all'acido
n satineren *n*
d Satinieren *n*

4651 SATIN PAPER — pap
f papier *m* satiné
e papel *m* satinado
i carta *f* rasata
n gesatineerd papier *n*, tafpapier *n*
d Atlaspapier *n*, satiniertes Papier *n*, Taftpapier *n*

4652 SATINING, CALENDERING — pap
f satinage *m*
e satinado *m*
i calandratura *f*, spianatura *f*
n satineren *n*
d Satinieren *n*

4653 SATURABLE — gen
f saturable adj
e saturable adj
i saturabile adj
n verzadigbaar adj
d sättigungsfähig adj

4654 SATURATED COLOR (US), SATURATED COLOUR (GB) — dec
f couleur *f* saturée
e color *m* saturado
i colore *m* saturo
n verzadigde kleur
d gesättigte Farbe *f*, Optimalfarbe *f*

4655 SATURATED COMPOUND — gen
f composé *m* saturé
e compuesto *m* saturado
i composto *m* saturo
n verzadigde verbinding
d gesättigte Verbindung *f*

4656 SATURATED HYDROCARBON — oic
f hydrocarbure *m* saturé
e hidrocarburo *m* saturado
i idrocarburo *m* saturo
n verzadigde koolwaterstof
d gesättigter Kohlenwasserstoff *m*

4657 SATURATED POOL BOILING — boi
f ébullition *f* à saturation
e ebullición *f* de saturación
i ebollizione *f* a saturazione
n kernkoken *n* in vat
d Sättigungssieden *n*

4658 SATURATED SOLUTION — gen
f solution *f* saturée
e solución *f* saturada
i soluzione *f* satura

n verzadigde oplossing
d gesättigte Lösung *f*

4659 SATURATED STEAM, WET STEAM — boi
f vapeur *f* saturée
e vapor *m* saturado
i vapore *m* saturo
n verzadigde stoom
d gesättigter Dampf *m*, Sattdampf *m*

4660 SATURATION — gen
f saturation *f*
e saturación *f*
i saturazione *f*
n verzadiging
d Sättigung *f*

4661 SAWDUST — mat
f sciure *f*
e serrín *m* de madera
i segatura *f*
n zaagmeel *n*, zaagsel *n*
d Sägemehl *n*

4662 SCALDING — bre
f nettoyage *m* en faisant bouillir
e hervor *m*
i sterilizzazione *f* per ebullizione
n uitkoken *n*
d Auskochen *n*

4663 SCALDING, SEETHING, SMOTHERING — tex
f échaudage *m*
e escaldadura *f*
i scottatura *f*
n broeien *n*
d Abbrühen *n*

4664 SCALE — boi
f incrustation *f*, tartre *m*
e incrustación *f*
i incrostazione *f*
n ketelsteen *n*
d Kesselstein *m*

4665 SCALE — met
f écaillage *m*
e cascarilla *f*, escama *f*
i scagliatura *f*
n oxydebladdering, oxydehuid
d Zunder *m*, Zunderschicht *f*

4666 SCALE FORMATION, INCRUSTATION — cor
f entartrage *m*, incrustation *f*
e incrustación *f*
i incrostazione *f*
n ketelsteenvorming
d Kesselsteinbildung *f*, Wassersteinansatz *m*

4667 SCALE REMOVAL — boi
f élimination *f* de tartre
e desincrustación *f*
i disincrostazione *f*
n afbikken *n* van ketelsteen
d Abklopfen *n* des Kesselsteins, Kesselsteinbeseitigung *f*

4668 SCALE-TIGHT FINISH — met
f finissage *m* résistant à l'écaillage
e acabado *m* resistente a la cascarilla
i finitura *f* resistente alla scagliatura
n oxydehuidvorming verhinderende bescherming
d zunderfreies Feinen *n*

4669 SCALE WAX — fat
f paraffine *f* brute en écaille
e parafina *f* bruta en escama
i paraffina *f* in scaglia
n schubbenparaffine
d Schuppenparaffin *n*

4670 SCALING, EXFOLIATING, PEELING OFF — dec
f écaillage *m*
e despellegamiento *m*
i screpolatura *f*
n afbladderen *n*
d Abblättern *n*, Abschälen *n*

4671 SCALPING, SLEDGING — brk
f broyage *m* préliminaire
e quebrantamiento *m* preliminar
i frantumazione *f* preliminare
n grof breken *n*
d Grobbrechen *n*

4672 SCARFING — cle
f décalaminage *m* au chalumeau
e escarpadura *f* con soplete
i scagliatura *f* alla fiamma
n afbranden *n*, schoonbranden *n*
d Abbrennen *n*, Abflammen *n*

4673 SCAVENGER — met
f éliminateur *m* d'impuretés
e eliminador *m* de impurezas
i eliminatore *m* d'impurezze
n reinigingsmiddel *n*
d Reinigungsmittel *n*

4674 SCAVENGING cle
f rinçage *m*
e enjuague *m*
i risciaquatura *f*
n spoelen *n*
d Spülen *n*

4675 SCENT, per
ODOR (US), ODOUR (GB),
SMELL
f odeur *f*
e olor *m*
i odore *m*
n geur, reuk
d Duft *m*, Geruch *m*

4676 SCENT, per
PERFUME
f parfum *m*
e perfume *m*
i profumo *m*
n odeur, parfum *n*
d Parfüm *n*

4677 SCHWEINFURT GREEN, dec
EMPEROR GREEN (US),
PARIS GREEN (GB)
f vert *m* de Scheele
e cardenillo *m* arsenioso, verde *m* de París
i verde *m* di Parigi
n Parijs groen *n*
d Parisergrün *n*, Schweinfurter Grün *n*

4678 SCISSION REACTION gen
f réaction *f* de séparation
e reacción *f* de separación
i reazione *f* di scissione
n splitsingsreactie
d Spaltungsreaktion *f*

4679 SCLEROPROTEIN, oic
ALBUMINOID
f albuminoïde *m*, scléroprotéine *f*
e albuminoide *m*, escleroproteína *f*
i albuminoide *m*, scleroproteina *f*
n albuminoïde *n*, scleroproteïne *n*
d Skleroprotein *n*

4680 SCORCH RETARDER, rub
ANTISCORCHER, ANTISCORCHING
AGENT
f antigrilleur *m*, retardateur *m* de grillage
e retardador *m* de la prevulcanización
i agente *m* antiscottatura
n vertrager van de aanvulcanisatie
d Vorvulkanisationsverzögerer *m*

4681 SCORCHED RUBBER, rub
BURNT RUBBER
f caoutchouc *m* brûlé
e caucho *m* quemado
i gomma *f* bruciata
n aangebrande rubber
d angebrannter Gummi *m*

4682 SCORCHING bur,tex
f roussissage *m*
e chamusquina *f*
i scottatura *f*
n schroeien *n*, zengen *n*
d Anbrennen *n*, Versengen *n*

4683 SCORCHING rub,tex
f grillage *m*
e prevulcanización *f*
i scottatura *f*
n aanvulcaniseren *n*
d Anbrennen *n*, Anvulkanisieren *n*

4684 SCORCHY ACCELERATOR rub
f accélérateur *m* précoce
e acelerante *m* con tendencia de quemar
i acceleratore *m* provocante scottatura
n versneller met neiging tot aanvulcanisatie
d Beschleuniger *m* mit anvulkanisierender Wirkung

4685 SCORIFICATION htt,min
f scorification *f*
e escorificación *f*
i scorificazione *f*
n louteren *n*
d Läutern *n*

4686 SCORING, dec,pla
DEVILLING
f rayage *m*
e raspado *m*
i rigatura *f*
n krassen *n*
d Kerben *n*

4687 SCOURING cle
f dérochage *m*, récurage *m*
e fregadura *f*
i strofinamento *m*
n schuren *n*
d Scheuern *n*

4688 SCOURING met
f décapage *m*
e decapaje *m*
i decapaggio *m*
n verwijderen *n* van de oxydelaag
d Entzundern *n*

4689 SCOURING nut
f vannage *m* mécanique
e aventamiento *m* mecánico
i ventilazione *f* meccanica
n mechanisch wannen *n*
d maschinelles Worfeln *n*

4690 SCOURING pol
f polissage *m*
e pulido *m*
i pulitura *f*
n gladschuren *n*
d Putzen *n*

4691 SCOURING tex
f dégraissage *m*, dessuintage *m*
e desgrase *m*, desuardado *m*
i sgrassaggio *m*
n ontvetting
d Entfettung *f*, Entschweissen *n*

4692 SCRAP gen
f déchets *pl*
e desechos *pl*
i rifiuti *pl*
n afval
d Abfälle *pl*

4693 SCRAP met
f chute *f*, ferraille *f*, mitraille *f*
e chatarra *f*
i rottami *pl*
n schrot *n*
d Schrott *n*

4694 SCRAP PAPER, pap
WASTE PAPER
f déchets *pl* de papier
e papel *m* de desechos, papel *m* para apuntes
i carta *f* straccia, cartaccia *f*
n oud papier *n*
d Altpapier *n*, Makulatur *f*

4695 SCRAP RUBBER rub
f déchets *pl* de caoutchouc
e desechos *pl* de caucho
i scarto *m* di gomma vulcanizzata e non vulcanizzata
n afvalrubber
d Gummiabfälle *pl*

4696 SCRAPING cle
f grattage *m*, raclage *m*
e rascadura *f*, rasqueteado *m*
i raschiatura *f*
n schrapen *n*, schrappen *n*
d Schaben *n*, Scharren *n*, Schrappen *n*

4697 SCRAPING, cer
FETTLING
f raclage *m*
e rascadura *f*
i raschiamento *m*, raschiatura *f*
n afschrappen *n*
d Abschroten *n*

4698 SCRATCH TEST tes
f essai *m* à rayure, essai *m* de rayage
e prueba *f* al rayado
i collaudo *m* per graffiatura
n krasproef
d Kratzprobe *f*, Ritzprobe *f*

4699 SCREEN ANALYSIS, ana
SIEVE ANALYSIS
f analyse *f* par tamisage
e análisis *f* por tamizado
i analisi *f* granulometrica
n zeefanalyse
d Siebanalyse *f*

4700 SCREEN PRINTING, tex
SERIGRAPHY
f impression *f* à la pellicule
e estampación *f* con estarcido
i stampaggio *m* alla pellicola
n zeefdruk
d Netzdruck *m*, Siebdruck *m*

4701 SCREENING, sap
SHIELDING
f blindage *m*
e blindaje *m*
i schermatura *f*
n afscherming
d Abschirmung *f*

4702 SCREENING, sor
SIEVING, SIFTING
f criblage *m*, tamisage *m*
e cribado *m*, tamizado *m*
i stacciatura *f*, vagliatura *f*
n zeven *n*, ziften *n*
d Sichten *n*, Sieben *n*

4703 SCREENINGS, sor
SHORTS, SIEVE RESTS
f débris *pl* de tamisage
e desperdicios *pl* del cribado, granzas *pl*
i residuo *m* sullo staccio
n zeefrest
d Siebrückstand *m*, Siebüberlauf *m*

4704 SCRUBBING cle,gas
f lavage *m* de gaz
e lavado *m* de gas
i lavaggio *m* di gas
n gas wassen *n*
d Gaswaschen *n*

4705 SCUM gen
f écume *f*
e espuma *f*
i schiuma *f*
n schuim
d Schaum *m*

4706 SCUM, bui
BLEMISH, EFFLORESCENCE
f carie *f*, efflorescence *f*, salpêtre *m*
e eflorescencia *f*
i efflorescenza *f* murale
n muurkanker, muursalpeter, muuruitslag
d Ausblühung *f*, Auswitterung *f*, Mauerfrass *m*, Salpeterfrass *m*

4707 SCUM (GB), met
DROSS (US)
f écume *f*
e espuma *f*
i feccia *f*, scoria *f* metallifera
n droes, schuim *n*
d Krätze *f*, Schaum *m*

4708 SCUM, bui
LAITANCE
f écume *f* de béton
e exudación *f* del hormigón, lechada *f* de cemento
i schiuma *f* di calcestruzzo
n betonschuim *n*
d Betonschaum *m*

4709 SCUM COVER iwt
f chapeau *m*
e cobertura *f* de espuma
i copertura *f* di schiuma
n schuimdek *n*
d Schaumdecke *f*

4710 SCUTCHING, tex
HACKLING, HECKLING
f peignage *m*
e peinado *m*, rastrillado *m*
i pettinatura *f*
n hekelen *n*
d Hecheln *n*

4711 SEA WATER, gen
SALT WATER
f eau *f* de mer
e agua *f* de mar
i acqua *f* di mare
n zeewater *n*
d Meerwasser *n*, Seewasser *n*

4712 SEAL hyg,sew
f barrage *m* liquide
e obturación *f* de agua
i sbarramento *m* liquido
n vloeistofversperring
d Flüssigkeitssperre *f* in Leitung

4713 SEALING vac
f étanchement *m* de pores
e procedimiento *m* para tapar poros
i sigillatura *f* di pori
n afdichten *n* van poriën
d Porenabdichtung *f*

4714 SEALING COMPOUND gen
f composition *f* d'étanchéité
e mezcla *f* para empaquetadura
i mescolanza *f* per chiusura stagna
n dichtingsmengsel *n*
d Dichtungsmasse *f*

4715 SEALING LIQUID iwt
f liquide *m* d'isolement
e líquido *m* obturador
i liquido *m* di chiusura
n afsluitvloeistof
d Absperrflüssigkeit *f*, Sperrflüssigkeit *f*

4716 SEARING, met
SLEEKING
f lissage *m*
e alisado *m*
i lisciatura *f*
n gladmaken *n*
d Glätten *n*

4717 SEASONING lea
f aprêtage *m* à l'albumen
e apresto *m* de albumina
i appretto *m* all'albume
n eiwitbehandeling
d Eiweissbehandlung *f*

4718 SEASONING pla
f maturation *f*
e maduración *f*
i maturazione *f*
n rijping
d Reifung *f*

4719 SEASONING, bui
WOOD SEASONING
f desséchement *m* de bois
e cura *f* de madera, desecación *f* de madera
i stagionatura *f* di legno
n drogen *n* van hout
d Holzauswitterung *f*

4720 SECOND BATH tan
f second bain *m*
e segundo baño *m*

i secondo bagno *m*
n reducerend bad *n*
d Reduktionsbad *n*

4721 SECOND MASHING bre
f seconde trempe *f*
e segunda mezcla *f*
i riscaldamento *m* della tempera ad una data temperatura
n tweede beslag *n*
d Aufmaischen *n*

4722 SECOND RIPENING, AFTERRIPENING, CHEMICAL RIPENING, DIGESTION pho
f maturation *f* chimique, maturation *f* en refonte
e maduración *f* química, segunda maduración *f*
i maturazione *f* chimica
n chemische rijping
d chemische Reifung *f*, Nachreifung *f*

4723 SECOND RUN OF WORT, AFTERWORT bre
f trempe *f* de lavage
e segundo mosto *m*
i secondo mosto *m*
n onderste wort, tweede wort
d Glatterwasser *n*, Nachwürze *f*

4724 SECOND RUNNING, AFTERRUN, LAST RUNNING dis
f queue *f* de distillation
e cola *f* de destilación
i coda *f* di distillazione
n naloop
d Nachlauf *m*

4725 SECONDARY FERMENTATION bre,fmi
f fermentation *f* secondaire
e fermentación *f* secundaria
i fermentazione *f* secondaria
n nagisting
d Nachgärung *f*

4726 SECONDARY PLASTICIZER, EXTENDER PLASTICIZER pla
f plastifiant *m* inerte
e plastificante *m* diluente
i plastificante *m* inerte
n secondair plastificeermiddel *n*
d Beschwermittel *n*

4727 SECONDARY REACTION gen
f réaction *f* secondaire
e reacción *f* secundaria
i reazione *f* secondaria
n nevenreaktie
d Nebenreaktion *f*

4728 SECONDARY RIPENING bre,fmi
f maturation *f* complémentaire
e maduración *f* complementaria
i maturazione *f* complementaria
n narijping
d Nachreifung *f*

4729 SECONDARY SEDIMENTATION, FINAL SETTLING iwt
f clarification *f* finale, décantation *f* secondaire
e clarificación *f* final
i sedimentazione *f* finale
n laatste klaring
d Nachklärung *f*

4730 SECONDARY YEAST bre,fmi
f levure *f* de fermentation secondaire
e levadura *f* de fermentación secundaria
i lievito *m* di fermentazione secondaria
n gist van de nafermentatie
d Nachgärungshefe *f*

4731 SEDIMENT bre
f dépôt *m*, sédiment *m*
e sedimento *m*
i sedimento *m*
n bezinksel *n*, sediment *n*
d Absatz *m*, Bodensatz *m*, Sediment *n*

4732 SEDIMENTARY SOLIDS iwt,sew
f matière *f* solide déposée
e materia *f* sólida depositada
i materia *f* solida di sedimentazio[ne]
n zinkstof *n*
d Sinkstoff *m*

4733 SEDIMENTATION, SETTLING cle
f sédimentation *f*
e sedimentación *f*
i sedimentazione *f*
n laten bezinken *n*, sedimentatie
d Absetzung *f*, Sedimentation *f*

4734 SEDIMENTATION ANALYSIS ana
f analyse *f* granulométrique par sédimentation
e análisis *f* granulométrica por sedimentación
i analisi *f* granulometrica per sedimentazione

n sedimentatieanalyse
d Sedimentationsanalyse *f*

4735 SEED CHARGE, PRECIPITANT, PRECIPITATING AGENT — gen
f agent *m* de précipitation, précipitant *m*
e agente *m* de precipitación, precipitante *m*
i agente *m* precipitante
n neerslagmiddel *n*
d Fällmittel *n*, Niederschlagsmittel *n*

4736 SEED YEAST — bre,fmi
f couche *f* médiane de levure
e levadura *f* de siembra
i lievito *m* di semenza, parte *f* centrale dello strato di lievito
n kerngist
d Kernhefe *f*, Samenhefe *f*

4737 SEEDINESS — dec
f rugosité *f*
e aspecto *m* granuloso
i puntinatura *f*
n ruwheid
d griesiges Aussehen *n*

4738 SEEPAGE WATER, LEAKAGE WATER — hyg,sew
f eau *f* de fuites
e agua *f* de infiltración
i acqua *f* di fughe
n lekwater *n*, sijpelwater *n*
d Leckwasser *n*, Sickerwasser *n*

4739 SEEPING — gen
f infiltration *f*, percolation *f*, suintement *m*
e percolación *f*, rezumamiento *m*
i percolazione *f*, stillicidio *m*
n doorsijpelen *n*
d Durchsickern *n*

4740 SEETHING, SCALDING, SMOTHERING — tex
f échaudage *m*
e escaldadura *f*
i scottatura *f*
n broeien *n*
d Abbrühen *n*

4741 SEGREGATED OILS — dec
f huiles *pl* séparées
e aceites *pl* segregados
i oli *pl* segregati
n afgesplitste oliën *pl*
d abgespaltene Öle *pl*

4742 SEGREGATION, LIQUATION — met
f liquation *f*, ségrégation *f*
e licuación *f*, segregación *f*
i liquazione *f*, segregazione *f*
n segregatie, zijgeren *n*
d Ausschwitzen *n*, Seigerung *f*

4743 SEIZING — met
f conglomération *f*
e conglomeración *f*
i conglomerazione *f*
n samenbakken *n*
d Zusammenbacken *n*

4744 SEIZING, GALLING — met
f écorchure *f*
e excoriación *f*
i scorticatura *f*
n aanvreten *n* door wrijving
d Fressen *n*

4745 SELECTIVE AERATION, DIFFERENTIAL AERATION — cor
f aération *f* sélective
e aireación *f* selectiva
i aerazione *f* selettiva
n selectieve aëratie
d unterschiedliche Belüftung *f*

4746 SELECTIVE COOLING, DIFFERENTIAL COOLING — clo
f refroidissement *m* sélectif
e enfriamiento *m* selectivo
i raffreddamento *m* selettivo
n selectieve koeling
d selektive Kühlung *f*

4747 SELECTIVE FERMENTATION — fmi
f fermentation *f* sélective
e fermentación *f* selectiva
i fermentazione *f* selettiva
n selectieve fermentatie
d selektive Fermentation *f*

4748 SELECTIVE FLOTATION, DIFFERENTIAL FLOTATION — met
f flottation *f* sélective
e flotación *f* selectiva
i flottazione *f* selettiva
n selectieve flotatie
d selektive Flotation *f*

4749 SELECTIVE HARDENING, DIFFERENTIAL HARDENING — pho
f durcissement *m* sélectif
e endurecimiento *m* selectivo
i indurimento *m* selettivo
n selectieve harding

d selektive Härtung *f*, Teilhärtung *f*

4750 SELENIUM TONING pho
f virage *m* au sélénium
e viraje *m* al selenio
i viraggio *m* al selenio
n seleniumkleuring
d Selentonung *f*

4751 SELF-ADHESIVE TAPE (GB), cpt
ADHESIVE TAPE (GB),
STICKING TAPE (US),
f ruban *m* adhésif, ruban *m* collant
e cinta *f* adhesiva
i nastro *m* adesivo
n plakband *n*
d Klebeband *n*, Klebstreifen *m*

4752 SELF-CLEANSING sew
f auto-épuration *f*
e autolimpieza *f*
i autodepurazione *f*
n zelfreiniging
d Selbstreinigung *f*

4753 SELF CORROSION, cor
LOCAL ACTION, LOCAL CORROSION
f action *f* locale
e acción *f* local
i azione *f* locale
n plaatselijke inwerking
d örtliche Einwirkung *f*

4754 SELF-CURING, rub
SELF-VULCANIZING
f autovulcanisant adj
e autovulcanizante adj
i autovulcanizzante adj
n zelfvulcaniserend adj
d selbstvulkanisierend adj

4755 SELF-HARDENING, met
AIR HARDENING
f trempe *f* à l'air
e temple *m* al aire
i tempera *f* all'aria
n harding aan de lucht
d Lufthärtung *f*

4756 SELF-SANITIZING PAINT, dec
GERM-REPELLING PAINT,
GERMICIDAL PAINT
f peinture *f* biocide
e pintura *f* germicida
i pittura *f* germicida
n biocide verf, kiemdodende verf
d biologische Selbstschutzfarbe *f*, biozide Farbe *f*, keimtötende Farbe *f*

4757 SELF-TONING PAPER pho
f papier *m* à autovirage
e papel *m* autovirador
i carta *f* autovirante
n zelfkleurend papier *n*
d selbsttonendes Papier *n*

4758 SEMI-WATERGAS gas
f gaz *m* mixte de gazogène
e gas *m* mixto de gasógeno
i gas *m* misto di gasogeno
n halfwatergas *n*
d Halbwassergas *n*

4759 SENEGAL GUM, glu
ACACIA GUM, ARABINE,
GUM ARABIC
f gomme *f* arabique, gomme *f* d'acacia
e goma *f* arábiga, goma *f* Senegal
i gomma *f* arabica, gomma *f* d'acacia
n acaciagom, Arabische gom, kordofangom, senegalgom
d Akaziengummi *m*, Arabin *n*, Gummiarabicum *n*, Senegalgummi

4760 SENSITIVENESS TO PERCUSSION exp
f sensibilité *f* à la percussion
e sensibilidad *f* a la percusión
i sensibilità *f* alla percussione
n slaggevoeligheid
d Schlagempfindlichkeit *f*

4761 SENSITIZATION, pho
SENSITIZING
f sensibilisation *f*
e sensibilización *f*
i sensibilizzazione *f*
n sensibilisatie
d Sensibilisierung *f*

4762 SEPARATION sep
f séparation *f*
e separación *f*
i separazione *f*
n scheiding
d Scheidung *f*, Trennung *f*

4763 SEPARATION sor
f classification *f*, triage *m*
e clasificación *f*, tría *f*
i cernita *f*, classificazione *f*
n classificeren *n*, sorteren *n*
d Klassieren *n*, Sichten *n*, Sortieren *n*

4764 SEPARATION OF MIXTURES mix,sep

f séparation *f* des constituents d'un mélange
e separación *f* de los constituentes de una mezcla
i separazione *f* dei costituenti d'una miscela
n ontmenging
d Entmischung *f*

4765 SEPIOLITE, MEERSCHAUM mat
f écume *f* de mer, sépiolite *f*
e espuma *f* de mar, sepiolita *f*
i schiuma *f* di mare, sepiolite *f*
n meerschuim *n*, sepioliet *n*
d Meerschaum *m*, Sepiolith *m*

4766 SEQUENCE ETCHING elc
f attaque *f* à l'acide composé
e ataque *m* al ácido compuesto
i attacco *m* all'acido composto
n samengestelde etsing
d Folgeätzung *f*

4767 SEQUESTERING AGENT, SEQUESTRANT, CHELATING AGENT, COMPLEXING AGENT gen
f agent *m* séquestrant complexant, chélatant *m*
e agente *m* secuestrador reactivo, agente *m* secuestrante, formador *m* de complejos
i agente *m* chelante, agente *m* sequestrante
n complexvormer
d chelatbildendes Reagenz *n*, Komplexbildner *m*

4768 SEQUESTRATION, CHELATE FORMATION, CHELATION, COMPLEXING ACTION oic
f action *f* complexante
e acción *f* quelacante, facultad *f* de formar complejos
i azione *f* complessante, azione *f* sequestrante, chelazione *f*
n chelateren *n*, complexvorming, klauwkoppeling, klauwverbinding
d Chelatbildung *f*, Chelierung *f*, Komplexbildung *f*

4769 SERIGRAPHY, SCREEN PRINTING tex
f impression *f* à la pellicule
e estampación *f* con estarcido
i stampaggio *m* alla pellicola
n zeefdruk
d Netzdruck *m*, Siebdruck *m*

4770 SERUM mcy
f sérum *m*
e suero *m*
i siero *m*
n serum *n*
d Serum *n*

4771 SESAME OIL, BENNE OIL fat
f huile *f* de sésame
e aceite *m* de sésamo
i olio *m* di sesamo
n sesamolie
d Sesamöl *n*

4772 SETTING OF CEMENT bui
f prise *f* du ciment
e fraguado *m* del cemento
i presa *f* del cemento
n binden *n* van cement
d Abbinden *n* von Zement

4773 SETTING OUT lea
f étirage *m*
e estriado *m*
i stiratura *f*
n strekken *n*
d Strecken *n*

4774 SETTING POINT lub
f point *m* de congélation
e punto *m* de fluidez crítica
i punto *m* di congelazione
n stolpunt *n*
d Stockpunkt *m*

4775 SETTING-POINT TEST lub
f essai *m* du point de congélation
e ensayo *m* del punto de fluidez crítica
i collaudo *m* del punto di congelazione
n stolpuntproef
d Stockpunktprobe *f*

4776 SETTLING, SEDIMENTATION cle
f sédimentation *f*
e sedimentación *f*
i sedimentazione *f*
n laten bezinken *n*, sedimentatie
d Absetzung *f*, Sedimentation *f*

4777 SETTLING TEST, SLUMP TEST bui
f essai *m* d'affaisement au cône d'Abrams
e ensayo *m* de aplastamiento, prueba *f* de trabajabilidad
i prova *f* di consistenza del cemento

n zettingsproef
d Kegelprüfverfahren *n*, Senkungsprobe *f*, Setzungsprobe *f*

4778 SEWAGE, WASTE WATER — iwt
f eau *f* résiduaire, eaux *pl* d'égout
e agua *f* residual, aguas *pl* cloacales, aguas *pl* sucias
i acqua *f* di rifiuto, fognatura *f*
n afvalwater *n*
d Abwasser *n*

4779 SEWAGE GAS, SLUDGE GAS — sew
f gaz *m* de boue, gaz *m* d'égout
e gas *m* cloacal, gas *m* de alcantarilla
i gas *m* di fogna
n rioolgas *n*
d Klärgas *n*

4780 SEWAGE SLUDGE — fer,iwt
f boue *f* des eaux d'égout
e fango *m* cloacal
i fango *m* delle acque di fogna
n rioolslib *n*
d Klärschlamm *m*

4781 SGRAFFITO, GRAFFITO — cer,dec
f sgraffite *m*
e esgrafiado *m*, obra *f* esgrafiada
i graffito *m*
n graffito *m*
d Sgraffitoarbeit *f*

4782 SHADE, HUE — dec
f teinte *f*, tonalité *f* chromatique
e tinte *m*, tonalidad *f* cromática
i tinta *f*, tonalità *f* cromatica
n kleurtoon
d Farbton *m*, Ton *m*

4783 SHADING, TINTING — dec
f mise *f* à la teinte, nuançage *m*
e matizadura *f*
i sfumatura *f*
n bijtinten *n*, op kleur brengen *n*
d Abtönen *n*, Nachtönen *n*, Nuanzieren *n*

4784 SHADOWING, SHIELDING — elc
f effet *m* d'écran
e efecto *m* de pantalla
i effetto *m* di schermo
n afscherming
d Abschirmung *f*, Blenden *n*

4785 SHAGREEN, GRAIN LEATHER — lea
f chagrin *m*
e cuero *m* granulado, cuero *m* chagrinado
i pelle *f* di pescecane
n chagrijnleder *n*, segrijnleder *n*, Turks leder *n*
d Körnerleder *n*, Narbenleder *n*

4786 SHAGREENING, GRAINING — lea
f crépissage *m*
e chagrinado *m*, granelación *f*
i spaccatura *f*
n chagrineren *n*
d Narben *n*

4787 SHAKING — mix
f secouage *m*
e sacudida *f*
i scottimento *m*
n schudden *n*
d Schütteln *n*

4788 SHAKING DOWN (US), STIRRING UP (GB) — mix
f agitation *f* du bain, brasure *f*
e agitación *f* del baño, braceado *m*
i agitazione *f* del bagno, rimescolata *f*
n omroeren *n*
d Rühren *n*

4789 SHALE — bui,mix
f schiste *m*
e esquisto *m*
i schisto *m*
n leisteen *n*
d Schiefer *m*, Schiefergestein *n*

4790 SHALE CLAY — min
f argile *f* de schiste
e arcilla *f* esquitosa
i argilla *f* di schisto
n leisteenklei
d Schieferton *m*

4791 SHALE OIL — fue
f huile *f* de schiste
e aceite *m* de esquisto
i olio *m* di schisto
n leisteenolie
d Schieferöl *n*

4792 SHALLOW HARDENING — met
f durcissement *m* superficiel

e endurecimiento *m* superficial
i indurimento *m* superficiale
n oppervlakkige harding
d Oberflächenhärtung *f*

4793 SHAPING pla,pre
f formage *m*, profilage *m*
e conformación *f*, perfilado *m*
i conformazione *f*, profilatura *f*
n profileren *n*, vormen *n*
d Formen *n*, Profilieren *n*

4794 SHARK LIVER OIL mcy,nut
f huile *f* de foie de requin
e aceite *m* de hígado de tiburón
i olio *m* di fegato di pescecane
n haaieleverolie
d Haifischtran *m*

4795 SHARP PAINT dec
f peinture *f* à séchage rapide
e pintura *f* magra
i pittura *f* ad essiccazione rapida
n sneldrogende verf, terpentijnrijke verf
d schnelltrocknende Farbe *f*

4796 SHARPENED LIME TREATMENT tan
f pelainage *m* renforcé
e pelambre *m* intenso
i macerazione *f* rinforzata
n geconcentreerd kalken *n*
d angeschärftes Äschern *n*

4797 SHARPENING tex
f élimination *f* d'oxygène
e eliminación *f* de oxígeno
i eliminazione *f* d'ossigeno
n zuurstofverwijdering
d Sauerstoffbeseitigung *f*

4798 SHATTERING POWER, BRISANCE, DETONATING VIOLENCE exp
f pouvoir *m* brisant
e brisancia *f*, efecto *m* rompedor
i effetto *m* dirompente, effetto *m* esplosivo
n brisante werking, brisantie
d Brisanz *f*

4799 SHATTERING TEST, BRISANCE TEST, VIOLENCE TEST exp
f essai *m* du pouvoir brisant
e prueba *f* de la brisancia
i collaudo *m* dell'effetto dirompente
n brisantieproef
d Brisanzprüfung *f*

4800 SHAVING lea
f dolage *m*, planage *m*
e adelgazado *m*
i raschiatura *f*
n schaven *n*
d Dollieren *n*

4801 SHAVING CREAM fat
f crème *f* à raser
e crema *f* de afeitar
i crema *f* per barba
n scheercrème
d Rasierseifencreme *f*

4802 SHAVING SOAP soa
f savon *m* à raser
e jabón *m* de afeitar
i sapone *m* da barba
n scheerzeep
d Rasierseife *f*

4803 SHEARINESS dec
f effet *m* gras
e efecto *m* opalescente
i lattescenza *f*
n vettigheid
d Fettigkeit *f*

4804 SHEATHING COMPOUND pla
f composé *m* pour gaines
e compuesto *m* para vainas
i composto *m* per guaine
n verbinding voor ommanteling
d Verbindung *f* für Armierung

4805 SHEEN, GLOSS, LUSTRE dec,tex
f brillant *m*, éclat *m*, luisance *f*, lustre *m*
e brillo *m*, lustre *m*
i lucentezza *f*, lustro *m*, splendore *m*
n glans
d Glanz *m*

4806 SHEET GLASS gla
f verre *m* à vitres
e vidrio *m* en hojas, vidrio *m* plano
i lastra *f* di vetro, vetro *m* in lastre
n vensterglas *n*
d Tafelglas *n*

4807 SHEET IRON met
f tôle *f* en fer
e chapa *f* de hierro
i lamiera *f* di ferro
n blik *n*, plaatijzer *n*
d Eisenblech *n*

4808 SHEET LEAD bui
f feuille *f* de plomb, plomb *m* laminé
e chapa *f* de plomo, plomo *m* en hojas
i piombo *m* in foglie
n bladlood *n*, loodfoelie
d Bleiblech *n*, Walzblei *n*

4809 SHEET POLYMER pla
f polymère *f* à branchement tridimensionnel
e polímero *m* de ramificación tridimensional
i polimero *m* a derivazione tridimensionale
n polymeer *n* met driedimensionale netvorming
d Polymer *n* mit dreidimensionaler Netzbildung

4810 SHEETING pla,rub
f laminage *m*, mise *f* en feuilles
e laminación *f*
i laminazione *f*
n uitwalsen *n*, walsen *n* van vellen
d Auswalzen *n*

4811 SHELL FLOUR FILLERS pla
f charge *f* en coquilles moulues
e carga *f* de cáscaras molidas
i carica *f* in gusci macinati
n vulstoffen *pl* uit gemalen doppen
d Füllstoffe *pl* aus gemahlenen Schalen

4812 SHELL LIME bui
f chaux *f* de coquilles
e cal *f* de conchas
i calce *f* di conchiglie
n schelpkalk
d Muschelkalk *m*

4813 SHELLAC, LAC dec
f gomme *f* laque
e goma *f* laca
i gomma *f* lacca
n schellak
d Schellack *m*

4814 SHERARDIZING coa
f shérardisation *f*
e sherardización *f*
i sherardizzazione *f*
n sherardiseren *n*
d Sherardisieren *n*

4815 SHIELDED-ARC WELDING wel
f soudure *m* à arc protégé
e soldadura *f* por arco protegido
i saldatura *f* ad arco schermato
n afgeschermd booglassen *n*
d Schutzgasschweissen *n*

4816 SHIELDING, SCREENING sap
f blindage *m*
e blindaje *m*
i schermatura *f*
n afscherming
d Abschirmung *f*

4817 SHIELDING, SHADOWING elc
f effet *m* d'écran
e efecto *m* de pantalla
i effetto *m* di schermo
n afscherming
d Abschirmung *f*, Blenden *n*

4818 SHIELDING GLASS nue
f verre *m* protecteur contre le rayonnement ionisant
e vidrio *m* protector contra la radiación ionizante
i vetro *m* protettore contro la radiazione ionizzante
n tegen ioniserende straling beschermend glas *n*
d gegen ionisierender Strahlung schützendes Glas *n*

4819 SHIVERING, PEELING OF GLAZE cer
f écaillage *m* de l'émail
e descamación *f* en el vidriado
i scheggiatura *f* dell'invetriatura
n schilferen *n* van de glazuur
d Abrollen *n* der Glasur

4820 SHOCK TEST tes
f essai *m* à chocs
e ensayo *m* de sacudidas
i collaudo *m* a scosse
n schokproef
d Stossprobe *f*

4821 SHODDY fer,tex
f laine *f* d'effilochage, laine *f* renaissance
e trapos *pl* de lana
i lana *f* di scarto
n wolafval
d Lumpenwolle *f*, Shoddy *m*

4822 SHORT-GROWN MALT bre
f malt *m* peu germé
e malta *f* poco germinada
i malto *m* poco germogliato

n kortmout *n*
d Kurzmalz *m*

4823 SHORT LIQUOR dye
f bain *m* de teinture à volume de liquide limité
e baño *m* de teñido con volumen de líquido limitado
i bagno *m* di tintura con volume di liquido limitato
n verfbad *n* met laag vloeistofvolume
d Färbebad *n* mit niedrigem Flüssigkeitsanteil

4824 SHORT-OIL VARNISH dec
f vernis *m* à faible teneur en huile
e barniz *m* con poco aceite
i vernice *f* a basso tenore d'olio
n magere vernis
d kurzöliger Klarlack *m*, magerer Klarlack *m*

4825 SHORT RESIDUE fue
f résidu *m* court, résidu *m* de distillation poussée
e residuo *m* corto, residuo *m* liviano
i residuo *m* corto
n van smeeroliefracties bevrijd residu *n*
d Vakuumrückstand *m*

4826 SHORT SHOT pla
f poids *m* injectable insuffisant
e peso *m* de un inyectado insuficiente
i peso *m* di un'iniezione insufficiente
n te klein vulgewicht *n*
d zu kleines Füllgewicht *n*

4827 SHORTERIZING, htt,met
FLAME HARDENING,
TORCH HARDENING
f trempe *f* au chalumeau
e temple *m* superficial a la llama
i tempera *f* alla fiamma
n vlamharding
d Flammenhärtung *f*, Oberflächenhärtung *f*

4828 SHORTS, sor
SIEVE RESTS, SCREENINGS
f débris *pl* de tamisage
e desperdicios *pl* del cribado, granzas *pl*
i residuo *m* sullo staccio
n zeefrest
d Siebrückstand *m*, Siebüberlauf *m*

4829 SHOT, met
SHOT METAL
f grenaille *f*
e granalla *f*
i graniglia *f*
n schroot *n*
d Schrot *m*

4830 SHOT (GB), pla
SHOT WEIGHT (US)
f capacité *f* par coup, poids *m* injectable
e peso *m* de un inyectado
i peso *m* di un'iniezione
n vulgewicht *n*
d Füllgewicht *n*

4831 SHOT BLASTING, cle
GRIT BLASTING
f grenaillage *m*
e rebarbado *m* por chorro de arenilla
i granigliatura *f*
n gritstralen *n*, gruisstralen *n*
d Abstrahlen *n*, Kiesstrahlen *n*, Schleuderstrahlen *n*

4832 SHOT PEENING cle
f grenaillage *m* à écrouissage
e golpo *m* con granalla
i pallinatura *f*
n staalstralen *n*
d Granalienstrahlen *n*, Strahlhämmerung *f*

4833 SHOWER PROOFING tex
f imprégnation *f* hydrofuge
e impregnación *f* hidrófuga
i impregnazione *f* idrofuga
n waterafstotend impregneren *n*
d wasserabstossendes Imprägnieren *n*

4834 SHREDDING oic
f déchiquetage *m*
e desmenuzamiento *m*, trituración *f*
i trinciatura *f*
n scheuren *n*
d Zerkleinerung *f*

4835 SHREDDING pap
f dilacération *f*
e desfibración *f*
i dilaceratura *f*
n lompen snijden *n*, vezelen *n*
d Zerfasern *n*

4836 SHRINK WRAPPING pla
f emballage *m* à rétrécissement

e embalaje *m* de encogemiento
i imballaggio *m* a ritiro
n krimpverpakking
d Schwundverpackung *f*

4837 SHRINKABLE FILM pla
f pellicule *f* rétrécissable
e película *f* de politeno enrogible
i pellicola *f* restringibile
n warmkrimpende film
d einlaufende Folie *f*

4838 SHRINKAGE, pla
CONTRACTION
f retrait *m*
e contracción *f*, retracción *f*
i ritiro *m*
n krimpen *n*
d Schrumpfen *n*, Schwinden *n*, Schwund *m*

4839 SHRINKAGE OF CLOTH tex
f rentrée *f* de tissu, rétrécissement *m* du tissu
e encogimiento *m* de la tela
i restringimento *m* del tessuto
n krimpen *n* van weefsel
d Eingehen *n* des Gewebes

4840 SHRINKAGE WATER cer
f eau *f* colloïdale
e agua *f* coloidal
i acqua *f* di ritiro
n krimpwater *n*
d Schwindwasser *n*

4841 SHRINKPROOFING tex
f traitement *m* irrétrécissable
e tratamiento *m* antiencogimiento
i trattamento *m* antiritiro
n krimpvrij maken *n*
d Krumpffreimachen *n*

4842 SICCATIVE, dec
DRIER
f siccatif *m*
e secante *m*, secativo *m*
i essiccativo *m*
n droogmiddel *n*, siccatief *n*
d Trockenmittel *n*, Sikkativ *n*

4843 SIDE CUT dis,fue
f fraction *f* latérale
e fracción *f* lateral
i frazione *f* laterale
n laterale fractie
d Seitenfraktion *f*

4844 SIDE REACTION gen
f réaction *f* secondaire
e reacción *f* secundaria
i reazione *f* secondaria
n nevenreactie
d Nebenreaktion *f*

4845 SIDESTREAM dis
f soutirage *m* latéral
e toma *f* lateral
i prelievo *m* laterale
n zijaftap
d Nebenstrom *m*, Seitenstrom *m*

4846 SIENNA dec
f terre *f* de Sienne
e tierra *f* de Siena
i terra *f* di Siena
n siënna, terrasiënna
d Sienaerde *f*

4847 SIEVE ANALYSIS, ana
SCREEN ANALYSIS
f analyse *f* par tamisage
e análisis *f* por tamizado
i analisi *f* granulometrica
n zeefanalyse
d Siebanalyse *f*

4848 SIEVING, sor
SIFTING, SCREENING
f criblage *m*, tamisage *m*
e cribado *m*, tamizado *m*
i stacciatura *f*, vagliatura *f*
n zeven *n*, ziften *n*
d Sichten *n*, Sieben *n*

4849 SIGHTING COLOURS (GB), dec
FUGITIVE TINTING COLORS (US)
f couleurs *pl* fugitives
e colores *pl* fugaces
i colori *pl* che sbiadiscono facilmente
n onbestendige kleurstoffen *pl*
d unbeständige Farbstoffe *pl*

4850 SILICA GLASS, gla
VITREOUS SILICA
f quartz *m* fondu
e cuarzo *m* fundido
i quarzo *m* fuso
n gesmolten kwarts *n*
d geschmolzener Quarz *m*

4851 SILICATE, nut,sap
LIQUID GLASS (US), SOLUBLE GLASS, WATER GLASS (GB)
f silicate *m* de sodium, verre *m* soluble
e silicato *m* sódico líquido, vidrio *m* soluble
i tetrasilicato *m* di sodio

n silicaat *n*, waterglas *n*
d Wasserglas *n*

4852 SILICATE CEMENT aic
f ciment *m* à verre soluble
e cemento *m* de vidrio soluble
i cemento *m* a tetrasilicato di sodio
n waterglaskit
d Wasserglaskitt *m*

4853 SILICATE PAINT dec
f peinture *f* au silicate, peinture *f* silicatée
e pintura *f* de silicato
i pittura *f* al silicato
n silicaatverf
d Silikatfarbe *f*

4854 SILICIFICATION min
f silicification *f*
e silicificación *f*
i silicificazione *f*
n silificatie, verkiezeling
d Silizifikation *f*, Verkieselung *f*

4855 SILICONIZING met
f formation *f* de siliciure
e siliconización *f*
i formazione *f* di siliciuro
n siliconiseren *n*
d Silizieren *n*

4856 SILK tex
f soie *f*
e seda *f*
i seta *f*
n zijde
d Seide *f*

4857 SILK DEGUMMING, SILK SCOURING tex
f dégommage *m* de la soie
e desgomación *f* de la seda
i sgommatura *f* della seta
n zijdeontgomming
d Seidendegummierung *f*, Seidenentbastung *f*, Seidenentschälung *f*

4858 SILK LOADING, SILK WEIGHTING tex
f chargement *m* de la soie
e cargado *m* de la seda
i carica *f* della seta
n zijdeverzwaring
d Seidenbeschwerung *f*

4859 SILK-SCREEN PRINTING PROCESS pri
f procédé *m* à l'écran de soie
e procedimiento *m* de impresión con estarcido de seda
i processo *m* di stampatura con schermo di seta
n gaasdrukproces *n*, zeefdrukproces *n*
d Siebdruckverfahren *n*

4860 SILKING dec
f éclat *m* soyeux
e aspecto *m* sedoso
i aspetto *m* di seta
n zijdeglans
d seidenartiges Aussehen *n*

4861 SILT, MUD, SLUDGE iwt
f boues *pl*, limon *m* épais
e fangos *pl*
i fanghi *pl*, melma *f*, mota *f*
n modder, slijk *n*
d Schlamm *m*, Schlick *m*

4862 SILTATION, SILTING iwt
f envasement *m*
e aterramiento *m*
i interrimento *m*
n dichtslibbing
d Verschlammung *f*

4863 SILVER BRAZING wel
f soudure *f* forte à l'argent
e platasoldadura *f*, soldadura *f* fuerte con plata
i brasatura *f* all'argento
n hardsolderen *n* met zilver
d Hartlötung *f* mit Silber

4864 SILVER BROMIDE PAPER, BROMIDE PAPER pho
f papier *m* au bromure d'argent
e papel *m* al bromuro de plata
i carta *f* al bromuro d'argento
n broomzilverpapier *n*
d Bromsilberpapier *n*

4865 SILVER PLATING coa,elc
f argenture
e plateado *m*
i argentatura *f*
n verzilveren *n*
d Versilbern *n*

4866 SILVER WHITE, FRENCH WHITE dec
f blanc *m* d'argent
e blanco *m* de plata
i bianco *m* d'argento
n zuiver loodwit *n*
d reinstes Bleiweiss *n*

4867 SILVERING coa
f argenture *f* chimique
e plateado *m* químico
i argentatura *f* chimica
n chemisch verzilveren *n*
d chemisches Versilbern *n*

4868 SINGEING, tex
GASSING
f flambage *m*, gazage *m*
e chamusquina *f*, gaseado *m*
i bruciatura *f*, gasatura *f*
n afzengen *n*, zengen *n*
d Flambieren *n*, Gasieren *n*, Sengen *n*, Versengen *n*

4869 SINGLE-EFFECT EVAPORATION eva
f évaporation *f* à effet simple
e evaporación *f* de efecto simple
i evaporazione *f* ad effetto semplice
n eenvoudige verdamping
d Einkörperverdampfung *f*, Einstufenverdampfung *f*

4870 SINGLE MASH PROCESS bre
f méthode *f* de brassage à une trempe
e sistema *m* de cocimiento de una mezcla
i sistema *m* d'impasto semplice
n enkelvoudig beslagproces *n*
d Einmaischverfahren *n*

4871 SINKAGE tex
f perte *f* de poids
e pérdida *f* de peso
i perdita *f* di peso
n gewichtsverlies *n*
d Gewichtsverlust *m*

4872 SINTERED CARBIDE, met
CEMENTED CARBIDE
f carbure *m* fritté
e carburo *m* sinterizado
i carburo *m* sinterizzato
n sintercarbid *n*
d Metallkarbid *n*, Sinterhartmetall *n*, Sinterkarbid *n*

4873 SINTERED GLASS, gla
FRITTED GLASS
f verre *m* fritté
e vidrio *m* fritado
i vetro *m* sinterizzato
n gesinterd glas *n*
d Sinterglas *n*

4874 SINTERING, htt,met
FRITTING
f frittage *m*
e sinterización *f*
i sinterizzazione *f*
n sinteren *n*
d Sintern *n*

4875 SIRUP (US), sin
SYRUP (GB)
f sirop *m*
e jarabe *m*
i sciroppo *m*
n stroop
d Sirup *m*

4876 SIZE dec
f colle *f* animale
e cola *f* animal
i colla *f* animale
n dierlijke lijm
d Tierleim *m*

4877 SIZE gen
f dimension *f*
e tamaño *m*
i dimensione *f*
n afmeting
d Mass *n*

4878 SIZE tex
f apprêt *m*
e apresto *m*
i agente *m* imbozzimatore
n slichtmiddel *n*
d Schlichtmittel *n*

4879 SIZE FRACTION, gen
PARTICLE-SIZE FRACTION
f fraction *f* granulométrique
e fracción *f* granulométrica
i frazione *f* granulometrica
n deeltjesfractie
d Kornfraktion *f*, Teilchenfraktion *f*

4880 SIZE REDUCTION, brk
BREAKING
f broyage *m*
e quebrantamiento *m*
i frantumazione *f*
n vermaling
d Zerkleinerung *f*

4881 SIZE WATER dec
f eau *f* de colle
e agua *f* de cola
i acqua *f* di colla
n lijmwater *n*
d Leimwasser *n*

4882 SIZING dec
f encollage *m*

e encoladura *f*
i incollatura *f*
n lijmen *n*
d Leimen *n*

4883 SIZING pap
f engommage *m*
e engomadura *f*
i incollatura *f*
n lijmen *n*
d Leimen *n*

4884 SIZING sor
f classification *f* volumique
e clasificación *f* volumétrica
i classificazione *f* volumetrica
n uitzeven *n*
d Sortieren *n* nach der Grösse

4885 SIZING tex
f apprêt *m*
e apresto *m*
i imbozzimatura *f*
n slichten *n*
d Schlichten *n*

4886 SIZING, tex
DRESSING, FINISHING
f apprêtage *m*
e apresto *m*
i apprettatura *f*
n appreteren *n*
d Appretur *f*

4887 SKEIN, tex
HANK
f échevan *m*
e madeja *f*
i matassa *f*
n streng
d Strähne *f*, Strang *m*

4888 SKIM MILK, dai
SKIMMED MILK
f lait *m* écrémé
e leche *f* desnatada
i latte *m* scremato, latte *m* sfiorato
n ondermelk, taptemelk
d Magermilch *f*

4889 SKIMMERS, bre
TAILINGS
f graines *pl* folles
e cebada *f* flotante
i orzo *m* galleggiante
n drijvende gerst
d Schwimmgerste *f*

4890 SKIMMING dai
f écrémage *m*
e desnatadura *f*
i scrematura *f*, sfioratura *f*
n afromen *n*, ontromen *n*
d Abrahmen *n*, Entrahmen *n*

4891 SKIMMING met
f écumage *m*
e espumadura *f*
i schiumatura *f*
n afschuimen *n*
d Abschäumen *n*, Entschäumen *n*

4892 SKIMMING (GB), met
DROSSING (US)
f décrassage *m*, écrémage *m*
e despumación *f*
i rimozione *f* della feccia
n afschuimen *n*
d Schlickern *n*

4893 SKIMMINGS bre
f mélange *m* de bière et levure
e mezcla *f* de levadura y cerveza
i schiuma *f* di birra e lievito
n afschuimsel *n*, bier-gistmengsel *n*
d Abgeschäumtes *n*

4894 SKIN DRYING met
f flambage *m*
e secado *m* a la llama
i fiammatura *f*
n afvlammen *n*, drogen *n* met de vlam
d Abbrennen *n*, Abflammen *n*

4895 SKINNING dec
f formation *f* de peaux
e formación *f* de costras, formación *f* de natas
i formazione *f* di pellicole
n velvorming
d Hautbildung *f*

4896 SKIVER lea
f peau *f* fendue
e cuero *m* rajado con cuchilla
i strato *m* superiore di crosta
n bovenste laag van splitleer
d oberste Schicht *f* von Spaltleder

4897 SLACK LIME, aic
SLAKED LIME, CHALK LIME
f chaux *f* éteinte
e cal *f* apagada
i calce *f* spenta
n gebluste kalk
d Äscherkalk *m*, Gerberkalk *m*, Löschkalk *m*

4898 SLACK MALT bre
f malt *m* trop humide
e malta *f* húmeda

i malto *m* che ha assorbito umidità
n kleffig mout *n*
d angezogenes Malz *n*

4899 SLACK WAX, fat
CRUDE WAX, PETROLEUM WAX
f paraffine *f* brute, paraffine *f* non-déshuilée
e cera *f* parafínica muy aceitosa, mezcla *f* de cera parafínica y de petróleo
i paraffina *f* greggia, paraffina *f* molle
n ruwe paraffine, uit aardolie verkregen paraffine
d Rohparaffin *n*

4900 SLAG, gen
CINDER, CLINKER
f laitier *m*, mâchefer *m*, scorie *f*
e escoria *f*
i scoria *f*
n sintel, slak
d Schlacke *f*

4901 SLAG BRICK, bui
BREEZE BRICK
f brique *f* en ciment et braise de houille
e ladrillo *m* de cemento y cenizas de coque
i mattone *m* di cemento e cenere di coke
n slakkensteen
d Schlackenstein *m*

4902 SLAG CEMENT bui
f ciment *m* de laitier
e cemento *m* de escoria
i cemento *m* di scoria
n hoogovencement *n*, slakkencement *n*
d Hochofenzement *m*, Schlackenzement *m*

4903 SLAG-OFF, htt,met
SLAGGING
f décrassage *m*, décrassement *m*
e desescorificación *f*
i scarico *m* delle scorie, scorificazione *f*
n afslakken *n*, ontslakken *n*
d Abschlacken *n*, Entschlackung *f*

4904 SLAG WOOL, clo
MINERAL WOOL
f laine *f* de laitier
e lana *f* de escorias
i lana *f* di scorie
n slakkenwol
d Schlackenwolle *f*

4905 SLAKING bui
f extinction *f*
e apagado *m*
i spegnimento *m*
n blussen *n*
d Löschung *f*

4906 SLATE bui,min
f ardoise *f*
e pizarra *f*
i ardesia *f*, lavagna *f*
n lei *n*
d Schiefer *m*

4907 SLEDGING, brk
SCALPING
f broyage *m* préliminaire
e quebrantamiento *m* preliminar
i frantumazione *f* preliminare
n grof breken *n*
d Grobbrechen *n*

4908 SLEEKING, met
SEARING
f lissage *m*
e alisado *m*
i lisciatura *f*
n gladmaken *n*
d Glätten *n*

4909 SLEEPINESS dec
f manque *m* de lustre
e aspecto *m* deslustrado
i appannatura *f*
n glansverlies *n*
d Glanzverlust *m*

4910 SLICING OF SUGAR BEETS sin
f découpage *m* de betteraves
e rebanado *m* de remolachas
i affettatura *f* di barbabietole
n snijden *n* van bieten
d Schnitzeln *n* von Rüben

4911 SLIME elc
f boue *f*
e limo *m*
i fanghi *pl* di deposizione
n slib *n*
d Rückstand *m*, Schlamm *m*

4912 SLIME min
f fines *pl*, limon *m*
e finos *pl*
i melma *f*
n slib *n*
d Schlamm *m*, Schlick *m*

4913 SLINGING cer
f lancement *m*
e eslingaje *m*
i lanciamento *m*
n slingeren *n*
d Schleudern *n*

4914 SLIP cer
f barbotine *f*, pâte *f* d'argile
e barro *m*
i pasta *f* d'argilla
n kleibrij, paat
d Schlicker *m*, Tonerdebrei *f*, Töpferbrei *f*

4915 SLIP ADDITIVE, SLIP AGENT pla
f agent *m* anti-adhésif
e agente *m* antiadhesivo
i agente *m* antiadesivo
n antiplakmiddel *n*
d Antihaftmittel *n*

4916 SLOP, DISTILLER'S WASH, SPENT MASH, STILLAGE, VINASSE nut
f drêches *pl*, vinasses *pl* de distillerie
e orujo *m*, vinazas *pl* de destilación
i burlanda *f*, trebbia *f*, vinacce *pl* distillate
n drab
d abgebrannte Würze *f*, Schlempe *f*, Spülicht *n*

4917 SLOP PADDING tex
f imprégnation *f* uniforme
e impregnación *f* uniforme
i impregnazione *f* uniforme
n gelijkmatige impregnering
d gleichmässige Imprägnierung *f*

4918 SLOT EXTRUSION, FLAT SHEET EXTRUSION pla
f extrusion *f* à filière plate
e extrusión *f* de laminas
i estrusione *f* di foglie piane
n foeliespuitmethode
d Breitbandspritzverfahren *n*

4919 SLOW ACCELERATOR, SLOW ACTION ACCELERATOR rub
f accélérateur *m* lent
e acelerador *m* lento
i accelerante *m* lento
n langzame versneller
d langsamer Beschleuniger *m*

4920 SLOW FERMENTATION, SLUGGISH FERMENTATION bre,fmi
f fermentation *f* lente
e fermentación *f* lenta
i fermentazione *f* lenta
n langzame gisting
d träge Gärung *f*

4921 SLOW SETTING CEMENT bui
f ciment *m* à prise lente
e cemento *m* de fraguado lento
i cemento *m* a presa lenta
n langzaam bindend cement *n*
d Langsambinder *m*

4922 SLUDGE, MUD, SILT iwt
f boues *pl*, limon *m* épais
e fangos *pl*
i fanghi *pl*, melma *f*, mota *f*
n modder, slijk *n*
d Schlamm *m*, Schlick *m*

4923 SLUDGE CONDITIONING, SLUDGE SEASONING iwt,sew
f amélioration *f* de boues, préparation *f* de boues, traitement *m* de boues
e tratamiento *m* de fangos
i trattamento *m* di fanghi
n slibverwerking
d Schlammaufarbeitung *f*

4924 SLUDGE GAS, SEWAGE GAS sew
f gaz *m* de boue, gaz *m* d'égout
e gas *m* cloacal, gas *m* de alcantarilla
i gas *m* di fogna
n rioolgas *n*
d Klärgas *n*

4925 SLUDGE REMOVAL, DESILTING, DESLUDGING iwt
f débourbage *m*, enlèvement *m* de boue, évacuation *f* de boue
e desenlodamiento *m*
i scarico *m* di fango
n ontslibbing
d Entschlammung *f*, Schlammabscheidung *f*

4926 SLUDGING iwt
f débourbage *m*
e separación *f* de fango
i separazione *f* di fango
n slibafscheiding
d Schlammabscheidung *f*

4927 SLUMP TEST, SETTLING TEST bui
f essai *m* d'affaisement au cône d'Abrams

e ensayo *m* de aplastamiento, prueba *f* de trabajabilidad
i prova *f* di consistenza del cemento
n zettingsproef
d Kegelprüfverfahren *n*, Senkungsprobe *f*, Setzungsprobe *f*

4928 SLURRY cer
f barbotine *f*, pâte *f*
e barbotina *f*, pasta *f*
i impasto *m* liquido, malta *f* liquida, pasta *f*
n dikvloeibare massa, slurrie
d Dickstoff *m*, Schlicker *m*, Tonschlamm *m*

4929 SLUSH MOLDING (US), SLUSH MOULDING (GB) pla
f moulage *m* de pâtes
e modelado *m* de pastas coquilla
i formatura *f* in conchiglia
n pastagietmethode, plastisolverwerking
d Plastisolverarbeitung *f*, Schalenguss *m*

4930 SMALT dec
f émail *m* de cobalt
e esmaltina *f*
i smaltino *m*
n smalt, smaltblauw *n*
d Kobaltblau *n*, Smalte *f*

4931 SMALT gla
f smalt *m*, verre *m* à oxyde de cobalt
e esmalte *m*
i smalto *m*
n kobaltglas *n*, smalt
d Blaufarbenglas *n*, Smalte *f*

4932 SMALTING cer
f azurage *m*
e teñido *m* de azul
i tintura *f* d'azzurro
n blauw verven *n*
d Blaufärben *n*

4933 SMELL, ODOR (US), ODOUR (GB), SCENT per
f odeur *f*
e olor *m*
i odore *m*
n geur, reuk
d Duft *m*, Geruch *m*

4934 SMELTING, MELTING gen
f fusion *f*
e fusión *f*
i fusione *f*
n smelten *n*
d Schmelzen *n*

4935 SMELTING FLUX ELECTROLYSIS elc
f électrolyse *f* à électrolyte fondu
e electrólisis *f* con electrólito fundido
i elettrolisi *f* ad elettrolito fuso
n elektrolyse met gesmolten elektrolyt
d Schmelzflusselektrolyse *f*

4936 SMOKE fue,gas
f fumée *f*
e humo *m*
i fumo *m*
n rook
d Qualm *m*, Rauch *m*

4937 SMOKE COMPOSITION exp
f artifice *m* fumant, mélange *m* à fumée
e composición *f* fumosa
i mescolanza *f* fumogena
n rook ontwikkelend mengsel *n*, rooksas
d Rauchsatz *m*

4938 SMOKE-DRIED MALT bre
f malt *m* fumé
e malta *f* ahumada
i malto *m* affumicato
n rookmout *n*
d Rauchmalz *n*

4939 SMOKE GAS fue
f gaz *m* de fumée
e gas *m* de humo
i gas *m* di fumo
n rookgas *n*
d Rauchgas *n*

4940 SMOKE POINT fue
f point *m* de formation de fumée
e punto *m* de formación de humo
i punto *m* di formazione di fumo
n walmpunt *n*
d Qualmpunkt *m*

4941 SMOKE TEST tes
f essai *m* de carneau
e ensayo *m* de conducto de humo
i collaudo *m* di condotto di fumo
n rookgangcontrole
d Rauchkanalprüfung *f*

4942 SMOKELESS POWDER exp
f poudre *f* sans fumée
e pólvora *f* sin humo
i polvere *f* senza fumo
n rookzwak buskruit *n*
d raucharmes Pulver *n*

4943 SMOKING nut
f fumaison *f*
e ahumado *m*
i affumicatura *f*
n roken *n*
d Räuchern *n*

4944 SMOKING, FUMIGATION agr
f fumigation *f*
e fumigación *f*
i fumigazione *f*
n roken *n*, uitroken *n*
d Ausräuchern *n*, Räuchern *n*

4945 SMOLDERING (US), SMOULDERING (GB) fue
f combustion *f* lente
e combustión *f* lenta
i combustione *f* lenta
n smeulen *n*
d Schwelen *n*

4946 SMOOTHING pap
f glaçage *m*, lissage *m*
e alisado *m*, satinación *f*
i appianatura *f*
n gladmaken *n*, strijken *n*
d Glätten *n*

4947 SMOOTHING ROLL pap
f cylindre *m* à glacer, cylindre *m* à lisser, cylindre *m* surfaceur
e cilindro *m* alisador, cilindro *m* de satinado
i cilindro *m* appianatore, cilindro *m* lucidatore
n strijkwals
d Glättwalze *f*

4948 SMOTHERING, SCALDING, SEETHING tex
f échaudage *m*
e escaldadura *f*
i scottatura *f*
n broeien *n*
d Abbrühen *n*

4949 SMUDGE, SOIL bui,hyg
f mélange *m* de noir de fumée et colle
e betún *m* de plomero
i mescolanza *f* di nero di fumo e colla
n roet-lijmmengsel *n*
d Russ-Leimmischung *f*

4950 SNAGGING pol
f ébarbage *m*
e desbastadura *f*
i sbavatura *f*
n poetsen *n*
d Fertigputzen *n*

4951 SNEEZE GAS gas,mit
f diphénylchlorarsine *m*, gaz *m* sternutatif
e difenilclorarsina *f*, gas *m* estornudatorio
i gas *m* starnutatorio
n niesgas *n*
d Reizgas *n*

4952 SOAKING met
f égalisation *f*
e resudación *f*, termodifusión *f*
i termodiffusione *f*
n egaliseren *n*, doorwarmen *n*
d Ausgleichen *n*, Durchwärmen *n*

4953 SOAKING tan
f reverdissage *m*
e maceración *f*
i macerazione *f*
n inweken *n*
d Einweichen *n*

4954 SOAKING, IMBIBITION pho
f imbibition *f*
e imbibición *f*
i imbibizione *f*
n doordrenken *n*, drenken *n*
d Durchtränken *n*, Imbibierung *f*, Tränken *n*

4955 SOAKING ZONE met
f zone *f* d'égalisation
e zona *f* de termodifusión
i zona *f* di termodiffusione
n doorwarmzone, opwarmzone
d Ausgleichzone *f*, Vorwärmzone *f*

4956 SOAP soa
f savon *m*
e jabón *m*
i sapone *m*
n zeep
d Seife *f*

4957 SOAP RESISTANCE dec
f résistance *f* au savon
e resistencia *f* al jabón
i resistenza *f* al sapone

n zeepvastheid
d Seifenbeständigkeit *f*,
Seifenfestheit *f*

4958 SOAPING AFTER-TREATMENT tex
f savonnage *m*
e enjabonadura *f*
i insaponatura *f*
n zeepbehandeling
d Seifebehandlung *f*

4959 SOAPSTONE, FRENCH CHALK (US), TALC (GB), TALCUM (GB) mat
f talc *m*
e piedra *f* de jabón, talco *m*
i talco *m*
n talk
d Talkum *n*

4960 SOAPSTONING (US), DUSTING (GB) coa
f talquage *m*
e entalcado *m*
i impolverizzazione *f* con talco
n met talk bepoederen *n*, talken *n*
d Talkumieren *n*

4961 SOAPY, SAPONACEOUS soa
f saponacé adj, savonneux adj
e jabonoso adj, saponáceo adj
i saponaceo adj, saponoso adj
n zeepachtig adj, zepig adj
d seifenartig adj, seifig adj

4962 SODA ASH sod
f soude *f* anhydre
e sosa *f* anhidra
i soda *f* anidra
n watervrije soda
d wasserfreie Soda *f*

4963 SODA EXTRACTION sod
f extraction *f* de soude
e extracción *f* de sosa
i estrazione *f* di soda
n sodaextractie
d Sodaextraktion *f*

4964 SODA HYPOSULFITE (US), SODIUM THIOSULPHATE (GB), ANTICHLORINE, GREEN MORDANT cle,tex
f antichlore *m*
e anticloro *m*
i anticloro *m*
n antichloor *n*, natriumthiosulfaat *n*
d Antichlor *n*, Natriumthiosulphat *n*

4965 SODA LIME bui
f chaux *f* sodée
e cal *f* sodada
i calce *f* sodata
n natronkalk
d Natronkalk *m*

4966 SODA-LIME GLASS gla
f verre *m* à chaux sodée
e vidrio *m* de cal sodada
i vetro *m* a calce sodata
n natronkalkglas *n*
d Natronkalkglas *n*

4967 SODA LYE gen
f lessive *f* de soude caustique
e lejía *f* de sosa cáustica
i lisciva *f* di soda caustica
n natronloog
d Natronlauge *f*

4968 SODA PROCESS pap
f procédé *m* à la soude
e procedimiento *m* de carbonato de sodio
i processo *m* al carbonato di sodio
n sodaontsluitingsproces *n*
d Soda-Kochverfahren *n*

4969 SODA RECOVERY pap
f récupération *f* de la soude
e recuperación *f* de la sosa
i ricupero *m* della soda
n sodaterugwinning
d Sodarückgewinnung *f*

4970 SODA RESISTANCE dec
f résistance *f* à la soude
e resistencia *f* a la sosa
i resistenza *f* alla soda
n sodavastheid
d Sodabeständigkeit *f*,
Sodafestheit *f*

4971 SODA WATER, AERATED WATER bre
f eau *f* de Seltz, eau *f* gazeuse
e agua *f* de Seltz, agua *f* gaseosa, gaseosa *f*
i acqua *f* di Seltz, acqua *f* gassosa
n sodawater *n*, spuitwater *n*
d kohlensaures Wasser *n*,
Sodawasser *n*

4972 SODA WOOD PULP (US), SODIUM WOOD PULP (GB) pap
f pâte *f* de bois à la soude
e pulpa *f* de madera a la sosa
i polpa *f* di legno trattato alla soda

n natronhoutpulp
d Natronzellstoff *m*, Sulfatzellstoff *m*

4973 SODIUM ETHYLATE, CAUSTIC ALCOHOL oic
f éthylate *m* de sodium
e etilato *m* de sodio
i etilato *m* di sodio
n natriumethylaat *n*
d Natriumäthylat *n*

4974 SODIUM HYDROXIDE, CAUSTIC SODA aic
f hydroxyde *m* de sodium
e sosa *f* cáustica
i soda *f* caustica
n bijtende soda
d Ätznatron *n*, Natriumhydroxyd *n*

4975 SODIUM RESINATE, RESIN SOAP soa
f résinate *m* de soude, savon *m* de résine
e resinato *m* de sodio
i resinato *m* di sodio
n harszeep, natriumresinaat *n*
d Harzseife *f*, Natriumresinat *n*

4976 SODIUM SOAP (US), CURD SOAP (GB), LAUNDRY SOAP (US) soa
f savon *m* dur
e jabón *m* duro
i sapone *m* duro
n harde zeep, kernzeep
d harte Seife *f*, Kernseife *f*

4977 SODIUM SULPHATE (GB), SALT CAKE (US) aic
f sulfate *m* de sodium commercial
e sulfato *m* de sodio comercial
i solfato *m* di sodio commerciale
n handelsnatriumsulfaat *n*
d technisches Natriumsulfat *n*

4978 SOFT FACING coa,met
f rechargement *m* doux
e recargue *m* suave
i rivestimento *m* dolce
n zachtmetaalbekleding
d Weichmetallauftrag *m*

4979 SOFT FLOW pla
f bonne plasticité *f*
e buen flujo *m*
i buona scorrevolezza *f*
n goed vloeivermogen *n*
d gute Fliessfähigkeit *f*

4980 SOFT RESIN pla
f résine *f* molle
e resina *f* blanda
i resina *f* molle
n zachte hars
d Weichharz *n*

4981 SOFT SOAP soa
f savon *m* mou, savon *m* vert
e jabón *m* blando
i sapone *m* molle
n zachte zeep
d Schmierseife *f*, weiche Seife *f*

4982 SOFT SOLDER wel
f soudure *f* à l'étain, soudure *f* tendre
e estañosoldadura *f*
i stagno *m* da saldare
n snelsoldeer *n*, zachtsoldeer *n*
d Weichlot *n*

4983 SOFT WATER hyg
f eau *f* douce
e agua *f* blanda
i acqua *f* dolce
n zacht water *n*
d weiches Wasser *n*

4984 SOFTENER gen
f agent *m* ramollissant, émollient *m*
e ablandador *m*
i agente *m* rammollente
n weekmaker
d Weichmacher *m*

4985 SOFTENING, LEAD IMPROVING met
f affinage *m* du plomb
e afino *m* del plomo
i affinatura *f* del piombo
n loodveredeling
d Bleivergütung *f*

4986 SOFTENING, WATER SOFTENING hyg,sew
f adoucissement *m* de l'eau
e ablandamiento *m* del agua
i addolcimento *m* dell'acqua
n waterontharding
d Wasserenthärtung *f*

4987 SOFTENING AGENT, WATER SOFTENER hyg
f adoucissant *m*
e agente *m* ablandante
i agente *m* d'addolcimento
n onthardingsmiddel *n*
d Enthärtungsmittel *n*

4988 SOFTENING POINT gen
f point *m* de ramollissement
e punto *m* de reblandecimiento
i punto *m* di rammollimento
n verwekingspunt *n*
d Erweichungspunkt *m*

4989 SOIL, bui,hyg
SMUDGE
f mélange *m* de noir de fumée et colle
e betún *m* de plomero
i mescolanza *f* di nero di fumo e colla
n roet-lijmmengsel *n*
d Russ-Leimmischung *f*

4990 SOIL ANALYSIS ana,agr
f analyse *f* du sol
e análisis *f* del suelo
i analisi *f* del suolo
n grondonderzoek *n*
d Bodenanalyse *f*, Bodenuntersuchung *f*

4991 SOIL CHEMISTRY agr
f chimie *f* du sol
e química *f* del suelo
i chimica *f* del suolo
n grondchemie
d Bodenchemie *f*

4992 SOIL FUMIGANT agr
f fumigant *m* du sol
e fumigante *m* del suelo
i fumigante *m* del suolo
n grondontsmettingsmiddel *n*
d Bodenentseuchungsmittel *n*

4993 SOIL RESIDUE, elc
RESIDUAL SOIL
f crasse *f* restante, souillure *f* restante
e residuo *m* de la suciedad, suciedad *f* restante
i residuo *m* di sporco
n resterend vuil *n*
d Restschmutz *m*

4994 SOLAR OIL, fat
DIESEL OIL, GAS OIL
f gas-oil *m*, huile *f* pour moteurs diesel
e gasoil *m*
i gasolio *m*, nafta *f* per motori diesel
n dieselolie
d Dieselöl *n*

4995 SOLATION col
f formation *f* de sols
e formación *f* de soles
i formazione *f* di soli
n solvorming
d Solbildung *f*

4996 SOLDER wel
f soudure *f*
e soldadura *f*
i saldatura *f*
n soldeer *n*
d Lot *n*, Lötmittel *n*

4997 SOLDERING wel
f soudage *m*, soudure *f*
e soldadura *f*
i saldatura *f*
n solderen *n*
d Löten *n*

4998 SOLDERING SOLUTION, wel
KILLED SPIRITS
f eau *f* à souder
e agua *f* de soldar
i acqua *f* per saldare
n soldeerwater *n*
d Lötwasser *n*

4999 SOLID PHASE RULE, gas
DISPERSE PHASE RULE
f règle *f* de la phase dispersée
e regla *f* de la fase dispersa
i regola *f* della fase dispersa
n disperse-fazeregel
d disperse Phasenregel *f*

5000 SOLID SOLUTION, gen
MIXED CRYSTAL, PRIMARY SOLUTION
f cristal *m* mixte, solution *f* solide
e cristal *m* mixto, solución *f* sólida
i cristallo *m* misto, soluzione *f* solida
n mengkristal *n*, vaste oplossing
d feste Lösung *f*, Mischkristall *m*

5001 SOLID STEEL, met
KILLED STEEL
f acier *m* calmé
e acero *m* reposado
i acciaio *m* calmo
n rustig staal *n*
d beruhigter Stahl *m*, ruhiger Stahl

5002 SOLIDIFICATION, gen
FREEZING
f solidification *f*
e solidificación *f*
i solidificazione *f*
n stolling
d Erstarrung *f*

5003 SOLIDIFICATION CONTRACTION, SOLIDIFICATION SHRINKAGE met,pla
f retrait *m* de solidification
e contracción *f* de solidificación
i ritiro *m* di solidificazione
n stollingskrimp
d Erstarrungsschrumpfung *f*, Erstarrungsschwund *m*

5004 SOLIDIFICATION POINT, FREEZING POINT gen
f point *m* de solidification
e punto *m* de solidificación
i punto *m* di solidificazione
n stolpunt *n*
d Erstarrungspunkt *m*

5005 SOLIDIFICATION RANGE gen
f intervalle *m* de solidification
e intervalo *m* de solidificación
i intervallo *m* di solidificazione
n stollingstraject *n*
d Erstarrungsbereich *m*

5006 SOLUBILITY gen
f solubilité *f*
e solubilidad *f*
i solubilità *f*
n oplosbaarheid
d Auflösungsvermögen *n*, Lösbarkeit *f*

5007 SOLUBILITY PRODUCT, IONIC PRODUCT elc
f produit *m* de solubilité, produit *m* ionique
e producto *m* de solubilidad, producto *m* iónico
i prodotto *m* di solubilità, prodotto *m* ionico
n ionisch produkt *n*, oplosbaarheidsprodukt *n*
d Ionenprodukt *n*, Löslichkeitsprodukt *n*

5008 SOLUBILIZER, HYDROTOPE soa
f agent *m* de solubilisation, tiers *m* solvant
e agente *m* solubilizante, fracción *f* disolvente
i agente *m* di solubilizzazione
n solubiliserend middel *n*
d Lösungsvermittler *m*

5009 SOLUBILIZING gen
f solubilisation *f*
e solubilización *f*
i solubilizzazione *f*
n oplosbaar maken *n*
d Solubilisation *f*

5010 SOLUBLE gen
f soluble adj
e soluble adj
i solubile adj
n oplosbaar adj
d auflösbar adj , löslich adj

5011 SOLUBLE GLASS, LIQUID GLASS, SILICATE, WATER GLASS aic
f silicate *m* de sodium, verre *m* soluble
e silicato *m* sódico líquido, vidrio *m* soluble
i tetrasilicato *m* di sodio
n silicaat *n*, waterglas *n*
d Silikat *n*, Wasserglas *n*

5012 SOLUBLE STARCH, AMYLODEXTRIN gla
f amidon *m* soluble
e almidón *m* soluble
i amido *m* solubile
n oplosbare stijfsel
d lösliche Stärke *f*

5013 SOLUTE gen
f corps *m* dissous
e soluto *m*, substancia *f* disuelta
i soluto *m*
n opgeloste stof
d gelöster Stoff *m*, Gelöstes *n*

5014 SOLUTION, DISSOLUTION gen
f dissolution *f*, solution *f*
e disolución *f*, solución *f*
i dissoluzione *f*, soluzione *f*
n oplossing
d Auflösung *f*, Lösung *f*

5015 SOLUTION, RUBBER ADHESIVE glu
f solution *f* de caoutchouc
e solución *f* de goma
i adesivo *m* di gomma
n rubberlijm, solutie
d Gummiklebstoff *m*, Klebkitt *m*

5016 SOLUTION RECLAIMING PROCESS, SOLVENT REGENERATING PROCESS rub
f procédé *m* de régénération par dissolution
e procedimiento *m* de regeneración por disolución

i processo *m* di rigenerazione in soluzione
n oplossingsregenereerproces *n*
d Lösungsregenerierverfahren *n*

5017 SOLUTION STRIPPING, CHEMICAL STRIPPING coa
f dépouillage *m* chimique
e eliminación *f* química
i degalvanizzazione *f* chimica
n chemische laagverwijdering
d chemische Entplattierung *f*

5018 SOLUTIZER PROCESS fue
f procédé *m* à dissolvant organique de mercaptans
e procedimiento *m* de disolvente orgánico de mercaptanos
i processo *m* a dissolvente organico di mercaptani
n proces *n* voor het oplossen van mercaptanen
d Verfahren *n* zur Auflösung von Merkaptanen

5019 SOLVATATION, SOLVATION col
f solvatation *f*, solvatisation *f*
e solvatación *f*
i solvatazione *f*
n solvatatie, solvatie
d Solvatation *f*, Solvation *f*

5020 SOLVAY PROCESS, AMMONIA-SODA PROCESS aic
f procédé *m* de la soude à l'ammoniaque, procédé *m* Solvay
e procedimiento *m* de sosa amoniacal
i processo *m* Solvay
n solvayproces *n*
d Ammoniaksodaverfahren *n*

5021 SOLVENT gen
f dissolvant *m*, solvant *m*
e disolvente *m*, solvente *m*
i dissolvente *m*, solvente *m*
n oplosmiddel *n*
d Lösungsmittel *n*

5022 SOLVENT, THINNER dec
f délayant *m*
e diluidor *m*, diluyente *m*
i diluente *m*
n verdunningsmiddel *n*
d Verdünner *m*, Verdünnungsmittel *n*

5023 SOLVENT CLEANING elc
f dégraissage *m* au solvant
e desengrase *m* por disolvente
i sgrassaggio *m* con solvente
n ontvetting door oplosmiddel
d Reinigung *f* mit Lösungsmittel

5024 SOLVENT DEGREASER cle
f dégraisseur *m*
e desengrasador *m*
i sgrassante *m*
n ontvetter, vetoplosmiddel *n*
d Fettlösungsmittel *n*, Reinigungsmittel *n*

5025 SOLVENT DYEING dye
f teinture *f* par solution
e teñido *m* por solución
i tintura *f* per soluzione
n oplossingsverven *n*
d Lösungsfärben *n*

5026 SOLVENT EXTRACTION fue
f extraction *f* par solvants
e extracción *f* por solventes
i estrazione *f* per solventi
n extractie met oplosmiddelen
d Lösungsmittelextraktion *f*

5027 SOLVENT NAPHTA fue
f solvant-naphta *m*
e nafta *f* disolvente
i benzina *f* solvente
n solventnafta
d Lösungsbenzol *n*, Schwerbenzol *n*, Solventnaphta *f*

5028 SOLVENT PLASTICIZER, GELATINIZER oic
f plastifiant *m* gélatinisant
e plastificante *m* gelatinizador
i plastificante *m* da gelatinizzare
n gelatinerende weekmaker
d gelatinierender Weichmacher *m*

5029 SOLVENT RELEASE dec
f lâchage *m* du solvant
e liberación *f* del solvente
i liberazione *f* del solvente, ritorno *m* del solvente
n loslaten *n* van het oplosmiddel
d Wiederabgeben *n* des Lösungsmittels

5030 SOOT fue
f suie *f*
e hollín *m*
i fuliggine *f*
n roet *n*
d Russ *m*

5031 SOOT BLACK, IMPINGEMENT BLACK dec

f noir *m* de combustion
pigment *m* noir déposé
e negro *m* de combustión, tizna *f*
i nero *m* di combustione,
nerofumo *m*
n roetzwart *n*, zwartsel *n*
d Farbruss *m*, Russschwarz *n*

5032 SORPTION abs
f sorption *f*
e sorción *f*
i
n sorptie
d Sorption *f*

5033 SORTING sor
f triage *m*
e separación *f*, tría *f*
i cernita *f*, separazione *f*
n sorteren *n*
d Sortieren *n*

5034 SOUPLING tex
f dégommage *m* partiel
e desengomado *m* parcial
i sgommatura *f* parziale
n gedeeltelijke ontgomming
d Teildegummierung *f*,
Teilentbastung *f*

5035 SOUR CRUDE fue
f huile *f* brute à excès de soufre
e crudo *m* sulfuroso
i olio *m* grezzo ad eccesso di
zolfo
n ruwe olie met overmaat zwavel
d Rohöl *n* mit Schwefelüberschuss

5036 SOUR GAS gas
f gaz *m* naturel acide
e gas *m* natural ácido
i gas *m* naturale acido
n zuur aardgas *n*
d saures Erdgas *n*

5037 SOUR GASOLINE fue
f essence *f* à excès de mercaptans
e gasolina *f* con exceso de
mercaptanos
i benzina *f* ad eccesso di
mercaptani
n mercaptaanrijke benzine
d merkaptanreiches Benzin *n*

5038 SOURING cle,tex
f vitriolage *m*
e acidulación *f*
i acidulatura *f*
n aanzuren *n*
d Ansäuern *n*

5039 SOURING (US), cer
AGEING (GB)
f mûrissage *m*
e curación *f*
i maturazione *f*,
omogenizzazione *f*
n homogenisatie
d Homogenisierung *f*

5040 SOY OIL, fat
SOYA BEAN OIL, BEAN OIL
f huile *f* de soya
e aceite *m* de soja
i olio *m* di soia
n sojaolie
d Sojaöl *n*

5041 SPALLING mat
f effrittement *m*
e machaqueo *m*
i sbriciolatura *f*, sgretolamento *m*
n verbrokkeling, verkruimeling
d Zerbröcklung *f*

5042 SPALLING, met
EXFOLIATING, FLAKING
f écaillage *m*
e desconchado *m*
i scagliatura *f*, screpolatura *f*
n afbladderen *n*
d Abblättern *n*, Häutung *f*

5043 SPANGLES, elc
FLOWERS
f fleurs *pl* de zinc
e floreado *m*
i fiori *pl* di zinco
n zinkbloemen *pl*
d Zinkblumen *pl*

5044 SPARGE LIQUOR bre
f dernières eaux *pl*, lavages *pl*
e aguas *pl* de lavado
i acque *pl* di lavaggio
n naspoeling
d Nachguss *m*

5045 SPARGE WATER bre
f eau *f* d'arrosage, eau *f* de
lavage
e agua *f* de lavado, agua *f* de riego
i acqua *f* d'annaffiamento,
acqua *f* di lavaggio
n sproeiwater *n*
d Anschwänzwasser *n*

5046 SPARGING bre
f arrosage *m*
e riego *m*, rociadura *f*
i annaffiamento *m*

n besproeien *n*
d Anschwänzen *n*, Überschwänzen *n*

5047 SPARK METALWORKING, ELECTRO-EROSION METALWORKING — met
f usinage *m* par électro-érosion, usinage *m* par étincelage
e trabajo *m* por electro-erosión
i erosione *m* mediante scintillio, lavorazione *f* a scintillamento
n elektro-erosieve behandeling
d Elektroerosionsbehandlung *f*

5048 SPATTER DASHED EFFECT, SPATTER EFFECT — dec
f effet *m* de peinture à la bruine
e efecto *m* de salpicadura
i effetto *m* di spruzzi
n spateffect *n*, spettereffect *n*
d Sprenkeleffekt *m*, Tüpfeleffekt *m*

5049 SPAVIN, CHAMOTTE, FIRECLAY, WARRANT — cer
f apyre *f*, argile *f* réfractaire, chamotte *f*
e arcilla *f* refractaria, chamota *f*
i argilla *f* refrattaria, terra *f* refrattaria
n chamotte, vuurklei
d feuerfester Ton *m*, Schamotte *f*

5050 SPEARMINT OIL — fat
f essence *f* de menthe crispée, essence *f* de menthe verte
e esencia *f* de menta crespa
i olio *m* di menta crespa
n kruizemuntolie
d Öl *n* der grünen Minze

5051 SPECTRUM ANALYSIS — ana
f analyse *f* du spectre
e análisis *f* del spectro
i analisi *f* dello spettro
n spectraalanalyse
d Spektralanalyse *f*

5052 SPENT ACID, WASTE ACID — gen
f acide *m* épuisé
e ácido *m* agotado
i acido *m* di rifiuto
n uitgeput zuur *n*
d Abfallsäure *f*

5053 SPENT AMMONIA LIQUOR, SPENT GAS LIQUOR, AMMONIA WASTE — gas
f eaux *pl* résiduaires ammoniacales
e agua *f* agotada amoniacal
i acqua *f* di rifiuto ammoniacale
n ammoniakhoudend afvalwater *n*
d Ammoniakabwasser *n*

5054 SPENT BARLEY, DREGS — bre
f drêches *pl* de malt
e bagazo *m* de malta, orujos *pl* de malta
i trebbie *pl* di malto
n moutdraf
d Malztreber *m*

5055 SPENT BEER, STALE BEER — bre
f bière *f* éventée
e cerveza *f* ventrada
i birra *f* svanita
n verschaald bier *n*
d abgestandenes Bier *n*

5056 SPENT LYE, TAIL LIQUOR — gen
f lessive *f* épuisée
e lejía *f* agotada
i sottolisciva *f*
n afgewerkte loog
d Abfallauge *f*, Ablauge *f*

5057 SPENT MASH, DISTILLER'S WASH, SLOP, STILLAGE, VINASSE — nut
f drêches *pl*, vinasses *pl* de distillerie
e orujo *m*, vinazas *pl* de destilación
i burlanda *f*, trebbia *f*, vinacce *pl* distillate
n drab
d abgebrannte Würze *f*, Schlempe *f* Spülicht *n*

5058 SPENT PICKLING LIQUOR — elc
f liqueur *f* de décapage épuisé
e lejía *f* de decapaje agotada
i lisciva *f* di rifiuto di decapaggio
n afgewerkte beitsvloeistof
d Abfallbeize *f*, Beizablauge *f*

5059 SPENT SULPHITE LIQUOR (GB), SULFITE WASTE LIQUOR (US) — sin
f lessive *f* résiduaire sulfitique
e lejía *f* residual sulfítica
i lisciva *f* di rifiuto solfitica
n afgewerkte sulfietvloeistof
d Sulfitablauge *f*

5060 SPERM OIL — fat
f huile *f* de spermaceti

e aceite *m* de sperma de ballena
i olio *m* di spermaceti
n spermacetiolie, spermolie
d Potratöl *n*, Spermöl *n*, Walratöl *n*

5061 SPERMACETI, CETACEUM fat
f blanc *m* de baleine, spermaceti *m*, sperme *m* de baleine
e blanco *m* de ballena, esperma *f* de ballena
i spermaceti *m*
n cetaceum *n*, spermaceti, walschot *n*
d Walrat *n*

5062 SPHERICAL YEAST fmi
f levure *f* sphérique
e levadura *f* esférica
i lievito *m* sferico
n bolvormige gist, gistbolletje *n*
d Kugelhefe *f*

5063 SPIKE OIL, LAVENDER OIL fat
f essence *f* de lavande
e esencia *f* de espliego
i olio *m* di lavanda, olio *m* di spigo
n lavendelolie
d Lavendelöl *n*

5064 SPIN HARDENING met
f endurcissement *m* par rotation
e endurecimiento *m* por rotación
i indurimento *m* per rotazione
n rotatieharding
d Drehungshärtung *f*

5065 SPIN WELDING pla,wel
f soudure *m* à chaud par rotation
e sellado *m* por rotación
i saldatura *f* per rotazione
n roterend lassen *n* van kunststof
d Drehsiegeln *n*

5066 SPINDLE OIL lub
f huile *f* minérale pour broches
e aceite *m* para husillos
i olio *m* per alberini
n spindelolie
d Spindelöl *n*

5067 SPINNING tex
f filage *m*, filature *f*
e hilado *m*, hilatura *f*
i filatura *f*
n spinnen *n*
d Spinnen *n*, Verspinnen *n*

5068 SPINNING DISK ATOMIZATION, soa CENTRAL DISK ATOMIZATION
f atomisation *f* à disque tournant
e atomización *f* por disco giratorio
i atomizzazione *f* mediante disco rotante
n verstuiving met draaiende schijf
d Zerstäubung *f* mittels rotierenden Tellers

5069 SPINNING DOPE (US), SPINNING OIL, SPINNING SOLUTION (GB) tex
f liquide *m* de filature
e líquido *m* de hilatura
i liquido *m* di filatura
n spinvloeistof
d Spinnflüssigkeit *f*

5070 SPINNING-OFF sin
f centrifugation *f*
e centrifugación *f*
i centrifugazione *f*
n centrifugeren *n*
d Zentrifugieren *n*

5071 SPIRIT dis
f esprit *m* de vin
e espíritu *m*
i spirito *m*
n spiritus
d Spiritus *m*

5072 SPIRIT LACQUER dec
f laque *f* à l'alcool
e laca *f* de espíritu
i lacca *f* a spirito
n spirituslak
d Spirituslack *m*

5073 SPIRIT OF CAMPHOR mcy
f alcool *m* camphré
e alcohol *m* alcanforado
i spirito *m* canforato
n kamferspiritus
d Kampferspiritus *m*

5074 SPIRITUOUS LIQUORS, ARDENT SPIRITS oic
f liqueurs *pl*, spiritueux *pl*
e bebidas *pl* espirituosas
i bevande *pl* alcooliche
n spiritualiën *pl*, sterke drank
d hochprozentige alkoholische Getränke *pl*, Spirituosen *pl*

5075 SPLASH COOLING clo
f refroidissement *m* à ruissellement

e enfriamiento *m* por salpicadura
i raffreddamento *m* ad annaffiamento
n sproeikoeling
d Berieselungskühlung *f*

5076 SPLIT lea
f croûte *f* d'une peau fendue
e cuero *m* hendido
i crosta *f*
n splitleer *n*
d Spaltleder *n*

5077 SPLITTING gen
f clivage *m*
e escisión *f*, ruptura *f*
i scissione *f*
n splijten *n*, splitsen *n*
d Abspaltung *f*, Spaltung *f*

5078 SPLITTING HYDROGENATION fat
f hydrogénation *f* destructive
e hidrogenación *f* de ruptura de cadena
i idrogenazione *f* destruttiva
n afbrekend hydrogeneren *n*
d spaltende Hydrierung *f*

5079 SPONGE IRON met
f éponge *f* de fer
e esponja *f* de hierro
i spugna *f* di ferro
n ijzerspons *n*
d Eisenschwamm *m*

5080 SPONGE PLATING, DOCTORING elc
f revêtement *m* électrolytique avec électrode à éponge
e revestimiento *m* galvánico con electrodo macrocelular
i rivestimento *m* galvanico con elettrodo macrocellulare
n tamponneren *n*
d Galvanisieren *n* mit Schwammelektrode

5081 SPONGE RUBBER rub
f caoutchouc *m* spongeux
e goma *f* esponjosa
i gomma *f* spugnosa
n sponsrubber
d Schwammgummi *m*

5082 SPONGINESS met
f spongiosité *f*
e espongosidad *f*
i spugnosità *f*
n sponsachtigheid
d Schwammigkeit *f*

5083 SPONGING AGENT, BLOWING AGENT, INFLATING AGENT rub
f agent *m* gonflant
e agente *m* esponjante, agente *m* hinchante
i agente *m* rigonfiante
n blaasmiddel *n*
d Blähmittel *n*, Treibmittel *n*

5084 SPONGY METAL met
f métal *m* spongieux
e metal *m* esponjoso
i metallo *m* spugnoso
n sponsachtig metaal *n*
d schwammartiges Metall *n*

5085 SPONGY PLATINUM, PLATINUM SPONGE met
f éponge *f* de platine
e esponja *f* de platino
i platino *m* spugnoso
n platinaspons *n*
d Platinschwamm *m*

5086 SPONTANEOUS CREAMING rub
f crémage *m* spontané
e desnate *m* espontáneo
i crematura *f* spontanea
n spontane oproming
d spontanes Aufrahmen *n*

5087 SPONTANEOUS EVAPORATION eva
f autovaporisation *f*, évaporation *f* spontanée
e autovaporización *f*, evaporación *f* espontánea
i autovaporizzazione *f*, evaporazione *f* spontanea
n zelfverdamping
d Selbstverdampfung *f*

5088 SPONTANEOUS FERMENTATION fmi
f fermentation *f* spontanée
e fermentación *f* espontánea
i fermentazione *f* spontanea
n eigen gisting
d Selbstgärung *f*

5089 SPONTANEOUS IGNITION fue
f combustion *f* spontanée
e combustión *f* espontánea
i combustione *f* spontanea
n zelfontbranding
d Selbstentzündung *f*

5090 SPOT TEST tes
f essai *m* à la goutte

e ensayo *m* a la gota
i collaudo *m* alla goccia
n druppelproef
d Tropfenreaktion *f*, Tüpfelprobe *f*

5091 SPOT TESTS tes
f essais *pl* à la touche
e ensayos *pl* al toque
i collaudi *pl* alla tocca
n druppelproeven *pl*
d Tüpfelanalyse *f*, Tüpfelproben *pl*

5092 SPOTTING dec
f apparition *f* de tachetures
e aparición *f* de manchas
i macchiatura
n vlekken *n*
d Fleckigwerden *n*

5093 SPRAY COOLING (AIR) cle
f atomisation *f* à froid, séchage *m* par l'air froid
e atomización *f* por enfriamiento, secado *m* por aire frío
i atomizzazione *f* a freddo, essiccazione *f* ad aria fredda
n koelen *n* met luchtstraal
d Kaltsprühung *f*

5094 SPRAY COOLING (WATER) clo
f refroidissement *m* par eau atomisée
e enfriamiento *m* por agua nebulizada
i raffreddamento *m* per spruzzo d'acqua
n afkoeling door besproeiing
d Spritzkühlung *f*, Zerstäubungskühlen *n*

5095 SPRAY DEVELOPMENT, JET DEVELOPMENT pho
f développement *m* sous jet
e revelado *m* de chorro
i sviluppo *m* con getto
n ontwikkelen *n* met sproeien
d Spritzentwicklung *f*

5096 SPRAY-DRIED RUBBER rub
f caoutchouc *m* en poudre
e látex *m* secado por pulverización
i lattice *m* secco polverizzato
n rubberpoeder *n* vervaardigd door latexverstuiving
d im Zerstäubungsverfahren hergestelltes Kautschukpulver *n*

5097 SPRAY DRYING soa
f séchage *m* par atomisation
e secado *m* por atomización
i essiccazione *f* per atomizzazione
n verstuivingsdrogen *n*
d Zerstäubungstrocknen *n*

5098 SPRAY DUST, SPRAY FUME dec
f brouillard *m* de pistolage, brume *f* de vaporisation
e niebla *f* de vaporización
i nebbia *f* di vaporizzazione
n spuitnevel, verfnevel
d Farbnebel *m*, Spritznebel *m*

5099 SPRAY DYEING dye
f teinture *f* par atomisation
e teñido *m* por atomización
i tintura *f* per atomizzazione
n verstuivingsverven *n*
d Zerstäubungsfärben *n*

5100 SPRAY PAINTING, PAINT SPRAYING dec
f pistolage *m*
e pintura *f* con pistola
i spruzzatura *f* di pittura
n verfspuiten *n*
d Spritzlackierung *f*

5101 SPRAY QUENCHING met
f trempe *f* à l'eau atomisée
e temple *m* con agua atomizada
i tempera *f* all'acqua atomizzata
n afschrikken *n* met verstoven water
d Abschrecken *n* mit zerstäubtem Wasser

5102 SPRAY RINSING, SPRAY SWILLING elc
f rinçage *m* au jet
e enjuague *m* al chorro
i sciacquatura *f* allo spruzzo
n spuitspoeling
d Spritzspülung *f*

5103 SPRAYED RUBBER rub
f latex *m* défloculé
e látex *m* secado en copos
i lattice *m* secco a fiocchi
n rubbervlokken *pl*
d Flockenkautschuk *m*

5104 SPRAYING coa
f atomisation *f*, pulvérisation *f*
e atomización *f*, pulverización *f*
i atomizzazione *f*, polverizzazione *f*
n verstuiving
d Zerstäubung *f*

5105 SPRAYING PAINT dec
f peinture *f* au pistolet

e pintura *f* de pistola
i pittura *f* alla pistola
n spuitverf
d Spritzfarbe *f*

5106 SPREADING rub
f enduisage *m*
e esparcimiento *m*
i spalmatura *f*
n nat berubberen *n*, strijken *n*
d Streichen *n*

5107 SPREADING CAPACITY, COVERAGE, OBSCURATION dec
f pouvoir *m* couvrant en surface
e extensión *f* especifica
i distendimento *m* specifico
n uitstrijkrendement *n*
d Ausgiebigkeit *f*

5108 SPREADING MIXTURE, COATING COMPOUND coa
f mélange *m* d'enduisage
e mezcla *f* de recubrimiento
i mescolanza *f* per ricoperto
n strijkmengsel *n*
d Streichmischung *f*, Überzugsmischung *f*

5109 SPROUTING, CHITTING bre
f piquage *m*
e apuntado *m*
i spuntatura *f*
n kiemverwijdering
d Spitzen *n*

5110 SPUN GLASS, GLASS WOOL gla
f coton *m* de verre, laine *f* de verre
e guata *f* de vidrio
i lana *f* di vetro
n glaswol
d Glaswolle *f*

5111 SPUTTERING coa
f pulvérisation *f* cathodique
e pulverización *f* catódica
i polverizzazione *f* catodica
n katodeverstuiving
d Katodenzerstäubung *f*

5112 SQUEEZE MOLDING (US), SQUEEZE MOULDING (GB) met,pre
f serrage *m* par pression
e apisonado *m* por presión
i stivatura *f* a pressione
n drukverdichting
d Druckverdichtung *f*

5113 SQUEEZING OUT pre
f pressurage *m*
e expresión *f*
i spremitura *f*
n uitpersen *n*
d Auspressen *n*, Ausquetschen *n*

5114 STABILITY IN STORAGE sto
f stabilité *f* au stockage
e estabilidad *f* de almacenaje
i stabilità *f* all'immagazzinamento
n bestendigheid bij opslag
d Lagerbeständigkeit *f*

5115 STABILITY TO LIGHT gen
f stabilité *f* à la lumière
e estabilidad *f* a la luz
i stabilità *f* alla luce
n lichtbestendigheid
d Lichtbeständigkeit *f*, Lichtechthe

5116 STABILIZATION, DESENSITIZATION exp
f flegmatisation *f*, phlegmatisation *f*
e flematización *f*, insensibilización *f*
i insensibilizzazione *f*
n flegmatiseren *n*, stabiliseren *n*
d Phlegmatisierung *f*

5117 STABILIZER cat
f agent *m* stabilisant
e agente *m* estabilisante
i agente *m* stabilizzatore
n stabiliserend middel *n*
d Stabilisierungsmittel *n*

5118 STABILIZING ANNEALING, STRESS-RELIEF ANNEALING htt,met
f recuit *m* de détente
e recocido *m* de atenuación de tensiones
i ricottura *f* di distensione
n herstelgloeien *n*
d spannungsfreies Glühen *n*

5119 STAGE DIGESTION OF SLUDGE iwt
f digestion *f* de boues échelonnée
e digestión *f* de fangos escalonada
i digestione *f* di fanghi a più stadi
n trapsgewijze slibverwerking
d mehrstufige Schlammfaulung *f*

5120 STAGE OF POLYMERIZATION oic
f étape *f* de polymérisation
e etapa *f* de polimerización
i gradino *m* di polimerizzazione, stadio *m* di polimerizzazione

n polymerisatietrap
d Polymerisationsstufe *f*

5121 STAGE OF PROCESS gen
f étape *f* d'un procédé
e etapa *f* de un procedimiento
i stadio *m* d'un processo
n procestrap
d Verfahrensstufe *f*

5122 STAGE VULCANIZING, rub
STEP-UP CURING
f vulcanisation *f* échelonnée
e vulcanización *f* escalonada
i vulcanizzazione *f* a gradini
n trapsgewijze vulcanisatie
d Stufenvulkanisation *f*

5123 STAIN gen
f tache *f*
e mancha *f*
i macchia *f*
n vlek
d Fleck *m*

5124 STAIN, dec
WOOD STAIN
f teinture *f*
e colorante *m* de la madera, tintura *f*
i tintura *f*
n houtbeits, kleurbeits
d Farbbeize *f*, Holzbeize *f*

5125 STAIN REMOVAL cle
f détachage *m*
e remoción *f* de manchas
i smacchiatura *f*
n ontvlekking
d Fleckentfernung *f*

5126 STAIN REMOVER cle
f détachant *m*, enlève-taches *m*
e removedor *m* de manchas
i smacchiatore *m*
n ontvlekkingsmiddel *n*
d Fleckentferner *m*

5127 STAIN SPOTS elc
f souillures *pl*
e mohos *pl*
i macchie *pl* da essudazione
n stippen *pl*
d Ausblühungen *pl*

5128 STAINER, dec
COLORING PIGMENT (US),
COLOURING PIGMENT (GB)
f pigment *m* colorant
e pigmento *m* de color
i pigmento *m* colorante
n pigmentkleurstof
d Farbpigment *n*

5129 STAINING, lea
BRUSH DYEING
f coloration *f* du cuir à la brosse
e coloración *f* del cuero con la brocha
i colorazione *f* del cuoio al pennello
n leder verven *n* met borstel
d Lederfärben *n* mittels Einbürsten

5130 STAINING, pap
GROUNDING
f teinture *f*
e tinción *f*
i tintura *f*
n gronderen *n*
d Foncieren *n*, Grundieren *n*

5131 STAINING METHOD bac
f procédé *m* de coloration
e procedimiento *m* de coloración
i processo *m* di colorazione
n kleurproces *n*
d Färbeverfahren *n*

5132 STAINING POWER dec
f pouvoir *m* colorant
e poder *m* colorante
i potere *m* colorante
n kleurkracht
d Färbkraft *f*

5133 STAINLESS met
f inoxydable adj, résistant adj à la corrosion
e corrosiorresistente adj, inoxidable adj
i inossidabile adj, resistente adj alla corrosione
n corrosievast adj, roestvast adj
d nichtrostend adj, rostfrei adj

5134 STAKING lea
f palissonnage *m*
e zurra *f*
i ragguagliamento *m*
n stoltsen *n*
d Stollen *n*

5135 STALE BEER, bre
SPENT BEER
f bière *f* éventée
e cerveza *f* ventrada
i birra *f* svanita
n verschaald bier *n*
d abgestandenes Bier *n*

5136 STALE SEWAGE iwt,sew
f eaux *pl* usées fades
e agua *f* usada añeja
i acqua *f* di rifiuto stantia
n aangerot afvalwater *n*, verschaald afvalwater *n*
d schales Abwasser *n*

5137 STAMPING brk,pap
f bocardage *m*
e bocarteo *m*
i battitura *f*
n stampen *n*
d Pochen *n*, Stampfen *n*

5138 STANDARD ELECTRODE POTENTIAL, NOBLE POTENTIAL elc
f tension *f* standard d'une électrode
e tensión *f* normal de un electrodo
i potenziale *m* normale d'un elettrodo
n normaalpotentiaal van een elektrode
d Normalpotential *n* einer Elektrode

5139 STANDARD SOLUTION gen
f solution *f* échelon, solution *f* titrée
e solución *f* patrón, solución *f* valorada
i soluzione *f* campione, soluzione *f* titolata
n normaaloplossing
d Standardlösung *f*, Titerflüssigkeit *f*

5140 STANDARD TITRIMETRIC SUBSTANCE ana
f substance *f* titrimétrique de référence
e substancia *f* patrón para valoraciones
i sostanza *f* campione per titolazioni
n oertiterstof
d Urtitersubstanz *f*

5141 STANDING BATH tex
f bain *m* de teinture à concentration constante
e baño *m* de teñido de concentración constante
i bagno *m* di tintura a concentrazione costante
n kleurbad *n* met constante concentratie
d Färbeflotte *f* konstanter Konzentration

5142 STANDING OFF gla
f refroidissement *m*, repos *m*
e dejación *f* en reposo
i raffreddamento *m* lento
n laten afkoelen *n*
d Abgehenlassen *n*, Abstehen *n*

5143 STANDOIL, BODIED OIL dec
f standolie *f*
e aceite *m* secante espesado
i olio *m* di lino cotto
n standolie
d Standöl *n*

5144 STANDOIL PAINT, DUTCH ENAMEL dec
f peinture *f* à la standolie
e pintura *f* al aceite secante
i pittura *f* all'olio di lino cotto
n standolieverf
d Standölfarbe *f*

5145 STANNATE OF SODIUM, PREPARING SALT tex
f sel *m* d'apprêt, stannate *m* de sodium
e estañato *m* sódico, sal *f* de apresto
i sale *m* d'appretto, stannato *m* di sodio
n natriumstannaat *n*, prepareerzout
d Präpariersalz *n*

5146 STANNIC SULFIDE (US), MOSAIC GOLD (GB) dec
f or *m* musif
e oro *m* musivo
i oro *m* musivo
n musiefgoud *n*, vals schelpgoud *n*
d Musivgold *n*, unechtes Muschelgold *n*, Zinn-IV-Sulfid *n*

5147 STANNIOL, TINFOIL coa
f étain *m* battu, feuille *f* d'étain, papier *m* d'argent
e estaño *m* en hojas, papel *m* estaño
i stagno *m* fogliato, stagnuola *f*
n bladtin *n*, staniol *n*, stanniool *n*, tinfoelie
d Blattzinn *n*, Stanniol *n*, Zinnfolie *f*

5148 STAR-ANISE OIL, ANISE-STAR OIL fat
f essence *f* d'anis étoilé, essence *f* de badiane
e esencia *f* de anís estrellado

i olio *m* d'anice stellato
n steranijsolie
d Sternanisöl *n*

5149 STAR FRAME DYEING dye
f teinture *f* à tourniquet porteur
e teñido *m* de torniquete portador
i tintura *f* con arganello portatore
n kruisraamverven *n*
d Kreuzrahmenfärbung *f*

5150 STAR STEAMING tex
f traitement *m* à la vapeur à tourniquet
e tratamiento *m* al vapor de torniquete
i trattamento *m* al vapore con arganello
n stoombehandeling met kruisraam
d Kreuzrahmendampfbehandlung *f*

5151 STARCH, AMYLUM oic
f amidon *m*
e almidón *m*
i amido *m*
n stijfsel
d Stärke *f*

5152 STARCH FERMENTATION, AMYLOLYTIC FERMENTATION fmi
f fermentation *f* amymolitique
e fermentación *f* amimolítica
i fermentazione *f* amimolitica
n stijfselfermentatie
d Stärkefermentation *f*

5153 STARCH GLUE glu
f colle *f* d'amidon
e cola *f* de almidón
i colla *f* d'amido
n stijfselpap
d Stärkekleister *m*

5154 STARCH GUM (GB), BRITISH GUM (US), DEXTRINE (GB) glu
f dextrine *f*
e dextrina *f*
i destrina *f*
n dextrine *n*
d Dextrin *n*, Stärkegummi *m*

5155 STARCH PASTE, FARINA tex
f apprêt *m* à base d'amidon
e apresto *m* de base de almidón
i appretto *m* a base d'amido
n pap
d Stärkekleister *m*

5156 STARCHING, STIFFENING cle
f amidonnage *m*
e almidonado *m*
i inamidatura *f*
n stijfselen *n*, stijven *n*
d Stärken *n*, Steifen *n*

5157 STARCHING BLUE, LAUNDRY BLUE, WASHING BLUE cle
f bleu *m* d'azur, bleu *m* d'empois
e azulete *m*
i turchinetto *m*
n blauwsel *n*
d Waschblau *n*

5158 STARTING BATH tex
f bain *m* initial
e baño *m* de salida
i bagno *m* d'avviamento
n beginbad *n*
d Anfangsbad *n*

5159 STEADY FLOW gen
f flux *m* stationnaire
e flujo *m* estacionario
i flusso *m* costante
n stationaire stroming
d stationäre Strömung *f*, stehende Strömung *f*

5160 STEAM boi
f vapeur *f*
e vapor *m*
i vapore *m*
n stoom
d Dampf *m*

5161 STEAM CLEANING cle
f épuration *f* à vapeur
e limpia *f* con vapor
i depurazione *f* con vapore
n reinigen *n* met stoom
d Dampfreinigung *f*

5162 STEAM CURING bui
f traitement *m* de béton à la vapeur
e curación *f* de hormigón al vapor
i trattamento *m* di calcestruzzo al vapore
n stoombehandeling van beton
d Betondampfbehandlung *f*

5163 STEAM CURING cer
f traitement *m* à la vapeur
e tratamiento *m* al vapor
i trattamento *m* al vapore
n stoombehandeling
d Dampfbehandlung *f*

5164 STEAM CURING, STEAM VULCANIZING rub
f vulcanisation *f* en vapeur
e vulcanización *f* a vapor
i vulcanizzazione *f* in vapore
n stoomvulcanisatie
d Dampfvulkanisation *f*

5165 STEAM DISTILLATION dis
f distillation *f* à vapeur d'eau
e destilación *f* con vapor de agua
i distillazione *f* con vapore d'acqua
n waterdampdestillatie
d Dampfdestillation *f*, Wasserdampfdestillation *f*

5166 STEAM DRYING dry
f séchage *m* à vapeur
e secado *m* con vapor
i essiccazione *f* con vapore
n stoomdroging
d Dampftrocknung *f*

5167 STEAM FINISHING pap
f humectation *f*
e humectación *f*
i umettazione *f*
n bevochtiging
d Benetzung *f*

5168 STEAM REFINING fat
f raffinage *m* par vapeur
e refinación *f* por vapor
i raffinazione *f* per vapore
n raffineren *n* met stoom
d Ausdämpfen *n*, Raffination *f* mittels Dampf

5169 STEAM TEMPERING, HOT PREPARATION cer
f traitement *m* de l'argile à la vapeur
e tratamiento *m* de la arcilla al vapor
i trattamento *m* dell'argilla con vapore
n stoombehandeling voor klei
d Dampfbehandlung *f* für Ton

5170 STEAMING, DECATIZING, ROLL BOILING tex
f décatissage *m*
e decatizado *m*, vaporización *f*
i decatissaggio *m*
n afglanzen *n*, decateren *n*
d Dekatieren *n*, Dekatur *f*, Glanzkrumpe *f*

5171 STEEL BALL PEENING (GB), CLOUDBURSTING (US) cle
f traitement *m* par jet de balles en acier
e tratamiento *m* por chorro con perdigones de acero
i trattamento *m* con pallottole d'acciaio
n kogelstralen *n*
d Kugelstrahlen *n*

5172 STEELING, ACIERAGE, IRON PLATING coa
f aciérage *m*
e aceración *f*, hierrado *m*
i acciaiatura *f*
n verijzering, verstaling
d Stählung *f*, Verstählung *f*

5173 STEEPED BARLEY bre
f orge *f* gonflée
e cebada *f* hinchada
i orzo *m* gonfiato
n geweekte gerst
d gequellte Gerste *f*

5174 STEEPING, MERCERIZATION tex
f mercerisage *m*
e mercerización *f*
i mercerizzazione *f*
n merceriseren *n*
d Mercerieren *n*, Mercerisieren *n*

5175 STEEPING WATER bre
f eau *f* de trempage
e agua *f* de remojo
i acqua *f* per macerazione
n weekwater *n*
d Weichwasser *n*

5176 STENCIL PRINTING tex
f impression *f* au pochoir
e estampación *f* con estarcido
i stampatura *f* allo stampino
n sjabloondruk
d Schablonendruck *m*

5177 STEP AERATION iwt,sew
f aération *f* repartie
e aireación *f* escalonada
i aerazione *f* a gradini
n trapsgewijze aëratie
d Stufenlüftung *f*

5178 STEP TITRATION ana
f titration *f* fractionnée
e valoración *f* fraccionada
i titolazione *f* frazionata
n trapsgewijze titratie
d Stufentitration *f*

5179 STEP-UP CURING, rub
STAGE VULCANIZING
f vulcanisation *f* échelonnée
e vulcanización *f* escalonada
i vulcanizzazione *f* a gradini
n trapsgewijze vulcanisatie
d Stufenvulkanisation

5180 STERIC HINDRANCE oic
f empêchement *m* stérique
e impedimiento *m* estérico
i impedimento *m* sterico
n sterische hindering
d sterische Hinderung *f*

5181 STERILE hyg,mcy
f stérile adj
e estéril adj
i sterile adj
n kiemvrij adj , steriel adj
d keimfrei adj , steril adj

5182 STERILIZATION bac,wit
f stérilisation *f*
e esterilización *f*
i sterilizzazione *f*
n sterilisatie, steriliseren *n*
d Sterilisation *f*, Sterilisieren *n*

5183 STICKINESS, dec
AFTERTACK
f tendance *f* au collant
e pegajosidad *f*
i appiccicatura *f*, collosità *f*
n kleverigheid, pikkigheid
d Klebrigkeit *f*

5184 STICKING PLASTER (US), mcy
ADHESIVE PLASTER (GB)
f emplâtre *m* adhésif, emplâtre *m* agglutinatif
e emplaste *m* adhesivo, esparadrapo *m*, parche *m* adhesivo
i cerotto *m*, impiastro *m* adesivo
n hechtpleister, plakpleister
d Heftpflaster *n*

5185 STICKING TAPE (US), cpt
ADHESIVE TAPE (GB),
SELF-ADHESIVE TAPE (GB)
f ruban *m* adhésif, ruban *m* collant
e cinta *f* adhesiva
i nastro *m* adesivo
n plakband *n*
d Klebeband *n*, Klebstreifen *m*

5186 STICKY, dec
TACKY
f poisseux adj
e pegajoso adj
i appiccicaticcio adj , attaccaticcio adj
n kleverig adj , pikkerig adj
d klebrig adj

5187 STIFF FLOW, pla
HARD FLOW
f basse fluidité
e mala fluidez *f*
i bassa fluidità *f*
n slecht vloeivermogen *n*
d harter Fluss *m*

5188 STIFFENER, rub
HARDENER
f durcisseur *m*, raidisseur *m*
e endurecedor *m*
i indurente *m*
n verhardingsmiddel *n*, verstijvingsmiddel *n*
d Härtungsmittel *n*, Versteifungsmittel *n*

5189 STIFFENING gen
f renforcement *m*
e refuerzo *m*
i rinforzo *m*
n versterking
d Verstärkung *f*

5190 STIFFENING, cle
STARCHING
f amidonnage *m*
e almidonado *m*
i inamidatura *f*
n stijfselen *n*, stijven *n*
d Stärken *n*, Steifen *n*

5191 STILLAGE, nut
DISTILLER'S WASH, SLOP, SPENT MASH, VINASSE
f drêches *pl*, vinasses *pl* de distillerie
e orujo *m*, vinazas *pl* de destilación
i burlanda *f*, trebbia *f*, vinacce *pl* distillate
n drab
d abgebrannte Würze *f*, Schlempe *f*, Spülicht *n*

5192 STILLINGIA OIL, dec
TALLOW TREE OIL
f huile *f* de stillinge
e aceite *m* de estilingia
i olio *m* di stillingia
n stillingiaolie
d Stillingiaöl *n*

5193 STINK DAMP (US), gas
HYDROGEN SULPHIDE (GB),
SULFURETTED HYDROGEN (US),
SULPHURETTED HYDROGEN (GB)
f hydrogène *m* sulfuré
e hidrógeno *m* sulfurado
i idrogeno *m* solforato
n zwavelwaterstof
d Schwefelwasserstoff *m*

5194 STIPPLING dec
f pochage *m*
e picado *m*
i picchiettatura *f*
n tamponneren *n*
d Tupfen *n*

5195 STIPPLING tex
f pointillage *m*
e punteado *m*
i punteggiamento *m*
n stippelen *n*
d Tüpfeln *n*

5196 STIPPLING PAINT dec
f peinture *f* à pocher
e pintura *f* de picado
i pittura *f* da picchiettare
n tamponneerverf
d Tupffarbe *f*

5197 STIRRING, mix
AGITATION
f agitation *f*
e agitación *f*
i agitazione *f*
n roeren *n*
d Rühren *n*

5198 STIRRING UP (GB), mix
SHAKING DOWN (US)
f agitation *f* du bain, brasure *f*
e agitación *f* del baño, braceado *m*
i agitazione *f* del bagno,
rimescolata *f*
n omroeren
d Rühren *n*

5199 STITCH WELDING, pla
TACK WELDING
f soudure *f* à la molette
e soldadura *f* discontinua
i saldatura *f* a tratti
n puntlassen *n*
d Heftschweissung *f*

5200 STOCK SOLUTION gen
f solution *f* de réserve, solution *f* mère
e solución *f* madre
i soluzione *f* di riserva
n voorraadsoplossing
d Vorratslösung *f*

5201 STONE CATCHING sin
f épierrage *m*
e colección *f* de piedras
i separazione *f* di pietre
n stenenvangst
d Steinfang *m*

5202 STONE WAX, fat
MONTAN WAX
f cire *f* de lignite
e cera *f* montana
i cera *f* montana
n montaanwas
d Montanwachs *n*

5203 STOPPER, dec
PUTTY
f mastic *m*
e masilla *f*
i mastice *m*
n stopverf
d Kitt *m*

5204 STOPPING bui,dec
f mastic *m* à reboucher
e masilla *f* para rellenar defectos, tapaporas *m*
i stucco *m*
n houtplastiek
d Spachtelkitt *m*

5205 STOPPING OFF elc
f isolation *f*
e aislamiento *m*
i isolante *m*
n afdekken *n*
d Abdecken *n*

5206 STORAGE sto
f dépôt *m*, emmagasinage *m*, emmagasinement *m*, stockage *m*
e almacenamiento *m*, depósito *m*
i deposito *m*, immagazzinaggio *m*, immagazzinamento *m*
n opslaan *n*
d Einlagerung *f*, Lagern *n*, Speichern *n*

5207 STORED MALT bre
f vieux malt *m*
e malta *f* vieja
i malto *m* vecchio
n oud mout *n*
d Altmalz *n*

5208 STORING, sto
HOLDING
f emmagasinage *m*, emmagasinement *m*

e almacenamiento *m*
i immagazzinamento *m*
n vergaren *n*,
verzamelen *n*
d Speichern *n*

5209 STOVING fou
f étuvage *m*
e estufado *m*
i essiccazione *f* in stufa
n drogen *n* in droogstoof
d Trocknen *n* im Trockenofen

5210 STOVING,
WET STOVING tex
f blanchiment *m* réducteur
e blanqueo *m* reductor
i candeggio *m* riduttore
n reducerend bleken *n*
d reduzierendes Bleichen *n*

5211 STOVING LACQUER dec
f laque *f* au feu
e laca *f* de fuego
i lacca *f* a fuoco
n vuurlak
d Einbrennlack *m*

5212 STRAIGHT-CHAIN
PARAFFIN fat
f paraffine *f* à chaîne droite
e parafina *f* de cadena recta
i paraffina *f* a catena lineare
n paraffine met onvertakte keten
d Paraffin *n* mit Geradkette

5213 STRAIGHT-RUN GASOLINE fue
f essence *f* de distillation
directe
e gasolina *f* de destilación
directa, gasolina *f* íntegra
i benzina *f* da distillazione
diretta
n destillatiebenzine, primaire
benzine
d Destillationsbenzin *n*,
Primärbenzin *n*

5214 STRAIN HARDENING,
WORK-HARDENING met
f écrouissage *m*
e endurecimiento *m* por
deformación en frío
i incrudimento *m* per deformazione
a freddo
n koudversteviging
d Kaltverfestigung *f*

5215 STRAINING cle,iwt
f criblage *m*
e coladura *f*
i colatura *f*
n ziften *n*
d Kolieren *n*, Seihen *n*, Sichten *n*

5216 STRAINING lea
f tendue *f*
e estirado *m*
i stringimento *m*
n spannen *n*
d Spannen *n*

5217 STRAINING CLOTH dye
f toile *f* de colature
e tela *f* coladora
i rete *f* per filtro
n zeefdoek
d Koliertuch *n*, Siebtuch *n*

5218 STRAW PAPER pap
f papier *m* de paille
e papel *m* de paja
i carta *f* di paglia
n stropapier *n*
d Strohpapier *n*

5219 STRAW PULP nut,pap
f pâte *f* de paille
e pulpa *f* de paja
i polpa *f* di paglia
n strostof *n*
d Strohstoff *m*, Strohzellstoff *m*

5220 STRAWBOARD pap
f carton *m* de paille
e cartón *m* de paja
i cartone *m* di paglia
n strokarton *n*
d Strohpappe *f*

5221 STREAMING POTENTIAL elc
f potentiel *m* de flux
e potencial *m* de flujo
i potenziale *m* di flusso
n stromingspotentiaal
d Strömungspotential *n*

5222 STRESS-RELIEF
ANNEALING,
STABILIZING ANNEALING htt,met
f recuit *m* de détente
e recocido *m* de atenuación de
tensiones
i ricottura *f* di distensione
n herstelgloeien *n*
d spannungsfreies Glühen *n*

5223 STRIKING elc
f amorçage *m*
e iniciación *f* de baño primario
i attivazione *f*

n aanbrengen *n* van basislaag
d Vorgalvanisieren *n*

5224 STRIKING HEAT, INITIAL TEMPERATURE — bre
f degré *m* initial, température *f* d'empâtage
e temperatura *f* inicial de mezcla
i temperatura *f* di miscela
n begintemperatuur van het beslag
d Einmaischtemperatur *f*

5225 STRIP HEATING — glu
f chauffage *m* à ruban
e calentamiento *m* de cinta
i riscaldamento *m* a nastro
n lintverwarming
d Streifenerwärmung *f*

5226 STRIPPABLE COATING, PEEL-OFF COATING — dec
f vernis *m* pelable
e barniz *m* pelable
i lacca *f* a spelamento
n afpelbare lak, striplak
d Abziehlack *m*

5227 STRIPPING — dec
f enlèvement *m* de la peinture
e remoción *f* de la pintura
i rimozione *f* della pittura
n verwijderen *n* van de verflaag
d Abkratzen *n* der Farbschicht

5228 STRIPPING — dis
f épuisement *m*
e agotamiento *m*, exhaustación *f*
i esaurimento *m*, strippaggio *m*
n uitdrijven *n*
d Abtrieb *m*, Austreiben *n*

5229 STRIPPING — elc
f dépouillage *m*
e eliminación *f* de la capa
i degalvanizzazione *f*
n verwijdering van de laag
d Entplattierung *f*

5230 STRIPPING — fue
f dégazolinage *m*, strippage *m*
e extracción *f* de fracciones livianas
i estrazione *f* di frazioni leggeri
n strippen *n*
d Strippen *n*

5231 STRIPPING — pho
f pelliculage *m* d'un film
e despegadura *f* de la emulsión
i distaccamento *m* dell'emulsione
n aftrekken *n* van de emulsie
d Abziehen *n* der Emulsion

5232 STRIPPING — tex
f élimination *f* des colorants
e eliminación *f* de los colorantes
i eliminazione *f* dei coloranti
n kleurstofverwijdering
d Farbstoffbeseitigung *f*

5233 STRIPPING COMPOUND — elc
f composé *m* de dépouillage
e compuesto *m* de despojamiento
i materiale *m* da strappamento
n scheidingsmiddel *n*
d Trennsubstanz *f*

5234 STRONG CLAY, FOUL CLAY, PLASTIC CLAY, PURE CLAY — bui
f argile *f* plastique
e arcilla *f* plástica
i galestro *m*
n zuivere klei
d reiner Ton *m*, reines Aluminiumsilikat *n*

5235 STRONG ELECTROLYTE — elc
f électrolyte *m* fort
e electrólito *m* fuerte
i elettrolito *m* forte
n sterke elektrolyt
d starker Elektrolyt *m*

5236 STRUCTURAL FORMULA, CONSTITUTIONAL FORMULA, GRAPHIC FORMULA — gen
f formule *f* de constitution
e fórmula *f* de constitución, fórmula *f* estructural
i formula *f* di struttura
n structuurformule
d Konstitutionsformel *f*, Strukturformel *f*

5237 STRUCTURAL VISCOSITY — dec
f viscosité *f* de structure
e viscosidad *f* estructural
i viscosità *f* strutturale
n structuurviscositeit
d Strukturviskosität *f*

5238 STUFFING — lea
f graissage *m*
e engrasación *f*
i trattamento *m* con grasso
n invetten *n*
d Einfetten *n*

5239 SUBERIC ACID — dye,mcy
f acide *m* subérique

e ácido *m* subérico
i acido *m* suberico
n kurkzuur *n*, suberinezuur *n*
d Korksäure *f*, Suberinsäure *f*

5240 SUBLIMATE gen
f sublimé *m*
e sublimado *m*
i sublimato *m*
n sublimaat *n*
d Sublimat *n*

5241 SUBLIMATION, gen
SUBLIMING
f sublimation *f*
e sublimación *f*
i sublimazione *f*
n sublimatie, sublimeren *n*
d Sublimation *f*, Sublimieren *n*

5242 SUBMERGED ARC-WELDING wel
f soudure *m* sous flux conducteur
e soldadura *f* sobre flujo electroconductor
i saldatura *f* ad arco sommerso
n booglassen *n* onder poederdak
d Unterpulverschweissen *n*

5243 SUBMERGED COMBUSTION bur
f combustion *f* noyée
e combustión *f* sumergida
i combustione *f* sommersa
n dompelverbranding
d Innenverbrennung *f* innerhalb der Flüssigkeit

5244 SUBSTANCE gen
f substance *f*
e substancia *f*, sustancia *f*
i sostanza *f*
n stof, substantie
d Stoff *m*, Substanz *f*

5245 SUBSTANTIVE DYE dye
f colorant *m* substantif
e colorante *m* directo
i colorante *m* sostantivo
n substantieve kleurstof, zoutkleurstof
d Direktfarbe *f*, substantiver Farbstoff *m*

5246 SUBSTANTIVE DYEING dye
f teinture *f* substantive
e teñido *m* directo
i tintura *f* sostantiva
n substantief verven *n*
d Direktfärbung *f*

5247 SUBTERRANEOUS fue
GASIFICATION
f gazéification *f* souterraine
e gasificación *f* subterránea
i gassificazione *f* sotterranea
n ondergrondse vergassing
d unterirdische Vergasung *f*

5248 SUCCINIC ACID, oic
AMBER ACID, ETHYLENE SUCCINIC ACID
f acide *m* d'ambre, acide *m* d'éthylène succinique, acide *m* succinique
e ácido *m* de ámbar, ácido *m* succínico
i acido *m* d'ambra, acido *m* succinico
n barnsteenzuur *n*, ethyleenbarnsteenzuur *n*
d Äthylenbernsteinsäure *f*, Bernsteinsäure *f*

5249 SUCKING OFF gen
f aspiration *f*
e aspiración *f*
i aspirazione *f*
n afzuigen *n*
d Absaugen *n*

5250 SUCRASE, sin
INVERTASE (GB), INVERTIN (US), SACCHARASE
f invertase *f*
e invertasa *f*
i invertasi *f*
n invertase, saccharase
d Invertase *f*, Invertin *n*, Saccharase *f*

5251 SUCROSE, sin
CANE SUGAR
f sucre *m* de canne
e azúcar *m* de caña
i zucchero *m* di canna
n rietsuiker
d Rohrzucker *m*

5252 SUCTION GAS fue
f gaz *m* pauvre d'aspiration
e gas *m* pobre de aspiración
i gas *m* povero d'aspirazione
n zuiggas *n*
d Sauggas *n*

5253 SUDS (US), soa
LATHER (GB)
f mousse *f* de savon
e espuma *f* de jabón
i schiuma *f* di sapone
n zeepschuim *n*
d Seifenschaum *m*

5254 SUFFOCATING GAS gas,mit
f gaz *m* suffocant
e gas *m* sofocante
i gas *m* soffocante
n verstikkend gas *n*
d Stickgas *n*

5255 SUGAR sin
f sucre *m*
e azúcar *m*
i zucchero *m*
n suiker
d Zucker *m*

5256 SUGAR BEET sin
f betterave *f*
e remolacha *f*
i barbabietola *f*
n suikerbiet
d Zuckerrübe *f*

5257 SUGAR CANE, sin
CANE
f canamelle *f*, cannamelle *f*, canne *f* à sucre
e caña *f* de azúcar, caña *f* dulce, caña *f* melar
i canna *f* da zucchero, cannamele *f*
n suikerriet *n*
d Zuckerrohr *n*

5258 SUGAR LIME, mcy
CALCIUM SUCRATE
f sucrate *m* de calcium
e sacarato *m* de calcio
i saccarato *m* di calcio
n calciumsacharaat *n*, suikerkalk
d Zuckerkalk *m*

5259 SUINT, fat
YOLK
f suin *m*, suint *m*
e juarda *f*, suarda *f*
i grasso *m* proveniente da sudore
n wolvet *n*
d Schafsschweiss *m*, Wollfett *n*

5260 SULFATE COOKING (US), pap
SULFATE PROCESS (US),
SULPHATE PROCESS (GB)
f procédé *m* de cuisson au sulfate
e procedimiento *m* de cocción al sulfato
i processo *m* di cottura al solfato
n sulfaatkookproces *n*
d Sulfatkochverfahren *n*

5261 SULFATE OF ALUMINIUM (US), mat
SULPHATE OF ALUMINIUM (GB),
FILTER ALUM
f sulfate *m* d'aluminium
e sulfato *m* de alumina
i solfato *m* d'alluminio
n aluminiumsulfaat *n*
d Aluminiumsulfat *n*, schwefelsauere Tonerde *f*, Tonerdesulfat *n*

5262 SULFATE PULP (US), pap
SULPHATE PULP (GB),
KRAFT PULP
f pâte *f* au sulphate
e pulpa *f* de kraft, pulpa *f* de sulfato
i polpa *f* a solfato
n sulfaatpulp
d Natronzellstoff *m*, Sulfatzellstoff *n*

5263 SULFATE TURPENTINE (US), oic
SULPHATE TURPENTINE (GB),
KRAFT TURPENTINE
f essence *f* de térébenthine résiduaire, essence *f* de térébenthine suédoise
e esencia *f* de trementina al sulfato
i trementina *f* al solfato
n sulfaatterpentijn
d schwedisches Terpentinöl *n*, Sulfatterpentinöl *n*, Zelluloseterpentinöl *n*

5264 SULFATING ROASTING (US), met
SULPHATING ROASTING (GB)
f grillage *m* sulfatant
e tostación *f* sulfatizante
i torrefazione *f* a solfatazione
n sulfaterend roosten *n*
d sulfatierendes Rösten *n*

5265 SULFIDE TONING (US), pho
SULPHIDE TONING (GB)
f virage *m* par sulfation
e viraje *m* por sulfatación
i viraggio *m* per solfatazione
n zwavelkleurig
d Schwefeltonung *f*

5266 SULFITATION (US), sin
SULPHITATION (GB)
f sulfitage *m*
e sulfitación *f*
i solfitazione *f*
n zwavelen *n*
d Schwefeln *n*

5267 SULFITE BLACK LIQUOR (US), pap
SULPHITE BLACK LIQUOR (GB)
f liqueur *f* noire
e líquido *m* residual
i lisciva *f* nera

n zwarte loog
d Schwarzlauge *f*

5268 SULFITE PULP (US), pap
SULPHITE PULP (GB)
f pâte *f* au sulfite
e pulpa *f* de sulfito
i polpa *f* al solfito
n sulfietcellulose
d Sulfitzellulose *f*

5269 SULFITE WASTE LIQUOR (US), sin
SPENT SULPHITE LIQUOR (GB)
f lessive *f* résiduaire sulfitique
e lejía *f* residual sulfítica
i lisciva *f* di rifiuto solfitica
n afgewerkte sulfietvloeistof
d Sulfitablauge *f*

5270 SULFONATION (US), oic
SULPHONATION (GB)
f sulfonation *f*
e sulfanación *f*
i solfonazione *f*
n sulfoneren *n*
d Sulfieren *n*, Sulfonieren *n*

5271 SULFUR BATH VULCANIZATION (US), vul
SULPHUR BATH VULCANIZATION (GB)
f vulcanisation *f* en bain de soufre
e vulcanización *f* en baño de azufre
i vulcanizzazione *f* in bagno di zolfo
n vulcanisatie in gesmolten zwavel
d Vulkanisation *f* in geschmolzenem Schwefel

5272 SULFUR BLOOMING (US), aic
SULPHUR BLOOMING (GB)
f efflorescence *f* du soufre
e eflorescencia *f* de azufre
i efflorescenza *f* di zolfo
n uitbloemen *n* van zwavel
d Ausschwefelung *f*

5273 SULFUR CEMENT (US), bui
SULPHUR CEMENT (GB),
METALLURGICAL CEMENT
f ciment *m* de fer, ciment *m* métallurgique
e cemento *m* de azufre
i cemento *m* di zolfo, mastice *m* di ferro
n ijzercement *n*
d Eisenkitt *m*, Schwefelkitt *m*, Schwefelzement *m*

5274 SULFUR CHLORIDE VULCANIZATION (US), rub
SULPHUR CHLORIDE VULCANIZATION (GB)
f vulcanisation *f* au chlorure de soufre
e vulcanización *f* al cloruro de azufre
i vulcanizzazione *f* al cloruro di zolfo
n chloorzwavelvulcanisatie
d Chlorschwefelvulkanisation *f*

5275 SULFURATED TURPENTINE OIL (US), fat
SULPHURATED TURPENTINE OIL (GB), HAARLEM OIL
f huile *f* de Haarlem
e aceite *m* de Harlem
i olio *m* di Haarlem
n haarlemmerolie
d Haarlemer Öl *n*

5276 SULFURETTED HYDROGEN (US), gas
SULPHURETTED HYDROGEN (GB),
HYDROGEN SULPHIDE (GB),
STINK DAMP (US)
f hydrogène *m* sulfuré
e hidrógeno *m* sulfurado
i idrogeno *m* solforato
n zwavelwaterstof
d Schwefelwasserstoff *m*

5277 SULFURIC ACID (US), sua
SULPHURIC ACID (GB)
f acide *m* sulfurique
e ácido *m* sulfúrico
i acido *m* solforico
n zwavelzuur *n*
d Schwefelsäure *f*

5278 SULFURING (US), hyg
SULPHUR FUMIGATION (GB)
f fumigation *f* à gaz sulfurant, sulfitage *m*
e fumigación *f* con azufre
i fumigazione *f* allo zolfo
n zwavelen *n*
d Ausräuchern *n*, Schwefeln *n*

5279 SULLING, elc
RUSTING
f oxydation *f* après décapage
e oxidación *f* después del decapaje
i ossidazione *f* dopo il decapaggio
n oxydatie na beitsen
d Oxydation *f* nach dem Beizen

5280 SUMACH WAX, JAPAN TALLOW, JAPAN WAX fat
f cire *f* du Japon
e cera *f* del Japón
i cera *f* del Giappone
n Japanse was
d Japanwachs *n*, Japantalg *m*

5281 SUN BLEACHING tex
f blanchiment *m* par la lumière solaire
e blanqueado *m* por la luz solar
i candeggio *m* per la luce solare
n bleken *n* aan zonlicht
d Sonnenbleichung *f*

5282 SUNFLOWER OIL fat
f huile *f* de tournesol
e aceite *m* de girasol
i olio *m* di girasole
n zonnebloemolie
d Sonnenblumenöl *n*

5283 SUPERACIDITY aic
f superacidité *f*, suracidité *f*
e hiperacidez *f*
i superacidità *f*
n peraciditeit
d Perazidität *f*, Superazidität *f*, Übersäuerung *f*

5284 SUPERCHLORINATION, EXCESS CHLORINATION, HIGH CHLORINATION aic
f chloration *f* à dose élevée, surchloration *f*
e sobrecloración *f*
i sovraclorazione *f*, superclorazione *f*
i overchloreren *n*
d Hochchlorung *f*, Überchlorung *f*, Überschusschlorung *f*

5285 SUPERCOOLING, SURFUSION, UNDERCOOLING clo
f surfusion *f*
e sobrefusión *f*
i surraffreddamento *m*
i onderkoeling
d Unterkühlung *f*

5286 SUPERFICIAL LAYER coa
f couche *f* superficielle
e capa *f* superficial
i strato *m* superficiale
n oppervlaklaag
d Oberflächenschicht *f*

5287 SUPERFINISHING, ULTRAFINISH pol
f superfinissage *m*
e superpulido *m*
i soprafinitura *f*
n superfijnen *n*
d Spiegelpolitur *f*, Superfeinen *n*

5288 SUPERHEATED STEAM boi
f vapeur *f* surchauffée
e vapor *m* recalentado
i vapore *m* surriscaldato
n droge stoom, oververhitte stoom
d Trockendampf *m*, überhitzter Dampf *m*

5289 SUPERHEATING boi
f surchauffage *m*
e sobrecalentamiento *m*
i surriscaldamento *m*
n oververhitting
d Überheizung *f*

5290 SUPERNATANT LIQUID gen
f liquide *m* surnageant
e líquido *m* sobrenadante
i liquido *m* sovrastante
n bovenstaande vloeistof
d obenstehende Flüssigkeit *f*

5291 SUPERSATURATION cle,hyg,tex
f sursaturation *f*
e supersaturación *f*
i soprassaturazione *f*
n oververzadiging
d Übersättigung *f*

5292 SUPERSULFATED CEMENT (US), SUPERSULPHATED CEMENT (GB) bui
f ciment *m* sursulfaté
e cemento *m* sulfático
i cemento *m* solfatico
n sulfaatcement *n*
d Sulfathüttenzement *m*, Sulfatzement *m*

5293 SURFACE ACTIVE AGENT, SURFACTANT, WETTING AGENT gen
f agent *m* mouillant, agent *m* tensio-actif
e agente *m* humector, agente *m* tensoactivo
i sostanza *f* tensio-attiva
n bevochtiger
d Netzmittel *n*

5294 SURFACE ACTIVITY gen
f activité *f* superficielle
e actividad *f* superficial
i attività *f* superficiale
n oppervlakactiviteit
d Oberflächenaktivität *f*

5295 SURFACE AERATION iwt,sew
f aération *f* par surface,
aération *f* superficielle
e aireación *f* superficial
i aerazione *f* superficiale
n oppervlakaëratie
d Oberflächenbelüftung *f*

5296 SURFACE BLUSH dec
f déteinte *f* superficielle
e descoloración *f* superficial
i annebbiamento *m* superficiale
n oppervlakverkleuring
d Oberflächenverfärbung *f*

5297 SURFACE BOILING, boi
LOCAL BOILING
f ébullition *f* locale
e ebullición *f* local
i ebollizione *f* locale
n lokaal koken *n*
d örtliches Sieden *n*,
Verdampfen *n* in unterkühlter
Flüssigkeit

5298 SURFACE COATING dec
MATERIALS,
PAINTING MATERIALS
f matériaux *pl* pour la peinture,
peintures *pl* et vernis *pl*
e materiales *pl* para pintar
i materiali *pl* per pitturare
n verfwaren *pl*
d Anstrichmittel *pl*, Farbwaren *pl*

5299 SURFACE COATING RESIN coa,pla
f revêtement *m* superficiel
plastique
e revestimiento *m* superficial
plástico
i rivestimento *m* superficiale
plastico
n kunstharsbedekking
d Kunststoffüberzug *m*

5300 SURFACE COMBUSTION, fue
FLAMELESS COMBUSTION
f combustion *f* sans flamme
e combustión *f* sin llama
i combustione *f* senza fiamma
n vlamloze verbranding
d flammenlose Verbrennung *f*

5301 SURFACE CONVERTING, met
CASE HARDENING
f trempe *f* de surface
e temple *m* de superficie
i tempera *f* di superficie
n inzetharding, oppervlakharding
d Einsatzhärtung *f*,
Oberflächenhärtung *f*

5302 SURFACE COOLING, clo
INDIRECT COOLING
f refroidissement *m* superficiel
e enfriamiento *m* superficial
i raffreddamento *m* superficiale
n oppervlakkoeling
d Oberflächenkühlung *f*

5303 SURFACE CORROSION cor
f corrosion *f* superficielle
e corrosión *f* superficial
i corrosione *f* superficiale
n oppervlakcorrosie
d Oberflächenkorrosion *f*

5304 SURFACE FERMENTA- bre,fmi
TION,
TOP FERMENTATION
f fermentation *f* haute
e fermentación *f* alta
i fermentazione *f* alta
n bovengisting
d Obergärung *f*

5305 SURFACE FINISHING pol
f finissage *m* superficiel
e acabado *m* superficial
i finitura *f* superficiale
n oppervlakpolijsten *n*
d Oberflächenpolierung *f*

5306 SURFACE GRINDING pol
f rectification *f* superficielle
e rectificación *f* superficial
i rettificazione *f* superficiale
n oppervlakslijpen *n*
d Oberflächenschleifen *n*

5307 SURFACE HARDENING met
f durcissement *m* superficiel
e endurecimiento *m* superficial
i indurimento *m* superficiale
n oppervlakharding
d Oberflächenhärtung *f*

5308 SURFACE HAZE dec
f trouble *m* superficiel
e enturbiamiento *m* superficial
i offuscamento *m* superficiale,
torbidità *f* superficiale
n oppervlakwaasvorming
d Oberflächentrübung *f*

5309 SURFACE IRRIGATION, sew
BROAD IRRIGATION,
OVERHEAD IRRIGATION
f irrigation *f* superficielle
e riego *m* con aguas cloacales,
riego *m* con aguas negras
i disperdimento *m* per spandimento
su terreni naturali permeabili,

irrigazione *f* superficiale
n sproeibevloeiing
d Oberflächenberieselung *f*, Rieselfeldreinigung *f*

5310 SURFACE OF LIQUID gen
f niveau *m* de liquide, surface *f* de liquide
e nivel *m* de líquido, superficie *f* de líquido
i livello *m* di liquido, superficie *f* di liquido
n vloeistofspiegel
d Flüssigkeitsspiegel *m*

5311 SURFACE SCUM, FLOATING LAYER iwt,sew
f boue *f* flottante, chapeau *m* d'écume
e fango *m* flotante, nata *f* espumosa
i cappellaccio *m*, fango *m* galleggiante
n schuimlaag
d Schwimmdecke *f*, Schwimmschlamm *m*

5312 SURFACE TEMPERING htt,met
f trempe *f* superficielle
e temple *m* superficial
i tempera *f* superficiale
n oppervlakontharding
d Oberflächenenthärtung *f*

5313 SURFACE YEAST, TOP FERMENTING YEAST bre,fmi
f levure *f* haute
e levadura *f* de fermentación alta
i lievito *m* per alta fermentazione
n bovengist
d Oberhefe *f*

5314 SURFUSION, SUPERCOOLING, UNDERCOOLING clo
f surfusion *f*
e sobrefusión *f*
i surraffreddamento *m*
n onderkoeling
d Unterkühlung *f*

5315 SURROGATE gen
f succédané *m*
e sucedáneo *m*
i succedaneo *m*, surrogato *m*
n surrogaat *n*
d Ersatz *m*

5316 SURROSION sor
f augmentation *f* du poids par corrosion
e aumento *m* del peso por corrosión
i aumento *m* del peso per corrosione
n gewichtstoeneming door corrosie
d Gewichtszunahme *f* durch Korrosion

5317 SUSPENDED MATTER iwt,sew
f matière *f* en suspension
e substancias *pl* en suspensión
i sostanze *pl* in sospensione
n zwevende deeltjes *pl*
d Schwebestoffe *pl*, Sinkstoffe *pl*

5318 SUSPENDING AGENT gen
f agent *m* de suspension
e agente *m* de suspensión, agente *m* suspendedor
i agente di sospensione
n suspensiemiddel *n*
d Suspensionsmittel *n*

5319 SUSPENSION gen
f suspension *f*
e suspensión *f*
i sospensione *f*
n suspensie
d Aufschlämmung *f*, Suspension *f*

5320 SUSPENSION BLACK, ANILINE BLACK, OXIDATION BLACK dye
f noir *m* d'aniline
e negro *m* de anilina
i nero *m* d'anilina
n anilinezwart *n*
d Anilinschwarz *n*

5321 SUSPENSION POLYMERIZATION oic
f polymérisation *f* en suspension
e polimerización *f* en suspensión
i polimerizzazione *f* in sospensione
n polymerisatie in suspensie
d Peripolymerisation *f*, Polymerisation *f* in Suspension

5322 SWEATING dec
f suintement *m*
e exudación *f*
i trasudamento *m*
n zweten *n*
d Schwitzen *n*

5323 SWEATING fue
f ressuage *m*
e resudación *f*
i sudata *f*, trasudazione *f*

n uitzweten *n*
d Ausschwitzen *n*, Schwitzen *n*

5324 SWEATING OUT, EXUDATION rub
f exsudation *f*
e exudación *f*
i essudazione *f*
n exsudatie, uitzweten *n*
d Ausschwitzen *n*, Exsudation *f*

5325 SWEET ALMOND OIL fat,soa
f huile *f* d'amandes
e aceite *m* de almendras
i olio *m* di mandorle
n amandelolie
d Mandelöl *n*

5326 SWEET OIL, OLIVE OIL fat,nut
f huile *f* d'olive
e aceite *m* de oliva
i olio *m* d'oliva
n olijfolie
d Olivenöl *n*

5327 SWEET WATER sin
f eau *f* sucrée
e agua *f* azucarada
i acqua *f* zuccherata
n suikerwater *n*
d Absüsser *n*, Absüssaft *m*

5328 SWEETENING fue
f adoucissement *m*, traitement *m* au plombite
e desmercaptanización *f*
i trattamento *m* al plombite
n plumbiteren *n*, stankvrij maken *n*
d Süssung *f*

5329 SWEETENING nut
f dolcification *f*
e dulcificación *f*
i addolcimento *m*
n zoet maken *n*, zoeten *n*
d Süssmachen *n*

5330 SWEETENING AGENT sin
f dolcifiant *m*
e dulcificante *m*
i dolcificante *m*
n zoetstof
d Süssstoff *m*

5331 SWEETENING OFF, SWEETENING OUT sin
f désucrage *m*, sucretage *m*
e desazucaramiento *m*
i ricupero *m* dello zucchero dalla melassa
n melasseontsuikering
d Absüssen *n*, Anschwänzen *n*

5332 SWELLING gen
f gonflement *m*
e hinchamiento *m*
i gonfiamento *m*
n zwellen *n*
d Aufquellen *n*, Quellen *n*

5333 SWELLING AGENT tex
f agent *m* de gonflement
e agente *m* de hinchamiento
i agente *m* di gonfiamento
n zwelmiddel *n*
d Quellmittel *n*, Schwellmittel *n*

5334 SWELLING CLAY col
f argile *f* colloïdale, argile *f* foisonnante
e arcilla *f* coloidal
i argilla *f* colloidale
n colloïdale klei
d Quellton *m*

5335 SYNTANS, SYNTHETIC TANNING AGENTS tan
f tanins *pl* synthétiques
e curtientes *pl* sintéticos
i concie *pl* sintetiche
n synthetische looistoffen *pl*
d synthetische Gerbstoffe *pl*

5336 SYNTHETIC FIBRE (GB), MAN-MADE FIBER (US) tex
f fibre *f* artificielle
e fibra *f* artificial
i fibra *f* artificiale
n kunstvezel
d Chemiefaser *f*

5337 SYNTHETIC RESIN, ARTIFICIAL RESIN pla
f résine *f* synthétique
e resina *f* sintética
i resina *f* sintetica
n kunsthars *n*
d Kunstharz *n*

5338 SYNTHETIC RUBBER (GB), ARTIFICIAL RUBBER, MAN-MADE RUBBER (US) rub
f caoutchouc *m* artificiel, caoutchouc *m* synthétique, élastomère *f* synthétique
e caucho *m* sintético, goma *f* artificial, goma *f* sintética
i gomma *f* sintetica
n kunstrubber, synthetische rubber
d Kunstgummi *m*, synthetischer Gummi *m*, synthetischer Kautschuk *m*

5339 SYNTHETIC WOOL (GB), ARTIFICIAL WOOL, MAN-MADE WOOL (US) tex
f laine *f* artificielle
e lana *f* artificial
i lana *f* artificiale
n kunstwol
d Zellwolle *f*

5340 SYRUP (GB), SIRUP (US) sin
f sirop *m*
e jarabe *m*
i sciroppo *m*
n stroop
d Sirup *m*

T

5341 TABLET mcy
f comprimé *m*, pastille *f*
e comprimido *m*, pastilla *f*, tableta *f*
i compressa *f*, pasticca *f*
n pastille, tablet *n*
d Pastille *f*, Tablette *f*

5342 TABLET, PELLET pla
f pastille *f* précomprimée
e pastilla *f* precomprimida
i pasticca *f* precompressa
n voorgeperst tablet *n*
d vorgepresste Tablette *f*

5343 TACK COAT coa,glu
f couche *f* de liaison
e capa *f* adhesiva
i rivestimento *m* adesivo
n hechtlaag, kleeflaag
d Klebeschicht *f*

5344 TACK WELDING, STITCH WELDING pla
f soudure *f* à la molette
e soldadura *f* discontinua
i saldatura *f* a tratti
n puntlassen *n*
d Heftschweissung *f*

5345 TACKIFIER glu
f agent *m* poisseux
e agente *m* de pegajosidad
i agente *m* d'appiccicosità
n kleverigmakend middel *n*
d Klebrigmacher *m*

5346 TACKY, STICKY dec
f poisseux adj
e pegajoso adj
i appiccicaticcio adj, attaccaticcio adj
n kleverig adj, pikkerig adj
d klebrig adj

5347 TAIL FRACTION dis,fue
f fraction *f* de queue
e fracción *f* de cola
i frazione *f* di coda
n staartfractie
d Endfraktion *f*

5348 TAIL LIQUOR sod
f liqueur *f* de queue, solution *f* finale
e lejía *f* de coda
i lisciva *f* finale di distillazione
n destillatieloog
d Chlorcalciumendlauge *f*, Destillationsendlauge *f*

5349 TAIL LIQUOR, SPENT LYE gen
f lessive *f* épuisée
e lejía *f* agotada
i sottolisciva *f*
n afgewerkte loog
d Abfallauge *f*, Ablauge *f*

5350 TAILING dye
f changement *m* de couleur progressif
e cambio *m* de color progresivo
i cambio *m* di colore progressivo
n geleidelijke kleurovergang
d allmähliche Farbübergang *m*

5351 TAILINGS bre
f déchets *pl* d'orge
e desperdicios *pl* de cebada
i scarto *m* dell'orzo
n gerstafval
d Gersteabfälle *pl*

5352 TAILINGS cer
f résidu *m* de broyage
e residuo *m* de molimiento
i residuo *m* di macinatura
n maalresten *pl*
d Mahlrückstand *m*

5353 TAILINGS min
f refus *pl* de flotation
e productos *pl* esteriles
i scarti *pl* di miniera
n ertsafval *n*
d Waschabgänge *pl*

5354 TAILINGS rub
f particules *pl* non-dévulcanisées dans le régénéré
e partículas *pl* no regeneradas en el regenerado
i particelle *pl* non devulcanizzate nel rigenerato
n niet-geregenereerde rubberdeeltjes *pl* in het regeneraat
d Gummirückstände *pl* im Regenerat

5355 TAILINGS, SKIMMERS bre
f graines *pl* folles

e cebada *f* flotante
i orzo *m* galleggiante
n drijvende gerst
d Schwimmgerste *f*

5356 TALC (GB), TALCUM (GB), FRENCH CHALK (US), SOAPSTONE — mat
f talc *m*
e piedra *f* de jabón, talco *m*
i talco *m*
n talk
d Talkum *n*

5357 TALL OIL, TALLOL, LIQUID RESIN — fat,pap
f huile *f* de Thal, résine *f* liquide
e líquido *m* resinoso aceitoso, resina *f* líquida
i olio *m* di sego
n tallolie
d Tallöl *n*

5358 TALLOW — fat
f suif *m*
e sebo *m*
i sego *m*
n talg, talk
d Talg *m*, Unschlitt *n*

5359 TALLOW FATTY ACID — fat
f acide *m* gras de suif
e ácido *m* graso de sebo
i acido *m* grasso di sego
n talgvetzuur *n*
d Talgfettsäure *f*

5360 TALLOW OIL — fat
f huile *f* de suif
e aceite *m* de sebo
i olio *m* di sego
n talgolie
d Talgöl *n*

5361 TALLOW TREE OIL, STILLINGIA OIL — dec
f huile *f* de stillinge
e aceite *m* de estilingia
i olio *m* di stillingia
n stillingiaolie
d Stillingiaöl *n*

5362 TALLOWING — fat
f graissage *m*
e engrase *m*
i trattamento *m* al sego
n invetten *n*
d Einfetten *n*

5363 TAMPING — cer
f serrage *m*
e apisonado *m*
i pigiatura *f*
n aanstampen *n*
d Verdichten *n*

5364 TAN LIQUOR — tan
f jus *m* tannant, jusée *f*
e licor *m* curtiente
i liquido *m* conciante
n looivocht *n*
d Gerbebrühe *f*, Lohbrühe *f*

5365 TANGENTIAL FLOW — mix
f flux *m* tangentiel
e flujo *m* tangencial
i flusso *m* tangenziale
n tangentiële stroming
d Tangentialströmung *f*

5366 TANK DEVELOPMENT — pho
f développement *m* chronométrique, développement *m* en cuve, développement *m* lent
e revelado *m* en tanque
i sviluppo *m* lento
n standontwikkeling, tankontwikkeling
d Standentwicklung *f*, Tankentwicklung *f*, Zeitentwicklung *f*

5367 TANK VOLTAGE — elc
f tension *f* de cuve
e tensión *f* de cuba
i tensione *f* di cella
n badspanning
d Badspannung *f*

5368 TANKAGE — fer
f issues *pl* de boucherie
e desperdicios *pl* de matanza, residuos *pl* de matadero
i residui *pl* di macelleria
n slachterijafval *n*
d getrocknete tierische Überreste *pl*

5369 TANNIC MORDANT DYE — dye
f colorant *m* mordant et tannant
e colorante *m* mordiente y curtiente
i colorante *m* mordente e conciante
n looiende beitskleurstof
d gerbender Beizenfarbstoff *m*

5370 TANNIN, QUERCITANNIC ACID — tan
f tanin *m*, tannin *m*
e tanino *m*

i tannino *m*
n looizuur *n*
d Gerbsäure *f*, Tannin *n*

5371 TANNING pho,tan
f tannage *m*
e curtido *m*, curtimiento *m*
i concia *f*
n looien *n*
d Gerbung *f*, Lohung *f*

5372 TANNING IN DRUMS tan
f tannage *m* en cuves
e curtido *m* en cubas
i concia *f* in tini
n looien *n* in vaten
d Küpengerbung *f*

5373 TAP WATER, CITY WATER, MUNICIPAL WATER hyg
f eau *f* du robinet, eau *f* potable
e agua *f* corriente, agua *f* potable
i acqua *f* di rubinetto, acqua *f* potabile
n drinkwater *n*, leidingwater *n*
d Leitungswasser *n*, Trinkwasser *n*

5374 TAPPING gen
f débouchage *m*
e sangrado *m*, sangría *f*
i spillatura *f*
n aftappen *n*
d Abstich *m*

5375 TAR mat
f goudron *m*
e alquitrán *m*, brea *f*
i catrame *m*
n teer
d Teer *m*

5376 TAR ACID mcy
f acide *m* de goudron
e ácido *m* de alquitrán
i acido *m* di catrame
n teerzuur *n*
d Teersäure *f*

5377 TAR BITUMEN bui
f goudron *m* butimineux
e mezcla *f* de betún y alquitrán
i catrame *m* bituminoso
n bitumenteer
d Teerbitumen *n*

5378 TAR BOARD, TAR PAPER pap
f carton *m* asphalté, papier *m* asphalté
e cartón *m* asfaltado
i cartone *m* catramato
n asfaltpapier *n*, teerpapier *n*
d Teerpapier *n*, Teerpappe *f*

5379 TAR OIL fat
f huile *f* de goudron
e aceite *m* de alquitrán
i olio *m* di catrame
n teerolie
d Teeröl *n*

5380 TAR SEPARATION fue
f séparation *f* de goudron
e desalquitranación *f*
i separazione *f* del catrame
n teerafscheiding
d Teerscheidung *f*

5381 TARING mea
f tarage *m*
e determinación *f* de la tara
i tarare v
n tarreren *n*
d Tarrieren *n*

5382 TARNISHING cor,met
f ternissure *f*
e empañadura *f*, pérdida *f* de brillo
i appannatura *f*, perdita *f* di brillo
n dofheid, matheid
d Mattigkeit *f*, Oberflächentrübung *f*

5383 TARRAGON OIL fat
f huile *f* d'estragon
e aceite *m* de estragón
i olio *m* di dragoncello, olio *m* di targone
n dragonolie
d Estragonöl *n*

5384 TARTARIC ACID cor
f acide *m* tartrique
e ácido *m* tartárico
i acido *m* tartarico
n wijnsteenzuur *n*
d Weinsäure *f*

5385 TAWING tan
f mégissage *m*
e curtido *m* en blanco
i concia *f* all'allume
n witlooien *n*
d Weissgerbung *f*

5386 TEAR GAS, LACHRYMATORY GAS (GB), LACRIMATORY GAS (US) mit
f gaz *m* lacrymogène

e gas *m* lacrimógeno
i gas *m* lacrimogeno
n traangas *n*
d Tränengas *n*

5387 TEAR TEST tes
f essai *m* de la résistance à la déchirure
e ensayo *m* de la resistencia al desgarramiento
i collaudo *m* della resistenza alla lacerazione
n scheursterkteproef
d Einreissfestigkeitsprüfung *f*

5388 TEARING, RUNNING dec
f coulage *m*, formation *f* de coulures
e derramiento *m*
i scolamento *m*
n vorming van lopers
d Abrinnen *n*, Läuferbildung *f*

5389 TEEMING, CASTING, POURING fou,pla
f coulée *f*
e colada *f*
i colata *f*
n gieten *n*
d Giessen *n*

5390 TELOMERIZATION oic
f télomérisation *f*
e telomerización *f*
i telomerizzazione *f*
n telomerisatie
d Telomerisation *f*

5391 TEMPER gla
f tension *f* résiduelle
e tensión *f* residual
i tensione *f* residua
n restspanning
d Restspannung *f*

5392 TEMPER, DISTEMPER dec
f détrempe *f*
e templa *f*
i tempera *f*
n lijmverf, temperaverf
d Leimfarbe *f*, Temperafarbe *f*

5393 TEMPERED GLASS, TOUGHENED GLASS gla
f verre *m* durci, verre *m* trempé
e vidrio *m* endurecido, vidrio *m* templado
i vetro *m* indurito, vetro *m* temprato
n gehard glas *n*
d gehärtetes Glas *n*

5394 TEMPERING cer
f délayage *m*
e amasado *m*
i preparazione *f* della pasta
n aanmaken *n*
d Anmachen *n*

5395 TEMPERING gla
f recuit *m*
e recocido *m*
i ricottura *f*
n langzaam koelen *n*, temperen *n*
d langsames Kühlen *n*

5396 TEMPERING htt,met
f revenu *m*
e revenido *m*
i rinvenimento *m*
n ontlaten *n*
d Anlassen *n*

5397 TEMPERING p'
f recuit *m*
e recocido *m*
i rinvenimento *m*
n temperen *n*
d Tempern *n*

5398 TEMPERING COLORS (US), TEMPERING COLOURS (GB) dec
f couleurs *pl* de trempe
e colores *pl* de temple
i colori *pl* di tempera
n aanloopkleuren *pl*, hardingskleuren *pl*
d Anlassfarben *pl*, Anlauffarben *pl*, Wärmetöne *pl*

5399 TEMPORARY HARDNESS hyg
f dureté *f* temporaire
e dureza *f* temporal
i durezza *f* temporanea
n tijdelijke hardheid
d Bikarbonathärte *f*, temporäre Härte *f*, vorübergehende Härte *f*

5400 TENACITY, TOUGHNESS gen
f ténacité *f*
e tenacidad *f*
i tenacità *f*
n taaiheid
d Zähigkeit *f*

5401 TENDERING tex
f affaiblissement *m* de la fibre

e ablandamiento *m* de la fibra
i affievolimento *m* della fibra
n aantasten *n* van de vezel
d Faserabbau *m*

5402 TENTERING tex
f élargissement *m*
e ensanchamiento *m*, estricado *m*, rameado *m*
i distendimento *m*
n spannen *n*
d Spannen *n*

5403 TERMOBACTERIA, PUTREFACTIVE BACTERIA fmi,iwt
f bactéries *pl* putrides
e bacterias *pl* de la putrefacción
i batteri *pl* della putrefazione
n rottingsbacteriën *pl*
d Fäulnisbakterien *pl*

5404 TERNE PLATE coa,met
f tôle *f* matte, tôle *f* terne
e chapa *f* emplomada
i lamiera *f* piombata
n verlood blik *n*
d Mattblech *n*, Mattweissblech *n*

5405 TERRA ALBA, LIGHT SPAR (US), NATURAL CALCIUM SULPHATE (GB) min
f blanc *m* brillant, blanc *m* de gypse, sulfate *m* de calcium naturel
e sulfato *m* de calcio natural
i terra *f* alba
n lenzine, lichtspaat *n*
d Federweiss *n*, Kalziumsulfat *n*, Leichtspat *m*, Lenzin *m*

5406 TERRACOTTA cer
f argile *f* cuite, terre *f* cuite
e terracota *f*
i terracotta *f*
n terracotta
d Kunstkeramik *f*, Terrakotta *f*

5407 TEST tes
f essai *m*
e ensayo *m*, prueba *f*
i collaudo *m*, prova *f*
n beproeving, proef
d Probe *f*, Prüfung *f*, Versuch *m*

5408 TEST PAPER ana,lab,tes
f papier *m* réactif
e papel *m* indicador, papel *m* reactivo
i carta *f* reagente
n reageerpapier *n*
d Reagenzpapier *n*

5409 TEST RUN dis,tes
f marche *f* d'essai
e operación *f* de prueba
i marcia *f* di prova
n proefloop
d Probelauf *m*

5410 TESTING tes
f épreuve *f*, essayage *m*
e prueba *f*, verificación *f*
i collaudo *m*, prova *f*
n beproeving, onderzoek *n*
d Prüfung *f*

5411 TEXTILE CHEMISTRY tex
f chimie *f* textile
e química *f* textil
i chimica *f* tessile
n textielchemie
d Textilchemie *f*

5412 TEXTILE FIBERS (US), TEXTILE FIBRES (GB) tex
f fibres *pl* textiles
e fibras *pl* textiles
i fibre *pl* tessili
n textielvezels *pl*
d Textilfaser *pl*

5413 TEXTILE LAMINATE, LAMINATED CLOTH, LAMINATED FABRIC tex
f tissu *m* laminé
e tela *f* laminada
i tessuto *m* laminato
n hardweefsel *n*
d Hartgewebe *n*

5414 TEXTILE PRINTING dye,pri
f impression *f* de tissus
e estampado *m* de telas
i stampaggio *m* di tessuti
n textieldrukkerij
d Textildruck *m*

5415 TEXTURE cer
f texture *f*
e textura *f*
i struttura *f*
n textuur
d Textur *f*

5416 TEXTURE FINISH, PLASTIC PAINT, RELIEF PAINT dec
f peinture *f* plastique, peinture *f* relief
e pintura *f* plástica
i pittura *f* plastica
n plastiekverf, reliefverf
d plastische Anstrichfarbe *f*

5417 THEOBROMA OIL, CACAO BUTTER, CACAONUT OIL	fat
f beurre *m* de cacao
e manteca *f* de cacao
i burro *m* di cacao
n cacaoboter
d Kakaobutter *f*

5418 THEORETICAL YIELD	gen
f rendement *m* théorique
e rendimiento *m* teórico
i rendimento *m* teorico
n berekende opbrengst, theoretisch rendement *n*
d berechnetes Rendement *n*, theoretische Ausbeute *f*

5419 THERMAL BARRIER	gla
f zone *f* de température très élevée
e zona *f* de temperatura muy alta
i zona *f* di temperatura molto alta
n hoge-temperatuurzone
d Hochtemperaturzone *f*

5420 THERMAL CONTRACTION	gen
f retrait *m* thermique
e contracción *f* térmica
i ritiro *m* termico
n thermische krimp
d thermischer Schwund *m*

5421 THERMAL CRACKING	fue
f craquage *m* thermique
e piezopirólisis *f* térmica
i piroscissione *f* termica
n thermisch kraken *n*
d Thermokrackung *f*

5422 THERMAL DECOMPOSITION	gen
f décomposition *f* thermique
e descomposición *f* térmica
i decomposizione *f* termica
n thermische ontleding
d thermische Zersetzung *f*

5423 THERMAL DIFFUSION	nue
f diffusion *f* thermique
e difusión *f* térmica
i diffusione *f* termica
n thermische diffusie
d Thermodiffusion *f*

5424 THERMAL DIFFUSION SEPARATION	fue,nue
f séparation *f* par diffusion thermique
e separación *f* por difusión térmica
i separazione *f* per diffusione term
n thermische diffusiescheiding
d thermische Diffusionsscheidung *f*

5425 THERMAL DISSOCIATION	gen
f dissociation *f* thermique
e disociación *f* térmica
i dissociazione *f* termica
n thermische dissociatie
d thermische Dissoziation *f*

5426 THERMAL ENDURANCE	gla
f résistance *f* à la chaleur
e resistencia *f* al calor
i resistenza *f* al calore
n hitteweerstandsvermogen *n*
d Hitzebeständigkeit *f*

5427 THERMAL REFORMING	fue
f réformation *f* thermique
e reformación *f* térmica
i riformazione *f* termica
n thermische reformatie
d Thermoreformation *f*

5428 THERMAL SEPARATION	sep
f séparation *f* thermique
e separación *f* térmica
i separazione *f* termica
n thermische scheiding
d Thermotrennung *f*

5429 THERMAL SHOCK	gla
f change *m* abrupt de température, choc *m* thermique
e choque *m* térmico
i sollecitazione *f* termica, urto *m* termico
n plotselinge temperatuurs-wisseling
d schroffer Temperaturwechsel *m*

5430 THERMAL SOFTENING, THERMAL WATER SOFTENING	hyg
f adoucissement *m* thermique
e ablandamiento *m* térmico
i addolcimento *m* termico
n thermische ontharding
d thermische Enthärtung *f*

5431 THERMAL TREATMENT	htt,met
f traitement *m* thermique
e tratamiento *m* térmico
i trattamento *m* termico
n warmtebehandeling
d Wärmebehandlung *f*

5432 THERMALLY ATTACKED RUBBER	rub
f caoutchouc *m* dégradé par la chale

e goma *f* degradada térmicamente
i gomma *f* degradata termicamente
n thermisch afgebroken rubber
d thermisch abgebauter Gummi *m*

5433 THERMITE wel
f thermit *m*, thermite *f*
e termita *f*
i termite *f*
n thermiet *n*
d Thermit *n*

5434 THERMITE PROCESS, ALUMINOTHERMIC PROCESS met
f procédé *m* aluminothermique, procédé *m* thermite
e procedimiento *m* aluminotérmico, procedimiento *m* termita
i processo *m* alluminotermico
n aluminothermisch proces *n*, thermietproces *n*
d aluminothermisches Verfahren *n*, Thermitverfahren *n*

5435 THERMITE WELDING, ALUMINOTHERMIC WELDING wel
f soudure *m* aluminothermique
e soldadura *f* aluminotérmica
i saldatura *f* alla termite
n thermietlassen *n*
d Thermitschweissung *f*

5436 THERMOCHEMISTRY gen
f thermochimie *f*
e termoquímica *f*
i termochimica *f*
n thermochemie
d Wärmechemie *f*

5437 THERMOCOLOR (US), THERMOMETRIC PAINT (GB) sap
f peinture *f* sensible à la chaleur
e pintura *f* sensible al calor
i pittura *f* sensibile al calore
n op warmte reagerende verf, temperatuurverf
d auf Wärme ansprechende Farbe *f*, Thermokolor *m*

5438 THERMODURIC (US), HEAT RESISTING gen
f résistant adj à la chaleur, résistant adj au feu
e resistente adj al calor, resistente adj al fuego
i resistente adj al calore, resistente adj al fuoco
n hittevast adj , vuurvast adj
d feuerfest adj , hitzebeständig adj

5439 THERMODYNAMIC POTENTIAL elc
f potentiel *m* thermodynamique
e potencial *m* termodinámico
i potenziale *m* termodinamico
n thermodynamische potentiaal
d thermodynamisches Potential *n*

5440 THERMOLABILE gen
f thermolabile adj
e termoinestable adj , termolábil adj
i termolabile adj
n niet bestand tegen matige warmte, thermolabiel adj
d nicht wärmebeständig adj

5441 THERMOPHILIC AEROBIC DIGESTION, HOT FERMENTATION fmi,iwt
f fermentation *f* à chaud
e fermentación *f* en caliente
i fermentazione *f* a caldo
n fermenteren *n* door broeien
d Heissvergärung *f*

5442 THERMOPLASTIC pla
f thermoplastique adj
e termoplástico adj
i termoplastico adj
n thermoplastisch adj
d thermoplastisch adj

5443 THERMOPLASTIC PUTTY bui
f mastic *m* thermoplastique
e mástique *m* termoplástico
i mastice *m* termoplastico
n thermoplastische kit
d thermoplastischer Kitt *m*

5444 THERMOPLASTIC RESIN pla
f résine *f* thermoplastique
e resina *f* termoplástica
i resina *f* termoplastica
n thermoplastische hars
d Thermoplast *n*

5445 THERMOSETTING pla
f thermodurcissant adj
e termoendurecente adj
i termoindurente adj
n thermohardend adj
d in Wärme aushärtend adj

5446 THERMOSETTING RESIN, HEAT-HARDENABLE RESIN, HOT SETTING RESIN pla
f résine *f* durcissable par chaleur
e resina *f* termoendurecente
i resina *f* termoindurente
n met warmte uithardbare hars,

thermohardende hars
d in Wärme aushärtendes Kunstharz *n*

5447 THICK JUICE sin
f jus *m* concentré
e zumo *m* concentrado
i succo *m* concentrato
n diksap *n*
d Dicksaft *m*

5448 THICK MASH bre
f trempe *f* épaisse
e mezcla *f* espesa
i tempera *f* densa
n dik beslag *n*
d Dickmaische *f*

5449 THICK MASH METHOD, DECOCTION METHOD bre
f méthode *f* par décoction
e método *m* por decocción
i metodo *m* per decozione
n afkookmethode, decoctiemethode
d Dekoktionsverfahren *n*

5450 THICKENED OIL, BASE OIL, OXIDIZED OIL fat,lub
f huile *f* épaissie, huile *f* oxydée, huile *f* soufflée
e aceite *m* oxidado
i olio *m* ossidato, olio *m* soffiato
n geblazen olie, geoxydeerde olie
d geblasenes Öl *n*, oxydiertes Öl *n*

5451 THICKENER, PUFFING AGENT dec
f matière *f* d'épaississement
e agente *m* espesante
i agente *m* d'addensamento
n verdikkingsmiddel *n*
d Verdickungsmittel *n*

5452 THICKENING dec
f épaississement *m*
e espesamiento *m*
i inspessimento *m*
n verdikken *n*
d Verdickung *f*

5453 THICKENING gen
f concentration *f*
e concentración *f*
i addensamento *m*
n indikken *n*
d Eindicken *n*

5454 THIN-FILM EVAPORATION eva
f évaporation *f* à couche mince
e evaporación *f* de capa delgada
i evaporazione *f* a strato sottile
n dunne-laagverdamping
d Dünnschichtverdampfung *f*

5455 THIN JUICE sin
f jus *m* sucré
e jugo *m* verde
i succo *m* verde
n dunsap *n*
d Dünnsaft *m*

5456 THINNER, DILUENT, DILUTING AGENT dec
f diluant *m*
e diluyente *m*
i diluente *m*
n verdunner, verdunningsmiddel *n*
d Streckmittel *n*, Verdünnungsmittel *n*

5457 THINNER, SOLVENT dec
f délayant *m*
e diluidor *m*, diluyente *m*
i diluente *m*
n verdunningsmiddel *n*
d Verdünner *m*, Verdünnungsmittel *n*

5458 THINNING OIL (US), DILUENT OIL (GB) dec
f huile *f* diluante
e aceite *m* diluyente
i olio *m* diluente
n verdunnende olie
d Verdünnungsöl *n*

5459 THIRTING cer
f découpage *m* de blocs d'argile
e corte *m* de bloques de arcilla
i taglio *m* di blocchi d'argilla
n snijden *n* van kleiblokken
d Schneiden *n* von Tonblocken

5460 THIXOTROPIC PAINT, GEL PAINT, JELLY PAINT dec
f peinture *f* thixotrope
e pintura *f* tixotrópica
i pittura *f* tixotropica
n thixotrope verf
d thixotrope Farbe *f*

5461 THIXOTROPY, FALSE BODY dec
f faux corps *m*, thixotropie *f*
e tixotropía *f*
i tixotropia *f*
n thixotropie
d falscher Körper *m*, Thixotropie *f*

5462 THOMAS PROCESS, met
BASIC BESSEMER PROCESS
f procédé *m* Bessemer basique, procédé *m* Thomas
e procedimiento *m* Thomas
i processo *m* al convertitore basico, processo *m* Thomas
n thomasproces *n*
d Thomas-Verfahren *n*

5463 THREE-COLOR PROCESS (US),pho
THREE-COLOUR PROCESS (GB)
f procédé *m* trichrome
e procedimiento *m* tricromo
i processo *m* tricromo
n driekleurenproces *n*
d Dreifarbenverfahren *n*

5464 THREE-DIMENSIONAL POLYMER oic
f polymère *m* à branchement tridimensionnel
e polímero *m* de ramificación tridimensional
i polimero *m* a diramazione tridimensionale
n polymeer *n* met driedimensionale netvorming
d Polymer *n* mit dreidimensionaler Verzweigung

5465 THREE-MASH METHOD bre
f méthode *f* de brassage à trois trempes
e método *m* de tres mezclas
i sistema *m* a tre tempere
n drievoudig-beslagproces *n*
d Dreimaischverfahren *n*

5466 THREE-STAGE COMPRESSION pre
f compression *f* en trois étages
e compresión *f* en tres etapas
i compressione *f* in tre stadi
n drietrapscompressie
d Dreistufenkompression *f*

5467 THROUGH, sor
DUFF, UNDERSIZE
f passé *m* de crible
e materia *f* cribada, materia *f* que pasa por la criba
i materia *f* passante, materia *f* vagliata
n zeefsel *n*, ziftsel *n*
d Siebdurchfall *m*, Unterkorn *n*

5468 THROUGH-CARBURIZING (US), htt,met
TOTAL CARBURIZING (GB)
f cémentation *f* totale
e cementación *f* total
i cementazione *f* totale
n door-en-doorcementeren *n*
d Durchzementierung *f*

5469 THROUGH-HARDENING (US), htt,met
TOTAL HARDENING (GB)
f endurcissement *m* total
e endurecimiento *m* total
i indurimento *m* totale
n door-en-doorharding
d Durchhärtung *f*

5470 THROWING cer
f tournassage *m*
e torneado *m* en la rueda de alfarero
i formazione *f* sul tornio del vasaio
n vorming op de draaischijf
d Formgebung *f* auf der Töpferscheibe

5471 THROWING tex
f filage *m* de la soie
e retorcimiento *m* de la seda
i ritorcitura *f* della seta
n fileren *n* van zijde
d Filieren *n* von Seide

5472 THROWING POWER elc
f pouvoir *m* couvrant, pouvoir *m* de pénétration
e poder *m* cubriente, poder *m* de penetración
i potere *m* ricoprente
n spreidend vermogen *n*
d Streuvermögen *n*

5473 THYME OIL fat
f essence *f* de thym
e esencia *f* de tomillo
i olio *m* di timo
n thymolie
d Thymianöl *n*

5474 TIE-GUM glu
f caoutchouc *m* d'ancrage
e goma *f* de fijación
i gomma *f* di fissaggio
n rubberhechtlaag
d Bindegummi *m*

5475 TIGHT CURING rub
f vulcanisation *f* complète
e vulcanización *f* completa
i vulcanizzazione *f* completa
n door-en-doorvulcanisatie
d Ausvulkanisation *f*, Durchvulkanisation *f*

5476 TIN DISEASE, cor,met
TIN PEST, TIN PLAGUE
f peste *f* de l'étain
e peste *f* del estaño
i peste *f* dello stagno
n tinpest
d Zinnpest *f*

5477 TIN-PLATE, coa,met
TINNED SHEET
f fer-blanc *m*
e hojala *f*
i banda *f* stagnata, latta *f*
n vertind blik *n*, witblik *n*
d Weissblech *n*, Zinnblech *n*

5478 TINCTORIAL POWER pho
f pouvoir *m* tinctorial
e poder *m* tintóreo
i potere *m* tintorio
n kleurkracht
d Farbkraft *f*

5479 TINCTURE mcy
f teinture *f*
e tintura *f*
i tintura *f*
n tinctuur
d Tinktur *f*

5480 TINFOIL, coa
STANNIOL
f étain *m* battu, feuille *f* d'étain, papier *m* d'argent
e estaño *m* en hojas, papel *m* estaño
i stagno *m* fogliato, stagnuola *f*
n bladtin *n*, staniol *n*, stanniool *n*, tinfoelie
d Blattzinn *n*, Stanniol *n*, Zinnfolie *f*

5481 TINMAN'S SOLDER wel
f soudure *f* de ferblantier
e soldadura *f* de hojalatero
i stagno *m* da lattoniere
n tinsoldeer *n*
d Weisslot *n*

5482 TINNING (GB), nut
CANNING (US)
f mise *f* en boîtes, mise *f* en conserves
e procedimiento *m* de conservar en latas
i conservazione *f* in scatola
n in blik conserveren *n*
d in Blechdosen Konservieren *n*

5483 TINT dec,dye
f nuance *f*, teinte *f*
e tinte *f*
i tinta *f*
n tint
d Farbton *m*, Nuance *f*

5484 TINTING dye
f nuançage *m*
e tintado *m*
i tingitura *f*
n bijkleuren *n*
d Abtönen *n*, Schattieren *n*

5485 TINTING, dec
SHADING
f mise *f* à la teinte, nuançage *m*
e matizadura *f*
i sfumatura *f*
n bijtinten *n*, op kleur brengen *n*
d Abtönen *n*, Nachtönen *n*, Nuanzieren *n*

5486 TISSUE PAPER pap
f papier *m* de soie, papier *m* mousseline
e papel *m* de seda
i carta *f* di seta, carta *f* velina
n vloeipapier *n*, zijdepapier *n*
d Seidenpapier *n*

5487 TITANIUM DIOXIDE, dec
TITANIUM WHITE
f bioxyde *m* de titanium, blanc *m* de titanium
e bióxido *m* de titanio, blanco *m* de titanio
i bianco *m* di titanio, biossido *m* di titanio
n titaanwit *n*
d Titanweiss *n*

5488 TITER (US), gen
NORMALITY (GB)
f normalité *f* de la solution
e normalidad *f* de la solución
i normalità *f* della soluzione
n normaliteit van een oplossing
d Normalität *f*, Titer *m*

5489 TITRATION ana
f titrage *m*
e valoración *f*
i titolazione *f*
n titreren *n*
d Titrieren *n*

5490 TOILET SOAP soa
f savon *m* de toilette
e jabón *m* de tocador
i sapone *m* da toeletta
n toiletzeep
d Toiletteseife *f*

5491 TONER dec
f agent *m* modificateur de la couleur
e agente *m* modificador del color
i agente *m* modificatore del colore
n kleurwijzigingsmiddel *n*
d Farbabänderungsmittel *n*

5492 TONER dec
f pigment *m* organique clair
e pigmento *m* orgánico claro
i pigmento *m* organico chiaro
n helder organisch pigment *n*
d klarer organischer Farbstoff *m*

5493 TONING pho
f virage *m*
e viraje *m*
i viraggio *m*
n kleuren *n*
d Tonen *n*

5494 TONING AND FIXING BATH, COMBINED BATH pho
f bain *m* viro-fixateur
e baño *m* de viraje y fijación
i bagno *m* di virofissaggio
n kleur-fixeerbad *n*
d Tonfixierbad *n*

5495 TOOTHPASTE hyg
f dentifrice *m*, pâte *f* dentifrice
e pasta *f* dentífrica
i dentifricio *m*, pasta *f* dentifricia
n tandpasta
d Zahnpaste *f*

5496 TOP, TOP PRODUCT dis
f produit *m* de tête
e producto *m* de cabeza
i prodotto *m* di testa
n fractie met laag kookpunt, topdestillaat *n*
d Kopfdestillat *n*, Kopfprodukt *n*

5497 TOP CHROMING, AFTERCHROME DYEING, BACK-CHROME DYEING tex
f chromatage *m*, teinture *f* postérieure au chrome
e cromado *m* secundario, postmordentado *m* crómico
i tintura *f* posteriore al cromo
n nachromeren *n*
d Nachchromieren *n*

5498 TOP FERMENTATION, SURFACE FERMENTATION bre,fmi
f fermentation *f* haute
e fermentación *f* alta
i fermentazione *f* alta
n bovengisting
d Obergärung *f*

5499 TOP FERMENTING YEAST, SURFACE YEAST bre,fmi
f levure *f* haute
e levadura *f* de fermentación alta
i lievito *m* per alta fermentazione
n bovengist
d Oberhefe *f*

5500 TOP FIRING cer
f grillage *m* à chauffage par la partie supérieure
e tostación *f* con calentamiento por la parte alta
i torrefazione *f* con fuoco alla bocca
n bovenvuurroosting
d Oberfeuerröstung *f*

5501 TOPPING dye
f correction *f* de la couleur
e corrección *f* del color
i correzione *f* del colore
n bijkleuren *n*, kleurcorrectie
d Farbkorrektion *f*

5502 TOPPING fue
f étêtage *m*, séparation *f* des produits de tête
e destilación *f* primaria
i distillazione *f* primaria
n aftoppen *n*, primaire destillatie
d Primärdestillation *f*, Toppen *n*

5503 TOPPING WITH GUM, PROOFING, RUBBERIZING coa,rub
f gommage *m*, recouvrement *m* de caoutchouc
e engomadura *f*, recubrimiento *m* con goma
i gommatura *f*, ricopertura *f* con gomma
n berubberen *n*, gommeren *n*
d Gummieren *n*

5504 TORCH CUTTING, FLAME CUTTING, GAS CUTTING met
f coupage *m* à l'autogène, oxycoupage *m*
e corte *m* por llama oxiacetilénica, oxicorte *m*
i taglio *m* per fiamma ossiacetilena
n autogeen snijden *n*
d Brennschneiden *n*

5505 TORCH HARDENING, htt,met
FLAME HARDENING,
SHORTERIZING
f trempe *f* au chalumeau
e temple *m* superficial a la llama
i tempera *f* alla fiamma
n vlamharding
d Flammenhärtung *f*, Oberflächenhärtung *f*

5506 TORRIFYING, bre
CURING, ROASTING
f torréfaction *f*
e tostación *f*
i torrefazione *f*
n roosting
d Röstung *f*

5507 TOSSING met
f écrémage *m*
e separación *f* de la espuma
i separazione *f* della melma
n vuilvangst
d Schlammfang *m*

5508 TOSSING, fou,met
KETTLE OXIDATION
f oxydation *f* par cuillerées
e oxidación *f* por cucharadas
i ossidazione *f* per cucchiaiate
n oxydatie door lepelen
d Oxydation *f* durch Löffeln

5509 TOTAL AIR-GAS MIXTURE fue,gas
f mélange *m* idéal air-gaz
e mezcla *f* ideal aire-gas
i miscela *f* ideale aria-gas
n ideaal lucht-gasmengsel *n*
d ideale Luft-Gasmischung *f*

5510 TOTAL CARBURIZING (GB), htt,met
THROUGH-CARBURIZING (US)
f cémentation *f* totale
e cementación *f* total
i cementazione *f* totale
n door-en-doorcementeren *n*
d Durchzementierung *f*

5511 TOTAL HARDENING (GB), htt,met
THROUGH-HARDENING (US)
f endurcissement *m* total
e endurecimiento *m* total
i indurimento *m* totale
n door-en-doorharding
d Durchhärtung *f*

5512 TOTAL SOLIDS, rub
DRY MATTER
f matières *pl* solides
e materia *f* seca
i materia *f* secca
n droge stof, indampresidu *n*
d Trockensubstanz *f*

5513 TOUGHENED GLASS, gla
TEMPERED GLASS
f verre *m* durci, verre *m* trempé
e vidrio *m* endurecido, vidrio *m* templado
i vetro *m* indurito, vetro *m* temprato
n gehard glas *n*
d gehärtetes Glas *n*

5514 TOUGHNESS, gen
TENACITY
f ténacité *f*
e tenacidad *f*
i tenacità *f*
n taaiheid
d Zähigkeit *f*

5515 TOWING, lea
CURRYING
f corroyage *m*
e tratamiento *m* con mezclas grasas
i trattamento *m* con mescolanze grasse
n touwen *n*
d Zurichtung *f*

5516 TOWN GAS, gas
ILLUMINATING GAS,
MANUFACTURED GAS
f gaz *m* d'éclairage, gaz *m* de ville
e gas *m* de ciudad, gas *m* del alumbrado
i gas *m* di città, gas *m* illuminante
n lichtgas *n*, stadsgas *n*
d Leuchtgas *n*, Stadtgas *n*

5517 TRACER, iso
INDICATOR
f indicateur *m*, traceur *m*
e indicador *m*, trazador *m*
i tracciante *m*
n indicator, tracer
d Indikator *m*, Leitisotop *n*, Tracer *m*

5518 TRACER COMPOSITION exp,mit
f composition *f* traceuse
e composición *f* trazadora
i composizione *f* tracciante
n rooksas
d Rauchspursatz *m*

5519 TRACKING, CREEP-PATH FORMATION pla
f formation *f* d'un passage conducteur
e formación *f* de un paso conductor
i formazione *f* d'un passaggio conduttore
n kruipwegvorming
d Kriechwegbildung *f*

5520 TRAIN OIL, FISH OIL fat
f huile *f* de baleine, huile *f* de poisson
e aceite *m* de ballena, aceite *m* de pescado
i olio *m* di balena, olio *m* di pesce
n traan, visolie
d Fischtran *m*, Tran *m*

5521 TRANSFER MOLDING (US), TRANSFER MOULDING (GB) pla
f moulage *m* par transfert
e moldeo *m* por transferencia
i stampatura *f* per trasferimento
n spuitgieten *n*
d Pressspritzen *n*, Spritzpressen *n*

5522 TRANSFERENCE NUMBER, TRANSPORT NUMBER elc
f nombre *m* de transport des ions
e número *m* de transporte de los iones
i numero *m* di trasporto degli ioni
n transportgetal *n*
d Überführungszahl *f*

5523 TRANSITION BOILING boi
f ébullition *f* transitoire
e ebullición *f* transitoria
i ebollizione *f* a transizione
n overgangskoken *n*
d Übergangssieden *n*

5524 TRANSITION POINT cry
f point *m* de transition
e punto *m* de transformación, punto *m* de transición
i punto *m* di transizione
n overgangspunt *n*
d Übergangspunkt *m*, Umwandlungspunkt *m*

5525 TRANSITION TINT dec
f teinte *f* de transition, teinte *f* passagère
e color *m* de transición
i colore *m* di transizione
n overgangskleur
d Übergangsfarbe

5526 TRANSLUCENT gla
f translucide adj
e translúcido adj
i traslucido adj
n doorschijnend adj
d durchscheinend adj

5527 TRANSPARENCY TEST pap,tes
f examen *m* de la transparence
e verificación *f* de la transparencia
i collaudo *m* della trasparenza
n doorzichtig adj , transparant adj
d Lichtdurchlässigkeitsmessung *f*

5528 TRANSPARENT gla,pap
f transparent adj
e transparente adj
i trasparente adj
n doorzichtig adj , transparant adj
d durchsichtig adj , transparent adj

5529 TRANSPARENT PARCHMENT (GB), GLASSINE (US), PERGAMYNE (GB) pap
f papier *m* cristal, papier *m* serpente, serpente *f*
e papel *m* cristal
i carta *f* glassene, carta *f* trasparente
n glaspapier *n*, kristalpapier *n*
d Glashaut *f*, Glassin *n*, Pergamin *n*

5530 TRANSPORT tra
f transport *m*
e transporte *m*
i trasporto *m*
n transport *n*, vervoer *n*
d Beförderung *f*, Transport *m*

5531 TRASH pap
f décrassage *m*
e recogida *f* de bagazo
i collezione *f* di ramaglia
n vuilvangst
d Schmutzfang *m*

5532 TREATED PIGMENT, COATED PIGMENT dec
f pigment *m* enrobé
e pigmento *m* revestido
i pigmento *m* rivestito
n omhuld pigment *n*, nabehandeld pigment *n*
d Hüllenpigment *n*, Kernpigment *n*

5533 TREATING WITH LIME bre
f chaulage *f*

e encalar v
i coprire v con calce
n met kalk bedekken *n*
d Einkalken *n*

5534 TREATMENT gen
f traitement *m*
e tratamiento *m*
i trattamento *m*
n behandeling, bewerking
d Bearbeitung *f*, Behandlung *f*

5535 TREE PUTTY, ANTISEPTIC ASPHALT MIXTURE rub
f mastic *m* antiseptique
e mezcla *f* antiséptica de asfalto
i mastice *m* a base d'asfalto antisettico
n antiseptisch asfaltmengsel *n*
d antiseptische Asphaltmischung *f*

5536 TREES AND NODULES elc
f arborescences *pl* et nodules
e arborescencias *pl* y nódulos *pl*
i arborescenze *pl* e noduli *pl*
n bomen *pl* en wratten *pl*
d Bäumchen *pl* und Knospen *pl*

5537 TRITURATION brk
f trituration *f*
e trituración *f*
i trituramento *m*, triturazione *f*
n fijnwrijven *n*
d Zerreiben *n*

5538 TRUE EXTRACT gen
f extrait *m* réel
e extracto *m* real
i estratto *m* reale
n werkelijk extractgehalte *n*
d wirklicher Extraktgehalt *m*

5539 TUMBLING, BARREL ROLLING, BARRELLING pol
f polissage *m* au tonneau
e pulido *m* en tambor
i pulitura *f* in tamburo
n trommelpolijsten *n*
d Trommelpolierung *f*

5540 TUMBLING (US), RATTLING (US), RUMBLING (GB) cle
f dessablage *m*
e desarenado *m*
i sterratura *f*
n poetsen *n*
d Putzen *n*, Scheuern *n*

5541 TUNA OIL mcy
f huile *f* de thon
e aceite *m* de atún
i olio *m* di tonno
n tonijnolie
d Thunfischöl *n*

5542 TUNG OIL, CHINA WOOD OIL fat
f huile *f* de bois de Chine, huile *f* de Tung
e aceite *m* de madera, aceite *m* de palo, aceite *m* de tung
i olio *m* di legno, olio *m* essiccante cinese
n houtolie, tungolie
d Holzöl *n*, Tungöl *n*

5543 TURBIDITY gen
f turbidité *f*
e turbidez *f*, turbiedad *f*
i torbidità *f*
n troebelheid
d Trübung *f*

5544 TURBIDITY POINT, CLOUD POINT fue
f point *m* de trouble
e punto *m* de enturbiamiento
i punto *m* d'intorbidimento
n troebelingspunt *n*
d Trübungspunkt *m*

5545 TURBULENT FLOW gen,mix
f courant *m* turbulent
e flujo *m* turbulento
i corrente *f* vorticosa
n turbulente stroming, wervelstroming
d turbulente Strömung *f*, Wirbelströmung *f*

5546 TURNING INTO CHEESE, CURDLING dia
f procédé *m* de manufacture du fromage
e procedimiento *m* de arrequesonar
i processo *m* di fabbricazione del formaggio
n kazen *n*
d Käsen *n*, Verkäsen *n*

5547 TURNING SOUR bre
f tourner v
e volverse v
i girare v
n omslaan *n*
d Umschlagen *n*

5548 TURPENTINE OIL fat
f essence *f* de térébenthine
e aguarrás *m*, esencia *f* de trementina
i olio *m* di trementina
n terpentijnolie
d Terpentinöl *n*

5549 TWIN HEAD PROCESS, DIRECT PROCESS, KASIKA PROCESS dec
f procédé *m* de pistolage à deux composants
e procedimiento *m* de chorreado con dos componentes
i processo *m* di spruzzo con due componenti
n tweecomponentenspuitproces *n*
d Zweikomponentenspritzverfahren *n*

5550 TWO-COLOR PROCESS (US), TWO-COLOUR PROCESS (GB) pho
f procédé *m* bicolore
e procedimiento *m* bicolor
i processo *m* a due colori, sistema *m* bicromico
n tweekleurenproces *n*
d Zweifarbenverfahren *n*

5551 TWO-DIMENSIONAL GAS gas
f film *m* gazeux
e película *f* gaseosa
i gas *m* bidimensionale
n gasfilm
d Gasfilm *m*

5552 TWO-MASH METHOD bre
f procédé *m* de décoction à deux trempes
e procedimiento *m* de dos decocciones
i processo *m* di decozione a due tempere
n tweevoudig beslagproces *n*
d Zweimaischverfahren *n*

5553 TWO-STAGE COMPRESSION pre
f compression *f* en deux étapes
e compresión *f* en dos etapas
i compressione *f* in due stadi
n tweetrapscompressie
d Zweistufenkompression *f*

U

5554 ULLAGE tra
f freinte *f*, vidange *m*
e merma *f*
i ammanco *m*, calo *m*
n vloeistofverlies *n*, wan
d Flüssigkeitsverlust *m*, Schwand *m*

5555 ULLAGES, bre
WASTE BEER
f bière *f* de reste
e cerveza *f* de resto
i resti *pl* di birra
n restbier *n*
d Restbier *n*

5556 ULMIC ACID, fer
HUMIC ACID
f acide *m* humique
e ácido *m* húmico
i acido *m* umico
n humuszuur *n*
d Huminsäure *f*, Humussäure *f*

5557 ULTIMATE ANALYSIS, ana
ELEMENTARY ANALYSIS,
ORGANIC ANALYSIS
f analyse *f* élémentaire
e análisis *f* elemental
i analisi *f* elementare
n elementaire analyse
d Elementaranalyse *f*,
Verbrennungsanalyse *f*

5558 ULTRA-ACCELERATOR rub
f ultra-accélérateur *m*
e ultra-acelerador *m*
i ultra-acceleratore *m*
n ultraversneller
d Ultrabeschleuniger *m*

5559 ULTRAFILTRATION fil
f ultrafiltration *f*
e ultrafiltración *f*
i ultrafiltrazione *f*
n ultrafiltratie
d Ultrafiltration *f*

5560 ULTRAFINISH, pol
SUPERFINISHING
f superfinissage *m*
e superpulido *m*
i soprafinitura *f*
n superfijnen *n*
d Spiegelpolitur *f*, Superfeinen *n*

5561 ULTRAMARINE, dec
ULTRAMARINE BLUE
f bleu *m* d'outremer
e azul *m* ultramarino
i azzurro *m* oltremarino
n bergblauw *n*, ultramarijn *n*
d Ultramarinblau *n*

5562 ULTRASONIC WELDING pla,wel
f soudure *f* ultrasonore
e soldadura *f* supersónica
i saldatura *f* ultrasonica
n ultrasoon lassen *n*
d Ultraschallschweissung *f*

5563 ULTRAVIOLET RADIATION hyg
f rayonnement *m* ultraviolet
e radiación *f* ultravioleta
i radiazione *f* ultravioletta
n ultravioletstraling
d Ultraviolettstrahlung *f*

5564 UNCOATED AREA, dec
BARE SPOT, HOLIDAY
f absence *f*, blanc *m*, manque *m*,
zone *f* non couverte
e superficie *f* pasada por alto
i zona *f* non coperta
n heiligedag
d blanke Stelle *f*, Jute *f*,
Jutlochstelle *f*

5565 UNCREASABLE tex
f infroissable adj
e inarrugable adj
i ingualcibile adj
n kreukvrij adj
d knitterfest adj

5566 UNCTUOUS mcy
f onctueux adj
e untuoso adj
i untuoso adj
n zalverig adj
d salbig adj

5567 UNCTUOUSNESS mcy
f onctuosité *f*
e untuosidad *f*
i untuosità *f*
n zalverigheid
d Salbungsvolle *f*

5568 UNCURED, rub
UNVULCANIZED
f non-vulcanisé adj

e no vulcanizado adj
i non vulcanizzato adj
n ongevulcaniseerd adj
d unvulkanisiert adj

5569 UNCURED COMPOUND (GB), GREEN COMPOUND (US) rub
f mélange *m* non-vulcanisé
e mezcla *f* no vulcanizada
i mescolanza *f* non vulcanizzata
n ongevulcaniseerd mengsel *n*
d unvulkanisiertes Gemisch *n*

5570 UNDERCOATING, INTERMEDIATE COATING, OVERCOATING dec
f main *f* intermédiaire, peinture *f* de sous-couche
e mano *f* intermedia
i mano *f* intermedia
n overgrondverf, voorlak
d Voranstrich *m*, Zwischenschicht *f*

5571 UNDERCOOLED GRAPHITE mat
f graphite *m* de surfusion
e grafito *m* de sobrefusión
i grafite *f* di segregazione, grafite *f* di sopraffusione
n onderkoelingsgrafiet *n*
d Unterkühlungsgraphit *m*

5572 UNDERCOOLING, SUPERCOOLING, SURFUSION clo
f surfusion *f*
e sobrefusión *f*
i surraffreddamento *m*
n onderkoeling
d Unterkühlung *f*

5573 UNDERCURED, UNDERVULCANIZED rub
f sous-vulcanisé adj
e subvulcanizado adj
i sottovulcanizzato adj
n ondergevulcaniseerd adj
d untervulkanisiert adj

5574 UNDERMODIFIED MALT bre
f malt *m* mal désagrégé
e malta *f* poco desagregada
i malto *m* poco disgregato
n onvoldoende opgelost mout *n*
d unterlöstes Malz *n*

5575 UNDERGLAZE COLOR (US), UNDERGLAZE COLOUR (GB) cer
f couleur *f* sous émail
e color *m* previo al esmaltado
i colore *m* sotto lo smalto
n onderglazuurverf
d Scharffeuerfarbe *f*, Unterglasurfarbe *f*

5576 UNDERPOLING met
f perchage *m* incomplet
e pertigado *m* insuficiente
i trattamento *m* al legno verde insufficiente
n onvolledig polen *n*
d unvollständige Polung *f*

5577 UNDERPRESSURE pre
f dépression *f*
e depresión *f*
i depressione *f*
n onderdruk
d Unterdruck *m*

5578 UNDERSATURATION gen
f sous-saturation *f*
e subsaturación *f*
i sottosaturazione *f*
n onderverzadiging
d Untersättigung *f*

5579 UNDERSINTERED htt,met
f fritté adj insuffisamment
e sinterizado adj insuficientamente
i sinterizzato adj insufficientamente
n onvoldoende gesinterd adj
d untersintert adj , zu schwach gesintert adj

5580 UNDERSIZE, DUFF, THROUGH sor
f passé *m* de crible
e materia *f* cribada, materia *f* que pasa por la criba
i materia *f* passante, materia *f* vagliata
n zeefsel *n*, ziftsel *n*
d Siebdurchfall *m*, Unterkorn *n*

5581 UNDERSTEEPING bre
f trempe *f* insuffisante
e remojado *m* insuficiente
i macerazione *f* insufficiente
n onvoldoende weking
d Unterweichung *f*

5582 UNFERMENTED bre
f non-fermenté adj
e no fermentado adj
i non fermentato adj
n ongegist adj
d ungegoren adj

5583 UNFERMENTED WORT bre
f moût *m* non-fermenté

e mosto *m* no fermentado
i mosto *m* non fermentato
n onvergiste wort
d unvergorene Würze *f*

5584 UNHAIRING, lea
DEPILATION
f dépilage *m*, épilage *m*
e depilación *f*
i depilazione *f*
n bloten *n*, ontharen *n*
d Enthaarung *f*

5585 UNIMOLECULAR LAYER, coa,fat
MONOMOLECULAR LAYER
f couche *f* monomoléculaire
e capa *f* monomolecular
i strato *m* monomolecolare
n monomoleculaire laag
d monomolekulare Schicht *f*

5586 UNINFLAMMABLE fue
f ininflammable adj
e ininflamable adj
i ininfiammabile adj
n onontvlambaar adj
d unentzündlich adj

5587 UNION PROCESS, bre
CLEANSING PROCESS
f procédé *m* de fermentation en tonnes
e procedimiento *m* de fermentación en barriles
i processo *m* di fermentazione in botte
n vatgistingsproces *n*
d Fassgärungsverfahren *n*

5588 UN-IONIZED elc
f non-ionisé adj
e no ionizado adj
i non ionizzato adj
n niet-geïoniseerd adj
d nicht ionisiert adj

5589 UNIT PROCESS gen
f procédé *m* unitaire
e procedimiento *m* unitario
i processo *m* fondamentale
n basisproces *n*, eenheidsproces *n*
d Einheitsverfahren *n*

5590 UNLOADED VULCANIZATE rub
(GB),
PURE GUM VULCANIZATE (GB)
f vulcanisat *m* sans charge
e vulcanizado *m* sin carga
i vulcanizzato *m* di gomma pura
n A-mengsel *n*, vulcanisaat *n* zonder vulstof
d Reinkautschukvulkanisat *n*

5591 UNMALTED GRAIN, bre
RAW GRAIN
f grain *m* cru
e grano *m* crudo
i grano *m* crudo
n ongemout graan *n*
d Rohfrucht *f*

5592 UNMODIFIED bre
f mal modifié adj
e no desagregado adj , no modificado adj
i non disgregato adj
n onopgelost adj
d unaufgelöst adj

5593 UNMODIFIED RESIN pla
f résine *f* non-modifiée
e resina *f* no modificada
i resina *f* non modificata
n niet-gemodificeerd hars *n*
d unmodifiziertes Harz *n*

5594 UNORGANIZED FERMENTS fmi
f enzymes *pl*, ferments *pl* sans structure
e enzimos *pl*, fermentos *pl* sin estructura
i enzimi *pl*, fermenti *pl* senza struttura
n enzymen *pl*, structuurloze fermenten *pl*
d Enzymen *pl*, strukturlose Fermente *pl*

5595 UNSATURATED gen
f insaturé adj , non-saturé adj
e no saturado adj
i non saturo adj
n onverzadigd adj
d ungesättigt adj

5596 UNSHIELDED ARC WELDING wel
f soudure *f* à arc non-protégé
e soldadura *f* con arco descubierto
i saldatura *f* ad arco nudo
n onbeschermd booglassen *n*
d ungeschütztes Lichtbogenschweissen *n*

5597 UNSLAKED LIME, aic
BURNT LIMESTONE, QUICKLIME
f chaux *f* vive
e cal *f* viva
i calce *f* viva
n ongebluste kalk
d Ätzkalk *m*, gebrannter Kalk *m*, ungelöschter Kalk *m*

5598 UNSTABLE gen
f instable adj
e inestable adj
i instabile adj
n onstabiel adj
d unbeständig adj

5599 UNTREATED WATER, NATURAL WATER hyg
f eau *f* naturelle
e agua *f* natural
i acqua *f* naturale
n ruw water *n*
d Rohwasser *n*

5600 UNVULCANIZED, UNCURED rub
f non-vulcanisé adj
e no vulcanizado adj
i non vulcanizzato adj
n ongevulcaniseerd adj
d unvulkanisiert adj

5601 UPPER CYLINDER LUBRICATION lub
f lubrification *f* de haut de cylindre
e lubricación *f* para la parte superior de los cilindros
i lubrificazione *f* a gocciolamento
n bovensmering
d Oberschmierung *f*

5602 UREA-FORMALDEHYDE RESINS pla
f résines *pl* à l'urée-formaldéhyde
e resinas *pl* de urea-formaldehido
i resine *pl* all'urea-formaldeide
n ureumformaldehydeharsen *pl*
d Harnstoff-Formaldehydharze *pl*, Karbamidharze *pl*

5603 UREA RESINS pla
f résines *pl* uréiques
e resinas *pl* ureicas
i resine *pl* uréiche
n ureumharsen *pl*
d Harnstoffharze *pl*

V

5604 VACCINATION mcy
f vaccination *f*
e vacunación *f*
i vaccinazione *f*
n inenting, vaccinatie
d Impfung *f*, Vakzination *f*

5605 VACCINE mcy
f vaccine *f*
e vacuna *f*
i vaccino *m*
n entstof, vaccine *n*
d Impfstoff *m*, Vaccin *n*, Vakzin *n*

5606 VACUUM gen
f vide *m*
e vacío *m*
i vuoto *m*
n luchtledigheid, vacuüm *n*
d Luftleere *f*, Vakuum *n*

5607 VACUUM COATING, VACUUM METALLIZING coa
f métallisation *f* sous vide
e vacuometalización *f*
i metalizzazione *f* sotto vuoto
n metalliseren *n* in vacuüm
d Aufdampfen *n* von Metallen in Vakuum, Vakuummetallisierung *f*

5608 VACUUM CRYSTALLIZATION cry
f cristallisation *f* sous vide
e vacuocristalización *f*
i cristallizzazione *f* sotto vuoto
n vacuümkristallisatie
d Vakuumkristallisation *f*

5609 VACUUM DEGASSING, VACUUM REFINING met
f dégazage *m* sous vide
e vacuodegasificación *f*
i degassamento *m* sotto vuoto
n ontgassen *n* in vacuüm
d Entgasung *f* in Vakuum

5610 VACUUM DISTILLATION dis
f distillation *f* sous vide
e vacuodestilización *f*
i distillazione *f* sotto vuoto
n vacuümdestillatie
d Vakuumdestillation *f*

5611 VACUUM DRYING dry
f séchage *m* sous vide
e vacuosecado *m*
i essiccazione *f* sotto vuoto
n vacuümdroging
d Vakuumtrocknung *f*

5612 VACUUM FILTRATION fil
f filtration *f* sous vide
e vacuofiltración *f*
i filtrazione *f* sotto vuoto
n vacuümfiltratie
d Vakuumfiltration *f*

5613 VACUUM FUSION, VACUUM MELTING met
f fusion *f* sous vide
e vacuofusión *f*
i fusione *f* sotto vuoto
n smelten *n* in vacuüm
d Vakuumschmelzen *n*

5614 VACUUM IMPREGNATION imp
f imprégnation *f* sous vide
e vacuoimpregnación *f*
i impregnazione *f* sotto vuoto
n onder vacuüm impregneren *n*
d Vakuumimprägnierung *f*

5615 VALERIAN OIL fat
f essence *f* de valériane
e esencia *f* de valeriano
i olio *m* di valeriano
n valeriaanolie
d Baldrianöl *n*

5616 VAPOR BLASTING (US), VAPOUR BLASTING (GB), LIQUID HONING pol
f honage *m* au jet de vapeur
e rectificación *f* al chorro de vapor
i rettificatura *f* al getto di vapore
n honen *n* met vloeistofstraal
d Honen *n* mit Dampfstrahl

5617 VAPOR BUBBLE GROWTH RATE (US), VAPOUR BUBBLE GROWTH RATE (GB) boi
f vitesse *f* de croissance de bulles de vapeur
e velocidad *f* de crecimiento de burbujas de vapor
i velocità *f* d'accrescimento di bolle di vapore
n groeisnelheid van dampbellen

d Wachstumgeschwindigkeit *f* von Dampfblasen

5618 VAPOR DEGREASING (US), cle
VAPOUR DEGREASING (GB)
f dégraissage *m* à vapeur de solvant
e desengrase *m* por vapor de disolvente
i sgrassaggio *m* mediante vapore di solvente
n ontvetting met verdampt oplosmiddel
d Entfettung *f* mit verflüssigtem Lösungsmittel

5619 VAPOR DENSITY (US), gen
VAPOUR DENSITY (GB)
f densité *f* de vapeur
e densidad *f* de vapor
i densità *f* di vapore
n dampdichtheid
d Dampfdichte *f*

5620 VAPOR-LIQUID EQUILIBRIUM dis
DISTILLATION (US),
VAPOUR-LIQUID EQUILIBRIUM DISTILLATION (GB)
f distillation *f* à équilibre liquide-vapeur
e destilación *f* con equilibrio líquido-vapor
i distillazione *f* con equilibrio liquido-vapore
n destillatie met vloeistof-dampevenwicht
d Destillation *f* mit Dampf-Flüssigkeitgleichgewicht

5621 VAPOR-PHASE CHLORINA- fue
TION (US),
VAPOUR-PHASE CHLORINATION (GB)
f chloruration *f* en phase vapeur
e cloruración *f* en fase vapor
i clorurazione *f* in fase vapore
n chloreren *n* in de dampfaze
d Dampfphasenchlorierung *f*

5622 VAPOR-PHASE CRACKING tes
(US),
VAPOUR-PHASE CRACKING (GB)
f craquage *m* en phase vapeur
e craqueo *m* en fase vapor, pirodesintegración *f* en fase vapor
i piroscissione *f* in fase vapore
n kraken *n* in de dampfaze
d Dampfphasenkrackung *f*

5623 VAPOR-PHASE DEPOSITION coa
(US),
VAPOUR-PHASE DEPOSITION (GB)
f déposition *f* de la phase vapeur
e deposición de la fase vapor
i deposizione *f* dalla fase vapore
n neerslaan *n* uit de dampfaze
d Niederschlagen *n* aus der Dampfphase

5624 VAPOR-PHASE HYDROGENA- fue
TION (US),
VAPOUR-PHASE HYDROGENATION (GB)
f hydrogénation *f* en phase vapeur
e hidrogenación *f* en fase vapor
i idrogenazione *f* in fase vapore
n hydreren *n* in de dampfaze
d Dampfphasenhydrierung *f*

5625 VAPOR PRESSURE (US), gen
VAPOUR PRESSURE (GB)
f pression *f* de vapeur
e presión *f* de vapor
i pressione *f* di vapore
n dampdruk
d Dampfdruck *m*

5626 VAPOR TENSION (US), gen
VAPOUR TENSION (GB)
f tension *f* de vapeur
e tensión *f* de vapor
i tensione *f* di vapore
n dampspanning
d Dampfspannung *f*

5627 VAPORIZING coa
f évaporation *f*, vaporisation *f*
e evaporación *f*, vaporización *f*
i evaporazione *f*, vaporizzazione *f*
n verdamping
d Verdampfung *f*, Verdunstung *f*

5628 VARNISH dec
f vernis
e barniz *m*
i vernice *f*
n vernis *n*
d Firnis *n*

5629 VARNISH STAIN dec
f vernis *m* teinté
e barniz *m* teñido
i vernice *f* tinta
n lakbeits
d Lackbeize *f*

5630 VARNISHING coa
f vernissage *m*
e barnizado *m*

i verniciatura *f*
n vernissen *n*
d Firnissen *n*

5631 VAT DYEING dye
f teinture *f* à la cuve
e teñido *m* en cuba
i tintura *f* in tino
n kuipverving
d Küpenfärbung *f*

5632 VAT DYESTUFFS dye
f colorants *pl* de cuve
e colorantes *pl* de cuba, colorantes *pl* de tina
i coloranti *pl* di tino
n kuipkleurstoffen *pl*
d Küpenfarbstoffe *pl*

5633 VAT PIGMENT dec
f pigment *m* de cuve
e pigmento *m* de cuba
i pigmento *m* di tino
n kuippigment *n*
d Küpenpigment *n*

5634 VATTED BEER bre
f bière *f* mûre
e cerveza *f* hecha
i birra *f* stagionata
n gelagerd bier *n*
d abgelagertes Bier *n*

5635 VEGETABLE COAL abs,fue
f noir *m* végétal
e negro *m* vegetal
i nero *m* vegetale
n absorptiekool, plantaardige kool
d Pflanzenkohle *f*

5636 VEGETABLE FIBER (US), VEGETABLE FIBRE (GB) tex
f fibre *f* végétale
e fibra *f* vegetal
i fibra *f* vegetale
n plantaardige vezel
d Pflanzenfaser *f*

5637 VEGETABLE GLUE glue
f colle *f* végétale
e cola *f* vegetal
i colla *f* vegetale
n plantelijm
d Pflanzenleim *m*

5638 VEGETABLE GLUTEN, GLUTEN, OSMAZOME glu
f gluten *m*, phytocolle *f*
e gluten *m*
i gluten *n*
n gluten *n*
d Gluten *n*, Tritizin *n*

5639 VEGETABLE OIL fat
f huile *f* végétale
e aceite *m* vegetal
i olio *m* vegetale
n planteolie
d Pflanzenöl *n*

5640 VEGETABLE TAN tan
f tanin *m* végétal, tannin *m* végétal
e curtiente *m* vegetal
i conciante *m* vegetale
n plantaardige looistof
d Pflanzengerbstoff *m*

5641 VEGETABLE TANNING tan
f tannage *m* végétal
e curtido *m* vegetal
i concia *f* vegetale
n looien *n* met plantaardige looistoffen
d Gerben *n* mit Pflanzengerbstoffen

5642 VEHICLE dec
f milieu *m* de suspension, véhicule *m*
e excipiente *m* líquido, vehículo *m*
i agente *m* legante, disperdente *m*, veicolo *m*
n bindmiddel *n*, vehikel *n*
d Bindemittel *n*, Farbstoffträger *m*, Lösungsmittel *n*

5643 VEHICLE mcy
f excipient *m*, véhicule *m*
e excipiente *m*, medio *m*, vehículo *m*
i eccipiente *m*
n excipiens *n*
d Arzneistoffträger *m*, Vehikulum *n*

5644 VELLUM PAPER pap
f papier *m* vélin
e papel *m* vitela
i carta *f* velina
n velijnpapier *n*
d Velinpapier *n*

5645 VELVETING rub
f suédage *m*, veloutage *m*
e aterciopelado *m*
i vellutatura *f*
n velveteren *n*
d Velvurisieren *n*

5646 VENTILATION clo
f ventilation *f*
e ventilación *f*
i ventilazione *f*
n ventileren *n*
d Lüftung *f*, Ventilation *f*

5647 VERBENA OIL fat
f essence *f* de verveine
e esencia *f* de verbena
i olio *m* di verbena
n verbenaolie
d Verbenaöl *n*

5648 VERDIGRIS, dec
AERUGO
f verdet *m*, vert-de-gris *m*
e cardenillo *m*, verdín *m*
i erugine *f*, verderame *m*
n kopergroen *n*
d Edelrost *m*, Grünspan *m*

5649 VERMILION, dec
VERMILLION, CINNABAR
f cinabre *m*, vermillon *m*
e bermellón *m*, cinabrio *m*
i cinabro *m*, vermiglione *m*
n cinnaber *n*, vermiljoen *n*
d Zinnoberrot *n*

5650 VERONESE GREEN, dec
VIRIDIAN GREEN
f viridine *f*
e verde *m* de Verona
i verde *m* di Verona
n Veronees groen *n*
d Cyprisch Grün *n*, Steingrün *n*,
Tiroler Grünerde *f*,
Veroneser Grün *n*

5651 VETIVER OIL fat
f essence *f* de vétiver
e esencia *f* de vetiver
i olio *m* di vetiver
n vetiverolie
d Vetiveröl *n*

5652 VIABLE YEAST fmi
f levure *f* à vitalité fermentative
e levadura *f* de vitalidad
fermentativa
i lievito *m* a vitalità
fermentativa
n fermenteerbare gist
d gärtüchtige Hefe *f*

5653 VIBRATION ger
f vibration *f*
e vibración *f*
i vibrazione *f*
n trillen *n*
d Schwingung *f*

5654 VIENNA CHALK, hyg,lub
FRENCH CHALK
f chaux *f* de Vienne
e cal *f* de Viena
i calce *f* di Vienna
n Weense kalk
d Wiener Kalk *m*

5655 VINASSE, nut
DISTILLER'S WASH, SLOP,
SPENT MASH, STILLAGE
f drêches *pl*, vinasses *pl* de distillerie
e orujo *m*, vinazas *pl* de destilación
i burlanda *f*, trebbia *f*,
vinacce *pl* distillate
n drab
d abgebrannte Würze *f*, Schlempe *f*,
Spülicht *n*

5656 VINE BLACK, dec
FRANKFORT BLACK
f noir *m* de fumée de pampres
e negro *m* de humo de sarmientes
de vid
i nero *m* di fumo di viticci
della vite
n rankenzwart *n*, wijngaardzwart *n*
d Frankfurter Schwarz *n*,
Rebenschwarz *n*,
Weinrebenschwarz *n*

5657 VINEGAR aam
f vinaigre *m*
e vinagre *m*
i aceto *m*
n azijn
d Essig *m*

5658 VINOUS FERMENTATION, oic
ALCOHOLIC FERMENTATION
f fermentation *f* alcoolique
e fermentación *f* alcohólica
i fermentazione *f* alcoolica
n alkoholische gisting
d alkoholische Gärung *f*

5659 VINYL RESINS pla
f résines *pl* vinyliques
e resinas *pl* vinílicas
i resine *pl* viniliche
n vinylharsen *pl*
d Vinylharze *pl*

5660 VIOLENCE TEST, exp
BRISANCE TEST,
SHATTERING TEST
f essai *m* du pouvoir brisant
e prueba *f* de la brisancia
i collaudo *m* dell'effetto
dirompente
n brisantieproef
d Brisanzprüfung *f*

5661 VIRUCIDE agr,mcy
f virucide *m*

e virucida *m*
i virucida *m*
n virus bestrijdend middel *n*
d virustötendes Mittel *n*

5662 VISCOSE tex
f viscose *f*
e viscosa *f*
i viscosa *f*
n viscose
n Natriumzellulosexanthogenat *n*, Viskose *f*

5663 VISCOSE PROCESS tex
f procédé *m* à la viscose
e procedimiento *m* de viscosa
i processo *m* alla viscosa
n viscoseproces *n*
d Viskoseverfahren *n*

5664 VISCOSE RAYON, VISCOSE SILK tex
f rayonne *f* de viscose
e rayón *m* de viscosa
i raion *m* di viscosa
n viscosezijde
d Viskoserayon *m*, Viskoseseide *f*

5665 VISCOSITY fat,lub
f viscosité *f*
e viscosidad *f*
i viscosità *f*
n viscositeit
d Viskosität *f*

5666 VISCOSITY BREAKING fue
f réduction *f* de viscosité
e reducción *f* de viscosidad
i riduzione *f* di viscosità
n viscositeitsafbraak
d Viskositätsabbau *m*

5667 VISCOSITY DEPRESSANT fat,lub
f agent *m* réducteur de la viscosité
e agente *m* reductor de la viscosidad
i agente *m* riducente della viscosità
n viscositeitsverlagend middel *n*
d viskositätsabbauendes Mittel *n*

5668 VISCOSITY INDEX IMPROVER lub
f améliorant *m* de l'indice de viscosité
e mejorador *m* del índice de viscosidad
i miglioratore *m* dell'indice di viscosità
n middel *n* ter verbetering van de viscositeitsindex
d Mittel *n* zur Viskositätsindex-verbesserung

5669 VISCOUS fat
f visqueux adj
e viscoso adj
i viscoso adj
n visceus adj
d viskos adj , Zähflüssig adj

5670 VISCOUS-ELASTIC MATERIAL pla
f matière *f* visco-élastique
e material *m* viscoelástico
i materiale *m* viscoelastico
n visceus-elastisch materiaal *n*
d visko-elastisches Material *n*

5671 VITAMINIZATION nut
f vitaminisation *f*
e vitaminización *f*
i vitaminizzazione *f*
n vitaminisering
d Vitaminisierung *f*

5672 VITREOUS, GLASSY gla
f vitreux adj
e vítreo adj
i vetroso adj
n glasachtig adj
d glasartig adj , glasig adj

5673 VITREOUS COATING coa
f revêtement *m* vitreux
e revestimiento *m* vidrioso
i rivestimento *m* vetroso
n verglazing
d Überglasung *f*

5674 VITREOUS SILICA, SILICA GLASS gla
f quartz *m* fondu
e cuarzo *m* fundido
i quarzo *m* fuso
n gesmolten kwarts *n*
d geschmolzener Quarz *m*

5675 VITRIFICATION coa
f vitrification *f*
e vitrificación *f*
i vetrificazione*f*
n verglazen *n*
d Verglasen *n*

5676 VOLATILE eva
f volatil adj
e volátil adj
i volatile adj
n etherisch adj , vluchtig adj
d ätherisch adj , flüchtig adj

5677 VOLATILE CONSTITUENT eva
f constituant *m* volatil
e constituente *m* volátil
i costituente *m* volatile
n vluchtig bestanddeel *n*
d flüchtiger Bestandteil *m*

5678 VOLATILE OILS, fat,per
ESSENTIAL OILS, ETHEREAL OILS
f huiles *pl* essentielles
e aceites *pl* esenciales
i oli *pl* essenziali
n etherische oliën *pl*
d ätherische Öle *pl*

5679 VOLATILE SOLVENT eva
f solvant *m* volatil
e solvente *m* volátil
i solvente *m* volatile
n vluchtig oplosmiddel *n*
d flüchtiges Lösungsmittel *n*

5680 VOLATILITY eva
f volatilité *f*
e volatilidad *f*
i volatilità *f*
n vluchtigheid
d Flüchtigkeit *f*

5681 VOLATILITY PRODUCT eva
f produit *m* de volatilité
e producto *m* de volatilidad
i prodotto *m* di volatilità
n vluchtigheidsprodukt *n*
d Flüchtigkeitsprodukt *n*

5682 VOLATILIZATION eva
f volatilisation *f*
e volatilización *f*
i volatilizzazione *f*
n vervluchtiging
d Verdunstung *f*, Verflüchtigung *f*

5683 VOLT EFFICIENCY elc
f rendement *m* en tension
e rendimiento *m* en tensión
i rendimento *m* di tensione
n spanningsrendement *n*
d Nutzspannung *f*

5684 VOLTOLIZATION fat
f voltolyse *f*
e voltólisis *f*
i voltolisi *f*
n voltolyse
d Voltolisieren *n*

5685 VOLUME ASSAYING, ana,min
VOLUMETRIC ASSAYING
f essayage *m* volumétrique
e ensayo *m* volumétrico
i assaggio *m* volumetrico
n volumetrisch essaaieren *n*
d volumetrische Erzprobe *f*

5686 VOLUMETRIC ANALYSIS ana
f analyse *f* volumétrique
e análisis *f* volumétrica
i analisi *f* volumetrica
n maatanalyse, volumetrische analyse
d Massanalyse *f*, volumetrische Analyse *f*

5687 VORTEX BOILING boi
f ébullition *f* tourbillonnaire
e ebullición *f* turbillonaria
i ebollizione *f* a vortice
n wervelingskoken *n*
d Wirbelsieden *n*

5688 VULCANITE, pla
EBONITE (GB), HARD RUBBER (US)
f ébonite *f*
e ebonita *f*
i ebanite *f*
n eboniet *n*
d Ebonit *m*, Hartgummi *m*

5689 VULCANIZED FIBER (US), elc
VULCANIZED FIBRE (GB)
f fibre *f* vulcanisée
e fibra *f* vulcanizada
i fibra *f* vulcanizzata
n fiber *n*, vulcaniet *n*
d Vulkanfiber *f*

5690 VULCANIZING, rub
CURING
f vulcanisation *f*
e vulcanización *f*
i vulcanizzazione *f*
n vulcaniseren *n*
d Vulkanisieren *n*

W

5691 WALL EFFECT dis
f effet *m* de paroi
e efecto *m* de pared
i effetto *m* di parete
n wandeffect *n*
d Randgängigkeit *f*, Wandeinfluss *m*

5692 WALL PAINT dec
f peinture *f* murale
e pintura *f* para paredes
i pittura *f* per parete
n muurverf
d Maueranstrich *m*, Mauerfarbe *f*

5693 WALLPAPER bui
f papier *m* de tenture, papier *m* peint
e empapelado *m*, papel *m* pintado
i carta *f* da parato
n behangselpapier *n*
d Tapete *f*

5694 WALNUT OIL fat
f huile *f* de noix
e aceite *m* de nuez, aceite *m* de nuez de leche
i olio *m* di noce
n noteolie, walnoteolie
d Walnussöl *n*

5695 WAR GAS, POISON GAS mit
f gaz *m* asphyxiant, gas *m* toxique
e gas *m* tóxico
i gas *m* aggressivo chimico
n gifgas *n*
d Giftgas *n*

5696 WARP PRINTING tex
f impression *f* de chaîne
e estampación *f* de urdimbre
i stampaggio *m* d'ordito
n scheringdruk
d Kettendruck *m*, Schinieren *n*

5697 WARRANT, CHAMOTTE, FIRECLAY, SPAVIN cer
f apyre *f*, argile *f* réfractaire, chamotte *f*
e arcilla *f* refractaria, chamota *f*
i argilla *f* refrattaria, terra *f* refrattaria
n chamotte, vuurklei
d feuerfester Ton *m*, Schamotte *f*

5698 WASH PRIMER, ETCHING PRIMER (GB), METAL CONDITIONS (US) dec
f peinture *f* pour couche primaire phosphante
e pintura *f* de mano primera fosfatizante
i pittura *f* di mano prima a base di fosfato
n hechtgrondverf voor metalen
d Haftgrundmittel *n*, reaktionsfähiger Haftgrund *m*, Metallwandler *m*

5699 WASH SIRUP (US), WASH SYRUP (GB), GREEN SYRUP (GB) sin
f eaux *pl* de deuxième et troisième jet
e jarabe *m* verde
i sciroppo *m* verde
n groene stroop
d Deckablauf *m*, Waschsirup *m*

5700 WASHABLE LEATHER lea
f cuir *m* lavable
e cuero *m* lavable
i cuoio *m* lavabile
n wasbaar leder *n*
d waschbares Leder *n*, Waschleder

5701 WASHED CLAY, MALM bui
f marne *f* artificielle
e marga *f* artificial, mezcla *f* de greda y arcilla
i marna *f* artificiale
n kunstmergel
d künstlicher Mergel *m*

5702 WASHING cle
f lavage *m*
e lavado *m*
i lavaggio *m*
n wassen *n*
d Waschen *n*

5703 WASHING AGENT, DETERGENT cle
f détergent *m*, détersif *m*
e detergente *m*, detersivo *m*
i detergente *m*, detersivo *m*
n detergens *n*, wasmiddel *n*
d Detergens *n*, Waschmittel *n*

5704 WASHING BLUE, LAUNDRY BLUE, STARCHING BLUE — cle
f bleu *m* d'azur, bleu *m* d'empois
e azulete *m*
i turchinetto *m*
n blauwsel *n*
d Waschblau *n*

5705 WASHING CLASSIFICATION, WET CLASSIFICATION — sor
f classification *f* à eau courante, classification *f* hydraulique
e clasificación *f* hidráulica, clasificación *f* por corriente de agua
i classificazione *f* a corrente d'acqua, classificazione *f* idraulica
n classificeren *n* met slibgoot
d Stromklassierung *f*

5706 WASHING FASTNESS, FASTNESS TO WASHING — tex
f solidité *f* au lavage
e resistencia *f* al lavado
i solidità *f* al lavaggio
n wasbestendigheid
d Waschechtheit *f*

5707 WASHING OFF, RINSING — cle
f rinçage *m*
e enjuague *m*
i risciacquamento *m*
n spoelen *n*
d Spülen *n*

5708 WASHING OIL, ABSORPTION OIL — cle,gas
f huile *f* lourde pour débenzolage du gaz
e aceite *m* de absorción
i olio *m* d'assorbimento
n wasolie
d Waschöl *n*

5709 WASHING OUT, ABLUTION — iwt
f ablution *f*
e ablución *f*
i abluzione *f*
n uitwassing
d Abspülung *f*, Abwaschung *f*, Auswaschung *f*

5710 WASHING PERFORMANCE, CLEANING EFFICIENCY, DETERGENT ACTION — cle
f action *f* détergente
e acción *f* detergente
i azione *f* detergente
n waswerking
d Waschwirkung *f*

5711 WASHING POWDER — cle
f poudre *f* à laver
e polvo *m* para lavar
i polvere *f* per lavare
n waspoeder *n*
d Waschpulver *n*

5712 WASHING SODA — cle
f soude *f* cristallisée
e sosa *f* cristal
i soda *f* cristallina
n kristalsoda
d Kristallsoda *f*

5713 WASTE, REFUSE — iwt
f déchets *pl*, refus *m*
e basuras *pl*, desechos *pl*, detritos *pl*
i cascame *m*, rifiuti *pl*, scarto *m*
n afvalstoffen *pl*
d Abfall *m*

5714 WASTE ACID, SPENT ACID — gen
f acide *m* épuisé
e ácido *m* agotado
i acido *m* di rifiuto
n uitgeput zuur *n*
d Abfallsäure *f*

5715 WASTE BEER, ULLAGES — bre
f bière *f* de reste
e cerveza *f* de resto
i resti *pl* di birra
n restbier *n*
d Restbier *n*

5716 WASTE DE-INKING — pap
f désencrage *m* de déchets
e extracción *f* de la tinta de desechos
i disinchiostratura *f* di rifiuti
n inktextractie uit afval
d Abfallentschwärzung *f*

5717 WASTE DISPOSAL — iwt
f élimination *f* de déchets
e eliminación *f* de desechos
i eliminazione *f* di rifiuti
n afvalafvoer, vuilopruiming
d Abfallbeseitigung *f*

5718 WASTE GASES, EXHAUST GASES — fue,gas,iwt
f gaz *pl* d'échappement

e gases *pl* de escape
i gas *pl* di scappamento
n afvoergassen *pl*, uitlaatgassen *pl*
d Abgase *pl*

5719 WASTE GRINDING iwt
f broyage *m* de déchets
e machacadura *f* de desechos
i macinazione *f* di rifiuti
n afvalvermaling
d Abfallzerkleinerung *f*

5720 WASTE LIQUOR, WASTE LYE gen
f lescive *f* résiduaire
e lejía *f* residual
i lisciva *f* residua
n restloog
d Ablauge *f*

5721 WASTE NEUTRALIZATION iwt
f neutralisation *f* de déchets
e neutralización *f* de desechos
i neutralizzazione *f* di rifiuti
n afvalneutralisatie
d Abfallneutralisierung *f*

5722 WASTE PAPER, SCRAP PAPER pap
f déchets *pl* de papier
e papel *m* de desechos, papel *m* para apuntes
i carta *f* straccia, cartaccia *f*
n oud papier *n*
d Altpapier *n*, Makulatur *f*

5723 WASTE SALTS min
f sels *pl* potassiques
e sales *pl* potásicas
i sali *pl* potassici
n kalizouten *pl*
d Abraumsalze *pl*, Kalisalze *pl*

5724 WASTE STEAM, DEAD STEAM, EXHAUST STEAM boi
f vapeur *f* d'échappement
e vapor *m* de escape
i vapore *m* di ritorno
n afgewerkte stoom
d Abdampf *m*

5725 WASTE UTILIZATION iwt
f utilisation *f* de déchets
e utilización *f* de desechos
i utilizzazione *f* di rifiuti
n afvalverwerking
d Abfallverwertung *f*

5726 WASTE WATER, SEWAGE iwt
f eau *f* résiduaire, eaux *pl* d'égout
e agua *f* residual, aguas *pl* cloacales, aguas *pl* sucias
i acqua *f* di rifiuto, fognatura *f*
n afvalwater *n*
d Abwasser *n*

5727 WASTE WATER CHLORINATION iwt
f chloruration *f* des eaux d'égout
e cloruración *f* del agua residual
i clorurazione *f* dell'acqua di fogna
n chloren *n* van afvalwater
d Abwasserchlorung *f*

5728 WASTE WATER PURIFYING iwt,sew
f purification *f* des eaux d'égout
e depuración *f* del agua residual
i epurazione *f* dell'acqua di fogna
n afvalwaterreiniging
d Abwasserreinigung *f*

5729 WASTE YEAST fmi
f levure *f* résiduelle
e levadura *f* usada
i lievito *m* scarto
n uitgewerkte gist
d Abfallhefe *f*, Nachzeug *m*

5730 WATER gen
f eau *f*
e agua *f*
i acqua *f*
n water *n*
d Wasser *n*

5731 WATER-BASED PAINT dec
f peinture *f* à émulsion
e pintura *f* de emulsión
i pittura *f* ad emulsione
n emulsieverf
d Emulsionsfarbe *f*

5732 WATER-CEMENT RATIO bui
f rapport *m* ciment-eau
e relación *f* cemento-agua
i rapporto *m* cemento-acqua
n cement-waterverhouding
d Zement-Wasserfaktor *m*

5733 WATER CONDITIONING, WATER TREATMENT hyg
f amélioration *f* de l'eau, conditionnement *m* de l'eau, correction *f* de l'eau
e tratamiento *m* del agua
i condizionamento *m* dell'acqua, potabilizzazione *f* dell'acqua, trattamento *m* dell'acqua

n waterzuivering
d Wasseraufbereitung *f*

5734 WATER CONTAMINATION, hyg
WATER POLLUTION
f contamination *f* des eaux, pollution *f* des eaux
e contaminación *f* del agua
i contaminazione *f* dell'acqua, inquinamento *m* dell'acqua
n waterverontreiniging, watervervuiling
d Wasserverunreinigung *f*

5735 WATER CONTENT gen
f teneur *f* en eau
e contenido *m* de agua
i contenuto *m* d'acqua, tenore *m* d'acqua
n watergehalte *n*
d Wassergehalt *m*

5736 WATER COOLING clo
f refroidissement *m* à l'eau
e enfriamiento *m* con agua
i raffreddamento *m* all'acqua
n koeling met water
d Wasserkühlung *f*

5737 WATER COOLING clo
f refroidissement *m* de l'eau
e enfriamiento *m* del agua
i raffreddamento *m* dell'acqua
n waterkoeling
d Wasserkühlung *f*

5738 WATER CURING, rub
WATER VULCANIZATION
f vulcanisation *f* à l'eau chaude
e vulcanización *f* en agua caliente
i vulcanizzazione *f* in acqua calda
n heet-watervulcanisatie
d Heisswasservulkanisation *f*

5739 WATER EXTRACT gen
f extrait *m* aqueux
e extracto *m* acuoso
i estratto *m* acqueo
n waterextract *n*
d Wasserextrakt *m*

5740 WATER FASTNESS, tex
FASTNESS TO WATER
f solidité *f* à l'eau
e resistencia *f* al agua
i solidità *f* all'acqua
n waterbestendigheid
d Wasserbeständigkeit *f*

5741 WATER GAS, gas
BLUE GAS, BLUE WATER GAS
f gaz *m* à l'eau, gaz *m* bleu
e gas *m* azul, gas *m* de agua
i gas *m* blu, gas *m* d'acqua
n blauw watergas *n*, watergas *n*
d Blauwassergas *n*, Koksgas *n*, Wassergas *n*

5742 WATER GLASS, aic
LIQUID GLASS, SILICATE, SOLUBLE GLASS
f silicate *m* de sodium, verre *m* soluble
e silicato *m* sódico líquido, vidrio *m* soluble
i tetrasilicato *m* di sodio
n silicaat *n*, waterglas *n*
d Silikat *n*, Wasserglas *n*

5743 WATER-HARDENING met
f durcissement *m* en eau
e endurecimiento *m* en agua
i indurimento *m* in acqua
n waterharding
d Wasserhärtung *f*

5744 WATER HARDNESS hyg
f dureté *f* de l'eau
e dureza *f* del agua
i durezza *f* dell'acqua
n hardheid van het water
d Wasserhärte *f*

5745 WATER-IN-OIL EMULSION fat
f émulsion *f* d'eau en huile
e emulsión *f* de agua en aceite
i emulsione *f* d'acqua in olio
n water-in-olie-emulsie
d Wasser-in-Öl-Emulsion *f*

5746 WATER INSOLUBLE gen
f insoluble adj dans l'eau
e hidroinsoluble adj
i insolubile adj in acqua
n onoplosbaar adj in water
d in Wasser nicht löslich adj

5747 WATER OF CRYSTALLIZATION, cry
CONSTITUTIONAL WATER
f eau *f* de cristallisation
e agua *f* de cristalización
i acqua *f* di cristallizzazione
n kristalwater *n*
d Konstitutionswasser *n*, Kristallwasser *n*

5748 WATER PAINT, DISTEMPER dec
f peinture *f* à l'eau
e aguada *f*
i colore *m* ad acqua
n waterverf
d Wasserfarbe *f*

5749 WATER-REPELLENT FINISHING tex
f apprêt *m* hydrofuge
e apresto *m* hidrófugo
i apprettatura *f* idrofuga
n waterafstotend maken *n*
d Wasserabstossendmachen *n*

5750 WATER-REPELLENT VARNISH coa,dec
f vernis *m* hydrofuge
e barniz *m* hidrófugo
i vernice *f* idrofuga
n vochtwerend vernis *n*, waterafstotend vernis *n*
d wasserabweisender Firnis *m*

5751 WATER ROLLING cle
f nettoyage *m* rotatif à l'eau de savonnage
e limpieza *f* giratoria con agua jabonosa
i pulitura *f* rotativa con acqua di sapone
n roterend spoelen *n* in zeepsop
d drehende Reinigung *f* mit Seifenwasser

5752 WATER SEWAGE DISPOSAL sew
f élimination *f* des eaux résiduaires, évacuation *f* des eaux d'égout
e eliminación *f* de las aguas cloacales
i eliminazione *f* dell'acqua di rifiuto
n afvalwaterlozing
d Abwasserbeseitigung *f*

5753 WATER SOFTENER, SOFTENING AGENT hyg
f adoucissant *m*
e agente *m* ablandante
i agente *m* d'addolcimento
n onthardingsmiddel *n*
d Enthärtungsmittel *n*

5754 WATER SOFTENING, SOFTENING hyg,sew
f adoucissement *m* de l'eau
e ablandamiento *m* del agua
i addolcimento *m* dell'acqua
n waterontharding
d Wasserenthärtung *f*

5755 WATER SOLUBILITY gen
f solubilité *f* dans l'eau
e hidrosolubilidad *f*
i solubilità *f* in acqua
n oplosbaarheid in water
d Wasserlöslichkeit *f*

5756 WATER STAIN dec
f teinture *f* à l'eau
e tintura *f* al agua
i tintura *f* all'acqua
n waterbeits
d Wasserbeize *f*

5757 WATER TOUGHENING met
f augmentation *f* de la ténacité par bain d'eau
e aumento *m* de la tenacidad con banõ de agua
i aumento *m* della tenacità con bagño d'acqua
n taaiheidsverhoging in waterbad
d Zähigkeitsverbesserung *f* in Wasserbad

5758 WATERMARK pap
f filigrane *m*
e filigrana *f*
i filigrana *f*
n watermerk *n*
d Wassermarke *f*, Wasserzeichen

5759 WATERMARKING pap
f application *f* du filigrane
e aplicación *f* de la filigrana
i applicazione *f* della filigrana
n aanbrengen *n* van een watermerk
d mit Wasserzeichen Versehen *n*

5760 WATERPROOF tex
f imperméable adj
e impermeable adj
i impermeabile adj
n ondoordringbaar adj voor water, waterproef adj
d wasserdicht adj , wasserfest adj , wasserundurchlässig adj

5761 WATERPROOFING tex
f imperméabilisation *f*
e impermeabilización *f*
i impermeabilizzazione *f*
n waterdicht maken *n*
d Wasserdichtmachen *n*

5762 WATERTIGHT sap
f étanche adj à l'eau
e estanco adj al agua
i stagno adj all'acqua
n waterdicht adj
d wasserdicht adj

5763 WATKINS DEVELOPMENT, FACTORIAL DEVELOPMENT pho
f développement *m* semi-automatique basé sur le coefficient de Watkins
e revelado *m* precalculado
i sviluppo *m* precalcolato
n van te voren berekende ontwikkeling
d Faktorenentwicklung *f*, vorberechnete Entwicklung *f*

5764 WAX fat
f cire *f*
e cera *f*
i cera *f*
n was
d Wachs *m*

5765 WAX FINISHING lea
f apprêt *m* de cire
e apresto *m* con cera
i apprettatura *f* con cera
n appreteren *n* met was
d Wachsappretur *f*

5766 WAX POLISH cle
f encaustique *f*
e cera *f* para lustrar, encáustico *m*
i cera *f* da lustrare
n boenwas
d Bohnerwachs *n*

5767 WAXING pap
f cirage *m*
e cerado *m*
i inceratura *f*
n paraffineren *n*
d Paraffinieren *n*

5768 WEAK BEER bre
f petite bière *f*
e cerveza *f* ligera
i birra *f* leggera
n licht bier *n*
d Dünnbier *n*

5769 WEAK ELECTROLYTE elc
f électrolyte *m* faible
e electrólito *m* débil
i elettrolito *m* debole
n zwakke elektrolyt
d schwacher Elektrolyt *m*

5770 WEAR gen
f usure *f*
e desgaste *m*, uso *m*
i logorio *m*, usura *f*
n sleet, slijtage
d Abnutzung *f*, Verschleiss *m*

5771 WEAR AND TEAR TEST tes
f essai *m* d'usure
e ensayo *m* de desgaste
i collaudo *m* d'usura
n levensduurbepaling
d Lebensdauerprüfung *f*

5772 WEAR BY FRICTION, ABRASION, ABRASIVE WEAR wea
f abrasion *f*, usure *f*
e abrasión *f*, desgaste *m* por abrasión
i abrasione *f*, usura *f*
n slijtage
d Abnutzung *f* durch Reibung, Verschleiss *m*

5773 WEAR RESISTANCE gen
f résistance *f* à l'usure
e resistencia *f* al uso
i resistenza *f* all'usura
n slijtvastheid
d Verschleissfestigkeit *f*

5774 WEAR TEST tes
f essai *m* d'usure
e ensayo *m* de desgaste, ensayo *m* de uso
i collaudo *m* della resistenza all'usura
n slijtageproef, slijtproef
d Abnutzungsprobe *f*, Verschleissprobe *f*

5775 WEATHERING cer
f désagrégation *f*
e disgregación *f*
i stagionamento *m*
n uit laten werken *n*
d Auswitterung *f*

5776 WEATHERING sto
f évaporation *f* de stockage
e evaporación *f* de almacenaje
i evaporazione *f* d'immagazzinamento
n opslagverdamping
d Lagerungsverdampfung *f*

5777 WEATHERING tex
f déteinte *f*
e palidecimiento *m*
i sbiadimento *m*
n verbleken *n*, verschieten *n*
d Verbleichen *n*, Verschiessen *n*

5778 WEATHERING OF GLASS gla
f effritement *m* du verre
e disgregación *f* del vidrio
i disintegrazione *f* del vetro

n glasverwering
d Glasverwitterung *f*

5779 WEATHERING TEST dec,pla tes
f essai *m* de vieillissement aux intempéries
e ensayo *m* de envejecimiento por intemperie
i collaudo *m* d'invecchiamento per intemperie
n weerbestendigheidsproef
d Bewitterungsprobe *f*

5780 WEATHERPROOF gen
f à l'épreuve des agents atmosphériques
e inatacable adj por los agentes atmosféricos
i resistente adj agli agenti atmosferici
n bestendig adj tegen atmosferische invloeden, weerbestendig adj
d wetterbeständig adj , wetterfest adj

5781 WEEDKILLER agr
HERBICIDE
f herbicide *m*
e herbicida *m*
i erbicida *m*
n onkruidverdelgend middel *n*
d Unkrautvertilgungsmittel *n*

5782 WEEP POINT dis
f point *m* de pleurage
e límite *m* superior de escape de líquido
i punto *m* di gocciolamento
n regengrens
d Regengrenze *f*

5783 WEIGHING gen
f pesée *f*
e pesada *f*
i pesata *f*
n wegen *n*
d Wägen *n*

5784 WEIGHT gen
f poids *m*
e peso *m*
i peso *m*
n gewicht *n*
d Gewicht *n*

5875 WEIGHTED CLOTH, tex
FILLED CLOTH
f tissu *m* chargé
e tela *f* cargada
i tessuto *m* caricato
n verzwaard weefsel *n*
d beschwertes Gewebe *n*

5786 WEIGHTING, tex
LOADING
f chargement *m*
e cargado *m*
i caricamento *m*
n verzwaren *n*
d Beschweren *n*

5787 WEIGHTING AGENT, tex
LOADING AGENT
f matière *f* de charge
e material *m* de carga
i materiale *m* di carica
n verzwaringsmiddel *n*
d Beschwerungsmittel *n*

5788 WELD wel
f soudage *m*, soudure *f*
e soldadura *f*
i saldatura *f*
n las, lasplaats
d Schweisse *f*, Schweissstelle *f*

5789 WELD, dye
LUTEOLIN
f lutéoline *f*
e luteolina *f*
i luteolina *f*
n luteoline
d Luteolin *n*

5790 WELD DECAY wel
f corrosion *f* intergranulaire de la soudure
e corrosión *f* intergranular de la soldadura
i corrosione *f* intergranulare della saldatura
n intergranulaire corrosie van de las
d intergranulare Korrosion *f* der Schweisse

5791 WELDING wel
f soudage *m*, soudure *f*
e soldadura *f*
i saldatura *f*
n lassen *n*
d Schweissen *n*

5792 WET ASSAYING ana
f essai *m* par voie humide
e ensayo *m* por vía húmeda
i assaggio *m* per via umida
n essaaieren *n* langs de natte weg
d Prüfen *n* auf nassem Wege

5793 WET CLASSIFICATION, WASHING CLASSIFICATION sor
f classification *f* à eau courante, classification *f* hydraulique
e clasificación *f* hidráulica, clasificación *f* por corriente de agua
i classificazione *f* idraulica, classificazione *f* a corrente d'acqua
n classificeren *n* met slibgoot
d Stromklassierung *f*

5794 WET ENAMELLING, DIP ENAMELLING coa
f émaillage *m* au trempé
e esmaltado *m* por inmersión
i smaltatura *f* ad immersione
n dompelemailleren *n*
d Nassemaillierung *f*

5795 WET FADING tex
f déteinte *f* en état humide
e palidecimiento *m* en estado húmedo
i sbiadimento *m* in stato umido
n nat verschieten *n*
d Nassverschiessen *n*

5796 WET FASTNESS TEST tex
f essai *m* de résistance à l'humidité
e ensayo *m* de resistencia a la humedad
i collaudo *m* di resistenza all'umidità
n vochtbestendigheidsproef
d Feuchtbeständigkeitsprobe *f*

5797 WET FINISHING tex
f apprêtage *m* liquide
e apresto *m* líquido
i apprettatura *f* liquida
n nat appreteren *n*
d Nassappretur *f*

5798 WET MILLING dec
f broyage *m* par voie humide
e molimiento *m* al líquido
i molitura *f* per via liquida
n nat malen *n*
d Nassvermahlung *f*

5799 WET SPINNING tex
f filage *m* dans un liquide
e hilatura *f* en un líquido
i filatura *f* in un liquido
n nat spinnen *n*
d Nassspinnen *n*

5800 WET STEAM, SATURATED STEAM boi
f vapeur *f* saturée
e vapor *m* saturado
i vapore *m* saturo
n verzadigde stoom
d gesättigter Dampf *m*, Sattdampf *m*

5801 WET-STEAM COOLING clo
f refroidissement *m* à vapeur humide
e enfriamiento *m* por vapor húmedo
i raffreddamento *m* a vapore umido
n koeling met vochtige stoom
d Nassdampfkühlung *f*

5802 WET STOVING, STOVING tex
f blanchiment *m* réducteur
e blanqueo *m* reductor
i candeggio *m* riduttore
n reducerend bleken *n*
d reduzierendes Bleichen *n*

5803 WET STRENGTH pap
f résistance *f* en état humide
e resistencia *f* en estado húmedo
i resistenza *f* in stato umido
n natte sterkte
d Nassfestigkeit *f*

5804 WETTABILITY gen
f mouillabilité *f*
e humectabilidad *f*, mojabilidad *f*
i inumidimento *m*
n bevloeibaarheid
d Benetzbarkeit *f*

5805 WETTING AGENT, SURFACE ACTIVE AGENT, SURFACTANT gen
f agent *m* mouillant, agent *m* tensio-actif
e agente *m* humector, agente *m* tensoactivo
i sostanza *f* tensio-attiva
n bevochtiger
d Netzmittel *n*

5806 WHALE OIL fat,mcy
f huile *f* de baleine
e aceite *m* de ballena
i olio *m* di balena
n walvistraan
d Walfischtran *m*

5807 WHITE BEER bre
f bière *f* blanche
e cerveza *f* blanca
i birra *f* di frumento
n witbier *n*
d Weissbier *n*

5808 WHITE BOILING sin
f cuite *f* à blanc
e cocción *f* para obtener azúcar blanco
i cottura *f* a bianco
n verwerken *n* op witte suiker
d Verarbeiten *n* auf Weisszucker

5809 WHITE CEMENT bui
f ciment *m* blanc
e cemento *m* blanco
i cemento *m* bianco
n witte cement
d Weisszement *m*

5810 WHITE DAMP (US), CARBON MONOXIDE (GB) gas
f oxyde *m* de carbone
e óxido *m* de carbono
i monossido *m* di carbonio, ossido *m* di carbonio
n koolmonoxyde *n*
d giftige Wetten *pl*, Kohlenoxyd *n*

5811 WHITE DISCHARGE tex
f corrosion *f* au blanc
e corrosión *f* al blanco
i corrosione *f* al bianco
n etsbeitsen *n*
d Ätzbeizen *n*

5812 WHITE FRENCH POLISH dec
f vernis *m* au tampon blanc
e barniz *m* de muñeca blanco
i vernice *f* a spirito bianca
n wit schellakpolitoer *n*
d weisse Schellackpolitur *f*

5813 WHITE LEAD, BASIC LEAD CARBONATE dec
f blanc *m* de céruse, céruse *f*, plomb *m* blanc
e albayalde *m*, blanco *m* de plomo, cerusa *f*
i biacca *f* di piombo, cerussa *f*
n basisch loodcarbonaat *n*, loodwit *n*
d Bleiweiss *n*, Deckweiss *n*

5814 WHITE LIME, WHITEWASH, FAT LIME dec
f chaux *f* à badigeonner, lait *m* de chaux
e jalbegue *m*, lechada *f* de cal
i calce *f* da imbianchino, latte *m* di calce
n witkalk
d Tünche *f*, Weisskalk *m*

5815 WHITE OIL fat
f huile *f* paraffinée
e aceite *m* blanco
i olio *m* bianco
n paraffineolie
d Weissöl *n*

5816 WHITE PASTE dec
f blanc *m* broyé en huile
e blanco *m* molido en aceite
i bianco *m* macinato in olio
n wit *n* in olie gemalen
d Ölweiss *n*

5817 WHITE PICKLING cle
f décapage *m* supplémentaire
e decapaje *m* suplementario
i decapaggio *m* supplementare
n nabeitsen *n*
d Nachbeizen *n*

5818 WHITE PRODUCTS fne
f produits *pl* de pétrole légers
e productos *pl* de petróleo ligeros
i prodotti *pl* del petrolio leggeri
n lichte olieprodukten *pl*
d leichte Ölprodukte *pl*

5819 WHITE REINFORCING FILLER, NON-BLACK REINFORCING FIL rub
f charge *f* renforçante blanche
e carga *f* reforzadora blanca
i carica *f* rinforzante bianca
n witte versterkende vulstof
d heller Verstärkerfüllstoff *m*

5820 WHITE RESIST tex
f réserve *f* blanche
e reserva *f* blanca
i riserva *f* bianca
n witte reserve
d weisse Reserve *f*

5821 WHITE RUST cor
f couche *f* d'oxyde de zinc
e capa *f* de óxido de cinc
i strato *m* d'ossido di zinco
n zinkoxydelaag
d Zinkoxydschicht *f*

5822 WHITE SHELLAC dec
f gomme *f* laque blanchie
e goma *f* laca blanqueada
i gomma *f* lacca candeggiata
n gebleekte schellak
d gebleichter Schellack *m*

5823 WHITE SPIRIT, MINERAL SPIRIT, MINERAL TURPENTINE dec
f essence *f* minérale, white-spirit

e aguarrás *m* mineral, trementina *f* mineral
i surrogato *m* d'acqua ragia
n lakbenzine
d Lackbenzin *n*, Schwerbenzin *n*, Terpentinersatz *m*, Testbenzin *n*

5824 WHITENING, WHITING — dec
f poudre *f* de craie
e polvo *m* de tiza
i polvere *f* di creta
n krijtpoeder *n*
d Schlämmkreide *f*

5825 WHITENING, WHITEWASH, LIMEWASH — dec
f badigeon *m*
e lechada *f* de cal
i latte *m* di calce
n witkalk
d Kalkmilch *f*

5826 WHITENING AGENT, IMPROVER — nut
f agent *m* blanchissant de farine
e blanqueador *m* de harina
i imbiancatore *m* di farina
n meelbleekmiddel *n*
d Mehlbleichmittel *n*

5827 WHITING MIXTURE — rub
f talc *m* pâteux
e pasta *f* de talco
i dispersione *f* antiadesiva di talco
n talkpap
d Talkumbrei *m*

5828 WHIZZING — dry
f séchage *m* par centrifugation
e secado *m* por centrifugación
i essiccazione *f* per centrifugazione
n centrifugerend drogen *n*
d Schleudertrocknung *f*

5829 WHOLE MALT — bre
f malt *m* entier
e malta *f* entera
i malto *m* intero
n volmout *n*
d Vollmalz *n*

5830 WHOLE RUBBER — rub
f caoutchouc *m* entier
e goma *f* entera
i tutta gomma *f*
n ingedampte latex
d Vollgummi *m*

5831 WIDE CUT — dis
f coupe *f* large
e separación *f* grosera
i taglio *m* largo
n fractie met groot kooktraject
d grober Schnitt *m*

5832 WIDE-CUT FUEL, AVIATION KEROSENE, AVIATION SPIRIT — fue
f essence *f* d'aviation
e gasolina *f* para aviación
i benzina *f* d'aviazione
n vliegtuigbenzine
d Fliegerbenzin *n*

5833 WIDE RANGE DISTILLATE — dis
f distillat *m* à grande gamme d'ébullition
e destilado *m* de gran gama de ebullición
i distillato *m* a gran gamma d'ebollizione
n destillaat *n* met groot kooktraject
d Destillat *m* mit grossem Siedebereich

5834 WILD BEER — bre
f bière *f* sauvage
e cerveza *f* salvaje
i birra *f* turbolenta
n wild bier *n*
d wildes Bier *n*

5835 WILD GAS (US), DAMP, FIREDAMP, PIT GAS (GB) — fue,gas
f grisou *m*, mofette *f*
e grisú *m*, mofeta *f*
i grisu *m*
n mijngas *n*
d Grubengas *n*, Schlagwetter *n*

5836 WILD GAS DETECTION (US), DAMP DETECTION, FIREDAMP DETECTION, PIT GAS DETECTION (GB) — sap
f détection *f* de grisou, détection *f* de mofette
e detección *f* de grisú, detección *f* de mofeta
i rivelazione *f* di grisu
n mijngasdetectie
d Grubengasanzeige *f*, Schlagwetteranzeige *f*

5837 WILD YEAST — fmi
f levure *f* sauvage
e levadura *f* salvaje
i lievito *m* selvaggio

n wilde gist
d wilde Hefe *f*

5838 WINCH DYEING, REEL DYEING dye
f teinture *f* à dévidoir
e teñido *m* con aspadera
i tintura *f* ad arcolaio
n verven *n* op haspel
d Haspelfärbung *f*

5839 WINE LEES, ARGOL, CRUDE CREAM OF TARTAR cor
f argol *m*, crème *f* de tartre brute
e crémore *m* tártaro crudo
i argol *m*, cremore *m* di tartaro crudo
n ruwe wijnsteen
d Argal *n*, roher Weinstein *m*

5840 WINTER BEER bre
f bière *f* bock
e cerveza *f* Bock
i birra *f* Bock
n bokbier *n*
d Bockbier *n*

5841 WINTERGREEN OIL, CHECKERBERRY OIL fat
f essence *f* de gaulthérie, essence *f* de wintergreen
e esencia *f* de gaulteria
i olio *m* di gaulteria
n wintergroenolie
d Wintergrünöl *n*

5842 WINTERING fat
f extraction *f* de la stéarine, frigélisation *f*
e extracción *f* de la estearina
i estrazione *f* della stearina
n stearine-extractie
d Entstearinieren *n*, Stearinextraktion *f*

5843 WIPING RESISTANCE dec
f résistance *f* au frottement
e resistencia *f* al fregamiento
i resistenza *f* alla sfregatura
n veegvastheid
d Wischfestigkeit *f*

5844 WIRE ROLLED GLASS, ARMOURED GLASS gla
f verre *m* armé, verre *m* métallifié
e vidrio *m* armado
i vetro *m* retinato
n gewapend glas *n*, draadglas *n*
d armiertes Glas *n*, Drahtglas *n*

5845 WOOD mat
f bois *m*
e madera *f*
i legno *m*
n hout *n*
d Holz *n*

5846 WOOD DISTILLATION dis
f distillation *f* du bois
e destilación *f* de la madera
i distillazione *f* del legno
n houtdestillatie
d Holzdestillation *f*

5847 WOOD FLOUR, WOOD MEAL exp,gen
f farine *f* de bois, sciure *f* de bois
e serrín *m*
i farina *f* di legno, segatura *f*
n houtmeel *n*, zaagsel *n*
d Holzmehl *n*, Sägemehl *n*

5848 WOOD GAS gas
f gaz *m* de bois
e gas *m* de madera
i gas *m* di legno
n houtgas *n*
d Holzgas *n*

5849 WOOD PAPER pap
f papier *m* de bois
e papel *m* de madera
i carta *f* di legno
n houtpapier *n*
d Holzpapier *n*

5850 WOOD PULP pap
f pâte *f* de bois
e pulpa *f* de madera
i pasta *f* di legno, polpa *f* di legno
n houtpulp, houtslijp *n*
d Holzschliff *m*, Holzstoff *m*

5851 WOOD SEASONING, SEASONING bui
f desséchement *m* de bois
e cura *f* de madera, desecación *f* de madera
i stagionatura *f* di legno
n drogen *n* van hout
d Holzauswitterung *f*

5852 WOOD SPIRIT, METHANOL, METHYL ALCOHOL dis
f alcool *m* méthylique, esprit *m* de bois, esprit *m* pyroxylique
e alcohol *m* metílico, espíritu *m* de madera, metanol *m*
i alcool *m* metilico, spirito *m* di legno
n houtgeest, methylalcohol
d Holzgeist *m*, Methylalkohol *m*

5853 WOOD STAIN, dec
STAIN
f teinture *f*
e colorante *m* de la madera, tintura *f*
i tintura *f*
n houtbeits, kleurbeits
d Farbbeize *f*, Holzbeize *f*

5854 WOOD SUGAR, dis
XYLOSE
f sucre *m* de bois, xylose *f*
e azúcar *m* de madera, xilosa *f*
i zucchero *m* di legno, xilosio *m*
n houtsuiker, xylose
d Holzzucker *m*, Xylose *f*

5855 WOOD TAR dis
f goudron *m* de bois
e alquitrán *m* de madera
i catrame *m* di legno
n houtteer
d Holzteer *m*

5856 WOOD TAR PITCH oic
f poix *f* de goudron de bois
e pez *f* de alquitrán de madera
i pece *f* di catrame di legno
n houtteerpek *n*
d Holzteerpech *n*, Schiffspech *n*, Schusterpech *n*

5857 WOOD WOOL gen
f laine *f* de bois, paille *f* de bois
e lana *f* de madera
i lana *f* di legno
n houtwol
d Holzwolle *f*

5858 WOODFREE PAPER pap
f papier *m* sans bois
e papel *m* sin madera
i carta *f* senza legno
n houtvrij papier *n*
d holzfreies Papier *n*

5859 WOODPULP PAPER, pap
CELLULOSE PAPER
f papier *m* de cellulose
e papel *m* de celulosa
i carta *f* di cellulosa
n cellulosepapier *n*
d Zellulosepapier *n*

5860 WOOL tex
f laine *f*
e lana *f*
i lana *f*
n wol
d Wolle *f*

5861 WOOL FAT, fat
LANOLIN
f graisse *f* de laine, lanoline *f*
e lanolina *f*
i lanolina *f*
n lanoline, wolvet *n*
d Lanolin *n*, Wollfett *n*

5862 WORK-HARDENING, met
STRAIN HARDENING
f écrouissage *m*
e endurecimiento *m* por deformación en frío
i incrudimento *m* per deformazione a freddo
n koudversteviging
d Kaltverfestigung *f*

5863 WORK LEAD, met
BASE BULLION
f plomb *m* d'oeuvre
e plomo *m* de obra
i piombo *m* argentifero, piombo *m* impuro
n ongezuiverd lood *n*, werklood *n*
d silberhaltiges Blei *n*, Werkblei *n*

5864 WORMSEED OIL, fat
CHENOPODIUM OIL,
SANTONICA OIL
f essence *f* de chénopodium
e esencia *f* de quenopodio
i olio *m* di chenopodio
n wormzaadolie
d Chenopodiumöl *n*, Wurmsamenöl *n*

5865 WORMWOOD OIL, fat
ABSINTHIUM OIL, ABSINTHOLE
f essence *f* d'absinthe
e esencia *f* de ajenjo
i olio *m* d'assenzio
n absinthol *n*, absintholie
d Wermutöl *n*

5866 WORT bre
f moût *m*
e mosto *m*
i mosto *m*
n wort
d Würze *f*

5867 WORT FROM THE LAUTER bre
TUB
f moût *m* clair
e mosto *m* claro
i mosto *m* chiaro
n heldere wort
d Läuterwürze *f*

5868 WORT TO BE PUMPED OFF bre
f préfiltre *m*
e prefiltro *m*
i prefiltro *m*
n voorloop
d Vorlauf *m*

5869 WRAPPED CURING, rub
WRAPPED VULCANIZATION
f vulcanisation *f* sous bandelage
e vulcanización *f* bajo vendaje
i vulcanizzazione *f* sotto benda
n in de katoen vulcaniseren *n*
d Wickelvulkanisation *f*

5870 WRINKLING, dec
BEADING, PUCKERING
f fripe *f*, frisage *m*
e arrugado *m*
i raggrinzamento *m*
n schroeien *n*
d Faltenbildung *f*, Runzelbildung *f*

X

5871 XANTHATION tex
f réaction *f* d'acide de xanthogène
e reacción *f* de ácido de xantógeno
i reazione *f* d'acido di xantogeno
n omzetting in xanthogenaat, xanthogeenzuurreactie
d Xanthogenierung *f*, Xanthogensäurereaktion *f*

5872 XENON LIQUEFACTION gas
f liquéfaction *f* de xénon
e licuación *f* de xenón
i liquefazione *f* di xeno
n vloeibaar maken *n* van xenon
d Xenonverflüssigung *f*

5873 XYLOSE, WOOD SUGAR dis
f sucre *m* de bois, xylose *f*
e azúcar *m* de madera, xilosa *f*
i zucchero *m* di legno, xilosio *m*
n houtsuiker, xylose
d Holzzucker *m*, Xylose *f*

Y

5874 YARN tex
f fil *m*, filé *m*
e hilado *m*, hilo *m*
i filato *m*
n garen *n*
d Garn *n*

5875 YARN COUNTS, YARN NUMBERS tex
f numéros *pl* des fils
e números *pl* del hilo
i titolazione *f* dei filati
n garennummers *pl*
d Garnnummern *pl*

5876 YARN DYEING, BUNDLE DYEING, HANK DYEING tex
f teinture *f* en écheveaux
e teñido *m* en suspensión de madejas
i tintura *f* delle matasse
n strengenverving
d Strähnenfärbung *f*

5877 YEAST bre,nut
f levure *f*
e levadura *f*
i lievito *m*
n gist
d Hefe *f*

5878 YEAST CELL fmi
f cellule *f* de levure
e célula *f* de levadura
i cellula *f* di lievito
n gistcel
d Hefezelle

5879 YEAST FOOD fmi
f aliment *m* de la levure
e nutrición *f* de la levadura
i alimento *m* per il lievito
n gistvoedsel *n*
d Hefenahrung *f*

5880 YELLOW BRASS, BRASS met
f laiton *m*
e latón *m*
i ottone *m*
n messing *n*
d Gelbguss *m*, Messing *m*

5881 YELLOW IRON OXIDE, FERRITE YELLOW dec
f jaune *m* d'oxyde de fer
e amarillo *m* de óxido de hierro
i giallo *m* d'ossido di ferro
n oxydegeel *n*, ijzeroxydegeel *n*
d Eisenoxydgelb *n*, gelbes Eisenoxyd *n*

5882 YELLOWING dec
f jaunissement *m*
e amarilleamiento *m*
i ingiallimento *m*
n vergelen *n*
d Vergilbung *f*

5883 YIELD gen
f rendement *m*
e rendimiento *m*
i rendimento *m*
n opbrengst, rendement *n*
d Ausbeute *f*, Rendement *n*

5884 YIELD POINT, YIELD VALUE pla
f seuil *m* de fluage
e límite *m* de fluidez
i limite *m* di fluidità
n vloeigrens
d Fliessgrenze *f*

5885 YLANG YLANG OIL, ANONA OIL, CANANGA OIL fat
f essence *f* d'ilang-ilang
e esencia *f* de cananga
i olio *m* di cananga odorata
n kanangaolie, ylang-ylangolie
d Anonaöl *n*, Kanangaöl *n*, Ylang-Ylangöl *n*

5886 YOLK, SUINT fat
f suin *m*, suint *m*
e juarda *f*, suarda *f*
i grasso *m* proveniente da sudore
n wolvet *n*
d Schafsschweiss *m*, Wollfett *n*

5887 YPERITE MUSTARD GAS gas,mit
f gaz *m* moutarde, ypérite *f*
e gas *m* mostaza, iperita *f*
i iprite *f*
n mosterdgas *n*, yperiet *n*
d Senfgas *n*, Yperit *m*

5888 YUFTEN, lea
JUCTEN, RUSSIAN LEATHER
f cuir *m* de Russie
e cuero *m* de Rusia, vaqueta *f* de Moscovia
i bulgaro *m*, cuoio *m* di Russia
n juchtleder *n*
d Juchtenleder *n*

Z

5889 ZAPON LACQUER dec
f laque *f* de zapon
e laca *f* de zapón
i lacca *f* di zapone
n zaponlak
d Zaponlack *m*

5890 ZINC BLOOM, HYDROZINCITE met
f fleurs *pl* de zinc, hydrozincite *f*, zincosine *f*
e hidrocincita *f*, zinconita *f*
i zinconite *f*
n hydrozinkiet *n*
d Zinkblüte *f*

5891 ZINC COATING coa
f zingage *m*
e cincado *m*
i zincatura *f*
n verzinken *n*
d Verzinken *n*

5892 ZINC DUST, BLUE DUST met
f poudre *f* de zinc
e polvo *m* de cinc
i polvere *f* di zinco
n zinkstof *n*
d Zinkstaub *m*

5893 ZINC OXIDE OINTMENT mcy
f onguent *m* à oxyde de zinc
e ungüento *m* de óxido de cinc
i unguento *m* di zinco
n zinkzalf
d Zinksalbe *f*

5894 ZINC PHOSPHATIZING coa
f revêtement *m* en phosphate de zinc
e revestimiento *m* de fosfato de cinc
i rivestimento *m* a fosfato di zinco
n zinkfosfaatbedekking
d Zinkphosphatschicht *f*

5895 ZINC REDISTILLATION dec,met
f distillation *f* sélective de zinc
e destilación *f* selectiva de cinc
i distillazione *f* selettiva di zinco
n selectieve zinkdestillatie
d selektive Zinkdestillation *f*

5896 ZINC SOAP soa
f savon *m* zincifère
e jabón *m* cincífero
i sapone *m* zincifero
n zinkzeep
d Zinkseife *f*

5897 ZINC WHITE dec
f blanc *m* de zinc
e blanco *m* de cinc
i bianco *m* di zinco
n zinkwit *n*
d Deckweiss *n*, Zinkweiss *n*

5898 ZINC YELLOW, CITRON YELLOW dec
f chromate *m* de zinc, jaune *m* de zinc
e amarillo *m* de cinc, amarillo *m* limón
i giallo *m* di zinco
n zinkchromaat *n*, zinkgeel *n*
d Zinkgelb *n*, Zitronengelb *n*

5899 ZINCATE TREATMENT coa
f traitement *m* au zincate
e tratamiento *m* al zincato
i trattamento *m* allo zincato
n zinkaatbehandeling
d Zinkatbehandlung *f*

5900 ZONE MELTING elc
f fusion *f* par zones
e fusión *f* fraccional, fusión *f* en zonas
i fusione *f* in zone
n zonesmelten *n*
d Zonenschmelzung *f*

FRANÇAIS

F

ESPAÑOL

E

ITALIANO

I

NEDERLANDS

NL

DEUTSCH

D